D1266572

McGRAW-HILL YEARBOOK OF
Science &
Technology

1997

McGRAW-HILL YEARBOOK OF
Science &
Technology

1997

Comprehensive coverage of recent events and research as compiled by the staff of the McGraw-Hill Encyclopedia of Science & Technology

McGraw-Hill

New York   San Francisco   Washington, D.C.   Auckland   Bogotá   Caracas   Lisbon   London   Madrid
Mexico City   Milan   Montreal   New Delhi   San Juan   Singapore   Sydney   Tokyo   Toronto

Hillsborough Community College LRC

McGRAW-HILL YEARBOOK OF SCIENCE & TECHNOLOGY
Copyright © 1996 by The McGraw-Hill Companies, Inc.
All rights reserved. Printed in the United States of America.
Except as permitted under the United States Copyright Act of 1976,
no part of this publication may be reproduced or distributed in any
form or by any means, or stored in a database or retrieval system,
without prior written permission of the publisher.

1 2 3 4 5 6 7 8 9 0   DOW/DOW   9 0 1 0 9 8 7 6

Library of Congress Cataloging in Publication data

McGraw-Hill yearbook of science and technology.
1962-   . New York, McGraw-Hill Book Co.

    v. illus. 26 cm.
    Vols. for 1962-   compiled by the staff of the
McGraw-Hill encyclopedia of science and technology.
    1. Science—Yearbooks.   2. Technology—
Yearbooks. 1. McGraw-Hill encyclopedia of
science and technology.
Q1.M13   505.8   62-12028

McGraw-Hill

*A Division of The **McGraw·Hill** Companies*

Printed on acid-free paper.

ISBN 045410-8
ISSN 0076-2016

International Editorial Advisory Board

Dr. Neil Bartlett
Professor of Chemistry
University of California, Berkeley

Dr. Richard H. Dalitz
Department of Theoretical Physics
Oxford University, England

Dr. Freeman J. Dyson
Institute for Advanced Study
Princeton, New Jersey

Dr. Leon Knopoff
Institute of Geophysics and Planetary Physics
University of California, Los Angeles

Dr. H. C. Longuet-Higgins
Royal Society Research Professor, Experimental Psychology
University of Sussex, Brighton, England

Dr. Alfred E. Ringwood
Professor of Geochemistry
Australian National University, Canberra

Dr. Arthur L. Schawlow
Professor of Physics
Stanford University

Dr. Koichi Shimoda
Department of Physics
Keio University, Tokyo

Dr. A. E. Siegman
Director, Edward L. Ginzton Laboratory
Professor of Electrical Engineering
Stanford University

Prof. N. S. Sutherland
Professor of Experimental Psychology
University of Sussex, Brighton, England

Dr. Hugo Theorell
Nobel Institute
Stockholm, Sweden

Lord Todd of Trumpington
Professor of Organic Chemistry
Cambridge University, England

Dr. George W. Wetherill
Director, Department of Terrestrial Magnetism
Carnegie Institution of Washington

Dr. E. O. Wilson
Professor of Zoology
Harvard University

Dr. Arnold M. Zwicky
Professor of Linguistics
Ohio State University

Editorial Staff

Sybil P. Parker, Editor in Chief

Katherine Moreau, Senior Editor
Jonathan Weil, Editor
Betty Richman, Editor
Glenon C. Butler, Editor
Patricia W. Albers, Editorial Administrator
Frances P. Licata, Editorial Assistant
Sonia Torres, Editorial Assistant

Ron Lane, Art Director
Vincent Piazza, Assistant Art Director
Angelika Fuellemann, Art Production Assistant

Joe Faulk, Editing Manager
Ruth W. Mannino, Senior Editing Supervisor

Thomas G. Kowalczyk, Production Manager
Suzanne W. B. Rapcavage, Senior Production Supervisor

Suppliers: North Market Street Graphics, Lancaster, Pennsylvania, generated the line art, and composed the pages in Times Roman, Helvetica Condensed Black, and Helvetica Condensed Bold.

The book was printed and bound by R. R. Donnelley & Sons Company, The Lakeside Press at Willard, Ohio.

Consulting Editors

Prof. R. McNeill Alexander. *Deputy Head. Department of Pure and Applied Biology, University of Leeds, England.* COMPARATIVE VERTEBRATE ANATOMY AND PHYSIOLOGY.

Prof. Eugene A. Avallone. *Consulting Engineer: Professor Emeritus of Mechanical Engineering: City College of the City University of New York.* MECHANICAL ENGINEERING.

A. E. Bailey. *Formerly, Superintendent of Electrical Science, National Physical Laboratory, London, England.* ELECTRICITY AND ELECTROMAGNETISM.

Prof. William P. Banks. *Chairman, Department of Psychology, Pomona College, Claremont, California.* PHYSIOLOGICAL AND EXPERIMENTAL PSYCHOLOGY.

Dr. Allen J. Bard. *Department of Chemistry, University of Texas, Austin.* PHYSICAL CHEMISTRY.

Dr. Alexander Baumgarten. *Director, Clinical Immunology Laboratory, Yale-New Haven Hospital, New Haven, Connecticut.* IMMUNOLOGY AND VIROLOGY.

Prof. Richard D. Berger. *Plant Pathology Department, University of Florida, Gainesville.* PLANT PATHOLOGY.

Prof. Gregory C. Beroza. *Department of Geophysics, Stanford University, California.* GEOPHYSICS.

Prof. S. H. Black. *Department of Medical Microbiology and Immunology, Texas A&M University, College Station.* MEDICAL MICROBIOLOGY.

Prof. Anjan Bose. *Director, School of Electrical Engineering and Computer Science, Washington State University, Pullman.* ELECTRIC POWER ENGINEERING.

Ronald Braff. *Principal Engineer. MITRE Corporation/Center for Advanced Aviation System Development, McClean, Virginia.* NAVIGATION.

Robert D. Briskman. *President, CD Radio, Inc., Washington, D.C.* TELECOMMUNICATIONS.

Michael H. Bruno. *Graphic Arts Consultant, Sarasota, Florida.* GRAPHIC ARTS.

Dr. John F. Clark. *Director, Graduate Studies and Professor, Space Systems Spaceport Graduate Center, Florida Institute of Technology, Satellite Beach.* SPACE TECHNOLOGY.

Ross A. Clark. *Executive Vice President, Search Energy, Inc., Calgary, Alberta, Canada.* PETROLEUM ENGINEERING.

Prof. David L. Cowan. *Chairman, Department of Physics and Astronomy, University of Missouri, Columbia.* CLASSICAL MECHANICS AND HEAT.

Dr. C. Chapin Cutler. *Ginzton Laboratory, Stanford University, California.* RADIO COMMUNICATIONS.

Dr. Jay S. Fein. *Division of Atmospheric Sciences, National Science Foundation, Arlington, Virginia.* METEOROLOGY AND CLIMATOLOGY.

Dr. William K. Ferrell. *Professor Emeritus, College of Forestry, Oregon State University, Corvallis.* FORESTRY.

Prof. Lawrence Grossman. *Department of Geophysical Science, University of Chicago, Illinois.* GEOCHEMISTRY.

Dr. Ralph E. Hoffman. *Associate Professor, Yale Psychiatric Institute, Yale University School of Medicine, New Haven, Connecticut.* PSYCHIATRY.

Prof. Stephen F. Jacobs. *Professor of Optical Sciences, University of Arizona, Tucson.* ELECTROMAGNETIC RADIATION AND OPTICS.

Dr. S. C. Jong. *Senior Staff Scientist and Program Director, Mycology and Protistology Program, American Type Culture Collection, Rockville, Maryland.* MYCOLOGY.

Prof. Karl E. Lonngren. *Department of Electrical and Computer Engineering, University of Iowa, Iowa City.* PHYSICAL ELECTRONICS.

Dr. Philip V. Lopresti. *Formerly, Engineering Research Center, AT&T, Princeton, New Jersey.* ELECTRONIC CIRCUITS.

Prof. Craig E. Lunte. *Department of Chemistry, The University of Kansas, Lawrence.* ANALYTICAL CHEMISTRY.

Dr. Michael L. McKinney. *Department of Geological Sciences, University of Tennessee, Knoxville.* INVERTEBRATE PALEONTOLOGY.

Dr. George L. Marchin. *Associate Professor of Microbiology and Immunology, Division of Biology, Kansas State University, Manhattan.* BACTERIOLOGY.

Prof. Melvin Marcus. *Department of Geography, Arizona State University, Tempe.* PHYSICAL GEOGRAPHY.

Dr. Henry F. Mayland. *Research Soil Scientist, Snake River Conservation Research Center, USDA-ARS, Kimberly, Idaho.* SOILS.

Dr. Orlando J. Miller. *Center for Molecular Medicine and Genetics, Wayne State University School of Medicine, Detroit, Michigan.* GENETICS AND EVOLUTION.

Prof. Conrad F. Newberry. *Department of Aerospace and Astronautics, Naval Postgraduate School, Monterey, California.* AERONAUTICAL ENGINEERING AND PROPULSION.

Consulting Editors (continued)

Dr. Gerald Palevsky. *Consulting Professional Engineer, Hastings-on-Hudson, New York.* CIVIL ENGINEERING.

Prof. Jay M. Pasachoff. *Director, Hopkins Observatory, Williams College, Williamstown, Massachusetts.* ASTRONOMY.

Prof. David J. Pegg. *Department of Physics and Astronomy, University of Tennessee, Knoxville.* ATOMIC, MOLECULAR, AND NUCLEAR PHYSICS.

Dr. William C. Peters. *Professor Emeritus, Mining and Geological Engineering, University of Arizona, Tucson.* MINING ENGINEERING.

Prof. W. D. Russell-Hunter. *Professor of Zoology, Department of Biology, Syracuse University, New York.* INVERTEBRATE ZOOLOGY.

Dr. Andrew P. Sage. *First American Bank Professor and Dean, School of Information Technology and Engineering, George Mason University, Fairfax, Virginia.* CONTROL AND INFORMATION SYSTEMS.

Dr. Alan Schiller. *Professor and Chair, Department of Pathology, Mount Sinai School of Medicine, New York, New York.* MEDICINE AND PATHOLOGY.

Mel Schwartz. *Materials Consultant, United Technologies Corporation, Stratford, Connecticut.* MATERIALS SCIENCE AND ENGINEERING.

Prof. Susan R. Singer. *Department of Biology, Carleton College, Northfield, Minnesota.* DEVELOPMENTAL BIOLOGY.

Prof. Marlin U. Thomas. *Head, School of Industrial Engineering, Purdue University, West Lafayette, Indiana.* INDUSTRIAL AND PRODUCTION ENGINEERING.

Prof. John F. Timoney. *Department of Veterinary Science, University of Kentucky, Lexington.* VETERINARY MEDICINE.

Prof. Romeo T. Toledo. *Department of Food Science and Technology, University of Georgia, Athens.* FOOD ENGINEERING.

Dr. Shirley Turner. *U.S. Department of Commerce, National Institute of Standards and Technology, Gaithersburg, Maryland.* GEOLOGY (MINERALOGY AND PETROLOGY).

Prof. Joan S. Valentine. *Department of Chemistry and Biochemistry, University of California, Los Angeles.* INORGANIC CHEMISTRY.

Dr. Blaire Van Valkenburgh. *Department of Biology, University of California, Los Angeles.* VERTEBRATE PALEONTOLOGY.

Prof. Frank M. White. *Department of Mechanical Engineering, University of Rhode Island, Kingston.* FLUID MECHANICS.

Prof. Richard G. Wiegert. *Institute of Ecology, University of Georgia, Athens.* ECOLOGY AND CONSERVATION.

Prof. Frank Wilczek. *Institute for Advanced Study, Princeton, New Jersey.* THEORETICAL PHYSICS.

Prof. W. A. Williams. *Department of Agronomy and Range Science, University of California, Davis.* AGRICULTURE.

Contributors

A list of contributors, their affiliations, and the titles of the articles they wrote appears in the back of this volume.

Preface

The 1997 *McGraw-Hill Yearbook of Science & Technology* continues a long tradition of presenting outstanding recent achievements in science and engineering. Thus it serves both as an annual review of what has occurred and as a supplement to the *McGraw-Hill Encyclopedia of Science & Technology,* updating the basic information in the seventh edition (1992) of the Encyclopedia. It also provides a preview of advances that are in the process of unfolding.

The Yearbook reports on topics that were judged by the consulting editors and the editorial staff as being among the most significant recent developments. Each article is written by one or more authors who are specialists on the subject being discussed.

The *McGraw-Hill Yearbook of Science & Technology* continues to provide librarians, students, teachers, the scientific community, and the general public with information needed to keep pace with scientific and technological progress throughout our rapidly changing world.

Sybil P. Parker
EDITOR IN CHIEF

A–Z

Aerodynamics

The aerodynamics of aircraft and missiles flying at high angles of attack is a subject of intensive research. Modern agile aircraft and missiles must perform well at very high incidence, near and beyond stall. Much of the past and present research and development efforts in aerodynamics are devoted to the investigations of the numerous flow phenomena that are accentuated as the angle of attack of the modern complex configurations increases toward stall and beyond. Modern fighters (such as the U.S. F-15, F-16, and F-18 and the Russian Mig 29 and Sukhoi 27), modern bombers (such as the B-2), supersonic and hypersonic transports (the Concorde and the proposed National Aerospace Plane), as well as the U.S. Shuttle spacecraft and the Russian *Burran* spacecraft have slender elongated shapes. The choice of slender configurations for these and other modern airplanes, spacecraft, and missiles is due to their improved operation at high angles of attack. The stall angle of attack of slender configurations can be delayed beyond 30°. These modern airplanes can maintain flying qualities and aerodynamic control capabilities for special maneuvers even beyond this high stall angle.

Separation and stall. It is well known that the maximum angle of attack of an airplane is limited by the stall characteristics of its lifting surfaces. The lifting surface stalls when the flow does not follow the surface contour. The flow is first accelerated over the leading-edge contour of the wing. After the flow passes the point of maximum velocity on the upper surface of the wing, the flow must decelerate. The maximum velocity on the forward part of the wing increases as the angle of attack is increased, so that the flow must be decelerated on the rear part of the wing. The strong deceleration of the flow causes separation of the viscous layer from the upper surface of the wing. When the separation covers most of the lifting surface, the effectiveness of the aerodynamic contour of the wing's airfoil is lost, lift is lost, and the wing stalls.

Such separated zones, due to increased angle of attack, may occur also in local areas on nonslender airplane configurations. Local separated zones may occur on highly deflected aerodynamic control surfaces, as well as on the extended flaps in landing configurations. In these cases, there is a local high-angle-of-attack effect of flow separation, while the airplane may be still flying at a regular angle of attack.

Methods of delaying separation. Extensive research on the separation phenomena is being conducted to develop methods for delaying the separation of the flow from critical aerodynamic control surfaces. Various methods for control of separation have been developed. Generally these methods are based on energizing the decelerating flow that is about to separate. The energy of jets and of vortical flows is utilized for this purpose. Thus, air jets are obtained by proper slot shapes in slotted flaps, enabling these slotted flaps to be effective even up to 60° deflection. Such large flap deflection results in more than doubling the lift coefficient of the wing, enabling considerable decrease in the landing and takeoff distances of heavily loaded airplanes. There are research programs on delaying separation and enhancing lift by jet blowing. Here, high-pressure gases are directed from the engine compressor and are blown as high-velocity jets through slots into regions where flow separation is imminent.

Another useful method for delaying separation over surfaces at increasing angles of attack is by utilizing the energy of vortices. Thus, so-called vortex generators are attached to the wing upper surface in regions where flow separation can occur. These vortex generators can be small plates, rectangular or triangular in shape, positioned at relatively high angles to the local flow. The flow over these generators rolls up into concentrated vortices. The rolled-up vortices interact with the retarded flow over the wing and energize it so as to delay the flow separation. Such vortex generators are used on many modern civil and military transport airplane wings as well as on wings of military fighters.

Research on the aerodynamics of configurations at high angle of attack makes significant contributions to many aerospace applications. First, it is applied to the aerodynamics of high-angle-of-attack flight of the agile slender military airplanes. In addition, local separated flows occur on most airplane configurations, from light general aviation aircraft to large transport airplanes as well as on the orbital shuttle in its landing configuration. Therefore, the understanding of high-angle-of-attack flow and the control of this flow with jets and vortices is applied to solve the flight problems of all these airplane configurations as their angle of attack increases toward stall.

Slender-body flow. High-angle-of-attack flight, at angles of attack of 30° and higher, is obtained with generally slender configurations of military fighters, such as the F-16, F-18, Mig 23, and Mig 29. The flying qualities of the slender configurations at increasing angles of attack are dominated by three-dimensional separated flow phenomena. In particular, their aerodynamic characteristics are determined by the dominant effects of the free vortices. These are shed from the leading edges of the lifting slender wing surfaces and from the leeward surfaces of the slender forebodies. A large part of the aerodynamic forces and moments at these high angles of attack is due to the induced flow generated by these vortices. The strength and position of these vortices depend on the shape of the planform and of the leading edges of the lifting surfaces, and on the geometry of the body.

These vortices can be generated and controlled by variation in the geometry of the wing planform. Vortices can be generated also by addition of small lift-generating surfaces on the body and by other means. Therefore, various modifications to the leading edges of the wing are used. These include a notch on the Mirage 2000, a sawtooth on the Kfir C2 and the Gripen JAS-39, and a claw on the Mig 23. The leading-edge extension (LEX) on wing planforms is used in many high-performance fighters. In many airplane designs various vortex generators are introduced for enhanced high-angle-of-attack capabilities. Another method for achieving enhanced high-angle-of-attack capabilities is the introduction of canard lifting surfaces ahead of the main wing. These canard surfaces, which can be either fixed or movable, generate strong vortices that interact and invigorate the main wing vortices at increasing angles of attack. This close-coupled canard configuration is very effective in achieving agile aircraft performance. The introduction of small fins, acting as vortex generators at strategic positions on the flight configuration, is used to affect the complete flow field.

Therefore, properly controlled vortex interactions may enable significant variation of the overall lift distribution on the aircraft with or without changes of its attitude.

Thus, controlled vortex interactions can be used as a form of direct lift control for flight at high angles of attack. The integration of devices for enhanced high-angle-of-attack capabilities is illustrated in **Fig. 1**, in the drawings of the F-18, the JAS-39, and the experimental high-angle-of-attack aircraft X-31.

Because wings for high-angle-of-attack operations need highly swept leading edges, the delta wing planform and its derivatives are most popular for agile high-angle-of-attack airplanes and missiles. Therefore, some airplanes have almost straight delta wings, like the Mirage III; some have shaped delta wings, like the Concorde; and some have double delta with leading-edge extension, like the F-16 and F-18. A canard-delta wing planform is used in the Griffen, Kfir C2, Lavi (Technology Demonstrator), X-31, Raffale, and Eurofighter aircraft.

Flow over delta wing. The type of flow seen over slender configurations at increasing angle of attack

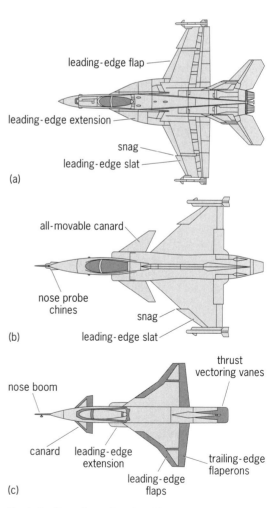

Fig. 1. Configurations of modern high-performance aircraft for high-angle-of-attack flight. (*a*) F-18. (*b*) Gripen JAS-39. (*c*) X-31A.

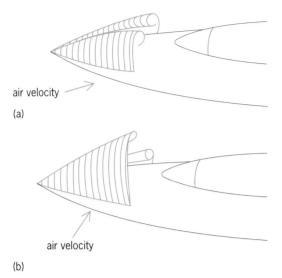

air velocity

(a)

air velocity

(b)

Fig. 2. Vortex separation on slender bodies. (*a*) Symmetric vortex separation at low angle of attack. (*b*) Asymmetric vortex separation at high angle of attack. (*After J. Rom. High Angle of Attack Aerodynamics, Springer-Verlag, 1992*)

is exemplified by the evolution of such flows over a delta wing. At low angles of attack, the flow remains attached to the surface. As the angle of attack increases, the flow separates by forming a shallow separation bubble attached to the surface of the wing. As the angle increases to moderate values the separation of the flow at the leading edge does not reattach. The separated free shear layers are rolled up into concentrated vortices that are fed by a vortex sheet attached to the leading edge. The strength of this concentrated vortex increases as the angle of attack increases. Then the phenomenon of vortex breakdown occurs. That is, the concentrated leading-edge vortices suddenly expand at a certain point along their axis. This vortex breakdown point starts back of the trailing edge and moves forward crossing the trailing edge and continuing toward the wing's tip. As the breakdown point crosses the trailing edge the lift of the wing begins to level off with increase of angle of attack. Eventually, as the breakdown moves toward the wing's tip, the lift falls off and the wing stalls. The breakdown of the leading-edge vortices is near the trailing edge of a delta wing. However, if the wing has even a small angle of yaw then this vortex breakdown is not symmetric. The resulting asymmetric lift causes a rolling moment around the longitudinal axis of the wing. *See* Vortex.

Wing rock. The onset of asymmetric vortex breakdown will cause a rolling motion. As the rolling angle increases the vortex on the down-moving wing may be reestablished, increasing the lift on that side, so that the roll may stop. Then the vortex on the up-moving wing may break down so that the roll motion reverses. This process can repeat itself, resulting in an oscillatory rocking motion that is known as wing rock. This rolling oscillation, which is generally only slightly damped, is experienced by a number of combat airplanes

while flying at high angles of attack. This wing rock impairs controllability of the airplane, particularly its capability in tracking and aiming. This limits the utilization of the full range of angle of attack for which the airplane is designed. In these cases the flight-control system must be designed or modified to compensate and eliminate these oscillations.

Sideslip at high angle of attack. Vortices are also generated by the separation of the flow from the nose of slender bodies. At low and medium angles of attack the separation and the vortices are symmetric with respect to the body axis (**Fig. 2*a***). At higher angle of attack the separation from the nose surfaces becomes asymmetric (Fig. 2*b*), so that side force and yawing moment are obtained while the body is still at zero sideslip. The situation becomes more interesting as the angle of attack is increased further. In this case the asymmetry of the vortex structure changes its direction followed by change in direction of the side force and yawing moment. The direction of the asymmetry can change again at still higher angles of attack. The yawing moment variation due to increase of angle of attack is clearly observed both in wind tunnel and flight data. The yawing moment causes the airplane to sideslip and requires large vertical surfaces for directional control. This may be the reason for the very large vertical tails found on the Russian Sukhoi 27 and the Mig 29. Both airplanes demonstrate excellent aerodynamic control in the performance of the so-called cobra maneuver.

Cobra maneuver. This is one of the most outstanding performances of high-angle-of-attack flying. It was demonstrated by a Sukhoi 27 at the 1990 Farnborough air show. The aircraft is flown at a velocity of close to 220 knots (110 m/s) at an initial altitude of 2000 ft (600 m). The power setting remains constant throughout the maneuver. The pilot has to deactivate the angle of attack limiter so as to be able to pull the full aft stick quickly. Then the aircraft pitches up to beyond vertical attitude (reaching an angle of 110–130°) while remaining at almost level flight. As the airplane reaches this maximum pitch angle in about 2–3 s, the pilot recovers by moving the stick forward; and the airplane levels to a regular flight attitude. At the end of this maneuver the aircraft slows down to about 80 knots (40 m/s) and is accelerated back in level flight (without loss of altitude throughout this maneuver) to higher flight velocity. This dynamic pitch up to beyond vertical attitude in level flight resembles the motions of a cobra head, which accounts for the name given the maneuver. This maneuver was also named the Pougachev maneuver for the first test pilot to perform it.

The success of the Cobra maneuver depends on the utilization of fast dynamic variation of angle of attack for obtaining high lift. It is well known that the lift of a pitching wing increases much above the steady-state value for fast pitch-up. This effect is known as dynamic stall. Therefore, the increase of the dynamic lift and pitching moment during the fast pitch-up of the Su Khoi 27 enabled transitory flight angle of more than 110°.

Another significant aerodynamic characteristic that is demonstrated in the cobra maneuver is the ability to control the yaw forces and moments of the aircraft at very high angles of attack. The elimination of right and left sideslip motions during the pitch-up are required in order to overcome the asymmetric side forces and moments experienced on slender bodies and wings at high angles of attack. This control is achieved by the large tail and rudder surfaces of the Sukhoi 27. The cobra maneuver was later demonstrated by the Mig 29, which also has large tail and rudder surfaces.

For background information SEE AERODYNAMIC FORCES; AERODYNAMICS; AIRFOIL; BOUNDARY-LAYER FLOW; FLIGHT CHARACTERISTICS; FLIGHT CONTROLS; MILITARY AIRCRAFT; VORTEX; WING in the McGraw-Hill Encyclopedia of Science & Technology.

Josef Rom

Bibliography. G. E. Erickson, High angle-of-attack aerodynamics, *Annu. Rev. Fluid Mech.,* 27:45–88, 1995; G. J. Hancock, *An Introduction to the Flight Dynamics of Rigid Aeroplanes,* 1995; J. Rom, *High Angle of Attack Aerodynamics,* 1992; Su-27 aerobatic routine highlights Farnborough show, *Aviat. Week Space Tech.,* 133(12):123, September 17, 1990.

Aerogel

Aerogels are materials obtained by removing the solvent from a gel with a method known as supercritical drying, that is, drying a material above the critical point of the solvent. Ever since the first report of aerogel synthesis in 1931, scientists and engineers have explored the potential of these materials, which possess unusual properties such as low thermal conductivity and high surface area, in diverse applications, including thermal insulation and catalysis. In particular, research has focused on the preparation of aerogels under ambient conditions, because their commercial viability depends on having production costs competitive with those of other available materials. The current challenge is to prepare aerogellike materials without using the high temperatures and pressures commonly encountered in supercritical drying. In other words, the trend is toward defining an aerogel in terms of its properties and not the way in which it is prepared.

Preparation and properties. An aerogel is a product of a gel, which is most commonly prepared by the sol-gel method (see **illus.**). In a sol-gel process, small particles are first formed in a liquid solution by polymerization of reactant molecules (precursors). The particles then connect to form a three-dimensional network. Thus, a gel can be visualized as consisting of many interconnecting strands that form an open structure, with the holes being filled with liquid. The conditions under which a gel is made determine the size of these particles and their interconnectivity, which in turn defines the microstructure of the gel. A gel containing very small holes (pores), which are desirable for many applications, can be prepared by a proper choice of the preparative variables.

The liquid filling the pores, usually the solvent used in the preparation, can be removed from a gel by drying. However, in conventional drying, a liquid-vapor interface develops as the liquid evaporates, and a significant shrinking of the gel occurs as a result of the accompanying capillary forces, yielding a material known as a xerogel. One way to circumvent this problem, as originally reported in 1931, is to bring the liquid above its critical point before removal, thus eliminating the liquid-vapor interface and consequently the capillary forces. This approach, referred to as supercritical drying, maintains the integrity of a gel network and leads to an extremely porous material. In fact, the term aerogel refers to a material that contains mostly air in its very open structure.

The openness of an aerogel can be appreciated by noting a few representative values (exact values

precursors   particles in solution (sols)   particles connecting to form a gel network

Diagram of a sol-gel process; OR represents functional (reactive) groups.

are dependent on specific samples and their preparations). A silica (SiO_2) aerogel, which is the most extensively studied, contains primary particles that are no larger than 50 nanometers; the characteristic diameter is 10 nm and the average pore size is about 20 nm. Because more than 95% of its volume consists of void spaces, such a material has a typical density of 0.1 g/cm^3 (which makes a solid 10 times less dense than water), and 1 g of it will provide a surface area as large as 1000 m^2 (11,000 ft^2; roughly the size of a baseball diamond).

Inorganic oxide aerogels other than silica can also be prepared by the sol-gel method. In general, gels for other oxides do not have as open a structure as a silica gel, because it is more difficult to control the chemistry of their more reactive precursors. But high surface area and high pore volume remain the common characteristics of aerogels. Recently, organic (carbon-based) aerogels have also been prepared from supercritical drying and subsequent thermal decomposition of gels that are formed by reacting resorcinol with formaldehyde and melamine with formaldehyde. Materials that are inorganic-organic hybrids represent another exciting area of research opportunities.

Applications. The low density of an aerogel and the variability of its density with preparation underlie its many unique properties that point to potential applications. For example, the index of refraction of silica aerogels is a function of density, and can be varied over a range that makes them useful as targets for high-energy particles. Silica aerogels are currently used as detectors in high-energy physics research.

The largest potential market for silica aerogels is thermal insulation. Again, because of the low density and small pore size of an aerogel, heat does not transport easily through it, either by conduction or by radiation. A typical value for the thermal conductivity of silica aerogels is 0.01 W/(m)(K) or 0.06 Btu/(h)(ft)(°F). (The thermal conductivity of vitreous silica is about 100 times larger at a comparable temperature.) Applications that could take advantage of this property include insulation for windows and walls, appliances such as refrigerators and water heaters, vacuum bottles, and shipping containers for temperature-sensitive materials. To enter these markets requires the ability to prepare materials at a competitive cost and, in the case of window insulation, to make them transparent (an aerogel consisting of nanosized particles scatters light). Another attractive feature of silica aerogels is that they do not seem to be an environmental hazard.

Aerogels might find applications as catalysts or catalyst supports. The high specific surface area of an aerogel provides a large interface at which catalytic reactions can take place (thus increasing the yields), and its open structure enables the easy diffusion of reactants and products. Indeed, many different aerogels have been found active and selective in catalyzing a wide variety of reactions. However, several practical obstacles (other than the cost considerations), including the low thermal conductivity, low density, and lack of mechanical stability, need to be overcome before catalytic aerogels can be used on a commercial scale.

Other potential applications of aerogels include transducers, electronic components, membranes for gas separations, and chemical sensors. The new organic aerogels have been found to be good conductors of electricity, and they could be used as capacitors.

Production without supercritical drying. The original preparation of aerogels under high temperature and pressure was tedious, hazardous, and not cost effective. Several developments have significantly simplified the process. The use of alkoxides as precursors eliminated the tedious step of exchanging water by repeated washing. The use of carbon dioxide as a supercritical drying agent enabled a safer drying step at a lower temperature (leading to materials known as carbogels). The current challenge is to produce materials that have the properties of aerogels without supercritical drying at all; in other words, to produce low-density xerogels.

A gel shrinks during evaporative drying because it cannot withstand the capillary forces. To prevent shrinkage, either a stiffer gel can be made or the capillary forces can be be minimized. Some level of control, albeit a qualitative one, over the stiffness of a gel can be achieved by changing the preparative variables. Specifically, the stiffness of a gel is governed by its microstructure, which is defined by the size of primary particles and their interconnectivity. Through a manipulation of the rates of hydrolysis and condensation, the two main classes of reactions in a sol-gel process, in principle it is possible to vary a gel's microstructure. For example, water and acid content used in the preparation can be changed. Drying-control chemical additives are available that, in a sol-gel solution, produce gels with uniform-sized pores. Such gels are stiffer because they are subject to a uniform, and not differential, stress distribution arising from capillary forces acting on uniform-sized pores.

Several approaches can minimize the capillary forces inside pores or eliminate them altogether. The original method of supercritical drying falls in this category. Freeze-drying follows the same concept of avoiding a liquid-vapor interface by freezing the liquid into a solid and then allowing the solid to sublime under vacuum. A potential problem with this method is that the solidification process itself may damage a gel network. Finally, exchanging the pore liquid with other liquids that have low surface tensions prior to drying is effective in maintaining the integrity of a gel network.

Surface modification is another promising approach for preparing aerogels under ambient conditions. It is applicable to another process that could contribute to gel shrinkage during drying. The surface of a gel is populated with hydroxyl groups

(—OH), which can interact in a condensation reaction. On a silica surface, such a reaction is as shown below. The formation of Si—O—Si linkages

$$2 \ Si — OH \longrightarrow Si — O — Si + H_2O$$

is believed to pull the network closer together, leading to shrinkage. Surface modification involves removing the hydroxyl groups from a surface by reacting them with another chemical species. The most common example is silylation, in which hydrogen (—H) in the OH group is replaced with an organosilyl group, such as —$Si(CH_3)_3$. Since methyl groups (—CH_3) do not condense with each other, a driving force for shrinkage is removed. Surface modification can also be used to alter the reactivity of a surface and its affinity toward water.

For background information SEE *CRITICAL PHENOMENA; GEL* in the McGraw-Hill Encyclopedia of Science & Technology.

Edmond I. Ko

Bibliography. C. M. Caruana, Aerogels set to take off, *Chem. Eng. Prog.,* pp. 11–17, June 1995; T. Heinrich, U. Klett, and J. Fricke, Aerogels—Nanoporous materials, pt. I: Sol-gel process and drying of gels, *J. Por. Mater.,* 1:7–17, 1995; G. M. Pajonk, Aerogel catalysts, *Appl. Catal.,* 72:217–266, 1991.

Agricultural soil and crop practices

Land-grown plants obtain water and nutrients from soil through uptake by roots. Plants need water and nutrients continuously throughout their growing season, but water by precipitation or irrigation and nutrients by fertilizer are added to soil only intermittently. Thus, at a given time, enough water and nutrients must be added to and stored in soil to sustain plants until the next additions. Also, roots must occupy as much soil as practical to reduce the potential for adverse effects due to limited water or nutrients. The volume needed depends on prevailing plant, soil, and climatic conditions.

Soil-root environment. Interactions among plant, soil, and climatic conditions create an extremely complex soil-root environment. No single plant root, soil, or climatic condition consistently affects plant productivity. However, soil compaction is an important factor because it affects the soil volume where water and nutrients can be stored and later extracted by roots. It also affects soil aeration, which provides for gaseous exchange within soil (oxygen flow to and carbon dioxide flow away from roots). Thus, compaction can greatly influence plant productivity.

Compaction changes a soil from a loose, porous condition to one that is more dense, with smaller spaces (pores) between individual particles. Compaction causes increases in soil bulk density and penetration resistance and decreases in pore size. Bulk density is a measure of dry soil mass (weight) per unit volume. Penetration resistance is a measure of the force needed to push a probe into a soil.

Extensive, deep plant rooting is very important in sandy soils that retain little water. Even in humid regions, short-term droughts on such soils may severely reduce plant growth and productivity. Extensive, deep rooting also is important for plant productivity without irrigation in subhumid and semiarid regions where precipitation often is infrequent. With limited precipitation, plants sometimes depend on stored soil water for survival and growth for several consecutive months. If water storage occurs in a greater soil volume and roots can extract that water, the potential for plant survival and growth increases. Unfortunately, surface or subsurface soil compaction often restricts water infiltration and root growth. Surface compaction impedes water infiltration, soil aeration, and seedling emergence. Subsurface compaction restricts water and root penetration, and may impede aeration.

Causes of soil compaction. Some soils are naturally dense or have a naturally compact layer that hinders root growth. Others become compacted because of human, animal, or equipment traffic; high-intensity rain; or irrigation. Any compaction may reduce plant productivity. Most induced compaction, except that caused by heavy-equipment traffic, is at or near the soil surface. Tillage to normal depths, freezing and thawing, swelling and shrinking due to wetting and drying, and biological activity often reduce such compaction. In contrast, compaction resulting from heavy-equipment traffic has become a major problem because both the depth and areal extent of compaction have increased as equipment weight and tire sizes have increased. Unless heavy-equipment traffic is controlled (restricted to specific zones), the entire soil surface often receives traffic during cultural operations for some plants. A typical production cycle for cotton and sugarbeet in some regions involves 10–15 cultural operations for plowing, planting, spraying, cultivating, and harvesting.

Equipment weights vary widely, ranging up to 16,000 kg (35,000 lb) for large tractors and 40,000 kg (88,000 lb) for loaded produce-hauling vehicles. Load per tire depends on the number of tires on the equipment, and tire sizes and inflation pressures affect contact pressure on the soil surface. The contact pressure in combination with the number of trips across the surface affects the depth to which compaction may occur. For example, corn roots grew to a depth of 90 cm (35 in.) with no traffic on the surface. One pass of traffic with a load resulting in 62 kilopascals (9 lb/in.^{2}) of contact pressure limited rooting to 70 cm (28 in.) depth. Fifteen passes of the same load limited rooting to a depth of 37 cm (15 in.). Root density in the top 20 cm (8 in.) of soil with no traffic was 5.7 mg/g (0.006 oz/oz). It was 5 mg/g (0.005 oz/oz) with one load pass, and 2 mg/g (0.002 oz/oz) with 15 passes. Soil water content and texture (sand, silt, and clay content) also affect the depth of compaction resulting from a given load.

Effects of soil compaction. Compaction affects water infiltration, root activity, and plant productivity.

Water infiltration and root activity. Soil water storage depends on water infiltration into soil. If a soil is compacted at or near the surface, water may simply flow across the surface.

Rapid water infiltration is desirable under many conditions but undesirable under others. On some porous soils, rapid infiltration results in water moving beyond the rooting depth of plants, thus reducing the amount of water potentially available. The water may also carry plant nutrients beyond the reach of roots, thus causing inefficient use of applied nutrients. In addition, the deeply carried nutrients may pollute underground water supplies. Intentional surface soil compaction with traffic or implements is a method of reducing excessive infiltration of irrigation water in some regions. Rapid infiltration into poorly drained soils may aggravate the condition by supplying too much water to the root zone. This excess can reduce soil aeration and cause poor plant growth and productivity. Surface or subsurface drainage systems often are used to remove excess water from such soils.

The ability of plant roots to grow in soil decreases as the level of compaction increases. Slight decreases in root growth may not adversely affect plant growth and productivity if adequate water and nutrients are available. However, major growth and productivity decreases can occur if reduced root growth limits the supply of water or nutrients to the plants.

Increased soil bulk density and penetration resistance are consequences of soil compaction. Soil water content affects penetration resistance but has little or no effect on bulk density of soils, except those with swelling clays. Penetration resistance generally decreases with increases in water content. The term soil strength often denotes the soil condition resulting from the above factors.

Roots readily penetrate loose, noncompacted soils with low strength. Under such conditions, roots easily extend to a depth typical for the plant, thus using water and nutrients from the entire root zone. As strength increases, root penetration decreases, with penetration being prevented at a strength of about 300 pascals (0.04 lb/in.^{2}). If roots cannot penetrate a compacted layer, they can obtain water and nutrients only from soil above this restrictive layer. Therefore, plants may suffer water stress and nutrient shortages, although additional water and nutrients are beneath the restrictive layer.

Plant productivity. The amount of plant material harvested represents plant productivity. The material may be the entire plant (for example, forage crops) or a specific plant part (for example, grain, fiber, or fruit). Regardless of the product harvested, compaction often strongly affects productivity. Compaction may hinder seedling establishment initially,

then limit the ability of roots to supply plants with water and nutrients. Compaction also affects soil aeration, which is essential for effective root activity.

A small root system usually can supply plants with adequate water and nutrients if these are readily available. Even with a compacted zone at a shallow depth, frequent additions can provide enough water and nutrients that plants do not become stressed. For example, productivity often is high for frequently irrigated plants, although the roots penetrate the soil only to a shallow depth. As availability decreases because of infrequent additions and continual extraction by roots, larger root systems generally are needed. If compaction restricts root development so that it reduces the amount of water and nutrients available to plants, productivity decreases. This situation sometimes occurs for precipitation-dependent crops. With limited precipitation or drought, low productivity or even crop failure is possible where compaction limits root growth to shallow depths. Under the same precipitation conditions but with rooting depth not limited by soil compaction, productivity usually is much greater.

Compaction alleviation and prevention. Compaction occurs in most soils if cultural operations are done when the soils are wet. Sandy soils are especially prone to compaction problems because they experience little natural loosening due to shrinking and swelling caused by drying and wetting. Besides natural loosening by freezing and thawing, wetting and drying, and biological activity, use of chisel plows and subsoilers can loosen compacted soils. Subsequent cultural operations should be done in a way that reduces the potential for recompaction. In extreme cases, use of deep plowing (to depths greater than 30 cm or 12 in.) and profile modification (mechanical mixing of the different soil layers) alleviates compaction. Performing cultural operations at optimum soil water contents, confining traffic to designated lanes, and reducing surface pressures (by using more and larger tires having lower inflation pressures) reduce the potential for compaction. Using crawler-type equipment and confining heavy loads (for example, produce-hauling traffic) to noncrop areas are also effective. Thus, careful management can reduce the potential for development of excessive compaction that could limit root growth and plant productivity on many soils.

For background information *SEE AGRICULTURAL SOIL AND CROP PRACTICES; IRRIGATION (AGRICULTURE)* in the McGraw-Hill Encyclopedia of Science & Technology.

Paul W. Unger

Bibliography. G. F. Arkin and H. M. Taylor (eds.), *Modifying the Root Environment to Reduce Crop Stress,* American Society of Agricultural Engineers, 1981; J. H. Taylor (ed.), Special issue on reduction of traffic-induced soil compaction, *Soil Tillage Res.,* 1987; P. W. Unger and T. C. Kaspar, Soil compaction and root growth: A review, *Agron. J.,* 86:759–766, 1994.

Air navigation

The continuing growth of aviation places increasing demands on airspace capacity, and emphasizes the need for the optimum utilization of the available airspace. A timely solution to the airspace management problem is required navigation performance (RNP). The International Civil Aviation Organization (ICAO) instructed three of its working panels to develop navigation performance standards for eventual worldwide implementation. These panels are the All Weather Operations Panel (AWOP), the Obstacle Clearance Panel (OCP), and the Review of the General Concept of Separation Panel (RGCSP). In 1994 the RGCSP completed its first required navigation performance standard.

Required navigation performance is an airspace-system function and not a navigation-sensor function, which means that the airspace requirements can be satisfied independently of the methods (that is, the sensors) by which they are achieved. Required navigation performance is therefore independent of technological advances. Thus it is quite different from the method used by regulating agencies at present, which specifies equipment to be carried on all aircraft for air navigation, and therefore constrains the optimum application and implementation of modern airborne equipment.

Aircraft safety criteria. Required navigation performance is a navigation requirement, and is only one factor in the determination of required aircraft separation minimums from other aircraft, ground terrain and obstacles, and the ocean. In addition to navigation performance, surveillance (conformance monitoring by an air-traffic management authority) and communications (intervention) are necessary to maintain aircraft separation. The risk of collision is therefore a function of navigation performance, aircraft exposure to other aircraft (traffic density), and the air-traffic management's ability to intervene to prevent a collision or maintain an acceptable level of navigation performance. The acceptable risk of a collision is referred to as the target level of safety. Once the separation criteria and the target level of safety are determined, a minimum level of performance can be set for the airspace-system parameters of navigation and air-traffic management intervention.

Flight mission phases. A flight mission can be partitioned into nine phases of flight: taxiing from the gate to the runway, takeoff, climb, en-route flight, descent, initial approach, final approach, landing, and finally taxiing from the runway to the gate. The All Weather Operations Panel developed required navigation performance for the approach, landing, and departure phases of flight, and the Obstacle Clearance Panel, for terminal-area operation (descent and initial approach); the Review of the General Concept of Separation Panel developed required navigation performance for the en-route phase of flight. For the approach and landing phases of flight the navigation function (that is, required navigation performance) is the principal factor determining the separation standard. For simplicity, the discussion below will describe required navigation performance in the context of the precision final approach phase of flight. In contrast to nonprecision approaches, a precision approach requires vertical in addition to lateral guidance.

Precision final approach. There are two basic sets of rules for flying in the national airspace, visual flight rules and instrument flight rules. Under conditions of poor visibility pilots must fly under instrument flight rules. In addition, when weather conditions cause poor visibility during final approach, a landing guidance system is required. A precision approach begins at the end of the initial approach phase of flight, which is about 5 nautical miles (9 km) from the runway threshold

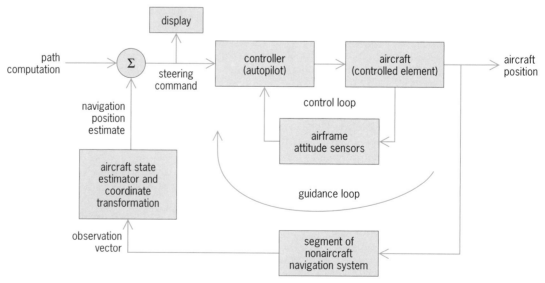

Fig. 1. Aircraft and navigation system model.

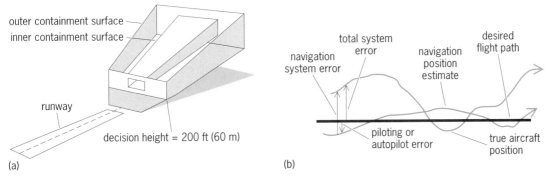

Fig. 2. Required navigation performance parameters. (*a*) Precision approach containment surfaces. (*b*) Total system error as vector sum of flight technical error and navigation sensor error.

and terminates at the minimum decision height of 200 ft (60 m), or higher at about 3000 ft (900 m) in front of runway threshold. The decision height is a height (above the runway elevation) below which a pilot must not descend if adequate visual reference has not been obtained. The pilot must be assured that there are adequate references to land by visual means or execute a missed-approach procedure at this height. A distinction is made between the decision height, which is determined by the obstacle clearance surface, and the weather minimum, which is determined by the visibility conditions. Approaches down to 200-ft (60-m) minimums are called category 1 approaches. If the weather minimum is less than the decision height, then a category 1 approach is not allowed.

Landing guidance systems. The total landing guidance system, comprising the aircraft, pilot or autopilot, and navigation system, is a control system wherein the navigation system is the feedback loop. The total system (**Fig. 1**) has an aircraft segment and a navigation system. A flight path is assigned to each approach phase of flight (**Fig. 2***a*). Around this flight path is specified an inner and an outer aircraft containment surface. Although not shown in Fig. 2*a*, each flight path is assigned an obstacle clearance surface, which is further from the desired flight path than the outer containment surface. The inner and outer containment surfaces contain the entire final approach down to the decision height. The difference between the actual flight path and the assigned flight path is the total system error (Fig. 2*b*). It has two error components, a navigation system error (NSE) and a flight technical error (FTE). The navigation system error reflects the imperfections of the navigation sensors and propagation effects to determine the aircraft's position, and the flight technical error is a measure of the aircraft's capabilities to track the desired flight path in the presence of steady winds and wind gusts. Guidance signals can be coupled to the aircraft in one of three modes: manual, flight director aiding, or autopilot. In the manual mode, the pilot maintains the aircraft's course by observing and correcting course deviations (the flight technical error) on the course deviation indicator or display.

This mode of flying is called manual flight. Its flight technical error is larger than that in either flight director aiding or autopilot modes.

Normal performance and containment surfaces. The central idea of the required navigation performance is to maintain the total system error less than the dimensions of the outer containment surface. In simple terms, the aircraft is inside the outer containment surface when the total system error is less than its dimensions, and it is outside this containment surface when total system error exceeds its dimensions. When the aircraft is outside the outer containment surface, it may be subject to collision hazards.

The purpose of the inner containment surface is to maintain a 95%-accuracy performance, which is the traditional aircraft navigation standard. Thus, the total system error must be less than the dimensions of the inner containment surface 95% of the time. Such performance ensures that the nominal aircraft tracking accuracy concentrates all but 5% of the lateral and all but 5% of the vertical path deviations about the center region of the containment surface.

Maintaining the total system error of the aircraft equal to or less than the dimensions of the inner containment surface 95% of the time, while keeping the total system error less than the dimensions of the outer containment surface, defines the normal (or fault-free) performance of an aircraft. Because under normal (fault-free) performance conditions, the total system error follows approximately a gaussian distribution, the role of the inner containment surface is now clear. It ensures that the tails of the total system error probability distribution are very small at values of the total system error exceeding the dimensions of the outer containment surface when this surface is chosen to satisfy operational requirements. This notion is critical to the understanding of the required navigation performance concept, for it permits rare normal and nonnormal total system error performance deviations to be truncated by the integrity monitors while permitting the airspace operational requirements to be achieved under normal total system error performance. Thus, the total system error of the aircraft and its navigation system must satisfy both the inner and outer containment surface dimensions.

Rare normal and nonnormal performance. Under normal performance conditions, the aircraft is said to be required navigation performance compliant (or to have required navigation performance capability). Rare normal navigation performance is defined when the total system error exceeds the dimensions of the outer containment surface. Rare normal events include, for example, severe wind gusts.

Other events can cause the total system error to exceed its dimensions. They are called nonnormal events. Nonnormal events include the aircraft flight-control system failures which degrade the aircraft's capability to track the flight-path commands and thus increase the flight technical error. They also include navigation system error equipment failures which induce large errors in the navigation position estimates (the aircraft location according to the landing guidance system). The outer containment surface can be viewed as a boundary which identifies those failure events that cause the total system error to exceed its dimensions.

Purpose of integrity monitors. With a rare normal total system error or a rare nonnormal equipment failure the aircraft total system error exceeds the dimensions of the outer containment surface, and the aircraft will unknowingly be in unprotected airspace. The object is to make these failures known to the pilot by monitoring those functions which are critical in maintaining total system error less than the dimensions of the outer containment surface. Flight-technical-error normal failures are rare and are easily identified; they are the excessive deviations shown on the pilot's course deviation display. By using this display as a monitor, the pilot applies corrective steering so that the aircraft's displacement from the desired course is within the outer containment surface. If the corrective steering is unsuccessful and the total system error exceeds the dimensions of the outer containment surface, the pilot performs a missed-approach maneuver. Equipment malfunctions, detected by individual executive monitors in the aircraft, generate a warning on the pilot's display. Similarly, any failures in the radio-navigation signals are detected by monitors on the ground (or in the space segment) which in turn shut down the transmitters that radiate the guidance signal and associated data to the aircraft. Turning off the guidance signal prevents erroneous information from being displayed to the pilot. These rare failures are called detectable failures. When they are detected the pilot leaves the containment surface knowingly, in a safe manner.

Finally, there are latent failures that are not detectable. These undetectable failures cause the aircraft's total system error to exceed the dimensions of the outer containment surface unknowingly and are the largest threat to flight safety. The following three sequences may give rise to such failures: (1) rare normal statistical fluctuations in the navigation guidance signals; (2) undetected failure of the equipment monitor, followed by an equipment failure; (3) equipment failure, followed by a monitor missed-detection failure; and (4) operational conditions which activate an unknown deterministic hardware or software design error.

Continuity versus integrity incidents. Monitoring systems are thus central to the required navigation performance compliance and are necessary to enunciate or remove the guidance when it is inaccurate because of failures. In general, a navigation or guidance equipment failure causes a monitor to generate a loss of required navigation performance capability warning to the pilot. Loss of required navigation performance capability alerts the pilot to the potential need to abort the approach and initiate a missed-approach procedure. Flight safety is maintained because the pilot leaves the containment surface following a missed-approach procedure by some preestablished safe flight path by using dead reckoning or secondary guidance. These detectable events are called continuity incidents, and the probability of a continuity incident is the continuity risk. Thus, the aircraft and the navigation system not only must be accurate but also must be reliable if an approach is to be completed successfully and economically. If the equipment fails without the pilot being alerted (an undetected failure), the guidance signal may not be truthful. Under such conditions, a navigation system may generate misleading information, which can lead the aircraft outside the outer containment surface without the pilot's knowledge. An integrity incident is induced whose probability of occurrence is called an integrity risk. Continuity incidents, therefore, are less hazardous than integrity incidents, because the pilot is forewarned about a critical failure, so that the maneuver to a missed approach point has a high pilot risk reduction factor of at least 1000:1. An integrity incident, however, has a much smaller pilot risk reduction factor in terms of the pilot's capability to eventually detect a latent error and fly to the missed approach point.

The purpose of the continuity risk requirement is to limit the frequency of aborted approaches. The integrity risk requirement is intended to limit the frequency of exposing the aircraft to conditions when it unknowingly violates the containment limit, the outer containment surface, and places the aircraft in a potentially hazardous environment.

Required navigation performance parameters. These results can be summarized by defining four required navigation parameters in terms of total system error, the inner and outer containment surfaces, and monitor warnings as follows:

1. Availability is the probability that all navigation and integrity functions are operational at the initiation of final approach.

2. Accuracy reliability refers to two probabilities: the probability, of 0.95, that the total system error is less than the dimensions of the inner containment surface, and the probability that the total

system error is less than the dimensions of the outer containment surface and that there are no warnings of loss of required navigation performance capability.

3. Continuity incident risk is the probability of loss of required navigation performance capability given that the system had required navigation performance capability at the initiation of the final approach.

4. Integrity risk is the probability that total system error exceeds the dimensions of the outer containment surface with no warning of loss of required navigation performance capability.

Outer containment surface. The purpose of the outer containment surface is to define acceptable navigation performance. Its dimensions are chosen to achieve operational benefits while simultaneously minimizing the number of missed approaches, which can result in revenue losses and aircraft arrival time delays. An example of an operational benefit is provided by an aircraft which can satisfy required navigation performance with an outer containment surface of small dimensions (and thus with an obstacle clearance surface of correspondingly smaller dimensions). Thus, the aircraft can have a lower decision height because the aircraft can fly between obstacles, which would not be possible with a larger containment surface under instrument flight rule conditions.

The pilot always knows when the aircraft is outside the outer containment surface because (1) large course deviations can be observed on the navigation display, and (2) monitors detect navigation system and aircraft flight-control system failures and warn the pilot that the system is no longer required-navigation-performance capable. Consequently, when there are no monitor warnings, the pilot knows that the aircraft is inside the outer containment surface. In effect, the dimensions of the outer containment surface determine the monitor limits.

The entire required navigation performance concept can be summarized in terms of total system error events and monitor warning events by using a Venn diagram. These events are three of the required navigation performance parameters described above: accuracy reliability (E_1), continuity risk (E_2), and integrity risk (E_3). The continuity and integrity risks are selected to satisfy the probability relation $P(E_1) + P(E_2) + P(E_3) = 1$. Provisional values for $P(E_2)$ and $P(E_3)$ are 10^{-4} and 6×10^{-7} per approach respectively. The accuracy reliability is therefore 0.9998994.

Risk assessment methodology. According to the present required navigation performance concept, continuity and integrity risks are allocated to only those navigation elements that generate navigation system errors (such as airborne sensors and the ground and space segments). Required navigation performance risk allocations are not assigned to those elements of the aircraft flight-control system that generate flight technical error failures (such as pilotage failures or autopilot failures). Although the normal flight technical error is included in the required navigational performance, the risk allocation to the aircraft flight-control system is part of a larger risk-assessment activity. Required navigation performance does not increase aircraft certification costs because expensive flight trials and extensive simulations are not required to confirm compliance with the condition that the aircraft be inside the outer containment surface. This is because the outer containment surface is related to the executive monitor limits of the navigation segments and aircraft flight-control system. If any one of the monitor limits is exceeded, the approach procedure reverts to the missed-approach procedure. If there are no monitor alerts, the system is required-navigation-performance compliant, and therefore the aircraft is inside the outer containment surface.

Pictorial description. A pictorial description of required navigation performance views the total system error as the radius of an error surface or bubble around the aircraft. The bubble, if smaller than the dimensions of the outer containment surface, will slide down the containment surface and roll out along the runway. When there is a failure, the total system error increases. The bubble then expands, causing it to get stuck, and a missed approach must be executed. In effect, required navigation performance formalizes for the aviation community what is implicitly intended by current navigation system monitor architectures, pilot displays, and obstacle clearance surfaces.

For background information SEE AIR NAVIGATION; AIR-TRAFFIC CONTROL; AUTOPILOT; RISK ANALYSIS in the McGraw-Hill Encyclopedia of Science & Technology.

Robert J. Kelly

Bibliography. R. J. Kelly and J. M. Davis, Required navigation performance (RNP) for precision approach and landing with GPS application, *Navigation*, 41(1):1–30, Spring 1994.

Air pollution

Airborne dust is a significant source of air pollution. The huge dust clouds generated by severe wind erosion are evidence that soils may produce significant quantities of dust that can hinder public transportation, reduce soil productivity, or be a nuisance. Dust also has an effect on human health, but the impact has not been quantified.

Geologic and climatic history may provide a guide to major changes in atmospheric dust: Recent studies have shown that atmospheric dust levels from Greenland ice cores reflecting deposits over the past 41,000 years varied as atmospheric circulations expanded to include new source regions. Upper atmospheric westerly winds from the United States are capable of transporting dust to central Greenland. About 24,000 years ago, dust levels were 1500 parts per billion compared to fewer than 100 ppb today.

Airborne dust. The deposition of airborne dust has been a factor in the formation of some very productive soils. The deep loess soils in western Iowa and eastern Washington are examples of wind deposits that occurred thousands of years before humans plowed the regions. These deposits are evidence that wind erosion has been a part of the geomorphological processes that shaped the Earth. However, dust also impacts air quality.

In the period 1860–1880, before the Great Plains in the United States were cultivated, blowing dust often limited visibility to a few feet during daylight hours. Overgrazing by native bison or domestic livestock may have contributed to the dust. Dust from wind erosion in the middle to late 1930s was a result of the combined hazards of drought, high winds, and application of inappropriate agricultural technology. Technology suitable for producing crops in the higher-rainfall regions of the eastern United States was not suitable for more arid regions. Currently, cultural practices that are more effective in arid regions, such as residue management and timely tillage, are widely used to control erosion.

The modern dryland farmer compensates for lower per-acre yields with larger fields and uses larger equipment to reduce production costs. Levels of dust from agriculture are mitigated by the use of larger equipment, which requires fewer tractor traffic zones. (The soil in traffic areas is pulverized, and may be more susceptible to wind erosion than soil from nontraffic areas.) In addition, with large equipment the farmer can till the fields when soil moisture conditions are better suited for forming nonerodible aggregates, thus reducing the number of tillage operations required to produce a crop. The fewer operations may further reduce dust emissions.

Measuring dust from agriculture. Estimated total quantities of fine dust in the atmosphere vary between 1.24 and 3.88 billion metric tons (1.36 and 4.27 billion tons) per year, with between 61 and 366 million metric tons (67 and 401 million tons) coming from wind erosion. The quantity of eroded material within a single dust cloud can be tremendous. Because of the high wind velocities, turbulent conditions, and low visibility, it is difficult to collect samples of airborne dust from within the dust cloud. Concentrations in the source area can be large. A dust haze in Mali varied from 26 to 13,735 micrograms per cubic meter. During intense storms, the concentrations exceeded 100,000 µg/m³.

The distance that dust can be transported by wind is an additional factor complicating determination of the effect of agricultural operations on dust emissions. Total-deposition atmospheric samples collected in Hawaii in March 1986 contained a considerable number of particles greater than 75 micrometers in diameter. Based on wind patterns for the previous 11 days, the particles were determined to have come from a major dust storm on March 15–17, 1986, in China.

The ominous dust clouds from major wind erosion events contain a mixture of particles of various sizes and compositions. With the development of field wind-erosion measuring equipment, it is possible to collect samples of the eroded material for size analysis. This equipment enables scientists to accurately measure soil erosion under natural field conditions and to verify wind-erosion prediction models. By collocating very fine particle samplers and erosion samplers, the wind-eroded material and concentrations of agricultural dust can be measured.

Health effects of dust. Dust from agriculture can be a hazard to human health. There are exposure limits, and some people are more sensitive than others. Prior to the use of air-conditioned tractor cabs, farmers who spent hundreds of hours on tractors were exposed to tremendous quantities of agricultural dust. During the Dust Bowl period in the United States, the Kansas State Board of Health reported 17 deaths traced to complications caused by dust: 14 deaths were attributed to pneumonia and 3 to suffocation when the victims were caught in the open during a dust storm. The magnitude and origin of dust are important in determining its impact on human health.

In a critical review of particle measurement methods in relation to compliance with ambient air-quality standards, particles in the nucleation range (<0.08 µm) are also termed ultrafine. Ultrafine particles are emitted directly from combustion sources or condense from cooled gases soon after emission. Their lifetime is usually less than 1 h because of their coagulation with larger particles. Particles or aggregates between 0.08 and 2 µm are considered to be in the accumulation range. They are finely ground dust, or they may result from the coagulation of smaller particles. Coarse particles larger than 2 or 3 µm result from grinding activities and are dominated by geological materials. Other sources of coarse particles include pollen, spores, and ground-up organic materials such as leaves or trash. Coarse particles between 2 and 10 µm may collect in the nasal passages and can contribute to upper respiratory conditions such as rhinitis, allergic reactions, and sinus infection. Particles larger than 10 µm are removed in the mouth or nose.

Geologic sources that contribute coarse particles follow a daily pattern of human activities, whereas the contributions from sources dominated by material 2–3 µm in diameter are homogeneously distributed over space and time. Weather conditions, such as strong inversions with the associated stagnant air, may influence the concentration of dust particles from each source.

Dust from agricultural activities and from natural erosion events contributes significant quantities of material to the atmosphere. Contributions of particles smaller than 10 µm (PM-10) do not always follow concentrations of larger particles. Depending on the type and strength of external forces, the fine particles may bond to larger particles. The resulting aggregate would be much heavier and would settle, unless the wind velocity increased.

Air-quality standards. In the United States, the first air-quality regulations were based on concentrations of total suspended particulates. The present regulations are based on particles smaller than 10 μm. Future air-quality regulations may be based on the concentration of particles smaller than 2.5 μm (PM-2.5). To ensure that future air-quality standards are more closely related to public health, which they are intended to protect, standards must be based on knowledge of the actual health effects, not on available contemporary monitoring technology.

To develop air-quality standards acceptable to all segments of society will be difficult. The intent of the standard should be to protect public health. To achieve this goal may require that the source of the particulate matter also be considered in establishing the acceptable standard. Of additional concern is the change from PM-10 to PM-2.5 as the standard. Establishing acceptable levels of inorganic dust will be a challenge because particles 2.5 μm in diameter are also associated with water droplets, organic hydrocarbons, and ammonium nitrate.

An air-quality paradigm that will be enforceable and meaningful should be a prerequisite for future standards. Past study results are varied, are sometimes contradictory, and have not always identified the threshold at which particulate matter no longer poses a health threat. An effective air-quality standard will be environmentally sound, protect public health, and provide a safe environment for the worker. Industrial concerns that factories have already been closed and jobs lost because of decisions based on incomplete or inaccurate data must be considered in developing an appropriate air-quality standard.

Particulate matter considerations. Neither the impact of particulate matter nor the importance of the source of the matter on human health has been adequately described. Until these issues are resolved, the impact of agriculture on air quality cannot be quantified. In addition, emissions from agriculture are much more variable than are emissions from industrial sources. Dust from agriculture may be a safety hazard for public transportation, contribute to off-site costs through additional maintenance, or be a nuisance, but the impact on human health has not yet been defined.

For background information SEE AIR POLLUTION; DUST STORM; EROSION; LOESS; PUBLIC HEALTH in the McGraw-Hill Encyclopedia of Science & Technology.

Donald W. Fryrear

Bibliography. Association News, Particulate matter: Linked to deaths but difficult to measure, *Environ. Manag.,* 1:182–183, 1995; P. R. Betzer et al., Long-range transport of giant mineral aerosol particles, *Nature,* 336(6199):568–571, 1988; J. C. Chow, Measurement methods to determine compliance with ambient air quality standards for suspended particles, *J. Air Waste Manag. Ass.,* 45:320–382, 1995; P. A. Mayewski et al., Changes in atmospheric circulation and ocean ice cover over the North Atlantic during the last 41,000 years, *Science,* 263:1747–1751, 1994.

Aircraft design

Design of an aircraft is an incredibly complex process involving many different skills, methods, and tools. It can be divided into three distinct phases: conceptual, preliminary, and detail design. Conceptual design deals with the initial development and optimal selection of the overall design approach, defining design basics such as the number and location of engines, the type of tails to be used, the location of passenger seating, and the method of landing gear retraction. Numerous design alternatives are studied, and the one judged to be best is ultimately selected.

Conceptual design. A typical conceptual design effort begins with specification of design requirements, which may include range, number of passengers, payload, and performance needs such as takeoff, landing, and rate of climb. These requirements are used to roughly estimate the vehicle weight and fuel weight required to meet the desired range, and to determine the required engine size and the wing and tail areas. These estimates are needed to make the aircraft design drawing.

The design drawing is a scaled description of the aircraft geometry, and includes the aircraft's wings, tails, fuselage, engines, fuel tanks, landing gear, crew station, passenger or payload area, major avionics, and additional subsystems and components as required. As the concept is developed, a wide variety of real-world constraints are considered, such as pilot vision, stability and control, maintenance, and safety.

The design is then analyzed for aerodynamics, weights, and propulsion, which are used to determine if the aircraft can carry enough fuel to meet the required range. If not, the aircraft must be redesigned and scaled upward in weight until range is met (**Fig. 1**). Also, aircraft performance calculations are made at this time and compared with requirements. Stability and control calculations are used to determine if the tails and control surfaces are the correct size for good handling characteristics.

Trade studies are conducted to optimize the design to reduce the weight and cost. Typical studies include those of wing loading, thrust-to-weight ratio, aspect ratio, and design sensitivities such as drag coefficient and specific fuel consumption. Analysis and trade study optimization results are then used to revise the design layout, and the process iterates.

Preliminary design. In this phase, the selected best design approach is studied and refined. Wind-tunnel tests of aerodynamics and stability are con-

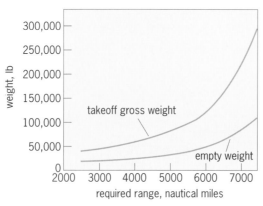

Fig. 1. Design trade study showing how the weight of an aircraft must increase to meet a greater range. 1 lb = 0.45 kg. 1 nautical mile = 1.85 km.

ducted, structural calculations are made, and other studies determine the design's characteristics. Design changes are hopefully limited to refinements resulting from optimization and solution of problems. Sometime during this phase, a so-called configuration freeze is declared, and the overall design is fixed.

Detail design. This phase proceeds with the actual design of every single part required to build the plane. Also, the tooling and manufacturing processes are designed, and many parts of the aircraft are built and tested. Detail design is very labor intensive, with hundreds or even thousands of designers involved. The final quality of a design, however, depends to a great extent on the development of a good overall design in the conceptual design phase.

Computer-aided design. The aircraft conceptual design process has experienced great advancement. Beginning in the 1970s, three-dimensional computer-aided design (CAD) software became available for aircraft design layout. Most major aerospace companies acquired commercial computer-aided design systems and began to use them for conceptual design, but because these programs were developed for detail design of production parts, they were not well suited for conceptual design.

In the late 1970s, a computer-aided design system called CDM was devised solely for aircraft conceptual design, featuring numerous capabilities strictly for the layout of a new aircraft design concept. Included are special commands for quick fuselage shaping including clearance of internal components, tail sizing, fuel-tank creation, and near-instant wing design from parametric inputs (area, aspect ratio, sweep, and so forth). CDM also allows the designer to later change a parameter such as wing sweep, and have the wing and all wing features including control surfaces, flaps, fuel tanks, and even wing structure automatically reshaped instantly to reflect the new wing geometry. Other built-in routines ensure compliance with design specifications such as landing-gear placement, cockpit geometry, and outside vision. Analysis routines allow immediate calculation of area distribu-

tion, wave drag, friction drag, aerodynamic center, weights, and other important results.

With such conceptual-design-specific routines, CDM greatly increased designer productivity, and was extensively used in the B-1B, X-31, and ATF (**Fig. 2**) conceptual design projects. Although never made available publicly, it is still in regular use at several major aircraft manufacturers, the National Aeronautics and Space Administration (NASA), and the U.S. Air Force and Navy.

Later, the CDM program was integrated with analysis and mission optimization programs to create a very powerful environment for complete design and optimization at the conceptual level.

Use of personal computers. As the power of personal computers advances, it is natural to move conceptual design activities onto them. A present-day personal computer is as powerful as a typical room-sized computer of the late 1960s, which was more than adequate to design the F-15 fighter, the 747, and the space shuttle. A highly integrated, user-friendly program based on classical design and analysis techniques from industry is in use at a number of companies and government agencies. It includes a three-dimensional computer-aided design module (**Fig. 3**); integrated analysis of aerodynamics, weights, and propulsion; mission sizing; range calculation; and a full complement of performance calculations including takeoff, landing, rate of climb, turn rate, and acceleration. It provides graphical output for drag polars, lift-to-drag ratios, thrust curves, flight envelope, range parameter, and other quantities. Optimization routines permit instant development of design carpet plots (discussed below) to allow selection of the lightest and cheapest design that meets all performance requirements. A simplified version of the program is in widespread use for aircraft conceptual design education.

Design optimization. This activity is crucial. For conceptual design, it is desirable to optimize the design geometry, including wing parameters such as area, sweep, and aspect ratio, as well as the engine size and type, fuselage shaping, tail geometry, and numerous other design features. In classical optimization, parametric methods are used. The design is repeatedly evaluated over a reasonable range of the design variables, and the resulting weights, costs, and performance results are used to pick the best combination of design variables. However, the number of cases that must be considered is (typically) equal to the number 5 raised to the power of the number of design variables. If, say, eight variables are considered crucial, then the design optimization must consider 390,625 separate cases, clearly an excessive workload.

In the past, aircraft designers have relied on the fact that the design optimization function is usually a so-called well-behaved function, and have used approximations, assumptions, and two-variable optimization techniques to reduce the workload, with reasonable success. Typically, a two-variable optimization graph called a carpet plot is used.

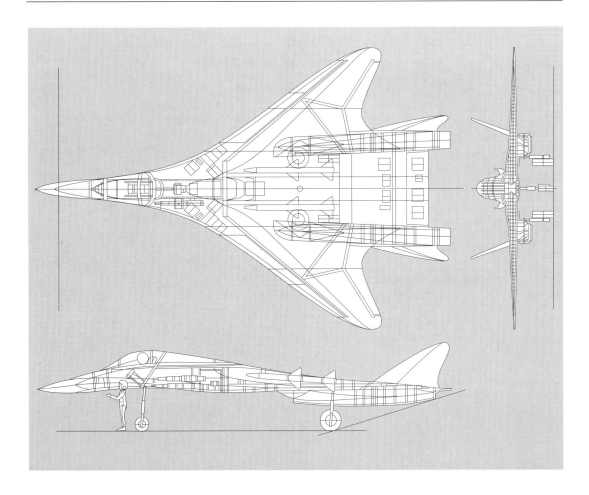

Fig. 2. Aircraft designed entirely by computer-aided design (CAD) in 1981. (*After D. P. Raymer, Aircraft Design: A Conceptual Approach, 2d ed., American Institute of Aeronautics and Astronautics, 1992*)

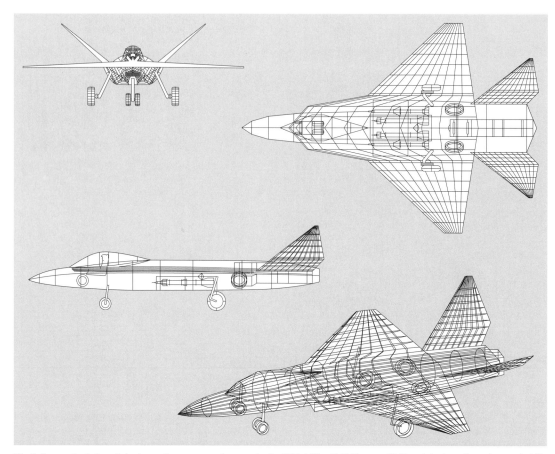

Fig. 3. Conceptual aircraft design using a personal computer in 1994. (*After D. P. Raymer, Notional design of an advanced strike fighter, AIAA Aircraft Eng. Cong., September 1995*)

However, with present-day computer power and optimization methods, true multivariable optimization is beginning to be used in early conceptual design. Such mathematical methods as decomposition, Latin squares, and steepest gradient allow design optimization of numerous variables by using workstations and even personal computers.

Knowledge-based engineering. This advanced technique offers a reduction of design time and cost with an increase in quality. Although mostly used so far on detail design, it will probably become common in conceptual design in the near future. With knowledge-based engineering, the computer-aided design program has built-in design rules and logic for creating routine parts such as wing ribs. Rather than laboriously constructing lines and surfaces, the designer just enters design requirements for the type of part required. Attachments and clearances to adjacent parts can be automated, and redesign work is greatly reduced because the computer can automatically adjust adjacent part geometries.

Computational fluid dynamics. This powerful tool in aircraft design involves the numerical evaluation of the entire airflow around an aircraft design, and offers far greater accuracy and flow-field understanding than did earlier aerodynamic analysis methods. Although the cost and setup time of computational fluid dynamics prevent its fullest use in early conceptual design, it is now quite common to run computational fluid dynamics once a baseline design has been selected. Thus early identification and solution of design problems is possible, resulting in a superior final design. A challenge for the future is to make computational fluid dynamics simple enough to use that the designers, not the computational fluid dynamics specialists, can quickly evaluate their latest design concept.

For background information SEE AIRCRAFT DE-SIGN; COMPUTER-AIDED DESIGN AND MANUFACTUR-ING; EXPERT SYSTEMS; OPTIMIZATION in the McGraw-Hill Encyclopedia of Science & Technology.

Daniel P. Raymer

Bibliography. P. Proctor, Boeing adopts "expert" design system, *Aviat. Week Space Tech.,* 142(17):27, April 24, 1995; D. P. Raymer, *Aircraft Design: A Conceptual Approach,* 2d ed., 1992; D. P. Raymer, *Aircraft Design on a Personal Computer,* SAE Pap. 951160, 1995; D. P. Raymer, Notional design of an advanced strike fighter, *AIAA Aircraft Eng. Cong.,* September 1995.

Animal evolution

The subphylum Vertebrata and two invertebrate subphyla, the Tunicata or tunicates and the Acrania (Cephalochordata) or amphioxus (lancelets), constitute the phylum Chordata. All chordates have several features in common, including pharyngeal gill slits, a dorsal tubular nerve cord, and an underlying stiffening rod termed the notochord. Amphioxus is generally considered to be the closest living relative of the vertebrates. **Figure 1** shows a living amphioxus (*Branchiostoma floridae*), which resembles a 2-in.-long [5-cm] jawless fish without paired fins or eyes. Although amphioxus shares many features with vertebrates, its body plan is much simpler. Unlike vertebrates, amphioxus has no skeleton and the notochord is retained in adults; the dorsal nerve cord has only a small anterior swelling (the cerebral vesicle) and lacks obvious divisions or segments such as forebrain, midbrain, and hindbrain; there are no optic or otic vesicles.

Fossil record. The common ancestor of amphioxus and the vertebrates has long been in question because there is no fossil record of intermediate forms. The first vertebrates appeared in the fossil record about 450 million years ago in the Ordovician. A fossil in Cambrian deposits, *Pikaia,* in which some paleontologists see indications of a segmental musculature and notochord resembling that of amphioxus, has been suggested as a possible ancestor of the vertebrates. However, this idea is far from generally accepted. Thus, attempts to reconstruct the vertebrate ancestor have been chiefly based on anatomical comparisons between both adult and embryonic vertebrates and amphioxus. The relatively simple body plan of amphioxus, however, has often made it difficult to discern homologies. For example, what part, if any, of the amphioxus nerve cord is homologous to the vertebrate brain is debated. One theory suggests that the cerebral vesicle is homologous to the entire vertebrate brain. Another proposes that the cerebral vesicle is equivalent to only the anterior portion of the hindbrain, the fore- and midbrain evolving only after the amphioxus and vertebrate lineages split. *SEE BRAIN.*

Developmental genes. A new technique made possible by molecular biology is helping to resolve such controversies by revealing previously cryptic homologies between distantly related organisms such as vertebrates and amphioxus. The expression patterns of developmental genes, that is, where and when in embryos and larvae the messenger ribonucleic acid (mRNA) for a particular gene is present, are used as new phenotypic characters. Developmental genes by definition code for proteins that are required for differentiation of the organism. Genes coding for proteins involved in so-called housekeeping functions such as cellular metabolism are not considered developmental genes even though they are expressed in embryos. Many developmental genes code for transcription factors [proteins that bind to deoxyribonucleic acid (DNA)] and turn on or off genes for other proteins.

Homeobox genes. One class of developmental genes is the homeobox genes, which code for homeodomain proteins. The homeobox is a 180-base-pair region that codes for the 60-amino-acid homeodomain, which binds to DNA. The name

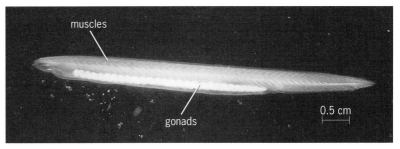

Fig. 1. Living amphioxus (anterior to the left). Chevron-shaped muscles run the length of the body and are visible though the epidermis dorsally. Segmental gonads are visible ventrally. (*Courtesy of M. D. Stokes*)

homeobox derives from the nomenclature of fruit fly (*Drosophila*) genetics. Mutations in some homeobox genes in fruit flies result in homeotic mutations in which anterior body parts are respecified to have a more posterior fate. For example, an antenna may be transformed into a leg. However, it is not understood precisely how mutations in homeobox genes cause such a transformation since the genes that homeodomain proteins bind to and regulate are largely unknown. Two features of homeobox genes make them useful tools for phylogenetic studies. First, the homeobox is highly conserved in evolution, and is thus useful in building gene trees which can give insights into how closely related living forms are but not into the nature of ancestral forms. Second, these genes have restricted spatiotemporal expression patterns in embryos and larvae, which are also evolutionarily conserved. A comparison of these expression patterns between living organisms can give insights into the structure of their common ancestor. The rationale is that if homologous genes of two organisms have restricted expression at about the same time in development within structures that are probably homologous, then the regions where those genes are expressed are also likely to be homologous. This technique works best when the organisms are not too distantly related and when the expression of several genes within a given structure can be compared. Applied to a comparison of the amphioxus and vertebrate nerve cords, it has provided some answers to the question of how the vertebrate brain evolved.

Hox and En genes. *Hox* and *engrailed* (*En*) are two classes of homeobox genes expressed in nerve cords of developing vertebrates and amphioxus. In mammals, there are 38 *Hox* genes and 2 *En* genes. The *Hox* genes are grouped into four clusters (a, b, c, d), each on a different chromosome. These clusters are thought to have arisen by duplication of a single cluster in the ancestral vertebrate. This idea is supported by the presence of just one such cluster in amphioxus. The vertebrate genes of the *Hoxb* cluster are expressed in an ordered series in the embryonic hindbrain and spinal cord. The hindbrain of vertebrates is transiently divided during development into seven segments called rhombomeres. Adjacent to the most posterior rhombomere (number 7) is the most anterior of the mesodermal somites. In the embryonic nerve cord, *Hoxb-1* is expressed only in

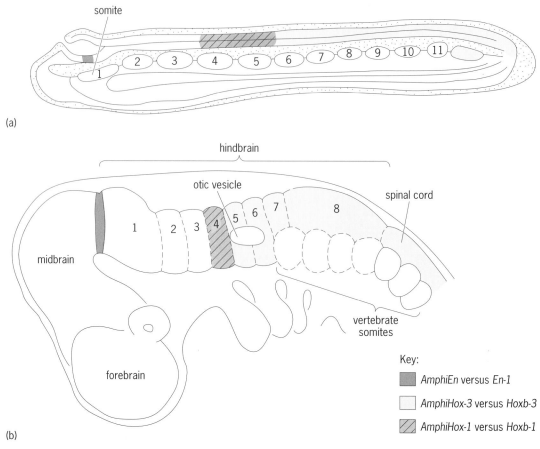

somite

(a)

hindbrain

otic vesicle

spinal cord

midbrain

1 2 3 4 5 6 7 8

forebrain

vertebrate
somites

Key:

▨ *AmphiEn* versus *En-1*

▢ *AmphiHox-3* versus *Hoxb-3*

▨ *AmphiHox-1* versus *Hoxb-1*

(b)

Fig. 2. Comparison of the expression domains of three homeobox genes in embryos (lateral view, anterior to the left). (*a*) Amphioxus. (*b*) Generalized vertebrate. Numbers indicate somites in amphioxus and segments of the hindbrain (rhombomeres) in the vertebrate.

rhombomere 4 (**Fig. 2**). The other *Hoxb* genes have nested expression domains with discrete anterior limits. The anterior limits of *Hoxb-2* and *Hoxb-3* expression are at the anterior boundaries of rhombomeres 3 and 5, respectively. A major expression domain of both *En* genes is located at the junction of the midbrain and hindbrain.

Figure 2 compares the expression domains of *Hoxb-1, Hoxb-3,* and *En* with those of their amphioxus homologs *AmphiHox-1, AmphiHox-3,* and *AmphiEn*. In the developing nerve cord, *AmphiHox-3* is expressed posterior to the level of the somite 4/5 boundary. Thus, rhombomere 5 in vertebrates corresponds to the amphioxus nerve cord at the level of somite 5. Correspondingly, *AmphiHox-1* is expressed in the amphioxus nerve cord in a stripe at the level of somites 4 and 5. Therefore, this region is equivalent to rhombomere 4 in the vertebrate brain. In addition, *AmphiEn* is expressed in a few cells within the cerebral vesicle, about midway between its anterior and posterior limits. This region is, therefore, equivalent to the vertebrate midbrain-hindbrain junction. Thus, a comparison of the expression patterns of three genes and their homologs between amphioxus and vertebrates shows that the amphioxus nerve cord does have an extensive homolog of the vertebrate hindbrain and, furthermore, places its anterior limit within the

cerebral vesicle. Therefore, at most only the anterior half of the cerebral vesicle remains for homologs of the vertebrate forebrain and midbrain.

Homeotic transformations. Additional support for these conclusions comes from perturbing the gene expression patterns with the teratogen all-trans retinoic acid. Treatment of early vertebrate embryos with excess retinoic acid results in severe craniofacial malformations. Depending on the embryonic stage being treated, the amount of retinoic acid, and the species, the forebrain and midbrain are reduced in size. In all vertebrates, there is a homeotic transformation in which the hindbrain anterior to the otic vesicle is respecified to a more posterior identity as shown by an anterior shift in the genetic expression patterns of *Hoxb-1* clusters. In mouse embryos, two phenotypes are observed: a segmented phenotype resulting from an application of retinoic acid after the somites have formed, and an unsegmented one resulting from the application of retinoic acid prior to the onset of somitic segmentation. In the segmented phenotype, rhombomere 4 expresses *Hoxb-1* as in the controls, but rhombomere 2 also expresses *Hoxb-1*. Thus, rhombomere 2 has apparently been converted to a rhombomere 4 identity. In the unsegmented phenotype, the hindbrain is shortened. Therefore, the entire portion anterior to the otic vesicle expresses

Hoxb-1, and has thus been converted to a rhombomere 4 identity. *SEE GENE.*

In some cases, amphioxus larvae are similarly affected by retinoic acid, causing the cerebral vesicle to become slightly shortened. The domain of *AmphiHox-1* expression, normally restricted to a stripe at about the level of somites 4 and 5, is extended anteriorly and displays both the segmented and unsegmented phenotypes. In the segmented phenotype, *AmphiHox-1* is expressed in all of the nerve cord anterior to about segment 6 except for a stripe at the level of somite 2; in the unsegmented phenotype, it is expressed in the entire nerve cord anterior to somite 6.

Implications. These results support the conclusion drawn from the expression domains of homeobox genes in normal larvae that amphioxus has an extensive homolog of the vertebrate hindbrain with its anterior limit within the cerebral vesicle. The vertebrate ancestor, therefore, probably possessed an extensive hindbrain, but at most had no more than a very rudimentary forebrain and midbrain.

For background information *SEE ANIMAL EVOLUTION; CEPHALOCHORDATA; GENE; MOLECULAR BIOLOGY; VERTEBRATA* in the McGraw-Hill Encyclopedia of Science & Technology.

Linda Z. Holland

Bibliography. E. M. De Robertis, G. Oliver, and C. V. E. Wright, Homeobox genes and the vertebrate body plan, *Science,* 263:46–52, 1990; C. Gans and R. G. Northcutt, Neural crest and the origin of the vertebrates: A new head, *Science,* 220:268–274, 1983; P. W. H. Holland et al., The molecular control of spatial patterning in amphioxus, *J. Mar. Biol. Ass. U.K.,* 74:49–60, 1994; H. Wood, G. Pall, and G. Morriss-Kay, Exposure to retinoic acid before or after the onset of somitogenesis reveals separate effects on rhombomeric segmentation and 3′ *HoxB* gene expression domains, *Development,* 120:2279–2285, 1994.

Animal phylogeny

Traditionally, zoologists divide the animal kingdom (Metazoa) into a small number of fundamentally important groups which are positioned above the level of phyla, and are sometimes called superphyla on the basis of their architecture. Simple small animals lack any body cavity and are acoelomate. Larger, more complex animals have different kinds of adult body cavities (usually termed the pseudocoelom and the coelom) which are also associated with profound embryological differences. One group is characterized by a primary body cavity (blastocoel) which originates within the hollow blastula embryonic stage and persists into adult life as the pseudocoelom (or at least some cavity reappearing in its place). This cavity supposedly occurs in animals collectively called the pseudocoelomates, or aschelminthes. The secondary or true body cavity, the coelom, arises in embryonic development after mesoderm has already formed, and is

defined as lying entirely within the mesoderm bounded by a mesothelial lining termed peritoneum. The higher animal groups such as annelids, echinoderms, and chordates are coelomates; so too (in theory at least) are the arthropod and mollusk groups, where the original pseudocoelom is dominant over a reduced coelom.

Origins of body cavities. There are two schools of thought concerning body cavities, each deriving from late-nineteenth-century embryological insights. Some perceive the ancestral form as an acoelomate, with pseudocoelomates as a derived sideline; coelomates are then a later development. Others have regarded the ancestor as being coelomate (the archecoelomate), with modern acoelomates and pseudocoelomates as regressions.

However, if a coelom is accepted as a useful phylogenetic character, all the traditional family trees require complicated arguments to allow the character to have arisen only once, because the modern coelom occurs as two types, giving the two coelomate superphyla: the Protostomia or Spiralia (annelids, mollusks, and arthropods), where the coelom derives from a split within a band of solid mesoderm and is termed a schizocoel; and the Deuterostomia (echinoderms, lophophorates, and chordates), where the cavity forms instead by three paired pouches budding off from the side of the gut wall, forming a tripartite enterocoel. The two superphyla thus become the two major higher branches of the animal kingdom's family tree (**Fig. 1**). Various theories take enterocoely or schizocoely to be the primitive mode of coelom formation; both versions require some phyla to have switched mode later, and both have difficulty with groups such as phoronids, brachiopods, or hemichordates, which have intermediate or entirely novel ways of forming a coelom.

If the coelom arose more than once, it is convergent, and a useless phylogenetic character; fundamental phylogenetic divisions should not rest upon it. Most analyses since the mid-1980s indicate that body cavities arose many times over by different routes, and are convergent and polyphyletic even within any one conventional category. Although functionally intriguing, they may be phylogenetically uninformative. This situation would force radical revision of existing trees of the animal kingdom.

Coelom. The archecoelomate version of metazoan phylogeny suggests that there is a regressive link between fully coelomate and pseudocoelomate and acoelomate conditions. This interpretation revolves around what a coelom is, since anything excluded becomes a blastocoel-related cavity by default. However, defining a coelom as lying within mesoderm raises new questions, since it is impossible to define mesoderm itself with clarity. This structure arises through invagination, or ingression from the ectoderm or endoderm, or from the blastopore region where the two primary germ layers meet. Some taxa have mesoderm

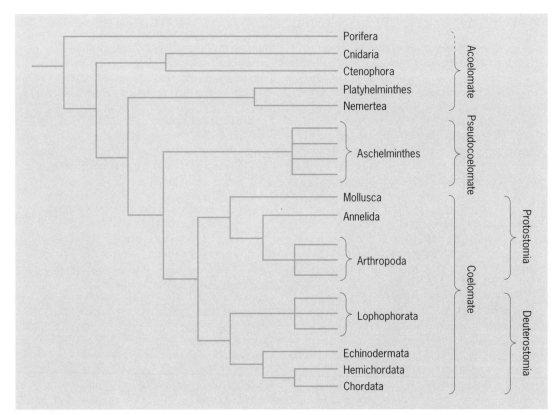

Fig. 1. Traditional pattern of the animal kingdom. Acoelomate and pseudocoelomate animals precede the more complex coelomates.

derived from both germ layers (most protostomes), and others (the main deuterostome groups and many pseudocoelomates) have only endomesoderm. Visceral and somatic mesoderms may have different patterns of gene expression, so should be regarded as two separate germ layers. Despite very active research in the area, there still seems to be no single molecular genetic marker for mesoderm, although a number of genes relating to partial mesoderm destiny are identified. It is therefore difficult to regard mesoderm as homologous between all phyla, in which case the intramesodermal cavities can hardly be so either.

Thus, in practice the two main types of cavity are often indistinguishable, structurally and at the gene expression levels, with the result that groups such as the priapulids, the bryozoans, and even the mollusks have been notoriously difficult to place on the traditional dichotomous schemes. The occurrence of many anomalies of body cavity formation within particular phyla, indicating the plasticity of higher-level features of body design, is another complication. Thus, the concept of a body cavity becomes multiply convergent; in particular, it may be misleading and unhelpful to perceive certain taxa with pseudocoeloms as being any more closely related to each other than they are to other taxa described as having coeloms.

These difficulties with traditional morphology may arise in part because of relative synchronicity in the emergence of most animal phyla, with bursts of radiation in the late Precambrian when many designs arose in parallel, giving a situation where original character states, and precise relations between groups, cannot be resolved. An independent data set is needed to test old morphological ideas. Molecular systematics and modern developmental biology may offer hope.

Molecular systematics. Ribosomal ribonucleic acid (rRNA) molecules are strongly conserved, and unlike morphological characters they can be compared among all living organisms; phylogenies can be constructed from similarities of nucleic acid sequences. The 18S rRNA has a very slow rate of change, and is particularly useful for phyletic relationships where connections at least 600 million years back are sought. The encoding ribosomal deoxyribonucleic acid (rDNA) is even more useful, lacking the problems of complex secondary structure. It has became amenable to analysis only recently with the amplifying effects of the polymerase chain reaction giving quick and repeated access to nuclear material from very much smaller and rarer animals.

Studies involving 18s rRNA, based on approximately 22 metazoans and 28s rRNA have been undertaken. Some of the analyses cannot clearly unite all the metazoans; those that do often do not clearly separate the groups with supposedly different kinds of body cavities. However, it is possible to state at this point that acoelomates appear to be basal, not secondarily regressive; therefore the acoelomate condition should be the ancestral one. The pseudocoelomate groups, still very little known, appear polyphyletic, and the key coelomate groups remain problematic: the coelomates usually group together and split from a common point, which might imply a unitary origin of the coelom. All molecular cladograms (phylogenetic trees) to date are impossible to resolve with confidence, and most do not support the protostome-deuterostome split (**Fig. 2**). However, the coelom might have evolved repeatedly during the late Precambrian radiation, and it remains doubtful whether molecules will ever resolve such a pattern at this chronological distance.

Some details of these molecular trees do begin to cast light on body cavity origins. Recently, most morphologists take the arthropods to be monophyletic and protostome, similar to annelids; their body cavity is seen as a fusion of unexpanded schizocoelomic cavities with a persistent primary

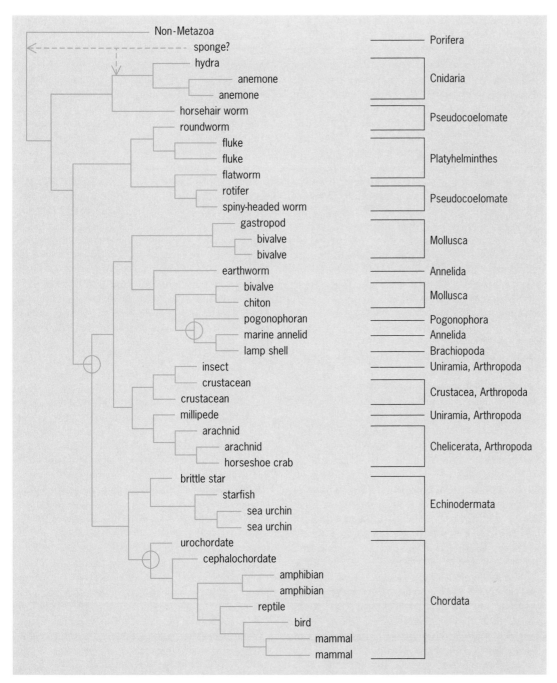

Fig. 2. Cladogram summarizing several recent ribosomal ribonucleic acid (rRNA) molecular cladograms of the animal phyla. Note the lack of resolution of the pseudocoelomates and of the protostome and deuterostome coelomates (open circles). Brachiopods and bryozoans lie among the schizocoelic groups, and arthropods also fail to group together as expected.

pseudocoelom to form a hemocoel. Molecular phylogenies have often had trouble supporting this picture, with some traditional arthropods lying apart from the remainder, or distant from annelids, or with accepted arthropod clades split up seemingly randomly. In contrast, the Brachiopoda, placed morphologically among the lophophorate phyla alongside the deuterostomes, appear in the middle of the traditional protostomes in molecular phylogenies. If the molecular data are correct, the most clear-cut of all morphological homologies is undermined—that the tripartite enterocoelic cavities of deuterostomes and their allies are good evidence of shared ancestry. There is some recent support for this anomalous brachiopod status from the fossil record of the halkierids, a group of Cambrian fossils. Furthermore, the preliminary molecular data on the lophophorate bryozoans indicate that they too should be placed among the protostomes.

However, molecular studies cannot be accepted uncritically. It is not known how molecular change relates to the passage of time; the rate of nucleotide substitution in DNA is highly and perhaps predictably variable (for example, it is much greater in endotherms than in ectotherms). There may also still be selectivity in choosing characters, as only a tiny part of the genome is analyzed, and information from a section that does not fit with preconceptions may be left out. Most importantly, there is still no consensus on exactly how to analyze homology in molecular data; the potentially huge data sets require more sophisticated analytical and statistical techniques.

Moreover, more information is required about variability within phyla before assessments can be made about variation between phyla. Using just a few, or often only one, species to represent a phylum is potentially very misleading; the species chosen have often been very aberrant, most notably in the case of *Drosophila,* which has proved almost useless for phylogenetic purposes, probably because it has evolved so fast.

Thus, molecules may lead to oversimplified conclusions; morphological characters may provide far more information, although it may be more difficult to quantify. But molecular evidence does yield independently derived phylogenetic trees or cladograms for comparison with those based on morphology of body cavities and associated characters. Molecules provide a means of bypassing some of the awkward assumptions made by morphologists, although it seems unlikely that they will unequivocally reveal the phylogenetic shape of the Metazoa.

Developmental mechanisms and body cavities. It is not known how morphological change corresponds to molecular change. Rates of evolution in the genome can be very different from rates of morphological change because so many genes have more than one phenotypic effect and interact with other genes. If developmental processes and morphological change depend mainly on a few regulator genes, then changes may be due to gene rearrangements (up to and including the chromosome scale) much more than to sequence mutations.

Therefore, new developments in the interaction of developmental biology and genetic and molecular studies also impinge on the analysis of body cavities and gross morphology. Beginning in the mid-1980s, the conservatism of mechanisms controlling developmental processes and the positional information available to dividing embryonic cells has been revealed. Indeed, the homeobox gene clusters may provide new homologies for the whole Metazoa, and the arrangement of these clusters in the genome may reveal new phylogenetic patterns. Many other genes and gene products are now known to affect tissue form; the relatively few molecules that determine cell-cell interactions and adhesion properties, conserved across large parts of the animal kingdom, may play key roles in determining morphology. It seems likely that genes controlling such major morphological patterns will soon be located, some of which may have a direct bearing on processes of body cavity formation.

Promising models have also appeared. One recent suggestion is that morphological evolution is generated initially by variations in intrinsic physical properties of cell aggregations, with the selection process favoring biochemical (and heritable) fixation of just a few viable morphologies. Thus, many developmental features including gastrulation, tissue layering, body cavity formation and segmentation could be explained. Also, important predictions—that major innovations occurred very early on in phylogeny, that they are likely to arise more than once, and that there would consequently be much functional redundancy among developmentally important genes—are possible. The most recent work in developmental biology supports these conclusions.

New technology. Techniques to test recent ideas are being developed: identifying and sequencing of genes, selectively splicing out or transplanting in relevant bits of genome, and monitoring cell interactions and resultant tissue morphology. In conjunction with labeling particular tissues by using new molecular markers, such techniques could be very powerful analytical tools. It may also be possible to recognize broader phylogenetically significant patterns in the genome, such as the sequence of genes on chromosomes or the patterns of introns within particular genes, features which would be much less susceptible to convergent change than either gross morphology or molecular sequence. This new technology should help to resolve the issue of how body cavities arose, whether they give any real phylogenetic information, and whether the animal kingdom should really be seen as a neat two-branched tree or as a rather unkempt meadow.

For background information SEE *ANIMAL EVOLUTION; ANIMAL MORPHOGENESIS; ANIMAL SYSTEMATICS; COELOM; PHYLOGENY* in the McGraw-Hill Encyclopedia of Science & Technology.

Pat Willmer

Bibliography. S. A. Newman, Generic physical mechanisms of tissue morphogenesis: A common basis for development and evolution, *J. Evol. Biol.,* 7:467–488, 1994; R. A. Raff, C. R. Marshall, and J. M. Turbeville, Using DNA sequences to unravel the Cambrian radiation of the animal phyla, *Annu. Rev. Ecol. Syst.,* 25:351–375, 1994; P. G. Willmer, *Invertebrate Relationships,* 1990.

Antibiotic

Any environment capable of supporting animal life can also support large populations of microbes. Consequently, host defense mechanisms are essential for all living organisms. Although lymphocytes and antibodies play vital roles in the acquired immunity of higher vertebrates to infection, recent studies have delineated an older arm of the immune system that operates by producing antibiotic peptides.

In contrast to the antibiotics produced by bacteria and fungi, which are mostly secondary metabolites, the antibiotic peptides of animals are encoded by genes and synthesized on ribosomes. Evidence for this system was initially obtained from two very different models involving mammalian white blood cells and the cell-free body fluids (hemolymph) of larval silk moths. More recently, the search for additional animal antibiotics has focused on the skin of frog; the blood, tongue, and windpipe of cattle; the small intestine of rats, mice, and pigs; and extracts of dragonflies, fruit flies, flesh flies, beetles, and horseshoe crabs. *See Immunological memory.*

Defensins. Certain white blood cells that normally circulate in the blood (neutrophils and monocytes) or reside in the tissues (macrophages) readily ingest bacteria and fungi. Most invading organisms are promptly killed and digested by these cells, thereby preventing initiation of serious disease. The cytoplasm of a human neutrophil contains more than 1000 minute granules, each 1–2 micrometers in diameter. Of the several types of granules that can be distinguished, about one-third are primary granules, which play a major role in killing ingested microbes. Between 30 and 50% of the total protein of primary granules consists of defensins, relatively small molecules comprising 29–35 amino acids. X-ray crystallography reveals that defensin molecules resemble tiny paper clips.

Defensins are also produced by specialized small intestinal cells (Paneth cells) whose secretions are believed to protect intestinal crypts from microbes that are swallowed with food or drink. Approximately 50 structurally distinct defensins have been identified from white blood cells or Paneth cells. Human neutrophils contain four different defensins (HNP 1–4), and human Paneth cells produce two other members of this family (HD 5 and HD 6).

Purified defensins kill a wide range of microbes in test tubes, and are thought to do the same within neutrophils or the intestinal crypts. Human defensins use their microbial target cells' electrochemical energy to damage the membrane or the target cell and form voltage-gated pores. These channels allow leakage of critical microbial constituents, including ions and metabolic intermediates. Defensins are more active against gram-positive bacteria than gram-negative bacteria, especially when they are tested in fluids that resemble serum in their ionic composition. Purified defensins can kill both yeast and mycelial phase fungi, and they directly inactivate certain enveloped viruses, including herpes simplex and influenza.

The potential of defensins as therapeutic agents has not been explored systematically, but it is probably limited by their complex structure (which makes their synthesis difficult) and by the presence of serum proteins, such as alpha-2 macroglobulin, that bind defensins. *See Influenza.*

Defensinlike peptides. At least three other families of defensinlike peptides also occur in animals: beta defensins, insect defensins, and big defensins. The sequences and structures of these peptides differ sufficiently from classical defensins to suggest that they are not closely related from an evolutionary perspective.

More than 20 different beta defensins have been identified. They are produced by epithelial cells that line the windpipe (trachea) and tongue of cattle, and multiple beta defensins are present in bovine white blood cells. The local production of beta defensins in bovine trachea and tongue increases after injury or exposure to bacterial components. Other beta defensins have been purified from the blood cells of chickens and turkeys, suggesting that this peptide family arose more than 150 million years ago, before the ancestors common to birds and mammals diverged.

Defensinlike molecules are also produced by many larval and adult insects in response to infection or injury. Members of this insect defensin family have been purified from various moths, flies, mosquitoes, dragonflies, and beetles. A potent member of this family, royalisin, is present in the royal jelly of honeybees (*Apis mellifera*). Recently, a relatively large defensinlike molecule (big defensin) was isolated from the white blood cells (hemocytes) of horseshoe crabs, a marine arthropod not much changed in external appearance from its fossilized, 650-million-year-old ancestors.

Horseshoe crab hemocytes also contain tachyplesins, exceedingly potent antibiotic peptides that are composed of 17–18 amino acids. They are stored in the cytoplasmic granules of cells that resemble the neutrophils of higher vertebrates. Protegrins, a family of five small (16–18 amino acids) antimicrobial peptides, occur in the granules of porcine leukocytes. Despite the many structural similarities between tachyplesins and protegrins, it is more likely that these families arose by convergent evolution than by descent from a common ancestral gene. The structural simplicity, potency, and broad antimicrobial spectrum of tachyplesins and protegrins makes them promising prototypes for developing novel antimicrobial agents.

Cecropins. Although insects possess neither lymphocytes nor antibodies, they respond rapidly and effectively to infection or injury by producing an array of peptide antibiotics. Cecropins A and B, the first antibiotic peptides characterized in insects, were purified from body fluid (hemolymph) obtained from larval giant silk moths (*Hyalophora cecropia*). Additional cecropins were found in many other insects, including fruit flies (*Drosophila*), flesh flies (*Sarcophaga*), and beetles.

High concentrations of cecropins appear in insect hemolymph within a few hours after the onset of bacterial infection, and contribute to the ability of immune hemolymph to kill many gram-positive and gram-negative bacteria. Cloning studies have revealed that expression of cecropin genes is regulated by signals that were later adopted by higher vertebrates to regulate the production of various acute phase proteins and antibodies. The purification of a cecropinlike antibiotic peptide from the small intestine of pigs suggests that such peptides may also contribute to host defense in vertebrates. *SEE IMMUNOLOGY.*

Cecropinlike peptides. Magainins are antimicrobial peptides composed of 21–27 amino acids. Magainins (and many other bioactive peptides) are stored in the granules of glandular cells found throughout the skin and intestinal tract of the African clawed frog (*Xenopus laevis*). The granular glands release these magainin-rich granules onto the skin after injury or in response to other stimuli, coating the surface with an antiseptic barrier that helps prevent infection.

Magainins kill a wide range of bacteria and fungi by permeabilizing their membranes after initially binding to acidic membrane phospholipids. A considerable effort is under way to develop analogs of magainins for use as topical antibiotics.

Antibiotic peptides. The roster of antibiotic peptides of animal origin is rapidly growing, and recently scores of additional molecules have been structurally characterized. Many of these peptides have structures that differ considerably from those of the peptides described above. For example, among the peptides purified from mammalian white blood cells have been several very proline-rich peptides (PR 39, prophenin, Bac5, and Bac7), a tryptophan-rich 13-amino-acid peptide (indolicidin), and a cyclic 12-amino-acid peptide (bactenecin dodecamer).

Several of the antimicrobial peptides have an ability to stimulate processes related to wound healing; to bind and detoxify bacterial lipopolysaccharide; and to synergize with other host defense systems (for example, antibody and complement) in protecting the host from infection. Several congeners of these antimicrobial peptides have been produced synthetically, and in some cases have outperformed their native counterparts.

For background information *SEE ANTIBIOTIC; ANTIMICROBIAL AGENTS; BLOOD; CLINICAL MICROBIOLOGY; INFECTION; PEPTIDE* in the McGraw-Hill Encyclopedia of Science & Technology.

Robert I. Lehrer

Bibliography. H. G. Boman, Peptide antibiotics and their role in innate immunity, *Annu. Rev. Immunol.*, 13:61–92, 1995; J. A. Hoffmann, C. A. Janeway, Jr., and S. Natori (eds.), *Phylogenetic Perspectives in Immunity: The Insect Host Defense*, 1994; R. I. Lehrer, A. K. Lichtenstein, and T. Ganz, Defensins: Antimicrobial and cytotoxic peptides of mammalian cells, *Annu. Rev. Immunol.*, 11:105–128, 1993; J. Marsh and J. A. Goode (eds.), *Antimicrobial Peptides,* Ciba Found. Symp. 186, 1994.

Arteriosclerosis

Arteriosclerotic plaques develop gradually over a period of years in susceptible human populations because of dietary, genetic, toxic, hormonal, and other factors. The atherogenic process is triggered by these factors, producing arteriosclerotic plaques of aorta, coronary, carotid, renal, and peripheral arteries.

The atherogenic process. The process begins with damage to the lining of arteries, deposition of cholesterol and other lipids in phagocytic foam cells of artery walls, degeneration of elastic fibers, and increased formation of smooth muscle cells and of extracellular proteoglycan matrix and collagen fibers. The resulting early fibrous and fibrolipid arteriosclerotic plaques form thickened, firm areas of the lining that narrow the lumen.

As arteriosclerosis progresses, these plaques enlarge and become calcified, further narrowing the lumen. Because of destruction of elastin fibers, the artery wall loses elasticity. Because of fibrosis and calcification, it becomes hardened, tough, thickened, and inelastic; hence the term arteriosclerosis (hardening of the artery). Some plaques also develop deposits of cholesterol crystals, other degenerated lipids, and amorphous proteinaceous material, producing atheromas, which are soft, swollen plaques that further narrow the lumen; hence the term atherosclerosis (hardening of arteries by atheromas). Finally, blood clots composed of fibrin, platelets, and erythrocytes adhere to the surface of advanced arteriosclerotic plaques, further narrowing or completely occluding the lumen. These mural blood clots become organized, develop new capillaries by angiogenesis, and become fibrotic, adding to the thickness of complex, advanced arteriosclerotic plaques.

Atherogenic factors. The etiological factors that enhance development of arteriosclerotic plaques include aging, a diet high in cholesterol, family history, exposure to toxins such as smoking, hormonal changes such as menopause or hypothyroidism, hypertension, diabetes, and kidney failure. In recent years, all these factors have been found to enhance atherogenesis by causing or contributing to hyperhomocysteinemia.

Elevation of blood levels of low-density lipoprotein and decreased levels of high-density lipoprotein also enhance atherogenesis. Genetic diseases (such as familial combined hyperlipidemia and dense low-density lipoprotein hypertriglyceridema) leading to highly elevated cholesterol and low-density lipoprotein levels accelerate the atherogenic process. Epidemiological studies have also associated increased risk of arteriosclerosis with elevated blood levels of small low density lipoprotein particles of increased density. A genetically determined fraction of low-density lipoprotein [lipoprotein (a)] is also associated with increased risk of arteriosclerosis.

Modification of the structure of low-density lipoprotein by chemical alteration or by exposure to cultured endothelial cells leads to increased uptake of altered lipoproteins by cultured macrophages to form foam cells. Modified lipoprotein contains oxidized lipids and protein constituents that result from a free-radical oxidation process. One type of oxidized lipid, oxycholesterol, is known to be highly atherogenic in animals, but highly purified cholesterol, protected from oxidation, is not atherogenic in animals.

Homocysteine. Analysis of the incidence of the genetic disease homocystinuria shows that increased blood levels of the amino acid homocysteine damage artery walls in affected children. This vascular damage leads to the formation of arteriosclerotic plaques which affect major organs of the body. Thrombosis of arteries to brain, heart, or kidney leads to early death from heart attack, stroke, or kidney failure. The causative effect of homocysteine in vascular damage, arteriosclerotic plaques, and thrombosis is proven by comparing the metabolic effects of three different enzyme deficiencies in different types of homocystinuria. The common factor in all cases is hyperhomocysteinemia (elevation of blood levels of homocysteine).

Epidemiological studies in humans indicate that hyperhomocysteinemia is a major risk factor for development of arteriosclerosis and consequent conditions, including heart attack, stroke, gangrene, and kidney failure. These studies also show that dietary deficiencies of vitamins B_6, B_{12}, and folic acid are associated with elevated homocysteine levels. Furthermore, increasing degrees of hyperhomocysteinemia are correlated with increased narrowing of arteries by arteriosclerotic plaques.

Effects in cells and tissues. Cultured skin cells from children with homocystinuria are very susceptible to the toxic effects of homocysteine because these cells are deficient in cystathionine synthase, the enzyme required for normal catabolism of homocysteine. Pyridoxine, the precursor of the coenzyme pyridoxal phosphate, reverses this toxicity by increasing the ability of these cells to dispose of excess homocysteine through cystathionine formation.

Exposure of cultured endothelial cells to excess homocysteine and copper ions causes cellular toxicity that is related to intracellular accumulation of hydrogen peroxide. Homocysteine is a highly effective chemical catalyst of oxidation reactions in the presence of ferric or cupric ions. In contrast to the toxic effect of homocysteine on cultured endothelial cells, homocysteine stimulates the growth of cultured smooth muscle cells by increased formation of cyclins that initiate cell division.

The highly reactive cyclic anhydride, homocysteine thiolactone, is synthesized in liver cells from methionine. Pure homocysteine thiolactone injected into normal mouse tissues causes an area of intense necrosis associated with calcium deposition, thrombosis within blood vessels, metaplastic squamous cells, and proliferation of small blood vessels, nerves, fibrous tissue, and glandular cells. When homocysteine thiolactone is applied to normal mouse skin, it causes ulceration, inflammation, dysplastic alteration of epidermal cells, and squamous cell carcinoma.

Reaction with lipoproteins. Homocysteine thiolactone reacts with amino groups of apoB protein of low-density lipoprotein to form aggregates of increased density that precipitate spontaneously. These aggregates are taken up by cultured macrophages both by membrane receptors and by phagocytosis. In macrophages the low-density lipoprotein of the aggregates is degraded to cholesterol, cholesterol esters, and other lipids, forming foam cells. The homocysteine of the aggregates is degraded and released by macrophages into surrounding cells and tissues of the intimal lining of arteries.

Homocysteine in the sulfhydryl form is a catalyst for the oxidation of low-density lipoprotein that occurs in the artery wall. The proteins of low-density lipoproteins are also oxidized to peptides and deposited in developing plaques. Homocysteine also reacts with nitric oxide, forming S-nitroso homocysteine and blocking the relaxing effect of nitric oxide on artery-wall muscle cells.

Vascular damage and arteriosclerotic plaques. Homocysteine in the thiolactone form reacts with thioretinaco of mitochondrial and microsomal membranes, converting it to thioco. Thioretinaco reacts with ozone and oxygen to form a thioretinaco ozonide disulfonium ion complex that binds adenosinetriphosphate (ATP) of mitochondrial membranes. The oxygen bound to thioretinaco ozonide is reduced to water with concomitant release of molecules of adenosinetriphosphate from the mitochondrion.

Homocysteine causes vascular damage by formation of thioco, which interferes with cellular oxidation, which is catalyzed by thioretinaco ozonide. As a result, free-radical oxygen molecules accumulate and react with unsaturated lipids and cholesterol of cellular membranes to form aldehydes and oxy-cholesterols. Free-radical oxygen also oxidizes the sulfur atom of thioretinamide to sulfite. Sulfite oxidase oxidizes sulfite to sulfate which is converted to phosphoadenosine phosphosulfate, forming sulfate esters of proteoglycosaminoglycan extracellular matrix. Proteins of lipoproteins such as ApoB are degraded by a similar oxidative process.

Conversion of thioretinaco ozonide to thioco by homocysteine thiolactone is a critical factor in increasing growth of smooth muscle cells of artery wall by activation of cyclins, insulinlike growth factor, and platelet-derived growth factor. Increased formation of smooth muscle cells initiates synthesis of collagen fibers, forming the fibrous component of arteriosclerotic plaques. The oxidative disturbances induced by homocysteine activate elastase of cellular lysosomes, leading to the degradation of elastica interna and elastic lamella of arteries and aorta that are characteristic of arteriosclerotic plaques. Inhibition of oxidative metabolism also leads to an influx of calcium into affected cells, causing calcium deposits within developing plaques.

In the developing plaque, homocysteine increases the binding of lipoprotein (a) to fibrin, enhances platelet aggregation at the site of damage, and promotes thrombosis. Advanced atheromas accumulate deposits of cholesterol crystals that result from degradation of lipoproteins within foamy macrophages and from binding of lipoproteins to sulfated proteoglycan matrix deposited in plaques by an increased production of smooth muscle cells.

For background information SEE ARTERIOSCLEROSIS; BLOOD VESSELS; CIRCULATORY DISORDERS; LIPID in the McGraw-Hill Encyclopedia of Science & Technology.

Kilmer S. McCully

Bibliography. K. S. McCully, Chemical pathology of homocysteine; I: Atherogenesis, *Ann. Clin. Lab. Sci.*, 23:477–493, 1993; K. S. McCully, Chemical pathology of homocysteine; II: Carcinogenesis and homocysteine thiolactone metabolism, *Ann. Clin. Lab. Sci.*, 24:27–59, 1994; K. S. McCully, Chemical pathology of homocysteine, III: Cellular functioning and aging, *Ann. Clin. Lab. Sci.*, 24:134–152, 1994; K. S. McCully and M. P. Vezeridis, Histopathological effects of homocysteine thiolactone on epithelial and stromal tissues, *Exp. Molec. Pathol.*, 51:159–170, 1989; M. Naruszewicz et al., Thiolation of low-density lipoprotein by homocysteine thiolactone causes increased aggregation and interaction with cultured macrophages, *Nutr. Metab. Cardiovasc. Dis.*, 4:70–77, 1994.

Atomic physics

The ability to fully ionize any atomic species in the periodic table, up to uranium (U^{92+}), in an electron-beam ion trap opens up new research opportunities for studies requiring very low energy, highly charged ions. Very heavy, highly charged, but slow ions can have up to several hundred kiloelectronvolts of total potential energy and essentially no kinetic energy. The interaction of such ions with surfaces, for example, causes nanometer-size defects via so-called Coulomb explosions, with evident possibilities for applications leading toward advances in nanotechnology. Precise atomic structure studies and studies of the reaction dynamics of very highly charged ions interacting with matter are of interest and can be performed. Calculations can be tested against precise measurements of transition energies and transition rates along isoelectronic sequences. Effects of relativity, quantum electrodynamics, and electron-nuclear overlap are most important in ions with the highest atomic number, Z. For example, the possibility of producing and confining hydrogenlike high-Z ions in an ion trap allows their hyperfine splitting to be measured by means of laser spectroscopy to high precision. This hyperfine splitting is proportional to the third power of the atomic number (Z^3). In addition, relativistic corrections increase the splitting for higher-Z ions, nearly doubling it for $Z = 82$ as compared to $Z = 1$. For this reason the transition frequency is high enough for a number of high-Z elements to be in the range of available laser sources. Furthermore, the physics of highly charged ions greatly affects conditions in hot plasmas, such as plasmas in controlled fusion devices, plasmas for x-ray lasers, and astrophysical x-ray sources. Thus, the new research capabilities which are currently available from an electron-beam ion trap or source lead to an expansion of the atomic database and, hopefully, the discovery of new phenomena relevant for different areas of physics and interdisciplinary research.

Electron-beam ion trap. The most successful production of slow, very highly charged ions has been achieved in the electron-beam ion trap and the electron-beam ion source. Here successive collisions between monoenergetic electrons and stationary atoms (or initially low-charge-state ions) at energies greater than the binding energies of the electrons to be removed are responsible for achieving the high degrees of ionization. This principle was first applied in the development of the electron-beam ion source and was implemented most successfully in an electron-beam ion trap. Here the very highest ion charge states possible, up to U^{92+}, have been produced. The impact ionization or excitation by successive electrons is efficiently achieved by causing the ions to be trapped in a compressed electron beam by the electron beam's space charge. The efficiency of the ionization depends predominantly on the electron-beam density and energy and the duration of ion confinement (up to few seconds) in the electron beam. The electron-beam compression is achieved by means of a high magnetic field. The ions are radially confined by the space-charge potential of the electron beam, and axially by a potential barrier of about 100 V applied to the top and bottom drift tubes, which are situated above and below the trapping region (**Fig. 1**).

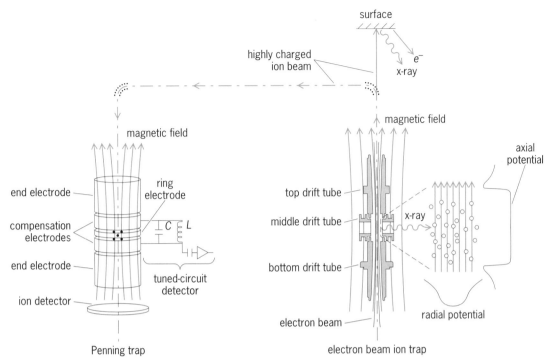

Fig. 1. Electron-beam ion trap operation. Ion extraction and the performance of internal electron-beam ion trap experiments (x-ray) and external experiments (surface) are shown, as well as the injection of ions, extracted from the electron-beam ion trap, into a secondary or Penning trap.

The trapping region is defined by a central drift tube, approximately 2 cm (0.8 in.) long and 1 cm (0.4 in.) in diameter. This length is the essential difference between an electron-beam ion trap and an electron-beam ion source, where the length of the center drift tube is about 1 m (3 ft).

The production of U^{92+} ions was made possible by increasing the electron-beam energy to 200 keV in a new trap called the super electron-beam ion trap. In addition, a so-called evaporative cooling technique was developed and found to be essential for the production of the very highly charged ions. The principle of the cooling scheme is to inject a small number of low-Z atoms (such as neon) into the region where the highly charged ions are trapped. The lower-charged lighter ions are pushed out of the trapping region via Coulomb collisions with the high-Z higher-charged ions. The electron-beam-heated high-Z ions lose their energy via collisions with escaping low-Z ions. The high-Z ions are therefore cooled and confined to the center of the electron beam. Typical operating parameters for the electron-beam ion trap and the super electron-beam ion trap are a 3-tesla magnetic field in which an electron beam carrying a current of about 150 mA is compressed until its current density reaches about 4000 A/cm². This produces a beam diameter of about 70 micrometers at electron-beam energies of 0.5–40 keV in the electron-beam ion trap or 5–200 keV in the super electron-beam ion trap. Neutral gas atoms can be injected from a side port through the magnetic fields into the center drift tube. Single or doubly charged ions, which are produced efficiently in an arc source, can be

injected axially into the trapping region. The ion extraction, which turns the electron-beam ion trap into an ion source, is simply done by lowering the potential of the axial trapping barrier on the top drift tube for a millisecond to eject the ions. The drift-tube potential multiplied by the ion charge state defines the energy of the ejected ions.

X-ray emission from trapped ions. The electron-beam ion trap is an ideal source for x-ray spectroscopy because of the fact that the x-ray emitting ions are essentially at rest and they form a line source which is easily adapted to different spectrometer geometries. An x-ray spectrum observed at the super electron-beam ion trap following the excitation of uranium with a 198-keV electron beam is shown in **Fig. 2**. The ions are trapped in the electron beam in an equilibrium ionization balance which is determined by the competing rates of ionization, radiative recombination, and charge exchange recombination with neutral atoms. The observed lines in the spectrum in Fig. 2 are due to radiative recombination and x-ray emission following the capture of an electron into a vacant shell of the uranium ion. Since the ions are stripped, capture can occur into shells with $n = 1, 2, 3, \ldots$, giving rise to x-ray emission at energies equal to the sum of the beam energy of 198 keV and the binding energy. The low-intensity peaks at 330 keV result from recombination into the open K shell of a small number of hydrogenlike and bare uranium ions present in the trap. This spectrum verifies the production of U^{92+} ions in the super electron-beam ion trap.

Ion-surface interactions. A bare U^{92+} ion carries a large amount of potential energy due to its degree

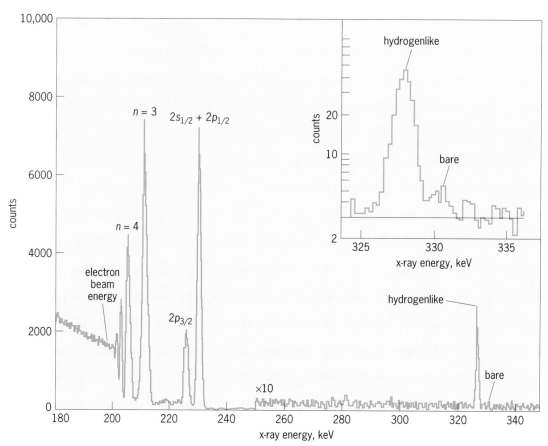

Fig. 2. X-ray emission spectrum following the ionization and excitation of uranium by a 198-keV electron beam. The inset shows details of peaks that result from recombination into the open K shell of hydrogenlike and bare uranium ions.

of ionization. The interaction of such slow, highly charged ions with surfaces causes a large number of electrons to be removed from the surface. These electrons (several hundred per ion) are partially captured in very highly excited states of the incident ion, leaving the core (inner shells) empty. The formation of these exotic atoms, which are called hollow atoms, and the response of the surface are currently being investigated.

It has been shown that the impact of a single highly charged ion on an insulator surface produces nanometer-sized defects (**Fig. 3**) and that the size of these defects can be controlled by changing the incident ion's charge state, and hence the potential energy it carries. Also, very highly charged ions are able to sputter off large numbers of ions and clusters from the surface (more than 1000 sputter ions per incident ion). These findings indicate the possibility that highly charged ions can be used to advance the knowledge of nanotechnology.

The combination of a high ionization state and good spatial resolution and energy definition of the ions produced by the electron-beam ion trap offers unique possibilities for material fabrication. These possibilities are based primarily on the ability of such ions to deposit energy at extremely high densities in a nonthermal process within nanometer-sized volumes. The continuing trend toward smaller device structures in the microelectronic industry,

beyond the current size of 0.25 μm in advanced devices, has created a growing need for enhanced lithographic tools and analytic techniques. The ability to manufacture these new devices is as limited by analytic capabilities needed to characterize the resulting structures as by the fabrication processes themselves. Focused low-emittance beams of highly charged ions, such as those produced with an electron-beam ion trap, could become critical tools in advanced device production. In particular, such beams have the potential for the development of

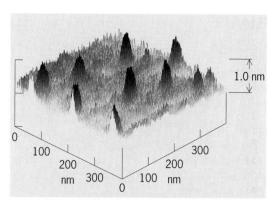

Fig. 3. Formation of defects by single highly charged ion (U^{70+}) impact on an insulator (mica), measured by performing atomic force microscope scans.

advanced methods of secondary ion mass spectrometry (SIMS).

Trapping of highly charged ions. The possibility of creating very highly charged ions in a laboratory allows such ions to be investigated under extreme and controlled conditions. Such investigation is most efficiently done by inserting them, under ultrahigh vacuum conditions (10^{-13} pascal or 10^{-15} torr), into an electromagnetic trap called the Penning trap (or RETRAP), situated at the electron-beam ion trap facility (Fig. 1). A few hundred thorium (Th^{80+}) ions can be confined, for example, up to several hours. Thus, precision measurements can be performed by exciting certain atomic transitions by using a laser (for example, the hyperfine transition in hydrogenlike ions), and their masses can be measured to very high accuracies.

This newly initiated research capability also allows one or more component plasmas to be formed inside the Penning trap. These plasmas, which consist of highly charged ions at low kinetic energies, are confined by the electromagnetic forces in the trap. Reduction of the ion's kinetic energies should lead to cooling and crystallization of the ion cloud. Matter in such a form is present in white dwarf stars. The ability to create it in a terrestrial laboratory and to study its thermodynamics will improve the understanding of stellar matter and perhaps help to establish the age of the universe.

For background information *SEE ATOMIC STRUCTURE AND SPECTRA; ION-SOLID INTERACTIONS; ION SOURCES; LASER SPECTROSCOPY; PARTICLE TRAP; PLASMA PHYSICS; SPUTTERING; X-RAY SPECTROMETRY* in the McGraw-Hill Encyclopedia of Science & Technology.

Dieter H. Schneider

Bibliography. J. P. Briand, Uranium ions stripped bare, *Phys. World*, 7(10):25–26, October 1994; I. G. Brown (ed.), *The Physics and Technology of Ion Sources*, 1989; R. Marrs, *Experimental Methods in the Physical Sciences*, vol. 29: *Electron Beam Ion Traps*, 1995; R. Marrs, P. Beiersdorfer, and D. Schneider, The electron beam ion trap, *Phys. Today*, 47(10):31–34, October 1994.

Automobile

In the rapidly evolving field of automotive electronics, new technologies are continually emerging. A trend toward up-integration that adds more functionality to circuit modules to reduce costs—such as combining engine management, cruise control, and transmission control on one power-train controller module—is evident.

Fuzzy logic. The IF (this is true)–THEN (do this) format of fuzzy logic rules is finding its place in a variety of automotive applications. Three typical situations in which fuzzy logic is appropriate for use in automobiles are (1) when environmental conditions are complicated and it is difficult to express them mathematically; (2) when the characteristics

of the system change with temperature or other parameters, despite the fact that the system to be controlled remains constant; and (3) when there are multiple control objectives.

Gear selection control in automatic transmissions is a widespread application of fuzzy logic. One system, for example, adds fuzzy logic to a conventional automatic transmission system by combining the driver's intentions, vehicle status, and road conditions into a final judgment to select from several modes: traversing even terrain, climbing a curvy hill, climbing a straight hill, and going downhill. For each mode, the system automatically determines throttle opening and gear position according to vehicle speed. Another system uses fuzzy logic instead of conventional techniques (see **illus.**). IF–THEN rules determine gear selection exclusively based on vehicle speed, throttle information, inclination, and other data. Fuzzy inference is used to determine when the person driving the car intends to decelerate. Other automotive fuzzy logic applications include systems for heating, ventilating, and air conditioning.

One-wheel braking. Several manufacturers are developing systems that use selective one-wheel braking to help a driver regain control of a car. These systems combine a network of sensors to monitor key dynamic variables such as wheel speeds, lateral acceleration, speed of the car's rotation on a vertical axis, and steering-wheel angle. When one of these variables goes out of a preprogrammed range, a computer automatically applies the brakes at one or more wheels. If, for example, the driver moves the steering wheel too aggressively for prevailing road conditions, resulting in a loss of rear traction, the car anticipates the spinout and automatically applies the brakes to one or more wheels without the driver stepping on the brake pedal (as must be done with antilock braking systems).

Navigation systems. Electronic systems to help drivers navigate have been available since the late 1980s. Advanced systems that rely on the Global Positioning System (GPS) as well as dead reckoning with map matching are now available. In one such system, a computer uses an electronic mapping database to figure a route. Guidance is supplied by positioning information from GPS satellites and from dead-reckoning calculations from the mapping database.

The driver enters into the system the details of the desired destination (an address, a road intersection, a place of public interest, or a previous destination stored in the system's computer memory). The system then calculates the most efficient route for the trip (unless the driver requests a scenic route) and displays a map on a screen with the route highlighted in color. Once the trip begins, the distance to and direction of each turning maneuver are displayed. In addition, a voice prompt advises when a turn is approaching. If a turn is missed or cannot be taken because of road construction, the system reroutes the trip at the touch of a single key.

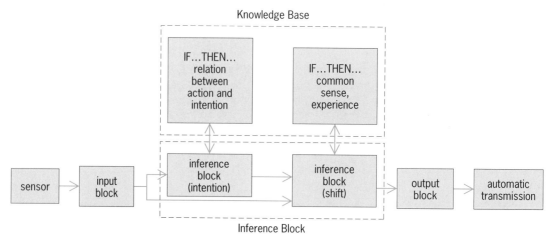

Block diagram of an automatic transmission control system that uses fuzzy inference to determine when the driver intends to decelerate.

Automated collision alert. Systems are now in use to assist police in locating stolen cars. But an even more important application for vehicle location is in the case of accidents where fast medical attention may be needed. Systems under development would sense a crash, locate its site, and automatically notify a medical service. Such systems need to be able to distinguish between an injury-inducing crash and normal vehicle operation. Accelerometers or electromechanical switches can detect large changes in a car's speed, and strain gages or accelerometers placed on various locations of a vehicle can act as chassis deformation detectors. Among the technologies that could be used to send news of the crash are those that use satellites in geosynchronous or low Earth orbit or that employ terrestrial communication systems. One notification system uses a three-axis accelerometer integrated with an eight-bit microcontroller to sense the crash, obtain data such as change in velocity, direction of impact, and indication of rollover, and then assemble the notification message to be transmitted by a little LEO (low-Earth-orbit) satellite.

Another such system has two buttons in the overhead console of the automobile, one with a two-truck icon and the other with an ambulance. When the driver pushes either button, the system uses the car's cellular phone to dial an emergency response center. An antenna in the car's trunk receives satellite signals that pinpoint its position within 100 ft (30 m). The center's equipment displays the type of alarm, the last recorded speed and direction of the vehicle, its serial number, and a call-back telephone number. An operator at the center can talk to the driver over the car's voice-activated phone and also can notify family members or friends.

Flat-tire warnings. Low tire pressure can increase tire wear and contribute to car accidents, but systems are available to warn drivers of the low-pressure situation. One such system has four small wheel modules and a radio receiver mounted under the car's dashboard. Each battery-powered module contains a piezoresistive pressure sensor and a 200-nanowatt pulse-code-modulated transmitter operating at 355 MHz. Each wheel's tire pressure is displayed and, if a tire is underinflated, its wheel identification box on the screen blinks. A similar system is available on automobiles equipped with extended mobility tires, which can run up to 200 mi (320 km) at 55 mi/h (88 km/h) at zero inflation pressure. Because these tires do not appear to be flat when they are without air, a low-tire-pressure warning system is needed.

Engine immobilizers. Automobile makers are looking increasingly to more effective theft-deterrent systems, and engine immobilizers are in high favor. One such system uses a low-frequency transponder as well as battery power to communicate with an on-board receiver when a motorist approaches the vehicle. In addition to activating the engine controller, the system can unlock the doors and initiate other electronic functions such as adjusting the driver's seat, the outside mirrors, and car cabin temperature to preset positions. A so-called smart-card version of the system requires no ignition key. A multistage switch is activated as soon as the motorist is seated in the car. Since the driver must be carrying the smart card, the engine cannot be turned on when the car is unattended.

Side-impact air bags. The effectiveness of air bags in protecting passengers in frontal collisions has been well demonstrated; thus the development of side-impact air bags was a logical next step. In side-impact collisions, the closeness of the occupants to the vehicle's sides can make a marked difference in the harm done. Consequently, impact sensing and air-bag deployment must be faster than for frontal collisions. A side-air-bag sensor has been developed that triggers in under 5 milliseconds. Monitoring the minimal gap between the vehicle's interior and its occupants requires extra, outlying sensors installed near the zone of initial impact. They could be mounted in the vehicle's periphery: on the B pillar or on cross members under the seat. The air bags themselves can be stowed in seats, back rests, headliners, roof rails, doors, or B pillars. Side-mounted sensors detect the impacts of large objects. A central control unit, to which the sensors are connected, controls the final decision to inflate the air bags.

Another system relies on a patented absolute-pressure sensor that measures change in external air pressure inside the door panel. A strong side impact is measured by the pressure sensor, which processes the data, and, if necessary, deploys the side air bag in 5–7 ms.

To avoid deploying an air bag for an empty passenger seat, a system has been developed that uses a capacitive measuring sensor to detect when that seat is occupied. The side air bag will not deploy if the seat is detected as unoccupied, even if it contains heavy objects or a rear-facing child seat.

Air bags under development are designed to inflate in various ways, depending on the circumstances. For example, sensors would be used to determine whether a driver or passenger is slumped or upright, centered or off-center, short or tall, and heavy or light. *See Mechatronics; Microengineering.*

Electromagnetic compatibility. Electromagnetic compatibility (EMC) has become a high priority in the design of electronic automotive modules as microcircuit technology has evolved with greater complexity, faster clock speeds, and smaller package size. To minimize the effect of electromagnetic compatibility due to faster rise and fall times, the design engineer must analyze the integrated-circuit package and associated decoupling capacitor. Choosing the right decoupling capacitor limits the amount of electromagnetic emissions radiated from the integrated circuits on the printed circuit board.

For background information *See Automobile; Electromagnetic compatibility; Fuzzy sets and systems; Satellite navigation systems* in the McGraw-Hill Encyclopedia of Science & Technology.

Ronald K. Jurgen

Bibliography. R. K. Jurgen (ed.), *Automotive Electronics Handbook,* 1995; R. K. Jurgen, The electronic motorist, *IEEE Spec.,* 32(3):37–41 and 44–48, March 1995; Society of Automotive Engineers, *Proceedings of the 1994 International Congress on Transportation Electronics,* 1994.

Aves

The question of bird origins has generally focused on the first known bird, the urvögel (*Archaeopteryx lithographica*), discovered in 1860–1861 in the fine-grained lithographic, Solnhofen limestone of Bavaria, dated at approximately 150 million years (m.y.) ago [Late Jurassic Period (Tithonian)]. Since the initial discovery of a single feather and a skeletal specimen with feathers, skeletons of six additional specimens have been recovered; the last two were discovered during the 1990s.

One specimen, discovered in 1876, is preserved with outstretched wings, indicating its intermediate status between reptiles and birds. The feathers and wings are those of a modern bird, as are most of the skull structures, the arrangement of the scapula and coracoid, and the rotation of the first toe to the rear of the foot as a hallux (an adaptation to aid in perching in trees, the rear toe opposing the front three). However, its reptilian link is vividly illustrated by the toothed jaws, the three clawed fingers of the hands, and the long reptilian tail from which a pair of tail feathers emanates from each vertebra.

Because *Archaeopteryx* was discovered just a year after the publication of Darwin's *On the Origin of Species* (1859), it became the focal point for all subsequent work on bird origins. Several important questions require resolution: (1) Which group of reptiles is ancestral to birds? (2) Did avian flight originate from the ground up (the cursorial theory), or from the trees down (the arboreal theory)? (3) What is the relationship of modern birds to the ancient, Mesozoic birds?

Bird ancestors. Many groups of reptiles have been envisioned as bird ancestors, ranging from lizards to pterosaurs. However, the most widely accepted hypotheses have focused on dinosaurs or on their antecedents, the basal archosaurs or thecodonts. The dinosaur theory was initiated by Thomas Huxley in 1868, and gained numerous proponents until the publication by Heilmann in 1926 of *The Origin of Birds,* which made a convincing argument that the dinosaurs were too specialized to have given rise to birds, and that birds arose from small, tree-dwelling archosaurs, the thecodonts. Since then, several other theories have related birds to the herbivorous dinosaurs, the ornithischians, and as a variation on Heilmann's theme to primitive crocodilomorphs, because of similarities in tooth structure.

The dinosaurian origin of birds was revived in 1973 by John Ostrom of Yale University. During the late 1960s Ostrom discovered the cursorial, predatory theropod dinosaur *Deinonychus* from the early Cretaceous of Montana, and in this dinosaur, identified many birdlike features. Although Ostrom's theory, which is in effect a refined and updated version of Huxley's dinosaurian origin of birds, has been widely accepted, the field is still open to debate. One question is whether the theropod dinosaurs acquired their birdlike features because they were closely related to birds, or by convergent evolution, that is, the acquisition of similar features in unrelated groups of organisms. In the latter case, dinosaurs would have come to look like ground-dwelling birds because they both evolved adaptations for a bipedal, cursorial way of life.

Evolution of avian flight. Avian flight evolved either from the ground up or from the trees down. Because dinosaurs are large, earth-bound reptiles, a dinosaurian origin of birds is inextricably linked to a cursorial origin of avian flight. Advocates of a theropod dinosaur ancestor imagine feathers evolving as a coating for warm-blooded (endothermic) dinosaurs; later, wing feathers elongated to capture insects. However, it is now thought that

Archaeopteryx and most dinosaurs were not warm blooded but more reptilelike in physiology. Also, it is almost impossible to model the origin of flight from a large, heavy, obligately bipedal, deep-bodied reptile with already foreshortened forelimbs (50% the length of the hindlimbs). Such a runner, fighting gravity all the way, would find it biophysically impossible to become transformed into a flier.

However, almost all groups of vertebrates have evolved flying forms, and in all cases from the trees. One can easily imagine how a small, quadrupedal, tree-dwelling reptile, jumping from branch to branch, would gain advantage from elongating featherlike scales, which would create drag and break the fall. In contrast, the same feathers that create drag in a falling animal would also create drag in a runner, but in the case of a runner drag would slow the animal down and be disadvantageous. Whatever the avian ancestor, birds surely evolved from the trees down, taking advantage of the energy provided by gravity. Thus, considerable doubt is cast on a dinosaurian origin of birds. There are a number of small, Late Triassic, arboreal thecodonts, and one, *Longisquama*, developed elongated featherlike scales.

Chronology. Another serious problem for a theropod dinosaur origin of birds is that birdlike dinosaurs occur primarily in the Cretaceous Period, ranging from some 40 to 75 m.y. after the occurrence of the first bird, *Archaeopteryx*. Indeed, the vast majority of dinosaurs with birdlike features occur in the latest Cretaceous. Advocates of a theropod dinosaur origin of birds state that the fossils have yet to be discovered; however, there are scores of well-preserved dinosaurs from these strata.

A recently described birdlike dinosaur that has gained much publicity and created considerable controversy is *Mononykus*, a turkey-sized creature from the latest Cretaceous deposits of the Gobi Desert of Mongolia. Most experts believe that it was a small, fleet-footed theropod dinosaur that developed a number of birdlike features through convergent evolution. Also, the flying reptiles, or pterosaurs (not related to birds), developed numerous birdlike features, including loss and deletion of bone; loss of teeth; hollow, pneumatic bones; a large orbit; similar cerebellum; fused thoracic vertebrae (notarium); a splintlike fibula; and birdlike tibia. Therefore, extreme caution must be exercised in making such comparisons.

Protoavis. A birdlike creature named *Protoavis* was described from the Late Triassic of Texas, and said to be the earliest bird, predating *Archaeopteryx* by millions of years. However, the material is somewhat fragmentary, its interpretation has been questioned, and there are no remains of feathers, the only truly diagnostic feature of early birds. It is possible that these remains represent an early thecodont that developed birdlike features. However, the answer to the riddle of *Protoavis* must await further discovery.

Cretaceous birds. Recently discovered Lower Cretaceous birds, especially from Spain and China, have provided new insight regarding knowledge of early bird evolution. Such forms as the Lower Cretaceous Spanish *Iberomesornis* and the Chinese *Sinornis* and *Cathayornis* show, along with other Cretaceous birds worldwide, that the dominant land birds of the Cretaceous were termed enantiornithines (opposite birds) because the three foot bones fused from proximal to distal, just the opposite of the direction of fusion in modern birds. Cretaceous bird fossils thought to represent modern orders have been restudied, and have been shown

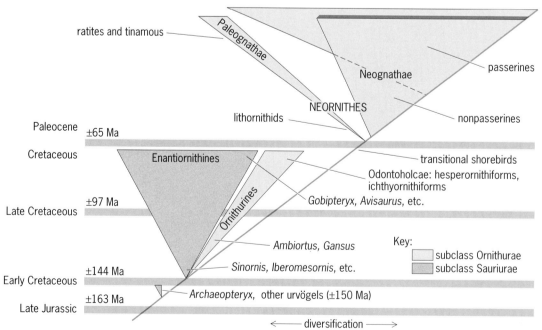

Evolutionary relationships among the various known avian lineages. (*After A. Feduccia, Explosive evolution in Tertiary birds and mammals, Science, 267:637, 1995*)

to belong to the opposite birds. These birds had a fully advanced flight apparatus but exhibited a primitive, *Archaeopteryx*-like, toothed skull and a primitive pelvic region. Opposite birds ranged from small sparrow-size birds to those the size of a vulture (see **illus.**).

Evolution of Tertiary birds. The traditional view of bird evolution has been one of the modern orders originating in the Cretaceous and gradually evolving to their present form over long periods of geologic time. However, with the discovery of the opposite birds and restudy of other fossils, it is clear that most modern birds evolved since the close of the Cretaceous, and therefore modern birds are truly of Tertiary origin. It is now understood that birds endured massive Late Cretaceous extinctions, including all of the opposite birds, as well as the toothed divers, the hesperornithiforms, and the volant, toothed, ternlike ichthyornithiforms.

Birds underwent a dramatic bottleneck at the Cretaceous-Tertiary boundary, represented by forms known as transitional shorebirds, which were probably the wellsprings of the modern avian radiation. Birds then closely paralleled mammals in a dramatic explosive adaptive radiation that probably produced most of the modern orders within a period of approximately 10 m.y. Hence almost all modern orders occur by the Eocene in both North America and Europe. Modern orders appeared by the Paleocene and Eocene, modern families by the late Eocene and early Oligocene, and modern genera by the Miocene. The second phase of explosive evolution produced the myriad passerine (songbirds) by the Late Tertiary. The first passerines are known from deposits of Oligocene time from the Northern Hemisphere, and these birds now constitute approximately 60% of living avian species, or 5700 of the 9700 living species of birds.

For background information *SEE ANIMAL EVOLUTION; AVES; CRETACEOUS; DINOSAUR* in the McGraw-Hill Encyclopedia of Science & Technology.

Alan Feduccia

Bibliography. A. Feduccia, Evidence from claw geometry indicating arboreal habits of Archaeopteryx, *Science,* 259:790–793, 1993; A. Feduccia, Explosive evolution in Tertiary birds and mammals, *Science,* 267:637–638, 1995; P. W. Houde, Paleognathous birds from the early Tertiary of the Northern Hemisphere, *Pub. Nuttall Ornith. Club,* vol. 22, 1988; D. S. Peters (ed.), Acta paleornithologica, *Courier Forsch. Senckenberg,* vol. 181, 1995.

Bioconjugation

Nearly all clinically approved drugs consist of low-molecular-weight agents that are derived from natural sources or are synthetically produced in the laboratory. Although such drugs have proven to be highly successful for the treatment of many common disorders such as bacterial infections, high blood pressure, and gastrointestinal ulcers, there is a great need for new therapeutic agents that are active against life-threatening diseases, such as cancer, heart disease, and major immunological and genetic disorders. With the advent of new interdisciplinary fields of research involving molecular biology, chemistry, enzymology, and other fields, it has been possible to create a new generation of clinically promising drugs that contain proteins, nucleic acid, or biocompatible polymers. Many of these agents are formed through the process of bioconjugation, in which two entities are joined together by using chemical or biological techniques. Bioconjugates include covalent polymer-protein and polymer-drug adducts, immunoconjugates consisting of cytotoxic agents attached to monoclonal antibodies, and noncovalent complexes formed between drugs and microcapsules.

Conjugation methodologies. Proteins are composed of a series of amino acids linked together through amide bonds. In addition, eukaryotic organisms often contain proteins that have carbohydrates appended to the amino-acid side-chain residues. Many techniques have been developed to utilize amino-acid side chains and protein carbohydrates for bioconjugate formation. Some of these chemical reactions are shown in the **illustration**. Typically, a bioconjugate is formed by reacting a protein with a bifunctional cross-linking reagent that is specific for functional groups present on each component of the conjugate. For example, lysines on the protein to be modified undergo reaction when they are combined with a second molecule that contains activated carboxyl groups. Conversely, it is possible to activate the carboxyl groups of glutamic acid and aspartic acid on a protein and then form a conjugate by adding an amine-containing molecule. Cysteines are particularly useful in conjugate formation, since they react rapidly with electrophilic reagents such as maleimides that can be introduced onto the molecule to be conjugated. A number of methods are available to modify protein tyrosine residues with radioactive iodide. Finally, the carbohydrates of proteins can be oxidized with reagents such as sodium periodate. The aldehyde groups generated undergo condensation reactions with amines or hydrazides.

The best method for conjugation inevitably depends on the particular properties of each component of the conjugate. It is often necessary to try several conjugation strategies in order to develop a modified protein with optimal biological properties.

Bioconjugates of monoclonal antibodies. Some properties of bioconjugates can be better understood by reviewing examples involving monoclonal antibody immunoconjugates. Monoclonal antibodies are produced in the laboratory by using a variety of techniques, and they differ from polyclonal antibodies in being homogeneous and specific. The most common medical use of monoclonal antibodies is for diagnostics. For example, in home pregnancy tests metal-labeled antibodies are used to detect indicator proteins in the urine.

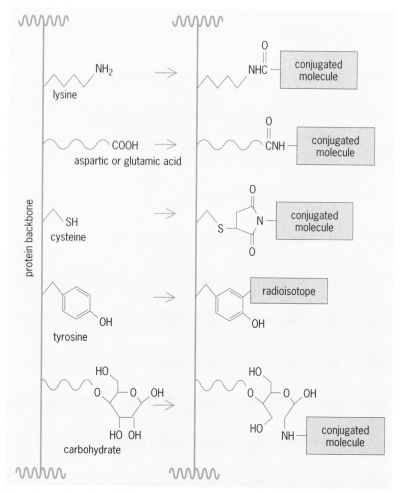

Bioconjugate formation utilizing protein amino-acid residues.

The interest in making bioconjugates of monoclonal antibodies stems from the finding that certain cell types, such as tumor cells, preferentially express antigens to which these antibodies can bind. This specificity has prompted research utilizing monoclonal antibodies for selective drug delivery.

Generally, three elements make up a monoclonal antibody conjugate: the antibody, the conjugated molecule, and the linker that joins them. Many common anticancer drugs, as well as a variety of potent toxin molecules, radionuclides, enzymes, and polymers, have been attached to monoclonal antibodies. The linkers are generally designed to form a stable bond between the monoclonal antibody and the targeted agent and, if necessary to release the conjugated molecule at the tumor site.

Anticancer drug delivery. Since the 1970s, scientists have been exploring the use of monoclonal antibodies for the delivery of the anticancer drug doxorubicin to tumor cells. This drug is of interest for targeted therapy because of its broad-spectrum antitumor activity, but unfortunately it exhibits significant cross-reactivity to normal cells.

In a new conjugation strategy for doxorubicin, the drug is linked to a monoclonal antibody known as BR96. This antibody recognizes antigens that are preferentially expressed on tumor cells. Once the BR96 antibody binds to antigens on tumor cells, it is internalized and then shuttled into acidic intracellular compartments. To take maximum advantage of this process, doxorubicin is attached by using a linker that releases active drug primarily under acidic conditions, such as are present inside the tumor cells that bound the bioconjugate. The BR96-doxorubicin conjugate has been tested against human lung, breast, and colorectal carcinoma tumors that were implanted in laboratory strains of mice and rats in which these human tumors could grow. The BR96-doxorubicin cured established tumors that had been resistant to doxorubicin treatment, without detectable toxic side effects. The next step is to use BR96-doxorubicin in clinical trials in humans.

The specificity displayed by monoclonal antibodies for tumor cells has enabled scientists to target molecules that are too toxic to be given systemically. For example, bioconjugates have been made by using the plant toxin ricin, which is one of the most toxic protein molecules known. A conjugate of ricin A-chain, a subunit of ricin, was formed by linking the toxin to an antibody that bound to human lymphoma cells. Individuals with advanced, bulky lymphoma tumors were treated with the ricin A-chain conjugate in a phase I clinical trial to determine whether the conjugate was toxic. Complete or partial tumor regressions were obtained in 27% of the patients receiving the conjugate. These results are particularly significant because the patients had large tumor burdens, and they had previously undergone multiple rounds of chemotherapy and radiation therapy without success. Clinical trials with other toxin molecules are under way.

Recently, molecular biologists have produced a new generation of bioconjugates in which the monoclonal antibody and toxin components are fused end to end by microorganisms. Such fusion proteins may prove to be superior to chemical conjugates because of their higher degrees of homogeneity.

Radioactive isotopes are often appended to antibodies for the detection and treatment of cancer and heart disease. For example, an antibody against B-cell lymphoma was radiolabeled with iodine-131 in a manner similar to that depicted in the illustration. Lymphoma patients were infused with the bioconjugate and then given autologous bone marrow reinfusions. Complete or partial responses were obtained in 18 of the 19 patients treated. It is worth noting that these patients had received an average of three previous therapeutic regimens prior to entering this clinical study.

One drawback in using monoclonal antibody conjugates for solid-tumor therapy is that the amount that actually gets into the tumor mass can be quite small and clearance from the blood is generally slow. As a result, some new approaches for monoclonal antibody-based therapies have been devised. Several laboratories are exploring the use of monoclonal antibody–enzyme conjugates for site-specific drug generation. In this approach, an enzyme is covalently attached to a monoclonal antibody that recognizes antigens on tumor cell surfaces. The conjugate is administered, and after allowing sufficient time for the conjugate to localize in the tumor mass and clear from the circulation, a nontoxic anticancer drug analog known as a prodrug is given. Upon contact with the targeted enzyme, the prodrug is converted into an active anticancer agent. Significant therapeutic effects have been obtained in preclinical tumor models with several different monoclonal antibody–enzyme–prodrug combinations. The conjugates used in this targeting approach have been made by using either chemical cross-linking reagents or fusion protein technology. Other multistep strategies for drug delivery involve localizing antibodies within tumor masses that can bind subsequently administered anticancer drugs and radionuclides, an approach that is undergoing clinical trials.

Bioconjugates of therapeutic proteins. Many highly active proteins have been found to have limited activity in the clinical setting because they are eliminated from the body very quickly or are rejected by the immune system. A number of strategies are being explored to improve the therapeutic uses of proteins through chemical modification. Polyethylene glycol is a biocompatible polymer that has been used extensively for protein modification, because it lowers immune responses against foreign proteins and increases protein biological half-life.

The enzyme adenosine deaminase has been given to individuals with severe combined immunodeficiency disease, even though blood clearance can be quite rapid. A polyethylene glycol–adenosine deaminase bioconjugate has been shown to clear from the body at a much slower rate. This modified protein has been approved by the U.S. Food and Drug Administration based on its efficacy in reversing the toxic effects of adenosine and deoxyadenosine accumulation in adenosine deaminase–deficient patients. Similar modification studies have been performed with the enzyme asparaginase. This enzyme is used to treat children with acute lymphoblastic leukemia, but is eventually rejected by the immune system since the protein is foreign. Polyethylene glycol–derived asparaginase has a prolonged plasma half-life and is better tolerated immunologically than the unmodified protein. Clinical studies have shown that the majority of children with acute lymphoblastic leukemia undergo partial or complete tumor remissions after treatment with polyethylene glycol–asparaginase. Another enzyme that has been subjected to polyethylene glycol modification is superoxide dismutase, which consumes highly reactive oxygen species at sites of injury and inflammation. Clinical studies have demonstrated that polyethylene glycol–superoxide dismutase is highly efficacious for treating patients with severe head injuries and for burn patients. In animal models for heart attacks, the polyethylene glycol–superoxide dismutase conjugate was much more active than the unmodified protein.

For background information *SEE CHEMOTHERAPY; MONOCLONAL ANTIBODIES; ONCOLOGY; OXYGEN TOXICITY* in the McGraw-Hill Encyclopedia of Science & Technology.

Peter Senter

Bibliography. O. W. Press et al., Radiolabeled-antibody therapy of B-cell lymphoma with autologous bone marrow support, *N. Eng. J. Med.*, 329:1219–1224, 1993; P. D. Senter et al., Generation of cytotoxic agents by targeted enzymes, *Bioconj. Chem.*, 4:3–9, 1993; P. A. Trail et al., Cure of xenografted human carcinomas by BR96-doxorubicin immunoconjugates, *Science*, 261:212–215, 1993; E. S. Vitetta, From basic science of B cells to biological missiles at the bedside, *J. Immunol.*, 153:1407–1420, 1994; S. S. Wong, *Chemistry of Protein Conjugation and Cross-Linking*, 1991; S. Zalipsky, Functionalized poly(ethylene glycol) for preparation of biologically relevant conjugates, *Bioconj. Chem.*, 6:150–165, 1995.

Bioelectronics

Molecular electronics is broadly defined as the encoding, manipulation, and retrieval of information at a molecular or macromolecular level. This approach contrasts with current semiconductor-based technology, in which these functions are accomplished via lithographic manipulation of bulk materials to generate integrated circuits. A key advantage of the molecular approach is the ability to design and fabricate devices from the bottom up, on an atom-by-atom basis. Lithography can never provide the level of control available through organic synthesis or genetic engineering. Bioelectronics is a subfield of molecular electronics that investigates the use of native as well as modified biological molecules in electronic or photonic (that is, light-activated) devices. Because natural selection processes have often solved problems of a similar nature to those that must be solved in harnessing organic compounds, and because self-assembly and genetic engineering provide sophisticated control and manipulation of large molecules, bioelectronics has shown considerable promise.

The two most commonly stated rationales for exploring molecular electronics are size and speed. A molecular computer could, in principle, be 1000 times smaller and 1000 times faster than a present-day semiconductor computer composed of a comparable number of logic gates. However, current projections suggest that semiconductor device sizes will approach the molecular domain around the year 2030. If size and speed were the only rationales for investigating molecular electronics, the field would have limited commercial potential. The opportunity to explore new architectures is one of the key aspects of molecular electronics that has prompted enthusiasm. For example, optical associative memories and three-dimensional memories can be implemented conveniently by using molecular electronics. Liquid-crystal display technology is a prime example of a hybrid molecular-semiconductor technology that has achieved widespread success.

Properties of bacteriorhodopsin. Many different bioelectronic devices are of interest, but this article concentrates on one approach that has achieved recent success because of a major international effort. The interest dates back to the early 1970s with the discovery of a bacterial protein that has unique photophysical properties. The protein, bacteriorhodopsin, is grown by a salt-loving bacterium that populates salt marshes. A light-absorbing group (the chromophore) embedded inside the protein matrix converts the light energy into a complex series of molecular events that store energy. Scientists using the protein for bioelectronic devices exploit the fact that this complex series of thermal reactions results in dramatic changes in the optical and electronic properties of the protein. The excellent holographic properties of the protein derive from the large change in refractive index that occurs following light activation. Furthermore, bacteriorhodopsin converts light into a refractive-index change with remarkable efficiency (approximately 65%). The size of the protein is 10 times smaller than the wavelength of light, which means that the resolution of a thin-film device is determined by the diffraction limit of the optical geometry rather than the graininess of the film. Also, the protein can absorb two photons simultaneously with an efficiency that far exceeds that of other materials. This capability allows the use of the protein to store information in three dimensions by using two-photon architectures. Finally, the protein was designed by nature to function under conditions of high temperature and intense light, a necessary requirement for a salt-marsh bacterial protein and a significant advantage for photonic device applications.

When the protein absorbs light in the native organism, it undergoes a complex photocycle which generates intermediates with absorption maxima spanning the entire visible region of the spectrum (**Fig. 1**). Most current devices operate at ambient or near-ambient temperature and utilize the following two states: the initial green-red absorbing state (bR) and the long-lived blue absorbing state (M), which can be stabilized at reduced temperatures, or by chemical or genetic modification of the protein. The forward reaction takes place only via light activation and is complete in about 50 microseconds. In contrast, the reverse reaction either can be light-activated or can occur thermally. The light-activated M → bR transition is a direct photochemical transformation using a photon of blue light. The thermal M → bR transition is highly sensitive to temperature, environment, genetic modification, and chromophore substitution. This sensitivity is exploited in many optical devices based on bacteriorhodopsin. Another reaction of importance is a photochemical

branching reaction from the O intermediate to form P. The O intermediate lives for about 8 milliseconds before thermally decaying back to form bR, but if O absorbs deep red light, it converts to form P. The P intermediate subsequently decays to form Q, a species that is unique in that the chromophore breaks the bond with the protein but is trapped inside the binding site. The Q intermediate is stable for extended periods of time (many years) but can be photochemically converted back to bR (as can P) by using blue light. This branching reaction provides for long-term data storage as discussed below.

Associative memories. Associative memories take an input data block (or image) and, independently of the central processor, scan the entire memory for the data block that matches the input. In some implementations, the memory will find the closest match if it cannot find a perfect match. Finally, the memory will return the data block in memory that satisfies the matching criteria, or it will return the address of the data block to permit access of contiguous data. Some memories will simply return a binary bit indicating whether the input data are present or not present. Because the human brain operates in a neural, associative mode, many computer scientists believe that the implementation of large-capacity associative memories will be required to achieve genuine artificial intelligence.

A salient example is the use of thin films of bacteriorhodopsin in holographic associative memories, as the photoactive components that provide real-time storage of the holograms. One widely used design (**Fig. 2**) includes both image feedback (which improves the accuracy of association) and thresholding (which helps to remove noise by filtering out stray signals below a given threshold intensity). The input image enters the system though a beam splitter at upper left and illuminates a spatial light modulator operating in the threshold mode. The light reaching the spatial light modulator is the superposition of all the images stored in the multiplexed holograms. Each image is weighted by the inner product between the recorded pattern from the previous iteration and itself. The pinhole array is designed so that the pinholes (diameter about 500 μm) are aligned precisely with the optical axes of the multiple Fourier-transform holographic reference images stored on the two protein films. The Fourier-transform holographic reference images enter from the middle left, and a separate Fourier-transform high-frequency-enhancement spatial light modulator is used to edge-enhance the images on bR film 1 in order to enhance the autocorrelation peak. The output image is a full reconstruction of the image stored on the Fourier-transform hologram that has the highest correlation with the input image (that is, produces the largest autocorrelation flux through its aligned pinhole). Thus, as the example in Fig. 2 shows, only a partial input image is required to generate a complete output image.

Although only four reference images are shown in Fig. 2, an optical associative memory can store many hundreds or thousands of images simultane-

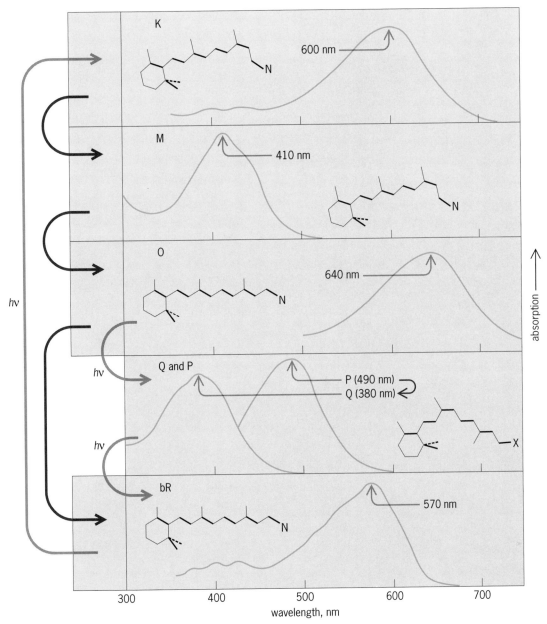

Fig. 1. Absorption spectra of selected intermediates in the photocycle of bacteriorhodopsin. Light gray arrows indicate light-activated processes (the symbol *h*v indicates a photon), and dark gray arrows indicate reactions that occur in the dark. The structures of the polyene chromophore are shown as inserts where N represents the nitrogen atom of the lysine which binds the chromophore to the protein and X represents either a lysine nitrogen (in P) or oxygen (in Q).

ously. This memory can also work on binary data by using redundant binary representation logic, and a small segment of data can be used to find which page has the largest association with the input segment.

The ability of the associative memory to rapidly change the holographic reference patterns via a single optical input while maintaining both feedback and thresholding increases its utility, and in conjunction with solid-state hardware, allows it to be integrated into hybrid computer architectures. The diffraction-limited performance of the protein films, coupled with the high write-erase speeds associated with the excellent quantum efficiencies of these films, represents a key element in the potential of this memory. The ability to modify the

protein by selectively replacing one amino acid with another provides significant flexibility in enhancing the properties of the protein.

Three-dimensional memories. The major impact of molecular electronics on computer hardware may eventually be in the area of volumetric memory. Three different types of protein-based volumetric memories are currently under investigation: holographic, simultaneous two-photon, and sequential one-photon. Since a holographic memory based on bacteriorhodopsin has already been described, the following discussion will be confined to the latter two architectures. These memories read and write information by using two orthogonal laser beams to address an irradiated volume (10–200 μm^3) within a much larger volume of a photochromic material.

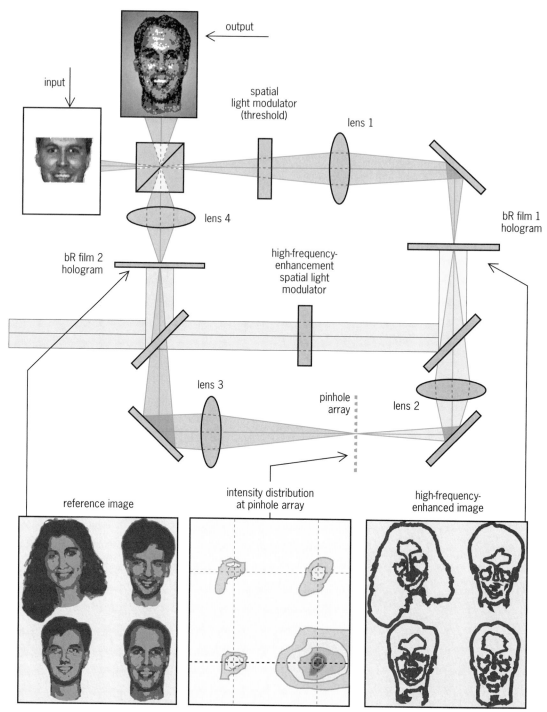

Fig. 2. Fourier-transform holographic associative memory with read-write Fourier-transform holographic reference planes using thin polymer films of bacteriorhodopsin to provide real-time storage of the holograms. A partial input image can select and regenerate the entire associated image stored on the reference hologram.

Either a simultaneous two-photon or a sequential one-photon process is used to initiate the photochemistry. The former process involves the unusual capability of some molecules to capture two photons simultaneously. The sequential one-photon process requires a material that undergoes a branching reaction, where the first photon activates a cyclical process and the second photon activates a branching reaction to form a stable photoproduct. The three-dimensional addressing capability of both memories

derives from the ability to adjust the location of the irradiated volume in three dimensions. In principle, an optical three-dimensional memory can store roughly three orders of magnitude more information in the same-size enclosure relative to a two-dimensional optical disk memory. In practice, optical limitations and issues of reliability lower the above ratio to values closer to 300. Nevertheless, a 300-fold improvement in storage capacity is significant. Furthermore, the two-photon or sequential one-photon approach makes parallel addressing of data possible, thus enhancing data read-write speeds and system bandwidth.

The simultaneous two-photon memory architecture has received a great deal of attention, and because bacteriorhodopsin exhibits both high efficiency in capturing two photons and a high yield of producing photoproduct after excitation, this material has been a popular memory medium. But more recent studies suggest that the branched-photocycle memory architecture may have greater

potential. This sequential one-photon architecture completely eliminates unwanted photochemistry outside the irradiated volume and provides for a particularly straightforward parallel architecture. The fact that the P and Q states can be generated only via a temporally separated pulse sequence provides a convenient method of storing data in three dimensions by using orthogonal laser excitation, with spatial light modulators to control the laser light. The process (**Fig. 3**) is based on sequences (1) and (2), where K, L, M, N, and O are all intermedi-

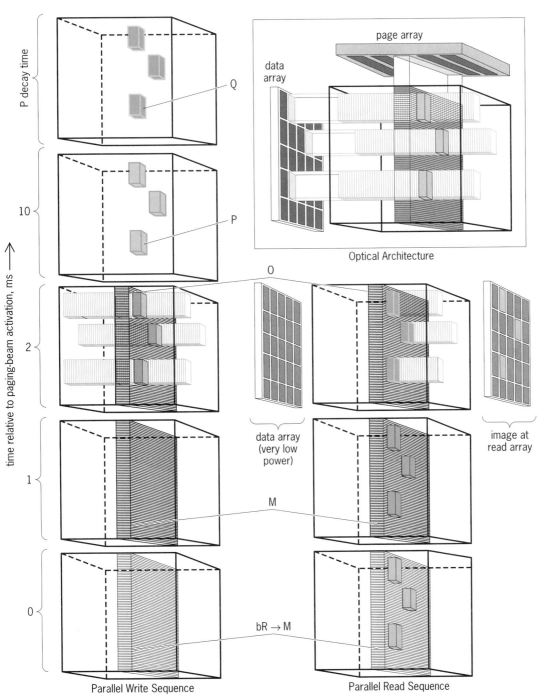

Fig. 3. Parallel write and read sequences associated with a branched-photocycle three-dimensional memory based on bacteriorhodopsin. The basic optical architecture is shown at upper right where spatial light modulators are used to control the laser light.

$$bR \text{ (state 0)} \xrightarrow{\text{photon 1}} K \longrightarrow L \longrightarrow M \longrightarrow N \longrightarrow$$
$$O \longrightarrow bR \quad \text{(paging)} \quad (1)$$
$$O \xrightarrow{\text{photon 2}} P \text{ (state 1)} \longrightarrow Q \text{ (state 1')}$$
$$\text{(write data)} \quad (2)$$

ates within the main photocycle, and P and Q are intermediates in the branching cycle (shown in Fig. 1, except for L and N).

A parallel write is accomplished by using an orthogonal optical architecture. Light activation is represented by using lines to represent the presence and the directionality of the light beam, and the vertical axis charts time relative to the firing of the paging laser. The timing is based on a memory cuvette at ambient temperature (20–30°C or 68–86°F). The paging beam with a wavelength of approximately 600 nanometers activates the photocycle of bacteriorhodopsin, and after a few milliseconds the O intermediate reaches near-maximum concentration. At this point, the data laser array is activated to photoselect the irradiated volume elements into which 1 bits are to be written. This process converts O to P in these, and only these, locations within the memory cube. After many minutes (the decay time is highly dependent upon temperature and the polymer matrix), the P state thermally decays to form the Q state. The latter is stable for months to years depending upon temperature and the polymer matrix. The bR state is assigned to binary state 0, and both P and Q to binary state 1. The entire write process is accomplished in about 10 ms, the time it takes the protein to complete the photocycle. If a 1024×1024 data array is used, then 1,048,576 data bits can be written, or about 105 kilobytes (one byte is represented by eight data bits and two error-correcting bits) within a 10-ms cycle. The overall write data is thus a throughput of 10^7 characters per second (10 megabytes per second), which is comparable to slow semiconductor memory. By using more than one memory cell, data write times improve proportionally. An entire page of memory can be erased by using blue light (which converts both P and Q back to bR).

The read process takes advantage of the fact that light around 680 nm is absorbed by only two intermediates in the photocycle of light-adapted bacteriorhodopsin, the primary photoproduct K and the relatively long-lived O intermediate (Fig. 1). A parallel read is accomplished by using a differential absorption process as shown in the right column of Fig. 3. The read sequence starts out in a fashion identical to that of the write process by activating the 600-nm paging beam. After 2 ms, the entire data array is turned on at a very low intensity (0.01% of the power used to write). A charge-injection-device array images the light passing through the data cube. Those elements that are in the binary 1 state (P or Q intermediates) do not

absorb the 680-nm beams, but those volumetric elements that started out in the binary 0 state (bR) absorb the 680-nm light because they have cycled into the O state. Because all of the volumetric elements outside the paged area are restricted to the bR, P, or Q states, the only significant absorption of the beam is associated with O states within the paged region. The charge-injection-device detector array is therefore observing the differential absorptivity of the paged region, and the paged region alone. Because the absorptivity of the O state within the paged region is more than 1000 times larger than the absorptivity of the remaining volume elements combined, a very weak beam can be used to generate a large differential signal. The read process is complete in about 10 ms, which gives a rate of 10 megabytes per second times the number of data cubes.

One of the commercial requirements of volumetric memory systems is the need for highly homogeneous memory media. Three space shuttle flights have been carried out to investigate the potential of manufacturing bacteriorhodopsin memory cubes in microgravity.

For background information SEE BIOELECTRON-ICS; COMPUTER STORAGE TECHNOLOGY; HOLOGRA-PHY; IMAGE PROCESSING; OPTICAL INFORMATION SYSTEMS; SPACE PROCESSING in the McGraw-Hill Encyclopedia of Science & Technology.

Robert R. Birge

Bibliography. R. R. Birge, Protein-based computers, *Sci. Amer.*, 272(3):90–95, March 1995; R. R. Birge (ed.), *Molecular and Biomolecular Electronics*, Adv. Chem. Ser. 240, 1994; D. Oesterhelt, D. C. Bräuchle, and N. Hampp, Bacteriorhodopsin: A biological material for information processing, *Quart. Rev. Biophys.*, 24:425–478, 1991; W. Stoeckenius, The purple membrane of salt-loving bacteria, *Sci. Amer.*, 234(6):38–44, June 1976.

Bioinorganic chemistry

Recent advances in bioorganometallic chemistry involve dimetal compounds with metal-metal bonds that exhibit antitumor activity, and organometallic compounds that can be used in immunoassays.

Antitumor chemistry. Metal-metal bonded complexes have been found that have potential as anticancer compounds.

Mononuclear platinum anticancer compounds. The medicinal properties of inorganic compounds are well known; metal compounds that exhibit anti-arthritic, antibacterial, anticancer, antidepressant, and antihypertensive properties are in routine clinical use. Interest in metal antitumor compounds stems from the extraordinary effectiveness of the platinum compound cisplatin [*cis*-DDP; structure (I)] and related complexes in the treatment of various deadly cancers.

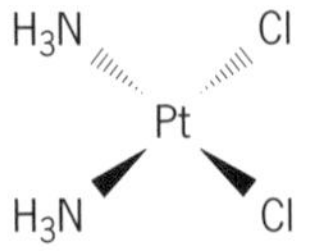

(I)

The introduction of cisplatin as an antitumor agent in the 1960s met with considerable opposition by the medical community. The replacement of highly toxic arsenic, antimony, and mercury antibacterial agents with organic compounds in the early 1900s ushered in a new era of medicinal chemistry that excluded heavy-metal compounds from therapeutic use. However, in the United States the Food and Drug Administration (FDA) approved cisplatin in the 1970s; this decision ultimately saved (and continues to save) the lives of many people suffering from testicular and ovarian cancers, as well as from tumors of the bladder, head, and neck.

Unfortunately, patients who receive cisplatin therapy experience severe side effects that require the administration of other drugs to alleviate the toxic side effects on normal cells of the kidneys, liver, and intestine. Second-generation drugs have been developed that exhibit high therapeutic activity with reduced toxicity; these include platinum(IV) [Pt(IV)] and platinum(II) [Pt(II)] compounds. Of this group, carboplatin [structure (II)] has recently been approved in the United States as a less toxic alternative to cisplatin in the treatment of ovarian cancer and small lung cancers, and iproplatin (III) is in

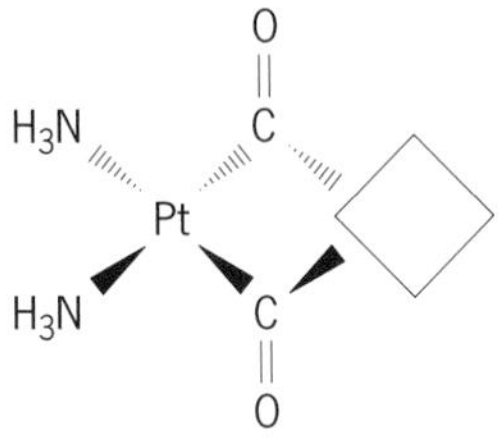

(II)

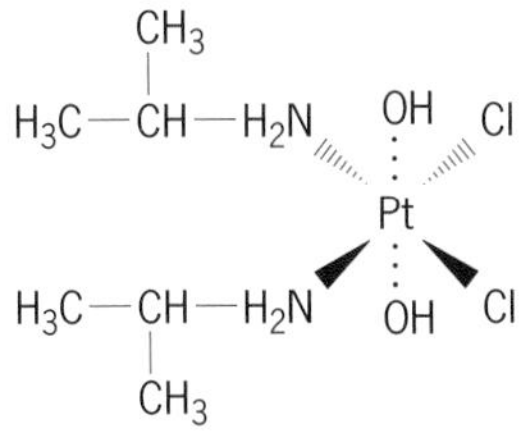

(III)

phase III of clinical trials. The challenge remains to develop alternative drugs whose biological activity is improved with concomitant decrease in toxicity.

Platinum binding to DNA. The mechanism by which cis-DDP [structure (I)] destroys tumor cells is still uncertain, but years of collective research have culminated in the conclusion that the active form of the drug is the water-solvated species that exhibits a strong preference for deoxyribonucleic acid (DNA) sequences containing guanosine (IV) nucleosides adjacent to a second guanosine or an adenosine (V), where R = deoxyribose.

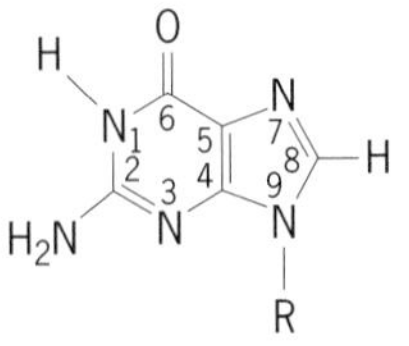

(IV)

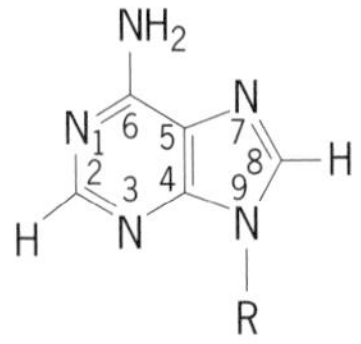

(V)

Modeling and x-ray crystallographic studies of cisplatin coordinated to DNA have verified the hypothesis that cisplatin binds to two adjacent guanine bases on the same DNA strand (intrastrand). This mode of binding and the changes it causes in the DNA helix are thought to be responsible for DNA inhibition, an event that leads to cell death.

Nonplatinum antitumor compounds. Many nonplatinum metal antitumor active compounds exhibit physical and chemical properties that are similar to cisplatin in that they are neutral molecules that possess labile cis leaving groups: these include compounds of titanium (Ti), vanadium (V), and molybdenum (Mo) of general formula $(Cp)_2MCl_2$ [Cp = cyclopentadienide $(C_5H_5^-)$], cis-Pd$(NH_2C_2H_2NH_2)$ $(NO_3)_2$, and Ti$(OC_2H_5)_2$(1-phenylbutane-1,3-dionato)$_2$ [budotitane]. Notable exceptions are the active cationic species cis-$[Ru(NH_3)_4Cl_2]^+$ and the trans isomer of RuCl$_2$(DMSO)$_4$ (DMSO = dimethylsulfoxide), which have not been found to be pharmacologically useful against cancer.

Dinuclear transition-metal compounds. One entirely different class of compounds whose structures and reactivities appear to defy most of the accepted guidelines for metal anticancer agents are dinuclear metal-metal bonded compounds of rhodium (Rh), ruthenium (Ru), and rhenium (Re). In particular, compounds of these metals with at least two bridging carboxylate ligands have been studied for their carcinostatic activity against various tumor cell lines with promising results. The basic lantern structures of these compounds, which are dimetal compounds containing a metal-metal bond and four bridging ligands (VI; R = CH$_3$, CH$_2$CH$_3$, CH$_2$CH$_2$CH$_3$,

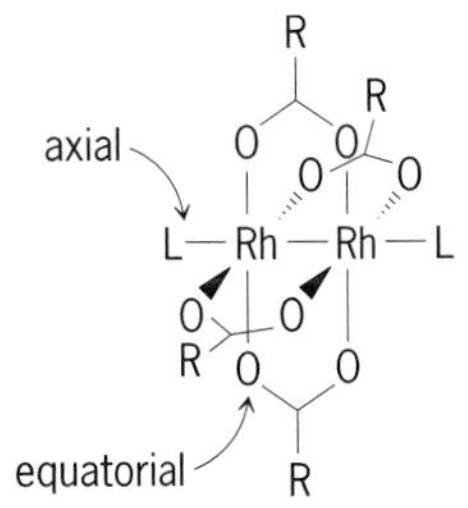

(VI)

CH$_2$OCH$_3$; L = donor solvent), allow for two possible types of binding sites, equatorial and axial.

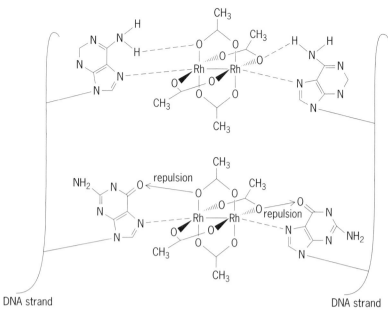

DNA strand   DNA strand

Fig. 1. Axial binding of deoxyribonucleic acid (DNA) to the Rh$_2$(O$_2$CR)$_4$ molecule.

Dirhodium tetracarboxylate compounds. Dirhodium tetracarboxylate compounds of the type Rh$_2$-(O$_2$CR)$_4$L$_2$ (VI) exhibit considerable antitumor activity in mice bearing the Erlich ascites tumor. In a similar fashion to *cis*-DDP, dirhodium tetraacetate was found to be a potent inhibitor of cellular DNA synthesis. Studies indicate that the lantern structure binds to denatured DNA, polyadenosine, and bovine serum albumin. Researchers have also concluded that dirhodium tetracarboxylates do not have an appreciable affinity for highly polymerized native calf thymus DNA, polyguanosine, or polycytidine. These DNA-binding preferences were taken as an indication of axial binding of DNA to the Rh$_2$(O$_2$CR)$_4$ molecule (**Fig. 1**) because steric repulsions between bridging carboxylate oxygen atoms and the O-6 atom of guanosine preclude its binding axially, whereas attractive hydrogen-bonding interactions favor axial interactions of adenosine with the dirhodium tetracarboxylates.

In contrast to the earlier studies, recent findings have established that guanine bases do in fact bind to Rh$_2$(O$_2$CH$_3$)$_4$, and in an unprecedented fashion. Results support an alternative substitution pathway for these compounds that involves displacement of equatorial bridging carboxylate ligands rather than mere axial substitution. Single-crystal x-ray studies have been performed on several compounds that possess novel bridging guanine and adenine groups. The discovery that a building block of DNA can act as a bridge between two metal centers is unprecedented and may be important in understanding how dinuclear complexes prevent DNA replication in tumors.

Molecular modeling is a valuable tool for the prediction of possible structures and relative energies of DNA-metal interactions. Recent computer simulations of DNA-dirhodium structures reveal surprisingly minor energy differences between a crystallographically known DNA-cisplatin structure and a DNA-Rh$_2$(O$_2$CCH$_3$)$_4$ structure (**Fig. 2**). The calculated minor energy difference of the DNA-metal interactions between the mononuclear Pt complex and the dinuclear Rh complex suggest that the interactions of Rh$_2$(O$_2$CCH$_3$)$_4$ with a guanosine-guanosine (GG) sequence are stable. The studies also reveal that the DNA backbones of both models are virtually indistinguishable, supporting the fact that little difference exists between the DNA interactions of the two different transition-metal complexes.

Diruthenium and dirhenium carboxylates. Since the discovery of the ability of dirhodium tetracarboxylates to inhibit cancer activity, related Ru$_2$ and Re$_2$ complexes have also been found to exhibit comparable antitumor activities. These compounds have been much less investigated than their dirhodium counterparts, but their biological mode of action of DNA inhibition is thought to be similar to the rhodium systems. The water-soluble diruthenium compounds Ru$_2$(O$_2$CR)$_4$Cl (R = CH$_3$, CH$_2$CH$_3$) are active in the leukemia P388 tumor system, whereas the dirhenium compound [Re$_2$(O$_2$CCH$_2$CH$_3$)$_4$][SO$_4$] exhibits antitumor behavior in mice bearing sarcoma 180 and leukemia P388. A related compound, *cis*-Re$_2$Br$_4$(O$_2$CCH$_3$)$_2$(H$_2$O)$_2$, the first compound with only two carboxylate [(RCOO)$^-$] ligands to be studied for antitumor behavior, showed considerable antitumor activity against B-16 melanoma and a bacteriostatic effect on the *Escherichia coli* strain W3350 (thy) similar to cisplatin. Like the dirhodium compounds, these dimetal systems selectively inhibit DNA synthesis with no effect on syntheses of ribonucleic acid (RNA) and protein. Although the rhenium anticancer compounds were found to gradually decompose in the test animals, they are virtually nontoxic, a situation that is in sharp contrast to all other metal complexes. Furthermore, rhenium itself appears to have little or no toxic

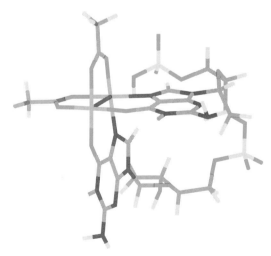

Fig. 2. Energy-minimized computer simulation of *cis*-Rh$_2$(O$_2$CCH$_3$){d(pGpG)}$_2$.

effects, no established poisonous levels having been reported.

Future of dimetal anticancer compounds. The main drawbacks to clinical use of the tetracarboxylate family of antitumor compounds (except Re) are their low solubilities and their severe toxicities due to rapid decomposition in the organism. The decomposition is related to the fact that carboxylate ligands are easily lost in redox reactions that eventually lead to the production of carbon dioxide (CO_2) and the metallic form of the metal. Dimetal compounds with robust bridging groups that serve to keep the metal centers intact and at the same time impart greater solubility or affinity for DNA are expected to lead to new generations of antitumor compounds. The use of ligands with intercalative properties (the ability to insert themselves between base pairs in DNA molecules) represents another challenge in the design of coordination compounds with antitumor potential.

Kim R. Dunbar; Kemal V. Catalan

Organometallics in immunoassay. Since the early 1960s, a rapid and significant development of immunoassay techniques has occurred. The first tracers used were exclusively radioactive [iodine-131(^{131}I), iodine-125(^{125}I), hydrogen-3(^{3}H), carbon-14(^{14}C)]. However, legal constraints and limitations attached to the use of such tracers have led since 1980 to the development of new, nonisotopic techniques that use either nonradioactive tracers such as enzymes or chemiluminescent or bioluminescent markers. Each of the currently available nonradioactive methods has its specific advantages but also its limitations, justifying a continued effort to explore new avenues, particularly in the area of the multi-immunoassay, which is currently one of the major challenges in the field. Unfortunately, the considerable developments in immunoassay techniques have not been accompanied by any major breakthroughs in the area of multi-immunoassays, despite there being a clear need for such techniques for economic reasons. The chief obstacle to the development of the multiassay is the difficulty of finding tracer combinations that produce simultaneously analyzable signals. A new and promising approach to this problem uses metal carbonyl complexes.

Carbonyl metallo-immunoassay. As part of the research focused on extending the uses of metal carbonyl complexes in the biomedical field, a nonisotopic immunoassay was developed, carbonyl metallo-immunoassay (CMIA). This technique uses metal carbonyl complexes as nonisotopic tracers and Fourier-transform infrared spectroscopy (FT-IR) as the detection method. Carbonyl metallo-immunoassay is a competitive type of immunoassay

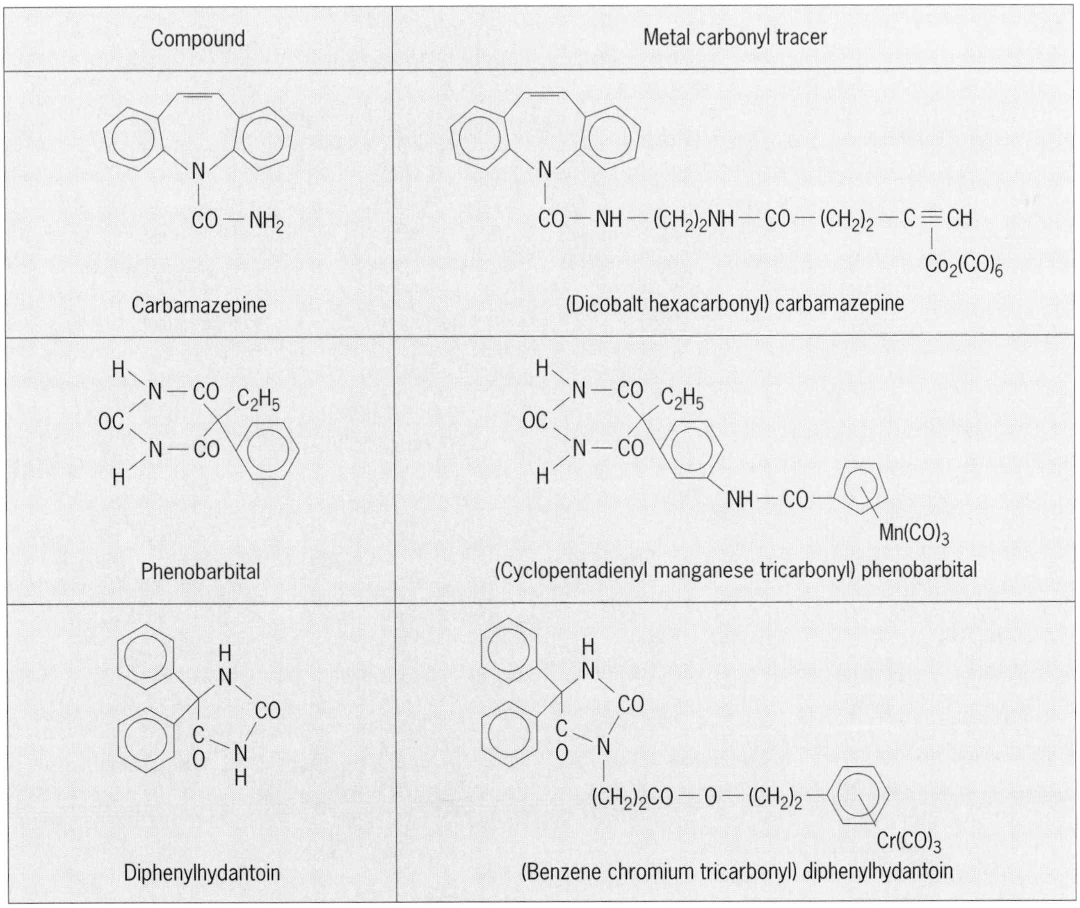

Compound	Metal carbonyl tracer
Carbamazepine	(Dicobalt hexacarbonyl) carbamazepine
Phenobarbital	(Cyclopentadienyl manganese tricarbonyl) phenobarbital
Diphenylhydantoin	(Benzene chromium tricarbonyl) diphenylhydantoin

Fig. 3. Analytes to be tested and their metal carbonyl tracers.

that involves three elements: the analyte to be assayed; specific antibodies of that analyte; and the tracer (formed from a molecule of the analyte), to which is attached a metal carbonyl moiety. During the incubation phase, there is competition for fixation to the specific antibody between a fixed quantity of tracer and variable quantities of the analyte being assayed. At the end of the incubation, the free and bound fractions of the tracer are separated by extraction of the free fraction by using an organic solvent that is immiscible with water (isopropyl ether, for example). Quantification of the tracer is then performed by FT-IR spectroscopic analysis.

Synthesis of tracers. One key step in preparing a carbonyl metallo-immunoassay is the synthesis of the organometallic tracers. In order to establish the feasibility of the approach, an attempt was made to monoimmunoassay drugs, specifically the three antiepileptic medications carbamazepine, phenobarbital, and diphenylhydantoin (DPH). Assaying is performed on a large scale in the hospital setting, because the difference between their effective and toxic doses is small.

Three tracers for the drugs were synthesized, each with a different metal carbonyl group (**Fig. 3**). For carbamazepine a dicobalt hexacarbonyl $[Co_2(CO)_6]$ fragment was attached to a triple bond; for phenobarbital the $(cyclopentadienyl)Mn(CO)_3$ moiety was used; and for diphenylhydantoin a complex with benzene chromium tricarbonyl was synthesized. These tracers retain good affinity for the specific antibodies of the substances to be assayed, and therefore are good candidates for use as tracers in carbonyl metallo-immunoassay.

The carbonyl metallo-immunoassay method relies on the unusual spectral properties of these metal-carbonyl complexes in infrared analysis: (1) the characteristic $\nu(CO)$ infrared vibration bands of the complexes lie in the 1850–2200 cm^{-1} region, where proteins and most organic molecules do not absorb; (2) these bands are three to six times more intense than any other band present in the infrared spectrum; and (3) the position and number of the bands vary according to the metal used and the symmetry of the metal carbonyl group $[M_x(CO)_y]$.

Tracer quantification. Quantification of the tracers is performed simply by measuring the absorbance value of one of the characteristic peaks of the complexes. In accordance with the Beer-Lambert law, this value is proportional to the quantity of tracer present in the medium. In general, the sharpest and most intense peak is the one that provides the best sensitivity. By using a small-volume (30-microliter) cell with a long optical path (2 cm), specially constructed for the purpose, a detection limit for (dicobalt hexacarbonyl) carbamazepine of 300 femtomoles (3×10^{-13} M) was obtained. This sensitivity range is comparable to that of the majority of immunological assay methods. The first step was the development of a single-assay carbonyl metallo-immunoassay for each of the three antiepileptics. It

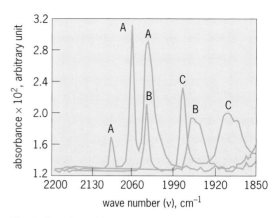

Fig. 4. Superimposition of the Fourier-transform infrared spectra (in the range 1850–2200 cm^{-1}) of three metal carbonyl tracers: (dicobalt hexacarbonyl) carbamazepine (A); (cyclopentadienyl manganese tricarbonyl) phenobarbitol (B); and (benzene chromium tricarbonyl) diphenylhydantoin (C).

was necessary to obtain titration and assay curves, starting from a series of infrared spectra, to allow quantification of the organometallic tracer present in each fraction. The curves obtained are completely in accord with those observed for classical methods such as radioactive immunoassay (which uses a radioactive tracer), confirming the validity of the approach.

Multi-immunoassay. One of the true novelties of organometallic tracers is the possibility for performing simultaneous quantitative analysis on a mixture of three (possibly even five) tracers. This analysis is performed starting from the infrared spectrum of the mixture in the 1850–2200 cm^{-1} range. In favorable cases, such as shown in **Fig. 4**, the analysis can be done by a simple stepwise calculation, as band overlap is only partial. In cases of greater overlap, more complex calculation methods must be employed. It has been demonstrated that the error is always less than 5%.

To assess the potential of the carbonyl metallo-immunoassay method in multi-immunoassay the three antiepileptics were chosen because they are often administered simultaneously, and simultaneous assay of these substances therefore answers a real need. The titration and assay curves obtained in both mono and multi carbonyl metallo-immunoassay were compared, and it was found that these curves can be superimposed. Thus it can be inferred that there is no interference between the different components of the assay, which in fact number nine (three substances to be assayed, three tracers, and three specific antibodies). The carbonyl metallo-immunoassay approach therefore is feasible. The multi-immunoassay possibilities of this method show great promise and may be one of its chief advantages. Further work is focused on the possibility of assaying four or even five analytes. An added advantage is that with current developments in the field of organometallic chemistry, synthesis of specific markers for any molecule, including proteins, may be possible in the near future, widening the application of the method even further. This novel method could represent a significant step forward in the field of nonisotopic immunoassay, with particular promise in the area of simultaneous multiimmunoassay.

For background information *SEE CHEMOTHERAPY; IMMUNOASSAY; METAL CARBONYL; ONCOLOGY; SPECTROPHOTOMETRIC ANALYSIS; X-RAY CRYSTALLOGRAPHY* in the McGraw-Hill Encyclopedia of Science & Technology.

G. Jaouen; A. Vessières

Bibliography. K. R. Dunbar et al., Structural evidence for a new metal-binding mode for guanine bases: Implications for the binding of dinuclear antitumor agents to DNA, *J. Amer. Chem. Soc.,* 116: 2201–2202, 1994; F. R. Hartley (ed.), *Chemistry of the Platinum Group Metals Recent Developments: Studies in Inorganic Chemistry 11,* 1991; G. Jaouen, A. Vessières, and I. S. Butler, Bioorganometallic chemistry: A future direction for transition metal organometallic chemistry?, *Acc. Chem. Res.,* 26:361–369, 1993; B. K. Keppler, Metal complexes as antitumor agents: The future role of inorganic chemistry in cancer therapy, *New J. Chem.,* 14:389–403, 1990; B. Rosenberg and L. VanCamp, Platinum compounds: A new class of antitumor compounds, *Nature,* 222:385–386, 1969; M. Salmain et al., Carbonyl metallo immunoassay (CMIA), a new type of non-radioisotopic immunoassay: Principles and applications to phenobarbital assay, *J. Immunol. Meth.,* 148:65–75, 1992; M. Salmain et al., A Fourier transform infrared spectroscopic method for the quantitative trace analysis of transition metal carbonyl labeled bioligands, *Anal. Chem.,* 63: 2323–2329, 1991; A. Vessières et al., An ultra-low volume, gold light-pipe cell for the IR analysis of dilute organic solutions, *Appl. Spectros.,* 44:1092–1094, 1990.

Biological pest control

Fungi are increasingly being utilized as agents for the biological control (biocontrol) of deleterious organisms. Targets include organisms responsible for spoilage of stored foods or fiber, weeds, harmful insects, and pathogens of agricultural crops. Many such pathogens are themselves fungi. *SEE FUNGI.*

The usual objective of biocontrol programs is a safe and economical reduction in the use of toxic or costly pesticides. Biological control may be used as the sole or primary weapon against harmful organisms, but more commonly is one of several components in integrated pest management. Integrated pest management also includes manipulation of cultural practices such as tillage, pruning, irrigation, fertilization, and crop rotation; development of epidemiological models; development of genetic resistance to pests; and judicious, restricted use of pesticides.

Fungi used as biocontrol agents may already exist in place, may be indigenous to the geographic area where they are deployed, or may be introduced. They may exercise control through predation, parasitism, antibiosis, or direct competition with the target organism. Pertinent to the use of fungi in biocontrol are surveys and experiments for selection of suitable agents; careful testing and regulation of candidate biocontrol agents; monitoring potential impacts on human and animal health; product development, including formulation, storage, and application of inoculum; and economic assessment from the standpoint of production and deployment.

The field of biological control, including fungi, is affected by developments in molecular genetics and genetic engineering. Most systems of biological control by fungi are still experimental, but commercial products are increasingly available.

Applications. Agriculture is the primary area for deployment of fungal biocontrol agents. The range of crops and cropping systems in which such agents are experimentally or commercially employed is considerable. However, because of the expense of production and application of the agent, commercial development has concentrated on high-value crops. Examples include the application of *Trichoderma* species to grape, strawberry, and raspberry blossoms to deter infection by another fungus, *Botrytis cinerea;* the control by several fungi of powdery mildew of vegetables grown in greenhouses; and use of *Gliocladium* to protect greenhouse-grown spruce seedlings from *Botrytis.* Biocontrol has also been used experimentally to protect field crops such as wheat, tobacco, and cotton.

Because crops that are stored represent substantial investment, considerable attention has been given to biological control of postharvest losses. Common plant-associated yeasts have been applied to stored apples and pears to prevent fruit rots. Conversely, protection of stored products can begin in the field; aflatoxin-producing strains of *Aspergillus* can be excluded from cottonseed (the source of cottonseed oil and an important ingredient in animal feeds) by inoculation of the crop with strains that do not produce the toxin.

There are a limited number of experimental and commercial applications of fungal biocontrol in public health and medicine. Cockroaches and mosquitoes are targets. Nonpathogenic strains of *Candida* have been incorporated into commercially available treatments for vaginitis.

Modes of action. Fungi can exert control over target pests in a variety of ways. The classic example of predation by fungi is the trapping of plant pathogenic nematodes (roundworms) by fungi in the genus *Dactylaria.* These fungi produce noose-like loops and other adhesive structures that trap the nematodes, which are then penetrated and digested. Although dramatic, predation is conceded to be a less effective control than parasitism. Some fungi, notably species in *Catenaria* and *Nematophthora,* are specialized parasites of nematodes. A wide variety of other fungi are opportunistic pathogens of nematodes and their eggs.

Instances of mycoparasitism (parasitism of one fungus by another) are frequent and are exploited for biological control. For example, the hyphae (threadlike structures that make up the body of a fungus) of *Pythium nunn* can coil around the hyphae of another fungus, or *P. nunn* can form infection structures on another fungus. Then the parasitized fungus is penetrated and digested. Fungi attacked by *P. nunn* include such important plant pathogens as *Phytophthora parasitica, Rhizoctonia solani,* and other *Pythium* species. In another type of mycoparasitism, *Sporodesmium sclerotivorum* specifically attacks sclerotia (multicelled resting and propagative structures) of several fungal plant pathogens.

Fungal plant pathogens can play a benevolent role when they are utilized as parasites of noxious weeds. Dodder (*Cuscuta*) is a leafless parasitic plant that attacks crop plants. *Colletotrichum gloeosporioides* f. sp. *cuscutae*, specialized to attack dodder, is used as a mycoherbicide for dodder control. *Colletotrichum gloeosporioides* f. sp. *aeschynomene* is used as a mycoherbicide on northern joint vetch (*Aeschynomene virginica*).

Insects are susceptible to control by fungal parasites. Entomopathogenic fungi in the genera *Beauveria* and *Metarhizium* have been used for control of codling moth, Colorado potato beetle, species of corn borer, spittle bugs, and other insect pests. *Lagenidium giganteum* and other fungi have been used for experimental control of mosquito larvae. SEE INSECT CONTROL, BIOLOGICAL.

In other instances, pests are controlled via antibiosis, that is, the production by organisms of compounds that are toxic to other organisms. The well-known instances of antibiosis involve compounds produced by bacteria, but several species in *Trichoderma* and *Gliocladium*, as well as other fungi, are thought to produce compounds inhibitory to fungal pathogens. Any given fungal biocontrol agent may display several modes of action against target organisms. For example, *T. harzianum* can act against another fungus via mycoparasitism or antibiosis, or by direct competition for nutrients.

Similar competition has been demonstrated in the protection of stored citrus fruit; yeast cells in high concentration can exclude *Penicillium* from small wounds in the fruit by utilizing nutrients that would otherwise be available to the pathogen. Competition for nutrients in soil microsites enhances the biocontrol efficacy of mycorrhizal fungi. These fungi establish a symbiotic relationship with plant roots. Their utilization of nutrients immediately proximal to roots restrains the growth of pathogenic fungi in that environment. Moreover, some mycorrhizae (ectomycorrhizae) also establish a physical barrier to pathogens by forming a coat or mantle around the roots. SEE MYCORRHIZAE.

Another mode of action is hypovirulence. Certain strains of the chestnut blight fungus, *Cryphonectria parasitica*, are infected with viruslike ribonucleic acid, which renders them less virulent than the uninfected strains. Moreover, under appropriate circumstances the hypovirulent strains can convert pathogenic strains to a hypovirulent condition. Exploitation of hypovirulence has controlled the spread of chestnut blight in European chestnut plantations.

Native versus introduced biocontrol agents. Introduction of nonindigenous species can have adverse environmental effects. Generally, the utilization of agents already in place, or at least geographically proximal, is regarded as lower risk than introduction of nonindigenous species. However, in some instances release of nonindigenous organisms has proven beneficial, especially in control of other nonindigenous organisms. For example, skeleton weed (*Chondrilla juncea*) has been controlled by the rust fungus *Puccinia chondrilla* in Australia, and the weed hamakua pamakani (*Ageratina*) by the smut fungus *Entyloma ageratinae* in Hawaii.

Alternatively, a potentially harmful organism may be controlled by indigenous microorganisms. *Sclerotinia sclerotiorum* (which causes white mold in beans) is found in the field with such ubiquitous fungi as *Cladosporium, Alternaria,* and *Epicoccum.* Experimental control of white mold has been obtained by repeated applications of *Epicoccum.* Certain species in *Trichoderma* and *Beauveria,* as well as many other fungi, have great potential as biocontrol agents and are already globally distributed.

Regulation, testing, and product development. The development of a biological control product is an arduous process. Candidate biocontrol organisms must be collected and tested against the target pest organisms. The candidate organism not only must be effective against the target pest(s) in laboratory and field tests but often must be tolerant of specific pesticides and adaptable to the cultural practices for the pertinent crop. Candidate organisms can be tested alone or in combination with other biocontrol agents. The organisms and other ingredients in the product must be assessed for adverse impacts on the environment and on human and animal health. Results are monitored by regulatory agencies and are subject to public scrutiny. Proper adjuvants such as nutrients for the biocontrol organism, buffers, and surfactants are necessary, as is careful definition of transport and storage conditions. Application procedures include sprays for field and greenhouse crops, dips for fruits and vegetables entering storage, injection into trees for canker control, coating of seeds for protection during germination and root development, and vectoring by bees for transfer of biocontrol agents to blossoms.

Biocontrol and biotechnology. New methodologies are transforming experimental biological control. One development is the generation of biocontrol strains tolerant to fungicides. Such strains, used in combination with low rates of fungicide, should enable suppression of pests with reduced application of pesticide. Parasexual (nonmeiotic) and sexual hybridization, protoplast fusion, and recombinant deoxyribonucleic acid (DNA) technology are tools for the creation of strains with improved biological control properties. Molecular genetics offers promise for conferring or enhancing production of crucial enzymes and antibiotics in biocontrol organisms. Conversely, pathways for production of undesirable compounds might be blocked or eliminated. Such methods as the polymerase chain reaction and DNA fingerprinting enable the tracking of released organisms in the environment and better assessment of their effectiveness in suppression of pests. *SEE BREEDING (PLANT).*

For background information *SEE BIOTECHNOLOGY; FUNGI; GENETIC ENGINEERING; MYCORRHIZAE; PLANT PATHOLOGY* in the McGraw-Hill Encyclopedia of Science & Technology.

Frank M. Dugan

Bibliography. I. Chet, *Biotechnology in Plant Disease Control,* 1993; R. J. Cook and K. F. Baker, *The Nature and Practice of Biological Control of Plant Pathogens,* 1983; D. O. Te Beest, X. B. Yang, and C. R. Cisar, The status of biological control of weeds with fungal pathogens, *Annu. Rev. Phytopathol.,* 30:637–657, 1992.

Biosensor

Many situations require the identification and quantification of almost any species in a sample regardless of the sample complexity. Examples are the determination of pesticides in soil, identification of carcinogens or mutagens in human fluids and tissues, and control of chemical reactions in large-scale industrial plant operations. Thus, there is growing interest in the development of new chemical sensing and biosensing schemes.

A chemical sensor consists of a recognition element that is typically immobilized on a platform and that specifically interacts with the species of interest, that is, the analyte. When the recognition element binds or associates with the target analyte, a transduction event occurs, resulting in a detectable change in the signal, which may be electrochemical, thermal, mass, or optical. A biosensor is a subcategory of chemical sensor in which a biological macromolecule (such as a protein, enzyme, or antibody) is used as the recognition element. An ideal biosensor would be inexpensive, small, simple to operate, sensitive, selective only for the analyte in question, reversible, rugged, and easily calibrated, and would exhibit rapid response to the analyte of interest. *SEE CAPILLARY ELECTROPHORESIS.*

Platform. In order to use a biological molecule as the recognition element, the species is generally immobilized on some substrate by using chemical attachment, physisorption, or entrapment procedures. These techniques are becoming increasingly desirable, because the biomolecule can be incorporated within a three-dimensional matrix without labor-intensive chemical modification, and the recognition element often maintains its native function and reactivity. Sol-gel processing methods have become an increasingly popular means of immobilizing recognition-sensor chemistries. This procedure makes it easy to form a porous, inorganic, glasslike material in which compounds such as biomolecules can be doped for chemical sensing purposes.

In the sol-gel process, a metal or semimetal (often silicon, titanium, or aluminum) alkoxide undergoes an acid or base hydrolysis. A common precursor is tetraethoxy orthosilicate $[(CH_3CH_2O)_4Si]$ or tetramethoxy orthosilicate $[(CH_3O)_4Si]$, which form silanols during hydrolysis. In a silanol structure, a silicon (Si) atom is bound to one or more hydroxyl (OH) groups. Condensation occurs between two resulting silanols, or a silanol and an alkoxide, to yield small colloidal particles. With time, polycondensation of these colloids results in Si-O-Si bonds, which in turn form the porous, three-dimensional, glasslike network.

A major advantage of this process is that the materials can be produced at room temperature, unlike conventional high-temperature glass processes that can damage temperature-sensitive dopants. Sol-gel encapsulation is also appealing, because there is no need to chemically modify the biological recognition element prior to its immobilization or entrapment. Thus, a sol-gel-immobilized biological recognition element can retain its conformation and reactivity. In covalent attachment schemes, biological recognition elements often orient in such a manner that their recognition sites become inaccessible to the analyte. Physisorption methods are random and often nonpermanent, and a significant fraction of the added biological recognition elements denature or lose their ability to recognize the analyte. In a sol-gel composite, the biological recognition element is generally added after the hydrolysis but before polycondensation is complete; thus it is essentially trapped within a porous, cagelike silicate network during the polycondensation process. Compared to other biosensing platforms, sol-gel-based materials have many positive features. For example, they are optically transparent and therefore suitable for common spectroscopic measurements; they are chemically inert, thermally stable, and easily processed; they can be molded into various sizes and shapes such as monoliths, thin films, powders, and fibers; and they are simple to produce.

Biorecognition element. Biological molecules are attractive as recognition elements because in nature their inherent function includes selective binding or transporting of chemical species. Therefore, biomolecules generally do not require any modifications to serve as recognition elements. Recently, much work has centered on building practical biosensors where proteins, antibodies, or enzymes serve as the biological recognition element. However, biological species are sometimes difficult to work with and are prone to lose functionality and denature. Fortunately, several researchers have shown that it is possible to stabilize biological recognition elements and simultaneously maintain their function (that is, molecular recognition) by using special sol-gel processing conditions. Also, because the sol-gel is inherently porous, the entrapped biological recognition elements remain active and accessible to the analyte. The large size of biological macromolecules such as proteins or enzymes also minimizes leaching of the biological recognition element from the silicate matrix. Biological dopants that have served as recognition elements include hemoglobin, myoglobin, glucose oxidase, urease, enzyme inhibitors, cyclodextrins, and antibodies.

Analyte. The target analyte to be detected may include species of biological, clinical, environmental, or social interest. For example, it may be necessary to determine pH, biologically relevant metals ions, gases, sugars, deoxyribonucleic acid (DNA), or environmentally harmful organic substances and metals in polluted soils or waters. Thus, development of a biosensor for a particular analyte requires identification of a biological recognition element that either can naturally identify or can be altered in such a way that it will recognize selectively the species of interest. For instance, myoglobin in its native form reacts nonspecifically with nitric oxide, because oxygen binds with it competitively. However, this reaction is specific to nitric oxide (with no interference from molecular oxygen) if the manganese myoglobin form is used as the recognition element.

Research and development. Examples of sol-gel-derived platforms used for biosensing include hemoglobin and myoglobin for the determination of molecular oxygen and carbon monoxide, and oxalate measurements with the entrapped enzyme oxalate oxidase [with nicotinamide adenine dinucleotide phosphate-reduced (NADH) as the enzyme cofactor].

A biosensor has been developed that uses pyoverdin (a natural fluorescent pigment) doped within a sol-gel powder for the continuous determination of iron(III) ions (Fe^{3+}) in tap water and human serum. Optical detection was used to follow the formation of a pyoverdin-Fe^{3+} complex. This particular sensor was reportedly selective, reversible, and stable for up to 1 year.

Glucose. Because of its clinical importance, glucose remains one of the most important of all analytes. As a result, considerable effort has focused on the use of glucose oxidase in a sol-gel composite for glucose detection and quantification. Several aspects of a functional glucose biosensor have been characterized and optimized. These properties include biosensor response time, long-term stability and storage conditions, detection limits, activity of the biological recognition element enzyme within various sol-gel matrices, and enzyme loading (concentration). Researchers have described a glucose biosensor formed by sandwiching glucose oxidase between two sol-gel-derived thin films. The sand-

wich configuration exhibited extremely promising results. Response times (90% of maximum) with the sol-gel glucose oxidase:sol-gel sandwich film sensor were as short as 30 s, and detection limits were on the order of 0.2 mM glucose. Working curves were linear from 5 to 35 mM glucose, detection precision was on the order of 5% relative standard deviation, and the sandwich film biosensors could be used repeatedly (see **illus.**). Also, the sandwich architecture was stable for well over 2 months, which compares favorably with previous immobilization of glucose oxidase.

Urea. Urease entrapment represents a much more difficult challenge; the urease molecule is very large (590,000 kilodaltons), and its functional form is a hexamer. By using a tetramethoxy orthosilicate–based sol-gel:enzyme:sol-gel sandwich architecture and a photometric detection scheme, scientists were able to quantify urea. This sandwich architecture simultaneously allows the analyte (urea) access to the enzyme (urease) and permits high levels of enzyme loading without any detectable leaching of the enzyme from the film. The analytical working curve of photometric signal versus urea concentration is linear over the physiological urea concentration found in human blood (2–18 mM), and detection limits are 0.5 mM urea. The response time

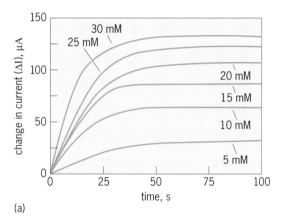

(a)

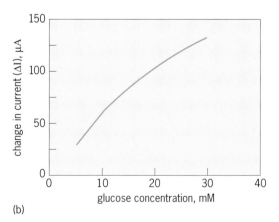

(b)

Typical experimental results using biosensors. (*a*) Current versus time profile after the addition of various concentrations of β-D-glucose. (*b*) Working curve: Amperometric response as a function of β-D-glucose concentration.

of the urea biosensor is on the order of 10 s, and it represents a hundredfold improvement over previous sol-gel-derived sensor schemes based on urease encapsulation.

Antibodies. Biorecognition is certainly not limited to metal-binding proteins and enzymatic reactions. Recent studies have shown that antibodies trapped within sol-gels promise new immunosensors. For example, the antibody antifluorescein has been entrapped within a sol-gel monolith and its affinity for its hapten studied. A significant population of the antibody retained its chemical affinity for fluorescein, and the antibody retained much of its intrinsic function. In these experiments, fluorescein simultaneously served as both the hapten and as the fluorescent probe. When fluorescein is not antibody bound, it exhibits a strong fluorescence signature at about 520 nanometers (green). The fluorescence is quenched significantly, and there is a distinct spectral shift when fluorescein is antibody bound. Thus, changes in the fluorescence spectrum, intensity, and excited-state intensity decay kinetics were used to probe the antifluorescein–fluorescein binding process within the sol-gel matrix. The results demonstrated that (1) an antibody can be trapped within a sol-gel glass matrix; (2) the analyte can diffuse into the porous sol-gel network; (3) the entrapped antibody remains functional, that is, it selectively binds its hapten; (4) the antibody retains a remarkable fraction of its affinity; and (5) the antibody within a hydrated, sol-gel-derived composite retains its affinity over time. Together these results bode well for the development of biosensing based on sol-gels and antibodies.

For background information SEE BIOELECTRONICS; COLLOID; ENZYME; GEL; MOLECULAR RECOGNITION in the McGraw-Hill Encyclopedia of Science & Technology.

Christine M. Ingersoll; Frank V. Bright

Bibliography. J. M. Barrero et al., Pyoverdin-doped sol-gel glass for the spectrofluorimetric determination of iron(III), *Analyst,* 120:431–434, 1995; A. M. Buckley and M. Greenblatt, The sol-gel preparation of silica gels, *J. Chem. Educ.,* 71(7):599–602, 1994; B. C. Dave et al., The sol-gel encapsulation methods for biosensors, *Anal. Chem.,* 66(22):1120A–1127A, 1994; U. Narang et al., Glucose biosensor based on a sol-gel-derived platform, *Anal. Chem.,* 66(19):3139–3144, 1994.

Bird migration

Migration is the seasonal movement of an animal to and from its breeding ground. Among birds, there are the true migrants, where all birds of the species migrate; the partial migrants, where only some birds of the species migrate; and the nonmigrants. Among the species that migrate, there is a great deal of variation in migratory patterns. Birds may migrate during the day or at night; they may migrate for relatively short distances or for long distances. Although some tropical species migrate merely a few kilometers as the seasons change, the Arctic tern migrates from its arctic or subarctic breeding grounds to winter in the Antarctic.

Flight muscles. Long-distance migrants may travel the entire distance without stopping, or they may break the journey into a number of shorter flights, with stopovers of up to several days between each section. Those birds that do perform long-distance migrations, however, perform sustained physical exercise at levels that are close to the maximum known among the vertebrates. It is necessary, therefore, for the flight muscles, which are responsible for generating the power required for the duration of the migration, to be adequately adapted for that purpose. In addition, the bird must carry sufficient fuel to provide the energy requirements of those muscles. These requirements are of particular significance for juvenile birds born during the summer, as the juveniles have to be sufficiently well developed to accompany their parents on the autumn migration to the wintering area.

One species that makes such migrations is the Svalbard population of barnacle geese (*Branta leucopsis*). During the last week of September, these birds fly in family groups approximately 1600 mi (2500 km) from their arctic breeding grounds along the west coast of the Svalbard archipelago (74–81°N) to winter at the Solway Firth (55°N) in southwest Scotland. They use flapping flight during the migration, expending energy at a very high rate. The goslings hatch during the first 2 weeks of July and leave the nest after approximately 24 h. They spend the next 12 weeks growing and preparing for the formidable migration that lies ahead.

Mass. An important aspect of locomotion in birds is that they have two independent locomotory systems, the wings and the legs. For the first 7 weeks of their lives, the goslings are unable to fly and use only their legs as they stay close to their parents and feed on the tundra. The leg muscles are approximately 13% of body mass in 3-week-old goslings (**Fig. 1**), whereas the pectoralis muscles (the major flight muscles) are a mere 1.5% of body mass. At 5 weeks of age, the relative mass of the leg muscles does not change much, but the pectoralis muscle mass more than doubles. By week 7 the pectoralis muscles are 14% of body mass, and the relative mass of the leg muscles has declined to approximately 9% of body mass. At this time, the fat content of the body of the goslings is low. During the next 4–5 weeks, the goslings accumulate fat as the fuel for the forthcoming migration and continue to increase their body mass. With their parents, they fly away from the breeding area to regions of lush vegetation, and they probably move progressively south as the colder weather approaches from the north. By the time they are ready to migrate, they have increased their body fat content by approximately 30-fold and the relative mass of the pectoral muscles has become approximately 17% of body mass. At the same time, the leg muscles are just

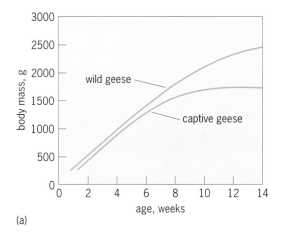

(a)

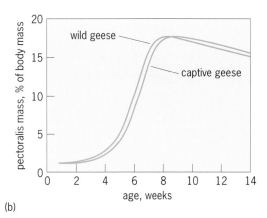

(b)

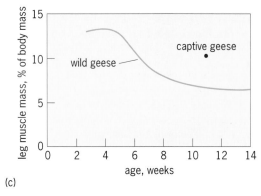

(c)

Fig. 1. Changes with age of body mass in populations of wild Svalbard and captive barnacle geese. (*a*) Total body mass. (*b*) Pectoral muscle mass. (*c*) Leg muscle mass. (*After C. M. Bishop et al., The morphological development of the locomotor and cardiac muscles of the migratory barnacle goose (Branta leucopsis), J. Zool., 1996*)

under 7% of body mass. Analysis indicates that, from the time the goslings are able to fly at 7 weeks of age, there is a direct relationship between body mass and the mass of the pectoral muscles. This relationship is also true for captive barnacle geese that have not migrated for several generations, although they do tend to have a lower body mass than the wild birds. Interestingly, in 11-week-old captive geese, which fly very little if at all, the mass of the leg muscles is, at approximately 10% of body mass, greater than that in wild geese at a comparable age.

Stamina. The flight muscles must have sufficient mass to enable them to generate the power required to make the birds of a given body mass airborne; in addition, they must be able to provide the lift and forward propulsion for the duration of the migratory flight. The flight muscles must have sufficient stamina. Stamina, or endurance, is the result of the production of energy from the oxidation of metabolic substrates such as glucose or fatty acids. Birds store fats rather than carbohydrates, because the energy derived from the oxidation of a given amount of fat is approximately 2.3 times that obtained from the oxidation of the same amount of carbohydrate. However, in order for the oxidation to occur, there must be sufficient amounts of oxygen delivered to the muscles by the respiratory and circulatory systems, and the enzymes responsible for the oxidation of the substrates must also be in sufficient quantity. An enzyme which is important for the oxidation of metabolic substrates is citrate synthase, and the relative activity of this enzyme can be used as an indicator of fitness.

The activity of citrate synthase in the pectoralis muscles is approximately 10 μmol substrate per minute per gram of tissue in 1-week-old birds, and does not change substantially until the goslings are beyond the age of 5 weeks, when there is a steep increase (**Fig. 2**). There is then a slower increase over the next 4–5 weeks to approximately 100 μmol substrate/(min)(g). It is interesting to note that there is a similar profile to the change in the concentration of plasma thyroxine and that both of these are similar to the relative growth of the pectoralis muscles. However, over the same period there is a continual decrease in the concentration of plasma growth hormone. Thus, it is possible that the increases in mass of the flight muscles and in their oxidative capacity are, at least partly, the result of an increase in the concentration of the thyroid hormone thyroxine against a background of a reduction in the concentration of growth hormone. Certainly, both of these variables are significantly affected in 6-week-old goslings by manipulating the plasma concentrations of thyroid hormones.

There is a large difference between the mass-specific activities of citrate synthase in wild and captive populations of 11–12-week-old geese. Unlike the mass of the pectoral muscles, there is a significant difference in the mass-specific activity of citrate synthase between the 11–12-week-old wild and captive populations of geese, when plotted against body mass. Also, in the captive geese the activity of citrate synthase actually decreases slightly after the goslings reach 7 weeks of age. Thus, the flight activity of the wild geese from the age of 7 weeks may be important in the maintenance and further development of the oxidative capacity of their flight muscles.

Citrate synthase activity in the heart of goslings from the age of 3 weeks until they migrate is approximately 80% of that in the pectoral muscles of 11–12-week-old geese before migration. In con-

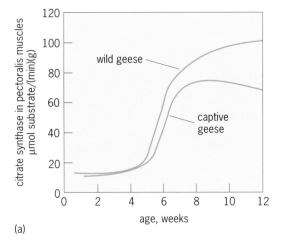

(a)

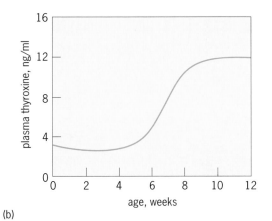

(b)

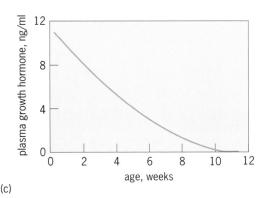

(c)

Fig. 2. Physiological changes with age in wild Svalbard and captive populations of barnacle geese. (*a*) Activity of citrate synthase in the pectoralis muscles. (*b, c*) Concentration of plasma thyroxine, *b*, and of plasma growth hormone, *c*, in the wild population. (*After C. M. Bishop et al., Development of metabolic enzyme activity in locomotor and cardiac muscles of the migratory barnacle goose, Amer. J. Physiol., 269:R64–R72,1995*)

trast, in a leg muscle of a 1-week-old gosling, the semimembranosus citrate synthase activity is approximately 50% of that in the pectoral muscle of an 11–12-week-old goose; this level is maintained until the bird reaches 7 weeks of age and begins to fly. At this time, there is a gradual decrease in citrate synthase activity, so that by the time the goose is

11–12 weeks of age and ready to migrate, citrate synthase activity in the semimembranosus muscle is approximately 20% of that in the pectoral muscle.

Implications. The development of the mass and the oxidative capacity of the flight muscles of barnacle geese appear to be genetically programmed in birds up to the age of 7 weeks, when they begin to fly. There is no difference between wild and captive birds in these respects, and thyroid hormones may be involved in these developments. Beyond the age of 7 weeks, the mass of the flight muscles is directly related to body mass in both groups of birds, whereas there is no such relationship for oxidative capacity. The implication is that flight activity (training) may be involved in maintaining muscle fitness in preparation for the migration. The development of the flight muscles is delayed until the birds are approximately 5 weeks old, when they begin to mature into the largest and some of the most oxidative muscles in the body.

For background information *SEE AVES; ENZYME; FLIGHT; MIGRATORY BEHAVIOR; MUSCULAR SYSTEM* in the McGraw-Hill Encyclopedia of Science & Technology.

P. J. Butler

Bibliography. P. J. Butler, Exercise in birds, *J. Exp. Biol.,* 160:233–262, 1991; E. Gwinner (ed.), *Bird Migration,* 1990.

Black hole

Recent advances in the study of black holes involve evidence for the existence of massive black holes in the centers of galaxies; and increased understanding of the apparent paradoxes associated with the phenomenon of black-hole evaporation, which may lead to their eventual resolution and to new insights into the physics of black holes and quantum gravity.

Evidence for massive black holes. Massive black holes have long been thought to exist in the cores of many galaxies that display energetic phenomena such as radio emission from relativistic jets and x-ray emission. These black holes should be surrounded by accretion disks through which material is funneled from the surroundings into the black hole. The efficient conversion of gravitational potential energy of this captured material into other forms of energy is the ultimate power source for the energetic phenomena seen in these galaxies. Although the black holes cannot be observed directly, their existence can be inferred from the behavior of material orbiting them in their accretion disks and beyond, and from the characteristics of radiation emitted from the infalling material from the accretion disk or radiation from background sources passing close to the black holes.

Gravitational redshift. A black hole can be characterized by its mass and its spin. Surrounding a black hole is an imaginary surface called the event horizon, inside of which not even light can escape. The radius of this surface, known as the Schwarzschild radius, is given by Eq. (1), where G is the gravita-

$$R_s = \frac{2GM}{c^2} \qquad (1)$$

tional constant, M is the mass of the black hole, and c is the speed of light. This equation was derived from newtonian physics in the eighteenth century (but with compensating errors due to limitations of the theory), as the radius at which the escape velocity equals the speed of light. The modern derivation, based on the principles of general relativity and quantum mechanics, gives the same answer. No information can pass through the event horizon to the outside world.

The energy, E_0, of a photon, or equivalently its frequency, leaving the vicinity of a black hole from outside the event horizon is reduced to energy E at the Earth according to Eq. (2), where R is the dis-

$$\frac{E}{E_0} = \sqrt{1 - \frac{R_s}{R}} \qquad (2)$$

tance from the black hole of the emitted photon. This effect, known as gravitational redshift, has been observed in the form of severe distortions in the spectral lines at optical and x-ray wavelengths suspected to arise from accretion disks. For example, emission from the $K\alpha$ line of iron at 6.7 keV has been observed by the Japanese *ASCA* satellite to have a very long wing that persists to below 5 keV. The wing is presumably caused by emission from various radii in an accretion disk ranging from about 3 to 10 times the Schwarzschild radius.

Orbital motion. Much of the evidence for black holes comes from observations of bodies in orbit around massive unseen objects. A body in a circular orbit of radius R around a central object of mass M will have an orbital speed given by Eq. (3), which is

$$V = \sqrt{\frac{GM}{R}} \qquad (3)$$

a variation of Kepler's third law. The mass of the Sun, for example, can be deduced from the orbital speed (or period) of the Earth and its orbital radius. In the search for black holes, examination of the velocities of stars as a function of radius from the centers of galaxies sometimes indicates high mass concentrations. However, it is difficult to obtain sufficiently high angular resolution to prove conclusively that the mass is not in the form of a dense star cluster. The Very Long Baseline Array (VLBA) is a radio telescope capable of achieving the high resolution and dynamic range needed to probe the cores of galaxies.

Telescope resolution. The angular resolution of any diffractive optical system is approximately equal to the reciprocal of its diameter in units of wavelength. Thus, the human eye has a resolution of about 1 arc-minute at a wavelength of 0.5 micrometer, and can therefore resolve or distinguish about 40 dots per centimeter (100 dots per inch) on a page held at arm's length. The Hubble Space Tele-

scope achieves its theoretical resolving power above the atmosphere of about 0.05 arc-second. In the radio range, wavelengths are much longer than in the visible range, for example, 1 cm rather than 0.5 μm. However, high resolution can be obtained by the method of interferometry, whereby the resolution of a very large aperture is achieved by an array of widely separated receiving elements or antennas. In one such array, the VLBA, ten 25-m-diameter (80-ft) antennas spread over the United States from Hawaii to St. Croix form an equivalent aperture of 8000 km (5000 mi) diameter. The resolution at 1.3-cm wavelength is therefore about 0.3 milli-arc-second. [If the human eye had such a resolution, newspaper print could be read at a distance of about 1000 km (600 mi).] Such an instrument is also well suited to track the relative motions of closely spaced objects. Motions as small as 1 micro-arc-second per year can be measured within a few years. If an astronaut on the Moon had radio beacons attached to the hand and fingernails, such a system could measure the rate of nail growth.

Natural masers. Only very bright objects can be discerned with such high resolution. Fortunately, very powerful beacons exist in the form of masers (an acronym for microwave amplification by stimulated emission of radiation), the microwave or radio version of the laser. Molecular clouds with trace amounts of water vapor can form one type of natural maser under certain circumstances. In the vicinity of a suitable energy source, the water vapor is pumped to high levels of excitation, which can lead to an excess population in energy levels that form the upper states of radio transitions. The cloud becomes an amplifier and produces beams of radiation with the equivalent intensity of a blackbody with a temperature of 10^{15} K. Because such a temperature is far beyond any reasonable physical temperature, the maser process is the only viable explanation for such a phenomenon. Such clouds appear as bright spots of very small angular size whose velocities along the line of sight can be inferred from the Doppler effect, since the rest frequency of the transition is known. When these masers can be detected, they are ideal probes of the dynamical properties of astrophysical systems. Water-vapor masers radiating at 1.3 cm (22 GHz) have been widely studied in this context.

Evidence for black hole in NGC4258. Strong evidence for the existence of black holes has been discovered by very high angular resolution VLBA observations of naturally occurring water-vapor masers in the center of the active galaxy NGC4258. This galaxy lies at a distance of about 2×10^7 light-years (1 light-year is approximately 10^{13} km or 6×10^{12} mi) in a cluster known as Canes Venatici, which is in the general direction of the Big Dipper. The full extent of the velocity range of the water-vapor maser in NGC4258 was discovered in 1992. The presence of emission at the systemic velocity of the galaxy, about 480 km/s (300 mi/s), was expected. However, there are also two clusters of high-velocity features offset from the velocity of the galaxy by ±1000 km/s (600 mi/s): a redshifted group near 1500 km/s (900 mi/s) and a blueshifted group near −500 km/s (−300 mi/s). The presence of this high-velocity emission indicates violent activity. The VLBA, with its angular resolution of 0.3 milli-arc-second and velocity resolution of 1 km/s (0.6 mi/s), revealed a disk viewed nearly edge-on in the center of the galaxy. The image delineated by the water-vapor masers appears as a highly elongated structure. At a distance of 2×10^7 light-years, an angular separation of 1 milli-arc-second corresponds to 0.1 light-year or about 10^{12} km (6×10^{11} mi). There is a smooth variation in velocities with position among the high-velocity features which precisely follows Kepler's law given by Eq. (3). The high-velocity features arise from the portion of the disk along a diametrical line that is perpendicular to the line of sight.

By application of Eq. (3) to the data in the **illustration**, the mass of the unseen central object that maintains the molecular disk in keplerian motion, as traced by the water-vapor masers, is estimated to be 3.5×10^7 times the mass of the Sun (the latter is 2×10^{30} kg or 4.4×10^{30} lb). More importantly, this mass must reside within the inner boundary of the disk, which has a radius of 0.4 light-year. If such a mass were evenly distributed in a spherical volume with a radius of 0.4 light-year, then the mass density would be 10^8 solar masses per cubic light-year. The Sun's nearest neighbor is 4 light-years distant, and thus the stellar density is in the solar vicinity of about 0.15 solar mass per cubic light-year. However, star clusters of density up to about 1000 stars per cubic light-year have been observed. These clusters are about 100,000 times less dense than the putative star cluster that might confine the molecular disk in NGC4258. Theoretical studies suggest that a star cluster of the density required in NGC4258 is not likely to survive for a significant fraction of the age of a galaxy. If the star cluster consisted of low-mass stars, then frequent collisions among them would disrupt the cluster. However, if it consisted mostly of massive stars, their powerful

gravitational interactions would expel member stars, leading to the evaporation of the cluster. Thus, it appears impossible to maintain a stellar cluster of the required density, and a black hole seems to be the only viable candidate to explain the central mass.

The Schwarzschild radius of an object with a mass of 3.5×10^7 solar masses is, from Eq. (1), 10^8 km (6×10^7 mi), about equal to the Sun-Earth distance. The molecular disk has a radius of about 40,000 times the Schwarzschild radius. Hence, whereas the mass is accurately determined, the direct relativistic effects normally associated with black holes are small. For example, the gravitational redshift, given by Eq. (2), is only about 4 km/s (2.5 mi/s) and the gravitational lensing effect is only about 0.1 micro-arc-second. Only the former effect is measurable.

Disk properties. The properties of the disk are precisely defined by the VLBA imaging. Although the disk is slightly warped, it is very thin. The ratio of its thickness to its radius (H/R) is less than 0.0025. If the disk is in hydrostatic equilibrium with gas pressure balancing gravity, the sound speed is $V(H/R)$. Hence, for H/R less than 0.0025 and $V = 1000$ km/s (600 mi/s), the sound speed must be less than 2.5 km/s (1.6 mi/s), which means the temperature must be less than 1000 K (1300°F). In addition, any magnetic field that contributes to the pressure must have a strength less than 0.3 gauss (3×10^{-5} tesla). The simple theory of viscous dissipation in keplerian disks shows that friction among particles due to random motions causes a gradual inward drift of material, with velocity of $\alpha V(H/R)^2$, where α is a friction parameter whose value is between 0 and 1. For $\alpha = 0.1$, the drift velocity is less than 6 m/s (20 ft/s), and the accretion rate into the black hole is less than about 10^{-4} solar mass per year. A 1% conversion efficiency of this mass to x-ray radiation would be sufficient to account for the observed x-ray source in the center of NGC4258 of 10^{34} W. The stability of the disk can be assessed by comparing the balance of gravity and pressure, as well as gravity and centrifugal force in response to small perturbations. The disk in NGC4258 has low enough mass (probably less than 10^5 solar masses) compared to the central (black hole) mass to be at least marginally stable.

The features in the systemic part of the spectrum show a systematic drift of about 9 km/(s)(yr) [6 mi/(s)(yr)], which is a measure of the centripetal acceleration in the disk, V^2/R. Since V is known, the measurement of acceleration yields an estimate of the linear radius of the disk. From the angular size determined from the VLBA observations, a distance estimate of $(2.0 \pm 0.3) \times 10^7$ light-years is obtained.

Prospects. The VLBA has extraordinary accuracy in measuring relative motions. At a distance of 2×10^7 light-years, the systemic features should move at a rate of 32 micro-arc-seconds per year. Although the rotation period of the inner edge of the disk is about 800 years, this motion can be measured accurately within a few years. Thus, a precise

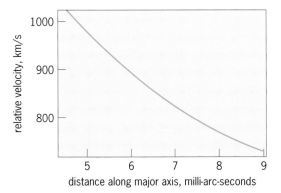

Magnitude of the line-of-sight velocities of high-velocity features of NGC4258 with respect to the systemic velocity of the galaxy as a function of angular offset from the center of the disk. Velocities lie near the curve given by Eq. (3) in the text, with mass of 3.5×10^7 solar masses. 1 km/s = 0.6 mi/s.

method of measuring the distance to the molecular disk and hence to the galaxy is provided.

About 500 galaxies have been surveyed for water-vapor masers, and about 12 masers have been discovered. Their distances range up to about 3×10^8 light-years. Careful imaging of these 12 could reveal evidence for other black holes.

James Moran

Problems of black-hole evaporation. Because of their remarkable properties, black holes play a key role in the search for a quantum-mechanical theory of gravity. The discovery of black-hole evaporation has created a growing tension between fundamental assumptions about gravity and quantum mechanics, in the form of an apparent loss of information during gravitational collapse. Recent developments have shed new light on the problem, either through a more careful application of known theories or by the introduction of new physical principles.

Black-hole radiance. According to classical physics, a black hole is a region of space from which nothing can emerge. Any particle that is located inside a surface, known as the horizon, is inevitably drawn farther into the black hole by the gravitational field. The area of the horizon, which for the simplest black holes is proportional to the square of the hole's mass, necessarily increases or stays constant, since matter can flow only inward through the horizon, not outward.

In the 1970s it was realized that quantum-mechanical effects dramatically alter this picture: particles are, in fact, emitted by the black hole, causing it to lose mass by evaporation. Although a proper explanation of this effect, known as Hawking radiation, would require a technical derivation, it can be made plausible as follows. According to the quantum theory, even regions of space that are normally perceived as empty are, in reality, filled with fluctuations; particle-antiparticle pairs are continually created and destroyed over very short time scales. Despite this activity, in the absence of gravitational or other fields, conservation of energy is sufficient to ensure that there is no net production of particles. But in the presence of a black hole this restriction is not applicable, as particles located inside the horizon can have negative energy. Thus there is the possibility of particle-antiparticle creation at the horizon, one particle flowing into the hole, one flowing out, without violating any conservation laws.

Upon detailed analysis, it is found that the spectrum of particles emitted is exactly the same as that from a thermal body of matter with a temperature that is inversely proportional to the mass of the black hole. The temperature is very small for sizable black holes; for example, a black hole whose mass equals that of the Sun has a temperature of 6×10^{-8} K. However, the evaporation process accelerates, since the temperature increases as the mass of the black hole decreases. But this effect is sufficiently weak that a solar-mass black hole would not have emitted a measurable fraction of its mass during a time equal to the present age of the universe.

The ultimate fate of an evaporating black hole is open to conjecture, as the radiance calculation depends on certain approximations which become unreliable once the black hole reaches the Planck mass, which is of the order of 10^{-5} g. Possibly, the black hole disappears completely, or some new effect causes the evaporation to terminate, leaving a stable remnant. A working theory of quantum gravity is needed to decide the issue.

Apparent information loss. A disturbing feature of the emitted radiation is its lack of information content. In particular, although it is possible to imagine forming a black hole in many different ways, from different kinds and distributions of matter, the radiation spectrum seems to depend only on the black hole's total mass, charge, and angular momentum. That this situation is unsettling may be appreciated by considering the case where the black hole evaporates completely. (The stable-remnant scenario has difficulties of its own, which will not be examined here.) This evaporation gives rise to a process by which two different configurations of matter can, by forming a black hole which subsequently evaporates, evolve into identical final configurations. In this sense, black holes seem to cause a loss of information. However, quantum mechanics has built into its very structure the existence of equations uniquely connecting the past with the future. If information loss really does occur, then quantum mechanics will have to be revised in order to be compatible with gravitational collapse.

Given the overwhelming successes of quantum mechanics, most physicists are loath to alter it in any substantial way, preferring to believe that the appearance of information loss is illusory. In order to see the problem more clearly, it is helpful to consider a less exotic process: the burning of a lump of coal. In this case, too, once the coal has been reduced to ash and radiation, it appears that information about the initial lump has been irretrievably lost. However, although this case is certainly true for all practical purposes, in principle, by making careful measurements, it is possible to recover all details of the lump's initial structure. The details are encoded in a complicated way in the correlations of the emitted radiation and the remaining ash.

Perhaps the information apparently lost in black-hole evaporation can be restored in a similar manner. Although the radiation has been characterized as being independent of the black hole's precise origin, this description is in fact known to be true only within the context of an approximate calculation. The original approximation neglected the fact that the radiation itself produces a gravitational field, which can then act back on the radiation. Later it was learned how to incorporate the gross, qualitative effects of this so-called back reaction by treating the radiation as a classical field with an attendant classical gravitational field. Although such an approach leads to many valuable insights, it is incapable of determining the precise correlations between the discrete quanta which compose the radiation, and which presumably carry the information, if the radiation is there at all. Any such determination requires that the radiation and gravitational field be treated as the quantum-mechanical variables that they are.

Recently, some steps were taken in this direction. Specifically, it was shown how to compute the modification to the rate of single-particle emission due to the effect of gravitational self-interaction. This approximation includes the force that a particle exerts upon itself, but neglects the forces between different particles. This truncation gives what is probably the dominant correction to the emission spectrum, and is able to provide reliable results in certain cases where previous methods fail completely. Still, the neglect of interparticle forces renders the multiparticle correlations trivial and removes the possibility of information recovery. Currently under way are attempts to refine the model so as to allow nontrivial correlations to emerge. It is hoped that these correlations will have sufficient structure to point the way toward a resolution of the puzzle.

Application of string theory. The approach just described is a relatively conservative one, as it proceeds by making systematic improvements to the standard approximation schemes, rather than by introducing new physical principles. Possibly, it will prove to be inadequate for this very reason. An alternative approach uses the properties of string theory, a theory that unifies gravity with the other known forces and posits that the fundamental building blocks of matter are extended strings rather than point particles. This class of ideas changes the character of the problem. In point-particle theories, the horizon sharply delimits the regions from which matter can or cannot escape from the black hole; a particle is either inside or outside the horizon. Thus it is extremely difficult to preserve information, for once a particle has fallen though the horizon there is no way that its information content can be communicated to the outside world. In string theory, however, there is a possibility which has no point-particle analog: a string can straddle the horizon, being both inside and outside. Because the inclusion of extended objects introduces many technical complications, a deeper understanding is required to ascertain whether this effect has a profound impact on the radiation.

It should also be pointed out that the notion of the horizon as a sharp boundary in particle theories exists only within a classical context. Once quantum effects are incorporated, the horizon becomes subject to fluctuations and ceases to exist at a well-defined location. But again, a quantitative understanding is lacking.

Black-hole entropy. Closely related to the problem of apparent information loss is the challenge of understanding what is known as black-hole entropy. In general, a body of matter in thermal equilibrium distributes its energy among a number of different internal states. Entropy is a measure of the number of such states, specifically its logarithm.

That a black hole should have an intrinsic entropy was in fact conjectured before the radiance phenomenon was discovered. This was possible because of a property remarked upon earlier: in classical gravity the total area of all black-hole horizons never decreases. Based on the similarity with the second law of thermodynamics, that entropy never decreases, it was argued that a black hole has an entropy proportional to the area of its horizon. The subsequent discovery of the radiance enabled the proportionality constant to be determined. Although the number of black-hole states is thereby known, their nature has never been elucidated. An attractive possibility is that the states are described by distinct configurations of the horizon, a discrete number of which might arise upon quantization. A solution to the information-loss puzzle can be envisioned along these lines, by having an infalling particle transfer its information content to the states of the horizon, which then emits correlated particles. In any event, it seems clear that further study of the entropy phenomenon will yield valuable insights into the workings of black holes and quantum gravity in general.

For background information SEE BLACK HOLE; ENTROPY; MASER; ORBITAL MOTION; QUANTUM GRAVITATION; RADIO ASTRONOMY; RADIO TELESCOPE; RELATIVITY; RESOLVING POWER (OPTICS); SUPERSTRING THEORY; X-RAY ASTRONOMY in the McGraw-Hill Encyclopedia of Science & Technology.

Per Kraus

Bibliography. S. W. Hawking, *A Brief History of Time: From the Big Bang to Black Holes,* 1988; J. Kormendy and D. Richstone, Inward bound: The search for massive black holes in galactic nuclei, *Annu. Rev. Astron. Astrophys.,* 33:581–624, 1995; M. Miyoshi et al., Evidence for a massive black hole from the high rotational velocities in a sub-parsec region of NGC4258, *Nature,* 373:127–129, 1995; Y. Tanaka et al., Gravitationally redshifted emission implying an accretion disk and massive black hole in the active galaxy MCG-6-30-15, *Nature,* 375:659–661, 1995; K. S. Thorne, *Black Holes and Time Warps: Einstein's Outrageous Legacy,* 1994.

Bone

Bone and most other vertebrate hard tissues have three main components: the fibrous protein collagen, a variant of calcium phosphate called apatite, and water. These are not materials that technologists would even begin to consider as starting blocks for making an engineering material. However, vertebrate mineralized tissues have quite respectable stiffness, and many of them are very tough. Toughness means being resistant to impacts, and not being made weak by any small flaws or notches. For example, glass fibers are strong but not tough. Different vertebrate mineralized tissues have greatly varying mechanical properties that are suitable for precisely the mechanical requirements imposed on them by natural selection. The stiffness of bone increases with the amount of mineral present, as can be fairly well understood from relationships derived from manufactured composite materials. However, how bone gets its toughness is much less well understood, even though toughness is just as important a property as stiffness.

Another feature of bone is that it avoids fracturing by being able to adapt to the loads put on it—the champion tennis player's serving arm is much more robust than the arm that merely throws the ball in the air. This adaptive property is of great importance for people designing orthopedic prostheses: bone responds to the presence of a prosthetic device by becoming understressed, and accordingly dissolving. Proper understanding of the signals that induce bone cells to produce more bone or to destroy bone is of great importance.

Toughness. The stress/strain curve of bone is superficially similar to that of metals: there is an initial straight region in which the strain is proportional to the stress, and then a so-called yield region where the curve bends over and is then almost flat, so that a small increase in stress will produce a large increase in strain (see **illus.**). If bone is to be tough, it is essential that there is this flattish part of the stress-strain curve, because it is here that much work has to be done to extend the bone to the point at which it breaks. In metals, this post-yield region is called the plastic region, and is caused by dislocations moving through the metal, allowing it to deform, but it is not associated with any fracture in the metal. In the post-yield region in bone (and other composite materials), the specimen develops thousands of tiny cracks. This microcracking can be heard by sensitive transducers (acoustic emission analysis). The interesting feature of this microcracking is that the cracks do not consolidate to form a large crack, which would run rather easily through the material. The study of microcracking in bone has been greatly helped by the development of the laser scanning confocal microscope, which allows optical slices to be taken though the depth of a specimen, and these optical slices can then be manipulated by a computer to give various three-dimensional views of the cracks.

Highly mineralized bone is almost unable to form microcracks, and it is brittle. For acoustical reasons, bones in the ear need to be very stiff and therefore must be highly mineralized. Thus, they are weak and brittle. However, as they are shielded from large loads, this brittleness is not important. Deer antlers, in contrast, are used for fighting, so require a high degree of toughness. Although adequately stiff, they are less mineralized than ordinary bone but are very tough, and they form a myriad of microcracks. It is not clear how these little cracks are prevented from joining together, but it is probably brought about by the microstructure of the bone. For instance, in a type of bone found in large animals such as cows, horses, and dinosaurs, sheets of more highly mineralized bone alternate with sheets of less highly mineralized bone. Cracks form in the highly mineralized layers, but are brought to a halt when they enter the poorly mineralized layers.

Toughening mechanisms are not restricted to the formation of microcracks. The tusk of the narwhal (an arctic whale) needs to be very tough to stand up to the testing knocks of rival males. It has a two-stage toughening mechanism. The tusk is made of dentine, which is like bone but with very uniformly oriented collagen fibers. Permeating the collagen fibers are little heaps of apatite which grow and fuse, leaving a poorly mineralized region where they abut. When tusk dentine is loaded with tension, these poorly mineralized interfaces slip a bit, at 45° to the load. No cracks are formed, but energy is used up in causing this slippage. If the load is increased, conventional microcracks appear and use up energy.

The formation of microcracks is advantageous because as they form they absorb energy that might otherwise drive big cracks forward. It seems that the microcracks that form in bone when it is loaded may slowly increase its resistance to impact. This is brought about by a process called stress shielding whereby the diffuse array of microcracks around the tip of a big crack reduces the stress at the tip itself. A damaged bone, therefore, may not have its strength reduced; it may even be increased. Although the bone is more compliant, it is less stiff than its original design. However, because bone is a living material, it can be remodeled, the microdamage removed, and the bone stiffened and restored to its pristine condition.

Adaptive remodeling. Bone can remodel adaptively in response to the loads put on it. Experiments to find out how this occurs are extremely difficult, but the situation is being improved by two quite unrelated scientific advances in computers and in molecular biology. The development of ever more powerful computers allows scientists to explore the effects of different remodeling algorithms on the behavior of bone. If bone is to change its shape adap-

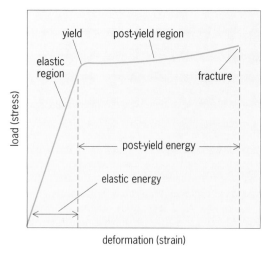

Stress-strain curve for a bone specimen loaded in tension. The energy absorbed up to any point on the curve is proportional to the area under the curve.

tively, a set of rules (an algorithm) is necessary to determine how the cells that lay down bone (osteoblasts) and the cells that destroy bone (osteoclasts) respond to the imposed loads. Until recently, the complex shape of bones made detailed analysis of the stresses in bone impossible, so the effect of remodeling could not be determined. However, finite element analysis is a computer technique that involves construction of a model of a complex shape, such as that of a bone, and the application of arbitrary loads to discover the resulting stresses and strains. Then, the model bone can be made to behave according to some algorithm, to find out its remodeling response and how remodeling affects the stress and strain. Finally, the experimentally observed response of the bone can be related to the responses of the computer model by using various algorithms. It is not known what algorithm, or set of algorithms, will most closely mimic the behavior of bone.

Recent experiments have studied the effects of strains on both living bone and cultured bone cells. After a brief period of loading, there is an immediate increase in the activity of the enzyme glucose-6-phosphate dehydrogenase in bone cells, followed within hours by increased ribonucleic acid (RNA) synthesis. Interestingly, individual bones in the skeleton have different responses to the presence or absence of factors that change bone mass. For example, the skull is loaded much less than the long bones during ordinary activity, yet it maintains its mass in the face of hormonal changes that induce loss in the more heavily loaded long bones. The mechanism for these regional differences in response to loading and hormones probably involves some location-dependent information, such as different cell-matrix interactions, allowing cells to alter the timing or amount of production of different cytokines (substances which act as local hormones).

The final goal of such studies is to identify the early genes in the cascade of events that culminate in altered bone mass. These experiments are leading toward attempts to mimic pharmacologically the effects of mechanical loading, so as to modify remodeling in living bone even more successfully than the finite element models.

For background information SEE BONE; PROSTHESIS; SKELETAL SYSTEM in the McGraw-Hill Encyclopedia of Science & Technology.

John D. Currey

Bibliography. F. Lyal and A. J. El Haj (eds.), *Biomechanics and Cells,* 1994; A. Odgaard and H. Weinans (eds.), *Bone Structure and Remodeling,* 1995; P. Zioupos and J. D. Currey, The extent of microcracking and the morphology of microcracks in damaged bone, *J. Mater. Sci.,* 29:978–986, 1994.

Brain

Recent research involving the brain has focused on the effects of early sensory experience on neural structural development and the connection of age-related changes with qualitative effects in cognition.

Early Sensory Experience and Neural Structural Development

Experimental neuroscience has determined that during the early stages of development, radical alterations of visual, somatosensory, or auditory experience alter patterns of neural circuitry. The capacity of early sensory experience to begin a cascade of downstream effects is of great interest because modern neuroscientists hold the fundamental assumption that human thoughts, feelings, and behavior come from the operations of neural anatomical circuitry. Thus, if an individual radically changes that circuitry and its operations, thoughts, feelings, and behavior will also change.

Sensory deprivation during early stages of brain development has been studied in visual, auditory, and somatosensory modalities. In various animal studies, early visual experience has been limited to monocular vision or exposure to lines in only one orientation; auditory experience, by plugging one ear; and somatosensory experience, by removal of whiskers or severing the nerve that transmits facial vibrissal sensory information to the brain.

Studies in developmental neurobiology have shown that by radically altering neural activity in sensory pathways of a normally developing brain early enough in its development neurons in distant locations (that otherwise would have been normal) will become abnormal in structure and function.

Types of structural changes. Different forms of early sensory deprivation produce different forms of anatomical abnormality. During fetal growth in rats, elimination of the neural signals from the facial whiskers causes both a reduction in the cerebral cortical map representing the facial whisker area and an excessive enlargement of its cortical map representing facial structures (the lower lip and jaw) that remain normally neurally active. In kittens, monocular deprivation interferes with the development of ocular dominance columns in the visual cortex as well as producing the types of abnormalities just described. Limiting the vision of newborn kittens to lines in one orientation also interferes with the development of cortical neurons that respond to lines in other orientations. In rare cases of humans who are born without eyes, or in studies using monkeys whose retinal activity has been eliminated from birth, novel cortical areas are formed, the size of the primary visual cortex is drastically reduced, and the number of callosal connections between secondary visual cortical areas is abnormally increased. Thus, experience-driven changes have been described for a variety of neural structures, ranging from synapses to axon arbors to cortical representational maps to entire cortical regions.

Mechanisms. During early stages of brain development, neural activity triggered by sensory experience may be thought of as a morphological agent. It is one of possibly several important neurobiological factors involved in shaping, specifying, selecting, and maintaining brain structure and function.

Absent, abnormal, or unbalanced sensory neural activity leads to unusual and abnormal structural and functional development. Abnormality in one site can propagate outward to distant sites, leading to altered and abnormal structure and function elsewhere. The cerebral cortex appears to be particularly vulnerable to such abnormal instructions from sensory systems and readily constructs abnormal representations which are enlarged, shrunken, absent, novel, underresponsive, or nonselectively overresponsive. For example, reduced activity in the peripheral receptors can lead to arrested or reduced growth in distant systems that receive input from that site, or to complex alterations of cerebral cortical maps in which some representations are reduced, while others are excessively enlarged.

Although it is not known whether the relative amount of activity or the pattern of activity within sensory systems is most critical, it is useful to keep in mind that "Neurons that fire together, wire together; those that don't, won't." Competition and cooperation are opposing and complementary forces that drive and shape neural structural development.

Critical periods. During development, there are critical periods during which the effects of deprivation of sensory experience are maximal. For example, if elimination of the neural signals from the facial whiskers of a rat had occurred in the neonatal instead of the fetal period, certain structural changes in the brain would not have occurred. Future research may find that each cortical system has its own timetable of vulnerability to such structural changes. In general, however, the earlier the onset and the greater the unbalanced or abnormal neural activity, the more abnormal will be the development of brain structures.

Future research. Many factors contribute to the formation of neural circuitry in the developing brain. In addition to the effects of early sensory experiences on neural circuitry formation, future research will examine the role of genetic constitution in minimizing or amplifying structural changes driven by sensory experiences. Also, because early sensory experiences affect structural development in sensory systems, it is reasonable to assume that they will also influence the neural structure of higher-order systems, including those involved in cognitive, attentional, memory, emotional, language, and social functions. Evidence to support this assumption will likely be provided by future research. Further, just as early sensory experience may trigger structural maldevelopment, abnormal neural activity initiated within an abnormally developed brain site may trigger structural abnormalities further downstream. Such effects may underlie some human developmental disorders. For example, in infantile autism, magnetic resonance imaging studies show that the cerebellum is abnormally developed, a finding corroborated by autopsy evidence, showing that the number of Purkinje neurons is substantially reduced in the cerebellum. The cause of these abnormalities in autism is as yet unknown, although genetic defect and viral damage are possible causative agents. A new theory suggests that in autism abnormal neural activity initiated within the damaged cerebellum triggers abnormalities in other brain structures which are not the original damaged site. SEE LANGUAGE PROCESSING.

Finally, to detect evidence of experience-driven brain changes, studies using mammals have employed radical methods of altering sensory experience. However, there is evidence from insects that, even under normal conditions, subtle differences in visual sensory input can significantly alter neural structure.

Eric Courchesne

Cognitive Function and Age-Related Changes

As humans age from the middle to the older adult years, changes in cognitive functioning are common. The most noticeable changes are a general slowing in many aspects of behavior and impairments in memory. Not only are motor responses slower (for example, elderly drivers brake more slowly when the traffic light turns red) but also cognitive responses are slower (for example, the elderly are slower to process and determine the meaning of a traffic light that has turned red). Memory for events in the distant past may be unchanged, but the ability to store and work with new information declines.

Tissue atrophy. The physical characteristics of the human brain also change with advancing age. There is a progressive loss of brain tissue with advancing age: compared to maximum brain size in young adulthood, 5% of the tissue has been lost by age 70, 10% by age 80, and 20% by age 90. Most human cognition functioning depends on a thin layer that forms the surface of the brain, that is, the cerebral cortex or neocortex. If the human cortex were spread out it would form a sheet about half a square meter in size. In order to fit into the skull, brain tissue is folded into a convoluted form of ridges and clefts. The ridges are called gyri, and the clefts are called sulci. Most of the loss in brain tissue occurs in the neocortex rather than structures located deep within the brain. The gyri atrophy, the sulci widen, and the ventricles, that is, the open spaces that carry cerebrospinal fluid to the brain, dilate. The cortex is conventionally divided into lobes: frontal, parietal, temporal, and occipital. The greatest loss of tissue is in the frontal lobe of the cortex, although parietal and temporal cortexes also show considerable loss.

Changes in blood flow. With age changes also occur in blood flow in the brain. The cerebral blood vessels become more coiled and tortuous. Their numbers decrease, particularly in deeper layers of the cortex where the greatest cell loss occurs. Even in healthy individuals, there are significant declines in the amount of blood flow, particularly in frontal, parietal, and parts of temporal cortex (see **illus.**).

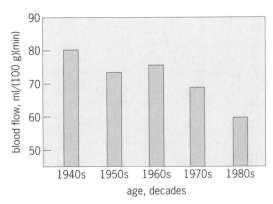

Blood flow in the cerebral cortex as a function of age. (*After T. G. Shaw et al., Cerebral blood flow changes in benign aging and cerebrovascular disease, Neurology, 34 (July): 855–862, 1980*)

However, it is still unknown whether brain tissue dies because of an impaired blood supply or because the blood supply is reduced as the number of surviving cells declines.

Neuronal loss. There is some evidence for loss of small interneurons that connect within a column of neurons or between adjacent columns. The neuronal loss is particularly evident in the hippocampus, that is, a structure which is folded into the inner or medial portion of the temporal lobe. The rest of the neocortex may have evolved from the hippocampus and surrounding structures. One technical problem is that neurons may not die but may, instead, shrink and be confused with glial cells. Glial cells provide support functions to neurons. These cells proliferate and increase in size in response to damage to neurons, and their numbers increase with advancing age.

Other changes. A number of other cellular abnormalities occur, including accumulation of the age pigment lipofuscin, plaques built around a core of amyloid protein, tangles of filaments, and vesicles filled with small, dense granules. These abnormalities are seen in pathologies such as Alzheimer's disease, but they are also found in normal, apparently healthy elderly adults. These abnormalities are also particularly common in the hippocampus and related areas.

Cognition. One theory of the relationship between brain change and cognitive function holds that because the changes in the brain are widespread, diffuse changes in cognition should be observed. Such theorists see general behavioral slowing as the fundamental cognitive change. Thus, many of the specific complaints of elderly adults, such as memory lapses, can be explained by slower and more erratic transmission of information through cortical networks that have suffered losses of neurons. Some theorists point to the evidence for extensive changes in areas such as the hippocampus and frontal lobes, predicting selective impairments in cognitive functions that are dependent on those areas. Both positions may be correct: there may be specific impairments superimposed on a general background of slowing.

Until recently, most of the evidence for selective impairments came from neuropsychology, that is, the study of the cognitive and behavioral effects of brain damage due to disease or injury. Individuals with damage to a particular brain area show certain behaviors. When elderly adults show similar behaviors, the tentative conclusion is that aging causes damage to that brain area. The clearest evidence is for the hippocampus and the frontal lobes.

Hippocampus. Damage to the hippocampus produces an inability to consolidate new memories, anterograde amnesia, which is similar to the memory difficulties in elderly adults but more severe. In both amnesia victims and older adults, short-term memory (for example, the ability to remember a phone number long enough to make a call) is spared but long-term memory (for example, the ability to recall in the afternoon items on a grocery list that were memorized that morning) is impaired. Studies of amnesia have found an important distinction between memory with and without awareness. Memory with awareness involves conscious recollection of the past, and is usually measured with explicit tests such as asking the person to recall a specific event; memory without awareness does not require conscious recollection and is usually inferred. Both amnesia victims and older adults show sparing of memory without awareness (implicit memory) and impaired memory with awareness (explicit memory). The sparing of implicit memory in older adults is not complete however. On tasks that appear to require a substantial contribution from the frontal lobes, older adults perform more poorly than amnesia victims. SEE MEMORY.

Frontal lobes. Damage to the frontal lobes affects executive, planning kinds of functions. It also affects memory performance, but rather than affecting the memory itself the effects are on the uses of memory: inferences from remembered information, placing the information in context, ordering the information in time, preventing irrelevant information from memory from interfering, and implementing appropriate strategies for encoding and retrieving information. Older adults perform poorly on neuropsychological tests to diagnose damage to the frontal lobes. One such test is the Wisconsin card-sorting test. The individual is given a deck of cards that vary in the number, shape, and color of items, and is asked to sort them into piles. The examiner has a rule, but the individual can learn the rule only by inference, that is, by placing a card and learning whether the placement was correct or not. The examiner changes the rule at intervals without telling the individual. A critical measure is perseveration, that is, how long the individual continues to use a rule that is no longer correct. Elderly adults and individuals with frontal lobe damage show much greater perseveration than young, normal adults. Curiously, individuals in both groups often say that they know they are sorting incorrectly after a rule shift, even though they continue to use the rule. Another such test is the Stroop color word test. This test requires an individual to name the color in which items are displayed. In the critical sections, the items are themselves names of colors that are inconsistent with the color in which they are presented; for example, the word red displayed in blue. The inconsistent words are particularly problematic for older adults and individuals with frontal lobe damage.

The neuropsychological evidence is consistent with impaired functioning of the hippocampus and frontal lobes in old age. However, the neuropsychological approach is not without difficulties. Brain damage is seldom confined to theoretically distinct brain areas, and most tests require processing in a number of areas. Moreover, older adults may show behavior analogous to individuals with frontal lobe damage not because of frontal damage, but because of damage to another area on which the frontal lobes are dependent.

Brain-imaging techniques. Functional brain imaging promises to avoid these problems and to substantially expand the understanding of aging and changes in brain functioning. In positron emission tomography (PET), small amounts of radioactively labeled oxygen are injected or inhaled. Radiation detectors can then identify the brain regions that are particularly active and therefore take up more oxygen. Thus, researchers can determine the brain systems involved in different cognitive activities in healthy, living individuals. Early studies revealed that activation was generally lower in elderly adults. Nonetheless, younger and older adults show the same increase in activation in response to a cognitive challenge, such as a difficult problem, or to increased motivation. The cortex is not less subject to arousal in older adults, in contrast to individuals with organic brain diseases who have lower activation levels that do not respond to challenge.

Preliminary studies of the brain systems underlying memory for faces and locations show that the same brain areas are activated in older and younger adults. There are, however, differences. For older adults (but not younger adults) when the task is to remember faces, brain areas for location memory are also activated. Thus the intriguing possibility is raised that in the older brain areas that carry out similar computations may be recruited to handle aspects of a task that can no longer be carried out by the primary area.

The most recent development in brain imaging is functional magnetic resonance imaging. Oxygen use in the brain is detected through the response of hemoglobin molecules to very rapid shifts in magnetic fields. Because no radioactive material is required, possible health risks to elderly volunteers are greatly reduced, and it is possible to test the same individuals repeatedly to explore longitudinal changes. SEE MAGNETIC RESONANCE IMAGING.

For background information SEE AGING; BRAIN; MOTIVATION; MEMORY; NEUROBIOLOGY in the McGraw-Hill Encyclopedia of Science & Technology.

Alan A. Hartley

Bibliography. F. I. M. Craik and T. A. Salthouse (eds.), *The Handbook of Aging and Cognition,* 1992; E. R. Kandel, J. H. Schwartz, and T. M. Jessell (eds.), *Principles of Neural Science,* 3d. ed., 1991; H. P. Killackey, Neocortical expansion: An attempt toward relating phylogeny and ontogeny, *J. Cog. Neurosci.,* 2:1–17, 1990; B. Kolb and I. Q. Whishaw, *Fundamentals of Human Neuropsychology,* 4th ed., 1995; M. I. Posner and M. E. Raichle, *Images of Mind,* 1994.

Breeding (plant)

The reduction or elimination of crop losses that result from pests and disease-causing organisms is a primary goal in agricultural production systems. Pests, such as insects and nematodes, cause damage by direct consumption of crop plants, whereas disease-causing organisms, such as bacteria, fungi, and viruses, incite symptoms that result in loss. Typical symptoms of crop diseases include rot, wilt, chlorosis, abnormal growth, and loss of vigor. Pests often transmit these disease-causing organisms.

Some of the damage caused by pests and diseases can be reduced by careful management practices. However, the recent trend toward reduced pesticide and fungicide use, which has resulted in some chemicals becoming unavailable, has made it significantly more difficult to manage pests and diseases that formerly were controllable. Aside from management practices, the primary method to control such organisms has been through the use of genetic resistance. SEE BIOLOGICAL PEST CONTROL.

Conventional breeding. Plant species evolved along with their pests and diseases, resulting in the development of genetic resistance or tolerance in native plant populations. Resistance is defined as the ability of a plant to slow or stop the spread of a pest or disease, whereas tolerance refers to the ability of a plant to produce a useful yield despite the action of pests or diseases. The goal of conventional breeding has been to exploit preexisting resistance (or tolerance) by sexual hybridization. The hybrid progeny will have exactly one-half of the genetic makeup of each parent. Although conventional breeding has been successfully applied to nearly all crops that exist today, there are several drawbacks: (1) only closely related (sexually compatible) parents can be crossed; (2) because the resulting progeny are a genetic mixture of the parents, prolonged backcrossing programs are needed to eliminate undesirable genes that accompany the resistance gene; (3) only genes present in the native population can be used. Thus, conventional breeding is inefficient when compared to the potential of improvement by genetic transformation.

Genetic transformation. Isolated genes that are inserted directly into the genome of an otherwise desirable plant (genetic transformation) can overcome all the drawbacks associated with conventional breeding. Transformation requires a plant regeneration system from which transformed cells ultimately can develop into whole plants. With only a few exceptions, genetic engineering systems utilize either of two approaches to insert desired genes into plants: a natural bacterial vector, *Agrobacterium,* is modified to insert the desired gene; or particle bombardment is used, whereby isolated genes are mechanically inserted into plant cells.

There are specific advantages to genetic transformation: genes from unrelated organisms, such as other plants, bacteria, fungi, and viruses, can be used; a single gene can be used and sexual recombination is not necessary, so that a resistance gene can be added to an otherwise desirable crop variety. Consequently, new types of resistance genes that do not exist in native populations can be inserted into plants, and years of conventional breeding are avoided. However, as disadvantages, relatively few resistance genes have been identified or prepared for use in plant transformation, and generally applicable transformation systems have not been developed for many important crops.

The resistance genes developed for use in plants generally can be categorized by the type of organism that they inhibit (viruses, fungi, bacteria, or insects). The application of genetic transformation for virus resistance is best understood.

Virus resistance. Because the viral genome is small compared to other organisms, it has been possible to utilize molecular genetics to identify, sequence, and clone specific genes with relative ease. The development of genes for virus resistance is based in the concept of pathogen-derived resistance, whereby genes taken from a disease-causing organism, then inserted and expressed in the host, cause resistance. For viruses, it has been found that genes that code for viral-coat protein, when inserted into plants, can inhibit infection and disease progression. This resistance phenomenon is called coat protein–mediated resistance. Unmodified copies of coat protein genes cause the plant cell to make an abundance of the protein and instill resistance. Modified (that is, disabled) forms of the genes that do not actually translate protein are shown to be better than the unmodified genes. Plants with such disabled genes should remain more vigorous because they are not forced to devote energy to produce the viral protein. Other genes from the viral genome, such as those that control virus movement and replication, also have been shown to be effective after transformation to instill resistance in plants.

The exact mechanisms by which coat protein–mediated resistance functions in plants are not well understood. For transformed plants that produce viral protein, the preexistence of coat protein in cells may disable the early infection process or signal a halt to the viral replication process. Fewer infection sites occur on inoculated transgenic plants, providing evidence of an early block in the infection process. Alternatively, the first stage of uncoating of the virus does not occur in transgenic plants, which is evidence for an early block to virus replication. A different explanation for coat protein–mediated resistance is that the presence of coat protein causes a shift in dynamic equilibrium away from viral uncoating and toward recoating, and this shift disables the process. It is also possible that an immune response occurs, such that antibodies are produced against the viral proteins. In plants with disabled viral genes, it is believed that resistance stems from mechanisms that operate at the ribonucleic acid (RNA) level (RNA-mediated resistance). The accumulation of transgenic RNA to a critical threshold level may activate a cellular system designed to identify and degrade aberrant transcripts; the system thus functions like an immune system.

With the possible exception of insect-resistant plants described below, transgenic virus-resistant plants are closest to the marketplace. For example, squash (*Cucurbita pepo*) was engineered with coat protein genes for three different viruses. Transgenic squash, highly resistant to the three most serious viral diseases, has been tested extensively in the field, and the release of this squash variety is awaiting final governmental approval.

Fungal and bacterial resistance. Strategies for fungal and bacterial resistance share many similarities. Genes isolated from fungi and bacteria that, in nature, are meant to combat other organisms have been isolated and introduced into plants. The most widely known class of genes code for fungal chitinase enzymes. Because the cell walls of many fungi are composed of chitin, plant cells that produce an adequate amount of chitinase are toxic to invading fungi. A broad class of genes that code for lytic peptides enable both fungal and bacterial resistance. For example, tobacco plants transformed with a gene called *Shiva-1* produce a lytic peptide and are resistant to a highly pathogenic strain of the bacterium *Pseudomonas solanacearum*.

Insect resistance. Insertion of genes that inhibit insects is of great interest, since direct feeding by various insects causes major damage to most agronomic crops. Control of many insects requires routine spraying with insecticide, which is expensive and often presents a biohazard. A solution to this problem uses the bacterium *Bacillus thuringiensis,* which produces a protein (known as Bt protein) that inhibits lepidopteran insects. When the *Bt* gene encoding this protein is inserted into plants, the plants produce the protein and are toxic to insects. This type of insect control is highly attractive, particularly since the protein is nontoxic to other types of animals and the use of pesticides is eliminated. The *Bt* gene can also be transferred into other lines by conventional breeding. Varieties of transgenic corn, cotton, and potato have been developed that carry the *Bt* gene and are highly resistant to several destructive insects. These varieties already have been evaluated in extensive field tests and are awaiting final governmental approval. Goals of continuing research are to modify the *Bt* gene so that the translated protein is toxic to a wider range of insect pests. *See Insect Control, Biological.*

Future directions. Transformation systems need to be improved, and the inventory of useful genes needs to be expanded. The identification of native resistance genes in resistant crop species and transference of these genes into susceptible species holds particular promise. Although the resistance genes are extremely difficult to identify in the large plant genome, new molecular techniques (such as transposon tagging) allow such genes to be found. A few native genes already have been isolated directly from resistant plant varieties with the goal of inserting them into susceptible varieties. For example, the flax L^6 gene, which confers resistance to the rust fungus *Melampsora lini,* has now been identified and cloned. The effect of these genes when placed into widely dissimilar crop types will be the focus of future research. Presumably, resistance genes that evolved in plants not only will have wide application but will allow molecular biologists to design experiments to better understand plant resistance mechanisms.

For background information *see Breeding (plant); Fungi; Fungistat and fungicide; Genetic engineering; Insect control, biological; Plant pathology* in the McGraw-Hill Encyclopedia of Science & Technology.

Dennis J. Gray

Bibliography. D. J. Gray and J. J. Finer, Design and operation of five particle guns for introduction of DNA into plant cells, *Plant Cell Tissue Org. Cult.,* 33:219, 1993; K. Lindsey, Genetic manipulation of crop plants, *J. Biotech.,* 26(1):1–28, 1992; A. Nejidat, W. G. Clark, and R. N. Beachy, Engineered resistance against plant virus diseases, *Physiologia Plantarum,* 80(4):662–668, 1990; B. J. Staskawicz et al., Molecular genetics of plant disease resistance, *Science,* 268:661–667, 1995.

Bridge

Modern cable-stayed bridges date from the mid-1950s. The basic design used inclined cables to provide intermediate supports for a bridge girder so that it could span a longer distance. Because long-span bridges were built mainly by steel companies, the early cable-stayed bridges were almost exclusively steel. Other distinctive features of these early structures were a minimum number of cables, large stiff girders, and lock coil–type cables (cables with S- and Z-shape outer wires that interlock when the cable is tensioned).

Theoretical advances and complex analyses performed with powerful computers, along with the introduction of new construction materials, have made possible the evolution of cable-stayed bridges into an array of elegant forms. As their designs have become more attractive both esthetically and economically, cable-stayed bridges have become more common.

Slenderness. Use of closely spaced cables allows the girder to be very slender. The local bending moment in the slender girder between the cables is significantly reduced so that the girder can be more flexible. The reduced stiffness also reduces the global bending moment, so that the girder stiffness can be reduced further.

By comparison, a normal girder bridge has a depth-to-span ratio of about 1:20, while for an early cable-stayed bridge the depth-to-span ratio is reduced to about 1:100. A modern cable-stayed bridge with closely spaced cables can decrease this ratio to 1:300, as in the ALRT Skytrain Bridge over the Fraser River in Vancouver, British Columbia, Canada (**Fig. 1**). With its girder only 1/300th of the span length between the towers, the bridge is extremely elegant.

Other elements such as the towers and tall piers are also made more slender when the actual behavior of the structure is determined to a greater extent by nonlinear analysis and buckling calculation. The number of cross struts in the portal-shaped towers is also greatly reduced.

Girder. Although steel is still being used in cable-stayed bridges, concrete cable-stayed bridges are becoming more popular. A steel bridge girder is lighter, but weight is not a disadvantage except for very long spans of more than 3000 ft (1000 m). An inclined cable can be effective only if it is tensioned to a certain force. This force is proportional to the weight of the bridge girder. Sometimes it may even be necessary to increase the weight to attain the required tension in the cable.

The advantages of the concrete girder are that its construction is less labor intensive; it can easily be cast into any desired shape; it is more economical; and concrete structures are relatively easier to maintain since they do not require painting at fixed intervals.

A concrete girder can be built by one of two methods: precast or cast-in-place. In the precast method, concrete girder segments are manufactured off site in a mold known as a casting bed and transported to the site for erection, where they are prestressed together by high-strength tendons and the joints between the segments are filled with epoxy. As the segments are lifted into place and stressed together, the cables are installed to support the completed portion of the bridge.

The cast-in-place method is similar to the precast method except that the mold is a specially designed piece of equipment known as a formtraveller. It is suspended onto the previously completed portion of the bridge girder, and fresh concrete is brought to the site and placed directly into the form in place, segment by segment. An advanced type of formtraveller developed for the Dames Point Bridge, Jacksonville, Florida (**Fig. 2**), used the permanent cable to provide additional support, and thus reduced weight by more than 50%. The development of this formtraveller has made the very flexible concrete girder easier to

Fig. 1. ALRT Skytrain Bridge over the Fraser River, Vancouver, British Columbia, Canada.

control during construction and its use more economical.

Composite girders, a combination of a steel grid with a concrete deck slab, have been used in several recent bridges. The deck of a full steel girder is usually an orthotropic plate, which requires a very labor intensive mode of construction, and is therefore very expensive; thus a full steel girder cannot compete economically with concrete girders. The composite girder, however, is competitively priced. Here the most expensive part of a full steel girder is replaced with less expensive concrete.

Towers. With few exceptions, the towers of modern cable-stayed bridges are made of concrete. Similar to the girder, a concrete tower is more economical, easier to build, and easier to maintain. The predominant shape is the rectangle because of its simplicity and attractiveness. Because the towers of a cable-stayed bridge are very large, a clean and simple form is usually preferable. The configuration of the towers may be a simple portal shape, diamond, A-shape, single-pole, or inverted-Y shape.

Cables. A cable usually consists of high-strength wires, straight or stranded into a specified pattern, and protected from weather in various ways. The early lock-coil-type cable is rarely used now. The most popular types are parallel wire cables and seven-wire strand cables.

A parallel wire cable consists of all galvanized wires running straight or twisted in a long pitch of about 7 ft (2 m). Each end of the cable has a steel socket that serves as the anchorage for the cable. The wires are extended into the socket, spread out, and hooked, flattened, or button-headed. The space

in the socket is then filled with zinc alloy or with epoxy, zinc powder, and small steel balls. The complete cable is then protected by extruding a layer of polyethylene over the wire bundle, which has been tied together by a spiral wire. Instead of extrusion, a polyethylene pipe may be used and the void between the wires and the pipe filled by cement grout after installation of the cable.

A seven-wire strand cable consists of a bundle of seven parallel wire strands, usually 0.6 in. (1.5 cm) in diameter. The strand bundle is placed into a polyethylene pipe, and the space is then grouted for protection. For additional protection the strands may be galvanized, coated with epoxy, or greased and sheathed.

The polyethylene pipe is usually supplied in black in order to resist ultraviolet radiation from the Sun; however, it absorbs heat in sunny weather, and may become too hot. Therefore, cables manufactured recently have been wrapped with a layer of polyvinyl fluoride tape in a lighter color, such as white or orange. The polyvinyl fluoride tape not only reduces overheating of the cable but also provides an additional layer of protection against the weather.

Aerodynamics. Cable-stayed bridge design has benefited from the exhaustive testing of the design of suspension bridges following the collapse of the First Tacoma Bridge in Washington in 1942. The basic criteria are the same. For vortex-induced vibrations and flutter, the analysis is similar to that for suspension bridges. However, several aerodynamic phenomena are peculiar to cable-stayed bridges: buffeting, wake vibration, and rain-wind vibration.

Fig. 2. Formtraveller as used in the construction of the Dames Point Bridge, Jacksonville, Florida.

Buffeting. This response to turbulent wind was initially regarded as unlikely to cause any problem. However, recent studies have found that if the frequencies of the bridge structure are close to the frequencies of a turbulent wind, the amplitude of the buffeting response can be very serious for a cable-stayed bridge, particularly during critical stages of construction. Consequently, tie-downs have been used to stabilize the structure.

Wake vibration. When two cables are placed next to each other horizontally, the wake created by the windward cable may excite the leeward cable. This type of response can be dampened by connecting the cable pair, either rigidly or by using an automobile shock absorber.

Rain-wind vibration. This phenomenon was discovered fairly recently. Some cables vibrate violently under the combination of a mild wind and light rain, but the vibration ceases if the rain becomes heavier or stops. The explanation is that a light rain creates a mantle of water surrounding the cable surface, and the wind deforms this mantle into an unstable shape, causing the vibration. The common method for suppressing this vibration is to tie the cables together so that they dampen each other.

Multispan configuration. A span is the bridge between two adjacent piers (supports). Cable-stayed bridges have been designed with either double spans (two adjacent cable-stayed spans) or triple spans. When a bridge has more than three spans, the middle spans become too soft and ineffi-

cient. Therefore, multispan cable-stayed bridges have been used for only very small spans. However, this situation may change as new approaches to stiffen the middle spans are being proposed.

Superlong spans. The maximum span of a cable-stayed bridge has been expanded significantly in recent years. The Duisburg-Neuenkamp Bridge in Germany had the world's longest span (1100 ft or 340 m) at its completion in 1970. In 1985 the record reached 1530 ft (465 m) with the construction of the Annacis Island Bridge in Vancouver, British Columbia, Canada. The 1970-ft-span (602-m) Yang Pu Bridge in Shanghai, China, was completed in 1994, and the 2808-ft-span (856-m) Normandy Bridge in France in 1995. The Tatara Bridge in Japan with a span of 2900 ft (890 m) was scheduled for completion in 1997. Even longer spans are being planned, and a cable-stayed bridge with a 6600-ft (2000-m) span will be possible in the near future.

For background information SEE AERODYNAMICS; BRIDGE; COMPOSITE BEAM; CONCRETE; CONCRETE BEAM; PRECAST CONCRETE; PRESTRESSED CONCRETE in the McGraw-Hill Encyclopedia of Science & Technology.

Man-Chung Tang

Bibliography. W. Podolny, Jr., and J. Scalzi, *Construction and Design of Cable-Stayed Bridges,* 2d ed., 1986.

Cancer (medicine)

The growth and division of somatic cells is regulated by multiple signals. Some are hormones or growth factors; other signals may be transmitted between cells in close contact. Each signal interacts with a corresponding receptor on the target cell. However, cancer is due to the emancipation of a single cell clone from growth control. Liberation from multifactorial control can occur only by multiple changes, and most human tumors arise after 5–7 successive mutations in the clonal and subclonal progeny of a single cell. Numerous genes have been shown to contribute to the malignant microevolution. They belong to three major groups: oncogenes, tumor suppressor genes, and destabilizing genes. SEE CELL (BIOLOGY).

Oncogenes. Oncogenes are normally involved in signaling circuits that trigger the entry of a cell into the cell cycle and prevent the elimination of excessively dividing cells by programmed cell death. They are highly conserved genes; that is, they have closely similar coding sequences in all vertebrates. Several of them can be traced down to invertebrates.

Categories. The oncogenes can be subdivided into five categories. Genes that encode normal growth factors can turn into cancer genes if they are activated to make their products in the wrong cell. If a cell that is normally driven by a growth factor produced by other cells switches on its own growth factor a cycle of self stimulation begins. Genes that encode receptors normally are triggered by specific growth factors. However, they may become cancer genes if mutations or other structural changes prompt them to emit a stimulatory signal without the triggering mechanism. Genes that encode proteins that enter a signal transduction pathway, whose normal function is to pass on the stimulatory impulse from a triggered receptor to the cell division machinery, may turn into cancer genes by mutations that activate them to emit signals in the absence of triggering. Genes encode proteins directly involved in the activation of cell division. These proteins include transcription factors that activate genes preparing the cell for the duplication of deoxyribonucleic acid (DNA) that must take place before the actual start of cell division. One oncogene in this category (a cyclin, D1) participates with other proteins in the triggering of the cell to pass a restriction point that normally prevents it from leaving the resting stage.

A separate category is represented by the *bcl-2* gene, which inhibits the cell from switching on the apoptotic mechanism of programmed cell death. Apoptosis is part of the normal homeostatic regulation that keeps the size of cell populations at a given level. Cells of the hemopoetic system and of many other tissues are continuously renewed. Their production must be counterbalanced by cell death. Lymphocytes, for example, are produced in much larger numbers than what is needed to provide the organism with immune responses against foreign invaders. Lymphocytes that are not called upon by the appropriate antigen within a short period of time are removed from the cell pool by apoptosis. This cleansing reaction prevents the obstruction of the system. Protection of the cells by illegitimate activation of the *bcl-2* gene (for example, as a result of chromosomal rearrangements) may increase the cell pool to a pathological size and ultimately lead to malignancy.

Mechanisms. Oncogenes that encode growth factors and receptors may become constitutively active, that is, independent of physiological triggering by ligands or signals, after they have undergone certain structural changes. Oncogenes that encode growth factors and transcription factors, and the *bcl-2* gene, are more frequently switched on by regulatory changes brought about by the insertion of a viral gene into the immediate neighborhood of the oncogene. Although the regulatory elements of the virus have evolved in competition with cellular regulatory signals, they may override the latter. Chromosomal rearrangements due to the exchange of terminal chromosomal fragments by reciprocal translocation may also activate oncogenes by juxtaposing them to highly active cellular genes. The frequency of chromosomal rearrangements may increase, for example, when cell division is chronically stimulated by microbial or chemical irritants that contribute to the carcinogenic processes. Chronically stimulated lymphocytes are particularly prone to such accidents. This outcome is understandable because their specific products, the antibodies made by the bone-marrow-derived B lymphocytes and the T-lymphocyte receptors involved in cell-mediated immunity are encoded by genes that normally diversify by physiological rearrangement of their DNA and are constitutively active.

Amplification of oncogenes is another activating mechanism that can contribute to the progression of tumors toward increased malignancy. It affects mainly oncogenes that encode receptors and transcription factors. Amplification raises the number of gene copies from the normal 2 per diploid cell to 30–50 or even more. In breast and lung cancer, and in neuroblastoma, amplification is correlated with increasing malignancy.

Tumor suppressor genes. These genes represent the second major category of genes that contribute to tumor development. In contrast to the oncogenes, they do so by their loss or functional inactivation.

The existence of genes that can counteract tumor development was first discovered by somatic hybridization between normal and malignant cells. It was found that the normal partner cell can suppress the tumorigenicity of the malignant partner. Reappearance of tumorigenicity was dependent on the loss of certain chromosomes from the normal parent. Later, nontumorigenic revertants could be isolated from transformed, tumorigenic cell cultures. The genes involved in the suppression of malignancy by somatic hybridization and reversion have not been defined except in a few experimental models. The most important developments came from the independent isolation of specific suppressor genes. The retinoblastoma and the *p53* gene were found to play particularly prominent roles.

Retinoblastoma gene. A mutated, functionally inactive form of the retinoblastoma gene is transmitted in the germ line of familial carriers of retinoblastoma, such as a malignant eye tumor of young children. The normal allele of the retinoblastoma gene is lost during somatic development, triggering the growth of the tumor. The retinoblastoma protein plays a pivotal role in the cell cycle clock apparatus. It permits the clock to control the expression of numerous genes that mediate the advancement of the cell through different phases of the growth cycle. Loss of retinoblastoma gene function deprives the cell of an important mechanism that regulates cell proliferation through the modulation of gene expression.

p53 gene. Another gene of major significance is a DNA-binding protein called *p53*. When expressed at a high level, it can stop DNA synthesis and thereby the growth cycle. DNA damage induces the stabilization of the protein, thereby raising its concentration. Cell division is arrested and the DNA can be repaired by enzymes. If DNA is successfully repaired, the concentration of the *p53* protein falls, and the cell resumes its DNA synthesis. If DNA damage is beyond repair, the apoptotic mechanism of programmed cell death is elicited and breaks down the DNA by an endolytic mechanism.

In *p53* knockout mice that carry an experimentally inactivated form of the gene, the lack of the guardian function renders the mice mutation prone. They are phenotypically normal at birth, but develop a variety of tumors later. A similar situation prevails in the human Li-Fraumeni syndrome, where a germ-line mutation of *p53* is transmitted, associated with the familial development of multiple tumors.

In the course of tumor progression, the loss of the *p53* function apparently facilitates the malignant growth of the cells. Alternatively, or in addition, it may act by preventing the apoptotic death that occurs frequently in oncogene-driven cells. Some mutants of *p53* can also directly contribute to the driving of cell division. Thus *p53* can act either as a suppressor gene or as an oncogene.

Destabilizing mutations. In the inherited *p53* mutation syndrome the loss of *p53* destabilizes the genome, because many other mutations, including those affecting oncogenes and suppressor genes, persist and if the cell is otherwise functional proliferate. In addition, several other destabilizing genes have been recently discovered, all involved with DNA repair. The *MSH2* gene, encodes a highly conserved DNA mismatch repair enzyme. Its loss is responsible for the familial cancer syndrome called hereditary nonpolyposis colon carcinoma, which promotes the appearance of colonic and other tumors. At least three other genes that can undergo destabilizing mutations have been recently identified. In their normal form, they are involved with DNA repair. Their germ-line mutations are associated with familial cancer syndromes.

Genetic analysis. Genetic analysis of the development of some major human tumors, including cancers of the large bowel, the lung, the prostate, the breast, and gliomas of the brain, indicate that genetic losses affecting known or presumptive suppressor genes are more common than oncogene activation events. Screening for suppressor gene losses is easier than the detection of oncogene activation, however, and it remains to be seen whether this is a spurious or a true difference. In familial cancer syndromes, genetic heterogeneity is created by the loss of some DNA repair mechanisms. Thus cancer incidence is increased by preventing the elimination of cell variants, many of which may carry oncogene or suppressor gene mutations.

For background information *SEE CANCER; CELL SENESCENCE AND DEATH; CHROMOSOME; GENE; MUTATION; ONCOGENE; TUMOR* in the McGraw-Hill Encyclopedia of Science & Technology.

George Klein

Bibliography. G. Klein and E. Klein, Evolution of tumours and the impact of molecular oncology, *Nature,* 315:190–195, 1985; P. Modrich, Mismatch repair, genetic stability, and cancer, *Science,* 266: 1959–1960, 1994; T. H. Rabbitts, Chromosomal translocations in human cancer, *Nature,* 372:143– 149, 1994; C. B. Thompson, Apoptosis in the pathogenesis and treatment of disease, *Science,* 267:1456– 1462, 1995; R. A. Weinberg, The retinoblastoma protein and cell cycle control, *Cell,* 81:323–330, 1995; R. A. Weinberg, Tumor suppressor genes, *Science,* 254:1138–1146, 1991.

Capillary electrophoresis

Recent advances in capillary electrophoresis involve methods for chiral separations and for separations using a microfabricated silicon chip.

Chiral separations. Chiral compounds possess asymmetry on a molecular scale, the classic example of molecular stereochemistry being a tetravalent carbon atom with four different substituents. Two forms or enantiomers of the molecule exist, differing in the spatial arrangement of the substituents such that the two forms are mirror images of each other. The molecule is said to be chiral, and although the enantiomers behave identically in a nonchiral environment, in the presence of other chiral species they may interact differently. Such enantioselectivity is important for the fate of chiral pharmaceuticals within an organism, which is full of chiral biomolecules. Thus, the enantiomers of a chiral drug can have differences in both their desired activity and unwanted side effects. These interactions have become better understood as analytical technology has advanced; recent moves to regulate chiral drugs have come about in large part because it is now feasible to separate many enantiomer pairs by using chromatography with chiral stationary phases. Recently, considerable progress in chiral analysis has been made by using capillary electrophoresis.

Chiral analysis. Stereoselective interactions between a chiral analyte and a single-isomer chiral selector form the basis of methods to analyze enantiomers. In chiral chromatography, the analyte is transported in a mobile phase, and allowed to interact with a chiral selector molecule immobilized on a support to form a chiral stationary phase. The stronger the binding between analyte and selector, the longer it takes for the analyte to pass through the column. Differences in enantiomer binding (**Fig. 1**) can result in a chiral separation. Despite the undoubted success of chiral chromatography, this technique has its limitations. In particular, many chiral stationary phases are enantioselective for just a few compounds or classes of compounds, and the development of a new chiral stationary phase is not a trivial undertaking.

Separation techniques. Capillary electrophoresis has been found to be useful for performing chiral separations. By using pure electrophoresis (that is, the movement of ions in solution in an electric field) it is not possible to separate enantiomer pairs; both have the same velocity or mobility (mobility is the velocity per unit field strength). However, by including a chiral selector in solution in the separation electrolyte, separations may be achieved on the basis of interactions between the chiral analyte and selector. One of the beauties of chiral capillary electrophoresis is that the selector need not be immobilized, and so a number of chiral compounds can be screened rapidly to determine their enantioselectivity just by dissolving them in the separation buffer, the buffer solution that fills the separation capillary during the analytical run. The most popular selectors are native and derivatized cyclodextrins (cyclic assemblages of chiral sugar molecules); crown ethers, chiral surfactants, some antibiotics such as vancomycin, and proteins such as albumin have also been used. One striking point is that the range of application of these selectors in capillary electrophoresis can be much wider than in high-performance liquid chromatography. The success of the cyclodextrins is particularly impressive, with a huge variety of different chiral compounds being separated. This success stems from the high efficiency and resolving power of capillary electrophoresis. A much smaller difference in the binding affinities of two enantiomers can give rise to a separation in capillary electrophoresis than in analytical-scale liquid chromatography. Another advantage of capillary electrophoresis is that interactions between the analyte and selector in dilute solution occur in a (reasonably) well-defined environment. Therefore, it should be possible to understand the separation on the basis of simple chemical equilibria, and conversely it is possible to obtain information on physicochemical properties such as binding constants based on separation measurements.

In a typical capillary-electrophoresis chiral separation, the separation buffer contains a chiral selector at millimolar concentrations. Since cyclodextrins are widely used as chiral selectors, an example of their use will be considered here. Except at very alkaline pH, the cyclodextrin itself is uncharged, and so will not migrate in the system under the influence of the electric field. If the enantiomers of a charged analyte are introduced into the capillary, and the field is applied, they will begin to move. While migrating they will interact with the cyclodextrin, just as they would if they were in solution in a beaker. With cyclodextrins, the primary mechanism of interaction is inclusion of the analytes into the hydrophobic internal cavity of the cyclodextrin. Just how well the analyte enantiomers fit and how strongly they are bound is affected by their three-dimensional structure. When an analyte enantiomer fits into the cyclodextrin, the entire complex will move under the effect of the electric field, because the analyte retains its charge; since the overall size of the complex is much bigger than the analyte alone, there will be more friction retarding its motion, and so the complex will move relatively slowly. Thus, the overall speed of each enantiomer through the capillary is related to the mobility of the free analyte, the mobility of the analyte/cyclodextrin complex, and the length of time that the analyte spends in the free and complexed states. The time factor is simply related to the strength of the binding holding the complex together. In the case of the cyclodextrin, the

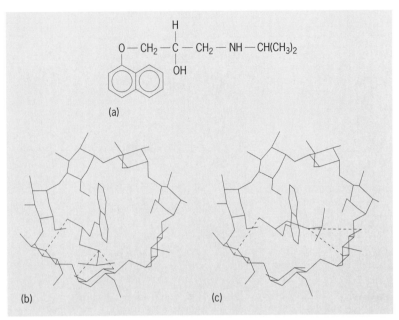

Fig. 1. Computer projections of the inclusion complexes of the two enantiomers of propranolol (*a*) with β-cyclodextrin from x-ray crystallographic data: (*b*) *d*-propanol complex and (*c*) *l*-propanol complex. Broken lines represent potential hydrogen bonds. The configurations represent the optimal orientation for each enantiomer on the basis of the highest degree of hydrogen bonding and complexation; differences in the strength of these complexes are the basis of the separation mechanism. (*After D. W. Armstrong et al., Separation of drug stereoisomers by the formation of β-cyclodextrin inclusion complexes, Science, 232:1132–1135, 1986*)

more strongly bound enantiomer would on average move more slowly through the capillary because it spends more time in the slower-moving complexed state. This difference in velocity results in a chiral separation as the analyte enantiomers move down the capillary toward the detector. Other combinations are possible. For example the analyte could be uncharged and the selector charged. In such a case the more tightly bound enantiomer would move more rapidly in the same direction as the selector.

Sample separations. **Figure 2***a* shows the separation of (6R)- and (6S)-leucovorin (a drug administered as a pair of stereoisomers to decrease the toxicity of cancer chemotherapy with methotrexate) and its metabolite 5-methyltetrahydrofolate. These compounds can be resolved by using γ-cyclodextrin as a chiral selector. With 0.08 *M* of γ-cyclodextrin in the separation buffer (upper trace) there is slight resolution of leucovorin but no separation of 5-methyltetrahydrofolate. With 0.25 *M* γ-cyclodextrin in the separation buffer (lower trace) both pairs of analyte stereoisomers are baseline resolved. The analyte peaks shown in Fig. 2*a* are very sharp. In Fig. 2*b*, the enantiomers of benzoin were separated by using human serum albumin as a chiral selector. Here the albumin is present in the separation buffer at a concentration of only 35 μ*M*

The great difference in the concentration of chiral selectors used in these separations reflects differences in the strength of binding between the analytes and selectors: leucovorin interacts weakly with the cyclodextrin, so high selector concentrations

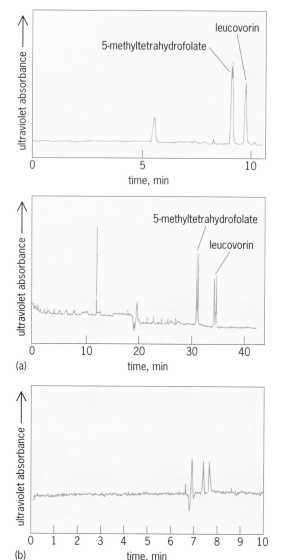

Fig. 2. Chiral separations. (*a*) Separation of the stereoisomers of leucovorin and 5-methyltetrahydrofolate by capillary electrophoresis, using a γ-cyclodextrin chiral selector. (*b*) Separation by using a protein chiral selector. The two peaks at 7.5 and 8 min are due to the separated enantiomers of benzoin, which interact differently with the 35 μ*M* of human serum albumin in the separation buffer. The baseline disturbance at 7 min is caused by the passage of the sample solvent through the detector.

must be used to achieve acceptable degrees of binding for separation to occur. Benzoin, however, binds strongly to albumin, and thus it is separated with a low selector concentration. The relationship between the analyte mobility and the strength of analyte-selector interaction has been used to probe enantioselective ligand-selector binding. Quantitative measurements of binding constants are possible in this way, as are investigations of drug-drug interactions in stereoselective protein binding. The advantage of using a separation technique such as capillary electrophoresis for these measurements is that a racemic compound may be used as a sample, whereas traditional techniques require separate enantioseparation and binding measurements.

Immobilized selectors. Despite the fact that chiral selectors used as buffer additives have been highly successful in capillary electrophoresis enantioseparations, there is still interest in systems where the selector is somehow immobilized in the capillary. A variety of approaches have been adopted including cross-linking of the selector in the capillary to form a gel, physical entrapment of the selector within a polymer network, and the use of chromatographic chiral stationary phases in electrochromatography. So far all these techniques have shown some promise, and there are a few advantages such as keeping the selector away from the detector window (which may be advantageous if the selector itself elicits a response from the detector, resulting in a large background signal). Nevertheless, for general analytical work the use of selectors in solution seems likely to remain the primary technique for capillary electrophoresis separations.

Prospects. Because of its wide applicability and relative simplicity, it is anticipated that capillary electrophoresis will replace chromatography for a number of chiral separation tasks. It is already being used for testing enantiomeric purity of bulk drugs or drug intermediates. Although it is more challenging to develop capillary electrophoresis methods for the analysis of chiral drugs and metabolites in biofluids, the potential to perform microscale bioanalysis by capillary electrophoresis in body compartments that were previously not amenable to analysis will provide impetus for further developments. The use of capillary electrophoresis to study binding affinities of enantiomers may also benefit from combination with its microanalytical abilities, permitting the design of experiments that use minute quantities of rare or expensive chiral selectors.

David K. Lloyd

Microchip separations. Capillary electrophoresis is a miniaturized separation technique in the sense that both sample and detection volumes are on the scale of nanoliters, and separations are performed in capillaries with a diameter of 100 micrometers or less. These small scales make capillary electrophoresis an ideal separation method to incorporate onto planar structures the size of an electronic microchip. A technology known as silicon micromachining, which utilizes microelectronic fabrication methods to make miniature, three-dimensional physical structures, can be adapted to the microfabrication of capillary electrophoresis systems on a chip. Networks of intersecting capillaries about half the diameter of a human hair are integrated onto a chip a few centimeters square to create a plumbing, microreactor, and separation system. In these microlabs, samples can be treated with various chemicals and then separated by using capillary electrophoresis, in order to determine the concentration of the targeted chemicals or biochemicals. The system functions without moving parts by using voltages to control the flow of fluids. The small dimensions permit reactions and analyses that can be performed in a matter of minutes, compared to times as long as hours with conventional methods. The eventual goal is a small, portable, chip-based system located in a hospital intensive care unit or emergency room, or carried into the field for automated, rapid analysis of environmental samples at the point of sample collection. Such a system represents a significant trend in biosensing, meeting a developing market demand for rapid, on-site, automated analyses. *SEE BIOSENSOR.*

Electroosmotic pumping. Electroosmotic flow in capillary electrophoresis is induced by the presence of an electrical double layer at the solution–capillary wall interface. This layer consists of fixed charge present on the walls because of surface chemistry (usually controlled by solution pH), with mobile counterions in solution creating a charged region within a few nanometers of the surface. An electric field applied to the capillary will cause the mobile charge to move, creating electroosmotic flow by dragging a solvent sheath along with the mobile counterions. The larger the width of the charged double layer the greater is the rate of electroosmotic flow, so that high fixed-charge density on the capillary walls or low concentrations of ions in solution will increase the overall solvent flow rate. Linear flow rates up to about 1 mm/s are easily achieved with this pumping method, although the so-called pumping force is sensitive to solution composition.

Electroosmotic flow is used to advantage in capillary zone electrophoresis to drive both cations and anions in the same net direction, because solvent flow in a fused silica or glass capillary is greater than the rate of ion migration induced by electrophoresis. This same pumping action can be applied in the microchip format to deliver sample and pump reagent streams with applied electric fields. In the design shown in **Fig. 3** a field could be applied between any two-channel endpoint reservoirs to control the flow of solvent within the capillaries and direct it between those two locations. More complex flow patterns required for mixing can be achieved by applying electric fields to several reservoirs at once, in order to accurately meter the flow from each channel into the main stream.

Each channel behaves as a resistor, so that the current flow in each channel and the potential at each intersection can be calculated by the application of Kirchhoff's laws for current flow in circuits. The corresponding flow of solvent (v) in each channel is then readily calculated from the electric field (E; potential gradient-unit length) along that channel and the electroosmotic mobility of the solvent (μ), as in the equation $v = \mu E$.

Electroosmotic pumping allows for the delivery of picoliter volumes of sample within the chip with 1% precision. Fluid can be directed around corners and across intersections without the need of pumps or valves, so that a fluidic system with essentially no moving parts is achieved.

Microfluidic systems. Microfabrication of capillaries on a planar substrate provides a unique feature

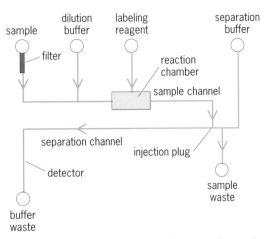

Fig. 3. Pattern of microchannels etched onto a planar substrate such as glass to form a complex sampling treatment and separation system.

in that complex networks of intersecting capillaries can be readily fabricated (Fig. 3). Mixing chambers, pre- and postseparation reactions, delivery of different reagents at different stages of a separation, or a combination of electrophoretic separation and other techniques such as liquid chromatography are among the combinations that may be fabricated on chip. In addition to the high level of automated sample processing that integration makes possible, the very short lengths and high heat capacity of planar substrates allow the application of very high electric fields of up to 2500 V/cm, compared to the 300–600 V/cm usually used in conventional capillary electrophoresis systems. These high electric fields can lead to very rapid separations on the 0.1–20-s time scale, without requiring complex and sophisticated equipment for sample injection.

On-chip separations. Most separations that have been demonstrated in conventional fused silica capillaries can also be performed in a chip format, by using a separation channel length of about 5 cm (2 in.) or less. In capillary zone electrophoresis the separation power depends principally on the voltage applied. Consequently, separations on chip can be as efficient as with conventional, meter-long, fused silica capillary as long as higher electric fields (up to 2500 V/cm) are applied to short channels on the chips. However, useful separations are hard to achieve with fields of more than 400 V/cm for deoxyribonucleic acid (DNA), so that serpentine channel paths may be required to maximize the length on chip. Demonstrated applications on chip include the separation of mixtures of six amino acids in less than 15 s, sequencing of up to 300 base DNA oligomers or of double-stranded DNA digested fragments in gel-filled columns on chip in 100–200 s, separation of metal cations, and separation of proteins or antibodies on chips with polymer-coated channel walls.

Unique chips for electrophoretic separation have been devised that can be prepared only by using microfabrication methods. A cyclic structure was prepared for separation of a mixture, in which the targeted component flows around in an endless loop, and the unwanted elements in the mixture are automatically directed out of the loop. This device yielded extraordinarily efficient separations equivalent to the separating power of 20–40 kV, while requiring only 2 kV. Theoretically, such a device could deliver the equivalent of more than a million volts of separating power.

On-chip chemical reactions. A key advantage of microlithographic fabrication of capillaries on chip is that there is essentially zero dead volume at capillary intersections, a feature that cannot be achieved with conventional capillary methods. The low dead volumes mean that very efficient mixing can occur, without sample dilution or degradation of the separation efficiency. In this way it is possible to fine-tune a series of separation steps to maximize performance. For example, a buffer appropriate for an initial preseparation reaction can be utilized at first, then a second buffer more effective in the separation step can be added when needed. Finally, another buffer or chemical reagent can be introduced by a channel at the end of the separation stage to optimize detection of the sample, perhaps causing a postseparation reaction, or adjusting pH to the optimum conditions for detection.

Figure 4 illustrates the separation of amino acids and hydrolyzed dansyl chloride, followed by postseparation reaction with the fluorogenic agent *ortho*-phthaldialdehyde at the end of the separation on chip. In liquid chromatography this is known as postcolumn reaction; it is a chemical reaction performed on the fly, so to speak, in the sample stream as it moves toward a detector. The *ortho*-phthaldialdehyde is readily delivered at a Y-shaped intersection by using electroosmotic pumping. For the example shown each 100-picoliter sample band passed the detector within about 90 ms, indicating the narrowly confined distribution of the sample plugs even after the chemical reaction. Such efficient on-chip chemical reactions demonstrate the usefulness of the chip for automated chemical processing and analysis of a sample.

On-chip antibody-based reactions offer a more complex type of sample reaction and analysis that is more directly related to clinical diagnostics. Antibodies are routinely used in clinical analysis because of their extremely high selectivity and sensitivity toward a variety of important drugs, biochemicals, bacteria, and viruses. Automation of antibody-based immunoassays in a microchip format involves electroosmotic pumping of sample and antibody into a reaction chamber, followed by injection, and then separation of the reaction products from the starting compounds. Such procedures illustrate the power of the chip for sophisticated sample processing and analysis, and indicate that a high degree of automation will prove possible.

Applications. The microchip format may prove to be a highly successful method of automating small,

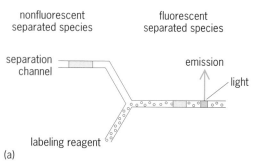

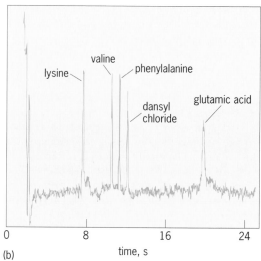

Fig. 4. (*a*) On-chip separations. Postseparation reactor created at an intersection between a separation and a reagent supply channel. (*b*) Separation of several amino acids and dansyl chloride followed by reaction to produce fluorescent products.

portable sample analyzers. Such systems, which could analyze complex samples and provide quantitative information about individual components of a sample, will begin to blur the distinction between sensors and instruments. Application of the microchip format to clinical diagnostics involving DNA probes, immunological reactions, or separations and analyses that are presently performed by using slab gel electrophoresis appear extremely promising. Other potential applications such as industrial process control or environmental monitoring also present realistic opportunities for portable or inexpensive automated systems. Questions related to the overall cost of the peripheral packaging that supports each chip, the level of reusability that is advisable (either from the perspective of cost or application), and the optimal methods for on-chip detection require resolution. However, the potential scope of applications for a fluidic system with no moving parts appears to be so large that positive resolution of these issues is inevitable.

For background information *SEE CHROMATOGRAPHY; CLATHRATE COMPOUNDS; ELECTROPHORESIS; ENANTIOMER; KIRCHHOFF'S LAWS OF ELECTRIC CIRCUITS; STEREOCHEMISTRY* in the McGraw-Hill Encyclopedia of Science & Technology.

D. Jed Harrison

Bibliography. P. R. Brown and E. Grushka (eds.), *Advances in Chromatography,* vol. 33, 1993; D. J. Harrison et al., Micromachining a miniaturized capillary electrophoresis-based chemical analysis system on a chip, *Science,* 261:895–897, 1993; R. Kuhn and S. Hofstetter-Kuhn, Chiral separations by capillary electrophoresis, *Chromatographia,* 34:505–512, 1992; S. F. Y. Li, *Capillary Electrophoresis: Principles, Practice and Applications,* 1993; A. Manz et al., Electroosmotic pumping and electrophoretic separations for miniaturized chemical analysis systems, *J. Micromech. Microeng.,* 4:257–265, 1994; M. M. Rogan, K. D. Altria, and D. M. Goodall, Enantioselective separations using capillary electrophoresis, *Chirality,* 6:25–40, 1994.

Carbon

In the traditional view, elemental carbon (C) occurs in two stable crystalline forms (allotropes)—diamond and graphite. In diamond, each carbon forms strong bonds to four nearest neighbors oriented (at angles of 109.4°) along the axes of a tetrahedron, yielding a solid that is exceptionally rigid in all directions. Each carbon atom in graphite forms strong bonds to only three other carbons in the same plane (all lying at angles of 120°). The individual planes in this so-called two-dimensional carbon interact only weakly with each other, which accounts for the well-known lubrication abilities. The valence electrons in diamond are strongly localized in strong single bonds formed by sp^3 hybridization, making it an exceptional electrical insulator. In contrast, graphite is a good electrical conductor because of the delocalization of electrons throughout the individual planes in the unhybridized p orbitals whereas the strong in-plane bonding results from the overlap of sp^2 orbitals. The familiar soot that accompanies hydrocarbon combustion processes is a disordered mixture containing both types of bonds between the carbon atoms.

Three-dimensional geometrical arrangements of the carbon atoms with sp^2 orbitals have recently been discovered; they are now taken to constitute the third form of carbon, fullerenes. Soccer-ball-shaped C_{60} is the archetype of this new class of carbon molecules, which have generated tremendous excitement since their discovery in 1985.

Another form of pure carbon was known to exist for some time, but only recently was it synthesized in macroscopic quantities. This fourth form of carbon is composed entirely of sp hybridized carbon atoms, arranged in a linear (or perhaps very slightly bent) geometry. Two resonance structures exist for describing the bonding in linear carbon: all double bonds (cumulene) or alternating single and triple bonds (polyyne).

Structure and bonding. Experimental data on the structural properties of small carbon molecules have, until recently, been extremely limited. Much of the present knowledge concerning these species derives from quantum chemistry calculations. The first such calculations, in 1959, predicted that carbon clusters smaller than C_{10} would form linear chain structures. For the odd-numbered clusters (C_3, C_5, C_7, and C_9), all electrons in the molecular orbitals are paired, resulting in a singlet electronic configuration for the ground (lowest) state. For the even clusters (C_4, C_6, and C_8), the two electrons in the highest occupied molecular orbital are unpaired, so that the ground-state configuration is a triplet. Carbon clusters larger than C_9 were predicted to exist as planar monocyclic rings.

Since the mid-1980s, a wealth of highly sophisticated theoretical predictions have appeared on structures and energetics for carbon clusters as large as C_{11}. More approximate treatments have been applied for cluster sizes between C_{11} and C_{20}. These studies have predicted subtleties in the structural behavior. For example, while the odd-numbered clusters smaller than C_{10} were also found to exist as linear chains, the even-numbered species have two structural isomers that are nearly equal in energy—the linear triplet chain and a singlet planar ring. The increased angle strain (relative to the ideal sp hybrid angle of 180°) of such small rings is essentially offset by the added stability that results from pairing the two outer electrons in going from a chain to a ring. For clusters larger than C_9, the more recent calculations support earlier predictions that planar monocyclic rings are the most stable structures.

The chemical bonding in these small carbon clusters is predicted to be principally cumulenic; that is, each atom in the cluster is sp hybridized and forms a double bond with each of its two nearest neighbors, analogous to the central carbon atom in the hydrocarbon allene ($H_2C=C=CH_2$). The outer atoms of the linear pure carbon chains are terminated by a pair of nonbonding electrons. Bonding diagrams for some representative carbon clusters are shown in the **illustration**. The average $C=C$ bond length for linear carbon molecules is approximately 0.128 nanometer. In contrast, the double-bond length in ethylene ($H_2C=CH_2$) is 0.134 nm, and the two $C=C$ bond lengths in allene are 0.131 nm, whereas the triple-bond length in acetylene ($HC\equiv CH$) is 0.121 nm.

Experimental studies. Carbon clusters and their ions have now been studied in the gas phase by using mass spectrometry, photoelectron spectroscopy, high-resolution optical and infrared spectroscopy, Coulomb explosion experiments, and gas-phase ion chromatography. Vibrational spectroscopy and electron spin resonance spectroscopy have been used to study carbon clusters trapped in solid neon or argon matrices at low temperatures. Carbon clusters have been observed in the gas surrounding carbon-rich stars, in the tails of comets, and in the interstellar medium by using visible, infrared, and far-infrared astronomy.

Although mass spectrometry is capable of extremely sensitive detection of carbon cluster

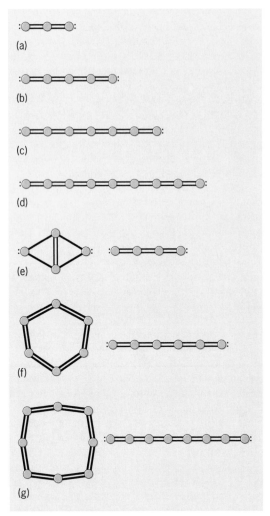

Low-energy structural isomers and electron configurations for carbon clusters: (a) C_3, (b) C_5, (c) C_7, (d) C_9, (e) C_4, (f) C_6, and (g) C_8. Only the linear isomers have been detected experimentally.

ions, it provides only limited information about the molecular structures. Photoelectron spectroscopy is also a highly sensitive technique that is used to study electronic configurations and to measure the frequencies of certain vibrational modes. However, difficulties in obtaining structural information arise because of the low resolution of this technique. Studies of carbon clusters trapped in cold solids are difficult to interpret for the same reason. Coulomb explosion experiments are ambiguous to interpret.

High-precision infrared laser spectroscopy provides the most detailed and reliable information about the molecular structures of gas-phase carbon molecules that can be obtained with presently available instrumentation. By analyzing the rotational transitions that accompany transitions between different vibrational energy levels, precise information can be obtained about bond lengths and bond angles, the electronic configuration of the cluster, and the dynamics of vibrational motions. Such experiments have recently been performed for carbon clusters as large as C_{13}.

Gas-phase ion chromatography is another useful tool that was recently devised to obtain structural information about cluster ions. Ion chromatography of carbon cluster ions has revealed a structural morphology for ions as large as C_{84}^+ that is far more complex than previously thought. For example, the linear chain structures were found to coexist with monocyclic rings for negatively charged ions as large as C_{20}^-. For larger carbon clusters, numerous structural isomers were present, including monocyclic and polycyclic rings, and fullerenes.

Linear C_{13}. High-resolution infrared spectroscopy has been used to characterize the linear carbon clusters C_3–C_7 and C_9, shown in the illustration. The rotational-vibrational spectra observed were consistent with theoretical predictions. Splittings of the absorption transitions for C_4 and C_6 indicated a triplet electronic state, and the absence of these splittings for the odd clusters confirmed a singlet configuration. At present, there is no direct experimental evidence for the cyclic isomers which are predicted to exist.

The infrared spectrum for linear C_{13} was recently discovered. Carbon clusters were produced by pulsed ultraviolet laser vaporization of graphite. In this method, carbon clusters initially begin to form in a laser-induced plasma that is swept through a narrow channel by a high-pressure pulse of helium. The gas mixture is then expanded as a supersonic molecular beam where the molecular degrees of freedom are rapidly cooled to temperatures as low as 10 K ($-440°$F). A tunable infrared diode laser spectrometer is used to measure the absorption transitions.

The spectrum is characteristic of a linear carbon cluster in a singlet electronic state, analogous to previous observations of C_3, C_5, C_7, and C_9, with an average bond length of 0.1277 nm. Analysis of the rotational transitions revealed that the spectrum could be assigned to the linear isomer of C_{13}, the largest carbon cluster to be characterized by high-resolution spectroscopy. This result was unexpected, because theoretical calculations predicted that carbon clusters of this size would exist predominantly as planar monocyclic rings. In fact, an earlier calculation found the cyclic isomer of C_{13} to be more stable than the linear form by 20–30 kcal/mol. After the observation of the linear C_{13} spectrum, a higher-level calculation revised the cyclic isomer to only 6.8 kcal/mol lower in energy. This study found that at the high temperature of carbon-cluster formation in laser vaporization sources the linear chain is actually the dominant isomer.

These observations, along with the results of gas-phase ion chromatography, indicate that an astounding variety of structural isomers of carbon clusters are present in high-temperature carbon-rich environments. Most previous attempts to model the clustering mechanism have assumed only cyclic isomers for clusters in the C_{10}–C_{20} size range. The reality, however, is that linear isomers of this size are present as well, and it is very likely that they play a critical role in the chemistry of these systems. For example, it was found that the polyyne form of carbon can be synthesized in a laser vaporization reactor, which can also be used to produce fullerenes. A gas-phase precursor added to the reactor stabilizes the polyynes by reacting with the terminal carbon atoms in the chain, and completely suppresses fullerene production. Thus a direct role for the linear form of carbon in the mechanism of fullerene synthesis is implied.

For background information *SEE ATOM CLUSTER; BOND ANGLE AND DISTANCE; CARBON; CHEMICAL BONDING; COULOMB EXPLOSION; MOLECULAR ORBITAL THEORY; POLYENE; SPECTROSCOPY; TRIPLET STATE* in the McGraw-Hill Encyclopedia of Science & Technology.

Alan Van Orden; Richard J. Saykally

Bibliography. B. Bleil, F.-M. Tao, and S. Kais, Structure and stability of C_{13} carbon clusters, *Chem. Phys. Lett.,* 229:491–494, 1994; T. F. Giesen et al., Infrared laser spectroscopy of the linear C_{13} carbon cluster, *Science,* 265:756–759, 1994; R. J. Lagow et al., Synthesis of linear acetylenic carbon: The "*sp*" carbon allotrope, *Science,* 267:362–367, 1995; G. Von Helden et al., Isomers of small carbon cluster anions—Linear chains with up to 20 atoms, *Science,* 259:1300–1302, 1993; W. Weltner, Jr., and R. J. Van Zee, Carbon molecules, ions, and clusters, *Chem. Rev.,* 89:1713–1747, 1989.

Cattle

Recombinant bovine somatotropins (rbSTs) are bacteria-made analogs of bovine growth hormone (bGH), a protein hormone secreted by the pituitary gland. These drugs promote milk production but have numerous side effects, including uterine disorders, mastitis, digestive disorders, a reduction in pregnancy rate, a decrease in gestation length, and a decrease in calf birth weight. The first marketed recombinant bovine somatotropin is sometribove. Its amino acid chain differs from the natural sequence in two positions. These differences may affect immunogenic responses in cows and milk consumers.

Milk production. The full increase in daily milk yield develops in 3–4 days after the initiation of recombinant bovine somatotropin treatment. The mean milk response was 11.0 lb/day (5.0 kg/day; 17% increase) in 29 commercial herds given sometribove for 12 weeks. The range was 5.3–16.8 lb/day (2.4–7.6 kg/day). The wide range of milk response may be partially attributable to management. Poor milk responses may be more consistently associated with poor management in certain countries, such as in Africa, and perhaps Russia, where poor conditions may reach extremes more frequently than in the United States and western Europe. Thus, poor management as a factor may apply to some extent in comparing regions, but it does not apply to individual herds within the United States and perhaps western Europe.

Nutrition. During recombinant bovine somatotropin treatment, the full increase in feed intake develops slowly relative to the milk increase, necessitating that body tissue be utilized to make the extra milk for a period of 8 weeks or more. A condition termed catabolic stress results, characterized by increased illness and infertility. Thus, catabolic stress enables the milk response but places the animal at risk for side effects. In one experiment, the combination of recombinant bovine somatotropin and inert fat synergistically affected milk production—a positive result—but increased tissue mobilization—an undesired effect from the point of view of preventive medicine.

The implementation of sound nutritional programs is advised; however, little or no consistent advantage has been found for increasing grain, the energy density of a total mixed ration, ruminally inert fat, protein, or ruminally undegraded protein. Until better nutritional programs are developed, the management program most likely to reduce or eliminate catabolic stress is to delay recombinant bovine somatotropin administration until the cow is in a strongly anabolic condition, that is, is restoring tissue, gaining in body weight and condition, and already pregnant. This anabolic start might be in month 5 or 6 of lactation rather than in month 3 as currently recommended. This safer application of the drug might increase its overall use, although for a shorter period. Also, administration during only the last half of lactation should allow the use of high-energy intakes that would otherwise increase the risk of the fat cow syndrome or the fatty liver syndrome (hepatic lipidosis) at the start of the next lactation.

Mastitis. An increased incidence of mastitis was argued to be related to the increase in milk production stimulated by sometribove. Mastitis is inflammation of the mammary gland usually associated with microbial infection. The average incidence of clinical mastitis in the United States and Europe is about 40 new cases per 100 cows at risk during a 10-month lactation, 20 cases in the first 2 months and another 20 in the last 8 months. Clinical mastitis is usually treated with antibiotics for 1–3 days; however, increased medication during recombinant bovine somatotropin usage may pose an indirect risk to consumers. In 1994, the Food and Drug Administration concluded that recombinant bovine somatotropin–induced mastitis is minor and managable; however, the quantitative comparisons are questionable.

The management hypothesis may be evaluated by regression analysis for mastitis: a low milk production in the control group indicates poor milk management, but a low control mastitis incidence reflects good mastitis management. The control mastitis incidences were 5–10% in significantly affected herds versus 16–26% in unaffected herds. The plot of the mastitis incidence response to sometribove against control incidence was significant and negative for the eight pivotal trials, disproving the hypothesis. Nevertheless, these eight trials served as the basis for the Food and Drug Administration's approval of the drug and allowance of the implied claim that good management positively influences the mastitis response to recombinant bovine somatotropin.

Other side effects. Recombinant bovine somatotropin side effects are the same set of diseases and disorders that attend the catabolic phase, usually during the first 2–3 months of natural lactation. Since all the side effects probably reflect catabolic stress, none is likely to be moderated in frequency or severity unless catabolic stress is reduced through a delayed start in drug administration until the cow is naturally in an anabolic state, or through the development of new nutritional programs that effectively reduce tissue utilization despite an earlier start.

Field experience. The Food and Drug Administration reported 96 complaints about sometribove during the first six months and another 710 during the second six, a total of 806 complaints of which 496 were deemed related to the drug. Clinical mastitis was named in 121 complaints relating to 2211 cows; other udder abnormalities, including swelling and abnormal milk (signs of mastitis), were named in another 73 complaints concerning 953 cows. Moreover, increased somatic cell counts, indicative of subclinical mastitis, were involved in 105 complaints affecting 3332 cows. Thus mastitic conditions accounted for 60% of complaints and involved 6496 cows.

The overall complaint level represents only 6% of farms using sometribove. This level is much smaller than the roughly 30% expected if administration started at day 63. When the drug was released in February 1994, farms using it treated most cows past day 63; the average would have been about day 184. Most cows past day 126 would have been in an anabolic condition and unlikely to suffer side effects. More cows were recruited at day 63, so that more cows in the treated group would become catabolic and develop side effects. Only during the drug's second year of use, when treatment will start at day 63 in most cows, will its full impact on health become clear.

Implications. Disclosure of side effects following approval of a recombinant bovine somatotropin drug has allowed independent review and recognition that some effects, notably mastitis, are statistically significant and clinically painful. The success of production promotants, such as recombinant bovine somatotropin, may depend on designing, testing, and revising new health management programs that control side effects.

For background information SEE DAIRY CATTLE PRODUCTION; MASTITIS; MILK in the McGraw-Hill Encyclopedia of Science & Technology.

David S. Kronfeld

Bibliography. D. E. Bauman, Bovine somatotropin: Review of an emerging animal technology, *J. Dairy Sci.,* 75:3432–3451, 1992; D. S. Kronfeld, Health management of dairy herds treated with bovine somatotropin, *J. Amer. Vet. Med. Ass.,* 204:116–130, 1994; Monsanto Company, *Technical Manual for Posilac,* 1993; P. Willeberg, An international perspective on bovine somatotropin and clinical mastitis, *J. Amer. Vet. Med. Ass.,* 205:538–541, 1994.

Cavitation

Many liquid flows involving rotating machinery and associated valves and piping have cavitation. Cavitation occurs when nuclei (microbubbles) enter a region where the local pressure in the liquid is equal to or less than the vapor pressure. Although the detailed mechanisms which cause cavitation in piping and valves are quite complex the phenomena are understood, and there exist empirical methods that are successful in predicting flow conditions when cavitation will occur. However, the efficient transfer of energy to or from a liquid by rotating machinery requires an unsteady cyclic operation that causes cavitation to be unsteady. The periodic formation of cavitation and the corresponding pressure fluctuations are detrimental to performance. They can lead to noise and vibration, material erosion, and performance breakdown. Blade surface, blade bubble, tip leakage vortex, and hub vortex forms of cavitation have all been observed.

In hydraulic terms, the cavitation performance of a machine is defined by the Thoma cavitation number, which is the ratio of net positive suction head, at the machine inlet and the operating head of the machine. The net positive suction head is defined as the total head (pressure plus velocity) on the suction side of the machine above vapor pressure. As the net positive suction head is lowered, there is a condition at which cavitation inception occurs. The value of net positive suction head which causes the operating head to drop by 3% is considered the performance design criterion for an allowable level of cavitation. However, because cavitation inception occurs at higher values of net positive suction head, levels of vibration, noise, and material erosion can be close to their maximum before the point at which the machine head drops by 3%.

Conventionally, studies of performance under cavitating conditions have been conducted by analyzing the relationship between inlet conditions and the machine's performance coupled with visual cavitation observations. A more comprehensive approach is now being developed to understand the fundamental mechanism of cavitation and to predict noise, vibration, and material erosion. Application of new instrumentation that measures the flow pressures and velocities, the radiated noise and machine vibrations, and the liquid quality have provided insight into cavitation effects on rotating machinery. These detailed measurements provide data for the emerging area of two-phase flow computational fluid dynamics (CFD).

Examples of cavitation. Some examples of recent work on cavitation in rotating machinery and valves will be discussed.

Cavitation performance testing. Cavitation model tests are used to determine the characteristic behavior of a prototype machine. It is essential to perform accurate tests with models that not only achieve dynamic similarity with the prototype but also reproduce bubble dynamic effects, that is, the behavior of bubbles and their interaction with the surrounding liquid. Both have a significant influence on cavitation inception and resulting cavitation patterns. As an example, it is important to perform tests with a model Francis turbine (an inward radial flow hydraulic turbine) at Froude similarity operating conditions (equal ratios of inertia force to gravitational force) to ensure similar pressure distributions. But, to guarantee similar cavitation patterns it is also necessary to include bubble dynamic effects by the control of the liquid quality. Liquid quality is a measure of cavitation susceptibility and is related to the size and number of nuclei.

Experimental studies have shown that cavitation patterns on a model Francis turbine are influenced by the liquid quality. **Figure 1** shows the influence of liquid quality on the relative efficiency of a Francis turbine as determined by using degassed water and water having microbubbles at the same operating condition.

In a flowing water system microbubbles are constantly being produced by the motion of the water and at the same time being dissolved in the water. Therefore, the liquid quality in a flow system is dynamic; that is, it has no constant value, only a statistical one, and its measurement is the motivation for the development of new instrumentation. Progress continues to be made on the development of two approaches. One approach characterizes the liquid quality by the liquid tension required for cavitation (that is, with the use of the venturi cavitation susceptibility meter and the ultrasonic method) and the other approach characterizes the liquid quality by the measurement of the microbubble distribution (that is, with the use of holography, light scattering, and the phase Doppler anemometer). The key issues are correlating these approaches with cavitation in the machine and utilizing this correlation to predict bubble dynamic scale effects, that is, the similarities between bubbles growing during a model-scale experiment and those at full scale.

In some cases, performance tests are conducted to determine cavitation inception. Experimental studies have been conducted to quantify the effects of bubble dynamics on cavitation inception by varying the liquid quality. Both acoustic and visual cavitation inception data were determined for the blade surface, traveling bubble, and tip vortex forms of cavitation for a range of liquid quality conditions. The liquid quality was determined from measurements with a cavitation susceptibility meter. Results clearly demonstrate a different influence of liquid

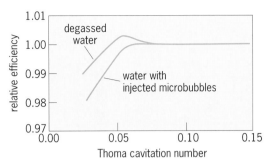

Fig. 1. Cavitation performance of a model of a Francis turbine using degassed water and water with injected microbubbles. (*After B. Gindroz, Institut de Machines Hydrauliques et de Mecanique des Fluides, Swiss Federal Institute of Technology*)

quality for each type of cavitation. In the case of blade surface cavitation, little variation in the cavitation inception index was found, even for a significant increase in liquid tension and reduction in the microbubble size distribution. Traveling bubble cavitation inception correlated directly with liquid tension measurements, but was independent of the microbubble size distribution. However, the tip vortex cavitation inception index decreased significantly for only a small increase in liquid tension. In addition, a strong dependence on the microbubble size distribution was noted.

Cavitation erosion. The collapse of the vaporous cavity can create pressures high enough to damage an adjacent surface. This material erosion is a key issue because, for economic reasons, most machinery is designed to operate near the condition that the net positive suction head causes the operating head to drop by 3%, and it therefore has cavitation. The material damage rate is determined by a complex combination of cavitation characteristics (impact pressure), material properties, and the exposure rate. Significant progress has been made in developing a prediction technique. The key to this prediction is knowing the spectrum of impact pressures on a solid surface. As a first step, an estimation of the collapsing cavity event rate must be made along with the impact pressure for each event. Many experiments have been designed to determine this impact pressure spectrum in a normalized, statistical distribution.

An emerging technology in this area is the application of nonmetallic coatings that are resistant to pitting. However, these coatings can fail internally because of the pressure waves created inside the coating by the impact pressure. Adhesion of protective coatings has been a problem that has limited their use in the past. *See* EROSION.

Cavitating inducers. In order to meet pump head requirements and yet be lightweight, rocket engine turbopumps must operate at high rotational speeds with low net positive suction head at the inlet. To satisfy these system requirements, a low-pressure rise, high-flow pump includes a cavitation resistance inducer upstream of the pump impeller. The inducer is essentially an axial flow pump with high-solidity blades, and its function is to increase the inlet head to prevent cavitation in the downstream impeller.

Experience with testing inducers shows that under most operating conditions cavitation occurs which results in significant blade and shaft vibrations and in blade erosion. Cavitating modes that are observed for a given flow condition as the inlet head is reduced are limited blade tip cavitation, blade surface cavitation, rotating or propagating cavitation, synchronous asymmetric cavitation, oscillating or chugging cavitation, and fully developed cavitation. Several continuing studies are aimed at avoiding the asymmetric, rotating, and oscillating cavitation forms that create excessive unsteady blade loads and shaft vibrations. This periodic formation and collapse of large cavities leads to the formation of so-called cloud cavitation that can cause more intense noise and has more potential for damage than fully developed cavitation.

Research has been conducted to investigate the dynamics of cavitating flows in inducers to understand the impact of cavity response on the rotor-dynamic forces. There is a strong coupling between the local dynamics of the cavitation and the global behavior of the flow. Inertial, damping, and compressibility effects on the dynamics of the cavities are important. As the inducer is whirling on a shaft, normal eccentricities result in the inducer being whirled in a circular motion in addition to its rotation. This whirling motion about the shaft rotation is usually at frequencies that are fractions of the main shaft frequencies. As the cavity dynamics change with time, pressure pulses are created that travel in the two-phase flow mixture at a local wave-propagation speed, that is, the speed of sound locally in the two-phase flow liquid. The effect of the whirl excitation on the cavitating flow is dependent upon the wave propagation speed and the cavity resonance behavior at natural frequencies associated with a characteristic length and local liquid velocity. These, in turn, lead to frequency-dependent rotor-dynamic forces which depend upon the void fraction of the cavitation and the mean flow properties. The dynamic response of the cavitation results in major deviations from the noncavitating flow solutions.

Prosthetic heart valves. One research area of current interest is cavitation in prosthetic heart valves. In artificial heart and ventricular assist devices, cavitation could cause valve structure erosion and blood damage. Tests conducted in a mock circulatory loop which models systemic capacitance and resistance show cavitation under normal physiologic conditions. An electrical ventricular assist device (EVAD) is used to power the loop which is filled with a transparent, glycerol-water solution that matches the kinematic viscosity of blood at high strain rates. **Figure 2** shows the transient cavitation when the inlet valve rapidly closes at the start of the cycle of the electrical ventricular assist device. Factors which may influence the cavitation are valve material, valve gap, closing velocity, and closing deceleration.

Two-phase flow computational fluid dynamics. Many new numerical methods are being developed to predict machine performance under cavitating conditions. Initial attempts to predict cavitation on hydrofoils used a free-streamline theory in which the flow boundary conditions are linearized (that is, with regard to incidence, camber, and so forth). A nonlinear theory followed which allowed boundary conditions at the exact location of the flow boundary. However, only fully numerical techniques can be utilized for the complex flows inside of a machine.

Recently, advances in computer performance have allowed the application of computational fluid dynamics to predict the complex three-dimensional flow in a machine. However, there are some difficulties in predicting where and what type of cavitation will occur. These difficulties are due in part both to the need to predict local fluid mechanics and to a lack of fundamental understanding of the mechanisms for the formation of cavities and their subsequent dynamics.

The application of the boundary elements method (BEM) has led to both two-dimensional and three-dimensional predictions of cavitation on hydrofoils (assuming steady cavities). This approach has also been applied to calculate the three-dimensional cavitation on a marine propeller by using the approximation of lifting surface theory. The main drawback of this method is that it does not include a nonuniform rotational flow.

Solutions to the three-dimensional Euler method which splits the velocity field into two parts, potential and rotational, have successfully predicted the cavitation patterns in an inducer. In this case, the cavitation model for the steady cavity is based on a pressure boundary condition.

In general, the modeling of cavitating flow on blades has taken two approaches. One approach seeks a solution only in the liquid domain, along with a description of the boundary cavity surface. The second approach solves the governing equations in the liquid-vapor domain by introducing a

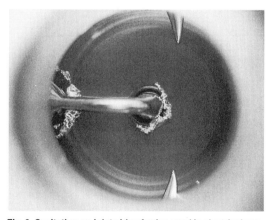

Fig. 2. Cavitation on inlet side of valve used in electrical ventricular assist device. (*D. Stinebring, Applied Research Laboratory and Bioengineering Program, Pennsylvania State University*)

pseudodensity which can vary between the liquid and vapor density extremes.

The all-encompassing approach of solving the Reynolds-averaged Navier-Stokes equations with cavitation bubbles in the flow inside a machine is making progress. This approach has been applied to the flow passage with some success. Needed insight into cavitation development and machine performance has been gained, but it is not yet possible to accurately address the three-dimensional incoming flow.

For background information SEE CAVITATION; DYNAMIC SIMILARITY; FROUDE NUMBER; HYDRAULIC TURBINE; PROSTHESIS in the McGraw-Hill Encyclopedia of Science & Technology.

Michael L. Billet

Bibliography. M. Billet et al., *Proceedings of the 21st International Towing Tank Conference—Cavitation Committee Report,* 1996; C. Brennen, *Cavitation and Bubble Dynamics,* 1995; P. Cooper, *Pumping Machinery, 1993,* 1993; T. O'Hern et al., *Cavitation and Gas-Liquid Flow in Fluid Machinery and Devices,* 1994.

Cell (biology)

A greater understanding of cellular proliferation and growth control has identified numerous positive and negative regulatory mechanisms which maintain homeostasis. Alterations of these processes have adverse consequences at both the cellular and organismal levels. Cellular proliferation can either be lost, as when a wound fails to heal, or become uncontrolled, as when a tumor forms. Between these two extremes lie many examples of how alterations in normal regulatory processes participate in the development and progression of disease as well as the normal aging process. Normal human and rodent cells in culture exhibit a finite life-span and cease proliferating, remaining metabolically active, a state termed cellular senescence. By contrast, cancer cells do not undergo senescence, and have an infinite life-span in culture. Evidence has been accumulating to support a genetic basis for cellular senescence and for the changes which alter this process to allow for the evolution of tumor cells.

Cellular senescence. Aging involves numerous changes in various organ systems, and the study of human aging has been further complicated by the vast number of genetic differences among aging individuals. Therefore, a simpler model system must be utilized to analyze age-related effects as distinguished from disease-induced phenomena.

Normal human cells have a limited division potential in culture. The observation of cellular senescence led to the proposal that the limited division potential of these cells in culture is an expression of aging or senescence at the cellular level. A variety of human cell types have been found to undergo senescence, including epidermal keratinocytes, smooth muscle cells, glial cells, endothelial cells, and T lymphocytes.

It has been demonstrated that the aging of cells in culture is related to aging of the organism as a whole. Fewer divisions in culture are observed in cells derived from shorter-lived animal species and individuals with genetic disorders that mimic premature aging (such as Hutchinson-Gilford and Werner's syndromes).

DNA synthesis inhibitor. The significance of the differences between young versus senescent cells is not fully understood because it is difficult to determine which changes are causal to, and which are the result of, senescence. However, it has become clear that cellular senescence is not a process of programmed cell death: these cells remain viable in culture for up to 3 years. The program of cellular senescence results in the inability of these metabolically active cells to initiate deoxyribonucleic acid (DNA) synthesis and progress through the cell cycle. No known combination of mitogens can stimulate DNA synthesis and cell division in senescent cells, unlike quiescent cells. It has been hypothesized that the end point of the senescence program is the buildup or specific production of an inhibitor which has the function of blocking the cell cycle machinery from initiating DNA synthesis.

Additional evidence for the DNA synthesis inhibitor was obtained from studies with senescent cell messenger ribonucleic acid (mRNA). Microinjection of mRNA isolated from senescent cells into young cells inhibited initiation of DNA synthesis. The result was the identification and isolation of several senescent cell-derived inhibitors of DNA synthesis (*SDI 1–3*). Only *SDI-1* mRNA levels were upregulated in nondividing cells (senescent and quiescent). Other studies have also identified the *SDI-1* gene products as a Cdk-interacting protein (CIP1/p21) and a tumor suppressor protein, p53-transactivated gene (*WAF1*).

The importance of *SDI 1* in cell survival is indicated by the fact that it has been impossible to detect mutations or deletions in this gene in a large number of immortal human cell lines. Collectively, these data suggest that *SDI 1* is required for growth arrest both in the normal cell cycle and at the end of the life-span in the laboratory, and it also controls cell growth to allow for DNA repair. The regulation and production of *SDI 1* is of paramount interest for the fields of cellular senescence and cancer biology, because loss of this function results in genetic catastrophe. SEE CANCER (MEDICINE).

Somatic cell hybrids. The second major approach to the study of cellular senescence is to examine the mechanisms by which cells escape senescence to become immortal. By understanding the genetic changes that allow for infinite division, identification of the growth regulatory genes and pathways that may be important in tumor formation is possible. The frequency with which cells escape senescence to become immortal varies among species. Spontaneous and induced immortalization occurs much more frequently for rodent cells than for human cells. Because human cells are resistant to spontaneous immortalization, and induced immortalization using various viral or chemical carcinogens occurs rarely, human cells are preferred for studies of cellular aging.

Fusions involving clonal populations of various immortal and normal human cells yielded hybrid cells with a limited doubling potential, providing evidence for a dominant genetic basis for senescence. Fusion of different immortal cell lines with highly differentiated cell types such as normal human T cells and endothelial cells also yielded hybrids with limited division potential. Collectively, these data support a genetic basis for aging and clearly indicate that cells escape senescence to become immortal because of recessive changes such as loss or inactivation of both copies of growth regulatory gene(s). They also suggest that the dominance of senescence is governed by common mechanisms in very different cell types.

By fusing various immortal human cell lines together, more than 40 different immortal human cell lines have been assigned to four complementation groups for indefinite division. The identification of only four complementation groups indicates that a few genes or gene pathways are altered to produce immortal cells, and that only a few primary genes are involved in the control of senescence. The complementation group studies greatly enhance the method of identifying genes involved in group-specific growth control, in an attempt to understand senescence and the changes responsible for immortalization.

Chromosome identification. Microcell-mediated chromosome transfer has been used in an attempt to identify chromosomes encoding senescence-related genes specifically altered in cell lines assigned to a given complementation group. This technique allows single normal human chromosomes to be introduced into immortal human cell lines assigned to the four complementation groups for indefinite division. By analyzing the growth potential of these microcell hybrids, the chromosome can be assessed for its ability to induce senescence in cell lines assigned to the same group, but not cell lines assigned to the other three groups.

Three chromosomes involved in normal growth control and senescence have been identified which suppress proliferation of specific cell lines. The analysis of additional chromosomes and the isolation of the senescence genes are ongoing.

Implications. The disruption of specific gene products can alter the proliferative capacity of a population of cells, and may result in immortalization. Because these genes clearly play regulatory roles, understanding their normal function is also likely to provide insight into the mechanisms underlying a number of diseases. This information could prove useful in the design of therapies for the treatment of disorders that result from altered growth regulation. For example, further characterization of the inhibitor of DNA synthesis may make this protein a target for manipulation in diseases such that the inhibitor could be introduced and upregulated in diseases of uncontrolled proliferation (such as tumors) or downregulated when proliferation is needed (such as for augmentation of wound healing). SEE TRANSPLANTATION BIOLOGY.

For background information SEE AGING; CANCER (MEDICINE); CELL SENESCENCE AND DEATH; CHROMOSOME; DEOXYRIBONUCLEIC ACID (DNA); GENE; TUMOR in the McGraw-Hill Encyclopedia of Science & Technology.

Patrick J. Hensler

Bibliography. E. Finch, Longevity, *Senescence and the Genome,* 1990; O. M. Pereira-Smith and J. Smith, Genetic analysis of indefinite division in human cell: Identification of four complementation groups, *Proc. Nat. Acad. Sci. USA,* 85:6042–6046, 1988; A. Noda et al., Cloning of senescent cell-derived inhibitors of DNA synthesis using an expression screen, *Exp. Cell Res.,* 211:90–98, 1994.

Cephalopoda

Coleoid cephalopod mollusks include the only living examples of large free-swimming invertebrate predators. They have a high biomass and purely carnivorous diet, are a major food source for higher predators such as whales and seals, and are the basis for commercially important fisheries.

These efficient predators possess a pair of large, powerful beaks (jaws or mandibles) located within a muscular buccal mass at the center of a ring of arms. The beaks are composed entirely of chitin, and their primary function is to bite and ingest food. The cephalopod jaw mechanism differs from those of arthropods or vertebrates in one major way: the beaks lack a linear jaw joint, and slide and rotate around an area rather than a fixed line of articulation.

Recent studies have shown links between beak shape and mandibular muscle volume, described the action of these muscles during biting, and interpreted the results in terms of adaptations to specific prey types.

Prey capture and diet. In coleoids, the mechanisms for prey capture (arms and tentacles) have become dissociated from those for killing and ingestion (beaks and poison). The way that cephalopods utilize their arms and buccal mass in dealing with prey is akin to that of primates, with arms and tentacles acting as hands to grab food and to manipulate it while feeding.

Squids generally prey on soft-bodied, highly mobile species such as fish and other squid, while cuttlefish also include crabs in their diet. In these decapod cephalopods (Decabrachia), the beaks are the prey-killing mechanism: fish, for example, are killed with a single bite which severs the spine just behind the cranium. In addition, cuttlefish paralyze crustacean prey by introducing toxins into the wound inflicted by the beaks.

In contrast to decapods, the preferred prey of benthic octopuses are mainly slow moving (crabs) or completely immobile (mussels, oysters). These prey rely to a great extent on armor to foil potential predators, and the crushing forces needed to deal with such defenses are beyond the capability of the chitinous cephalopod beaks. Instead, Octopoda (Octobrachia) utilize poison and enzymes to attack their prey. They paralyze prey with toxins and then drill a hole in the shell with the radula (a filelike ribbon studded with toothlike structures) and salivary papilla (a nipplelike eminence), through which they introduce a mixture of organic compounds. This chemical attack causes, among other things, bivalves to relax the muscles which hold their shells closed and crabs to go into the molt cycle. Thus, the octopus gains access to the flesh of the prey with minimal physical effort. Differences in beak shape and muscle anatomy are expected to reflect these dietary distinctions.

Beak shape. The front portion of the beak consists of a raised hood and wing area (**Fig. 1**) that encloses the anterior part of the mandibular muscles. The tip of the biting surfaces is called the rostrum. The side of the beaks is termed the lateral wall, and where the two lateral walls meet is the crest. The crest and lateral wall area is entirely covered on its outer surface by muscle (**Fig. 2**).

The octopuses found in shelf waters have short, parrotlike, blunt beaks, while beaks of squids and cuttlefish have more elongated tips. Cephalopod species can be determined on the basis of beak shape alone, and this method is often used to identify undigested beaks found in seal and whale stomachs.

A functional analysis of beak shape suggests that in decapods the pointed upper beak rostrum confers the benefits of both high bite strength (small surface area at the tip) and speed of movement (low mechanical advantage). The elongated form has therefore evolved as a stabbing tool with which to deliver a fast killing bite to a mobile, struggling prey. In contrast, the prey-handling strategy followed by octopuses requires no such fast kill, and the blunt, chopping beaks are used only in cutting up the food items into pieces small enough to swallow.

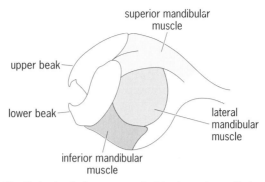

Fig. 2. Anatomical arrangement of beaks and mandibular muscles within the buccal mass. (*After A. J. Kear, Morphology and function of the mandibular muscles in some coleoid cephalopods, J. Mar. Biol. Ass. U.K., 74:801–822, 1994*)

This stabbing bite of the decapods has resulted in other differences in shape: in order to accommodate the pointed upper rostrum, the crest of the lower beak must shorten in length from the octopod form (or the upper beak tip would punch a hole in the lower when the beak was closed). This change requires a simultaneous increase in breadth and surface relief if the area available for muscle attachment is to be kept constant. One such surface feature is the development of a ridge or fold on the lateral wall, as in *Ancistrocheirus*. Other differences in beak morphology are related to the layout of the mandibular muscles and their action during biting.

Bite cycle. The cycle of beak movements during biting comprises an opening phase; fully open; a closing phase; closed with the upper beak pulled deeper back inside the lower (retracted); and a small reopening movement to return the beaks to their resting position (closed but not retracted) [**Fig. 3**]. The muscles responsible for beak movement are the superior mandibular muscle, a pair of lateral mandibular muscles, and the inferior mandibular muscle (Fig. 2). The superior mandibular muscle is the largest.

Studies of the beak movements and the action of the mandibular muscles lead to the following interpretation of the bite cycle:

1. Closing movements. The large superior mandibular muscle provides the force of the bite and contributes the largest component of the closing motion. Simultaneously, the inferior mandibular muscle acts to retract the upper beak, causing shearing action, and dorsal portions of the lateral mandibular muscles flex the upper beak walls outward, probably to accommodate the backward sweep of the radula during closing.

2. Opening movements. To open the beaks, the ventral portions of the lateral mandibular muscles pull the rear lateral walls of the two beaks toward each other. This pulling action moves the lower beak back, relative to the upper beak, causing the rostra of the beaks to move away from each other in an opening motion.

Muscle differences between species. The buccal mass as a whole represents a greater proportion of body size in juveniles (2.21–4.34% of body weight)

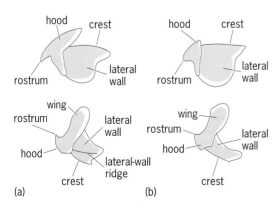

Fig. 1. Side view of disarticulated cephalopod beaks: (*a*) Decapod. (*b*) Octopod.

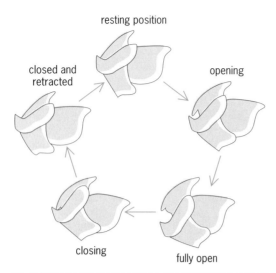

resting position

closed and retracted

opening

closing

fully open

Fig. 3. Side view of decapod beaks, showing the full cycle of movements during a bite. Food is punctured as the beaks close, and the mouthful is sheared off as the upper beak retracts inside the lower. (*After A. J. Kear, Morphology and function of the mandibular muscles in some coleoid cephalopods, J. Mar. Biol. Ass. U.K., 74:801–822, 1994*)

than in maturing or adult animals (0.40–1.55%), indicating that in hatchlings it may play as important a role in prey capture as the arms and tentacles.

In comparing different groups of cephalopods, the buccal mass weighs more in squid and cuttlefish (0.65–2.03% of body weight) than in octopuses of similar age (0.40–0.77%). This weight difference is accounted for by a much larger superior mandibular muscle in the decapods, which is a reflection of their powerful killing bite.

Since beak shape is related both to the surface area where the mandibular muscles attach and to the volume that they occupy, there is a large upper beak hood in all decapods to accommodate this larger superior mandibular muscle. Some species (for example, *Ancistrocheirus*) also can alter the angle by which the wings meet the lateral wall to house a larger volume of superior mandibular muscle.

Conversely, in octopuses the large lateral wall of the upper beak results in a larger lateral mandibular muscle than that of decapods. The area enclosed by the lateral wall is also larger in the former. Both of these features are adaptations associated with the need to accommodate a bigger radula and salivary papilla when drilling hard shells.

Evolution of buccal mass. Because of their soft-bodied nature, the fossil record of coleoids is sparse, but some information can be gleaned about ancestral feeding mechanisms. For example, Jurassic coleoids had hooks instead of suckers on the arms, no tentacles, and a primitive beak of a straight-edged form with no pointed tip to the rostrum.

Decapods use their tentacles to capture small or fast prey and their arms for capturing slower items. In the several modern squid species which possess them, tentacle and arm hooks have been interpreted as an adaptation for catching soft, fleshy prey. Species without hooks pursue a different

strategy: the toothed suckers found in other squid can grip the smooth carapace of soft-bodied crustaceans as well as fleshy prey, while octopuses have broad, flat suckers suitable for handling hard shells. Holes drilled by octopuses have been found in shells dating back to the Pliocene (1.64–5.2 million years ago).

Ancestral coleoids therefore probably consumed slow-moving fleshy prey. The tentacles and stabbing beaks of modern decapods represent later specializations for dealing with more mobile prey types. Similarly, the chopping beaks and utilization of poison by octopods demonstrate a change in diet with the adoption of a more benthic life-style. Further investigation of fossil material may reveal exactly how and when these changes took place.

For background information *SEE CEPHALOPODA; DECAPODA; OCTOPODA; PLIOCENE* in the McGraw-Hill Encyclopedia of Science & Technology.

Amanda J. Kear

Bibliography. P. R. Boyle (ed.), *Cephalopod Life Cycles,* vol. 2, 1987; R. G. Bromley, Predation habits of octopus past and present and a new ichnospecies, *Oichnus ovalis, Bull. Geol. Soc. Denmark,* 40:167–173, 1993; M. R. Clarke and E. R. Trueman (eds.), *The Mollusca,* vol. 12; *Palaeontology and Neontology of Cephalopods,* 1988; A. J. Kear, Morphology and function of the mandibular muscles in some coleoid cephalopods, *J. Mar. Biol. Ass. U.K.,* 74:801–822, 1994.

Cetacea

Modern cetaceans (whales, dolphins, and porpoises) are fully aquatic mammals whose ancestors lived on land. Until recently, fossils of the transition from land to water were too incomplete to show how this transition took place, especially for such important behaviors as locomotion. In 1994, discoveries of the earliest fossil cetaceans from Pakistan shed light on the issue: *Ambulocetus natans* (the walking whale) is the missing link between the land ancestors of whales and their swimming descendants. It is one of the best examples of the fossil record presenting direct evidence about a major morphological transition in evolution.

Cetaceans breathe air and produce milk to nurse their live-born young. Many have remnants of hair on their faces and small rods of bones in their abdominal wall that are a leftover of the hindlimbs. These are a few of the most conspicuous differences between cetaceans and fish, and they indicate that cetaceans are more closely related to land mammals. Ancestral mammals lived on land for more than 100 million years before cetaceans originated, implying also that cetaceans had ancestors that lived on land. Anatomical facts, such as the presence of lungs, confirm this suggestion.

Origin. For more than a century, scientists have tried to determine who the closest living relatives of cetaceans were. In 1994, molecular systematists documented resemblances in the deoxyribonucleic acid (DNA) of cetaceans and Artiodactyla (the order including camels, pigs, deer, and cows), suggesting that these are close relatives.

Paleontologists have also tried to find the group of land mammals most closely related to cetaceans. More than a decade ago, they found convincing similarities between cetaceans and a group called Mesonychia (or Mesonychidae). Mesonychia are extinct and their DNA cannot be analyzed, but fossil representatives are found throughout the Northern Hemisphere from the Paleocene through the Oligocene. Mesonychia were very diverse, ranging in size from that of a weasel to that of a brown bear. The skeletons of some Mesonychia suggest that they may have lived like modern wolves or hyenas, as active predators that chased prey on land. The dentitions of Mesonychia vary greatly in size but are similar in shape, and very different from dentitions of most other mammals. The only animals with similar teeth are fossil cetaceans; in fact, teeth were the main reason for thinking that the two groups were related. However, prior to recent fossil cetacean finds the skeletons of mesonychids and the oldest known cetaceans seemed to bear little resemblance to one another. *See Mammalia*.

The origin of cetaceans can be traced to the early Eocene, about 55 million years ago. At that time, the northern edge of the Indian plate was inhabited by several ancestral cetaceans, such as *Pakicetus* and *Ichthyolestes*. The teeth of these mammals are very similar to those of Mesonychia, but aspects of the ear link *Pakicetus* with cetaceans. Unfortunately, the only fossils known for *Pakicetus* and *Ichthyolestes* are teeth and skull fragments, and these fossils do not indicate clearly how these earliest cetaceans lived. The oldest complete skeletons of fossil cetaceans are represented by *Basilosaurus* from the late Eocene of Egypt. *Basilosaurus* was an enormous animal, more than 30 ft (10 m) long, with tiny hindlimbs. Unlike Mesonychia, it is inconceivable that the hindlimbs were used for locomotion on land, and *Basilosaurus* fossils reveal little concerning the transition from locomotion on land to water. Scientists think that *Basilosaurus* may have used its hindlimbs for embrasure while mating, in a manner similar to the claspers of sharks.

Ambulocetus. Fossils of a new species of whale, *A. natans,* found in Eocene sediments of the Kuldana Formation of Punjab, Pakistan, solved the puzzle in 1994. A considerable portion of the skeleton of *A. natans* was found, including the bones of the forelimb and of the foot. *Ambulocetus* is the only member of the Cetacea known to have had feet suitable for locomotion on land. *Ambulocetus* displayed several characters that were similar to those of the terrestrial Mesonychia and others that were similar to those of aquatic cetaceans.

Ambulocetus was about the size of a male southern sea lion (*Otaria,* 660 lb, or 300 kg), but it did not look like a sea lion (**Fig. 1**). It had a long snout, and the eyes were set high on top of the head, as in a crocodile. Of the forelimb, only parts from the elbow down were found: the bones were short and powerful, and the five fingers had nails in the shape of a hoof, convex from side to side as well as along their length. Several hindlimb elements were found, including the femur, knee, and many foot bones. The femur is also short and stout but much longer than the forelimb. The foot is especially impressive; it has four robust toes that are nearly twice as long as the fingers. *See Pinnipedia*.

Ambulocetus is similar to *Pakicetus* and the mesonychians in the shape of its teeth. All have reduced areas on the molars for crushing and cutting (so-called basins and crests), and the teeth mainly consist of a number of blunt elevations (cusps). Upper molars have three large cusps, lower molars two.

The presence and shape of large hindlimbs establish a clear similarity to land mammals, and details of their form, such as the hooves and the large size of the third and fourth digits, are similar to those of mesonychians. The skeleton of *Ambulocetus* looks unlike that of any known cetacean, but fortunately the elements that most clearly identify it as a cetacean were found. These fossil elements are the bones that encase parts of the hearing organ, the

Fig. 1. Reconstruction of *Ambulocetus natans.* (*Courtesy of Marion Lipka*)

middle ear. One, the tympanic, has a greatly enlarged medial lip (involucrum) and a thin crest on the lateral side (sigmoid process). In addition, the lower jaw has an enlarged foramen for the entrance of arteries and nerves (mandibular foramen). In modern cetaceans, the mandibular foramen is very large and contains a fat pad that transmits sounds for underwater hearing. The involucrum, sigmoid process, and enlarged mandibular foramen are characteristic of cetaceans and are unknown in any other vertebrate. All three characteristics occur in *Ambulocetus,* and the first two also occur in the only braincase known for *Pakicetus.*

The swimming behavior of *Ambulocetus* can be inferred from its skeleton. The size and shape of the vertebrae suggest that the animal had very strong back muscles that could flex the spine up and down through the water. Modern cetaceans swim in this manner, with a horizontal tail fluke that is moved up and down by the back muscles. Only a few elements of the tail of *Ambulocetus* are known, and these suggest that the animal had a long tail and probably no tail fluke. Instead, it probably held its large hindlimbs behind the body for swimming, moving them up and down with its back and trunk muscles. Modern sea otters (*Enhydra*) swim in this manner, undulating their entire body dorsoventrally. In some sense, then, *Ambulocetus* combined the swimming techniques of its ancestors, land mammals that swam by dog-paddling, and of its fully aquatic descendants, cetaceans that use tail oscillations. Like the former, *Ambulocetus* used its hindlimbs for propulsion, but like the latter, the motion of the hindlimbs resulted from contraction of the back and trunk muscles.

Ambulocetus was found in beds with oysters and marine snails (gastropods), indicating that it lived in shallow-water seas. Its teeth and jaws are very sturdy, implying that it fed on large animals, and its body proportions suggest that although it could locomote on land it was not very agile. The dorsal position of the eyes indicates that it hunted from a submerged position, watching for the approach of terrestrial prey to the water's edge. Thus it would appear that *Ambulocetus* ambushed prey close to the water's edge in ways very similar to modern crocodiles.

Rhodocetus. In addition to *Ambulocetus* a second new fossil cetacean was discovered in 1994. *Rhodocetus kasrani* was found in middle Eocene sediments from central Pakistan, making it slightly younger than *Ambulocetus.* Moreover, it was found in rocks that indicate it lived in deeper seas, off shore. The hands and feet of *Rhodocetus* were not found, but scientists did unearth a skull, parts of the vertebral column, and some elements of the hindlimbs. These finds suggest that *Rhodocetus* had limbs that were shorter than those of *Ambulocetus* and that probably could not support its body weight. The sacrum of *Rhodocetus* consists of unfused vertebrae, just as in modern cetaceans and

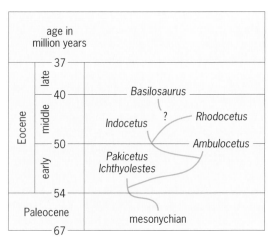

Fig. 2. Phylogenetic relations of some of the earliest cetaceans and their geologic ages.

unlike mammals that locomote well on land. The eyes of *Rhodocetus* are on the side of its head, and it was probably a pursuit hunter of fish in the water, like modern seals. However, *Rhodocetus* had much longer limbs than late Eocene cetaceans from Egypt, such as *Basilosaurus.*

Implications. Conceivably, *Rhodocetus* spent part of its life on land, and may have lived like a modern seal or sea lion.

Contemporaneous with *Rhodocetus* lived yet another whale, *Indocetus,* for which first skeletal material has also been found. *Indocetus* may be intermediate between *Ambulocetus* and *Rhodocetus* (**Fig. 2**).

Thus, the cetacean fossil record is quickly improving. More fossils of *Pakicetus* and *Ichthyolestes* are being found, and the new finds may finally reveal what these oldest cetaceans looked like. Fossils of these primitive cetaceans will also make it possible to compare the bones of artiodactyls and cetaceans in detail to determine if these two groups are as closely related as the molecular data suggest. In the past, comparisons of these fossils were not possible because the known fossil cetaceans had accrued so many specializations that they did not resemble any modern mammal very closely.

The discovery of *Ambulocetus* proves that intermediate forms occur everywhere in evolution. However, the fossil record is incomplete because these intermediates are only rarely preserved.

For background information *SEE* Cetacea; MAMMALIA in the McGraw-Hill Encyclopedia of Science & Technology.

J. G. M. Thewissen

Bibliography. P. D. Gingerich et al., New whale from the Eocene of Pakistan and the origin of cetacean swimming, *Nature,* 368:842–844, 1994; S. J. Gould, Hooking Leviathan by its past, *Nat. Hist.,* 94(5):8–15, 1994; J. G. M. Thewissen, S. T. Hussain, and M. Arif, Fossil evidence for the origin of aquatic locomotion in archaeocete whales, *Science,* 263:210–212, 1994.

Charge-coupled devices

Charge-coupled devices (CCDs) have become essential tools for modern astronomical research. Recent developments of these array imagers have led to substantial improvements in sensitivity, allowing more to be learned about the universe.

Most knowledge of the universe comes from recording electromagnetic radiation of one form or another. Astronomers have always relied upon optical images to observe the sky, and the majority of current optical astronomy is performed by using charge-coupled devices as light sensors. Silicon charge-coupled devices have also found an important application as x-ray sensors, usually on board satellites, to study celestial x-ray sources. X-ray photons liberate charges in proportion to their energy (wavelength), and the charge-coupled device therefore acts as a unique imaging spectrometer, providing spatial and spectral information in one image. This article emphasizes the use of charge-coupled devices for optical astronomy; however, many of the device refinements and technical developments are utilized in both fields of use, as well as in other scientific imaging applications. In fact, imaging arrays using charge-coupled devices have extended the range of wavelengths that can be efficiently measured and therefore complement other modern astronomical techniques (for example, gamma-ray, ultraviolet, and infrared observations). Since the first charge-coupled device was demonstrated in 1969, the original concept has been refined considerably, and these imaging arrays now find applications that include home video camcorders, sophisticated telescopes, and many aerospace, military, and scientific projects. Astronomers were among the first to recognize the advantages offered by the silicon charge-coupled device. This article discusses how the device has been developed and used for modern astronomy.

Operation. The basic charged-coupled device is constructed as shown in **Fig. 1.** Semiconductor fabrication techniques are used to construct a monolithic silicon device that comprises a two-dimensional array of picture elements (pixels). Photons of light falling on the surface are directly converted to electrical charges, which are collected and stored at each pixel. By correctly controlling electrical signals (clock waveforms) this photocharge can be moved from one element to another; hence the name charge-coupled device. When charge has been manipulated to an output corner it is then converted to a voltage, fed to a buffer transistor, and externally amplified before being measured, usually by a computer-linked electronics system.

This straightforward construction and direct measurement process yields many of the performance benefits of this type of sensor: high quantum efficiency (conversion of photons to electrons), up to almost 100%; wide spectral range (the ability to detect a wide range of colors, from ultraviolet to infrared), spanning wavelengths from 300 to 1100 nanometers; wide dynamic range (the ability to measure small and large signals at the same time), up to 10,000:1; low readout noise, meaning that very small signals can be measured with good signal-to-noise ratios; direct readout of the signal to an electronic computer system; very good linearity (meaning that the measured electrical signal is proportional to incoming light level); good spatial resolution (many small pixels), for example, 4096 pixels, each 15 micrometers square; and excellent stability (allowing repeatable measurements to be made). A typical scientific charge-coupled device can have 2048×2048 pixels, each 15 μm^2, about ⅙ the size of a human hair. The total picture resolution considerably exceeds that of even the highest-definition television screen.

Application to astronomy. For the reasons presented above, charge-coupled devices provide a powerful tool for the recording of optical images. They do have a smaller area than photographic plates, but this disadvantage is far outweighed by their increased sensitivity. Furthermore, several observatories are engaged in projects to utilize

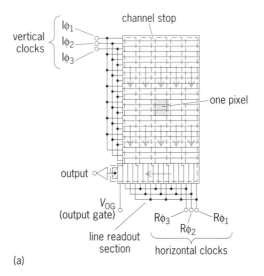

(a)

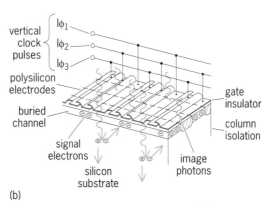

(b)

Fig. 1. Typical imaging charge-coupled device. (*a*) Two-dimensional array of pixels. For simplicity, only a few pixels are shown. In this example, each pixel is defined by a set of three electrodes, which provide vertical clock pulses $I\phi_1$, $I\phi_2$, and $I\phi_3$. (*b*) Section of device. (*EEV*)

mosaics of charge-coupled-device arrays so as to achieve a larger available image detection area. For those few applications where a high-speed time response is desirable, intensified photon-counting detectors (including photomultipliers) have a role to play. However, the majority of quantitative astrophysical research is now carried out using charge-coupled-device sensors, in order to take advantage of their linearity, stability, and wide dynamic range.

The most direct use of the device is at the prime focus of a telescope. Normally a fast beam (small f/number) is produced which concentrates the light well onto the detector and provides high sensitivity by matching the pixel size to the resolved size of astronomical objects. Signal levels, especially from the sky background, can be quite high, and so the high dynamic range of the charge-coupled device is helpful in the recording of faint superimposed astronomical objects. Such direct imaging is used to study clusters of stars and the structure of galaxies, and to observe very faint (distant) galaxies and other celestial objects. The ability to record the faintest observable object allows cosmologists to analyze the universe more deeply than ever before, since faint objects are generally farther away and were created earlier in time.

However, in many cases charge-coupled devices offer more advantages when used as spectroscopic detectors on telescopes. In this case the light collected by a telescope is passed through a spectrograph (using a diffraction grating or prism). The spectrograph disperses the light into a spread of wavelengths, which are then projected onto the two-dimensional detector. The sensor allows the recording of many wavelengths simultaneously from one or more objects. In this case, the signal levels are usually low, and the low readout noise of the charge-coupled device allows very faint spectra to be recorded. The analysis of such spectral information is a long-established tool of astrophysics; accurate information about temperatures, velocities, and hence distances (of stars or galaxies) can be derived.

Performance evolution. Early devices demonstrated these basic principles quite well, but still suffered from some drawbacks. Although scientific researchers used such devices to great advantage, instrument scientists and manufacturers have improved the basic charge-coupled device in several ways, particularly since the mid-1980s. Astronomers have played a strong role in this work since sensitive (efficient) detection of faint light (from distant celestial objects) is crucial to this field of research. The most advanced charge-coupled devices have much larger active areas, higher light sensitivity, and improved performance at the lowest light levels than their predecessors.

The availability of a large sensitive area is particularly useful, since it allows the simultaneous recording from a wide area of sky, or of a wide range of wavelengths when used on a spectrograph. Semiconductor processing techniques allow the manufacture of only modest-sized silicon devices. For example, a

charge-coupled device with 2048×2048 pixels (each $15\ \mu m^2$) occupies 30×30 millimeters (1.2×1.2 in.), and is one of the largest semiconductors made. Astronomers have created mosaics of such sensors in order to construct even larger effective sensor arrays. A matrix of such large charge-coupled devices can form an advanced sensor mosaic, 12×12 cm (5×5 in.) or larger, which begins to match the size of traditional photographic plates, but with vastly improved sensitivity and performance.

Design and manufacturing techniques have also been refined so that modern devices can record the maximum amount of incoming light. Current sensors can approach 100% efficiency, compared with the photographic plate, which achieves only 1–2% efficiency (and has poorer linearity of response). Indeed, many amateur astronomers with charge-coupled devices on small telescopes can now routinely capture dramatic images like the ones the very largest telescopes struggled to record photographically in the mid-1980s. The increased sensitivity also spans a wide wavelength range, which allows even more spectroscopic data to be obtained.

One fundamental limitation to the performance of charge-coupled devices has always been the difficulty of measuring very small amounts of electrical charge, corresponding to only a few photons of light. Advances in design now allow the measurement of one or two electrons of charge from each pixel. (The so-called readout noise is quoted as $2\ e^-$ root-mean-square, where e^- is the electron charge). Coupled with the very high optical sensitivity, the best charge-coupled devices can now record, and measure, almost single photons of light. Thus astronomical measurements are limited only by the faintness of distant objects, and by the length of time available to collect their images.

In summary, if basic charge-coupled devices from around 1985 (512×512 pixels, 50% response, $10\text{-}e^-$ noise levels) are compared with 1995 sensors (4096×4096 pixels, 100% response, $2\text{-}e^-$ readout noise), the combination of these factors indicates that current charge-coupled-device arrays are almost 1000 times more powerful.

Observations. A substantial amount of astronomical data has already been collected by using charge-coupled devices. The Hubble Space Telescope, furnished with charge-coupled-device optical sensors, provides a powerful tool. In particular, its ability to record images with exceptional clarity allows it to be uniquely sensitive to faint point sources of illumination such as distant stars or galaxies. Hubble Space Telescope observations have been used to measure the brightness variations of so-called Cepheid variables, and these observations help give important information about the distances of other galaxies, and hence about the scale of the universe (through the Hubble constant). The sensitivity, and precision, of the charge-coupled device allows the detection of more distant Cepheids, and hence extends the range of measurements out into the universe. *SEE UNIVERSE.*

Fig. 2. Image of Jupiter–Shoemaker-Levy collision taken in July 1994 with the 4.2-m (165-in.) William Herschel Telescope. This image illustrates excellent imaging quality obtained (from a ground-based telescope) with a high-quality charge-coupled device. Impacts of fragment G (dark spot in upper left) and fragment H (top center) can be seen, as well as the red spot of Jupiter (light oval in upper right). (*Royal Greenwich Observatory*)

Ground-based telescopes are also used to determine such cosmological parameters using charge-coupled devices. Several large telescopes around the world utilize a mosaic of charge-coupled devices to record significant areas of sky (many square degrees) during each set of exposures. When certain stars explode (type 1a supernovae), it is possible to calculate their distance by comparing their measured brightness with their expected standard brightness. By mapping the sky and looking for new bright objects, supernovae can be detected in distant galaxies. The charge-coupled device is particularly attractive for this application because large images (for example, 4096 × 4096 pixels, yielding 32 megabytes of data) can be recorded electronically very directly. Multiple frames (yielding, say, 1 gigabyte of data per night) can be compared by using a fast data-processing computer so that detections can be made before the supernova fades. Thus the Hubble constant, which is one of the major unknowns of modern astronomy, can be determined in another way.

The July 1994 collision of Comet Shoemaker-Levy with the planet Jupiter provided another exciting opportunity to use charge-coupled devices for the study of a unique event. Although this event has now passed, the large body of data obtained is still being analyzed, and new results continue to provide better understanding of the original collision. **Figure 2** shows a direct image taken with a 1024 × 1024 charge-coupled device on the William Herschel 4.2-m (165-in.) telescope at the international La Palma Observatory in the Canary Islands. The excellent imaging quality of the telescope-detector combination allows high-resolution images to be recorded, and the wide dynamic range of the detec-

tor allows excellent recording of a greater range of brightness than can be reproduced in this figure. Spectra have been recorded using a similar detector array on the Cassegrain spectrograph of the Isaac Newton 2.5-m (98-in.) telescope (also at La Palma).

For background information *SEE ASTRONOMICAL SPECTROSCOPY; CHARGE-COUPLED DEVICES; OPTICAL TELESCOPE; SATELLITE ASTRONOMY; X-RAY TELESCOPE* in the McGraw-Hill Encyclopedia of Science & Technology.

Paul R. Jorden

Bibliography. S. B. Howell (ed.), *Astronomical CCD Observing and Reduction Techniques,* 1992; G. H. Jacoby (ed.), *CCDs in Astronomy,* 1990; P. R. Jorden, New detectors for new astronomy, *Phys. World,* 7(5):40–45; May 1994; I. S. Maclean, *Electronic and Computer-Aided Astronomy,* 1989; A. Wells and K. Pounds, CCDs map the x-ray sky, *Phys. World,* 6(5):32–37, May 1993.

Chemical force microscopy

Intermolecular forces are responsible for a wide variety of phenomena in condensed phases extending from lubrication at macroscopic length scales, through micelle and membrane self-assembly on a mesoscopic scale, to molecular recognition and binding at the nanoscale. Development of a fundamental understanding of such important phenomena, regardless of the length scale, requires detailed knowledge of the magnitude and range of intermolecular forces. For example, macroscopic measurements of friction and adhesion between surfaces are influenced by complex factors such as surface roughness and adsorbed contaminants. Microscopic studies of these forces should, however, be interpretable in terms of fundamental chemical forces such as van der Waals, hydrogen bonding, and electrostatic interactions. In addition, an understanding of intermolecular forces in condensed phases is of significance to nanoscale chemistry, where noncovalent interactions are important to the manipulation, assembly, and stability of new nanostructures and to imaging functional groups on surfaces. A new technique that directly probes interactions between functional groups and maps spatially their distribution on surfaces is chemical force microscopy.

Components. The chemical force microscope is based upon well-defined modification of an atomic force microscope probe tip with an organic monolayer that terminates with specific chemical functional groups (**Fig. 1**). By modifying probe tips with functional groups, the force microscope becomes sensitive to specific molecular interactions between these groups and those on a sample surface. This conceptually simple elaboration of the force microscope enables researchers to probe binding interactions between functional groups on the tip and

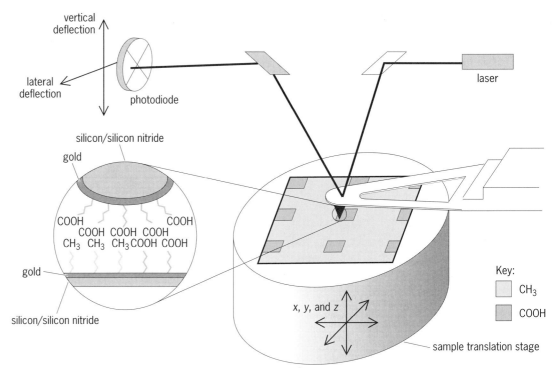

Fig. 1. Chemical force microscope. A laser beam is reflected from the back side of the cantilever-tip assembly into a photodiode to measure normal and lateral forces. The inset shows the interactions between functional groups on the tip and sample surfaces.

surface, and to use this information to map the distribution of different functional groups on a sample surface.

In general, a force microscope consists of several key components, including (1) an integrated cantilever–probe tip assembly, (2) a mechanism to detect the normal and lateral deflection of the cantilever, and (3) an x,y,z sample translation stage (Fig. 1). The cantilever–probe tip assemblies in conventional force microscopes are fabricated from silicon nitride (Si_3N_4) or silicon (Si). These inorganic surfaces, which contain uncontrolled amounts of oxide or adsorbates, can be modified with well-defined organic monolayers by using the technique of self-assembly. One successful method involves self-assembly of functionalized organic thiols onto the surfaces of gold-coated Si_3N_4 probe tips. Stable and rugged monolayers of alkyl thiols or disulfides containing a variety of terminal groups can be readily prepared, and they enable systematic studies of the interactions between basic chemical groups on the probe tip and similarly modified gold substrates.

Binding interactions between functional groups. The adhesive or binding interaction between different functional groups has been determined from force versus cantilever displacement curves. In these measurements the deflection of the cantilever is recorded as the sample approaches, contacts, and is then withdrawn from the probe tip along the vertical axis in Fig. 1. The observed can-

tilever deflection is converted into a force by using the cantilever spring constant.

Typical force-displacement curves can be obtained in ethanol solution by using tips and samples that were functionalized with monolayers terminating in either methyl (CH_3) or carboxyl ($COOH$) groups. These measurements were carried out in ethanol solution rather than air to eliminate contributions from large capillary forces that arise from adsorbed water and other impurities. The hysteresis in the force displacement curves (that is, approach versus withdrawal) corresponds to the adhesion or binding between functional groups on the tip and sample surface. The magnitude of the adhesive interactions between tip/sample functional groups decreases in the following order: COOH/COOH > CH_3/CH_3 > CH_3/COOH. This observed trend in adhesive force agrees with qualitative expectations that interaction between hydrogen-bonding COOH groups will be greater than non-hydrogen-bonding CH_3 groups. The uncertainties in the adhesive interactions between different functional groups have also been characterized by recording multiple force curves. This analysis yields mean adhesive forces of 2.3 ± 0.8, 1.0 ± 0.4, and 0.3 ± 0.2 nanonewtons for the interactions between COOH/COOH, CH_3/CH_3, and CH_3/COOH groups, respectively. Significantly, since the mean value for each type of interaction is outside the uncertainty range for the other interactions, these results show that it is possible to differentiate reproducibly between chemically distinct functional groups by measuring the adhesion or binding force.

In addition, it is possible to assess the energetics of the different intermolecular interactions from the experimentally observed forces by using established models of adhesion mechanics. For example, in the case of COOH-terminated monolayers on the tip and sample surfaces, the surface free energy is 4.5 millijoules per square meter. This value together with that determined for CH_3-terminated monolayers and CH_3/COOH adhesion force also enables calculation of the CH_3/COOH interface free energy: 5.8 mJ/m^2. The large value of the interface free energy readily explains the ordering of intermolecular adhesive forces; that is, the large and unfavorable interface free energy dominates the COOH and CH_3 surface free energies in ethanol and results in a smaller adhesive interaction for CH_3/COOH versus CH_3/CH_3. This approach can be extended to other functional group pairs and different solvent systems; therefore, such studies should provide a wealth of thermodynamic data for predicting, for example, binding.

Friction measurements. It is also interesting to consider whether friction and adhesive forces correlate directly with each other because microscopically both forces originate from the breaking of intermolecular interactions. This idea has been studied by measuring the friction force between CH_3- and COOH-terminated monolayers attached to the tip and sample surfaces. The friction force is measured by recording the lateral deflection of the tip cantilever, while the sample is scanned in a forward-backward cycle. The resulting curve is called a friction loop. It has been determined that the friction force increases linearly with the applied load for each combination of modified tip and surface, and that for a fixed applied load the friction force decreases as follows: COOH/COOH > CH_3/CH_3 > COOH/CH_3. Notably, the trend in the magnitudes of the friction forces is the same as that observed for the adhesion forces: COOH/COOH-terminated tips/samples yield large friction and adhesion

forces, while COOH/CH_3 functionality yields the lowest friction and the smallest adhesion forces. These results confirm the suggestion that the friction and adhesion forces between structurally similar but chemically distinct monolayers correlate directly with each other.

Chemically sensitive imaging. The differences in friction can also be exploited to produce lateral force images (that is, x-y maps of friction) of surfaces containing more than one type of functional group. For example, by using a photochemical method it is possible to produce monolayers having 10-micrometer-square regions that terminate with COOH groups and repeat every 30 μm in a square pattern; the regions of the monolayer surrounding these hydrophilic squares terminate with hydrophobic CH_3 groups. **Figure 2** shows topography and lateral force images of the patterned monolayers recorded by using tips modified with monolayers terminating in either COOH or CH_3 groups. The surface exhibits a flat topography across the CH_3- and COOH-terminated regions of the sample. Although several small adventitious particles are detected on the sample surface, no chemical information is revealed. However, a friction map of the same area shown in Fig. 2a provides chemical information about the surface. Friction maps recorded with the COOH-terminated tips (Fig. 2b) exhibit high friction (light color) in the square area of the sample that contains the COOH-terminated monolayer, and low friction in the CH_3-terminated regions. Images recorded with CH_3-terminated tips exhibit a reversal in the friction contrast: low friction (dark color) in the square area of the sample that contains the COOH-terminated monolayer, and higher friction over the surrounding CH_3-terminated regions (Fig. 2c). The reversal in image (friction) contrast occurs only with changes in the probe tip functionality, and thus it can be concluded that it is possible to image with sensitivity to chemical functional groups. Furthermore, this chemically sensitive imaging is predictable in that friction forces correlate with adhesive interactions between

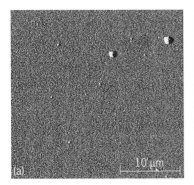 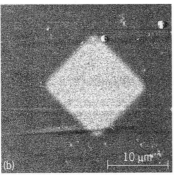 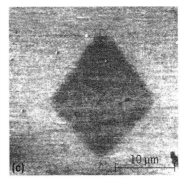

Fig. 2. Chemical force microscopy images on one square of a photopatterned sample. (*a*) Topography. (*b*) Friction force when using a tip modified with a COOH-terminated monolayer. (*c*) Friction force when using a tip modified with a CH_3-terminated monolayer. The COOH-terminated region of the sample is in the centers of the three images. Lighter shading corresponds to higher friction.

functional groups, and the relative magnitudes of these latter interactions follow directly from chemical intuition.

There are also a number of prospects for the future of chemical force microscopy. For example, a host of different types of molecular interactions can be studied, since the modification of probe tips with monolayers provides a general way to introduce chemical functionality. Basic thermodynamic information can also be extracted from the analysis of adhesion data obtained in different solvent media, and should prove useful to chemists and biologists interested in binding and assembly. Lastly, chemical force microscopy imaging of polymer and biomolecular systems could lead to new insights into the spatial distribution of functional groups that are critical to the structure and assembly of these macromolecules.

For background information *SEE INTERMOLECULAR FORCES; MICELLE; MONOMOLECULAR FILM; SCANNING TUNNELING MICROSCOPE* in the McGraw-Hill Encyclopedia of Science & Technology.

Charles M. Lieber

Bibliography. L. H. Dubois and R. G. Nuzzo, Synthesis, structure, and properties of organic surfaces, *Annu. Rev. Phys. Chem.,* 43:437–461, 1992; C. D. Frisbie et al., Functional group imaging by chemical force microscopy, *Science,* 265:2071–2074, 1994; S. M. Hues et al., Scanning probe microscopy of thin films, *MRS Bull.,* 18:41–49, 1993; A. Noy et al., Chemical force microscopy: Exploiting chemically-modified tips to quantify adhesion, friction and functional group distributions in molecular assemblies, *J. Amer. Chem. Soc.,* 117:7943–7951, 1995.

Chromosome

In many different types of organisms, the determination of sex is under the control of a single gene or a small number of genes. In many species, the two members of the pair of homologous chromosomes on which the sex-determining genes are located are identical in appearance. However, in other species, one of these chromosomes is very different from its homolog. This modified chromosome is called the Y, and the other chromosome of the pair is called the X. Individuals carrying an X and a Y chromosome are males; those carrying two X chromosomes are females (in some species, such as butterflies and birds, the females have different sex chromosomes, called Z and W, and the males have two Z chromosomes). The changes that eventually result in different sex chromosomes can be limited to a small segment of the Y (or W) chromosome, or they can affect the entire chromosome.

Since the discovery of sex chromosomes and their correlation with sex determination, the selective forces that underlie these changes have been the object of considerable speculation. A general scenario assumes that the most primitive mechanism for the determination of separate sexes is a simple difference in the alleles (or forms) present in a single gene; one sex would be *Aa* and the other *AA*. The *a* allele is restricted to one sex, and mutations that accumulate on the chromosome in the vicinity of this allele also tend to be restricted to the same sex. Mutations occurring in more distal genes on the chromosome would show the same tendency if recombination were reduced between these genes and the *a* gene. The vast majority of mutations reduce or abolish the normal function of genes, causing the sex-limited region of the *a*-bearing chromosome to become more and more dysfunctional. This region (or in some cases the whole chromosome) eventually would degenerate into inactive, constitutive (permanent) heterochromatin (specialized chromosome material which remains tightly coiled even in the nondividing nucleus).

Evolution of Drosophila sex chromosomes. The fruit fly *Drosophila* provides a unique opportunity for the study of the evolution of sex chromosomes. Recent experimental evidence supports the notion that, in the absence of recombination of linked genes during meiosis, mutations accumulate on the sex-limited member of a pair of homologous chromosomes. The experiment used different *Drosophila* lines. In one set of lines, two autosomes (chromosomes in the nucleus that are not the sex chromosomes) were always transmitted to the next generation by the males; therefore, because of the absence of crossing over during meiosis in this sex, these chromosomes were never subjected to recombination. In control lines, the opportunity for the transmission of recombinant autosomes to the next generation was the same as in natural populations. After as few as 35 generations, a dramatic reduction in fitness, resulting from the accumulation of harmful mutations, was registered for those individuals carrying the nonrecombinant autosomes. Other recent experiments with *Drosophila* have shown that many of the mutant alleles that accumulate on the sex-limited chromosome (Y or W) result from the insertion of transposable elements into or near the genes in question.

Ancestral elements. The basic chromosomal complement of the genus *Drosophila* is six chromosome arms, designated A through F (**Fig. 1**). During the course of evolution, rearrangements of these arms have resulted in different chromosome configurations in different species. Nevertheless, the genetic content of the arms has been sufficiently conserved to allow geneticists to determine from which ancestral arm they are derived. For example, in *D. melanogaster,* the X chromosome is the ancestral A element while the two major autosomes were formed by the fusion of the B and C arms and the D and E arms. In *D. pseudoobscura,* the X chromosome has two arms and consists of the fusion of the ancestral A and D arms. Females of the closely related species *D. miranda* exhibit the same chromosomal complement, but *D. miranda* males have an odd number of chromosomes with one member

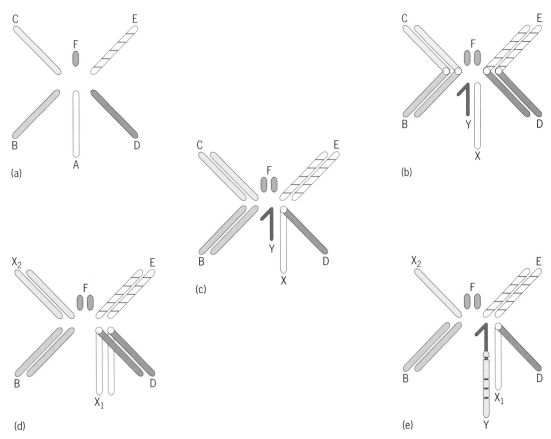

Fig. 1. Chromosome arrangements in four species of the genus *Drosophila*. (*a*) Basic array of the six elements that are found in different configurations in these species. (*b*) Male of *D. melanogaster*. (*c*) Male of *D. pseudoobscura*. (*d*) Female of *D. pseudoobscura*. (*e*) Male of *D. miranda*.

of a pair of rods (chromosome 3 of *D. pseudoobscura* or the ancestral C element) apparently missing. This chromosome is present in two doses in females and only one dose in males; because this is characteristic of X chromosomes, it is called X2. Instead of being completely condensed, as in *D. pseudoobscura*, the Y chromosome of *D. miranda* males contains a considerable portion that has the normal appearance of other chromosomes. Clearly, the *D. miranda* Y chromosome arose by a fusion of the Y and chromosome 3 of *D. pseudoobscura*.

Mutations. Because of the absence of recombination in *D. miranda* males, the chromosome 3 material attached to the Y chromosome is limited to the male sex; therefore, it is expected that this neo-Y chromosome would accumulate mutations as well as transposable elements. A preliminary molecular study showed that repetitive deoxyribonucleic acid (DNA) sequences appear to invade this chromosome. Specific supportive evidence was obtained by using in-place hybridization to chromosomes. A repetitive element called ISY3 is much more prevalent on the chromosome material translocated onto the neo-Y chromosome than on its homolog, the X2 chromosome. A novel transposable element, called TRIM, occurs twice as frequently in males as in females, with the extra copies located to the new portion of the neo-Y chromosome. The invasion by insertional elements of a chromosomal region that

lacks crossing over is greatly facilitated if their frequency of excision is sufficiently low.

Transposable elements can cause mutations by inserting in the regulatory or transcribed regions of genes. Recently, it has been demonstrated that the inactivation of resident genes by repetitive elements can occur without altering the inherent functional aspects of these genes. On chromosome 3 of *D. pseudoobscura*, and on the X2 of *D. miranda*, a cluster of four genes encodes the larval cuticle proteins. These genes are expressed normally while their counterparts on the neo-Y are not active, presumably because of the insertion into the cluster of several repetitive elements including copies of the TRIM transposon and ISY3 repetitive sequence. When the repetitive elements were removed by means of molecular cloning techniques, the genes of the cluster were found to be quite active. Thus, the invasion of a sex-limited region of the genome by transposable elements leads to changes that repress gene activity and eventually transform functional regions of the chromosome into genetically inert heterochromatin.

John C. Lucchesi

Evolution of mammalian sex chromosomes. In mammals, females have two X chromosomes, and males a single X and a Y. Meiosis ensures that half the sperm formed in testis carries an X, and half a Y, so that the sex ratio of offspring is more or less equal. The mammalian Y is male determining because XXY individuals are male and XO female. Testis determination is the first step in mammalian sex determination, and male characteristics are subsequently controlled by testicular hormones. The testis is determined by a Y-borne testis determining factor, identified a few years ago as the *SRY* gene on the Y chromosome. Other genes on the Y are active in the testis and may have a role in spermatogenesis. However, the Y seems to be something of a chromosomal wasteland, containing few active genes and much noncoding (junk) DNA.

In contrast, the X contains 2000 or 3000 genes. These appear to be a rather average mix of housekeeping genes and genes for specialized functions, with no role in sex determination. A remarkable chromosome-wide system inactivates one of the two X chromosomes in a female cell, ensuring equal dosage of X-linked gene products in males and females.

X and Y chromosomes are morphologically and genetically so distinct that it might seem preposterous to suggest that they evolved from an ordinary homologous pair of chromosomes. However, recent evidence indicates that the Y is essentially a broken-down X, and its few genes are relics of this degradation (**Fig. 2**).

Human and mouse sex chromosomes. Although the human X and Y are very different, they do pair at male meiosis. Crossing-over is confined to the very tips of the X and Y within a homologous region. The tiny pseudoautosomal region on the short arm of the chromosome contains seven active genes.

The bulk of the Y does not recombine with the X. The bottom half consists entirely of a few highly repeated sequences and literally glows in the dark when stained with fluorochromes. The rest has been cloned and exhaustively searched for active genes. At the last count there were eight active genes and several degraded and nonfunctional pseudogenes, some present in multiple copies. Most genes (even including *SRY* itself) and pseudogenes are related to genes on the X.

The mouse X chromosome contains almost the same set of genes as the human X chromosome, in line with a venerable prediction that mammalian X chromosomes are conserved to maintain the X inactivation system. The mouse Y chromosome is small and heterochromatic. It contains genes homologous to most of the human Y-borne genes, including *SRY,* and again there are related genes on the X. However, there are striking differences in the activity, numbers, and even the presence or absence of genes on the human and mouse Y, and the gene

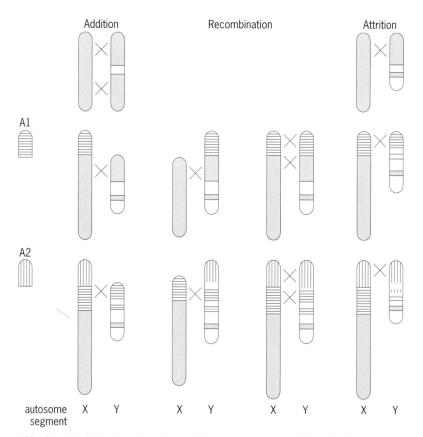

Fig. 2. Addition-attrition hypothesis for the origin of mammalian sex chromosomes. The original sex chromosomes (top) were homologous except at a sex-determining locus (white) and underwent homologous pairing (crosses). The Y chromosome was progressively degraded, leaving few functional genes (horizontal lines on a background of noncoding deoxyribonucleic acid), and a small paired pseudoautosomal region at the terminus. Addition of an autosomal segment A1 to the X was followed by recombination within the pseudoautosomal, adding the segment to the Y and producing enlarged sex chromosomes paired over the added region. The enlarged Y was further degraded, leaving a few functional genes (vertical lines) and a minimal pseudoautosomal. The cycle of addition and attrition was repeated with autosomal region A2.

contents of the pseudoautosomal regions do not overlap at all. Such inconsistency is unprecedented in other regions of the genome.

The identity of the pseudoautosomal region, and the relationship between X- and Y-borne genes in the differentiated regions, supports the contention that the X and Y chromosomes were once identical.

Sex chromosomes of divergent mammals. Some unexpected insights into the differentiation of mammalian X and Y chromosomes have been gained by studying the sex chromosome variation among the most distantly related mammal groups, marsupials and monotremes, which diverged from placental (eutherian) mammals approximately 130 and 170 million years ago.

Marsupials (for example, kangaroos) and monotremes (for example, platypus) have distinct X and Y chromosomes, but their sizes, pairing relationships, and gene contents differ from those of eutherian mammals in revealing ways. Marsupials have a smaller basic X and a tiny Y, which do not appear to recombine. Monotremes have large X and Y chromosomes, which pair over much of the Y.

Gene-mapping studies show that two-thirds of the human X chromosome is shared by the marsupial and monotreme X chromosome, identifying a 170-million-year-old conserved mammalian X chromosome. However, genes from the rest of the human X chromosome are clustered on three other chromosomes in both marsupials and monotremes, suggesting that a region was recently added in bits to the eutherian X. These comparative studies identify the evolutionary origins of different regions of the human X chromosome.

The marsupial Y is testis determining, and contains *SRY* as well as another gene shared with the mouse Y. Mammalian Y chromosomes are therefore likely to have a common evolutionary origin. However, most of the X-Y shared human genes fall within the recently added region, implying that autosomal regions were added to the placental Y as well as to the X, probably by an addition to an ancient pseudoautosomal region of one partially differentiated sex chromosome, and then by recombination onto the other. The X and Y have both grown incrementally, while the Y has been progressively degraded (Fig. 2). Thus some genes on the human Y chromosome derived from an ancient X chromosome, and others from recent autosomal additions.

Degradation and evolution of male-specific function. All genes on the Y chromosome probably started out with a partner on the original or added region of the X chromosome, and most have been degraded or deleted. Sequence comparisons show that genes on the Y evolve rapidly. The loss of genes from the differential region of the Y is undoubtedly related to its isolation from recombination, which means that mutations and deletions can no longer be expunged. Loss of Y chromosomes having the fewest mutations could occur by chance, or a degraded Y could be selected if it contained a new favorable mutation.

Only genes with a selectable function are likely to withstand the degradation of the Y chromosome, and the only critical functions open to Y-borne genes are in male sex determination and spermatogenesis. These male-specific functions may have been acquired quite recently by genes isolated on the Y. For instance, putative spermatogenesis genes seem to have evolved from genes which are ubiquitously expressed in both sexes, and are therefore likely to have general housekeeping functions. The *SRY* gene appears to be a relic of an active X-linked gene located in the central nervous system, and may act indirectly by inhibiting a related gene. Even this critical gene has been amplified in some rodent species and lost in others, its function apparently taken over by another, unknown gene.

For background information *SEE CHROMOSOME; CROSSING-OVER (GENETICS); GENE; GENETIC MAPPING; GENETICS; HUMAN GENETICS; MEIOSIS; MUTATION* in the McGraw-Hill Encyclopedia of Science & Technology.

Jennifer A. Marshall Graves

Bibliography. B. Charlesworth, The evolution of sex chromosomes, *Science,* 251:1030–1033, 1991; J. A. M. Graves, The origin and function of the mammalian Y chromosome and Y-borne genes: An evolving understanding, *BioEssays,* 17:311–320, 1995; S. Ohno, *Sex Chromosomes and Sex-Linked Genes,* 1967; W. R. Rice, Degeneration of a nonrecombining chromosome, *Science,* 263:230–232, 1994; M. Steinemann and S. Steinemann, Degenerating Y chromosome of *Drosophila miranda:* A trap for retrotransposons, *Proc. Nat. Acad. Sci. USA,* 89:7591–7595, 1992; M. Steinemann, S. Steinemann, and F. Lottspeich, How Y chromosomes become genetically inert, *Proc. Nat. Acad. Sci. USA,* 90:5737–5741, 1993.

Chrysotile

There is a continuing interest in the study of asbestos because of the health hazards associated with these fibrous materials. The term asbestos includes a range of different fibrous minerals with different structures and chemical compositions. The most important commercially is chrysotile. Information about the structure of the very fine fibers can be obtained by using the complementary techniques of transmission electron microscopy and x-ray diffraction; the use of high-resolution imagery has revealed that chrysotile has fivefold symmetry. It is the first time such symmetry has been observed in a natural crystalline material.

Chrysotile structure. Chrysotile is one of the serpentine group of minerals, all of which have the approximate formula $Mg_3Si_2O_5(OH)_4$. The structure consists of layers made up of two component parts joined together. One part is a sheet of linked tetrahedra of oxygen (O) atoms with a silicon (Si) atom at the center of each tetrahedron. These tetrahedra are arranged in a pseudohexagonal network.

The other part is a sheet of linked hydroxyl (OH) octahedra with a magnesium (Mg) atom at the center of each octahedron. Two out of every three (OH) groups on one side of this sheet are replaced by apical oxygens of the silicate (SiO_4) tetrahedra, thus bonding the two parts together strongly to form one composite layer (**Fig. 1**).

What makes the serpentine minerals so interesting is that the two layers have different dimensions, and the mismatch between tetrahedral and octahedral components is compensated for, at least partially, in different ways in the different forms of serpentine. The tetrahedral part has the smaller dimensions. In chrysotile, the fibrous form of serpentine, sheets curl into cylinders (usually about the x axis), either stacked concentrically or as spirals. These form fibers that are typically about 20 nanometers in diameter and often millimeters or centimeters long. In a second form of serpentine, lizardite, the sheets are flat, not curved, and there is some distortion in the tetrahedral part of the sheet. In the third form, antigorite, sheets curve about the y axis in an alternating-wave structure.

The unusual cylindrical structure of chrysotile was modeled from x-ray diffraction patterns. Much later, the cylindrical structure was confirmed by direct imaging, using transmission electron microscopy. The nature of the crystal symmetry around the fiber axis could not be derived directly from early experimental work, although modeling showed that structural units should repeat exactly five times around a circumference. Recent work has provided direct evidence for fivefold symmetry in chrysotile.

Evidence for fivefold symmetry. The occurrence of fivefold symmetry on a local scale has been well established in quasicrystals of intermetallic phases. Until recently, however, fivefold symmetry for a complete crystal has not been reported; symmetry axes previously recognized for organic and inorganic phases include two-, three-, four- and sixfold axes. Work reported since 1993 has shown that the unusual curved structure of chrysotile coupled with the particular unit cell parameters of the material must lead to a fivefold symmetry about the fiber axis.

One type of evidence for fivefold symmetry is found in images of chrysotile recently obtained by transmission electron microscopy. Chrysotile is very sensitive to the electron beam, degrading to amorphous material in seconds. At a particular stage of this breakdown images (**Fig. 2**) from some chrysotile samples show 15 radial spokes where the concentric layers are still visible, separated by radial bands of amorphous material. The reason for these 15 bands lies in the fact that 2π multiplied by the layer spacing is exactly equal to five times the repeat of the structural unit around a circumference. Thus, if consecu-

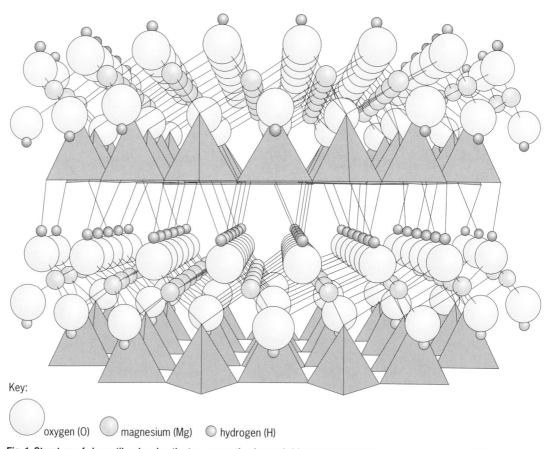

Key:

oxygen (O)   magnesium (Mg)   hydrogen (H)

Fig. 1. Structure of chrysotile, showing the two serpentine layers (without curvature). The pyramids represent SiO_4 tetrahedra. (*After B. A. Cressey and E. J. W. Whittaker, Five-fold symmetry in chrysotile asbestos revealed by transmission electron microscopy, Mineral. Mag., 57:729–732, 1993*)

tive layers are in register with one another at all, then they must be in step five times around the circumference. In this condition, (OH) groups on the outside surface of one layer line up with O atoms on the inside of the next layer outward 15 times per circumference (three times per unit cell), enabling strong hydrogen bonds to form between layers at these 15 radial positions. Because of this stronger bonding, the layer structure remains visible for longer at these 15 positions, providing visual evidence for the fivefold symmetry of chrysotile.

A second type of evidence for fivefold symmetry is derived from diffraction work. The overall symmetry of a crystal is commonly determined from analysis of x-ray or electron diffraction patterns. Because chrysotile fibers are very small, difficult to align accurately, and beam sensitive, it has not been possible to obtain diffraction patterns down the fiber axis of individual crystals by normal methods. To obtain information about the arrangement of atoms in this direction, the structure of chrysotile was modeled in the past by using optical diffraction masks consisting of arrays of dots. By ordering the dots in one direction, it was shown that fivefold symmetry was possible. Such optical transforms showed a tenfold symmetry, reflecting the underlying fivefold symmetry of the model.

Recently, scientists in France and Italy have modeled several possible arrangements of the chrysotile structure and derived diffraction patterns for the arrangements by using computational methods. They demonstrated that regardless of whether successive concentric layers are stacked in register with one another (certain rows of lattice points lining up radially) or not, fivefold symmetry is still evident in the simulated diffraction patterns, as it is also if layers are spirally stacked rather than concentrically, or are elliptically deformed. Further, by using transmission electron microscopy, these workers also obtained a high-resolution image from a fiber oriented with its axis exactly parallel to the beam. They produced a numerical Fourier transform from it that resembled one of their simulated diffraction patterns from a slightly disordered structure showing 10 intensity enhancements, thereby demonstrating fivefold symmetry in a chrysotile fiber.

Polygonal fibers—a related structure. Some years ago, diffraction studies suggested that certain types of serpentine fibers have structures closely related to those of chrysotile, but with subtle differences that could be interpreted as resulting from an arrangement of flat layers stacked around the fiber axes. This structure was confirmed by transmission electron microscopy; this type of fiber is still tubular, but polygonal in section rather than circular. Strictly speaking, it cannot be called chrysotile, which, by definition, must consist of curved layers, as it is made of flat, lizardite-type sheets stacked in sectors. However, the structures of these two fiber types are very closely related, and polygonal fibers always occur in close association with the more common cylindrical fibers. Geological evidence and observations using

Fig. 2. Group of chrysotile fibers in cross section in the transmission electron microscope. The fibrous crystals appear circular in cross section because they consist of rolled-up layers (like rolls of carpet) or concentric layers (like the growth rings of trees). Under certain conditions the fiber layers appear to break up into 15 dark spokes, reflecting the underlying five-fold symmetry of the crystal. (*From B. A. Cressey and E. J. W. Whittaker, Five-fold symmetry in chrysotile asbestos revealed by transmission electron microscopy, Mineral. Mag., 57:729–732, 1993*)

transmission electron microscopy suggest that the polygonal form probably develops from the cylindrical form by recrystallization.

Several scientists have noted that complete polygonal fiber sections almost always have 15, or less commonly 30, sectors. If cylindrical fibers with hydrogen bonding between successive layers in 15 radial bands were to recrystallize to form polygonal fibers with 15 sectors, the degree of interlayer bonding between the flat sheets would be greatly extended, and this should be a more stable arrangement, especially for large-diameter fibers.

The observation that polygonal fibers generally have 15 or 30 sectors led one British worker to propose a geometrical model for the growth of equal polygonal sectors. In this model, five extra b-repeats are incorporated per circumference of each successive layer outward (as in cylindrical chrysotile), by adding whole numbers of extra $Mg(O,OH)_6$ octahedra to the new layer in each sector. A consideration of the geometry and crystallography shows that only arrangements with 15 or 30 equally developed sectors can do this with minimal misfit.

Other scientists in France and Italy have proposed an alternative model for the formation of polygonal serpentine fibers, assuming recrystallization from chrysotile. This model requires 15 partial dislocations per circle and would result in a cyclic distribution of twins and different layer stacking patterns to produce the most energetically stable arrangement. This model suggests that fivefold symmetry would be preserved, at least approximately, as chrysotile is transformed into polygonal serpentine.

For background information SEE ASBESTOS; CRYSTAL STRUCTURE; CRYSTALLOGRAPHY; ELECTRON MICROSCOPE; HYDROGEN BOND; SERPENTINE; X-RAY DIFFRACTION in the McGraw-Hill Encyclopedia of Science & Technology.

Barbara A. Cressey

Bibliography. A. Baronnet, M. Mellini, and B. Devouard, Sectors in polygonal serpentine: A model based on dislocations, *Phys. Chem. Minerals,* 21:330–343, 1994; B. A. Cressey and E. J. W. Whittaker, Five-fold symmetry in chrysotile asbestos revealed by transmission electron microscopy, *Mineral. Mag.,* 57:729–732, 1993; B. A. Cressey, G. Cressey, and R. J. Cernik, Structural variations in chrysotile asbestos fibers revealed by synchrotron x-ray diffraction and high-resolution transmission electron microscopy, *Canad. Mineral.,* 32:257–270, 1994; B. Devouard and A. Baronnet, Axial diffraction of curved lattices: Geometrical and numerical modeling; application to chrysotile, *Eur. J. Mineral.,* 7:835–846, 1995.

Coal, low-rank

Significant research and development in North America, Europe, and the countries of the Pacific Rim (such as Japan, China, Australia, and Indonesia) is focused on beneficiating poor-quality lignites to improve their economic value and to allow abundant regional reserves to be used more efficiently. Lignite is a solid carbonaceous material generally categorized as a low-rank coal because of its low degree of coalification. Low-rank coals are found in young geologic settings and are typically characterized by relatively high-moisture contents, low densities, and simple open chemical structures. Generally, the term low-rank coal includes peat, lignite, brown coals, and subbituminous coals. Low-rank coals are found throughout the world; they represent a large, relatively untapped fuel resource in North America, Europe, and throughout the Pacific Rim. Unfortunately, the low calorific value and the high-moisture and high-ash contents of the as-mined low-rank coals severely limit the efficient, economical use of these resources.

Coal rank. Historically, most processing research and development of low-rank coals has focused on producing synthetic natural gas and synthetic diesel fuel (coal oil). The German efforts prior to and during World War II are still the basis for most synthetic fuel production technologies. These processes use a combination of heat, pressure, and the addition of hydrogen to alter the physical and chemical structure of the raw coal so as to produce the desirable hydrocarbon products. Low-rank coals have a hydrogen-to-carbon ratio of about 1:1. Hydrocarbon liquids have a hydrogen-to-carbon composition of about 4:1 (for example, methane, CH_4). As the degree of coalification increases, the coal rank increases, the percentage of carbon increases, and the percentage of hydrogen decreases. Thus, as coal

rank increases, the hydrogen-to-carbon ratio shifts further from that of hydrocarbon liquids or gases, so that the higher-rank coals are less amenable to conversion than the lower-rank coals. Since high moisture content characterizes low-rank coals, the thermal efficiency of conversion is lower than for the higher-rank coals because of the heat required to evaporate a greater percentage of moisture. All beneficiation processes for low-rank coals must overcome the increased economic cost caused by the lower carbon density of the low-rank coals, which requires that more tons be processed than for higher-rank coals to yield the same amount of carbon.

Upgraded fuels. Efforts to make low-rank coals more useful have shifted toward producing a solid synthetic fuel that can be used like a higher-rank coal rather than converting solid low-rank coal into a liquid or gaseous hydrocarbon. A number of government agencies and private concerns have sponsored extensive projects for upgrading low-rank coals since the mid-1970s, with the goals being to produce higher-ranked coal products for general industrial and utility fuel use, reduce transportation costs, increase electrical energy conversion efficiencies, and reduce solid waste and air pollutants. Most of the recently developed upgrading involves partial pyrolysis or mild gasification that heats the raw coal in an inert atmosphere, causing thermal, chemical, and physical alterations within the coal to produce a chemically superior fuel product. These upgraded products from low-rank coal have lower equilibrium moisture capacities (inherent or natural moisture levels) and reduced contents of oxygen functional groups (usually hydroxyl, carbonyl, and carboxyl hydrocarbon radicals); and they have more condensed chemical structures with a higher percentage of aromatic carbon moieties. Additionally, the upgraded low-rank coals produced from the western United States subbituminous coals have shown a potential to significantly reduce emissions of sulfur dioxide, nitrogen oxides, and other hazardous air pollutants (such as arsenic and mercury) when substituted for eastern United States bituminous coals. Unfortunately, these upgraded low-rank coals typically are susceptible to spontaneous combustion, and tend to be very dusty when used in conventional bulk coal handling systems. These characteristics will require changes in the industrial and utility infrastructure to use these fuels effectively; thus, upgraded low-rank coals should not be considered as direct substitutes for higher-rank coals.

The possible efficiencies that could be gained in transportation of these upgraded fuels and conversion to electricity, heat, and other products offer significant economic advantages to industrial and utility users. As a result, there is continued interest in the development of technologies for upgrading low-rank coal. Several of these technologies are under development on a demonstration scale which produces 20–60 tons (18–54 metric tons) per hour of production. Processes for upgrading low-rank coal target operating conditions of moderate reaction temperatures, 300–500°C (570–930°F), and low pressures to partially pyrolyze and thermally alter the low-rank coals. All of the upgrading processes under development for low-rank coals dehydrate the low-rank coal feedstock, and further alter the chemical structure by decarbonylating, decarboxylating, devolatilizing, and carbonizing the feedstocks to various degrees. Of the processes under development to produce upgraded fuel from low-rank coal, none directly hydrogenate the solid material. The upgrading reactions generally result in a solid material with increased carbon and reduced hydrogen. Dehydration removes chemically bound water and condenses the coal structure. Decarboxylation removes chemically bound carbon dioxide. Decarbonylation is similar to decarboxylation, but it removes chemically bound carbon monoxide and condenses the coal structure.

Upgrading technologies. All the processes for upgrading low-rank coals cause a thermal decomposition of the coal's physical structure and result in a product with smaller particles; the smaller particles cause process developers and potential product users to be concerned about dust problems that might occur with conventional bulk handling. Research has been focused on various techniques to address the issues of spontaneous combustion and product dustiness; these techniques include briquetting or pelletizing, the use of chemical treatments or additives, and other processing to alter the chemical structures of the upgraded coal further.

Some of the technologies under development actually operate at a temperature high enough to substantially devolatilize the raw low-rank coal, producing coal gases, the condensable portion of which can be cooled to yield liquid coal tars, and the remaining (noncondensable) portion of which can provide a low-to-medium-Btu gaseous fuel (that is, 100–600 Btu/ft^3). These upgrading technologies typically use the liquid coal tars evolved from the process as a binding agent to briquette or pelletize the product, in order to make it more useful. The noncondensable gaseous fuel is normally consumed on site for process heat, since it is not pipeline quality and not suited to substitute directly for natural gas. One technology produces a coal-oil slurry by further processing the liquid coal tars to reduce their viscosity and mixing them with the devolatilized coal char to make a coal-oil mixture.

Before technologies for upgrading low-rank coal can be fully commercialized, solutions to the spontaneous combustion problem must be resolved, and potential users must find it acceptable that upgraded low-rank coals have different characteristics and features than higher-rank coals. The most likely group to accept technologies for upgrading low-rank coals and to sponsor the conversion to these new fuels are the industrial fuel users. They can take advantage of the enhanced characteristics almost immediately; are generally smaller users, making any necessary modifications more easily;

and are not dominated by fuel costs as are utility operations, which can afford substantial modifications to use fuels that provide the cheapest Btu's possible. Because of the large global reserves of low-rank coals, most of which are recoverable by low-cost surface-mining techniques and an expanding fuel demand in economically developing areas, interest in research and development for upgrading low-rank coals will continue.

For background information SEE COAL; COAL GASIFICATION; LIGNITE; ORE DRESSING; PYROLYSIS; SYNTHETIC FUEL in the McGraw-Hill Encyclopedia of Science & Technology.

Ray Sheldon

Bibliography. G. H. Gronhovd et al., *Low Rank Coal Technology*, 1982; K. L. Smith et al., *The Structure and Reaction Processes of Coal*, 1994; S. C. Tsai, *Fundamentals of Coal Beneficiation and Utilization*, 1982.

Combinatorial chemistry

Combinatorial chemistry is an approach to synthetic organic chemistry whereby large numbers of structurally related molecules are generated in a short time. Compound libraries made in this manner can be evaluated for biological activity, such as inhibiting the action of an enzyme or receptor. Combinatorial chemistry techniques are being adapted by the pharmaceutical industry to decrease the time needed to proceed from the identification of a molecular target for a particular disease (generally an enzyme or receptor) to the initiation of clinical trials in humans.

A combinatorial library is a collection of molecules made from all possible combinations of a given set of building blocks. Libraries of peptides were the first synthesized, but therapeutic limitations of this class of molecules have spurred the development of small-molecule libraries. Because combinatorial chemistry allows the rapid synthesis and selection of molecules with novel properties, its methods will play an important role in the pharmaceutical industry.

Drugs. A drug is a chemical substance used in the treatment, cure, or prevention of disease. Some drugs, such as insulin and human growth hormone, are high-molecular-mass proteins. However, most are classified as small molecules, that is, organic compounds of less than 1000 daltons.

Biochemical inhibitors. Often, small-molecule drugs work by interfering with an enzyme or receptor that is essential for a particular disease. The antibiotic penicillin irreversibly inhibits an enzyme that bacteria use to make their cell walls. When the normal functioning of this enzyme is blocked, bacterial growth and reproduction are halted. The analgesic ibuprofen works by inhibiting the enzyme cyclooxygenase. Normally, this enzyme catalyzes a key step in the biosynthesis of prostaglandins, molecules that cause pain and swelling after tissue injury. Blockage of this pathway forms the basis for most nonprescription pain medications.

Molecular targets for disease. Recent advances in molecular biology have made it possible to precisely define the biochemical mechanisms for many diseases. For example, the specific mutations in cellular deoxyribonucleic acid (DNA) responsible for many kinds of cancer have been determined. These mutations can create a cell with its machinery for growth and division permanently switched on, causing the ceaseless cellular division characteristic of cancer. Small-molecule drugs might be able to block the interactions that cause this uncontrolled cell division. In the case of acquired immune deficiency syndrome (AIDS), several key enzymes in the life cycle of the human immunodeficiency virus (HIV) have been identified as potential targets for therapeutic intervention. The HIV reverse transcriptase and the HIV protease have been investigated the most thoroughly.

Drug discovery process. Once a biological interaction has been identified as important for a particular disease, the next step in the development of a drug is the initial identification of inhibitory molecules. Researchers evaluate many thousands of compounds to identify just a few lead compounds with modest potency. Often these compounds are natural products, collected from all over the Earth and derived from plants, animals, or fermentation. A pharmaceutical company also makes extensive use of its collection of synthetic compounds, built up over years of research. (Although the use of computers can accelerate some aspects of this process, it is still not possible to design a ligand to inhibit the action of a given receptor or enzyme by giving the structure of the target.)

After the identification of a lead compound, many analogs are synthesized to increase the desired biological activity. This process is repetitive and is partially based upon trial and error. The final goal of this optimization process is to obtain an organic molecule that exhibits high affinity and specificity for the target of interest and is reasonably inexpensive to mass-produce. A final drug candidate also must have low toxicity, should have a long circulating half-life in the blood, and ideally is absorbed through the intestine (to allow for oral administration). The extensive optimization and biological evaluations that are needed just to bring a candidate to clinical trials make this process extremely time consuming and expensive.

Combinatorial chemical libraries. The development of combinatorial chemistry has been driven by the continual identification of new pharmaceutical targets and the need for new lead compounds and drug candidates for these targets. Synthesis of chemical combinatorial libraries allows the rapid generation of large numbers of structurally related but dissimilar compounds to expedite either the lead discovery or the optimization process.

Inspiration from immunological diversity. The intellectual inspiration for the development of combinatorial libraries is the diversity of the immune system. To recognize foreign invaders, nature has evolved an immune system that produces more than 100 million different antibodies. This collection of diversity is sufficient to recognize nearly any foreign invader, suggesting that, given enough structurally diverse molecules, members of the set that have a particular desired property can be found.

Library concept. The idea behind a combinatorial library is very simple. A number of sets of different building blocks are brought together in all possible combinations to yield a large number of diverse yet related molecules. If a given class of target molecules is to be synthesized from three sets of building blocks (A, B, and C), and if 20 derivatives of each building block are used (A^1–A^{20}, B^1–B^{20}, C^1–C^{20}), an exhaustive combinatorial library would include $20 \times 20 \times 20$, or 8000, different compounds ($A^1B^1C^1, A^1B^1C^2, \ldots, A^{20}B^{20}C^{20}$). If lead compounds for a particular target are identified from an initial library, a more focused library can be synthesized to optimize the properties of the initial lead.

Solid support. Chemical combinatorial libraries are generally synthesized by using a polymeric support material, frequently small polystyrene beads. The compound being made is covalently tethered to the support during synthesis and is cleaved from the support at the completion of the synthesis. There are several important differences between the solid-phase method and traditional solution-based organic synthesis. The solid-phase method allows for easy purifications between chemical steps, because unreacted reagents (anything that is not covalently bound) may simply be rinsed away from the support by using a fine filter. However, during a solid-phase synthesis the analysis of a synthetic intermediate is often complicated by the tether to the polymeric support. Finally, the bond-forming reactions used in a particular synthesis sequence must proceed in high yield for a variety of building-block derivatives; otherwise the desired compound will not be the major component of the final product mixture.

Peptide libraries. Peptides are molecules that have widespread and important actions in biological systems. Since the 1960s, general methods have been developed for the (noncombinatorial) chemical synthesis of peptides on solid supports. For these reasons, peptides were the focus of initial library efforts. The disconnection of a generic tripeptide into three building-block sets, each an amino acid, is shown in reaction (1). Peptides, being oligomeric

Tripeptide

$$(1)$$

compounds, are somewhat unique in that the same kind of building block (in this case an amino acid) is used to introduce diversity at each successive position in the target molecule. The actual synthesis of a combinatorial library can be performed by either the spatially separate or the split-and-mix approach. The use of both methods for the synthesis of a combinatorial library using two building-block sets with three building blocks per set gives 3×3, or 9, different compounds in the final library. Both strategies also are used for the synthesis of nonpeptide small-molecule libraries.

Spatially separate approach. Each possible compound (A^1B^1, A^1B^2, A^1B^3, A^2B^1, ..., A^3B^3) is prepared in a separate vessel. An evaluation of biological activity provides direct information about the activity of each member in the library. The number of compounds synthesized can become quite large. Management of large numbers is greatly simplified by racks that hold 96 compounds at a time; more than 10,000 spatially separate compounds have been synthesized at one time by using this method. Recent advances in robotic instrumentation allow for the biological evaluation of more than 10,000 compounds per day.

Split-and-mix approach. This approach aims to substantially reduce the number of reaction vessels at the cost of increasing the assay complexity or reducing the information gleaned from a library. The synthesis of a nine-member library from two building-block sets, each with three members (A^1, A^2, A^3 and B^1, B^2, B^3), is shown in the **illustration**. The support is divided evenly into three flasks. Building-block A^1 is then coupled to the support in the first flask, A^2 in the second, and A^3 in the third. The support from all three flasks is then thoroughly mixed and repartitioned into three flasks for reaction with B^1, B^2, and B^3. The first flask now contains equal amounts of the three library members that incorporate B^1, the second contains equal amounts of the three compounds that incorporate B^2, and the third contains equal amounts of the three derivatives that incorporate B^3. An assay of these three pools indicates which has the highest biological activity, and consequently which of the three derivatives at B is the most active. A second library of three separate compounds is then synthesized by varying A while fixing the B position to the highest-activity building block as determined in the first biological assay. A second assay indicates which derivative at A is the most active, in the hope of getting one high-activity compound. The number of possible compounds in the library increases dramatically as the number of building-block sets and the number of compounds per set increase.

Benefits and limitations. Peptide libraries have been used for the rapid identification of molecules that bind to specific antibodies, ligands that mimic the natural opiates in humans, and compounds that inhibit platelet aggregation. These and many other successes clearly establish the power of the combinatorial method. However, certain limitations of

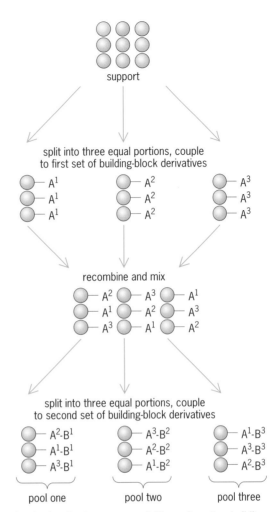

Synthesis of a two-component library from two building-block sets, each with three members (A^1, A^2, A^3 and B^1, B^2, B^3), by the split-and-mix method. Three different building-block derivatives are used at each point of variability, giving 3×3, or 9, different compounds in the completed library.

peptides as drugs have become apparent. Peptides generally cannot be orally administered, and even if introduced directly into the bloodstream are cleared rapidly. These problems have spurred the development of synthetic strategies for the construction of combinatorial libraries of nonpeptide small molecules.

Small-molecule libraries. Several requirements must be met for the successful synthesis and use of a combinatorial organic library. First, the bond-forming reactions that are used must proceed efficiently, with many different building blocks for each set. Second, the building blocks for a particular combinatorial library must be commercially available or readily synthesized. If substantial effort is spent on the time-consuming synthesis of starting materials, the purpose of combinatorial synthesis is defeated. Third, a method to identify compounds with the desired activity must be available. The identification process depends on the strategy used for the construction of a particular combinatorial library. It can be as simple as maintaining each different library element as a spatially discrete compound, or as complex as decoding a chemical tag of the synthetic history of a compound.

1,4-Benzodiazepines. Combinatorial libraries of many classes of molecules have been synthesized and evaluated. One of the most studied classes of molecules for combinatorial library synthesis are the 1,4-benzodiazepines. The best-known benzodiazepine is the anxiolytic diazepam. However, benzodiazepine derivatives have been discovered that act as anticoagulants, HIV reverse transcriptase inhibitors, ras farnesyl transferase inhibitors, and opioid antagonists. The broad biological activity and favorable pharmacokinetic properties of this class of compounds make it a natural target for the synthesis of a combinatorial library for lead identification. The possible dislocation of a generic 1,4-benzodiazepine derivative into structural fragments is shown in reaction (2), where R = functional

1,4-Benzodiazepine
derivative

$$(2)$$

groups and Bpoc = protecting group. Diversity is provided by variation of the three building-block sets; R^A is introduced by an acid chloride, R^B by an amino acid, and R^C by an alkylating agent. Millions of benzodiazepine derivatives may be synthesized by this route, since well over 100 variants of each building block are available.

Results. Benzodiazepine libraries have been used to identify lead compounds for several important therapeutic targets. Derivatives have been found that act to inhibit tyrosine phosphorylation by the enzyme known as pp60[c-src]. A phosphorylated tyrosine in certain proteins acts as a molecular "on" switch, spurring the host cell to divide. Benzodiazepine derivatives have also been identified that inhibit interaction between an antibody and an individual's DNA. This autoimmune reaction has been implicated in systemic lupus erythematosus. These compounds are not yet viable for the treatment of disease. However, they represent success in the second step of the drug discovery process, obtaining a small-molecule inhibitor that works in a test tube.

Future directions. Ongoing research in combinatorial chemistry involves development of chemical reactions that may be used to construct a library; optimization of library design to minimize time, cost, and effort; selection of an optimal set of building blocks for a particular library; and application of combinatorial methods to materials science.

Synthetic methodology. Successful synthesis of a combinatorial library requires bond-forming reactions that proceed in high yield for a variety of building-block derivatives. The adaptation of chemical reactions for use in solid-phase chemistry creates an ever-expanding "toolbox" of reactions for constructing a combinatorial library. Another area of interest is the development of analytical methods to monitor the progress of solid-phase reactions.

Library design. Many competing factors must be considered in the design and construction of a combinatorial library. A researcher wants to identify and optimize drug candidates in the shortest amount of time, at the lowest cost, and by making the fewest compounds. These goals often conflict with each other. Powerful techniques based on group theory and multivariate analysis are emerging to simplify library synthesis and analysis. These methods make it possible to determine the activity of individual members of a library when the library members are synthesized and analyzed as mixtures of many derivatives.

Reagent selection. Another area of current interest is the issue of building-block selection, because the building blocks chosen for a particular library will vary with its purpose. A library constructed to identify lead compounds will be as diverse as possible to maximize the chances of finding an inhibitor. Likewise, the design of a library for optimization will take advantage of prior knowledge of the functional groups that give the desired activity. Computational methods based on the comparison of molecular structure have been developed to assess diversity. Neural networks can make it possible to bias the building-block derivatives used for a particular library. Therefore, structural elements known to be advantageous for a particular receptor can be incorporated into library elements.

Materials science. The combinatorial approach has already had a significant effect on the fields of molecular biology (affinity selection, directed evolution, complementary deoxyribonucleic acid libraries, and bacteriophage libraries) and organic chemistry (combinatorial libraries for drug discovery and catalysis development). In these areas it is difficult or impossible to design a molecule with a specific function or property. A new area of interest is the extension of combinatorial methods to materials science. Researchers are interested in combinatorial methods to discover high-transition-temperature (T_c) superconductors, liquid crystals for better flat-panel displays, and materials to construct thin-film batteries.

For background information SEE AMINO ACIDS; ORGANIC CHEMICAL SYNTHESIS; PEPTIDE; PHARMACEUTICAL CHEMISTRY in the McGraw-Hill Encyclopedia of Science & Technology.

Matthew J. Plunkett; Jonathan A. Ellman

Bibliography. B. A. Bunin, M. J. Plunkett, and J. A. Ellman, Synthesis and evaluation of 1,4-benzodiazepine libraries, *Meth. Enzymol.,* vol. 267, 1996; E. M. Gordon et al., Application of combinatorial technologies to drug discovery, *J. Med. Chem.*, 37:1233–1251, 1994; L. A. Thompson and J. A. Ellman, Synthesis and applications of small-molecule libraries, *Chem. Rev.*, 96:555–600, 1996.

Communications satellite

Satellite-delivered personal and business telephony using small, hand-held receivers is a direct outgrowth of the evolution of space-based consumer communications services. Emerging personal satellite communications systems are the result of several converging technology trends, specifically advances in solid-state circuitry, miniaturization, and digital transmission. Concurrently, an increase in launch capabilities and reliability combined with decreasing weight of spacecraft and payload components has accelerated the growth and diversification of communications via satellite. This article provides a brief overview of the evolution of satellite personal communications systems and services, concluding with a review of current proposals and programs to provide a global, instantaneous telephone service employing hand-held telephone units similar to those available in areas of comprehensive terrestrial cellular coverage.

Development of satellite communications. The space age generally is considered to have begun with the launch in October 1957 of *Sputnik 1* by the former Soviet Union. Less than 10 years later, in April 1965, the age of commercial communications via satellite was initiated by the orbiting of *INTELSAT 1,* popularly known as *Early Bird,* by the United States for the International Telecommunications Satellite Organization (INTELSAT). This global satellite system, created under the Communications Satellite Act of 1962, was offered to the nations of the world on a nondiscriminatory basis. The Soviet Union followed suit with its Intersputnik system, offered to nonaligned and Soviet-bloc nations.

Since then, communications via satellite has evolved rapidly to include long-haul telephony and facsimile services between continents and across oceans. Sophisticated video, voice, and data business services utilizing dedicated and shared private satellite networks numbering, in some cases, in the thousands of receive sites are common. Commercial communications satellite capabilities include delivery of television and entertainment programming to cable head-ends, direct-to-home television services, navigation, location, search and rescue, and other services.

Personal communications systems. The providers of satellite services consistently have moved to reduce the weight and complexity of the ground segment, generally considered to include the satellite user's receiving unit and antenna, uplink equipment, power source, and associated electronics. At the end of 1995, Inmarsat, a global mobile satellite services organization, was offering international voice, facsimile, and data connectivity using briefcase-sized mobile satellite user terminals.

The next step is the general introduction by about the year 2000 of satellite services predicated on the development of small, transportable personal telephone units similar to the hand-held cordless telephones now employed by cellular users worldwide. Several proposals for the provision of this service, which would permit voice and data communications in the most remote areas of the Earth, are being pursued by international consortia. While this effort has been under way, various regulatory agencies on the national and international level have worked to establish frameworks for service provision and acceptance. Technological, regulatory, and operational bottlenecks have been resolved.

Big LEO systems. In 1995 the U.S. Federal Communications Commission (FCC) approved the plans of several consortia to provide global mobile hand-held telephony and other services by using satellite personal communications systems in low Earth orbit (LEO; see **table**). These potential service providers are known generically as the Big LEOs.

The proposals are vastly different, and evolutionary in nature. Ongoing advances in various technology areas have resulted in several system redesigns seeking to capitalize on various orbital configurations, available frequency and spectrum resources, and proprietary manufacturing and other advantages.

The Globalstar plan calls for a 48-satellite system plus spares, which will provide resource monitoring, wireless data, voice services, and paging and messaging. The satellites, placed in eight inclined planes, will work as simple repeaters rather than as a linked satellite network. Each orbital plane will include six satellites, 763 nautical miles (1414 km) above the Earth.

Plans for the Iridium system call for a 66-satellite constellation to provide satellite-based cellular personal and business communications to any location on Earth (see **illus.**). Eleven satellites will fill each of six polar orbits 421 nmi (780 km) above the Earth. Each satellite will have direct links to adjacent satellites, permitting calls to hop from satellite to satellite. Iridium satellites will project a grid of cells onto the Earth's surface, similar to the cell structure evident in a terrestrial cellular telephone network. Through these intersatellite links, users will be handed off from one beam or satellite to another, in much the same way that a traveling cellular telephone user seamlessly moves from one cell site to another.

The Odyssey plan calls for a medium-Earth orbit, 12-satellite system. Each satellite is to be equipped with a multibeam antenna pattern that separates coverage into 19 contiguous parts, each covering an area 500 miles square. Odyssey will extend cellular phone services to clients who need service beyond the distance capabilities of regular terrestrial service providers.

The proposed Ellipso system is a 16-satellite configuration which is unique in that it is the first

Satellite personal communication systems

System name	Number of satellites	Orbital parameters	Types of service
Globalstar	56, including 8 spares	8 planes of 6 at 52° inclination, 763 nmi (1414 km) altitude	Voice, fax, data, paging, positioning
Iridium	66 plus spares	6 planes of 11 at 86.4° inclination, 421 nmi (780 km) altitude	Voice, data, paging
Odyssey	12	3 planes of 4 at 50° inclination, 5590 nmi (10,354 km) altitude	Voice, data, fax, paging, personal computer data
Ellipso	16 plus spares	2 planes of 5 at 116.5° inclination, 4236 nmi (7846 km) apogee, 280 nmi (520 km) perigee; 6 in circular equatorial orbit at 4353 nmi (8062 km) altitude	Voice, data, fax, paging
Orbcomm	36, including spares	2 planes of 2 at 70° inclination, 4 planes of 8 at 45° inclination, 418 nmi (775 km) altitude	Tracking, data acquisition and monitoring, messaging, search and rescue
Starsys	24	6 planes of 4 at 53° inclination, 540 nmi (1000 km) altitude	Data, positioning
Vitasat	2 plus 1 spare	1 plane of 2 at 82° inclination, 540 nmi (1000 km) altitude	Electronic mail, data
ICO Global Communications	12	2 planes of 5 plus 2 spares, 5615 nmi (10,400 km) altitude	Voice, data, fax, paging

planned commercial satellite system to use an elliptical orbit.

Little LEO systems. In addition to the Big LEO proposals, several companies have applied to the FCC for authority to construct and operate similar systems that would provide messaging, data, and location services but would not include voice capabilities (see table). These potential service providers are referred to as Little LEOs and use satellites substantially smaller than those of Big LEOs.

Orbcomm, a proposed 36-satellite constellation, is the first Little LEO to provide two-way messaging and position location to compact terminals. The constellation will fly 418 nmi (775 km) above the Earth. Plans call for four satellites in orbits inclined at 70°, with up to 32 additional spacecraft, 8 each in 4 orbits inclined at 45°.

The first two experimental Orbcomm satellites were launched in early 1995. After initial problems in orbit, the satellites were able to perform various tests validating the system concept. Commercial service on Orbcomm began on February 1, 1996.

The Starsys constellation will comprise 24 satellites orbiting 540 nmi (1000 km) above the Earth. Starsys will provide two-way communication and position determination service throughout the world. Subscribers, by using portable or mobile ter-

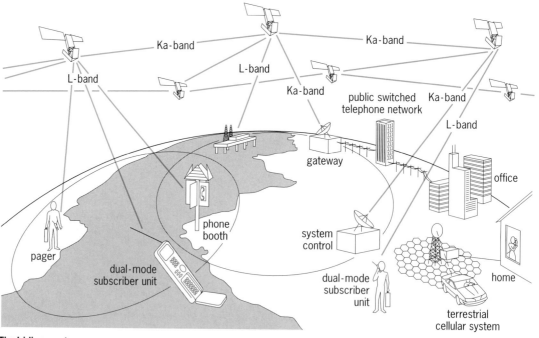

The Iridium system.

minals, can receive or transmit messages of up to 32 characters.

The planned VITA (Volunteers in Technical Assistance) system is composed of three satellites: two in orbit and one on-ground spare. This constellation would provide store-and-forward services, with satellites picking up data in one location, storing it temporarily in the satellite's memory, and down-linking it to correspondents at another. The system would not offer real-time messaging.

Other satellite systems. The planned ICO Global Communications system (see table) will comprise 12 S-band satellites (10 operational and 2 spares) in intermediate circular orbits of about 5600 nmi (10,400 km).

The *MSAT 2* satellite, launched in April 1995, is a large geosynchronous (GEO) spacecraft designed to provide satellite-based mobile voice, data, and messaging services in the United States. The operators of *MSAT 2* and those of the identical *MSAT 1* spacecraft, which provides services to Canada, are working to provide seamless and redundant service across the North American continent. Communications are provided for land-mobile, maritime, aeronautical, and fixed-site applications, with a focus on land-mobile cars and trucks. Mobile terminals are similar to a standard cellular telephone.

For background information SEE COMMUNICATIONS SATELLITE; MOBILE RADIO in the McGraw-Hill Encyclopedia of Science & Technology.

Scott Chase

Bibliography. R. Cochetti, *The Mobile Satellite Handbook,* 1994; H. Hudson, *Communication Satellites: Their Development and Impact,* 1990; J. L. McLucas, *Space Commerce,* 1991; W. L. Pritchard, H. G. Suyderhoud, and R. A. Nelson, *Satellite Communication Systems Engineering,* 2d ed., 1993.

Composite material

Recent advances in composite materials involve the development of microinfiltrated macrolaminated composites and the technique of reinforcing concrete with carbon fibers.

Microinfiltrated macrolaminated composites. Ceramics offer attractive properties, including good high-temperature strength and resistance to wear and oxidation. However, the major limitation to their use in structural applications is their inherent low fracture toughness, that is, the tendency to break (fracture) and produce low values. Ceramics have low fracture toughness, whereas steels and superalloys have better fracture toughness. Methods currently being used to improve the toughness of ceramics involve incorporation of reinforcing whiskers and fibers; inclusion of a phase that undergoes transformation within the stress field associated with a crack; and cermet technology (ceramic/metal composites), in which the tough metallic component absorbs energy. Improved toughness is attributed to various mechanisms, including crack branching, transformation-induced residual stresses, crack bridging, and energy absorption by plastic flow.

Another possible approach to improving the toughness of ceramics is via laminated construction. Sophisticated coating techniques have been used to produce laminated microstructures of ceramics and metals, but the cost of producing a bulk-laminated composite of useful size is prohibitive. Fabrication of microinfiltrated macrolaminated composites offers an economically feasible approach to produce a variety of cermet/metallic and ceramic/metallic bulk-laminated composites (**Fig. 1**). The basic architecture of a microinfiltrated macrolaminated composite is a double-layer structure. One layer consists of a soft, ductile material having a low modulus of elasticity, low strength, and high toughness; the second layer consists of a hard, brittle material having a high modulus of elasticity and low toughness. This double layer is repeated as many times as necessary to form the bulk composite; in addition, the brittle material is infiltrated with the ductile constituent.

Stress and strain behavior. The potential benefits of the architecture of microinfiltrated macrolaminated composites can be better understood by considering the nature of damage development in a typical cermet (tungsten carbide–cobalt, or WC-Co) and a tungsten heavy alloy (tungsten-nickel-iron, or W-Ni-Fe). Damage in the cermet occurs by a process of both brittle and ductile fracture. In particular, the early stages of damage involve the brittle fracture of carbides (transgranular microfracture) or of carbide contiguous interfaces (intergranular microfracture). Although these sites constitute the bulk of the resulting fracture surface, they represent little of the total energy expended in fracture.

Instead, most of the resistance to failure (or fracture toughness in the instance of a cracked sample) stems from plastic flow in and rupture of the matrix ligaments, effectively bridging the macroscopic crack plane. However, the capability of this mechanism to strengthen and toughen the composite is constrained by limitations on plastic-zone development. Limitations are imposed by the carbide spacing, which effectively controls the height of the crack-tip plastic zone and confines it to approximately an in-plane (of the crack) strip. Under dynamic loading, tungsten heavy alloys behave in a similar manner, with microcracking dominating at contiguous particle interfaces.

Conventional composite microstructures inhibit the maximum functionality of the intrinsic toughness of the matrix phase. Thus, under the same nominal stress intensity, the plastic zone in such a composite is considerably different from a zone at the tip of a crack residing within a bulk tough matrix. In a bulk matrix, the equilibrium plastic zone would be more extensive on either side of the crack plane, and the added energy-absorbing capability would result in correspondingly greater toughness.

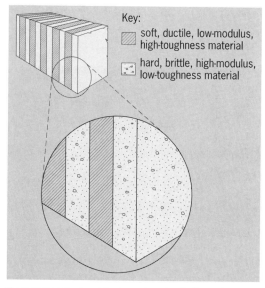

Key:

soft, ductile, low-modulus, high-toughness material

hard, brittle, high-modulus, low-toughness material

Fig. 1. Conceptual architecture of a microinfiltrated macrolaminated composite.

The architecture of the microinfiltrated macrolaminated composite offers a compromise to the conventional composite microstructure by providing repeated alternating layers of bulk tough metallic and brittle cermet and ceramic materials. Any crack introduced into the brittle constituent will, upon entering the metallic interlayer, be subjected to a potential crack-stopping higher toughness.

Process concepts. In any laminate-fabrication approach, the resulting composite mechanical properties will follow the rule of mixtures (a measure of strength properties) only if there is good adhesion between layers. One way to improve adhesion is to set the phases into alternating layers by vapor deposition, which also results in extremely fine grain sizes. Another approach is to deposit the layers at a high temperature to allow some interdiffusion between layers. Similar results can be achieved by heat-treating a laminate to promote some interdiffusion.

Good adhesion between layers is difficult to achieve with materials that undergo little or no interdiffusivity. For this situation, the best layer-bonding method is to have one layer penetrate into the other on a microscale (like tiny claws penetrating a surface). This concept is used in artificial bone implants, where a thin outer layer of porosity in the implant allows the ingrowth of bone tissue, and produces strong mechanical bonding between the implant and the natural bone. This type of structure provides good mechanical bonding between layers.

None of the coating processes discussed above can produce laminates in which one of the constituents will penetrate the other. However, fabrication of microinfiltrated macrolaminated composite materials offers a processing route to economical manufacture of large, bulk-laminated composites in which an interpenetrating microstructure is achievable. Use of both ceramic and metal powders in the process opens the door to a potentially wide range of material combinations. The same fundamental processing steps can be used to produce a wide variety of microinfiltrated macrolaminated composites, including ceramic/metal, metal/intermetallic, and ceramic/intermetallic systems.

Composition and fabrication. Various combinations of composite properties, including hardness, strength, ductility, and fracture toughness, are possible by varying the laminate layer thicknesses. Microinfiltrated macrolaminated composites, when used in bulk form, are expected to have properties far superior to those of the individual monolithic constituents of the composite.

Alternative processing routes for microinfiltrated macrolaminated composites are available if the composite constituents have some solubility for one another, such as in the tungsten carbide-cobalt/cobalt (WC-Co/Co) and tungsten-nickel-iron heavy alloy/nickel (W-Ni-Fe/Ni) systems. In general, to obtain a large composite of this type, the best approach is to use the tape-casting process (well known in the ceramic industry) to produce large, thin tapes (sheets) of the material. Tape casting consists of mixing powders with a suitable binder to form a slurry, and then casting the tape. The material can be cut into required sizes in the green condition (when it is easier to cut) or after sintering. A material in green condition has been put together initially but not fully consolidated or fired (sintered); it also has not reached its full density. Consolidation (compaction) helps the material achieve its full density by removing gas voids, porosity, and so forth. An attractive alternative for producing sheets consists of rolling the powder material followed by sintering to the required percent of theoretical density, retaining a certain level of interconnected porosity.

The fabrication steps used to make a W-Ni-Fe/Ni microinfiltrated macrolaminated composite illustrate the process. Tungsten powder is tape cast, and the sheets are sintered to 60–70% of theoretical density, which produces totally interconnected porosity. Small amounts of nickel can be used together with tungsten to promote activated sintering and to produce a porous tungsten sheet that can be handled safely. Sheet thickness ranges 0.04–0.4 in. (1–10 mm).

An 80:20 ratio of nickel and iron powders also is tape cast and sintered to full density. Sheets of porous tungsten and fully dense Ni-Fe alloy material are laid up in alternate layers and heated to about 2687°F (1475°C), which is about 72°F (40°C) above the melting point of the Ni-Fe alloy. The molten Ni-Fe alloy infiltrates the porous tungsten sheets and takes into solution a fraction of the tungsten. A thin layer of the liquid phase can be retained between the tungsten sheets after the infiltration process is complete. This retention is possible if the tungsten sheet contains porosity levels that are lower than the volume of liquid formed by melting the Ni-Fe sheets.

Difficulties in using this technique can be minimized by avoiding excessive liquid-phase formation, which can cause slumping of the composite, and by avoiding pore formation at the layer interfaces.

An alternative method for producing a similar composite involves the formation of W-Ni-Fe tape castings from elemental powders. The nickel content is slightly lower than that normally used in tungsten heavy alloys. Fully dense W-Ni-Fe sheets are produced via sintering, and thin nickel foils and W-Ni-Fe sheets are stacked in alternate layers and then heated to the liquid-phase sintering temperature. If the nickel layers are too thin, very fast diffusion kinetics can result in homogenization. In contrast, excessive liquid formation and slumping occur if the nickel layers are too thick.

Solid-state bonding provides an alternative method for avoiding the difficulties associated with controlling liquid-phase sintering. Tape casting and composite lay-up (assembly) are the same. Sintering is carried out at a temperature of 2550°F (1400°C). Solid-state bonding requires the use of a pressure assist in case the sheets are not perfectly flat.

Similar developmental work also is being conducted to fabricate a WC-Co/Co microinfiltrated macrolaminated composite. Alternate sheets of sintered WC-Co and pure cobalt are hot pressed in a graphite die at a temperature of 2280°F (1250°C) and a pressure of 5945 lb/in.^{2} (41 megapascals).

It is possible to produce a compositional gradient in the soft, interpenetrating ductile layer. The softer phase must have some solubility for the hard ceramiclike material, and the thickness of the softer layer should be great enough to prevent total compositional homogenization over the entire thickness. Thus, there are potentially many alloy possibilities in microinfiltrated macrolaminated composites for materials that have mutual solubility in one another.

In contrast, many material combinations, such as ceramic/metal, have very little or no solubility in one another. Producing microinfiltrated macrolaminated composites from these materials presents a major challenge. Processing of such materials is conceptually similar to that described previously. For example, alternating layers of porous aluminum oxide (Al_2O_3) tape castings and nickel sheets are heated to a temperature below the melting point of nickel but high enough that the nickel is extremely malleable. The laminate is subjected to a pressure sufficiently high to extrude thin ligaments of nickel into the pores of the Al_2O_3. Conventional hot pressing of the layered stack in graphite dies does not generate enough pressure to extrude the nickel into the fine-pore structure of the Al_2O_3. Thus, a pressing technique in which very high pressures can be applied quickly to the stack of material is required.

An alternative route to fabricate ceramic/metal microinfiltrated macrolaminated composites is to form comixed porous or dense sheets of Al_2O_3 and nickel. It is possible to produce a comixed Al_2O_3/Ni sheet in which both phases are interconnected by selecting the proper powder volume fraction (V_f)

and relative particle size ratio. By conventionally hot-pressing a layered stack of comixed sheets and either nickel sheets or foils, the interconnected nickel ligaments present in the comixed sheets easily bond to the nickel layer, yielding the proper microstructural architecture. The same processing steps can be used to produce other material combinations consisting of a hard, brittle phase interpenetrated by a soft, ductile material.

Applications. The design of microinfiltrated macrolaminated composites could prove advantageous in applications involving impact or ballistic penetration. The metallic interlayer would function to hold together damaged portions of the brittle constituent. Thus the development of ceramic armor having multihit capability becomes possible. Current metal/ceramic composite armor is layered on an extremely macroscopic scale. An approach involving microinfiltrated macrolaminated composites permits optimization of layer frequency and thickness to resist specific threats, as in antipersonnel weapons or armored vehicles.

In addition, high-temperature composites, consisting of a ceramic and intermetallic compound [such as aluminum oxide/nickel aluminide (Ni_3Al) plus boron] could be fabricated in the form of microinfiltrated macrolaminated composites to yield combinations with new properties. Ultrahigh-temperature composites incorporating high-temperature ceramics and ductile niobium is another potential application in order to obtain various combinations of high wear resistance and toughness. The hard ceramic outer layer would provide a wear-resistant surface, with toughness being provided by the ductile material. A similar concept could lead to a new generation of cutting tools, in which the outermost cutting layer could be made of ultrahard materials suitably layered with a soft, ductile constituent for toughness.

Other potential applications are in heat-and-oxidation-resistant, low-density structural components and aerospace parts having low density, high strength and modulus, and good fracture toughness.

Mel Schwartz

Reinforcement of concrete with carbon fibers.

The reinforcement of concrete with carbon fibers is a recently developed technique for the construction of structures such as buildings, bridges, and tunnels. Advantages of carbon-fiber reinforcement include outstanding electromagnetic shielding, high resistance to corrosive environments, light weight, and high mechanical property strength. Two methods are used to achieve the reinforcement: mixing carbon fibers, either chopped or mat, directly into cement (carbon fiber–reinforced concrete); and using carbon fiber/plastic composites in rod or tape form for strengthening the concrete.

Carbon fiber–reinforced concrete. This material was first used for the Al Shaheed monument in Baghdad, Iraq, and the Ark Hills office building (known as the Ark Tower) in Tokyo, Japan. The Al Shaheed monument, with twin domes, 130-ft (40-m) height, and 150-ft (45-m) base diameter, required a lightweight material with enough mechanical strength and durability to withstand severe weather conditions. For this application, cladding tile panels of carbon fiber–reinforced concrete with an area of about 108,000 ft² (10,000 m²) was used.

The Ark Tower building, with 37 stories, has curtain walls of carbon fiber–reinforced concrete with a total area of about 340,000 ft² (32,000 m²). This design yielded a remarkable reduction in wall weight. The material was easily handled with a small lift, which permitted a shortened construction period.

The success of these projects led to the application of carbon fibers in various types of structures. Chopped carbon fibers of general-purpose grade were employed; an addition of only 2% by volume of carbon fibers doubled the tensile strength and changed the fracture mode from brittle to ductile.

Many fundamental and engineering studies involving carbon fiber–reinforced concrete have been carried out by using not only chopped carbon fibers based on polyacrylonitrile and pitch but also mats (papers) and fabrics of continuous carbon fibers. The addition of small amounts of carbon fibers into the cement mortar increases the mechanical strength by a factor of more than 5. The addition of carbon fiber also increases toughness of the concrete. Carbon fiber–reinforced concrete has a high durability and high dimensional stability in a high-temperature and high-humidity environment.

Carbon fiber–reinforced concrete has been used in various portions of buildings, such as for curtain walls, waterproof roofs, and construction of floors having chemical stability and low noise level. This material has also been found to be effective for electromagnetic shielding in modern buildings (sometimes known as intelligent buildings); such shielding blocks out electrical and magnetic noise from the outside that might impede steady operation of the building's functions by computers. Recently, carbon fiber–reinforced concrete was successfully used in the demagnetization function of piles in the Yokohama Bay area in Japan.

An example of a new application of carbon fibers is the construction of a frame for placing concrete. A thin mortar plate reinforced by carbon fiber mesh and chopped carbon fibers is used for the frames; some advantages of this technique are ease in forming by permitting bending at the site of construction (**Fig. 2**), and ease in transportation and storage because of properties such as light weight, low thickness, and chemical and mechanical stability. This method also saves labor, as the frame does not have to be removed.

Carbon fiber–reinforced concrete injected as a paste has also been used to strengthen the walls in tunneling with an automatic drilling tool. It has been reported that this method was used successfully at a depth of about 240 ft (73 m).

Reinforcement by carbon fiber/plastic composites. Two novel ways for reinforcing the concrete by using carbon fibers have been developed. One uses a rod

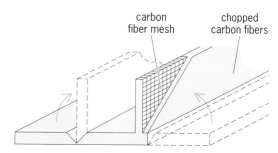

Fig. 2. Use of carbon fiber–reinforced concrete in the form for placing concrete.

of carbon fiber/plastic composites instead of steel wires in the concrete, and the other uses sheets (tapes) of unidirectionally aligned carbon fibers with a polymer matrix.

Rods consisting of a few carbon fiber/plastic composite strands have been tested for reinforcing a small concrete bridge directly exposed to erosive sea wind in Ishikawa Prefecture facing the Japan Sea. In the past the bridge had to be reconstructed at least once a year because of rapid erosion of the steel wires. With the use of rods of carbon fibers to reinforce the concrete, no defects in the bridge have been observed even after a period of 6 years. This type of reinforced concrete has been used successfully in a pedestrian bridge located in the city of Funabashi, Chiba Prefecture, Japan, since 1989.

The prepreg sheets of carbon fibers have been used for reinforcing the piers (columns) of highways and chimneys. After the severe earthquake in the Kobe area (January 17, 1995) and also because of a large increase in traffic in the Tokyo area, the urgent necessity of reinforcing columns of highways and pillars in buildings became apparent. In this method of reinforcement the tapes of carbon fiber prepregs are initially glued to the surface of the pier along its axis after the surface of concrete has been completed, and then either the same prepreg tapes or carbon fiber strands of 1200–2000 fibers are wound perpendicular to the pier axis. In the case of chimneys eroded by highly acidic exhausts, the advantages of light weight and ease of repair from the exterior of the chimney have been recognized; about 20 repairs of this kind have been done in Japan. The largest chimney repaired by this method is 330 ft (100 m) high.

As for highway columns, most have square cross sections and special attention was required at the corners so that the carbon fibers would not break. Many new technologies are being developed and introduced for such repairs. Examples are a remotely controlled system to check the corrosion in large concrete constructions, and use of an electrical current to harden the plastic matrix after forming.

For background information *SEE* CERMET; COMPOSITE MATERIAL; GRAPHITE; REINFORCED CONCRETE; SINTERING; STRESS AND STRAIN in the McGraw-Hill Encyclopedia of Science & Technology.

M. Inagaki

Bibliography. M. Inagaki, Research and development on carbon/ceramic composites in Japan, *Carbon,* 29(3):187–195, 1991; A. I. Isayev and M. Modic, Self-reinforced melt processible polymer composite: Extrusion, compression, and injection molding, *Polym. Compos.,* 8(3):158–175, 1987; K. Kobayashi et al., *Mater. Des.,* vol. 9, no. 10, 1988; A. Richardson et al., The production and properties of poly(arylether ketone) (PEEK) rods oriented by drawing through a conical die, *Polym. Eng. Sci.,* 25(6):355–361, 1985.

Computer

Although every computer, and indeed every object in the physical world, is described by the laws of quantum physics, in all ordinary present-day computers the storage and switching functions, involving the aggregate action of myriad individual quantum systems, actually obey quite accurately the laws of classical physics. That is, a conventional electronic switch passes from its "on" to its "off" state in a deterministic way under the influence of definite forces, in accord with Newton's laws. But in a quantum computer, the time evolution of the state of the individual switching elements of the computer is itself governed by the laws of quantum mechanics. A quantum computer does not presently exist, but may once the working components of a computer have been shrunk to atomic dimensions. Physicists and computer scientists have been working out its properties.

Classical and quantum computers. In an ordinary computer, at any instant in time the machine exists in a particular discrete state of its bits, which is some number in binary representation (for example, 0111000101 . . .). This state evolves in time according to some program of instructions through a sequence of other discrete states. After a successful run, the state of the computer contains the answer to some arithmetic calculation. Implicit in this description is the notion that the time evolution of the computer is governed by a classical, newtonian law (albeit a very complicated one).

The state of a quantum computer is described by a wave function, Ψ, which might consist of a coherent linear superposition of bit states, for example, $\Psi = x|0111000101\ldots\rangle + y|0101000011\ldots\rangle +\cdots$. The information which this wave function contains is that, if the state of the computer is observed, it will be in the first binary state with probability $|x|^2$, in the second with probability $|y|^2$, and so on. But the wave function also contains additional information. If the state of the computer is unobserved, the various superposed computations will all evolve forward in time simultaneously. In the course of this time evolution, two or more computations may arrive at the same bit state, in which case they will interfere with one another. The interference may be either constructive or destructive, according to the relative phases of the complex coefficients x, y, and so forth.

Quantum computer implementation. Two-state quantum-mechanical systems which might serve as an individual bit in a quantum computer (called a qubit) are well known in physics. Examples are the ground state and first excited state of an atom, the zero- and one-photon states of an electromagnetic cavity, and the spin-up and spin-down states of a particle with a total angular momentum $\hbar/2$ (where $\hbar$ is Planck's constant divided by 2π), called a spin-½ particle. It is also well known, at least at the few-qubit level, how to implement operations which serve as the elementary logic gates of a quantum computer; they correspond to techniques used in the science of spectroscopy. For example, the inverter or NOT gate is implemented by using a technique known as tipping-pulse spectroscopy. A tipping pulse is a burst of alternating-current radiation (for example, a burst of radio-frequency magnetic field in the case of nuclear magnetic resonance spectroscopy, when the two-level system is the spin state of an atomic nucleus) which is tuned to be in resonance with the transition energy between the lower and upper states of the qubit. The tipping pulse evolves the wave function of the qubit in time from an initial state $\Psi_i = x_i|0\rangle + y_i|1\rangle$ to a final state $\Psi_f = x_f|0\rangle + y_f|1\rangle$, where the final coefficients, x_f and y_f, are related to the initial coefficients, x_i and y_i, by a rotation by an angle $\theta = \Omega T/2$ (**Fig. 1**). Here, Ω, the Rabi frequency, depends on the physical parameters of the qubit, and T is the duration of the tipping pulse. If the spectroscopist tunes T such that the tipping angle, θ, equals π radians or 180°, then the usual inverter or NOT gate is obtained: if the initial state is $|0\rangle$, then the final is $|1\rangle$, and vice versa (apart from phases). This quantum inverter has a number of nonclassical properties: it will properly invert each component of the linear superposition represented by the wave function, and if the tipping angle is set to be different from π radians, an operation having no classical analog is obtained. (For example, for angle $\pi/2$ radians = 90°, an operation that has been named square-root of NOT is obtained.)

Physicists have extended the spectroscopies which can be performed on individual qubits, leading to techniques of double-resonance spectroscopy which

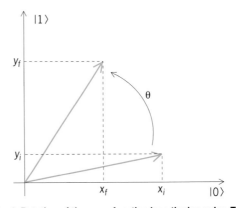

Fig. 1. Rotation of the wave function by a tipping pulse. The coefficients *x* and *y* can be complex.

implement simple logical operations on pairs of qubits. For example, a protocol using a sequence of alternating-current pulses called ENDOR (electron-nucleus double resonance), developed in the 1950s to manipulate the state of silicon-29 nuclei coupled to donor electron spins in crystalline silicon, implements an exclusive-or (XOR) gate on the state of the two qubits. **Figure 2a** shows the truth table for executing this XOR, and Fig. 2b shows the symbol for the quantum XOR gate. Unlike an ordinary XOR, this gate has two output wires rather than one. These wires actually represent the states of individual elementary particles, which cannot be created and destroyed during the tipping-pulse operations. These wires also reflect the fact that quantum mechanics implements a kind of reversible logic, which can always be run backward from the output to obtain the inputs. Mathematically, only the XOR and the (generalized) NOT gates are needed as elementary building blocks for quantum computation; any n-qubit computation can be built up as a sequence of these operations applied successively to individual qubits or pairs of qubits. For example, the logical AND operation is constructed as shown in Fig. 2c with three XORs and four $\pm\pi/4$ single-qubit tipping pulses. The physical implementation of such quantum circuits is difficult, since it requires that interactions be turned on and off between specified pairs of two-level quantum systems; recent proposals for doing this involve, for example, passage of atoms in a beam in and out of cavities in which these interactions take place. But there is no established best technique for building up these networks.

Application to factoring integers. If such networks of quantum operations can be built up on a large scale, they would have the capacity to solve certain kinds of mathematical problems much more efficiently than any conventional computer. The outstanding example of this possibility is P. Shor's quantum algorithm for factoring integers into products of prime numbers. If prime factoring were easy, a very popular form of cryptography known as public-key cryptography would become obsolete, because it relies on the fact that it is presently very hard to determine from the publicly known encoding key, which is a large composite integer, the prime factors of that integer which are used for the decoding (and are known only to the decoder). Shor's quantum factoring procedure begins with the observation that the function $f(x) = c^x(\bmod N)$, for any integer c which has no prime factors in common with N, is periodic in x with some period r [that is, $f(x + r) = f(x)$], and that r contains information about the prime factors of N. If N is itself prime, then $r = N - 1$, but if N is composite, then r is less than $N - 1$. Since $f(0) = 1$, it follows that $f(r) = 1$, or $c^r = 1 + kN$ for some integer k. If r is even, then this expression can be factored as in the equation below, showing that $c^{r/2} \pm 1$ is likely to

$$(c^{r/2} + 1)(c^{r/2} - 1) = kN$$

have prime factors in common with N. The problem with this approach to prime factoring is that r is of order N, so that in a classical computation r would have to be found by an exhaustive search of all the numbers between 0 and N, a very inefficient pro-

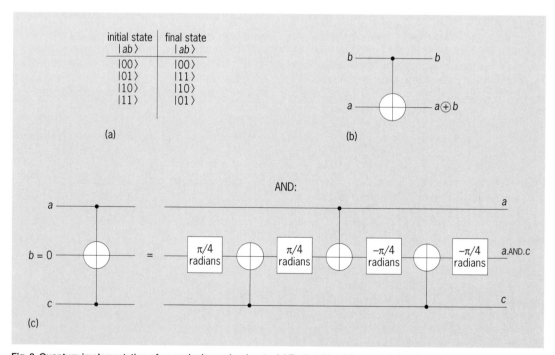

Fig. 2. Quantum implementation of an exclusive-or (XOR) gate. (a) Truth table of the XOR, giving the evolution of different components of the two-qubit wave function from the initial to the final time. (b) Symbol for the quantum XOR gate. (The a qubit ends up in the sum of the values of the original states of a and b, modulo 2.) (c) Construction of the reversible AND gate, using XORs and tipping pulses on individual qubits (tipping angle $\Omega T/2 = \pm\pi/4$ radians = $\pm 45°$). The result of the AND operation on a and c is placed in the work bit b.

cess. But quantum computation can extract r efficiently by performing operations in parallel. Shor's procedure is to form a quantum superposition of all possible inputs x to the function, compute all the function values $f(x)$ in parallel, and then, by performing a Fourier transform on the result, extract the period r in much the same way as the period of a crystal is determined by x-ray diffraction: constructive interference of the different pathways occurs only when the phases are equal in every period. **Figure 3** shows this computational procedure, which requires exponentially fewer steps (that is, magnetic resonance operations) than any known classical factoring procedure.

Technological requirements. Despite the incomparable mathematical power inherent in quantum-mechanical operations, it will be a long time before an operating quantum computer will be built. As mentioned above, it is difficult to execute accurately a long sequence of quantum gate operations on a large set of qubits; the technology of experimental spectroscopy has not advanced to the point where

operations of this complexity have been performed. Although there appears to be no reason in principle why such machines could not be built, there is a fundamental reason why constructing a quantum computer will require very advanced technology: Virtually any stray interaction with its environment (absorption of a single photon, or of a single quantum of vibrational energy) will cause the qubit to dephase. In dephasing, the definite phases which enable different time evolution pathways (that is, different computations) to interfere constructively or destructively (as in Fig. 3) are lost; so this process destroys the essence of quantum computation. To take one example, the longest known dephasing times for electron states in semiconductors are about 1 nanosecond, an inadequate length of time for completing most useful quantum computations. Still, other qubits have longer, more potentially useful dephasing times: about 0.1 s for positive indium (In^+) ions in a single-ion vacuum trap, and many hours for the nuclear spins of xenon-129 in the gas phase. Qubits of this sort will have to be used, or the dephasing of other qubits will have to be greatly

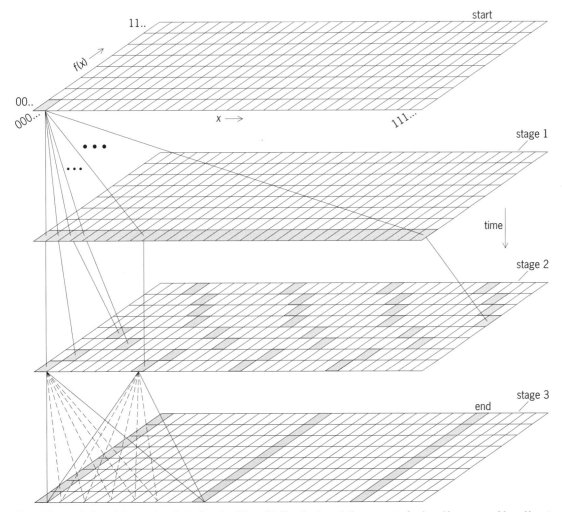

Fig. 3. Time evolution of the quantum factoring algorithm of P. Shor. In stage 1, the computer is placed in superposition of input states x. In stage 2, $f(x)$ is evaluated. In stage 3, the Fourier transform is taken. Because of the periodicity of $f(x)$, constructive interference results only when the analog of the Bragg condition is satisfied (solid paths); most paths (broken lines) combine destructively.

reduced by very stringent isolation techniques, before significant progress toward building a quantum computer can take place.

For background information SEE CRYPTOGRAPHY; DIGITAL COMPUTER; FOURIER SERIES AND INTEGRALS; LOGIC CIRCUITS; NONRELATIVISTIC QUANTUM THEORY; NUMBER THEORY; QUANTUM MECHANICS; X-RAY DIFFRACTION in the McGraw-Hill Encyclopedia of Science & Technology.

David P. DiVincenzo

Bibliography. C. H. Bennett and R. Landauer, Physical limits of computation, *Sci. Amer.*, 253(1):48–56, July 1985; G. Brassard, Cryptography column: Quantum computing, the end of classical cryptography?, *SIGACT News*, 25(4):15–19, 1994; J. Brown, A quantum revolution for computing, *New Sci.*, 133(1944):21–24, 1994; D. Deutsch, Quantum computation, *Phys. World*, 5(6):57–61, 1992; IEEE Computer Society, *Proceedings of the 35th Annual Symposium on the Foundations of Computer Science*, 1994.

Computer-based systems engineering

Both complex physical systems and sophisticated software systems include distributed processing and databases, internal communication systems, heterogeneous processing components, and sophisticated human-machine interfaces. Processing components by themselves can compose a system, or they may be embedded in a physical system such as an automobile, aircraft, or medical diagnostic system. Both the encompassing system and the processing system are known as a computer-based system (CBS). Other computer-based systems include the San Francisco Bay Area Rapid Transit (BART), telecommunications systems, airline reservations systems, weapons systems, manufacturing automation systems, and the New York Stock Exchange. These systems are inherently difficult to develop as they are large and technically complex, and usually involve a cooperative effort by a number of organizations. It is important that the parts and interfaces be consistent, and that the system-level aspects such as security, performance, and safety requirements be met. Security and reliability are frequently required to protect finances, privacy, and property. Safety is a concern, for example, for transportation systems, nuclear power plants, radiation therapy machines, and air-traffic control, where processing errors can result in the loss of human life. In real-time systems, meeting timing requirements is as important as meeting functional requirements. Delayed performance can cause catastrophic results, as in digital flight control or nuclear generator control. Dependability, performance, and such are emergent system properties; that is, they depend on a number of parts or subsystems, including the communication system, and cannot be guaranteed at the component level.

Integration of information technologies into physical and human activities dramatically in-creases the interdependencies among components, actors, and processes, generating complex dynamics not taken into account in previous generations of systems. To deal with this complexity, a new profession and discipline in the engineering of computer-based systems has been advocated.

Need for system-level discipline. The systems engineer of computer-based systems develops a system within a system; the properties of the former have pervasive effects throughout the larger system. The computer-based system consists of all components necessary to capture, process, transfer, store, display, and manage information. Components include software, processors, networks, buses, firmware, application-specific integrated circuits (ASICs), storage devices, and humans (who also process information). Embedded computer-based systems interact with the physical environment through sensors and actuators, and also interact with external computer-based systems. The systems engineer of computer-based systems has to have a thorough understanding of the system in which the computer-based system is embedded, for example, an automobile, medical diagnostic system, or stock exchange.

Computer-based system resources are frequently geographically dispersed and under the control of different organizations. System-level engineering is needed to define a holistic architectural approach and to avoid system-level disasters. Systems engineering of computer-based systems requires knowledge of distributed computer system design and development, including in-depth knowledge concerning computer hardware and software disciplines. Unfortunately, many systems engineers of large physical systems do not have this knowledge, resulting in computer hardware engineers and software engineers performing their tasks without an integrated system-level approach to computer-system requirements definition and allocation, architecture, security, reliability, safety, and fault tolerance.

The systems engineering of computer-based systems is analogous to traditional systems engineering, but the focus and necessary skills are different. The concerns of computer-based systems engineering include definition of the requirements of the computer-based system; decisions concerning processes and methods; design decisions concerning the distributed architecture of the computer-based system; allocation of resources to component developers and management of the coordinated process; allocation of functions and data to resources of the computer-based system (processors, software, datastores, displays, and humans); computer-based system strategies with respect to safety, security, and fault tolerance; global system management strategy; definition of system services; performance allocations (timing, sizing, and availability); testing (of component, integration, and interoperability with the external environment); logistics support (including maintenance and training); and implementation of the computer-based system within the existing environment.

Differences between disciplines. Engineering of computer-based systems is an interdisciplinary field that includes systems engineering, electrical engineering, software engineering, communications, and human factors. To clarify the differences between systems engineering, software engineering, and computer-based systems engineering, the way in which these disciplines work together to develop a sophisticated vehicle will be examined. The systems engineer has primary responsibility for specifying and allocating requirements for the vehicle. From the start of the program, the computer-based systems engineer works with the systems engineer on system-level requirements that relate to the distributed processing system. One requirement is that the vehicle operator be able to navigate. The systems engineer and the computer-based systems engineer make decisions concerning the system boundary. They decide which navigation functions the system automates and which navigation functions the vehicle operator performs. They also make design decisions, such as that the system will include maps and that the Global Positioning System will provide location coordinates. To architect the computer-based system, the computer-based systems engineer must understand the Global Positioning System and how the operator will use it, together with the maps, to navigate. Another requirement is that the system control speed to avoid hitting another vehicle. To satisfy this requirement, the systems engineer and computer-based systems engineer must perform trade-offs and decide on various vehicle components. They decide that a proximity sensor is needed to provide data on nearby vehicles.

The computer-based systems engineer must also determine what is done in software, hardware, and firmware. These decisions affect performance and maintainability, as hardware is fast but software is changeable. This engineer must also make decisions with respect to resources and allocation. These decisions determine the types of processors, communication links, and storage devices that are needed, such as signal processors for processing data from the proximity sensor and a CD-ROM (compact-disk read-only memory) for storing the map database. Based on dependability needs, the computer-based systems engineer chooses fault-tolerance features such as dual buses, standby processors, fault-tolerant operating systems, and software recovery blocks. The computer-based systems engineer also optimizes the number of standard parts in the computer-based system (a practice known as commonality), as well as the cost of the computer-based system. Standard software and hardware parts and standard interfaces reduce development and maintenance costs.

After considering various trade-offs, including processing allocation, the systems engineer and computer-based systems engineer define the subsystems and interfaces. Interfaces should be minimal for ease of change and good maintainability. Subsystems for the vehicle include the vehicle body, engines, a fuel system, a navigation system, a guidance and control system, a proximity sensor, and an acceleration and braking system.

Software engineers work with computer-based systems engineers to define software requirements, and allocate them to processing resources. In the above example, some software will be allocated to the signal processor, and other software will be allocated to data processors. Such allocation decisions can be difficult to make and can seriously affect performance. Tools that simulate software functionality running dynamically on different hardware configurations support these trade-offs.

Problems in developing systems. Most corporations and university programs are organized around a component paradigm, an obsolete paradigm that would work only for nondistributed, nonembedded computer systems, if at all. In this paradigm, software engineering and computer hardware engineering are not well integrated, and systems engineers allocate software requirements to subsystems prior to having a detailed understanding of the distributed system requirements as a whole. Frequently, the resulting system has overly complex interfaces and poor performance. Thus the architecture must be changed after the system is already built, resulting in an architecture that is difficult to maintain.

One reason why the development of computer-based systems is difficult is that few system, software, and hardware engineering tools are integrated. For example, if design tools in these disciplines were integrated, it would be much easier to simulate software functionality running on selected hardware and to provide analysis of processing and communication resources. Such tool integration has begun, and will make it possible to perform architectural trade-offs more easily.

Communication between engineering disciplines is also a problem. Systems engineers frequently have an engineering background, and use the vernacular of their domain of expertise, for example, aircraft, medical systems, or nuclear-power generation. Software engineers frequently have educational training in mathematics and computer science. They are used to dealing in abstractions, rather than hard science, and often do not have extensive domain expertise. Thus, the two disciplines have difficulty communicating. Most systems engineers do not understand what software engineers need to perform their tasks, and software engineers are not able to explain this need adequately. The computer-based systems engineer, having an electrical engineering as well as a software background and some domain expertise, would be able to improve communication.

Technical support. The development of distributed computer systems requires an engineering discipline that integrates systems engineering, electrical engineering, software engineering, communications, and human factors. This integrated knowledge is necessary to deal with the complexities of distributed systems and to design the architecture of maintainable, reliable systems with good performance.

To advance this interdisciplinary field, the Computer Society of the Institute of Electrical and Electronics Engineers (IEEE) formed the Technical Committee on the Engineering of Computer-Based Systems. This committee facilitates and encourages research in the field, and serves as a forum for exchange of ideas and for collection and dissemination of knowledge among interested practitioners, researchers, educators, and students. The committee also encourages development of an interdisciplinary academic field to train computer-based systems engineers, and is establishing a framework for education and training. Promotion and institution of standards for computer-based systems is also an important committee activity.

For background information SEE DISTRIBUTED SYSTEMS (COMPUTERS); EMBEDDED SYSTEMS; FAULT-TOLERANT SYSTEMS; REAL-TIME SYSTEMS; SOFTWARE ENGINEERING; SYSTEMS ENGINEERING in the McGraw-Hill Encyclopedia of Science & Technology.

Stephanie M. White

Bibliography. H. Lawson (ed.), *Proceedings of the 1994 Tutorial and Workshop on Systems Engineering of Computer-Based Systems*, 1994; N. G. Leveson and C. S. Turner, An investigation of the Therac-25 accidents, *Computer*, pp. 18–41, July 1993; B. Melhart and J. Rozenblit (eds.), *Proceedings of the 1995 IEEE International Symposium and Workshop on Systems Engineering of Computer-Based Systems*, 1995; B. Thorne and M. Voss (eds.), *Proceedings of the 1996 IEEE International Symposium and Workshop on Engineering of Computer-Based Systems*, 1996; J. E. Tomayko, G. Goos, and H. Hartmanis (eds.), *Proceedings of SEI 1991 Conference on Software Engineering Education*, Lecture Notes in Computer Science, no. 536, 1991; S. White et al., Systems engineering of computer-based systems, *IEEE Comput.*, pp. 54–65, November 1993.

Computer storage technology

The use of computer memory equipment presents a very complex picture, with 14 generically different kinds of digital memory being employed. For relatively slow but inexpensive archival storage, tape and disk are widely used. Silicon-based dynamic random access memories (DRAMs) satisfy the bulk of the very high density (64–256 megabit), high-speed (35-nanosecond) applications. Static random access memories (SRAMs), including those on gallium arsenide (GaAs), are in use for faster (few-nanosecond) but lower-density devices (not more than 256 kilobits). Finally, a wide variety of devices have nonvolatile applications in which memory is retained when power is interrupted. The use of nonvolatile devices is less widespread than the other types, and is dominated by electrically erasable programmable read-only memories (EEPROMs), magnetic core memory, bubble memory, and some radiation-hard embodiments [such as silicon-on-sapphire, plated wire, or complementary metal-oxide silicon (CMOS), with battery backup]. In 1995, devices based upon ferroelectric thin films began to be used in place of many of the devices listed above.

DRAMs. The bit density of DRAMs has doubled every few years since the 1970s. In a typical memory chip, most of the area is taken up by the capacitors; the resistors and transistors utilize relatively little space. Consequently, recent developments have emphasized complicated geometries to increase the surface area available for the capacitors, involving the cutting of steep trenches with large depth-to-width ratios in each dielectric layer, using three-dimensional stacking of capacitor layers, and making the surface wavy instead of perfectly flat (corrugation). Some progress was also made in replacing the standard capacitor material (silicon dioxide) with substances having higher dielectric constants. Since the capacitance of a layer is proportional to the product of its area and its dielectric constant, doubling the dielectric constant is equivalent to doubling the area. Tantalum oxide (Ta_2O_5) appeared until recently to be the material of choice. However, its relative permittivity (dielectric constant) is only about 20, and the prospect of using more complex ferroelectric oxides, such as $PbZr_{1-x}Ti_xO_3$ (PZT) or $Ba_{1-x}Sr_xTiO_3$ (barium strontium titanate or BST) with relative permittivities of 800–1300 has obvious appeal. (Ferroelectric is a misnomer, as the materials contain no iron. This term arose because ferroelectrics have a net polarization that can be switched in an electric field, just as ferromagnets have a net magnetization that can be reversed in a magnetic field.)

BST-based DRAMs are produced at 64- and 256-Mb densities. The materials are utilized in the form of thin films of fine-grained ceramics (with thicknesses of 25–150 nanometers and grain diameters of 40–100 nm). Both sputtering and metal-organic decomposition (MOD) processing is used, with excellent results. PZT is also used in prototype DRAMs, but this material has some disadvantages with respect to BST, including the volatility and toxicity of lead, relatively poor performance at very high frequencies (at a 100-MHz internal clock rate, a DRAM capacitor must have low loss and frequency-independent dielectric constant), higher leakage currents, and generally poor electrical characteristics when fabricated in the very thin films (thinner than 100 nm) required for DRAM capacitors. Undoped PZT typically has dc leakage currents of 1–10 mA/m^2 at 150 nm thickness, about 10 times too great for DRAM device requirements. This can be improved via donor doping, but the surface-layer characteristics and dielectric-electrode interface parameters make BST the preferred material.

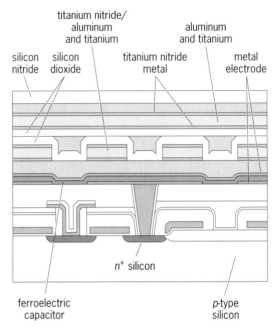

titanium nitride/
aluminum
and titanium

aluminum
and titanium

silicon
nitride

silicon
dioxide

titanium nitride
metal

metal
electrode

n^+ silicon

ferroelectric
capacitor

p-type
silicon

Fig. 1. Planarized 256-Mb DRAM structure with ferroelectric capacitor. (*After J. F. Scott, Ferroelectric memories, Phys. World, pp. 46–50, February 1995*)

With the use of high-permittivity (ferroelectric) capacitors, chips up to 256 Mb can be completely planarized (**Fig. 1**). (At 100 nm thickness, BST will provide 120 femtofarads per square micrometer, thus satisfying the 30 fF per 0.5 × 0.5-μm cell requirement of most chip designers.) This saves approximately 200 processing steps, and should improve yield and reduce costs. For the 1-gigabit chip, to be introduced around 2001, high-permittivity films such as BST must be used in addition to stacking and trenching.

Nonvolatile devices. In contrast with their use in DRAMs, ferroelectrics in nonvolatile devices provide the actual storage medium itself. Ferroelectrics are crystals in which the electric polarization (dipoles) can exist in two or more thermodynamically stable configurations, and in which application of an external voltage permits switching between these configurations. In a computer memory, these two positive and negative electrical states can be used to store the 1 and 0 of the binary boolean algebra used. This physical mechanism is analogous to the use of positive or negative magnetic spin domains in ferrite core memories; but ferroelectrics have the advantage that their switching is voltage driven, rather than current driven, so that they use much less power than magnetic memory devices. In nonvolatile memories the high dielectric constant of most ferroelectrics is irrelevant; in fact, it can actually be a disadvantage. Consequently, the ideal ferroelectric material for nonvolatile RAM applications (in which it is the storage medium) is generally different from the ideal ferroelectric material for DRAM applications (in which it is simply a capacitor). Ferroelectric memories offer great improvements in erase and rewrite speeds (less

than 100 ns, compared with milliseconds for EEPROMs) and much longer endurance (10^{13} rewrites, compared with 10^5 for EEPROMs). At present, the best ferroelectric material for nonvolatile memory applications is $SrBi_2Ta_2O_9$. This layer-structure perovskite (**Fig. 2**) exhibits superior retention (shelf life), little if any fatigue (degradation with repetitive use), and excellent performance in 95-nm thicknesses. In addition, since it is lead free, worker safety or toxic waste disposal issues are eliminated. (An alternative material is $SrBi_2NbTaO_9$.)

Ferroelectric FETs. The nonvolatile ferroelectric memories discussed above employ a transistor pass-gate circuitry and a destructive read-out oper-

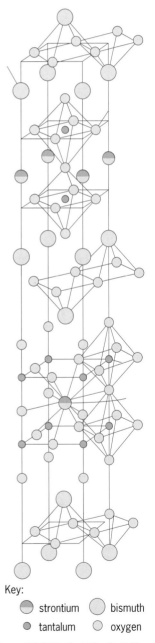

Key:
◑ strontium   ◯ bismuth
● tantalum   ○ oxygen

Fig. 2. Structure of $SrBi_2Ta_2O_9$. (*After C. A. Paz de Araujo et al., Fatigue-free ferroelectric capacitors with platinum electrodes, Nature, 374:627–629, 1995*)

ation. In order to read the stored information in each cell, it is necessary to switch it, compare the displacement current with that from a reference cell, and thereby determine whether the cell polarizations are the same. Subsequently the addressed cell must be rewritten in its original state. For applications in which there are many reads (more than 10^{12}) but very few writes (less than 10^5), such a destructive read-out (DRO) device will suffer unnecessary fatigue. For such applications a nondestructive read-out (NDRO) operation would be a significant improvement. Ferroelectrics ($BaMgF_4$ and $LiNbO_3$) have also been used in prototype NDRO devices. Such a device is basically a silicon field-effect transistor (FET) in which the metal gate has been replaced with a ferroelectric thin film (**Fig. 3**). The source-drain current magnitude is increased or reduced when the polarization of the ferroelectric gate is switched. When the gate has negative polarization, down toward the silicon substrate, the electrons in the source-drain current are repelled, taking a less direct path to the drain and thereby reducing the current. For positive gate polarization the source-drain current is increased. Hence, by simply monitoring the source-drain current, it is possible to read the polarization of each cell, and thus the 1 or 0 stored in each bit.

Although this NDRO ferroelectric FET is clearly an improved device in comparison with the DRO pass-gate transistor ferroelectric memory, it is more difficult to fabricate. The pass-gate device has the ferroelectric film in direct contact with a metal electrode (usually platinum on a titanium adhesion layer); there is no direct ferroelectric-silicon or ferroelectric–gallium arsenide (GaAs) junction. However, with the ferroelectric FET the dielectric is directly on silicon. In this situation the electrical properties of the interface may be dominated by surface states, which are electron levels at the interface that can trap electrons. The trap density for such states (that is, the number of such traps per cubic centimeter) in a typical ferroelectric oxide is approximately 1000 times greater than for ordinary silicon devices. This can preclude successful operation, since the result of applying a voltage to the ferroelectric will be to fill or empty the traps

and not to reverse the lattice polarization. To minimize this problem, a buffer layer, usually of silicon dioxide (SiO_2), is put between the ferroelectric and the silicon. However, the use of such a buffer layer is not without disadvantage. Since much of the applied source-gate voltage will drop across the buffer layer and not across the ferroelectric, a larger applied voltage will be required for switching, making it more difficult to fabricate devices operating at the 2.5- or 3.0-V standard for high-density silicon or gallium arsenide chips, respectively. These complications have thus far delayed full commercial development of ferroelectric FET nondestructive read-out nonvolatile memories. However, excellent prototypes have been fabricated with an output conductance of 0.49 microsiemens at 2.0-V gate voltage, amplification factor of 178, and 1.15-V threshold operation.

Applications. Most applications of thin-film ferroelectrics have involved silicon devices, but full integration of ferroelectric barium strontium titanate into gallium arsenide devices was made in 1993 in the form of bypass capacitors for five different 0.8–2.3-GHz microwave monolithic integrated circuits (MMICs) for television and digital telephones. Thus, gallium arsenide technology (which lacks EEPROMs) may soon benefit significantly from the introduction of these new materials. A 500-picofarad BST capacitor was also utilized in an 8-bit silicon microprocessor in 1993. As a result, considerable experience has been gained in the processing compatibility of these substances prior to their introduction into 64- or 256-Mb DRAMs and 256-kb ferroelectric nonvolatile RAMs.

For background information *SEE COMPUTER STORAGE TECHNOLOGY; FERROELECTRICS; INTEGRATED CIRCUITS; PEROVSKITE; SEMICONDUCTOR MEMORIES; TRANSISTOR* in the McGraw-Hill Encyclopedia of Science & Technology.

James F. Scott

Bibliography. C. A. Paz de Araujo et al., Fatigue-free ferroelectric capacitors with platinum electrodes, *Nature,* 374:627–629, 1995; J. F. Scott, Ferroelectric memories, *Phys. World,* pp. 46–50, February 1995; J. F. Scott and C. A. Paz de Araujo, Ferroelectric memories, *Science,* 246:1400–1405, 1989; J. F. Scott, C. A. Paz de Araujo, and L. D. McMillan, Ferroelectric memories, *Condens. Matter News,* 1:16–21, 1992.

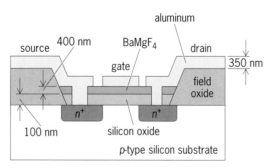

Fig. 3. Ferroelectric field-effect transistor in which a ferroelectric gate permits nondestructive read-out (NDRO) operation. (After J. F. Scott, Ferroelectric memories, Phys. World, pp. 46–50, February 1995)

Conductor (electricity)

The significance of the thermal behavior of an electrical conductor is that the current-carrying capacity of the conductor often depends on the maximum temperature that it attains and the period for which that temperature is held. In the case of an insulated conductor, the temperature and its duration may cause chemical and mechanical degradation of the insulation, leading to electrical breakdown of the circuit. With a bare conductor, higher temperatures may cause excessive sag, leading to a dangerous reduction in the clearance to ground, grounded objects, or other energized circuits. Also, extended periods of elevated temperatures may cause annealing of the material of the conductor, with a consequent reduction in its mechanical strength.

The current-carrying capacity (rating) of an insulated conductor is based on the temperature of the outer surface of the conductor, hence on the inner surface of the insulation. The rating of a bare conductor is usually based on the average temperature of the conductor, or it is assumed that the conductor is isothermal. This assumption is reasonable for a solid conductor but not for a conductor consisting of many wires stranded together in concentric layers. In this case, the restrictions to the radial heat paths, where the wires in adjacent layers cross one another, result in an effective radial thermal conductivity which is about $\frac{1}{100}$ of that for a solid conductor. In other words, the temperature difference between the center and the surface of a stranded cylindrical conductor is about 100 times that for a solid conductor having the same diameter. This temperature difference does not usually exceed about 10% of the temperature rise of the surface above ambient, and amounts to a few degrees Celsius. The apparent area of each wire-wire contact is very small, of the order of 1 millimeter squared. The area of the metal-metal contacts (**Fig. 1**) is about $\frac{1}{1000}$ of this value, the remainder of the area being occupied by air gaps. Since these air gaps are only about 1 micrometer thick, most of the heat flows to the surface of the conductor across the air gaps, rather than across the metal-metal contacts. The heat transferred to the surface of the conductor by radiation, conduction, and convection across the air voids between the wires is negligible.

Steady-state heating. The balance between the heat gains and the heat losses determines the relationship between the current and the temperature of a conductor. Two terms are neglected, because they occur sporadically and randomly. These are the heat lost by mass transfer during condensation and rain, and the heat gain due to electrical corona on polluted or wet conductors at high voltage. The heat gains comprise the rate of generation of resistive (joule) heat (P_J); the rate at which heat is generated by the magnetic field in a steel-cored conductor, in the form of hysteresis and eddy-current losses (P_M); and the total rate at which the conductor receives heat from the Sun (P_S). The heat loss rates comprise those for radiated heat (P_R) and convected heat (P_C). For a unit length of conductor, these quantities satisfy Eq. (1).

$$P_J + P_M + P_S = P_R + P_C \qquad (1)$$

This appears to be a simple equation, but it can be quite complex to solve. For a solution, either the current or the temperature of the conductor must be assumed. If a value is assumed for the temperature of the surface of the conductor, and, moreover, if the ambient temperature is known or assumed, then the temperature rise of the surface of the conductor is the difference between the surface and ambient temperatures. The radiated heat rate can then be calculated as the product of the difference between the fourth powers of these two temperatures, the surface area, the Stefan-Boltzmann constant, and the emissivity of the surface. The emissivity depends on the condition of the surface, and can vary from about 0.1 when the surface is bright to almost 1.0 when it is very weathered.

The rate of convected heat depends on whether the conductor is in still air (natural convection), or whether it is exposed outdoors (forced convection). In still air, the convected heat depends mainly on the diameter of the conductor and the temperature rise of its surface above the ambient temperature. In moving air, the rate of convected heat also depends on the speed and direction of the wind. When the wind speed is less than about 1 m/s (3.3 ft/s), it may be necessary to take account of both natural convection and forced convection.

The next step is to calculate the solar heat received by the conductor. This amount varies during the day, and is, of course, zero at night. The solar heat depends on the intensity of the solar beam, the altitude of the Sun, the diffuse radiation from the sky dome, the radiation reflected from the ground, the inclination of the conductor to the horizontal, and the azimuths of the Sun and the conductor. The heat absorbed by the conductor depends on the nature of its surface: the absorptivity varies from about 0.3 with a bright surface to almost 1.0 with a very weathered surface.

Having found the magnetic and solar heat gain rates and the radiative and convective heat loss rates, the resistive heat gain rate can be found by subtraction. Because this rate is the product of the square of the current and the resistance of the conductor, the current can then be determined. Although this seems a complicated way to find the rated current, it should be realized that some simplifications have already been made; for example, it is assumed that the wind does not fluctuate in speed and direction, and that it is nonturbulent. For design purposes, the calculations are simplified by assuming standard values for the more important parameters: for example, crosswind speed, 0.6 m/s (2 ft/s); intensity of total solar radiation, 1000 W/m^2 (93 W/ft^2); ambient temperature, 35°C (95°F); conductor surface temperature, 80°C (176°F); and

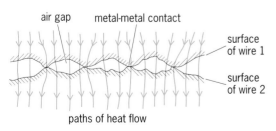

Fig. 1. Conduction of heat between contacting metallic surfaces.

emissivity = absorptivity = 0.5. It is obvious that this combination of parameters will rarely occur, so that this type of rating is inherently conservative; that is, for a certain design current, the conductor temperature will be less than the design temperature for most of the time. This situation can be partially rectified by allowing for the variation of the solar radiation and the ambient temperature with the seasons. Although there is no solar heating of the conductor at night, it is notable that the wind speed often falls to near calm, resulting from the absence of solar heating of the ground.

It has been assumed that the current is uniform across the cross section of the conductor. This assumption is almost true for a monometallic conductor carrying a direct current, although there will be a small radial nonuniformity, because of the radial temperature distribution. A monometallic conductor carrying an alternating current experiences a current distribution which increases toward the surface. This skin effect, which increases with increasing frequency, is due to the internal inductance of the conductor. When conductors, such as busbars, are close together, mutual inductance between them causes marked nonuniformity of the current density, known as the proximity effect. Many transmission-line conductors have a steel core to support a large part of the mechanical tension. The magnetic flux that is induced in this core gives rise to a redistribution of current within the conductor, known as the transformer effect. In a stranded steel-cored aluminum conductor with three layers of aluminum wires, the current density in the center layer increases by about 30% at the expense of the inner and outer layers. All these nonuniformities of the current will affect the temperature distribution within the conductor.

In addition to the radial temperature gradient within a conductor carrying a current, there will also be circumferential and axial temperature distributions. The circumferential distribution results from the natural or forced convective flow around the surface and the effect of other close conductors or surfaces. The axial temperature distribution is caused by heat conduction to end connectors, spatial variations in wind speed and direction, and the effect of moving clouds on the solar heat gain.

Dynamic heating. Higher current can be carried by a conductor for a short period without overheating. This is useful in an emergency, for example, when a transmission line is out of service, and other lines have to compensate by carrying higher current than normal. Many electricity supply utilities employ 10- or 20-min conductor ratings to meet such contingencies. When the current is increased, the temperature of the conductor increases at a rate that depends on the total heat transfer coefficient and the heat capacity, that is, the product of the mass per unit length (m) and the specific heat (c). The heat equation now becomes Eq. (2), where

$$P_J + P_M + P_S = P_R + P_C + mc(d\theta/dt) \qquad (2)$$

$d\theta/dt$ is the rate of change of the mean temperature rise of the conductor (above the ambient temperature), which may be positive or negative.

It is common practice to base the temperature rise after the change in the current, or one or more of the meteorological parameters, on the thermal time constant, which is the time duration for the temperature rise of the conductor above ambient to reach 63.2% of its final value. Strictly speaking, this assumes that the radiative heat loss is negligible compared with the convective loss, and that the rate of increase of the temperature rise above ambient decreases exponentially. This assumption is generally not true for a conductor in still air, particularly if the emissivity of the surface is high. It is often assumed that the thermal time constant of the cooling of a conductor following a decrease in the current or in a meteorological parameter is identical to that for the heating following an increase in the current or in a meteorological parameter. However, the time constant during heating is greater than that during cooling, because the resistive heat gain during heating increases as the temperature of the conductor increases.

Probabilistic heating. A further increase in the rating of a conductor can be obtained by taking account of the statistical variation of the electrical loads in the power system and the meteorological parameters, such as wind speed and direction, intensity of total solar radiation, and ambient temperature. The temperature history of the conductor can be measured or calculated, assuming mutual independence between the parameters. For a known statistical variation of current, the risk of exceeding the maximum permissible temperature of the conductor can then be determined. **Figure 2** shows the measured statistical variations (probability density functions) of the surface temperature rise (above the ambient temperature) of an exposed horizontal 90-m (295-ft) span of a conductor having a core of 7 steel wires, each of 3.5-mm (0.14-in.) diameter, surrounded by 54 aluminum wires, each of 3.5-mm (0.14-in.) diameter, in three layers, carrying a constant 50-Hz current of 1500 A for day and night over a period of 32 months. It is seen that there is greater probability of lower temperature rises during the day, because of the higher wind speeds resulting from solar heating of the ground. At night, higher temperature rises are more likely as the wind speed falls.

Adiabatic heating. When the current in a conductor increases rapidly to a high value, for example, following a short circuit or a lightning strike, the temperature increases at a fast rate. The total heat losses, and any external heat gain, may then be negligible compared with the resistive heat gain, which increases with the square of the current. The temperature of the conductor then depends on the current density and its rate of change, and the variation with temperature of the thermophysical properties of the conductor material, such as the specific heat, the thermal conductivity, and the mass density.

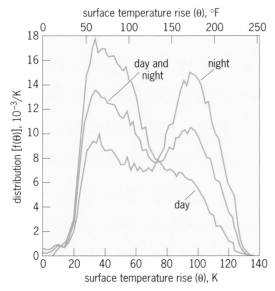

Fig. 2. Statistical distribution of the average surface temperature rise (above ambient temperature) of an exposed conductor carrying a constant current. (*After V. T. Morgan, Statistical distribution of the temperature rise of an overhead-line conductor carrying a constant current, Elec. Power Sys. Res., 24:237–243, 1992*)

If the cross-sectional area of the conductor is small, the temperature may reach the fusion temperature, and the whole conductor may melt isothermally, or part of it may rupture because of the electromagnetic force and the reduction in tensile strength. A thin wire may form unduloids (geometric figures formed by the revolution of a wavy line around a straight line parallel to its axis of symmetry), which may separate into a chain of beads because of surface tension. With even higher current, or with smaller cross section, the temperature of the conductor may reach the boiling temperature, and the conductor vaporizes. The vapor channel has a high resistance, so that the current is cut off, or the vapor is expelled by the electromagnetic force. This phenomenon is known as an exploding wire.

For background information SEE CONDUCTION (HEAT); CONDUCTOR (ELECTRICITY); CONVECTION (HEAT); EDDY CURRENT; JOULE'S LAW; MAGNETIC HYSTERESIS; SKIN EFFECT; TRANSMISSION LINES in the McGraw-Hill Encyclopedia of Science & Technology.

Vincent T. Morgan

Bibliography. V. T. Morgan, The overall convective heat transfer from smooth circular cylinders, *Adv. Heat Trans.*, 11:199–264, 1975; V. T. Morgan, The radial temperature distribution and effective radial thermal conductivity in bare solid and stranded conductors, *IEEE Trans. Power Deliv.*, PWRD-5:1443–1452, 1990; V. T. Morgan, Statistical distribution of the temperature rise of an overhead-line conductor carrying a constant current, *Elec. Power Sys. Res.*, 24:237–243, 1992; V. T. Morgan, *Thermal Behaviour of Electrical Conductors*, 1991.

Corrosion

Aircraft are designed to carry passengers, cargo, stores, and crew to destinations safely and at a reasonable cost. Since their advent early in the twentieth century, significant strides have been made in developing aircraft that can carry passengers, cargo, and crew for greater distances in shorter times and with greater efficiency. As with all technological developments, these improvements have come at a price in terms of both money and human lives. In 1928, the design of aircraft was changed significantly by the crash in Germany of an aircraft that resulted in the death of all on board. The crash was found to have resulted from the failure of a critical component by fatigue (the application of cyclic loads to a component or system of the aircraft). Since then, a great deal of attention has been focused on the factors that impact the integrity and durability of aircraft, and in particular on four principal time-related failure modes that may occur in the metallic components used in aircraft. These four failure modes are corrosion, creep, fatigue, and wear. Also, interactions of these four means of degradation have been dealt with to some extent. This article focuses on the issue of corrosion of aging aircraft.

Aircraft environments. Naturally, aircraft would be expected to age since components that make up the aircraft are subjected to forces that result from the varying loads imposed on the structural members from flight, landing, and so forth. In addition, portions of the aircraft are exposed to varying and high temperatures as well as interactions between components. The former results in creep deformation and the latter result in wear phenomena. Chemical environments also impact the integrity and durability of the components. The potential for corrosion is evident from a consideration of the possible chemical environments that the inside and outside components of aircraft encounter throughout their operation. Some aircraft fly through rain, chemically contaminated environments (smog for example), snow, and salt-laden air (near or over oceans and salt-water bodies such as the Great Salt Lake). In addition, aircraft landing gear and the lower parts of the fuselage are exposed to runway fluids and elements of the environment such as salt and salt water. Some interior aircraft components are exposed to complex chemical environments from moisture condensation from human and animal breathing, galley spills, lavatory spills, fuel, and cargo fluid spills. The chemical environments to which various components and systems on aircraft are subjected vary considerably.

Conditions for corrosion. Aircraft are primarily constructed out of metallic materials such as aluminum, titanium, or iron alloys. Some propulsion components are manufactured from nickel-base superalloys as well. Electronic and avionics components are made of metals, polymers, and ceramics in various combinations. All these materials are subject to interaction with the chemical environment.

When certain electrochemical conditions are fulfilled, corrosion can occur. The primary such conditions are (1) the availability of both an anode and a cathode, (2) the presence of an electrolyte (a chemical, usually a liquid or gas), (3) a connection between the anode and the cathode (which usually occurs through the electrolyte), and (4) a flow of electrons.

Even though the aluminum and titanium alloys have a protective layer of oxide, this layer can be broken by the loads on aircraft, by wear, and so forth. This situation permits the environment to contact the materials and start the corrosion processes. Corrosion occurs in several main types, including general attack, pitting, exfoliation, filiform, stress-corrosion cracking, fretting, intergranular attack, oxidation, and erosion. Several of these types involve interaction of the by-products of the chemical reactions that occur, and these chemical species may diffuse and may embrittle the host material.

Effects of corrosion. Corrosion may give rise to three major technical problems related to the safety and integrity of critical aircraft components (aside from its possible adverse effects on aircraft appearance). These problems are as follows:

1. A reduction of cross-sectional area may significantly elevate the stress on the component.

2. Localized discontinuities may create a local elevation of stress, and lead to the nucleation of cracks. The cracks may propagate, and their rate of propagation may be affected by the chemical environment.

3. The chemical species that diffuse into the material may cause embrittlement that changes the toughness and strength of the material.

Each of these processes may have a significant effect on the integrity of the components that are viewed as critical in terms of the magnitude of stress placed upon them. These processes also may change the properties of critical components as well as the assessment of critical locations.

Response to corrosion problem. The problem of corrosion has been addressed through the development of a corrosion prevention and control program for each aircraft, and to a degree this effort has been successful. However, a long history of failures since the 1960s has shown that aircraft corrosion has not been dealt with adequately, because in part of a lack of emphasis on this problem in both aircraft design and the education of engineers. Since the 1980s, more attention has been focused on aircraft corrosion in order to prevent accidents and incidents and also to decrease corrosion-related maintenance costs of aging aircraft.

Corrosion-related accident. On April 28, 1988, near Hawaii, the crown section of the fuselage of a 737 aircraft was ripped off because of a sequence of events that led to the generation of multiple fatigue cracks at the rivet holes. One of the major factors responsible for this situation was the occurrence of corrosion between sections of the fuselage skin either concomitant with or after the breakdown of adhesive bonds holding the skin together. Not only was this accident significant in itself, resulting in the loss of one life, but it also served as a clear signal that something was not receiving adequate attention in the design, maintenance, and inspection of aircraft. Furthermore, the principle that was developed after the first fatigue failure of an aircraft, namely, the safety-by-inspection principle, was now in question. This obviously had significant safety and durability implications.

Response to accident. After the Hawaii accident, a series of fleet surveys resulted in significant findings. The most notable finding was the extent of corrosion in aircraft of all types, and in many places on aircraft where it was not expected. This applied not only to commercial aircraft but to military aircraft as well. Another key finding was the detection of corrosion in structural members that were not viewed as critical with regard to fatigue or strength (mainly residual strength). However, with corrosion they now became critical components and critical areas with regard to safety. This finding prompted aircraft manufacturers, operators (including the military and the National Aeronautics and Space Administration), regulatory authorities, and the technical community to develop improved procedures to deal with corrosion, especially on aircraft that had been used for some years and had either exceeded or were close to exceeding their lifetimes based on the original design criteria. Operators were utilizing aircraft much longer than expected, and so additional procedures were required to deal with the potential for increased corrosion with age of the aircraft. Research and development was also accelerated into aspects of corrosion that had not previously received enough attention. These areas included (1) evaluation of means to characterize corrosion damage and develop procedures that assess its significance on aircraft integrity, (2) development of better means of nondestructive inspection to detect corrosion, (3) characterization of the chemical environments that impact the potential for corrosion to occur in aircraft components, and (4) evaluation of the components of aircraft that are most likely to experience corrosion and the development of an archive of information on these components. Numerous additional research and development activities are under way to assure that corrosion will not significantly affect the high standard of aircraft safety.

For background information SEE AIRCRAFT DESIGN; CORROSION; METAL, MECHANICAL PROPERTIES OF in the McGraw-Hill Encyclopedia of Science & Technology.

David W. Hoeppner

Bibliography. National Transportation Safety Board, *NTSB Aircraft Accident Report, Aloha Airlines, Flight 243, Boeing 737-200, N73711, Near Maui, Hawaii, April 28, 1988,* NTSB/AAR-89/03, 1989; W. Wallace, D. W. Hoeppner, and P. V. Kandachar, *AGARD Corrosion Handbook,* vol. 1:

Aircraft Corrosion: Causes and Case Histories, AGARD-AG-278, 1985.

Cosmic background radiation

Cosmic background radiation is remnant light from the big bang that has been traveling through the universe for more than 10^{10} years. Its discovery by A. Penzias and R. Wilson in 1964 was compelling evidence for the big bang model. Nearly 30 years later, a map showing the tiny variations in the intensity of cosmic background radiation across the sky was obtained for the first time by using instruments aboard the *Cosmic Background Explorer* (*COBE*) satellite. The map is a snapshot of the universe when it was 10^5 times younger than it is now. The snapshot can be used to further test the big bang model and discriminate among competing theories of the origin and evolution of large-scale structures in the universe.

Cosmic microwave background. According to the big bang picture, cosmic background radiation originated in an early epoch when the universe was 10^3 times hotter and 10^9 times denser than it is now, long before there were any planets, stars, or galaxies. All matter was compressed and heated into an ionized plasma of free electrons, protons, and nuclei. Light quanta or photons scattered and rescattered from the free electrons as the universe expanded and the plasma cooled. After the first 10^5 years or so, the gas cooled sufficiently that electrons and nuclei could combine to form a neutral gas of atoms. With no free electrons remaining, the photons were suddenly free to stream through the universe. Over the subsequent 10^{10} years, the continued expansion of the universe caused the photons to redshift and cool from a value of nearly 10,000 K down to 2.726 ± 0.010 K.

The hot-big-bang picture predicts that the average intensity of the radiation should vary with wavelength in the same way as an ideal blackbody. Observations by spectrophotometers aboard *COBE* and recent rocket experiments have validated the prediction with exquisite accuracy. Consequently, the tiny variations in intensity across the sky can be reliably interpreted as a direct image of the infant universe, just as the photons were last scattering from the ionized plasma.

COBE observations. The *COBE* satellite used a series of differential microwave radiometers (**Fig. 1**) to detect variations in the cosmic background intensity. Each radiometer included a pair of antenna horns pointing in two directions in the sky 60° apart. Each horn received radiation from a patch in the sky spanning roughly 10°. The radiometer electronically subtracted the intensity of the signals from the two horns and recorded the difference. The satellite traveled in orbit around the Earth and spun on its axis so that, over the course of 4 years, the whole sky was covered several times over. The satellite included three sets of radiometers operating at different wavelengths to dis-

criminate the blackbody spectrum of the cosmic background from other sources of radiation. The record of intensity differences was then converted into a cosmic background intensity map of the sky (**Fig. 2**). The shadings highlight what are really minuscule differences in intensity, less than 0.002% on average.

Significance of COBE results. The *COBE* map shows that the universe was extraordinarily homogeneous a few hundred thousand years after the big bang. Yet, surveys of the distribution of nearby galaxies show that the universe is highly inhomogeneous now, with matter clumped into large-scale structures such as galaxies, galaxy clusters, and enormous sheets of galaxies separated from one another by vast voids.

The simplest explanation of how the universe evolved from highly homogeneous to highly inhomogeneous, consistent with the big bang picture, is that gravity is the cause. Gravity draws matter into the slightly overdense regions and away from the slightly underdense ones. In this way, the degree of inhomogeneity is progressively amplified by gravity. To explain the degree of inhomogeneity seen today, however, there must be sufficient inhomogeneity at the 10^5-year mark. According to theoretical calculations, the *COBE* result of 0.002% intensity variation is just enough. Hence, *COBE* is important new evidence supporting the big bang picture and the notion that gravity plays the essential role in making the large-scale structure of the universe.

The understanding of this evolution is complete only if there is an explanation of how the initial inhomogeneities were created. The big bang picture offers no information. In fact, according to the big bang model, there was not sufficient time after only 10^5 years for light or matter or any physical forces to propagate across the expansive patches of high and low intensity shown in Fig. 2. Hence, the patches cannot have been created by any ordinary dynamical processes in a big bang universe.

The realization that the big bang model is an incomplete description of the universe has led to several alternative cosmological models. One leading candidate is the inflationary universe model.

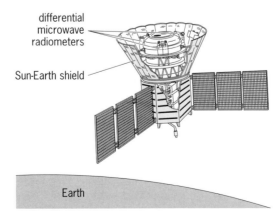

differential microwave radiometers

Sun-Earth shield

Earth

Fig. 1. Cutaway of the *COBE* satellite, showing the differential radiometer antennas. (*G. Smoot*)

According to this model, the universe occupied a microscopic region in the first instants after creation, much tinier than predicted by the big bang model. The region was sufficiently compact that light and other physical processes could propagate back and forth across the entire patch, smoothing out the initial distribution of matter and energy. At the same time, quantum effects, similar to those that cause fluctuations in the position and momentum of an electron orbiting a nucleus, created random fluctuations in energy throughout the small patch. Then, according to the inflationary model, the universe underwent a brief epoch of extraordinarily rapid expansion (inflation) during which the universe stretched at least 10^{100} times faster than the standard big bang picture. The tiny patch grew to cosmic proportions. When inflation stretched the patch, it also stretched the quantum-generated inhomogeneities. A key prediction is that the intensity difference between two patches in the sky is nearly independent of the angle between the patches. A pattern with this property is called nearly scale invariant.

An important conclusion from *COBE* was that the map of cosmic background intensity is consistent with a nearly scale-invariant pattern. However, other cosmological models are also in agreement. The 10° resolution of the *COBE* radiometers washed out intensity variations on finer angular scales that could discriminate among the models.

Cosmic fingerprint. The goal after *COBE* is to obtain a high-resolution, all-sky map of the cosmic background intensity. The map shows the intensity, $I(\theta,\phi)$, where θ and ϕ are the two angles which label sky position in spherical coordinates. The function $I(\theta,\phi)$ can be written as a discrete sum of spherical harmonic functions, $Y_{lm}(\theta,\phi)$. The decomposition into spherical harmonics is similar to a Fourier decomposition into sinusoidal functions, except that $I(\theta,\phi)$ depends on angles rather than position

or time. Because there are two angles, the spherical harmonic functions are parameterized by two integers, l and m, with $-l \le m \le l$. The coefficients in the decomposition, a_{lm}, depend on the orientation of the coordinate system, whereas the combination $C_l = (2l + 1)^{-1} \Sigma_l |a_{lm}|^2$ does not. The C_l's are called multipoles.

A plot of the multipoles, usually shown as $l(l + 1)C_l$ versus l, is a coordinate-independent fingerprint of the intensity pattern, the key graph that is used to test decisively cosmological models. Roughly, multipole C_l is the average difference in intensity between two points in the sky separated by the angle π/l radians or $180°/l$. Different cosmological models predict different dependencies on angle.

The fingerprint predicted by the inflationary model of the universe is shown in **Fig. 3.** The leftmost side ($l = 2$) is the average intensity difference between points separated by nearly 90°. Scanning toward the right shows how the difference changes for smaller and smaller angles. The key features are a plateau at small l and a series of peaks beginning at $l \sim 200$. The plateau is characteristic of a model that predicts a nearly scale-invariant spectrum of initial inhomogeneities. The peaks beginning at $l \sim 200$, corresponding to intensity differences between points separated by roughly 1°, are characteristic if the models also predict that the energy density of the universe is at the critical value.

Figure 3 also displays results from a wide variety of cosmic background experiments. *COBE,* the first, is the only space-satellite experiment at present. The others include ground-based and high-altitude balloon-borne experiments at the South Pole, the Canary Islands, and Saskatoon, Saskatchewan, Canada. Each experiment is sensitive to a certain range of angles or multipoles, depending on the experimental design. At present, the inflationary model of the universe fits consistently with the data,

Fig. 2. Projection of the full sky showing the variations in the cosmic background intensity measured by *COBE.*

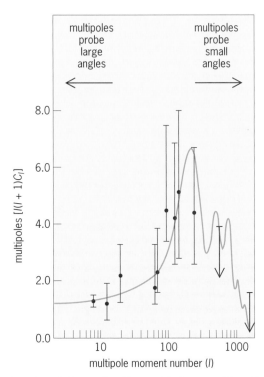

Fig. 3. Cosmic fingerprint of how multipoles vary with multipole moment number (l), as predicted by the inflationary universe model compared to measurements by *COBE* (leftmost data point) and subsequent balloon- and ground-based experiments through 1994.

and fits better than competing models. However, the present experimental uncertainties are large. All but the two rightmost measurements are claimed detections of intensity variation, but the amplitude is poorly resolved. The two rightmost measurements, representing the highest-angular-resolution experiments available, are shown as upper bounds because they are unable to detect any intensity variations so far. Substantial improvement is needed before the fingerprint curve is finely resolved and definitive conclusions can be reached.

Prospects. Great improvements in experimental resolution are anticipated. There are plans to launch new satellite experiments which can resolve much finer angles than *COBE*. There are plans for long-duration balloon flights that will circumnavigate Antarctica for several weeks and support larger, higher-resolution experiments. There are also plans for very large ground based experiments at the South Pole, using large arrays of radio interferometers to measure patches of the sky with resolutions of a few arc-minutes.

The highly precise cosmic fingerprint obtained from these experiments may validate one of the current models, such as the inflationary universe theory, making it possible to trace the evolution of the universe back to the first instants after creation. Alternatively, the fingerprint may force consideration of totally unanticipated ideas. In addition, the shapes and positions of peaks and dips in the fingerprint are highly sensitive measures of cosmic

parameters, such as the cosmic energy density, expansion rate, and density of ordinary and dark matter. Finally, when joined with the ongoing projects to measure the redshifts of millions of galaxies and obtain a three-dimensional map of the universe, the cosmic background measurements will reveal the long-sought secrets of how large-scale structure in the universe was formed.

For background information *SEE BIG BANG THEORY; COSMIC BACKGROUND RADIATION; COSMOLOGY; INFLATIONARY UNIVERSE COSMOLOGY; SPHERICAL HARMONICS; UNIVERSE* in the McGraw-Hill Encyclopedia of Science & Technology.

Paul J. Steinhardt

Bibliography. E. W. Kolb and R. Peccei (eds.), *Particle and Nuclear Astrophysics and Cosmology in the Next Millennium*, 1995; J. P. Ostriker (ed.), *Some Unsolved Problems of Cosmology*, 1995; G. Smoot and K. Davidson, *Wrinkles in Time*, 1993.

Cosmology

The density of matter in the universe has been a major concern in the field of cosmology. This density determines the future evolution and ultimate fate of the universe. Observations of the motions of stars and galaxies indicate that most of the matter is invisible, the so-called dark matter, and theoretical arguments suggest that the universe has exactly the critical density, at which its expansion will gradually slow to a halt but never reverse. The first section of this article discusses searches for dark matter through observations of the phenomenon of gravitational lensing. The second section examines the relations between the geometry, evolution, and density of the universe, and discusses an alternative version of the inflationary universe cosmology that allows for a subcritical universe, in which the density of matter is less than the critical density.

Dark matter in the universe. Astronomers studying the motions of stars and galaxies throughout the universe have long noticed a strange discrepancy. Stars move around galaxies as if pulled by 10 times the matter that can be seen. Galaxies move around other galaxies in clusters at speeds implying up to 30 times the visible mass of the cluster. The force of gravity at many size scales seems to exceed the amount of visible matter by more than an order of magnitude. The curious implication of these results, that most matter in the universe is invisible, has led physicists and astronomers to hunt for the dark matter. However, a search for an invisible entity is intrinsically difficult. Theorists have proposed viable constituents ranging from black holes 10^6 times more massive than the Sun to swarms of exotic subatomic particles 10^{-8} the weight of a hydrogen atom. Many such proposals, if verified, would offer further, testable predictions about the origin and fate of the universe. *SEE BLACK HOLE.*

Although several independent lines of evidence have made the big bang a well-established theory of cosmogony, knowledge of the fate of the universe awaits a measurement of its overall density. If there is insufficient matter to cause a reversal of the expansion initiated in the big bang, the universe will continue to expand, eventually (in about 10^{11} years) cooling into blackness. But given a high-enough density, the expansion will instead stop and the universe recollapse, ending finally in a big crunch. The critical density is the value between these two destinies, where the expansion halts but never quite reverses. Although to many such a finely tuned cosmos seems unlikely, some very powerful theories (inflationary models) suggest that the universe has exactly the critical density. Other inflationary models, discussed below, do not require this. Searches for dark matter are designed to help decide between these competing theories and their predictions.

Gravitational lensing. Because dark matter, by definition, cannot be directly observed, several searches have exploited the predicted gravitational effects of its mass on the observed light from nearby stars. The fundamental idea is not new; A. Einstein first predicted that mass can bend light. Experimental confirmation of this aspect of his general theory of relativity came during a 1919 total solar eclipse, when the images of distant stars passing near the limb of the Sun were displaced by the predicted amount. Extragalactic examples of this gravitational lensing have also been seen. Lensing amplifies the apparent brightness of the distant light source by bending light rays toward the observer's line of sight. Although the relative positions of the three elements of a gravitational lens (observer, lensing mass, and background lensed light source) critically affect the brightness amplification factor, the apparent relative motions of extragalactic objects such as galaxies and quasars are so slow that observable changes in brightness are not expected in a human lifetime.

If the dark matter pervasive in galaxies is composed of normal atomic (baryonic) matter in clumps such as planets, stars, or black holes, then these masses can also be expected to produce gravitational lensing locally. Because the motions of stars suggest that large amounts of dark matter diffused in a halo around the Milky Way Galaxy, occasionally a star in the Galaxy should appear to brighten as a dark compact object crosses the line of sight from the Earth to the star (**Fig. 1**). Because the mass of a typical star is at most 10^{-11} the mass of a typical galaxy, the strength and duration of the lensing are much smaller than for the extragalactic case. Lensing by compact objects such as stars and planets is thus called microlensing. To produce detectable brightening, the alignment of observer, lensing mass, and lensed star must be less than a micro-arcsecond, about the angular size of a dime 2×10^6 mi (3×10^6 km) distant. A wide variety of galactic models predict that such alignments in the Galaxy should be extremely rare.

Detection of microlensing. To detect these rare and often weak microlensing events, millions of stars must be repeatedly observed. Such observation requires sensitive imaging that covers a wide area of sky, accurately measuring the brightness of hundreds of thousands of stars each night. Measurements of each star must be compared nightly to search for brightness variability. Many types of stars are known to vary for reasons unrelated to microlensing, and are certain to populate the observed field. Such stars include eclipsing binary star systems, pulsating giant stars, and rare episodic outbursts from more exotic systems such as dwarf novae and x-ray binaries. A large number of variable stars are being discovered in ongoing searches for galactic microlensing, providing a valuable database for understanding the physically revealing phases in the lives of stars. But when the goal is the observation of dark matter, other varying stars primarily act as contaminants. Fortunately, the brightness changes expected from microlensing can be distinguished with only a little extra effort.

Before a change in brightness of a star can be confidently identified as microlensing, it must pass a battery of tests. First, since an alignment occurs only in passing, the event must not repeat. Second, the brightening and subsequent dimming over time (the light curve) must follow a well-defined symmetric curve. By convention, the maximum change in brightness must exceed 30% for the event to be considered a detection. Third, the light curve must be achromatic; the shape and amplitude of the variations must be the same when observed through filters of different colors. This last test requires the

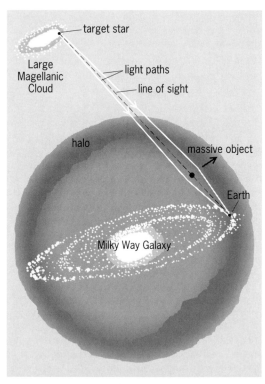

Fig. 1. Massive object in the halo of the Milky Way Galaxy traversing the line of sight to a distant target star, bending and focusing its light.

observer to monitor every star in two different colors throughout the event.

MACHO, EROS, and OGLE experiments. Three teams of astronomers are trying to acquire a large number of microlensing events to further understanding of dark matter in the Milky Way Galaxy. These experiments have been given the acronyms MACHO (Massive Compact Halo Objects search), OGLE (Optical Gravitational Lensing Experiment), and EROS (Experience de Recherche d'Objets Sombres, or Research Experiment on Dark Objects).

The MACHO team uses an electronic camera called a charge-coupled device (CCD), related to the light-sensitive imaging devices found in video cameras. The difference is that the team's camera is much larger and more sensitive. With eight CCD arrays, each with 2048 picture elements on a side, more than 3.3×10^7 separate numbers are generated for each image. This enormous camera is used with an Australian 50-in. (1.3-m) telescope, and focused repeatedly on the Large and Small Magellanic Clouds, dwarf galaxies orbiting the Milky Way Galaxy. Every clear night, several million stars in the Large Magellanic Cloud are observed, and their brightness compared to previous measurements. Two exposures, through blue and red filters, are used to ensure that every candidate lensing event is achromatic. When the Large and Small Magellanic Clouds have swung low in the southern sky, the MACHO team also images the bulge, a spheroidal region of the Milky Way Galaxy toward its center. The nightly imaging of the MACHO experiment means that lensing objects heavier than Jupiter can be detected. *SEE CHARGE-COUPLED DEVICES.*

The EROS observations are performed in two ways. One method uses computer analysis of large-area photographic plates. These plates are also most sensitive to dark bodies more massive than Jupiter. A second method points a camera with 16 CCDs toward the Large Magellanic Cloud, taking more rapid exposures. The data should be sensitive to lensing masses as small as the Earth's.

The OGLE team uses a single 2048×2048 CCD to image several fields toward the galactic bulge, also measuring several million stars each night.

Microlensing observations. All these groups have observed very strong candidates for microlensing events. As of December 1995, about a dozen had been seen toward the Large Magellanic Cloud, and nearly 100 toward the center of the Milky Way Galaxy. The large number of lensing events toward the bulge was a surprise, and perhaps indicates a larger stellar density there than was thought. This may be evidence that the Milky Way has a barlike structure that has been seen in some external galaxies. Most predictions of the number of detectable lensing events had underestimated the effect of lensing stars far from the midpoint of the sight line to the source star. Although a lensing mass is most effective halfway to the background source, somewhat heavier objects near the source star can also amplify its light.

The results so far suggest lens masses in the range of one to a few tenths the mass of the Sun. These masses typically represent objects much heavier than Jupiter, and are most likely to be small faint stars that are common throughout the Milky Way Galaxy. Although it now seems likely that microlensing is caused by ordinary low-mass stars, much remains to be gleaned from ongoing studies. Estimates of the amount of baryonic dark matter in compact form in the Milky Way Galaxy await more accurate determinations of the search efficiency, and of the percentage of lensing events that can be seen by these experiments as a function of mass and distance. Many more lensing events along several sight lines are necessary to provide better statistics.

Paul J. Green

Subcritical universes and inflation. In the standard big bang model of the universe, it is assumed that the universe is expanding and that on very large scales, after the local clumpiness from mass clustered into galaxies and clusters of galaxies has been smoothed over, the universe is very nearly homogeneous and isotropic. Homogeneity is the property that the universe looks the same to observers at different positions but at the same cosmic time. Isotropy is the property that the universe looks the same in different directions. In general relativity these requirements of spatial homogeneity and isotropy define a narrow class of cosmological models that satisfy the Einstein equations, known as Friedmann-Robertson-Walker universes. A Friedmann-Robertson-Walker universe consists of an underlying three-dimensional space satisfying the requirements of homogeneity and isotropy that expands with increasing time. This expansion is described by a scale factor $a(t)$, where t is cosmic time. At equal cosmic time the physical distance d between two points separated by a comoving distance r in the underlying three-dimensional space is given by Eq. (1). Consequently, distances between

$$d = a(t) \cdot r \qquad (1)$$

points at rest with respect to the expansion of the universe, or the Hubble flow, increase in proportion to $a(t)$.

Geometry of the universe. The requirements of homogeneity and isotropy allow three distinct possible geometries for the underlying three-dimensional space for a Friedmann-Robertson-Walker universe. The simplest possibility is ordinary three-dimensional euclidean space, characterized by the absence of curvature. In the absence of curvature, a sphere of radius r has an area equal to $4\pi r^2$. With positive curvature this area would be smaller; with negative curvature it would be larger. Another consequence of the absence of curvature is that a vector parallel-transported around a closed curve preserves its original orientation. This property does not hold in a curved space. An example is the surface of the Earth, which is curved. If a vector lying parallel to the surface of the Earth is transported halfway around the Equator and then in a great arc of the same length through the North Pole and back to its starting point, the vector will undergo a half-rotation, despite the fact that the vector was not allowed to rotate relative to the Earth's surface during its journey.

The surface of the Earth, as any two-sphere, is a two-dimensional space with positive curvature. Thus arises the second possibility for the spatial three-dimensional geometry, that of a three-dimensional sphere. Just as a two-dimensional sphere may be thought of as an embedding in ordinary three-dimensional euclidean space, a three-dimensional sphere may be thought of as an embedding in four-dimensional euclidean space consisting of those points a fixed distance R from a fixed point. However, it may also, perhaps more properly, be thought of in a more intrinsic manner, independent of any embedding into a space of higher dimension. The area of a two-dimensional sphere of radius r on the three-sphere is $4\pi R^2 \sin^2 (r/R)$, indicating positive spatial curvature because this area is smaller than the value $4\pi r^2$ for a flat euclidean geometry. The third possibility is a hyperbolic geometry, for which a sphere of radius r has the area $4\pi R^2 \sinh^2 (r/R)$, which is larger than that in flat space. Hyperbolic space cannot be embedded in a euclidean space of higher dimension.

Another property that distinguishes these three geometries is the behavior of parallel lines (or geodesics) lying in the same plane. In euclidean space, given a line L and a point P not on L, in the plane containing L and P there exists precisely one line L' that passes through P but never intersects L (**Fig. 2**a). The line L' is parallel to L and may be constructed by drawing another line L'' through P and perpendicular to L, so that L' is the unique line through P that is perpendicular to L''. If this construction is repeated on the surface of a sphere (where great arcs are considered to be lines), it is found that every line through P intersects L, because initially parallel lines curve toward each other (Fig. 1b). By contrast, in hyperbolic space there exists a whole family of lines through P that do not intersect L (Fig. 1c) because in such space initially parallel lines curve away from each other.

Euclidean space, also known as flat space, is infinite and lacks an intrinsic scale. By contrast, both the spherical and hyperbolic geometries are characterized by an intrinsic scale, R. Locally, on length scales much shorter than R, the spherical and hyperbolic geometries differ little from euclidean space, but over larger distances the differences become substantial. Because of its finite volume, a spherical universe is also known as a closed universe, and a hyperbolic universe is also known as an open universe. It is clear that locally the geometry of the universe is close to euclidean. Therefore, if the universe is open or closed, rather than flat, then $a(t) \cdot R$ must now be very large.

Evolution of the universe. An intimate connection exists between the evolution of the scale factor $a(t)$

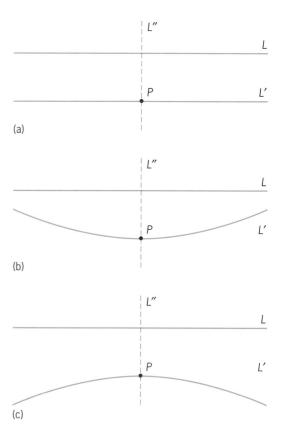

Fig. 2. Behavior of parallel lines in possible geometries of the universe. (*a*) Euclidean or flat space. (*b*) Spherical geometry. (*c*) Hyperbolic geometry.

and the geometry of the universe. The Hubble constant (which is not really constant), or expansion rate, is given by Eq. (2), where $\dot{a}$ is the time deriva-

$$H(t) = \frac{\dot{a}(t)}{a(t)} \qquad (2)$$

tive of a. In the case of an open universe, the quantity R may be absorbed into a rescaling of a, so that R may be taken to equal 1. The scale factor $a(t)$ of the universe evolves according to Eq. (3), where

$$H^2(t) = \frac{8\pi G}{3}\rho(t) - \frac{k}{a^2(t)} \qquad (3)$$

$k = -1, 0$, or $+1$, depending on whether the universe is open, flat, or closed, respectively; $\rho(t)$ is the mean density at time t; and G is Newton's gravitational constant. The term $k/a^2(t)$ reflects the influence of curvature on the expansion rate of the universe. A flat (critical) universe requires that the present mean density of the universe ρ_0 be exactly equal to the critical density, given by Eq. (4), where H_0 is the

$$\rho_{\text{crit}} = \left(\frac{3}{8\pi G}\right)H_0^2 \qquad (4)$$

present value of the Hubble constant. (The subscript zero is used to indicate the present values of cosmological parameters.) Thus, for a flat universe the dimensionless cosmological density parameter $\Omega_0 = \rho_0/\rho_{\text{crit}}$ is equal to exactly 1. It follows that a subcritical universe with Ω_0 less than 1 is hyperbolic, and that a supercritical universe with Ω_0 greater than 1 is closed. *SEE UNIVERSE.*

Measurement of density parameter. Much observational effort has been devoted to determining the density parameter, Ω_0, but no definitive conclusion has yet been reached. As discussed above, there seems to be a consensus that the universe contains more matter than just the luminous or the baryonic matter, and that an additional weakly interacting dark-matter component must exist. But most observations also infer a mean density less than the critical density, thus implying that the universe is open (hyperbolic). However, most measurements of Ω_0 are sensitive only to clustered matter, and an alternative explanation for low measurements of Ω_0 is an additional unclustered component of the energy density, arising for example from a nonzero cosmological constant.

Dicke coincidence and inflation. One argument in favor of $\Omega_0 = 1$, the Dicke coincidence, considers the consequences of supposing that Ω_0 today is not equal to 1. Extrapolating the present equation of state for matter or radiation back to much earlier times leads to the conclusion that ρ scales with a in proportion to a^{-3} or a^{-4}. This implies that the contribution from curvature at very early times (which scales as a^{-2}) must have been negligible, and at very early times Ω must have been exceedingly close to 1. Therefore, for Ω_0 to be away from 1 today would require that Ω be tuned to near 1 but not equal to 1 with astounding precision at some very early time.

With inflation, however, this line of reasoning breaks down. Inflation is based on an equation of state in which the density at extremely early times changes very slowly compared to the expansion of the universe; in other words, ρ is approximately constant. During inflation, Ω evolves toward 1 rather than away from 1.

In many cases, inflation gives a flat universe with $\Omega_0 = 1$. In the standard slow-roll inflation (that is, new or chaotic inflation), inhomogeneities in the initial conditions prior to inflation may be described as spatial variations in Ω, which decay during inflation as Ω flows toward 1. Provided that there is enough inflation to erase these initial homogeneities, inflation predicts a flat universe with $\Omega_0 = 1$.

Alternative inflationary scenarios. However, it is quite simple to obtain values of Ω_0 less than 1 within the context of inflation by introducing negative spatial curvature during inflation. Bubble nucleation through quantum tunneling introduces negative spatial curvature during inflation in a very precise way. Instead of a potential for the scalar field responsible for inflation, known as the inflaton field, that decreases monotonically toward the true minimum, this potential may have the form sketched in **Fig. 3**, in which there is a false minimum. Initially, the inflaton field gets stuck in this false minimum at $\phi = \phi_F$, during an initial epoch of old inflation, which smooths out whatever initial homogeneities may have existed prior to inflation.

During this first epoch of inflation the horizon and smoothness problems, which were two of the original motivations for inflation, are solved, and the space-time geometry rapidly approaches that of de Sitter space. Then through quantum tunneling the inflaton field decays out of the false minimum onto a part of the potential gently sloped toward the true minimum of the potential at $\phi = 0$. This quantum tunneling is a process of bubble nucleation.

A space-time diagram for this process is sketched schematically in **Fig. 4**. Time flows upward and the horizontal direction is the radial coordinate. The angular coordinates have been suppressed. Each point in the diagram represents a two-dimensional sphere. The solid lines indicate surfaces on which the inflaton field is constant. The lower part of the diagram (below the horizontal line) indicates the classically forbidden part of the process, in which a critical bubble materializes through quantum tunneling. Far toward the bottom and to the right the inflaton field is in the false vacuum. At later times (above the horizontal line) the critical bubble expands classically, at a speed rapidly approaching the speed of light. The straight diagonal line emanating from M, which represents the forward light cone of M, may roughly be considered to lie in the middle of a thick bubble wall, which expands roughly at the speed of light. The region above this light cone is the interior of the bubble. Inside the bubble the inflaton field slowly rolls toward the true minimum, just as in new or chaotic inflation.

The crucial point is that the geometry of the bubble interior is that of an open universe. The light cone emanating from M corresponds to time $t = 0$ in these open coordinates, and time increases as one moves inward toward the bubble interior. Viewed by using the open coordinates, the bubble exterior corresponds to what happens before the beginning of time, or before the supposed big bang. But this is nothing more than an artifact of this particular choice of coordinates. In this model, there is nothing singular about the surface defined by $t = 0$ in the open coordinates. No physical quantity diverges as this surface is approached, much in the same way that the horizon of a Schwarzschild black hole is nonsingular despite the fact that it appears to be singular in the usual black-hole coordinates.

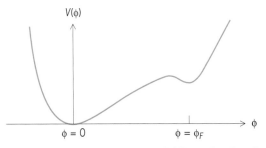

Fig. 3. Potential for the inflaton field [$V(\phi)$] as a function of order parameter (ϕ) in an inflationary scenario that gives rise to a subcritical universe.

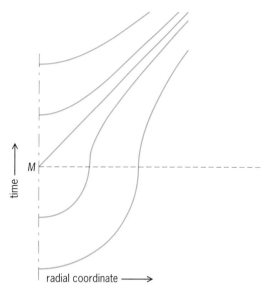

Fig. 4. Space-time diagram for bubble nucleation through quantum tunneling.

Inside the bubble the inflaton field is constant on surfaces of constant negative curvature (that is, surfaces of constant time t in the open coordinates). Because the value of the inflaton field determines the time until reheating (that is, until the end of inflation), the surfaces on which the inflaton field is constant are the surfaces that later evolve into surfaces of constant density. One feature of these expanding bubbles somewhat contrary to everyday intuition is that these bubbles have no well-defined center. They expand in a manner so symmetric that every observer perceives his or her position to coincide with the center of the bubble. The only invariant measure of position inside the bubble is the time t in the open coordinates.

Inside the bubble is a shortened epoch of slow-roll inflation, which gives Ω_0 somewhere midway between 0 and 1 today. At $t = 0$ the density parameter Ω vanishes, but as t inside the bubble increases, Ω increases, approaching unity. After the end of inflation, Ω decreases, flowing away from unity. The shortened epoch of slow-roll inflation inside the bubble results in Ω_0 being less than 1 today. The present value of Ω_0 is determined by the amount of inflation inside the bubble and the reheat temperature. If there is too much inflation, Ω_0 today is almost 1; if there is too little, Ω_0 today is almost 0, corresponding to an almost empty universe now. Just the right amount of inflation inside the bubble (to within some 10% accuracy) results in Ω_0 being in an interesting range about midway between 0 and 1. For this scenario of single-bubble inflation to work, it is essential that the bubble nucleation rate be very small, so that another expanding bubble does not collide with the bubble that contains the observable universe. Fortunately, a small bubble nucleation rate is easily achieved.

This inflationary scenario retains all of the beautiful properties of inflation, namely, the solution of the horizon, smoothness, and monopole problems, and a mechanism for generating primordial density predictions. But unlike more conventional inflationary scenarios, it does not predict a flat universe with $\Omega_0 = 1$. The Dicke coincidence problem is circumvented; rather mild tuning of the features of the potential give interesting values of Ω_0 between 0 and 1. Thus, although inflationary theories may offer a beautiful explanation for a flat universe, flatness is not an inexorable prediction of inflation. In fact, if preliminary indications that the universe is open turn out to be correct, inflation will be needed more badly than ever to avoid the extreme fine tuning suggested by the Dicke coincidence argument.

For background information SEE BIG BANG THEORY; COSMOLOGY; GRAVITATIONAL LENS; INFLATIONARY UNIVERSE COSMOLOGY; LIGHT CURVES; RIEMANNIAN GEOMETRY; UNIVERSE in the McGraw-Hill Encyclopedia of Science & Technology.

Martin A. Bucher

Bibliography. E. Aubourg et al., Evidence for gravitational microlensing by dark objects in the galactic halo, *Nature*, 365:623–625, 1993; M. W. Browne, Dark matter hints at new shape for Galaxy, *New York Times*, p. C1, April 18, 1995; M. Bucher, A. S. Goldhaber, and N. Turok, An open universe from inflation, *Phys. Rev. D*, 52:3314–3337, 1995; J. R. Gott III, Creation of open universes from de Sitter space, *Nature*, 295:304–307, 1982; S. W. Hawking and W. Israel (eds.), *General Relativity*, 1979; D. N. Schramm, Dark matter and the origin of cosmic structure, *Sky Telesc.*, 88(4):28–35, October 1994.

Cover crops

Cover crops are relatively short-term crops grown during times of the year when annual grain or fiber crops are not normally grown, or when a perennial crop is being established. Reasons for planting cover crops include (1) providing soil cover and erosion control during otherwise noncrop periods; (2) immobilizing soluble nutrients (nitrates) to prevent loss by leaching; (3) for legumes, converting atmospheric nitrogen to biomass nitrogen, which can mineralize in the soil and become available for the next grain crop; (4) adding organic matter to soils and improving aggregation; and (5) providing cover and protection to seedlings of perennial crops during establishment. Except for the last example, cover crops are not normally harvested and sold. However, in some instances cover crops may provide short-term forage grazing for livestock. Cover crops are frequently grown during winter months, between annual summer grain crops.

Species. Numerous species are grown as cover crops. Some of the more common species used in temperate regions are listed in the **table**. In more tropical regions the genera *Cajonus* (pigeon pea), *Crotalaria* (crotalaria species), and *Vigna* (beans, grams, cowpea) are also used. In addition and frequently unintentionally, numerous annual weeds serve some functions of a cover crop if left uncontrolled.

The species of cover crop chosen depends primarily on the purpose for which it is grown. Winter cover crops are usually selected to grow between summer annual grain crops to provide soil erosion control, reduce nitrate leaching, or improve various soil properties.

In temperate regions, rye is often used. In northern regions, if a legume cover crop is selected to reduce the amount of fertilizer nitrogen needed for the following grain crop, hairy vetch (see table) is commonly used. In regions with less severe winters, other legume crops include crimson clover and subterranean clover.

Varieties of *Brassica* species, winter wheat, and rye are sometimes used when livestock will graze on the cover crop. Also, if the following summer crop (such as sorghum or sunflowers) is normally planted in late spring or early summer or, in warmer climates, in the early spring, several short-season annual grain legumes may be used (for example, lentils, peas, or beans). These legumes are planted 50–90 days before the planting date of the summer grain crop. For winter grain crops such as wheat, it may be possible to grow a cover crop after harvest; often tropical or subtropical species (such as cowpea or mung bean) are planted.

In most cases no produce is harvested from the cover crop. The cover crop is usually killed by tillage or by spraying with herbicides before it reaches maturity and produces viable seed. This practice has a serious economic disadvantage, because new seed must be purchased and planted each year. To reduce this cost, species or varieties that produce seed before they are destroyed are sought. This endeavor is sometimes successful with certain bean and pea species and a few clovers and medics. For example, mung might be grown to maturity after wheat harvest. Also, the cover crop is sometimes grazed to partially offset the cost of annual reseeding.

Use with perennial crops. Cover crops are often used to provide protection when seeding and establishing a perennial crop. A common example is the use of oats as a cover crop for the establishment of alfalfa, various clovers, or perennial grasses. This practice is common in humid temperate regions. The oats mature early in the summer and can be harvested for grain, providing some income during the year that the perennial crop is being established. By the time the perennial crop is established, it can withstand adverse ambient conditions. In such situations the cover crop provides protection from cold, winds, and excessive radiant drying for the slowly emerging perennial species. The cover crop selected must not exhibit aggressive growth such that it provides excessive competition for the perennial being established. Oats fill this need better than most other small-grain crops.

In regions where water availability is usually the primary factor limiting plant growth, the use of a cover crop when seeding a perennial species is not advantageous. The rapidly growing cover crop will usually deplete the surface soil of available water within a few weeks after planting, thereby reducing the germination and establishment of the perennial crop. Therefore, an alternative system is often used in which the cover crop is grown alone the first year, which usually permits the harvest of grain from this crop. Wheat and barley, rather than oats, are the preferred choices. Then, the perennial crop is directly seeded into the standing stubble of the first crop, using no-till technology. Weeds are controlled with suitable herbicides. The standing stubble provides protection from adverse weather that the newly emerging perennial crop needs, but in a noncompetitive manner. This technology is frequently used not only for seeding cropland back to perennial grasses but also in establishing perennial vegetation in critical situations in dry regions, such as roadsides, deep gullies, and reclaimed mined land. This procedure is much less expensive and more effective than covering areas seeded to perennials with a straw mulch, or hydromulch, a method commonly used on disturbed lands in earlier years. In hydromulching, grass seed and various mulching materials are mixed with water, and the resulting slurry is sprayed onto the areas to be seeded. After drying, the mulching material remains in place, protecting the grass seedlings until they are established.

Controlling soil erosion. Many research projects have demonstrated the effectiveness of cover crops in controlling soil erosion. In the midwestern and southeastern regions of the United States, winter and spring precipitation and spring thaws often result in severe soil erosion from soybean or cotton

Species used as cover crops in the United States	
Scientific name	Common name
Avena sativa	Oats
Beta vulgaris ssp. *vulgaris*	Garden beet, mangel
Brassica juncea	Mustards
B. napus var. *napobrassica*	Rutabaga, swede
B. rapa var. *rapa*	Turnip
Daucus carota ssp. *sativus*	Carrot
Hordeum vulgare	Barley
Lathyrus sativus	Chickling vetch (pea)
L. sylvestris	Flat pea
L. tingitanus	Tangier pea
Lens Miller culinaris	Lentil
Lolium multiflorum	Annual (Italian) ryegrass
Lotus corniculatus var. *corniculatus*	Bird's-foot trefoil
Lupinus albus	White lupine
Medicago lupulina	Black medic
M. sativa ssp. *sativa*	Alfalfa
Pisum sativum ssp. *sativum*	Peas, Austrian winter pea
Secale cereale	Rye
Trifolium alexandrinum	Berseem clover
T. incarnatum	Crimson clover
T. pratense	Red clover
T. repens	White clover
T. subterraneum	Subterranean clover
Vicia faba	Fava (horse) bean
V. villosa ssp. *villosa*	Hairy vetch

stubble or from land tilled in the fall season. In such situations, planting winter cover crops in September or October is usually highly effective in reducing soil loss by erosion during the winter and spring, even though dry matter production of the cover crop may be only a few hundred kilograms per hectare. The Natural Resources Conservation Service (formerly the Soil Conservation Service) of the U.S. Department of Agriculture has recognized the effectiveness of this practice, and frequently recommends the use of winter cover crops to meet the conservation compliance requirements for highly erodible land as dictated by recent legislation. Rye, hairy vetch, and crimson clover are the cover crops most commonly used. These crops are likewise used in the eastern and northeastern United States. For example, hairy vetch and rye are commonly grown as cover crops in association with corn-soybean rotations to protect quality of water entering Chesapeake Bay.

Reduction of nitrate leaching. In other areas, there is major concern about reducing nitrate leaching into ground water. The usual solution to this problem is to reduce fertilizer nitrogen inputs to the extent possible and to use practices that reduce nitrate-nitrogen levels in the soil during the period when leaching is most likely, that is, in the spring and early summer. These objectives can be met to some extent by use of legume cover crops. Such cover crops take up and immobilize residual nitrate-nitrogen left in the soil after harvest of the previous crop; fix atmospheric nitrogen and thereby reduce fertilizer requirement; and decompose and release during midsummer nitrogen immobilized in the cover crop, at the time when nitrogen requirement by the summer grain crop is greatest. When corn, sorghum, or cotton is the summer crop, hairy vetch is frequently the most suitable winter cover crop. However, there is also some potential for growing short-season seed legumes for 50–90 days before planting the summer grain crop. Species such as field pea or Austrian winter pea, fava bean, and lentils may be used for this purpose.

Use with winter grains. Cover crops are infrequently used with winter grains such as wheat, because wheat is generally harvested in summer (May to August), and the remaining stubble provides adequate soil erosion protection. Also, in most areas of the United States hot and relatively dry weather at that time of the year provides essentially no potential for nitrate leaching. However, in southern regions there is potential to grow short-season legumes to maturity after harvesting wheat. For example, in eastern Oklahoma and Texas, mung beans are sometimes used for this purpose. In drier regions where wheat is often grown after summer fallow, few good cover crop species are available. In northern regions, early-season legumes (peas, beans, lentils) may be grown for about 60 days in early spring of the fallow year. Presumably there will be sufficient precipitation later in the summer to replace the soil water used by the cover crop.

A similar procedure using short-season cover crops is sometimes used in northern regions such as the Northern Great Plains. In this variant, the cover crop is seeded in the fall in narrow strips (0.3–1 m or 1–3 ft) spaced 10–30 m (30–90 ft) apart across the field. The species used are barley, oats, or flax (sometimes in combination). Compared to solid seeding of the cover crop, this arrangement greatly reduces the amount of water removed from the soil by reducing leaf area and subsequent transpiration loss. In addition, the cover-crop strips act as miniature snow barriers, causing snow drifts to develop on the leeward side. This snow cover reduces evaporation losses from the frozen soil during winter (sublimation) and increases soil water storage by trapping snow which may potentially infiltrate the soil during thaw. Thus, it is not uncommon for water content to be increased in the upper 2–8 cm (1–3 in.) of the soil by spring where these barriers are used.

Results. Use of cover crops often improves several soil physical properties such as aggregation, porosity, infiltration rate, and soil organic matter content. Magnitude of this effect varies widely but is usually dependent upon the quantity of organic carbon returned to the soil in the cover crop. In general, soil characteristics are enhanced because more organic residues and additional carbon are generally returned to the soil by growing a cover crop. Thus, if cover crops are used for several years, it is likely that soil organic matter will be increased and corresponding improvements in soil physical properties will be observed. The improved soil physical properties will in turn reduce water runoff and soil loss by erosion. Over a period of years, the reduced erosion and improved soil physical properties will be reflected in greater crop growth and greater return of organic carbon in the residues of these crops. It has been demonstrated in the southern Piedmont region of the United States that a high level of productivity can be restored in 3 years to the severely eroded soils there by use of cover crops in association with no-till techniques.

For background information SEE AGRICULTURAL SOIL AND CROP PRACTICES; COVER CROPS in the McGraw-Hill Encyclopedia of Science & Technology.

J. F. Power

Bibliography. W. L. Hargrove (ed.), *Cover Crops for Clean Water,* Soil and Water Conservation Society, 1991; W. L. Hargrove (ed.), *Cropping Strategies for Efficient Use of Nitrogen and Water,* Amer. Soc. Agron. Spec. Publ. 51, 1988; J. F. Power (ed.), *The Role of Legumes in Conservation Tillage Systems,* Soil and Water Conservation Society, 1987.

Crab

Recent innovative applications of biotelemetry are providing remarkable insights into the daily lives of various animals, including crabs, in coastal ecosystems. For example, small battery-powered transmitters that give off ultrasonic signals are attached to the backs of crabs moving freely in marine and estuarine habitats. The signals are detected by a hand-held hydrophone and receiver. Thus, ecologists can track the movement of the crabs to determine not only their specific habitat but also details of their physiology and behavior, including molting, feeding, and fighting. Biotelemetry data are yielding information about the ecology of various crab species that are harvested in major commercial fisheries or that are important predators in the coastal zone. Because limited visibility under water prevents direct observations of the movements and behavior of crabs, such crucial data were previously impossible to acquire.

Ultrasonic telemetry. Solid-state electronics have allowed scientists and electronics companies to make telemetry tags that are small enough to use on large crabs without altering their normal behavior. Although radiotelemetry has been used successfully to track animals in air and fresh water, ultrasonic telemetry is used to locate crabs in estuaries or the ocean, where saltwater rapidly weakens radio signals. The tag is usually less than 5% of the animal's weight, so that it will not interfere with normal movement and behavior. For ultrasonic telemetry, the size of the tag is determined primarily by the size of the battery, which determines the signal power and transmitting life, and the size of the sound-generating piezoelectric ring, which sets the frequency range of the signal. These features of tag size must be optimally matched with the particular species' movement patterns and habitat characteristics. Ultrasonic telemetry tags for crab research have a battery life of 2 weeks to more than 1 year. The tags range in size from 0.3 to 0.6 in. (8 to 16 mm) in diameter and 2 to 4 in. (40 to 90 mm) in length, with a weight in water of 0.1 to 0.5 oz (3 to 15 g). These tags send a 75-kHz signal which can be adjusted to produce a particular rate or pattern of beeps that can be used to track several animals at a time; alternatively, the frequency can be adjusted around 75 kHz to allow distinction of individual tags.

The receiving range of these ultrasonic tags is highly dependent upon the conditions of the water, bottom, and other features of the habitat; typical ranges are 330–3300 ft (100–1000 m). Temperature discontinuities (thermoclines or marked thermal gradients) or other discontinuities in water density (haloclines, pycnoclines) in the water column can cause the sound waves to reflect into the bottom, thus rapidly diminishing the receiving distance. Certain habitat features can also interfere with the signal, especially features that contain gas bubbles, such as flotation structures in some algae or oxygen bubbles that often form on the surface of sea-grass blades carrying out photosynthesis. Animals which move into burrows, dens, or caves of solid rock may also present tracking problems. In addition, other underwater noise, especially from snapping shrimp (family Alpheidae) or from boat propellers, can mask or confuse signal reception. Nevertheless,

maintaining a zigzag pattern of tracking generally makes it easy to locate small ultrasonic telemetry tags to within 3–7 ft (1–2 m) in shallow 3–100-ft (1–30-m) waters.

Because the transmitters are comparatively expensive, ultrasonic telemetry commits the research to a strategy of intensive tracking and analysis of data from relatively few individual animals. This strategy contrasts with nontelemetry studies using traditional mark-recapture methods, in which thousands of animals are typically tagged in hopes of returns for 5–10% of the sample. Moreover, telemetry provides extensive information on the details of variation in movements and habitat use throughout the tracking period; whereas standard mark-recapture methods give information only at the point of release and recovery, with no additional information in between. Depending on the research objective and the pattern of movement, crabs in telemetry studies have either been tracked continuously or relocated periodically at intervals of 12–24 h to produce detailed maps of individual movement that are digitized for computer analysis.

Movement and habitat use. Using ultrasonic telemetry to track crabs has also yielded interesting comparative data on speed and patterns of movement for several species. In the swimming crab family (Portunidae), adult blue crabs (*Callinectes sapidus*) in Chesapeake Bay meander at about 40–50 ft (12–15 m) per hour while feeding, then have sudden rapid movement at 1300–2600 ft (400–800 m) per hour directionally along the estuarine axis before settling down to a meandering pace in a new location. The tropical swimming crab, *Scylla serrata*, exhibits a similar pattern of movement, meandering at about 30–60 ft (10–19 m) per hour, and rapidly (220 ft or 70 m per hour) moving directionally into the current within muddy estuaries of South Africa. In the spider crab family (Majidae), snow crabs (*Chionoecetes opilio*) in the Gulf of Saint Lawrence move slowly at speeds of about 3–7 ft (1–2 m) per hour; *Maja squinado* moves at 1 ft (0.4 m) per hour while meandering to feed on rocky outcrops in summer, and 0.6 ft (0.2 m) per hour on the soft bottom floor during the winter. In the family Cancridae, the Dungeness crab (*Cancer magister*) off British Columbia moves at intermediate speeds averaging 36 ft (11 m) per hour in males and 56 ft (17 m) per hour in females, with bursts of higher speed up to 125 ft (38 m) per hour, and faster movement during summer than during winter months.

Habitat. In combination with other biological data, ultrasonic telemetry provides information on habitat utilization and seasonal changes in movement patterns for different life stages of these species. For example, during the summer juvenile blue crabs use shallow waters near the shoreline and remain within subestuaries, while large adult blue crabs feed in channel waters of subestuaries and freely exchange with the mainstem Chesapeake Bay. In the fall, adult *C. sapidus* migrate 6–120 mi (10–200 km) into deep water of the bay, with females over-wintering in high salinities at the mouth of the subestuary and males remaining in lower-salinity zones. Similarly, juvenile *M. squinado* off Spain are primarily restricted to an algal-covered rocky bottom, but after maturation in late summer the large adults migrate 3–60 mi (5–100 km) offshore on soft bottoms at speeds of 3 ft (0.9 m) per hour to overwinter in deep water. Canadian snow crabs also undergo a seasonal migration of moving tens of kilometers from deeper waters during winter to shallow nearshore waters during spring to mate and hatch their eggs before moving back into deeper waters. During progressive growth and maturation, Dungeness crabs of the northeast Pacific shift their habitat from shallow waters of estuaries and bays as small juveniles, to channels as larger immature crabs, and then to deeper water of the outer coast or large sounds as adults. These various ontogenetic and annual changes in movement patterns and resource utilization are adaptive for exploiting habitats that provide refuge from predators to small juveniles, afford optimal temperatures for growth and reproduction during seasonal cycles, and provide egg-hatching sites that aid in offshore larval transport and hence a reduction in plankton predation.

Behavior. Biotelemetry also provides data on key physiological functions and behaviors. Blue crabs in Chesapeake Bay are being studied with new, sophisticated ultrasonic transmitters that signal molting, feeding, and agonistic behavior.

Molting tags use a small reed switch that is glued to the back of the blue crab along with a small magnet that is inserted into a tube next to the switch and connected to a springy line attached to the bottom of the crab. When the crab molts, it splits open its carapace at the rear and backs out of the old shell, causing the magnet to be pulled out of the tube and away from the reed switch, thus changing the pulse rate of the telemetry signal. These molting tags show that immature male blue crabs alter their movement pattern just prior to molting, when they move up into small salt creeks and select the shallow edges of marshes as a microhabitat for molting.

Feeding tags have small electrodes inserted through holes that are drilled through the carapace at the site of attachment of the mandibular muscle. When the muscle contracts to cause biting action, the electrical signal of the muscle's myopotential is detected by the tag and transmitted in addition to the regular location signal. The tag allows determination of the time and location of blue crab feeding, and of the number of bites a crab takes to consume its prey. For example, small clams (*Macoma balthica*), which are the primary prey of blue crabs, require about 200 bites in order to be completely consumed by a crab. The biotelemetry pattern of feeding shows that blue crabs feed about four to seven times per day, with significant peaks in feeding activity near dawn and dusk.

Agonistic display tags use reed switches on the upper segments of the blue crab's chelae and small magnets attached on the distal segment of the claw.

During a resting posture, the claws are folded and the magnets are positioned away from the reed switches. When the crab is threatened, exhibiting the stereotypical extension of the claws, the magnets are brought in proximity to the reed switches and the telemetry signal changes. Data from agonistic display tags indicate that blue crabs fight frequently, with bimodal peaks in claw-spread threats during the early morning and late afternoon that are associated with the periods and locations of peak feeding.

Combined biotelemetry. In addition to telemetry of single behaviors or physiological variables, newer multichannel tags allow the transmission of two or more behaviors in the same organism. Crab movement and the location and timing sequence of these behaviors can be recorded by a computer for rapid data storage and analysis. Combined biotelemetry of feeding and agonistic threat displays shows that blue crabs are attracted to feed on experimental patches of clams, but that increased local abundances of feeding crabs result in high levels of fighting that reduce foraging efficiency and stimulate blue crabs to move away from the feeding sites. Other combinations of muscle activity and appendage movement can be readily telemetered to determine various behaviors by using these multichannel tags.

This new biotelemetry technology can also provide information on environmental variables of water depth, temperature, salinity, and light intensity that an animal encounters. Furthermore, these techniques can be applied readily to other groups, including lobsters, lithodid crabs, horseshoe crabs, fish, mollusks, reptiles, and mammals. SEE XIPHO-SURIDA.

For background information SEE BIOTELEMETRY; CRAB; CRUSTACEA in the McGraw-Hill Encyclopedia of Science & Technology.

Anson H. Hines

Bibliography. B. J. Hill, Activity, track and speed of movement of the crab *Scylla serrata* in an estuary, *Mar. Biol.,* 47:135–141, 1978; A. H. Hines et al., Movement patterns and migrations in crabs: Telemetry of juvenile and adult behavior in *Callinectes sapidus* and *Maja squinado, J. Mar. Biol. Ass. U.K.,* 75:27–42, 1995; T. G. Wolcott and A. H. Hines, Ultrasonic biotelemetry of muscle activity from free-ranging marine animals: A new method for studying foraging by blue crabs (*Callinectes sapidus*), *Biol. Bull.,* 176:50–56, 1989; T. G. Wolcott and A. H. Hines, Ultrasonic telemetry of small-scale movements and microhabitat selection by molting blue crabs (*Callinectes sapidus*), *Bull. Mar. Sci.,* 46:83–94, 1990.

Crinoidea

Crinoids are often regarded as living fossils because they were abundant and diverse in Paleozoic seas and because modern species have retained the same basic body forms. Most extant crinoids are unstalked feather stars (comatulids), a group that occurs com-

Fig. 1. Deep-sea stalked crinoid with arms held in the feed-ing-fan posture. The water current flows from left to right.

abyssal zones of the deep sea. No stalked crinoid can swim; indeed, most individuals probably spend their entire adult lives either cemented to the same substratum or attached to a single piece of rock by their basal structures called cirri. It has recently been shown that some of the species which attach with cirri (family Isocrinidae) can change attachment sites. For example, *Endoxocrinus parrae,* a slope species in the Caribbean Sea, sometimes releases its grip on the substratum and moves slowly across the bottom with a rowing motion of the arms. Also, these same stalked crinoids move the individuals arms of their feeding crowns in response to various kinds of mechanical stimuli, including silt, sand particles, adjacent sea anemones, and small crustaceans. They respond to all types of irritants with a single effective swatting behavior that utilizes a single band of muscles antagonized by mutable collagenous ligaments. Study of the stimuli that elicit behaviors in modern crinoids should elucidate the selective pressures that may have favored the evolution of arm musculature in articulated crinoids some 200 million years ago.

Previously, all known behaviors of stalked crinoids were associated with their habit of filtering food particles from the water; virtually all such behaviors are slow and deliberate. Stalked crinoids feed passively by holding their many arms in the shape of a parabolic dish, which the current orients like an umbrella in the wind (**Fig. 1**). Food is collected on the downstream side of this umbrella by

monly in both shallow and deep water, particularly in the tropics. Comatulid crinoids lose their attachment stalks as juveniles and are therefore able to crawl, climb, and sometimes swim as adults. Stalked crinoids are sedentary animals with very limited mobility. There are now fewer than 80 species of living stalked crinoids, all restricted to the bathyal and

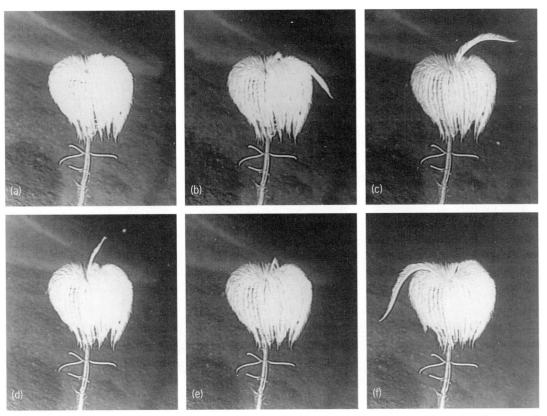

Fig. 2. Video sequence (*a–f*) showing a stalked crinoid, *Cenocrinus asterius,* waving various arms in response to sand particles being dropped on its crown. (*From C. M. Young and R. H. Emson, Rapid arm movements in stalked crinoids, Biol. Bull., 188:89–97, 1995*)

tiny tube feet in the side branches (pinnules) of each arm. Once collected, the food is transferred in mucus to the centrally located mouth by means of ciliary tracts. In still water, crinoids abandon their feeding posture and assume a flowerlike pose.

Arm movement. Crinoids can hold their arms erect for extended periods of time while feeding. It is thought that they maintain this posture by means of ligaments made of collagen (catch connective tissue), the stiffness of which can be regulated by the nervous system. The crinoid stalk is a flexible structure constructed of a stack of calcareous disk-shaped ossicles held together by ligaments. Despite having no muscles, the stalk can alter its orientation with respect to the current. These stalk movements have been shown to depend on ligaments of catch connective tissue.

By using a high-resolution close-up video camera mounted on the outside of a research submersible, the details of rapid arm movement in stalked crinoids

have been documented. Crinoids wave individual arms up and down during both feeding and nonfeeding periods. An arm flexure occurs when longitudinal muscles lying within the upper (oral) side of the arm contract, causing the arm to bend at all points where ossicles articulate. There are no antagonistic muscles on the aboral side of the arm to bring the arm back into position; instead, the arm is returned to position by the resiliency of the same longitudinal ligaments that maintain the feeding posture. Microscopic examination of these ligaments reveals that they are bounded by cells similar in appearance to the juxta-ligamental cells that regulate the viscosity of catch connective tissue in other echinoderms. The upward (effective) stroke of arm waving is faster than the downward (recovery) stroke. This difference in speed is caused by the different speeds of muscular contraction and elastic recoil of ligaments.

A single wave of the arm may last for 2–21 s, depending on how long the longitudinal muscles in

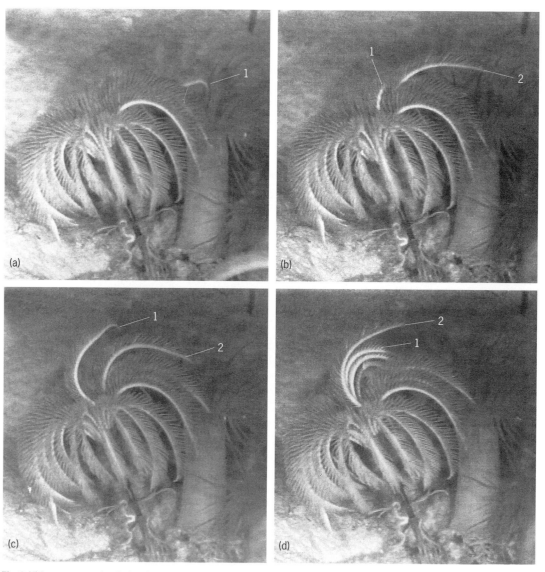

Fig. 3. Video sequence (*a–d*) showing a stalked crinoid, *Endoxocrinus parrae*, escaping from the tentacles of a sea anemone by flexing its arms. Changes in the positions of 1 and 2 arms can be seen.

the arm remain contracted. An arm may flex through an arc as large as 180° from its original position. The nerve fibers that control the arm muscles measure 2.5–3.75 micrometers in diameter, several times larger than the diameter of typical echinoderm nerves. These large nerves permit crinoids to respond to stimuli with a reaction time of less than 0.5 s. Rapid and dramatic arm movements were not previously expected in these passive and sedentary animals; nevertheless, crinoids are clearly equipped to perform such movements quickly and efficiently.

Function. In the food-limited deep-sea habitats where stalked crinoids live, their waving of individual arms up and down would be a major waste of valuable energy unless it had an important function. During the past several years, biologists have had numerous opportunities to observe stalked crinoids living on the steep slopes of the Bahamas. The arm-waving behavior of crinoids appears to have many different functions, some being responses to irritating or dangerous organisms and others involving cleaning of the delicate feeding crown.

Crinoid-crustacean encounters. When crinoids were illuminated with the bright lights of a submersible, the incidence of arm waving increased with time; the longer the animals were illuminated, the more frequently they waved their arms. Initially this reaction was thought to be a response to the lights themselves, but careful analysis of the videotapes demonstrated that the crinoids were actually responding to the individual strikes of small crustaceans attracted to the lights of the submersibles. Longer periods of illumination attracted more crustaceans, resulting in more crustacean-crinoid encounters. Consequently, the number of swimming crustaceans visible on the video screen was correlated with the total number of animals waving arms and also with the average number of times each individual crinoid waved its arms. Small crustaceans often swam directly toward the oral side of the crinoid arm, where the ciliated food groove conveys food-laden mucus toward the mouth. Whenever a crustacean struck the food-groove region of an arm, the crinoid immediately flicked that particular arm, invariably driving the crustacean away. It is suspected that under natural circumstances crustaceans attempt to steal the concentrated organic particles and mucus present in the food groove and that the arm-waving behavior helps crinoids minimize losses. On one occasion, a crustacean remained attached to the food groove of a crinoid for several seconds. The crinoid waved the affected arm repeatedly until the crustacean was shaken off.

Other stimuli. Arm-waving behavior was also observed in response to other kinds of mechanical stimuli. When fine sediment was picked up with the manipulator arm of the submersible and dropped on the crowns of stalked crinoids, the animals responded by waving many different arms until the water was clear. This response was also elicited by larger sand particles falling on their crowns (**Fig. 2**). During one experiment, a large piece of flocculent

material, probably a decomposed piece of algae that had drifted down the water column, became lodged on the oral side of an arm. The crinoid waved the affected arm as well as an adjacent arm repeatedly until all the material had been dislodged.

Defense mechanisms. Crinoids living in proximity to sea anemones sometimes become entangled in the anemone tentacles, which are laden with stinging nematocyst cells. The crinoid frees itself by pulling away the entangled arm, using the same muscle that is used for arm-waving behavior (**Fig. 3**).

Many organisms that spend their entire lives in one place have the ability to withdraw rapidly into tubes or burrows as a defense against predators. Stalked crinoids were not previously known either to need or to exhibit such escape behaviors.

For background information SEE CRINOIDEA; CRUSTACEA; ECHINODERMATA in the McGraw-Hill Encyclopedia of Science & Technology.

Craig M. Young

Bibliography. C. G. Messing et al., Relocation movement in a stalked crinoid (Echinodermata), *Bull. Mar. Sci.,* 42:480–487, 1988; I. C. Wilkie, R. H. Emson, and C. M. Young, Smart collagen in sea lilies, *Nature,* 366:519–520, 1993; I. C. Wilkie, R. H. Emson, and C. M. Young, Variable tensility of the ligaments in the stalk of a sea-lily, *Comp. Biochem. Physiol.,* 109A:633–641, 1994; C. M. Young and R. H. Emson, Rapid arm movements in stalked crinoids, *Biol. Bull.,* 188:89–97, 1995.

Crop rotation

Crop rotation is the sequential growth of different crop species on the same land. The practice has been used for centuries because early agriculturists experienced low yields with continuous monocropping, and they found that rotations were necessary to maintain productivity. Increased availability of nitrogen fertilizers, herbicides, and pesticides, as well as higher short-term profit, caused farmers to abandon extended rotations following World War II, but interest in crop rotation has been renewed by a greater awareness of both on-site and off-site impacts of soil and crop management practices.

Crop rotations as known today can be traced to eighteenth-century Norfolk, England. Turnip, barley, clover, and wheat were typically grown in a 4-year sequence. It was thought that yields were improved because each crop obtained its nutrients from a different part of the soil.

English customs were adopted in the United States. Crop rotation and animal manure were used extensively by progressive farmers in an attempt to restore productivity to levels observed when soils were first tilled. George Washington wrote that a good rotation for Long Island consisted of corn with manure, oats or flax, wheat with four to six pounds of clover and one quart of timothy, and meadow or pasture. With slight modification, this rotation was also popular in Pennsylvania, but in

Virginia farmers used wheat, corn, potatoes, peas, rye, clover, and buckwheat with animal manure.

Scientific reasons for using rotations were not known until the midnineteenth century, when researchers discovered that legumes were able to fix nitrogen from the atmosphere and that some of this nitrogen was available for succeeding crops. Thus, crop rotation remained popular into the early twentieth century.

Local use of crop rotation was highly dependent on the amount and cost of new land. This practice was not widely used in the Corn Belt (where soils were extremely fertile), even though it increased yields at several locations. Crop rotation in the southern United States revolved around cotton, tobacco, or rice, and included corn, wheat, oats, peanut, cowpea, and crimson clover in 2- or 3-year combinations.

Legume rotations were deemphasized after World War II, except by farmers who needed forage for livestock. The availability of inexpensive nitrogen fertilizers, herbicides, and pesticides as well as improved crop varieties and the replacement of draft animals with tractors reduced the perceived need for and use of extended rotations by farmers and researchers. Abandonment of extended rotations has had environmental consequences, including decreased soil organic matter content; degraded soil structure; increased soil erosion and sedimentation of streams, lakes, and reservoirs; increased external inputs into farming; and increased surface- and ground-water contamination.

Increased crop yield may be one of the most practical reasons for crop rotation. Simply changing from continuous corn to a corn-soybean rotation generally increases yields 5–20%. Rotation can frequently eliminate corn-yield decreases when no-till practices are used to reduce soil erosion.

Currently, about 20% of the corn in the United States is in continuous monoculture, while the rest is grown in a 2-year rotation with soybean or in short (2- or 3-year) rotations with cotton, dry beans, alfalfa, or other crops. Nutrient cycling, water availability, soil structure, weed control, disease suppression, insect pressure, and soil microbial activity are among the factors affected favorably by crop rotation, but many factors, processes, and mechanisms responsible for increased yield remain unknown.

Water use. In low rainfall areas, crop rotations help farmers adapt to seasonal rainfall patterns by providing complementary root systems that improve overall water-use efficiency. For example, by using small grains, grasses, deep-rooted crops, and a minimum amount of summer fallow, soil-water loss by deep percolation can be reduced and development of saline seeps can be minimized. The sustainability is increased by mimicking natural ecosystems.

Nutrients. Crop sequence influences nutrient use efficiencies and can be used to reduce nitrogen loss. Including alfalfa in rotations can help scavenge soil profiles for nitrogen, but if inorganic sources are not available, alfalfa can symbiotically fix nitrogen from the air. Phosphorus availability is not directly affected by crop rotation, but returning nonharvested portions to the soil can reduce stratification. Potassium and micronutrients can be indirectly affected by crop rotation, if plant root development and function are improved or if mycorrhizal communities become established. Crop rotation can also increase availability of micronutrients such as iron, copper, and zinc.

Soil quality. The need to reduce negative on- and off-site impacts of agricultural practices has increased interest in crop rotation. Soil quality, which can be evaluated by assessing productivity, resistance to degradation, and environmental buffering against pesticide and nutrient loss, provides one measure.

Abandonment of extended rotations has degraded soil structure as measured by aggregate stability, bulk density, erosion, and water infiltration rate. Crops such as bahiagrass have been shown to improve soil quality in compacted soils by creating biopores to depths that were not previously affected by tillage.

Rotations that return the most crop residue to the soil usually reduce soil bulk density, although traffic patterns, tillage, and sampling techniques can complicate the assessments. Rotations that involve long periods of sod, pasture, or hay generally increase soil organic matter. Thus, resistance to water and wind erosion is increased, and often aggregate stability, and water entry and retention—factors which enhance soil quality—are improved.

Pests and disease. Pest control is enhanced by growing different crops in rotation. Pests controlled by crop rotation must be deprived of an inoculum source in the field, have a fairly narrow range of hosts, and be incapable of surviving long periods without a living host. Examples include soil- and root-dwelling nematodes, soil-borne pathogens, and vegetatively propagated weeds such as nutsedge. Rotation does not control highly mobile pests, which can invade from adjacent fields or other areas.

Temporal and spatial diversity achieved through rotation can reduce weed population density and biomass production. Rotation helps control weeds because they thrive and increase in crops with similar growth habits. With rotations, opportunities for growth and reproduction change annually.

Insects that have specific or at least narrow host ranges, and are incapable of extended migration are particularly susceptible to control by crop rotation. For example, northern corn rootworm becomes an economic problem about 30% of the time with corn grown as a monocrop, but in a corn-soybean rotation, economic thresholds are reached less than 1% of the time. However, as such 2-year rotations have increased, the insect is gradually adapting its diapause or resting period. For insect species such as black cutworms, crop rotation is worse than monoculture since alternate crops attract the moths.

Crop rotation is often justified as a method for preventing fungal diseases, but this argument is not accepted by everyone. Depending upon the specific disease, rotation may or may not be effective. Crop rotations were demonstrated to be very effective for controlling root knot and cyst nematodes in tobacco and soybean in North Carolina during the 1950s and 1960s. Rotation research was deemphasized as priorities shifted to developing resistant cultivars and using nematicides. However, as resistant cultivars have been developed and grown, nematode races have become modified and now make some resistant cultivars less effective.

Biological diversity. Rotations increase temporal and spatial diversity. Temporal diversity is important for breaking pest cycles, reducing soil erosion, and increasing yields. Spatial diversity has repeatedly been shown to influence the abundance of wildlife, but this effect is less well known. One example is the impact of increased crop diversity on the ring-neck pheasant population. Crop rotation and other conservation practices have provided more grass or forage areas where the pheasants can nest. Row crops generally do not provide sufficient ground cover for nesting in early summer or late spring. In fact, most wildlife species that rely on agricultural habitat for survival sustain their populations much more effectively on landscapes with diverse rotations than on those with monocultures or continuous row crops, because the diversity provides alternative food sources.

Bioenergy. The use of bioenergy crops in rotations is receiving increased emphasis for at least three reasons: petroleum energy reserves are finite; selected bioenergy products are relatively clean and less polluting than petroleum counterparts; and farmers view the energy market as very large and as having significant economic potential. Traditional energy plants such as trees are not suitable for crop rotation schemes because of their semipermanency, but crops such as switchgrass or reed canary grass may become more so. Opportunities for twenty-first-century crop rotations will undoubtedly increase as technological advances in ethanol manufacture are made.

For background information SEE AGRICULTURAL SOIL AND CROP PRACTICES; ECOSYSTEM; MYCORRHIZAE; SOIL ECOLOGY in the McGraw-Hill Encyclopedia of Science & Technology.

Douglas L. Karlen

Bibliography. C. A. Francis, C. B. Flora, and L. D. King (eds.), *Sustainable Agriculture in Temperate Zones*, 1994; D. L. Karlen et al., Crop rotations for the 21st century, *Adv. Agron.*, 53:1–45, 1994.

Cytochemistry

Analytical chemistry has been directed toward determination of ever smaller amounts of material. This trend has opened up new applications such as chemical analysis of single cells. Recently, electro-

analytical methods have found an important application in monitoring chemical signals sent by cells.

Chemical messengers. Chemicals are frequently used as messengers when cells communicate with each other. A pancreatic β cell, for instance, upon sensing an increase in glucose concentration in the blood secretes the hormone insulin; the insulin then enters the bloodstream and makes its way to receptors on target cells. The insulin causes an increase in glucose consumption, which eventually reduces the glucose levels. In effect, insulin brings the message "glucose is available, begin using it" to target cells. A host of hormones such as insulin are used to regulate many physiological processes, including growth, development, metabolism, and stress response. Neurons also secrete chemicals, known as neurotransmitters, to communicate with each other. Neurotransmitters may be simple inorganic compounds such as nitric oxide (NO), low-molecular-weight organic compounds such as dopamine, or small peptides such as galanin. Understanding the secretion and chemical communication process is fundamental to understanding the physiological and mental processes that are orchestrated by intercellular communication. In addition, many diseases are associated with defects or irregularities in secretion. For example, type II diabetes is associated with insulin secretion that is insufficient to bring glucose levels back to normal. Improved methods for monitoring chemical signals have profound implications for progress in these important areas.

Most experiments aimed at monitoring chemical messengers have relied upon measurements of chemicals released from large groups of cells. Although important information is gained from this approach, the information is limited to the average of all cells sampled. Recently, electrochemical methods have been developed that allow chemical secretions from single cells to be monitored in real time. They are rapidly advancing the study of secretion and chemical communication.

Electrochemical monitoring of single cells. Two electrochemical methods, amperometry and voltammetry, have recently been used for monitoring secretion from single cells. In amperometry, the sensing electrode is poised at a constant potential that is sufficient to cause compounds contacting the electrode surface to be oxidized; that is, the compounds transfer electrons to the electrode, resulting in a current that is monitored. The charge passed (Q) can be related to the number of moles oxidized (N) according to Faraday's law, $Q = nFN$, where n is the number of electrons transferred per oxidation and F is Faraday's constant (96,485 coulombs/equivalent). In voltammetry, the electrode potential is swept in a triangular waveform around the potential for oxidation and reduction of the compounds of interest. The current recorded during the potential scan, known as a cyclic voltammogram, is a complex function of the concentration of analyte, rate of electron transfer, rate of mass transport to the electrode, and electrical artifacts. Cyclic voltammograms are characteristic for a chemical, and are useful for verifying the identity of compounds detected.

To monitor secretion at a single cell, microelectrodes with dimensions similar to, or smaller than, the cell are fabricated. A useful electrode material is carbon fiber of 1–10 micrometers in diameter. The fiber is coated with insulator around the edge so that only its tip is exposed to solution, resulting in a disk-shaped electrode. The electrode is then carefully positioned near a single cell that has been isolated in culture. **Figure 1** shows a carbon-fiber electrode with 1-μm tip diameter positioned near a differentiated PC12 cell. PC12 cells secrete dopamine and norepinephrine, two easily oxidized compounds that are used as chemical messengers. The PC12 cell is part of a tumor cell line that will differentiate, that is, grow fingerlike projections, when treated with nerve growth factor. The projections are functionally similar to axons in neurons, and these cells are frequently used as models of neurons. The electrode in Fig. 1 is positioned near a varicosity, a small bulge on the process, which is believed to be a site of secretion.

Once the electrode is positioned, secretion is initiated by an appropriate stimulus. Highly dependent on the cell type used, the stimulus may be a chemical that interacts with receptors on or inside the cell, or it may be an electrical pulse. The current trace that is recorded is a direct measure of the chemicals being secreted. **Figure 2** illustrates a typical current trace resulting from stimulation of a single PC12 cell with nicotine. The current spikes are indicative of exocytosis, the secretion mechanism used by these cells. In this process, vesicles that contain small amounts of hormone fuse with the plasma membrane. The fusion of vesicle membrane and cell membrane results in a pore that allows the interior of the vesicle to come in contact with the extracellular environment without degrading the integrity of the cell. The pore expands, allowing hormone inside to escape as a small packet, resulting in detection of a current spike. Individual spikes correspond to detection of single exocytosis events. The area under these spikes has units of charge, and can be used to determine the total amount released during the exocytosis events

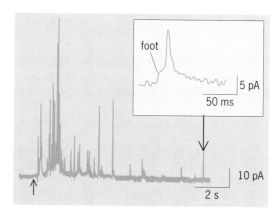

Fig. 2. Amperometric recording of catecholamine secretion from a single adrenal chromaffin cell, illustrating measurement of secretion. The cell was stimulated by applying nicotine for 10 s, beginning at the arrow. The inset shows enlargement of a single spike that has a so-called foot.

by using Faraday's law. Amounts detected per release event in PC12 cells average about 0.3 attomole. Although exocytosis has long been accepted as the mode of release in these cells, the electrochemical methods have provided the first direct measurement of the hormone release at the level of single exocytosis events.

Studies of catecholamine secretion. The ability to directly quantify the amount of material released has led to some surprising results. For example, not all of the catecholamine in an adrenal chromaffin cell vesicle is released during an exocytosis event. This finding was demonstrated by comparing spike areas at different extracellular pH values. At pH 8.4, spike areas were almost 50% greater than spike areas at pH 7.4 (normal extracellular pH). Thus, it was concluded that at normal pH a significant portion of the catecholamine remained in the vesicle or was released too slowly to be detected. Previously it had been believed that the entire hormone content was rapidly extruded during exocytosis. The lack of complete release has been attributed to the complex nature of the extrusion of hormone from the vesicle. Catecholamines are stored in vesicles at high concentrations (about 0.3 M) in association with a matrix that is composed, at least in part, of the water-soluble protein chromogranin A. In order to escape from the cell, the catecholamine must dissociate from chromogranin A. Presumably, the normal extracellular environment is not appropriate for driving complete dissociation of catecholamine from the matrix. The significance of this apparent inefficiency in release remains to be elucidated.

Spike shape is another useful parameter provided by the electrochemical techniques. The shape of spikes is determined by the dynamics of exocytosis, and therefore provides valuable insight into the secretion events. For instance, in some cases a small increase in current, or foot, is observed prior to the abrupt upstroke of the spike (Fig. 2b). This foot has been attributed to leakage of the hormone out of the fusion pore prior to the complete opening of the vesicle. Other studies on mast cells have shown

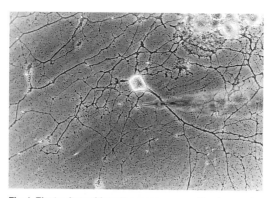

Fig. 1. Electrode positioned to measure secretion from a single varicosity on a differentiated PC12 cell.

that the fusion pore opening is sometimes transient, and is reversible (that is, the fusion pore can reclose without complete exocytosis).

Chemically modified electrodes. Most of the work so far on studying secretion at single cells has focused on release of epinephrine and norepinephrine from adrenal chromaffin cells. These two compounds are easily oxidized at carbon electrodes and thus are readily detected. Release of these compounds, and of their relatives dopamine and serotonin, has also been measured at neurons, PC12 cells, and mast cells. The secretion of other electroactive compounds, such as histamine and small tryptophan-containing peptides, such as melanocyte-stimulating hormone, have also been measured at single cells. Unfortunately, many important chemical messengers are not as easy to detect even though their oxidation or reduction is thermodynamically favored. Thus efforts have been made to use chemically modified microelectrodes for single-cell monitoring. In this approach, the surface of the electrode is modified in such a way that the rate of oxidation of the analytes is enhanced. Typically, a catalyst is immobilized on the electrode surface to mediate transfer of electrons from analyte to electrode. In other cases, modifications may be used to increase selectivity. For example, if an anion is a possible interference in the experiment, a cation-exchange film may be coated onto the electrode to exclude anions from the electrode surface. The modifications that are necessary will depend on the goals of the experiment and the complexity of the cellular environment. However, in all cases the electrode must be small to allow precise positioning and minimize background noise, and it must have a fast response time to allow the fast chemical changes to be followed.

One example of using a chemically modified electrode is for monitoring insulin secretion from pancreatic β cells. In this case, a composite of ruthenium oxide and cyanoruthenate is electrochemically deposited on the electrode. This film is a potent catalyst for the oxidation of disulfide bonds such as those in insulin. **Figure 3** represents detection of insulin from single β cells following three different stimuli. The difference in the timing of exocytosis events is indicative of the difference in the coupling of stimulus to secretion. For example, with glucose there is a delay before the spikes begin. This latency is attributed to the need for glucose to be metabolized and for build-up of the concentration of intracellular messengers, such as adenosinetriphosphate (ATP), to levels sufficient to cause depolarization of the cell and ultimately secretion. In addition, secretion continues for some time after the glucose stimulus, indicating the persistence of the intracellular messenger. In contrast, tolbutamide and potassium ion (K^+) act by directly depolarizing the cell. In the case of such stimulation, the spikes start almost immediately and end soon after the stimulus is removed. The use of this electrode is starting to yield other insights into the

insulin secretion process. For example, in contrast to adrenal chromaffin cells the spikes from insulin cells always rise smoothly. The suggestion is that insulin cannot leak out of the fusion pore during exocytosis. A likely explanation is that insulin, which is stored as a solid inside vesicles, is slow to dissolve from the granule and therefore does not readily leak out from the vesicle during fusion pore formation.

Chemically modified electrodes have also been used to detect release of nitric oxide from endothelial cells. For nitric oxide detection, the base electrode is a carbon fiber 2–6 μm long with a tip diameter of 0.5 μm. The electrodes are modified by electrochemically depositing a polymeric porphyrin that contains nickel as the central metal. The electrode is then covered with a cation-exchanging polymer by dip coating. The polymer greatly reduces interferences from anionic substances. For this application, the most important interference is the nitrite ion (NO_2^-), which can also be detected at the porphyrinic-nickel–modified electrodes. Electrodes prepared in this fashion exhibit remarkable analytical characteristics. For nitric oxide, the linear dynamic range is more than four orders of magnitude, the detection limit is a concentration of 10 nanomolar, and the response time is 10 ms. The electrode does not detect NO_2^- at concentrations up to 20 micromolar, indicating excellent selectivity for nitric oxide. These analytical characteristics make the electrode well suited to the single-cell application.

This chemically modified electrode has been used to detect nitric oxide secretion from endothelial cells. Endothelial cells control and communicate with muscle cells by releasing endothelium-derived relaxing factor. (Nitric oxide is responsible for the activity of endothelium-derived relaxing factor.) The endothelial cells can be stimulated to produce nitric oxide by several agents, including ATP, histamine, and bradykinin; these agents act by binding to cell

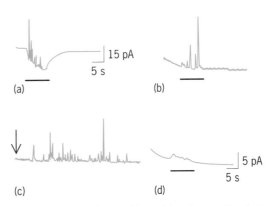

Fig. 3. Amperometric recordings of insulin secretion from single pancreatic β cells using various stimulations: (*a*) 200 micromolar tolbutamide, (*b*) 64 millimolar potassium ion, (*c*) 16 m*M* glucose, and (*d*) blank. In *a* and *b* the bar indicates the duration of the stimulus, and in *c* the stimulus was applied for 30 s prior to the point indicated by the arrow. The differences in the patterns and timing of exocytosis are attributed to differences in the mechanism of stimulation.

surface receptors. In one experiment, nitric oxide sensors were placed at the surface of a single endothelial cell and another was implanted inside a smooth muscle cell. Nitric oxide secretion was found to occur as a smooth curve after stimulation of the endothelial cells. Nitric oxide is intracellularly produced on demand by the action of nitric oxide synthetase on arginine. It then diffuses out of the cell to target cells. Therefore, the time course of release measured with the nitric oxide sensor is a smooth curve rather than the sharp spikes observed with exocytosis. A quantitative examination of the release curves has revealed that the arrival of nitric oxide at the electrode surface is delayed because of chemical reactions that consume the nitric oxide. By using the electrode, it was also possible to detect the arrival of nitric oxide inside the muscle cell. The time resolution of these single-cell measurements has provided an opportunity to study the kinetics governing the release and production of nitric oxide.

Prospects. The use of microelectrodes has allowed secretion of as few as 100,000 molecules to be detected with millisecond time resolution from single endocrine and neuronal cells. These methods have allowed scientists to listen to the chemical conversations of single cells with unprecedented clarity. The continued used of modified electrodes will increase the number of compounds that can be monitored by this approach. The power of these methods will be amplified as they are combined with other methods compatible with single-cell analysis. For example, combination with chemical microscopy, capillary electrophoresis, capillary chromatography, or electrophysiology will provide new details about how cells secrete and about the message implied by secretion.

For background information SEE CATALYST; ELECTROCHEMICAL TECHNIQUES; ELECTRODE; HORMONE; pH in the McGraw-Hill Encyclopedia of Science & Technology.

Robert T. Kennedy

Bibliography. R. H. Chow, L. von Ruden, and E. Neher, Delay in vesicle fusion revealed by electrochemical monitoring of single secretory events in adrenal chromaffin cells, *Nature,* 356:60–63, 1992; T. Malinski and Z. Taha, Nitric oxide release from a single cell measured in situ by a porphyrinic-based microsensor, *Nature,* 358:676–678, 1992; S. G. Weber (ed.), *Trends Anal. Chem.,* vol. 14, no. 4, 1995; R. M. Wightman et al., Resolved catecholamine concentration spikes correspond to vesicular release from individual chromaffin cells, *Proc. Nat. Acad. Sci. USA,* 88:10754–10758, 1991.

Cytolysis

The inability to transplant tissues between members of the same species (unless they are identical twins) has been apparent for more than a hundred years. In the early 1900s, it was suggested that transplant rejection might be an immunological phenomenon. But until the 1950s, the only known immunological effector mechanisms were antibodies. Antibodies can bind to cells recognized as foreign; a series of proteins referred to collectively as complement bind to the antibodies and damage the underlying cell membrane, leading to cell death. However, most attempts to demonstrate a role for antibody and complement in graft rejection have failed. Because no other known immune mechanism could account for the phenomena associated with graft rejection, many scientists were reluctant to accept graft rejection as immune mediated.

Cell-mediated cytotoxicity. This idea changed dramatically when André Govaerts found in 1960 that lymphocytes collected from lymph nodes draining a graft site could, when coincubated in the laboratory with cells taken from the original graft donor, destroy the graft cells in a matter of hours. Thus, the field of cell-mediated cytotoxicity (CMC) emerged. Cell-mediated cytoxicity is a major mechanism used by the immune system to rid the body of cells infected with intracellular viruses and bacteria, as well as cancer cells. Graft rejection is of course an unintended side effect.

Exploration of the cellular basis of cell-mediated cytotoxicity ultimately established the existence of certain CD8[+] T cells as the effector cells, the cytotoxic T lymphocytes. Establishing the mechanism(s) by which cytotoxic T lymphocytes destroy other cells then became a major focus for researchers studying cell-mediated cytotoxicity. Any reasonable mechanism that could be imagined was tested. Chief among these was complement, but complement components could not be demonstrated in association with cytotoxic T lymphocytes, or in the membrane of cells destroyed by them.

Perforin and granzymes. A major advance came in 1981 with the discovery of perforin, a complement-like molecule stored in cytoplasmic granules in cytotoxic T lymphocytes. When cytotoxic T lymphocytes bind to target cells through the surface receptor for antigen, the cytoplasmic granules move toward the interface between the cytotoxic T lymphocytes and the target cell, and their contents are released into the intercellular space. Once released, perforin assembles into pore structures, which insert into the target cell membrane and cause irreversible damage. The resulting membrane pores are similar to, although somewhat larger than, pores created by classical complement. It is not obvious how the cytotoxic T lymphocytes escape damage from the released granule contents.

Highly purified perforin causes cells to die by necrosis. In combination with other granule components, such as the granzymes (granule-associated proteolytic enzymes), perforin can induce apoptotic cell death, such as occurs with cells killed by cytotoxic T lymphocytes. Perforin itself may not be directly cytolytic, but may simply facilitate the entry of granzymes into the target cell. Granzymes themselves are thus proposed as the apoptosis-inducing agent. This point remains controversial.

Perforin-related mechanisms. During the initial period after it was first reported, it seemed that perforin and granzymes would account for all the phenomena associated with cell-mediated cytotoxicity. However, within a few years this notion was challenged. First, several highly potent sources of cytotoxic T lymphocytes, such as those found in the peritoneal cavity after rejection of an ascites allograft, have essentially no perforin or other granule contents. Second, killing by cytotoxic T lymphocytes can often be observed in the absence of calcium ions, where both the release of perforin and the assembly of perforin into pore structures are blocked. Thus, several laboratories proceeded with the creation of perforin-less mice, using the technique of gene disruption by homologous recombination to search for alternative lytic pathways. The initial reports describing such mice appeared in 1994. Perforin-less mice are healthy, both physiologically and reproductively, and have anatomically and cytologically normal immune systems. Although generally maintained in ultraclean facilities, their immune response to many pathogens, particularly bacterial pathogens, is quite good.

Fas molecule. A second lytic pathway in cytotoxic T lymphocytes was reported based on the Fas molecule, which is a member of the tumor necrosis factor (TNF-α) family of cell-surface receptor molecules. Cytotoxic T lymphocytes express a surface ligand for Fas, and when engaging a target cell through their antigen receptor, expression of the Fas ligand is greatly upregulated. This ligand can then engage Fas on the target cell surface and, quite independently of degranulation products, induce apoptotic target cell death. The Fas pathway appears to fit the characteristics of the putative perforin-independent lytic pathway proposed by several laboratories. The kinetics of apoptosis and cell death induced by Fas are quite similar to those seen in degranulation-induced death. Under artificially induced cell-mediated cytotoxicity assays, both perforin and Fas cause target cell death in 3–4 h at low (for example, 1:1) cytotoxic T lymphocyte to target cell ratios. It appears that Fas and perforin together can account for all of the acute cytolytic activity in cytotoxic T lymphocytes.

Tumor necrosis factor. In addition to the perforin and Fas lytic pathways, cytotoxic T lymphocytes also have a surface form of tumor necrosis factor α that can deliver a lethal signal to target cells displaying a tumor necrosis factor α receptor. Given the ubiquity of expression of the receptor for tumor necrosis factor α in mammals, particularly during an inflammatory response, this is a potentially important lytic pathway. Under laboratory conditions, killing target cells via the tumor necrosis factor pathway is considerably slower (18–24 h compared to 3–4 h for Fas and perforin). However, for reactions taking place over 7–10 days this difference may not be meaningful.

Immunity. The immune status of perforin-less mice was initially tested by immunizing them with lymphocytic choriomeningitis virus, which is thought to be cleared almost exclusively by CD8$^+$ cytotoxic T lymphocytes. Perforin-less mice were found to be unable to clear a lymphocytic choriomeningitis infection. When lymph node cells from infected perforin-less mice were tested for their ability to kill infected target cells in culture, it seemed initially that they were unable to do so. However it was subsequently realized that the targets used in these experiments were Fas$^-$; if (and only if) the target cell is Fas$^+$, lysis by perforin-less cytotoxic T lymphocytes is rapid and effective. The perforin-dependent cytolytic pathway thus seems to be absolutely required to clear this virus in the living organism. Some interesting questions about the Fas cytolytic pathway are thus raised. Perforin-less mice do have an intact Fas-Fas ligand system. Laboratory-controlled comparisons of the Fas and perforin pathways suggest they are comparable in regard to allowing cytotoxic T lymphocytes to kill target cells. Given that the Fas molecule is quite ubiquitous, especially in tissues at an inflammatory site, researchers are investigating why perforin-less mice do not use the Fas lytic system to clear a lymphocytic choriomeningitis infection.

Although perforin-less mice are unable to clear the infection, they appear to develop serious immunopathological consequences. The virus itself is completely harmless in mice; mice without functional immune systems are not affected. Yet about one-half of perforin-less mice infected with lymphocytic choriomeningitis become sick and die. They are found to have numerous hemorrhagic foci on organs and tissues throughout the body, and the hematocrit is greatly reduced. This effect can be completely eliminated by reducing the number of CD8 T cells in these mice prior to infection. Unlike normal mice, in which the CD8 cells used to clear the virus from the system disappear after a few days (along with the virus), in perforin-less mice, both the virus and the CD8 cells persist for a long time. Apparently, these cells cause ongoing immune damage. Whether this damage is caused by the Fas lytic system, or by cytokines released from activated CD8 cells, remains to be determined.

The response of perforin-less mice to infection with other intracellular parasites is currently being investigated. It has been shown that perforin-less mice are able to clear infections by *Listeria monocytogenes*: mice that have overcome a primary *Listeria* infection can overcome subsequent higher doses more readily than uninfected perforin-less mice, indicating development of some sort of immunological memory. This protection appears to be mediated by CD8 T cells. Spleen cells from *Listeria*-infected perforin-less mice are cytotoxic toward *Listeria*-infected hepatocytes in culture. This cytotoxicity is presumably either Fas-mediated or mediated by tumor necrosis factor. To what extent this laboratory manifestation of cytotoxicity is responsible for the protection seen in living organisms is not yet clear.

Allograft rejection. Recent results involving perforin-less mice have raised some important questions about the role of cell-mediated cytotoxicity in allograft rejection. These mice are inherently deficient in one of its three known mechanisms. By using allogeneic cells and tissues lacking both the Fas antigen and the tumor necrosis factor receptor (TNF-R) for transplantation into perforin-less mice, it is possible to create a transplant situation where none of the mechanisms can operate: perforin is missing in the host; Fas and tumor necrosis factor receptor are absent on the transplanted cells. In many instances, there is no apparent effect on the rate at which the allografts are rejected. However, elimination of CD8 T cells from perforin-less mice prior to transplantation profoundly reduces the ability to reject Fas⁻, TNF-R⁻ allografts. The inflammation-promoting properties of CD8 T cells may thus be more important than their cytotoxic properties for allograft rejection. Although highly potent cytotoxic T cells are clearly produced as a result of allograft rejection, there is some question as to whether the cytolytic function of these cells really causes allograft rejection. This issue is important to resolve.

For background information SEE CELLULAR IMMUNOLOGY; CYTOLYSIS; HISTOCOMPATABILITY; TISSUE CULTURE; TISSUE TYPING; TRANSPLANTATION BIOLOGY in the McGraw-Hill Encyclopedia of Science & Technology.

William R. Clark

Bibliography. D. Kägi et al., Cytotoxicity mediated by T cells and natural killer cells is greatly impaired in perforin-deficient mice, *Nature*, 369:31–39, 1994; A. Ratner and W. R. Clark, Role of TNF-α in acute CTL-mediated lysis, *J. Immunol.*, 150:4303–4310, 1993; E. Rouvier, M. F. Luciani, and P. Golstein, Fas involvement in Ca2⁺-independent T cell-mediated cytotoxicity, *J. Exp. Med.*, 177:195–201, 1993; C. M. Walsh et al., Immune function in mice lacking the perforin gene, *Proc. Nat. Acad. Sci. USA*, 91: 10854–10862, 1994.

Data communications

The race to build faster computing devices and high-speed applications is placing new demands on computer networks; they are running out of bandwidth. New applications, faster user devices and servers, and an increasing numbers of users are pushing networks to their performance limits. In response to these challenges, asynchronous transfer mode (ATM) switching has emerged as a technology with the potential to alleviate the bottlenecks in local-area networks (LANs) and wide-area networks (WANs) and act as a technology model to move networking capabilities into the future.

Development of ATM. ATM originated in the Consultative Committee on International Telephony and Telegraphy (CCITT). The CCITT is an international body, working under the auspices of the United Nations and the International Telecommunications Union, which sets standards for the world's telephony and other public networks.

In the early 1980s, the CCITT developed standards for the Narrowband Integrated Services Digital Network (N-ISDN), to allow a limited capability for public networks to carry digital data traffic. By the mid-1980s, public telecommunications providers saw the need for higher-bandwidth data transport, due in large part to the widespread adoption of digital communications equipment and the growing economic importance of data traffic. Study Group XVII of the CCITT began working on a successor to N-ISDN, known as Broadband ISDN (B-ISDN). The goal of this effort was to define, essentially, the eventual replacement for the entire public network infrastructure. A key objective was to define a single switching technology that would allow all types of traffic to be carried on the same network. In this way, many of the problems of disjoint network fabrics for each type of traffic could be avoided, and a technology developed that could scale far beyond the limits of then current technologies, in recognition of the exponential growth in data traffic rates. Public telecommunications providers could upgrade their existing infrastructure of transmission, switching, and multiplexing systems and easily move to an all-digital network.

The CCITT examined many different technologies to meet these goals. Variants of both time-division multiplex (TDM) technologies, synchronous transfer mode (STM) and packet switching, were examined. In 1988, the CCITT decided to base the development of B-ISDN on ATM, which was formalized by a set of recommendations. The term ATM is now sometimes used as a synonym for B-ISDN. However, B-ISDN is just one of the possible services that can use ATM technology.

ATM technology. ATM is one of the general classes of packet technologies that relay traffic to a selected network destination via an address contained within a packet. In general, packet switching means data transport over packet-switched networks that use variable-length addressed packets that incorporate extensive error checking for use over noisy analog transmission facilities. In ATM, all data are transferred in fixed-length 53-byte cells. Each cell has a 5-byte header that identifies the cell's route through the network and a 48-byte payload containing user data. These user data in turn carry any headers or trailers required by higher-level protocols. Using fixed-length cells provides a method of predicting and guaranteeing bandwidth for applications that need it, whereas variable-length packets can cause delays at switches.

The operation of an ATM switch is conceptually simple. The header of each cell contains a virtual connection (VC) identifier, consisting of a virtual path identifier (VPI) and a virtual channel identifier (VCI). On each incoming link, an arriving cell's VC identifier uniquely determines a new VC identifier to be placed in the cell header and the outgoing link over which to transmit the cell. In the case of a multicast connection (in which one cell can be sent to one or more destinations in a single operation), the VC identifier maps to a set of new VC identifiers and outgoing links.

The simplicity of the algorithm for forwarding ATM cells lends itself to efficient hardware implementation, which is the only way that the required switching speeds may be obtained. The VC routing information that the hardware uses is updated by switch-control software when a connection is established. In the case of a multiple-switch ATM LAN, the control software on the switches must cooperate to route the connection through the network, but again only at the time the connection is established; during the transfer of cells over the connection, only the switch hardware is involved.

ATM allows the definition and recognition of individual communications by virtue of the label field inside each ATM cell header; in this respect, it resembles conventional packet transfer modes. Like packet-switching techniques, ATM can provide a communication with a bit rate that is individually tailored to the actual need, including time-variant bit rates.

The term asynchronous in the name of the new transfer mode refers to the fact that, in the context of multiplexed transmission, cells allocated to the same connection may exhibit an irregular recurrence pattern as cells are filled according to the actual demand (**Fig. 1**). In the synchronous transfer mode (STM; **Fig. 2**), a data unit associated with a given channel is identified by its position in the transmission frame. In ATM a data unit or cell associated with a specific virtual channel may occur at essentially any position. The flexibility of bit-rate allocation to a connection in STM is restricted because of predefined channel bit rates and the rigid structure of conventional transmission frames. These normally will not permit individual structuring of the payload or will permit only a limited selection of channel mixes at the

corresponding interface at the time that a connection is established.

In ATM-based networks the multiplexing and switching of cells are independent of the actual application. So the same piece of equipment in principle can handle a low-bit-rate connection as well as a high-bit-rate connection, either of stream or burst nature. Dynamic bandwidth allocation on demand with a fine degree of granularity is provided. Thus, the definition of high-speed channel bit rates is now, in contrast to the situation in an STM environment, a secondary function of the network.

Benefits of ATM. These include scalability, statistical multiplexing, traffic integration, and granularity.

Scalability. This is one of the most important properties of ATM. ATM is scalable in speed, size, distance, and traffic type. The essential factors contributing to the scalability are a switched-based architecture and a common cell structure across ATM components. Users across a wide variety of system types can access ATM networks irrespective of media types and applications. Within the limitations imposed by the bandwidth of the physical link or connections, an arbitrary bit rate can be allocated to the user which remains assigned for the entire connection. Simultaneously, from a network-systems perspective, the capacity of a node or switch can be increased to meet new load requirements by interconnecting several collocated switches (scaling) and distributing the switching bandwidth over the multiple physical switches. As the network load increases and as more users need to be connected, more switch ports and switches can be added.

Statistical multiplexing. ATM is attempting to solve the unused bucket problem of STM by statistically multiplexing several connections on the same link based on their traffic characteristics. In other words, if a large number of connections are very bursty, then all of them may be assigned to the same link in the hope that statistically they will not all burst at the same time; and, if some of them do burst simul-

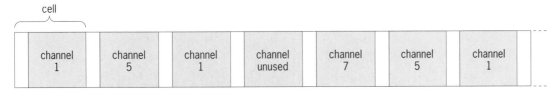

Fig. 1. Typical allocation of cells in asynchronous transfer mode (ATM).

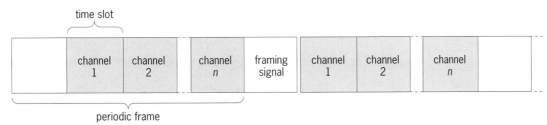

Fig. 2. Structure of transmission frames in synchronous transfer mode (STM).

taneously, that there is sufficient elasticity that the burst can be buffered and put in subsequently available free cells. This characteristic is called statistical multiplexing, and it allows the sum of the peak bandwidth requirements of all connections on a link to even exceed the aggregate available bandwidth of the link under certain conditions of discipline. This capability was impossible on an STM network, and it is the main distinction of an ATM network.

Traffic integration. ATM networks can carry integrated traffic because they use small fixed-size cells. The small fixed size helps to overcome the problem of uncertain delay experienced by packet-switched networks with their variable-length packets. This is accomplished by dedicating resources for a brief cell time to each source, which allows data from different sources to appear as if they are transmitted concurrently and transparently. Since different types of transmissions (voice, video, and data) can be interleaved in these small cells, a cell carrying delay-sensitive data need wait only a very short length of time. ATM networks will also implement various traffic-shaping and policing management protocols to ensure that all nodes receive the amount of bandwidth they require. In ATM, the true integration of traffic takes place at the ATM layer as cells from different ATM Adaptation Layer (AAL) classes, signaling, and management data are mixed in the same cell format.

Granularity. ATM allows a network to be engineered for the set of applications rather than forcing applications to fit the network. Many network technologies have difficulty dealing with anything that does not fit the limited granularity of the digital hierarchy. ATM allows the user to deliver traffic at rates and degrees of burstiness compatible with the applications running, not at the rates convenient to the network. ATM provides the mechanism to attain the ultrahigh-speed transmission necessary for emerging multimedia applications because (1) it is a simple, very fast switching and routing process based on call address, and (2) unlike the X.25 protocol, it does no processing in the network above the most rudimentary cell level, thereby simplifying and speeding the handling of traffic.

For background information *SEE DATA COMMUNICATIONS; INTEGRATED SERVICES DIGITAL NETWORK (ISDN); LOCAL-AREA NETWORKS; PACKET SWITCHING; WIDE-AREA NETWORKS* in the McGraw-Hill Encyclopedia of Science & Technology.

Gregory E. Federline

Bibliography. G. E. Federline, ATM: What to expect in 1995, *Telecommun. Amer. Ed.,* pp. 69–70, January 1995; *Proceedings of LANWAN Asia,* November 1993; *Proceedings of UNIX EXPO,* 1995.

Data compression

The digital representation of analog audio and video signals requires sampling and quantization. It is often desirable to reduce the number of bits needed for this representation, a procedure known as compression. The first section of this article discusses general techniques of image and video compression, particularly those that are combined in the MPEG-2 standard for high-quality video. The second section discusses trellis-coded quantization, a powerful compression technique.

Image and Video Compression

Image and video are visual perceptions. They can be represented in digital forms after appropriate discretization and quantization. Thus, an image or picture can be represented by an array of pixels or pels, and each pel can be represented by a certain number of bits, typically 8. Video is a sequence of images displayed at a certain number of frames per second, typically 30. For color images, each pel can be represented by three primary components, red (R), green (G), and blue (B), and each requires the same number of bits for representation. Consequently, the total number of bits required to represent a typical color image with a resolution of 512×512 pels is approximately 6.29 megabits. Similarly for a videocassette-recording (VCR) quality video with a resolution of 240×360 pels at 30 frames per second, the required bit rate is approximately 62.2 megabits per second. As a reference, the storage capacity and access rate for the widely used CD-ROM (compact-disk read-only memory) are 5200 megabits and 1.2 megabits per second respectively. Without image or video compression, the above-mentioned CD-ROM could store only 84 s of VCR-quality video and would take 72 min to retrieve it. Assuming that a typical television channel can carry a digitally modulated signal of 18 megabits per second, the same uncompressed VCR-quality video would require approximately 3.5 channels to broadcast with quality much inferior to that of regular television. However, because of abundant redundancy among pels within a frame and also between frames, a great deal of image and video compression is possible.

MPEG-2. Numerous image and video compression techniques have been proposed. At present, no single technique can achieve a high degree of compression and also work in all situations. Consequently, a modern high-compression scheme is usually a combination of several complementary techniques. One outstanding example is the international standard for high-quality video called MPEG-2 (developed by the Moving Pictures Experts Group of the International Standards Organization). MPEG-2 is a generic coding standard that supports a variety of applications by means of profiles and levels. A profile defines a number of technical features and functionalities, such as signal-to-noise ratio and spatial scalabilities, required by a cluster of similar applications. A level is a set of constraints imposed on parameters, such as picture resolution, within a profile. MPEG-2 can compress high-quality television to about 9 megabits per second and high-definition television (HDTV) to

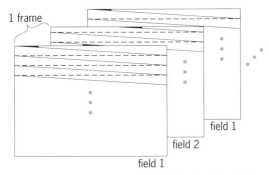

Fig. 1. Division of a frame into two fields by interlaced video.

about 18 megabits per second. Another video coding standard, called H.262, is a low-delay version of MPEG-2 suitable for two-way visual communication. MPEG-2 can compress either progressive or interlaced video with several different spatial resolutions. The progressive video is based on frames, and each frame contains a certain number of scan lines. The interlaced video divides each frame into fields 1 and 2, which consist of odd- and even-numbered lines respectively (**Fig. 1**). The interlaced video was introduced in order to reduce the bandwidth of a television signal without sacrificing too much of its visual quality. Each pel is represented by luminance (Y) and two chrominance (U and V) components which can be converted from RGB. The reason for using YUV is that it requires fewer bits than RGB because the chrominance components can be subsampled without losing much visual quality. The major elements of MPEG-2 are discrete cosine transformation (DCT), quantization, variable-length coding, and motion estimation.

Discrete cosine transform. This is a mathematical technique that transforms a block of pels, typically 8×8, into a block of frequency-domain coefficients of the same dimension. It is like Fourier-series expansion in two dimensions consisting of a constant plus various frequency components of the block in both horizontal and vertical directions. Because human vision is not sensitive to the high-frequency (rapidly changing) content of a visual pattern, these high-frequency domain coefficients can be discarded (thus achieving compression) without degrading the picture quality. The discrete cosine transform is a computational intensive operation, typically requiring several hundred million multiplication and accumulation operations per second for high-quality video. Because of recent advances in microelectronics and fast techniques of computing the discrete cosine transform, it is no longer a bottleneck in achieving cost-effective implementation of video compression. For interlaced video, the computation of the discrete cosine transform is somewhat complicated depending on whether the compression is done on a frame or field basis. To illustrate this, a block of 16×16 pels is shown in **Fig. 2***a*, where the o's and e's represent pels in the odd- and even-numbered lines, respectively. For the frame-based coding, the discrete

cosine transform should be performed on the 8×8 blocks shown in Fig. 2*b*. For the field-based coding, since the o's and e's belong to different fields, the discrete cosine transform should be performed on the 8×8 blocks containing the o or e pels exclusively, as shown in Fig. 2*c*.

Quantization. After the discrete cosine transform, the coefficient values can be quantized to a small number of levels which in turn can be represented by binary bits. A finer quantization would produce better image quality but require more bits. In contrast, a coarse quantization would produce inferior image quality but require fewer bits. In most appli-

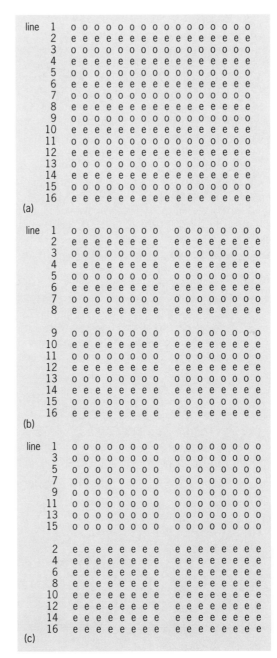

Fig. 2. Computation of discrete cosine transform in interlaced video. (*a*) Original block of pels. (*b*) Blocks are transformed in frame-based coding. (*c*) Blocks are transformed in field-based coding.

Table 1. Example of the Huffman code

Symbol	Probability	Huffman code	Fixed-length code
a	0.5	0	00
b	0.25	10	01
c	0.125	110	10
d	0.125	111	11

cations, the number of levels in a quantizer changes dynamically during the coding according to a certain strategy so as to maintain a constant bit rate.

Variable-length coding. This is a lossless coding technique which is completely reversible. The inverse operation of variable-length coding is called variable-length decoding (VLD). By using the statistical properties of a given set of symbols, variable-length coding assigns a shorter code to a more frequently occurring symbol and vice versa. Various variable-length coding techniques have been proposed. Because of its simplicity and high efficiency, the Huffman code has been adopted in many video coding standards, including MPEG-2.

A simple example of the Huffman code for a set of four symbols with their probability of occurrence is shown in **Table 1**, together with its fixed-length code. For instantaneous decodability of a variable-length code, it is necessary (and also sufficient) that no code word be a prefix of some other code words. In Table 1, only one bit has been assigned to the most frequently occurring symbol, a. The average word length for the fixed-length code is 2. However, for the variable-length code, it is the sum of word length of each symbol weighted by its probability of occurrence, which is $0.5 \times 1 + 0.25 \times 2 + 0.125 \times 3 + 0.125 \times 3 = 1.75$. The resultant compression ratio is 1.14:1 for this simple case. In general, a compression ratio ranging from 1.5:1 to 4.0:1 can be expected, depending on symbol statistics.

The implementation of variable-length coding is very straightforward. However, some ingenuity is required for the implementation of variable-length decoding.

Motion estimation. Interframe redundancy can be removed by predictive coding with motion estimation. Motion estimation is done by comparing a block of pixels, typically 16×16, in a current frame with blocks in a previously coded frame within a search window. The best-matched block is used as the prediction. If the prediction is perfect, no bits need to be transmitted. Otherwise, the prediction error is coded via the discrete cosine transform, quantization, and variable-length coding. Meanwhile the location of the matched block is also coded via variable-length coding. This approach requires far fewer bits than the original block. Motion estimation is the most computation-intensive and challenging part of video compression. It can consume as much as 75% of total signal processing power devoted to compression and decompression, and it must be done in real time. Generally speaking, motion estimation techniques can be divided into two major categories, full search and fast search. The full search method is to compare the current block with every block in the search window. It would require about 1.1×10^{10} absolute difference calculations per second for high-quality video with a search range from −16 to +15 pels. Nevertheless, because of its regular data structure, very large scale integrated-circuit (VLSI) implementations of full-search motion estimation with varying degrees of capability have been reported. Because a full search requires too much computation, it is highly desirable to devise a fast-search technique that would deliver comparable performance. A large number of fast-search techniques have been proposed and some have been used in practice.

One of the reasons why MPEG-2 can achieve high-quality video with high compression is the use of bidirectional motion estimation with a field-frame option. A sequence of frames is first formed into a group of pictures (GOP; **Fig. 3**). The first frame, I, uses intrapicture coding; that is, only spa-

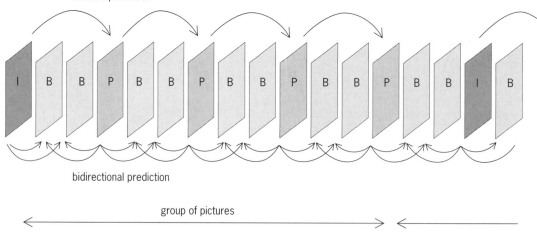

Fig. 3. Group of pictures in bidirectional motion estimation.

tial redundancy within the frame is removed. P is the predicted frame, with motion estimation based on either the I frame or the previous P frame. Between the I and P frames, there can be a number of B frames (for bidirectional coding), typically two as shown in Fig. 3. Motion estimation for the B frames can be obtained from one of the following three cases depending on which one is more accurate: the previous I or P frame, the following P frame, or the average of both. Motion estimation from the following frame would provide much more accurate information for the uncovered areas created by moving objects. Because of the bidirectional nature of the motion estimation, the coding sequence is different from the source sequence. Also, a few-frame coding delay would occur, which is not desirable for simultaneous two-way visual communication. For MPEG-2, motion estimation can be done on either a frame basis or a field basis, or on a combination of both. This is extremely useful because, for a fast-moving scene, prediction from a field is more accurate than from a frame and the reverse is true for a stationary scene. After a full-pel motion estimation is completed, a half-pel motion estimation can be carried out to increase its accuracy.

By properly combining the major compression techniques, MPEG-2 can achieve a compression ratio up to about 50:1 with very good visual quality. A higher compression ratio is possible; however, the visual quality will be greatly degraded. Research on very high video compression, of the order of 200:1, continues. Techniques such as object-oriented segmentation, model-based coding, and fractal coding, which are fundamentally different from those used in MPEG-2, have been intensively investigated for future applications.

Ming Liou

Trellis-Coded Quantization

To store and process sound, images, and video data in computers, it is necessary to convert the analog waveform to a digital representation. Analog-to-digital conversion consists of two stages, sampling and quantization. Sampling the waveform creates a discrete time signal (also known as a time series), where the value of each sample can be any real number. Quantizing maps the value of the sample to one of N levels. The level can then be represented by $\log_2 N$ bits. Typically, for compact-disk quality digital audio, the sampling rate is approximately 44,100 samples per second, and each sample is stored as 16 bits. This precision gives 65,536 (2^{16}) possible levels. Digital video is usually stored with 8 bits per pixel ($2^8 = 256$ levels) for each of the three color components, red, green, and blue.

For transmission over many channels, or for long-term storage, it is often desirable to compress the audio or video so that fewer bits are needed than for the normal digital representation. The easiest way to do this is to requantize each sample into a smaller number of levels. Although easy to implement and cost effective, this method usually results in unacceptable quality. Among the many techniques studied for compression, trellis-coded quantization has emerged as one of the most powerful. Its computational requirements are modest, and it can achieve quality that is relatively close to the theoretical limits. Because scalar quantization is embedded in trellis-coded quantization, it will be briefly discussed.

Scalar quantization. **Figure 4** shows the input-output relationship of a scalar quantizer. The sample's input value is given by the **x** axis, and the output value by the **y** axis. For a range of input values, the output value is the same. In this example, there are 8 output levels; therefore, 3 bits per sample are required. The number of input levels depends on the application. When quantizing a sampled analog waveform, each input value can be any real number; for requantizing a digital value, the input values must be from some finite set of numbers. Using audio as an example, there are over 65,000 possible input values between x_0 and x_8. After quantization, the number of bits required to represent the sample has been reduced, but so has the dynamic range.

The scalar quantization design problem becomes one of selecting the end points, x_k, for the input regions, and the output value, y_k, for each region. This selection is done by assuming that the probability density function is known for the input signal. The distortion between the input and output values is calculated. The Lloyd-Max algorithm is then used to adjust the end points and output values so that the distortion is minimized.

The use of a scalar quantizer may be illustrated by the values for x_k and y_k in **Table 2**. These values are typical for the quantization of a gray-scale image, originally represented by 8 bits per pixel. If five input pixels are 68, 100, 110, 133, and 180, then the corresponding output pixels are 79, 111, 111, 143, and 175. The total distortion, ρ, caused by quantizing the five pixels is given by Eq. (1), and is

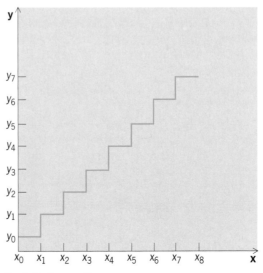

Fig. 4. Scalar quantizer.

$$\rho = (68-79)^2 + (100-111)^2 + (110-111)^2$$
$$+ (133-143)^2 + (180-175)^2 = 368 \quad (1)$$

also known as the squared error. The average distortion per pixel (mean-squared error) is 368/5 = 73.6. After quantization, only 3 bits per pixel ($2^3 = 8$ levels) are required, but the values have been distorted.

Trellis coders. Although trellis coders have been designed and implemented for a number of years, interest in them has increased since 1990. This is primarily due to a discovery in the related field of digital modulation theory. In this field, one of a number of signals is transmitted over a noisy channel during a time interval known as the symbol period. The receiver then attempts to determine which signal was transmitted during that interval. Usually, the decisions are made independently for each interval. In 1982, G. Ungerboeck showed that by expanding the number of possibly transmitted signals by a factor of two, while using the previously received signal to select a subset of the signals to test, performance could be significantly increased.

A similar approach is applied to trellis quantization. Instead of using a quantizer with N levels, a quantizer with $2N$ levels is used, leading to better performance. This larger quantizer is partitioned into a number of smaller quantizers (subquantizers) which will be discussed below. To maintain the same output bit rate, only some of the subquantizers, containing a total of N levels, are used at any time. At each sample time, the encoder is in one of a finite number of states. Each state determines the set of quantizers that can be chosen to encode the input sample.

The key to understanding this process is a trellis diagram, shown in **Fig. 5**a. There are four states in the trellis, labeled s_0, s_1, s_2, and s_3. There are two branches leaving each state and two entering. Three time intervals, corresponding to three input samples, are shown. An 8-level scalar quantizer has been partitioned into four quantizers, each containing two possible output levels. Figure 5b shows the partitioning, with the four quantizers labeled D_0 through D_3.

In a further example, the trellis-coded quantization encoding procedure for the five pixels in the

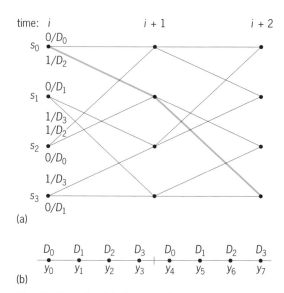

(a)

(b)

Fig. 5. Trellis coding. (a) Diagram of four-state trellis. (b) Partition of an eight-level scalar quantizer into four sets.

previous example will be shown in order to look at how the input pixel values determine a path through the trellis. The first two pixels, with values 68 and 100, are used. The encoder always starts in state s_0. In this state, only reproduction levels from quantizer D_0 or D_2 can be used. From Table 2, the closest value for the first pixel is 79, which belongs to D_2. On the trellis diagram, the state moves from s_0 to s_1 along the heavy path. One bit, with a value of 1, is used to indicate which path was taken, and another bit, a zero, is used to select the smaller value in quantizer D_2. The next pixel is now input. Its value is 100, and in state s_1 only quantizers D_1 or D_3 can be used. The closest value is 111, the smaller level in D_3. The state changes to s_3, again moving along the heavy path.

Although 3 bits are required to specify 8 levels, this trellis-coded quantization system has a bit rate of 2 bits per sample since only 4 levels can be used at a given state. One bit per sample is used to specify the path through the trellis, and the other bit is used to determine which of the two reproduction levels is used. Performance will be close to that of the scalar quantizer at 3 bits per sample. Trellis coders with larger numbers of states can provide even better performance, at the cost of greater computational complexity.

Encoding procedure. Trellis-coded quantization obtains its performance primarily because it seeks to minimize the total quantization distortion for the entire sequence, compared to minimizing it sample by sample. The encoding process for an input sequence of K samples consists of two steps. The first step computes a distortion value, known as the branch metric, for each branch in the trellis. This step is done for all of the input samples. Because each branch has a quantizer associated with it, the distortion between the input pixel and the two output levels in the quantizer are calculated, and the smaller distortion becomes the

Table 2. Input range end points and output levels for a scalar quantizer

Input end points		Output levels		Subquantizer*
x_0	0.0	y_0	15.0	D_0
x_1	31.0	y_1	47.0	D_1
x_2	63.0	y_2	79.0	D_2
x_3	95.0	y_3	111.0	D_3
x_4	127.0	y_4	143.0	D_0
x_5	159.0	y_5	175.0	D_1
x_6	191.0	y_6	207.0	D_2
x_7	223.0	y_7	239.0	D_3
x_8	255.0			

* The last column gives the subquantizer to which the particular output level belongs.

branch metric, $b^i_{j,k}$. The superscript i denotes the time, subscript j denotes the state the branch originates from, and k determines which of the two branches is used. When this is completed, a trellis of length K has been constructed. The second step in the encoding process is a search of the trellis to find the output sequence that results in the minimum cumulative distortion. The search is done by using the Viterbi algorithm, used to decode convolutional codes in communication systems.

Figure 6 shows the complete branch-metric labeling of the trellis for the five input pixels used in the first example. To compute the branch metrics for the input pixel, 68, the first step is to determine which of the two values in each quantizer, D_i, is the best for the input pixel: the scalar quantizer that results in the smallest distortion is chosen. For D_0, the possible values in Table 2 are 15 and 143. The former is closer to 68, so it is used. Similarly, the best value in D_1 is 47. **Table 3** shows these results for all four quantizers. After calculating the distortion values for each quantizer, each branch is given a value based on which quantizer is used on the branch. The branch from state s_0 to s_0 uses D_0, as shown in Fig. 5a. It is given the value of 2809.

The Viterbi algorithm is a recursive method used to determine the maximum estimate, based on the data, of a sequence of states for a discrete-time Markov process. Rather than giving a mathematical description of the algorithm, an example will be used to show how the algorithm works. The vector $\mathbf{s}$ = $(s^0, s^1, \cdots, s^i)$ is defined to be the state sequence from time zero to time i, and $\hat{\mathbf{s}}(s^i_j)$ to be the survivor

path to state s_j at time i. The survivor path is the state sequence with the smallest distortion that ends in the given state. Examples of these are given below. The quantity $\Gamma(s^i_j)$ is defined to be the corresponding distortion for each survivor path.

In the case of the labeled trellis in Fig. 6, the steps taken by the Viterbi algorithm are shown in **Table 4** for the same five input samples. Because the algorithm always starts in state s_0, the initial survivor path for this state is the one-element vector (s_0). The survivor paths for the other three states are undefined and denoted by (?). The (cumulative) distortion for the path to s_0 is initialized to zero, while the initial distortions for the other three states are infinite.

For each step in the recursion, that is, transitioning from time i to time $i + 1$, the algorithm determines the best path to each of the states. This is done in two steps, by using the best paths to each state at the previous time step and the branch metrics. First, the distortion for following the trellis from state s_{j1} to state s_j is calculated, using Eq. (2).

$$\Gamma(s^{i+1}_j, s^i_{j_1}) \doteq \Gamma(s^i_{j_1}) + b^i_{j_1,k} \qquad (2)$$

The last term is the branch metric for the branch used to move between the two states. Eq. (2) is computed for every possible state transition. The four-state trellis has eight possible transitions among the states. Second, for each s^{i+1}_j, the new distortion for the best survivor path is found according to Eq. (3).

$$\Gamma(s^{i+1}_j) = \min_{s_{j_1}} \Gamma(s^{i+1}_j, s^i_{j_1}) \qquad (3)$$

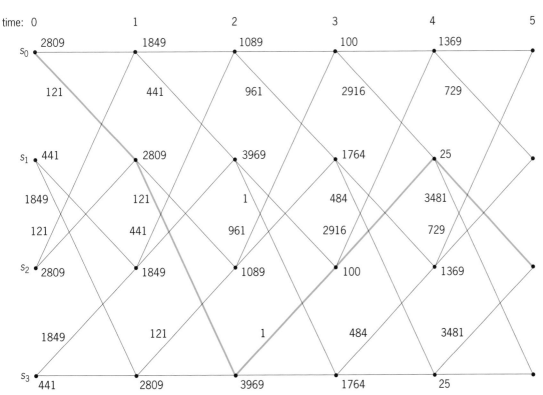

Fig. 6. Five time steps of the trellis with all the branch metrics.

Table 3. Computation of branch metrics for the first time step

Subquantizer	Best output value	Distortion
D_0	15.0	$(68 - 15)^2 = 2809$
D_1	47.0	$(68 - 47)^2 = 441$
D_2	79.0	$(68 - 79)^2 = 121$
D_3	111.0	$(68 - 111)^2 = 1849$

For example, the best way to reach s_0^1 is from s_0^0. The branch metric is 2809, and the previous distortion is zero. The new value of the distortion is shown in Table 4, as is the survivor path. Similarly, to get to s_1^1, the best path comes from s_0^0. The branch metric is 121, and adding this to the previous distortion of zero gives a distortion of 121. At time 1, states s_2 and s_3 cannot be reached, so their distortions are still infinite and their survivor paths are denoted by (?,?). These states can be reached from time 2 on.

After all the samples have been encoded, the survivor path with the smallest distortion is selected. It ends in state s_2, yielding a total distortion of 368. In Fig. 6, it is shown by the heavy line. Two bits are stored per sample. The first one indicates if the lower or upper branch is taken for each sample, and the second bit selects the reproduction level in the appropriate quantizer, D_i. The path selected by this example gives output pixels of 79, 111, 111, 143, and 175. These pixels are the same as those of the scalar quantizer, which needs 3 bits per sample.

The decoding is very simple. As each sample is processed, the decoder follows the path through the trellis, using the 1 bit per sample that specifies the path. The other bit per sample is used to choose the reproduction level which is then output. Configurations of trellis coding can be designed to obtain bit rates below 1 bit per sample. The simplicity of the decoder makes this method attractive for applications such as broadcasting, where there are many decoders but only a few encoders.

Codebook design. There are three main issues involved in the design of a trellis-coded quantization system. The first is determining the trellis, specifically the number of states and the branching structure. The second issue is choosing the scalar quantizer to be embedded in the trellis. The third issue is how the quantizer is partitioned and assigned to the branches.

Initial trellis-coded quantization systems adopted the trellises used by G. Ungerboeck for trellis-coded modulation because the results obtained are close to the best theoretically possible. Ungerboeck designed trellises containing 4–256 states by using a number of results from rate-distortion theory and heuristic design rules. An important consequence of his work is that close to optimal performance can be obtained by doubling the number of signals (or doubling the number of quantizer levels in trellis-coded quantization). Although performance improves with the number of states, the gain is less for large-state trellises. Because more states require a more complex implementation, the computational ability of the hardware will often limit the number of states.

Originally, trellis-coded quantization used scalar quantizers designed with the Lloyd-Max algorithm. Because quantizers were already designed for common probability density functions such as gaussian and laplacian, it is necessary to determine only the desired number of reproduction levels. This method works very well for higher bit rates, that is, 2 or more bits per sample, but at bit rates around 1 bit per sample, performance is not as good. Better results can be obtained by using a training sequence to design the quantizer. Although the details are somewhat involved, the basic idea is to use a typical input sequence in the design. An initial quantizer is chosen, and samples from the training sequence are coded with it. The samples mapped to a particular quantization level are then used to refine the level. The procedure is iterated a few times to obtain the new quantizer.

The last issue in the design is the partitioning of the quantizer and the assignment of quantizer values to the branches of the trellis. Here also trellis-coded modulation design rules are used. A set partitioning rule is employed. It essentially states that the quantizer should be divided into two sets (subquantizers) such that the maximum distance between the two sets is minimized. Each set is divided into two more sets using the same criterion, and the process is repeated until the required number of sets is obtained. These sets of quantizers form a binary tree where each set has one parent and two children. In the examples discussed previ-

Table 4. Survivor paths to each state with their associated cumulative distortion

Time, i	0	1	2	3	4	5
$\hat{s}(s_0^i)$	(s_0)	(s_0, s_0)	(s_0, s_0, s_0)	(s_0, s_1, s_2, s_0)	$(s_0, s_1, s_3, s_2, s_0)$	$(s_0, s_0, s_1, s_3, s_2, s_0)$
$\hat{s}(s_1^i)$	(?)	(s_0, s_1)	(s_0, s_0, s_1)	(s_0, s_1, s_2, s_1)	$(s_0, s_1, s_3, s_2, s_1)$	$(s_0, s_1, s_3, s_2, s_0, s_1)$
$\hat{s}(s_2^i)$	(?)	(?, ?)	(s_0, s_1, s_2)	(s_0, s_1, s_3, s_2)	$(s_0, s_0, s_1, s_3, s_2)$	$(s_0, s_1, s_3, s_2, s_1, s_2)$
$\hat{s}(s_3^i)$	(?)	(?, ?)	(s_0, s_1, s_3)	(s_0, s_0, s_1, s_3)	$(s_0, s_1, s_2, s_1, s_3)$	$(s_0, s_1, s_3, s_2, s_1, s_3)$
$\Gamma(s_0^i)$	0	2809	4658	3891	3159	4464
$\Gamma(s_1^i)$	∞	121	3250	4019	343	3888
$\Gamma(s_2^i)$	∞	∞	2930	243	3735	368
$\Gamma(s_3^i)$	∞	∞	242	3251	4503	3824

ously, the eight-level quantizer was divided into two sets. The first set contains (y_0, y_2, y_4, y_6), and the second one contains (y_1, y_3, y_5, y_7). Each set is partitioned to yield $D_0 = (y_0, y_4)$, $D_2 = (y_2, y_6)$, $D_1 = (y_1, y_5)$, and $D_3 = (y_3, y_7)$.

The assignment of quantizers to the branches is done so that the quantizers used when leaving or entering a particular state contain half the total number of levels. Additionally, the two quantizers should be derived from the same parent quantizer. It is also desirable that all quantizers be used equally often. These constraints help assign the quantizers to the branches. For trellises with a large number of states, the rules can be augmented by using numerical optimization algorithms.

For input sequences where the samples are uncorrelated, performance is quite good. However, for the commonly occurring case where there is correlation among the input samples, a change must be made. Instead of inputting each sample to the trellis-coded quantization system, a prediction of the sample is made based on the previous samples. The prediction is subtracted from the sample, and the difference is sent into the trellis-coded quantization system. At the decoder, the decoded value of the previous sample is added to the decoded difference value to get the present sample. The first sample is coded without prediction to start the process. Performance is now dependent not only on the trellis-coded quantization but also on the quality of the prediction. Still, for many practical applications, the performance is as good as any other known method.

Video compression applications. Because of the very high bit rate of digital video, one of the major applications of trellis-coded quantization is video compression. A digital image with resolution similar to conventional television has 720 pixels per scan line and 480 scan lines, for a total of 345,600 pixels. At 30 frames per second, this translates into more than 10^6 pixels per second. Color video uses 24 bits per pixel, giving an aggregate uncompressed bit rate of almost 250 megabits per second. HDTV will have a bit rate at least four times greater.

Some of the emerging video compression algorithms use digital filter banks followed by subsampling to filter each video frame into a number of spatial frequency subbands. Each subband typically has different statistical properties. The lowest-frequency subband contains a low-resolution version of the input image, while the higher-frequency subbands contain edge information. Most pixels in these subbands have values close to zero except for those representing the edges. Trellis-coded quantization, and an improved variation, entropy-constrained trellis-coded quantization, have been successfully applied to code these subbands.

For background information *SEE ANALOG-TO-DIGITAL CONVERTER; COMPACT DISK; INFORMATION THEORY; INTEGRAL TRANSFORM; TELEVISION; TELEVISION STANDARDS* in the McGraw-Hill Encyclopedia of Science & Technology.

Robert E. Van Dyck

Bibliography. T. R. Fischer and M. Wang, Entropy-constrained trellis coded quantization, *IEEE Trans. Info. Theory,* 38:415–425, 1992; G. D. Forney, Jr., The Viterbi algorithm, *Proc. IEEE,* 61:268–278, 1973; A. Gersho and R. M. Gray, *Vector Quantization and Signal Compression,* 1992; International Standards Organization, *Generic Coding of Moving Pictures and Associated Audio,* Committee Draft ISO/IEC 13818-2, Recommendation H.262, ISO/IEC JTC1/SC29, WG11/602, 1993; M. W. Marcellin and T. R. Fischer, Trellis coded quantization of memoryless and Gauss-Markov sources, *IEEE Trans. Comm.,* 38:82–93, 1990.

Decapoda (Crustacea)

The most characteristic type of movement of shrimps and lobsters is the tail-flip escape response. In the tail flip, the abdomen is rapidly flexed and re-extended to produce a strong propulsive force that propels the animal backward through the water at high speed. Crustaceans ranging in size and habit from large bottom-dwelling clawed lobsters and crawfish to small planktonic shrimps and mysids show this response, which may be evoked by a variety of noxious or threatening stimuli but probably most commonly by the approach of potential predators. Such stimuli may, in some instances, elicit only a single tail flip; in other circumstances, the response may be a sequence of escape swimming powered by a repeated series of tail flips. The characteristics (timing, direction, speed, duration) of the resulting movement are important in determining whether the individual will survive the encounter or fall prey to the attacker.

Biomechanics. The generation of propulsive forces in the tail flip has been compared to rowing with a single oar, with movement resulting from a combination of a reactive force (added mass) and a resistive force (drag). In addition, a hydrodynamic squeeze force, that is, a type of jet propulsion, is produced as the abdomen is pressed against the cephalothorax toward the end of flexion. Thrust generation is enhanced by an increase in cross-sectional area of the posterior end of the abdomen caused by spreading of the segments of the telson (tail fan) as flexion of the abdomen begins. The flexion of the tail has a tendency to rotate the animal in a somersault and to generate useful directional movement, the balance between these behaviors depending on the relative masses of the abdomen and the cephalothorax. An optimal relationship between these two elements can be predicted, on the basis of theoretical hydrodynamic analysis of the forces involved, in which effective directional movements are maximized while useless rotational movements are reduced.

Although basically similar, the tail-flip responses of different crustaceans show characteristic features related to their different sizes and morphologies. In larger decapods, such as the clawed lobsters and crawfish, the center of mass lies within the cephalothorax, anterior to the point of flexion. The thrust generated by the relatively small abdomen acts against the inertia of the mass of the heavy, armored cephalothorax together with its appendages, restricting the possible range of movements that can be generated by the tail flip. In smaller decapod crustaceans, such as the brown shrimp *Crangon crangon,* and in mysids, such as *Praunus flexuosus,* the cephalothorax and abdomen are more nearly equal in mass, and the point of flexion is close to the center of mass which lies within the first abdominal segment. Rotational forces produced by the simultaneous movements of the cephalothorax and abdomen effectively cancel each other. Also, in these species a head fan is formed by the antennal scales that normally lie on either side of the rostrum. As flexion begins, these scales extend laterally, and the stiff setae that fringe their margins spread out, increasing their surface area, and balancing the similarly spread tail fan in area and in generation of thrust.

Neural organization. In all groups of crustaceans, the tail-flip response is mediated by giant nerve fibers situated in the ventral nerve cord. Details of its neuronal basis have been studied most extensively in the crayfish *Procambarus clarkii.* Here, two pairs of fibers, the medial and lateral giants, are responsible for the initiation of the first flexion of the abdomen in response to a sufficiently strong stimulus. Subsequent tail flips in a series do not involve giant fibers but are generated by a neuronal network. The form of the first flexion depends upon which of the giant fibers is activated. Visual or mechanical stimuli received anteriorly activate selectively the medial giant fibers to produce a tail flip in which all the abdominal segments flex. The result is an escape trajectory which carries the animal directly backward. In contrast, posterior stimuli, which activate the lateral giant fibers, cause only the anterior abdominal segments to flex, producing an upward forward-pitching trajectory. In the Norway lobster (*Nephrops norvegicus*) stimulation of the medial giant fibers produces a wave of abdominal flexion traveling forward sequentially from the most posterior segment, resulting in a flat backward trajectory; when the lateral giant fibers are stimulated, the initial point of flexion occurs at the anterior end of the abdomen and the wave of flexion travels backward, producing an upward and backward trajectory.

Speed and orientation. The effectiveness of the tail-flip response and tail-flip swimming as a means of evading capture by natural predators, such as fishes, is determined by factors such as the timing of the response, its orientation, the speed of movement produced, and the overall distance traveled.

The maximum velocity generated during escape swimming varies over only a relatively small range 2–9.2 ft/s (0.6–2.8 m/s) among a variety of crustaceans from mysids to lobsters that range in adult size from 0.4 to 10.6 in. (9.5 to 270 mm) in body length, although this range represents a 30-fold difference in velocity expressed as body lengths per second. Velocity, however, is not the most important factor influencing effectiveness of the response. Timing in relation to the developing threat, determined by the threshold for activation of the response, is critical: reacting too soon allows the predator to adjust its attack; reacting too late allows the crustacean to be caught. Thresholds for activation of the tail-flip response tend to be high, in common with those for other types of startle reaction.

Orientation or trajectory of the response is important. Although orientation or trajectory is largely constrained by the direction in which the the propulsive force is generated, it can also be influenced by separate steering actions. For some crustaceans, such as the Norway lobster, actions may involve redirection of the propulsive force by slight rotations of the abdomen; for others, the abdominal appendages are involved. Equilibrium organs (the statocysts) provide the animal with positional information that assists in controlling these adjustments. Additionally, smaller decapods and other small crustaceans such as mysids are capable of directing the propulsive force of the tail flip into a different plane than that of their initial orientation in relation to the predator. This change in orientation is brought about by a rotation of the whole body about its longitudinal axis by up to 90°. In the mysid *P. flexuosus,* this rotation is complete within 5 milliseconds of stimulation. The forces needed to achieve reorientation are generated by asymmetrical movements of the left and right elements of the head and tail fans as they open out.

Effectiveness. Quantitative assessments of the effectiveness of the tail-flip response in actual encounters between crustaceans and their natural predators are rare. In one study carried out in laboratory aquaria, the mysid *Neomysis integer* showed an escape success rate of 75% from attacks by one of its predators, the 15-spined stickleback, *Spinachia spinachia.* In a study of encounters between shrimp *C. crangon* and juvenile cod *Gadus morhua,* escape success was shown to be dependent on the relative sizes of predator and prey: the greater the size differential, the greater the chance of capture and ingestion at the first encounter; the smaller the differential, the greater the chance of escape. In common with mysids, shrimps can redirect the tail flip by means of a lateral rotation of the body so that the initial trajectory of the response is roughly at 90° to the initial orientation. Steering may then lead to a change in direction as further tail flips in a swimming sequence follow. With good visibility, juvenile cod continue to chase shrimp that escape the first encounter. They may be successful in catching the prey in subsequent attacks, although the chances of this happening are reduced if sediment is available in which the shrimp can conceal themselves. Thus, escape swimming is most effective when the immediate environment provides a refuge from further pursuit.

Exploitation. Tail-flip swimming in crustaceans may also be stimulated by the approach of unnatural aggression, such as trawl-fishing gear. Norway lobsters, for example, respond to the approach of a trawl by swimming up in the water. Those facing the fishing gear show the upwardly directed response mediated by the medial giant fibers; those facing away from the gear show the shallower backwardly directed response mediated by the lateral giant fibers and then swim away from the gear in the direction of towing. However, provided commercial trawls are operated at speeds that exceed the average swimming speed of the lobsters, both groups are likely to be caught. Other features of tail-flip swimming are exploited by humans. For example, the limited vertical movement generated in the response provides a basis for the incorporation of a horizontal panel of netting into commercial trawl nets used for catching fish and Norway lobsters. This panel effectively separates the crustaceans into the lower compartment, and the fish, such as haddock and whiting, swimming higher in the water column, into the upper compartment. Thus, knowledge of the characteristics of tail-flip swimming behavior can provide a strong biological basis for the design of crustacean fishing gear.

For background information SEE CRUSTACEA; DECAPODA (CRUSTACEA); MYSIDACEA in the McGraw-Hill Encyclopedia of Science & Technology.

Alan D. Ansell

Bibliography. T. L. Daniel and E. Meyhöfer, Size limits in escape locomotion of carridean shrimp, *J. Exp. Biol.*, 143:245–265, 1989; D. M. Neil and A. D. Ansell, The orientation of tail-flip escape swimming in decapod and mysid crustaceans, *J. Mar. Biol. Ass. U.K.*, 75:55–70, 1995; P. L. Newland and C. J. Chapman, The swimming and orientation behaviour of the Norway lobster, *Nephrops norvegicus, Fish. Res.*, 8:63–80, 1989; D. C. Sandeman and H. L. Atwood (eds.), *The Biology of Crustacea*, vol. 4, 1982.

Digital camera

Digital still photography is playing an increasingly important role in the work of commercial and industrial organizations. Whether it is for recording a car being insured or photographing merchandise for an advertisement or catalog, digital photography is becoming the medium of choice; it eliminates the need of film processing and related ecological concerns, and it makes digital files that are immediately available for printing processes. Replacing a film or print scanner, digital still cameras are being used in more and more applications.

Evolution. Original analog still electronic cameras were introduced in 1981, but true digital still cameras did not appear until the late 1980s. Although the first publicly announced (1981) electronic still camera was considered to be a digital camera because of its use of the floppy disk, in reality it was an analog camera with images recorded in the same form as video tape signals are recorded on linear tape.

The first actual digital camera was a limited-resolution black-and-white camera. That original simple digital still camera served a number of needs for recording photographs and storing the images digitally in the camera.

In 1990, more impressive digital still cameras appeared, using regular 35mm camera mechanisms. The exposing plane of the linear film was replaced by a charge-coupled device (CCD) that served as an electronic light recording device. These higher-resolution cameras were suitable for photojournalism and were also used in law enforcement and industrial applications. These original digital cameras led to development of three types of digital cameras: area-array cameras, linear-array cameras, and aperture cameras.

Area-array cameras. In these cameras photos are made by using a charge-coupled device that can be compared to the flat capturing surface of a piece of film. Instead of silver halide crystals collecting photons of light energy, a series of either rectangular or square sensor elements collect light energy. These sensor elements might be compared to buckets collecting rainwater in a downpour. The photons, now represented as electrical charges in each sensor element (that is, picture element or pixel), are moved or pushed out of the sensor element and converted to a voltage. This voltage then becomes a digital value that represents a precise moment.

Two different types of area arrays are used for inexpensive still digital cameras. The same charge-coupled device that is used in video camcorders is used in some cameras. In these video sensors, two successive field images are made for each picture frame. Each field contains half of the alternate line information of a complete video frame of information. In video camera charge-coupled devices the images are made in $\frac{1}{60}$th of a second.

A full-frame charge-coupled device used in both lower- and higher-level area-array cameras records the image in one moment, and this image is then transferred from an electronic form to a voltage form and finally to a digital form within the camera.

Color information is recorded in two ways: The most common way is by placing filters over individual pixels. Another approach is to use a black-and-white charge-coupled-device sensor and place color filters in front of the lens, with one exposure being made through a red filter, another through a green, and a third through a blue. Although both the individual filtered pixels and the separate exposure of the charge-coupled device produce color images, the greatest amount of color fidelity is achieved by the individual exposure of the black-and-white charge-coupled device. To see the final color image, the three images—red, green, and blue—must be viewed or printed together.

Besides having enough pixels to accurately represent details of an original scene, another important element of the individual charge-coupled device is its ability to render shades of color. A minimum requirement for good color is a sensor that will produce eight bits of color from each picture element. When joined together with the eight bits of colors from the other two values, more than 16 million colors will be available if the equipment used for viewing or printing the images can accommodate such.

Cameras with filters over each individual pixel have the ability to take photographs of subjects in motion or when motion can be stopped with electronic flash equipment. When a filter must be placed over the black-and-white sensor with one exposure for each of the three colors, no action pictures can be made since any movement of the object between the exposures would produce multiple images, but of different colors.

The newest filter technology, using liquid-crystal color filters, allows a monochrome sensor to instantly record three colors with electronic flash or action subjects.

Linear-array cameras. Charge-coupled devices are available not just as area arrays but also as linear arrays—a row of individual sensor elements or three rows of linear arrays together, one for recording red, one for green, and one for blue image information. The linear array can be moved across the equivalent of an exposure plane where an image can be captured one line or three lines of color at a time.

An entire series of cameras of this type are used extensively for commercial photography where products are photographed without live models. Linear-array cameras have exposure times of up to several minutes in duration, requiring that an image be stationary and that the lighting not be changed during the time of the exposure.

The resulting images from these linear-array sensors can be very high resolution, up to several thousand pixels in both horizontal and vertical directions. Linear-array cameras or camera backs provide images that can be enlarged or that are suited for reproduction as large magazine illustrations.

The area-array cameras may be compared to reflected or transmitted light scanners in which a linear array moves past a photograph or film to produce a scanned digital image for computer use.

Aperture cameras. Because of the lack of availability of low-cost, high-resolution area-array sensors, several companies have developed a generation of digital cameras known as aperture cameras. Here a smaller charge-coupled-device sensor is used and moved in a pattern resembling tiles on a wall to produce a series of images that are joined together to form a single color image.

With high-quality images, the aperture camera is used primarily for commercial product photography. Because a number of images must be made of the same scene, it is important that lighting be a constant value and that no movement occur in the camera or the subject.

Digital imaging. Most photographers who have used digital cameras find that the color rendition is more accurate than in images made on traditional photographic film—especially when individual exposures are made on a black-and-white area array or on linear-array systems using high-quality filters.

The ability to render a subject with many light-level values is another factor that makes digital cameras superior to traditional film cameras. Although film has a limited range of light levels, digital cameras have a much broader range.

An important consideration is that only a limited number of digital cameras can make action photographs of subjects. Producing an action photograph is only now beginning to reach the level of quality that 35mm films have traditionally provided to photographers. The difference is that the cost of the digital camera to produce the same result is considerably higher than that of the traditional photographic film camera. The increase in cost can often be compensated for by the fact that the image is digitized, and in many applications the need for a scanner is eliminated.

With the advent of high-definition television and its need for higher-resolution charge-coupled-device chips for cameras and camcorders, it is possible that the price of higher-resolution still digital cameras may be reduced. As the demand for high-definition television increases, the market for digital cameras will expand.

New approaches to bringing all of the elements of the digital camera to a single piece of silicon have been demonstrated. Workers at both Edinboro University of Pennsylvania and the U.S. National Aeronautics and Space Administration have demonstrated the viability of this approach to camera construction. This approach could be the breakthrough needed for an economically priced digital camera for the general public in the foreseeable future.

The digital camera is presently available for the specialized user, and will become even more interesting as the information highway develops. This technology will provide a convenient means of sending business and personal photographs in full color to remote locations.

For background information SEE CAMERA; CHARGE-COUPLED DEVICES; TELEVISION in the McGraw-Hill Encyclopedia of Science & Technology.

J. Larish

Bibliography. J. Larish, *Digital Photography: Pictures of Tomorrow,* 1994.

Dinosaur

Fossil evidence of dinosaurs includes, in order of decreasing abundance, footprints, skeletal remains, eggs, nests, and coprolites (fossil dung). However, paleontologists have traditionally focused attention on skeletal remains or body fossils, and regarded footprints and other forms of evidence, known as trace fossils, as rare and rather insignificant. Recent research has revealed that dinosaur footprints are particularly abundant, and much more useful for interpreting behavior than previously supposed: unlike bones, the residue of death, footprints clearly represent the activity of animals during life. Dinosaur tracks can reveal many aspects of paleobiology, ranging from individual behavior (speed, locomotion, and gait) to social behavior (herding) and ecology (population structure, relative abundance, diversity, habitat preference, and paleogeographic range). Tracks are also useful in applied geologic fields such as global biostratigraphic correlation, and interpretation of sedimentological conditions in ancient depositional environments. It is also evident that the distribution of tracks is a function of how sedimentary rocks accumulate, and suggestions that tracks are more likely to be preserved during periods of high sea level are apparently correct.

Modern dinosaur track research began in the late 1970s and early 1980s with a debate regarding the speeds attained by these animals. At this time, formulas for calculating speed from trackways were proposed. Based on trackways preserved in Cretaceous rocks in Texas, theropods (bipedal carnivorous dinosaurs) were estimated to have run at a rate of 40 km/h (25 mi/h), the same speed as an Olympic sprinter. However, trackways of large quadrupedal dinosaurs indicate much slower speeds with no evidence for anything more than moderate walking speeds of about 5 km/h (3 mi/h). Such evidence fits well with expectations, because bipedal dinosaurs, including theropods, are generally considered to be cursorial (that is, adapted for running), whereas most quadrupedal forms are described as mediportal or graviportal (that is, having significant weight-bearing adaptations). The rarity of trackways of running dinosaurs is also a function of probability. Assuming that dinosaurs spent less than 1% of their time running, especially on soft ground, there is a very low probability of finding trackways indicative of such activity.

Classification. Naming of dinosaur tracks is a difficult problem because of uncertainty about the identity of the track maker. However, in recent years much progress has been made in correlating new track discoveries with track makers. The first convincing examples of *Tyrannosaurus* tracks were found in strata of appropriate age (Late Cretaceous, Maastrictian) in New Mexico and named *Tyrannosauripus* (**Fig. 1**). Similarly, the first substantial assemblage of ceratopsian (horned dinosaur) tracks was found in Maastrichtian strata from Colorado, and named *Ceratopsipes.*

Preservation. Quality of preservation of dinosaur tracks is also a field that is being seriously studied. Incomplete brontosaur trackways, consisting only of front footprints, had previously been cited as evidence of swimming behavior. The sauropod was thought to have walked on its front feet while pro-

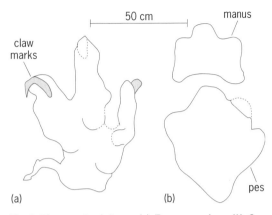

Fig. 1. Dinosaur track types. (*a*) *Tyrannosauripus.* (*b*) *Ceratopsipes.*

pelling itself with its tail through the water. Such interpretations have proved incorrect, and instead are simply the result of poor preservation of walking traces, as underprints or undertracks, because of the deeper penetration of the small front foot (manus) which in some cases exerts more force than the broad pes (**Fig. 2**). Such tracks have also been referred to as ghost prints, and result from the distortion of feet on underlayers below the surface over which track makers walked. Such evidence indicates that an understanding of how tracks are preserved is needed before interpretation.

Trackway interpretations. Brontosaur trackways have also been differentiated on the basis of the distance between right and left footprints as narrow gauge and wide gauge: the narrow forms are largely confined to the Jurassic; wide-gauge trackways are dominant in the Cretaceous. Such patterns

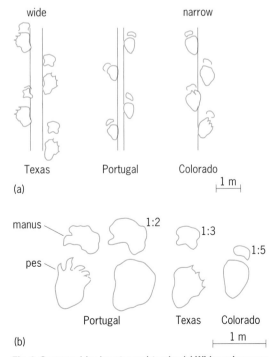

Fig. 2. Sauropod (or brontosaur) tracks. (*a*) Wide and narrow gauge types. (*b*) Variation in the relative size of hind and front feet. Ratios 1:2, 1:3, and 1:5 indicate the area of manus tracks relative to pes footprints. 1 m = 3.3 ft.

probably reflect the evolution and distribution of different brontosaur families during these two periods. Recent work also shows significant differences among brontosaur species in the relative size of manus and pes tracks, a phenomenon known as heteropody. Such distinctive morphological differences allow further discrimination of brontosaur track types.

Trackway evidence for social behavior or herding is provided by surfaces with multiple parallel trackways of the same type. Such evidence is particularly common among large herbivorous brontosaurs and ornithopods (duck-billed dinosaurs), and suggests that throughout their history both groups often engaged in gregarious activity; there is limited trackway evidence for this activity among small ornithopods, theropods, and ceratopsians. Some groups of parallel trackways were made by animals in a narrow size range, suggesting that gregarious species sometimes separated into distinct age groups, as in some modern mammals.

Many dinosaur track assemblages consist of dozens of trackways that include a wide range of tracks of similar shapes but different sizes. These assemblages are thought to represent a herd. In combination with estimated skeletal growth rates, the tracks can be used to assess the relative ages of individuals. This kind of information on the number and size of associated individuals is rarely obtained from the bone record, especially at single sites. The track evidence provides very little evidence of the activity of small, hatchling dinosaurs, but generally provides much more abundant evidence of intermediate-sized juvenile and subadult dinosaurs than can be obtained from the bone record. Thus, data can be easily accumulated on size distributions within dinosaur populations.

A fossil trackway site usually includes only a limited diversity of distinct dinosaur track types, often only 2–5, but in some cases 10 or more. Interestingly, these same footprint types often recur in characteristic proportions in track assemblages (ichnocoenoses). Such ichnocoenoses correlate with particular sedimentary environments, such as floodplains or lake margins, shedding light on ancient ecology (or paleoecology). This evidence of repeated patterns in the proportions of track makers in given sedimentary environments suggests that the track record, at least in part, genuinely reflects the relative abundance of species and the composition of ancient animal communities in a particular environment or habitat. Such inferences are supported by similar patterns in track assemblages produced by modern animals.

Recurrent trace fossil assemblages have been used to define ichnofacies that reflect both the nature of the ancient environment and the biota that inhabited it. Although not recognized until recently, there are a number of distinctive vertebrate ichnofacies. The *Brontopodus* ichnofacies describes the recurrent association of multiple sauropod track assemblages with low-latitude car-bonate platform facies (that is, environments rich in lime mud). An ichnofacies characterized by large Cretaceous ornithopod tracks is associated with humid, well-vegetated coastal plain deposits.

Biostratigraphy. Fossil footprints have also been applied to the problem of biostratigraphy, that is, the correlation of strata. The applied biostratigraphic practice of using fossil footprints for correlation, known as palichnostratigraphy (ancient track stratigraphy) and introduced by German track specialists in the 1970s, has been shown to have applications on a global scale.

An increasing number of recent discoveries have demonstrated that tracks from the same aged strata in different regions are often remarkably similar. Traditionally, it has been accepted that in the late Paleozoic and early Mesozoic, when the continents were coalesced as the supercontinent Pangaea, there were few barriers to migration. Thus, animal communities were cosmopolitan, creating a situation where both tracks and skeletal remains were similar over a large area. Recent work extends this type of palichnostratigraphy into a significant part of the late Mesozoic, and has resulted in several track correlations between North America, Europe, and Asia that postdate the Early Jurassic breakup of Pangaea, and so provide evidence of continued faunal interchange.

The distribution of tracks in various sedimentary formations throughout geologic time is also significant. Distinctive track types are found not only in strata of a particular age and type but in some cases in specific layers over regionally extensive layers on the order of tens of thousands of square miles. Such extensive track layers have been called megatracksites. Most evidence suggests that such megatracksites are associated with changes in sea level, specifically sea rises. Such circumstances allow for the buildup, or accumulation, of coastal plain environments. In such environments, large areas of wetlands and wet substrates are created, providing the ideal medium for registering many tracks.

Recent studies suggest that dinosaur tracks are the only fossil record available in many formations. Such phenomena as megatracksites demonstrate that the number of animals represented in the track record may be several orders of magnitude greater than individuals represented by skeletal remains. Moreover, tracks show recurrent ichnofacies associations, indicative of habitat preferences. Such associations have not been recognized or defined to any degree in the skeletal record. Thus, footprints are an integral part not only of the fossil record of behavior but also of the sedimentological or geologic record. As such they add much to developing a better understanding of the behavior of dinosaurian track makers and their global distribution in both space and time.

For background information *SEE DINOSAUR; FOSSIL; GEOLOGY; PALEONTOLOGY; STRATIGRAPHY* in the McGraw-Hill Encyclopedia of Science & Technology.

Martin G. Lockley

Bibliography. R. McN. Alexander, *The Dynamics of Dinosaurs and Other Extinct Giants,* 1989; D. D. Gillete and M. G. Lockley, *Dinosaur Tracks and Traces,* 1989; M. G. Lockley, *Tracking Dinosaurs,* 1991; M. G. Lockley and A. P. Hunt, *Dinosaur Tracks and Other Fossil Footprints of the Western United States,* 1995; R. A. Thulborn, *Dinosaur Tracks,* 1990.

Earth crust

Destructive earthquakes are constant reminders that the Earth's crust consists of mobile plates whose relentless motions cause active deformation where two plates interact. Geodesy is the science concerned with the measurement of the size, shape, and motion of the Earth and its surface. Recent advances in space geodetic methods permit direct observation of the plate motions that occur at rates of up to a few inches per year. It is also possible to measure in great detail the deformation patterns between, during, and following individual earthquakes along a crustal fault. Studies of the distribution of deformation along a plate boundary zone can identify regions of particular seismic risk for earthquake hazards. Even though most crustal deformation and earthquakes occur along plate boundaries, geodetic studies of the North American midcontinent show some evidence for localized deformation within the plate. These areas of intraplate deformation are susceptible to earthquakes. *SEE EARTHQUAKE.*

Space geodetic methods. Until the mid-1980s, geodetic measurements of crustal motions relied on terrestrial methods, of which triangulation, trilateration, and spirit leveling were the most commonly used. In these geodetic methods, surveyors repeatedly measure angles, distances, or elevation differences between benchmarks set in the ground. Changes of these measurements with time indicate deformation of the crust. In the mid-nineteenth century the first triangulation measurements were made in many parts of the world, and were replaced in the 1960s by precise trilateration measurements. Reobserving these old networks with modern methods provides valuable information on deformation over many decades. Even today, spirit leveling provides the most precise measurements of vertical deformation over distances of up to 100 mi (160 km), but it is very time intensive and costly.

Space geodetic methods rely on extremely precise measurements of signals from extraterrestrial objects or Earth-orbiting satellites. Space geodetic methods are rapidly replacing traditional methods. Very long baseline interferometry (VLBI) and the Global Positioning System (GPS) made studies of motions on a global scale possible. GPS has become the (affordable) tool of choice for studies of regional deformation along active plate boundaries. Synthetic-aperture-radar (SAR) interferometry is the latest and potentially the most effective addition to the space-geodetic methods for studying crustal deformation. It is noteworthy that all these technologies were developed for other scientific or technological applications such as radio astronomy (VLBI), vehicle navigation (GPS), and topographic mapping (synthetic-aperture radar).

Unlike terrestrial methods, space geodetic methods do not require line of sight between stations in a network, they do not depend on good weather and daylight, and measurement errors accumulate much more slowly with increasing distance.

VLBI uses faint signals emitted from stellar sources located at the edges of the universe that are received at radio telescopes distributed on the Earth's surface. The signals are compared from two or more radiotelescopes to determine the time delay between reception of the signals. This time delay depends on the angle between the signal source and the baseline vector between two stations. If signals from several sources at multiple stations are included, the baseline vectors between the stations can be computed with a precision of a few centimeters. Repeating these measurements every few months allows the detection of changes in the relative station positions that are due to crustal deformation, on the order of centimeters per year. The most powerful aspect of this method lies in its ability to measure these precise positions over distances of thousands of kilometers. However, high costs and improvements in GPS methods for long-baseline measurements have resulted in a reduction of VLBI measurements for crustal deformation studies.

Receivers equipped to measure signals from GPS satellites operated by the U.S. Department of Defense allow the determination of three-dimensional positions accurate to several meters in a few seconds. Through careful treatment of the data and the use of the carrier phase signal transmitted at two frequencies, the precision of relative station positions improves to the level of few millimeters. This is about five orders of magnitude more precise than the original specifications for the GPS navigation system. The principle of GPS geodesy is based on measuring the distance or range to several GPS satellites from the time it takes for a signal to be transmitted from the satellites that orbit the Earth at ~12,000 mi (20,000 km) altitude to a receiver. If the range to four or more satellites can be measured at the same time and if the position of each satellite is known, the three-dimensional station positions in an Earth-centered reference frame can be computed. Satellite orbit information is part of the signal broadcast by the GPS satellites. A total of 24 GPS satellites (plus some spares) ensure that four or more satellites are in view at any time anywhere in the world. To achieve the precision necessary for most crustal deformation studies, it is necessary to carefully model the effects of signal propagation delays in the ionosphere and the troposphere, to cancel out common error sources by differencing data from at least two receivers, and to average data over many hours or days. For distances longer than about 60 mi (100 km), it is also

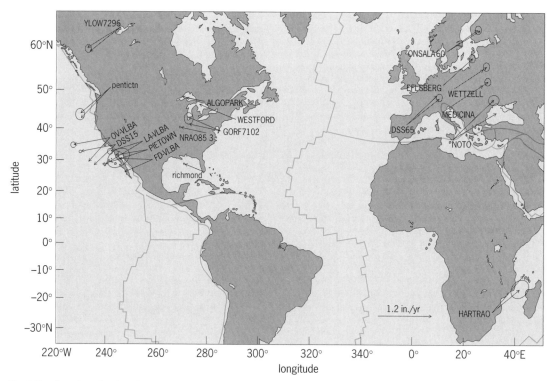

Fig. 1. Very long baseline interferometry (VLBI) station velocities in Europe and North America show the active opening of the Atlantic Ocean. The arrows represent vectors. The arrows without the 3-sigma error ellipses represent estimates of motions averaged over the last 3 million years. 1 in. = 2.54 cm. (*After C. Ma, J. W. Ryan, and D. S. Caprette, NASA Space Geodesy Program: GSFC VLBI Data Analysis, 1993, NASA Tech. Mem. 104605, 1993*)

necessary to compute a more precise position of the GPS satellites. For this purpose, data from a worldwide net of continuously running GPS sites are used. With determination of the position of these global stations to within less than an inch, improved orbit tracks for the GPS satellites can be computed.

The latest revolution in crustal deformation studies, using synthetic-aperture-radar interferometry, made its entry in 1993. By using interferometric images of satellite radar data collected by the *Earth Resource Satellite-1* (*ERS 1*), before and after the 1992 Landers earthquake in California, scientists were able to image the deformation surrounding the rupture in very fine detail. For this purpose it is necessary to successfully interfere the phase information of reflected radio signals (echo) from observations taken several weeks or months apart. If the reflecting area has changed its relative position, a phase shift will be revealed in the interferogram. Obviously it is necessary to carefully correct for the effects of the difference in the orbital tracks when the two images were taken and to remove the effects of varying topography. Any change in surface characteristics, such as vegetation growth, causes a loss of coherence, and it is not possible to interfere the phase data. In the resulting interferogram, each fringe corresponds to one cycle of about 1 in. (2.5 cm) of motion away from or toward the *ERS-1* satellite (half the 56-mm wavelength of the synthetic-aperture radar). The ultimate precision of

synthetic-aperture-radar interferometry is at the millimeter level, with dense spatial sampling of 300 ft (100 m) per pixel.

Crustal plate motions. The Earth's crust consists of about a dozen plates that continuously move at rates of less than one to several centimeters per year, driven by slow convection in the Earth's mantle below. For example, the San Andreas fault system in California represents part of the boundary between the Pacific and North American plates that move side by side at about 2 in. (5 cm) per year. Along the western and northern margins of the Pacific plate, the oceanic crust is being shoved below the neighboring plates in subduction zones. There is another type of plate boundary in the middle of the Atlantic Ocean. Here new crust forms from lava cooling along the Mid-Atlantic ridge where the North American and Eurasian plates continuously diverge from one another. The Himalaya represent another classic example of active plate tectonics where collision of the Indian plate and the Eurasian plate continues to compress and deform the crust along the plate boundary.

Until recently, knowledge of this jostling plate puzzle depended mostly on geologic evidence, such as the nature and structure of uplifted mountain ranges and the morphology of the ocean floor. Geologists estimated the rates of deformation averaged over several million years from the magnetic signature of past magnetic pole reversals that are frozen into oceanic crust along mid-ocean ridges.

Measurements of the ongoing motion of the large crustal plates were not possible until the development of space geodetic methods.

Figure 1 shows the motions of radio telescopes in North America and Europe and their 3-sigma error ellipses computed from data collected between 1979 and 1992 by VLBI. The results show that the two plates are moving apart at high rates. Measurements with VLBI have made it possible to visualize the shifting of the Earth's crust. More recently, GPS measurements have achieved comparable precision on intercontinental baselines. Also shown are vectors (arrows without ellipses in Fig. 1) that represent averaged motions over the last 3 million years. The velocities of stations within the plates averaged over the last few years agree well with the plate displacements estimated for the last few million years. Thus plate tectonics is a very steady process where plates shift their paths and speed only after many million years. Furthermore, these measurements indicate that plate interiors deform very little.

However, it can be seen that the velocities from VLBI in the western United States do not agree very well with the geologic rigid plate displacements. Near the boundary between two tectonic plates, in this case the Pacific and North American plates, the plates do not behave as rigid blocks, and deformation is not steady. There observations indicate a complex cycle of slow strain buildup and sudden release through earthquakes.

Crustal deformation. Relative plate motion at many plate boundaries does not occur along a single fault, with plates steadily sliding past one another. Rather, deformation is distributed across a broad zone and occurs intermittently, often in the form of deadly earthquakes. Along most of the San Andreas fault system in California, deformation spreads across a 60-mi-wide (100-km) zone that contains several active faults. Another example of an active plate boundary is the collision zone between the Indian and Eurasian plates, greater than 600 mi (1000 km) wide. A complex system of mountain ranges (including the high Himalaya), faults, and folds has accommodated the 40-million-year-old head-on collision of the two continents.

In what was probably the first measurement of plate boundary deformation, surveyors recognized that the great 1906 San Francisco earthquakes had significantly distorted the regional northern California triangulation network. Remeasurement of

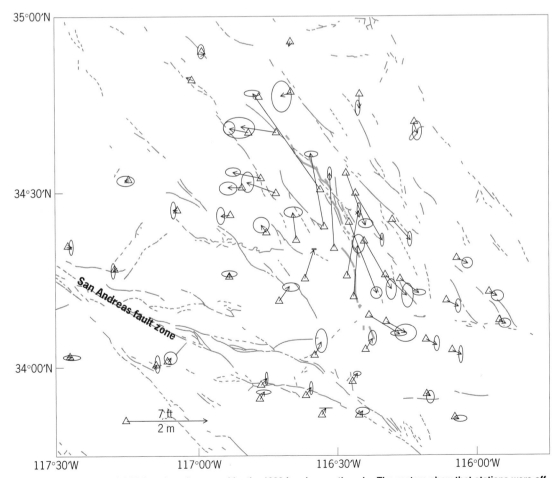

Fig. 2. Displacements of GPS benchmarks caused by the 1992 Landers earthquake. The vectors show that stations were offset from their positions before the earthquake by up to 10 ft (3 m) close to the ruptured fault (colored fault lines) and were affected at distances exceeding 60 mi (100 km). (*After J. Freymueller, N. E. King, and P. Segall, The co-seismic slip distribution of the Landers earthquake, Bull. Seismol. Soc. Amer., 84(3):646–659, 1994*)

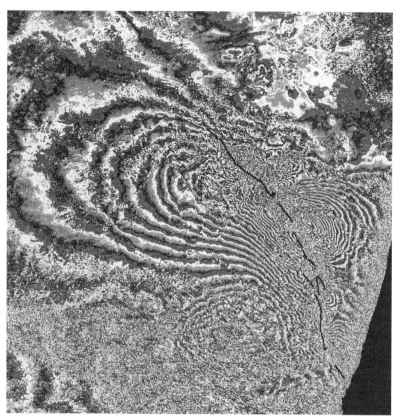

Fig. 3. Synthetic-aperture-radar interferometry image of the 1992 Landers earthquake deformation derived from *ERS 1* satellite radar data collected before and after the earthquake. Each fringe (band) in the image represents a change of range between the satellite and the Earth's surface of 1.2 in. (2.8 cm) caused by the earthquake. The earthquake rupture is shown as black lines. (*From D. Massonnet et al., The displacement field of the Landers earthquake mapped by radar interferometry, Nature, 364(6433):138–142, 1993*)

angles between the monuments undertaken to repair the network showed that the land to the west of the San Andreas Fault moved northward by about 20 ft (6 m) during the earthquake. With the development of highly precise electronic distance measurement devices (trilateration) in the 1960s, it became possible to measure the slow deformation that builds up stresses between earthquakes. By the late 1980s, after 20 years of time-consuming and costly measurements (conducted in large part by the U.S. Geological Survey), a broad zone of elastic strain across the San Andreas fault system was revealed.

Many scientists believe that the development of the GPS led to a revolution in crustal deformation studies in the early 1990s. Evaluation of the first results of a large number of crustal deformation studies conducted with GPS along plate boundaries all over the world was undertaken only recently. These studies reveal the detailed motions of the shifting plates and the complex and variable deformation at their boundaries.

In southern California, data collected by GPS and VLBI from 1985 until 1992 not only confirm the results of the earlier trilateration measurements but also show that the Los Angeles and Ventura basins are being slowly compressed at about 0.2 in. (5 mm) per year in a north-south direction. A bend in the great San Andreas fault system causes this squeeze of the metropolitan Los Angeles region. The Northridge earthquake in 1994 was a direct consequence and was a costly reminder of this compression.

Although the deformation measured between earthquakes occurs at very slow rates, an earthquake can suddenly move and deform an area by several meters. The largest earthquake to occur in California since the mid-1960s was the magnitude-7.3 Landers earthquake in the Mojave Desert in 1992. It broke a 50-mi-long (80-km) sequence of segmented faults, with displacements across the vertical fault plane reaching 15 ft (5 m). Motions were primarily horizontal; viewed from one side of the fault, the other side moved to the right by that amount.

More interesting to geophysicists than the displacements at the fault itself, however, was how the earthquake deformed the surrounding region over distances of 100 mi (160 km) or so. Soon after the earthquake, GPS data collected at many stations throughout the area that had also been surveyed in previous years revealed a fascinating pattern of distributed deformation. **Figure 2** shows horizontal displacements that reached up to several meters near the fault. These data allowed geophysicists to determine both the size of the fault plane below the Earth's surface and the amounts and distribution of slip on the fault to a depth of 7 mi (12 km).

The Landers earthquake marks a watershed in earthquake geodesy. For the first time, scientists succeeded in measuring crustal deformation by using radar images from space. An interferogram created from images taken before and after the earthquake revealed a fringed pattern caused by the deformation. Each fringe (band) in **Fig. 3** represents about 1.2 in. (2.8 cm) of change in distance between the Earth's surface and the *ERS-1* satellite. Since the Landers earthquake, researchers successfully used the method to image several other earthquakes and deforming volcanoes. Although still in its infancy, synthetic-aperture-radar interferometry promises to revolutionize crustal deformation research.

The measurements of deformation from the 1906 San Francisco earthquake support the hypothesis of an earthquake cycle consisting of occasional earthquakes and gradual elastic strain buildup in between. By using the improved resolution of modern geodetic tools, scientists can test more complex models that suggest that deformation following a large earthquake accelerates in the nearby region. This acceleration of deformation is caused by flow of hot and somewhat viscous rock below about 6 mi (10 km) and by accelerated creep of neighboring fault segments in response to the earthquake rupture. Some of this deformation occurs within a few weeks following an earthquake, but postearthquake anomalies may persist for decades. Thus, different processes may play a role at vastly different time scales. To fully understand these patterns, measurement precision of ~0.05 in. (1 mm) may be required. Partly for this reason, efforts are being made to use

GPS receivers to continuously collect data 24 h a day.

In the past few years, it has been recognized that even though most deformation of the Earth's crust occurs at plate boundaries, there is evidence that plate interiors are not always stable. In the heart of the North American continent near New Madrid, Missouri, three great earthquakes occurred in the early nineteenth century. A comparison of recent GPS data with historic triangulation measurements revealed that strains there may be as high as one-third of that found along the San Andreas fault system. Thus localized zones of active deformation may exist within plates that were previously considered to be rigid and safe from earthquakes. Similar monitoring efforts are now being undertaken to characterize active deformation on the Indian subcontinent and other regions where some evidence exists that the plates are internally deforming. These measurements may identify regions where no earthquakes occurred since settlement by humans but that are at risk of earthquake damage in the future.

Within less than a decade, space geodetic measurements have confirmed the relentless motion of the crustal plates, allowed the determination of the complex deformation patterns from several earthquakes, confirmed the existence of anomalous strain transients in the months and years following an earthquake, and identified unexpected levels of deformation far away from plate boundaries. Future research promises more significant advances in understanding the Earth's crust and its deformation. Integrated GPS networks and a number of synthetic-aperture-radar interferometry satellites will permit close monitoring of this deformation and better understanding of its causes and consequences.

For background information SEE EARTH CRUST; EARTH DEFORMATIONS AND VIBRATIONS; EARTHQUAKE; GEODESY; INTERFEROMETRY; MID-OCEANIC RIDGE; PLATE TECTONICS; REMOTE SENSING; SATELLITE NAVIGATION SYSTEMS in the McGraw-Hill Encyclopedia of Science & Technology.

Roland Bürgmann

Bibliography. B. H. Hager, R. W. King, and M. H. Murray, Measurements of crustal deformation using the Global Positioning System, *Annu. Rev. Earth Planet. Sci.*, 19:351–382, 1991; L. Liu, M. Zoback, and P. Segall, Rapid intraplate strain accumulation in the New Madrid seismic zone, *Science*, 257:1666–1669, 1992; D. Massonnet et al., The displacement field of the Landers earthquake mapped by radar interferometry, *Nature*, 364(6433):138–142, 1993.

Earthquake

Earthquake prediction has long been a goal of geophysicists. In practice, useful prediction has proved difficult for several reasons. The buildup of stress within the Earth is very gradual, taking tens to hundreds of years. The final stress that causes failure is like the straw that broke the camel's back. Moreover, unlike weather prediction, where the air is readily observed, even shallow earthquakes begin at a depths greater than several kilometers, where the cost of monitoring is prohibitive. In addition, there has not been enough time to establish a phenomenological base such as exists for weather. For example, no large earthquake has occurred along the northwest Pacific coast of the United States since the arrival of Europeans and the creation of written records. Finally, the physics of rock failure is much more poorly understood than the physics of clouds and air. Better understanding of rock physics may determine whether earthquakes are in fact predictable, and may make it possible to delineate observable processes that precede them. Although the goal of earthquake prediction has not been achieved, progress has been made in understanding the nature of fault failure. Research has revealed that hot water under high pressure deep below the surface has extensive chemical and mechanical effects on the behavior of rocks within fault zones. Consequently, laboratory experiments are now focused on understanding the behavior of rocks under such conditions. SEE EARTH CRUST.

Exhumed fault zones. Recent large earthquakes in California have started at depths from 3 to 17 km (2 to 10 mi). At such depths, conditions (involving extremely high rock and fluid pressures and temperatures to 400°C, or 750°F) are quite different from those near the surface. Under such conditions, rocks react chemically with water and undergo slow spatially continuous deformation known as creep. The familiar ground-up rock (gouge) formed within near-surface fault zones has little relevance to these processes.

Fortunately, geologically recent changes in the orientation of the San Andreas fault zone east of Los Angeles have resulted in the uplift and exposure of older parts of the fault that were several kilometers deep when active slip was occurring. Microscopic observations of rock samples have revealed that the fault zone consists of a extremely sheared core that is only a tenth to a few tenths of a meter wide within a zone of damaged rocks about 100 m (330 ft) wide. Essentially all the slip (the relative displacement) on the fault, probably comprising many kilometers, occurred within the narrow highly sheared core.

The pore fluid, hot saline water, had extensive chemical interaction with this highly sheared core. The core material consists of alteration products of the original minerals, resulting in hydrous minerals, including clays and newly deposited material in small veins. The core rock appears to have undergone repeated episodes of rupture and of growth of mineral grains by the addition of layers of atoms to their surfaces. Rapid slip during earthquakes opened veins and cracks, which later compacted and filled with precipitated minerals. When open, the veins acted as a conduit for fluids within the core of the fault zone. Mineral precipitation tended to seal leaks from the core of the fault zones to the surrounding damaged zone.

The core of the fault zone results from a feedback where rock becomes weaker as it is ground up, thus concentrating strain. The final grain size, typically less than 10 micrometers, in the core of the fault zone is small enough that its high surface area affects chemical and mechanical properties. Obviously, the material becomes highly reactive. In addition, small grains are slightly more soluble than larger grains, and thus dissolve to reprecipitate on the surfaces of the larger grains. This process, known as Ostwald ripening, limits the extent to which grinding reduces grain size. It also implies that the grains behave ductilely to some extent, because pressure solution creep, where material dissolves from stressed grain-grain contacts and precipitates on unstressed surfaces, has a similar dependence on the reactivity of the surfaces of small grains.

Weak faults. Measurements of the heat flowing out of the Earth provide indirect information on the conditions within fault zones, because friction during slip on a fault zone generates heat. Eventually this heat diffuses to the surface, where it can be measured. (In practice, the increase in temperature with depth and the thermal conductivity of the rocks—the amount of heat transferred per time across a unit surface area in the presence of a unit temperature change per unit distance—are measured and the heat flow is obtained from the product of these quantities.) However, excess heat flow as expected from this reasoning is not observed along the San Andreas Fault in California.

Thus an upper limit is placed on the shear stress (force per area) on the fault while it is slipping. Mathematically, the heat per area generated by friction is the slip distance times the shear stress. The slip distances and rates are well known from geological observations along the San Andreas Fault. Thus, the shear stress is obtained by modeling the predicted excess heat flow at the surface from a given stress at depth and comparing the result with observations. In the actual case, no excess heat flow is evident from the data, and only an upper limit is obtained from models where the predicted excess is small enough that it would escape recognition among background heat-flow variations from other causes. Only average properties at depth are obtained, because the measured surface heat flow averages over the time (on the order of a million years) that it takes for heat to reach the surface, and over a lateral extent comparable to the depth of the heat source. Thus, local regions of the fault could have shear stresses that are considerably higher or lower. Such studies have limited the shear stress on the fault to less than 20 megapascals (about 200 times atmospheric pressure).

This amount of stress is much less than expected for frictional failure of rocks. The shear stress at failure is the effective pressure (force per area normal to the fault surface) times the coefficient of friction, which is typically about 0.7 for broken rock.

Returning to the stress state on fault planes, the effective pressure is approximately the lithospheric pressure, that is, the weight of the overlying rocks ($\rho_r gZ$; here ρ_r is rock density, g is the acceleration of gravity, and Z is depth minus the pore fluid pressure). If well-connected pore space exists, fluid pressure is hydrostatically related to the weight of the overlying water ($\rho_w gZ$; here ρ_w is water density). The effective stress thus is expected to be ($\rho_r - \rho_w)gZ$. For example, if the densities differ by 1600 kg/m^2, a shear stress for failure for the upper limit of 20 MPa (140×10^6 lb/in.^{2}) is reached at a depth of only 1.7 km (1 mi). A shear stress of 340 MPa (2.4×10^9 lb/in.^{2}) is needed for failure at the 20-km (12-mi) depth to the base of the seismic zone.

In detail, failure is more complicated and depends on the previous history of preseismic frictional creep on a fault rather than just the instantaneous stress; this relationship has been observed in many laboratory experiments. Observational and inferential evidence for the dependence of failure on history is extensive: triggering of one earthquake by another is often delayed from the time of the sudden change of stress; there is no resolvable correlation between the times of earthquake occurrence and peak daily stresses associated with earth tides. The dependence of failure on history has been extensively studied, because an understanding of this effect might lead to prediction and because preseismic creep within a nucleation zone, if detected, may give a warning. Nevertheless, it is clear that friction on faults is much lower than expected.

High fluid pressure. Explanations for this observation maintain that the fluid pressure is actually near the weight of the overlying rocks and that the effective stress is small, at least at the time of an earthquake. (This process can be illustrated by shaking and turning over a soda can on a smooth surface. The excess pressure of the carbonated liquid separates the lip of the can from the surface, allowing easy sliding.) Such high-pressure water tends to leak upward toward the surface and also into any regions near the fault zone where the fluid pressure is essentially lower. This explanation requires a seal to form between the fault zone and its surroundings. Such seals are in fact observed in exhumed fault zones. The physics of their formation is relatively simple. The solubility of commonly dissolved rock constituents, such as silicon dioxide, decreases with fluid pressure at conditions relevant to crustal faults. Precipitation occurs where the fluid flows from high pressure to low pressure. Clogging and sealing occur, because the greatest rate of change in pressure and hence precipitation occurs where the flow path is already constricted. Self-sealing of fault zones has also been simulated in laboratory experiments at high temperatures. Sealing also occurs when fluid percolates upward to regions of lower temperature, where silicon dioxide is less soluble. Quartz (crystalline silicon dioxide) veins, which sometimes contain gold deposits, are often formed in this way.

A source of the high-pressure water is needed, as surface water clearly cannot flow down and directly generate high pressures. In some areas, fault zones are underlain by sedimentary rocks from which water is being expelled by compaction under the pressure generated by the weight of the overlying rocks. High fluid pressures and water expulsion have been observed during drilling of oil wells in sedimentary basins (they are a source of danger, because the well may blow out), and also landward of oceanic trenches, where sediments are dragged down to great depths by the subsiding tectonic plate. Veins and the lack of open pore space in slightly metamorphosed rocks formed from sediments indicate that high fluid pressures initiated cracking and that fluid expulsion from sediments is common. Water is also expelled from low-grade metamorphic rocks when a change in conditions (typically a temperature increase) causes water-bearing minerals to break down and release water. This process may be occurring at present beneath bends in the San Andreas Fault where there is a component of horizontal convergence perpendicular to the fault. One side of the fault moves over the other, driving it downward. An example is the Loma Prieta bend on the San Andreas Fault near San Francisco, California, which produced a magnitude-7 earthquake in 1989.

In addition, fault zones often act as conduits for the egress of water. High pressures of ascending water within fault zones are inferred where branch veins extend from the fault zone into the surrounding rock. In fact, mineral deposits formed by precipitation along such exhumed faults are so common that mining terminology is traditionally used to describe the geometry of faults.

An interaction between earthquakes and the egress of water is observed in exhumed faults. Earthquakes produce cracking that dilates the fault zone, decreasing fluid pressure and also making the fault zone act as a connected conduit. Between earthquakes, creep may compact fault zones and thus increase the fault pressure, and the connected pore space may close, either by this process or by mineral precipitation. It has been proposed that the plane of the fault zone thus organizes into a lattice of sealed compartments, each a few hundred meters across. Shortly before an earthquake, preseismic creep may rupture the seals between compartments, allowing rapid fluid flow and perhaps triggering the earthquake by changes in effective stress.

Such fluid flow is potentially observable at the surface. A promising method involves the short-term variations of the Earth's magnetic field such as those observed before the 1989 Loma Prieta earthquake. Physically, water flow through a porous material generates an electric current; ions of one charge are preferentially concentrated near the grain surfaces where flow is slow, while ions of the other charge are constrained within the faster-flowing fluid away from the grain surfaces. The electric currents generate subtle magnetic fields that

can be observed with sophisticated instruments. So far, little generalization has been possible because only the initial Loma Prieta data set is available. The general difficulty in studying processes prior to large earthquakes lies in such earthquakes being so rare that even systematic monitoring yields few good data sets.

Orientation of fault zones relative to stress. Measurement of stress in deep boreholes and inferences of stress from earthquake mechanisms and geological deformation have indicated that the San Andreas Fault is weaker than the surrounding rock. The formalism for describing the orientation of stress in three dimensions is somewhat more complicated than the stress on a single plane.

The stress within a solid is defined by the orientation of three perpendicular axes. The force per area resolved on the planes perpendicular to these axes is purely perpendicular to the planes, with no resolved shear stress acting along their surfaces. The three axes are ranked by the size of the normal stresses to give a least, intermediate, and most compressive axis. Both shear stress and normal stress act along all other planes. The maximum shear stress occurs in the planes defined by the intermediate axis and the directions halfway between the other two axes. This plane, however, is not the preferred plane of slip in a homogeneous solid, because failure depends on both the normal stress through the coefficient of friction and the shear stress. Typically a plane defined by the intermediate axis and a direction 60° from the most compressive axis is the preferred plane of slip.

The San Andreas Fault, however, is nearly perpendicular to the regional axis of maximum compression. This orientation is unexpected in terms of the above discussion because little shear stress exists to drive slip on this plane and because the normal stress is expected to be especially high on this plane. For this situation to occur, the strength of the fault plane needs to be much weaker than that of the surrounding rock. That is, faulting in the preferred plane 60° from normal to the most compressive axis is expected in the surrounding rock. Such secondary faulting is in fact observed near the exhumed parts of the San Andreas Fault in southern California and from earthquakes in rocks near the fault.

Stresses within the weak core of the fault zone behave similarly to those in a very viscous fluid between rigid parallel planes. Mechanical equilibrium requires that the shear stress and the normal stress be continuous on the plane of the edge of the fault zone. The fluid condition relaxes stresses, so the plane of maximum shear stress is the fault plane and the minimum compressive stress is the resolved normal stress outside the fault plane minus the outside shear stress. This condition is relevant to the formation of veins, which typically form in the plane normal to the least compressive stress when the fluid pressure exceeds that of stress. First, the fluid pressure required to form veins is much greater within the fault zone than within the surrounding rock. Second, high-pressure veins that escape from the fault zone may readily fracture the surrounding rock but with a different orientation. This change in orientation is in fact observed near exhumed faults.

Earthquake cycle in hard rocks. Compaction of sedimentary rocks and dehydration of low-grade metamorphic rocks are not an applicable mechanism for high fluid pressures in all geological environments. Hard rocks, for example, granite, provide no source of high-pressure water. A means of producing high fluid pressures in faults cutting such rocks has been proposed on the basis of studies of exhumed faults and laboratory experiments that indicate fault zones should self-seal. Starting the cycle just after an earthquake, extensive pore space created by cracking during slip decreases the fluid pressure to low pressure such as in the surrounding rocks, allowing some water to leak in. During the interseismic level, ductile creep compacts the pore space, increasing the fluid pressure, because the same amount of fluid is now in a smaller pore space. Partway through the cycle, the fluid pressure reaches that of the surrounding rock and the fluid begins to leak out slowly. Eventually, the increased fluid pressure and shear stress buildup on the fault causes failure, which produces pore space, thus continuing the cycle. This mechanism does not require a source of high-pressure water, because the leakage in and out of the fault may balance through a cycle.

However, a conceptual difficulty arises with this mechanism, because earthquakes are believed to begin with slow preseismic creep within a localized nucleation zone on part of the fault rather than with sudden failure everywhere over the fault. Such preseismic creep should create pore space, decrease fluid pressure, and thus strengthen the fault plane. Relatively small amounts of new pore space can stabilize preseismic creep and stabilize potential earthquakes. Such stabilization is particularly efficient for slow creep within nucleation zones, because the heat generated by friction on the fault has time to diffuse away. Conversely, heat does not have time to diffuse away during an earthquake. This heat expands the pore fluid, increasing its pressure and offsetting or even overwhelming the pressure decrease associated with new pore space. The interaction of slow preseismic creep and fluid pressures is still poorly understood.

For background information *SEE EARTH TIDES; EARTHQUAKE; FAULT AND FAULT STRUCTURES; GEO-ELECTRICITY* in the McGraw-Hill Encyclopedia of Science & Technology.

Norman H. Sleep

Bibliography. F. M. Chester, J. P. Evans, and R. L. Biegel, Internal structure and weakening mechanisms of the San Andreas fault, *J. Geophys. Res.*, 98:771–786, 1993; S. Hickman, R. Sibson, and R. Bruhn, Introduction to special section: Mechanical involvement of fluids in faulting, *J. Geophys. Res.*, 100:12831–12840, 1995.

Echinodermata

Echinoderm tube feet, the external appendages of the water-vascular system, are also probably the most advanced hydraulic organs of the animal kingdom. They can take part in locomotion, burrowing, feeding, sensory perception, or respiration. In some echinoderm groups, a single type of tube foot fulfills different functions; in others, different types of tube feet coexist, each with its own function.

Starfishes possess several hundred tube feet that are located on the oral surface of their arms where, depending on the species, they are arranged in either two rows or four. Each tube foot is connected to an ampulla, a muscular sac located within the starfish arm. The tube feet consist of an extensible cylindrical stem topped by a somewhat wider flat specialized disk (see **illus.**). Starfishes use their tube feet for locomotion and feeding.

During locomotion, the tube feet either act as a system of levers when the animal is on a horizontal surface, or form a traction system when the animal is on a vertical surface. In both types of locomotion, the stem and the disk together work as a functional unit: the stem allows the tube foot to lengthen, flex, and retract, and the disk allows it to adhere to the substratum. Traditionally, starfish tube feet are viewed as suckered organs in which adhesion is due partly to suction (mechanical attachment) and partly to adhesives (chemical attachment). Suction has often been regarded as the primary means of tube-foot adhesion in starfishes. However, recent studies concluded that suction is a secondary adjunct to the adhesion established by secretions. This latter method is less well understood.

The stem. Tube feet are hollow organs consisting of a central, fluid-filled, water-vascular lumen surrounded by a very thin wall. The tube-foot wall is

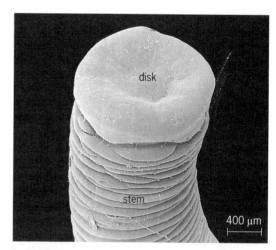

Scanning electron micrograph of outer aspect of a locomotory tube foot of the starfish *Marthasterias glacialis*. The tube foot has been preserved while retracted; hence the wrinkled aspect of the stem. (*After P. Flammang, S. Demeuleneare, and M. Jangoux, The role of podial secretions in adhesion in two species of seastars (Echinodermata), Biol. Bull., 187:35–47, 1994*)

made up of different layers: an inner mesothelium, a connective tissue layer, a nerve plexus, and an outer epidermis. However, these layers are differently organized in the stem and in the disk.

Within the stem, the mesothelium mainly consists of longitudinally arranged muscular cells that together form the tube-foot retractor muscle. The connective tissue layer is made up of two concentric cylindrical sheaths comprising numerous collagen fibers. In the outer sheath, the fibers are oriented longitudinally, and in the inner sheath they are arranged as a crossed-fiber helical array. Such an array consists of connective tissue fibers that wrap the tube foot in both left- and right-hand helices. The nerve plexus is poorly developed, and the epidermis is of the unspecialized type.

The adhesive disk. Within the disk, the mesothelium is greatly reduced, the connective tissue forms a thick supportive plate, the nerve plexus is well developed, and the epidermis is thickened and specialized compared to the stem general epidermis. Adhesion is brought about by secretions from the cells of the disk epidermis. These cells belong to four different categories: nonciliated secretory cells of two different types (NCS1 and NCS2), ciliated secretory cells (CS cells), nonsecretory ciliated cells (NSC cells), and support cells. A two-layered glycocalyx, the cuticle, covers the epidermal cells of the disk.

Nonciliated secretory cells (NCS1 and NCS2) are flask shaped; each cell body sends out a long apical process that reaches the apex of the tube foot. The cytoplasm of both the cell body and the apical process is filled with densely packed membrane-bound secretory granules. At the end of the apical processes, the granules are extruded through a secretory pore. NCS1 and NCS2 cell secretory granules differ in shape, size, and ultrastructure of their contents. Both types of granules presumably contain associations of proteins and glycans, although the exact composition of these compounds remains unknown.

Ciliated secretory cells have the same shape and size as nonciliated secretory cells. The entire cytoplasm of the cell is filled with small, spherical membrane-bound secretory granules of unknown content. The apex of the ciliated secretory cells bears a subcuticular cilium. Nonsecretory ciliated cells are narrow and have a centrally located nucleus. Their characteristic feature is a single short cilium whose apex protrudes into the outer medium. Both secretory ciliated cells and nonsecretory ciliated cells terminate basally within the nerve plexus. This, together with the fact that they both bear a short and rigid cilium, is indicative of both neural origin and sensory function.

Adhesive mechanism. The different epidermal cells of starfish tube-foot disks are involved in both adhesion and de-adhesion. The two types of nonciliated secretory cells are considered to be adhesive because they are the only secretory cells that extruded some of their secretory granules during observations of tube feet firmly attached to a substrate. Their secretions, a protein-polysaccharide complex, form an adhesive layer joining the tube-foot cuticle to the substratum, presumably through electrostatic interactions. The significance of two types of adhesive cells in the disk epidermis of starfish tube feet can be explained by the fact that only one type of adhesive cell (NCS1 cells) is used in locomotion on horizontal substrata, whereas both types are necessary during locomotion on vertical substrata or for anchorage, when a stronger adhesive bond is required.

The de-adhesion is due to the material enclosed in the granules of ciliated secretory cells. Indeed, there must be a detachment mechanism that is under control of the animal and allows the tube feet to easily become detached from the substratum. The ciliated secretory cells are the best candidates for this function, as they appear to release some of their secretory granules after tube feet are voluntarily detached from the substratum. The composition and mode of action of this de-adhesive secretion remain unknown. One possibility is that the de-adhesive material interferes with the electrostatic bonds between the adhesive layer and the cuticle; another, that it acts as an enzyme, releasing by lysis the cuticle from the epidermis.

These secretions are presumably controlled by stimulation of the two types of ciliated receptor cells which interact with specific secretory cells via the nerve plexus.

Locomotion. Starfish locomotion requires the formation of an adhesive bond between the tube foot and a substratum, as well as a great amount of mobility of the tube feet. Mobility of starfish tube feet can be explained by combinations of three basic movements: protraction, flexion, and retraction. These movements result from the antagonist action within the hydraulic pressure of the water-vascular fluid and tube-foot retractor muscle. The protraction of the tube foot is caused by the contraction of its associated ampulla. The elongation process is made possible by the properties of the stem-connective tissue inner sheath that prevents dilatation of the tube-foot wall: thus, the total amount of hydrostatic pressure is exerted on the connective tissue plate at the distal end of the tube foot. Retraction is due to the contraction of the tube-foot retractor muscle; the water-vascular fluid is then driven back in the ampulla. The retractor muscle is also involved in bending the tube foot toward one direction or another; it seems that each sector of the muscular cylinder can contract independently from the others and, therefore, act on the hydraulic skeleton to produce a flexion.

Starfish locomotion on horizontal surfaces seems to be a stepping process. The retracted tube foot is oriented in the direction of movement, protracted, and when fully extended swung back through an angle of about 90° with the disk attaching, not necessarily firmly, to the substratum. Because cuticle-protruding cilia of nonsecretory ciliated cells are the first to contact the substratum, it is likely that their stimulation triggers the release of adhesive material by NCS1 cells. When the disk is attached, the pendulumlike motion of the tube foot has the effect of thrusting the base of the foot forward relative to the disk; the sum of the forward thrusts of all the stepping feet determines the movement of the animal as a whole. At the end of the backswing, the disk detaches because of the release of a de-adhesive secretion by ciliated secretory cells (the secretion of their granules would be induced by a stimulation of their subcuticular cilia). The tube foot is then retracted and reoriented, and in this way acts as a strut used as a lever.

In locomotion on an angled surface or against a current, the tube feet do not function as levers but by traction. After protraction of a tube foot, the disk attaches firmly to the substratum. As in the previous type of locomotion, when the disk contacts the substratum, the cilia of nonsecretory ciliated cells are stimulated, leading to the release of the secretory granules by the nonciliated secretory cells. This time, a combination of the secretory materials from both NCS1 and NCS2 cells is required to achieve a stronger adhesive bond. Once the tube foot is firmly anchored to the substratum, it retracts, the combined action of all the attached feet pulling the animal along. Detachment of the disk from the substratum is then brought about by the secretion of ciliated secretory cell secretory granules.

Other marine invertebrates. Adhesion is a way of life in the sea. Indeed, marine representatives of the bacteria, of all the lower plants from unicellular algae to macroalgae, and of all the animal phyla can attach to foreign surfaces, including other organisms. Three forms of adhesion may be distinguished: (1) transitory adhesion involving the secretion of a viscous material and permitting simultaneous adhesion and movement along the substratum (for example, the foot secretions as trails of some mollusks); (2) permanent adhesion involving the secretion of a cement (for example, the attachment of barnacles on rocks); and (3) temporary adhesion allowing an organism to attach strongly but momentarily to the substratum.

Among the adhesive organs of marine invertebrates, duogland organs, enclosing two types of secretory cells (cells releasing an adhesive secretion and cells releasing a deadhesive secretion) and involved in temporary attachment, are morphologically the closest to the starfish tube-foot adhesive system. These adhesive organs are frequently found in small invertebrates inhabiting the interstitial environment (for example, in Turbellaria, Gastrotricha, Nematoda, and Polychaeta). A duogland adhesive system has also been described, recently, in the captacula (that is, the food-collecting tentacles) of scaphopod mollusks. In every species studied, the adhesive organs contain two types of closely associated secretory cells. Cells of the first type are specialized epidermal cells similar to starfish nonciliated secretory cells, and they release an adhesive material. Cells of the second type are derived from sensory nerve cells and resemble starfish ciliated secretory cells; they are de-adhesive in function.

However, the temporary adhesion mechanism and complex chemical nature of the adhesive and de-adhesive used are largely unknown. Indeed, the small size of duogland organs in most marine invertebrates has precluded any biomechanical or biochemical studies on temporary adhesion. The study of macroinvertebrate adhesive systems therefore is a promising and obligatory method of understanding temporary adhesion. Among all macrobenthic organisms, echinoderms appear to have exploited temporary adhesion the most efficiently through their tube-foot adhesive system. Future studies should permit isolation and purification of both adhesive and de-adhesive substances, and define their physicochemical characteristics and structures. Such information, together with the current morphological data, would provide the necessary basis to understand how echinoderm adhesive systems are used in temporary adhesion.

For background information SEE BIOPHYSICS; ECHINODERMATA in the McGraw-Hill Encyclopedia of Science & Technology.

Patrick Flammang

Bibliography. P. Flammang, S. Demeuleneare, and M. Jangoux, The role of podial secretions in adhesion in two species of seastars (Echinodermata), *Biol. Bull.*, 187:35–47, 1994; M. Jangoux and J. M. Lawrence (eds.), *Echinoderm Studies*, vol. 5, 1996; R. S. McCurley and W. M. Kier, The functional morphology of starfish tube feet: The role of a crossed-fiber helical array in movement, *Biol. Bull.*, 188:197–209, 1995; L. A. Thomas and C. O. Hermans, Adhesive interactions between the tube feet of a starfish, *Leptasterias hexactis*, and substrata, *Biol. Bull.*, 169:675–688, 1985.

Electric power systems

In recent years, voltage instability of large electric power systems has caused blackouts and brownouts. Voltage instability and collapse have caused major blackouts in France (1978), Belgium (1982), Sweden (1983), western France (1987), Tokyo, Japan (1987), and Israel (1995). There have been many lesser incidents in the United States and elsewhere. In order to avoid instability, power companies have been forced to limit power imports and to use more expensive generating plants.

Voltage stability. Power-system voltage stability, sometimes called load stability, is associated with the load dynamics in large power systems. By contrast, electromechanical stability is associated with maintaining synchronism of interconnected synchronous generators, sometimes called generator stability. Active power balance governs synchronous stability, while reactive power balance governs voltage stability.

Power companies try to keep voltages within about ±5% of rated voltage. Voltage instability may result in either high or low voltage, but the major concern is low voltage during heavy load conditions such as very hot or very cold weather. Voltage collapse is a condition of either abnormally low voltage (brownout) or zero voltage (blackout).

Transient voltage stability. There are two types of voltage instability. Transient voltage instability occurs within a few seconds. For example, a transmission-line short circuit occurs that slows down induction motors. In a fraction of a second, circuit breakers isolate the faulted transmission line, but the supply system is now weakened, with remaining lines overloaded. The motors draw a lot of current to reaccelerate. As a consequence, voltage sags and the motors may stall. Stalled motors that are not disconnected draw about six times more current than during normal operation, causing further voltage sags and stalling of other motors. The load area may then suffer a voltage collapse.

Longer-term voltage stability. Longer-term voltage instability occurs over 0.5–30 min. Following outage of transmission lines or generators, voltages in load areas sag. Load reduces during voltage sags, and this load reduction or load relief stabilizes voltage. However, some load-restoring mechanisms and other voltage-instability mechanisms operate over the 0.5–30-min time frame. Longer-term voltage instability is currently of greatest interest, and was responsible for the blackouts mentioned earlier.

Reactive power. Active power, measured in watts, kilowatts, or megawatts, is associated with current that is in phase with the voltage 60-Hz waveform. Reactive power is associated with the electric or magnetic fields in transmission lines, transformers, motors, capacitors, and other equipment. For example, the alternating-current sinusoidal waveform of current drawn by an induction motor lags the applied voltage waveform due to so-called magnetizing-current requirements. At no load, energized motors and transformers require magnetizing current that lags applied voltage by 90°; multiplying this magnetizing current by the voltage gives reactive power. For a loaded motor, the current waveform lags the voltage waveform by about 35°.

However, capacitors draw current that leads voltage by 90°, resulting in reactive power with the opposite sign. The lagging or leading angle is termed the power-factor angle, and the power factor is the cosine of this angle. Rather than dealing with voltage and current waveforms, engineers use effective or root-mean-square values of voltage and current. Related to the power-factor angle, complex numbers and phasors are used as shown in **Fig. 1**. A phasor is a rotating vector that describes the voltage or current magnitude and phase relationship. Although reactive power is mathematically imaginary, it cannot be ignored.

Capacitors produce or generate reactive power, and inductive equipment consumes or absorbs reac-

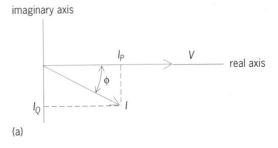

(a)

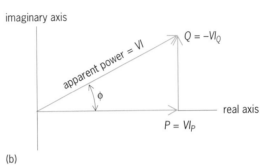

(b)

Fig. 1. Voltage and current phasors, and active and reactive power. (a) Current phasor, *I*, lagging the reference voltage phasor, *V*, by the power factor angle, ϕ, as for an induction motor. The current phasor has component *I*ₚ in phase with the voltage phasor, and component *I*_Q in quadrature with the voltage phasor. (b) Corresponding active power, *P*, and reactive power, *Q*, using the convention that the motor is consuming positive active and reactive power.

tive power. For high active power loading, transmission lines consume inductive reactive power.

Transmission. A key aspect of voltage stability is the need to provide reactive power to support voltages. It is possible to transmit active power long distances, but it is not possible to transmit reactive power long distances, especially during heavy load conditions. Large reactive power losses occur along the transmission path because of the series inductance of transmission lines and transformers. These losses are proportional to the square of the line or transformer current. Also, reactive power transmission requires the voltage at the sending or generation end to be significantly higher than the voltage at the receiving or load end.

Compensation. To avoid transmitting reactive power, utilities and industrial customers apply reactive power compensation near the points of need. Industrial customers install shunt capacitor banks to provide for the motor magnetizing current. (Shunt signifies connections between the three phases, or between the phases and ground.) Thus the overall power factor is improved and utility charges for low power factor are avoided. Similarly, utilities use capacitor banks on the transmission and distribution networks during heavy load conditions.

Capacitor banks are inexpensive, have very low losses, and are widely used to support voltage. They are, however, destabilizing because the reactive power output is proportional to the square of the voltage. During voltage decay, output goes down when it is needed most.

The other important source of reactive power and voltage control is automatic excitation (field-winding) control of synchronous generators. The generator voltage is continuously controlled to a desired value. In addition to active power production, generators provide reactive power to compensate for loads with lagging power factors and reactive power losses of the transmission network.

Load-restoration mechanisms. Immediately following an outage, all loads are voltage sensitive. Motors act like an impedance (resistive) load, with a drop in power proportional to the square of the voltage. Within a few seconds, however, inertial dynamics (Newton's second law applied to rotation) adjust the motor speed and torque to meet the mechanical power demand regardless of the voltage. Electronically controlled loads such as adjustable speed drives and computer power supplies operate even faster.

In a much slower time frame, tap-changing transformers regulate voltages at substations near loads. Most bulk power delivery transformers incorporate automatic underload tap changing. A typical transformer converts 115-kV transmission voltage to 12.5-kV distribution voltage, and delivers around 20 MW of power. Following a drop in voltage, tap-changing transformers have an initial delay time of 0.5–2 min, and then 5–10 s of mechanism time between steps. A ±10% voltage tap range is common.

An even slower load-restoration mechanism is the automatic or manual control of loads that must deliver a certain amount of energy. Electric heating loads are constant-energy loads of this type; the power delivered is proportional to the square of the voltage, and the energy is proportional to the time that the heater is on. Electric space heating, electric water heating, ovens, and cooktops are either thermostatically controlled or operator controlled. Following a drop in voltage, heaters must stay on longer to meet the energy demand. Thus more heaters in a load area will be on at a given time, reducing the natural load diversity of an area. Thermostatically controlled electric space heating loads are important in wintertime voltage stability situations.

Current limiting at generators. A second important mechanism for longer-term voltage instability occurs at the generating plants. Generators have many seconds or a few minutes of time-overload capability in the field and armature windings. This time-overload capability, along with the initially voltage-sensitive loads, allows stable operation for perhaps several minutes following severe outages. Field-winding time-overload limits are usually reached first, and are typically controlled automatically by reducing synchronous generator field excitation voltage and current. Thus the generator reactive power output is reduced, increasing the reactive power demand on other generators. With continued voltage decay, the armature windings rapidly reach their time-overload limit. Armature current limiting (usually performed manually by

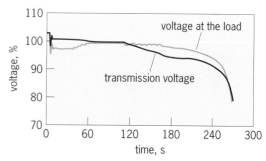

Fig. 2. Plot of voltage collapse. Both the voltage at the load that is regulated by a tap-changing transformer and the transmission voltage are shown. (*After C. W. Taylor, Power System Voltage Stability, McGraw-Hill, 1994*)

power plant operators) requires a more drastic reduction in generator reactive power, leading to voltage collapse.

Simulation. **Figure 2** demonstrates the process of voltage collapse by time simulation. Following a transmission-line outage and synchronizing oscillations, voltages are almost steady in the 10–40-s time period. Restoration of voltage at the load by tap changing then starts, and the tap steps can be seen. After 60 s the voltage at the load is restored to its initial 100% level, and another period of steady voltages ensues. The transmission voltage, however, is below its initial value of 102.7%, and generators are above their continuous-field current limits. Starting at about 110 s, voltages start to decay because of field-current limiting at nearby generators. Further tap changing occurs, preventing load relief (Fig. 2). Voltages collapse as the field-current limiting is enforced and reactive power transmission from remote generators is unsuccessful.

Voltage instability countermeasures. Utility engineers must find cost-effective solutions to prevent voltage instability and to allow high power transfers. For example, utilities install capacitor banks to allow generators to operate near unity power factor with a large reserve of automatically controlled reactive power capability for emergencies. The likelihood of field-current limiting is thus reduced.

For infrequently occurring disturbances, some utilities install undervoltage load shedding. This load shedding is analogous to underfrequency load shedding that most utilities installed following the 1965 blackout in northeastern North America.

In the future, utilities may economically ensure voltage stability by direct control of customer load over the so-called information superhighway. During emergencies, air conditioners, water heaters, space heaters, and other noncritical loads can be rapidly turned off until other stabilizing measures can be taken. SEE INFORMATION SUPERHIGHWAY.

For background information SEE ALTERNATING-CURRENT CIRCUIT THEORY; ELECTRIC POWER SYSTEMS; STATIC VAR COMPENSATOR; VOLT-AMPERE in the McGraw-Hill Encyclopedia of Science & Technology.

Carson W. Taylor

Bibliography. C. Concordia (ed.), Special issue on voltage stability and collapse, *Int. J. Elec. Power Ener. Sys.*, vol. 15, no. 4, 1993; O. I. Elgerd, *Basic Electric Power Engineering*, 1977; C. W. Taylor, *Power System Voltage Stability*, 1994.

Electric power transmission

The United States has more than 100 million electric power customers, more than 2300 generators with a total generation capacity of about 700 gigawatts (1 GW equals 10^9 W), and 147,000 mi (237,000 km) of high-voltage alternating-current (ac) and 2600 mi (4200 km) of direct-current (dc) transmission lines. Most of this is operated as one huge, interconnected system connecting Florida with Manitoba and Nova Scotia with the Rockies. In effect, this may well be the most complex machine ever built and operated by humans. However, the complexity of the system is hidden from the average person, who is used to having an unlimited supply of power. The function of the transmission system is to share generation resources over large regions. Thus, it is the key to satisfying the instantaneous demand for power whenever a customer turns on a switch.

Power-system control objectives. In contrast to the telephone system or a gas pipeline, the delivery of power cannot be definitely associated with a given producer of power because the electrons flow where they want to flow. Electric energy must also be consumed as soon as it is generated or converted into other forms of energy like kinetic energy in, for example, pumped hydroelectric storage plants if intended for later use. The so-called electric woodshed is missing, with the exception of batteries and superconducting coils. Thus the control of power systems is unique. Some other control objectives are as follows:

1. The system has to be operated with near-perfect frequency because otherwise clocks would not keep time correctly.

2. Voltages must be kept within narrow limits to avoid burning out light bulbs in a few hours and still provide good illumination of the workplace.

3. The customers should not suffer a blackout because of a loss of any single generator or transmission line.

The power system is always bordering on being unstable because system losses are kept at a minimum to achieve the highest possible efficiency of the system. To provide reliable power, the system operators must therefore also carefully consider the dynamics of the system. Unusually hot or cold weather can put severe stresses on the system. Blackouts, either preplanned or unplanned, would be the result of insufficient generation. Many blackouts are a result of cascading failures. That is, one failure leads to another, and so forth. The transmission system is used to manage these kinds of system stresses and also to share resources to meet differ-

ing seasonal, weekly, and daily variations in regional power demands. Thus, utilities whose operation is based on hydroelectric power will sell power when there is a surplus and buy power in the dry season. Transmission lines are also used to supply, in effect, coal by wire from power plants at the coal fields instead of shipping the coal by train to generators at load centers. The transmission lines connecting the generators to the loads are therefore often rather long, but long lines have poor load-carrying capacity. *See* ELECTRIC POWER SYSTEMS.

Power transfer. For ac lines, the power transfer of a transmission system (**Fig. 1**) is proportional to the sending- and receiving-end voltages, inversely proportional to the series impedance, and a function of the electrical angle across the line (**Fig. 2**). No active power will flow if the electrical angle is zero. The maximum power will flow when the angle is 90°. The ohmic resistance of the line impedance thermally limits the line's load-carrying capability, but the major component of the impedance is an inductive reactance. This measures the magnetic field arising from the current in the conductors. This reactance consumes so-called reactive power because it does not perform any work; only so-called active power is converted to real work. Figure 1 also shows shunt impedances to ground. These are capacitive reactances which are charged and discharged from the voltage applied to the line. They can normally be ignored except for very long lines and for transmission cables.

Power transfers can be controlled by changing the voltages, the impedance, or the electrical angle, but ac systems cannot operate with too large an angle across the line. The maximum usable power angle may be 30–40°. Thus the required stability margin to maintain power-system reliability is provided. Also, all but the generator voltages are fixed. Additional components are therefore needed to make the transmission system controllable.

Voltages can be controlled by means of on-load tap changers on transformers or by compensating the reactive-power portion of the loads. If the load is inductive, shunt capacitors are used for power-factor corrections; if it is capacitive, shunt reactors are used.

The impedance can be controlled by inserting a capacitor or reactor in series with the line. Capacitors will reduce the impedance of the line and increase the power flow, and series reactors are used to reduce power flows.

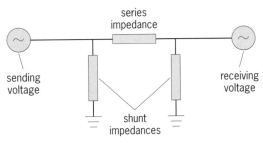

Fig. 1. Elementary two-machine transmission system.

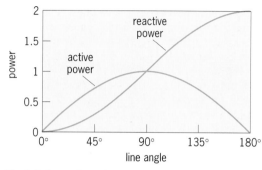

Fig. 2. Active and reactive power as functions of line angle. Values of power are normalized to the value at an angle of 90°.

For phase-angle control, special transformers, so-called phase-angle regulators (PAR), are used.

When active power flows through a line, reactive power is also consumed by the line. Figure 2 shows how the reactive power varies as a function of the line angle. If the sending- and receiving-end voltages have the same amplitude then the same amount of reactive power has to be supplied from each end of the line. The reactive power must be supplied or the system will collapse. Thus, the operators must balance not only the active power generation to load, but also the generation and consumption of reactive power. This can be a problem also when the load is low because then the combined capacitance of all lines may generate too much reactive power and cause overvoltages. Such overvoltages are controlled by shunt reactors.

Thyristor-based equipment. Often, possibly as a result of numerous power sales and purchases between utilities, some lines will carry more than the desired load and other lines will be underutilized. Series compensation or phase-angle regulators are the most effective means for balancing the line loadings to make the power flow where the operators want it to flow. Such compensation can be fixed or switched by using conventional breakers. However, a high frequency of operations is often needed, and then mechanical switches are not adequate. High-speed control is also frequently needed. Large-power semiconductor switches called thyristors can be used to solve these problems. The thyristor is a *pnpn* semiconducting device with alternating *p*- and *n*-doped semiconducting layers. They can be turned on but not off by applying a current to a control gate. Only the so-called gate-turn-off (GTO) thyristor can also be turned off. One device can occupy the entire surface of a wafer 5 in. (125 mm) in diameter. It can block 8000 V and conduct 4000 A, but for high-voltage switch applications many thyristors must be connected in series and parallel. Thyristors are also found in most homes in simple dimmer switches.

Thyristor switches became available in the 1970s when they became sufficiently powerful to replace mercury-arc valves in high-voltage dc transmission systems. They were first applied in static-var control systems (SVC); thyristor-controlled reactors were

often combined with fixed or thyristor-switched capacitors for reactive power control used primarily for improved power quality by reducing the effects of flicker from arc furnaces and so forth. A number of new types of thyristor-based equipment are now manufactured or under development. They are becoming known as flexible ac transmission system (FACTS) devices or systems as a result of a major development effort initiated by the Electric Power Research Institute (EPRI) beginning in 1988. However, the first flexible ac transmission system application of a static-var control system took place as early as 1978. A new type of var compensator, often referred to as a static compensator (STATCOM), has now been developed and is installed for trial operation in Tennessee. This system is based on what is called voltage source converter technology, which requires the use of gate-turn-off thyristors.

Thyristor-controlled or switched-series capacitor equipment is also available. A 208-megavar, 3000-A, three-phase thyristor-controlled series capacitor (TCSC) system has been installed for trial operation in the Pacific AC Intertie system, a 500-kV line in Oregon (**Fig. 3**). Development of what is called a unified power flow controller (UPFC) has also begun. This device can control voltages, phase angles, and series impedance all at the same time, and thus promises to become a universal compensator. The unified power flow controller requires the use of semiconductor devices which can be turned off by means of gate control and is very similar to what is used in static compensator systems.

The flexible ac transmission system is based on technologies which for more than 25 years have been used in high-voltage dc systems, so in that sense it is a mature technology. However, new semiconductor devices are also under development. One is the so-called metal-oxide-semiconductor controlled thyristor (MCT), which is expected to become the ideal high-power semiconductor switch. There is also a search for new semiconductors based on such materials as silicon carbide and diamond, which

Fig. 3. A 208-megavar, three-phase thyristor-controlled series capacitor bank installed in a 500-kV line in Oregon. (*General Electric Company*)

could revolutionize the technology. Thus the field would be opened to new and innovative circuit solutions in future flexible ac transmission systems. The development of flexible ac transmission systems is driven by the need to utilize the existing transmission systems to their limits because it is almost impossible to get approvals for building new lines. It will also be driven by the need to deal with transmission access requests from independent power producers. Flexible ac transmission systems are part of an emerging technology expected to play a significant role in future systems.

For background information *SEE ALTERNATING CURRENT; ALTERNATING-CURRENT CIRCUIT THEORY; ELECTRIC POWER SYSTEMS; SEMICONDUCTOR RECTIFIER; STATIC VAR COMPENSATOR; TRANSFORMER; TRANSMISSION LINES; VOLT-AMPERE* in the McGraw-Hill Encyclopedia of Science & Technology.

Stig L. Nilsson

Bibliography. P. Kundur, *Power System Stability and Control*, 1994; North American Reliability Council, *Transmission Transfer Capability*, 1995.

Electromagnetic field

It is important to understand the electromagnetic fields associated with the electric power transmission and distribution system. Historically, it has been important for determining that systems operated efficiently and reliably. For example, electromagnetic field analysis has been used to determine equivalent transmission-line parameters for system power flow, loss, and stability studies. It has also been used in studies of lightning protection and insulation failure. Another concern related to the electromagnetic environment has been safety from the hazards of electrical shock. Such shocks may occur both by direct contact with the power system and by indirect contact with systems that are electromagnetically coupled to the power system. Electromagnetic coupling can also cause interference with wire communication systems such as telephone and railroad signaling circuits.

Over the years, power-line voltages have been increased in order to improve the efficiency of power transmission over long distances. One consequence is a concern for the effects of power lines on the environment. For example, electromagnetic interference (EMI) to broadcast communications and audible noise from a corona (that is, electrical discharges at the conductor surface) on high-voltage power lines have been issues since the 1930s. Recent concern involves possible health effects from exposure to the extremely low frequency (ELF) electric and magnetic fields associated with power lines.

This article will review recent research on ELF electric and magnetic fields near overhead power lines, in particular, the health-effects controversy and recent advances in calculation, measurement, and mitigation of these fields.

Definitions. Before further discussion, some definitions of electrical terms will be made by analogy to a water-pumping system, in which a pump forces water continuously through a pipe and back to the pump again. The pressure generated by the pump is analogous to the voltage of a simple electrical system, and the amount of water passing through the pipe per unit of time is analogous to the current passing though the resistor in the same electrical system. Voltage is measured in volts and current in amperes.

In most power systems, the voltage and current reverse direction periodically in time. Each pair of reversals is defined as a cycle, and the number of cycles per second is defined as the frequency of the system. The power system in North America is operated at a frequency of 60 cycles per second, or 60 hertz.

Simple models for calculating the ELF electric and magnetic fields of a power line have been developed. The electric field (measured in volts per meter) can be calculated from a knowledge of the power-line voltage and geometry. Generally, as voltage increases, the electric field increases. The magnetic field can be calculated from a knowledge of the power-line current and geometry. Generally, as the current becomes larger, the magnetic field becomes larger; thus, the magnetic field is not directly related to voltage. Magnetic field is here used as equivalent to the magnetic flux density. It is measured in milligauss (mG) in the United States and in microtesla (μT) in Europe; 0.1 μT = 1.0 mG.

Health-effects controversy. In the early 1970s, the electric field was the major concern when possible health effects of electric power lines were considered, because calculations of electric current induced in humans under typical transmission lines showed that the electric field was the dominant contributor. Since 1979, with the publication of a landmark epidemiological study, concern has been gradually shifting to magnetic fields. In this original study, a substitute for the magnetic field (rather than the magnetic field itself) was found to have a weak but statistically significant association with childhood leukemia. This substitute is called a wire code, a formula based on the power-line wire size, voltage level, and distance from the residence under study. It was used by early researchers as a substitute for making magnetic-field measurements in residences.

An association does not necessarily imply cause and effect, and in fact no compelling evidence for cause and effect has been found. Subsequent epidemiological studies, although of better quality, have failed to either establish or reject an association between magnetic fields and health outcomes. Nevertheless, they show a consistent association between health outcomes and the magnetic-field substitute. This consistency is what drives the continued interest and research in this area. None of these epidemiological studies has found a statistically significant association between the electric field and any health outcome.

In recent years, there has been a great deal of research in this area and vigorous debate about whether there are any health hazards of exposure to ELF magnetic fields at levels typically found near power lines. Although the consensus among scientists is that there are biological effects at some level of magnetic-field exposure, there is no consensus as to the minimum level or to whether there is sufficient evidence for the existence of health hazards or any basis for exposure regulations.

Calculating power-line fields. Power-line voltages are constant to within roughly 10%. Thus, the ELF electric field from a given line can be calculated with reasonable accuracy at any time. The behavior of the current and hence the magnetic field, however, is very different. Both are dependent upon power-system configuration and operating conditions. For example, on a given line the current can vary over a day by a factor of more than 4. Significant seasonal variation is also observed. These variations occur because of shifts in the demand for electric power due to daily patterns of use and seasonal weather changes.

Thus, the magnetic fields of a power line may vary considerably over time; as a consequence, it is difficult to define a magnetic-field level that characterizes the line. It is, in fact, appropriate to describe the field statistically. Examples are the median level (the level exceeded 50% of the time) and the practical maximum (the level exceeded only 10% of the time). The median is preferred because it corresponds most closely with magnetic-field values reported in health-effects studies. Care must be used in defining levels because attempts to regulate magnetic fields are complicated by the question of how to define the magnetic field at a point near a power line by a single number.

Measurement of power-line fields. Recent research has focused on the design of instruments and protocols for measurement of ELF electric and magnetic fields. Meters used for measurement of electric fields include the free-body meter, the ground-reference meter, and the electrooptic meter. Meters used for measurement of magnetic fields at levels typically encountered near power lines include the coil-probe meter and the fluxgate.

Mitigation of magnetic fields. A power line consists of a set of current-carrying conductors. The magnetic fields from these conductors depend on the quantity R_a, defined as the distance from the measurement point to the origin of a set of coordinates near the center of the conductor set. For typical power lines and operating conditions, the magnetic fields decay as $1/(R_a)^2$. The magnetic fields can be reduced if the spacing between wires is reduced or the distance R_a is increased. Additional reduction is possible if power lines are designed in certain ways. For example, two power lines may be operated in parallel. If the currents on the two lines are split equally and the wires arranged in an optimum way, the magnetic fields will decay as $(1/R_a)^3$, which results in a smaller magnetic field. This type

of power line, called a double-circuit low-reactance line, is in common use. Measurements on an operating line of this type show that its magnetic fields are typically one-third those of a conventional line carrying the same power.

Methods involving shielding or field cancellation have also been used. Shielding or cancellation of electric fields is relatively easy, and even structures such as wood houses act as shields. However, the shielding or cancellation of magnetic fields is a more difficult problem. The classical method is to use magnetic material which shields by a process called flux shunting. In this method the magnetic fields are pulled away from the area to be shielded and directed into the shield. A related method uses conducting materials in which induced eddy currents partially cancel fields. One special case of this technique is the use of a separate loop of wire mounted near a power line to partially cancel the magnetic fields of the power line. Although this method has been demonstrated on an operating line, it is not yet commonly used.

Problems with field reduction. The reduction of magnetic fields is not necessarily done without penalty. It can be shown, for example, that many power lines designed for reduced magnetic fields have larger electric fields at the conductor surface, and hence more corona and related audible noise and electromagnetic interference. Worker safety can also be compromised if smaller wire spacing is used to reduce magnetic fields. Reduced clearances increase the possibility of shock if maintenance is conducted while the power line is energized. For these reasons, design modifications must conform to the National Electric Safety Code.

For background information SEE ELECTRIC FIELD; ELECTRIC POWER SYSTEMS; ELECTRICAL INTERFERENCE; ELECTROMAGNETIC COMPATIBILITY; EPIDEMIOLOGY; INDUCTIVE COORDINATION; MAGNETIC FIELD; TRANSMISSION LINES in the McGraw-Hill Encyclopedia of Science & Technology.

Robert G. Olsen

Bibliography. Electric Power Research Institute, *Transmission Line Reference Book, 345 kV and Above,* 2d ed., 1982; Institute of Electrical and Electronics Engineers, *IEEE Recommended Practice for Instrumentation: Specifications for Magnetic Flux Density and Electric Field Strength Meters, 10 Hz to 3 kHz,* IEEE Std. 1308-1994; National Institute of Environmental Health Sciences and U.S. Department of Energy, *Questions and Answers About EMF: Electric and Magnetic Fields Associated with the Use of Electric Power,* DOE/EE-0040, 1995; G. Taubes, Fields of fear, *Atl. Month.,* pp. 94–108, November 1994.

Electrooptic materials

Materials that change color as a result of an applied potential are called electrochromic. These materials function by an electrochemical process; that is, the oxidation state of the material is changed. Most of the materials studied have been metal oxides or conducting organic polymers. The devices are relatively complicated as they require two electrodes and an electrolyte, which has to be sandwiched between two sheets of glass. Electrochromic devices have undergone a large amount of developmental work, but they suffer from certain disadvantages: they require a relatively large power input, they are slow to react, and the color changes are relatively limited. Thus their applications for displays are limited, and their use for large windows is very expensive.

A varied group of organic and organometallic complexes have been shown to possess novel electrooptic properties. These materials change their intensities of absorbance in the visible spectrum when an electrical potential is applied across them. It is thought that this effect results from novel solid-state changes within the materials.

The aim of research is to develop materials that will lighten or darken when a potential is applied across them in the solid state. Such technology could be used to coat glass that would allow the regulation of light or heat entering through a window. The implications for energy saving are large. One advantage of these materials is that a very low power input is required to bring about this effect; thus another application could be as part of an energy-efficient long-term display device.

The materials are organic conductors, and a number of studies have been made of their semiconducting and superconducting properties. The materials consist of stacks of planar, aromatic donors or acceptors [as shown in structures (I)–(III)] with closed-

(I)   (II)

(III)

shell counterions. The stacks are partially charged, and the overlap of the aromatic systems results in pseudo-one-dimensional conduction, where the conductivity in the plane of the stacks can be as high as that for graphite. Although these types of materials have been known since the mid-1940s, the complexes that demonstrate these electrooptic properties are novel because their counterions are noncentrosymmetric.

The complexes that have been studied have either tetrathiafulvalene [(I); S = sulfur], metal

maleonitriles [(II); M = transition metal, CN(NC) = cyano group], or tetracyanoquinadimethane (III) as the planar aromatic donors or acceptors. In these structures, R represents an organic group.

Growth of materials. The materials grow on most substrates as fine, dense, black crystals. In this form they are of little use for optical purposes. When the complexes are grown onto indium tin oxide, which is an optically transparent conductor, they form amorphous, translucent materials. The planar molecules that form the stacks are dissolved in a nonaqueous solvent such as acetonitrile or ethanol. The counterions that will be incorporated in the complex are added to the solution as the electrolyte. For example, when tetrathiafulvalene (TTF) is used as the substrate, the glass coated with indium tin oxide is made into the anode in the electrolytic cell and the reaction shown below occurs; X^- represents a counterion.

$$TTF + X^- - e^- \longrightarrow [TTF\ X] \downarrow$$

Because a nonaqueous solvent is used, the TTF^+ formed is poorly solvated in solution and so forms a complex with the counterion (X^-) from the electrolyte on the electrode surface. The choice of the solvent is crucial: too polar and the complex will not precipitate; too nonpolar and many electrolytes will not dissolve.

Optical properties. The **illustration** shows the visible spectrum for a sample of a typical material and the effect of an applied potential. The absorbance decreases across the entire spectrum. The material is observed to become lighter, rather than changing color. This effect is observed only when the closed-shell counterion of the complex is noncentrosymmetric. All complexes containing symmetrical ions, such as the tetrafluoroborate (BF_4^-), do not change their optical properties in the visible region when a potential is applied across them. It is thought that these novel electrooptic properties arise from physical changes occurring within the crystal structure. The noncentrosymmetric ions have a dipole moment at the center of volume, and these ions rotate within the crystal lattice to become aligned with the potential field. Since the optical properties of these materials depend upon a charge-transfer process between the stacks and the counterions, a rotation of the latter will result in a change in interaction and hence a change in the intensity of the light absorbed.

The difference between this electrochromic phenomenon and the electrooptic effects described here is that the latter are brought about by a physical rather than a chemical change. Little current should be required to bring about these changes, and any device based on such materials would have a low power input.

Another advantage of these materials is that once a potential has been applied, the intensity of light absorbance is not reversed when the potential field is removed. If these materials were used in a long-term display device, the power requirements

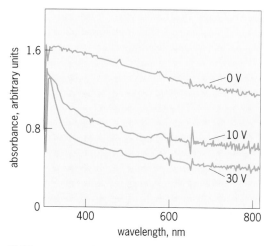

Visible spectrum of tetrathiafulvalene mandalate complex grown onto glass coated with indium tin oxide when a potential is applied across the material in the solid state.

would be very low, as the potential would have to be applied only to change the material and not to keep it in its altered state. The electrooptic effect can be reversed only by heating the material (thus randomizing the dipole orientations) or by applying a high potential field in the opposite direction. The change in the intensity of the absorbance is also dependent on the direction in which the potential field is applied. For most of the films studied, the change in absorbance when the potential is applied parallel to the film is greater than when it is applied perpendicular to the film. This result suggests that the orientation of the counterion with respect to the planar moiety is critical. Experiments in which the material is electrically polarized in an electrolyte solution show similar optical changes without the flow of a faradic current. Thus it is apparent that the effect is not due to a chemical change in the material or heating of the sample.

One factor that has hampered the optimization of these new materials has been the large number of variables that affect the speed of rotation and the magnitude of the optical change. Several molecular and bulk properties of the materials can be varied independently to change the nature of the electrooptic effect. These properties include the size and dipole moment of the counterion, the size of the cavity in which the ion rotates, the electronic structure of the planar moiety, the substrate onto which the material is grown, and the conditions of electrodeposition.

Size and dipole moment. The size of the ion with respect to the cavity in which it resides will affect the magnitude of the energy barrier to rotation. It may be expected that the larger the dipole moment, the greater will be the effect of a given potential difference. Initial results show that the smaller the counterion, the smaller the potential that is required to bring about the optical changes in the materials. This relationship is to be expected as there will be a smaller steric barrier to the rotation of the ion. The

dipole moment of a single ion cannot be measured, and so nonempirical molecular orbital calculations have been carried out to determine the size and dipole moments of the ions. The effect of the magnitude of the dipole moment is not yet apparent. Preliminary observations show no direct correlation between the dipole moment and the rate, the magnitude, or the onset potential of the electrooptic effect—probably because steric factors dominate the reorientation process.

Size of cavity. The size of the cavity in which the ion rotates can be changed by altering the R groups in structures (I)–(III). Bulky groups will tend to expand the size of the cavity and will also change the interaction with the counterions. However, only certain groups can be used, as very bulky groups will not allow the stacking of the aromatic systems; hence the materials will not conduct and will not grow on the electrode surface.

Electronic structure of planar moiety. The donor and acceptor qualities of the planar moieties can be changed by substituting the sulfur atoms for heavier atoms such as selenium and tellurium. The properties such as color and electrical conductivity can also be changed by increasing the degree of conjugation of the aromatic system. For the metal maleonitrile complexes, the transition metal, M, also affects the color and magnitude of the change in absorbance.

Substrate. The crystal structure and surface topology of the electrode will affect the structure and degree of crystallinity of the material that grows on it.

Electrodeposition conditions. The morphology of the material will be dependent upon the electrode potential, the applied current density, the concentration of electrolyte and electroactive species, the solvent, and the temperature of the deposition process. All these factors will affect the structure of the deposit, and hence vary the size of the cavity and the interaction of the planar moiety with the counterion.

Electrical properties. In addition to novel electrooptic effects, these materials have variations in their electrical properties associated with the changes in optical absorbance. It has been found that the capacitance and conductivity of the materials change by several orders of magnitude upon the application of a potential field. The large changes in the conductivity (often four to five orders of magnitude) can result only from a change in overlap of the aromatic moieties in the stacks. It is well known that the application of an electric field to any material has two results; first there is an instantaneous polarization of the electronic charge clouds; second there is a much slower change in the polarization because of a reorientation of any constitutional electric dipoles. The analogy with simple electronic circuits containing resistors and capacitors is well established, and the measured changes in capacitance and resistance correlate well with the potentials and time scales observed in the optical measurements. Thus it can be concluded that the cause of the electrooptic effect also changes the electrical properties of the material from those typical of a semiconductor to those of a conductor. Researchers have proposed that the rotation of the counterions within the lattice causes simultaneous structural changes in the stacks. This idea would aid in the explanation of the optical effects, because such a process would be expected to be relatively slow and would be more difficult to reverse.

For background information *SEE COORDINATION COMPLEXES; ELECTROOPTICS; ORGANIC CONDUCTOR; SOLID-STATE CHEMISTRY* in the McGraw-Hill Encyclopedia of Science & Technology.

Andrew P. Abbott

Bibliography. A. P. Abbott, P. R. Jenkins, and N. S. Khan, *J. Chem. Soc. Chem. Commun.,* pp. 1935–1936, 1994; P. M. S. Monk, R. J. Mortimer, and D. R. Rosseinsky, *Electrochromism: Fundamentals and Applications,* 1985.

Environmental fluid mechanics

The definition of environmental is that which deals with the aggregate of surrounding things, conditions, and influences. Essentially, the word can invoke a variety of meanings dealing with nature and its exploitation by humans. The environment on Earth is intimately tied to the fluid motion of air (atmosphere), water (oceans), and species concentrations (air quality). In fact, the very existence of the human race depends upon its abilities to cope within Earth's environmental fluid systems. Only recently has a close examination begun of the role and influence of the human race on the environment, and its consequences.

The fluid motion of the environment has been studied by meteorologists, oceanologists, and geologists for many years. Although weather and ocean-current forecasts are still of major concern, the motion of the environment is also the main carrier of pollutants. More recently, biologists and engineers have joined in studies examining the effect of pollutants on humans and the environment, and the means for environmental restoration. The impact of pollutants on drinking-water quality has become of special importance in the study of ground-water flow. It is also well known that lake levels are significantly influenced by climatic change, a relationship that has become of some concern in view of the global climatic changes that may result from the greenhouse effect (whereby the Earth's average temperature increases because of increasing concentrations of carbon dioxide in the atmosphere).

Scales of motion. Environmental fluid mechanics basically deals with the study of the atmosphere, oceans, lakes, streams, surface and subsurface water flows (hydrology), building exterior and interior air flows, and pollution transport within all the above categories. Such motions occur over a wide range of scales, which accounts in large part for the difficulties associated with understanding fluid motion within the environment. In order to impart motion (or inertia) to the atmosphere and oceans, internal and external forces must develop. Global external forces consist of gravity, Coriolis and centrifugal forces, and electric and magnetic fields (to a lesser extent). The internal forces of pressure and friction are created at the local level, that is, on a much smaller spatial scale; these influences also have different time scales. However, the winds and currents arise as a result of the sum of all these external and internal forces.

Global motion is generally referred to as synoptic-scale, or macroscopic, motion; regional- or intermediate-scale motion is known as mesoscale; local motion is commonly referred to as microscale motion. Interestingly, humans live in the microscale associated with atmospheric motion, that is, the boundary layer of air which extends about 1 km (0.6 mi) above the Earth's surface. The terrain of the Earth as well as ocean surface conditions significantly affect both the microscale and mesoscale motion of the atmosphere, and hence weather conditions.

El Niño, a name given to the periodic flow of warm waters along the western coast of South America, disrupts the coastal ocean and the upwelling of cold waters, producing large amounts of precipitation, along with widespread destruction of plankton, fish, and sea birds (which prey on the fish). Major El Niño events have occurred in 1925, 1941, 1957–1958, 1972–1973, 1982–1983, and 1992. Only recently has it been determined that the events are caused by changes in the surface winds over the western tropical Pacific, which periodically release and drive warm waters eastward to the South American continent.

The study of air pollution falls within the category of environmental fluid mechanics, because the air within the lower atmosphere steers (or advects) and diffuses pollutants. Atmospheric winds near the Earth's surface are generally turbulent and gusty, which helps to clear polluted areas; the velocity varies with altitude, local stability (level of turbulence), and roughness of the terrain. However, when the winds are calm, stagnant conditions can prevent pollutants from being cleared from a city, resulting in high levels of bad air quality and smog. Of particular importance on the mesoscale level is acid rain (where rainfall removes sulfates and nitrates within the atmosphere), which has resulted in serious environmental damage. Likewise, mixing of pollutants into the upper atmosphere can cause long-term changes in the ozone layer (even though the causes, such as propellants within spray cans, may have been generated within the microscale layer). The recent explosion of Mount Pinatubo in the Philippines resulted in the discharge of many tons of particulates into the Earth's atmosphere; these particulates in turn acted as seed nuclei for precipitation, and were the cause of much of the flooding and climatic changes over the following few years.

Governing equations. The foundations of environmental fluid mechanics lie in the same conservation principles as those for fluid mechanics, that is, the conservation of mass, momentum (velocity), energy (heat), and species concentration (for example, water, humidity, other gases, and aerosols). The differences lie principally in the formulations of the source and sink terms within the governing equations, and the scales of motion. These conservation principles form a coupled set of relations, or governing equations, which must be satisfied simultaneously. These governing equations consist of nonlinear, independent partial differential equations which describe the advection and diffusion of velocity, temperature, and species concentration, plus one scalar equation for the conservation of mass. In general, environmental fluids are considered to be newtonian, making it possible to utilize Newton's postulate (in 1686) for viscosity whereby fluid stresses vary linearly with fluid strain. This is conveniently handled by relating the stresses to the velocities of deformation which are expressed in terms of space derivatives of the velocity components themselves. The resulting equation of motion for atmospheric flow is generally referred to in fluid mechanics as the Navier-Stokes equation for incompressible flow.

Driving mechanisms of flow. The mechanisms which drive the flow patterns in the atmosphere and oceans are vastly different. The atmosphere is thermodynamically driven, with the major source of energy coming from solar radiation. This short-wave radiation traverses the air and becomes partially absorbed by the land and oceans, which reemit the radiation at longer wavelengths. This long-wave radiation heats the atmosphere from below, creating convection currents in the atmosphere.

In the oceans, periodic gravitational forces of the Sun and Moon generate tides; in addition, the ocean surface is affected by wind stress that drives most of the ocean currents. Local differences between the air and sea temperatures generate heat fluxes, evaporation, and precipitation, which ultimately act as thermodynamical forces that create or modify wind-driven currents.

Environmental layers. Fortunately, not every term in the Navier-Stokes equation is important in all layers of the environment. The horizontal component of the motion is usually the most significant in most cases; this horizontal motion is subjected to maximum frictional forces at atmosphere-ocean interfaces. This frictional force causes the formation of a boundary layer in which the velocity of air at the surface of the Earth is zero (relative to the Earth), and the velocity at the surface of the ocean is a minimum equal to the surface velocity of the water. The ocean current is primarily generated by the wind; hence, the water velocity at the surface is a maximum and decreases in depth, again as a result of frictional forces. In both instances, frictional forces cause strong velocity gradients and vorticity (rotation) within the boundary layer.

Figure 1 shows the velocity distribution in the atmosphere and ocean.

The thickness of the atmospheric boundary layer varies with the wind speed, degree of turbulence, and type of surface. For atmospheric flows, the layer is on the order of 1 km (0.6 mi) thick; within the ocean, it may be 30 m (100 ft) thick. Beyond this layer, the environmental flow is typically considered to be viscous free (without turbulent shear), or inviscid. Because there are no shear stresses, the motion of the inviscid layer is governed only by the advection, pressure, and body-force terms. In atmospheric flows, the rotation of the Earth strongly influences this layer of flow, generally referred to as the geostrophic layer. (The motion is called macroscopic scale because it deals with the largest portions of the troposphere and the oceans.) Just above the surface, the mean velocities are small and the advection terms and the Coriolis force (which depends on the velocity) are negligible compared to the shear forces (viscous terms), which appear to be constant in this inner layer. However, within the outer, or Ekman, layer advection is still negligible and the viscous forces are small; this part of the boundary layer is in equilibrium with the Coriolis, pressure, and Reynolds stresses (turbulence). The **table** shows typical scales of length, velocity, and time for both atmospheric and oceanic motions. Oceanic motions are slower and more confined, and tend to evolve more slowly, than atmospheric motions.

Relative importance of terms. The key to being able to obtain solutions to the Navier-Stokes equation lies in determining which terms can be neglected in certain specific applications. For convenience, problems can be classified on the basis of the order of importance of the terms in the equations utilizing nondimensional numbers based on various ratios of values. The Rossby number is the ratio of the advection (or inertia) forces to the Coriolis force, $Ro = V/L\Omega$, where V is the velocity, Ω is the Earth's angular velocity, and L is a specified reference length. When the Rossby number is much less than 1, the inertia forces become insignificant, implying that these types of flows are more geostrophic. The importance of the viscous to Coriolis forces is defined by the Ekman number, $Ek = \mu/\rho L^2 \Omega$, where ρ is density and μ is viscosity. The ratio of inertia to viscous forces is referred to as the Reynolds number, $Re = \rho VL/\mu$. The product of the Ekman and Reynolds numbers therefore yields the Rossby number. Thus, when the Rossby number is large and the Ekman number small, the motion is geostrophic; when the Rossby number is small and the Ekman number large, an Ekman-type boundary layer develops. As the Reynolds number increases, the ratio of the flow velocity to viscosity increases (that is, the advection terms become more important than the viscous terms), with the flow eventually becoming turbulent. Boundary-layer flow near the Earth's surface is nearly always turbulent.

Measurements. Because of the scales of motion and time associated with the environment, and the somewhat random nature of the fluid motion, it is difficult to conduct full-scale, extensive experimentation. Likewise, some quantities (such as vorticity or vertical velocity) resist direct observations. It is necessary to rely on the availability of past measurements and reports (as sparse as they may be) to establish patterns, especially for climate studies. However, some things can be measured with confidence.

Both pressure and temperature can be measured directly in the atmosphere and ocean with conventional instruments. In the ocean, the pressure is a much better measure than depth; depth is typically calculated from measured pressures obtained from instruments lowered into the sea. In the atmosphere, ground precipitation, radiative heat fluxes, and moisture content can be accurately measured. Likewise, the salinity of the ocean can be determined from electrical conductivity, and the levels monitored at shore stations. Unfortunately, many of these scalar quantities must be combined to obtain products and quantities that can be used to gain physical insight into the motion. Concentration samples are generally collected at receptor sites over long periods of time. The data are then examined to determine specific concentration levels and particulate sizes; this information is used to determine isopleth (concentration) levels and exposures over various atmospheric conditions. Occasionally inert tracer gases are released to determine wind directions as well as atmospheric diffusion (turbulence levels) and trajectories.

Vector quantities such as horizontal winds and currents are typically measured by using anemometers and current meters. Fixed instruments such as anemometers, which sit atop buildings and towers, and current meters attached to mooring lines at fixed depths offer fine temporal readings but are too expensive to adequately cover a large spatial area. Instruments are routinely deployed on drifting platforms in the ocean, and balloons are released in the atmosphere. (However, such measurements are mixed in time and space.) Although vertical speeds can be measured, the signals are usually below the level of the ambient turbulence and within the sen-

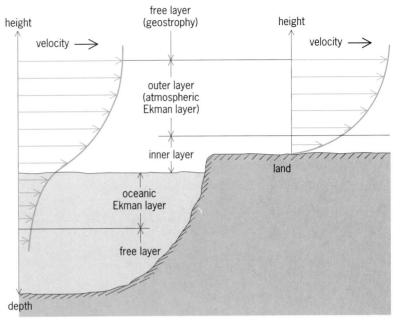

Fig. 1. Velocity distribution in the atmosphere and ocean. (*After S. Eskinazi, Fluid Mechanics and Thermodynamics of Our Environment, Academic Press, 1975*)

sitivity error of the instrument. Measuring the three-dimensional velocity components simultaneously and obtaining meaningful three-dimensional heat fluxes is difficult, and essentially relegated to small-scale laboratory experiments.

Recent advances utilizing satellite imagery, Doppler radar, acoustic sounding, and lidar (laser) have made it possible to obtain highly detailed data, including turbulence information, over much broader spatial distances. The use of Doppler radar has yielded three-dimensional velocity data and rotational characteristics within thunderstorms which can be used reliably to predict the onset of tornadoes. Although they are expensive, the National Weather Service is attempting to place more Doppler radar units around the country, particularly at airports.

Modeling. There are two types of modeling strategies: physical models and mathematical models. Physical models are small-scale (laboratory)

Length, velocity, and time scales in the Earth's atmosphere and oceans

Phenomenon	Length scale, km (mi)	Velocity scale, m/s (mi/h)	Time scale
Atmosphere			
Sea breeze	5–50 (3–30)	1–10 (2–20)	12 h
Mountain waves	10–100 (6–60)	1–20 (2–40)	Days
Weather patterns	100–500 (60–300)	1–50 (2–100)	Days–weeks
Prevailing winds	Global	5–50 (10–100)	Seasons–years
Climatic variations	Global	1–50 (2–100)	Decades
Ocean			
Internal waves	1–20 (0.6–12)	0.05–0.5 (0.1–1)	Minutes–hours
Coastal upwelling	1–10 (0.6–6)	0.1–1 (0.2–2)	Several days
Large eddies, fronts	10–200 (6–120)	0.1–1 (0.2–2)	Days–weeks
Major currents	50–500 (30–300)	0.5–2 (1–4)	Weeks–seasons
Large-scale gyres	Basin scale	0.01–0.1 (0.02–0.2)	Decades

SOURCE: B. Cushman-Roisin, *Introduction to Geophysical Fluid Dynamics*, Prentice Hall, 1994.

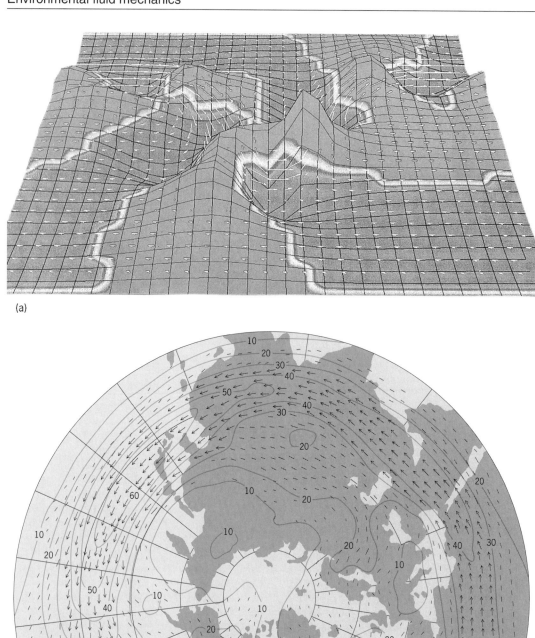

(a)

(b)

Fig. 2. Fluid-flow simulations. (*a*) Stability analysis for a flow path with a narrow trough (*after D. Adamec, Large-scale ocean modeling, NAS Technical Summaries, March 1989–February 1990, NASA Ames Research Center, 1990*). (*b*) Monthly mean winds over the Northern Hemisphere (north of 15°N latitude) for January 1991 at the 300-millibar (30-kPa) pressure level; contours of equal wind speed are shown, labeled in m/s (1 m/s = 2.2 mi/h); wind vectors are shown by arrows, except in regions of low speed or sparse data; parallels and meridians are at intervals of 15° in latitude and longitude (*after B. Cushman-Roisin, Introduction to Geophysical Fluid Dynamics, Prentice Hall, 1994*).

mockups which can be measured under variable conditions with precise instrumentation. Such modeling techniques are effective in examining wind effects on buildings and species concentrations within city canyons (flow over buildings). Generally, a large wind tunnel is needed to produce correct atmospheric parameters (such as Reynolds number) and velocity profiles. Mathematical models (algebraic and calculus based) can be broken down further into either analytical models, in which an exact analytical solution exists, or numerical models, whereby approximate numerical solutions are obtained by utilizing computers. By far the most interesting and widely used models are the numerical models. The reason for their popularity is that it is possible to model more of the actual physics of the flow, that is, solve the Navier-Stokes equation, rather than make assumptions and eliminate key components of the physics. Although the Navier-Stokes equation and other equations of the same form are highly nonlinear, such partial differential equations can now be solved with some measure of confidence and reliability. In many instances involving environmental fluid flow, the use of supercomputers is required.

Numerical methods. Several broad classes of solution techniques can be employed to solve the various derivatives and terms of the Navier-Stokes equation. The most common and widely used numerical methods are finite-difference schemes (which use a form of truncated Taylor-series expansion); finite-element schemes (which minimize the error between the actual and approximate solutions by using a local basis, or shape, function); spectral methods (in which dependent variables are transformed to wave-number space by using a global basis function, such as the Fourier transform); pseudo-spectral methods (which use truncated spectral series to approximate derivatives); interpolation techniques (whereby polynomials are used to approximate the dependent variables in one or more spatial directions); boundary-element methods (where a Green's function is employed along with discretization of only the boundary, by using finite elements, to model the Laplace and Poisson equations); and particle methods (which use Lagrangian particles whose trajectories are calculated within a conventional eulerian grid). Such numerical models depend strongly on boundary and initial conditions; care must be exercised to correctly initialize and specify all variables at the boundaries of the computational model. All the aforementioned schemes except the particle methods require knowledge of properties such as viscosity, dispersion coefficients, and thermal conductivity; particle methods require no constitutive models for particle viscosity or thermal conductivity, but do require a large number of particles for an accurate description of the flow field. The most widely used modeling approaches continue to be finite-difference, finite-element, and interpolation schemes, especially for mesoscale and synoptic-scale simulations.

Capabilities. The continuing rapid increase in computational hardware capabilities has made it possible to model much more complicated problems, that is, include more physics (or mathematical terms) in the governing equations. Simulations of environmental fluid flow over microscale and mesoscale regions without simplifications of the equations of motion are now fairly common. Arrays consisting of millions of nodes can be calculated within a few hours on supercomputers, and three-dimensional graphical displays can be generated on work stations. By using satellite, radar, and conventional surface observations as input data to meteorological models, reasonably accurate local forecasts can be made for up to 3–12 h. Advances in numerical techniques as well as computer hardware will continue, making it possible to perform more detailed calculations over broader expanses with reasonable accuracy over longer forecast periods.

Examples. An example of large-scale ocean modeling is shown in **Fig. 2***a*. Three-dimensional quasi-geostrophic and primitive-equation ocean models were used to simulate ocean mesoscale variability in the midlatitudes. The results of a flow-stability analysis for a flow path with a narrow trough are shown. The initial flow is indicated by the white arrows, while the three-dimensional surface and shading indicate the magnitude of the instability (the tendency to turbulent, chaotic motion). The instability with the largest magnitude occurs just downstream of the trough crest.

Figure 2*b* shows monthly winds over the Northern Hemisphere for January 1991 at the 300-millibar (30-kPa) pressure level (a height of approximately 9 km or 30,000 ft). The jet stream is clearly visible around the 45°N parallel, except over the eastern North Pacific and eastern North Atlantic (because of blocking effects).

For background information SEE AIR POLLUTION; BOUNDARY-LAYER FLOW; DYNAMIC METEOROLOGY; FINITE ELEMENT METHOD; FLUID-FLOW PRINCIPLES; INTERPOLATION; METEOROLOGICAL INSTRUMENTATION; NAVIER-STOKES EQUATIONS; NUMERICAL ANALYSIS; OCEAN CIRCULATION; SIMULATION; WEATHER FORECASTING AND PREDICTION in the McGraw-Hill Encyclopedia of Science & Technology.

Darrell W. Pepper

Bibliography. B. Cushman-Roisin, *Introduction to Geophysical Fluid Dynamics,* 1994; S. Eskinazi, *Fluid Mechanics and Thermodynamics of Our Environment,* 1975; *NAS Technical Summaries, March 1989–February 1990,* NASA Ames Research Center, 1990; R. A. Pielkie, *Mesoscale Meteorological Modeling,* 1984; P. Zannetti, *Air Pollution Modeling,* 1990.

Environmental geophysics

Geophysics is the earth science discipline that is based upon measuring physical and chemical properties and processes from a distance. These measurements may be performed in the laboratory, in boreholes, between boreholes, from the surface of the Earth, from vehicles (including boats and aircraft), and from satellites. Most of the measurements are performed noninvasively, that is, without the necessity of sampling or drilling. Geophysical measurements often require processing, modeling, and interpretation to derive other desired quantities. Earthquakes, volcanoes, and other natural hazards; exploration for minerals, ground water, and petroleum resources; and generic subsurface characterization have been investigated with geophysical methods for many decades.

Development of the discipline. Environmental geophysics is a much more recent evolution of the application of geophysics, beginning with water-pollution monitoring about 1970. It is principally concerned with the use of geophysical techniques for site characterization, monitoring of remediation or containment of contaminants, and verification. Since the mid-1980s, the growth in environmental geophysics has been explosive, including the formation of a new professional society, the Environmental and Engineering Geophysical Society; convening of annual conferences; the development of expert systems; and publication of books.

Geophysics uses instruments that passively detect changes in the Earth's natural magnetic, electric, thermal, stress, radiation, and gravity fields, as well as instruments that actively stimulate the Earth with electric, electromagnetic, seismic, and other forces in order to to measure the resulting responses. Whether active or passive, these measurements indicate changes in physical and chemical properties and processes within the Earth. In the 1970s, the instruments and techniques that were used in environmental geophysics were mostly adapted from mining geophysics, but in the 1980s instruments and techniques began to be developed specifically for shallow subsurface environmental and geotechnical characterization problems.

Site characterization. Adequate site characterization of geological, geochemical, hydrological, and anthropogenic features is necessary to understand and efficiently to mitigate environmental problems. The need to minimize further waste migration and to maximize personnel safety mandates that initial site investigations be performed with noninvasive airborne or surface geophysical methods. Selective drilling, coring or sampling, and borehole or hole-to-hole geophysical methods should then be used to maximize the return from the invasive holes. No single technique meets all site characterization requirements, necessitating the use of a variety of monitoring methods. Once all available data are acquired at a site, characterization proceeds by finding the model of the site that is consistent with the data, and any inconsistencies are investigated—a process sometimes called integrated interpretation or data fusion. The resulting site characterization model is then used to describe the severity of the environmental problem, estimate the risk and success of various courses of mitigation, and guide the eventual mitigation effort. During mitigation, continued geophysical site characterization is performed periodically. Such ongoing activity detects mitigation actions that could ultimately prove detrimental unless rapidly forestalled (such as pump and treat activity that would unintentionally spread contamination further). Once mitigation is complete, comparison of a postmitigation site characterization with a premitigation site characterization baseline provides a method for assessing the efficacy or failure of the mitigation effort, that is, quality control.

Methods of characterization. Most environmental problems are site specific. The same source contaminant chemical behaves in different ways at different sites, depending upon the local conditions. Movement and eventual equilibrium conditions for the contaminant are strongly influenced by the geology (for example, sandy soil, mineralogical clay, or fractured bedrock), hydrology (such as shallow or deep water tables, perched aquifers, or infiltration rate), geochemistry (including factors such as naturally occurring heavy metals, reactive clay minerals, or natural organic concentration), and preexisting or current anthropogenic features (such as utility trenches, presence of other wastes, or agricultural practices).

Site characterization by drilling or coring is often hazardous because of the presence of punctured drums, breached clay barriers, drilling-waste disposal, and so forth; thus characterization by such methods frequently misses the problem entirely or acquires information out of context because of the limited number of samples provided. Much safer and more broadly representative data can be acquired by first using noninvasive geophysical methods.

Magnetic techniques. Perturbations in the Earth's natural magnetic field near magnetic objects such as iron drums or barrels are measured by magnetic techniques. Large nearby concentrations of iron such as fences, utilities, culverts, vehicles, or buildings may interfere with the technique. High-iron-content soils, such as greensands, basalts, or red hematitic soils, may be sufficiently magnetic to hide objects detectable under other soil conditions.

Electromagnetic techniques. Such techniques generate low-frequency time-varying magnetic fields to induce electrical currents in the Earth (and are the principal methods employed by most metal detectors). The electrical conductivity of the Earth is proportional to the secondary magnetic field generated from the induced currents. By measuring and mapping the changes in electrical conductivity, electromagnetic induction techniques may directly locate plumes of inorganic contaminants, clay lenses, metallic objects such as buried drums or wires, and inhomogeneities in geology such as fractures. Electromagnetic induction techniques may be ineffective in areas with many good conductors, such as fences, pipelines, and utility cables. These methods acquire data very quickly over large areas, whereas electrical resistivity methods are preferred for acquisition of depth information.

Electrical resistivity techniques. These methods use electrodes in contact with the ground to measure electrical resistivity (the reciprocal of conductivity). The depth of investigation is a function of the electrode spacing and geometry; larger spacings permit acquisition of data from greater depths. By measuring and mapping the changes in electrical resistivity, these techniques observe the same things as the electromagnetic induction methods noted above. A variation of the technique known as complex resistivity measures the electrical properties of the ground as a function of frequency, mapping electrochemically reactive areas, with the ability to detect clay-organic reactions and thus map the presence of reactive organic contaminants.

Ground-penetrating radar. This technique is used to measure changes in the propagation of radio-frequency electromagnetic energy in the ground. Such changes typically occur from changes in water content and bulk density. Ground-penetrating radar is a sensitive indicator of soil stratigraphy and bedrock fracturing, and it is an excellent way to map the water table. It may sometimes directly detect organic contaminants either by changes in scattering properties (the texture of the radar record) or by dielectric contrast (such as oil or gasoline floating on water). This type of radar works well in high-resistivity environments such as dry or fresh-water saturated coarse sand or granite. Low-resistivity salt water and mineralogical clays such as montmorillonite severely limit the depth of penetration and effectiveness of ground-penetrating radar. However, it is the highest-resolution geophysical technique, best used for detailed mapping of geological heterogeneity (the single biggest cause of failure in mitigation of contaminated materials).

Seismic techniques. These methods measure changes in the propagation of low-frequency elastic compressional or shear energy (seismic waves) in the ground. They image the subsurface by using reflected or refracted seismic waves; they are sensitive to changes in density and water content; and they are useful in defining subsurface geological and hydrological structure. Although seismic techniques cannot detect contaminants directly, they can locate trenches or other disturbed burial zones in the ground that may contain contaminants. In urban environments, high noise or difficult coupling (such as through concrete or asphalt) may make their use too difficult. Seismic and radar techniques are complementary, as seismic techniques work well in mineralogical clay soils where radar ones do not; radar works well in loosely compacted sandy soil where seismic does not. Each is sensitive to different sources of noise and interference.

Gravity techniques. These methods measure minute variations in the gravitational field of the Earth. Such variations are interpreted in terms of changes in density and porosity in the ground. Microgravity techniques may be useful in locating trenches, voids, and incipient subsidence problems, but they cannot directly detect contaminants.

Radiometric techniques. These techniques measure the radiation emitted from the decay of radioactive isotopes. Radioactive contaminants may be masked by high background levels of natural radioactivity or by roughly a meter of overlying soil cover, depending upon the type and strength of the source. Radiometric techniques are generally useful only at radioactive waste disposal sites; however, they may be useful in locating natural radioactive hazards (such as radon gas sources), early radium processing plants, mining mill tailings, and similar problems.

Combined techniques. Some problems, such as locating buried steel barrels, are easily performed with one technique (magnetometry in the case of steel barrels). However, if the barrels are plastic or are suspected of leaking organic chemical contaminants, the problem is more complicated, possibly requiring ground-penetrating radar, complex resistivity, or geochemical soil gas sniffers. Drilling would not be advisable because of the hazard involved in puncturing otherwise intact barrels and because of the possible remobilization and downward migration of the organic contaminants that are denser than water.

Noninvasive environmental geophysical techniques cannot directly detect contaminants at the parts-per-billion level of regulatory concern. Also, the results of interpretation of many noninvasive measurements are often not unique; for example, several possible models will fit the data. Thus, there is a need to acquire data by invasive probing, drilling, coring, or trenching. Soil and water sampling and analyses are then performed. Care must be taken in locating the intrusive sampling points so as to be representative and yet not provide new migration pathways for site contaminants or water. Invasive sampling points are located by using the data from the noninvasive techniques to best meet these goals. To minimize the number of invasive points required, borehole geophysical logging and hole-to-hole geophysical methods should be used to acquire data between invasion points and to maximize the return for the invasive risk.

For background information SEE BOREHOLE LOGGING; EXPERT SYSTEMS; GEOELECTRICITY; GEOMAGNETISM; GEOPHYSICS in the McGraw-Hill Encyclopedia of Science & Technology.

Gary R. Olhoeft

Bibliography. J. R. Boulding, *Practical Handbook of Soil, Vadose Zone, and Ground-Water Contamination: Assessment, Prevention and Remediation,* 1995; J. E. Lucius et al., *Properties and Hazards of 108 Selected Substances,* USGS Open-File Rep. 92-527, 1992; G. R. Olhoeft, *Geophysics Advisor Expert System Version 2.0,* USGS Open-File Rep. 92-526, 1992; S. H. Ward (ed.), *Geotechnical and Environmental Geophysics,* 3 vols., 1990.

Equine morbillivirus

The equine morbillivirus is a recently discovered virus that caused an outbreak of lethal pneumonia in horses in Queensland, Australia, in 1994. It is also known to have infected two humans (one fatally) who came in close contact with the horses during this outbreak. The disease, previously called acute equine respiratory syndrome but now known as equine morbillivirus pneumonia, appears to be completely new. It has not reemerged in Australia since 1994, although in October 1995 another fatal human infection was detected and traced back to exposure in 1994. Medical and veterinary authorities have been monitoring for infection since the first outbreak. At this point, only 22 horses and 3 humans are considered to have been infected.

Epidemiology. In September 1994, a very severe outbreak of disease occurred among thoroughbred horses at a stable in the Brisbane suburb of Hendra. Shortly before, the horse trainer had become sick and died after intensive hospital care. At the same time, his stablehand developed fever and muscle pains. Both the stable workers and the horses at the stable had contact with a pregnant mare that soon died from a very severe respiratory disease. The mare had come 2 weeks previously from a nearby pasture where there was no history of respiratory disease in other horses.

An immediate standstill order on horse movements in the area was imposed, and the stables were placed under quarantine to prevent any spread. All infected horses were destroyed. Within a few weeks, a survey indicated that the virus had not spread in horses or humans and, since no further cases had developed, the restrictions were lifted. Australia has been considered free of the disease since then.

However, in September 1995 a horse stud owner in Mackay, 497 mi (800 km) north of Brisbane, became infected with equine morbillivirus and died from encephalitis in October. It was reported that he had had close contact with two horses which had died on his property 1 year previously, and he had been involved in conducting an autopsy on them. Archival tissues from these horses tested positive for equine morbillivirus. Active surveillance of the other horses and animals on the farm and in surrounding areas has failed to demonstrate any other infections. Moreover, no link has been established between the Hendra outbreak of 1994 and the cases in Mackay, which would be the most likely explanation for these occurrences. Thus, it is possible that there may have been two separate foci of disease.

Characteristics. The equine virus is assigned to the genus *Morbillivirus,* one of three genera in the family Paramyxoviridae, because of its physical structure and gene sequence. A human strain of the virus, identical in appearance and gene sequence, was isolated from the kidney tissue of the trainer.

This new virus is the first morbillivirus to be detected that infects humans since measles virus was described in the tenth century. It is the first to infect horses and to cause disease in both horses and humans. The closest relatives of the virus are measles virus (humans), distemper virus (dogs), rinderpest virus (cattle), and phocine distemper viruses (seals and dolphins). However, the equine virus is quite distinct. A comparison of the gene sequence of one of the main proteins of the virus, the matrix protein, with that of the other Paramyxoviridae shows it to differ 95% from other genera in the family, and it is 50% different from the other morbilliviruses. (The other morbilliviruses are about 20% different from each other.) In time, the equine virus may prove to be the progenitor virus for the group or even be assigned to a new genus.

Virus particles vary in size (from approximately 40 to 600 nanometers) and in shape. They are covered by an envelope which has surface projections, giving the appearance of a fringe. The viral genetic material is ribonucleic acid (RNA); inside the virus are RNA-protein complexes known as nucleocapsids which have a herringbone structure, characteristic of the morbilliviruses.

Tests are also available to detect antibody to the equine morbillivirus; they can be used alone or in combination. The most specific test available is the serum neutralization test, where live virus is reacted with test serum and then checked to see if it is still infectious for cell cultures. If antibody is present, the virus is neutralized. The virus does not appear to react with antibodies against other members of the group, for example, measles virus.

Clinical signs. After exposure, horses show few, if any, signs of infection for a week or more. However, they subsequently develop an increased temperature and heart and respiratory rate, collapse, and die within 1–2 days. Other symptoms include staggering, coughing, and head pressing. At terminal stages, a frothy nasal discharge can occur. In humans, the clinical signs of infection are not so clear-cut, perhaps because there have been so few cases. The stable hand from Hendra suffered from muscle pains, headaches, lethargy, and vertigo but no respiratory signs. During the preliminary stages of infection, the trainer experienced similar symptoms, but then developed a severe and progressive respiratory disease before dying from cardiac complications. It has been conjectured that both individuals had been directly exposed to infectious material in the nasal discharge from the mare.

A number of animals have been tested experimentally to see if they are susceptible to the virus. A disease can be induced in cats which is very similar to the horse pneumonia; guinea pigs also become seriously ill and die. However, the results from a survey of 500 cats in Queensland show that they are not naturally infected. Although the limited evidence from the outbreak at Hendra suggests that the virus is not highly contagious, in Australia it is agreed that it should be handled in the laboratory only under the highest level of biocontainment conditions because of its highly virulent nature.

Diagnosis. The outstanding postmortem finding is heavy, fluid-filled lungs. In field cases there may be froth tinged with blood in the airways. The virus grows in the cells which line the small blood vessels in the lungs and other organs, and damages them. Infected cells coalesce and form characteristic giant cells in the blood vessel walls. Consequently, thrombosis and hemorrhage occur, and the lungs fill with fluid, occluding the airways.

Horses which experience any undiagnosed severe respiratory infection should be tested for the equine morbillivirus. In humans, the situation is less clear, although individuals who develop undiagnosed severe respiratory disease and, possibly, unexplained meningitis should be tested for the virus, especially if there has been an association with horses.

For background information *SEE ANIMAL VIRUS; PARAMYXOVIRUS; RESPIRATORY SYSTEM DISORDERS* in the McGraw-Hill Encyclopedia of Science & Technology.

Keith Murray

Bibliography. K. Murray et al., A morbillivirus that caused fatal disease in horses and humans, *Science,* 268:94–97, 1995; L. A. Selvey et al., Infection of humans and horses by a newly described morbillivirus, *Med. J. Austral.,* 162:642–645, 1995; H. A. Westbury et al., Equine morbillivirus pneumonia: Susceptibility of laboratory animals to the virus, *Austral. Vet. J.,* 72:278–279, 1995.

Erosion

Two-phase flows of fluids or gases with suspended solid particles can be found in many applications such as pneumatic conveying, food processing, surface blasting, fluidized beds, and industrial, marine, and aviation gas-turbine engines. Whether the presence of the particles in the flow is induced or whether they are unavoidable, their higher inertia causes their trajectories to deviate from the flow streamlines. As a result they impact the surfaces bounding the flow field, and cause them to erode. The erosion produces changes in the impacted surface characteristics and its geometry, which adversely affect the performance level and the life of the equipment. Extensive research efforts have been directed at developing material and coatings that are resistant to erosion. Design efforts to eliminate or reduce erosion in fluid-solid flow interactions require a multidisciplinary approach. Experimental and analytical studies of material erosion behavior and particle dynamics continue to develop an understanding of the factors affecting particle-surface impacts and the associated surface erosion. Experimental studies of surface erosion by impinging solid particles indicate that this erosion is strongly dependent on the particle-impact velocities and angles, the sur-

rounding temperature, material properties of the abrasive particles and the impacted surfaces, and the angularity or roundness of the particles.

Test facilities. Early erosion experiments were conducted in sand-blasting test facilities, in which a small jet of particle-laden flow was aimed at a stationary target. Erosion tests performed in such facilities provided information on the comparative erosion behavior of different materials. However, because of the spreading and mixing of the jet flow, the erosion data obtained at a given test condition actually encompassed a range of particle impingement velocities and impact angles. More recently, erosion tunnels have been designed to obtain test results at uniform particle-impacting conditions that are representative of the environments of systems operating with particulate flows.

Such an erosion test facility (**Fig. 1**) consists of a particle feeder, a main air supply pipe, a combustor, a particle preheater, a particle injector, an acceleration tunnel, a test section, and an exhaust tank. A measured amount of abrasive grit of a given mixture of constituents is placed into the particle feeder. The particles are fed into a secondary air source and blown up to the particle preheater, and then to the injector, where they mix with the main air supply, which is heated by the combustor. The particles are then accelerated by the high-velocity air in a constant-area steam-cooled duct, and impact the specimen in the test section. The particulate flow is then mixed with the coolant and dumped in the exhaust tank. Erosion tests can be conducted in

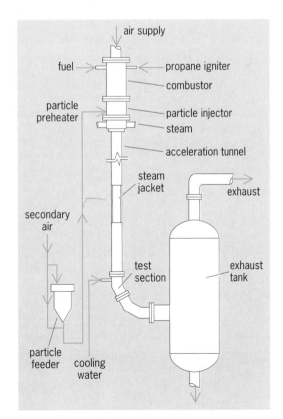

Fig. 1. Erosion test facility.

the tunnel over a range of impact velocities and impingement angles and temperatures.

The particle velocity is controlled by varying the tunnel air flow, while the particle impingement angle is controlled through the target sample rotation relative to the flow. The test temperature is varied by heating the flow stream, which in turn heats the erodent particles and sample to the desired temperature. Because the tunnel geometry is uninterrupted from the acceleration tunnel into the test section, the particle-laden flow is channeled over the specimen and the aerodynamics of the fluid passing over the sample are preserved. Numerous fan compressor and turbine blade materials and coatings have been tested in such a tunnel to determine their resistance to erosion in particulate flow environments under various flow conditions.

Test results. The experiments are conducted by placing a target sample at predetermined angle, and measuring the change in the sample weight before and after erosion. The erosion test results are usually expressed in terms of the erosion mass parameter (milligrams per gram of impacting particles), or they are converted to an erosion volume parameter (cubic centimeters per gram of impacting particles).

Figure 2 shows typical variation of the erosion rate with particle impingement angles for ductile and brittle materials. The brittle-material erosion rate continuously increases with the impingement angle up to its maximum at normal impingement. However, the erosion rate of ductile materials initially increases until it reaches a maximum at an impingement angle between 20 and 35°, depending on the material. It then decreases with further increase in the impingement angle to a residual value at normal impact.

During the early experiments it was proposed that the erosion loss is proportional to the kinetic energy of the erodent particles, and that consequently the erosion rate (ε) is proportional to the square of the impacting particle velocities (V). However, erosion experimental results indicated that the

velocity exponent (n) in the erosion equation, shown below, can vary widely depending on the target

$$\varepsilon \propto V^n$$

material and that it also depends on the impingement angle and temperature. The value of the velocity exponent can be determined from logarithmic plots of the erosion rate versus impact velocity, which produce a straight line whose slope gives the velocity exponent at a given impact condition. Velocity exponent values as high as 4 and as low as 0.68 have been reported for different materials over the range of tested temperatures.

In general, the erosion rate increases with increased temperature. Experimental evidence indicates that the erosion rate at 1200°F (650°C) can be double that at room temperature. It is therefore important that the erosion experiments at high temperatures be conducted in a carefully controlled environment and that the target is heated to a steady temperature in the hot stream before obtaining the results.

Suspended-particle dynamics. When a system operates with suspended particles in its flow field, the deviation of the particle trajectories from the flow streamlines depends on their inertia. In general, the smaller lighter particles tend to follow the streamlines since they can accelerate quickly and approach the local gas velocity. They have few surface impacts at shallow angles. The larger heavier particle trajectories, however, can be very different from the flow streamlines. These particles suffer repeated impacts with the various surfaces, which dominate their trajectories. Because the magnitude of the surface erosion by particle impacts is a function of the impact velocity, impingement angle, and frequency of impacts, it is strongly affected by the particle size distribution.

Particle trajectory studies are conducted to determine the distribution of the particle impact conditions over the various surfaces of systems operating with particulate flows. These data, when combined with the experimental results of target material erosion, can be used to predict the intensity and pattern of the associated surface erosion. In the particle trajectory calculations, the equations of particle motion under the influence of particle-fluid forces of interaction are integrated throughout the flow field until the particles impact a solid surface. The continuation of particle trajectory analysis requires models for particle-surface interactions.

Particle restitution studies. Particle restitution tests are used to determine the particle rebounding conditions after surface impacts. The experiments are conducted in tunnels that are similar to the previously described erosion test facility, with a window in the test section to measure the pre- and post-collision particle velocities very close to the target sample, using laser doppler velocimetry.

Turbomachinery blade erosion. The prediction of particle trajectories and the associated blade erosion in turbomachines is an example of the combined analytical and experimental efforts to study

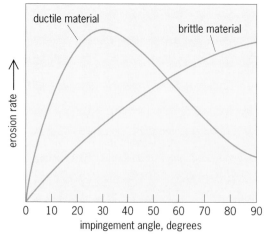

Fig. 2. Variation of erosion rate with impingement angle.

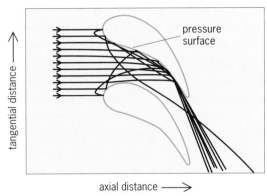

Fig. 3. Sample particle trajectory projections in a gas turbine.

erosion by suspended particles in gas flows. It requires modeling of a complex three-dimensional flow field between the blade passages, and of the cumulative influence of the repeated particle interactions with the stationary and rotating blade rows. The trajectories provide the distribution of the particle impact conditions on the blade surface, which is combined with the blade-material erosion test results to determine the intensity and pattern of blade erosion. The resulting turbomachinery performance deterioration is associated with the increased blade-tip clearances and airfoil surface roughness in gas-turbine engines, and to the blunting of the blade leading edge and reduced chord length in fans and compressors.

A sample of the particle trajectories simulation results through an axial flow turbine stator is shown in **Fig. 3**. The particle trajectories and how they impact the turbine blades, especially the pressure surface, are demonstrated. The particles also migrate in the radial direction as they are centrifuged because of the circumferential velocities they acquire from the blade surface impacts.

For background information *SEE FLOW MEASUREMENT; FLUIDIZED-BED COMBUSTION; GAS TURBINE; PARTICULATES* in the McGraw-Hill Encyclopedia of Science & Technology.

Awatef Hamed

Bibliography. I. Finnie, J. Wolak, and I. Kabil, Erosion of metals by solid particles, *J. Mater.*, 2:682–700, 1967; A. Hamed, and T. P. Kuhn, Effects of variational particle restitution characteristics on turbomachinery erosion, *J. Eng. Gas Turb. Power*, 117:432–440, 1995; A. Hamed, W. Tabakoff, and R. Wenglarz, *Particulate Flow and Blade Erosion*, von Karman Instit. Fluid Mech. lect. ser. 1988-08, June 1988; W. Tabakoff, M. Metwally, and A. Hamed, High-temperature coatings for protection against turbine deterioration, *J. Eng. Gas Turb. Power*, 117:146–151, 1995.

Evolution

Because of human-induced changes to the Earth, there is a growing suspicion that the current rate of species extinction exceeds historical background rates calibrated from fossil data. It has been suggested that the Earth is entering a phase of mass extinction that will rival the major episodes of extinction documented in geologic history. Thus, it is imperative to monitor changes in current levels of global species diversity, and to better understand linkages between biodiversity and physical changes to the Earth. However, a possible impediment to determining these linkages is the probability that the biological effects of many physical phenomena will not be detectable over the short term but will require thousands of years, if not more, to play out. The fossil record provides unique opportunities to overcome this problem by permitting exploration of relationships between global biodiversity and physical changes to the Earth over periods of time much longer than those observable with present-day biotas. Moreover, some human-induced changes to the Earth (for example, possible global warming) have natural analogs during Earth history, thereby providing an opportunity to measure directly their effects on life in the past.

Calibration of global biodiversity. Paleontologists have sought to calibrate global biodiversity, and changes thereof, throughout the history of life. Much of this effort has focused on marine faunas of the Phanerozoic Eon (the most recent 540 million years of geologic time), which constitute the majority of organisms preserved in the fossil record. The most effective calibrations involve assembly and analysis of composite databases that include information gleaned from published descriptions of taxa present in fossiliferous rocks of various ages worldwide. From these databases, global diversity curves are constructed through geologic time. Among the highlights of diversification in the Paleozoic Era (the earliest of the three Phanerozoic eras) were an initial diversity increase during the Cambrian Period, characterized by the first appearances of most fundamental animal designs (the Cambrian explosion); a broad-based proliferation during the Ordovician Period of groups that would eventually dominate marine settings through the balance of the Phanerozoic (the Ordovician radiation); and major episodes of extinction (mass extinctions), recognized in global compilations as dramatic, short-term diversity declines that occurred during the Late Ordovician, Late Devonian, and Late Permian periods (see **illus.**). Subsequent to the Paleozoic Era, global biodiversity increased further in the Mesozoic and Cenozoic Eras, interrupted for brief periods by additional mass extinctions. *SEE FOSSIL; PALEONTOLOGY.*

Selectivity. A major challenge confronting geologists is to determine the mechanisms responsible for episodes of large-scale diversification and extinction. Among post-Paleozoic biological events, the Late Cretaceous mass extinction of dinosaurs and other organisms has received considerable attention, in part because of its possible linkage to the impact(s) of large comets or asteroids. Paleontologists have recognized the importance of dissecting the global Late Cretaceous signal, to determine possible geographic, environmental, taxonomic, or other patterns of selectivity associated with it. Recognition of selectivity, or the lack thereof, helps to narrow the search for causes of extinction, because certain mechanisms would be expected to impart unique selectivity signatures. In the case of the Late Cretaceous event, recent analyses of extinction patterns among marine bivalve mollusks suggest that little selectivity was exhibited; the mechanism(s) responsible for the extinction was of such a catastrophic nature globally that there was no special protection afforded bivalves that exhibited certain kinds of life habits or that lived in particular habitats or geographic regions.

Ordovician radiation and mountain building. Analyses of selectivity should be also be viewed as essential components of attempts to account for major episodes of diversification. In parallel with changes in global biodiversity that have characterized the Phanerozoic, it is now recognized that the Earth has undergone a variety of changes in its physical and geochemical constitution. During the Paleozoic Era, one of the more apparent physical changes was a worldwide increase in the amount of mountain-building activity (orogeny) during the Cambrian and Ordovician periods, in association with an increased number of collisions among tectonic plates.

Quantification of global orogenic activity throughout the Paleozoic Era yields a time-series graph resembling that of Paleozoic global biodiversity; both orogeny and biodiversity increased substantially during the Ordovician Period (see illus.). Although this coincidence hints at a possible relationship between orogeny and diversification, a far more definitive test has involved assessment of geographic patterns of selectivity among diversifying Ordovician groups with respect to probable centers of orogeny. These geographic analyses indicate a significant tendency throughout the Ordovician for diversity to be concentrated in regions associated with orogeny, strengthening the case for a link between orogeny and diversification at that time. Possible explanations for this kind of linkage include increased primary productivity and thus food supply, generated by an influx of nutrients associated with the erosion of newly uplifted regions; wholesale changes to marine bottom substrates, also related to the influx of eroded materials; and a greater likelihood of geographic speciation because of increased fracturing of habitats that are associated with heightened tectonic activity.

Phanerozoic diversification and volcanism. Paleontologists have long sought to establish linkages between volcanic activity and extinction. However, it has recently been suggested that heightened submarine volcanism was associated with the two most significant episodes of global diversification in the history of life, which occurred during the early Paleozoic and the Mesozoic. It is thought that increased volcanism at those times induced elevations of global temperatures because of concomitant increases in atmospheric levels of carbon dioxide, which along with additional volcanic effects may have augmented global nutrient and energy levels. Thus, in turn, the subsistence and diversification of organisms with relatively high metabolic rates may have become possible. To some degree, this hypothesis dovetails with the proposed relationship between diversification and mountain building.

Devonian mass extinction and land plants. As linkages between physical and biological activity continue to be explored, it has become apparent that this relationship is two way: not only do physical changes to the Earth affect life but the evolution of life also affects the physical Earth, suggesting that the dichotomy recognized between physical and biological processes is to some degree artificial. In fact, in some instances a feedback loop may exist between biological and physical transitions. For example, a recent hypothesis links the Late Devonian marine mass extinction and the rise of land plants. During the Devonian, the fossil record indicates a substantial increase in the diversity, maximum size, and environmental and geographic extent of vascular land plants. Because of accompanying increases in root biomass, it has been suggested that there was a dramatic increase in soil formation, which over the short term elevated the import of nutrients to the oceans, thereby stimulating algal blooms in semirestricted marine settings and inducing anoxic conditions on sea floors. Evidence for this increase is provided by the high concentration in the Devonian rock record of

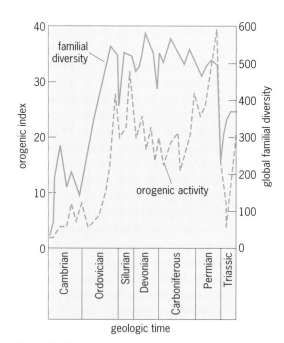

Global biodiversity (at the taxonomic level of family) and orogenic activity during the Paleozoic Era and the Triassic Period of the Mesozoic Era. (*After V. E. Khain and K. B. Seslavinskii, Historical Geotectonics—Paleozoic, Nedra, 1991; J. J. Sepkoski, Jr., Ten years in the library: New data confirm paleontological patterns, Paleobiology, 19:43–51, 1993*)

black shales, which are indicative of anoxic conditions. It has been suggested, in turn, that the marine mass extinctions are associated temporally with these black shale events, thereby implicating land plants indirectly in the marine crisis.

The possibility of a linkage between land-plant diversification and the marine mass extinction could be tested further by evaluating geographic patterns of extinction to determine if the organisms that became extinct were concentrated in regions characterized by black shales. Regardless of the outcome of such research, the set of linkages hypothesized for the Devonian provides an excellent reminder of the probable complexities of linkages between the physical and biological worlds, as well as the terrestrial and marine realms. *SEE ANIMAL EVOLUTION.*

Prospects. Although the science of paleontology is historical in scope, its relevance to life in the future is evident. Paleontological analyses are entering a new era of sophistication in which additional, previously unappreciated linkages between biological and physical process will become established. The result should be a much-improved sense of the long-term prospects for life on Earth.

For background information *SEE ANIMAL EVOLUTION; EXTINCTION (BIOLOGY); FOSSIL; OROGENY; PALEOBOTANY; PALEONTOLOGY; PLANT EVOLUTION* in the Mcgraw-Hill Encyclopedia of Science & Technology.

Arnold I. Miller

Bibliography. T. J. Algeo et al., Late Devonian oceanic anoxic events and biotic crises: "Rooted" in the evolution of vascular land plants?, *GSA Today,* 5(46):64–66, 1995; D. Jablonski and D. M. Raup, Selectivity of end-Cretaceous marine bivalve extinctions, *Science,* 268:389–391, 1995; V. E. Khain and K. B. Seslavinskii, Global rhythms of the Phanerozoic endogenic activity of the earth, *Stratig. Geol. Correl.,* 2:520–541, 1994; A. I. Miller and S. Mao, Association of orogenic activity with the Ordovician radiation of marine life, *Geology,* 23:305–308, 1995; G. J. Vermeij, Economics, volcanoes, and Phanerozoic revolutions, *Paleobiology,* 21:125–152, 1995.

Extinction (biology)

Biological extinction occurs when the last member of a plant or animal species dies, and the species as a living entity ceases to exist. Extinctions have occurred throughout geological history; indeed, most species that have ever existed are now extinct. However, the disappearance of a species from the fossil record does not always correspond with the actual point of its extinction. A famous example of such a discrepancy is the coelacanth, a fish whose most recent fossil occurs in rocks more than 65 million years old. It was presumed to be extinct, but in 1938 a living specimen was found off the coast of Madagascar. The apparent extinction of the coelacanth was due to the lack of fossil preservation, not

the demise of the species. This problem of pinpointing the precise moment of extinction in the often incomplete fossil record has long been a concern for biostratigraphers and other paleontologists. In recent years it has fueled controversies surrounding mass extinction events, particularly the end-Cretaceous mass extinction 65 million years ago (Ma), in which the dinosaurs and other distinctive forms of Mesozoic life perished.

Mass extinctions. It was recognized very early in the history of paleontology that there were episodes in the geological past during which many species seemed to become extinct simultaneously. Charles Lyell and many other nineteenth-century geologists (Charles Darwin among them) attributed these apparent mass extinctions to imperfections in the stratigraphic record. They argued that the stratigraphic ranges of the species in question were merely truncated by a gap in the rock record; if strata representing the missing interval could be found, the true pattern of extinction would be more gradual.

Although this explanation holds true in many instances of abrupt faunal turnover, especially in local stratigraphic sections, decades of ever more detailed stratigraphic studies and improved methods of absolute age dating of strata have convinced most paleontologists that mass extinctions are not geological artifacts. For major events such as end-Permian (245 Ma) and end-Cretaceous (65 Ma) extinctions, there is not enough of the record missing to account for all the disappearances through normal background extinction.

Although most modern paleontologists agree that mass extinctions are real phenomena, there is still much disagreement over the degree to which the vagaries of sedimentation and fossilization obscure the true pattern of extinction through time. The concern of modern paleontologists is the reverse of that addressed by earlier workers: imperfections in the record can blur the timing of extinctions and other events, making truly abrupt events appear gradual. Thus, attempts to link extinctions with specific causes (climate change, volcanic eruptions, or asteroid impacts) become confounded because species can appear to have become extinct before, or survived beyond, the event which supposedly caused their extinction. Both types of artifacts, spurious range truncation and extension, are topics of active research and debate among students of mass extinction.

Signor-Lipps effect. The odds are strongly against any individual organism becoming preserved as a fossil. After death, the tissues of most organisms are consumed by scavengers or broken down by fungi and bacteria. Even decay-resistant structures such as bones or shells are usually destroyed and their mineral constituents recycled in the environment. Those potential fossils that survive long enough to become buried may be obliterated afterward by dissolution or recrystallization within the sediments. Finally, the fossil-bearing sedimentary rocks can themselves be destroyed by erosion or metamorphism.

Because preservation is such an unlikely event, the odds that the remains of the last surviving member of a species will be preserved (and later collected by a paleontologist) are effectively zero. By this logic, for any given species its apparent point of extinction, that is, its last appearance in the fossil record, will tend to be stratigraphically below its true point of extinction (excluding the possibility of redeposition). The discrepancy will be greatest for those species that are uncommon as fossils, either because they were not readily preservable (that is, they lacked durable skeletons or lived in areas of low sedimentation) or because they were numerically rare in life. In the case of mass extinctions, where a wide range of species become extinct simultaneously, the fossil record would show a gradual decline in number of species, with uncommon species disappearing first and abundant species persisting almost up to the true extinction horizon. This phenomenon is termed the Signor-Lipps effect, after the paleontologists who first described it mathematically.

The Signor-Lipps effect has been invoked to explain the apparent extinction of many dinosaur species well before the putative asteroid impact that marked the end of the Cretaceous Period. In the sedimentary rocks of western North America, where both dinosaur bones and the Cretaceous-Tertiary boundary layer are preserved, the stratigraphically highest dinosaur bone yet found lay several tens of centimeters below the boundary layer, and the last representatives of most species are even lower in the strata. Although some paleontologists have used this stratigraphic sequencing to argue that dinosaurs were long gone before the impact occurred and that their extinction was due to unrelated causes, others have argued that the Signor-Lipps effect is a more likely explanation.

Two approaches have been used to discern the true pattern at extinction boundaries despite the Signor-Lipps effect. The first, and conceptually the simplest, approach has been to collect more fossils from the extinction interval. Ammonites, cephalopods with coiled shells that died out at the end of the Cretaceous, were initially thought to represent a counterexample to the instantaneous, impact-caused extinction model because they showed a gradual decline in numbers of species prior to their final extinction. However, intensive collecting and restudy of Upper Cretaceous rocks, particularly in Spain and France, have shown that many forms thought to have become extinct earlier actually survived right up to the impact layer.

Another, more sophisticated approach has been to mathematically estimate confidence intervals, or error bars, for the point of extinction of fossil species. The several methods for estimating confidence levels differ in their computational details and mathematical assumptions, but all are based on the frequency of occurrence of a fossil species within its known stratigraphic range and can be set at any degree of statistical certainty. A species that occurred only sporadically (for example, one speci-

men on average every 33 ft or 10 m of stratigraphic section) would have broader error bars (more uncertainty) regarding the timing of its extinction than one that occurred more frequently (about every 6 ft or 2 m). If the error bars for a group of species showing a gradual extinction pattern extend to an event horizon (such as the Cretaceous-Tertiary boundary layer), the gradual pattern can be reasonably attributed to the Signor-Lipps effect. If the error bars do not reach the horizon, the gradual pattern is likely to be real.

Reworking and bioturbation of fossils. Fossils already preserved in layers of sediment are occasionally exhumed by erosion and redeposited in sediments of lesser age. This process is called reworking. In the deep sea and other areas of slow sedimentation, worms and other burrowing organisms can churn sediments, causing fossils from different layers to be mingled together. This biologically produced disruption of sedimentary layering is called bioturbation. Both reworking and bioturbation can extend the apparent stratigraphic ranges of species beyond their true points of extinction, and thus present problems for paleontologists.

Microfossils, because of their small size, are particularly vulnerable to displacement by reworking and bioturbation. The possible reworking of Cretaceous foraminifera (marine microorganisms with calcium carbonate shells) into post-Cretaceous sediments has contributed to a debate among micropaleontologists over the abruptness of foraminiferal extinctions at the end of the Cretaceous. The gradual pattern of extinction observed at some localities may be an artifact of reworking and bioturbation, and many of the apparent survivors of the extinction event may have been reworked from lower strata.

Reworked fossils can often be distinguished from younger fossils by the quality of their preservation; they are more likely to be abraded or discolored, for example, but this is not a completely reliable indication. Unabraded dinosaur teeth found in Paleocene sediments have been cited as evidence that some dinosaurs survived the end-Cretaceous extinction event, but subsequent studies have shown that teeth can experience significant reworking and transport without showing signs of abrasion. Thus, the dinosaur teeth may have been reworked, despite their pristine condition.

A promising technique for distinguishing reworked from undisturbed fossils involves analyzing the isotopic composition of fossil shell material. During growth, the different isotopes of elements such as carbon (^{12}C and ^{13}C) and strontium (^{86}Sr and ^{87}Sr) are incorporated into the shells of most organisms in approximately the same ratios that they exist in the environment. Because isotopic ratios in the environment have fluctuated through geologic time, shells formed at different times will bear different isotopic ratios. Barring later contamination or recrystallization, which can alter the ratios, the distinctive isotopic ratios of different-aged fossils can allow reworked fossils to be recognized, even if they are otherwise indistinguishable from unreworked fossils.

The problems of reworking and bioturbation can also be approached mathematically by using diffusion equations and other models to estimate the probability that a fossil or other particle has been displaced a given stratigraphic distance. These models have been applied primarily to the microfossil record in deep-sea sediments. The potential for upward transport is highly dependent on the thickness of the sediment layer reworked by water currents or bioturbated by organisms. The chance of finding a particle declines exponentially above the horizon of the its origin.

The simplest of the models, the Berger and Heath model, predicts that extinct microfossil species will occur in sediment cores at measurable abundances (0.1% of their abundance while extant) up to 24 in. (60 cm) above their true extinction horizon, assuming a bioturbated layer thickness of 4 in. (10 cm). In some cases, bioturbation models can be mathematically inverted to unmix the fossil record in a sediment core, thus restoring the original stratigraphic positions of species' extinctions and abundance fluctuations. Such analyses are, of course, highly subject to the model's underlying assumptions.

For background information SEE CRETACEOUS; DINOSAUR; EXTINCTION (BIOLOGY); FORAMINIFERIDA; FOSSIL; PERMIAN in the McGraw-Hill Encyclopedia of Science & Technology.

Alan H. Cutler

Bibliography. D. E. Briggs and P. R. Crowthes (eds.), *Paleobiology: A Synthesis,* 1992; A. H. Cutler, Notes from the underground, *The Sciences,* 35(1):36–40, 1995; S. M. Kidwell and A. K. Behrensmeyer, Paleontological Society, *Short Courses in Paleontology,* vol. 6: *Taphonomic Approaches to Time Resolution in Fossil Assemblages,* 1993.

Feline immunodeficiency virus

First identified in 1987, feline immunodeficiency virus (FIV) is a pathogen of domestic cats that belongs to the lentivirus group of retroviruses. Its structure and biology closely resemble those of the human immunodeficiency virus (HIV), the causative agent of human acquired immune deficiency syndrome (AIDS). Similar but distinct viruses demonstrated in wild felids (for example, lions and panthers) do not appear to contribute to the circulation of feline immunodeficiency virus in domestic cats. Like HIV-infected humans, cats infected with feline immunodeficiency virus develop a slow but progressive impairment of the immune system which leads to the development of a variety of superinfections and tumors. Neuropathological manifestations are also a prominent feature of infection. The clinical result is called feline acquired immune deficiency syndrome (FAIDS) because of its many similarities to human AIDS.

Feline immunodeficiency virus should not be confused with feline leukemia virus, another widespread retrovirus of cats that primarily induces tumor growth and may also cause severe immunosuppression.

Epizoology. The main means of transmission is believed to be through bites; feline immunodeficiency virus is recovered readily from the saliva of infected cats. Infection is usually acquired after 1 year of age and is two or more times more frequent in male cats, which have a greater propensity for biting than females. Prevalence is high in places where cats are numerous and have an aggressive behavior because they are left free to roam outside. Transmission to offspring also occurs but is frequent only when mothers become infected during pregnancy or immediately after parturition (when mothers have long-term infections, such transmission rarely occurs, probably as a result of the higher levels of antiviral immunity). The offspring become infected within the uterus, resulting in abortions or in birth of subnormal-weight and immunodeficient kittens; or after birth, by ingestion of colostrum and milk (materials that harbor the virus) and possibly by maternal grooming. Alternative modes of transmission appear to be less effective; sexual transmission has never been clearly documented. Experimentally, feline immunodeficiency virus is readily transmitted by any parenteral route that brings the virus directly into the cat tissues whereas oral, vaginal, and rectal routes require large doses of virus and skin scarification has proved unsuccessful. Thus, considering the fragility of feline immunodeficiency virus, the risk of transmission by unprotected surfaces is negligible. In addition, cats maintained in stable households for prolonged periods rarely bite each other; therefore, noninfected cats living in close contact with infected ones usually remain infection free. Infection rates range from 1% or less in healthy cats in the United States and central European countries to 30% or more in sick cats in Japan and Australia.

On the basis of divergences in the nucleotide sequence of the viral genome, feline immunodeficiency virus has recently been grouped into at least four subtypes, and it is likely that additional subtypes will be recognized. The limited epidemiological data available indicate that the relative prevalences of circulating subtypes may vary considerably worldwide. Although it is apparent that future vaccines for feline immunodeficiency virus will have to be multivalent because of the high degree of divergence in the genes, it has yet to be determined whether the subtypes differ in routes of transmission or result in different clinical outcomes. In the laboratory, cats infected with one subtype can be readily infected by another subtype and genetic recombination can occur between the two viral strains, but the significance of such phenomena outside the experimental setting remains to be determined.

Infection. Information about life cycle of the virus in the infected cat is limited and derives mostly from cats infected experimentally with virus doses considerably higher than those involved in natural transmission. In these animals, the virus begins to appear in lymphoid tissues shortly after inoculation, and dissemination occurs very rapidly throughout the body. In one study of the primary stages of infection, at 1 week after infection, the highest concentrations of virus were detected in the bone marrow, thymus, and lymph nodes, while lower levels were found in the kidney, lung, liver, and other tissues. Initially, T lymphocytes were preferentially infected, together with an unidentified population of mononuclear cells and a few macrophages, but at later stages of the acute disease phase the proportion of infected macrophages increased dramatically. In other investigations, an early peak of viral particles in the blood was detected immediately before the onset of acute-phase symptoms. Plasma virus remained at high levels throughout the primary stage of infection and then declined but remained easily demonstrable. The central nervous system also represents an early target. Infectious feline immunodeficiency virus can be observed in the cerebrospinal fluid and brain tissue during the first weeks after infection.

Immune response. Following infection, immune responses to feline immunodeficiency virus are prompt, vigorous, and sustained. Antibodies to the major proteins and glycoproteins usually become detectable in serum within 2–6 weeks and remain high throughout the animal's life-span. Some of these antibodies have the capacity to inhibit (neutralize) infectivity of feline immunodeficiency virus for cells in the test tube. These antibodies have received considerable attention because neutralizing antibodies represent a hallmark of protection in most antiviral vaccines. They can be classified as broadly reactive or strain specific depending on whether they are effective at blocking any strain of feline immunodeficiency virus or only the specific strain (or closely related strains) that has induced their production. Broadly neutralizing antibodies detected using feline kidney cell cultures reach high concentration in infected cats but apparently do not exert significant protective effects. In contrast, strain-restricted neutralizing antibodies detectable in lymphoid cells appear to possess at least some protective activity in infected cats: neonatal kittens that were nursed by immunized female cats and received high titers of transplacental and colostral antiviral antibodies were effectively protected from experimentally administered feline immunodeficiency virus.

Although broadly neutralizing antibodies appear to be mostly directed against the third variable region of the surface glycoprotein of the mature virus particle, the epitopes (corresponding to parts of the antigen) recognized by strain-specific neutralizing antibodies are still unknown. It has recently become evident, however, that strain-specific neutralizing antibodies effectively block the infectivity of laboratory-adapted strains of feline immunodeficiency virus but have little or no effect on strains obtained from field cats. A similar phenomenon observed in HIV type 1 has raised considerable discussion and concern for the possible implications in the evaluation of anti-HIV vaccines under consideration.

Cell-mediated immune responses are still poorly characterized. Although helper and cytotoxic T lymphocytes specific for feline immunodeficiency virus have been described in infected cats, the levels of reactivity were generally low and the viral antigens recognized have yet to be characterized. The relative contributions of cell-mediated and humoral immune responses in determining baseline resistance to feline immunodeficiency virus also remain unknown.

The immune response developed by cats probably contributes to keeping the infection under control, as suggested by the prolonged clinical latency, but fails to entirely eliminate the virus. Once established, feline immunodeficiency virus usually persists for the lifetime of the cat. Persisting infectious virus is recoverable from 100% of seropositive cats and viral loads are generally high, with a tendency to increase during late stages of infection when immunodepression and other clinical signs become evident. Mechanisms of persistence have not been intensively investigated. Possible viral factors are the ability of feline immunodeficiency virus to invade and hide within macrophages and other immunocompetent cells, low sensitivity to antibody-mediated neutralization, and the tendency of viral antigens to mutate.

Induced disease. The deterioration in the immune system is clearly evident at the level of blood lymphocyte counts. As in HIV infection, infection by feline immunodeficiency virus is characterized by a fall of $CD4^+$ T lymphocytes that is already visible in the early postinfection weeks and becomes progressively more profound. Other immunological alterations described in infected cats include increased blood levels of gamma globulin, decreased lymphocyte responses to interleukin 2 and mitogens (a compound that stimulates cells to undergo mitosis), diminished antibody responsiveness, and reduced production of cytokines. Neurological function abnormalities also develop early in the course of infection, and include alterations of the sleep pattern and delayed reflexes.

Clinical manifestations are similar to those caused by HIV, and as a consequence clinical staging closely reflects the classification systems used for HIV-induced pathology (see **table**). Primary acute-phase infection is often accompanied by a mild illness akin to the initial mononucleosislike illness of HIV infection, but generally it goes unnoticed by cat owners. A generally long clinically silent period follows, and then a slow deterioration of the health status occurs, with an increased susceptibility to intercurrent infections usually being the most obvious manifestation, or at least that first noticed by owners. The opportunistic illnesses described in seropositive cats for feline immunodeficiency virus are numerous and extremely variable in individual animals; they may be episodic, although with increasing time postinfection they tend to become permanent. The onset of full-blown feline acquired immune deficiency syndrome requires at least 3–5 years, but infection is ultimately fatal.

Coinfection with the feline leukemia virus has resulted in more severe forms of disease and shortened survival time compared to infection with feline immunodeficiency virus alone, but the mechanisms of disease potentiation have remained elusive. In contrast, attempts to establish whether other incidental infections and immune stimuli function as cofactors for infection with feline immunodeficiency virus and enhance disease progression have given conflicting results.

As clinical and histopathological changes are nonspecific, diagnosis relies on laboratory methods. However, in veterinary practice identification of infection is easy; demonstration of antibodies to feline immunodeficiency virus in serum or saliva is unequivocal evidence for ongoing infection.

Prevention and control. Although effective vaccines exist against the feline leukemia virus, there are no effective vaccines against feline immunodeficiency virus. The only means of protecting cats from infection is to keep them indoors to avoid fights with infected cats.

The development of vaccines is encountering the same obstacles that are hampering the preparation of effective anti-HIV vaccines, including existence of several virus subtypes, great genetic and antigenic plasticity of the virus, and lack of clear correlation between protection and induction of neutralizing antibody. Many vaccination experiments have failed to induce significant protection or have paradoxically accelerated infection, upon challenge. Vaccines composed of proteins of feline immunodeficiency virus and synthetic peptides have generally induced the development of broadly reactive neutralizing antibodies but have uniformly failed to protect against subsequent infection. Similar results have been obtained with experimental vaccines for the simian immunodeficiency virus (SIV), indicating that great caution should be used in deciding about candidate anti-HIV vaccine testing in humans. More encouraging results have been obtained by immunizing cats with inactivated whole virus or infected cell vaccines, although the protection afforded did not extend to strains of feline immunodeficiency virus antigenically distinct from those used in vaccine preparation, and vaccinated cats resisted challenge by parenteral routes but remained susceptible to mucosal challenge. Results indicate that vaccinal prevention against feline immunodeficiency virus is indeed achievable but difficult to generate. The development of an effective vaccine would have far-reaching implications, since the same strategies should be applicable to the production of vaccines against human AIDS.

Several anti-AIDS drugs have been seen to inhibit growth of feline immunodeficiency virus and also of cellular damage induced by it in the test tube at concentrations similar to those required for HIV. Unfortunately, drug-resistant mutants emerge as rapidly as with HIV. Some substances have also been shown to exert an anti–feline immunodeficiency virus effect when administered to infected cats, although effectiveness was generally lower than might have been expected from laboratory studies. Their usefulness in preventing or delaying feline acquired immune deficiency syndrome in infected cats is still uncertain. Clinical improvements in infected cats after treatment with reverse transcriptase inhibitors have been noted, but other cases did not respond to therapy. Prophylactic use of antiviral compounds, that is, treatment initiated just before or immediately after exposure to the virus, has also proved only marginally beneficial.

Clinical stages of infection with feline immunodeficiency virus

Stage	Manifestations	Duration	Response to symptomatic treatment
Primary infection	Most frequently unnoticed; usually mild constitutional symptoms (fever, dullness, loss of appetite, diarrhea, conjunctivitis, enlarged lymph nodes) are transiently present	Weeks to months	Yes
Asymptomatic carrier	Complete clinical normalcy apart from moderate depletion of blood CD4+ T cells	Years	—
Persistent generalized lymphadenopathy	Long-lasting generalized enlargement of lymph nodes, vague constitutional signs (recurrent fever, loss of appetite), vague behavioral changes; circulating CD4+ T cells continue to slowly decline	Months to years	—
AIDS-related complex	Enlarged lymph nodes; fever; mild chronic secondary infections of mouth, skin, upper respiratory, digestive, and urinary tracts; neurologic abnormalities; slight weight loss; extremely low levels of blood CD4+ T cells	Months to years	Yes
Feline acquired immune deficiency syndrome	Severe often multiple chronic secondary infections, tumors, neurologic disorders, wasting; terminally, excessively low levels of blood CD4+ T cells	Months	No

A model for AIDS studies. Feline immunodeficiency virus is actively investigated because of its importance in veterinary medicine and as one of the most versatile animal models for AIDS. Advantages include safety to laboratory personnel (feline immunodeficiency virus does not infect humans) and existence of large reservoirs of naturally infected hosts. Efforts are focused on immuno- and neuropathogenetic mechanisms, development of protective vaccines, and testing of therapies. For example, the similar susceptibilities of feline immunodeficiency virus and HIV to antivirals provide a potentially useful system for preclinical evaluation of therapies in humans.

For background information SEE ACQUIRED IMMUNE DEFICIENCY SYNDROME (AIDS); ANIMAL VIRUS; ANTIBODY; FELINE LEUKEMIA; IMMUNITY; VIRUS INFECTION, LATENT, PERSISTENT, SLOW in the McGraw-Hill Encyclopedia of Science & Technology.

Mauro Bendinelli

Bibliography. M. Bendinelli et al., Feline immunodeficiency virus: An interesting model for AIDS studies and an important cat pathogen, *Clin. Microbiol. Rev.,* 8:87–120, 1995; J. H. Elder and T. R. Phillips, Feline immunodeficiency virus as a model for development of molecular approaches to intervention strategies against lentivirus infections, *Adv. Virus Res.,* 45:228–248, 1995; M. J. Hosie et al., Protection against homologous but not heterologous challenge induced by inactivated feline immunodeficiency virus vaccines, *J. Virol.,* 69:1253–1255, 1995; J. A. Levy (ed.), *The Retroviruses,* vol. 2, 1995.

Fluid flow

Advances in the science and technology of fluid flow involve the development of heat sinks that make use of fluid flow in microchannels, and of devices that utilize the electrorheological effect and flows.

Heat Sinks with Microchannels

Because of the demand for miniature high-heat-flux devices for the cooling of electronics, many investigations of single-phase and two-phase heat sinks with micro- or miniature channels have been recently reported. The combination of the microscale heat-transfer phenomena and high heat flux results in the various nonconventional transfer mechanisms encountered in micro- and miniature heat-sink experiments.

Heat-sink types and evaluation. Small-size heat sinks can be divided into several categories depending on their dimensions, number of phases of working fluid, and design complexity. The terms micro and miniature refer mainly to the characteristic cross-sectional dimensions of a device. Channels and grooves with sizes that can be more conveniently expressed in micrometers than in millimeters are usually called microchannels. The term miniature is more appropriate for devices with an inner diameter or opening of more than 1 mm; however, miniature heat sinks can have microgrooves on the inner surface.

Micro and miniature heat sinks can be operated as a single- or two-phase flow with the fluid transport usually initiated by a mechanical pump. However, in micro heat pipes, which are closed two-phase devices, the fluid circulation is maintained by capillary forces. Two-phase heat sinks can work with boiling of the liquid or thin-film evaporation at the enhanced inner surfaces. The most important characteristics of heat sinks are heated length and external thickness; uniformity of temperature over the surface; effective heat transfer coefficients (or thermal resistance); maximum attainable heat fluxes on the surface; pressure drop and mass flow rate; stability of operation; and operating temperature versus absolute pressure range.

Microchannel single-phase heat sinks. A schematic of a microchannel heat sink is shown in **Fig. 1**. The hydraulic diameter, D, of microchannels usually exceeds 10 μm; however, it can be as small as 1 μm. For a smaller hydraulic diameter the ratio of the perimeter to the cross-sectional area of the channel increases and the role of the solid-liquid interface becomes more important. Therefore, phenomena at this interface can be more important for a microchannel than for a conventional-sized channel.

For microchannels of extremely small diameters, continuum fluid mechanics cannot provide appropriate predictions of heat transfer and fluid flow because it assumes that the molecular spacing and mean free path of the fluid particles, λ, are at least two orders of magnitude smaller than the system's size, which is not always true for microchannels. Several flow regimes are possible in microchannels, which can be distinguished based on the value of the Knudsen number, $Kn = \lambda/D$. The continuum regime ($0 < Kn < 0.01$) is usually described by the no-slip Navier-Stokes equations (wherein the velocity of the liquid at the wall is set equal to zero). For $0.01 < Kn < 0.1$, the Navier-Stokes equations with slip-flow boundary conditions should be used, and the flow is referred to as the slip-flow regime. For $0.1 < Kn < 3$, the Navier-Stokes equations are not applicable because of flow rarefaction (this is the transition regime), and the full Boltzmann transport equation should be used. The Boltzmann transport equation is based on the kinetic theory and describes the flow of particles that can collide with one another and also with solid walls. Finally, for $Kn > 3$, the free molecular

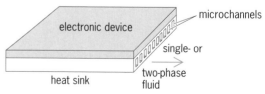

Fig. 1. Microchannel heat sink.

flow regime occurs, with only rare collisions, and the so-called collisionless Boltzmann equation (without the terms describing collisions of the particles) is applicable since the flow is sufficiently rarefied.

Some important features of fluid flow and heat transfer in microchannels observed by various investigators will be discussed.

Influence of surface roughness. The experimental results concerning the role of the roughness in microtubes differ from those common for macrotubes. Increased roughness appears to cause larger pressure drops for both laminar and turbulent micro-scale flow, and also decreases turbulent heat-transfer effectiveness. Therefore, unlike conventional-sized heat sinks, ultrasmooth inner surfaces might be advantageous for microchanneled heat sinks.

Friction coefficients. Experimental investigations with microchannels found that the friction factor–Reynolds number characterizing the flow friction was 53 rather than 64 for laminar flow in conventional-sized channels and the friction factor was up to 30% lower than those usually observed for turbulent flow.

Slip-flow boundary conditions. The no-slip solution of the Navier-Stokes equations fails to adequately model the momentum transfer from the fluid to the channel wall for the Knudsen number larger than 0.01. In that case, velocity slip and temperature jump (the temperature of liquid at the wall differs from that of the wall) boundary conditions must be applied at the wall. One of the promising alternative methods for describing the high-Knudsen-number flow regime (Kn > 1) is the direct simulation Monte Carlo (DSMC) technique. Direct simulation Monte Carlo is a statistical numerical technique which traces trajectories of many particles in chosen cells. It accounts for reflection of particles from a solid surface and collisions between particles.

Transition to turbulence. The Reynolds numbers, which characterize the flow regime (Re = $\bar{u}D/\nu$, where $\bar{u}$ is the mean velocity and ν is the kinematic viscosity), can be two times as high at the heat sink outlet as at the inlet because of the variation of the liquid temperature along the heated microchannel. Therefore, zones with laminar, transition, and fully turbulent flows can exist along the heated microchannel. The transition phenomena in microchannels cannot be explained in terms of Reynolds or Reynolds and Grashof numbers, as is possible for macrochannels.

Variation of properties. Nonlinear pressure distributions in uniform microchannels have been measured for many working fluids such as nitrogen and helium. Because the temperature drops across microchanneled heat sinks can be very significant (dozens of degrees centigrade) and the pressure drops can be as much as 10^5 pascals (1 atm) or higher, the effects of compressibility and variation of thermophysical properties, influencing the pressure distribution, should be taken into consideration in both numerical modeling and experimental data analysis.

Microchannel single-phase heat sinks operate with high pressure drops even though their lengths are usually less than 10 mm. Isothermal heat sinks with low thermal and hydraulic resistances and those which can be long enough and capable of withstanding high heat fluxes can be of more practical use in various technical applications. Therefore, two-phase (liquid-vapor) micro and miniature heat sinks have also been extensively studied.

Two-phase heat sinks. Although two-phase heat sinks provide a more uniform temperature distribution along the length of the channels, the mechanisms of heat transfer and those leading to dryout phenomena in two-phase micro and miniature channels have not been fully investigated because of their complexity. The multiple physical phenomena observed are due to the two-phase nature of the fluid flow.

Regimes of boiling in small-size channels. The nature of bubble formation in microchannels is not completely understood. For example, it has been reported that when wall superheat is in the fully developed boiling region, small bubble strings immediately emerge in the outlet chamber as the heated liquid flows out of the microchannels. However, no bubbles have been directly observed in microchannels even for cases of fully developed nucleate boiling heat fluxes. Evaporating liquid microlayers on the inner walls of microchannels with two-phase flow are believed to play an important role in enhancing the overall heat transfer rate.

Comparative performance of heat sinks. A comparative investigation of two-phase minichannel and

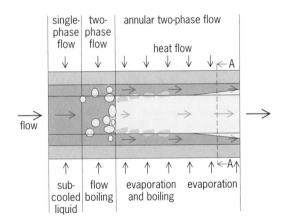

(a)

(b)

Fig. 2. Two-phase miniature heat sink with axial microgrooves. (*a*) Longitudinal cross section. (*b*) Transverse cross section at A-A.

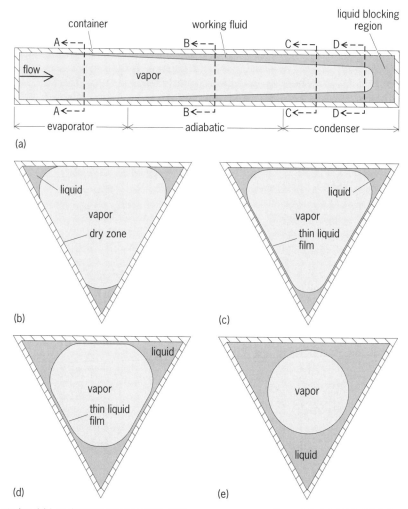

Fig. 3. Micro heat pipe. (*a*) Longitudinal cross section. (*b*) Transverse cross sections at A-A, (*c*) B-B, (*d*) C-C, and (*e*) D-D.

microchannel heat sinks revealed that minichannels operate with significantly smaller pressure drops than microchannels with boiling and the overall temperature drops are about the same for high heat fluxes.

Evaporative heat sinks with microgrooves. Enhancement of the inner surfaces of the two-phase miniature heat sink can be extremely useful for high-heat-flux conditions. One of the possible enhanced designs of a heat sink has small axial grooves on the inner surface of a rectangular channel (**Fig. 2**). These microgrooves provide significant heat-transfer enhancement during evaporation and boiling of the liquid and accelerate rewetting of the inner surface after occasional local dryout. Moreover, since the liquid can flow along the axial grooves under the influence of the capillary pressure and the pressure gradient along the flow channel, the heated length can be larger for this configuration than in the case of miniature channels with smooth walls.

With moderate heat fluxes, the impact of evaporation on the heat transfer prevails over the impact of boiling over the heated length because of intensive evaporation from ultrathin liquid films attached to the liquid-vapor menisci. Other designs

of two-phase heat sinks with enhanced surfaces are also possible.

Although micro and miniature heat sinks and the related fundamental physical phenomena need to be studied further, the existing experimental results are very impressive: the heat fluxes achieved range from 100 to 10,000 W/cm^2.

Micro heat pipes. Heat pipes are autonomous, closed two-phase devices with vaporization and condensation of the working fluid which effectively transport heat with very small temperature gradients. Heat is transported because of the capillary-driven vapor-liquid circulation initiated by the temperature drop along the heat pipe. Therefore, heat pipes do not need a mechanical pump for fluid circulation. Micro heat pipes have been proposed to maintain a uniform temperature of computer chips by distributing heat over their surface. They can be manufactured directly in the silicon wafer of the electronic chip. Typically, micro heat pipes have convex cross sections with sharp inner corners (for example, a polygon), as shown in **Fig. 3**. The axial transport of liquid in a micro heat pipe takes place in the corners, where most of the liquid resides, because of the capillary pressure gradient arising when one part of the micro heat pipe is heated, which triggers the fluid circulation.

Because the micro heat pipe hydraulic diameter can be as small as 10 μm, the vapor continuum limitation can be encountered in the micro heat pipe while it is also subject to the operating limits of conventional heat pipes such as capillary limit. The heat-transport capability of a heat pipe operating under the rarefied or free molecular vapor condition is very limited, and a large temperature gradient exists along its length. As a result, for the Knudsen number greater than 0.01, the micro heat pipe will lose its advantages as an effective cooling method. The maximum heat flux based on the conventional formulation is of the order of 10 W/cm^2.

Amir Faghri

Electrorheological Effect and Flows

The electrorheological effect is recognized as the phenomenon of a rapid reversible change in mechanical (rheological) properties of dielectric suspensions of fine particles in a nonconducting oil in the presence of strong electric fields. It was discovered in 1947 by W. Winslow, who observed a change in the effective viscosity (fluidity) of dispersions. An electrorheological fluid consists of a carrier medium (any nonconducting oil) with excellent insulation capability and a filler (particles) with a different dielectric constant dispersed in this medium. Chlorinated paraffin, silicone oil, and mineral oils are the most common carrier fluids. The filler consists of suspended solid particles of 0.1–100 μm in diameter. The dispersed phase may be either organic material such as microfine powders of soybean casein or starch cellulose or inorganic material such as micropowder mica, silica gel (barium titanate), various polymers such as phenolic resin, or metallic powders. These micropowders are used either untreated or after surface treatment to improve their dispersability. Although a simple electrorheological fluid such as cornstarch in cooking oil can be easily mixed, it is quite difficult to manufacture an electrorheological fluid which meets commercial expectations.

Fluid behavior. Electrorheological fluids fall into the category of smart materials, which can adjust their physical properties to changing conditions in a broad sense. Although these fluids have been manufactured since 1947, they have gradually become commercially viable only since the mid-1980s as a result of the advances made in creating new fluids and, above all, the advent of microcomputers. Electrorheological fluids remain in the liquid state in the absence of an electric current. When subjected to a current, they almost instantaneously turn into a gel-like solid, and revert to the liquid state upon the removal of the current in time spans smaller than a millisecond. Morever, the consistency of the gelling can be adjusted by varying the field strength, that is, the voltage. An intermediate state between a complete solid and a complete liquid can be instantaneously and smoothly selected by applying the voltage required, as electrorheological fluids are extremely sensitive to the intensity of the current. Real-time application of the precise voltage required and its variation by minute amounts in devices employing these fluids recently became a reality because of the availability of microcomputers. In most applications, the combination of electrorheological fluid and microcomputer provides the enormous advantage of eliminating most moving parts in machinery, leading to simpler machine designs which respond faster and more precisely to electric signals than existing technologies. It also opens the way for revolutionary new technologies in a number of industries, for instance, in designing new generations of remote-control valves, brakes, clutches, shock absorbers, self-adjusting steering systems, automatic transmissions, engine and machine-tool mounts, other hydraulic devices, and robotic manipulators. Electrorheological fluids could be used in helicopter rotor blades made of a material which would automatically compensate for temperature variations by self-adjusting its stiffness as the stresses on it vary, and in suspensions for electron microscopes to eliminate blurring vibrations.

Structural changes. The observed constitutive behavior can be explained by analyzing the internal structural changes in suspensions. The effect of an electric field on an electrorheological fluid in a condenser gap can be directly observed by high-speed photography and speckle interferometry. The particles of the solid phase, initially uniformly distributed in the dielectric medium, start to move sharply toward the electrodes, oscillate between them, and form separate aggregates upon establishing a current. They proceed to coalesce rapidly and form a multitude of bridges connecting unlike charged electrodes, resulting in the formation of a solid anisotropic structure.

Structural changes in the presence of an electric field, accompanied by the formation of individual aggregates, are typical of a number of other dispersions based on nonpolar fluids, and are well known. However, unlike electrorheological fluids these materials form structures with separate particles weakly connected to each other, which easily disintegrate even when a small load is applied. But the internal structural transformation in an electrorheological fluid because of the electrorheological effect cannot be easily destroyed, and is capable of carrying considerable shearing loads depending on the temperature, the composition of the fluids, and so forth.

Dielectric polarization. Structural changes which occur in an electrorheological fluid in the presence of an electric field are based on the polarization of components, conduction, and coagulation. A thorough study of dielectric polarization activated in electrorheological fluids by the presence of high-voltage potential fields shows that specific features of dielectric polarization of disperse systems are determined by processes in electric double layers, hydrate layers, and absorption layers. Free and bound charges accumulate at an interface in the presence of an electric field, producing macrostructural (or migration) polarization. In polarizable media, the occurrence of chain aggregates is usually associated with polarization of an ionic atmosphere, while intrinsic polarization of materials in these cases has no appreciable influence. In nonpolar media, particularly in hydrocarbons, where charge concentration is very low, the structure of the disperse-phase particles (Maxwell-Wagner polarization) and hydration of particles with associated film polarization is quite important.

In the case of electrical contact between particles, exchange of charges results in a decrease of the force of polarization, attraction, and even a change of sign. Charge decrease due to exchange between particles occurs in a finite time, which depends on the resistance and capacitance of the electric contact. In low-frequency electric fields, charges are neutralized with time and a conductance path is formed which weakens the electrorheological effect. One way or another, a conductance increase of the particles increases the attractive polarization forces until free charges move within the limits of a particle and charge transfer between them is, in fact, absent. These processes in intermediate layers are often more dependent on electric properties at an interphase than on bulk characteristics of the particles and medium (their permittivity). Confirmation was obtained by experiments conducted with ceramics (barium titanate, $BaTiO_3$) and Seignette's salt, with a dielectric constant equal to 10^3, which did not show an appreciable enhancement of the electrorheological effect. These results have made it possible to treat the electrorheological effect as a phenomenon fully dependent on the electric field at the interphase, lending impetus in recent years to the synthesis and study of highly efficient electrorheological fluids by modifying the interphase and the surface properties of the particles.

Dispersed fillers. A dispersed filler in many cases contains water, which causes phenomena resulting in the deterioration of the fluid's performance. Water causes the dielectric breakdown voltage to drop sharply and the electrode's metal to ionize. It promotes ionization with temperature increase, resulting in higher ionic currents. Water can easily be diffused or vaporized with a temperature rise, which unfortunately causes the fluid's performance to become unstable.

Dispersed fillers are classified with respect to their electric characteristics into the following groups: normal dielectric materials, ferroelectric materials, semiconductors, electron conductors, and ion conductors. The dielectric materials may be considered to have characteristics similar to the ion conductors because of the small water, formic acid, and aniline contents.

When an electric field is applied to a semiconducting or conducting filler, the electrons within the filler shift, because of the electric field, causing the filler to polarize. In the case of a dielectric or ferroelectric material, orientation polarization occurs within the filler, if it is thought of as bulk material. In addition, space charges that are present in the interface also polarize. Called interfacial polarization in the case of an ion conductor, this polarization presumably occurs when ion pairs at the surface ionize. Interfacial polarization due to the surface ions is likely to occur particularly when the filler contains water and other ionic materials.

Improved fluid systems. Several new fluid systems have been reported to solve problems for aqueous electrorheological fluids. These systems can be roughly classified into two groups: systems that are improved aqueous fluids with better stability due to pretreatment, such as zeolite-based electrorheological fluids, and totally nonaqueous fluids using electron polarization. Among nonaqueous fluids are those with three types of dispersed phase: inorganic, ferroelectric microparticles; organic, semiconductor microparticles, represented by condensation-polymerized aromatic compounds; and insulating film-coated metallic microparticles. It has been reported that fluid systems in which microparticles of calcium titanate, an inorganic ferroelectric material, are dispersed in a naphthene-based liquid compound, show excellent durability. Various organic semiconductors, chlorinated aromatic liquids containing 25–35 wt % microparticles of poly-naphthalene-quinone or poly-anthracene-quinone and no water, were found to show high viscosity and shear stress with an applied voltage. A superior electrorheological effect was manifested by the fluid systems in which aluminum particles, whose surface was coated with an oxide film less than 1 μm thick, were dispersed in tri-2-ethythexyl-trimellitate.

The use of a solid phase consisting of light, featherlike fibers, in particular phosphorocarbon based on a cellulose precursor, has been proposed. In such element-containing fibers, required electric properties of a surface are determined by the temperature of preparation. Thus the electrorheological fluids can be stable against sedimentation while remaining highly sensitive to the electric field, as compared to electrorheological fluids based on the same material in powderlike form. In recent years, considerable advances have been made in theoretical description and mathematical modeling of the electrorheological effect, in particular with the use of the relation between conductivity and dielectric losses for solid and liquid phases.

At present, no ideal highly effective general-purpose electrorheological fluid exists. There are several problems to be solved before a smart fluid can be used effectively in any field. These problems are viscosity increase with shear stress, base viscosity, filler particle size, high voltage, time response, and upper limits on working temperature.

Applications. One of the most important reasons for an optimistic faith in the future of electrorheological fluids is their promising use in energy conservation. However, no estimates of their economic impact are presently available, because that depends on the particular design of electrorheological devices, conditions of their use at production sites, and the skills of designers.

Smart fluids are expected to replace many conventional technologies or cause new technologies to be developed in wide-ranging fields, including vibration-control systems, mainly for automotive applications; vibroprotection systems for large machines; power-transmission systems, such as clutches; flow-rate control systems; control systems for airplane wings; and micromachine systems.

Dielectric motors, liquid-power generators, and motion converters are, in principle, novel energy-saving devices consuming a minimum of electric energy. Indeed, despite the high voltages required (of the order of kilovolts), the power consumption is only about 0.1–1 W because of the semiconducting properties of the working media (a current of about 10^4 A). For the same reason, the application of an electric field allows energy savings in the cooling of objects or effectively transferring heat to other objects.

Application of available technological procedures using electrorheological fluids also indirectly saves energy because of a decrease in expenditures for the manufacture of mechanical fixtures such as clamps and grips, or magnetic and vacuum fixtures which possess limited capability (they are applicable only to magnetic materials or unperforated workpieces). Also, the use of electrorheological fluids makes it possible to achieve new levels of quality in the execution of many production operations that were previously attainable only at a high cost.

For background information SEE BOLTZMANN TRANSPORT EQUATION; BOUNDARY-LAYER FLOW; COLLOID; CONVECTION (HEAT); DIELECTRIC MATERIALS; ELECTRIC INSULATOR; FERROELECTRICS; FLUID FLOW; GAS DYNAMICS; HEAT TRANSFER; KNUDSEN NUMBER; NAVIER-STOKES EQUATIONS; PERMITTIVITY; POLARIZATION OF DIELECTRICS; REYNOLDS NUMBER; RHEOLOGY in the McGraw-Hill Encyclopedia of Science & Technology.

Dennis A. Siginer; E. V. Korobko

Bibliography. A. Faghri, *Heat Pipe Science and Technology,* 1995; G. F. Hewitt (ed.), *Proceedings of the 10th International Heat Transfer Conference, Brighton, U.K.,* 1994; *Proceedings of the 30th AIAA Thermophysics Conference,* 1995; R. Tao (ed.), *Electrorheological Fluids: Proceedings of the International Conference, Carbondale, 1991,* 1992; D. B. Tuckerman and R. F. Pease, High-performance heat sinking for VLSI, *IEEE Electron. Device Lett.,* 2:126–129, 1981; U.S. Department of Energy, *Electrorheological (ER) Fluids: A Research Needs Assessment Final Report,* DE-AC02-91ER30172, May 1993; W. M. Winslow, Induced fibration of suspensions, *J. Appl. Phys.* 20:1137–1140, 1949.

Food engineering

Recent advances in food engineering include packaging systems for perishable foods and robotic water-jet slicing systems.

Packaging for perishable foods. Packaging systems for perishable foods are usually designed to be maintained and distributed in an environment that is below ambient temperatures.

Half of all foods moving through distribution channels in the United States, Canada, and western Europe are chilled foods. Consumers value these foods for quality, convenience, nutritional value, and the perception that the fresher the food the more healthful it is. Chilled foods generally are minimally processed; that is, they may be trimmed, cleaned, comminuted, and perhaps heated or subjected to other mild preservation treatment to retain quality. Examples of minimal processes are washing precut vegetables with chlorinated water, thermal pasteurization of high-acid sauces, aseptic packaging, temperature reduction, and reduced-oxygen packaging. Each process, individually or in combination, has contributed to the extension of shelf life of perishable foods.

Controlled atmosphere packaging. Receiving much attention since the mid-1980s has been controlled/modified atmosphere/vacuum packaging, which had its origins many years earlier in warehouse storage of apples and truck shipment of lettuce. Early manifestations of the same basic packaging technology were found in vacuum packaging of cured meats and cheeses and vacuum shrink packaging of primal cuts of red meat such as beef, both of which are now mainstream chilled food preservation procedures. Since the mid-1980s reduced-oxygen packaging has been increasingly used for prepared foods, retail poultry, sandwiches, precut produce, and even beverages. In 1995, the volume of food prepared and packaged under reduced-oxygen modified atmospheres exceeded that of frozen or canned foods.

In controlled atmosphere packaging, the gaseous environment of the food is totally controlled to some predetermined level throughout distribution. The term modified atmosphere indicates that the gaseous environment is initially altered to a specified level, and subsequently changes over time as a result of product or microbiological respiration, gas transmission through the package structure, temperature, and so forth. With respiring products such as vegetables or fruit, initial atmospheric modification often eventually results in achieving a desirable optimum equilibrium between the surrounding atmosphere and the product.

Most of the recent research and development has centered on respiring foods such as red meats and precut vegetables and fruits and prepared (that is, precooked) main course or side dish foods. The preservation effects of modified-atmosphere packaging derive from several interacting variables. The most important of these is temperature reduction, which decreases the rates of enzymatic and biochemical deterioration and significantly retards the rate of growth of microorganisms. Reduction in the number of deteriorative vectors by processing, sanitation, and other unit operations adds to the preservation effect of refrigeration. Reduction of oxygen reduces the rate of oxidative deteriorations such as development of lipid rancidity, enzymatic browning, and growth of aerobic microorganisms responsible for most of the food spoilage. Increase in carbon dioxide also reduces the rate of growth of aerobic microorganisms. Increasing the water vapor concentration retards water loss from the food and also suppresses deteriorative aerobic respiratory reactions. The results of altering the environments surrounding and within foods are significantly enhanced by temperature reduction; for example, the solubility of carbon dioxide is inversely proportional to temperature.

Respiratory anaerobiosis. Although reduced oxygen can deliver major quality-retention benefits, a total absence of oxygen can lead to adverse results. In low-acid foods, and particularly those processed or packaged so that competitive aerobic spoilage microorganisms are suppressed, respiratory anaerobiosis (anoxia) resulting from a lack of oxygen can permit the outgrowth of anaerobic pathogenic microorganisms if the temperatures are not maintained below 36°F (3°C). Their growth can lead to the production of toxins, which can represent a major public health hazard. Respiratory anaerobiosis in respiring vegetables and fruit leads to the production of compounds with undesirable flavors and colors. Although these substances are not generally hazardous to health, they can have economic consequences. One mechanism to obviate the problems, or to retard them, is to engineer the package structures to permit higher gas (and oxygen) transmission into the package from the exterior air.

Physical openings. Probably the easiest way to address the respiratory anaerobiosis issue is to puncture the package to ensure that oxygen enters. But this practice generally permits a gaseous exchange that delivers an air environment within the package; thus it is counter to the idea of product content preservation. Packages employing this concept have been engineered with microperforations in which the package plastic has a multiplicity of specific tiny openings through which air can enter. Any physical opening permits gases to pass in the desired ratio of one oxygen to one carbon dioxide (that of normal plant tissue respiration). However, the opening also allows water vapor to pass, usually out, and microorganisms to enter. These latter two are undesirable consequences in packaged vegetables and fruit.

Microporous materials. A variation on actual physical openings is the use of microporous materials such as mineral-filled polyolefin films. By incorporating gross insoluble particulates such as calcium carbonate, which disrupt the continuity of the plastic film, a path is created through which air from the exterior can readily pass into the package. By controlling the ratio of mineral content to the total area through which the air can pass, the quantity of air entering a package can be controlled.

Microbiocidal films. Commercial polyolefin films containing small quantities of silver salts with allegedly microbiocidal properties have been tested. Although silver salts are effective antimicrobial agents, their application in package materials is restricted by the need to ensure contact between the salt and the microorganisms; this is extremely difficult because of the irregular shapes of foods.

High-gas-permeability-rate films. Respiratory anaerobiosis in fresh produce may also be avoided by increasing the surface-to-volume ratio of the plastic package. Increased package surface area will increase total oxygen permeation, thus minimizing the problem of anaerobiosis.

Respiratory anaerobiosis may also be minimized by engineering the plastics themselves to have the desired gas permeabilities. In past plastic packaging, materials were generally designed to reduce oxygen permeability. Highly plasticized polyvinyl chloride, which is highly gas permeable, has been used extensively for wrapping fresh red meats; it permits oxygen to enter the package so as to help retain the red oxymyoglobin color. Plasticized polyvinyl chloride is very clear, but it is relatively expensive, is mechanically difficult to handle because of a lack of stiffness, and has been a focus of environmental concern.

Low-density polyethylenes have traditionally been regarded as high-gas-permeability-rate films, but their inherent rates are too low to avoid respiratory or microbiological anaerobiosis. Enhancing the plastic with ethylene vinyl acetate generally increases the gas permeability rate to a desirable value, but as the content of ethylene vinyl acetate increases so does its stickiness, with consequent declines in the ability to commercially machine the material.

Lowering the density of polyethylene increases the gas permeability, and so the newer ultralow-density polyethylene resins in coextrusion or film form for lamination are being used as constituents of plastic film package structures. The gas permeabilities of these materials can be controlled.

One problem with any plastic structure is that although its gas permeability increases with temperature, these changes are not nearly as rapid as those for gas consumption or production by the respiring product itself. Avoidance of anaerobiosis carries with it the objective of obviation under elevated temperatures that are not uncommon in commercial distribution. At least two different developmental structures are being tested: In one, an added branch chain of the polyolefin polymer changes with temperature to significantly increase its gas permeability. In the other, slits in a multilayer plastic structure are engineered so that one of the plastic components is distorted by elevated temperature more than the other, with the result that the total structure curls and the slit opens to a diameter directly proportional to the temperature increase. It appears that this thermocouple analog is reversible.

High-gas-permeability plastic structures must be tailored to the specific product contents. Different film is required for each commodity. Further, from a microbiological safety perspective, none of the plastics even approaches the oxygen permeabilities required to obviate the growth of anaerobic pathogenic microorganisms in hermetically sealed packages. Therefore, the hazard and risk of microbiological anaerobiosis is not overcome by known alterations in the package material structures.

Oxygen scavengers. In recent years, oxygen scavengers have been incorporated inside packages to remove internal package oxygen and thus extend shelf life. Oxygen removal may create conditions conducive to the generation of anaerobic pathogenic microbiological problems or respiratory anaerobiosis. To achieve oxygen reduction, usually sealed sachets containing either a ferrous salt or ascorbic acid, both of which remove oxygen as they undergo oxidation, are introduced into a package. Oxygen scavenging is successful in situations where zero oxygen poses no public health hazard from *Clostridium botulinum,* such as in packaging of cured meats or bakery goods. Nevertheless, oxygen scavengers are occasionally incorporated into modified atmosphere packages of fresh pastas and bulk packages of fresh poultry. Research and development is focused on incorporating oxygen scavenging compounds directly into the packaging material structure, avoiding the separate and more expensive sachet that can cause problems if consumed.

Aaron L. Brody

Robotic water-jet slicing of foods. A very thin jet of water at very high velocity can slice through biological tissue and even through hard materials such as rocks or metal. The use of high-velocity water jets to cut rocks and ceramic materials has been under development since the 1970s. The system is attractive because of worker safety, minimal mechanical breakdown of equipment, no problem with dulling of cutting blades, and reduced noise and dust. Experimental applications of water-jet cutting of foods was first described in the 1980s. The system has been used to trim beef chucks of fat and connective tissue; to slice potatoes for potato chips; to remove tops from carrots, and to top, tail, and skin onions. Water jets are particularly effective in slicing onions for onion rings without inducing tearing of the eyes in workers. More recently, robotic water-jet cutting systems have been shown to be commercially attractive for precise trimming of meat to exacting specifications for portions with the desired shape and weight for the food service industry.

Water-jet cutting. Water-jet slicers consist of high-pressure water at around 50,000 lb/in.^{2} (3.5 megapascals) fed to a diamond or sapphire nozzle. Water emits from the nozzle as a thin jet with a diameter of about 0.008 in. (0.2 mm) and a velocity three to five times the speed of sound. This jet is very powerful and can easily slice through about 1.6 in. (4 cm) of flesh and connective tissue. Bones may shatter under a high-velocity water jet; therefore, the system is used for only boneless meat. Because the water jet exceeds the speed of sound, it generates a shock wave that is manifested by a very high intensity noise. Noise propagation is avoided by using deflector shields to change the velocity and direction of the jet as it emerges under the cut material. The rate of cutting depends upon the volume of water emitting from the nozzle, which is dependent upon pressure for a given orifice size. High-speed cuts in different directions that are needed to produce the desired shape may be achieved by using multiple nozzles, each driven by a single-axis robot or by using a single nozzle driven by a dual-axis robot. Multiple nozzles require higher power requirements for the intensifier, which generates the high-pressure water.

Robotic slicing. The robotic slicing system consists of robotic arms that move the water-jet wands; a machine vision system that measures the product shape and contours; a computer that computes the extent of trimming necessary to remove excess product relative to specifications for the desired product and that signals movement of the robotic arm; and a system for conveying the product through the measurement and cutting stations. In one system used for poultry, boneless breasts are positioned flat on a solid belt that conveys the product into a measuring station. A laser line is directed across the meat and is deflected by the contours of the product, producing a wavy line. The wavy line is detected by a camera that maps the product contour in 0.01-in.-square (0.25-mm) grids. Each grid is defined by a pixel that is given an arbitrary weight. The computer then analyzes the number of pixels needed to deliver the shape and weight specifications and directs the robotic arm to trim the excess product.

Because the measuring conveyor belt must be solid to eliminate cavities under the product and prevent measurement errors, and because water jets must have free space under the product, a transferring mechanism is needed to precisely transfer the product from the measuring to the slicing conveyor. All conveyor movements and robot arm movements are computer controlled to be in synchrony.

Future versions of these robotic cutting systems may incorporate the use of color cameras to enable differentiation between fat and muscle and dark-colored flesh that might be bruised or oxidized, providing the proper information for trimming fat and other undesirable tissue. As the robot movement and differentiation capability becomes more sensitive, these systems will find use for applications other than meat and poultry.

For background information SEE COMPUTER VISION; FOOD ENGINEERING; FOOD MANUFACTURING; POLYMER; ROBOTICS in the McGraw-Hill Encyclopedia of Science & Technology.

James Hewell; Romeo T. Toledo

Bibliography. R. Becker and G. Gray, Evaluation of a water jet cutting system for slicing potatoes, *J. Food Sci.,* 57:132–135, 1992; A. L. Brody, *Controlled/Modified Atmosphere/Vacuum Packaging of*

Foods, 1989; A. L. Brody, *Modified Atmosphere Food Packaging*, 1994; P. Demetrakakes, Cutting the fat with electric eyes, *Food Process.*, 55(6):78, June 1994; W. Heiland, R. Konstance, and J. Craig, Robotic high pressure water jet cutting of chuck slices, *J. Food Process. Eng.*, 12:131–136, 1990; J. Hewell, *Compuscan 3000, Portioning Control System*, Cantrell Machine Co., Gainesville, Georgia, 1995; R. T. Parry, *Principles and Applications of Modified Atmosphere Packaging of Food*, 1992; G. Robertson, *Food Packaging*, 1993.

Food microbiology

Microbial inactivation by pulsed electric fields is a new technology for food preservation that subjects food to only a small increase in temperature and minimally influences food components. This technology involves subjecting food placed between two electrodes to a short burst of high-voltage electricity. Food at ambient or subambient temperature is treated with pulses of microsecond duration, minimizing energy consumption by heating. This new technology has the potential to improve energy usage economically and efficiently, as well as to provide consumers with microbiologically safe, minimally processed, nutritious foods with freshlike quality.

Foods are commonly heat processed to increase shelf life and remove safety hazards by inactivating spoilage and pathogenic microorganisms. However, heat treatment can have adverse effects on the flavor, taste, and nutrient content of foods. Use of high-intensity pulsed electric fields is an attractive alternative to traditional pasteurization methods.

Generation of pulsed electric fields. The generation of pulsed electric fields requires a pulsed power supply and a treatment chamber. In the pulsed power supply, common utility line voltage is converted to a high voltage, and at the same time electric energy at low levels is collected over an extended period and stored in a capacitor. That same energy can then be discharged almost instantaneously (in a millionth of a second) at very high levels of power. The discharge is accomplished by using state-of-the-art high-voltage switches, which operate reliably at high power and with a high rate of repetition. The switches can be selected from gas spark gaps, vacuum spark gaps, solid-state switches, thyratrons, and high-vacuum tubes. The experimental repetitive high-voltage pulser used for microbial inactivation was designed with a wide range and adjustable operating voltage, repetition rate, pulse duration, and pulse energy. The treatment chamber is used to transfer the high-voltage pulse into a high-intensity pulsed electric field in the food.

Treatment chambers commonly are of three types: parallel plate static, parallel plate continuous, and coaxial cylinder continuous. In the parallel plate static chamber, two disk electrodes are held in position by an insulating spacer that also forms an enclosure containing the food. The electrodes are fabricated of stainless steel or other conductive materials, and the spacer and lids are made of polysulfone or other insulating materials. A practical design for pasteurization of fluid foods using a pulsed electric field allows liquid food to flow continuously through the treatment chamber. The simple flow-through chamber is a modified static chamber with baffled flow channels inside. The coaxial cylindrical configuration of a continuous chamber can provide a well-defined field distribution. Both the high-voltage electrode and the grounded electrode in a pulsed electric field system may contain circulating cooling fluid to control electrode temperature. The gap between the coaxial electrodes in a coaxial cylinder continuous system is determined by the diameter of the inner electrode.

A continuous pulsed electric field system consists of a high-voltage pulse generator, a continuous treatment chamber, a fluid-food pump, a heat exchanger, and a control and data acquisition system with a computer. The pulse generator is a repetitive capacitor discharge modulator. The continuous treatment chamber is used to transfer pulsed high voltage into a high-intensity pulsed electric field. Typical electric fields and pulse durations are 20–80 kV/cm and 0.5–10 microseconds. The temperature of the food treated in the pulsed electric field is rapidly reduced to a refrigerated temperature for aseptic filling. The flow rate during the treatment is controlled by a variable-speed pump. Electrical and flow parameters are selected so that each unit volume of food is subjected to the necessary number of pulses to create the desired inactivation of microbial cells.

Inactivation of microorganisms. Under pulsed electric fields, microbial inactivation is caused by irreversible pore formation and destruction of the semipermeable barrier of the cell membrane. Pulsed electric fields are successful in the inactivation of fluid-suspended microorganisms such as *Escherichia coli*, *Staphylococcus aureus*, *Lactobacillus delbruekii*, *Bacillus subtilis*, *Salmonella enteritidis*, *Klebsiella pneumoniae*, *Pseudomonas aeruginosa*, *Listeria monocytogenes*, *Candida albicans*, and *Saccharomyces cerevisiae*. The kinetics of microbial inactivation in fluid media are a function of field intensity, treatment time (pulse duration and number of pulses), and a model constant determined by the microorganism and its physiological status. Other parameters affecting the inactivation of microorganisms by the pulsed electric field are applied pulse wave shape, ionic concentration of the food, temperature of the food, microbial cell concentration, and growth stage of the cell.

Applied peak electric field intensity is one of the important factors influencing microbial inactivation. With a fixed number of applied pulses, the rate of inactivation increases with an increase in the electric field intensity. After pulsed electric field treatment of *S. cerevisiae* in apple juice, the structure of the yeast cells is changed. The **illustration** shows the transmission electron micrograph of *S. cerevisiae* before and after application of the pulsed electric field. Massive damage, in the form of a hole on the cell wall, is caused by the application of pulsed electric fields.

Pulsed electric fields may be applied in the form of exponentially decaying, square-wave, oscillatory, or bipolar pulses. Oscillatory pulses are least efficient in inactivating *S. cerevisiae* when tested at an electric field strength of 40 kV/cm and an oscillating frequency of 100 kHz. The energy efficiency of square-wave pulses is greater than that of exponentially decaying pulses. The bipolar pulses are more lethal to *E. coli* than monopolar pulses at an electric field intensity of 40 kV/cm.

Pulsed electric field treatment combined with a moderate temperature exhibits a synergistic effect on the inactivation of microorganisms. This synergistic effect reflects the temperature-related phase transition of phospholipid molecules from a rigid gel structure to a liquid-crystalline structure and the associated reduction in bilayer thickness. Cell membranes are more susceptible to electric fields at relatively high temperatures, which are, however, below the traditional heat treatment temperatures.

Food processing. The pulsed electric field treatment system is suitable for the pasteurization of fruit juices, dairy products, eggs, and other fluid foods. For example, apple juice from concentrate has been subjected to this type of treatment in the continuous system at a processing temperature of 95°F (35°C). After the pulsed-electric-field processing, apple juice is aseptically filled into packages for shelf-life studies. The shelf life of apple juice from concentrate treated with the pulsed electric field is at least 4 weeks at room temperature. There is no apparent change in the physical and chemical properties of the juice. A sensory panel found no significant difference between the treated and nontreated juice.

Fresh-squeezed apple juice, raw skim milk, beaten eggs, and pea soup have also been treated by pulsed electric fields in controlled treatment temperatures (below 113°F or 45°C). No apparent change has been found in the physical and chemical properties of the products. The shelf life of the products treated with pulsed electric fields is at least 2 weeks at refrigeration temperature (39–43°F or 4–6°C). The sensory panel in the triangular test found no significant differences between the fresh food and the treated product. In the acceptance test, a panel generally preferred the treated foods to the thermal-treated products.

However, application of the pulsed electric field method is restricted to food products that can withstand high electric fields. Homogeneous fluids are ideal for continuous pulsed electric field treatment. Nonfluid foods can also be processed by pulsed electric fields in a batch mode, as long as dielectric breakdown in the foods is prevented. Air bubbles in the food must be removed; otherwise localized arcing will occur in the high electric fields. In general, the pulsed electric field method is not suitable when solid food products containing air bubbles are placed in the treatment chamber. Another limitation of the method is the size of particulates in fluid foods. The largest particle must be smaller than the gap in the flow channels.

For background information SEE CAPACITOR; ELECTRIC FIELD; ELECTRIC SWITCH; FOOD ENGINEERING; FOOD MICROBIOLOGY; GAS TUBE in the McGraw-Hill Encyclopedia of Science & Technology.

Gustavo V. Barbosa-Cánovas;
Barry G. Swanson; Bai-Lin Qin

Bibliography. G. V. Barbosa-Cánovas and J. Welti-Chanes (eds.), *Preservation of Foods by Moisture Control: Fundamentals and Applications*, June 1995; H. Hülsheger, J. Potel, and E.-G. Niemann, Electric

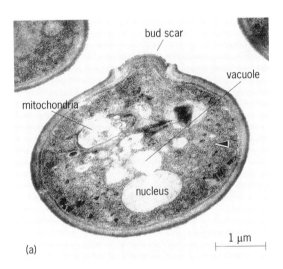

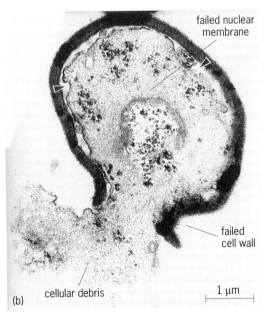

Transmission electron micrographs showing *Saccharomyces cerevisiae* before and after application of a pulsed electric field. (*a*) An untreated cell. The arrow indicates ribosomes. (*b*) The cell after being treated with 40 kV/cm electric fields. The arrows indicate shrinkage of cytoplasmic material from the cell wall.

field effects on bacteria and yeast cells, *Radiat. Environ. Biophys.*, 22:149–162, 1983; B. L. Qin et al., Inactivation of microorganisms by pulsed electric fields of different voltage waveforms, *IEEE Trans. Dielec. Elec. Insulat.*, 1(6):1047–1057, 1994.

Forest canopy

Until recently, most knowledge about forest ecosystems was based upon observations at ground level. This was due to the logistic difficulties of reaching the canopy, which has created an obvious limitation in the perception of forests because the majority of the organisms and plant biomass exist well above ground level. Forest canopies have been called the last frontier. Until recently, very little information had been collected from in-place tree crowns. Most information about forest canopies was obtained by examining the crowns of felled trees. Recent innovative and creative techniques have facilitated canopy studies, and have led to the development of new perspectives on this important biological system. Access techniques include the use of special hardware and ropes, the construction of aerial walkways and platforms, and even such devices as dirigibles and construction cranes. Thus scientists can collect information and formulate hypotheses for this above-ground three-dimensional system that were not previously possible.

In addition, over the past hundred years most ideas about forest ecology were developed in temperate regions; tropical forests remained relatively remote and inaccessible to most scientists. In contrast, much recent research in forest canopies has been pioneered in the tropics. In 1978, a method of climbing into tropical tree canopies by using ropes and climbing hardware was described and the use of single rope techniques signaled a rapid expansion of canopy research worldwide.

Importance. There are a number of important reasons why canopy access has become a research priority. First, as rainforests continue to dwindle, the urgency of surveying the biodiversity in their tree crowns challenges many researchers. There are reputedly numerous undiscovered organisms throughout forest canopies; some may contain important economic and medicinal products. By virtue of their complexity, it is believed that tropical tree canopies house the largest diversity of terrestrial organisms. Temperate deciduous forest canopies contain a lower diversity of plants (for example vascular epiphytes are almost nonexistent) and fewer invertebrates. In oak or maple forests of the American northeast, insects are relatively diverse during the temperate summer season but decline to negligible numbers during the winter months. Second, canopy processes are essential to life on our planet; canopy organisms are integral to the maintenance of rainforest ecosystems, and the canopy is a major site of productivity in terms of photosynthesis and turnover of carbon dioxide.

Since the mid-1980s, the amount of information on canopy biology has greatly expanded. Achievements of note include the quantification of the diversity and abundance of invertebrates in the canopy, the accurate measurements of photosynthesis in this environment where foliage is extremely dense, the temporal and spatial measurements of insect damage to leaves, and the measurements of nutrient cycling from the atmosphere into the canopy and through to the understory. The development of access techniques has stimulated collaborative research where measurements of atmospheric components, respiration, leaf mortality, and tree-growth architecture can be surveyed simultaneously and modeled at the ecosystem level. Future research priorities include intercontinental comparisons of canopy processes, further studies of the behavior and population dynamics of the more cryptic groups of organisms (for example, mites, leaf-sucking insects, miners, stem-borers, birds with large home ranges). Of greatest priority is the ability to integrate canopy information into quantitative results that are useful in the conservation and management of canopy environments.

There are many challenges to conducting scientific research in forest canopies. Scientists face enormous limitations when attempting to work in a large, aerial three-dimensional space, such as in the detection and sampling of organisms in a heterogeneous environment where human mobility is curtailed and accounting for the differences in longevity and seasonality of different organisms. As with the expansion of coral reef ecology in the 1970s with the advent of the self-contained underwater breathing apparatus (SCUBA), canopy biology will benefit from new sampling protocols that will quantify the spatial, temporal, and substrate heterogeneity of the canopy environment.

Components. The three major components of forest canopies, each requiring a different method of investigation by scientists, are sessile organisms (trees, vines, epiphytes and epiphylls, some insects), mobile organisms (birds, some insects, and mammals), and the ecological processes whereby these organisms interact, such as herbivory and nutrient cycling.

Sessile organisms. Studies of sessile organisms pose fewer logistic difficulties for measurement than other canopy components because these organisms can usually be counted. Sessile organisms include the obvious organisms such as trees and vines, as well as smaller species such as tiny mosses or lichens, and even some insects (for example, scales) that remain in one place throughout most of their life cycle. Trees are the largest sessile organisms in forest canopies, composing the major substrate in the canopy ecosystem. Tree species, with their varying architecture, limb strength, surface chemistry and texture, play fundamental roles in shaping the canopy community. Epiphytes (also called air plants) live on the surfaces of trees, as do many epiphylls (tiny air plants such as lichens that live on leaf surfaces). In addition, trees serve as both shelter and food for many mobile organisms, and most canopy processes are directly dependent upon the trees.

The architecture of trees is far more complex in the tropical forests than in the temperate regions, and over time as tree crowns expand the communities within them increase in complexity. For example, patches of leaves that are heterogeneous in their age structure, foliage quality, and distribution throughout the crown attract different populations of insects. In turn the location of predators is affected, and also possibly the canopy processes within different portions of the crowns. Thus, all sessile organisms, from the smallest moss up to a large emergent fig tree, make up the substrate, the shelter, and often the food supply for the canopy ecosystem.

Sessile organisms such as trees are currently the focus of extensive investigation; scientists are interested in tree survival as part of forest conservation. Long-term plots to measure seedling germination, sapling survival, and eventual growth into canopy trees are being established in tropical forests worldwide, including Australia, Panama, Ecuador, and Malaysia. After many years of measuring individual trees in the forests, scientists learn how fast different species grow and die. Some trees exhibit very unusual means of survival. For example, the suicide tree (*Tachigalia versicolor*) in Panama has a unique pattern of shedding its seed only once as an adult and then dying immediately afterward. Whether this pattern is a means to open up a gap for its offspring or a result of reproduction stressing the tree severely, or some other phenomenon, is not yet understood.

Mobile organisms. Mobile organisms in the canopy pose great challenges of measurement for scientists, but exciting progress has been made. Canopy access has led to the discovery of unexpected canopy foraging among some mammals previously considered ground dwellers [species of mice (*Peromyscus*)]. Similarly, studies of birds are beginning to reveal complex patterns whereby species exhibit preference for certain vertical regions of the canopy, or for specific food trees. Such patterns have obvious implications for the niche theory, although the population dynamics of most canopy organisms require additional study to better quantify the niches and interactions involved. Studies of invertebrates in tropical forest canopies have perhaps created more controversy than any other aspect of canopy research. Fogging experiments have raised speculation that the number of species on Earth may be as high as 30 million, rather than the 1–2 million previously estimated. The enormous variability in space and time within a tropical rainforest canopy, however, as well as the obvious limitations of sampling in this region, make estimates of numbers of mobile canopy insects very difficult.

In Australia, a certain beetle (*Novocastria nothofagi*) utilizes its mobility to take advantage of young leaves for its diet, and also has eluded canopy scientists for many years. Antarctic beech trees (*Nothofagus moorei*), which dominate the cool temperate rainforests in the montane regions of Australia, undergo a major leaf flush from September to November. The beetle larvae feed on the young leaves as they emerge, moving throughout different canopy levels as the buds burst. Damage from the voracious larvae results in foliage losses of over 50% to some tree canopies, but scientists failed to observe the ephemeral culprit for many years. Soon after leaf emergence, the beetle larvae disappear by dropping from the canopy into the soil for metamorphosis, only to emerge several weeks later as mature beetles. In contrast, many sucking insects (for example, lerps) remain relatively immobile on leaf surfaces, and can be counted over time, making their populations easier to measure.

Processes. Canopy processes represent the machinery that link all of the organisms. They include reproductive biology, herbivory (consumption of foliage), nutrient cycling, and photosynthesis. Different canopy methods have enabled scientists to study different processes, but logistics still present a formidable limitation. For example, studies of reproductive biology require measurements of flowers and their pollinators. These components are often located in the uppermost branches of tall trees, and are relatively inaccessible by access methods such as ropes, ladders, or tree platforms, all of which require large branches for safe operation. In Malaysia, scientists built very tall ladders into the upper canopies of dipterocarp trees to observe tiny thrips pollinating the flowers.

Access to tree crowns has also stimulated interest in canopy nutrient cycling, particularly with reference to epiphytes. Many bromeliads, with their cuplike architecture, serve as sinks for the deposition of wind-blown fine litter, nutrient particles, and water vapor. Within these cups high up in the trees reside a variety of organisms, including frogs, invertebrates, and other aquatic life.

Herbivory of foliage by insects and some mammals is a canopy process that directly affects the mechanism of energy production by trees. Extensive removal of foliage can lead to reduction in photosynthesis or to mortality of entire trees, or both. How insects select for specific leaf material, how they navigate through tree crowns to locate their food, and how leaf area removal affects different plants are important questions relating to the health of the entire forest. Many insect herbivores in forest canopies are nocturnal feeders, making the logistics of research even more daunting.

Ecological challenges. Now that techniques have been designed that enable scientists to access forest canopies, the real challenges lie ahead. Canopy organisms, both mobile and sessile, must be mapped and measured. Canopy processes must be sampled with respect to the complex differences in light, height, tree species, and seasonality. Establishing rigorous sampling techniques and conducting long-term research that requires the scientist to dangle from a rope are ambitious objectives that constitute the current priorities of forest canopy research.

The pressures of human populations on tropical rainforests provide an added incentive for scientists to understand the dynamics of forest canopies. The next several years will be critical, as scientists attempt to classify the biodiversity and ecology of forest canopies before habitat fragmentation takes its toll. Hopefully, canopy access techniques will continue to provide valuable information about the many organisms and processes that occur in forest canopies, thus facilitating the implementation of sound forest conservation practices.

For background information SEE FOREST AND FORESTRY; FOREST ECOSYSTEM; FOREST MANAGE-MENT; FOREST SOIL in the McGraw-Hill Encyclopedia of Science & Technology.

Margaret Lowman

Bibliography. M. D. Lowman and M. Moffett, The ecology of tropical rain forest canopies, *Trends Ecol. Evol.*, 8:104–107, 1993; M. D. Lowman and N. Nadkarni (eds.), *Forest Canopies*, 1995.

Forest fires

Forest fires have sculpted the historic landscapes of the western forests of the United States, and they will continue to be important in the future. Over the last century, western forests and fire were managed inadequately, primarily because the fire-control paradigm that prevailed was based on the mistaken assumption that all fire on the landscape could be successfully suppressed. A broader and more rational paradigm of fire management is replacing fire control.

Fire regimes. Fire in the landscapes of the West occurred in a wide range of frequencies and intensities and affected the biota in myriad ways. In some ecosystems fire was frequent (every 5–15 years) but of low intensity and did not damage mature trees; in others it was an intense but infrequent occurrence that sometimes killed most of the trees. One way to understand future fire-management options is to categorize these ecosystems into a general set of fire regimes—combinations of historical fire frequency, intensity, and extent that may have had similar ecosystem effects.

Low-severity fire regimes evolved with frequent low-intensity fires. These forests are characterized by trees that are adapted to this type of fire regime; typically such trees have high crowns and thick bark, which are excellent adaptations to resist frequent surface fires. Fires often killed small portions of the cambium (the layer responsible for secondary growth and for generating new cells), so counting annual growth rings healing over the dead part of the cambial circumference can produce a chronology of all past fires. Some trees have recorded up to 30 different low-intensity fires at fire-return intervals of fewer than 10 years. In con-trast, high-severity fire regimes evolved with infrequent but generally intense fires. These forests are characterized either by trees that have little adaptation to fire (because it is infrequent) or by trees that are adapted to intense fires by their ability to aggressively colonize burned sites. Between the low- and high-severity fire regimes are the moderate-severity fire regimes, which have a high variability in both frequency and intensity.

Ponderosa pine (Pinus ponderosa) forests. These forests are the classic example of a low-severity fire regime. The prehistoric pine forests were stable and sustainable, in part because frequent fires of low intensity maintained open forests with low fuel levels. These fires burned lightly around the bases of fire-resistant large pines, doing little harm to the thick-barked mature trees. Fire-return intervals of 5–15 years were typical, and these fires kept the forest understory clean of small trees [pine and more shade-tolerant associates, such as Douglas-fir (*Pseudotsuga menziesii*)] and tall shrubs. Most of the grasses and shrubs were adapted to these fires by sprouting. Insects and diseases were held in check, wildlife habitat was stable, and the forests were of high esthetic quality.

However, fire exclusion practices result in another scenario. One example is the ecological change in a typical ponderosa pine forest where Douglas-fir is a shade-tolerant component. In 1909, before effective fire suppression was practiced, there was an overstory of ponderosa pine and a diverse understory of low shrubs and herbs. Historic fire frequency was probably every 10–15 years, and minor logging operations had occurred. By 1948, successful fire exclusion practices resulted in a tree jungle, with associated increases in potential for catastrophic fire. By the 1990s, almost 50 years later, the situation was even worse.

Subalpine fir (Abies lasiocarpa) forests. These forests exemplify the high-severity fire regime. Most studies have shown fire-return intervals of 75–250 years. Most postfire stands had a high component of lodgepole pine (*P. contorta*), which maintained a crown seed bank within serotinous cones. Serotinous cones hold viable seed but remain closed in the tree crown until a fire passes through. The cones then open and release seed into a competition-free ashbed on the forest floor, regenerating the area the next spring. If fire returns to the site within the next two centuries, another generation of lodgepole pine is the result, but if not, the forest eventually becomes dominated by subalpine fir. This forest type is common in the vicinity of Yellowstone National Park, and the new lodgepole pine forests regenerating after the massive 1988 fires that burned across the park are replacing forests that regenerated after similar fires in the early 1700s.

Multicanopied stands. A trend toward multi-canopied stands occurs in most western conifer forests that are protected from fire. But the effect of this trend differs significantly, depending on the fire regime. In subalpine fir forests, it is a natural progression that is well within the natural range of variation for this forest ecosystem. The Yellowstone fires were characteristic of fires that burn in these high-elevation forest ecosystems. In ponderosa pine forests, multicanopied stands rarely spread widely across the landscape because the frequent fires maintain an open forest with a single fire-resistant canopy.

Fire management. Twentieth-century management, for all its good intentions, has created serious problems for the landscape. The problem is most obvious in the low-elevation, dry, low-severity fire regimes containing ponderosa pine, Douglas-fir, and mixed-conifer forests. The species composition and structure of these forests have been radically altered by fire exclusion and logging. Forests have lost their larger, fire-tolerant pines through selective harvesting, and effective fire suppression has allowed pines to be replaced by many smaller, less fire-tolerant trees, usually Douglas-fir, grand fir (*A. grandis*), or white fir (*A. concolor*). The forests have insect and disease problems that appear to be much worse than those that existed in the past. Pines stressed from understory competition are attacked by bark beetles. Defoliating insects, which focus on the firs, have more available food, so that their outbreaks have become longer and more intense. Through fire exclusion, a forest has been created that no longer supports the frequent low-intensity fires of antiquity. The thick stands of small, dense trees serve as fuel ladders that carry wildfire into the canopy of the forest, and the friendly flame has been replaced by the firestorm. These problems constitute a decline in forest health.

Solutions have focused on the low-severity fire regimes where forest health has been most altered by twentieth-century management. However, there is a significant social component to the issue of improving forest health, and the spectrum of vested interests ranges from those of the environmental community to those of the timber community. Both groups agree that reducing tree density and favoring fire-tolerant trees are appropriate to the solution, but they differ substantially in how to achieve that goal. The environmentalists emphasize reintroduction of low-intensity fire, as it is a more natural process than tree cutting. The timber community emphasizes removal of timber as an alternative solution, with less emphasis on restoring fire to the ecosystem. The first step in a sustainable solution is to recognize that the primary goal of any treatment is to leave a more sustainable disturbance-tolerant forest to improve forest health. If this measure is applied to the drier forests of the West, tens of millions of acres might eventually be treated. The massive restoration projected for these forests will require many of the methods now in use, including timber harvest and prescribed fire (see **illus.**).

One promising sign is that fire control is being replaced by fire management, a concept emphasizing both use and control of fire. Control and prevention of fires are still extremely important parts of fire

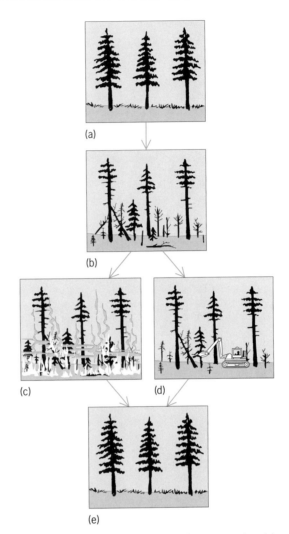

(a)

(b)

(c)   (d)

(e)

Rationale for correcting a history of fire suppression: (*a*) Natural open ponderosa pine forest with grassy floor. (*b*) Dense growth of small trees and accumulation of deadfall resulting from fire suppression and grazing. (*c*) Prescribed fire and (*d*) mechanical thinning remove fuels and reduce fire danger, leading to (*e*) restoration of the historic balance of the forest. (*After J. Phillips and D. Lambert, The crisis in our forests, Sunset Mag., pp. 87–92, July 1995*)

management. Wildfires burning during the dry season will not help to restore forest health. Programs such as Smokey Bear will continue to be an essential part of the fire-management picture, but prevention alone will only exacerbate the current problems. Fire-management education must become broader than the traditional "prevent wildfire" statement.

Prescribed fires. Prescribed fire, ignited by managers under controlled conditions, is intended to play a larger role in meeting land management objectives. Fires are ignited under prescribed conditions of fuels, weather, and topography in order to keep fire behavior, usually indexed by the length of the flames, within certain limits. Burning under moist conditions (early spring and late fall for much of the West) allows fuel cleanup to occur without harm to the larger trees, as flame lengths can be maintained below 1 m (3 ft). Smaller trees are killed, and fuel loads are successfully reduced. The practice of prescribed natural fires and of lightning-caused fires that are allowed to burn under defined conditions with monitoring will continue in remote park and wilderness areas, but most of these areas contain higher-elevation forests where forest health problems are not as severe as elsewhere.

Other tools. Among the emerging tools for fire managers are computer programs that track the spread and behavior of wildland fires. One program, BEHAVE, predicts the behavior of fires given inputs of a fuel model (standard grass, shrub, timber, logging slash models, or locally developed fuel models), weather, and slope conditions. The rate of spread and fire-line intensity (scaled to flame length) are the major products of the model. Similar fire behavior prediction models are incorporated with geographic information systems (GIS) in a new model called FARSITE to show geographical fire spread over a landscape, given a data stream of fire weather that can change over time.

Prescribed fire will be an important tool in forest restoration, but it cannot work alone for several reasons. Many trees that were the product of early fire exclusion, those that would have been killed by recurring fires while still very small, are now large enough that prescribed fires hot enough to remove them may be too hot for older, pre-fire-exclusion trees to survive. If these larger younger trees are left, crown densities may remain high enough to sustain crown fires, although high surface fire intensities would be required to initiate crowning. Even low-intensity but long-smoldering duff fires, a result of forest-floor buildup, may be damaging to residual stands. (Duff is the partly decayed organic matter on the forest floor.)

The smoke from prescribed fires might exceed air-quality standards such that only small areas of the landscape could be treated. Across much of the West, the natural fire regimes must have generated substantial haze, and reintroducing fire may increase that haze over more populous landscapes. In the drier forest types, however, prescribed fire will eventually create fire-safe landscapes where wildfires can be more easily controlled, so a trade-off of wildfire versus prescribed-fire smoke exists. Nevertheless, at the scale that prescribed fire might be contemplated, smoke problems will be a major limitation.

Prospects. There is the question of whether timber harvesting alone can succeed. Probably not, unless substantial amounts of smaller debris are removed by pile burns or by chipping operations. Entry can be accomplished with much less ground disruption than in previous decades, but some landscapes may not be capable of sustaining necessary new roads without environmental damage. Most operations need to be low thinnings, that is, careful removal of the smaller trees while leaving the bigger, more economically valuable ones that may be classed as lower-vigor trees. Across the West, older sawmills not capable of utilizing small trees will be replaced by computer-intensive modern mills that can cut small-diameter logs efficiently. Today, much of the smaller material can be removed from the forest with little environmental damage, and commercial utilization allows these investments in forest health to pay for themselves.

This new vision of fire management integrates land management goals with all fire strategies, ranging from fire prevention to allowing natural fires to burn in some park and wilderness areas. Fire is a natural process that will be present in some form—prescribed or wild—over future decades. Integrating use of fire with fire control in the drier forests of the West will ensure wise stewardship of these natural resources.

For background information SEE DENDRO-CHRONOLOGY; FOREST AND FORESTRY; FOREST FIRE CONTROL; FOREST MANAGEMENT in the McGraw-Hill Encyclopedia of Science & Technology.

James K. Agee

Bibliography. J. K. Agee, *Fire Ecology of Pacific Northwest Forests,* 1993; S. J. Pyne, *World Fire,* 1995.

Forest soil

Soil organic layers on forest floors contain a large array of arthropod species (insects, mites, spiders, and others) arranged in elaborate food webs. Together with other invertebrate animals, such as earthworms and nematodes, forest-floor arthropods are important agents affecting processes such as plant-litter decomposition, cycling of important nutrients, and creation of soil structure. Microbes (bacteria and fungi) are the main agents of decomposition and nutrient transformation, but these processes are influenced by the interaction of the microbes with soil animals. In feeding habits, forest-soil arthropods range through those subsisting on living plant parts (roots), saprovores (feeders on dead organic matter), fungivores (fungal feeders), and predators on other animals.

Nearly all forest arthropods live in the soil at some time in their life histories. Permanent residents include many of the ubiquitous mites and collembolans (springtails). Many insect species, such as ants and termites, are periodic soil inhabitants, spending their life histories in the soil but emerging as flying adults to disperse. Others (many species of flies, for instance) are temporary residents, completing their juvenile stages in the soil but living aboveground as adults. Transient species, such as caterpillars feeding in the forest canopy, enter the soil to pupate or to overwinter. This array of strategies serves to connect food webs in the soil with those in the forest canopy. Caterpillars descending from treetops to pupate in the soil are a major seasonal food item for spiders and beetles on the forest floor. SEE FOREST CANOPY.

Functional classification. Forest-soil arthropods are divided into microarthropods and macroarthropods, based on body size. The numerous microarthropods include mites, collembolans, and other tiny insects small enough to move freely through the natural pore spaces in forest soil [about 0.04 in. (1 mm) or less in width]. Macroarthropods, such as larger beetles and spiders, millipedes, and centipedes, must create their own passageways (biopores). The separation into these two groups is one of convenience, based in part upon methods used for collection and study. Hand collecting and sorting suffice for macroarthropods, but microarthropods require extraction by heat from soil samples with Berlese funnels, or by flotation in salt solutions.

Biodiversity. Forest-floor microarthropods are a numerous, diverse assemblage, rich in species. Numbers may exceed 500,000 individuals per square meter of forest floor. Many of the mites and collembolans in temperate and tropical forests are poorly known taxonomically (that is, they have not yet been given scientific names). Immature stages are not well known, and information is lacking about their use of food resources. Many ant species are yet to be discovered in the tropics. Forest-floor spiders in North America and ground beetles and millipedes in the southern Appalachian mountain forests are also species-rich groups. Together, these arthropods are an enormous resource for biodiversity.

Trophic structure and food webs. The basis of the food webs involving forest-floor arthropods is the rich mat of decomposing organic matter. Each year several hundred grams of dead leaf material and smaller amounts of woody material cover each square meter of forest floor. These materials are attacked by microbes. Saprophagous microarthropods and millipedes feed upon this mixture, some ingesting bits of organic litter (macrophytophages), some ingesting only fungal hyphae (microphytophages), and some ingesting both (panphytophages). A further input of energy and nutrients to these food webs comes from living roots. Some arthropods feed upon roots directly, but others feed upon microbes in the rhizosphere (the zone of influence of the root) or find their prey there. The importance of the root resource for soil food webs is poorly known, and this resource is currently under study using observation windows or transparent tubes (minirhizotrons) in forest soils.

Some forest-soil arthropods are opportunistic feeders and defy classification into simple food chains. Mites and collembolans, for example, switch their feeding habits to include nematodes when those worms become abundant. Ants that feed upon seeds or fungus may begin to eat mites or collembolans. For many arthropods, dietary preferences are not known. An interesting question is how so many species manage to coexist, given their proximity in the forest-floor habitat. New research approaches measure the activity of enzymes in guts of forest-floor mites in an attempt to group species into feeding guilds. Species in one major group of mites, the oribatids, contain a wide variety of enzymes, including amylase, pectinase, cellulase, and other oligosaccharides. Whether these enzymes are produced by the mites themselves or by microbes in their guts is unknown.

Influences on decomposition processes. The decomposition of dead organic matter on the forest floor is the work of microbes, bacteria, and especially fungi. However, the process is influenced by soil arthropods, and the actual rates of decomposition are the product of interactions between microbes and soil fauna. Through their feeding activities, arthropods fragment the dead organic litter (comminution). In that way, a greater surface area is made available for microbial attack and for leaching of nutrients from the substrate. Animal activities mix the decomposing material with mineral soil, making it more available to a variety of microbial species. Arthropods introduce fungal spores into decomposing organic matter. Their feces may act as a further substrate for microbial activity.

The influences of soil arthropods on decomposition rates have been measured by using experimental procedures. Arthropods may be excluded from decomposing litter either by chemical procedures (insecticides) or by enclosing leaf litter in fine mesh bags on the forest floor. Such exclusion of arthropods has been found to cause delays in the decomposition process. Nutrient releases from decaying organic matter also are slower when arthropods are excluded. Arthropod activities are more important in the decomposition of recalcitrant leaf-litter species, which may contain large amounts of lignin, polyphenols, tannins, or lipid cuticles. Furthermore, arthropods appear to be more important in later stages of organic matter decomposition, after the more labile fractions have been decomposed.

Effects on soil structure. Forest soil is the product of interactions between physical and biological processes. Root exudates and their activities, and the importance of earthworms, have been thoroughly documented as biological factors contributing to the aggregate structure in forest soils. Macroarthropods make significant contributions to soil structure by their tunneling activities, and their fecal pellets contribute to soil aggregates. Microarthropods, as noted above, may be limited by the pore size in soils, but microscopic study of soil sections reveals that fecal pellets of mites and collembolans may also contribute to aggregate structure on a smaller scale. Termites are major movers of soil in tropical regions, as earthworms are in temperate forests. Ants, however, may move more soil on a worldwide basis than do termites and earthworms combined.

Effects of environmental factors. Soil processes vary with latitude. In general, rates of decomposition, soil respiration, and nutrient dynamics increase from polar regions to the tropics. The numerical abundance of soil arthropods, however, seems more closely related to the thickness of the litter layer than to atmospheric temperature. Microarthropods appear to be more abundant in southern temperate forests than in tropical forests, where leaf-litter production is continual and little accumulation occurs because of rapid decomposition. Microarthropods living in forest soils exist in an atmosphere saturated with moisture (the atmosphere of pore spaces). Macroarthropods living on the surface of the forest floor, such as millipedes, are also dependent upon atmospheric moisture, and they suffer desiccation easily. During droughts or seasonal dry spells, forest-soil arthropods congregate in moist areas beneath logs or stones.

Effects of management practices. Populations of forest-floor arthropods show differential responses to forest management techniques. In extensive modifications of tropical and temperate forests, many arthropod species become diminished in numbers or are lost from the forest. Intense harvests, such as clear-cutting, lead to drier, warmer forest floors and reduced populations of many arthropod species. Burning of the forest floor may cause transitory depressions in populations of some arthropods. However, populations of others, such as the collembolans, may increase following burning. Practices which change the abundance or distribution of coarse woody debris on the forest floor can lead to shifts in arthropod species abundance.

Forest management practices that protect the diversity of the arthropods include those which leave the forest floor relatively undisturbed. Practices which leave some vegetation in place reduce extremes of temperature. Decomposing wood left on the forest floor is an important structural feature for both microarthropods and macroarthropods.

Influences on soil quality. Soil quality is an elusive but important concept in assigning relative values to different soil conditions. Important considerations include promotion of biological activity, mediation of water flow, and maintenance of soil organic matter. Forest-soil arthropods make contributions in these areas. They have a large, biological diversity and participate in complex food webs. Moreover, they participate in the dynamics of soil organic matter. Macroarthropods, like earthworms, may affect the movement of water through soils. Some forest-soil arthropods have been proposed as indicator organisms for overall soil health, a concept better developed in Europe than in North America.

For background information SEE ARTHROPODA; COLLEMBOLA; FOOD WEB; FOREST CANOPY; FOREST ECOSYSTEM; FOREST SOIL; FUNGI in the McGraw-Hill Encyclopedia of Science & Technology.

D. A. Crossley, Jr.

Bibliography. D. C. Coleman and D. A. Crossley, Jr., *Fundamentals of Soil Ecology*, 1996; D. L. Dindal, *Soil Biology Guide*, 1990; B. Hölldobler and E. O. Wilson, *The Ants*, 1990; K. Killham, *Soil Ecology*, 1994; T. R. Seastedt, The role of microarthropods in decomposition and mineralization processes, *Annu. Rev. Entomol.*, 29:25–46, 1988.

Fossil

The Ediacaran biota is an ancient community of marine organisms, which included the earliest animals. Ediacaran fossils, some as long as 3.2 ft (1 m) occur in sedimentary rocks deposited before the Cambrian Period, which began approximately 541 million years ago. A new discovery of these fossils from a mountain range near the town of Caborca in northwestern Sonora, Mexico, sheds considerable light on the origins of the Ediacaran biota. This find at a field site about 60 mi south of Mexico's border with Arizona is important both because Ediacaran soft body fossils have not previously been described from this part of the world and because the fossils are the oldest Ediacaran specimens known.

Fossil locality. The fossils occur in the Clemente formation, a widespread pre-Cambrian (late Proterozoic) stratum in this part of Sonora, which occurs low in the sequence of layered rocks that constitute the Caborca stratigraphic section. The most distinctive horizon, composed of yellowish-weathering dolomite, is the Clemente formation oolite (a sedimentary rock composed of small ooliths, spherical grains formed by mineral accretion around a tiny nucleus).

The Clemente formation had previously been prospected for ancient fossils without success. Scraps of fossillike objects were recovered from the formation between 1982 and 1990, but definitive evidence of ancient life proved elusive. In 1995, the fossils were discovered in a low outcrop invisible on air photos or from more than a few meters away on the ground, and in an area of desert pavement that would not ordinarily be prospected for fossils.

The fossils occur in fine sandstones and siltstones that lie stratigraphically below the Clemente formation oolite. These sediments cannot be directly dated by current methods of radiometric (radioisotopic) analysis. However, analysis of the ratio of heavy carbon to light carbon in the dolomite and limestone layers both above and below the fossil horizon, and analysis of trace fossils occurring in the Clemente formation above the oolite, indicate an age of approximately 600 million years. This great age (the oldest reported for a confirmed occurrence of the Ediacaran biota) is further attested to by the fact that several major unconformities (gaps in the stratigraphic sequence) are present between the horizon of the new discovery and the overlying, trilobite-bearing Cambrian strata of the Caborca stratigraphic sequence. One unconformity occurs at the upper boundary of the Clemente formation oolite: an erosive destruction of the strata that had been lain down is indicated by a conglomerate layer. The duration in geologic time of this unconformity or period of nondeposition is not known, but it could amount to millions of years.

Classification. Among fossils recovered were both body fossils (actual remains of creatures) and trace fossils (crawling and burrowing tracks). The body fossils are represented by at least three unnamed genera and four previously unknown species. Two types of trace fossils are known, including one new trace fossil species and one trace fossil species that is widespread in younger sedimentary rocks.

The body fossils consist of three discoidal fossils and a third fossil consisting of juxtaposed elongate parallel tubes. Discoidal fossils in the Ediacaran biota are conventionally referred to as medusoids to indicate their possible relationship to the medusae (jellyfish) of modern cnidarians. The degree to which the fossil forms are actually related to cindarians such as jellyfish, however, is a topic of considerable controversy. Some paleontologists consider the fossils to represent animals belonging to extant phyla such as Annelida (segmented worms) and Cnidaria (jellyfish and sea anemones); some have assigned them to an independent phylum of animals; others have assigned them to the protoctists (eukaryotes which are not plants, animals, or fungi); and others have argued that they represent an entirely distinct and now extinct kingdom of organisms. The unlikely suggestion has even been made that the Ediacaran organisms were lichens.

Anatomy and physiology. Because the Ediacaran creatures apparently lacked mouths and guts, it has been suggested that they created their own food with the help of photosynthetic or chemosymbiotic internal, microbial symbionts. Recent evidence supports this explanation, which has the advantage of not being dependent on the taxonomic placement of the organisms because many types of eukaryotic organisms can undergo these types of symbioses.

The largest of the discoidal Mexican fossils is preserved as a protuberance on the base of a sandstone bed. This protuberance occurs as a very high profile, domal central cone approximately 0.82 in. (2.1 cm) in diameter and 0.2 in. (7 mm) in height. This cone or dome is smooth but displays a distinct wrinkle where it meets the relatively flat outer ring. The flat outer ring bears fine, radially oriented tubular structures.

The organism represented by this fossil apparently rested on the sea floor in a shallow depression. The outer ring rested almost flat on the sediment surface, and was probably used as a light-gathering organ if the creature was indeed photosymbiotic. The radially oriented tubular structures are reminiscent of the tentacles of a cnidarian, but because they appear to have been sand-filled in life it is unlikely that they could have functioned as food-capturing organs in the manner of modern anemones.

Another specimen is a circular to elliptical fossil 0.4 m (1 cm) in greatest dimension with a preserved relief of 0.063 in. (1.6 mm). It is apparently the fossil of a nonmineralized conical structure that was the semirigid basal attachment disk of an early cnidarian. The conical object has been laterally compressed and flattened by compaction of enclosing sediment.

The third medusoid fossil is an incomplete specimen of a form consisting of radiating sand-filled tubes. These radiating, sand-filled cylinders are up to 0.75 in. (19 mm) in length. The surfaces of some of the tubes are slightly undulose. A second order of cylinders is located between the larger, first-order cylinders. Some of the first-order tubes terminate in bulbous expansions. Small pustules occur at the ends of the most distally swollen first-order cylinders.

The fourth body fossil is incomplete, and consists of an impression of six parallel and adjacent tubular cylinders, varying from 0.051 to 0.078 in. (1.3 to 2.0 mm) in width. The cylinders run approximately straight for 0.39 to 0.59 in. (1.0 to 1.5 cm), then turn in the same direction, making an arc of about 35° over a distance of about 0.2 in. (7 mm). In their last 0.059–0.79 in. (1.5–2.0 mm), each cylinder separates from the others, tapers distally, and comes to a point. These pointed ends of tubes form a serrated edge to the fossil.

Trace fossils. Two trace fossil types occur with the body fossils described above; in some cases both trace and body fossils occur on the same piece of rock. The first trace fossil species is represented by specimens preserved as elongate, low-relief teardrop-shaped convex bulges on the base of thin sandstone beds. In one example three individual traces are oriented with their long axes (the long axis of the teardrop shape) along the same line.

Also occurring in this biota are trace fossils belonging to the species *Palaeophycus tubularis.* This species is represented by eight specimens of straight to slightly sinuous horizontal cylindrical burrows up to 0.02–0.091 in. (0.6–2.3 mm) in diameter. The burrows are lined by a very thin clay layer. Three specimens are preserved as concave impressions on the base of sandstone beds, four specimens as convex protrusions on the base of sandstone beds, and one specimen as a convex protrusion at the top of a sandstone bed. The longest burrow, a sinuous trackway, is 3 in. (7 cm) in length.

Interpretation. Although, as mentioned above, the affinities of the body fossils are controversial, the trace fossils are clearly the products of the activities of ancient animals. These trace fossils are the most ancient known in the Mexican Proterozoic (late pre-Cambrian) sequence. The organisms represented by these trace fossils, and the associated body fossils, lived in a shallow-water marine habitat above fair-weather wave base, as indicated by both the stratigraphic proximity of the Clemente formation oolite and the presence of small-scale cross-bedding (inclined laminations in sandstones indicative of the motion of sediment by the force of flowing water) in the siltstones and fine sandstones of the fossiliferous interval.

This discovery suggests that the organisms of the Ediacaran biota began their evolutionary radiation in shallow water. These organisms developed a fairly high level of diversity at an early stage, and, judging from the presence of trace fossils, animals were definitely part of the biota. Unless older equally convincing specimens of the Ediacaran biota are found elsewhere, it can be argued from this discovery that the Ediacaran biota first evolved on or near the North American craton.

For background information SEE ANNELIDA; CAMBRIAN; COELENTERATA; EDIACARIAN FAUNA; PALEONTOLOGY; STRATIGRAPHY in the McGraw-Hill Encyclopedia of Science & Technology.

Mark A. S. McMenamin

Fuel cell

Fuel-cell technology is currently considered a promising answer to many of the environmental problems caused by conventional combustion technologies. By using electrochemical reactions rather than combustion, fuel cells can generate electric power from fuels more efficiently than heat engines such as gas turbines and internal combustion engines, while almost entirely avoiding the production of pollutants that arise from combustion. Although the most apparent linkages between fuel-cell technologies and the environment are their efficient and clean operation, the prospects for environmental benefits that derive from more subtle synergisms, namely, opportunities for cogeneration and compatibility with renewable sources of energy, are even more compelling. This article outlines the main roles that fuel cells can play in facilitating the environmentally sustainable use of energy for electric power generation and transportation.

Applications. Fuel cells were first employed in a practical application in the mid-1960s on-board the National Aeronautics and Space Administration's *Gemini* spacecraft, and have been approaching commercial application for stationary power production and transportation. Research and development efforts have led to several important advances in the performance, economy, and reliability of fuel-cell systems, leading to several demonstrations of the viability of the technology. A 200-kW phosphoric acid fuel-cell system for production of electric power from natural gas has been commercialized, and about 75 units have been purchased and installed by utilities and other users, yielding excellent performance results with respect to efficiency, emissions, and reliability. The first molten carbonate fuel-cell systems are under construction, and have the capacity to yield even higher efficiency than phosphoric acid fuel cells at lower commercial costs. For transportation applications, several fuel-cell vehicles operating on either hydrogen or methanol have been demonstrated, and developers including major auto manufacturers are proceeding toward commercial fuel-cell vehicles.

Technological advances. A variety of technological advances have enabled this rapid progress toward the widespread practical application of what was once an exotic technology limited to aerospace applications. For example, polymer electrolyte fuel cells have become much more attractive for transportation applications in the near term, owing in part to rapid progress toward operating them on fuels other than pure hydrogen. Developers are now operating fuel processors designed to accept various fuels, including those that are already widely available such as gasoline, and convert them into a hydrogen-rich gas suitable for consumption by the fuel cell. Carbon monoxide, a performance-reducing so-called catalyst poison produced when carbon-containing fuels are processed, can now be largely eliminated by selective oxidation over a suitable catalyst before reaching the fuel cell, with any residual carbon monoxide being, in effect, burned off the anode catalyst with a small amount of added oxygen. Increasing the power density of the fuel-cell system, which has also been central to the development effort, is being achieved in part through major advances in polymer electrolyte membrane technology, such as the development of much thinner membranes with lower internal resistance and correspondingly higher power. Concurrently, the need for expensive materials has been dramatically reduced through more efficient material use and replacement by inexpensive materials. For example, the requirement for platinum catalyst has been reduced by more than an order of magnitude from what was a prohibitively expensive quantity to what is now an amount comparable to that found in the catalytic converters of automobiles. Other engineering and design improvements, such as the replacement of relatively expensive milled graphite with metal and plastic components, have helped to further reduce the cost of materials and manufacturing.

The next ten or fifteen years is generally seen as the time period over which fuel cells will become able to start contributing substantially to worldwide energy conversion requirements. In so doing, fuel cells offer the opportunity to shift toward a more sustainable energy path in four main ways: (1) by generating power and providing transportation at greater efficiency than is possible with conventional combustion technologies; (2) by drastically reducing emissions of energy-related air pollutants; (3) by greatly expanding the capacity for energy savings through cogeneration of heat and power; and (4) by facilitating the transition to renewable sources of energy.

Efficiency. Unlike conventional combustion technologies, which generate electric power via the intermediate production of heat, fuel cells convert the chemical energy of fuels directly into electricity. The thermodynamic efficiency of heat engines such as gas turbines and internal combustion engines is subject to the Carnot limit, which reflects the second-law constraint that a high-entropy form of energy such as heat cannot be entirely converted into a low-entropy form such as mechanical or electrical energy. Because the Carnot limit is less severe at higher operating temperatures, the most efficient power plants have relied on large-scale designs that reduce heat-loss inefficiencies and economically exploit advanced materials that can reliably endure high temperatures in industrial settings. Because fuel cells are not Carnot limited, they can attain higher efficiencies than combustion-based power plants, even at small scales. **Table 1** compares the thermodynamic efficiencies of fuel-cell and combustion power plants, at small and large scales, using natural gas and coal.

Table 1. Efficiencies of fuel-cell and conventional power systems

Stationary power production		
Fuel	Efficiency of fuel-cell power plant, %	Efficiency of combustion power plant, %
Natural gas (~2 MWs)	50*	35[†]
Natural gas (~hundreds of megawatts)[‡]	65	55[§]
Gasified coal[‡]	52	44

Vehicles		
Vehicle power plant	Fuel	Relative efficiency[¶]
Internal combustion engine	Gasoline	1.00
Fuel cell	Gasoline	1.85
Fuel cell	Methanol	2.37
Fuel cell	Hydrogen	2.76

* Nearly commercial system.
[†] Commercial diesel-engine system.
[‡] Projections.
[§] Gas-turbine system employing steam-bottoming cycle.
[¶] Taking into account inefficiencies associated with on-board processing.

In vehicles, the efficiency benefits of fuel-cell systems are much greater still. A fuel cell has high efficiency at its peak power and even higher efficiency at lower power, whereas internal combustion engines achieve their maximum efficiency near their peak power and are substantially less efficient when operating at lower power. Because vehicles operate at low power most of the time (for example, during idling or cruising at a constant speed), a fuel-cell vehicle achieves an efficiency improvement that is much greater than is suggested merely by its higher efficiency at peak power. Table 1 compares the efficiency of an internal combustion engine vehicle operating on gasoline to that of fuel-cell vehicles having comparable performance operating on various fuels. The fuel-cell vehicle's superior fuel economy offers the possibility of dramatically reducing the amount of energy used for transportation applications, which in the United States accounts for roughly one-third of primary energy consumption.

Emissions of air pollutants. Fossil-fuel combustion is overwhelmingly the major contributor to air pollution through the release of compounds formed from fuel-bound contaminants (such as sulfur oxides), the generation of partial combustion products (such as carbon monoxide, unburned hydrocarbons, particulates, and volatile organic compounds), and the production of other unwanted combustion products (such as nitrogen oxides). However, fuel-cell systems produce absolutely no local pollutants (emissions that cause local environmental problems such as smog and acid rain, as opposed to greenhouse gases such as carbon dioxide) when fueled with hydrogen, and when fueled with other fuels produce only the very minor emissions that result from the processing of the fuel prior to its consumption in the fuel cell. **Table 2** shows the emissions resulting from fuel-cell power plants and fuel-cell buses, with comparisons to the federal standards for emissions levels for natural gas–fired power plants and heavy vehicles, respectively. In both cases, the measured emissions are markedly lower (in some cases by several orders of magnitude) than the standards, and significantly lower than can be achieved by using conventional combustion technologies. Because of the rapidly growing demand for energy and transportation services worldwide, an inherently clean technology such as fuel cells will almost certainly be necessary to prevent pollution levels from rising dramatically.

Congeneration. Although fuel-cell power plants generate electric power more efficiently than combustion-based power plants, a much greater potential for energy savings lies in the ability to use fuel-cell technologies to greatly expand the cogeneration of electric power and heat beyond what is possible with conventional combustion technologies. The combustion technologies currently used for electric power generation produce large amounts of heat as a by-product, but typically this heat is simply discarded since no local demand for heat exists at today's power plants, which are often remotely located and built very large (several hundred megawatts) in order to improve efficiency and take advantage of scale economies. At the same time, more than 20% of the United States' primary energy consumption goes toward water heating and space conditioning in residential and commercial buildings, with still more energy consumed in industrial settings for process heat and steam generation, applications for which the current practices of burn-

Table 2. Emissions of regulated pollutants

Fuel-cell power plants	Particulates, lb/MBtu*	Nitrogen oxides (NO_x), lb/MBtu*	Fuel-cell buses	Carbon monoxide (CO), g/hp-h[†]	Nitrogen oxides (NO_x), g/hp-h[†]
Emissions limits	0.030	0.300	Federal vehicle standards	15.0	4.0
200-kW plant	0	0.020	Hydrogen-fueled bus	0.0	0.0
2-MW plant	0	0.00006	Methanol-fueled bus	0.2	<0.002

* 1 lb/MBtu = 4.3×10^{-10} kg/J.
[†] 1 g/hp-h = 3.7×10^{-10} kg/J.

ing fossil fuels and consuming electricity would be unnecessary if there were a convenient supply of heat at these sites. Because fuel cells are highly efficient even at very small scales, and because they are clean, quiet sources of power that can be sited almost anywhere, they are ideally suited for on-site cogeneration. Because the varied fuel-cell technologies generate heat over a wide range of operating temperatures, fuel-cell systems sited in buildings and factories are able to simultaneously generate electric power for the utility grid as well as satisfy local water heating, space heating, space cooling (by using heat-driven absorption coolers), and industrial heat needs, enabling dramatic reductions in the burning of fossil fuels and the consumption of electricity for these applications.

Compatibility with renewable energy. Fuel cells can facilitate the transition to renewable sources of energy by making economical use of transportation fuels produced from renewable resources, and by providing a promising technology for production of power from renewable biomass resources.

Transportation fuels. The much higher fuel economy of fuel-cell vehicles not only reduces energy consumption but also makes possible the cost-effective use of fuels that would be prohibitively expensive for use in internal combustion engine vehicles. Although the first fuel-cell vehicles to be commercialized may be fueled with readily available fossil fuels (such as gasoline or diesel), over time the necessary infrastructure can evolve for delivering more desirable fuels such as hydrogen or a hydrogen carrier (for example, methanol) which is easily and efficiently processed on board the vehicle.

Hydrogen can be produced electrolytically from renewable power sources such as wind or photovoltaic electricity. Even though current estimates for the future cost of producing hydrogen electrolytically from renewable sources are high in terms of dollars per unit of energy, the cost per mile of driving is still comparable to the cost of gasoline per mile for a combustion vehicle because of the much higher fuel-cell vehicle efficiency. Thus, the fuel would be no less affordable, yet would generate no emissions of local pollutants or carbon dioxide (CO_2).

Alternatively, hydrogen can be produced at lower cost from natural gas at large scales with commercial technology and from coal and biomass via processes that begin with thermochemical gasification. Biomass is a renewable energy source, and the net life-cycle emissions of carbon dioxide from

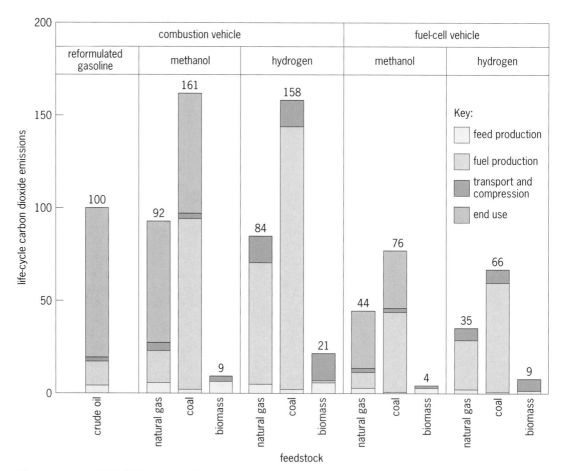

Life-cycle carbon dioxide (CO_2) emissions for fuel-cell vehicles compared to emissions from comparable internal combustion engine vehicles fueled with methanol and hydrogen derived from natural gas, coal, and biomass. For reference, the carbon dioxide emissions for an internal combustion engine vehicle fueled with reformulated gasoline are also shown. Emissions are given relative to the total emission from this engine, which is set equal to 100.

biomass-derived fuels are extremely low because sustainably harvested biomass releases no more carbon dioxide as it is converted to energy than that absorbed from the atmosphere during its growth. The only emissions come from the relatively minor contribution from fossil fuels used for biomass production (planting, fertilizing, harvesting, transporting, processing, and so forth). For hydrogen derived from natural gas, coal, and biomass, respectively, the life-cycle carbon dioxide emissions per mile for fuel-cell vehicles would be 35, 66, and 9% as large as for gasoline combustion vehicles of similar performance. The corresponding life-cycle emissions for methanol derived from the same three feedstocks would be 44, 76, and 4% as large as for gasoline internal combustion vehicles respectively.

Thus, although fuel-cell vehicles make possible significant reductions in carbon dioxide emissions even when operated on fuels derived from fossil sources, truly drastic reductions occur when fuels are derived from renewable sources such as biomass. In the **illustration**, carbon dioxide emissions of fuel-cell vehicles are compared with those of combustion vehicles, and the emissions are broken down into those generated in extraction or production of the energy feedstock, in production of the fuel from the feedstock, in transport and additional processing, and in final use of the fuel.

The potential biomass resource is vast. Fast-growing trees and perennial grasses can be grown for energy on dedicated farms in a manner inherently more environmentally friendly than growing annual food crops. Even though the overall photosynthetic efficiency is low (for example, what is considered a good yield of 15 dry metric tons/per hectare/per year or 7 tons per acre per year corresponds to an annual average photosynthetic efficiency of just 0.5%), the overall process of growing biomass, converting it to hydrogen, and using the hydrogen to power fuel-cell vehicles is relatively energy efficient. This fuel cycle will support nearly seven times as many vehicle-kilometers of travel per hectare than the present commercial process of making ethanol from grain for use in combustion vehicles. In fact, to run the entire expected worldwide fleet of 10^9 cars in the year 2020 on biomass-derived hydrogen would require only 600,000–700,000 km^2 (230,000–270,000 mi^2) of land, approximately the area of Texas.

Stationary power production. In the case of stationary power production as well, fuel cells are potentially well suited for use with biomass resources. Biomass gasifiers, which thermochemically convert solid biomass feedstocks into a gaseous fuel, can be integrated with fuel-cell systems to efficiently produce electric power. This coupling of technologies borrows heavily from the previous coupling of the coal-gasifier and fuel-cell technologies, and from the coupling of the biomass-gasifier and gas-turbine technologies, which are approaching commercial readiness and are currently regarded as the leading candidates for the generation of electricity from biomass.

There are also engineering advantages. Biomass is more reactive than coal, making biomass gasification potentially less costly and more easily integrated with the downstream fuel-cell system. The much lower levels of sulfur and heavy metals make the cleanup of biomass-derived fuel gas simpler and less costly. Compared to biomass-gasifier and gas-turbine systems, the main advantage of biomass-gasifier and fuel-cell systems is that they would be at least as efficient even at scales as small as a few megawatts. This small size enables fuel requirements to be met entirely by nearby sources of biomass, keeping the cost of transporting biomass low and significantly expanding the biomass resources that can be economically used for generating power.

For background information SEE AIR-POLLUTION CONTROL; BIOMASS; COGENERATION; ENERGY SOURCES; FUEL CELL; INTERNAL COMBUSTION ENGINE in the McGraw-Hill Encyclopedia of Science & Technology.

Sivan Kartha

Bibliography. L. Blomen and M. Muwerga, *Fuel Cell Systems*, 1993; S. Kartha and P. Grimes, Fuel cells: Energy conversion for the next century, *Phys. Today*, 47(11):54–61, November 1994; S. S. Penner et al., Commercialization of fuel cells, *Energy: Int. J.*, 20:331–470, 1995; R. H. Williams et al., Methanol and hydrogen from biomass for transportation, *Ener. Sustain. Dev.*, 1(5):18–32, 1995.

Functional analysis diagrams

Functional analysis is performed in systems engineering, software systems engineering, and business process reengineering as part of the design process. The design process typically involves the steps of requirements definition and analysis, functional analysis, physical or resource definition, and operational analysis. This last step involves the union of functions with resources to determine if the requirements are met.

Functional analysis addresses the activities that the system, software, or organization must perform to achieve its desired outputs, that is, the transformations necessary to turn the available inputs into the desired outputs. Additional considerations include the flow of data or items between functions, the processing instructions that are available to guide the transformations, and the control logic that dictates the activation and termination of functions. Functional analysis diagrams have been developed to capture some or all of these concepts.

The concept of examining the logical architecture via functional analysis prior to and concurrent with the development of the physical architecture has become a well-accepted principle in the related fields of systems engineering, software systems engineering, and business process reengineering. Experience has shown that rushing to a physical representation of the system or organization will likely produce a set of resources that cannot produce some of the outputs desired by the system.

Elements of diagrams. There are four elements to be addressed by any specific functional analysis approach. First, the functions are represented as a hierarchical decomposition, in which there is a top-level function for the system or organization. This top-level function is partitioned into a set of subfunctions that use the same inputs and produce the same outputs as the function itself. Each of these subfunctions can be partitioned further, with the decomposition process continuing as often as it is useful.

Second, functional analysis diagrams can represent the flow of data or items among the functions within any portion of the functional decomposition. As the first and subsequent functional decompositions are examined, it is common for one function to produce outputs that are not useful outside the boundaries of the system or organizations. These outputs are required by other functions in order to produce the needed and expected external outputs.

Processing instructions are a third element that appears in some functional analysis diagrams. These instructions contain the needed information for the functions to transform the inputs to the outputs.

The fourth element is the control flow that sequences the termination and activation of the functions so that the process is both efficient and effective. This information answers the following sample questions: (1) Can these functions work serially or must they be processed concurrently? (2) Are these functions activated once or a series of times? (3) What circumstances dictate that one function rather than another be activated? Included in this control flow are the completion criteria that dictate that a function has finished, and the activation logic that dictates when a function begins.

Functional analysis approaches. The major functional analysis approaches are structured analysis, dataflow and control flow diagrams, function-flow block diagrams and N^2 charts, and behavior diagrams.

Structured analysis. The structured analysis and design technique (SADT) was developed from 1969 to 1973. SADT is a graphical modeling language and a comprehensive methodology for developing models. In the 1970s, the U.S. Air Force incorporated SADT into its Integrated Computer-Aided Manufacturing (ICAM) program as the definition language for manufacturing systems, yielding the phrase IDEF (for ICAM definition). $IDEF_0$ focuses on functional models of a system; $IDEF_1$ is a model of the information needed to support the functions of a system; while $IDEF_{1x}$ is a semantic data model using relational theory and an entity-relationship modeling technique. Many other IDEF representations exist.

Within $IDEF_0$, a function or activity is represented by a box, described by a verb-noun phrase and numbered to provide context within the model. Inputs enter from the left of the box, controls enter from the top, mechanisms (or resources) enter from the bottom, and outputs leave from the right. A flow of material or data is represented by an arrow or arc that is labeled by a noun phrase. The label is connected to the arrow by an attached line, unless the association is obvious.

An $IDEF_0$ model has a purpose and viewpoint and comprises two or more diagrams. The A0 page is the context diagram and establishes the boundaries of the system or organization being modeled. The A0 page (**Fig. 1**) defines the decomposition of the A0, or top-level, function by two to six functions for display reasons. The decomposition of a parent function (A0 in this case) preserves the inputs, controls, outputs, and mechanisms of the parent. There can be no more, no less, and no differences. Every function must have a control. An input is optional. Boxes are usually placed diagonally. Arcs are, like functions, decomposable. Feedback is modeled by having an output from a higher-numbered function on a page flow upstream as a control, input, or mechanism to a lower-numbered function.

$IDEF_0$ models can also address the interaction of the system with other systems. This interaction is modeled on the A1 page, which takes the A0 function and places it in context with other systems or organizations. This representation is often critical to understanding the relationship of the system being addressed to its outside world and establishing the origination of inputs and controls and the destinations of outputs. In order to emphasize the readability and understandability of the models, the $IDEF_0$ community has been very strict in establishing criteria for constructing $IDEF_0$ diagrams correctly. In fact, the National Institute of Standards and Technology has released a standard for $IDEF_0$ and $IDEF_1$.

Dataflow diagrams. Dataflow diagrams (DFDs), dating from the early 1970s, are one of the original diagramming techniques used in the software and information systems communities. The basic constructs of DFDs are the function or activity, dataflow, store, and terminator.

The circle is the most standard representation for a function, defined as a verb phase. Arcs again represent the flow of data or information between functions, or to and from stores. Double-headed arcs are allowed; these represent dialog between two functions, for example, a query and a response. The labels for an arc are noun phrases and are placed near each arrow. Branches and joins are allowed and are depicted as forks.

A new concept is introduced, the store or buffer, a set of data packets at rest. Again, there are several legal representations of a store. A store can be represented as a noun phrase between two diagonal lines that are not connected, connected at one end by a straight line, or connected at both ends by semicircles.

The final syntactical elements of dataflow diagrams are terminators, or other systems. In fact, context diagrams that show the interaction between the external systems, or two terminators, and the system

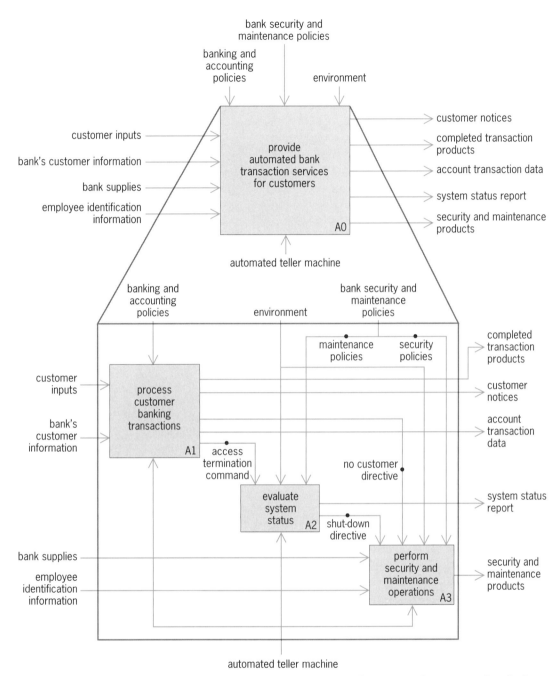

Fig. 1. Functional decomposition in an IDEF$_0$ model, showing the preservation of inputs, controls, outputs, and mechanisms.

being designed or analyzed are standard in DFDs (**Fig. 2**). Terminators are shown in boxes.

A DFD model is typically a set of "leveled" DFD diagrams showing the flow of data between functions as well as the hierarchical decomposition of the functions. No standard has developed within DFDs, as many practitioners have modified the basic DFDs as described above to suit their needs. For example, continuous dataflow is often represented as a double arrow. The flow of events is represented by a broken line, and the continuous flow of events by a broken line with a double arrow.

Control-flow diagrams. Sometimes control-flow diagrams (CFDs) are used in conjunction with DFDs, either as separate but parallel diagrams or superim-posed on DFDs. Control flow is information that is transmitted between functions or between a function and the outside and that determines how the functional processes must operate under specific changes in the operating modes. These operating modes may dictate that certain functions are present or absent, or change the ways in which these functions perform. This control flow is typically shown as broken arcs; therefore, control flow is not used in conjunction with the expanded distinctions of event flows in DFDs because control and event flows would both be represented by broken lines.

Function-flow block diagrams. This was the original approach to functional decomposition in systems engineering, dating to the late 1960s and early 1970s.

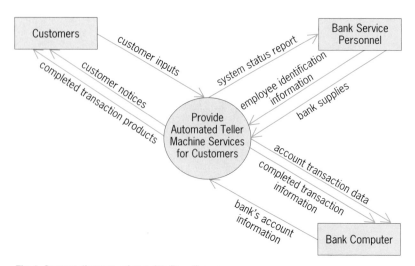

Fig. 2. Context diagram using a dataflow diagram.

Function-flow block diagrams (FFBDs) show the functions at a given level, and a control structure that dictates the order in which the functions can be executed. Functions are typically shown in boxes with their associated number.

In the original, or basic, FFBD syntax (**Fig. 3**), four types of control structure were allowed: series, concurrent, selection, and multiple-exit function. A set of functions defined in a series control structure must be executed in that order; the second cannot begin until the first is finished, and so forth. Control passes from left to right along the arc shown from

outside (depicted by a function in a box with broken top and bottom lines) and activates the first function. When the first function has been completed (that is, its exit criterion has been satisfied), control passes out of the right face of the function and into the second function.

The concurrent structure allows multiple functions to be working in parallel; thus it is sometimes called parallel. However, the concurrent structure should not be confused with the concepts of parallel in electric circuits or redundant systems. Essentially, control is activated on all lines exiting the first AND node, and control cannot be closed at the second AND node until all functions on each control line are completed.

A selection structure and a multiple-exit function achieve essentially the same purpose: the possibility of activating one of several functions. The multiple-exit function achieves this end by having a function placed at the fork to make the selection process explicit. When the selection function has been completed, one of the several emanating control lines is activated. Once all of the functions on the activated line have finished execution, control passes through the closing OR node. The exit criteria for the multiple-exit function are labeled. (An exit criterion exists for every single-exit function but does not appear on the diagram.)

For the selection construct, which is an exclusive or, the first OR node passes control to one of the exiting control lines in a manner that is unspecified on

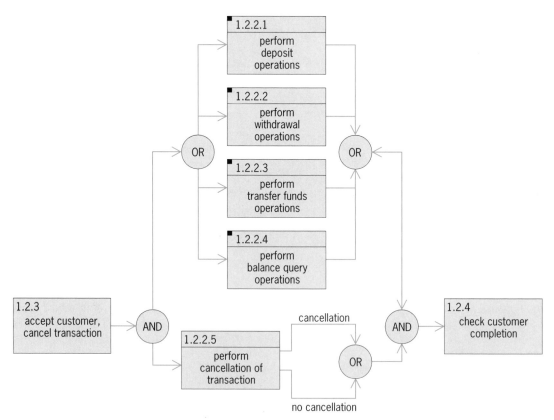

Fig. 3. Concurrent, selection, and multiple-exit functions in a function-flow block diagram.

the diagram. This control line stays active until the set of functions on that control line is completed; control then passes through the second OR node.

Additional control structures have been added to FFBDs to form enhanced FFBDs: iteration, looping, and replication. Iteration involves the repetition of a set of functions as often as needed to satisfy some domain set; this domain set must be defined based upon a number or an interval. Looping provides a similar control structure, but in this case it is possible to exit the loop if the appropriate criterion has been satisfied. Finally, replication involves repeating the same function concurrently by using identical resources.

N² charts. N² charts were created with FFBDs to depict the data or items that are the inputs and outputs of the functions in the functional architecture. The charts are called N² because for a set of N functions the chart contains N^2 boxes to show the flow of items within (or internal to) the N functions. The N functions are placed along the diagonal. Items flowing from function i to function j are defined in the i,j box. Additional boxes along the top and down the right are needed to add the flow of external items into and out of the set of N functions, respectively. N² charts provide information comparable to $IDEF_0$ and DFDs.

Behavior diagrams. Behavior diagrams originated as part of the Distributed Computer Design System of the Department of Defense. System behavior is described through a progressive hierarchical decomposition of a time sequence of functions and their inputs and outputs. Functions are represented as verb phrases inside boxes. A control structure is represented by lines that flow vertically, from top to bottom, through the boxes. The control lines have only one entry path into a function but may have multiple exit control paths. Input and output items are represented in boxes with rounded corners; their entry to and exit from functions is depicted by arcs that enter and exit the boxes, respectively. Specific control structures for sequence, selection, iteration, looping, concurrency, and replication have been defined within behavior diagrams, just as they have been in FFBDs.

For background information SEE BLOCK DIAGRAM; DATAFLOW SYSTEMS; SOFTWARE ENGINEERING; SYSTEMS ENGINEERING in the McGraw-Hill Encyclopedia of Science & Technology.

Dennis M. Buede

Bibliography. D. J. Hatley and I. A. Pirbhai, *Strategies for Real-Time System Specification,* 1988; R. H. Thayer and M. Dorfman (eds.), *System and Software Requirements Engineering,* 1990; E. Yourdon, *Modern Structured Analysis,* 1989.

Fungal genetics

The fungi have been exploited for millennia in a range of traditional biotechnological processes. In the last decade their potential for commercial exploitation has been enhanced by the emerging recombinant deoxyribonucleic acid (DNA) technology of the biopharmaceutical industries. The ability to precisely and stably alter the genetic makeup of a fungal species through the rational manipulation of endogenous genetic information, or by the introduction of foreign (heterologous) genetic information, has underpinned these developments. At the academic level, much effort has been devoted to the genetic engineering of one particular fungal species, the yeast *Saccharomyces cerevisiae.* Much of what has been learned with this unicellular fungus is now being transferred to other yeast species and to the more complex fungal species: the filamentous fungi.

The successful genetic engineering of any fungal species requires understanding the organization and expression of the endogenous genes of the chosen species and efficiently introducing heterologous or modified endogenous genes without altering the overall genetic integrity of the fungal cell. The number of fungal species that have been subjected to the new sophisticated genetic engineering methods is relatively limited.

Gene transfer systems. The genetic transformation of a fungal cell with exogenous DNA requires a means of transferring the DNA across its cell wall and membrane and into the nucleus. The strategy originally developed involved the enzymatic removal of the cell wall, to generate osmotically sensitive spheroplasts which could then take up the DNA, and regeneration of the cell wall to create a viable recombinant cell. This procedure has been replaced by electroporation-based methods which use a pulse of high voltage to permeablize the fungal cell to exogenous DNA. Electroporation not only facilitates the genetic manipulation of a number of fungal species recalcitrant to spheroplast-based methods but also improves the efficiency of transformation for the more manipulatable fungi, particularly the yeasts.

Once the DNA has entered the fungal nucleus it either becomes integrated into the host chromosome (integrative transformation) or remains as an autonomously replicating molecule. Its fate can be determined by the nature of the DNA molecule introduced. For example, a linear DNA molecule which is unable to self-replicate will become integrated into one of the host cell's chromosomes. In yeasts, this integration is largely via the process of homologous recombination, the DNA integrating at a chromosomal location containing an identical DNA sequence. Thus, the DNA can be precisely targeted to a specific chromosomal locus. This degree of control has been exploited in the rational genetic engineering of the yeast genome; for example, it allows the replacement of a gene in the chromosome with a modified form of the same gene without disturbing the surrounding DNA sequences.

For *S. cerevisiae* there exists an extensive array of autonomously replicating circular DNA molecules that can be used to introduce target DNA into living cells in single or multiple copies. These vectors are based on bacterial plasmids and carry a gene that can be used to select for cells carrying the plasmid (for example, a gene encoding resistance to an antibiotic) and a sequence that confers the property of autonomous replication (an origin of replication). A centromere sequence, which enhances the mitotic stability of the plasmid, can also be added, but it reduces the number of copies of the plasmid in the cell from 20–50 to 1–2 copies per cell. Autonomously replicating linear plasmids have also been developed and are maintained as linear molecules by the inclusion of telomere sequences (the natural ends of chromosomes). Perhaps the most widely exploited linear plasmids are the yeast artificial chromosomes which are used to isolate large fragments (up to 1 million base pairs) of human DNA and are a pivotal tool for human genome researchers. Vector systems, analogous to those developed for *S. cerevisiae,* have also been developed for at least a dozen other fungal species.

Expressing foreign genes. One of the most economically significant applications of recombinant DNA technology to fungi has been to engineer cells to efficiently express a gene that encodes a protein of potential value as a biopharmaceutical. Fungal cells (particularly yeasts) offer real benefits over what a bacterial cell expression system can provide, because they are able to express and process complex mammalian proteins authentically.

Saccharomyces cerevisiae, with an ease of genetic manipulation combined with favorable fermentation properties, has been instrumental in the development of fungal cell expression systems. These expression systems are designed to produce large quantities of a recoverable recombinant protein that is both authentic and cheap to produce. There are two basic strategies for achieving this: expressing the protein in the cytoplasm of the fungal cell (intracellular expression) and directing the protein into the secretory pathway and out of the cell (extracellular expression). The latter route is necessary if the recombinant protein requires posttranslational modifications such as the addition of sugars (glycosylation) or the formation of disulfide bridges between cysteine residues.

Genes chosen for expression in *S. cerevisiae* must be complementary DNAs (cDNAs) because the yeast RNA-splicing machinery is unable to excise introns from messenger ribonucleic acids (mRNAs) other than its own. The cDNA is linked to a DNA promoter sequence that is efficiently recognized by the host RNA polymerase to generate high-level transcription of the target cDNA into mRNA. Ideally, the ability of the RNA polymerase to bind the promoter sequence should be controllable by simple changes to the growth conditions of the yeast cell. The most widely utilized promoter in this context is the *GAL1,10* promoter which is efficiently recognized by RNA polymerase in cells grown on galactose as the sole carbon source, yet is not recognized when glucose is used as the carbon source. Regulated promoters are of value when expressing a protein that is toxic to yeast. A DNA sequence necessary for terminating transcription is also introduced after the target cDNA to ensure the correctly sized mRNA is made.

Optimizing the transcription of the target cDNA as well as the process of translation of the mRNA into protein must be accomplished. This action involves both the noncoding mRNA sequences, which flank the region decoded by the cell's ribosomes, and the cellular parameters which control the turnover of the mRNA.

In order to target a protein to the fungal secretory pathway, it must contain a short (15–30 amino acids) hydrophobic polypeptide sequence at its amino terminus. This signal sequence is removed as the protein is translocated across the endoplasmic reticulum (ER) membrane and into the secretory pathway. Once the protein crosses the endoplasmic reticulum membrane it is subjected to a range of posttranslational modifications and folds into its final conformation. Finally the protein is transferred from the membrane-bound secretory pathway, via secretory vesicles, to the outer surface of the cell. More than 50 different recombinant proteins have been successfully secreted from *S. cerevisiae.* However, *S. cerevisiae* is not ideal for expressing secretory proteins: it is unable to carry out some of the more complex posttranslational modifications, and the efficiency of extracellular expression is low because the secretory pathway is readily saturated, leading to an accumulation of aberrantly folded and modified recombinant proteins in the endoplasmic reticulum.

Expressing foreign genes in other fungi. Variable levels of intracellular expression, coupled with the low efficiency of extracellular secretion of recombinant proteins from *S. cerevisiae,* have led to efforts to identify and exploit other fungal species as potential expression hosts. Two groups of fungi have proven fruitful in this context: the methylotrophic yeasts and the filamentous fungi.

Methylotrophic yeasts, which utilize methanol as their sole carbon source, have been extensively studied as expression hosts; particularly successful in this context are *Pichia pastoris* and *Hansenula polymorpha.* Although their plasmid-based expression systems are rudimentary, they are cheap to ferment in bulk and can be grown to very high cell densities. High levels of expression of mammalian proteins can be achieved in methylotrophic yeasts by using the methanol-induced alcohol oxidase (*MOX*) gene promoter, which is extremely effective and tightly regulated. Indeed, *P. pastoris* is rapidly becoming the fungal cell of choice for the intracellular expression of recombinant proteins.

A number of other fungal species have been used to express mammalian proteins although with limited success, including the yeasts *Kluyveromyces lactis* and *Yarrowia lipolytica.* The filamentous fungi have also been used and have the advantage over the yeast expression systems in that they naturally secrete a wide range of endogenous proteins with high efficiency. Gene transfer and expression systems exist for several species of filamentous fungi, the most advanced being for *Aspergillus niger.* Although a number of different proteins have been successfully expressed in filamentous fungi, few have been generated in yields sufficient for commercial exploitation. Therefore, it is unlikely that this group of fungi will replace the yeast-based expression systems.

For background information SEE FUNGAL GENETICS; FUNGI; GENE ACTION; GENETIC ENGINEERING; RECOMBINATION (GENETICS); SACCHAROMYCETALES; YEAST in the McGraw-Hill Encyclopedia of Science & Technology.

Mick F. Tuite

Bibliography. R. G. Buckholz and M. A. G. Gleeson, Yeast systems for the commercial production of heterologous proteins, *BioTechnology,* 9:1067–1072, 1991; J. R. Kinghorn and G. Turner, *Applied Molecular Genetics of Filamentous Fungi,* 1992; M. A. Romanos, C. A. Scorer, and J. J. Clare, Foreign gene expression in yeast: A review, *Yeast,* 8:423–488, 1992; A. E. Wheals, A. H. Rose, and J. S. Harrison (eds.), *The Yeasts,* vol. 6, 1995.

Fungi

Recent developments concerning fungi involve advances in biotechnology and in interpretation of the phylogeny and evolution of these organisms.

Fungal Biotechnology

Fungi are ubiquitous organisms which can be broadly divided on the basis of their morphologies into filamentous forms, unicellular budding yeasts, and mushrooms or toadstools. These organisms have been used since the dawn of civilization as a food, as a source of alcohol, and to produce the carbon dioxide required to leaven bread. During the twentieth century, fungi have been used more widely in industry, first to produce biochemicals such as citric acid and then antibiotics, notably penicillin. Previously, the use of fungi in biotechnology was relatively limited. However, there is increasing interest in the application of these diverse organisms in all areas of biotechnology, a trend which will continue as a result of the increasing sophistication of fungal molecular genetics.

Medical biotechnology. The preeminent medical use of fungi is as a source of antibiotics. Penicillin and its semisynthetic derivatives continue to be the most important fungal-produced antibiotics. Others include the cephalosporins, and of lesser importance the fusidanes and the antifungal griseofulvin. Antitumour and antiviral agents are also produced by fungi. For example, lentinan, a polysaccharide produced by the edible mushroom *Lentinus edodes,* has been used in the treatment of stomach cancers and also as an antihistamine.

Of increasing interest and importance are fungal products with immunomodulatory properties. In particular, the advent of the cyclosporins has led to a dramatic improvement in the success rate of human organ transplants. Produced by *Trichoderma polysporum* and *Cylindrocarpon lucidum,* these compounds are nonpolar polypeptides which also show antifungal activity. Cyclosporin A has a powerful immunosuppressant effect and appears to act by preventing both antibody and cell-mediated immunity. Gliotoxin, formerly regarded only as a mycotoxin, is also being evaluated for its immuno-regulatory properties. Ergot alkaloids have been used since antiquity to hasten uterine contraction during childbirth; produced by the black sclerotia of the fungus *Claviceps pupurea* (which overwinters on the head of rye) these fungal products cause the often fatal condition called Saint Anthony's fire, which is associated with the appearance of swollen blisters all over the body and even the loss of whole limbs.

Fungi are increasingly being used to transform biochemicals into new medically useful products. The most successful example is the use of *Rhizopus arrhizus* to produce and modify the structure of steroids. Another example of such transformation is the use of *Aspergillus sclerotiorum* to improve the effectiveness of lucanthone in the treatment of schistosomiasis (bilharziasis), a tropical disease caused by the blood fluke *Schistosoma.* The ability of fungi to transform chemicals in this manner provides an efficient and economical means of obtaining specific compounds which would be impossible or too expensive to produce either chemically or by extraction from animal organs.

Although most fungal products used in medicine are produced by filamentous fungi, mushrooms are increasingly being evaluated as sources of medically important compounds. Species of *Grifola,* for example, produce antitumor glucans which can be used to reduce blood pressure and cholesterol levels. SEE NUTRACEUTICALS.

Industrial biotechnology. Fungi are currently being evaluated for use in the paper, fuel, and mineral processing industries. For example, white rot fungi, such as *L. edodes,* can be used to reduce the lignin content of wood pulp (biopulping), a process which is potentially less energy demanding and polluting than the standard physicochemical approaches. Complete delignification is not essential, as the production of even partially delignified pulp reduces energy costs.

The ability of fungi to degrade lignin is also used in the biosolubilization of coal. The wood-rotting fungus *Polyporus versicolor,* for example, releases black, oily droplets when grown on low-grade coal. Under optimal conditions, complete coal solubilization takes place in about a week, and results in a solution containing a complex mixture of primary solutes which can be burnt, gasified, or used as a chemical feedstock. Although limited fungal solubilization of hard coals has been achieved, this biotechnology seems best suited to low-grade coals such as lignite.

The lignin contained in plant material, such as elephant grass, can also be gasified by filamentous fungi, such as *Penicillium* species, to a mixture of methane and hydrocarbons, emphasizing the possibility of a wide range of renewable lignin-rich plants as energy sources. Finally, yeasts such as *Candida lipolytica* have also been used to remove paraffins and waxes from petroleum to produce low-boiling-point kerosenes for jet aviation fuel. SEE COAL, LOW-RANK.

A number of biotechnological companies are currently evaluating the application of fungi in various aspects of mineral biotechnology. Fungi can be used to release aluminum from aluminosilicates, and to liberate metals (for example, zinc and copper) from metal filter dusts and fly ash. The mycelium of filamentous fungi such as *A. ochraceus* also takes up uranium when incubated with uranium ores, suggesting a potential use in uranium bioleaching. Finally, in the textile industry filamentous fungi can be used to make fabrics. Such mycelial fabrics also appear to promote wound healing, and may eventually find a medical use as wound dressings.

Environmental biotechnology. Emphasis is being placed on the use of white rot lignin-decomposing fungi in the process of bioremediation, that is, the removal of pollutants from waste waters and soils. The utility of fungi, most notably *Phanerochaete chrysosporium*, for this purpose is based on their ability to produce nonspecific ligninases, which break down not only lignins but also recalcitrant pollutants such as trinitrotoluene (TNT), polycyclic aromatic hydrocarbons, and various chlorinated lignin compounds produced by the paper industry.

Most filamentous fungi can also adsorb metal ions and insoluble particulates from solution. Thus, they can be used to remove precious metals from wastes, and pollutants from effluents. One commercial product is used, with greater efficiency than ion exchange resins, to remove uranium from solution. In many cases, the metal can be removed and the biomass reused for a variable number of times; waste fungal mycelium from penicillin or citric acid manufacture can be used in this way, thereby utilizing a material that would otherwise have to be disposed of at some cost.

Agricultural biotechnology. In the past, the impact of fungi on agriculture largely focused on their role as plant pathogens. However, fungi are increasingly being evaluated as biocontrol agents for controlling pathogenic fungi, nematodes, insects, and weeds. Some commercially available mycoinsecticides contain *Verticillium lecanii*, and are used in greenhouses to control whitefly on cucumbers and tomatoes. A mycoherbicide containing chlamydospores of *Phytophthora palmivora* is used commercially to control strangleweed. Waste fungal mycelium from fermentation industry can also be used as a fertilizer and soil conditioner, and has found a novel use in Switzerland for revegetating bare ski slopes. SEE BIOLOGICAL PEST CONTROL.

Fungi can also be employed as inoculants to improve crop growth. Particular attention has focused on the inoculation of mycorrhizal fungi to help the plant take up water, phosphorus, or other nutrients. A commercial product based on the fungus *Hymenoscyphus ericae* is used to increase yields of blueberries in New Zealand. Finally, crop growth might be improved by addition of so-called biofertilizers. Species of *Penicillium*, for example, are being used to solubilize insoluble soil phosphates, thereby providing soluble phosphorus to plants growing in phosphorus-deficient soils. SEE MYCORRHIZAE.

The possibility of growing crops for gasification and fuel use has already been considered. A prime example of the use of fungi to produce renewable energy is the production of alcohol by growing the yeast *Saccharomyces cerevisiae* on sugar cane. The potential also exists for alcohol production from starch-rich root products and food wastes.

Food biotechnology. Recent developments in food biotechnology include the application of fungi as protein sources. Originally devoted to yeast production, this technology has recently produced a high-protein, low-fat material called mycoprotein. Mycoprotein is produced from the filamentous fungus *Fusarium gramanearum*. The filamentous nature of the product gives the material bite (the texture and feel of meat when being chewed), allowing it to be effectively used as a meat substitute. Finally, an increasingly wide variety of often exotic edible mushrooms is now available for public consumption.

Biodeterioration and biodegradation. Most natural and many synthesized products are used as substrates by fungi. The nonspecific ligninases of white rot fungi which can be effectively used in biotechnology also allow fungi to attack the wood used in buildings. The most notorious of these wood-decomposing fungi is *Serpula lacrymans*, the cause of dry rot. Other rots include those produced by the cellar fungus (*Coniophora puteana*) and the white pore or mine fungus (*Fibroporia vaillantii*). Wood rots are traditionally prevented by using fungicides; however, biocontrol agents are also being evaluated for this purpose. For example, the decay of creosote-treated wooden poles by *L. lepideus* can be effectively prevented by inoculation with the antagonistic, saprophytic fungi, such as species of *Scytalidium* and *Trichoderma*.

Molds also bring about extensive deterioration of foods, even during refrigeration. Fungal growth on stored grains and foodstuffs (such as peanuts) can lead to the production of mycotoxins such as aflatoxin. These toxins are associated with cancers in animals and possibly also humans. Mycotoxin contamination can be prevented by employing storage conditions which exclude fungal growth. Again, biological control is being seriously considered. For example, species of the bacterium *Flavobacterium* and the fungus *A. parasiticus* can be effectively used to degrade aflatoxins produced by molds that grow on contaminated foodstuffs.

Future developments. Fungi are being extensively used or evaluated in all areas of biotechnology. This trend is likely to increase because of recent developments in the genetic manipulation of industrial fungi. Such developments already include the production of heterologous proteins such as human alpha interferon and human growth hormone by yeasts.

The use of yeasts to produce fuel cells provides a final example of the continuous development of fungal biotechnology. Although yeast-based batteries currently generate only sufficient electricity to power digital watches or calculators, future developments may open up exciting new possibilities.

Milton Wainwright

Phylogeny and Evolution

Recent new estimates suggest that 1.5 million species of fungi exist; however, of this number only 5% have been inventoried as species known and described. Although the fungi are of great consequence agronomically, bioindustrially, medically, and biologically, fungal phylogeny and evolution among the fungi and between these and other organisms is not well understood. Interpretation of fungal phylogeny and evolution based on fungal fossils is difficult since fossil records are fragmentary. In the past, phylogenetic speculation of the fungi was based mainly on comparative analyses of morphological, ontogenetical, and biochemical data sets. In the 1980s, the birth of molecular phylogenetics based on the neutral theory of molecular evolution extended studies on relationships, phylogeny, and evolution of organisms. In the early 1990s, the invention of a polymerase chain reaction gave rise to new debates on fungal phylogeny and evolution from a molecular and morphological perspective.

Molecular phylogenetic analysis. Ribosomal ribonucleic acid (rRNA) sequence comparisons offer a means for estimating phylogenetic relationships. Initially, comparative analyses based on 5S rRNA sequence comparisons improved understanding of fungal phylogeny and evolution. Because there are only 120 nucleotides available for comparison, resolution is limited. In recent years, the molecular phylogenetic analysis of the fungi shifted to the small nuclear (18S) and large (23S to 28S) rRNA genes. Phylogenetic analyses among distantly related taxa using nuclear 18S ribosomal deoxyribonucleic (rDNA) sequence divergence are almost all well resolved and statistically supported.

Circumscription. Evidence from the 18S rDNA sequence put an end to a debate concerning the composition of the kingdom Fungi and whether chytrids were the true fungi, and confirmed the extent of true fungi. The true fungi are chytridiomycetes, zygomycetes, ascomycetes, and basidiomycetes. They are hyphal, have cell walls throughout most or all of their life cycle, and are exclusively absorptive in their nutrition. These fungal groups form a monophyletic group distinguished from cel-

lular slime molds (Acrasiomycetes), plasmodial slime molds (Myxomycetes), and oomycetes (Oomycota) [**Fig. 1**]. In this phylogenetic tree, the two groups of slime molds diverged separately, prior to the terminal radiation of eukaryotes. This result is consistent with many differences between slime molds and fungi in form, function, and life cycle. The oomycetes form a cluster with brown algae and diatoms, whose cells contain chlorophylls *a* and *c*. This result is also consistent with principal phenotypic characters. A recent 18S rDNA phylogeny indicated that the hyphochytriomycetes form a monophyletic group with the oomycetes and heterokont algae.

The four true fungal groups are also characterized by chitinous cell walls and an alpha aminoadipic acid–lysine biosynthetic pathway. However, the oomycetes, hyphochytrids (Hyphochytriomycetes), labyrinthulids (Labyrinthulomycetes), thraustochytrids (Thraustochytriaceae), plasmodiphoromycetes (Plasmodiphoromycetes), and slime molds, which contain cellulose and have a diaminopimelic acid–lysine pathway, are not members of the kingdom Fungi. Among these, the oomycetes, hyphochytrids, labyrinthulids, and thraustochytrids are called the pseudofungi or protistlike fungi. The slime molds are peculiar organisms and share the features of animals and fungi. Protozoologists prefer to place them in the Mycetozoa of the kingdom Protozoa, although they have usually been studied by mycologists rather than protozoologists. The Mycetozoa ingest nutrients by phagocytosis, have tubular mitochondrial cristae like most protozoa, and have no cell walls in the trophic phase. Ultrastructure, wall chemistry, feeding mode, and rDNA phylogeny of these organisms are also evidence that the Mycetozoa are associated with the kingdom Protozoa rather than the Fungi.

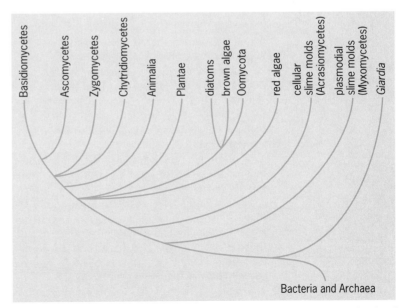

Fig. 1. Phylogenetic tree showing relationships of eukaryotes, including the true fungi and pseudofungi. *Giardia* represents an early diverging eukaryote. (*After D. L. Hawksworth, ed., Ascomycete Systematics: Problems and Perspectives in the Nineties,* Plenum Press, *1994*)

Evolutionary relationships between the three kingdoms of eukaryotes, fungi, and plants and animals are still controversial because of lack of solid fossil evidence. Recent 18S rDNA phylogenies and amino acid sequencing of elongation factor [a protein involved in the translation of messenger ribonucleic acid (mRNA)] show that the closest relatives to fungi are animals not plants. A phylogenetic analysis based on 23 different protein species from the three kingdoms by three different methods (maximum-likelihood method, neighbor-joining method, and maximum-parsimony method) also shows that the kingdom Animalia is closely related to Fungi and is distantly related to the kingdom Plantae. Among the ideas concerning the boundaries between the Fungi and other organisms is a proposal that the kingdom Fungi evolved from a choanoflagellate. However, the full evidence from both molecules and morphology is lacking.

Phylogenetic divergence of lower fungi. The four divisions of fungi—the Chytridiomycota (chytridiomycetes), Zygomycota (zygomycetes), Ascomycota (ascomycetes), and Basidiomycota (basidiomycetes)—now can be recognized as sharing a number of important characteristics. In the lower fungi (Chytridiomycota and Zygomycota) anastomoses (networks of branches) and cross walls are rare. In contrast, anastomoses and cross walls (frequently with clamp connections in basidiomycetes) predominate in the higher fungi (Ascomycota, Basidiomycota, and related forms). The molecular phylogenies support this framework. Nuclear 18S rDNA sequence data available for phylogenetic analyses are limited in the lower fungi which comprise the divisions Chytridiomycota, with posteriorly monoflagellate zoospores, and Zygomycota, with sexual production of zygospores. Evolutionary analyses of 18S sequences led to the inference that the lower fungi branched earlier than the higher fungi within the kingdom Fungi. A recent molecular phylogeny deduced from 18S rDNA sequence places the four taxa in the Entomophthorales (primarily parasites of insects) in two clusters: *Basidiobolus ranarum* in one cluster containing the aerobic chytrid *Chytridium confervae*, the anaerobic rumen chytrids *Spizellomyces acumninatus* and *Neocallimatix* species, and the remaining three entomophthoralean fungi *Conidiobolus coronatus*, *Entomophthora muscae*, and *Zoophthora radicans* in another cluster containing the typical mucoralean fungus *Mucor racemosus*. Apart from the other three chytrids the aerobic chytrid, *Blastocladiella emersonii*, formed the eariest branch within the fungi. The mycorrhizal fungus *Glomus etunicatuum* is basal to ascomycetes and basidiomycetes, and the trichomycete *Smittium culisetae* was placed close to the divergence of Entomophthorales from the chytrid-*Glomus*-ascomycete-basidiomycete clade. The analysis also strongly suggests the difficulty of clear separation between Chytridiomycota and Zygomycota, loss of flagella on several lineages, and great phylogenetic divergence among the chytrids.

Phylogenetic divergence of higher fungi. Phylogenetic trees from 18S rDNA sequence data clearly indicate the existence of the two divisions (Ascomycota and Basidiomycota) among the higher fungi. The former is characterized by the ascus as a meiosporangium producing ascospores (sexual endospores), whereas the latter is characterized by the basidium as a meiosporangium producing basidiospores (sexual exospores). Both divisions appear to be monophyletic but their origin is ambiguous and controversial. In most of the higher fungi the asexual process involves the reproduction of asexual spores (for example, conidia). The sexual state of a higher fungus is now termed the teleomorph, the asexual state the anamorph, and the whole state the holomorph.

Ascomycotina. The Ascomycota is composed of three major lineages: the archiascomycetes (basal or early ascomycetes), ascomycetous yeasts (hemiascomycetes), and filamentous ascomycetes (euascomycetes). The archiascomycetes diverged prior to the separation of other two major lineages as sister groups. The archiascomycetes, which may not be monophyletic, include diverse fungal taxa, that is, the plant parasites *Taphrina deformans*, *T. populina*, *T. wiesneri*, *Protomyces inouyei*, and *Pr. lactucae-debilis*, the mitotic, saprobic yeast *Saitoella complicata*, the fission yeast *Schizosaccharomyces pombe*, and the human lung pathogen *Pneumocystis carinii*. Among these, *Pn. carinii* was considered to be a protozoan for many years. However, the molecular phylogenies support its placement, and its meiospore characters seem ascomycetous. A lack of common characteristics that are needed to define the Archiascomycetes as a new class suggests that *Taphrina* and relatives, *Schizosaccharomyces*, and *Pneumocystis* may have radiated rapidly. The hemiascomycetes include fungi that typically grow as yeasts and produce naked asci singly, such as *Saccharomyces cerevisiae* and the anamorphic yeast *Candida albicans*.

The euascomycetes, with comparatively well-developed fruiting bodies, compose the plectomycetes, pyrenomycetes, loculoascomycetes, laboubeniomycetes, and discomycetes. The plectomycetes with closed fruiting bodies (cleistothecia) and pyrenomycetes with flask-shaped fruiting bodies (perithecia) constitute a monophyletic group. The plectomycetes include *Eremascus albus*, *Ascosphaera apis*, the human pathogens causing deep mycoses, *Aspergillus fumigatus*, and *Penicillium notatum*. One member of the pyrenomycetes is *Neurospora crassa*. The class Loculoascomycetes (fissitunicate ascomycetes) is not monophyletic but the order Pleosporales appears as a monophyletic group that includes the families Pleosporaceae and Lophiostomataceae. One lineage of the laboulbeniomycetes (obligate parasites of arthropods) lies outside the other ascomycetes with oval-shaped fruiting bodies among loculoascomycetes and discomyctes, where taxon sampling is still incomplete.

The discomycetes (apothecial ascomycetes) may not be monophyletic.

Basidiomycotina. The 18S rDNA sequence-based phylogenies divide the Basidiomycota into three major lineages. The first major lineage (Ustilaginales smuts) is composed of the smut fungi (Ustilaginales), represented by *Ustilago maydis, U. hordei,* and *Tilletia caries,* including the strictly anamorphic yeast *Sympodiomycopsis paphiopedili.* The type species *Graphiola phoenicis* and *G. cylindrica,* both of which parasitize palm leaves in the tropical and subtropical regions of the world, are grouped together with the smut fungi. The small samples of Ustiliaginales smuts seem to form a monophyletic group, which may be basal to other basidiomycetes. The second major lineage (simple septate basidiomycetes) includes the smutlike, teliospore-forming yeast species, which are characterized by simple septal pores and no xylose in the cell wall. *Mixia osmundae,* a *Taphrina*-like parasitic fungus previously in the Ascomycota, and the rust fungi *Cronartium ribicola* and *Peridermium harknessii* are included in this lineage. The hymenomycetes represent the third major lineage and include filobasidiaceous yeast species, *Cystofilobasidium capitatum, Mrakia frigida, Filobasidium floriforme,* and *Filobasidiella neoformans* (anamorph: *Cryptococcus neoformans*), the ballistospore-forming yeast species (anamorph: *Bullera alba*), and the arthroconidium-forming yeast species. Members of the heterobasidiomycetes *Tremella, Dacrymyces,* and *Auricularia* and selected hymenomycetous taxa represented by *Athelia bombacina, Coprinus cinereus, Boletus satanas,* and *Spongipellis unicolor* are also included in this lineage.

Under the dual nomenclatural system, fungi showing the structures associated with sexual reproduction are classified in the Ascomycota or Basidiomycota, but those that lack evidence of sexual reproduction are classified in Deuteromycota or Deuteromycotina. However, the molecular characters (both 5S rRNA and 18S rRNA sequences) do not support the placement of the deuteromycetes as higher taxa. They are phylogenetically assigned to the Ascomycota or Basidiomycota. This placement may suggest that asexual organisms do not contribute to the major events of evolution.

Lichens. The lichen life style is the symbiotic association of fungi with the photosynthetic members such as the green algae *Trebouxia, Pseudotrebouxia,* and *Trentepohlia,* and the cyanobacterium *Nostoc.* It is found in various representatives of the higher fungi, both the Ascomycota and Basidiomycota. However, there are no lichenized hemiascomycetes, plectomycetes, and laboulbeniomycetes. Lichen-forming fungi constitute about 18,000 species, suggesting that only 50–70% of the world's species are currently known. Lichens occur in a variety of habitats from the Arctic and the Antarctic and all regions in between: on soil and bare rock, on tree trunks, on frozen substrata in the polar regions, and on plant leaves particu-larly in the tropics. This is a reflection of fungal diversity.

A recent phylogenetic analysis, based on 18S rDNA sequences, suggests multiple origins of lichen symbioses in fungi. At least five independent origins of the lichen habit in disparate groups of the Ascomycota and Basidiomycota are suggested. Within the Basidiomycota, the phylogenetic hypothesis supports three independent origins of the lichen habit indicated by *Multiclavula, Omphalina,* and *Dictyonema,* each corresponding to groups supported by morphological characters. Within the Ascomycota, it supports at least two independent origins of the lichen habit represented by members of the order Lecanorales, and *Arthonia radiata* and allied species of the order Arthoniales. This molecular phylogenetic analysis concludes that neither mutualism nor parasitism should be construed as an endpoint in symbiont evolution because lichen associations arose from parasitic, mycorrhizal, or free-living saprobic fungi. The concept of lichen is ecologically but not phylogenetically meaningful.

Fungus radiations and the geological record. The absolute timing of origin of fungal groups (**Fig. 2**) has been estimated from molecular and fungal fossil evidence. The percentage of nucleotide substitution that lineages have accumulated since divergence from a common ancestor was used to estimate the relative timing of origin of fungal lineages. Another indication was the calibration points from the fossil record, such as the appearance of fossilized fungal clamp connections. Clamp connections in fungal filaments first appear in the fossil record in woody tissue from a 290-million-year-old Carboniferous fern.

From these estimates it was shown that the chytridiomycetes split from a lineage of terrestrial fungi that gave rise to the zygomycetes, ascomycetes, and basidiomycetes approximately 550 million years ago (Ma). After plants invaded the land about 440 Ma, the ascomycetes split from the basidiomycetes. The three major lineages in the Ascomycota were established perhaps during the coal age, 330–310 Ma. Mushrooms, many ascomycetous yeasts represented by the true yeast *Kluyveromyces lactis,* baker's yeast (*S. cerevisiae*), and common molds in the economically important anamorphic genera (*Aspergillus* and *Penicillium*) may have evolved after the origin of angiosperm plants and during the last 200 million years. The holobasidiomyctes, including mushrooms, radiated in the Cretaceous approximately 130 Ma. The remains of two gilled mushrooms were recently discovered in Turonian (90–94-million-year-old; mid-Cretaceous) amber from central New Jersey. These fossils provide another calibration point for basidiomycete molecular clocks and are consistent with current conclusions.

More taxa and more sequence data are required to elucidate the phylogenetic relationships among fungi, from the lower to the higher in the light of

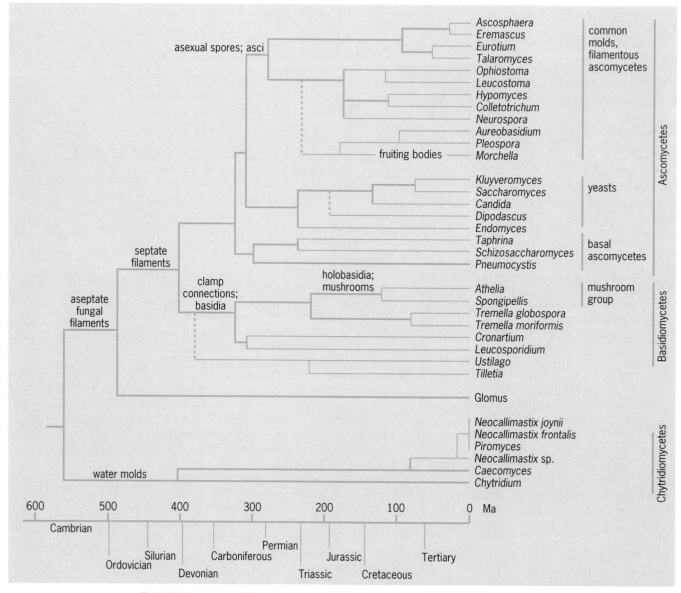

Fig. 2. Phylogenetic tree illustrating the estimated fungal divergence times superimposed on the 18S ribosomal deoxyribonucleic acid sequence-based phylogeny. Branch lengths on the tree are proportional to the average percent of nucleotide substitutions, corrected for lineage-specific differences in substitution rates, indicating that the phylogeny was in conflict with branch length. (*From M. L. Berbee and J. W. Taylor, Dating the evolutionary radiations of the true fungi, Can. J. Bot., 71:1114–1127, 1993*)

the fungal species diversity. Knowledge of fungal molecular phylogenetics is advancing rapidly, and it has been suggested that the fungal taxonomic framework be revised.

For background information *SEE ANTIBIOTIC; ASCOMYCOTINA; BASIDIOMYCOTINA; BIOTECHNOLOGY; CLASSIFICATION, BIOLOGICAL; FUNGAL GENETICS; FUNGI, FUNGISTAT AND FUNGICIDE; INDUSTRIAL MICROBIOLOGY; LICHENES; MYCOLOGY; PALEOBOTANY; PLANT KINGDOM; PLANT PHYLOGENY; SOIL ECOLOGY; YEAST; ZYGOMYCOTINA* in the McGraw-Hill Encyclopedia of Science & Technology.

Junta Sugiyama

Bibliography. M. L. Berbee and J. W. Taylor, Dating the evolutionary radiations of the true fungi, *Can. J. Bot.*, 71:1114–1127, 1993; T. D. Bruns et al., Fungal molecular systematics, *Annu. Rev. Ecol. Syst.*, 22:525–564, 1991; R. L. Metzenberg, The impact of molecular biology on mycology, *Mycol. Res.*, 95:9–13, 1991; T. Nagahama et al., Phylogenetic divergence of the entomophthoralean fungi: Evidence from nuclear 18S ribosomal RNA gene sequences, *Mycologia*, 87:203–209, 1995; H. Nishida and J. Sugiyama, Archiascomycetes: Detection of a major new lineage within the Ascomycota, *Mycoscience*, 35:361–366, 1994; J. F. T. Spencer and D. M. Spencer (eds.), *Yeast Technology*, 1990; M. Wainwright, *An Introduction to Fungal Biotechnology*, 1992; M. Wainwright, Novel uses for fungi in biotechnology, *Chem. Indus.*, pp. 31–34, January 1991.

Gene

Animal development could not proceed without a mechanism for telling different parts of the embryo their correct position with respect to the rest of the body. The coordinate system that signals to one cell its position within the developing head and to another its position within the developing tail arises from the expression of a collection of genes collectively known as the HOX cluster.

The HOX genes were originally identified in the fruit fly, *Drosophila melanogaster.* Mutations in the genes of the HOX cluster, known as the homeotic genes in *Drosophila,* result in homeotic or part-for-part conversions of body structures. For example, in one kind of mutant, structures normally found only in the thorax, the legs, will also grow from the head in place of the antennae. Isolation and sequencing of the genes responsible for these phenotypes reveals that they all contain within their coding regions a conserved deoxyribonucleic acid (DNA) sequence, the homeobox. The homeobox codes for a protein motif, the homeodomain, that allows a protein to bind DNA. The proteins encoded by homeobox genes function during development by binding to other genes and turning these targets on and off.

Genetic mapping. Mapping the HOX genes within *Drosophila*'s genome reveals that the homeotic genes are all arranged on a single chromosome. Researchers examining the location in the embryo of homeotic gene expression found that it parallels the order of the genes along the chromosome (see **illus.**). The *labial* (*lab*) gene on one end of the cluster is expressed at the anterior end of the embryo. The next gene in the cluster, *proboscipedia* (*pd*), is expressed to the right of *lab*.

The ordered expression of the HOX cluster genes clarifies position within the fruit fly embryo; that is, the expression of each gene tags a portion of the fly with a specific address relative to the head and tail. The homeotic phenotypes that originally evoked interest in these genes result from specific changes in the expression of the homeotic genes. A

fly sprouting legs instead of antennae on its head suffers from misexpression of a thoracic homeotic gene, *Antennapedia* (*Antp*), in the head region of the embryo, leading cells in the head to develop as if they were in the thorax.

HOX clusters have recently been identified in many animal species, including nematodes (roundworms), shrimps, amphioxi (a group closely related to vertebrates), hydras, mice, and humans. Over the past several years, HOX clusters have become versatile tools for analyzing evolutionary and developmental questions in animal systems. This versatility is possible for three reasons: First, the HOX cluster appears to be ubiquitous among animal species, allowing researchers to compare clusters between both closely and distantly related organisms; HOX gene function can be usefully compared between flies and butterflies, or between flies and humans. Second, the molecular details of how the HOX cluster performs its basic role of providing positional information have been intensely studied in *Drosophila,* and that information can often be generalized to other systems. Third, the HOX cluster genes are vital in development, making them likely candidates for genes that, upon alteration, could result in significant evolutionary change.

Studies of HOX genes within a single species as well as comparisons of HOX clusters between different species have provided numerous insights into the evolution and development of animals. It also has become clear that HOX genes can have roles beyond simply providing positional information.

HOX clusters and vertebrate evolution. Higher vertebrates' genomes contain more than one HOX cluster; this unique feature among animal taxa may have implications for how novel structures such as the vertebrate jaw evolved. Mice, humans, and birds all possess four different types of HOX clusters, labeled HOXa–HOXd. Researchers are examining the HOX cluster in the genomes of lower vertebrates and closely related, nonvertebrate taxa (such as amphioxus) to determine approximately when in evolutionary history the HOX cluster duplications occurred. If the duplication occurred before vertebrates arose as a separate taxa, closely related species such as amphioxus would also contain more than one HOX cluster.

Researchers are instead discovering that the amphioxus genome contains only a single HOX cluster, indicating that duplication of the HOX clusters occurred in the vertebrate line after the last common ancestor between amphioxus and vertebrates. Other work in the lamprey, one of the most primitive fishes, indicates that the lamprey genome has at least two and possibly more HOX clusters, strengthening the correlation between having a HOX cluster duplication and being a vertebrate.

The duplication of the HOX cluster and other genes in the vertebrate ancestor may have aided the evolution of novel structures by providing extra genes that could be coopted for new functions. Once

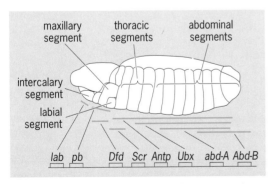

***Drosophila* embryo, 10 h old, shown with the approximate positions of epidermal expression of the HOM-C genes. (*After W. McGinnis and R. Krumlauf, Homeobox genes and axial patterning, Cell, 68:283–302, 1992*)**

the HOX cluster was duplicated, two copies of each HOX gene existed and would initially have been redundant, each regulating the same targets in the same way. Mutations that changed the function of one gene in a pair would not harm the organism because an extra copy was available. In this way duplicated genes could be recruited for new roles during development, leading to novel structures such as limbs and jaws. HOX genes, which act by regulating other genes, may have been especially important in evolution; a change in one HOX gene could affect a large number of targets. Duplication of the HOX genes probably did not directly cause the evolution of the vertebrates but did provide redundant genetic information that could be put to new uses.

HOX genes and limb development. Limbs begin as buds of tissue that elongate, elaborate, and differentiate during development. During the early stages of limb development, HOX genes are expressed within the developing limb bud in an anterior to posterior arrangement that mirrors the placement of the genes along the chromosome.

The ways in which the HOX genes influence limb development are complex. Initial data suggested the HOX genes might be acting in a similar fashion to their counterparts in *Drosophila,* specifying the location of the developing digit in reference to the anterior and posterior end of the limb bud. Experiments with chick embryos in which *Hoxd-11,* normally expressed only in the posterior part of the limb bud, was artificially turned on in the anterior portion resulted in the first digit (digit I) developing with an appearance similar to digit II, including an additional segment. Digit I appeared to develop as if it were located at the position of digit II, in a fashion analogous to the legs growing in place of the antennae in an *Antp* mutant fly.

Although this development resembles a homeotic conversion, recent evidence contradicts a simple positional information role for HOXd genes in limb development. The early patterns of gene expression do follow an anterior to posterior pattern, but later in development all the HOXd genes are expressed throughout the limb bud, and mutations that knock out, or eliminate, a single HOXd gene affect all digits, indicating HOXd genes are necessary for the normal development of all digits. If HOXd genes provided only positional information, mutations in these genes would affect only subsets of the digits.

Thus, the function of the HOXd genes in vertebrate limbs appears to have been modified beyond simply providing positional information, an example of how evolution has tinkered with the functions of initially redundant genes.

HOX genes and insect development. Data from research on insect development has also suggested other roles for HOX genes. Recent comparisons of wing development and homeotic gene expression among *Drosophila,* butterflies, and wingless insects suggest that the targets for specific homeotic genes have changed in different insect species, and that these changes are responsible for some of the morphological and evolutionary differences between those species.

Drosophila possess a single pair of wings on one thoracic segment (T2) and structures known as halteres on the next segment (T3). Insects with a single pair of wings are thought to be derived from insects with two pairs, such as dragonflies and butterflies. Mutations in the homeotic gene *Ultrabithorax* (*Ubx*) that prevent *Ubx* expression in T3 during development result in the halteres being replaced by a second pair of wings that appear identical to those on T2. This phenotype was originally thought to represent a reversion to a four-winged ancestor, with expression of *Ubx* in T3 preventing full development of wings. Butterflies possess two morphologically distinct pairs of wings arising from thoracic segments T2 and T3. According to the above theory *Ubx* is not expressed in segment T3 of butterfly embryos, thus allowing full expansion of the second pair of wings.

However, researchers discovered that *Ubx* is expressed in segment T3 of butterfly embryos, indicating that *Ubx* expression in that segment is an ancient condition, and that the morphological differences between the T3 structures of *Drosophila* and butterflies arise from differences in the genes *Ubx* regulates in each species, and not from a change in the expression of *Ubx* itself.

Another area of research that may have an answer in the HOX genes is the mechanism by which the winged insects initially reduced their number of wings. Fossil evidence shows that early winged insects had wings attached to all thoracic and abdominal segments, a vast extravagance when compared to the one or two pairs seen in modern insects. A comparison between *Drosophila* and a wingless, closely related insect, *Thermobia,* has shown that the pattern of homeotic gene expression is highly conserved in both. The loss of wings is, therefore, not due to a difference in the patterns of homeotic gene expression. Instead, over evolutionary time different HOX genes were recruited to suppress wing formation in the segments in which those genes are expressed.

For example, mutants lacking the particular homeotic gene *Sex-combs reduced* (*Scr*), which is expressed in the normally wingless thoracic segment T1, show expression of some wing-specific genes in the T1 segment during embryo development. These patches of expressing cells do not form normal, adult wings, probably because other genes necessary for wing formation are still off. Nevertheless, the abnormal expression of even a subset of wing genes indicates that *Scr* plays a role in preventing wing formation.

Genes and evolution. Other areas of HOX cluster research include comparing HOX gene expression between different arthropod groups and correlating morphological changes with changes in HOX patterns; studying the effects of turning off specific

HOX genes in transgenic mice and examining the effects on brain and skeletal development; and examining the patterns of gene duplication and deletion within HOX clusters of different species—some organisms possess a greater or lesser number of homeobox genes in their HOX clusters than are present in *Drosophila*.

For background information SEE CHROMOSOME; DEVELOPMENTAL GENETICS; EMBRYOLOGY; EMBRYONIC DIFFERENTIATION; GENE; GENE ACTION; MUTATION; PATTERN FORMATION (BIOLOGY) in the McGraw-Hill Encyclopedia of Science & Technology.

Kyle A. Serikawa

Bibliography. S. B. Carroll, Homeotic genes and the evolution of arthropods and chordates, *Nature*, 376:479–485, 1995; P. W. H. Holland et al., Gene duplications and the origins of vertebrate development, *Develop. Supp.*, pp. 125–133, 1994; B. A. Morgan and C. Tabin, Hox genes and growth: Early and late roles in limb bud morphogenesis, *Develop. Supp.*, pp. 181–186, 1994; M. P. Scott, Intimations of a creature, *Cell*, 79:1121–1124, 1994.

Genetic disease

Most purebred domestic animal stocks evolved from relatively small gene pools, each establishing an original breed. The common practices of inbreeding and linebreeding (a less concentrated form of inbreeding) descendants of the foundation stock have promoted genetic mutations which inevitably have led to the increased transmission and recognition of genetic diseases. These inherited conditions should be distinguished from congenital disorders, which are present at birth and may be hidden, only to be discovered later on in life. Thus, although genetic defects are congenital because they are present at birth, not all congenital defects are of genetic origin.

Definitions. Genes are dominant when they are manifested or expressed in the heterozygous as well as homozygous state. In contrast, in recessive genes the effects of the gene are hidden (nonexpressed) in heterozygotes and revealed only in homozygotes.

A homozygous individual, or homozygote, carries two of either the dominant or recessive gene of a pair of alleles. A heterozygous individual, or heterozygote, carries both the dominant and recessive genes of a pair of alleles. Thus, homozygotes have two like genes (homozygous dominant or homozygous recessive), and heterozygotes are not alike because they carry one dominant and one recessive gene of a pair of alleles.

Another commonly used term is incomplete dominance. This form of genetic expression is the second most frequent after recessivity, and refers to the variable penetration or expressivity of a particular dominant gene. Related individuals can be normal or can express the trait in a mild, moderate, or severe form depending on the degree of gene penetrance in each individual.

A carrier is an individual heterozygous for some recessive character or trait. Carrier states can be silent for one or more generations and then reappear. Examples include coat color variations and the hemophilia gene (where the carriers are female).

Hemophilia. Autosomal traits involve genes that are carried on the body chromosomes or autosomes and affect both sexes. Sex- or X-chromosome-linked traits, by contrast, involve genes carried on the X chromosome of females and expressed primarily in males. Hemophilia is a common example of a sex-linked disorder that is also recessive. In this situation, the female heterozygote carries the gene on one X chromosome while the other is normal. The female does not exhibit any clinical signs of the carrier state of this genetic defect. If the female is mated with a normal male, 50% of her daughters will inherit the hemophilia gene and be heterozygous carriers; the other 50% will be genetically normal. Similarly, 50% of her sons will inherit the hemophilia gene and be affected hemophiliacs. They will not be homozygous for the hemophilia gene, however, because males have only one X chromosome. Thus, the affected hemophilic male is a hemizygote; the other half of the male offspring from this mating will be genetically normal.

A heterozygous female carrier of hemophilia could also be mated with a hemizygous hemophilic male that survives to reproductive age. Although rare, this situation occurs in inbred and linebred families of dogs or in situations where a relatively mild form of hemophilia is present. The expected progeny from such a mating will include normal males that inherit the normal X chromosome from their mother; hemophilic males that inherit the

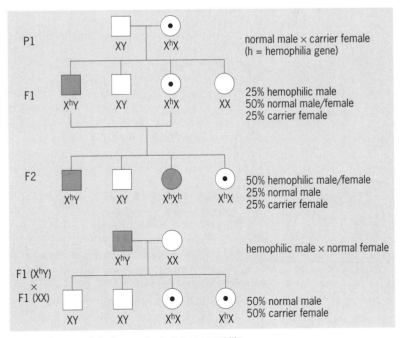

Progeny for a sex-linked recessive trait for hemophilia.

hemophilic X chromosome from their mother; carrier females that inherit the hemophilia gene from their father but the normal gene from their mother; and female hemophiliacs that inherit an affected gene from each parent. To sum up the situation with the various genotypes in hemophilia, carrier females are heterozygotes, affected males are hemizygotes, and affected female hemophiliacs are homozygotes (see **illus.**).

Genetic screening. At least 400 disorders of genetic basis or predisposition have been recognized in dogs. Large-scale screening programs for the identification of affected and carrier animals have been in place for about 30 years, and are an effective mechanism to discover and eventually control the frequency of specific genetic defects within the population at large. Screening programs of this type have been used successfully in humans for many years (for example, for Tay-Sachs disease, phenylketonuria) and also have been applied to a variety of animal species. Recently, ethical issues have surfaced concerning the advisability of routinely screening asymptomatic humans for inherited diseases. To some extent, these same issues apply in the genetic screening of domestic animal populations.

Depending on the mode of inheritance, different approaches may need to be applied for the detection and control of genetic diseases. It is advantageous to be able to select against heterozygotes (carriers) rather than have to eliminate affected individuals from a breeding program once the condition has already appeared. Control and elimination of the disease by testing is feasible and reliable in cases where a specific genetic probe has been identified or a biochemical marker of the trait is expressed in heterozygous carrier animals (for example, as measured in blood, urine, or saliva tests or by electrocardiogram, skin biopsy, eye examination, or hair analysis). Recently, genetic probes have been developed for progressive retinal atrophy, that is, an inherited eye disease of Irish Setters that causes night blindness; pyruvate kinase enzyme deficiency of the red blood cells of Basenjis, which produces chronic anemia; and phosphofructokinase enzyme deficiency of the red blood cells and muscles of English Springer Spaniels, which causes chronic anemia and enlargement of the spleen from recurring episodes of exercise-induced lysis of red blood cells and muscle fatigue. In addition, it is desirable to evaluate breeding stock for overall soundness, reproductive health, and performance.

Cardiovascular malformations. An estimated 0.7% of canine hospital admissions involve anomalies of the cardiovascular system; the incidence is lower in cats and domestic farm animals. Ventricular septal defects, in which a hole or weakening is present between the tissue (septum) dividing the two main chambers (ventricles) of the heart, are more common in horses and cattle than atrial septal defects, in which the septum dividing the upper chambers (atria) above the ventricles is defective. One of the most common cardiovascular anomalies of the dog is patent ductus arteriosus. In this condition, the blood vessel remnant joining the aorta and pulmonary artery fails to close properly at birth, thereby shunting blood away from the lungs and causing cyanosis (bluish color of tissues from oxygen deprivation), shortness of breath, and stunted growth. It is a polygenic trait in Miniature and Toy Poodles. Valvular pulmonic stenosis is hereditary in Beagles, and is caused by failure of the pulmonic valve of the heart to open properly. Persistent right aortic arch is a developmental problem in which one of the fetal blood vessels near the heart fails to close off as it should during embryonic development. It is inherited in German Shepherds, and is also seen in cattle, horses, and cats. Subaortic stenosis is a condition where the outflow opening to allow blood to flow from the heart into the aorta is too small or tight (stenotic), causing difficulty with poor circulation and respiratory embarrassment. It occurs in Newfoundlands, Golden Retrievers, Boxers, and German Shepherds. In the Newfoundland breed, subaortic stenosis is inherited as an autosomal dominant trait with variable expression determined by modifying genes. In swine, aortic stenosis is also an inherited trait. Tetralogy of Fallot is caused by a four-way (tetra-) developmental anomaly of the heart and great blood vessels that is usually incompatible with life. It is hereditary in the Keeshond, and is also seen in cattle, horses, and cats. For each anatomical form of hereditary congenital heart disease, the degree of severity is positively correlated with the gene dosage.

Hematologic disorders. The inherited bleeding disorders are quite prevalent in a variety of domestic animal species and are similar to the analogous human diseases. All of these conditions have autosomal inheritance with the exception of the two types of hemophilia. Affected animals exhibit a tendency to bleed abnormally and may have life-threatening hemorrhage after injury or surgery. Inherited dysfunction of blood platelets or thrombocytes occurs in Basset Hound, Spitz, and Otterhound dogs, Simmental cattle, swine, and cats. The blood-clotting defects are caused by deficiency or dysfunction of the blood plasma coagulation proteins. The most commonly recognized are von Willebrand's disease (seen in more than 60 breeds of dogs, several breeds of cats, quarterhorses, and Poland-China swine) and the two forms of hemophilia. Von Willebrand's disease results from an abnormality of the adhesive blood protein called von Willebrand factor. Its role is to assist platelets in sticking to wounds and sealing leaking or severed blood vessels. Of the hemophilias, hemophilia A is more common than hemophilia B (Christmas disease), and has been recognized in most dog breeds (especially German Shepherds), many cat breeds, standardbred and thoroughbred horses, Hereford cattle, and sheep. Hemophilia B occurs in about 20 dog breeds and in cats.

Cyclic hematopoiesis (cyclic neutropenia) is an autosomal recessive trait of gray Collie dogs and is linked to their coat color. The disorder is characterized by periodic lowering of neutrophils, a type of white blood cell important to control infections. Affected dogs develop chronic, treatment-resistant infections. Leukocyte adhesion defect produces white blood cells (leukocytes) that have lost or never developed the ability to react normally in fending off infections. It is inherited in cattle, Irish Setters, and Doberman Pinschers.

Eye and ear disorders. Inherited diseases of the retina of the eye include Collie eye anomaly, a recessive trait with variable expression caused by selection for the breed's very narrow head which produces malformed eyes; progressive retinal atrophy or night blindness of many dog breeds and cats; retinal dysplasia, a malformation of the retina of Labrador Retrievers that also have skeletal dysplasia (shortening) of the forelegs; and optic nerve hypoplasia of Miniature Poodles where the nerve going from the eye to the brain is too small. Other eye diseases include abnormal inward rolling of the eyelids and Vogt-Koyanagi-Harada syndrome in Akitas (an autoimmune syndrome where the eyes, blood, and other tissues are progressively destroyed, leading to blindness and death); persistent pupillary membranes in Basenjis where the membrane forming the iris of the eye persists and blocks vision; glaucoma or abnormally high blood pressure within the fluid of the eyes of Basset Hounds and American Cocker Spaniels; and cataracts (opacity of the lens of the eyes) of several types which occur in many breeds and can progress to blindness. Deafness is associated with a white coat color and blue eyes, and is commonly seen in white Boxers, Dalmatians, white Bull Terriers, Scottish Terriers, Border Collies, Old English Sheepdogs, blue merle Shetland Sheepdogs, English Setters, and Fox Terriers.

Musculoskeletal disorders. Cattle commonly inherit dyschondroplasia, an abnormally developed cartilage causing deformed limbs and poor growth. Dwarfism is seen in German Shepherds and Alaskan malamutes. The latter also have a red blood cell deformation (stomatocytosis) associated with the dwarfism. Hereford and Aberdeen Angus cattle exhibit specific types of dwarfism. Muscular dystrophy is inherited as a sex-linked disorder of Golden Retrievers, Irish Terriers, Samoyeds, and Belgian Shepherds. It produces signs such as poor growth, weakness, abnormal gait, difficulty in eating and swallowing, and muscle wasting (atrophy). Labrador Retrievers have an inherited deficiency of a specific muscle fiber (type II), which produces similar signs of muscular dystrophy. Double muscling disease is an autosomal recessive disorder of cattle causing bunching up and cramping of muscles.

Hip dysplasia (partial dislocation from looseness or laxity of the hip joint) is a prevalent polygenic trait recognized in many dog breeds. It causes a characteristic rear leg gait (shuffle with legs placed further forward under the body or splayed sideways) and can progress to painful arthritic changes in the hip joint sockets.

Cleft lip or palate occurs commonly in dogs, cats, horses, and cattle. Undershot or overshot jaws are common anomalies that would decrease in frequency if breeders removed affected stock from the gene pool. In some dog breeds, malocclusion of the teeth and jaws is the normal pattern (for example, Bulldogs, Boxers).

Nervous system disorders. Congenital hydrocephalus (an accumulation of fluid within the ventricles of the brain) with muscle disease is seen in Hereford cattle; hydrocephalus alone occurs in many dog breeds, but especially in the Chihuahua, Pekingese, Poodle, and Boston Terrier. Contracted flexor tendons is an autosomal recessive condition of newborn foals where the tendons tighten up and prevent the limbs from straightening out. Limber leg is a simple recessive disease of Jersey cattle which makes them appear double-jointed; mule foot occurs in Holstein and other cattle where the hoof is not cloven; and the development of multiple toes also occurs in cattle. Cerebellar cortical atrophy is a condition where the cerebellum of the brain is malformed and degenerates, causing incoordination and eventual incapacity. It is an incompletely dominant trait in Holstein, Aberdeen Angus, and Charolais cattle; sheep; and dogs. Hereditary neuraxial edema (swelling) of the brain occurs in Hereford cattle, causing them to stumble and succumb, and Weaver disease (bovine progressive degenerative myeloencephalopathy) is a disorder of Brown Swiss cattle that causes them to weave about. Spastic paresis and lethal spasm cause attacks of paralysis (paresis) in Jersey and Hereford cattle and are simple autosomal recessive traits.

Mannosidosis is a simple autosomal recessive deficiency of the enzyme alpha-mannosidase in cats and cattle, which produces an uncoordinated weak cat or calf that eventually dies. The condition has had a considerable economic impact in several cattle breeds and necessitated extensive screening of breeding stock. GM_1 gangliosidosis (storage disease) is a specific enzyme deficiency which causes lipids (fats) to accumulate in nerve cells. It occurs in Siamese cats, Portuguese Water Dogs, and Friesian cattle, and is a simple autosomal recessive deficiency of the enzyme beta-galactosidase.

Hereditary ataxia (incoordination) is a recessive disorder of Fox Terriers. Wobbler's syndrome is a recessive disorder of Doberman Pinschers resulting from cervical vertebral instability which pinches off the spinal cord at the top of the neck and causes a characteristic high-stepping gait. English Pointers have an autosomal recessive self-mutilation disorder called peripheral sensory neuropathy or acral mutilation in which they lack pain sensation and progressively gnaw at their own limbs. Spinal dysraphism is a disorder of Weimaraners and Rhodesian Ridgebacks where the spinal cord is deformed; it is characterized by a hopping gait which can lead to syringomyelia (cavities within the spinal cord). Hereditary spinal muscular atrophy of Brittany Spaniels is an autosomal dominant disease of motor nerves characterized by weakness and muscle wasting. Affected dogs have a typical gait, and the disorder progresses to dangling of the head and a drooping, paralyzed tail. Severely affected animals become paralyzed and die by 3–4 months of age.

Familial amaurotic idiocy occurs as a simple recessive trait in German Shorthaired Pointers and English Setters, and causes stupor, seizures, visual loss, and depression. Scotty cramp is a form of recurrent leg cramping of Scottish Terriers because of abnormal metabolism of serotonin (a cellular signal component). It is a common, simple autosomal recessive trait. Epileptic seizures are another common inherited disorder seen in many dog breeds including Beagles, Belgian Tervurens, Labrador Retrievers, German Shepherds, Irish Setters, Vizslas, and American Cocker Spaniels; in Swedish Red cattle (autosomal recessive) and Brown Swiss cattle (autosomal dominant). Narcolepsy can occur in Doberman Pinschers and Labrador Retrievers in which affected dogs suddenly fall asleep and collapse. Episodes can occur during periods of activity and last for varying lengths of time.

Disorders of the skin. Sebaceous adenitis is an autosomal recessive disease of sebaceous (sweat) glands characterized by reactive inflammation of the tissue and autoimmune self-destructive wasting of these glands. Progressive hair loss occurs and is poorly responsive to treatment. The condition is seen most often in Standard Poodles, Akitas, Vizslas, and Samoyeds.

Disorders of reproduction. Cryptorchidism (hidden testicle or testicles) is seen commonly in swine, horses, and dogs. The existence of only one testicle also occurs in dog breeds. Gonadal hypoplasia (small reproductive organs or gonads) is inherited in Swedish Highland cattle. Hermaphroditism, a true bisexual condition, occurs in dog breeds such as the Alaskan Malamute and English Cocker Spaniel. Rectovaginal constriction is a simple autosomal recessive disorder of Jersey cattle where both rectum and vagina are constricted in size; affected calves fail to grow. Narrowing of the vagina by a circular ring of tissue also occurs in various dog breeds and prevents natural breeding and conception.

Gastrointestinal tract disorders. Gastric torsion, commonly called bloat, is a life-threatening disorder of certain large dog breeds, where the stomach produces excessive gas and enlarges severely enough to twist upon itself and rapidly cause shock and death unless the pressure can be quickly relieved. Breeds at greatest risk are Great Danes, Weimaraners, Saint Bernards, Gordon and Irish Setters, and Mastiffs.

Urinary tract disorders. Renal dysplasia (malformed kidneys) and renal hypoplasia (small kidneys) are commonly inherited urinary tract disorders, and are seen in numerous dog breeds including the Chow-Chow, Lhasa Apso, Shih Tzu, Poodle, Doberman Pinscher, Beagle, Miniature and Standard Dachshund, Soft-Coated Wheaton Terrier, and Miniature Schnauzer. Bladder stones of various types are also commonly seen in some breeds, such as the Dalmatian, which has a biochemical defect in purine metabolism, leading to the formation of uric acid (urate) stones. Cystine and urate stones occur in the English Bulldog, and calcium oxalate stones in the Miniature Schnauzer.

Inborn errors of metabolism. Most inborn errors of metabolism are metabolic blocks caused by mutations in genes coding for specific enzymes of intermediary metabolism. The resultant disorders are usually autosomal recessive traits. Bedlington Terriers, West Highland White Terriers, and Doberman Pinschers have an abnormality of copper metabolism called copper storage disease. It results from an inability to utilize and store copper properly, and causes a bronze discoloration of the skin and progressive liver failure from the copper that accumulates there. Other examples of metabolic blocks include vitamin B_{12} responsive malabsorption of Giant Schnauzers that causes a lethal form of pernicious anemia from the deficiency of vitamin B_{12} and chronic diarrhea; cystinuria, an accumulation of a specific amino acid in the urine; and renal Fanconi syndrome of Basenjis, in which carbohydrate (glycogen) is released into the urine causing progressive renal failure; gyrate atrophy, a degeneration of a portion of the brain in cats, caused by a defect in the enzyme ornithine aminotransferase; mannosidosis; tyrosinemia type II, a defect in the enzyme tyrosine aminotransferase that causes accumulation of a specific amino acid in the blood in dogs; mucopolysaccharidosis VI (accumulation of a type of carbohydrate) in Siamese cats, in which the head is small with a broad, flat face, the cornea of the eyes is opaque, joints crackle and are painful, and the sternum is concave (the clinical severity of the disorder depends on the degree of mutation in the gene for the enzyme arylsulfatase B); GM_1 gangliosidosis which is widespread and exhibits severe skeletal deformities, visual impairment, depression, tremors, paralysis, large colorful granules in white blood cells, and high levels of the accumulated metabolite dermatan sulfate in the urine; and globoid cell leukodystrophy, abnormal, white globoid brain cells, that develop in Cairn and West Highland White Terriers (Krabbe's disease). Krabbe's disease produces tremors, paralysis, muscle wasting, mental changes, and impaired vision. Bladder stones, discussed above, often occur in dog breeds having an underlying block in amino acid metabolism.

Breeding. Even though a particular genetic defect may initially have been recognized in a specific line or family within a breed, all important breeding stock of the breed should be screened for the defect or undergo planned test matings because of a similar genetic background. Otherwise, the frequency of genetic defects inevitably will increase, and will have a negative impact on the health and longevity of many domestic animal species.

For background information SEE EYE DISORDERS; GASTROINTESTINAL TRACT DISORDERS; GENETIC MAPPING; GENETICS; HEART DISORDERS; HEMATOLOGIC DISORDERS; HEMOPHILIA; METABOLIC DISORDERS; MUSCULAR SYSTEM DISORDERS; NERVOUS SYSTEM DISORDERS; REPRODUCTIVE SYSTEM DISORDERS; SKELETAL SYSTEM DISORDERS; SKIN DISORDERS; URINARY TRACT DISORDERS in the McGraw-Hill Encyclopedia of Science & Technology.

W. Jean Dodds

Bibliography. *Association of Veterinarians for Animal Rights, Canine Consumer Report: A Guide to Hereditary and Congenital Diseases in Purebred Dogs*, 1994; W. J. Dodds, Estimating disease prevalence with health surveys and genetic screening, *Adv. Vet. Sci. Comp. Med.*, 39:29–96, 1995; N. Frost, Ethical implications of screening asymptomatic individuals, *FASEB J.*, 6:2813–2817, 1992; D. F. Patterson et al., Research on genetic diseases: Reciprocal benefits to animals and man, *J. Amer. Vet. Med. Assoc.*, 193:1131–1144, 1988; C. A. Smith (ed.), New hope for overcoming canine inherited disease, *J. Amer. Vet. Med. Assoc.*, 204:41–46, 1994.

Glacial epoch

Recent research has suggested that the tropical Americas and adjacent oceans may have cooled by 5°C (9°F) during the last glacial maximum, which occurred about 18,000 radiocarbon years ago. (The term radiocarbon years implies a value based on formulaically converted radiocarbon data, which may vary by several thousand years from the real age.) This finding contradicts the long-held belief that the tropics did not cool significantly. The new evidence is based on a paleotemperature record derived from the concentrations of atmospheric noble gases dissolved in ground water.

Glacial temperatures in the tropics. Whether temperatures in the tropics were lower during the last glacial period is an ongoing controversy in paleoclimate research. In the framework of the CLIMAP (Climate Long-Range Investigation, Mapping and Prediction) project in the 1970s, a global map of glacial sea surface temperatures was reconstructed based on faunal abundances in ocean sediment cores. The CLIMAP paleotemperature records indicate that tropical sea surface temperatures during the last glacial maximum were at most 2°C (4°F) lower than today. In the mid-1980s, this finding was confirmed by a method that uses oxygen isotope ratios ($^{18}O/^{16}O$) of foraminifera in ocean sediments as a paleothermometer. However, evidence from the tropical Americas shows that the altitude of snow lines and characteristic vegetation zones was about 1000 m (3300 ft) lower during the last glacial maximum as compared to today. These findings

indicate a significant cooling of the tropics at high altitudes (between 2500 and 5000 m or 8200 and 16,000 ft), most likely of the order of 5°C (9°F) or more. Low-elevation records of vegetation changes also point to significantly lower glacial temperatures. However, their quantitative interpretation appears to be more difficult than at higher elevation. An increased vertical temperature gradient (lapse rate) during the last glacial maximum could possibly reconcile the different degrees of cooling of tropical ocean surfaces and high-elevation continents. However, no process has yet been identified that could significantly change the lapse rate of the tropical atmosphere.

The application of new techniques to reconstruct sea surface temperatures has yielded conflicting results. Strontium/calcium (Sr/Ca) and oxygen isotope ratios obtained from corals at Barbados indicate a temperature change consistent with the continental records (5°C or 9°F). However, a technique that uses the concentration ratio of certain long-chain organic molecules (alkenones with 37 carbon atoms) in sediments of the equatorial Atlantic as a paleothermometer seems to confirm the glacial sea surface temperature reconstructions based on faunal abundances and oxygen isotopes.

The new continental paleotemperature record derived from atmospheric noble gases dissolved in glacial ground waters is located in northeastern Brazil at low elevation (400 m or 1300 ft) close to the coast (500 km or 300 mi); therefore it fills the gap between high-altitude continental and oceanic records.

Noble-gas thermometer. The noble-gas thermometer is based on the temperature dependency of the solubility of noble gases (neon, argon, krypton, and xenon; also known as the inert gases) in

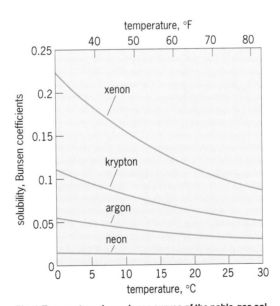

Fig. 1. Temperature dependence curves of the noble-gas solubilities in fresh water expressed as Bunsen coefficients (volume of dissolved gas under standard pressure and temperature per volume of water when equilibrated with the gas at standard pressure).

water (**Fig. 1**). Although water percolates through the unsaturated soil zone, gases are continuously exchanged with air. The last equilibration takes place at the water table where the water enters the saturated zone, typically a few meters to several tens of meters below the surface. As a result, noble gases in ground water reflect the temperature at the water table. In most cases, the measured noble-gas concentrations exceed the expected ones due to solubility equilibrium at the water table. This additional component, termed excess air, is likely caused by fluctuations of the water table trapping small air bubbles that are then partially or totally dissolved under increased hydrostatic pressure. Some of this excess air may subsequently be lost by a secondary gas exchange across the water table. The noble-gas composition of a ground-water sample is therefore influenced by three unknown parameters: (1) temperature at the water table, (2) amount of excess air, and (3) degree of loss of excess air by secondary gas exchange. These parameters are being determined from the four measured noble-gas concentrations in an iterative procedure by optimizing the agreement of the four noble-gas temperatures.

It has been demonstrated that in suitable ground-water flow systems (aquifers) the measured noble-gas concentrations, after subtraction of excess air, closely reflect the mean annual soil temperature at the water table.

Aquifers as archives of climate. Confined aquifers, for example, a sandstone layer embedded in clays, appear to be the best paleoclimate archives (**Fig. 2**). They often contain high-quality potable water, and therefore can be conveniently sampled by existing supply or observation wells. Frequently, percolating rainwater recharges confined aquifers near the surface (Fig. 2), and then follows a highly permeable sandstone layer into the deeper subsurface. As a consequence, the age of the water in the aquifer increases as a function of distance from the recharge area in the direction of flow. Typical flow velocities in confined aquifers are of the order of 1 m (3 ft) per year. Theoretically, a 100-km-long (62-mi) aquifer should yield a 100,000-year paleoclimate record. The only reliable tool for dating glacial ground water, the radiocarbon technique, limits the accessible time scale to about 30,000 years. Ground-water radiocarbon ages are characterized by a typical uncertainty of a few thousand years because of the complex carbon hydrochemistry in aquifers. In addition, small-scale mixing processes smooth the recorded climate signal. However, model calculations and several pilot studies have shown that the last glacial-to-interglacial climate transition is often well preserved in the aquifer.

The disadvantages of this technique compared to others used to reconstruct continental climate records, such as vegetation changes as recorded in lake sediments, are its limited time resolution and dating uncertainty. The advantages are that it is based on a relatively simple physical principle and that the derived temperature reflects (multi) annual mean temperatures with a high precision (0.5–0.8°C or 0.9–1.4°F). Noble-gas paleotemperature records have been reconstructed for several sites in Europe, North America, and southern Africa.

Paleotemperatures for Brazil. The first noble-gas temperature record in the tropics was derived recently from a confined aquifer in the Maranhão Basin at 41.5°W longitude and 7°S latitude in the

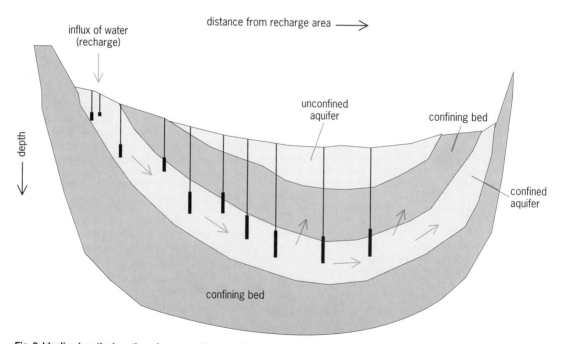

Fig. 2. Idealized vertical section of an unconfined aquifer and a confined aquifer separated by confining beds. The thick black bars indicate wells to be sampled to derive a noble-gas temperature record. Arrows indicate the direction of water movement.

central part of the semiarid Piaui province in north-eastern Brazil. The average noble-gas temperature of the water samples characterized by radiocarbon ages of less than ≈10,000 years and therefore recharged at temperatures very similar to present ones (29.6 ± 0.3°C or 85 ± 0.5°F) agrees well with the mean annual soil temperature in the area (29.1 ± 1.2°C or 84 ± 2.2°F). This agreement indicates that the aquifer is a suitable recorder of noble-gas temperatures. The difference between the average noble-gas temperatures for the present interglacial (radiocarbon age less than ≈10,000 years) and the last glacial maximum (radiocarbon age more than ≈10,000 years) is 5.4 ± 0.6°C (9.7 ± 1.1°F).

Paleoclimatic implications. The estimate of a 5°C (9°F) cooler climate at the last glacial maximum is consistent with the lowering of the snow lines and vegetation zones at high elevation in the tropical Americas, and with recently obtained ice-core records in Peru. It appears that the glacial vertical temperature gradient (lapse rate) was similar to the current temperature gradient and that a different lapse rate cannot be used to solve the puzzle. Temperature changes of the order of 5–8°C (9–14°F) have been found at many sites in the Americas between a latitudes of 40°N and S. Two of these records (Texas, 29°N, and New Mexico, 38°N) were also derived from noble gases in ground water. The combined evidence suggests that during the last glacial maximum the latitudinal temperature gradient had not been increased but that the tropical and subtropical Americas were uniformly cooler. Cooler glacial continental tropics are also consistent with the Sr/Ca and oxygen isotope record derived from corals in Barbados. If the hypothesis of cooler tropics can be extended to the adjacent oceans, the current understanding of the glacial world and the climate system of the Earth will have to be reassessed. Atmospheric general circulation models simulating glacial climate, for example, frequently used the CLIMAP sea surface temperatures as a boundary condition, and consequently they would have to be revised. Current ideas of latitudinal heat transport by the ocean or the atmosphere would have to be reconsidered. However, the serious disagreement between continental and coral records, on one hand, and sea surface temperature records based on faunal abundances and oxygen isotope ratios, as well as alkenones, on the other, persists and will require further research.

For background information SEE AQUIFER; FORAMINIFERIDA; GEOLOGIC THERMOMETRY; GLACIAL EPOCH; INERT GASES; PALEOCLIMATOLOGY; RADIOCARBON DATING in the McGraw-Hill Encyclopedia of Science & Technology.

Martin Stute

Bibliography. CLIMAP project members, Seasonal reconstruction of the Earth's surface at the last glacial maximum, *Geol. Soc. Amer.*, Map and Chart Series, vol. 36, 1981; D. Rind and D. Peteet, Terrestrial conditions at the last glacial maximum and CLIMAP sea-surface temperature estimates: Are they consistent?, *Quart. Res.*, 24:1–22, 1985; M. Stute et al., Cooling of tropical Brazil (5°C) during the last glacial maximum, *Science*, 269:379–383, 1995; P. K. Swart et al., Climate change in continental isotopic records, *Geophys. Monogr.*, 78:89–100, 1993.

Hepatitis

Hepatitis is an inflammation of the liver. It can lead to serious illness accompanied by fever, jaundice, gastrointestinal symptoms (nausea and vomiting), and in some cases death. Hepatitis is an ancient disease: the "catarrhal jaundice" reported by Hippocrates (460–377 B.C.) was almost certainly a description of the clinical symptoms of hepatitis. SEE INFLAMMATION.

Although there are many possible causes of hepatitis, including drugs, toxins, bacteria, protozoa, and fungi, viruses are the major cause, and are responsible for several hundred million cases worldwide. The clinical symptoms, histopathological lesions, and course of illness for acute viral hepatitis are similar, regardless of the specific agent causing the disease. Therefore, laboratory tests are necessary for identifying the specific virus. Several viruses, including those of the herpes virus family, yellow fever virus, rubella virus, and enteroviruses, can cause liver inflammation, but hepatitis is not the primary disease caused by these agents. Until recently, two types of viral hepatitis were recog-

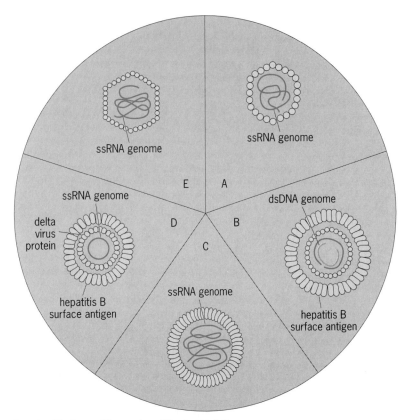

Structural features of human hepatitis viruses A–E.

nized: infectious hepatitis or short-incubation hepatitis, now known to be caused by hepatitis A virus, and serum hepatitis or long-incubation hepatitis, now known to be caused by hepatitis B virus. Currently, five viruses, designated hepatitis A, B, C, D, and E, are considered the primary agents of viral hepatitis in humans, and account for most of the diagnosed cases.

Following the identification of hepatitis A and B viruses, it was evident that there were additional primary hepatitis viruses, and the term non-A, non-B hepatitis was used to describe these unidentified viruses and to point out that, at the time, their diagnosis was one of exclusion. The non-A, non-B hepatitis viruses were separated into parenterally transmitted (not through the digestive tract) and enterically transmitted (intestinal or through the digestive tract), and eventually, two different viruses were identified, the parenteral form being named hepatitis C and the enteric form hepatitis E. The hepatitis delta agent, or hepatitis D virus, was discovered as an agent closely associated with hepatitis B virus. It is likely that additional human hepatitis viruses will be identified as more information on the basic characteristics of the viruses is attained and as better diagnostic tests are developed and applied.

Although these five hepatitis viruses cause similar acute disease, they are distinct from each other, and have different biological characteristics and modes of transmission. Some of the structural features of the viruses are shown in the **illustration**. Characteristics of the viruses and their associated

disease are summarized in the **table**. Several characteristics including final taxonomic placement, are currently unclear and subject to change as more information becomes available.

Hepatitis A. This virus has a single-stranded RNA genome (ssRNA genome) and was demonstrated by electron microscopy in 1973 in stool samples of individuals with acute hepatitis. Most cases of hepatitis caused by hepatitis A virus are self-limiting, and mild or asymptomatic infections, often without jaundice, are relatively common. Primates, especially chimpanzees and marmosets, can be experimentally infected with the virus, and may possibly be infected naturally. However, humans are the major natural host for hepatitis A virus. Fecal material is the primary source of contamination, and transmission is mainly by ingestion. Person-to-person transmission is common, and is frequently seen within households and among children in nurseries or daycare centers. Food-borne and water-borne outbreaks are also common, and the virus can be passed from shellfish harvested from sewage-contaminated waters, especially if they are consumed uncooked. Dating from the early 1970s, hepatitis A virus infections have steadily decreased in the United States, probably because of improved sanitation and educational programs. In less developed countries, however, serologic evidence suggests that up to 90% of adults may have been exposed to the virus. Probably many of these cases are mild or asymptomatic during childhood. The virus is very resistant to the usual procedures for

Human hepatitis viruses and their epidemiological and clinical features

Characteristic	Hepatitis A	Hepatitis B	Hepatitis C	Hepatitis D	Hepatitis E
Family	Picornaviridae	Hepadnaviridae	Flaviviridae	Unclassified	Caliciviridae
Prevalence	High	High	Moderate	Low, regional	Regional
Percentage of total viral hepatitis	50%	40%	5%	<1%	<1%
Age preference	Children, young adults	All ages	All ages	All ages	Young adults
Onset	Sudden	Slow	Slow	Variable	Variable
Incubation period	15–45 days (avg. 25–30)	7–180 days (avg. 60–90)	15–160 days (avg. 50)	28–45 days	14–56 days (avg. 35–42)
Acute disease	Mild to moderate	Moderate	Mild to moderate	Severe	Moderate (severe in pregnancy)
Fulminant disease	Infrequent	Infrequent	Infrequent	Frequent	In pregnancy
Chronic disease	No	Yes (10%)	Yes (> 50%)	Yes (50–70%)	No
Carrier state	No	Yes	Yes	Yes	No
Liver cancer	No	Yes	Yes	No ?	No
Main route of transmission	Fecal-oral	Parenteral , (blood, sexual, perinatal)	Parenteral (blood, sexual)	Parenteral (blood, sexual)	Fecal-oral
Location					
Stool	Yes	No	No ?	No	Yes
Blood	Yes	Yes	Yes	Yes	?
Saliva, semen	Rare (saliva)	Yes	?	Yes	?
Control and prevention	Hygiene, sanitation, vaccine, immunoglobulin	Public health, vaccine, interferon, immunoglobulin	Public health, interferon	Public health, hepatitis B vaccine, interferon	Hygiene, sanitation

disinfection, and remains infectious in the environment for long periods. As there is no specific antiviral drug available for persons with acute hepatitis A, supportive care, including rest and proper nutrition, is the recommended treatment. Anti-hepatitis A immunoglobulin is approximately 80–90% effective in preventing clinical illness when given prior to exposure or early during the incubation period. A recently developed formalin-inactivated vaccine appears to be highly immunogenic and effective against hepatitis A virus, and its use may reduce the incidence of infection and disease.

Hepatitis B. The identification of hepatitis B virus was accomplished when a new antigen, known as Australia antigen, was found in the blood of an Australian aborigine in 1963. This antigen was associated with a form of hepatitis, and examination of samples from individuals with Australia antigen showed the presence of a spherical or filamentous form identified as hepatitis B surface antigen, and Dane particles identified as the actual hepatitis B virus. The virion has several unusual features which differentiate it from other viruses: its genome is circular deoxyribonucleic acid (DNA) and is partially double stranded (dsDNA). It uses a reverse transcriptase and a ribonucleic acid (RNA) intermediate to replicate itself. The virus is not related to any other known human virus, and is classified in the family Hepadnaviridae. However, similar viruses of animals, especially one infecting the woodchuck, provide useful animal models for the disease in humans.

Hepatitis B virus is a primary cause of acute and chronic hepatitis, and of hepatocellular cancer. In the United States, up to 0.5% of the population may be chronic carriers of the virus. Sexual transmission is common; about one-half of these infections are thought to be sexual in origin. Worldwide, as many as 300–350 million individuals may have chronic hepatitis B virus infection, and these infected individuals are the major viral reservoir. They are also at a high risk of developing serious health problems, and more than 1 million individuals die from the disease annually. Hepatocellular carcinoma is about 10–300 times more frequent in carriers of hepatitis B virus than in noncarriers, depending upon the specific population. In some parts of the world the number of carriers is very high. Although hepatitis B virus is involved in only 1–2% of the hepatocellular cancer cases in North and South America, in Eastern Asia and sub-Saharan Africa it is estimated that between 5 and 20% of the population are carriers of the virus and have an increased risk of developing liver cancer. The probability of developing chronic hepatitis B virus seems to depend largely upon the age at which an individual is infected. Chronic hepatitis may develop in as many as 90% of babies infected at birth, about 20–25% of those infected at 1–5 years of age, and about 5–10% of older children and adults.

Hepatitis B virus is transmitted mainly by exposure to infectious blood and other body fluids. The virus is thought to be stable for several days on environmental surfaces, and so indirect inoculation can occur from inanimate objects. The risk of transfusion with contaminated blood, once a major risk, has been reduced to very low levels through proper testing of the blood supply. The sharing of contaminated needles during intravenous drug use, accidental needle sticks, injuries from sharp instruments, and sexual and perinatal transmission remain major public health problems. The incubation period of the virus is variable, ranging between 1 week and about 5 months. Humans are the primary natural host, although subhuman primates, especially the chimpanzee, can be infected experimentally.

The initial hepatitis B vaccine was a purified, inactivated suspension of the surface antigen harvested from human plasma. Later, a recombinant vaccine was produced by inserting a plasmid containing the gene for the hepatitis B surface antigen protein into yeast cells. Both vaccines have been effective in prevention of disease, with protection lasting for at least 10 years. The duration of protection and its role in reducing liver cancer are current areas of active research. Although vaccines were used initially in selected high-risk groups, many countries, including the United States, now recommend routine vaccination of infants.

Hepatitis C. This virus causes acute and chronic hepatitis and cirrhosis and is a major contributor to hepatocellular carcinoma throughout the world. A very high percentage of persons with hepatitis C virus infection become chronically infected, and approximately one-half of these are likely to develop chronic liver disease or cirrhosis. In some countries, for example Japan, hepatitis C virus may be as important as hepatitis B virus as a precursor to liver cancer. The primary means of transmission of the virus is by contaminated blood. By the mid-1970s, it was obvious that a large number of cases of posttransfusion hepatitis were not caused by hepatitis B virus. At that time, it was possible to accurately identify only hepatitis A and hepatitis B viruses, so the name non-A, non-B hepatitis was applied to the unidentified viral agent(s) as a diagnosis of exclusion. During the next 15 years, standard techniques of virology failed to identify the virus. Finally, in 1989 molecular biological techniques accomplished the cloning of the virus, and it was identified and named hepatitis C virus.

The virion was shown to be a small enveloped particle with a single-stranded RNA genome. It is thought to be present as one of two forms: intact whole virus particles or nucleocapsids (nucleic acid and surrounding protein) without the host-derived lipid envelope. The virus seems most similar to members of the family Flaviviridae, although its final classification is subject to further studies. Hepatitis C virus was recognized as the major cause of posttransfusion hepatitis, responsible for 80–90% of the cases in the United States, Asia, and Europe prior to anti–hepatitis C virus screening. Routine screening of blood donors for hepatitis C virus began in the United States during the middle of 1990. Currently, the main risk factor is through abuse of injection drugs. Interferon alpha has been approved for treating chronic hepatitis C, although its benefits seem transient. The antiviral drug ribavirin has demonstrated the ability to improve liver enzyme levels in individuals with chronic hepatitis C during treatment, but its long-term value is unclear. Current research suggests that a combination of interferon and antiviral drug(s) is most likely to prove useful in treatment of the disease. The observations that individuals can apparently be reinfected by hepatitis C virus, and that the virus may readily undergo mutation, may make development of an effective vaccine more difficult.

Hepatitis D. In 1977 an antigen was noted in hepatocytes from severe cases of individuals infected with hepatitis B virus and was also found inside viruslike particles that had a coat of hepatitis B surface antigen. This particle was later named the delta agent, or hepatitis D virus. Hepatitis D virus is a defective or incomplete virus that requires the presence of hepatitis B virus as a helper for transmission. In fact, its small, circular RNA genome is actually enclosed in a coat of the hepatitis B surface antigen. Transmission of hepatitis D virus is similar to that of hepatitis B virus, although sexual transmission may be less efficient than for hepatitis B virus. Hepatitis D virus can be acquired either simultaneously with hepatitis B virus as a coinfection or as a superinfection of chronic hepatitis B virus carriers. The combination of both viruses leads to more severe illness and a higher incidence of chronic liver disease with cirrhosis than is seen with hepatitis B virus alone. Laboratory diagnosis is accomplished by demonstration of the hepatitis D antigen or antibodies to the antigen in the patient's blood.

The geographic distribution of hepatitis D virus is irregular or patchy. It is most frequently found in Italy, Africa, the Middle East, and South America, and is endemic in persons with hepatitis B virus infection in those areas. In nonendemic areas (such as the United States and Europe), it is more common in individuals frequently exposed to blood and blood products, especially hemophiliacs and drug addicts. There is currently no specific means of preventing hepatitis D infection, but because the virus is dependent upon the presence of hepatitis B virus to provide its coat, prevention of hepatitis B virus infection also prevents infection with hepatitis D virus. Thus, public health measures to reduce the risk of transmission of hepatitis B virus and the vaccine against hepatitis B virus are also effective against the transmission of hepatitis D virus.

Hepatitis E. This virus is a major cause of enterically transmitted hepatitis. Hepatitis E virus is transmitted mainly by the fecal-oral route, especially by drinking contaminated waters. This virus was first noted in 1955, during an epidemic in New Delhi, India. Initially, the virus was thought to be hepatitis A virus. Molecular cloning techniques have shown the virus to be spherical, nonenveloped and to have a single-stranded RNA genome. Hepatitis E virus has not been shown to cause chronic infections. However, pregnant women are at an increased risk to the virus, and may have a mortality rate as high as 20%. It is unclear if immune serum globulin provides protection, and there is currently no approved treatment. Based upon its physical properties and morphology, hepatitis E virus is considered similar to caliciviruses which infect the intestines, but may be reclassified as more is learned about the virus. Hepatitis E virus is apparently widespread, and will likely be found worldwide as better diagnostic assays are developed. Research efforts are aimed at developing an effective vaccine, possibly for use in combination with the hepatitis A virus vaccine.

Prevention and control. The primary means of combating viral hepatitis is by prevention of exposure to the virus and immunization against those viruses for which a vaccine is available. Preventing exposure depends largely upon breaking the lines of transmission of the virus. For those types which are blood- and secretion-borne or are sexually transmitted (hepatitis B, C, and D viruses), proper testing of the blood supply has provided a powerful tool for eliminating exposure via transfusion, and awareness of the risk of using contaminated needles helps against transmission by intravenous drug use. The proper use of condoms is an important means of preventing sexual transmission of these viruses. Education and prevention programs for the human immunodeficiency virus (HIV) are equally useful for preventing the spread of these hepatitis viruses. For the hepatitis viruses which are transmitted in contaminated food or water (hepatitis A and E viruses), proper water treatment and food handling are the primary tools for preventing exposure. Thus, good personal hygiene and public health measures are the first line of defense against hepatitis viruses, and are very successful when properly applied. Administration of immunoglobulin prior to or very soon after exposure is most useful in preventing disease or reducing its severity for hepatitis A and B viruses, but less certain for the other hepatitis viruses. Safe and effective vaccines are currently available for hepatitis A and B viruses, and may become available for the other hepatitis viruses.

For background information SEE ACQUIRED IMMUNE DEFICIENCY SYNDROME (AIDS); ANIMAL VIRUS; ANTIGEN; HEPATITIS; IMMUNOGLOBULIN; INFLAMMATION; LIVER in the McGraw-Hill Encyclopedia of Science & Technology.

John M. Quarles

Bibliography. E. E. Mast and M. J. Alter, Epidemiology of viral hepatitis: An overview, *Sem. Virol.*, 4:273–283, 1993; P. R. Murray et al., *Medical Microbiology*, 2d ed., 1994; K. J. Ryan (ed.), *Sherris Medical Microbiology: An Introduction to Infectious Diseases*, 3d ed., 1994; D. O. White and F. J. Fenner, *Medical Virology*, 4th ed., 1994.

Hybrid materials

A hybrid material is a combination of two or more different material systems that results from the attachment of one material to another. Polymets are a new class of hybrid materials composed of metals and polymers. Recent research has demonstrated the feasibility of extruding metal-polymer composites (polymets) of aluminum (Al) with poly(etherether ketone) and of aluminum with a liquid-crystal polymer. Extrusion through either a high-shear 90° die (**Fig. 1**) or a converging conical die improves the properties of these aluminum-polymer composites. The improvement in properties is believed to be the result of texture development, that is, the bulk and molecular orientation of the polymer in the extrusion direction. The yield strengths of these metal-polymer composites were found to be 13–18% greater than that predicted by the rule of mixtures, and their specific yield strengths 14–21% greater than that of the aluminum control specimen.

Material processing. Numerous researchers have found that the properties of polymers such as polyethylene, polypropylene, and poly(etherether ketone) have been substantially improved by extrusion and drawing through conical converging dies. Nearly a threefold increase in the modulus of elasticity for poly(etherether ketone) after drawing through a die reduction of 3:1 (the area reduction of a material to one-third its original size prior to being drawn through a die) at 310°C (590°F) has been reported. Such an improvement has been attributed to the molecular alignment produced by the drawing operation. Several studies have indicated that thermotropic, liquid-crystal copolyesters that exhibit rigid rodlike morphologies in the melt are readily drawn into fibers. A copolyester is an aromatic thermoplastic polyester that is melt processable. Another study found that fibers produced from copolyesters of 2,6-naphthyl and 1,4-phenyl exhibit a hierarchical fibril structure, that is, an orderly level of structure—micro (smaller) and macro (larger). Diameters of these fibers range from 5 micrometers (macrofibrils) to 0.05 μm (microfibrils), depending upon local shear stresses. Other research revealed that under certain conditions thermotropic, liquid-crystal polymers (LCP) form high-modulus and high-strength filaments when deformed in a flexible polymer matrix at high strain rates.

Recent research involved polymets that were prepared from commercially pure aluminum (99.99% pure) and a wholly aromatic, thermotropic, liquid-crystal copolyester. In these materials the aluminum forms the matrix; the polymer is the second-phase reinforcement. The processing involved extensive extrusion deformation in order to align the polymeric material directionally. In addition, a commercially pure aluminum alloy was prepared in an identical manner to that used to produce the polymets, and it was used as a control specimen.

The first step in the processing was communition of the pellets of the liquid-crystal copolyester, 3–4 mm (0.12–0.16 in.) in diameter, which were screened to −20 mesh. The metal-polymer blends were prepared by mixing −240/+325 mesh, commercially pure aluminum powder, and powdered liquid-crystal copolyester (5–20 vol %) in a rotating V-cone blender for 1 h. The powder blends were sealed in a fully annealed aluminum can. The canned powders were vacuum degassed at 300°C (570°F) for approximately 1 h. The materials were then hot-extruded on a 200-ton extrusion press at 300°C (570°F), by using an extrusion die with an angle of 45° and area reduction 10.7/1. The polymets were evaluated after the can material was removed (scalped). Portions of the materials were sealed in cans, vacuum degassed, and re-extruded for effective area reductions of 114/1 and 1225/1.

Microstructure and mechanical properties. In order to better understand the deformation behavior of the aluminum-polymer hybrid materials, fractured tensile specimens were examined by using scanning electron microscopy. The fracture surfaces of the tensile specimens that were tested were examined by using a scanning electron microscope operated at 30 kV in the secondary electron emission mode.

Standard tensile tests were performed on the consolidated alloys in order to evaluate their response at ambient temperature. Several observations involving the microstructures were made:

The polymer phase elongates and thins in the extrusion direction. The observed breadth of the polymeric phase in polymets extruded at 10.7/1 (to 1/10.7 of its original size) is between 20 and 50 μm, and the breadth of those extruded at 114/1 (to 1/114 of the original size) is less than 10 μm. The breadth of the polymer phase in polymets extruded at area reductions of 1225/1 is between 0.5 and 3 μm.

Similar to the material that is 95 vol % Al + 5% LCP aromatic polyester, the polymer phase in polymets with 10 and 20 vol % LCP aromatic polyester elongates and thins in the extrusion direction. However, the higher polymer content promotes polymer segregation, and isolated aluminum powder particles often appear as "islands" in a "sea" of polymer. During extrusion, the polymer inhibits

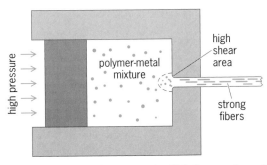

Fig. 1. Polymet extrusion process in which fibers are formed in place at the melt temperature.

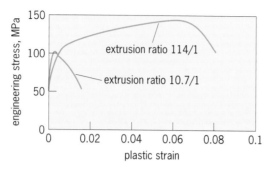

Fig. 2. Stress-strain diagram for the polymet 95 vol % aluminum + 5% liquid-crystal copolyester extruded at 10.7/1 and 114/1. Engineering stress is the ratio of the applied load to the original cross-sectional area.

and often prevents these isolated aluminum powder particles from bonding with the matrix.

Tensile properties. The strength and ductility of the polymets decreases with increasing polymer content. For polymets extruded at 10.7/1, the yield strength decreases from 100.9 megapascals (14.63×10^3 lb/in.^{2}) to 55.9 MPa (8.10×10^3 lb/in.^{2}) as the vol % liquid-crystal copolyester is increased from 5% to 10%. Ductility also decreases; tensile elongation declines from 1.6% to 1.2%.

Thus, the properties of the polymets are affected by the extrusion ratio: high extrusion ratios decrease the yield strength, increase ultimate tensile strength, and improve ductility of the polymets.

The yield strength of aluminum extruded at 10.7/1 and 114/1 decreased from 115.7 MPa (16.27×10^3 lb/in.^{2}) to 83.0 MPa (12×10^3 lb/in.^{2}). Ductility of the aluminum increased from 18.3% to 27.0%; however, the ultimate tensile strength of the aluminum remained virtually unchanged, that is, between 136 and 138 MPa (15.72 and 20×10^3 lb/in.^{2}).

The stress-strain curves for the polymet 95% Al + 5% LCP aromatic polyester extruded at 10.7/1 and 114/1 are shown in **Fig. 2**. The yield strength of this polymet decreased from 100.9 MPa (14.63×10^3 lb/in.^{2}) to 83.9 MPa (12.13×10^3 lb/in.^{2}) when extruded at reduction ratios of 10.7/1 and 114/1, respectively; however, the ultimate tensile strength increased from 104 MPa (15.08×10^3 lb/in.^{2}) to 144 MPa (20.88×10^3 lb/in.^{2}). The elongation of 95% A + 5% LCP aromatic polyester increased significantly with extrusion ratio from 1.6 to 7.5.

An approximation of the energy required to cause tensile failure was obtained by measuring the area under the stress-strain curve. The energy expended to cause tensile failure of a polymet is affected by the extrusion ratio: high extrusion ratios increase expended energy. The energy absorbed by the composite 95% Al + 5% LCP aromatic polyester extruded at 10.7/1 and 114/1 increased from 1.3 millijoule mm^3 to 10 mJ/mm^3, that is, nearly 670%.

The yield strengths of the polymets decreased with increased extrusion ratio. Interestingly, there was a concomitant decrease in the yield strength of the aluminum control specimen.

More work and further microstructural analysis are required to understand the reasons for the var-

ious increases and decreases in properties. Possible causes could be processes involving matrix recovery, recrystallization, and grain growth.

In order to more accurately assess the effect of the extrusion ratio on polymet yield strength, the decrease in matrix yield strength (as observed in the aluminum control sample) must be taken into account.

The normalized yield strengths of the polymets increase with extrusion ratio. The average normalized yield strength of the polymets extruded at 114/1 is 26.7% greater than those extruded at 10.7/1. The normalized yield strength of the polymet 95% Al + 5% LCP aromatic polyester increased 15.3%, that is, from 100.9 MPa (14.63×10^3 lb/in.^{2}) to 116.3 MPa (16.86×10^3 lb/in.^{2}). By inference, the increase in normalized yield strengths of the polymets is attributable to morphological and crystallographic changes in the polymeric phase.

Fracture behavior. Representative fractographs of polymets extruded at 114/1 reflect polymer films, typically 0.5 μm thick and more than 10 μm wide, which are clearly visible. The polymer films appear to emanate from grain boundaries or prior particle boundaries. This type of fracture morphology may indicate poor matrix cohesion, and result in lower ductility and reduced transverse strength. There was insufficient 1225/1 extruded material produced to determine tensile properties, but the very fine fibrils found in the microstructure of the 5% LCP aromatic polyester material appear encouraging.

Fractographs of the polymet 95% Al + 5% LCP aromatic polyester clearly show fibrils emanating from matrix dimples. The fibrils are approximately 1.5 μm in diameter, and the dimple size and spacing are similar to those of the aluminum control specimen.

The research of the group that studied the aluminum-polymer polymets yielded a number of observations: (1) formation of in-place polymer fibrils and films result from extrusion processing; (2) the ultimate tensile strength and ductility of Al–LCP aromatic polyester polymets is improved by increasing the extrusion ratio from 10.7/1 to 114/1; (3) the yield strengths of the polymets decline by increasing the extrusion ratio, but the normalized yield strengths of the polymets increase an average of 26.7%; (4) increases in the ultimate tensile strengths and normalized yield strengths of the polymets with extrusion ratio are attributable to formation of polymer fibrils and films; and (5) the energy required to produce tensile failure is significantly enhanced by increasing the extrusion ratio from 10.7/1 to 114/1.

For background information SEE COMPOSITE MATERIAL; LIQUID CRYSTALS; STRESS AND STRAIN in the McGraw-Hill Encyclopedia of Science & Technology.

Mel Schwartz

Bibliography. R. Bhagat et al. (eds.), *Metal and Ceramic Matrix Composites: Processing, Modeling and Mechanical Behavior*, 1990; A. I. Isayev and M. Modic, Self-reinforced melt processible polymer composites: Extrusion, compression, and injection molding, *Polym. Compos.,* 8(3):158–175, 1987; A. Richardson et al., The production and properties of poly(arylether ketone) (PEEK) rods oriented by drawing through a conical die, *Polym. Eng. Sci.,* 25(6):355–361, 1985; L. C. Sawyer and M. Jaffe, The structure of thermotropic copolyester, *J. Mater. Sci.,* 21:1837–1913, 1986; A. E. Zachariades and J. A. Logan, The preparation of oriented morphologies of the thermotropic aromatic copolyester of poly(ethylene terephalate) and 80 mole percent *p*-acetoxybenzoic acid, *Polym. Eng. Sci.,* 23(15): 797–802, 1983.

Immunological memory

In 1781, the Faroe Islands were hit by an epidemic of measles that decimated the population. The population of this tiny island group, isolated except for the occasional visit by Danish trading ships, had never been immunized to measles. In 1846, another outbreak occurred, again brought by a contaminated ship. This time two physicians were sent to the aid of the islands. Dr. P. L. Panum, a brilliant observer, worked out the incubation period of the virus and suggested several protective measures. Most important were his observation that people older than 65 did not become ill and suggestion that they were protected because of their exposure during the 1781 epidemic. Because neither their children nor close relatives were protected, there could have been no measles on the islands for 65 years. Their immune systems had remembered the previous exposure to measles and could mount a response that was faster and stronger than the immune responses of the younger islanders.

This capacity to generate a fast, strong, and superbly protective response at the second encounter with a pathogen is the hallmark of immunological memory. It is the reason that children get sick much more often than adults, and it is also the reason that vaccination works. There are two relevant features of this protective response: (1) the differences that make the memory immune response more efficient than the first; and (2) the mechanisms that maintain those differences for years.

Protective response. The secondary response to infection is usually faster and stronger than the primary response. It is also often tailored to the particular pathogen involved, whereas the primary is more generic (see **illus.**).

Speed. The time lag from the moment a pathogen enters the body until it is finally cleared by the immune system is due to the many steps that are required for the development of an efficient immune response. First, the immune cells must become aware that a pathogen has entered. This is the function of special antigen-presenting cells that capture some of the pathogens and take the pieces to local lymph nodes where they present the antigens to circulating immune cells. Second, among the millions of immune cells, those few that are specific for the pathogen must become activated to divide and make an army of effector cells with the same specificity. It takes several rounds of cell division before there are enough activated effector cells to deal with the pathogen. The activated effector cells must then either leave the lymph nodes and circulate to the infected areas (if they are killer cells) or begin secreting antibodies (if they are antibody-producing cells). At this stage the infection is almost over, but it nevertheless takes time for the killer cells to kill the last infected cell (for example, during a virus infection) and the antibodies to clear the last pathogen (for example, a bacterium).

Many of these steps are faster the second time. After the first exposure, circulating antibodies in the blood bind to the entering pathogen and either destroy it completely or, if that fails, cover it so that the antigen-presenting cells can more easily and quickly pick it up. Because of the cell division in the first response, the number of specific immune cells is also higher the second time, and it therefore takes less time to generate enough effectors to fight off the pathogen and to produce a threshold level of antibody. There is also some evidence that the activation requirements of memory cells may be less stringent than those of inexperienced cells, thus allowing a faster response by a greater number of cells.

Strength. The increased number of cells available to respond in a secondary response contribute to an increase in response strength. However, there is one additional feature: during the primary response, the B cells making antibodies go through a series of mutations from which the highest-affinity antibody producers are selected. Thus, the antibodies in a secondary response are not only generated faster and to higher levels but also consist of antibodies having a greater binding capacity to the pathogen.

Class. The immune system tailors the response to the type of pathogen and the organ that is affected. For example, it normally generates killer cells to clear viruses, and immunoglobulin E to clear worms. The beginning of a primary response tends to be specific for the pathogen but rather generic as to the class of antibody or effector cell. Generally, for example, immunoglobulin M antibodies are involved, regardless of the pathogen. Later on in the primary response and during the secondary response, the class of response switches and new types of antibodies are made, such as immunoglobulin G, or immunoglobulin A. Different classes of antibodies have different functions. For example, immunoglobulin A antibodies are very good at protecting against infections through respiratory and intestinal mucosa. They are the primary type of antibody made after immunization with the oral polio vaccine. Immunoglobulin E antibodies are useful at clearing worms, and are most often found during the late primary and secondary responses to worm infections.

Although the mechanisms that control these decisions are not yet understood, there is emerging evidence that different tissues affect the class of response. The eye, for example, which protects itself to a large degree by producing lysozyme in tears, also produces substances to ensure the production of immunoglobulin, the main antibody found in tears. The gut seems to do the same. The three features of speed, strength, and effector class can work together so efficiently that a previously exposed or vaccinated individual may never even feel ill during subsequent exposures to the same pathogen.

Maintaining memory. There are currently two main theoretical points of view. One postulates that immune cells switch during a primary response to a long-lasting memory state whose maintenance requires no additional input, the other that memory cells need to be restimulated periodically. In other words, one view says that the immune system remembers and the other that it must be reminded (see illus.).

Long-lasting memory state. Virgin and memory cells have different life-spans. The newly developed virgin immune cell is short lived. It circulates through the body for a few weeks and then dies to make room for new cells. However, it changes in two important ways during the primary response: first, it reproduces itself, making large numbers of identical daughter cells. Second, the daughter cells have a long life-span and can circulate for years, in the absence of any further stimulation, waiting for the next invasion by the same pathogen. Therefore, the increased speed and strength of the secondary response are both results of the increased frequency of antigen-specific cells, and the persistence of these differences is due to the change in their intrinsic life-span.

The advantage of this model is that the mechanism is simple. The disadvantages are that it is diffi-

cult to see how the immune system maintains a fairly constant number of cells, if they expand each time there is an infection; or, if it has a way to hold to a constant number, by what mechanism it maintains the slots for old memory cells under the onslaught of constantly expanding memory cells to new pathogens.

Restimulation of memory cells. This view sees the maintenance of memory as a dynamic process; virgin and memory cells have similar life-spans and memory is the result of periodic restimulation by antigen. In this model, the pool of memory cells is constantly contracting as cells die and expanding as they are restimulated.

There are three ways in which a pathogen can restimulate memory cells. Some pathogens (for example, herpes viruses) persist in the body at very low levels for very long periods. In essence, there is a constant dynamic balance between the pathogen's attempt to expand and the immune system's counterattack. Although the pathogen cannot cause overt symptoms during this period, the interactions act as a persistent restimulation of the immune system. Other pathogens do not persist in the body but do persist in the environment, and every time they reinfect the host they serve to restimulate the memory cells. For example, parents are boosted by the diseases their children bring home from school. In addition, a third mechanism of restimulation has been constructed by the immune system itself to cover those cases in which the pathogen might not persist or return often enough to be effective at maintaining memory.

The immune system keeps a library of the antigens it has seen. The antigens are complexed to antibodies bound to the surface of specialized cells in spleen and lymph nodes, called follicular den-

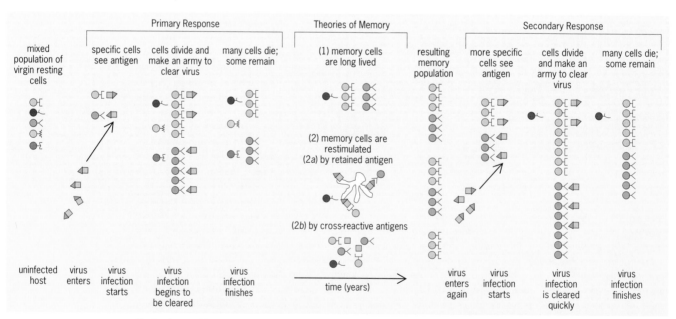

Mechanisms of immunological memory that can maintain the differences between primary and secondary repsonse to a pathogen for years: the immune system either (1) remembers (memory cells are long lived) or (2) must be reminded (memory cells are restimulated).

dritic cells, and have been found to last at least 1 year. Periodically, bits of the retained antigen are released from this depot, and the immune system reacts as though it were seeing the real pathogen. Thus, the immune system is periodically restimulated even when the dangerous pathogens are absent.

Another source of antigen can help to maintain memory. Although immune cells are each quite specific for particular antigens, their ability to discriminate differences is not perfect. Different antigens that share enough common elements to be able to stimulate the same set of immune cells are called cross reactive. Some immune cells specific for influenza, for example, might also be able to recognize polio or hepatitis virus. Thus, infections with new pathogens can also induce cross-reactive memory to old ones.

Because of such cross-reactions, an individual's previous immunological history influences the response to new antigens. For example, a primary response to influenza might consist of virgin cells specific for flu and a few memory cells for polio that also cross-react with flu. Memory cells respond more easily than virgin cells, and therefore should have an advantage in the competition for expansion. The effect of such competition between memory and naïve cells is often seen in responses to flu, where the response to a new strain (such as Hong Kong flu) can be overwhelmingly composed of memory cells responding to antigens that are also present on a previously observed strain (such as Puerto Rican flu). Thus, as an individual ages, the system becomes dominated by memory cells that respond to current environmental pathogens while retaining memory of old, less frequently encountered pathogens.

The advantage of this model is that it easily explains how memory cells can persist over time, despite competition from new virgin cells and from memory cells responding to other antigens. The disadvantage is that the maintenance of memory to old, infrequently encountered pathogens may not be very efficient. However, from an evolutionary point of view, memory that lasts beyond the child-rearing years may be superfluous.

Vaccines. The controversy between the two views of memory maintenance, one that memory cells have intrinsic longevity and the other that they require periodic restimulation, is not resolved, and experiments are under way to test the different hypotheses. This question is not only an interesting theoretical problem but also has relevance for the rational design of effective vaccines. For example, if follicular dendritic cells and their antigen and antibody complexes are major contributions to memory maintenance, then it becomes essential to design vaccines able to induce strong antibody responses. If cross-reactions contribute to memory, the design should incorporate cross-reactions to environmentally common antigens. With a better understanding of the complexity of the immune system, scientists should be able to induce protective memory against the many pathogens living in our environment.

For background information SEE ANTIBODY; ANTIGEN; ANTIGEN-ANTIBODY REACTION; CELLULAR IMMUNOLOGY; IMMUNITY; INFLUENZA in the McGraw-Hill Encyclopedia of Science & Technology.

Polly Matzinger; Francesca Di Rosa

Bibliography. C. A. Janeway, Jr., and P. Travers, *Immunobiology: The Immune System in Health and Disease,* 1994; D. Male et al., *Advanced Immunology,* 2d ed., 1994.

Immunology

Young mammals use their mothers' antibodies for defense while their own immune systems develop. The fetal immune system is normally sheltered from bacteria, viruses, fungi, and parasites, and therefore does not make antibodies. However, the mother is not sheltered, and the antibodies that are formed during pregnancy reflect the pathogens to which she is exposed. Those same pathogens are likely to await the newborn. Thus, a mother's immunologic experience is directly relevant to the needs of her offspring: she makes the antibodies they are most likely to need. The major antibody type in blood is immunoglobulin G (IgG). Maternal IgG antibodies are transmitted to fetal or newborn mammals and provide immunity during the first months after birth. Humans are born with IgG which has been transported across the placenta. The IgG molecules that are transported are a cross section of the mother's antibodies, and are not selected on the basis of the antigens they recognize. After birth, immunoglobulin A from breast milk can provide local defense in the gastrointestinal tract but does not enter the infant's blood.

Small amounts of IgG are transported from the maternal blood to the fetus as early as the twelfth week of pregnancy. However, the IgG concentration in fetal serum remains low until the period between the twenty-second and twenty-sixth week, when it rises approximately tenfold to approach the neonatal level. Fetal IgG synthesis is relatively low, and most IgG in the blood of newborns is maternal in origin. At birth, the IgG concentration in fetal blood often exceeds the maternal concentration, indicating active transport. IgG is transported in preference to immunoglobulin A (IgA), immunoglobulin M (IgM), or other serum proteins. There are four subclasses of human IgG: IgG_1, IgG_2, IgG_3, and IgG_4. Each subclass is transported to the fetus; however, IgG_2 is transmitted somewhat less efficiently than the other subclasses.

The placental barrier. The placenta is the interface between the pregnant woman and the fetus. Within the placenta, the maternal and fetal bloodstreams are close but separate. The mother's blood fills the spaces between finely branched chorionic villi (which extend from the outer membrane of the embryonic sac) on the fetal side of the placenta. The outermost layer of the chorionic villi, in contact with maternal blood, is called the syncytiotrophoblast. The apical surface of this layer is convoluted into a dense array of fingerlike microvilli, creating a large surface area 75–183 ft^2 (7–17 m^2) over which molecules can be transported. Beneath this layer, the cytotrophoblast forms a complete cell layer in early pregnancy but becomes discontinuous as the surface area of the chorionic villi expands. The cores of the villi contain fetal blood vessels and loosely packed cells (the stroma), including macrophages (Hofbauer cells). The cellular barriers that separate the blood of the mother and fetus are the syncytiotrophoblast and the endothelial cells that line the fetal capillaries, and early in gestation the cytotrophoblast.

Maternal IgG is moved across the placental barrier. In contrast, soluble immune complexes formed by maternal IgG and paternally derived fetal antigens are mostly trapped within the placenta. Both transport and trapping appear to involve receptors (Fc gamma receptors) for the Fc region of IgG in the placenta. First, Fc receptors on syncytiotrophoblast and endothelial cells may transport IgG from maternal blood in the intervillous spaces to the fetal capillaries. Second, Fc gamma receptors on Hofbauer cells and on the endothelium may bind immune complexes and prevent their deposition in fetal tissues. If the antibody IgG is considered as a letter Y, the Fc region contains part of the hinge of IgG and domains C_H2 and C_H3, the tail. This part of the molecule is the same in all immunoglobulins of the same subclass, regardless of what antigens the antibodies bind. For this reason, Fc gamma receptors are able to bind IgG molecules of all specificities.

Route of IgG transport. The route of IgG transport has been investigated by studying the localization of maternal IgG in the placenta and by visualizing labeled IgG taken up by isolated placental tissue. IgG is detected at the surface of and within the syncytiotrophoblast during the first trimester and at term. During the first trimester, IgG penetrates only the syncytiotrophoblast, perhaps unable to cross the nearly complete cytotrophoblast layer. By term, IgG is also present in the stroma, endothelial cells, and capillaries (but not cytotrophoblast). At the ultrastructural level, IgG is detected in coated pits, coated vesicles, noncoated vesicles, tubulovesicular bodies, and multivesicular bodies in the syncytiotrophoblast. Coated vesicles isolated from term placenta also contain IgG. IgG is located in vesicles and multivesicular bodies within endothelial cells, but not in the paracellular clefts between them.

These observations and timed uptake studies suggest the following transport route: First, IgG enters the syncytiotrophoblast by endocytosis at coated pits. Next, coated vesicles deliver IgG to tubulovesicular bodies (which may be early endosomes). IgG is then sorted to noncoated vesicles in the basal cytoplasm that deliver it to the basal plasma membrane. IgG is released into the intercellular spaces of the stroma, and finally enters the fetal circulation by crossing the capillary endothelia, again by a vesicular pathway.

IgG binding proteins in placenta. IgG binding sites have been detected in syncytiotrophoblast, fetal capillary endothelium, and Hofbauer cells. Two approaches have been used to characterize placental Fc gamma receptors. Proteins were solubilized from placental membranes and those that bound IgG were isolated. Two have been identified as placental alkaline phosphatase and annexin II. It remains unclear whether the ability to bind IgG is related to the biological functions of these molecules. More recently, antibody and nucleic acid probes have been used to look in placenta for Fc gamma receptors that have already been studied in other tissues.

Distribution of Fc gamma receptors. Fc gamma receptors on white blood cells mediate several of the effector functions of the immune system. These functions include phagocytosis of antibody-coated pathogens by monocytes, macrophages, and neutrophils, and antibody-dependent cell-mediated cytotoxicity by monocytes, macrophages, neutrophils, and natural killer cells. Monoclonal antibodies (mAbs) raised against leukocyte Fc gamma receptors have been used to detect Fc gamma receptor I (CD64), Fc gamma receptor II (CD32), and Fc gamma receptor III (CD16) in placenta. Additionally, messenger ribonucleic acids (mRNAs) for these three classes of Fc gamma receptors are detected in placenta.

Fc gamma receptor III and II are present in the syncytiotrophoblast and capillary endothelium, respectively. Because these cell layers separate the maternal and fetal blood, Fc gamma receptor II and III may function in the transport of maternal IgG. Alternatively, these receptors together with Fc gamma receptor I, II and III on Hofbauer cells may sequester immune complexes, preventing their transmission to the fetus.

Distribution of neonatal Fc receptors in placenta. The neonatal Fc receptor was originally purified from neonatal rat intestine. This receptor has a high affinity for monomeric IgG. In contrast to humans, rats and mice receive relatively little maternal IgG during gestation compared with that acquired after birth. They obtain most maternal IgG from milk, by transport across the intestinal epithelium. Both prenatal and postnatal transmission of IgG are carried out by the two subunits of the neonatal Fc receptor, beta 2 microglobulin and an integral membrane protein that resembles the alpha chains of class I major histocompatibility complex proteins.

Human placenta contains neonatal Fc receptors. Neonatal Fc receptor RNA is detected in the syncytiotrophoblast of first trimester and term placenta by in-situ hybridization, and the neonatal Fc receptor protein is detected in the same location by anti–neonatal Fc receptor alpha chain antibodies. This pattern of expression is consistent with a role for human neonatal Fc receptor in transport of maternal IgG to the fetal circulation across the placental syncytiotrophoblast, but not the endothelium. Syncytiotrophoblast lacks classical class I major histocompatability complex proteins and is reported to contain no beta 2 microglobulin, but recent reevaluation suggests that beta 2 microglobulin is present, although nonuniformly distributed. Thus, functional neonatal receptors could be assembled in beta 2 microglobulin–positive regions of syncytiotrophoblast. Human neonatal Fc receptor binds IgG optimally below pH 6. Although it does not bind at the neutral pH of maternal blood, it could bind at the acidic pH within endosomes.

Model of IgG transport. A model for the mechanism of IgG transport is as follows. Maternal IgG is delivered by fluid-phase endocytosis or by another protein (potentially Fc gamma receptor III) to endosomes in syncytiotrophoblast, where it binds neonatal Fc receptor. Vesicles that bud from endosomes would then carry IgG bound to neonatal Fc receptor to the basal plasma membrane, where IgG would be released at neutral pH. IgG released on the stromal side of the syncytiotrophoblast must cross the capillary endothelium in order to enter the fetal circulation. This process is transcellular, and does not occur by way of the paracellular clefts between endothelial cells. Fc gamma receptor II is the only Fc gamma receptor that has been identified in the fetal vessel endothelium, and may therefore mediate IgG transport across endothelial cells.

Medical implications. The clinical importance of maternal IgG for the immunologic defense of the newborn is well established. Most IgG is transmitted to the fetus late in the third trimester of pregnancy; therefore, very premature infants have low serum IgG and are especially vulnerable to perinatal infection. A pooled IgG preparation, intravenous gamma globulin, is sometimes given to these infants. Children of women with abnormally low serum IgG also have low IgG at birth and are at risk of infection. Placental transmission has been exploited in such cases by administering intravenous gamma globulin to the mother, thereby increasing the amount of antibody available to the fetus.

However, not all antibodies are beneficial. As the following examples indicate, maternal IgG antibodies resulting from isoimmunization by fetal erythrocytes, leukocytes, or platelets can harm the fetus and newborn. Antirhesus antibodies that cross the placenta cause lysis and stimulate phagocytosis of fetal erythrocytes, resulting in rhesus hemolytic disease. Less commonly, IgG anti-ABO antibodies cause anemia. A decrease in neonatal white blood cells and the number of platelets in the blood is caused by isoimmunization with fetal leukocytes and platelets, respectively. Additionally, transmission of autoantibodies in systemic lupus erythematosus, myasthenia gravis, Graves' disease, and other autoimmune diseases of the mother can be damaging to the fetus.

For background information SEE ANTIBODY; ANTIGEN; ANTIGEN-ANTIBODY REACTION; AUTOIMMUNITY; BLOOD; CELLULAR IMMUNOLOGY; IMMUNOGLOBULIN; PLACENTATION; PREGNANCY in the McGraw-Hill Encyclopedia of Science & Technology.

Neil E. Simister; Craig M. Story

Bibliography. P. M. Johnson and P. J. Brown, Review article: Fc gamma receptors in the human placenta, *Placenta*, 2:355–370, 1981; H. Metzger (ed.), *Fc Receptors and the Action of Antibodies*, 1990; F. Saji, M. Koyama, and N. Matsuzaki, Current topic: Human placental Fc receptors, *Placenta*, 15:453–466, 1994.

Industrial engineering

An important aspect of industrial engineering is the concept of product usability, sometimes referred to as ease of use or user friendliness. This concept enters into product design and is related directly to the quality and reliability of the product and indirectly to the productivity of the work force. Products that are well designed and usable can increase worker productivity by decreasing the time needed to learn the product or the time needed to execute tasks with the product.

Components of product usability. Consumer surveys indicate that the usability of the product is important to the perception of product quality and reliability. Designing usable products will reduce the number of products returned to the vendor and increase product sales. Cost-benefit analyses have been used to determine the cost of designing usable products and the resultant benefit to a company. These studies show that for each dollar spent on usability, 2 to 100 dollars will be recovered because of increased sales or increased productivity. Reducing the learning time, execution time, or errors can have a large effect on productivity, especially if the task is performed repetitively. Customer surveys show that usability is an important component of quality. In these surveys, quality is often broken down into six components (listed in order of importance): reliability, durability, ease of maintenance, usability, trusted or brand name, and price. Ease of maintenance and usability both relate to product usability. It can be argued that reliability also possesses a component of usability. If a product is too difficult to use, the customer may think that it is broken because it appears not to work properly. Thus, the customer would return the product to the vendor, not because it was unreliable but because it did not work the way the customer thought it should.

Since quality is often the main determinant in a customer's decision to buy the product, products that are highly usable are more likely to be purchased. Usability is especially important for computer products, both hardware and software, because poorly designed products can be almost impossible to use. Software products often emphasize their user friendliness.

The product's usability can be measured in terms of several usability components. These include the time required to perform a task (the execution time), learnability, mental workload (the mental effort required to perform a task), consistency in the design, and errors.

Execution time. The ability to execute tasks quickly is a sign of good usability. Execution time is dependent on the number of steps that must be performed by the user. Steps can include either mental or physical operations. Decision making, matching items in memory, and retrieving information from memory are examples of mental operations. Physical operations can include keystrokes, button pushes, rotations of knobs, moving a joystick, eye movements, or speech. Execution times can be reduced and usability increased by eliminating steps, utilizing more efficient physical operations, or laying out the keys, knobs, and buttons so that hand-movement times can be reduced. Designers must balance the elimination of steps, which increases the usability according to this execution time component, with the other usability components. For example, several keystrokes could be eliminated by having a special key that performs the subtask with a single keystroke. Although the execution time would be reduced, undesirable consequences could be an increase in learning time, more errors due to hitting the key inadvertently, an increase in the number of mental steps needed, and an increase in the hand-travel time (more keys would be needed). A designer has to make decisions about acceptable trade-offs with the different usability components.

Learnability. Usable products utilize simple operations that can be learned quickly. Learning time is related to the number of steps, methods, or selection rules that must be learned by the user. A method is a sequence of steps needed to accomplish a goal or subgoal for a task or subtask. A selection rule is an if-then rule for determining which method, from a set of alternative methods, is the most appropriate to use in the context of the task. Once again, usability trade-offs occur. An example is the telephone. A simple telephone may be easy to use because it has few features, few tasks, and few methods to learn for the tasks. A more complicated telephone, with the feature of storing phone numbers and then using one-button automatic dialing, saves execution time because of the elimination of digits (six key presses could be eliminated for a normal seven-digit number), but would be more difficult to learn about. The user would have to learn the methods to store the number and the selection rules for when to use the one-button dialing compared to the normal dialing method. Different consumers may have different needs in terms of reduced learning time or reduced execution time.

Mental workload. The designer of a usable product will try to incorporate the clues of how to use the product within the physical design instead of relying on the user's memory. This tactic is often referred to as placing the knowledge in the world instead of the head. Any design that relies on a person's memory for performing the task will increase the mental workload of the task. In developing human information-processing theories, scientists have assumed that information is stored temporarily in people's working memories. The capacity of an individual's working memory is limited; experiments have shown that people can store only three to nine items in working memory at one time. If people are required to remember more items, mental capacity is exceeded and information may be lost from working memory. The lost information may make the product impossible to use.

Consistency in design. Two kinds of design consistencies are usually considered: internal and external. Internal refers to the consistency for tasks within a product. If the user has to accomplish several subgoals to perform a task, the design is consistent if similar methods and steps within methods are used to accomplish the subgoals. Internal consistency can be measured by recording the steps required by a given method and counting the number of steps that are the same for other methods. External consistency refers to the overlap of methods and steps between different products. Computer interface designs have style guides for how the interaction should be designed by software producers. The style guides try to ensure that external consistency is achieved by using similar methods across different tasks. As an example, editing a picture (by cutting and pasting) and editing text will use similar methods (even though they are different products) by following the style guide for cutting and pasting. Designing consistent interfaces reduces the learning time, because the same methods, steps, and selection rules do not have to be learned a second time.

Errors. The errors can be categorized according to the first four components of usability. For instance, errors will be caused if a task requires so many steps that the user cannot execute the steps in the proper order (execution time); if a task has too many methods, selection rules, or steps to learn (learnability); if the information required to perform the task exceeds the capacity of working memory (mental workload); or if the steps in the task are so inconsistent with other steps that they are difficult to remember (consistency in the design).

Theories of usability. The task of designing usable interfaces is often theory driven. Although many theories exist for guiding the design, they can be broken down into four main approaches: cognitive, anthropomorphic, predictive-modeling, and empirical.

Cognitive. For the cognitive approach, theories in cognitive science and cognitive psychology are applied to the interface design to make the information display compatible with how people think and solve problems. Cognitive theories state how humans perceive, store, and retrieve information from working memory and long-term memory, manipulate that information to make decisions and solve problems, and carry out responses. These theories state that people learn through metaphors and analogies and that problem solving is enhanced through the manipulation of images. Icons and graphics are used to make products more usable by being compatible with human cognitive abilities. An important contribution of cognitive theories involves the practice of designing interfaces around a metaphor in which an old, familiar situation is applied to a new situation. As an example, most computer interfaces use a desktop metaphor. Familiar items such as file folders, trash cans, and scissors are used to insinuate a desktop metaphor in the user's head. The user can apply knowledge of how to manipulate items on a real-world desktop to understand how to manipulate items on a computer interface.

Anthropomorphic. The anthropomorphic approach applies human qualities to nonhuman entities. Under this approach, the designer uses the process of human-human communication as a model for human-product interaction. Several humanlike qualities have been designed into the interface to make products more usable: natural language, voice communication, help messages, tutoring, and friendliness.

Another important set of concepts involved in this approach is affordances and constraints. An affordance occurs when a person can look at an object and understand how to use it. Several objects have affordances associated with them: buttons are for pushing, knobs are for turning, objects with hand grips that fit the fingers are for grasping (for example, joysticks), and handles are for pulling. When an affordance is violated, the product is not very easy to use. An example is a door with a handle that is used for pushing instead of pulling.

Constraints are used so that items cannot be used in the wrong way. An example of a constraint on computer interfaces is "graying out" commands that are inoperable in certain situations. When a command word is grayed out, the word appears fainter on the display, and clicking on the command with the mouse button will have no effect. Users are constrained to only those commands that are operable at the time.

Predictive-modeling. This approach tries to predict performance of humans interacting with products similar to the predictions that engineering models make for physical systems. Models have been developed to specify the cognitive steps that must be performed to execute a task. One class of models, the GOMS (goals, operators, methods, and selection rules) models, is based upon theories of human information processing. When the usability of products is modeled, the cognitive steps are specified for performing a task. Each step, as validated from experimental studies, requires a certain time to execute. By specifying the steps and assign-ing time values to the steps, the user execution times for different product designs can be predicted. Learning time can also be calculated from the number of steps that the user must learn. Both of these measures are related to product usability. These predictive models have been shown to be extremely accurate when compared to empirical studies on usability.

Empirical. Under the empirical approach, usability questions are settled by running experiments on people to see which design is the best. Many companies, especially those producing computers or consumer electronics, maintain laboratories in which to test the usability of their products. Formal and informal experimentation can be used to test usability. Formal experiments utilize a controlled laboratory setting in which the different conditions can be manipulated without the possible confounding of uncontrolled factors. Usability is measured most often by counting the errors or recording the time needed to execute a task. Often, performance on different products or features will be compared to determine which can be executed and learned fastest with the fewest errors. In a product design cycle, formal experimentation is often used to test users on early mock-ups, later on the product prototypes, and then on the final product. The advantage of formal experimentation is that clear cause-and-effect relationships can be established.

Informal experiments use videotapes, surveys, and questionnaires as ways of providing product designers with ideas on how the product can be improved. The surveys and questionnaires can address users' perceptions of the product, including what they like and dislike about it. Clear cause-and-effect relationships cannot be established, but the information received from the questionnaires could be very valuable for alerting the product designers to potential problems in usability. Users can be videotaped in natural settings as they perform various tasks on the product. The videotapes can be especially useful for identifying errors from which the users cannot recover. These kinds of errors are likely to cause the customer to return the product. Informal experiments are especially important at the beginning and end of the design process. At the conception of the product, questionnaires can assess the need for a new product or the need to redesign an existing product. Once the product is placed on the market, product surveys can assess how the product is being received by users.

For background information SEE COGNITION; HUMAN-FACTORS ENGINEERING; INDUSTRIAL ENGINEERING; INDUSTRIAL HEALTH AND SAFETY; INFORMATION PROCESSING (PSYCHOLOGY) in the McGraw-Hill Encyclopedia of Science & Technology.

Ray Eberts

Bibliography. R. Eberts, *User Interface Design*, 1994; D. Norman, *The Psychology of Everyday Things*, 1988.

Infection

By the early 1950s, deoxyribonucleic acid (DNA) was identified as the genetic material that forms a blueprint for life which encodes all the information required for a single cell to develop into a complex organism. Until recently, however, it was not realized that DNA also functions to alert white blood cells to the presence of infection. Several types of white blood cells are able to detect a simple yet characteristic pattern in DNA molecules of bacteria and viruses that distinguishes these molecules as being foreign. In response to this DNA signal, white blood cells rapidly become activated, and within a few hours start secreting chemical messengers called cytokines which activate a broad range of immune defense mechanisms to fight the infection. It is hoped that this pathway of immune activation can be used to prepare better vaccines, to trigger the immune system to fight some types of infections better, and to destroy tumors.

DNA in vertebrates and bacteria. The genetic information in DNA is contained within the sequence of four bases (letters): adenine (A), cytosine (C), guanine (G), and thymidine (T). Combinations of these letters specify the unique sequences that make up each gene. One combination of two bases is surprisingly uncommon in the DNA of vertebrates; base AC is followed by a G only about one-third as often as would be expected. This combination is called a CpG, p indicating that the bases are linked by a phosphodiester chemical bond which forms the backbone linking the individual bases. The frequency of CpG in bacterial DNA is essentially the same as that of any other two base combinations. However, this underrepresentation of CpG in vertebrate DNA, termed CpG suppression, has been unexplained and the subject of much speculation.

Another difference in vertebrate DNA is that about 80% of the CpGs present are modified by the addition of a chemical cap, called a methyl group, on the cytosine. Only the cytosines that are followed by a G are capped in this manner; cytosines followed by other bases are not methylated. Taken together, CpG suppression and methylation distinguish microbial DNA from that of vertebrates; unmethylated CpGs are present at the expected frequency of 1 in 16 bases in bacterial DNA, but are about 1/15 as common in vertebrate DNA.

Immune defenses. The best-understood immune defense mechanisms recognize very specific patterns in the proteins on the surface of bacteria and viruses, and make antibodies against these patterns. These antibodies have to be very precise to distinguish foreign proteins from those of the host. To make high-quality protective antibodies against a new virus or bacteria that the immune system has never before encountered takes about 7–10 days. These types of immune defenses are termed acquired immunity because they develop in response to an infection (or immunization).

Although acquired immunity is extremely important in fighting off infections, fortunately other immune defenses (termed innate immunity) act much more rapidly. Bacteria can double in number every 20 min and viruses proliferate even more rapidly. If this growth rate were to remain unchecked for even a day, organisms would be completely consumed by the countless microbes. Skin provides a simple type of innate immunity by keeping a barrier of dead cells against the environment. One type of T cell defends the human body against infection by mycobacteria, which can cause tuberculosis. Another form of innate immunity is provided by certain pattern-recognition immune proteins, such as mannose-binding protein, that are able to recognize and bind to structural features common in infectious organisms but not in human tissues. When bound, these defense proteins mark the invading organisms for attack by specialized white blood cells and trigger other, more sophisticated defenses.

The activation of the immune system whereby bacterial DNA can be distinguished from self DNA depends on the detection of unmethylated CpGs, since bacterial DNA that has been modified by adding methyl groups to its CpGs is no longer stimulatory. Thus, in theory if the immune system were able to recognize DNA containing high levels of unmethylated CpG, it could safely assume that the outer defenses had been breached, and that an immediate counterattack was needed. Best of all, such a defense would be directed not by the bacterial surface proteins, which show enormous variation, but rather by the very genetic material, which is much less variable.

Several kinds of white blood cells are activated by CpG DNA, and within 30 min begin to produce and secrete a kind of chemical messenger called a cytokine. There are several types of cytokines, including more than a dozen interleukins (IL) each specialized for activating particular protective immune responses, and several types of interferons, so named because of their ability to interfere with the growth of viruses. CpG DNA activates B cells to produce IL-6 and IL-12, T cells to produce IL-6, natural killer cells to produce interferon γ, and monocytes to produce a complex combination of IL, interferons, and other protective compounds. This pattern of cytokines activates a coordinated set of immune responses that includes humoral immunity (antibodies) and cellular immunity (T cells and natural killer cells) which destroy foreign cells or infected host cells.

The effect of CpG DNA on B cells may help to promote the preferential activation of protective cells. In the absence of infection, when a B-cell antigen receptor binds an antigen for which it is specific, the cell is either inactivated or deleted by undergoing a kind of suicide called apoptosis. Typically, more than 95% of lymphocytes never participate in an immune response but die by apoptosis, the physiologic mechanism limiting the risk of excessive immune activation, or immune responses to self tissues (autoimmunity).

For B cells to avoid apoptosis, a second costimulatory signal is needed along with the antigen. Such signals are typically provided by specialized T cells that help the B cells to make antibodies. CpG DNA can also provide a costimulatory signal, and stimulates B cells to make antibody even in the absence of T cells. CpG DNA also protects white blood cells against elimination. Lymphocytes exposed to CpG DNA are protected against apoptosis, and thereby tend to increase the magnitude and duration of immune responses.

CpG DNA and disease. Like any other immune defense, there are likely to be certain circumstances in which immune activation by CpG DNA may have harmful instead of protective effects. Because bacterial DNA normally is present on the skin, and in the mouth and intestines, in theory, CpG DNA might contribute to conditions such as eczema, gum inflammation, or some inflammatory diseases of the intestine.

When bacteria enter the bloodstream, a condition called the sepsis syndrome can result in which bacterial products such as endotoxin trigger the excessive production of host cytokines and other inflammatory chemicals. Although small amounts of these compounds elicit protective immune responses, large amounts overwhelm the system and can lead to massive organ failure and death. When mice were injected with bacterial DNA in combination with a sublethal dose of endotoxin, 75% of the mice died, suggesting that bacterial DNA may be one of the microbial products that can lead to sepsis.

Systemic lupus erythematosus is an autoimmune disease in which the immune system attacks the body's own tissues, instead of defending them against infection. Although the causes of lupus are poorly understood, bacterial infection has been suspected as a triggering factor. The studies described above suggest that CpG motifs in bacterial DNA could promote the development of lupus by causing activation of B cells and other lymphocytes, overexpression of immune stimulatory cytokines such as IL-6, and resistance to apoptosis, thereby potentially allowing the survival of autoimmune cells. For example, DNA-specific B cells in systemic lupus erythematosus may be triggered by the concurrent binding of bacterial DNA to membrane lg, and costimulatory signals provided by CpG motifs.

Aside from the possible role of exogenous bacterial DNA, individuals with lupus are exposed to another source of CpG DNA within their own cells. Lymphocytes in these individuals have a lower level of the enzyme responsible for maintaining DNA methylation. As a result, individuals with lupus have elevated levels of hypomethylated CpG DNA in their cells and in their bloodstream.

Evolution. The question remains: if bacteria can evolve much faster than humans why was the immune defense of humans avoided by methylating bacterial CpGs, or suppressing their number? It may be important to consider that this immune defense will be activated only if bacteria have broken through the outer defense of the skin and are growing within tissues where the white blood cells are patroling. Bacteria on skin and in intestines cause no harm, and may even be helpful by preventing other, harmful disease-causing bacteria from gaining a foothold. Indeed, it may be in the evolutionary interests of these bacteria to keep humans alive and healthy so that they have a host.

For viruses the situation is different. Viruses lack the machinery to proliferate, and must do so by infecting a host cell, taking over its machinery, and making more copies of itself. Thus, viruses would be under great evolutionary pressure to evade the CpG defense. Almost all kinds of virus have very low levels of CpGs, indicating that they have indeed sought to avoid activating the antiviral aspects of the CpG defenses. The human immunodeficiency virus (HIV), which causes acquired immune deficiency syndrome (AIDS), has about ½ the predicted level of CpGs, raising the interesting question of whether treatment with CpG DNA might help individuals with AIDS more effectively to fight off the virus. It is noteworthy that viruses that infect bacteria, which would not be expected to benefit from CpG suppression, have the expected frequency of CpGs. Thus, the explanation for viral CpG suppression likely resides in the fact that they replicate in eukaryotic hosts rather than in some other factor such as the size of their genomes.

Applications. The remarkable immune-activating properties of CpG DNA suggest several possible therapeutic applications. For example, CpG DNA might be added to vaccines to increase their effectiveness. Individuals with an inadequate immune response to certain types of infections may be helped by boosting their level of immune activation with CpG DNA. Finally, studies in mice with cancer suggest that treatment with CpG DNA can lead to an antitumor response that, in some cases, is curative, and within the next few years, research studies will determine whether the immune-activating properties of this DNA sequence are just a laboratory curiosity or the basis for a new class of drugs.

For background information SEE ACQUIRED IMMUNE DEFICIENCY SYNDROME (AIDS); ANTIGEN-ANTIBODY REACTION; AUTO IMMUNITY; DEOXYRIBONUCLEIC ACID (DNA); IMMUNITY; INFECTION in the McGraw-Hill Encyclopedia of Science & Technology.

Arthur M. Krieg

Bibliography. A. P. Bird, CpG islands as gene markers in the vertebrate nucleus, *Trends Genet.*, 3:342, 1987; A. M. Krieg, CpG DNA: A pathogenic factor in systemic lupus erythematosus?—Clinical immunotherapeutics, *J. Clin. Immunol.*, 15(6): 284–292, 1995; A. M. Krieg et al., B cell activation by oligodeoxynucleotides that contain stem-loop motifs, *Nature*, 374:546, 1995.

Inflammation

New insights into the inflammatory response should result in more precise therapeutic inventions designed to block inflammation and accompanying tissue injury. Much information exists describing interactions between phagocytic cells (neutrophils, monocytes) and the endothelium. The recognition that the inflamed endothelium is different from the normal endothelium suggests that interventional agents may be targeted to only the activated endothelium. In order for the inflammatory response to commence, both leukocytes and the endothelium must be activated.

Adhesive interactions. Activation of phagocytic cells is best defined by increased surface expression of molecules such as beta 2 integrin, CD11b/CD18 (also known as Mac-1) that cause adhesion to cell surfaces. In addition, the heterodimeric beta 2 integrin, CD11a/CD18 (LFA-1), may undergo a change as a result of cell activation, resulting in higher affinity of binding for its so-called counterreceptor on endothelial cells. The counterreceptor for both LFA-1 and Mac-1 is the endothelial intercellular adhesion molecule 1 (ICAM-1), which is expressed on endothelial cells but is upregulated when these cells come into contact with agonists such as tumor necrosis factor alpha (TNF alpha), interleukin-1, or bacterial lipopolysaccharide, which induce gene activation and cause expression of messenger ribonucleic acid (mRNA) coding for intercellular adhesion molecule 1. Another example of adhesive interactions between phagocytic cells and the activated endothelium involves a glycoprotein, very late arising antigen 4, which like the beta 2 integrins can undergo conformation change resulting in increased reactivity with its endothelial counter-receptor, vascular cell adhesion molecule 1. Because very late arising antigen 4 does not exist on neutrophils but is expressed on monocytes and eosinophils, the endothelial expression of its counterreceptor, vascular cell adhesion molecule 1 may specify a restricted type of cellular inflammatory response.

Two other prominent endothelial adhesion molecules are E-selectin and P-selectin, which are normally not expressed on the endothelium. E-selectin is upregulated by the same agonists that cause expression of endothelial intercellular adhesion molecule 1, and P-selectin, being stored in Weible-Palade granules of endothelial cells, is rapidly (and transiently) expressed on endothelial cells activated with histamine or the complement activation product, C5a. Glycoconjugates reactive with P- and E-selectins are normally expressed on phagocytic cells, especially neutrophils. Because neither P- nor E-selectin on endothelial cells is normally expressed, their ligands on leukocytes are functionally silent.

The third selectin, L-selectin, is present on all leukocytes. It seems to play a significant role in directing leukocyte traffic, although its counter-receptor on endothelial cells has not been defined. In the case of lymphocytes, L-selectin was originally described on the lymphocyte homing receptor, because lymphocyte L-selectin appears to react with counterreceptors or high endothelial venular cells. These special endothelial ligands for lymphocyte L-selectin may function chiefly to direct lymphocyte recirculation from blood to lymph nodes and ultimately back to blood via lymphatic channels.

Blocking strategies. Several different clinical strategies for blocking the inflammatory response have been proposed. One strategy involves blockading of the endothelial adhesion molecules, such as endothelial intercellular adhesion molecule 1 by monoclonal antibodies. Theoretically, if the role of adhesion molecules in a given type of inflammatory response (for example, rheumatoid arthritis, psoriasis, ischemia-reperfusion injury, or organ transplantation) were known, blockading by the antibody approach could be effected, although immune responses to the antibodies pose a substantial obstacle. Other strategies under consideration involve the use of peptides that mimic and, therefore compete with the binding site on an adhesion molecule or the use of specific glycoconjugates. To what extent these strategies will gain clinical acceptance remains to be determined. A primary problem is to define the most appropriate candidate adhesion molecule. For example, although much is known about adhesion molecules that are involved in lymphocyte recirculation and trafficking to specific lymphoid organs (for example, Peyers patches in the small intestine and peripheral lymph nodes), whether the same complex of adhesion molecules is at play in lymphocyte-rich inflammatory responses (for example, in psoriatic lesions or in allografted organs) remains to be seen.

Cytokines. Another area of major advance in the field of inflammation is the understanding of cytokines (also collectively known as interleukins). The number of cytokines is bewildering, because most of these mediators were originally designed by their immune response–enhancing function. However, their potential as inflammatory mediators will soon become evident. The first cytokines, interleukin-1 and tumor necrosis factor alpha, were quickly found to have a plethora of inflammation-promoting functions. These early-response cytokines have a key function in activating endothelial cells that initiate upregulation of endothelial intercellular adhesion molecule 1, E-selectin, and vascular cell adhesion molecule 1. In experimental studies, blocking of either cytokine markedly suppresses the inflammatory system by reducing upregulation of endothelial adhesion molecules. Blockade has been achieved by the use of antibodies or naturally produced antagonist molecules. Two examples include soluble tumor necrosis factor alpha receptor 1 and interleukin-1 receptor antagonist. The former functions like an antibody, intercepting tumor necrosis factor alpha before it can attach to its cellular receptor, and the latter interacts with the receptor, thereby competing with naturally produced interleukin-1. Both cytokine inhibitors have been effective in experimental inflammatory models. Their effectiveness in human inflammatory conditions has yet to be demonstrated.

Another promising area of cytokine biology involves the regulatory cytokines which demonstrate anti-inflammatory activities. Those cytokines currently include interleukin-4, interleukin-10, and interleukin-13. Although their mechanism of action is not well understood, these regulatory cytokines seem to suppress the production of proinflammatory cytokines (interleukin-1 and tumor necrosis factor alpha) and other factors, such as nitric oxide synthase in phagocytic cells. In the case of interleukin-4 and interleukin-10, several experimental models have shown the anti-inflammatory functions. The ability of interleukin-4 and interleukin-10 to reduce upregulation of endothelial adhesion molecules represents an important linkage.

Production of oxidants. Finally, researchers have some understanding of why and how phagocytic cells recruited into an inflammatory site have damaging effects via their production of oxidants. Resting phagocytic cells demonstrate little, if any, oxidant production, but activated phagocytic cells can produce large quantities of oxidants, with two main pathways of oxidant production being identified. In the first pathway, nicotinamide adenine dinucleotide phosphate oxidase exists in a quiescent form in neutrophils, monocytes, and macrophages. Cell activation leads to translocation of at least two cytoplasmic subunits, activating the assemblage of nicotinamide adenine dinucleotide phosphate oxidase on the cell membrane. This enzyme ultimately transfers single electrons to molecular oxygen to generate a series of products, and finally water. A major function of this pathway is microbicidal activity, much of which occurs via neutrophil enzymes such as myeloperoxidase reacting with hydrogen peroxide in the presence of a halide such as chlorine (Cl^-) to form the powerful oxidant hypochlorous acid. The importance of nicotinamide adenine dinucleotide phosphate oxidase is seen in individuals with a genetically based defect (chronic granulomatosis of childhood) in one of the subunits for assembly of this enzyme complex, the result being deficient microbicidal activity for bacteria.

The second major enzyme linked to inflammatory injury is inducible nitric oxide synthase. This enzyme can be induced in monocytes and macrophages by contact with bacterial lipopolysaccharide, tumor necrosis factor alpha, or interferon-gamma, and appears to play a major role in containment of intracellular pathogens. Although production of nitric oxide by endothelial cells is important in regulating vascular tone and preventing excessive vasoconstriction leading to hypertension, its generation by monocytes or macrophages appears to have the potential for the induction of tissue injury. Thus, blockade of intracellular nitric oxide synthase may be a strategy for reducing tissue injury during the inflammatory response.

For background information *SEE ARTHRITIS; BLOOD; INFLAMMATION; INTERLEUKIN; METABOLIC DISORDERS; MONOCLONAL ANTIBODIES; RHEUMATISM* in the McGraw-Hill Encyclopedia of Science & Technology.

Peter A. Ward

Bibliography. S. M. Albelda, C. W. Smith, and P. A. Ward, Adhesion molecules and inflammatory injury, *Faseb. J.,* 8:504–512, 1994; J. I. Gallin (ed.), *Inflammation: Basic Principles and Clinical Correlates,* 2d ed., 1992; M. S. Mulligan et al., Role of endothelial-leukocyte adhesion molecule 1 (ELAM-1) in neutrophil-mediated lung injury in rats, *J. Clin. Invest.,* 88:1396–1406, 1991; R. R. Ruffalo, Jr., and M. A. Hollinger (eds.), *Inflammation: Mediators and Pathways,* 1995; T. P. Shanley et al., Regulatory effects of intrisic IL-10 in IgG immune complex-induced lung injury, *J. Immunol.,* 154:3454–3460, 1995.

Influenza

Influenza viruses are negative-strand segmented ribonucleic acid (RNA) viruses of types A, B, and C and belong to the family Orthomyxoviridae. However, only influenza A viruses cause pandemics of disease. Both of the surface antigens of the influenza type A viruses undergo two types of variation: antigenic drift and shift. Antigenic drift involves minor antigenic changes in the hemagglutinin (HA) and neuraminidase (NA), and antigenic shift involves major antigenic changes in these molecules resulting from replacement of gene segment(s).

Antigenic drift results in the annual epidemics of influenza, and occurs by accumulation of a series of point mutations resulting in amino acid substitutions in antigenic sites at the membrane distal region of the hemagglutinin. These substitutions prevent binding of antibodies induced by a previous infection, allowing the virus to infect the host.

Antigenic shifts cause pandemics of influenza in humans at irregular intervals. Since 1933, when influenza virus was first isolated, pandemics have occurred in 1957 when the H2N2 subtypes (Asian influenza) replaced the H1N1 subtype, in 1968 when the Hong Kong (H3N2) virus appeared, and in 1977 when the H1N1 virus reappeared. Pandemic strains of influenza A viruses also emerge in lower animals and birds.

Reservoirs. Influenza A viruses infect a variety of animals, including humans, pigs, horses, sea mammals, and birds. There is convincing evidence that all 14 hemagglutinin subtypes of influenza A viruses are perpetuated in the aquatic bird populations of the world, being especially prevalent in ducks, shorebirds, and gulls; there is no evidence that influenza viruses persist for extended periods in individual animals. Thus some mechanism has evolved for maintaining influenza viruses in aquatic avian species.

The occurrence of influenza A infections among wild ducks in nature is important because their annual migrations put them in contact with large numbers of other animal species that could potentially become infected. Infections caused by most strains of influenza virus are asymptomatic in aquatic birds, indicating that avian viruses are highly adapted to their natural hosts. In wild ducks, influenza viruses replicate preferentially in the cells lining the intestinal tract, and are excreted in high concentrations in the feces. Influenza viruses have been isolated from unconcentrated lake water, indicating that waterfowl are able to transmit influenza viruses both to other ducks and to other domestic and wild birds by depositing feces in their water supply.

Evolutionary pathways. Studies on the ecology of influenza viruses have led to the hypothesis that all mammalian influenza viruses derive from the avian influenza reservoir. Support for this theory comes from phylogenetic analyses of influenza A RNA sequences from a variety of hosts, geographical regions, and subtypes. Phylogenetic analyses of the nucleoprotein gene show that avian influenza viruses have evolved into five host-specific lineages: an ancient equine lineage, which has not been isolated in more than 16 years, a recent equine lineage, a lineage in gulls, one in swine, and one in humans. The human and classical swine viruses have a genetic sister group relationship, which shows that they evolved from a common origin. The ancestor of the human and classical swine virus was evidently an intact avian virus which, like the influenza virus currently circulating in pigs in Europe, derived all of its eight genes from avian sources. The avian influenza viruses can be separated into two clads, one in the Americas and the other in Europe and Asia.

A surprising discovery from phylogenetic analyses was that avian influenza viruses, unlike human strains, are not highly variable. In fact, influenza viruses in aquatic birds appear to be in evolutionary stasis, with no evidence of net evolution over the past 60 years. Nucleotide changes have continued to occur at a similar rate in avian and mammalian influenza viruses; although there has been no net accumulation of amino acid changes in the avian viruses, all eight mammalian influenza gene segments continue to accumulate changes in amino acids. The high level of genetic conservation suggests that avian viruses are approaching or have reached an adaptive optimum, where further nucleotide changes provide no selective advantage, and that the source of genes for past pandemic influenza viruses exists unchanged in the aquatic bird reservoir.

Overall, the most important implication of phylogenetic studies is that the ancestral viruses which caused the Spanish influenza in 1918, as well as the viruses that provided gene segments for the Asian (1957) and Hong Kong (1968) pandemics, are still circulating in wild birds, with few or no mutational changes.

Interspecies pandemics. The available evidence indicates that each of the human pandemics of influenza which occurred in the twentieth century were derived from avian influenza viruses either after reassortment with the currently circulating human strain or by direct transfer. Genetic and biochemical studies have indicated that the 1957 Asian H2N2 strain obtained its HA, NA, and PB1 genes from an avian virus and the remaining five genes from the preceding human H1N1 strain. The Hong Kong 1968 (H3N2) strain contained HA and PB1 genes recently transmitted from an avian donor; its NA and five other genes were from the H2N2 strain circulating in early 1968. The Russian (1977) strain was a reintroduction of a human influenza virus that had been absent for 27 years but had been stored in the frozen state. It is unclear whether this was a laboratory accident or whether the virus was frozen in nature; however, the evidence argues against it having been reintroduced from an avian reservoir.

To fully explain the origin of human pandemics, epidemiologists need to understand where the reassortment between human and avian influenza viruses occurs. Reassortment requires the simultaneous infection of a host animal with both human and avian influenza viruses. The pig is the leading contender for this role of intermediate host. Swine are the only abundant domesticated mammalian species that serve as common hosts for both the human and avian influenza viruses. Humans occasionally contract influenza viruses from pigs: a young soldier died at Fort Dix of swine influenza in 1976, triggering the dubious swine influenza vaccine program. At that time, it was not known that swine influenza virus infects about 10% of persons occupationally exposed to swine and that occasional fatalities occur. A recent example involved a young pregnant woman who attended an agricultural display of swine and subsequently died of swine influenza. These events are rare, and the viruses have not yet spread widely in humans. There is only indirect evidence that humans can be directly infected with avian influenza virus. However, pigs in Europe were infected with avian influenza virus in 1979, and these viruses may transmit to humans. Thus, there is a considerable body of circumstantial evidence supporting the notion that pigs may be the mixing vessel for human influenza.

The majority of human influenza pandemics since about 1850 have originated in China, raising the possibility that southern China may be an influenza epicenter. Unlike the temperate or subarctic regions of the world where human influenza is a winter disease, influenza occurs year round in the tropical and subtropical regions of China. Influenza A viruses of all subtypes are present throughout the year in ducks. Pigs are a common source of protein in China. The high concentration of people, pigs, and ducks in this region of the world provides conditions for interspecies transmission and genetic exchange between influenza viruses. To date, the knowledge about pandemic influenza viruses in southern China is anecdotal. The occurrence of influenza viruses in humans after direct transmission of a host-range mutant is uncertain. There is precedent for direct transmission of influenza viruses to horses, seals, and domestic poultry, and indirect evidence from phylogenetic studies suggests that the catastrophic Spanish influenza of 1918 which killed at least 20 million persons worldwide may have possessed all gene segments of an avian influenza virus.

Future pandemics. The transmission of European swine influenza virus to humans needs to be monitored closely, because this H1N1 virus contains all eight genes from an avian influenza virus. It is in an optimal position to reassort with human influenza viruses, and such reassortants were detected in children in The Netherlands in 1993. This situation is reminiscent of the hypothesized origin of the Asian and Hong Kong pandemics. The continuing circulation of descendants of the Russian strain that are immunologically related to the viruses in European pigs reduces the likelihood of direct transmission at this time.

With the realization that a reservoir of all known influenza A subtypes exist in aquatic birds, it must be accepted that influenza is not an eradicable disease and that this disease must be thought of in terms of prevention and control. Phylogenetic and epidemiological evidence indicates that previous human influenza pandemics originated from avian reservoirs with pigs serving as the intermediate hosts. If southern China is an epicenter, it is conceivable that pandemics could be prevented by changing agricultural practices to minimize contact of pigs, people, and ducks.

Present knowledge offers the possibility of preventing pandemic influenza and reducing the likelihood of catastrophic outbreaks due to genetic shift. However, it is possible that there will be another human pandemic of influenza, and so it is necessary to develop an international plan to deal with what could be a catastrophy of global proportions. SEE VIROLOGY.

For background information SEE ANIMAL VIRUS; ANTIGEN; AVES; IMMUNITY; INFLUENZA; MUTATION in the McGraw-Hill Encyclopedia of Science & Technology.

Robert G. Webster

Bibliography. R. M. Krug (ed.), *The Influenza Viruses*, 1989; R. G. Webster et al., Evolution and ecology of influenza A viruses, *Microbiol. Rev.*, 56:152–179, 1992.

Information superhighway

The terms national information infrastructure (NII) and information superhighway describe a next-generation telecommunications and information industry. Sometimes these terms are used without reference to a specific vision of the future. The NII concept grows out of a desire to rationalize the communications and computer technologies that are currently being developed. Other terms that are used to describe parts of this evolution are convergence and digitalization.

Technologies and systems. A number of technologies and systems are evolving into the NII. These include the Internet, asynchronous transfer mode (ATM), cable systems, long-distance and local telephone companies, personal computers, digital television, and wireless.

Internet. A networked collection of more than a million computers, the Internet is probably the best model for the future NII. The Internet connects hundreds of different types of computers and computer programs from thousands of sources. Key to the success of the Internet is a set of standards that define the interconnection of computers and networks and permit applications, such as electronic mail (e-mail) and electronic publishing, to exchange and share information. The Internet spans multiple vendors, national boundaries, and regulatory regimes to provide valuable services.

Asynchronous transfer mode. ATM is a packet-based technology for use in high-speed networks. SEE DATA COMMUNICATIONS.

Cable systems. Cable operators are replacing coaxial cable with fiber optics, expanding bandwidth, and moving to include digital modulation and two-way capabilities in their architectures.

Long-distance telephone companies. These firms have installed nationwide fiber-optic networks that expand capacity manyfold and lower the cost of moving bits across the country.

Local telephone companies. These companies are upgrading their local loop facilities and are connecting fiber-optic lines to the premises of larger customers. They are also exploring various technologies that expand the communications capabilities of the household.

Personal computers. The processing speed, memory, and communications capabilities of personal computers continually increase. Current personal computers have the capabilities of the supercomputers of a decade ago. The capabilities of personal computer software increase every year.

Digital television. This technology is in the process of coming into general use. The proposed standard for advanced television would transmit television pictures far superior to traditional television using a 2×10^7 bit-per-second information stream.

Wireless. The cellular industry and the newly created personal communications service (PCS) industry are deploying digital public mobile telecommunications services.

Standards. These disparate elements are evolving separately, but they possess great complementarities. A common standard for data transfer would allow a consumer to connect to a data service by using the telephone network, the cable network, or a cellular connection. Common standards allow software developed for one application to be used for another application without expensive reprogramming.

Thus, these developments cannot provide their maximum benefit to consumers in isolation. Rather, common standards and applications programs are needed to allow consumers to exploit the capabilities of hardware systems. The need for common standards is sometimes referred to as harmonization.

Naïve definition. Understanding the NII can be facilitated by envisioning an expanded Internet, one that has high-speed connections to the computers or local-area networks (LANs) in all offices and similar connections to many homes. Such an improved Internet would have two major advantages. First, many services, such as electronic mail, would become more valuable because of their wider availability. Similarly, the incentives to provide services on the Internet would expand as the potential market expanded. Second, the use of higher-performance communications would make existing applications more responsive and more valuable, and would permit sophisticated applications to be feasible everywhere, not just at research centers.

The current Internet supports many valuable services, but it still reflects its origins in the computer science research community. Many, perhaps most, people in the industry work with computers connected to the Internet. In contrast, only a few lawyers work daily on computers connected to the Internet (although their numbers are continually increasing). Thus, many firms in the computer industry make sales and support information available on the Internet because they can easily reach their intended audience. Some computer industry publications are available on the Internet, and societies for computer professionals provide services to their members over the Internet. As the NII grows, services catering to legal professionals will expand as more lawyers are connected to the network. A similar process will occur for other professional groups.

Increasing the bit rate on Internet connections will permit new services to be offered. At present, a user with a low-data-rate communications link cannot rapidly browse through a document with graphics. That is, users cannot browse a catalog, an encyclopedia, or a collection of medical images over the network nearly as easily as they could with the hard-copy version. The Internet and attached personal computers are still a poor substitute for a sales catalog or an illustrated car repair manual.

The service expansions made possible by improved infrastructure will extend far beyond improvements in the services now offered over the Internet. The Internet can now carry voice and audio in a reasonably convenient fashion. The NII will be able to carry switched video services in much the same fashion.

Technical definition. An engineering task force, the Cross Industry Working Team (XIWT), has developed a view of the NII as composed of three layers of functions: bitways, services, and applications. Bitways are the so-called data pipes that carry information from one place to another and the software that controls this transmission. Bitways are made of coaxial cable, copper wire, fiber optics, and radio links. Services are the building blocks for higher-level applications. Examples of services are file transfer, directory support, and financial transactions. Applications, the highest layer in the architectural model, perform the tasks desired by the user. Applications are performed by software that controls the presentation of the information to the user and calls on services to get the necessary information and provide the supporting transactions. Applications include a shopper browsing a remote catalog, a radiologist viewing an x-ray while conferring with a physician, and a student viewing a video clip.

Perspectives. The NII can be considered to be three different activities: (1) an organizing paradigm, (2) a standards process, and (3) a series of research and demonstration projects.

Organizing paradigm. One great benefit of the NII vision lies in the organizing concept it provides for designers and users of a wide mix of systems. Consumers will benefit if the hardware and software they purchase can work together in a seamless fashion. Articulating this compatibility goal in advance and anticipating the problems in achieving it are inevitable consequences of the NII discussion.

Standards process. Standards need to be developed to support NII activities. This process has already begun with the work on next-generation Internet (IPng) standards that explicitly take into account the possibility that every television receiver might need to be treated as a host on the Internet. If standards for NII activities are developed that meet needs for performance and interconnection, they will have a good probability of being accepted. Such standards pave the way for wider NII activities.

Research and demonstration projects. Demonstration and pilot projects will test the NII concepts. Although the NII will be built as a part of the commercial and personal infrastructure (computers and LANs purchased by businesses and governmental organizations for their own use; residential purchase of computers, software, and telecommunications service; and construction of bitways by telecommunications service providers), demonstration projects will be put in place in advance of commercial deployment of specific elements of the NII. Such demonstration and research projects form an important part of government participation in the development of the NII.

The concept of the NII forces a beneficial examination of the future of telecommunications and information services and facilitates the development of better systems. Although some form of the NII and its worldwide analog, the global information infrastructure (GII), is almost certainly inevitable, the pace and ease of moving toward the

NII-GII vision depend upon current decisions on standards and system design. The discussion of the NII helps ensure that today's decisions promote the development of a compatible and efficient NII.

For background information SEE DATA COMMUNICATIONS; DIGITAL COMPUTER; INTEGRATED SERVICES DIGITAL NETWORK (ISDN); LOCAL-AREA NETWORKS; OPTICAL COMMUNICATIONS; TELEPHONE SERVICE; TELEVISION; WIDE-AREA NETWORKS in the McGraw-Hill Encyclopedia of Science & Technology.

Charles L. Jackson

Bibliography. Cross Industry Working Team, *An Architectural Framework for the National Information Infrastructure,* 1994; National Telecommunications and Information Administration, *The National Information Infrastructure: Agenda for Action,* 1993; J. S. Patterson and W. L. Smith, The North Carolina information highway, *IEEE Netw.,* 8(6):12–17, November–December 1994.

Infrared radiation

Infrared detectors operating in the very long wavelength infrared region (that is, wavelengths of 6–18 micrometers) are of great interest to a variety of ground-based and space-based applications such as night vision, early-warning systems, navigation, flight-control systems, weather monitoring, and astronomy (infrared detectors are used in the focal planes of telescopes). In addition, infrared detectors in this spectral region can be used for pollution monitoring and for monitoring the relative-humidity profiles and the distribution of different constituents in the atmosphere, such as ozone (O_3), carbonmonoxide (CO), and nitrous oxide (N_2O). Indeed, most of the absorption lines of gas molecules lie in the infrared spectral region. Applications include monitoring global atmospheric temperature profiles to an accuracy of 1 K (1.8°F) and the depth of the Earth's atmosphere to an accuracy of 1 km (0.6 mi) by measuring the temperature dependence of the absorption signatures of the carbon dioxide molecules, as well as relative-humidity profiles and the distribution of minor constituents in the atmosphere. Such studies are already being planned for the Earth Observing System (EOS) of the National Aeronautics and Space Administration (NASA). This spectral region is also rich in information vital to the understanding of the composition, structure, and energy balance of molecular clouds and star-forming regions of the Milky Way Galaxy. Thus, there is great commercial, scientific, and academic interest in infrared detectors operating in this wavelength region.

Long-wavelength infrared detectors. It is customary to make infrared detectors in this spectral span by utilizing the interband absorption of narrow-band-gap semiconductors, such as indium antimonide (InSb) and mercury cadmium telluride ($Hg_{1-x}Cd_xTe$). Infrared radiation is absorbed by the photosensitive material when an incoming photon has sufficient energy (equal to $h\nu$, where h is Planck's constant and ν is the frequency of the incoming photon) to photoexcite an electron from the valence band to the conduction band. In a detector structure, these photoexcited carriers are collected by applying an electric field, thereby producing a photocurrent or a photovoltage. Because the absorbed photon energy is greater than the energy of the band gap, both electrons and the holes are created. Thus, the semiconductor does not need to be doped. (These detectors are called intrinsic detectors.) Large two-dimensional arrays of indium antimonide (512×512 pixels) and mercury cadmium-telluride (128×128 pixels) detectors have been demonstrated up to cutoff wavelengths of 5 and 11 micrometers, respectively. In indium antimonide, since the band gap is fixed, these detectors cannot operate at longer wavelengths. In contrast, mercury cadmium telluride can be made into narrow-band-gap materials by varying the alloy composition. However, the long-wavelength, large mercury cadmium telluride arrays are highly nonuniform in composition and doping, resulting in large nonuniformities in spectral response and sensitivity. Furthermore, there are large numbers of trap centers throughout the band gap and charged surface states, which produce substantial and detrimental parasitic currents, resulting in higher $1/f$ noise. In addition, such narrow-band-gap materials are more difficult to grow and process into devices than wide-band-gap semiconductors. Thus, there is research into utilizing the artificial low-effective-band-gap structures made of wide-band-gap semiconductors such as gallium arsenide (GaAs), which are easy to grow and process into devices. The basic advantages of the gallium arsenide–based quantum-well infrared photodetectors (QWIPs), namely the highly mature gallium arsenide growth and processing technologies, become more important at longer wavelengths where the narrow-band-gap materials become more difficult to work with.

Multiple quantum wells. The idea of using multiple quantum-well structures to detect infrared radiation can be understood by using the basic principles of quantum mechanics. The quantum well is equivalent to the particle-in-a-box problem in quantum mechanics, which can be solved by the time-independent Schrödinger equation. The solutions of this problem are the eigenvalues that define the energy levels inside the well in which the particle is confined. The positions of the energy levels are primarily determined by the well dimensions (height and width). Therefore, by tailoring the quantum-well structure, the separation between the allowed energy levels can be adjusted so that the infrared photons can induce an intersubband transition between the ground state and the first excited state. The lattice-matched gallium arsenide–aluminum gallium arsenide materials system ($GaAs/Al_xGa_{1-x}As$, where x is the molar ratio between aluminum and gallium), is a good candidate in which to create such a potential well, since the band gap of aluminum gallium arsenide is higher than that of gallium arsenide and can be changed continuously by varying the molar ratio, x (and hence the depth of the well or height of the barrier). Carriers (electrons for n-type material and holes for p-type material) can be introduced by doping the gallium arsenide well. These carriers will occupy the ground state of the quantum wells at low temperatures.

QWIP operation. The first QWIP was based on an intersubband transition between two bound quantum-well states (that is, the ground state and the first excited state were inside the well). The intersubband absorption excites an electron from the ground state to the first excited state, where it can tunnel out to the continuum (the continuous energy levels above the quantum well) in the presence of an external electric field, thereby producing a photocurrent. By reducing the quantum well width it is possible to push the second bound state (the first excited state) into the continuum, resulting in a strong bound-to-continuum intersubband absorption (see **illus.**). The major advantage of the bound-to-continuum QWIP is that the photoexcited electron can escape from the quantum well to the continuum transport states without being required to tunnel through the barrier (see illus.). As a result, the bias required to efficiently collect the photoelectrons can be reduced dramatically, thereby lowering the dark current. Because the photoelectrons do not have to tunnel through the barriers, the aluminum gallium arsenide barrier thickness of the bound-to-continuum QWIP can now be increased without reducing the photoelectron collection efficiency. The performance

of QWIPs has been improved further by utilizing the bound-to-quasibound intersubband absorption (occurring when the first excited state is in resonance with the top of the barrier). This transition maximizes the intersubband absorption while maintaining the excellent electron transport.

Dark current. In addition to the photocurrent, all detectors, including QWIPs, produce a parasitic current called a dark current, which must be minimized to achieve high performance. In QWIPs, the dark current originates from three different mechanisms (see illus.). The first such process is quantum-mechanical tunneling from well to well through the aluminum gallium arsenide barriers (sequential tunneling). This process is independent of temperature. Sequential tunneling dominates the dark current at very low temperatures (less than 30 K or −405°F). The second mechanism is thermally assisted tunneling, which involves a thermal excitation and tunneling through the tip of the barrier into the continuum energy levels. This process governs the dark current at medium temperatures. The third mechanism is classical thermionic emission and dominates the dark current at higher temperatures (greater than 45 K or −378°F). At higher temperatures the last mechanism is the major source of dark current, and the thermal generation rate is determined by the lifetime of the carriers and the well doping density.

Increasing the barrier width from a few tens of nanometers to 50 nm can reduce the ground-state sequential tunneling by an order of magnitude. Furthermore, constructing bound-to-quasibound QWIPs, in which the energy barrier for thermionic

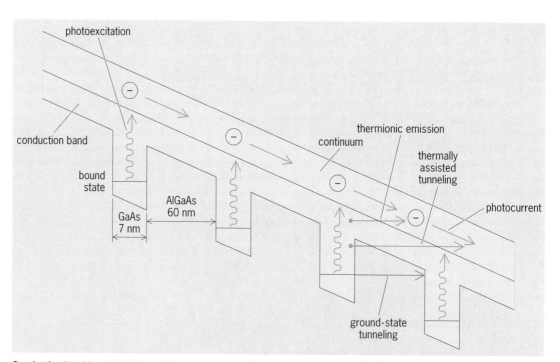

Conduction band in a bound-to-continuum quantum-well infrared photodetector (QWIP) with an electric field. Infrared photon absorption can photoexcite electrons from the quantum-well ground state into the continuum, causing a photocurrent. Dark-current mechanisms—ground-state tunneling, thermally assisted tunneling, and thermionic emission—are also shown. (*After S. D. Gunapala et al., 9 μm cutoff 256 × 256 GaAs/Al$_x$Ga$_{1-x}$As quantum well infrared photodetector focal plane array camera, IEEE Trans. Electron. Devices, 1996*)

emission is equal to the photoionization energy, reduces the dark current from thermionic emission by an order of magnitude relative to that in bound-to-continuum QWIPs, where the thermionic barrier is 10–15 meV less than the photoionization energy. Because of this high performance and the excellent uniformity of gallium arsenide–based QWIPs, large two-dimensional imaging arrays of 128×128 and 256×256 pixels have been demonstrated up to a cutoff wavelength of 15 μm.

Light coupling to QWIPs. QWIPs do not absorb radiation incident normal to the surface because the light polarization must have an electric field component normal to the superlattice (the growth direction) to be absorbed by the confined carriers. For imaging, it is necessary to be able to couple light uniformly to two-dimensional arrays of these detectors. Efficient light coupling to QWIPs in a large array in the focal plane of an imaging camera is achieved by fabrication of linear or two-dimensional gratings, which diffract the incoming light, on top of each detector.

Many more passes of infrared light inside the detector structure can be obtained by incorporating a randomly roughened reflecting surface on top of the detectors, and a factor-of-8 enhancement in QWIP responsivity compared to 45°-illumination geometry has been demonstrated by using single-element detectors. The random structure on top of the detector prevents the light from being diffracted normally backward after the second bounce, as happens in the case of a two-dimensional grating. After each bounce, light is scattered at a different random angle, and the only chance for light to escape out of the detector is when it is reflected toward the surface within the critical angle of the normal. For the galium arsenide–air interface this angle is about 17°, defining a very narrow escape cone for the trapped light.

The reflector was designed with two levels of scattering surfaces located at quarter-wavelength separations. The area of the top unetched level is equal to the area of the etched level, a quarter-wavelength deep. Therefore, the normally reflected light intensities from the top and bottom surfaces of a random reflector are equal and 180° out of phase, thus maximizing the destructive interference at normal reflection (and hence lowering the light leakage through the escape cone). The random structures were fabricated on the detectors by using standard photolithography and selective dry etching. The advantage of the photolithographic process over a completely random process is the ability to accurately control the feature size and preserve the pixel-to-pixel uniformity which is a prerequisite for high-sensitivity imaging focal-plane arrays.

QWIP performance. Figures of merit such as responsivity and detectivity are commonly used to compare the performance of detectors. Responsivity is the ratio of the signal current to the incident radiation power on the detector. Detectivity is the

signal-to-noise ratio normalized to unit area and unit bandwidth. The primary noise source in QWIPs is the shot noise produced by the dark current. Therefore, unlike the narrow-band-gap detectors, in which the noise is dominated by temperature-independent processes at low temperatures, QWIP performance can be further improved by cooling to cryogenic temperatures.

Exceptionally rapid progress has been made in the performance (detectivity) of very long wavelength QWIPs, starting with bound-to-bound QWIPs, which had relatively poor sensitivity, and culminating in high-performance bound-to-quasicontinuum QWIPs with random reflectors. The achieved detectivities are more than sufficient to demonstrate large two-dimensional imaging arrays (128×128 pixels or larger) at very long wavelengths, presently not possible with intrinsic narrow-band-gap detectors. Because of the pixel-to-pixel nonuniformities, the detectivity of a single pixel is not sufficient to describe the performance of a large imaging array. Noise induced by nonuniformity has to be taken into account for complete evaluation. The general figure of merit to describe the performance of a large imaging array is the noise-equivalent temperature difference (NEΔT), which includes the spatial noise originating from pixel-to-pixel nonuniformities. The noise-equivalent temperature difference is the minimum temperature difference across the target that would produce a signal-to-noise ratio of unity. Because of superior material quality and the high uniformity associated with the gallium arsenide–aluminum gallium arsenide materials system, arrays of very long wavelength QWIPs can be fabricated with very low noise equivalent temperature difference.

These initial arrays gave excellent images with 99.98% of the pixels working, demonstrating the high yield of gallium arsenide technology. The uncorrected photocurrent uniformity (standard deviation/mean) of the 256×256 focal-plane array (65,536 pixels) was only a few percent. The uniformity after two-point correction was reduced to 0.05%. The first hand-held long-wavelength infrared camera was demonstrated by installing this focal-plane array into a commercially available hand-held infrared camera. The measured mean noise-equivalent temperature difference of the focal-plane array was 26 mK at an operating temperature of 70 K (–334°F) and bias of –1 V for a 300 K (80°F) background. This excellent yield uniformity is due to the excellent gallium arsenide growth uniformity and the mature gallium arsenide processing technology.

For background information *SEE INFRARED RADIATION; NONRELATIVISTIC QUANTUM THEORY; PHOTOCONDUCTIVITY; REFLECTION OF ELECTROMAGNETIC RADIATION; SEMICONDUCTOR; SEMICONDUCTOR HETEROSTRUCTURES* in the McGraw-Hill Encyclopedia of Science & Technology.

Sarath D. Gunapala

Bibliography. M. Francombe and J. Vossen (eds.), *Physics of Thin Films*, vol. 21: *Homojunction and*

Quantum-Well Infrared Detectors, 1995; B. F. Levine, Quantum-well infrared photodectectors, *J. Appl. Phys.,* 74:R1–R81, 1993; *Proceedings of Innovative Long Wavelength Infrared Detector Workshop,* 1990; V. Swamirathan et al. (eds.), *Proceedings of 2d International Symposium on 2–20 μm Wavelength Infrared Detectors and Arrays: Physics and Applications, 1994,* 1995.

Insect control, biological

Recent research on the biological control of insect pests has focused on neem tree (*Azadirachta indica*) and *Bacillus thuringiensis* insecticides, and on baculoviruses as components of integrated pest management.

Neem Tree Pesticides

The neem tree (*A. indica*) is a broad-leaved evergreen related to mahogany. It probably originated in Burma, but now thrives in India and has been planted in Africa, Saudi Arabia, the Philippines, South America, and the United States. Mature trees produce yellowish oval fruits (0.5–1 in.) (1.4–2.4 cm) long. Inside the fruits are brown, almondlike seed kernels covered with a hard white shell. Seed kernels can be isolated by drying the fruit and removing the white shell. Pesticidal substances are present in all parts of the neem tree including the leaves, but the greatest concentrations are found in the oily seed kernels.

Seed oil can be removed from the kernels either by a hydraulic press or by solvent extraction. Solvent extracts of pulverized seed kernels or of the expressed oil have the greatest commercial potential in Western markets.

Neem oil. Neem seed oil contains 4 major and at least 20 minor biologically active compounds. The major components—azadirachtin, salannin, nimbin, and nimbidin—and many of the minor active compounds belong to a chemical class called the limonoids. These are bitter-tasting complex molecules composed only of carbon, hydrogen, and oxygen. The large multiringed carbon structure is balanced by oxygen in the form of alcohol, ether, ester, and epoxide groups. The presence of oxygen tends to make the limonoids more soluble in water, methanol, or ethanol than in hexane, gasoline, or other similar solvents.

Although neem oil has many insecticidal components, most of its insecticidal action is due to azadirachtin. Azadirachtin is a mixture of at least nine closely related chemical and structural isomers. Azadirachtins A and B are present in greatest quantity. Although there is some variation, about 83% of natural azadirachtin is azadirachtin A, and about 16% azadirachtin B. The concentration of azadirachtin in neem seed kernels is extremely variable; the maximum concentration is 1%.

Neem insecticides are available only as extracted formulations. Several different types of solvent are used to extract azadirachtin and the limonoids from neem seed or neem oil to produce a number of proprietary insecticide products.

Mechanism of action. Azadirachtin acts as an antifeedant and as a pseudosteroidal insect growth regulator, disrupting molting and metamorphosis. In some insects it disrupts mating and sexual communication. The mating disruption may be hormonal or an indirect toxic effect. Azadirachtin can also prevent swallowing and decrease gut motility. It is a chitin synthesis inhibitor; however, it is weaker than synthetic substances such as diflubenzuron.

Two kinds of antifeedant effects are seen with neem—primary and secondary. The desert locust (*Schistocerca gregaria*) has gustatory sensors for azadirachtin and probably experiences a primary antifeedant effect; that is, it does not like the taste and refuses to eat it. Many insects either cannot detect neem by taste or do not find it repulsive. However, after they have eaten some of the treated foliage, they suddenly stop feeding. Exposed insects refuse to eat even when transferred to untreated food. Topical application to vulnerable larvae can also produce this secondary antifeedant effect. The secondary effect is probably due to azadirachtin, and may be hormonally controlled. Even when larvae are able to eat, neem can cause decreased growth due to impaired secretion of digestive enzymes such as trypsin.

A number of other adverse biological effects are shown by neem, including sterilization of insect eggs in rare cases, reduction of adult life-span, and reduced fecundity. The major effects, however, are antifeedant activity and hormonal disruption, leading to formation of permanent larvae and insects frozen in larval-pupal, nymphal-pupal, nymphal-adult, and pupal-adult transitions.

The growth-regulating properties of neem are mostly due to broad antihormonal effects of azadirachtin. Azadirachtin reduces body concentrations (titers) of the molting hormone ecdysone in a number of insects. Juvenile hormone concentrations are also reduced in some cases. The exact mechanism has not yet been resolved, but disturbance of the brain and neuroendocrine system is likely.

Azadirachtin and neem. For many insects, most of the hormonal and antifeedant effects of neem are due to azadirachtin; however, the activity of neem varies with the insect and the crop. For example, pesticidal action on the variegated cutworm (*Peridroma saucia*) and on the milkweed bug (*Oncopeltus fasciatus*) correlates with the amount of azadirachtin present, but this correlation may not apply to the strawberry aphid (*Chaetosiphon fragaefolii*).

Although most of the pesticidal activity in neem is due to azadirachtin, the oil contains volatile sulfides such as di-*n*-propyl sulfide, which is larvacidal to the yellow fever mosquito (*Aedes aegypti*), tobacco budworm (*Heliothis virescens*), and corn earworm (*Helicoverpa zea*). Also, for mite control, pentane extracts of neem seed are more effective than ethanol extracts containing azadirachtin. Feeding experiments of a second larval instar diamondback moth (*Plutella xylostella*) show that pure synthetic azadirachtin is about three to four times less effective than extracts of neem having the same azadirachtin concentration. Either natural synergists of azadirachtin are present in the oil, or the neem oil helps with absorption.

Toxicity. Because neem is toxic only by ingestion, only pest insects feeding on treated plants are killed. Parasitoids and predators attacking pests that have eaten neem-treated foliage generally show only slight symptoms. Also, neem oil is practically nontoxic to mammals, and is used in India to make soap and toothpaste. For instance, the oral LD50 (the amount of toxin necessary to kill half a test population: the greater the LD50, the less toxic the material) of neem oil in rats is greater than 12,000 mg/kg. The oral LD50 of one formulation is about 4500 mg/kg. Both neem oil and azadirachtin-enriched neem formulations are less toxic than salt (sodium chloride), which has an oral LD50 in rats of 3750 mg/kg.

Insects affected. Although azadirachtin affects at least 200 different insect species in several different orders, the best results are obtained with immature stages of insects that undergo complete metamorphosis. Caterpillars are the most susceptible, especially those with a limited host plant range. However, good results are seen with applications to early larval stages of beetles. Leafminers are also very sensitive to neem and azadirachtin formulations.

Chewing insects such as caterpillars are often better controlled and with lower concentrations of azadirachtin than are sucking insects such as aphids. Neem extracts are not very effective against mealybugs and scales, and variable results are obtained with other homopterans (sucking insects). Leaf- and planthoppers, however, experience the antifeedant effect, and immature whiteflies are vulnerable to growth disruption. Thrips are not very sensitive to neem extracts and azadirachtin, but repeated applications can stop growth between the first and second nymphal stage.

Azadirachtin is less useful for aphids than caterpillars, beetle larvae, leafminers, and similar pests. Aphids are generally not repelled by neem treatments; contact with dried residues has little effect, and aphid control comes only after the insect starts feeding. Some aphids are more resistant than others, and persistence plus repeated applications are required.

Systemic effects. Neem extracts containing azadirachtin can be applied as foliar sprays or as a soil drench. When used as a soil drench, the azadirachtin is taken up by plant roots and transported to leaves and other organs. The systemic effect works better in some plants than in others. However, leaves and stems of wheat, beans, barley, rice, sugarcane, tomatoes, cotton, and chrysan-themums have been protected from susceptible insects.

Advantages and drawbacks. Because of its low mammalian toxicity, greenhouses, gardens, and fields may be reentered as soon as sprays of neem extracts dry. Because azadirachtin and other limonoids do not persist in the environment, cumulative food chain toxicity of the dichlorodiphenyltrichloroethane (DDT) type is unlikely. Because of low toxicity to beneficial insects, neem is compatible with biological control and other elements of integrated pest management. For instance, neem extracts are compatible with parasitoids, predators, pyrethrins, *B. thuringiensis,* soaps, and viruses, but may not be compatible with some entomophagous fungi. Finally, insects are not yet resistant to azadirachtin. Integration of neem products into a multicomponent integrated pest management program may be the best insurance against insect resistance.

From the grower's point of view, the greatest drawback to neem is its lack of persistence. Under typical field conditions foliar applications of neem extracts last 4–8 days. Rainfall washes the material off foliage, ending the treatment effect. Neem works best in warm climates, and effectiveness is reduced in cool weather. Also, neem extracts work slowly, and thus treatment must be applied early in an insect's life cycle to avoid unacceptable plant damage.

William Quarles

Bacillus thuringiensis Insecticide

The annual global agrochemical market is worth more than $27,000 million. Of this only 1% is from biologically based products, the majority of which are microbial insecticides based on the bacterium *B. thuringiensis.* This gram-positive soil bacterium is characterized by its ability to produce crystalline inclusions during sporulation. These inclusions consist of a protein that exhibits a highly specific insecticidal activity. The use of *B. thuringiensis* as a microbial insecticide, which dates back to the 1950s, initially had little impact on the crop protection market as a whole. However, with new advances in genetic manipulation and insecticide formulation and application, coupled with increasing public concern over chemical insecticides and the growth in use of integrated pest management, the utilization of *B. thuringiensis* as an insecticide is set to fulfill its potential.

Mode of action and specificity. The *B. thuringiensis* insecticide is usually applied to the crop plant, where it is ingested by the pest insect as it consumes the foliage, grain, or fruit. The crystalline inclusion is ingested along with the rest of the bacterium, after which a combination of the alkaline pH and the proteolytic activity of the insect gut fluids dissolves the crystal, thus activating the toxic polypeptide fragment known as the δ-endotoxin. The δ-endotoxin initiates the sequence that eventually causes the insect to die of a lethal septicemia. The activated toxin penetrates the peritrophic membrane (a tubular chitinous sheath) of the gut wall creating small pores in the lining cells. An inflow of water results, causing the breakup of the gut wall and ultimately insect death.

Identification of this general mechanism of action of the δ-endotoxin has been important for identifying the key processes responsible for the specificity of the crystal proteins, and hence the specificity of *B. thuringiensis* insecticides. The most obvious factors that could influence the host range of a crystal protein are (1) differences in larval gut affecting the solubilization of the protoxin and (2) the presence of specific toxin-binding sites (receptors) in the gut of different insects. Research has shown that the second factor is probably the more important, at least among different species of lepidoptera, where variation in midgut cell receptors seems to have a major influence on toxin specificity.

A knowledge of the factors influencing host specificity and the potential ability to modify it is important in the use of *B. thuringiensis* as an insecticide. In the past one of the problems with its use as an insecticide was the limited range of pests against which each of its crystal proteins could be applied. However, recent developments have made it possible to broaden the host range of a *B. thuringiensis* insecticide. The key discovery was the confirmation that the crystal proteins of the δ-endotoxin are single gene products, which led to the observation that very similar gene products can have very different host specificities. A difference of three amino acids is sufficient to account for differences in specificity between the larvicidal protein of *B. thuringiensis aizawai* that is toxic to both lepidoptera and diptera and the monospecific lepidopteran toxin from *B. thuringiensis berliner*. Such similarity lends support to the notion that most of these strains are evolutionarily related.

Selection and manipulation of strains. New strains and isolates of *B. thuringiensis* are being continually found and evaluated for use as insecticides. It is important that the search for new isolates not be neglected, because these isolates are the source of genetic material required for future exploitation. Traditionally, useful strains have been identified through a selection procedure that assesses potency relative to a standard strain (HD-1 strain of H-type *B. thuringiensis kurstaki*). Strain selection has rarely been used beyond this initial screening, although recent work suggests that it is possible to use induced selection pressure as a means of producing *B. thuringiensis* isolates with favorable characteristics such as resistance to ultraviolet light. However, current emphasis is on development through the techniques of genetic manipulation. The techniques of protoplast transformation, transduction, and conjugation have been used to vector genetic material into *B. thuringiensis*. These methods provide opportunities not only for determining mechanisms of action but also for improving formulation, toxicity, and host

range. Such an approach has been utilized to transfer crystal genes responsible for toxicity in lepidoptera into beetle-active strains, thus generating new hybrid clones that are active against both insect orders. A similar approach has produced a strain that is active against both lepidoptera and diptera. Toxicity has been enhanced by genetic manipulation of the active agent against pests such as the fall armyworm (*Spodoptera frugiperda*), a pest of corn, and the gypsy moth (*Lymantria dispar*), a forest pest. It is anticipated that other genetically manipulated *B. thuringiensis* insecticides will become available in the near future.

Fermentation, formulation, and application. One of the great advantages of using *B. thuringiensis* as an insecticide is the ease with which it can be produced. This is achieved through the use of conventional bulk fermentation techniques, a well-developed technology. Small improvements in production techniques can invariably be made, but any changes are unlikely to drastically reduce production costs unless the use of genetic manipulation techniques identifies completely new, reliable methods of production, perhaps for instance by the insertion of genes coding for easy-to-grow organisms such as *Pseudomonas*.

The formulation of an insecticide is crucial for the ease with which it can be applied and its retention, uptake, and persistence. In the past the poor persistence of *B. thuringiensis* has proved a major obstacle to its success as an insecticide. The bacterial spore of this bacillus is inactivated by ultraviolet light, which means that the insecticide does not persist on the foliage; hence frequent, regular applications are required. The commercial viability of a *B. thuringiensis* insecticide is thus decreased relative to more persistent chemicals. Traditionally, ultraviolet protectants have been added, either directly or during tank mixing; however, more innovative methods have become available that are proving successful. *Bacillus thuringiensis* spores and crystals have been encapsulated in a starch matrix containing ultraviolet screens; and most recently the endotoxins have been encapsulated in killed pseudomonad cells, offering both a flexible delivery system as well as enhanced persistence.

Bacillus thuringiensis insecticides can be applied with conventional pesticide application equipment. Field results published since 1985 are certainly encouraging for the future of *B. thuringiensis* as an insecticide, because satisfactory levels of control have been obtained for a wide range of pests in a variety of situations including stored products, field and horticultural crops, and forestry. Where problems have been encountered, the reasons for failure are most often associated with a poor understanding of spray technology. Factors such as spray coverage, droplet deposition and size, loss of toxin crystals, and contact rates can have a major impact the efficacy of a *B. thuringiensis* insecticide. Efforts directed toward a better understanding of spray technology and deposition on foliage have improved the success of this insecticide in forestry, and similar studies in other systems are expected to yield important benefits.

Prospects. Resistance to *B. thuringiensis* insecticides has been recognized since about 1988, and it has been demonstrated in at least five insect species. The inevitability of this resistance, especially as its use becomes more widespread, has led to increased research on deployment strategies that might delay or prevent its evolution. The use of *B. thuringiensis* in the context of integrated pest management programs, where it is only one of a number of control measures, should also delay the development of resistance. SEE PLANT (BREEDING).

David R. Dent

Baculoviruses

As a safer alternative to widespread chemical applications, viruses are increasingly being assessed for their potential to control insect pests, either in their natural form or with genetic modifications of the genome. Recent research has centered on one group, the baculoviruses, although other viruses such as the entomopoxviruses and iridescent viruses have also been considered for both their pest control potential and their capacity for genetic modification.

Baculoviruses are deoxyribonucleic acid (DNA) viruses found only in invertebrates and primarily in insects. Their limited distribution and host range are key factors in their development as pest control agents because they present less risk to nontarget species than alternative forms of control. Their use as pest control agents, particularly for caterpillar pests, goes back many decades. Although baculoviruses have been used successfully in a wide range of crops, their use, especially in agriculture, has not reached its full potential. A key reason for this is that the virus needs time to develop an infection, which makes it slower acting than the synthetic chemical insecticides with which they tend to be compared. Thus, current research using genetic engineering is being directed at developing faster-acting baculoviruses.

Genetic modification. Genetic modification of baculoviruses is possible as a result of the development of the baculovirus expression system. There are two main groups of baculoviruses: the nucleopolyhedroviruses and the granuloviruses, both of which are occluded (that is, the DNA-containing virus particles are packaged, either singly or multiply, in a proteinaceous coat which enables the virus to persist outside the host). The protein forming this coat in nucleopolyhedroviruses (polyhedrin) is expressed very late in the infection cycle and has been shown to be nonessential for virus replication. It is also under the control of a strong promoter which results in the synthesis of large quantities of protein, making it an attractive site for the incorporation of genes encoding foreign proteins. Understanding the mechanisms of the expression system has also meant that it is possible to introduce genes with the goal of altering the biological activity of the virus.

The current strategy for improving the efficacy of baculovirus insecticides is focused on increasing their speed of action. Single genes have been incorporated into the genome with the aim of poisoning the insect or disrupting its development, resulting in more rapid mortality or cessation of feeding. Introduced genes have thus fallen into two categories: insect-selective toxins or insect hormones and enzymes. However, a third option has been brought about by the discovery that certain baculovirus genes prolong the life of their host. The best-studied example of this type of gene is one which codes for the enzyme ecdysteroid UDP-glucosyltransferase, which inhibits the molting of an infected host. The full role of this gene in the insect and baculovirus interaction is unknown, but one effect of its deletion is a reduction in feeding and in the time taken for the virus to kill its host. The deletion of this gene has also been used in combination with the addition of genes from the other two categories to enhance final efficacy. However, not all these approaches have been successful: some have produced no effect at all; others have resulted in significant enhancement. Currently, the most effective genetically modified constructs have been those which utilize insect-selective toxins derived from arthropods.

Field testing. Field testing of these genetically modified viruses has lagged behind their development in the laboratory. The approach taken for their release has been cautious. The first release of a genetically modified baculovirus was carried out in the United Kingdom in 1986 and involved the alfalfa looper, *Autographa californica* nucleopolyhedrovirus (AcNPV). *Autographa californica* nucleopolyhedrovirus is the type species for nucleopolyhedroviruses, and its genome has been sequenced in its entirety, allowing a high degree of precision in its modification. The virus contained only a small, nonfunctional modification, that is, the addition of a noncoding 80-base pair synthetic oligonucleotide marker in the genome. This was the first stage in a step-wise program of confined field releases in which progressive changes were made to the baculovirus genome; the second release involved a virus in which the polyhedrin gene had been removed. The latter was followed by a release in which a functional marker gene (the beta-galactosidase gene from *Escherischia coli*) had been introduced. The final release was of a virus expressing a functional gene which enhanced its insecticidal properties.

The first field test of a baculovirus genetically modified to have enhanced insecticidal properties was carried out in the United Kingdom in 1993. This trial involved the release of *A. californica* nucleopolyhedrovirus which expressed an insect-selective toxin (AaHIT) from the scorpion, *Androctonus australis*. Laboratory assays showed that expression of the toxin paralyzed the larvae, resulting in a reduction in the median time to death in cabbage looper, *Trichoplusia ni,* larvae by approximately 25%. The field test compared three concentrations of the recombinant virus and its parent clone on cabbages infested with *T. ni* larvae. Leaf area analysis at the end of the trial showed a significant reduction in crop damage in the recombinant treated plots as compared to those treated with the wild-type virus, indicating that the strategy of increasing speed of kill may be a viable approach to developing novel crop protection agents.

In natural situations the normal cycle of baculovirus infection is for a susceptible larva to become infected while feeding. In most species little or no virus is released from the insect until after it dies, at which point the cuticle ruptures, releasing large numbers of virus polyhedra which then contaminate the environment. If an insect dies early enough in the larval period there is also the potential for a secondary cycle of infection in any remaining susceptible larvae. One interesting aspect of the 1993 trial highlighted differences in the effects that the two viruses (natural and modified) had on their hosts, which resulted in differences in secondary cycling. Because the genetically modified viruses killed their hosts more rapidly, the resulting virus yield was considerably reduced (compared to the wild type). The larvae paralyzed by the scorpion toxin also had a tendency to fall off the cabbages, thereby removing the inoculum they produced from any remaining susceptible hosts which were feeding on the plant. The combination of these factors appeared to result in reduced secondary cycling in plots treated with the genetically modified virus. Later field trials in the United Kingdom confirmed this finding, demonstrating that transmission of the faster-acting virus expressing an insect-selective scorpion toxin gene is considerably lower than for the wild type. These factors have interesting repercussions for the risk assessment of genetically modified baculoviruses and possibly other genetically modified biopesticides.

Risk assessment. Genetically modified baculoviruses are pathogens, and therefore their release is seen as a potential risk to nontarget species, either by the direct effects of the pathogen or by the inserted gene moving into another host. Thus, in addition to the development of more effective baculoviruses, research has also centered on risk assessment. Current work on this topic is taking a two-pronged approach: either assessing whether risks actually do exist or investigating strategies of biocontainment. The possession of a protein coat means that baculoviruses have the capacity to persist for considerable periods of time in habitats where they are protected from the degrading effects of ultraviolet radiation. For wild-type baculoviruses, this is undoubtedly an asset, as it provides the means by which the virus can survive between generations and be dispersed by passive vectors such as birds and predatory arthropods. The spread of baculoviruses from release points has been well documented. However, for a genetically modified baculovirus, this persistent capacity is not necessarily seen as beneficial. Thus, research into biocontainment has mainly involved looking at mechanisms whereby persistence is reduced so that genetically modified baculoviruses will be lost rapidly from the environment.

Reducing persistence has been achieved by deleting genes which are not essential for baculovirus replication, such as polyhedrin, *p10* and *pp34.* Polyhedrin minus viruses were released in the early field tests in the United Kingdom; the gene deletion reduced environmental persistence significantly as predicted, but this reduction was so severe that polyhedrin minus viruses were considered to be too unstable to be useful as a pest control agent. Other deletion mutants have yet to be released in the field. An intermediate option has been field-tested recently in the United States, that is, the co-occluded virus. In co-occluded viruses each polyhedra contains a mixture of wild-type (polyhedrin positive) and polyhedrin minus virus particles. Laboratory studies indicated that the polyhedrin minus virus would not persist on passage from one insect to the next, resulting in a rapid loss in the environment. If the polyhedrin minus virus also contained a biologically active gene, this would also die out. The field release supported this; data collected over 3 years showed a decline in the proportion of polyhedrin minus virus in recycling virus populations. However, the maintenance of conditions for co-occlusion requires very high concentrations of virus which may not be economically viable. It may also be argued that for certain types of modification (for example, accelerating speed of kill), this type of approach is superfluous as the engineered virus is at a selective disadvantage compared to the parent wild-type virus and will die out naturally after its release without further alterations.

For background information SEE BIOLOGICAL PEST CONTROL; FOREST PEST CONTROL; GENETIC ENGINEERING; INSECT CONTROL, BIOLOGICAL; INSECTICIDE; PESTICIDE in the McGraw-Hill Encyclopedia of Science & Technology.

Jennifer S. Cory

Bibliography. H. D. Burgess, *Microbial Control of Pests and Diseases,* 1981; J. S. Cory et al., Field trial of a genetically improved baculovirus insecticide, *Nature,* 370:138–140, 1994; J. C. Fry and M. J. Day (eds.), *Genetically Engineered and Other Microorganisms,* 1992; D. G. Jones (ed.), *Exploitation of Microorganisms,* 1993; National Research Council, *Neem: A Tree for Solving Global Problems,* 1992; H. Schmutterer, Properties and potential of natural pesticides from the neem tree, *Azadirachta indica, Annu. Rev. Entomol.,* 35:271–297, 1990.

Insect physiology

Unlike vertebrates, invertebrates do not produce antibodies or antigen-specific receptors (T-cell receptors) in their immune systems. However, invertebrate and vertebrate immune systems share processes which reflect a common origin in the primordial immune systems that existed before the divergence of chordates and arthropods. The outlines of primordial immunity are becoming clearer as more is learned about invertebrate immunity and features shared in common with vertebrates. Primordial immunity probably included both cellular and humoral responses. Primordial blood cells were able to recognize nonself and engulf or wall off foreign invaders. Antimicrobial peptides were synthesized, following gene activation by the Rel family of transcription factors, which are important components of both vertebrate and invertebrate immunity. Nonspecific antimicrobial peptides constitute an important part of invertebrate and vertebrate immunity. Insect systems may be studied to reveal shared fundamental properties of immunity or unexpected processes in vertebrates.

Cell-mediated immunity. Invertebrate blood-cell (hemocyte) types are not as well characterized as vertebrate ones, and there are large differences between species. However, most invertebrates have fewer hemocyte cell types than are found in vertebrate blood. Descriptive classification has been performed in several model systems, and monoclonal antibodies have been produced against different hemocyte types in *Manduca sexta,* the tobacco hornworm moth; unfortunately, it is not possible to do genetic work with this insect. In *Drosophila,* there are conflicting hemocyte classification systems. It is possible to identify genes that are expressed in hemocytes, which should provide the basis for hemocyte classification.

Phagocytosis and encapsulation are the two main cell-mediated immune responses in invertebrates. Microbes and small particles are phagocytized just as in vertebrates, while larger objects including parasites are encapsulated, a process similar to granuloma formation in vertebrates. Encapsulation begins when hemocytes flatten themselves against the foreign surface, as though they would phagocytize it if they could reach completely around. Layers of flattened cells are surrounded by other cells, until mature capsules are chemically cross-linked to form an impermeable surface.

Phagocytosis and encapsulation follow cellular recognition of foreign surfaces as nonself. Vertebrate macrophages are thought to use the sugar mannose and scavenger receptors to accomplish this task. A scavenger receptor has recently been identified in the fruit fly. This receptor and related proteins presumably serve similar functions on hemocytes as on macrophages. Thus, these two cell-mediated immune responses are much the same, and certainly must have constituted an important part of primordial immunity.

It is not surprising to find vertebrates and invertebrates sharing a similar cellular mechanism to recognize nonself. This mechanism is a central requirement for immunity, and once a functional system developed, it was conserved through evolution. Although vertebrates later acquired a more sophisticated and specific system involving immunoglobulin superfamily genes (and a corollary mechanism to prevent autoimmunity), they retained the functional, relatively nonspecific primordial mechanisms.

Some caution must be exercised in evaluating shared immune mechanisms. Tissue rejection is a well-known T-cell-mediated immune response in vertebrates. Heterograft rejection has been demonstrated between inbred lines of cockroaches (a usually long-lived insect), but these results are difficult to interpret since heterografts are accepted by other invertebrates and T cells do not exist in invertebrates. Tissue rejection is therefore not likely to have been a part of primordial immunity.

Humoral immunity. Two invertebrate humoral immune responses are still poorly understood. Nodule formation, somewhat similar to clotting, is the aggregation of microbes into gelatinous masses. Phenoloxidase may also be activated in a cell-free manner to kill invading organisms. More is known about a third humoral response, the synthesis of antimicrobial peptides which kill prokaryotic cells. Lysozyme is a well-known protein which is found in many vertebrate and invertebrate species, and often acts in digestion as well as in immune protection. Other antimicrobial peptides are known in different species; this is an active area of research and progress is rapid. While all of these antimicrobial peptides are nonspecific, some have wider ranges of activity (for example, gram positive versus gram negative) than others. A new antifungal peptide named drosomycin has recently been identified in *Drosophila.* This peptide shows no activity against bacteria, even though it is induced like the others by bacterial infection.

In vertebrates, a bewildering array of antibacterial peptides have been isolated from the skins of frogs and toads. There is little sequence conservation between these different peptides, obscuring evolutionary relationships. However, the existence of antimicrobial peptides in such widely different species demonstrates the primordial origin of this humoral response. In mammals, besides lysozyme and intracellular antimicrobials such as defensins, antimicrobial peptides have been found in bovine airway, porcine gut, and human wound fluid. Interestingly, the epithelial expression of antibacterials in insects is still somewhat unsettled because of the thinness of this tissue and its close apposition to cuticle, making histochemistry difficult.

Cytokines. There is evidence for intercellular signaling in invertebrates, which is accomplished by cytokines in vertebrates. Gentle abrasion of silkmoth cuticle induced only local cecropin (an antimicrobial peptide) expression, while stronger abrasion resulted in cecropin expression in fat body throughout the animal. Local injury of *Drosophila* larvae and adults results in antibacterial peptide induction in distant sites. These results are difficult to explain without soluble messengers traveling from the site of injury, but no cytokine-like proteins have been isolated from insects yet. In snails and other invertebrates, anti–interleukin 1 antibody cross-reacts with hemolymph components, but these components have not yet been identified.

Transcription factors. Insect antibacterial peptide expression is primarily controlled at the transcriptional level, another mechanism shared by vertebrates and invertebrates. A transcription factor regulating immune gene induction in the moth *Hyalophora cecropia* (cecropia immune factor) was found to be similar to NF-κB, a mammalian transcription factor important in regulating many genes in inflammation and the immune response. NF-κB consists of Rel proteins (Rel refers to a family of transcription factors including oncogenes and a gene from the reticular endotheliosis virus, which gave the family its name). NF-κB's activity is regulated by an inhibitory factor, IκB, which holds NF-κB in the cytoplasm. Upon activation, IκB is degraded, while active NF-κB moves into the nucleus.

Similar Rel factors have been identified in the flesh fly and in *Drosophila*. Dorsallike immune factor binds to the cecropin promoter and activates cecropin transcription in cell transfection assays. This factor is clearly a key element in antibacterial peptide induction, and the similarities between dorsallike immune factor and NF-κB in function (and presumably regulation) are compelling. Another Rel protein, Dorsal, plays a role in immunity; it binds to the diptericin promoter and activates gene transcription in cell transfection assays. Recently, a third *Drosophila* Rel protein, the gene *Relish,* was discovered. *Relish* contains IκB-like domains, similar to several mammalian NF-κB proteins. One of the NF-κB domains of these proteins has been shown to have IκB-like activity, and the inhibitory domain must be proteolytically removed to generate active NF-κB. Whether this is also the case with *Relish* is not yet known.

It is striking that Rel factors are key regulators of immune responses in both vertebrates and invertebrates. Even though effector pathways have diverged—the activation of inflammatory genes and immunoglobulins in vertebrates versus antimicrobial peptides in invertebrates—both groups use similar transcription factors to induce gene synthesis in response to infection. This clearly shows that Rel factor involvement in immunity is primordial in origin. SEE INFLAMMATION.

For background information SEE ANIMAL VIRUS; ANTIBODY; ANTIGEN; IMMUNITY; PHAGOCYTOSIS in the McGraw-Hill Encyclopedia of Science & Technology.

Mitchell Dushay

Bibliography. N. E. Beckage et al. (eds.), *Parasites and Pathogens of Insects,* 1993; S. Cociancich et al., The inducible antibacterial peptides of insects, *Parasitol. Today,* 10:132–139, 1994; A. P. Gupta (ed.), *Immunology of Insects and Other Arthropods,* 1991.

Irrigation (agriculture)

Recent inventories suggest that about 15–17% of the world's cultivated cropland is irrigated. About 40% of these acres were developed after 1960 to meet the food and fiber needs of the world's population, which grew rapidly after World War II in response to medical and hygienic advances. The lack of significant new water resources and the cost of developing present ones have caused irrigated acreage to plateau at around 6×10^8 acres (2.4×10^8 hectares) since the late 1980s. Recent advances in irrigation include the further development of sprinkler technology.

Impact on global harvest. The increase in irrigated acres from about 1960 through the 1980s, coupled with advanced agronomic practices and improved crop varieties, became known as the green revolution, and provided a brief period of global plenty. Despite the large increases in world population, food surpluses rose to some of their highest levels in recent history. Famine-prone nations such as India became exporters of grain. Slowing of irrigation development since 1980 has been an important factor in the renewed decline in world food reserves.

Earth's irrigated acres (one-sixth of all cropland) provide about one-third of the annual harvest, making irrigation more than twice as productive as rain-fed agriculture. The monetary value of irrigation's harvest is even greater than one-third the total. More importantly, one-third of the world's food crop harvest comes from a mere 1.25×10^8 acres (5×10^7 ha) of irrigated land, despite the fact that irrigated agriculture generally occurs on relatively poor soils.

Irrigated agriculture's efficiency and elevated monetary return stem mainly from two factors: First, irrigation generally results in higher-quality commodities because of its ability to better regulate inputs and prevent stresses than rain-fed agriculture. Second, the ability to reduce risk of stress, disease, and input inefficiencies often allows commercial production of crops having higher intrinsic monetary value.

In fact, on a global scale, most irrigated lands contribute positively to agricultural sustainability. Arid soils typically have low organic matter contents and high base saturation (a high ratio of adsorbed nutrient cations to hydrogen cation). Thus, arid soils seldom require liming to maintain favorable pH levels. They usually need less potassium fertilizer and lower soil-applied herbicide application rates for comparable weed control. The arid climate associated with most of irrigated agriculture not only maximizes photosynthesis but also in many cases translates to lower disease and pest pressures, often reducing required pesticide inputs for profitable production.

Another important aspect of irrigation is the role it plays in food security. As populations continue to increase, not only does the amount of production needed to meet human nutrition and fiber requirements increase, but so does the need for reliable production levels. Irrigated agriculture has proven less subject to the variability of climate and water supply than rain-fed agriculture. In properly managed irrigated schemes the impact of a short-term or even season-long drought can be mitigated by use of the water stored in reservoirs and groundwater aquifers. In rain-fed agriculture, droughts commonly cause serious yield reduction and sometimes complete crop failure.

Impact on environment. If all the world's irrigated land were taken out of production, it would require the cultivation of new land equal to more than 36 times the farmed area of Iowa to replace irrigated agriculture's contribution to Earth's harvest (a conservative estimate based on average production). Even if the land currently irrigated were left in rain-fed production, the extent of new land required to fill the production void would not be significantly reduced, for two reasons: First, the level of food and fiber production on land currently irrigated would be much lower than average rain-fed production, since most irrigation occurs in arid environments. Second, the world's choicest arable land is already in production. The productivity of the remaining lands, with their natural soil-climate associations, would not equal the existing mean rain-fed cropland base. In this way the use of irrigated agriculture currently allows more than 1.2×10^9 acres (4.8×10^8 ha) of Earth's rainforests, grasslands, ranges, and marshes to remain in undisturbed natural ecosystems.

In addition, irrigated agriculture's efficiency and concentration in arid and semiarid ecosystems have minimized species loss by limiting the need to clear more densely speciated environments (for example, rainforests) for agricultural development.

The most frequent criticism of irrigation focuses on disruption of natural ecosystems and species (or even cultural) displacement. Yet these outcomes also result from land management for rain-fed agricultural development, and usually occur on a larger scale because of the lower efficiencies. Critics of irrigation development frequently point to elevation of water tables and soil salinization as inherent problems. In reality, these problems usually result not from inherent failures of irrigated agriculture but from improper irrigation development. Modern irrigation design recognizes the need for adequate natural or artificial soil drainage and for application of adequate amounts of water to overcome salt accumulation.

It is worth noting that development of water resources has often had significant benefits beyond environmental ones. Water reservoirs and waterways provide hydropower, flood control, recreational resources, transportation, and fishing resources.

R. E. Sojka

Sprinkler technology. Sprinkler irrigation was developed to overcome many of the inherent limitations of earlier irrigation methods that relied on the soil surface to convey and distribute water to crops. Sprinkler application devices have evolved from simple orifices to a variety of nozzles designed to provide flexible and efficient water distribution over a wide range of soil and topographical conditions. Even though sprinkler irrigation requires energy to pressurize the system, the additional energy costs are largely offset by improved crop quality, reduced water applications, and labor savings. The gross amount of water applied is reduced because the applications are more uniform. Improved application uniformity allows smaller amounts of water to be applied more frequently, thus maintaining more favorable moisture conditions in the root zone so as to decrease water stress and improve the quality and yield of crops.

Technology development. Water droplets from sprinklers are subject to wind drift and evaporation. Contrary to the general perception that evaporation losses are significantly increased with sprinklers, recent energy budget studies have shown that maximum evaporation losses probably do not exceed 5% of the applied water. The effects of wind drift and distortion of application uniformity are more significant and have encouraged the development of improved nozzles that create larger droplets, which are less susceptible to wind drift. Size and spatial distribution of water droplets can be measured with laser technology; such studies are used to design and select sprinklers best suited for the operating conditions at a site.

Microprocessor technology is being used to improve the reliability and versatility of large mechanically moved irrigation systems. With improved control capabilities, producers are adopting innovative crop production practices that improve crop quality while conserving valuable water resources. Judicious application to meet crop water needs reduces leaching, which can transport valuable nutrients below the root zone as well as degrade water quality of the leachate that may eventually become part of the groundwater supply. Sprinklers may also be used in conjunction with drip irrigation systems to leach accumulated salts below the root zone during the nongrowing season or to germinate sensitive shallow-rooted crops.

Application of chemicals. Sprinkler equipment has been adapted to apply chemicals necessary for crop production through the irrigation system, a practice called chemigation. Fertilizers, herbicides, insecticides, fungicides, and growth regulators can be injected into the water as it is applied to the crop. Fertilizers that are susceptible to leaching, such as nitrate nitrogen, can be more effective and less likely to be leached when applied through the irrigation system as the crop needs it. Herbicides are the most frequently applied pesticide, and water applied simultaneously with the herbicide is usually necessary to activate the chemical. Fungicides applied with irrigation water have been demonstrated to be effective in controlling soil-borne diseases in a variety of vegetable crops. The advantages of chemigation are improved control of timing and uniformity of chemical applications, resulting in reduced chemical application amounts and costs. Food quality is improved by more timely applications of chemicals, while the environmental hazards of leached nutrients and chemicals degrading ground-water quality are reduced.

Low-pressure and low-volume sprinklers. New sprinkler applicators requiring lower pressures have been developed to combat rising energy costs. Assuming equal application uniformity and irrigation performance, low-pressure sprinklers require less energy than earlier high-pressure ones. For row crops, an irrigation system known as the low-energy precision application was developed to make more efficient use of limited water supplies as well as to maximize effectiveness of natural precipitation. Low-pressure nozzles are suspended from the overhead pipes of center-pivot or linear-move systems to a height of 15–18 in. (38–46 cm) above the soil surface. Small dikes every 4–6 ft (1.2–1.8 m) are created in each furrow to catch the applied water and allow it to infiltrate where applied.

Low-volume sprinklers that operate at low pressures, such as plastic microspray nozzles, have been developed primarily for use on tree crops in soils with low water-holding capacities and with limited ability to spread water by capillary action. The advantages of microsprays are low cost, a smaller volume of water delivered at lower pressures, and the ability to apply fertilizers and chemicals around the trees for maximum effectiveness.

Modification of microclimate. In fruit-producing areas susceptible to freezing temperatures, sprinklers can be used to modify the microclimate and protect the fruit crop from frost. When water is applied at a very low rate during freezing temperatures, heat is released when the water freezes, preventing damage to the plant tissue itself. Good management is required so as to operate sprinklers to achieve protection without causing ice damage to the trees.

Another example of microclimate modification is using sprinklers for evaporative cooling of tree or vegetable crops. When temperatures are higher than desirable, crops can be sprinkled frequently to create a cooler environment, which delays bloom and bud formation, thus reducing the likelihood of freeze damage as well as improving fruit yield. During critical growth periods, extremely high temperatures can cause detrimental stress in the plants. Water applied with sprinklers evaporates, cooling the temperature in the vicinity of the crop and reducing the water and temperature stress on the crop.

Wastewater reuse. Although the concept of using sprinklers to spread wastewater on cropland is not new, there is increased interest in using sprinkler irrigation as an environmentally sound disposal method that recognizes the nutrient value of wastewater as a resource. Especially in water-short areas, wastewater is reclaimed and used for irrigating cash crops as well as public areas such as parks and golf courses. Sprinkler irrigation systems can also apply sludge slurry to reclaim old mined or severely nutrient-deficient areas. The main benefits of sludge application are the release of nutrients for crop growth and the improved physical condition of the soil from increased organic matter. The main difficulty is the inability to match the available nutrients with the crop's nutritional needs, which can cause a detrimental buildup of certain nutrients and heavy metals. Recent work has focused on determining long-term, environmentally sound disposal rates.

Prospects. The trend for increased sprinkler irrigation is likely to continue because of labor savings and the ability to apply water more uniformly. Efforts to improve the sprinkler's versatility will continue as long as there are economic incentives to conserve water, energy, and nutrient resources.

For background information *SEE ECOSYSTEM; IRRIGATION (AGRICULTURE); SOIL CHEMISTRY; WATER CONSERVATION* in the McGraw-Hill Encyclopedia of Science & Technology.

Gerald W. Buchleiter

Bibliography. P. H. Gleick, *Water in Crisis,* 1993; G. J. Hoffman, T. A. Howell, and K. H. Solomon (eds.), *Management of Farm Irrigation Systems,* 1990; M. A. Hunst and B. V. Powers (eds.), *Agricultural Statistics,* 1993; W. M. Lyle and J. P. Bordovsky, Low energy precision application irrigation systems, *Trans. Amer. Soc. Agr. Eng.,* 24(5):1241–1245, 1981; D. Tribe, *Feeding and Greening the World,* 1994.

Language processing

Only humans communicate through speech: both talking (speech production) and listening (speech perception) involve the brain as much as the tongue and the ear. It is not known which neural circuits in the human brain interpret changes in sound pressure at the eardrums to distinguish speech and, more importantly, differentiate the sounds as a request, a demand, or a statement, and which circuits turn an idea into a meaningful utterance, either spoken or signed. Neuroscientists have not determined how many different brain areas are involved, what each does (or should do), and in what order each task is performed. Practically, no cognitive scientist could possibly build a robot with speech capabilities equal to that of a human if it had to be done on the basis of current knowledge about speech and the brain. *SEE BRAIN.*

The role of the brain. The human brain is a bilaterally symmetrical structure, interconnected by two main bridges of neurons. Neuropsychological research has demonstrated that an adult with left hemisphere damage is likely to have some difficulty in speaking (or signing) and in making sense of what is said or signed (a condition called aphasia). Moreover, if the bioengineer's specifications for an ideal speech processor require that the machine be able to appreciate the various nuances of meaning in the way an utterance is spoken, rather than in what is actually said, then the right hemisphere must be intact.

Developmental neuropsychologists and neurobiologists have investigated the way the brain handles speech changes with age. For example, young children do not process speech in exactly the same manner as adults. Furthermore, damage to the left hemisphere of the brain in a young child is much less devastating to normal speech development than the equivalent damage to the adult brain. Indeed, children with only one hemisphere can learn to hear and talk normally, and it does not seem to matter which hemisphere remains. Being left-handed also may mitigate against the detrimental consequences of left hemisphere damage, presumably because the right hemisphere plays a greater role in language processing.

Speaking occurs under voluntary (and therefore cortical) control including the inferior motor cortex in the zone of the lip, tongue, and larynx representations, and the supplementary motor cortex. The most recent data indicate that speech perception and production require the concerted action of anterior and posterior cortical areas within both hemispheres and various subcortical structures.

The discovery that large parts of the brain are involved in speech should not be surprising given the complexity of recovering phonetic structure (vowels and consonants) and eventually words from an essentially continuous acoustic stream. For speech reception, the representation of a word must be derived from the acoustic waveform and its various transformations along the auditory pathway; for speech production, the representation of a word must be realized in terms of a set of motor commands to the speech apparatus. In either case, any given acoustic segment typically contains information for more than one phonetic segment, and information for any given phonetic segment is typically spread across a series of acoustic segments. This lack of a one-to-one correspondence between acoustic and phonetic segments makes understanding what someone says and uttering a reply in response an extremely complex act which the brain carries out almost automatically at remarkable speeds. Individuals apparently find it easy to perceive voices never before heard in the face of different speaking rates, styles, and accents; and yet, they are incapable of accurately perceiving or producing nonspeech sounds at the same rate for any extended period of time.

Morphology of the brain. Because generally left hemisphere damage leads to severe speech-related problems in humans, neuropsychologists sought the speech areas of the brain by zeroing in on the gross morphological features that differentiated left and right hemispheres. The two sides of the human brain, much like the human face, are different despite their overall similarity. Close to the junction of the temporal and parietal lobes, a region called the planum temporale is longer and more curved in the left than the right hemisphere as if it is covering a greater expanse of cortex. The percentage of individuals with this asymmetry correlates significantly with the number who cannot speak if their left hemisphere is numbed with sodium amytal but have no problems if their right hemisphere is put to sleep. Clearly, the left hemisphere is essential for speech production. However, although numbing the left hemisphere arrests speech if an individual attempts to count, recite the alphabet, or tell a story, it does not prevent the person from understanding spoken language. Thus, the right hemisphere alone is capable of speech perception, and the brain areas involved in language comprehension are more broadly distributed than are the areas involved in language production.

Until recently, most of the evidence for speech and the role of the brain was based on language-related difficulties of individuals with brain damage following stroke and surgical removals or separation of the two cerebral hemispheres to relieve epileptic seizures. It was thought that speech was the sole province of two discrete areas in the dominant left hemisphere: production, under control of the motor speech center (Broca's area) that controls tongue and mouth movements; and perception, the responsibility of the sensory speech center (Wernicke's area).

The possible involvement of the right hemisphere in speech processing remains controversial, ranging from the view that normally it has no role whatsoever but can take over if the speech areas in the left hemisphere are compromised (especially early in life) to the view that it is critical to the understanding of the more figurative, metaphorical, and pragmatic aspects of language. In this latter view, an intact right hemisphere is essential for understanding idioms, metaphors, proverbs, indirect requests, and certain types of jokes as well as the nuances of meaning in the prosody, intonations, and emotional modulations of speech.

Brain imagery. Because gross anatomical differences between the hemispheres are not sufficient to account for functional differences, a more direct source of evidence on how and when different parts of the brain are engaged during speech has come from images of the intact brain taken as a person listens to or produces utterances. Since about 1980, techniques have been developed for constructing various images of brain structures (computed axial tomography [CAT]; magnetic resonance imaging [MRI]) and how they function (event-related brain potentials [ERPs]; functional magnetic resonance imaging [fMRI], magnetoencephalography [MEG], and positron emission tomography [PET]). There are techniques for measuring the electrical activity (event-related brain potential), the magnetic activity (magnetoencephalography), and the changes in cerebral blood flow or cellular metabolism (positron emission tomography, functional magnetic resonance imaging). Structurally, the three-dimensional magnetic resonance imaging image provides detailed static images of the brain. Of the functional approaches both event-related brain potential and magnetoencephalography provide recordings with temporal resolution on order of milliseconds. *SEE MAGNETIC RESONANCE IMAGING.*

In more general terms, the brain-imaging data from brain-damaged individuals as well as from intact individuals have challenged the assumption that there is a one-to-one mapping between structure and function. Thus, although there is clearly some segregation of responsibility for various speech operations that psychologists and linguists have proposed, there is no single brain center for either speech perception or speech production; no single region is responsible for construing meaning and following the rules of grammar. It appears that even the apparent specialization of the left hemisphere for speech may be a side effect of its ability to deal with very rapid transitions in the acoustic signal. Position emission tomography data show that the left hemisphere is more active than the right during the processing of acoustic transitions on the order of 50 milliseconds or less for both speech and nonspeech sounds. However, this type of analysis is a prerequisite for speech perception, particularly for hearing consonants rather than vowels. The right hemisphere by contrast is better equipped to process the envelope of the acoustic signal, including aspects of its duration and intensity.

Positron emission tomography studies further show that the primary auditory cortex contributes to the early acoustic analysis of all incoming auditory signals (speech and noise) whereas the superior temporal gyrus in both hemispheres appears to be responsible for higher-order auditory signal analyses. Speech, but not noise, also involves more anterior regions of the superior temporal gyrus. Pitch discriminations involve brain foci in the right prefrontal cortex. By contrast, phonetic discriminations involve regions in the left hemisphere, including Broca's area and the superior parietal area. Some of these same areas appear to be involved in speech production. Positron emission tomography blood flow data from the brain of intact individuals during speech production have implicated Broca's area and the lower motor cortex regions for the larynx, tongue, and face. Electrical stimulation also has revealed a basal temporal language area.

Implications. These new techniques have revealed that both speech perception and speech production are complex processes requiring a distributed network of systems in the brain with the broad involvement of the left cerebral hemisphere, parts of the right hemisphere, and various subcortical structures (thalamus, basal ganglia, cerebellum). These data also have demonstrated that the global similarity of human brains belies their striking variability in both structure and function.

For background information *SEE BRAIN; HEMISPHERIC LATERALITY; MEDICAL ULTRASONIC TOMOGRAPHY; PSYCHOLOGY, PHYSIOLOGICAL AND EXPERIMENTAL; SPEECH PERCEPTION* in the McGraw-Hill Encyclopedia of Science & Technology.

Marta Kutas

Bibliography. S. E. Blumstein, Impairments of speech production and speech perception in aphasia, *Phil. Trans. Roy. Soc. London B*, 346(1315): 29–36, 1994; B. M. Mazoyer et al., The cortical representation of speech, *J. Cog. Neurosci.*, 5(4):467–479, 1993; S. L. Miller, T. V. Delaney, and P. Tallal, Speech and other central auditory processes: Insights from cognitive neuroscience, *Curr. Opin. Neurobiol.*, 5(2): 198–204, 1995; R. J. Zatorre et al., Lateralization of phonetic and pitch discrimination in speech processing, *Science*, 256:846–849, 1992.

Laser

Recent laser technology advances involve a very broad range of areas. This article discusses the effects on matter of ultrahigh-intensity, ultrashort-pulse lasers; the quantum cascade laser, a new type of semiconductor laser based on tunneling and quantum confinement; and free-electron lasers, in which high-intensity radiation is generated by the interaction of an electron beam with a periodic magnetic field called a wiggler.

Laser-Matter Interactions at Ultrahigh Intensities

Apart from the most extreme conditions that existed during the first few minutes after the big bang (and are found in certain astrophysical objects even today) never has matter been exposed to such high light intensities as it is when illuminated with ultrashort-pulse lasers. Laser pulses can routinely be produced lasting less than a picosecond and reaching intensities exceeding 10^{19} W/cm^2 at a focus. The electric fields associated with such laser beams can be so high as to ionize a heavy atom such as uranium to U^{82+} in less than 1 ps. These lasers are, therefore, beginning to have a revolutionary impact on atomic physics, plasma physics, and accelerator technology.

Plasma formation. When an ultraintense laser pulse is focused in a medium, the medium is rapidly ionized to produce a plasma. An intense laser pulse can ionize an atom even when the photon energy is much less than the ionization energy of the atom. This is possible because an atom can absorb many photons at once. Multiphoton ionization of single atoms has been studied extensively for many years. However, researchers are only now beginning to appreciate the unique properties (and therefore applications) of a medium ionized in this manner. Among the properties of a gaseous medium ionized via multiphoton ionization are uniquely determined charge states, electron density and temperature, spatial size, and extremely large transient

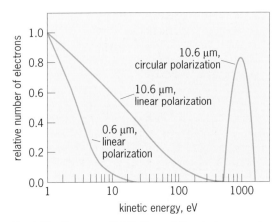

Fig. 1. Kinetic energy distributions of the electron popula-tion produced by pulsed laser ionization of xenon gas at the wavelengths and polarizations indicated. (*After P. B. Corkum, N. H. Burnett, and F. Brunel, Above-threshold ion-ization in the long-wavelength limit, Phys. Rev. Lett., 62:1259–1262, 1989, and R. R. Freeman et al., Above-threshold ionization with subpicosecond laser pulses, Phys. Rev. Lett., 59:1092–1095, 1987*)

magnetic fields. Experimental measurements of the threshold appearance intensity, the minimum laser intensity for the production of a particular ionic state, may be compared with the calculated thresh-old intensity by using the barrier suppression model. In this model the intense electric field of the laser suppresses the atomic potential barrier so that a bound electron can escape this barrier and become free. The various ionic states of noble gases have an appearance intensity that is predicted rather well by the barrier suppression model. These data clearly show that a plasma which predominantly contains a specific ionic species can be generated by choosing the laser intensity. When a gaseous medium is thus ionized, the resulting initial elec-tron distribution function can be varied over a very wide range of energies by changing the laser wave length or polarization (**Fig. 1**).

One immediate application of this control over charge state and electron temperature of the plasma is to induce lasing action in the extreme ultraviolet region of the electromagnetic spec-trum. In one approach to realizing such a laser, known as the recombination scheme, a dense but cold plasma of fully stripped ions is created via multiphoton ionization. Collisional recombination preferentially populates the high-quantum-number states of the hydrogenlike ion. This process, cou-pled with the faster radiative decay of the lower levels, leads to population inversions between lev-els with principal quantum numbers up to 4. Las-ing at 13.5 nanometers in the far ultraviolet by recombination in a hydrogenlike lithium plasma produced by an ultrashort laser pulse has been demonstrated.

A short intense laser pulse that forms a plasma via multiphoton ionization undergoes dramatic spectral changes as a result of the time-dependent decrease in the plasma's refractive index. This nonlinear effect is called self-phase modulation. At some point dur-ing the intensity rise of the laser pulse a plasma begins to form, lowering the medium's refractive index and consequently blueshifting a portion of the laser pulse. This time-dependent frequency change is called a chirp. Because the laser pulse propagates through the medium at close to the speed of light, the ionization front it creates also propagates rela-tivistically. Such ionization fronts have been used not only to chirp laser pulses but also for compressing intense ultrashort pulses.

High harmonic radiation. When ultrashort laser pulses were first focused onto high-density gas-jet targets, a surprising result was observed. The interac-tion generated extremely high odd-order harmonics of the incident laser wavelength. The harmonic spec-trum thus generated has two characteristic features: The conversion efficiency falls off rapidly with increasing order, but only up to about the eleventh harmonic. Then comes a plateau, where the conver-sion efficiency (roughly 10^{-8} to 10^{-9}) falls off rather slowly (**Fig. 2**). Experiments with extremely short laser pulses have approached or reached the 135th harmonic of 1.05-micrometer incident radiation. These very high harmonics show a cutoff at photon energies of about three times the electron quiver or oscillatory energy in the electric field of the laser.

Solid-density plasmas. When an intense, short laser pulse hits a solid target, the electron-ion colli-sion times are much shorter than the pulse dura-tion. Therefore, the laser pulse duration determines the time scale for temperature changes in the plasma formed from the solid. The expansion of such a solid-density plasma, which erases the den-sity discontinuity at the solid surface, occurs on a still longer time scale. Therefore the optical radia-tion is in effect interacting with the still-solid target. Thus, in such experiments a short burst of contin-uum x-ray radiation is seen as well as discrete x-ray emission lines. The x-ray burst from a solid target can be a few picoseconds in duration. Although the radiation is incoherent, it can have a peak power

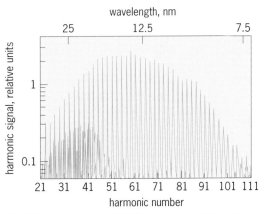

Fig. 2. Power spectrum of high harmonic radiation from a pulse of 0.8-μm laser light focused to a peak intensity of 10^{15} W/cm^2 in a neon gas jet. (*After J. J. Macklin, J. D. Kmetec, and C. L. Gordon, III, High-order harmonic generation using intense femtosecond pulses, Phys. Rev. Lett., 70:766–769, 1993*)

many orders of magnitude higher than can be obtained in the 1–10 keV range from synchrotron light sources. Such short, intense x-ray pulses offer the possibility of picosecond resolution for many kinds of experiments in chemistry and biology. If the terawatt lasers with kilohertz repetition rates that are now being contemplated are employed, the x-rays from such high-density plasma should find commercial applications in x-ray lithography and x-ray imaging.

At 10^{18} W/cm² the light pressure of an ultrashort laser pulse is 0.3 gigabar (30 terapascals). The light pressure can exceed the thermal pressure of the low-temperature plasma, whose density is still close to that of the solid from which it sprang. Thus, the light pressure can both impede the plasma's expansion and push the plasma aside. The radiation pressure of an ultraintense pulse plays a crucial role in the fast ignitor scheme, a new concept for achieving energy from laser fusion. In this scheme the radiation pressure of a short laser pulse bores a hole through the plasma atmosphere of an imploding fusion pellet so that yet another ultrashort pulse can propagate through the hole and reach the solid surface. At the surface of the pellet the energy of the ultrashort pulse is converted into a beam of megaelectronvolt electrons. These hot electrons in turn deposit their energy in the compressed deuterium-tritium fuel, igniting the fusion reactions.

Collective effects. A plasma is a quasineutral collection of electrons and ions. In addition to producing such a plasma via multiphoton ionization, an intense burst of photons from a laser pulse can strongly perturb the plasma. Electrons are displaced from the ions, thereby setting up an electric field. This electric field in turn acts as a restoring force causing these initially displaced electrons to oscillate about the ions, thus setting up oscillations or plasma waves.

Of particular interest are relativistic plasma waves that have the largest useful phase velocities, very nearly the speed of light. They are useful because the longitudinal electric field arising from the displaced electrons can be very high for such waves and may be used to accelerate externally injected electrons to high energies in a very short distance. There are three ways of exciting such a relativistic plasma wave. The first is called wakefield generation. Here, a short, intense laser pulse, only half a plasma wavelength long, disturbs the plasma electrons and leaves behind a wake of plasma oscillations, which oscillates at the plasma frequency, much like the wake of a motorboat. The second method is called beat-wave excitation. A beat pattern created by two copropagating laser pulses slightly differing in frequency can be thought of as a series of short laser pulses, which resonantly build up a large plasma oscillation if the beat frequency matches the plasma frequency. The third method is called the Raman forward scattering (RFS). Raman forward scattering is a parametric instability where the laser photons decay into rela-

tivistically propagating plasmons and Stokes or anti-Stokes photons that are down or up shifted in frequency by the plasma frequency. Until very recently, the ultraintense laser pulses needed to excite plasma wake fields interesting enough for particle accelerator applications were not easily available. However, remarkable results have been obtained that demonstrate ultrahigh-gradient particle acceleration by using the beat-wave and Raman forward scattering techniques.

In 1993, by using a two-frequency carbon dioxide (CO_2) laser, researchers were able to excite a plasma beat wave that had a large density modulation and therefore, according to Gauss' law, a large electric field. By injecting 2-MeV electrons from an external linear accelerator into a plasma beat wave approximately 1 cm long, they observed up to 30-MeV electrons. These experiments implied an acceleration gradient of approximately 2.8 GeV/m, in contrast to current benchmark gradient of approximately 20 MeV/m in radio-frequency linear accelerators, opening up possibilities of making compact particle accelerators.

If the beat-wave technique has set the record for the highest-gradient acceleration of externally injected particles, the record for the highest-gradient acceleration of plasma electrons that are self-trapped by a relativistically moving plasma wave goes to the Raman forward scattering scheme. The maximum possible electric field of such a wave is obtained when the relativistic wave breaks. This so-called wave-breaking limit is characterized by sudden acceleration of copious amounts of electrons from the plasma itself. Experiments demonstrated electron acceleration from rest to 44 MeV in just 350 μm or an average energy gain greater than 100 GeV/m via such a wave-breaking process. This acceleration gradient represents by far the highest collective-wave field ever produced in the laboratory.

Relativistic effects. When the oscillating velocity of an electron in the laser field becomes nearly the speed of light, relativistic effects become important. One such effect is relativistic guiding. As in any nonlinear optical medium, self-focusing of a laser pulse in a plasma arises because the refractive index becomes higher where the intensity of the radiation is greatest, causing the wavefront to become concave. Relativistic self-focusing or guiding is caused by the reduction in plasma frequency (and therefore an increase in the refractive index) by the relativistic mass increase of plasma electrons oscillating in the laser field.

Relativistic guiding has an important application in collective plasma accelerators. The theoretically maximum energy gain in a plasma accelerator is possible only if the plasma wave exists over a distance much greater than the Rayleigh range of a focused laser beam. (The Rayleigh range is the distance on either side of the best focus where the intensity of the laser beam drops by a factor of two because of diffraction.) Relativistic guiding is one way to maintain the large laser intensities needed to excite a large plasma wave.

Another important relativistic effect is the generation of odd harmonics when laser pulses are incident upon a sharp interface between a vacuum and a solid-density plasma. As mentioned above, such pulses can exert gigabar pressures, creating large-density oscillations and relativistic electron velocities at the plasma surface, which lead to efficient harmonic generation.

Chandrashekhar Joshi

Quantum Cascade Laser

The quantum cascade laser is a semiconductor laser involving only one type of carrier and based on two fundamental phenomena of quantum mechanics, tunneling and quantum confinement. Although the basic concept was proposed in 1971, an actual device was demonstrated only in 1994.

Because of their small size (typically the size of a pinhead), high reliability, and ease of use, conventional semiconductor diode lasers play a dominant role in telecommunications and consumer electronics in applications such as optical fiber communications and compact-disk players. In these lasers, the light originates from the recombination of negative and positive charges (electrons and holes) injected into a small region of the structure (the active region) by means of a *pn* junction diode. In this process electrons from the filled energy states in the conduction band make a transition to empty states in the valence band (holes), emitting a photon whose energy equals the energy separation between the two bands. The energy separation between the two bands, referred to as the energy bandgap, thus determines the lasing wavelength. As an example, infrared diode lasers emitting at a wavelength of 10 μm require much smaller bandgap semiconductors than near-infrared lasers, emitting around 1 μm, used for compact-disk players and optical communications. Unfortunately, these small-bandgap semiconductors, such as mercury cadmium telluride (HgCdTe) or lead tin telluride (PbSnTe) alloys, are hard to process and are prone to the formation of defects.

The quantum cascade (QC) laser is based on a completely different approach. In a quantum cascade laser, electrons make transitions between bound states created by quantum confinement in ultrathin alternating layers of semiconductor materials. Because these ultrathin layers, called quantum wells, are comparable in size to the electron's de Broglie wavelength, they restrict the electron motion perpendicular to the plane of the layer. Because of this effect, called quantum confinement, the electron can no longer have an arbitrary energy perpendicular to the layer, but must have one of the few energies that are now allowed. These special values of the energies are called bound states, and the electron can jump from one state to another only by discrete steps. The spacing between the steps depends on the width of the well, and

increases as the well size is decreased. This laser is freed from so-called bandgap slavery; that is, the emission wavelength depends now on the layer thicknesses and not on the bandgap of the constituent materials. *SEE INFRARED RADIATION.*

In a laser, electrons make transitions between two states, emitting photons of light. The electrons stay in one of these states for a limited time which is characteristic of this state and is called the lifetime of this state. To obtain lasing action, a laser must satisfy the condition of population inversion; that is, the population (number of electrons) of the upper state must be larger than the population of the lower state. This special situation can be obtained if the lifetime of an electron in the upper state is longer than the lifetime of this electron in the lower state. Indeed, the longer the lifetime of a state, the larger the population on this state. In other words, the lower state must empty faster than it fills. In a normal solid state or gas laser, this right set of lifetimes, resulting from the natural chemical and electronic properties of the material, is obtained by choosing the right pair of states. In contrast, in a quantum cascade laser, these lifetimes are obtained by careful design of the structure.

Design of the structure. The material chosen for the first quantum cascade lasers is an alloy of indium gallium arsenide for the well material and aluminum indium arsenide for the barrier material. These materials may be said to be sprayed one monolayer (approximately 0.25 nanometer thick) at a time by using an ultrahigh-vacuum evaporation technique called molecular beam epitaxy (MBE). This technique allows for the very sharp interfaces and the thickness control to about a monolayer that are needed for the growth of a quantum cascade laser. Such an accuracy is difficult to comprehend. For comparison, one monolayer is about 1/1000 the thickness of the finest line on the best microchips manufactured. Such thin layers are necessary to observe the electron's quantum confinement effects.

The overall structure of the quantum cascade laser is a repetition of 25 identical cells. Each cell (**Fig. 3***a*) consists of an active region followed by a electron injector. The active region is the part of the structure where the photon emission takes place. It consists of a series of three quantum wells that, under an appropriate electrical bias, form an energy ladder of three states. Lasing occurs between the level 3 created by the first (and narrowest) well and the level 2 created by the adjacent well. From level 2, electrons are first siphoned out into level 1 and from there escape into the electron injector from where they can be injected into the next period. Population inversion is maintained between levels 3 and 2 because level 3 is designed to have a much longer lifetime than level 2.

All the states depicted in Fig. 3*a* are truly bound only in the growth direction; that is, electrons can still move freely in the plane of the layers. For this

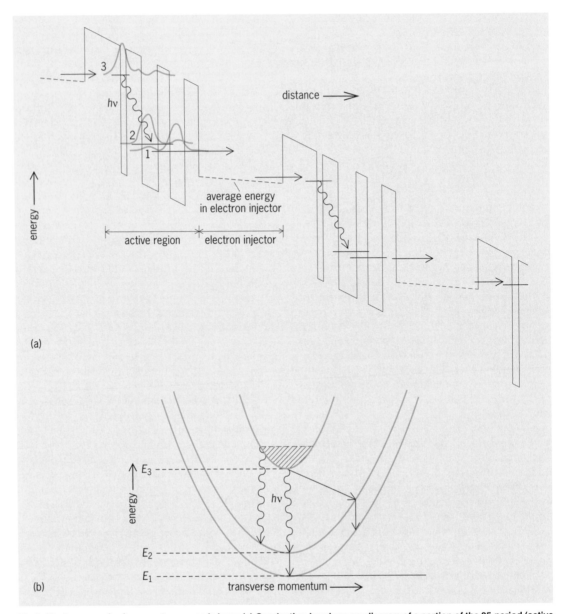

Fig. 3. Electron energies in a quantum cascade laser. (*a*) Conduction-band energy diagram of a portion of the 25-period (active region plus electron injector) cell of the laser. Wavy line represents emission of photons of energy *h*ν, where *h* is Planck's constant and ν is the corresponding frequency. (*b*) Plot of dispersion, that is, energy versus transverse momentum (momentum parallel to the plane of layers) of the electron for states 1, 2, and 3. The wavy lines represent emission of photons from state 3, the straight lines the emission of optical phonons. The minimum energies of electrons in states 1, 2, and 3 are E_1, E_2, and E_3, respectively.

reason, an electron in a excited state can emit an optical phonon and fall into a lower state (Fig. 3*b*). Fortunately, the probability for such an event decreases strongly when the energy separation between the two states is increased well above the optical phonon energy. The structure is therefore designed such that the energy spacing between levels 3 and 2 (300 meV) is much larger than the optical phonon energy of approximately 34 meV in this material system, while maintaining the spacing between levels 2 and 1 close to the optical phonon energy. In this way, the lifetime of level 3 is much longer than that of level 2 because it takes much longer to emit an optical phonon from level 3. Additionally, a careful examination of the probability

distributions of the electrons in levels 3 and 2 in Fig. 3 shows that these distributions are much less over-lapped between levels 3 and 2 as compared to levels 2 and 1, further enhancing this ratio of lifetimes.

The overall band diagram of the structure (Fig. 3*a*) is tilted by an applied electric field. This field is generated by the electrical bias applied to the device and serves to drive the electrons across the structure. It is essential for the good performance of the device that the electrons follow a well-defined path when cascading down the structure. To this end, a region called the electron injector was designed next to the active region. This region is formed by alternating ultrathin layers of barrier and well materials to create what is called a digital alloy. In a way similar to the simulation of shades of gray in a printed image by varying the densities of black dots on white paper, the electron does not feel the individual layers but only the average energy of the well and barriers. This average energy changes with position so as to almost completely cancel the effect of the applied field in the injector region (Fig. 3*a*). Because the bottom of the electron injector is almost flat, electrons are collected from the active region, and allowed to lose their energy and therefore cool down. From the bottom of the injector, the electrons must be injected with a high efficiency into the upper state of the lasing transition (state 3 in Fig. 3*a*), because only the electrons staying in state 3 will participate in the lasing. Electrons injected above or below this energy would be wasted and simply increase the current passing through the structure. This efficient injection is obtained by finely tuning the composition of the injector in such a way that, under applied bias, it is nearly flat and at the same energy as state 3 of the next active region. In this way, electrons can literally be funneled into this state by a process called resonant tunneling, in which the presence of state 3 at the same energy as the injector increases dramatically the probability of an electron crossing the barrier that separates the injector from the next well.

Features. The first such structure fabricated lased at a wavelength of 4.26 micrometers, in the midinfrared part of the electromagnetic spectrum. A signature of lasing action is shown in **Fig. 4**, where the spectrum of the laser as a function of wavelength is displayed for increasing values of current. For currents smaller than 600 mA, the spectrum is broad, characteristic of spontaneous emission. As the current increases, the spectrum narrows regularly as the gain increases. Then suddenly, at a current of 850 mA, the gain has risen to the point where it equals the cavity losses. The threshold for lasing action is reached: the spectrum exhibits a dramatic line narrowing, and the power increases by several order of magnitude, from microwatts to milliwatt levels. These early devices operated only in the pulse mode (that is, the current was injected in short bursts) and at cryogenic temperatures of 10 K (−263°C, −442°F).

Perspectives. Since this initial demonstration, much progress has been made. Advantage has been taken from the freedom from bandgap slavery and lasers have been designed and operated at wavelengths in the 4–5- and 8–10-μm regions by using the same heterostructure material. These structures operate in the continuous-wave mode up to temperatures of 100 K (−173°C, −280°F) and in the pulse mode up to 230 K (−43°C, −46°F). These structures already outperform in optical power the existing semiconductor technology for this range of wavelengths, which is based on lead salt crystals.

The quantum cascade laser is likely to become a major contender in the race for an inexpensive, high-operating-temperature midinfrared source for

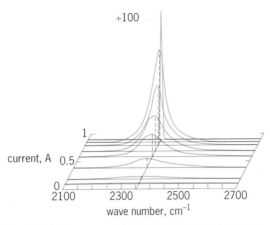

Fig. 4. Emission spectrum of the quantum cascade laser at various drive currents. The strong line narrowing and the large increase of the optical power at currents above 850 mA demonstrates laser action. The emission wavelength is λ = 4.26 μm.

pollution monitoring, optical remote sensing, or collision avoidance radar. In a more distant perspective, development of new materials with larger band discontinuities (energy difference between the band edge in the two semiconductor materials, which is equal to 0.52 eV in current structures) would allow structures to be designed with larger photon energies, for operation at shorter wavelengths in the near-infrared or even visible end of the spectrum.

Jérôme Faist; Frederico Capasso

Free-Electron Lasers

In a free-electron laser, radiation of high intensity is generated by the interaction of a relativistic electron beam and a spatially periodic magnetic field, known as a wiggler (or an undulator). In contrast with conventional lasers, which rely on radiation emitted by bound electrons as they undergo transitions between discrete atomic energy levels, the electrons in a free-electron laser are free. One of the most notable attributes of a free-electron laser is its tunability: it is possible to change the wavelength of a free-electron laser continuously by changing either the energy of the electron beam or the wavelength of the wiggler magnetic field, or both. The free-electron laser can be tuned over a very wide range of wavelengths, making it a useful device for several applications in physics, chemistry, and biology, in academic research as well as industry. Since its discovery in the 1970s, free-electron laser research has matured to the point that several user facilities provide radiation spanning several decades in wavelengths (from less than 1 μm to 1 millimeter) for use by biomedical practitioners, chemists, and condensed-matter physicists.

Basic principles. **Figure 5** is a schematic representation of a free-electron laser device. An electron beam from an accelerator enters a cavity containing a periodic magnetic field and a seed radiation field. The accelerated charged particles

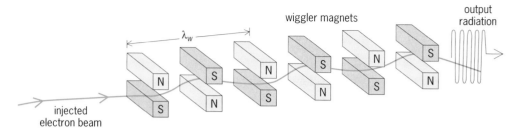

Fig. 5. Free-electron laser amplifier. N and S are north and south poles of magnets. Electrons in the beam undergo periodic oscillations with wavelength λ_w equal to the period of the wiggler magnetic field. (*After C. W. Roberson and P. Sprangle, A review of free electron lasers, Phys. Fluids, 1:3–42, 1989*)

emit radiation according to the laws of electromagnetism. The radiation emitted can be of two types, spontaneous and induced, in analogy with the emission from a conventional laser. The spontaneous emission is the random superposition of the radiation fields due to the individual electrons in the electron beam and is hence incoherent, with an intensity proportional to the number of electrons (or the total current of the electron beam). From the point of view of free-electron-laser operation, the more interesting part of the radiation is the induced emission that occurs once the interaction of the electron beam with the radiation field leads to the organization of the electrons into bunches spaced approximately one radiation wavelength apart. These bunches of electrons emit coherent radiation that is synchronized in phase, with the consequence that the intensity of the coherent emission scales as the square of the number of electrons (or the square of the total current of the electron beam).

Free-electron lasers can be configured as oscillators or amplifiers. In an oscillator configuration, the wiggler has partially reflecting mirrors at its two ends, and the radiation intensity is built up by many passes of the radiation over the electron beam. In an amplifier, a strong seed radiation field provided by an external source is amplified by single pass in the high-gain regime.

In contrast with conventional lasers which can be explained only by quantum mechanical principles, the behavior of a free-electron laser can be understood entirely in classical terms by using the laws of relativistic mechanics and electromagnetism. (Although the earliest theoretical treatments of free-electron-laser dynamics made use of quantum mechanics, Planck's constant, a ubiquitous presence in quantum-mechanical calculations, was seen to cancel out of the formula for the laser wavelength or the optical gain, strongly suggesting that the free-electron-laser mechanism is intrinsically classical.) Because of the Lorentz force imposed by the periodic wiggler magnetic field, every electron in the electron beam undergoes periodic oscillations, indicated qualitatively by the wavelike trajectory in Fig. 5, with wavelength λ_w equal to the period of the wiggler field. The wavelength λ of the electromagnetic radiation obtained

at the end of the wiggler is given by Eq. (1). Here, $\gamma = (1 - v^2/c^2)^{-1/2}$ is the relativistic factor, where v is

$$\lambda = \frac{\lambda_w\left(\dfrac{1 + a_w^2}{2}\right)}{2\gamma^2} \qquad (1)$$

the electron beam velocity and c is the speed of light; and a_w is the wiggler parameter given by Eq. (2). In Eq. (2) e is the magnitude of the electron charge,

$$a_w = \frac{eB_w\lambda_w}{2\pi mc^2} \qquad (2)$$

B_w is the magnitude of the wiggler field, and mc^2 is the rest energy of the electron. Equation (1) reveals the tunability that is one of the most attractive properties of a free-electron laser: by changing γ and λ_w continuously, it is possible to scan continuously over a range of radiation wavelengths. Furthermore, because the free-electron-laser mechanism involves conversion of the kinetic energy of the electron beam into radiation energy, the availability of high-power accelerators offers the potential, in principle, of generating radiation of high power.

Efficiency enhancement. The free-electron laser was originally conceived with a constant-parameter wiggler, that is, a wiggler with constant wiggler wavelength λ_w and wiggler parameter a_w. However, such devices have stringent limitations on the efficiency of energy extraction. The fraction, η, of the initial electron beam energy that is converted to radiation energy is typically less than 10% with constant-parameter wigglers. The physical reason for this limitation can be understood from Eq. (1), which essentially represents a resonance condition between the electrons and the phase velocity of the so-called ponderomotive wave that grows out of the interaction between the electron beam and the radiation field. When this resonance condition is obeyed, the electrons can transfer energy to the radiation field, but in doing so they fall out of resonance because their energy decreases. When the resonance is lost, no further energy transfer can occur from the electron beam to the radiation field. For fixed wavelength, one way in which the resonance condition of Eq. (1) can be preserved along the length of the wiggler as the electron beam energy (proportional to γ) decreases is by decreasing the wiggler parameter a_w as a function of distance along the wiggler. This is essentially the idea underlying efficiency enhancement using a so-called tapered wiggler. Some tapered devices have generated microwaves with efficiency $\eta = 34\%$.

Spectral purity. In a conventional laser, induced emission is well known to produce sharp wavelengths. However, the spectral purity of a free-electron laser is hampered by the excitation of the so-called sideband instability, which is caused by the resonant coupling of the main signal with other parasitic wavelengths, mediated by the periodic oscillations of electrons in the ponderomotive wave. The excitation of sidebands can cause a conspicuous degradation of the spectrum generated by the free-electron laser. **Figure 6a** shows the output of a 10-μm free-electron laser amplifier using an untapered wiggler, as seen in a computer simulation. The main signal, centered at zero frequency, is flanked by numerous parasitic frequency components which are excited by trapped-electron instabilities. (Similar parasitic excitations are also seen in free-electron-laser oscillator configurations.) It has been shown theoretically as well as experimentally that it is possible to eliminate parasitic modes in oscillators, when necessary, by passive optical techniques such as the placement of a filter that absorbs the wavelength of a sideband.

Recently, it has been realized that strong tapering of a free-electron-laser amplifier can achieve simultaneously the goals of high extraction efficiencies and strong sideband suppression. Figure 6b shows the spectrum of the 10-μm free-electron laser amplifier, obtained from a computer simulation, with a strongly tapered wiggler in which the wiggler parameter a_w is decreased linearly with distance along the wiggler. The near elimination of the parasitic sideband modes is accompanied by an enhancement in the amplitude of the primary signal in Fig. 6b relative to the amplitude in Fig. 6a by nearly a factor of 10. The extraction efficiency, η, in such a strongly tapered system is predicted to exceed 50%.

Optical guiding. It is well known that light spreads by diffraction. Perhaps the simplest example of diffraction is the spreading of the light beam from a flashlight which illuminates a wider area, but with lower intensity, as the distance from the flashlight increases. The radiation from a conventional laser or a free-electron laser can exhibit a similar behavior. For a fixed energy output, this increase in the spot size means a loss of intensity in the optical beam. Because long wigglers are often necessary to obtain large gain, diffractive losses impose constraints on the maximum realizable gain. One of the beneficial effects in a free-electron laser, known as optical guiding, is that the electron beam, under some conditions, acts as an optical fiber with a refractive index slightly greater than unity. In other words, the optical beam is spontaneously focused toward the axis of the wiggler, as when a light beam in air falls on a convex glass lens. This guiding effect reduces the diffractive losses, and allows the propagation of the optical beam over wigglers that are several meters in length. Other very interesting ramifications of the analogy between an electron beam and an optical fiber include the possibility of generating solitary optical pulses that can propagate long distances without changing shape.

For background information *SEE LASER; NONLINEAR OPTICS; NUCLEAR FUSION; OPTICAL PULSES; PARTICLE ACCELERATOR; RAMAN EFFECT; SEMICONDUCTOR HETEROSTRUCTURES; SYNCHROTRON RADIATION; TUNNELING IN SOLIDS* in the McGraw-Hill Encyclopedia of Science & Technology.

Amitava Bhattacharjee

Bibliography. C. A. Brau, *Free-Electron Lasers*, 1990; M. Everett et al., Trapped electron acceleration by a laser-driven relativistic plasma wave, *Nature*, 368:527–529, 1994; J. Faist et al., Quantum cascade lasers, *Science*, 264:553–556, 1994; J. Faist et al., A vertical transition quantum cascade laser with Bragg-confined excited state, *Appl. Phys. Lett.*, 66:538–540, 1995; C. Joshi and P. B. Corkum, Interactions of ultraintense laser light with matter, *Phys. Today*, 48(1):36–43, January 1995; T. C. Marshall, *Free-Electron Lasers*, 1995; M. D. Perry and G. Mourou, Terawatt to petawatt subpicosecond lasers, *Science*, 264:917–924, 1994; C. W. Roberson and P. Sprangle, A review of free electron lasers, *Phys. Fluids*, 1:3–42, 1989; C. Sirtori et al., Quantum cascade laser with plasmon-enhanced waveguide operating at 8.4 μm wavelength, *Appl. Phys. Lett.*, 66:3242–3244, 1995.

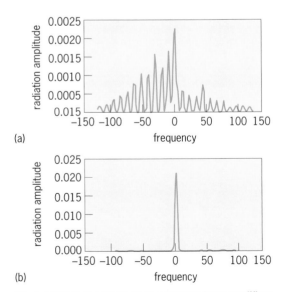

Fig. 6. Frequency spectra of free-electron laser amplifiers from computer simulations. (*a*) Untapered amplifier. (*b*) Strongly tapered amplifier. The radiation amplitude and the frequency are both dimensionless quantities in these simulations. The zero of the frequency scale is set at the primary carrier frequency.

Launch vehicle

Two important vehicles under development are the X-33 single-stage-to-orbit (SSTO) reusable launch vehicle (RLV) and the X-34 small launch vehicle.

X-33 Advanced-Technology Demonstrator

The next generation of United States launch vehicles must dramatically lower the cost of space access. Many promising space missions and experiments are now grounded because of overwhelming launch costs, and only the highest-priority payloads are being launched. An all-rocket single-stage-to-orbit fully reusable launch vehicle appears to be the best blend of near-term achievable technology and affordability for low-cost routine space access. Unlike the space shuttle, it will not expend stages during its ascent to orbit, thus simplifying ground and flight operations and processing requirements. The goal of the X-33 advanced-technology demonstrator is to mature the technologies essential for a single-stage-to-orbit next-generation reusable launch system capable of reliably serving national space transportation needs at substantially reduced costs.

Primary objectives. The X-33 is an experimental single-stage-to-orbit rocket proof-of-concept demonstrator. Its primary objectives are to (1) mature the technologies required for the next-generation system through hardware and software demonstrations in relevant environments, (2) demonstrate the capability to achieve low development and operational cost and rapid launch turnaround times through repeated flight testing of the demonstrator, and (3) reduce business and technical risks to encourage significant private investment in the commercial development and operation of the next-generation system.

Requirements. The overarching requirement is that the X-33 system, subsystems, and major components be designed and tested to ensure their traceability (technology and general design similarity) and scalability (directly scalable weights, margins, loads, design, fabrication methods, and testing approaches) to a full-scale single-stage-to-orbit rocket system. The X-33 will demonstrate the critical technologies for single-stage-to-orbit rockets in realistic operational environments. To the extent practical, the X-33 will be tested in the ascent and reentry flight environments of a full-scale single-stage-to-orbit rocket. Critical characteristics of single-stage-to-orbit systems, such as the structural-thermal concept, aircraftlike operations and maintenance concepts, flight dynamics, flight loads, ascent and entry environments, mass fraction, fabrication methods, and so forth, will be incorporated into the X-33 system.

In addition, the X-33 will focus on operational issues critical to the development of reliable inexpensive reusable space transportation, an order of magnitude better than the current fleet of space shuttles and expendable launch vehicles. It will incorporate more advanced materials with weights and margins equivalent to those required by a single-stage-to-orbit rocket. The X-33 ground-support and flight-control systems will be designed to accomplish operations and supportability goals which are key to less expensive system operations.

The operability and performance demonstrated by the X-33 will provide the necessary data to establish the detailed requirements for a future operational single-stage-to-orbit system. At a minimum, the X-33 will be an autonomous, suborbital, experimental, single-stage rocket flight vehicle.

Concepts. Three basic classes of X-33 are being investigated. The geometric scale of the X-33 will be approximately one-half that of the next-generation system.

A vertical-takeoff, horizontal-landing, lifting-body configuration (**Fig. 1**) relies on its outer-moldline shape to provide aerodynamic lift during reentry and landing in a gliding mode. Another unique aspect of this configuration is the use of a linear aerospike main-engine system integrated with the lifting-body shape. This engine is an alternative to the main engine used in the early stages of space shuttle development. An aerospike is an altitude-compensating nozzle which uses an external expansion surface (that is, a bell-type nozzle reversed inside out) combined with changing ambient air pressure to maintain an optimal expansion ratio. Thus the system can gain greater engine performance compared to conventional bell-type nozzles.

A second alternative is the vertical-takeoff, vertical-landing configuration (**Fig. 2**). After completing reentry in a nose-down attitude, the vehicle performs a rotation maneuver to a tail-down orientation in preparation for a propulsive landing. This maneuver has been demonstrated in the DC-X program conducted by the Ballistic Missile Defense Organization from 1993 to 1995. The DC-X was a quarter-scale (40 ft or 12 m in height) demonstrator of a next-generation system which was intended to prove that a cryogenically fueled rocket could be operated more like an aircraft.

The vertical-takeoff, horizontal-landing, winged-body configuration (**Fig. 3**) requires a delta-shape wing for aerodynamic lift during reentry and landing. This vehicle reenters and lands in a glide mode similar to that of the space shuttle.

All the concepts require approximately 500 pound-force (2200 newtons) of vacuum thrust for

Fig. 1. X-33 vertical-takeoff, horizontal-landing, lifting-body configuration. (*NASA*)

Fig. 2. X-33 vertical-takeoff, vertical-landing configuration. (*NASA*)

main propulsion. Thus the X-33 will be able to reach a minimum velocity of Mach 15, which is required in order to accurately simulate the aerodynamic and aerothermodynamic loads which will be experienced on the next-generation system.

Current activities. A competitive X-33 concept definition and design activity combined with ongoing technology developments and demonstration is currently under way. This effort, initiated in 1995, is scheduled to culminate in the selection of the X-33 concept and industry team in 1996. A wide range of technology candidates will be demonstrated to a level of maturity sufficient to reduce the number of alternatives, enabling the design and development of a cost-effective, large-scale technology demonstrator.

Flight testing. The X-33 will serve as the flight testbed for large-scale elements of critical reusable launch-vehicle technologies. The vehicle will be flight-tested by using an incremental envelope expansion process, whereby the vehicle flight envelope (the maximum velocity, altitude, heating loads, and so forth) is systematically expanded, based on results of previous flights; and it will demonstrate aircraftlike operations (minimal ground-crew size, short turnaround times, and so forth). Testing will start in relatively benign flight environments (for example, subsonic), moving to hypersonic flight as confidence in the system is built. Flight testing will be accomplished at an appropriate test range. Edwards Air Force Base in Mojave, California, White Sands Missile Range in Las Cruces, New Mexico, and Cape Canaveral Air Force Station in Florida are the primary launch points being evaluated. The X-33 may require up to 1000 mi (1600 km) between launch and landing points in order to adequately simulate the next-generation system's ascent and reentry profiles. Existing or temporary facilities will be utilized in order to maximize system flexibility while minimizing costs.

Key technology demonstrations. The following represent the critical technologies that the X-33 must demonstrate in order to adequately reduce the technical risk to development of a next-generation system. These technologies will be demonstrated in a combination of X-33 flight testing to verify system performance and operability in combined environments and parallel ground testing to more rigorously understand individual system responses to extended life-cycle and failure-mode tests.

Vehicle structural and thermal technologies. These technologies encompass reusable cryogenic tank systems, graphite-composite structures, and thermal protection systems. The efforts focus on the operability and integration of the load-carrying airframe structure, cryogenic insulation (as required), reentry thermal protection material, and associated health management for the next-generation system. (In this context, health management is the capability to detect faults in vehicle systems on board and in flight, to reconfigure around them, and to provide data to the ground crew for maintenance.)

The reusable rocket must return from orbit with its cryogenic propellant tanks, presenting complex thermal-structural challenges. Issues associated with life-cycle effects on the integrated tank system, including the tank wall, cryogenic insulation, and thermal protection system, must be addressed. Aluminum-lithium alloys and graphite-composite tank materials are being considered. To significantly reduce structural mass and alleviate fatigue and corrosion concerns, nonpressurized airframe structures will be constructed of graphite composite, drawing on current aircraft and rocket designs. These include both low- and high-temperature composite materials. Thermal protection system candidates for large areas of the vehicle (the wings, fuselage, tail, and so forth) involve both ceramic and metallic concepts. Leading-edge, nose-cone, and control-surface material candidates include advanced carbon-carbon and ceramic matrix composites.

Propulsion technologies. Propulsion-technology efforts will demonstrate the operational and performance characteristics of candidate engine and main propulsion systems and define and establish a set of derived requirements for an operable propulsion system. Key goals for the next-generation propulsion system are increased robustness (insensitivity of engine performance to external factors), greater operability, and higher thrust-to-weight ratio (70–80) than those of the space-shuttle main engines; and an affordable development program with acceptable risk. Demonstrations of ground-based engine and main propulsion systems will provide a testbed for demonstration of operability and performance targets.

Fig. 3. X-33 vertical-takeoff, horizontal-landing, winged-body configuration. (*NASA*)

Candidate engine systems currently identified for reusable vehicles are based on liquid-oxygen and liquid-hydrogen engines. Engine concepts being evaluated include a space-shuttle-main-engine (SSME)–derived engine, a linear aerospike engine, and an engine derived from the core stage engine of the Russian Energia launch vehicle. Key technologies for these concepts include oxygen-rich compatible materials, modular combustion chamber development, and advanced ceramic matrix composite materials in component designs.

Operations technologies. Short turnaround times, small ground crews, and airline-type maintenance procedures are critical to achieving dramatic reductions in operations costs. Automated operations technologies to be applied involve automated checkout, vehicle health management and monitoring systems, autonomous flight controls, and so-called smart avionics and guidance navigation. Incorporation of process enhancements such as one-time flight certification, hazardous materials elimination, ground-scheduling systems, and a philosophy of reliability-centered maintenance and minimum operations between flights will contribute to an aircraftlike operations process.

Private-sector development. The ultimate goal of the X-33 is to mature the technologies to a degree that will allow private-sector development and operation of the next-generation system. In order to promote this commercial goal, the National Aeronautics and Space Administration (NASA) has given industry the lead in development, with government laboratories contributing critical research and development expertise and facilities. Based on the results of the X-33 program, a decision can be made on the feasibility of a commercially developed and operated space-launch system.

Stephen Cook

X-34 Small Launch Vehicle

X-34 is the official designation for a joint government-industry program to develop a small, mostly reusable orbital launch vehicle. The origin of the program can be traced to an initiative by Orbital Sciences Corporation (OSC) in 1993 to develop a less expensive replacement for its Pegasus expendable small launch vehicle. Early trade studies concluded that a partially reusable configuration (fully reusable booster combined with a smaller expendable orbital vehicle) resulted in a system with minimum life-cycle cost for this size class when development costs were balanced with flight operations cost and the recurring cost of expendable hardware. OSC approached NASA Marshall Space Flight Center to inquire about possible NASA interest in participating in the development of the system. The NASA Reusable Launch Vehicle Program, prepared in response to President Clinton's

August 5, 1994, Presidential Decision Directive, includes the accelerated development and flight testing of a small reusable booster. In March 1995, NASA announced that it had selected an OSC-Rockwell team as the government's private sector partner for the X-34 cooperative agreement.

Description of system. The X-34 system is composed of a subsonic launch carrier aircraft, a reusable suborbital booster, and an expendable orbital vehicle. The launch carrier aircraft is one of NASA's two 747 shuttle carrier aircraft. Air launch was selected to reduce the cost-performance ratio of the system as well as to mitigate uncertainties in the future certification and operational regulation environment, for which X-34 is the pathfinder for the entire reusable launch vehicle program.

Although the overall reusable launch vehicle program strives to achieve operational flexibility approaching that of commercial airliners, that is, safe operations without the traditional rocket firing range approach (fly anywhere, anytime), uncertainty over the technical and political feasibility of this goal makes it impossible for the industrial partners to invest in the X-34 unless the system is capable of operating profitably either within or away from established rocketry ranges. By allowing launch of the booster several hundred miles from the recovery airfield, air launch provides operational flexibility to partially offset this uncertainty.

A special adapter structure (**Fig. 4**) interfaces the existing shuttle orbiter attachment points on the 747 shuttle carrier aircraft to the smaller X-34 booster. This structure provides approximately 12° of relative incidence angle just prior to separation, while keeping the booster at a low drag incidence angle of 2° during the potentially lengthy flight to (and, in case of launch, abort from) the launch point.

The booster vehicle (**Fig. 5**) weighs about 22,000 lb (10,000 kg) dry and 120,000 lb (55,000 kg) at separation from the launch carrier aircraft, and is approximately 83 ft (25 m) long with a 48-ft (15-m) wingspan. Wing area is approximately 1100 ft^2 (100 m^2). The structure is entirely composed of graphite composite, with a stainless-steel-lined liquid oxygen tank. A single 200,000-lbf (900,000-N) thrust class liquid-oxygen and kerosine liquid engine powers the booster. The large payload bay fully encloses the orbital vehicle as well as the orbital payload cantilevered off it. A structural cradle carries the orbital vehicle's launch loads, allowing a lightweight orbital vehicle structure.

The orbital vehicle is composed of a main 15,000-lbf (65,000-N) thrust liquid-oxygen and kerosine propulsive stage (OV-1) with a smaller pressure-fed bipropellant orbital trim stage (OV-2) for final injection accuracy. The orbital vehicle is designed for minimum recurring cost while offering acceptable orbit injection accuracy.

Flight operations. Although separate launch and recovery airfields may be used during the flight-test program, it is expected that for commercial operations a single airfield will be used for both takeoff and recovery, requiring the launch carrier aircraft to fly about 800 nautical miles (1500 km) to the launch point (**Fig. 6**). The booster and orbiter's liquid oxygen tanks are topped off 1 h prior to takeoff of the launch carrier aircraft. The insulation on the tanks is sized to eliminate the need for in-flight liquid oxygen replenishment.

After reaching the launch point, the booster is elevated to the separation incidence angle and the launch carrier aircraft initiates a descent with reduced thrust to match the effective lift-to-drag ratios of both airframes for safe unpowered separation. (For the unpowered booster this ratio is probably less than 7.) After vertical and lateral separation is ensured, the booster engine ignites, accelerating the booster to a velocity of approximately 9300 ft/s (2.8 km/s) and an altitude of 375,000 ft (115 km) 155 s after launch. After booster burnout, the payload bay door opens to allow ejection of the orbital vehicle with its attached orbital payload. After a brief separation maneuver, the OV-1 engine ignites, accelerating to orbital velocity. The OV-2 engine is then used to either circularize the orbit or initiate a fuel-efficient Hohmann transfer to a higher orbit.

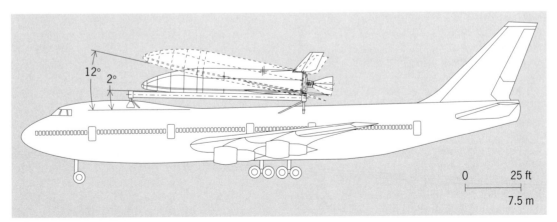

Fig. 4. X-34 booster vehicle on 747 launch carrier aircraft. The two-position adapter structure configures X-34 for low-drag flight to drop point or for high-drag incidence angle for separation.

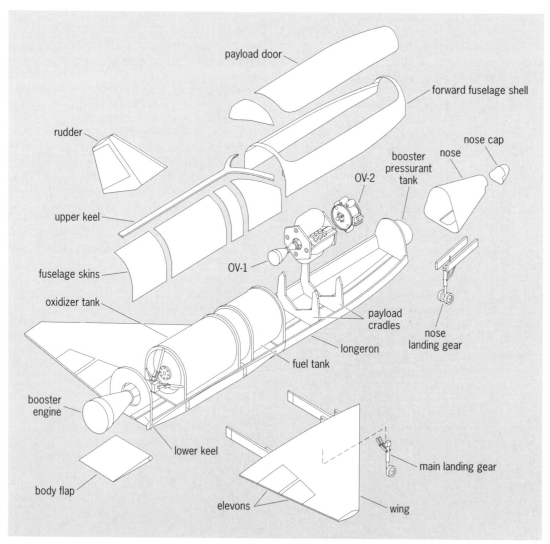

Fig. 5. X-34 booster internal arrangement showing the two elements (OV-1 and OV-2) of the orbital vehicle.

Meanwhile, the booster payload bay door is closed and a fixed attitude established for reentry. [Booster apogee occurs at an altitude of 500,000 ft (150 km) and a velocity of about 8500 ft/s (2.6 km/s), 286 s after launch.] Although the reentry velocity is a fraction of that of the shuttle, its reentry flight path angle is much steeper, resulting in a maximum temperature of the thermal protection system (about 2200°F or 1200°C) comparable to that of the shuttle orbiter (with a peak leading edge temperature of about 3000°F or 1650°C), but with shorter duration, which reduces the thermal soakback problem. (Thermal soakback is a phenomenon whereby, because of the lag in propagation of temperature changes through insulating materials, the maximum temperature of the thermally protected structure may be reached a certain time after the outside of the protective coating experiences its maximum temperature.) Significant advances in the durability of the thermal protection system as well as cost and weight reductions below those of the shuttle are possible by the use of a toughened tile material, AETB-12 (for a lumina-enhanced

thermal barrier), in the nose cap and leading edges, and extensive use of advanced, high-temperature, weatherproofed blankets over the majority of the structure: the X-34 uses only about 300 tiles, compared to the tens of thousands in the shuttle orbiter. For the first time, the thermal protection system will cover a carbon composite primary structure.

After reentry, the booster flies a decelerating glide trajectory similar to that of the shuttle, although at lower approach speeds (approximately 130 knots or 67 m/s) due to its lower wing loading. A split rudder is used to modulate drag during final approach, and the differential Global Positioning System and a radar altimeter are used for navigation during the automatic landing phase. Plans call for turnaround of the booster (including ground processing of health-monitoring data recorded in flight, which is used in lieu of detailed inspections and tests to qualify the vehicle for the next flight) and preparation and integration of the orbital vehicle and payload to require 15 8-h shifts by a multidisciplinary team of between 15 and 25 people.

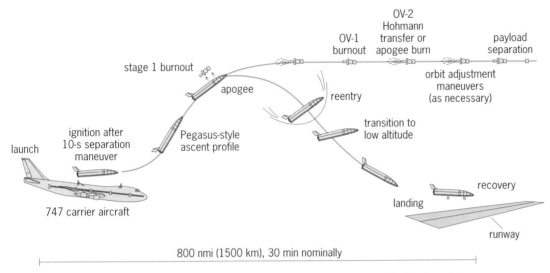

Fig. 6. Typical X-34 orbital mission profile. The exact distance from launch to landing is mission dependent.

With the exception of the main booster engine, which is expected to last only 20 flights, no routine refurbishments are planned during the booster's nominal 50-flight lifetime. Although there is no other life-limited element in the booster, the economic design of the system is based on the conservative assumption of having to replace an entire booster, either from attrition or from equivalent unplanned maintenance costs, every 20 flights.

For background information *SEE ATMOSPHERIC ENTRY; COMPOSITE MATERIAL; ROCKET PROPULSION; SATELLITE (SPACECRAFT); SPACE FLIGHT; SPACE SHUTTLE; SPACE TECHNOLOGY* in the McGraw-Hill Encyclopedia of Science & Technology.

Antonio L. Elias

Bibliography. J. R. Asker, X-34 to be acid test for Space Commerce, *Aviat. Week Space Technol.*, 142(14):44–53, April 3, 1995; A. L. Elias, D. Hays, and J. Kennedy, *Pioneering Industry/Government Partnerships: X-34*, AIAA 1995 Space Programs Technol. Conf., Pap. 95-3777, September 1995; National Aeronautics and Space Administration, *NASA Access to Space Study*, January 1994; *Presidential Space Policy Directive NTSC-4*, August 1994.

Lentil

Since its domestication in the Near East, lentil (*Lens culinaris Medikus*) has held a prominent place in cropping sequences in semiarid regions of the world, where it has provided local populations with an important source of dietary protein. However, the lentil crop is often relegated to the most marginal areas, where land races (genetically diverse cultivated forms evolved from a natural population of plant species) are usually grown without the benefit of fertilization, herbicides, pest control chemicals, or irrigation. Consequently, yields have remained constant or declined. Lentil is popular in those areas because it is a traditional food and it appears to be one of only a few crops that can be successfully grown. Lentil crops produced on such poor and stony soils generally need to be harvested by hand pulling, an extremely labor-intensive operation. Attempts to mechanize lentil harvest have had limited success in traditional production areas because stony soil and the short stature of the plant make direct cutting difficult.

India is by far the largest producer, followed by Turkey, Canada, and Syria; Ethiopia, Morocco, Chile, and Argentina are also major producers. Middle Eastern countries such as Egypt, Jordan, Iraq, and Lebanon are major consumers of lentil but not major producers. In addition, Canada and the United States are major lentil producers; however, the crop is mostly grown for export. Because of a recent emphasis on growing legumes, Australia may emerge as a major producer and exporter in the near future.

Domestication. The Near East and Asia Minor are the accepted centers of origin of cultivated lentils. *Lens culinaris* subspecies *orientalis*, which closely resembles the cultivated subspecies *culinaris*, is widely accepted as the progenitor species. *Lens culinaris* subspecies *orientalis* can be found over an extended geographical range from the Near East to Afghanistan. The species is found in rocky and stony habitats with very little soil and in association with other annual legumes such as the medics and annual grasses. The conclusion that the cultivated lentil originated in the Near East arc from *L. culinaris* subspecies *orientalis* is based on discoveries of carbonized remains of apparent cultivated lentils in the same region over which this subspecies is widely distributed. Such carbonized remains appear in early Neolithic settlements that date back to 7000–6000 B.C.

Taxonomy. Cultivated lentil belongs to the genus *Lens*, which is associated with other genera of the Vicieae tribe, comprising *Lens, Vicia, Pisum, Lath-*

yrus, and *Vavilovia.* The primary genepool of *L. culinaris* comprises subspecies *culinaris,* which is cultivated, and its presumed wild progenitor subspecies *orientalis.* Three other wild *Lens* species are recognized in the secondary gene pool: *L. odemensis, L. nigricans,* and *L. ervoides.*

Characteristics. Stipule shape distinguishes the wild *Lens* species. In contrast, the stipules of *L. culinaris* subspecies *orientalis* are shaped like the head of a lance (lanceolate), similar to those of *L. culinaris* subspecies *culinaris* and *L. ervoides. Lens nigricans* and *L. odemensis* have semihastate (hastate meaning shaped like an arrowhead with divergent barbs) or dentate (toothlike) stipules. The stipules of *L. nigricans* are oriented parallel to the stem while those of *L. odemensis* are oriented perpendicular to the stem.

All *Lens* species are herbaceous annuals with slender stems and branches (**Fig. 1**). Plant height usually ranges from 10 to 12 in. (25 to 30 cm), but may vary from 6 to 30 in. (15 to 75 cm) depending on genotype and environmental conditions. Plants have a slender taproot with fibrous lateral roots.

The flowers are borne singly or in multiples on stalks (peduncles) that originate from the upper nodes of the plant. Each peduncle normally bears from one to three, and rarely four flowers. The flowers are complete and have a typical butterfly-like structure (**Fig. 2**). Each individual flower has five sepals (leaves composing the calyx) and a corolla (the petals of a flower) consisting of a large

Fig. 1. Typical lentil plant, showing branch structure and podding habit. (*From F. J. Muehlbauer et al., Production and breeding of lentil, Adv. Agron., 54:283–332, 1995***)**

petal, two internal smaller petals, and an inner central petal structure. There are 10 stamens and a flat glabrous ovary that normally contains one or two ovules. The style is covered with hairs on the inner side and usually develops at a right angle to the ovary. The stigma is generally flattened on the outer side.

Seeds are lens shaped and weigh between 0.001 and 0.005 oz (20 and 80 mg). Seed diameter ranges from 0.08 to 0.3 in. (2 to 8 mm) and the seed coat may be light green or greenish red, gray, tan, brown, or black. Purple and black mottling and speckling of seeds is also common in some cultivars and accessions. Lentil genotypes are sometimes grouped as macrosperma, with large seeds that range from 0.24 to 0.31 in. (6 to 8 mm) in diameter. The macrosperma types are common to the Mediterranean basin and in the Western Hemisphere, and the microsperma types, which are less than 0.24 in. (6 mm), in diameter predominate throughout the Indian subcontinent and in parts of the Near East.

Uses. A major food use of lentil is as dhal (lentils with the seed coat removed and the seed split). It is a principal ingredient in soups and other dishes prepared on the Indian subcontinent. In addition, whole lentil seeds are often ground into flour and added to cereal flour in the preparation of breads and other baked products. Lentils are also used in dishes containing rice and cereal grains. When combined with the cereal grains, lentil provides a nutritionally well-balanced diet. Lentils have relatively large concentrations of lysine which compensate for the minimal concentrations in the cereal grains, and the cereal grain compensates for the minimal concentrations of sulfur-containing amino acids in lentil.

Feed for livestock is an important use for the crop residues. In the traditional production areas of the Middle East, West Asia, North Africa, Ethiopia, and the subcontinent of India, residues from threshing of the lentil crop are essential for livestock feeding. In some years, the residues have commanded prices equal to or greater than that of the grain.

Production. Lentil crops are most often grown in rotation with cereals, either wheat or barley. In the United States, land intended for lentil is usually rough-tilled in the fall and harrowed in the spring after application of soil-incorporated herbicides. Lentils are planted as early as possible in the spring with the same equipment that is used to plant cereals. Seeding rates of 60–75 lb/acre (65–85 kg/ha) are generally considered adequate. Soil temperatures above 43°F (6°C) are needed for good germination and seedling growth.

Lentil harvesting is completely mechanized in the United States. The most common method is to windrow the crop with a swather and, after a drying period of 7–10 days, to combine the crop directly from the windrows.

Nitrogen fixation. Effectively nodulated lentils seldom respond to an application of inorganic nitro-

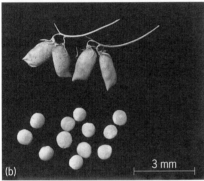

Fig. 2. Lentil flowers and pods and seeds. (a) Typical papilionaceous structure. (b) Pods, usually containing one or two seeds. (From F. J. Muehlbauer et al., Production and breeding of lentil, Adv. Agron., 54:283–332, 1995)

gen fertilizer. Inoculation with an appropriate strain of the bacterium *Rhizobium leguminosarum* is necessary when lentils are seeded into fields for the first time or after a lapse of several years. Special care should be taken when using fungicide seed dressings potentially toxic to *Rhizobium*. The nitrogen-hunger phase, which is often experienced by grain legumes when crops are seeded into cool, wet soil before significant symbiotic nitrogen fixation begins, can be avoided by the application of a small starter dose of 9–23 lb/acre (10–25 kg/ha) inorganic nitrogen fertilizer adjacent to but not in contact with the seeds.

Constraints to production. Lentils are attacked by a variety of insects during crop development and in storage. The principal soil-borne pests include seedcorn maggots, wireworms, cutworms, and larvae of weevils that attack the seeds and developing seedlings soon after planting. After the plant emerges, a variety of insects feed on the leaves, stems, and flowers; these pests include thrips, aphids, leaf weevils, lepidopterous larvae, and grasshoppers. The most important insect pests of the pods and seeds include lygus bugs, bruchid beetles, and lepidopteran pod borers. Bruchid beetles are also major storage pests, but they do not occur in the United States or Canada.

The serious disease problems of lentil include the root rot and wilt complex caused by *Pythium, Rhizoctonia, Sclerotinia,* and *Fusarium* species. Research is under way toward selection for resistance to the various components of the root rot and

wilt complex. Rust, anthracnose, and Ascochyta blight are important diseases of lentil in many countries, especially in wetter regions or during years with heavy rainfall. Also, several aphid-transmitted viruses can cause serious damage. Seed quality can be adversely affected by infections of *Ascochyta, Botrytis, Fusarium,* and *Phoma.*

Germplasm. Lentil improvement programs that maintain germplasm stores are maintained at a number of locations including the U.S. Department of Agriculture–Agricultural Research Service Regional Plant Introduction Station located at Pullman, Washington. Large collections are also in Aleppo, Syria, and New Delhi, India. These germplasm collections comprise the land races, and in the case of the larger collections, the related wild species.

Breeding. Lentil breeding is a recent endeavor when compared with efforts devoted to cereals or some of the other food legumes. Land races still occupy most of the area planted to lentil in the major producing countries. Moreover, most of the cultivars released have been derived from selection within heterogeneous land races, and are not the result of hybridization.

The methods of breeding lentil are similar to those used in breeding other self-pollinated crops and include pure line selection or hybridization followed by the bulk-population breeding method of handling the populations. However, the pedigree breeding method is seldom used.

Lentil breeding programs throughout the world have similar objectives, with larger and more stable seed yields being the most important. Breeding for adaptation to stress environments, especially to drought, and resistance to diseases and insects are also major objectives. Priorities and breeding goals usually differ between regions depending on specific problems and special considerations related to farmers' needs and consumer demands. In the developing countries, for example, one of the major breeding goals is the development of genotypes that are suitable for mechanical harvesting. Moreover, improving the amount of crop residue produced by lentils is important because of its value as animal feed and as a residue for the control of soil erosion. In the major exporting countries, increased seed yield, improved disease resistance, and improved seed quality are principal breeding goals. Breeding efforts have resulted in the development and release of cultivars which have distinct advantages over previously grown land races. Seven lentil varieties have been developed and released in the United States. Brewer, Redchief, and Spanish Brown occupy significant acreages.

Crop improvement. Research on improvement of the lentil crop was minimal until the establishment in 1978 of the International Center for Agricultural Research in the Dry Areas in Aleppo, Syria. In addition, national programs actively conducting research on all aspects of the lentil crop have been established in all of the major producing countries.

A gene map for lentil has been developed and contains more than 100 loci; new loci are being added from research at a number of laboratories worldwide. The goal is to produce a gene map which will enable geneticists and breeders to use genetic markers in the development of better germplasm and cultivars.

Wild species, closely related to the cultivated lentil, are being used by geneticists and breeders to expand the genetic base. Traits considered to be available in the wild species include drought and cold tolerance and disease resistance. Cold-tolerant lentils would make fall planting possible, enabling better utilization of available moisture for crop growth during the fall and early spring when evapotranspiration is low.

For background information SEE AGRICULTURAL SOIL AND CROP PRACTICES; BREEDING (PLANT); LENTIL; SOIL ECOLOGY in the McGraw-Hill Encyclopedia of Science & Technology.

Fred J. Muehlbauer; Walter J. Kaiser

Bibliography. F. J. Muehlbauer et al., Production and breeding of lentil, *Adv. Agron.*, 54:283–332, 1995; G. Ladizinsky, Wild lentils, *Crit. Rev. Plant Sci.*, 12:169–184, 1993; S. O'Brien (ed.), *Genetic Maps*, 1993.

Magnetic resonance imaging

Magnetic resonance imaging (MRI), an established anatomical imaging modality of clinical value, is now being used to noninvasively identify and map regional brain activation and functional transients within activated sites. Functional MRI uses conventional MRI technology and equipment to image the intrinsic hemodynamic changes which are constantly occurring in human cognitive functions such as vision, motor skills, language, memory, and indeed all mental processes. Because it is a noninvasive imaging modality, unlike positron emission tomography or single-photon emission tomography, which use injected radio-labeled tracer compounds, MRI can acquire functional images from an individual in any plane or volume at comparatively high resolutions and then overlay the functional centers of activation onto the underlying cerebral anatomy, also imaged with an MRI scanner. Since its inception in 1992, functional MRI is rapidly evolving beyond the localization of visual, motor, and somatosensory responses to activation patterns (paradigms) in human volunteers. Mapping of the brain prior to the surgical removal of tumors, localization of epileptic foci, and analysis of brain function altered by pathologies such as stroke are now possible. Given the large number of clinical MRI scanners operating worldwide, functional MRI will give rise to routine clinical assessment of brain and tumor function, in addition to the anatomical imaging role of present-day MRI.

Functional mapping. Changes in MRI intensity can be induced by alterations in the number of hydrogen nuclei within a pixel (the proton density), proton relaxation rates, proton diffusion, blood flow, as well as by variations in the local magnetic field surrounding a tissue. Conventional magnetic resonance images, which display excellent soft tissue contrast, derive that contrast from proton density and the relaxation rates that describe the reorientation of proton motions in the magnetic field surrounding them.

Several tissue contrast mechanisms have been described that are sensitive enough to image the effects of functional stimulation in brain. In the most popular of these functional MRI methods, image contrast is produced by the brain's hemodynamic response to neuronal activity or to a challenge to the regional blood flow. Neuronal activity increases local metabolic demands and induces an autoregulated increase in local blood flow, which occurs in the small arteries feeding the neurons. The delivery of increased blood flow alters the local magnetic field homogeneity, and is observable as small (2–15%) changes in the magnetic resonance image intensity.

Image intensity. The magnetic resonance image intensity is strongly influenced by the microvascular oxygenation state of the tissue. In this blood-oxygenation-level-dependent approach, the rate of decay of the proton phase coherence is a measure of the local magnetic field homogeneity and is modulated by the presence and oxygenation state of blood hemoglobin. The iron in blood hemoglobin is an efficient magnetic-field-altering intravascular contrast agent inherent in all perfused tissues. Hemoglobin is a unique local indicator of vascular function because blood contains largely oxyhemoglobin (oxygenated hemoglobin). Oxygenated iron is diamagnetic; that is, it has a small magnetic-field-altering effect. Therefore, it does not greatly affect tissue homogeneity or image intensity. However, tissue metabolism requires oxygen and converts oxyhemoglobin into deoxyhemoglobin, which contains a more paramagnetic species of iron. The presence of the paramagnetic iron oxygenation state disturbs the local magnetic field and leads to the greater rate of magnetic resonance signal decay. Increases in the concentration of deoxyhemoglobin in blood lead to a decrease in image intensity, and increases in the delivery of arterial blood, composed of oxygenated hemoglobin, will cause magnetic resonance signal increases. An increase in oxygenated arterially delivered blood in response to neuronal activation results in more oxygenated iron in the capillary and venous vascular beds. The result is a relatively longer regional magnetic field homogeneity throughout the surrounding tissue and an increase in the observable image intensity. This concept applies in all perfused tissues in the body, including tumors, and has led to the concept of functional imaging of tumor and organ oxygenation.

The hemodynamically induced magnetic-field homogeneity changes in the primary motor cortex that occur during finger exercise can be visualized (see **illus.**). Magnetic resonance images are acquired during a simple finger opposition exercise, where the thumb is touched to each of the fingers on one or both hands for 20 s, followed by 20 s of rest. During exercise, rapid neuronal activation occurring in the motor cortex causes a surplus in the amount of oxygenated hemoglobin delivered to the activated tissue. The larger amount of the diamagnetic oxyhemoglobin is apparent from the magnetic resonance images acquired during exercise as an increased signal in the regions of the cortex that are being activated (illus. *b*). Images acquired at rest do not show these signal changes. The increased signal can then be overlaid onto an image to complete the functional MRI map (illus. *c*). Magnetic resonance evaluations can be done in single-slice, multislice, or true three-dimensional volume modes.

Functional MRI studies have demonstrated functional changes within the primary cerebral cortex in response to simple activation tasks. During simple motor movements, functional increases in local magnetic field homogeneity (blood oxygenation level dependent) were found in the contralateral primary motor cortex, the supplementary motor area, the premotor cortex, the supplementary motor area, the premotor cortex of both hemispheres, and the contralateral somatosensory cortex. Work on cortical motor control and activation has also focused on cerebellar activation in subjects performing a series of alternating wrist flexion and extension movements against constant inertial loads, and three different spatial patterns of activation have been observed.

Requirements. The resources needed for a functional evaluation are surprisingly few and can be performed on most MRI systems operating today. The evaluation shown in the illustration consisted of acquiring a series of 128 fast-scan images over 192 s from four slices during consecutive intervals of exercise, rest, exercise, and rest. Magnetic resonance signal intensity changes were then examined for each of the pixels in each of the images. Those

regions in the images which altered their magnetic resonance intensity at the frequency of the paradigm (those with a 40-s period) were then mapped to form the functional image corresponding to that activation paradigm (illus. *b*). A similar paradigm can be constructed to blow puffs of air onto each of the fingers of both hands to map the corresponding sensory centers known to be near the motor centers. Both functional maps can be shown together from data acquired from two exams from the same volunteer, a process requiring less than 15 min.

Advantages. The high spatial detail and good temporal resolution of the acquired noninvasive functional magnetic resonance images offer researchers the best look yet at local dynamic and regional hemodynamic events occurring during brain function. These blood-oxygen-level-dependent functional MRI methods have significant advantages over other functional imaging methods such as positron emission tomography because tissue oxygenation or magnetic field homogeneity changes can be imaged continuously before and during activation of functional centers. The blood-oxygen-level-dependent magnetic field homogeneity changes are observed directly within seconds, allowing observation of time courses, activation delays, and time-resolved phase effects. Each volunteer serves as both control and subject; activation can be directly observed and reproduced for a single individual. Both functional and underlying anatomical images can be acquired from each subject by using the same equipment and often by using the same MRI sequences during the same medical evaluation. The chances of misregistration, which severely limits all other functional mapping methods, are greatly reduced. No other functional mapping method can image the underlying anatomy. Finally, the observed blood-oxygen-level-dependent effects seen in activated regions are not functions of the local cortical geometry or orientation, as in electromagnetic methods such as electrocardiography.

Reliability. The results of recent research are widely varying and add to the potential of functional MRI as a superior noninvasive brain-mapping tool.

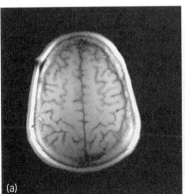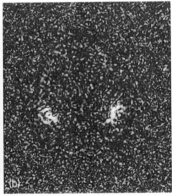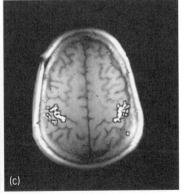

Visualization of the local magnetic field homogeneity differences of the human brain employed in blood-oxygenation-level-dependent functional MRI. (*a*) Magnetic resonance image in an axial slice. (*b*) Map of those pixels whose image intensities vary at the exercise activation period (40 s). Only those areas which change signal during exercise are seen. (*c*) Overlay of activation onto anatomy, producing the functional map.

At present, most studies have focused on the reliability and robustness of the methods used. Signal changes in functional MRI of human brain activation not only arise from the brain's response to a stimulus, reflecting associated adjustments of cerebral blood flow and oxygen consumption, but also strongly depend on the blood-oxygen-level-dependent MRI technique chosen and the actual experimental setting. Many magnetic resonance parameters exist that can influence the size and magnitude of the magnetic resonance signal strength, for example, static field homogeneity of the magnet, magnetic resonance pulse sequence and signal type, field-gradient strengths, field-gradient waveforms, radiofrequency receiver bandwidth, and resulting voxel size. Because a local signal increase during functional activation may reflect a regional change in cerebral blood flow or deoxyhemoglobin concentration or both, proper localization of these different contributions is important. Experimental strategies are often evaluated that either discriminate MRI effects in large vessels from those in cortical microvasculature or separate changes in blood-flow velocity from those in blood oxygenation in reported studies of the human visual and motor cortex.

Applications. Clinical functional MRI applications are rapidly moving toward routine noninvasive mapping of distortions of the functional motor and somatosensory cortex and other cortical regions as a result of brain tumors, trauma, cerebral aneurysms, arteriovenous malformations, and stroke. Nonsurgical applications include improved functional mapping of brains subjected to degenerative conditions such as multiple sclerosis, Alzheimer's disease, Parkinson's disease, epilepsy, and other diseases causing neuronal loss. Preoperative mapping with functional MRI will mature rapidly, with the development of paradigms that will produce a motor strip map in a single medical evaluation lasting a few minutes. The acquisition of total brain volume data will become routine as newer and faster MRI techniques develop. Whole-body MRI systems with higher magnetic field strengths are being installed at major medical centers and will be capable of observing much smaller activation-induced magnetic field homogeneity changes from smaller regions of tissue. *SEE MULTIPLE SCLEROSIS.*

Epilepsy. When functional MRI has been used together with positron emission tomography and single photon emission tomography, new opportunities for noninvasive brain investigation have appeared. The use of integrated imaging techniques allows investigations of brain abnormality in individuals with chronic epilepsy. With the demonstration of the utility of noninvasive brain mapping apparent in seizure and motion disorders, the proper use and interpretation of the findings provided by these new technologies will be a major challenge to epilepsy programs. This challenge has been addressed in recent studies by using functional MRI to map any cortical activation that would occur during focal seizures.

It is now widely considered that functional MRI can provide new insights into the dynamic events that occur in the epileptic brain and their relationship to brain structure, especially when functional MRI is coupled with high-resolution magnetic resonance images of the hippocampus. Typically, left and right hippocampi were symmetrical in the control subjects from magnetic resonance images; however, for individuals with epilepsy the hippocampus was significantly smaller on the side of the seizure focus. Moreover, the left-right hippocampal ratio was significantly smaller on the side of the seizure focus and significantly differentiated the control subjects from each epileptic group. The left temporal lobe was significantly smaller than the right in control subjects. The temporal lobes of epileptics were smaller on the side of the seizure focus, compared to the temporal lobes in the control subjects. MRI hippocampal measurements were compared to hippocampal neuronal densities obtained postoperatively.

Physiological stress. Mapping the autoregulated response of any organ (or tumor) to some controlled physiological stress is also possible. Tumors can be distinguished from normal tissue by the altered response of the tumor to mild bouts of hyperoxia (breathing 100% oxygen for a few seconds) or of hypercarbia (breathing small amounts of carbon dioxide). The same is true for the brain. Cerebral response to breathing altered gas mixtures can identify stroke from hypoperfused and from normal brain. The sensitivity of functional MRI to changes in cerebral blood oxygenation has been introduced for monitoring autoregulation in the human brain under vasodilatory stress. Following the administration of acetazolamide, an injected potent vasodilator, signal intensities of images sensitive to magnetic field homogeneity increased in cortical and subcortical gray matter and to a lesser extent in white matter. The result reflects an increase in oxygenation in the veins stemming from an increase in cerebral perfusion with oxygen consumption remaining constant. Recording local tissue oxygenation by functional MRI will enhance understanding of the modulation of vasomotor tone and cerebral perfusion in normal and pathological states. Furthermore, this technique may prove valuable for assessing the cerebrovascular reserve capacity in individuals with carotid artery occlusive disease.

Activation mapping. This method charts areas of the brain associated with mental tasking of image rotation, mental calculation tasking, color selection, picture naming, verb fluency, sentence processing, and word association in normal and deaf volunteers. Activation mapping has occurred along with studies of very brief visual stimuli (second and subsecond activations) and onset variability. When human volunteers were exposed to brief sudden photic stimulation, an early subsecond response occurred leading to a reduction of the magnetic resonance signal with a slower response (1–2 s after stimulus) resulting in the usual blood-oxygen-level-dependent signal increases. The fast negative response may be attributed to increased oxygen consumption which would occur immediately upon increased metabolism and before local blood flow increases, followed by a slower vascular response with overcompensation in blood oxygenation. These same methods have produced a series of new observations involving learning patterns in different languages, selective learning semantic analysis of word associations, and the recognition of activated signal alterations in different Brodmann areas of the cerebral cortex. Most importantly, and unlike other positron emission tomography and single-photon emission tomography methods, functional MRI can map graded activations during a single medical evaluation, in order to elucidate how the human brain thinks harder and if functional MRI can detect decreased neuronal activity. SEE LANGUAGE PROCESSING.

For background information SEE COMPUTERIZED TOMOGRAPHY; MEDICAL IMAGING; NUCLEAR MAGNETIC RESONANCE; SEIZURE DISORDERS in the McGraw-Hill Encyclopedia of Science & Technology.

Michael E. Moseley

Bibliography. S. W. Atlas (ed.), *Magnetic Resonance Imaging of the Brain and Spine,* 2d ed., 1995; R. R. Edelman, J. R. Hesselink, and M. Zlatkin (eds.), *Clinical Magnetic Resonance Imaging,* 2d ed., 1996; C. Jack et al., Sensory motor cortex: Correlation of presurgical mapping with functional MRI and invasive cortical mapping, *Radiology,* 190:85–92, 1994; R. I. Kuzniecky, *Magnetic Resonance in Epilepsy,* 1994.

Mammalia

One of the most fundamental aspects of the biology of mammals is their warm-blooded or endothermic nature: mammals produce enough heat by internal, metabolic means to maintain a high and constant body temperature over a wide range of ambient conditions. Although the metabolic costs of endothermy are high, it provides two significant advantages over ectothermy, or cold-bloodedness: a stable body temperature that allows mammals to remain active even under cold or nocturnal conditions, and an increased aerobic capacity that enables mammals to sustain levels of activity well beyond the scope of ectotherms. These two factors, homeothermy and stamina, are largely responsible for the present success of mammals in the terrestrial environment; therefore the evolutionary origins of endothermy are of considerable interest.

Despite its profound significance, endothermy leaves a very poor fossil record, and has traditionally been virtually impossible to demonstrate in extinct forms. Previous studies had to rely extensively on indirect lines of evidence and remained inconclusive. Recently, however, the realization

that the respiratory turbinate bones of mammals (**Fig. 1**) are strongly correlated with endothermic processes, and that traces of turbinate bones can be found in fossil forms, has led to new insights into the origins of mammalian endothermy.

Endothermy and the fossil record. Physiologically, endothermy is achieved through greatly elevated rates of oxygen consumption. Even at rest, the metabolic rates of mammals are typically about 6–10 times greater than those of (ectothermic) reptiles of the same body mass and temperature; and mammalian field metabolic rates (which incorporate the much higher activity levels of mammals) exceed those of similar-sized reptiles by as much as 16–40 times. To support such high levels of oxygen consumption, mammals have made profound structural and functional modifications to enhance the uptake, transport, and delivery of oxygen. For example, compared with reptiles mammals have greatly expanded lung volumes and higher ventilation rates, and fully separated pulmonary and systemic circulatory systems. Mammalian blood volume, oxygen-carrying capacity, and cardiac output are also greatly increased, as are mitochondrial density and enzymatic activity levels. Similar modifications also occur in birds, which, like mammals, are fully endothermic. Unfortunately, these key features of endothermy are never found preserved in fossils of any kind.

Previous investigations of the possible presence of endothermy in extinct vertebrates relied principally on hypothetical correlations of metabolic rate with a variety of circumstantial criteria, such as predator-prey ratios, fossilized trackways, and paleoclimatological inferences, or on correlations with avian or mammalian posture, bone histology, or brain size. However, close scrutiny of these arguments has revealed that virtually all are equivocal at best. For example, the implications of modern predator-prey ratios are ambiguous, and the preservation of true community structure in fossil assemblages is unreliable. The distribution of histological bone types among known endotherms and ectotherms is inconsistent, and the functional relationship between bone histology and metabolic rates is unclear. Like-

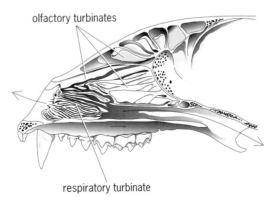

Fig. 1. Right saggital section of the skull of a raccoon, *Procyon lotor.* Arrows indicate the path of respired air. (*After W. J. Hillenius, The evolution of nasal turbinates and mammalian endothermy, Paleobiology, 18:17–29, 1992*)

wise, the association with the thermoregulatory status of a number of mammalian features often cited to support endothermy in fossil forms, such as upright posture, large brain size, and the reduction of lumbar ribs, is obscure. Previously, without a preservable osteological characteristic that demonstrates a clear and exclusive functional relationship to endothermy, the origins of endothermy remained unresolved.

This situation recently changed with the demonstration that the respiratory turbinate bones of mammals are tightly correlated with the maintenance of high lung ventilation rates and endothermy. Traces of turbinate bones have been found preserved in the fossil ancestors of mammals, and these structures thus represent the first direct morphological indicator of endothermy that has been observed in the fossil record. *See Dinosaur.*

Turbinates. The correlation between respiratory turbinates and endothermy rests on both morphological and functional considerations. Turbinate bones are thin but highly complex structures that occupy the nasal cavity of mammals (Fig. 1). Two distinct types of turbinates occur in virtually all mammals. A large number of olfactory turbinates are typically located in the dorsal and posterior portions of the nasal chamber. These turbinates are covered with sensory (olfactory) epithelium and serve to enhance the sense of smell. Respiratory turbinates are located in the anterior portion of the nasal cavity and are covered exclusively with moist mucociliated epithelium. Often very large, elaborately scrolled or branched structures, these turbinates are positioned directly in the path of respired air, whereas the olfactory turbinates are located outside (above or behind) the respiratory airflow. A similar arrangement of complex respiratory and olfactory turbinates also occurs in birds, but not in reptiles. Reptiles have much simpler cartilaginous nasal folds that support only olfactory tissues; folds or turbinates with a respiratory function are completely absent in these animals.

Embryological and developmental studies have shown that the reptilian nasal folds and the olfactory turbinates of mammals and birds are most likely homologous (and thus probably represent a primitive characteristic for amniotes, with no particular bearing on thermoregulatory status), whereas the respiratory turbinates are new morphological structures, derived independently by endothermic birds and mammals.

Only the respiratory turbinates have a strong functional association with endothermy. In both mammals and birds, endothermy is tightly linked to

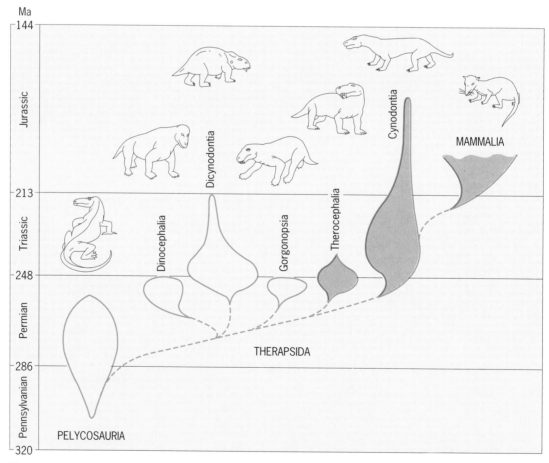

Fig. 2. Stratigraphic ranges of the main groups of mammallike reptiles. Shading indicates the groups in which evidence of respiratory turbinates is found, and in which endothermy may have evolved. (*After W. J. Hillenius, Turbinate in therapsids: Evidence for Late Permian origins of mammalian endothermy, Evolution, 48:207–229, 1994*)

high levels of oxygen consumption and rapid, continuous lung ventilation. Typical lung ventilation rates of resting birds and mammals are at least 3.5–5 times greater than those of similar-sized reptiles, and estimates for avian and mammalian ventilation rates in the field exceed reptilian rates by at least 20–30 times. By necessity, air in the lungs is always fully saturated, and unless this respiratory water is reclaimed the high ventilation rates of endotherms could lead to potentially disastrous desiccation rates.

The respiratory turbinates of mammals recover a significant portion of the water vapor contained in exhaled air through countercurrent exchange of heat and water between respired air and the moist epithelial linings of the turbinates. Water recovery occurs in two alternating stages. During inhalation, cool external air absorbs heat and moisture from the turbinal linings; by the time it reaches the trachea, inhaled air has been warmed to core body temperature and is fully saturated with moisture. This process not only prevents undue heat loss and desiccation in the lungs but also cools the turbinal epithelia and creates a thermal gradient along the turbinates. Upon exhalation, this process is reversed: warm, moist air from the lungs is cooled as it returns through the turbinal complex. As a result of this cooling, the air becomes supersaturated, and excess water vapor condenses on the turbinal surfaces, where it can be reclaimed and recycled. Over time, a substantial amount of water (proportional to the degree of cooling of exhaled air) can thus be saved, rather than lost to the environment. Without a turbinal water recovery mechanism, the respiratory water loss rates of mammals would likely exceed tolerable levels, and continuously high rates of oxidative metabolism, and endothermy itself, would probably be unsustainable. These observations suggest that the respiratory turbinates of mammals are intimately associated with high mammalian lung ventilation rates, and most likely evolved together with the development of elevated aerobic capacity and endothermy.

Turbinates in the fossil record. The fossil record of the ancestors of mammals, the synapsid or mammallike reptiles, is quite extensive and reasonably complete (**Fig. 2**). Early representatives of this lineage, the pelycosaurs, differed relatively little from basal amniotes, and probably remained essentially ectothermic. But advanced synapsids, the therapsids, gradually acquired so many mammalian characteristics, such as erect posture, complex dentition, and a bony secondary palate, that this group has long been suspected of being mammallike in both physiology and thermoregulatory status. New evidence for the presence of respiratory turbinates among certain therapsids confirms these suspicions.

The fragility of the turbinal bones usually precludes their preservation in fossilized specimens, but their presence and function are readily revealed by the distinctive pattern of ridges by which the turbinates attach to the walls of the nasal cavity.

Studies of the nasal cavities of mammallike reptiles indicate that olfactory turbinates were present throughout this lineage, but respiratory turbinates first appear in two groups of advanced therapsids, the therocephalians and cynodonts. Thus the evolution of mammalian oxygen consumption may have begun as early as the late Permian, about 260 million years ago. At present, the earliest record of respiratory turbinates occurs among primitive therocephalians, such as the South African form *Glanosuchus*, which had not yet acquired a bony secondary palate. In these animals, the essentially reptilian palatal configuration limited the size of the respiratory nasal chamber, and most likely its capacity to modify respired air. This combination of features suggests that the respiration rates of primitive therocephalians had expanded only moderately beyond ancestral ectothermic rates. Later, the respiratory passage of both therocephalians and cynodonts expanded gradually, and by the Early Triassic both groups had independently acquired a full bony secondary palate. Thus the respiratory capacity of these animals probably also gradually increased. Full mammalian endothermy was probably not achieved until the first mammals appeared, 220–210 million years ago, and thus may have taken as much as 40–50 million years to develop.

Two main theories for the evolution of endothermy have been proposed. According to the homeothermic model, high resting metabolic rates were developed to facilitate metabolically based thermoregulation and a stable body temperature. However, metabolic heat production at most levels intermediate between those of living reptiles and mammals is insufficient to establish endothermic homeothermy, and thus provides little thermoregulatory benefit while substantially increasing the animal's energy budget. Given the apparently slow development of mammalian metabolic rates, it is unlikely that homeothermy was a significant selective factor in this process. Instead, the gradual evolution of endothermy is more consistent with the aerobic capacity model, which invokes the advantages of an expanded capacity for sustained, aerobically supported activity. Here, even modest increments of the metabolic rates of therocephalians and cynodonts were probably tightly associated with expansion of endurance and of normal, sustainable activity levels.

For background *SEE FOSSIL; MAMMALIA; METABOLISM; RESPIRATORY SYSTEM* in the McGraw-Hill Encyclopedia of Science & Technology.

Willem J. Hillenius

Bibliography. A. F. Bennett, The evolution of activity capacity, *J. Exper. Biol.*, 160:1–23, 1991; W. J. Hillenius, Turbinates in therapsids: Evidence for Late Permian origins of mammalian endothermy, *Evolution*, 48:207–229, 1994; N. Hotton et al. (eds.), *The Ecology and Biology of Mammal-like Reptiles*, 1986; R. C. Schroter and N. V. Watkins, Respiratory heat exchange in mammals, *Respirat. Physiol.*, 78:357–368, 1989.

Manufacturing engineering

Recent developments in manufacturing engineering involve a new discipline known as factory physics, which seeks to explain essential types of manufacturing behavior, and rapid prototyping, which permits quick production of design prototypes.

Factory Physics

Manufacturing facilities present complex logistical problems such as how to schedule work, how to control inventories, how to schedule and allocate the work force, how to monitor and ensure quality, and how to maintain equipment. In modern factories, most of these problems are addressed through computerized planning systems. Unfortunately, although computers have grown increasingly powerful in recent years, many fundamental aspects of manufacturing behavior are still poorly understood. As a result, there is considerable discussion about what types of planning systems are effective in different kinds of manufacturing systems. A new field known as factory physics has evolved that explains essential types of manufacturing behavior and thereby provides a foundation for better methods of designing and managing production systems.

Science of manufacturing. Despite a widely publicized scientific management movement in the early twentieth century, a consistent science of manufacturing has never been developed. Although researchers have studied specific pieces—the physics of metal cutting and methods for scheduling certain types of factories, for example—the basic laws describing the behavior of manufacturing systems have remained elusive. As a result, manufacturing designers and engineers have had to rely on experience, common sense, and popular trends instead of well-founded principles.

As the pace of technological change and the intensity of global competition have intensified, the lack of a science of manufacturing has become increasingly apparent. Firms relying on familiar traditional techniques have frequently found themselves overtaken by competitors that could supply products with higher quality, lower cost, or more responsive delivery times. The pressures associated with this problem have stimulated manufacturing research, leading to the active compilation and development of the principles of manufacturing behavior. The term factory physics is used to describe this emerging science of manufacturing.

Performance measures. A primary objective of virtually all manufacturing enterprises is to generate profit. Although many things affect profit, three of the most critical performance measures at the plant level are throughput, work in process, and cycle time. Throughput, the average rate at which a product is produced, directly affects revenue by fueling sales. Work in process, the amount of inventory (raw materials and semifinished product) held on average in the facility, affects the cost of business because it represents money tied up in materials and therefore not available for other purposes. Cycle time, the average amount of time required to manufacture a product, affects customer service by influencing how rapidly customer orders can be filled.

Although manufacturing systems, and hence the performance measures, involve a host of physical processes (cutting, grinding, soldering, lifting, conveyoring, and so on), it is the integration of these processes that most critically determines the performance of the overall system. The key to good manufacturing management, therefore, is understanding how to influence the physical and human elements in a manufacturing system to achieve positive performance measures. This requires solid intuition on the part of managers, so that they can determine which activities have a large impact on performance measures. Providing a framework with which to develop this needed intuition is a major goal of the field of factory physics.

Operations view. To develop intuition about the relationships between processes and performance measures in a manufacturing facility, it is useful to adopt the operations view, in which parts and processes are viewed in generic terms and the focus is placed on the flow of material through the system. In a sense, the operations view is an intermediate perspective. It is higher than the process perspective, which focuses on matters such as the physics of metal cutting and is too detailed to incorporate plantwide performance measures. It is lower than the corporate perspective, which focuses on the overall financial performance of the firm and is too general to incorporate the specifics of manufacturing processes.

The operations view looks at parts (for example, printed circuit boards, automobile frames, and steel slabs) as entities that flow through processes (for example, circuitizing, assembling, and galvanizing), and is concerned with what factors affect this flow. Because the processes influence the flow, the operations view is linked to the process perspective. Because the flow translates into throughput and hence profit, it is linked to the corporate perspective as well. But by providing a general framework in which to examine material flow, the operations view makes it possible to sift through the myriad

Performance measures for a simple production line			
Work in process	Throughput	Cycle time	Throughput × cycle time
1	0.25	4	1
2	0.50	4	2
3	0.75	4	3
4	1.00	4	4
5	1.00	5	5
6	1.00	6	6
7	1.00	7	7
8	1.00	8	8
9	1.00	9	9
10	1.00	10	10

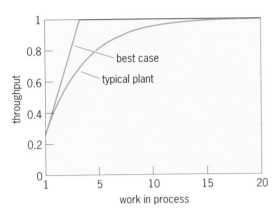

Fig. 1. Throughput plotted as a function of work in process, showing best-case and typical plant values.

process for those details that have significant performance implications.

Simple production lines. An example of a simple production line is one that produces large pennies for use in Fourth of July parades. It consists of four processes: head stamping, which stamps the head design on a penny blank; tail stamping, which stamps on the tail design; rimming, which places a rim around the penny; and deburring, which removes any sharp burrs. Each operation requires 1 min. The first concern in developing intuition concerning the basic manufacturing behavior for this process is to characterize the three performance measures—throughput, work in process, and cycle time—and their relationship to one another.

In order to examine these relationships, a thought experiment is performed in which the work-in-process level in the line is held constant and the other two measures are observed. For instance, for a work-in-process level of one, one penny blank is released into the front of the line and is completely finished before another is released. Since each penny will take 4 min to finish, the throughput is one penny every 4 min, or 0.25 penny/min; and the cycle time is 4 min.

When the work-in-process level is increased to two, two blanks are released into the system; then another blank is released each time a finished penny exits the line. Although the second blank must wait for 1 min to get into the stamp 1 station (because it was released into the line simultaneously with the first blank), this effect is transient and does not occur after the start of the experiment. In the long run, the pennies will follow one another through the line, each taking 4 min, resulting in an output of two pennies every 4 min. Hence, throughput is 0.5 penny/min and cycle time is 4 min.

Increasing the work-in-process level to three causes throughput to rise to three pennies every 4 min, or 0.75 penny/min. Again, after an initial transient period in which the second and third blanks wait at stamp 1, there is no waiting at any station; hence each penny requires 4 min to complete. Cycle time is still 4 min.

When the work-in-process level is increased to four, something special happens: 3 min after the four blanks are released to the line (when the first blank has reached the last station), each of the four processes has one penny to work on. From that point onward, all the processes are constantly busy. One penny finishes at the last station every minute, so the throughput is 1 penny/min, which is the maximum output the line can achieve. In addition, since each machine completes its penny at exactly the same time, no penny must wait at a process before beginning work. Therefore 4 min is the minimum value possible for cycle time. This special work-in-process level, which results in both maximum throughput and minimum cycle time, is called the critical work in process. In a balanced line (that is, all the machines require the same amount of time) made up of single machine stations, the critical work in process will always equal the number of stations, because each station requires one job to remain busy. In lines with unbalanced capacities at the stations (for example, where some of the stations consist of multiple machines in parallel), the critical work in process may be less than the total number of machines in the system.

If the work-in-process level is increased above the critical level, for example to five, then waiting (or queueing) begins to occur. With five pennies in the line, one must always be waiting at the front of the first process, because there are only four machines in the line. Each penny will wait an additional minute before beginning work in the line. The result will be that while throughput is 1 penny/min, as was the case when work-in-process level was four pennies, cycle time is 5 min because of waiting time. Increasing the work-in-process level even more will not increase throughput, since throughput of 1 penny/min is the capacity of the line, but it will increase the cycle time by causing even more waiting.

The **table** summarizes the throughputs and cycle times that result from work-in-process levels between 1 and 10. Throughput and cycle time are plotted as a function of work in process in **Figs. 1** and **2**. The so-called best-case curve represents a system with absolutely regular processing times. If the best case were achievable, it is clear that the

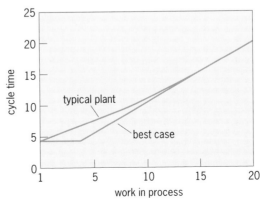

Fig. 2. Cycle time plotted as a function of work in process, showing best-case and typical plant values.

optimal strategy would be to regulate work in process to hold it right at the critical level of 4 pennies, thus maximizing throughput while minimizing cycle time. Lower work-in-process levels will cause a loss of throughput, and hence revenue; higher work-in-process levels will inflate cycle time with no increase in throughput. Unfortunately, the best-case scenario is virtually impossible to achieve in real-world manufacturing. Still, consideration of such an ideal case provides a baseline from which to judge actual performance and also yields a fundamental result.

Basic manufacturing law. The last column in the table gives the product of throughput and cycle time. In every case this product equals the work in process. Indeed, this property is a highly general law of manufacturing, which can be written as Eq. (1).

$$\text{Work in process} = \text{throughput} \times \text{cycle time} \quad (1)$$

This very useful law implies that if two of the performance measures are known, the third can be determined. For instance, if for a given production line or process the throughput is 500 parts per month and the average work-in-process level is 1000 parts, cycle time can be computed as work in process divided by throughput, or $1000 \div 500 = 2$ months.

Impact of variability. Equation (2) is implied by

$$\text{Throughput} = \frac{\text{work in process}}{\text{cycle time}} \quad (2)$$

the law stated in Eq. (1). Thus it is possible to achieve the same throughput with either a small value for work in process and a small cycle time, or a large value for work in process and a large cycle time, since throughput is determined by the ratio. Clearly, a small work in process (for low cost) and a small cycle time (for responsive deliveries to customers) would be preferable. The main factor that prevents most manufacturing systems from being able to achieve small values of work in process and cycle time is variability.

Variability is anything that prevents the manufacturing system from running in a uniform, clock-like manner. Machine failures, operator outages, delays to adjust equipment, quality problems, and many other factors can make the processing occur in a fluctuating manner and hence can drive up the work in process and the cycle time required to achieve a given throughput. In Figs. 1 and 2 this effect is illustrated by the curves labeled "typical plant." The typical plant requires much more work in process to attain the same level of throughput as the best-case plant. Alternatively, the typical plant achieves a lower throughput for the same work-in-process level as the best-case plant.

The challenge of improving an existing manufacturing facility may be viewed as trying to move a plant from the typical plant behavior in Figs. 1 and 2 closer to the best-case behavior. In general, anything that reduces the variability of the times required to process parts on the machines will help. But the impacts of variability can be subtle; therefore evaluating the impact of specific reduction policies requires additional factory physics relationships. A manufacturing manager who has developed sound intuition about these relationships is well prepared to concentrate the plant's resources on policies that will have the greatest impact on profit. As such, factory physics is a powerful framework for diagnosing problems in a manufacturing system and systematically achieving continual improvement of manufacturing environments.

Prospects. Factory physics is still in its infancy. Although researchers have worked on formal problems related to manufacturing logistics for almost 100 years, only in recent years have attempts at integration into a scientific framework been made. It can be expected that the discipline of factory physics will develop as most sciences do, according to the process of conjecture and refutation. That is, from time to time someone will conjecture a new framework (or partial framework) for the science of manufacturing. If it gains acceptance, researchers will use it to do modeling and empirical studies of manufacturing systems. Eventually, however, someone will refute the conjecture (for example, find it inadequate under some set of circumstances). This refutation will lead to a new and presumably better conjecture, and work will continue.

In a mature scientific field such as general physics, major refutations can take decades to arise. However, because efforts to place manufacturing practice into some kind of scientific framework are still new, it is likely that conjectures and refutations will take place rapidly in factory physics. Thus the field of manufacturing research will be exciting, and significant changes in the way manufacturing facilities are designed and managed are likely in the not-too-distant future.

Wallace J. Hopp; Mark L. Spearman

Rapid Prototyping

A number of methods are available for rapid production of prototypes. Whatever the technique used, the ability to rapidly produce prototypes of new designs is increasingly important to the world's manufacturers. This ability can remove months from new product development and mean the difference between profit and loss.

Once a three-dimensional solid model of a new product is available in a computer-aided-design (CAD) system, the next phase of the design process can begin. This phase is aided by generating a prototype or facsimile of the product. The prototype does not need to be made of the same material as the final product, but it should have many of the same physical characteristics.

There are several ways in which a prototype of the new product can be generated. All the techniques require that the database of the solid model in the computer-aided-design system be converted into a form that is usable by the prototyping machine. Most rapid prototyping systems produce a model by solidifying a medium of some kind, for example, a photosensitive polymer, a thermoplastic substance, a ceramic powder, or another formless material.

Stereolithography. One of the most popular prototyping techniques is called stereolithography. The stereolithography process was developed in the United States, and equipment has been commercially available since 1989. This process converts a three-dimensional computer-aided-design model into a number of horizontal slices, which are then used to create the model from photosensitive resin. First, the computer-aided-design system converts the boundaries of the solid model into a series of triangles. The data for each triangle consist of the coordinates of the three vertices and a normal vector, indicating how each triangle is oriented in space. The resulting file is called the stereolithography file or .STL file. The second step involves converting the triangles into data representing horizontal slices of the solid. This conversion is accomplished by proprietary software within the stereolithography equipment. The slice data are then used to guide a laser that cures a layer of photosensitive resin contained in a vat. The curing process is done by playing the laser beam back and forth on the surface of the photopolymer. Any place that the beam strikes will solidify and become part of the prototype. Once a layer has been drawn, it is lowered a small amount into the resin vat. The next layer is then built upon the last, resulting in a stack of slices that is the model. Depending on the resin used, the model may have to be cleaned and postcured. Some machining of the model may also be necessary to remove any supporting structures that were added to prevent drooping during the building process. Exposure of the resin can take 8 h or more. It is an automated process that is often performed overnight.

Laminated object manufacturing. This technique was developed in the United States. The laminated object manufacturing system also uses the .STL files and generates slices of the model internally. The model is built up from thin layers of paper. The slice data are used to control a laser to cut the outline of the part into the paper. Each layer of paper is laminated to the last automatically by using a heat-sensitive polymer. After a new layer is laminated to the stack, the height of the stack is measured. This measurement is used to generate the next slice. In this way, the thickness of the paper does not have to be absolutely accurate. The sections of each piece of paper that are not part of the model are hatched and left in place. They form a support structure for the model which is easily broken away when the model is complete. An advantage of the laminated object manufacturing system is that only the periphery of a layer needs to be cut. There is no need to have each section completely filled in as with stereolithography or other methods.

Solider. In a system called solider, developed in Israel, the negative of the slice data is photographically transferred to a piece of glass, resulting in the area around the model being black. Then, as with stereolithography, a layer of photosensitive resin is exposed to ultraviolet light by using the glass as a mask. The resin protected by the black areas of the mask are left uncured. These areas are removed and replaced by wax to provide a support structure for the next layer. Once the model is complete, it is heated to remove the supporting wax. This process is potentially faster than stereolithography, since the entire photopolymer surface is exposed to ultraviolet light through the mask. A disadvantage of the solider system is that a machining operation is required to remove excess wax and make sure that each layer is level.

Selective laser sintering. This method, developed in the United States, is similar to stereolithography, except that the photosensitive resin is replaced by a heat-fusible powder, composed of thermoplastics and wax. The heat for fusing the powder again comes from a laser. The selective laser sintering model is formed on a table that descends through a cylinder. First, a layer of the powder is placed on the table. Then, the laser draws the model data on the surface of the powder, much as the stereolithography system exposes photosensitive resin, melting and fusing the powder that will become the model. Each layer is drawn on top of the previous one. The unfused powder is left to provide a supporting structure for subsequent layers, and is removed once the model is complete.

Fused deposition modeling. This system, developed in the United States, resembles a flatbed X-Y plotter. Instead of a pen, there is an extruder head from which a stream of thermoplastic material or wax exits. The prototype model is drawn by using this head, and each layer of the model is built upon the last one. No supporting structure is normally used. The material cools quickly to form a solid and allows some overhanging without supports.

Ballistic particle manufacturing. This process became available in July 1995. The model is built on a platform that moves in the z axis. A five-axis robotic ejection head is used to generate and direct millions of microscopic particles of thermoplastic material that will make up the model. The system automatically determines the need for support structures and builds them with perforations for easy removal. The robotic head contains a piezoelectric jetting system that shoots droplets of molten thermoplastic approximately 75 micrometers in diameter at a rate of 12,000/s. The system offers the advantage of small size and relatively low cost. It operates in a standard office environment with no need for special venting equipment.

Direct shell production casting. A new technique called direct shell production casting became available commercially in August 1995. This process produces ceramic molds, such as those used for investment casting, directly from computer-aided-design files. It is similar to selective laser sintering in that a thin layer of fine alumina powder is spread over a working surface. But unlike selective laser sintering, a printhead injects tiny drops of silica binder onto the powder surface from 128 ink jets.

The droplets solidify the powder on contact, and the unbound powder remains as support for the following layers. Thus, when the process is complete, the casting shell remains buried in a block of loose ceramic powder. The ceramic powder is removed, and the finished shell is fired to remove moisture and to preheat it. The shell is then filled with molten metal. After cooling, the shell is broken away to reveal the completed casting.

Other methods. Not all new products require three-dimensional prototyping capabilities. For example, sheet-metal parts might be basically two-dimensional, with a uniform thickness in the third dimension. Prototypes of these parts can be made by using a high-power laser operating in two dimensions—much as an X-Y plotter. Although such parts will cost more than stamped components, the savings in time and tooling expenses may justify the higher costs.

Prototypes for sheet-metal objects may not need to be made from metal. It may be sufficient to evaluate parts made from plastic or cardboard to verify proper geometry and hole placement. In these cases, prototype parts can be made by using a low-power laser (50 W or less).

Some products contain components that cannot utilize the newer rapid prototyping techniques. But that is not to say that they cannot benefit from rapid prototyping. For example, a new part may be too large to fit in a prototyping machine. Or it may be necessary to have the prototype made from the same material as the final product. In these cases, numerically controlled machines may be used.

For background information SEE COMPUTER-AIDED DESIGN AND MANUFACTURING; MANUFACTURING ENGINEERING; NUMERICAL CONTROL; QUEUEING THEORY in the McGraw-Hill Encyclopedia of Science & Technology.

Walter H. Hoppe

Bibliography. R. Askin and C. Standridge, *Modeling and Analysis of Manufacturing Systems*, 1993; J. Buzacott and J. Shanthikumar, *Stochastic Models of Manufacturing Systems*, 1993; W. Hopp and M. Spearman, *Factory Physics: The Foundations of Manufacturing Management*, 1996; W. H. Hoppe, Rapid prototyping goes 2-D, *Mach. Des.*, pp. 148–149, March 9, 1995; Y. Uziel, Art to part in 10 days, *Mach. Des.*, pp. 56–60, August 10, 1995; C. E. Young, Getting up to speed: Rapid prototyping and tooling—Where are the technologies?, *Rapid News*, 3(2):6–8, 1995.

Marine mammal medicine

There are several groups of marine mammals: the order Cetacea includes dolphins and whales; Pinnipedia includes seals, sea lions, and walruses; Sirenia includes the manatee. Sea otters and polar bears are also classified as marine mammals.

Exotic animal veterinarians treat marine mammals in oceanariums or, if the animal is stranded, on the ocean shore. However, animals in oceanariums receive much more extensive treatment. The most common animals in marine parks are the bottle-nosed dolphin (*Tursiops truncatus*) and the California sea lion (*Zalophus californianus*). Several parks have harbor seals (*Phoca vitulina*), walrus (*Odobenus rosmarus*), and additional species of whales and dolphins.

Husbandry. Husbandry is the most important aspect of marine mammal medicine. The marine park must create an aquatic environment that is suitable for the health and care of these animals. Water quality is of prime importance to good animal health. The two types of water systems are open and closed. The open system pulls water from the ocean through the holding pools and then returns it. The water may or may not be filtered or chemically treated. A park that uses a closed system may be located inland. This system uses artificial seawater that is constantly filtered and treated chemically to clean and refresh it.

The food given to the animals must be clean and disease free. Most marine parks use frozen fish and closely monitor its quality and calorie content. Vitamin tablets are placed in a small amount of the fish to replace the nutrients that are lost during the thawing process.

One of the most important aspects of husbandry is observation; animal keepers and trainers are in close contact with each animal and are quick to recognize signs of possible health problems. In addition, the animals in the collection are medically checked routinely. If an animal is suspected of being ill, further diagnostic procedures are followed.

Preventive medicine measures include the administration of vitamins and vaccinations for disease prevention in some species. Pinnipeds receive medication to prevent heartworms, a mosquito-borne disease. Fecal samples are examined regularly for parasites.

Medicine. The physical examination of a captive marine mammal includes checking body condition, skin, weight, all body orifices, teeth, mouth, and blowhole. This procedure is easier with trained than with untrained animals. The most important diagnostic procedure is obtaining a blood sample. In the dolphin or whale, blood is usually taken from the fluke, although the pects or dorsal fin may also be used. Blood samples from the sea lion can be taken from the rear flippers or from the gluteal vein lateral to the caudal spine. In walruses and seals, the sample may be taken midline between the caudal vertebrae in the venous sinus or, in trained or restrained animals, from the rear flippers. Blood sampling from the manatee is taken on the ventral surface of the pect between the radius and ulna. Polar bears and otters are bled similarly to canines (from the front leg, rear leg, or neck).

Many of the dolphins and whales perform husbandry behaviors and are trained to present the ventral side of their flukes for blood sampling. Some dolphins and whales will remain motionless for a fecal sampling, and a few female animals will urinate into a cup. Untrained animals, however, must be restrained to collect a blood sample. A stretcher is used for a dolphin, a squeeze cage for a sea lion; sedation may be necessary for polar bears. Hematological analysis includes white blood cell count, red blood cell count, hemoglobin, packed cell volume, differential, and other parameters. The serum is harvested from each sample, and 20–30 different chemistries are checked to evaluate the health of internal organs. Progesterone or estrogen levels are also studied in the females. If a urine sample is available, a complete urinalysis is run. Bacterial cultures may be taken from the blowhole or nares; fecal samples may be examined for parasites; and tests may be given to observe the types of cells present in the animal.

After a careful review of the laboratory data, a plan is formulated for treatment and further diagnostic tests (if necessary). If an enteric problem is suspected, an endoscope can be passed down the esophagus into the stomach so that any internal lesions can be visually observed. This procedure is relatively simple in dolphins; however, pinnipeds must be restrained by either physical or chemical methods. While the scope is in place, samples of stomach content can be collected along with a biopsy sample, or foreign bodies can be removed. Fecal cytology can be taken from any animal by passing a tube into the colon and extracting a sample. With proper sterile technique, the cultures obtained from the blowhole (dolphin) or external nares (sea lion) and the fecal sample can be tested for antibiotic sensitivity. Thus, appropriate medication can be administered.

In some cases, x-rays are taken. Powerful equipment is required because of the size and density of the animals. The availability of portable x-ray equipment is helpful because it is easier to move the machine than the animals. Sonography is also used extensively in marine mammal medicine because the machines are small and portable. This procedure is very useful for diagnosing internal lesions. It is also very helpful during the early stages of pregnancy. Some marine mammal veterinarians use a thermogram camera to measure heat emissions from the body. Inflamed areas that are not visible to the eye will be readily visible on the screen. One advantage of thermography is that the subject can be some distance from the camera so that restraining the animal is not required.

Once the veterinarian has the test results, a plan of treatment can be determined. However, initial treatment usually begins before many of the cell culture, cytology, and pathology reports are received.

Therapy. After diagnosis, appropriate medications are administered, usually orally by placing medication in food. However, if the animal is not eating, the medication must be administered by injection. In some cases, supportive measures may be taken such as force feeding or tube feeding. The animal is always rechecked after 2 or 3 days to

determine if it is responding to the treatment. If there is little or no improvement, a reevaluation of the diagnosis and therapy is necessary.

If an animal has external lesions, the areas must be cleaned and antibiotic therapy started.

Surgery (for removal of an infected uterus, tusk, or tooth, or a foreign body) requires the use of anesthesia, either local (for minor procedures such as small-growth removal or skin biopsies) or general (for major procedures). The veterinarian must always weigh the benefits of the surgery against the risk of general anesthesia in any exotic species.

Reproduction. Marine mammals reproduce well in oceanariums, although in some parks the breeding is controlled either surgically (by castration) or with deferent hormones. Many killer whales (*Orcinus orca*), false killer whales (*Pseudorca crassidens*), beluga whales (*Delphinapterus leuca*), bottle-nosed dolphins (*Tursiops truncatus*) and pinnipeds have been born and raised in these environments. Several second and third generations have also been successfully raised. There have been a few walrus births. Rarely is a difficult birth encountered. In such a case, the veterinarian assists the birth mechanically or with labor-enhancing drugs. Occasionally, a cesarean section is performed. Neonates may need assistance. Different milk formulas for the different species of marine mammals are given by nursing bottles or administered by stomach tubes. Sometimes, a mother may be milked of the first milk to give the newborn essential antibodies. If the first milk is not available, gamma globulin can be harvested from the mother's blood serum and given orally.

Emergency treatment. All stranded animals must be considered an emergency situation. Stranded animals have come ashore or, having become incapacitated in some way, cannot function normally and are in immediate danger of dying. Most animals need more extensive care than can be given at the stranding site, and need to be transported to an oceanarium. Treatment at the stranding site is basic emergency first aid. Some animals may have only minor problems, such as ropes around the pects of a manatee that do not endanger the limb. These animals can be treated and released, but an animal in danger of losing the limb needs to be transported to a marine park for further treatment.

Stranded cetacea are by far the most critical of beached animals; more than 90% of beached cetacea die. First-aid measures include providing shade to reduce extreme sunburn, administering fresh water (orally), and administering antibiotics and cortisone by injection to relieve stress. The animal should be transported to a marine park as soon as possible for further evaluation and treatment.

Many pinnipeds become stranded on the shores of the western United States. Most of these animals are heavily parasitized. Supportive therapy and administration of parasiticides to rid the animals of lung worms and intestinal worms have saved many pinnipeds. The animals are released back to the wild after regaining their strength.

Many manatees in the southern United States become stranded. Most are injured by boat collision, some are entrapped in flood control gates, and others become entangled in crab-trap lines. In the winter, many become debilitated from cold exposure. These animals are picked up and brought to the different marine parks associated with the Marine Mammal Stranding Network. Their injuries range from minor propeller wounds to fractures of the vertebrae, collapsed lungs, and large areas of tissue loss. All the animals undergo extensive evaluation. A large majority of manatees can be restored to health and released in the wild.

Polar bears. Although the polar bear is a land animal, it is classified as a marine mammal because it spends most of its life in or around the sea. Bears are treated in the same way as other large dangerous carnivores. Minor treatment procedures such as deworming medications can be given orally. If hands-on treatment is required, the bear must be sedated. Sedatives may be administered orally or by remote injection, for example, dart or pole syringe. Blood sampling and other procedures can then be accomplished.

For background information SEE BEAR; CETACEA; MEDICAL IMAGING; OTTER; PINNIPEDIA; SIRENIA in the McGraw-Hill Encyclopedia of Science & Technology.

Deke Beusse

Bibliography. L. A. Dierauf, *CRC Handbook of Marine Mammal Medicine: Health, Disease, and Rehabilitation,* 1991; J. R. Geraci and V. J. Lounsbury, *Marine Mammals Ashore: A Field Guide for Strandings,* 1993.

Marine natural products

Many marine plants and animals contain secondary metabolites, known as marine natural products, that are uniquely identified with the respective organisms. About 7500 marine natural products have been reported. They are not uniformly distributed among the various phyla of marine plants and animals; in the invertebrates they are found more often in species that are sessile or slow-moving and lack a shell, a hard exoskeleton, spines, or other physical defenses against predation. For example, many natural products have been isolated from soft corals, while few have been reported from hard corals, where the living tissues are protected by a calcium carbonate skeleton. An analysis of the distribution of marine natural products leads to the conclusion that they often function to defend the producing organism against predation by making it less palatable or more toxic than alternative foods. In addition, there is increasing evidence that the sexual reproduction of marine organisms, particularly brown algae, is mediated by chemicals that enable the male gametes (sperm) to locate the correct female gametes (eggs) when both have been released into the water column.

Toxins. Studies of marine toxins have featured prominently in the development of marine natural product chemistry not only because of the incidence of accidental poisonings but also because of their value in elucidating biochemical mechanisms. Tetrodotoxin, which is responsible for occasional fatalities in those who eat pufferfish (fugu), is used as a molecular probe to study the sodium channel of excitable membranes. Like many toxins that affect the seafood industry, tetrodotoxin is of dietary origin, and it is produced by *Shewanella alga,* a bacterium that is a symbiont or epiphyte of an alga consumed by pufferfish. Saxitoxin, the causative agent of paralytic shellfish poisoning, is produced by dinoflagellates (microscopic algae) that are eaten by filter-feeding shellfish, which concentrate the poison. Although tetrodotoxin and saxitoxin are relatively small molecules, the compounds associated with ciguatera (seafood poisoning) are large molecules of the polyether class that have been traced to the dinoflagellate *Gambierdiscus toxicus.* The largest of these toxins, maitotoxin, has a molecular weight of 3422 and may be the most lethal nonprotein toxin.

Pharmaceuticals. Among all natural sources, marine sponges are considered the most prolific producers of biologically active natural products. Okadaic acid, which is a potent protein phosphatase inhibitor, was first isolated from sponges but was later found to be produced by dinoflagellates. A similar situation may exist for halichondrin B, a sponge metabolite that is being studied for anti-cancer activity. Most sponge metabolites, however, are produced by the sponges from which they are isolated. The anti-inflammatory agents manoalide and scalaradial, which inhibit the enzyme phospholipase A_2, are produced by *Luffariella variabilis* and *Cacospongia* spp., respectively. Many other anti-inflammatory agents are produced by sponges. Discodermolide, from a deep-water sponge, is a very powerful immunosuppressive agent that has the potential to suppress organ rejection after transplant surgery. A large proportion of sponge metabolites are reported to possess cytotoxic, antifungal, or antimicrobial properties; but few, if any, are currently being considered as candidates for drug development, despite their often unique chemical structures. However, the value of new chemical structures should not be underestimated; for example, the antiviral drug Ara-A (Vidabarine) is a direct descendant of the arabinosyl nucleosides isolated from the sponge *Tethya crypta* in 1950.

Other sessile invertebrates that have yielded potential drugs include bryozoans, soft corals, gorgonians (sea whips and sea fans), and tunicates (sea squirts). Bryostatin 1 is a complex metabolite from the bryozoan *Bugula neritina* that has undergone clinical trials as an anticancer agent. The same is true for didemnin B, a metabolite of tunicates of the genus *Trididemnum.* Anti-inflammatory agents have been found in both gorgonians and soft corals, with the pseudopterosins, such as pseudopterosin A from the gorgonian *Pseudopterogorgia elisabethae*, receiving the most attention.

Halogenated compounds. In contrast to their terrestrial counterparts, many marine natural products contain one or more halogen atoms. In some compounds, for example the monoterpene from the red alga *Plocamium cartilagineum*, the combined mass of the bromine (Br) and chlorine (Cl) atoms is more than half of the molecular weight of the compound. This large mass is hardly surprising since seawater contains a very high concentration of chloride ion and a lesser, though significant, concentration of bromide ion. Based on the relative concentrations of halide ions, more chlorinated than brominated natural products might be expected. In fact, the opposite is true, because in order to incorporate halogens into organic compounds the halide ion must be oxidized by a haloperoxidase enzyme, and less energy is required to oxidize a bromide ion than to oxidize a chloride ion.

The red algae are the major producers of halogenated and particularly polyhalogenated natural products. Brominated metabolites have also been found in bacteria, green algae, sponges, tunicates, bryozoans, mollusks, various marine worms, and a few coelenterates. Many halogenated marine natural products are feeding deterrents that can protect the producer from most predators. However, specialist predators are not affected by the halogenated compounds in their diet. Sea hares are shelless mollusks that specialize in consuming those red algae that contain large concentrations of halogenated metabolites. The sea hares not only tolerate compounds that deter most other predators but use the same compounds for their own chemical defense.

For background information *SEE ALGAE; BRYOZOA; GORGONACEA; HALOGEN ELEMENTS; TOXIN; TUNICATA* in the McGraw-Hill Encyclopedia of Science & Technology.

D. John Faulkner

Bibliography. D. J. Faulkner, Biomedical uses for natural marine chemicals, *Oceanus,* 35(1):29–35, 1992; D. J. Faulkner, Marine natural products, *Nat. Prod. Rep.,* 10(5):497, 1993; V. J. Paul, *Ecological Roles of Marine Natural Products,* 1992; Special edition on marine natural products chemistry, *Chem. Rev.,* 93(5):1671–1944, 1993.

Mass spectrometry

Since the 1970s, the measurement of the masses of atoms and molecules has grown from a specialty within nuclear physics to a widely used tool in science and industry. The recent advent of single-ion measurements in Penning traps is dramatically extending the capabilities of mass spectrometers, increasing the accuracy of mass measurements by factors of 10 to 1000 (see **table**). In addition to improving upon traditional uses of mass spectrometry, the new accuracy is generating many new applications in chemistry, metrology, and fundamental physics.

Single-ion mass spectrometer. The most accurate mass spectrometer in the world is the single-ion Penning-trap mass spectrometer at the Massachusetts Institute of Technology. This machine combines several advanced technologies in order to compare masses with an accuracy of one part in 10^{10}. Each measurement involves isolating a single charged atom or molecule, trapping it in the center of an evacuated space, and precisely monitoring its motion. The vacuum must be good enough for the ion to remain in the trap for many hours without colliding with anything. The ion trap sits in the center of a superconducting magnet, several times stronger (8.5 tesla) than the magnets used to lift automobiles, but so stable that the hourly change in the field is insufficient to visibly deflect a compass needle. The whole apparatus is cooled by liquid helium to 4° above absolute zero.

Figure 1 shows the essential parts of the spectrometer. Positive ions are created inside the trap by letting a tiny amount of gas into the vacuum chamber and using an electron beam to strip an electron from one of the atoms (or molecules) of the gas. The resulting ion is confined vertically by the positive voltages on the end caps, and confined horizontally by the magnetic field, which bends the horizontal motion into a circle. (If more than one ion is made, either the extras are driven out by radio waves or the trap is emptied and a new ion is made.)

The horizontal motion of the trapped ion is a circular orbit revolving at the ion's cyclotron frequency, equal to the product of the ion's charge and the magnetic field divided by 2π times the ion's mass. Because the charge is always an integral number of electron charges and the magnetic field is extremely stable, the cyclotron frequency provides an accurate measure of the ion's mass. Mass comparisons are made by making alternate measurements of the cyclotron frequencies of two different ions several times during a night, and fitting the data to a smooth curve that characterizes the field drift (**Fig. 2**). The atomic mass of an ion is determined by directly or indirectly comparing that ion to carbon-12, which is defined to have an atomic mass of exactly 12 u. (The international unified atomic mass unit, abbreviated u, is also called a dalton or an amu.)

The use of single ions greatly improves the accuracy of mass measurements since the repulsive forces between multiple ions shift their cyclotron frequencies, but detecting one ion several millimeters (about ¼ in.) from the nearest surface presents a difficult and subtle challenge. As the ion moves inside the trap, it generates tiny currents (of the order of 10^{-14} A) in the electrodes which are detected by a superconducting resonant circuit coupled to a SQUID (superconducting quantum interference device) preamplifier. However, any detector coupled to the ion's cyclotron motion inevitably causes some tiny shift in the cyclotron frequency. The spec-

Atomic mass table, comparing values from single-ion spectrometer with accepted values*

Atom	Mass from single-ion Penning-trap mass spectrometer, u	Accepted mass values, u[†]
^{1}H	1.007 825 031 6 (5)	1.007 825 035 0 (120)
n	1.008 664 923 5 (23)	1.008 664 919 0 (140)
^{2}H	2.014 101 777 9 (5)	2.014 101 779 0 (240)
^{13}C	13.003 354 838 1 (10)	13.003 354 826 0 (170)
^{14}N	14.003 074 004 0 (12)	14.003 074 002 0 (260)
^{15}N	15.000 108 897 7 (11)	15.000 108 970 0 (400)
^{16}O	15.994 914 619 5 (21)	15.994 914 630 0 (500)
^{20}Ne	19.992 440 175 4 (23)	19.992 435 600 0 (22000)
^{28}Si	27.976 926 532 4 (20)	27.976 927 100 0 (7000)
^{40}Ar	39.962 383 122 0 (33)	39.962 383 700 0 (14000)

* The numbers in parentheses indicate the uncertainty in the last digits.
† From A. H. Wapstra and G. H. Audi, The 1983 atomic mass evaluation, *Nucl. Phys.*, A432:1–54, 1985.

trometer circumvents this problem by using carefully timed pulses of spatially varying electric fields to move the ion vertically in a way that is partially synchronized with its (horizontal) cyclotron motion. The detector senses only this vertical motion, allowing an ideal combination: the ion revolves freely most of the time, with the detector only occasionally monitoring its position.

Applications. Penning-trap mass spectrometers are a classic case of a tool developed for basic research finding unanticipated uses. Originally developed for the study of electrons, they are now a standard tool used in hundreds of laboratories.

Chemical analysis. Many of the current applications are in analytical chemistry. Chemical experiments usually involve loading the trap with large molecules which break up during ionization. The weights and relative abundances of the ionized fragments serve to identify or determine the structure of the original molecules. Mass spectrometry has made major contributions to several branches of chemistry including analysis of protein structures, dynamic studies of gas reactions, and detection of trace pollutants.

Replacing the kilogram. The international unit of mass, the kilogram, is defined as the mass of a platinum-iridium cylinder that is stored in a vault in Sèvres, France. This artifact standard has many disadvantages. To reduce the chance of damage, the prototype kilogram has been compared to secondary standards only three times this century. A greater drawback is that its long-term drift is simply unknown. Several teams of scientists are working to redefine the kilogram as a fixed number of atomic mass units. Standard masses would then be made from nearly perfect crystals of silicon containing a known number of atoms. The spacing between atoms in silicon has been measured to within a few parts in 10^8, but until recently the atomic mass of silicon was not known with enough accuracy to make a good standard. The mass measurement with the single-ion Penning-trap mass spectrometer, accurate to one part in 10^{10}, has completely eliminated this problem.

Fine-structure constant. The fine-structure constant is a dimensionless ratio, approximately $\frac{1}{137}$, which is proportional to the square of the electron charge divided by the product of Planck's constant and the speed of light. It is arguably the most important dimensionless ratio in physics. It determines the relative strength of all electromagnetic interactions as well as the so-called weak interactions of nuclear and elementary particle physics. The effects of the fine-structure constant show up in settings as diverse as giant particle accelerators and the coldest of cryogenic refrigerators. Furthermore, the fine-structure constant can be measured in a variety of seemingly unrelated ways, allowing accurate checks of theories ranging from quantum electrodynamics to superconductivity.

Several of the better measurements of the fine-structure constant are based on the Rydberg constant, which is very accurately known from laser spectroscopy of hydrogen. The Rydberg constant is equal to the product of the square of the fine-structure constant, the electron mass, and the speed of light, divided by twice Planck's constant. Because the speed of light is known exactly, a measurement of the electron mass divided by Planck's constant is effectively a measure of the fine-structure constant. However, this ratio is hard to measure directly. Much better measurements of the ratio of a parti-

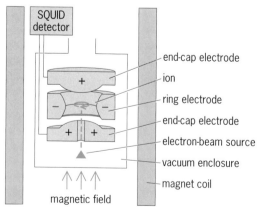

Fig. 1. Single-ion Penning-trap mass spectrometer. The end-cap voltages are typically 10 V higher than the ring.

cle's mass to Planck's constant have been made on neutral particles such as neutrons or cesium atoms. Highly accurate Penning-trap measurements of mass ratios such as the ratio of the electron mass to the neutron mass and the ratio of the electron mass to the cesium-atom mass are making it possible for these values of the mass-to-Planck's-constant ratio to contribute to the knowledge of the fine-structure constant.

Masses of gamma rays, neutrinos, and chemical bonds. Accurate mass spectrometers can be used to determine the energy released in nuclear and chemical reactions by measuring the mass difference between the initial and final states and by using Einstein's relation, which states that the energy release equals the product of the mass difference and the square of the speed of light. Consequently, mass measurements have long played a central role in nuclear physics, especially for the calibration of gamma-ray energies, where uncertainties have recently been reduced by a factor of 100. The unprecedented accuracy of Penning-trap spectrometers is now making it possible to measure much smaller mass differences such as the (possible) rest mass of the neutrino, or the masses corresponding to chemical binding energies.

The rest mass of the electron neutrino is one of the leading mysteries of contemporary physics. Neutrinos with nonzero rest mass could supply the missing dark matter in the universe, account for the missing solar neutrino flux, and point the way to new physics beyond the standard model of elementary particles. Measurements indicate that the neutrino mass is less than about 7 eV (divided by the square of the speed of light), but the best fit to the data actually yields the doubtful result that the square of the mass is negative. The most accurate measurements of the neutrino mass come from studying the decay of tritium into helium-3, an electron, and an antineutrino, as in the reaction below.

$$^3\text{H} \longrightarrow {}^3\text{He}^+ + e^- + \bar{\nu}$$

The neutrino mass (if any) must be less than the helium-tritium mass difference minus the maximum kinetic energy of the electron (divided by the square of the speed of light). Penning-trap spectrometers have measured the ^{3}H-^{3}He mass difference to within 2 eV, and even more accurate measurements are expected. *See Cosmology.*

Current understanding of chemistry depends heavily on knowledge of the binding energies of molecules, radicals, and ions, but these energies have never been measured for many of the most reactive species because they are too difficult to produce in quantity. At present, the best mass spectrometers are barely accurate enough to detect the existence of chemical bonds, but improvements to the single-ion Penning-trap spectrometer are expected to enhance its accuracy by as much as a factor of 30. This new accuracy, combined with the ability of Penning traps to use single ions, may soon open an entire new field of chemistry by allowing

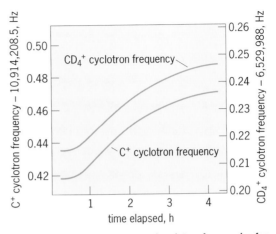

Fig. 2. Polynomial fit for measured cyclotron frequencies for C$^+$ and CD$_4^+$ ions as a function of time. The curves characterize magnetic field drift. Different vertical scales are used for the two ions. These data provide a value for the atomic mass of deuterium.

measurements of binding energies for cluster ions (such as H$_{13}^+$), doubly charged molecules, negative ions, and other reactive species.

For background information *see* Chemical bonding; Fundamental constants; Gamma rays; Mass spectrometry; Mass spectroscope; Neutrino; Particle trap; Physical measurement; Rydberg constant; SQUID in the McGraw-Hill Encyclopedia of Science & Technology.

Fred L. Palmer; David E. Pritchard

Bibliography. E. A. Cornell et al., Mode coupling in a Penning trap, *Phys. Rev.*, A41:312–315, 1990; F. DiFilippo et al., Accurate masses for fundamental metrology, *Phys. Rev. Lett.*, 73:1481–1484, 1994; R. S. VanDyck, Jr., D. L. Farnham, and P. B. Schwinberg, High precision mass ratio measurements in a Penning trap, *J. Mod. Opt.*, 39(2):243–255, 1992; A. H. Wapstra and G. Audi, The 1983 atomic mass evaluation, *Nucl. Phys.*, A432:1–54, 1985.

Matter in strong fields

The primary reason for the stability of matter is that electrons are bound in atoms by electric fields that are far larger than any field which can be normally produced in the laboratory. For example, a field of 10^5 volts per centimeter is extraordinarily difficult to achieve safely in the laboratory, but the electron of a hydrogen atom is held to its proton with a field 10^3 to 10^4 times larger. Thus, normal matter is immune to the application of even large electric or magnetic fields, and special methods are needed to separate electrons from their nuclei.

Photoelectric effect. One of these methods is based on the photoelectric effect, first explained by A. Einstein in 1905. The electrons of a particular atom are exposed to light with sufficiently high frequency, or short-enough wavelength, that the light photon is absorbed by the electron. In this manner the electron gains enough energy to liberate it from the nucleus. This is purely a quantum effect, relying upon the fact that light photons carry energy proportional to their frequency, and that these photons can disappear, giving up all their energy to the electrons. If the frequency were not high enough, the energy absorbed by the electrons would be too small to liberate them. This explanation resolves the seemingly counterintuitive observation that irradiating the atom with light of increasingly high intensity but of insufficiently high frequency can never result in the liberation of electrons.

Rydberg atoms. A more recent method is the application of laboratory-sized fields to atoms whose electrons are in excited states. The excited states of an atom that electrons can occupy are labeled by the quantum number n, where n is an integer that ranges in value from 1 (corresponding to the ground state where atoms remain unless otherwise perturbed) to infinity (corresponding to ionization, where the electron leaves the nucleus). As n increases, the strength of an external electric field required to distort and ultimately strip the electron from the nucleus decreases as n^4. Thus, a field of 10^3 V/cm will have very little effect on an atom in its ground state, but an atom that has been excited to $n > 40$ will become highly distorted in this field and ultimately lose its electron. Atoms with their outer electrons in high-n states are often called Rydberg atoms.

This method is also applied to the study of the effects of magnetic fields on atoms. A significant area of investigation in astrophysics concerns the properties of atoms exposed to enormous fields on the surface of collapsed stars. The fields on these stars can exceed 10^8 tesla (10^{12} gauss), roughly 10^6 times greater than the largest fields produced in the laboratory. The laboratory study of atomic behavior in these huge magnetic fields is made possible by the application of the same technique used in the study of the effects of intense electric fields on atoms. In this case, the effect of a magnetic field increases as n^4. Instead of trying to reproduce the huge fields of the star in the laboratory, physicists excite the atoms into high-n states and apply modest magnetic fields. The physical principles are essentially the same in either case, but the promotion of electrons within the atom to high-n levels is far easier.

Laser excitation. The excitation of atoms into states with high-enough values of n that laboratory fields can affect them has been made possible by the advent of the tunable, high-powered dye laser. Prior to this, the only method of excitation was nonselective; that is, the atoms were excited in such a way that the electrons of individual atoms ended up not in a single state but in a variety of n states.

The operation of the tunable, high-powered dye laser is based on a version of the quantum effect discussed above. Laser light has an extremely narrow distribution of frequencies, and can therefore be tuned so that the photon energy is precisely equal to the energy difference between the electrons in their ground state and the energy of some higher n state. This energy purity of the light from the laser makes possible the preparation of a large quantity of atoms with their electrons excited to a given, single n state. Thus, the results of experiments that apply electric or magnetic fields to these excited atoms can be directly related to understanding how an atom in its ground state interacts with much larger fields.

Hidden states. In the case of electric fields, it has been shown that when atoms in excited states are placed in large fields, a series of states essentially "hide" behind the nucleus, opposite to the field direction. These states are quite stable even though the field is strong enough that, naively, the electron should be pulled away from the nucleus almost instantaneously. This wholly unexpected effect was confirmed by detailed computer calculations.

Orbit replication. In the case of Rydberg atoms placed in large magnetic fields, a remarkable result was obtained: the electrons tend to move in orbits that not only are stable but have a strong tendency to replicate themselves. Even more remarkable was the observation that this stability and orbit replication extend to energies far above the ionization energy, where the electron should have simply left the nucleus. This is evidence of an entirely new quantum system, consisting of an electron, a nucleus, and a magnetic field. This system evidently has its own symmetries, quantum states, and classical analogs, each different from the cases in which an electron is in either the atom or the magnetic field alone.

Chaotic behavior. This quantum system provides the simplest example of an effect that often occurs in physics but is almost always too difficult to completely understand or calculate: two competing forces act on an object (in this case an electron) with essentially the same magnitude. Under this condition, the object responds in a manner quite different from its behavior when only one of the forces is present. This field of research was instrumental in originating the study of chaos. The excited atom in a large magnetic field remains one of the principal experimental grounds for chaos theory.

High-power laser experiments. Lasers have been constructed with such extremely high powers that the light is best thought of as an oscillating electric field, with strengths that are comparable to the field of the nucleus on the electron, even in the ground state. These lasers produce their high power by emitting a modest amount of energy in an extremely short time. A typical high-power laser will emit 0.1 joule of energy in a pulse with a duration of less than 10^{-13} s. When focused onto an atom, the resultant electric fields are greater than 10^8 V/cm.

Multiphoton ionization. Under these conditions, a far more complicated stripping of the atoms occurs than in the simple photoelectric effect discussed above. Instead of requiring that the electron absorb only one photon with enough energy to take the electron from $n = 1$ to $n = \infty$, the huge fields of these lasers cause the electron to absorb multiple photons to reach the ionization limit, essentially stripping the atom independent of the frequency of the light. Experiments have shown that atoms can be stripped of their electrons by lasers emitting light at frequencies for which the electron must absorb more than 12 photons simultaneously in order for the energy state to jump from $n = 1$ to $n = \infty$.

Above-threshold ionization. An extreme example of the application to atoms of electric fields that are equal to or greater than the nuclear fields is above-threshold ionization (ATI). Here the intensity of the laser light causes an electron to continue absorbing photons from the field, even after it has been promoted to an energy state above $n = \infty$. In one of the most dramatic examples of two forces acting on an electron simultaneously, the electron experiences such a strong field that it absorbs enough energy to be stripped, but before it can escape the field of the nucleus it goes on absorbing more photons. The result is not the simple threshold law of ionization proposed by Einstein but a series of thresholds, each separated by the photon energy. This effect is a direct result of the field of the nucleus and the field of the laser controlling the motion of the electron as it is stripped away from the atom. Above-threshold ionization is used as a mechanism for the study of high-field effects because it is quite easy to observe with lasers of sufficiently high intensity.

For background information *SEE ATOMIC STRUCTURE AND SPECTRA; CHAOS; LASER; LASER SPECTROSCOPY; NEUTRON STAR; OPTICAL PULSES; PHOTOEMISSION; QUANTUM MECHANICS; RYDBERG ATOM* in the McGraw-Hill Encyclopedia of Science & Technology.

Richard R. Freeman

Bibliography. G. N. Gibson, R. R. Freeman, and T. J. McIlrath, New questions in the multiphoton ionization of atoms, *Opt. Photon. News*, 3(12):22, December 1992; U. Mohideen et al., High intensity above-threshold ionization of He, *Phys. Rev. Lett.*, 71:509–512, 1993; Optical Society of America, *Proceedings of Short Wavelength*, V: *Physics with Intense Laser Pulses*, 1994.

Mechatronics

Mechatronics is a branch of engineering that incorporates the ideas of mechanical and electronic engineering into a whole. It covers those areas of engineering concerned with the increasing integration of mechanical, electronic, and software engineering into a production process. The closeness of this integration particularly characterizes mechatronics, and often demands that the designer choose between different ways—mechanical, electronic, or software—to realize the required functionality of the system. This unification of traditionally competing domains of engineering lies at the heart of the new developments.

Mechatronic systems include robotic manufacturing systems, vehicles, and many items of domestic equipment. In some cases, the mechatronic system provides facilities similar to a previous product, for example, an electronically controlled domestic appliance such as a washing machine. The advantages are improved control and reliability stemming from the replacement of some mechanical parts that are subject to wear by electronic elements that are more reliable. Additional features, for example a wider range of operating programs, are often incorporated into the mechatronic design. At a higher level, the application of mechatronic principles to product design results in the development of products with distinctly more advanced features. For example, the automatic camera incorporates exposure control, which was previously a manual function.

At an even higher level, mechatronic methods can provide advances in areas not even possible without the advent of electronic techniques. The control by an engine-management system of the automobile engine provides an example whereby the efficiency and pollution of the engine are controlled.

Generalized mechatronic system. From the description above, it is clear that a mechatronic system incorporates many features of a control system in which energy flows are regulated, but that in addition there is a requirement to interface with the user and the outside world, which implies the manipulation of information. A block diagram of a generalized system, which emphasizes the division between the information and energy domains of the system, is shown in the **illustration**. Within these domains the central role of the sensors and actuators, which interface with the fundamental mechanism that is being controlled, is critical.

A mechatronic system has several of the following features: complexity, but with a high level of systems integration; reallocation of functionality between the mechanical, electronic, and software domains; provision of higher performance; processing that is embedded and often distributed; low-level responsibility for performance that is automatic, with the user concentrating on higher levels; sophisticated

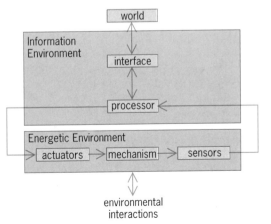

Generalized mechatronic system.

use of sensors to replace human sensing; and systems operation that is generally transparent to the user.

Examples. To illustrate these concepts, three examples will be discussed: a manufacturing system, the automobile, and the automatic camera.

Manufacturing system. It was in connection with manufacturing systems that the word mechatronics first gained wide use in the early 1980s. Because of the intrinsic complexity of most manufacturing systems, they illustrate an important feature that most large mechatronic systems possess, that of hierarchy. At its lowest level the individual machine tool is itself a mechatronic system. It has mechanical, electronic, and software features that are closely integrated into the total machine. Within it are sensors, actuators, and feedback systems that provide the essential internal elements of the system. The machine tool also possesses an internal communications network to coordinate the behavior of the machine and to enable it to communicate upward to the next level in the hierarchy.

The next level in the hierarchy is usually a so-called island of automation, with several machines connected together to provide a complete manufacturing cell. These machines are connected electronically through a local-area network (LAN), which enables the complete system to function as an integrated whole. At this and higher levels the software considerations dominate the design of the system.

At the highest level, the complete factory may be regarded as a totally integrated mechatronic system. The islands of automation are interconnected, and often further connected to other systems concerned with, for example, the managerial and financial functions of the company. At this level, a broadband network employing the manufacturing applications protocol (MAP) may be used.

Automobile. The modern automobile is a sophisticated mechatronic system. Not only is its engine frequently controlled by a microprocessor-based system, but the entire power train from the engine to the road wheels is subject to such control.

The control of the chassis systems, which include the steering, braking, and suspension systems, is also based on mechatronic principles. The first developments in this area were antilock braking systems; subsequent developments include active suspension systems and steering systems which improve the handling characteristics and vehicle safety. Electronics is also used to control the safety and comfort of the passenger accommodation, and in this area air crash bags have probably been the most extensively employed application.

All these developments have required the invention and development of sensors and actuators that are appropriate for use in ever more inexpensive classes of automobiles. Sensors are applied in areas related to the engine (engine speed and torque, fuel mixture, coolant temperature, oil temperature, emission composition, and ignition conditions), the chassis (gearbox and clutch operation, suspension, braking conditions, speed over ground, wheel speed, and steering), and the body (security systems, crash systems, ice warning, and environment control). SEE AUTOMOBILE.

Automatic camera. For the autofocus camera, each of the fundamental system elements—lens, flash system, and body—contains its own sensing, processing, and actuation elements, and thus may be considered as a mechatronic system in its own right. These systems are connected by a communication network inside the camera, enabling it to assume responsibility for its operation and leaving the user free to concentrate on picture composition. Thus, the lens subsystem controls focus, aperture, and zoom; the flash (if needed) setting is controlled; and the body subsystem controls the film advance and rewind. The user need only select the high-level procedures and compose the picture.

Enabling technology. The advance of silicon-based integrated circuit technology has provided the critical impetus to the advance of mechatronics. Originally, the silicon-integrated circuit was developed for military use. It incorporated on a single silicon chip many components which had previously been connected in a system. This microelectronic approach responded to a need to reduce weight, improve reliability, and reduce power requirements in, for example, aeronautical and space applications. The cost was initially of secondary significance. However, as the technology advanced during the 1970s and 1980s the range of applications to nonmilitary systems increased dramatically with the finding that the cost of the integrated microelectronic solution was often significantly lower than traditional approaches. Thus, cost effectiveness combined with reduced size and the added functionality that could be achieved became the driving force in the adoption of microelectronics in an ever wider range of civilian industrial and consumer products. In portable equipment the low power consumption and weight advantages were also exploited; reliability was an advantage as well.

In this way, the processing elements of modern mechatronic systems were adopted, but the cost and technical specifications of other parts of the system, particularly the sensors and actuators, lagged behind. Newly developed integrated microminiature components using silicon and other technologies are now being substituted for more traditional devices. An example is the miniature accelerometer based on silicon technology that is used in many automobile crash-bag systems. SEE MICROENGINEERING.

Prospects. The advances of the past three decades in microelectronic technology have not yet run their course. The complexity of individual microcircuits is still increasing, and the cost of an individual circuit at the level of a logic gate on a chip continues to decline. The incorporation of a greater proportion of the electronics on a single chip of silicon can be foreseen, and this advance will open up new application areas as system costs come within range of commercial exploitation.

New sensors, particularly microminiature types, are moving into a phase of rapid development, and these are likely to open up new fields where the intrinsic small size of the device will be an advantage. Advances in medical mechatronics are a likely development.

Some areas for development are said to be safety critical. The already wide exploitation of mechatronic principles in the automotive industry is an example, and future developments in the medical field another. Recent advances in the design and testability of microelectronic components will likely permit the progress of mechatronic approaches to design, using such microelectronic components in ever more critical application areas.

For background information SEE AUTOMOBILE; CAMERA; COMPUTER-INTEGRATED MANUFACTURING; EMBEDDED SYSTEMS; FLEXIBLE MANUFACTURING SYSTEM; INTEGRATED CIRCUITS; MEDICAL CONTROL SYSTEMS; NUMERICAL CONTROL; ROBOTICS in the McGraw-Hill Encyclopedia of Science & Technology.

A. P. Dorey

Bibliography. D. A. Bradley et al., *Mechatronics: Electronics in Products and Processes*, 1991; A. P. Dorey and D. A. Bradley, Measurement science and technology—Essential fundamentals of mechatronics, *Meas. Sci. Tech.*, 5:1415–1428, 1994; A. P. Dorey and J. H. Moore, *Advances in Actuators*, 1995; D. Tomkinson and J. D. Horne, *Mechatronics*, 1996.

Medical geography

Medical geography is the study of relationships between human health and geographic location through the identification of disease patterns and their possible causes. Disease clustering is an emerging field of medical geography that plays an increasing role in the analysis of public health problems.

Disease clusters arise whenever there is an excess of cases in space (space clustering), in time (time clustering), or in both (space-time clustering). A fourth kind of clustering, called interaction, occurs when nearby cases appear at about the same time and may be caused by transient exposure to a spatially localized toxin or by an infectious agent. For example, a spatial cluster of cholera cases in the vicinity of the Broad Street pump in London was investigated by Dr. John Snow in 1854. An example of a time cluster is the excess of flu cases that occurs every fall.

One of the earliest examples of the use of medical geography to determine the causes underlying a disease cluster is Snow's investigation of the cholera epidemic in London. Snow mapped the locations of cholera deaths (see **illus.**), and observed that many of them occurred near a water pump on Broad Street. He removed the pump's

handle, effectively ending the epidemic, which is now known to have been caused by consumption of the well's sewage-contaminated water.

More recent examples are notable for the controversy they have generated. Reports of clusters of childhood leukemia and other cancers in the vicinity of nuclear facilities in Sellafield, England, and other locations led to a flurry of studies that began in the late 1980s and are continuing. Some of the studies found an excess of cancers, and controversy arose regarding the biological mechanisms that might explain the apparent clusters. In general, levels of ionizing radiation and the release of isotopes (both possible carcinogens) by the nuclear facilities were not large enough to explain the excess cancers.

Another example involves a cluster of childhood leukemia near a hazardous waste disposal facility in Woburn, Massachusetts, in 1986. A statistically significant relationship was found between childhood leukemia incidence and consumption of well water contaminated by trichloroethylene, which appeared to originate from the waste disposal facility. This case, and others like it, prompted several state health departments to initiate active screening of their cancer registries for geographic disease clusters.

Public health context. Since the early 1980s public health agencies have investigated an increasing number of alleged clusters of adverse health events such as leukemia, cancers of the brain and central nervous system, sudden infant death syndrome, and birth defects. In 1990, the Centers for Disease Control of the U.S. Department of Health and Human Services formulated guidelines for a protocol for investigating disease clusters, from initial contact and response through a full epidemiologic study.

During stage 1, initial contact and response, the cluster is brought to the attention of the health agency, usually by a concerned non-health professional who notices a local excess of disease. The cases in the cluster are identified, and their characteristics (for example, places of residence, symptoms, and date of onset) are recorded. Residents in the area of the suspected cluster are often deeply concerned, and the health agency must maintain communication with the public and reply to inquiries throughout the course of the investigation. Stage 2, cluster assessment, seeks to determine how unusual the aggregation of cases is. Statistical methods are used to quantify the cluster and to guide the future course of the investigation. Many diseases, such as cancer, occur at some basic rate within a population; and the initial problem is to determine whether the cases in the cluster exceed this background rate. The cluster investigation usually is halted when an excess of cases does not exist. When an excess does exist, the investigation continues in an effort to identify a common cause underlying the cluster, such as exposure to an environmental carcinogen. In stage 3, case evaluation, the initial diagnoses are verified. Although some practitioners prefer to verify cases before undertaking the statistical analysis, the Centers for Disease Control recommends that case verification be postponed until the preliminary statistical analysis is completed. The final stages involve the design and implementation of study to investigate the origin of the disease in order to evaluate hypotheses that might explain its distribution in a population, and hence the alleged cluster.

Kinds of clusters. Potential causes of a cluster are not always known, and the clusters themselves may be poorly defined. In other instances, the cluster is well defined and clearly associated with a common exposure. Clusters can be described as perceived or true and are distinguished by health outcome, potential exposure, exposure-health link, and statistical significance.

Perceived clusters often reflect the environmental health concerns of the non-health professionals reporting the cluster. The health outcomes—the medical conditions which describe the cases in the cluster—are often poorly defined with cases characterized by unrelated illnesses (for example, em-

physema, breast cancer, pneumonia) or symptoms—when such cases are identifiable. A common exposure is absent or not obvious, and there may be many possible explanations for the reported health outcomes. A hypothesis describing the distribution of the disease in a population which would explain the health outcome may be lacking, and an excess of cases may not be found under the statistical analysis. Some perceived clusters arise because of chance, so that an excess of cases can exist but have independent underlying causes. A parallel is found in the so-called Texas sharpshooter problem, in which a sharpshooter fires bullets at a barn and then draws a circle around the tightest cluster as a testimony to his or her accuracy. Perceived clusters arise in an analogous fashion whenever a chance grouping of cases is put forward as a true cluster.

True clusters have a common cause or exposure underlying the cases, and explain fewer than 5% of reported clusters. They have a definable health outcome with a clinical diagnosis, such as leukemia or cancer of the brain and central nervous system, or

Key: × pump
      • deaths from cholera

0   50   100 yd

One of the earliest examples of a disease map, created by John Snow based on his investigation of the cholera epidemic in London in September 1854. (*After E. R. Tufte, The Visual Display of Quantitative Information, Graphics Press, 1983*)

are typified by a suite of related symptoms. A potential exposure (such as to benzene) may be suspected as the common cause underlying the cases. Further, a plausible biological mechanism relating the exposure to the reported health outcomes is known or hypothesized (for example, the link between benzene exposure and leukemia). Finally, the pattern of cases itself is unusual and is statistically significant.

Disease maps. Disease maps provide visual summaries of geographic disease patterns. Two kinds of data, involving rates and case locations, can be mapped. Disease rates summarize the number of health events that occur in a given area over a defined time span, and consist of a numerator and a denominator. The numerator is the number of cases or deaths, and the denominator is the size of the population at risk. Rates are often displayed as choropleth maps that color the areas based on the magnitude of the rate in each area. Rates are said to be unstable when the denominators are small or vary greatly from one area to another, and may be stabilized by using statistical techniques such as bayesian estimation models or by aggregating geographic regions. Many countries and many states in the United States maintain registries for certain diseases, such as cancer, which are used to construct atlases composed of disease maps. These maps provide visual summaries of geographic disease patterns, and are used for disease surveillance in an attempt to identify places with high disease rates. Rates may not be useful when the study area is small and the size of the denominator is small or unknown; in these instances point maps are used to display the locations of individual health events, for example, residences of women diagnosed with breast cancer in a neighborhood.

Advances. The term spatially referenced data is applicable when the geographic locations and values of observations are known. In the 1990s the use of geographic information systems (GIS) to manage and display spatially referenced public health data has greatly increased. These systems provide a wealth of information and permit the rapid construction of disease maps. Thus, the increased availability of spatially referenced health and exposure data is changing the ways in which disease clustering and health surveillance is conducted. Recent research has focused on how improved estimates of human exposure to toxic and carcinogenic compounds may be used to identify links between the environment and human health.

In the long term two research issues will probably be paramount. The first is the identification, within geographic information systems, of meaningful data targeting specific public health issues; and the second is the quantification of statistically unusual patterns on maps. Until recently, difficulties in obtaining spatially referenced information made medical geography relatively poor in data, and techniques for selecting data addressing meaningful epidemiologic hypotheses from data-rich

geographic information systems will need to be developed. Maps produced by these systems will be used to identify and evaluate possible relationships between the environment and health, and improved techniques are needed for determining whether these relationships are statistically unusual or can best be explained by chance alone.

For background information *see* CARTOGRAPHY; EPIDEMIOLOGY; ESTIMATION THEORY; PUBLIC HEALTH in the McGraw-Hill Encyclopedia of Science & Technology.

Geoffrey M. Jacquez

Bibliography. Centers for Disease Control, Guidelines for investigating clusters of health events, *Morbid. Mortal. Week. Rep.,* 39(RR-11):1–23, 1990.

Memory

Scientific and practical concerns have led to recent research on developmental aspects of memory. From birth, memory systems are able to pick up, store, and use information from the environment; this capability seems especially true for the acquisition of general information and motor skills. However, because the ability to remember events depends upon other types of memory, it develops more slowly. These findings are especially important when the memories of children are used in legal situations. The malleable nature of adult memories of childhood has also been examined.

Types of memory. Human memories contain a vast amount of information, including personal facts, such as activities; general knowledge, such as the spelling of words; and the knowledge needed to perform tasks such as tying shoelaces. Autobiographical memories, linked to particular times and locations, are called episodic memories. More general knowledge about the world such as the meaning of words, the use of language, and other information that is not associated with a particular time and place is known as semantic memory. Memory for motor skill and operations is known as procedural memory. Procedural knowledge often is not available for conscious recall. For example, an individual who knows how to ride a bike will have a difficult time explaining how to do it. Semantic and episodic memories can be expressed in words. Episodic memories depend on semantic information because personal knowledge such as where an individual went to school would make little sense if that individual did not know the meaning of the word school.

Semantic and procedural memory in infants. A useful model of semantic memory is a network of associated concepts. Evidence exists that individuals begin to learn associations very early in life. One series of studies tested the ability of infants to remember procedural knowledge. Each infant was placed in a crib with an overhead mobile that was attached to the infant's ankle. When the infant kicked, the mobile would move. After the infant had learned the connection between kicking and activating the mobile, the infant was removed from the crib. Some time later, the infant was placed back in the crib, and the time to relearn the connection between kicking and moving the mobile was measured. Older infants took less time to relearn. These studies showed that procedural memory is fragile at first but develops rapidly: 2-month-old infants forgot very quickly and showed no procedural memory after only 3 days, whereas 6-month-old infants showed memory up to 3 weeks later.

Episodic memory and infantile amnesia. Unlike procedural and semantic memory, episodic memory has not been detected in the first years of life. Most adults have no episodic memories from the first years of life. This lack of memory is known as infantile amnesia. The earliest autobiographical memories mark the offset of infantile amnesia. Estimates of the age of the offset of infantile amnesia have varied from 2 to 6 years. One method to estimate the offset of infantile amnesia is to ask individuals about events such as the birth of a sibling or a hospitalization. Very few people can recall anything about the birth of a sibling when that birth occurred before they were 3 years old. The same is true for personal information associated with a historical event, such as the assassination of President Kennedy. Adults who were between the ages of 1 and 7 at the time of the assassination were asked questions about where they were and what they were doing when they heard that the President had been shot. Adults and many older children had clear memories, whereas more than half of those who were younger than 5 had no memories.

Although 2½-year-old children can recall events that took place 6 months earlier, infants do not form lasting episodic memories for a number of reasons. Brain structures, such as the hippocampus, which are important for the formation of episodic memories are not fully developed. When young children do form memories, continued development of the hippocampus may overwrite or make those memories unavailable. Understanding and language are also lacking in infants. Infants are able to experience and interact with their surroundings, but they are only beginning to develop the ability to understand those experiences. Memories are at their best when the meaning of events instead of the raw perceptual details can be recalled. Because infants are unable to attach meanings to many of the events they experience, they are unlikely to be able to recall those experiences.

Children as eyewitnesses. Recent awareness of child abuse has focused attention on the reliability of children as witnesses. Increasing concern over the prevalence of abuse and a willingness to prosecute the perpetrators has resulted in an increasing number of children being called upon to testify in court. In the United States prior to the 1980s, many states required corroborating evidence before the testimony of children would be allowed in court. Later, this requirement was abandoned, making the testimony of children even more important. There is good evidence that the testimony of children can be useful to the courts. However, children are more easily misled than adults, and sometimes confuse fact and fantasy.

Hundreds of studies have shown that adults can also be misled. For example, subjects are shown an event such as a car running a stop sign and hitting a pedestrian. Some subjects then are given reading materials with misleading information about the event, for example, reference to a yield sign instead of the stop sign that was in the visual presentation. When compared with subjects who did not read the misleading information, these subjects are more likely to say they saw a yield sign instead of a stop sign. Studies using similar methods have found that young children are more susceptible to misleading information than adults. When forming memories of events, children often focus on the perceptual features (the details) of the events rather than the gist or meaning of the event. Because memories for perceptual features are weaker than memories based on meaning, children may forget the original information (the stop sign) and therefore be more willing to accept the misleading information (the yield sign) than adults.

Fact and fantasy. Source monitoring refers to the ability to attribute a memory to a particular source. Recent studies suggest that information about the source of memories is stored separately from the memory. For this reason, individuals do not always remember the source of their memories, and so they often confuse memories from different sources. For example, someone who sees a stop sign and then reads about a yield sign may remember both bits of information but may not remember which sign was read about and which sign was seen. A particular kind of source monitoring, one that is particularly difficult for children, is reality monitoring. Reality monitoring refers to the ability to distinguish memories for real experiences from memories for things that are imagined. For example, 6- and 9-year-olds were asked to say or imagine saying a set of words. Six-year-olds had a harder time than the 9-year-olds remembering which words they had imagined saying and which words they actually had said. In another study, children either watched actors performing actions (such as touching their nose) or imagined the actors performing the actions. When compared with adults, children were much less accurate at recalling which actions they had seen and which actions they had imagined.

Even children who talk as though they can distinguish reality from fantasy often act as though they cannot. For example, children were more likely to investigate a box in which they imagined a friendly creature such as a rabbit than a box in which they imagined a monster. Also, children who were asked to imagine a monster in a black box said they knew there was no monster, but they refused to be left alone with the box. Even though the children knew the boxes were empty, their act of imagination had created an available image of the creature. Having a mentally available image increases the subjective likelihood that the events are real.

Adult memories of childhood. Sigmund Freud believed that humans store all experiences but that some memories are repressed so that there is no longer conscious memory of them. According to Freudian theory, repression occurs when a memory is pushed from consciousness because it is too discomforting. However, these repressed memories can be influential, and hidden memories from childhood can become disruptive many years later. The way to control the disruptive influence of repressed childhood memories is to bring them back to consciousness and confront the conflicts that are inherent in them. Some psychotherapists have adopted this position, and attempt to help their clients remember events that they have no conscious knowledge of and that are alleged to have happened years or decades earlier. However, there is no solid scientific support for the repression folklore, and there is increasing evidence that attempting to dig out such memories can lead to the creation of false memories.

In one demonstration, a family member told participants of a memory about a time when a particular participant was 5 years old and was lost in a shopping mall and ultimately rescued. On the basis of this suggestion, many other participants created a false memory for being lost and rescued. Similar procedures have been used to induce people to create memories for such things as going to the hospital for a suspected ear infection, knocking over a punch bowl at a wedding, and getting their hand caught in a mousetrap. False memories created through suggestion are exceedingly difficult to tell from real memories.

For background information SEE AMNESIA; CONSCIOUSNESS; MEMORY in the McGraw-Hill Encyclopedia of Science & Technology.

Charles G. Manning; Elizabeth F. Loftus

Bibliography. S. J. Ceci and M. Bruck, Suggestibility of the child witness: A historical review and synthesis, *Psychol. Bull.*, 113:403–439, 1993; E. F. Loftus and K. Ketcham, *The Myth of Repressed Memory*, 1994; A. Searleman and D. Herrmann, *Memory for a Broader Perspective*, 1994.

Metal coatings

Thermal spray processes are utilized to apply protective coatings for the prevention of wear, oxidation, corrosion, and adverse effects of high temperatures. Thermal spray has been used since the late 1950s to apply ceramic coatings for aerospace and other advanced applications where monolithic ceramics were either too costly or technically impractical. The use of thermally sprayed ceramics is increasing, with applications in net-shape manufacturing, printing, paper manufacturing, aircraft engines, orthopedic implants, electronics, and other areas.

Thermal spray. This broad term refers to a group of processes that utilize a heat source and atomization jets to melt and propel materials toward a prepared substrate. Thermal spray processes are typically classified under the subsets of flame spray, electric-arc spray, and plasma-arc spray. Each of these subsets is divided into a multitude of variants.

Metallizing, or wire flame spray, is the oldest form of thermal spray, in use for more than 80 years. The invention of the plasma-arc spray in the late 1950s, driven largely by aerospace applications, brought thermal spray out of the garage-type job shops and into advanced-materials processing and coatings applications. Plasma-arc spray is the most flexible of thermal spray processes in terms of materials that can be sprayed; and it produces the highest-quality coatings. Electric-arc spray followed in the early 1960s, bringing high-quality, high-rate spraying of metals to thermal spray. Not until the early 1990s was electric-arc spray accepted as an alternative to plasma-spray coatings.

Materials considered to be sprayable are those with a definite melting point, that is, materials that do not dissociate, sublime, or the like. Thus, most metals, metal alloys, intermetallics, cermets, polymers, and all forms of ceramics, including oxides, carbides, nitrides, and silicides, are sprayable.

Devices. The name of each device (gun) used for thermal spraying is derived from the type of heat source used to melt the feedstock material. The flame spray gun (**Fig. 1**) utilizes an oxygen and fuel gas mixture (oxy-fuel), typically oxy-acetylene, to melt feedstock in the form of powder, wire, or rod. With the electric-arc spray gun (**Fig. 2**), an arc is drawn between two feedstock wires to cause melting. Compressed air is used in most flame and electric-arc guns to atomize and propel the feedstock.

Plasma is a highly energized, ionized gas stream used to melt powdered feedstock. Arc temperatures are 20,000 K (35,000°F) within the gun and 10,000 K (18,000°F) at 1 in. (25 mm) from the exit. Plasma gases expand rapidly because of heating and are accelerated through the nozzle of the plasma-arc spray gun (**Fig. 3**), propelling the molten powder onto the substrate or part to be coated. The high temperatures associated with plasma are well suited for spraying refractory materials such as ceramics.

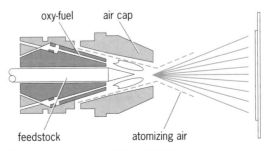

Fig. 1. Components of a flame spray gun.

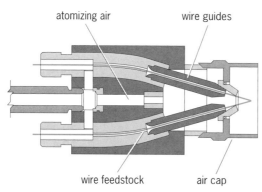

Fig. 2. Components of an electric-arc spray gun.

The mainstay of plasma-arc spray is atmospheric (air) plasma spray. In all of these thermal spray processes, feeding mechanisms maintain a continuous supply of material to the heat source.

Stabilized zirconia. Yttria-stabilized zirconia represents the bulk of all sprayed ceramics. This material is used primarily for thermal barrier coatings in aircraft, rocket, and reciprocating engines. Thermal barrier coatings are applied to engine components to lower substrate temperatures so that combustion gas temperatures can be higher, thereby increasing engine power and efficiency and lowering emissions. Stabilized zirconia is unique for its high coefficient of thermal expansion and low thermal conductivity. The high coefficient of thermal expansion correlates well with the base metals to which zirconia is commonly applied, reducing stresses that are induced by differential expansion. Plasma-sprayed, yttria-stabilized zirconia is also reasonably resistant to thermal fatigue and chemical attack.

Other elements used to stabilize zirconia include the oxides of magnesium, calcium, and cerium. Phase stabilization is used to mitigate the large volume change which zirconia undergoes during heating to and cooling from service temperatures. The phase transformation from low-temperature monoclinic to high-temperature tetragonal can be arrested by the inclusion of stabilizing components such as yttria. Fully stabilized zirconia maintains a cubic structure throughout heating. Partially stabilized zirconia, which has both cubic and tetragonal phases, is reported to be tougher and to have a better match of coefficient of thermal expansion with engine materials.

MCRALYs. Metallic coatings are used between the ceramic coating and substrate both to enhance bonding and to provide a barrier that prevents substrate oxidation and corrosion. As a class, these materials are denoted by the term MCRALY, which is derived from the components: a base metal (M), chromium (CR), aluminum (AL), and yttria (Y). The M component is iron (FE), cobalt (CO), or nickel (NI), singly or in combination. The coatings are then called FECRALY, COCRALY, NICOCRALY, and so forth.

Coating techniques. Thermal spraying as a technology is characterized by considerable flexibility in materials choices, high deposition rates, high thickness capability, low capital investment, and minimal environmental impact. Across the spectrum of application techniques, characteristics such as microstructure, porosity, grain size, and chemistry are controllable.

Recent advances in thermal spraying have focused on controls. Process control includes both feedstock materials and processing parameters. Historically, thermal spray coatings have been applied at a confidence level of around 70%. The confidence level is a composite of errors imposed by current aerospace materials specifications (currently the industry standard), gun-manufacturing tolerances, processing parameter control, and other variables. The thermal spray industry is evolving, driven by influences outside its historical moorings in aerospace applications. Automotive, medical, biomedical, electronic, and other emerging applications are demanding a higher level of confidence, of an order greater than 99%. These industries produce products in millions of units, whereas more conventional thermal spray applications have been limited to a few thousand. Where coatings are used for critical service in the larger markets, reliability and liability are of greater concern. In response to these new demands, adaptive process control is being applied to thermal spray processing. Conventional methodology utilizes control of process inputs; adaptive control derives data from process outputs that are then used as the point of control. This approach, when coupled with recent advances in instrumentation and attention to feedstock specification, makes attainable the goal of greater than 99% reliability.

Applications. Common use of thermally sprayed yttria-stabilized zirconia involves net-shape manufacturing. Oxygen sensors for automobile emission control systems are manufactured by applying the coatings of yttria-stabilized zirconia to removable mandrels. When the mandrel is removed, a freestanding shape is left.

Net- and near-net-shape techniques facilitate the fabrication of parts that is not practicable by other means. Freestanding net-shape ion engines have been manufactured from tungsten by using inert-chamber, plasma-arc spray. Plasma spraying in an argon atmosphere eliminates oxidation of reactive materials, such as tungsten. Precision alumina tubes [0.03 in. (0.75 mm) wall thickness, 3 in. (75 mm) diameter, and 48 in. (1.2 m) length] would be difficult, if not impossible, to fabricate by casting and grinding; however, they have been successfully fabricated by thermal-spray net-shape techniques. To the other extreme, multilayer ceramic tubes with 0.04-in. (1-mm) inside diameter have been made to join blood vessels.

Ceramic coatings are applied to medical instruments used for endoscopic and other forms of minimal invasive surgery. Plasma spray is used to apply aluminum titanate and other materials to absorb

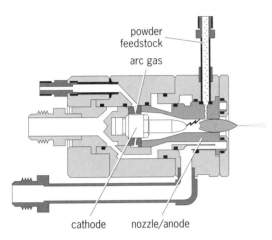

powder
feedstock

arc gas

cathode   nozzle/anode

Fig. 3. Components of a plasma-arc spray gun.

carbon dioxide laser light, which is used for cutting in many surgical procedures. The function of the coating is to prevent reflection of the laser light. Alumina compounds are also used for dielectric coatings on bipolar scissors, which simultaneously cut and cauterize tissue, facilitating near-bloodless surgery.

Plasma-sprayed chromia is now in common use as a coating for anilox (engraved) printing rolls. The coating is laser-engraved with a pattern of inverted pyramidal pockets (cells), in a range of 400–1200 cells per lineal inch. The engraved roll is used to pick up and transfer ink. Chromia is used for its ink-release properties and resistance to wear from the doctor blade.

Hydroxyapatite and porous titanium are used as coatings on implant prostheses to enhance bone attachment. Hydroxyapatite is a hydroxyl form of calcium phosphate, an essentially artificial bone matter. Since hydroxyapatite resembles bone matter, newly forming bone cells attach to and react favorably with the implant, speeding the healing and recovery process. Porous titanium improves bone attachment by providing greater surface area to which newly forming cells can attach. Bone ingrowth into the porosity helps provide increased mechanical strength to the bone-implant interface. Both materials are typically applied by plasma spraying. Titanium is sprayed within an inert chamber to prevent oxidation. Tooth implants, hip joints, and knee joints are typical examples of coated prostheses.

Automotive applications are one of the rapid growth areas for thermal spray. Piston rings, transmission syncro rings (parts of transmission-shifting mechanisms), body sail panels, and many other drive-train and chassis components are coated by using thermal spray. Air plasma spray is used to apply a wear-resistant coating of molybdenum/nickel-chromium-boron to piston rings. Molybdenum oxide, a solid-state lubricant, forms in service and reduces friction. Elemental molybdenum is also used on synchro rings for the same reasons. The body panel formed by the joining of the roof line to

the rear quarter panel is the sail panel. Historically the joint was filled with molten lead to cover the weld seam. Because of governmental regulations, the use of lead is no longer permitted. Electric-arc-sprayed silicon bronze is applied to the weld seam, thus reducing health and environmental concerns and increasing productivity.

For background information SEE *ADAPTIVE CONTROL; CRYSTAL STRUCTURE; METAL COATINGS; PROCESS CONTROL* in the McGraw-Hill Encyclopedia of Science & Technology.

Daryl E. Crawmer

Metal injection molding

Injection molding is one of the most productive techniques for shaping materials. Until recently, injection molding was restricted to thermoplastic polymers (polymers that melt on heating). However, metals have property advantages over polymers. They are stronger, stiffer, are electrically and thermally conductive, can be magnetic, and are more wear and heat resistant. The concept of metal injection molding combines metal powders and a thermoplastic binder to allow shaping of complex objects in the high-productivity manufacturing setting associated with injection molding. After shaping, the metal powders are sintered to densities nearly close to those listed in handbooks; that is, they have very few pores. Accordingly, the process delivers materials with metallic properties in an efficient manner.

Derivation from powder metallurgy. Metal injection molding, and the related process of ceramic injection molding, is a derivative of powder metallurgy. Metal powders can be shaped in a semifluid state (powders that are poured into containers take on the shape of the container), but after heating to high temperatures the particles bond into a strong, coherent mass. In ceramics this is analogous to the manufacture of clay pottery, where shaping occurs with a water-clay system at the potter's wheel; but after kiln firing, the structure is strong and rigid. For metal injection molding, a fine metallic powder is mixed with a wax-polymer binder that allows shaping in a thermoplastic injection molding machine. After shaping, the wax-polymer binder is removed, usually by heat, and then the structure is sintered in a manner similar to traditional powder metallurgy or ceramics firing.

A key aspect of the process is identifying a fine powder, whose particles are typically 20 micrometers or smaller (less than 0.001 in.) in diameter, that can be sintered. Binders are used to shape these fine powders and to hold them in place until they are bonded in a sintering furnace. The overall process is similar to other powder-binder forming processes such as tape casting, slip casting, pressure filtration, transfer molding, extrusion, and freeze firing. Although the technology is relatively new and was widely commercialized only in the late

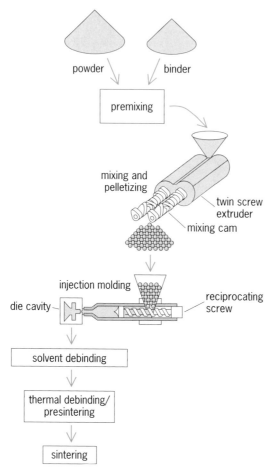

Metal injection molding process.

1980s, products made by using metal injection molding are now in widespread use. Some examples include orthodontic brackets for straightening teeth, porous filters for hot waste water, computer disk drive magnets, small gears for electric hand tools and toothbrushes, sport shoe cleats, surgical tools such as scalpels, electrical connectors, handgun components, and microwave filters for high-frequency computer chips.

Basic process. Plastic products formed by injection molding are in widespread use because of their low cost and high shape complexity. Metal injection molding builds on that shaping flexibility, but relies on a high metal particle content in the plastic to form a metallic shape. The process is outlined in the **illustration**. It begins by mixing selected powders and binders. The powders are fine in order to facilitate sintering, often with particles between 0.1 and 20 μm and a nearly spherical shape. One of the most commonly used powders is formed by chemical vapor decomposition of an iron pentacarbonyl molecule. The particles in this carbonyl iron powder are usually about 5 μm in size. The binders are thermoplastic mixtures of waxes, polymers, oils, lubricants, and surfactants.

The metal injection molding process is practiced with many variations, yet basically they all are similar. Fine powders are used to aid sintering densifi-

cation. Many of the most popular alloys include steels, iron-nickel, low-alloy steels, tool steels, and stainless steels; pure iron is also used. A favorite binder system consists of 65% paraffin wax, 30% polypropylene or polyethylene, and 5% stearic acid. This binder melts at about 150°C (300°F). The amount of binder is about 40% by volume of the mixture; for a steel-binder mix, that corresponds to about 6% by weight of binder.

It is desirable to attain a high relative content of metal particles in the powder-binder mixture, while maintaining a low mixture viscosity to aid in shaping. Sufficient binder is needed to fill all voids between particles and to lubricate the particles during molding. The most successful molding is attained with mixtures whose viscosities are similar to that of toothpaste. That viscosity depends on the inherent binder composition, as well as the mixture temperature, shear rate, solids content, and type of surface wetting agents (such as stearic acid) included in the binder. At too high a solid-to-binder ratio there is insufficient binder to fill all void space between the particles, and the feedstock is impossible to mold.

After the powder-binder mixture is completely combined, it is injection-molded into the desired shape by heating the mixture and, when hot, feeding it under pressure into a die cavity. The liquid polymer makes the mixture sufficiently fluid so that it can be shaped very rapidly. Cooling channels in the die cause the polymer to solidify, thereby preserving the molded shape.

Molding equipment. The equipment used for shaping the compact is the same as is used for polymer injection molding. It consists of a set of clamped dies that is filled through a gate from a heated barrel. A motor-driven rotating screw stirs the powder-binder feedstock to maintain a homogeneous mixture, and generates the pressure needed to fill the dies. The molten powder-binder mixture is forced forward to fill the cold dies in less than a second. Molding pressures are dependent on the die geometry, binder, and powder characteristics, and can be as high as 60 megapascals (9000 lb/in.^{2}). Pressure is maintained on the feedstock during cooling to minimize the formation of shrinkage voids. After cooling, the dies open, the compact is ejected, and the cycle is repeated.

Debinding. This molded part is used directly as molded for certain applications, including small magnets and frangible bullets. However, in most instances the binder is removed from the molded compact by debinding, which is done in several ways, including thermal, solvent, and capillary debinding. Thermal debinding is used frequently; here the compact is heated slowly to 600°C (1112°F) in air to decompose the binder. Simultaneous oxidation of the powder provides handling strength to the compact. An alternative is to immerse the compact in a solvent that dissolves one of the binder constituents, leaving some polymer behind to hold the particles in place.

Sintering. The next step is sintering, which can be incorporated directly onto a thermal debinding cycle. Sintering provides strong interparticle bonds and densifies the structure. Isotropic powder packing allows for predictable and uniform shrinkage, so the original compact is oversized as appropriate for the final compact dimensions. Sintering is usually performed in a protective atmosphere (nitrogen and hydrogen) or in a vacuum. Usually, sintering is done at a temperature high enough so that the space originally occupied by the binder is eliminated by compact shrinkage. For steels and stainless steels these temperatures are often in the 2050–2640°F (1120–1350°C) range for periods of 30–120 min. Densification makes the sintered compact substantially smaller than the initial die cavity, often by about 12–18%. To compensate for this shrinkage, the die cavity is oversized, and in this regard computer-aided design is a powerful tool in making possible the desired final shape with specified dimensions.

After sintering, the resulting compact has a uniform microstructure and is quite strong; it often exhibits mechanical properties superior to those obtainable by other processes. The compact can be heat treated as required by the design specifications.

Typical product. A typical component fabricated by metal injection molding is a trigger guard for a shotgun; the guard is the curved piece that surrounds the trigger on the underside of the barrel. The guard is fabricated from a low-alloy iron-nickel steel, usually with a final weight of 40 g (1.6 oz). To make the alloy, a mixture of powders containing particles of 5-μm-diameter carbonyl iron and 8-μm-diameter carbonyl nickel is used. The mixture is combined with wax and polyethylene to form feedstock that can be molded at 58% solid by volume. During molding, the nozzle temperature in the molding machine is 175°C (347°F) with a die temperature of 40°C (100°F). The maximum pressure applied during mold filling is 20 MPa (3000 lb/in.^{2}), with a pressure of 8 MPa (1200 lb/in.^{2}) sustained during mold cooling. In fabricating the trigger guard, the mold filling time is relatively rapid, about 0.5 s, but the mold cooling time is 18 s; the total cycle time is 37 s between parts.

Applications. Metal injection molding is applied in the production of high-performance components that have complex shapes. The most successful applications involve components that have complex shapes and also must be inexpensive to produce while providing high relative performance. Generally, metal injection molding can be used for all shapes that can be formed by plastic injection molding, especially for very small complex geometries. Improved productivity in metal injection molding requires multicavity dies, which can be quite large. In a few operations, up to 40 cavities can be accommodated in a single mold set. Most optimization software for injection molding compares the cost of machining multiple cavities with increased production rates in order to determine which combination of machining and production costs gives the lowest final part price.

Many applications have been developed for metal injection molding. They include components for use in diverse fields ranging from surgical tools to microelectronic packaging. Recently, automotive components such as sensor parts in the air-bag actuator mechanism have moved into production. The technology has been extended to utilize a wide range of materials, including steels, stainless steels, nickel, copper, cobalt alloys, tungsten alloys, niobium alloys, nickel-base superalloys, and intermetallics. Additionally, most ceramics can be processed in the same manner; these include silica, alumina, zirconia, silicon nitride, silicon carbide, aluminum nitride, cemented carbides, and various electronic ceramics.

Although metal injection molding is generally viable for all shapes that can be formed by plastic injection molding, it is not cost competitive for relatively simple or asymmetric geometries. Large components require larger molding and sintering devices that are more difficult to control. Accordingly, metal injection molding is largely applied to smaller shapes with masses below 100 g (4 oz).

The formation of thick cross sections remains a problem with metal injection molding because binder removal times vary with the square of the section thickness. Although there is no upper limit on section thickness, most are less than 25 mm (1 in. thick), and the maximum lateral dimension is less than 100 mm (4 in.). Sections as thin as 0.25 mm (0.01 in.) have been formed by metal injection molding, but they create new problems with rapid cooling in the die. Dimensional tolerances usually range from ±0.1 to 0.3% of section thickness, and in a few instances ±0.05% is attainable.

As the engineering community better appreciates its favorable attributes, metal injection molding applications will grow. Although the current industry is relatively small, the anticipated growth is impressive. One area of keen interest is in the processing of titanium alloys for biomedical applications. Another is in the co-molding of different materials, that is, forming part of a component from one material and then another portion from a second material. This option has merit for forming corrosion barriers, wear surfaces, and electrical interconnections in ceramics. Other major growth areas include high-performance magnets, technical ceramics, materials associated with heat dissipation in electronic circuits, hard materials, ultra high strength metals, and biocompatible materials.

For background information SEE MATERIALS SCIENCE AND ENGINEERING; PLASTICS PROCESSING; POWDER METALLURGY; SINTERING in the McGraw-Hill Encyclopedia of Science & Technology.

Randall M. German

Bibliography. P. H. Booker, J. Gaspervich, and R. M. German (eds.), *Powder Injection Molding Symposium—1992,* 1992; R. M. German, *Powder Injection Molding,* 1990.

Microbial mining

The use of bacterial processes in mining of mineral ores—specifically for copper, gold, and uranium—has become commercially competitive in recent years. Microbial mining is expanding at a rapid rate worldwide. Such rapid growth precludes an accurate estimate of the ultimate limitation of this technology, but assuredly, microbial mining has become commonplace.

It is estimated that more than 25% of current world copper production is from bacterial acid leaching of ores too low grade for physical extraction processes to be economical. Although bioleaching of copper on a commercial scale occurred at the Rio Tinto mine in Spain in the 1700s, its mechanism was not recognized until the 1950s. The bacteria oxidize sulfur in the ore to sulfuric acid, which then releases the mineral cations. Bacterial metabolism has also been used to extract soluble uranium(VI) from uranium ores. However, the scale of global uranium mining has been radically reduced in recent years. In contrast, microbial bioextraction of gold from gold-containing arsenopyrite ores is being used on an increasing scale in North and South America, Africa, and Australia. Microbes are also useful in bioremediation (metabolism) of organic and inorganic pollutants that are generated by the mining processes.

Mineral mining occurs in an environment seemingly inhospitable for living organisms; yet here bacteria grow and thrive. Through their metabolic activities, bacteria extract valuable minerals in production processes that are less expensive, less energy intensive, and more environmentally favorable than traditional physical and chemical processes.

Copper. Acid solubilization of copper from sulfidic minerals occurs naturally. It was recognized in Roman times, but leaching was first used on a commercial scale in eighteenth-century Spain. Only in the late 1950s was this process recognized as microbial in origin, and the bacteria responsible isolated and characterized. The primary bacterial species involved is *Thiobacillus ferrooxidans,* although other bacterial species (*T. thiooxidans* and *Leptospirillum*) are also involved. Because of the huge amounts of ore processed and the relatively low value of copper, heap leaching is the preferred method. Two related processes are dump leaching, which involves piling of fractured low-grade ore in huge piles (up to 1.1×10^9 tons or 10^9 metric tons) along slopes or valleys; and heap leaching, in which smaller piles (sometimes 1.1×10^5 tons or 1×10^5 metric tons, with pipes laid in the heaps during construction for aeration) are layered on plastic-lined concrete foundations. In dump leaching, the ore piles can reach tens of meters in height, and the time for complete recovery of copper can be 20 years; in heap leaching, the ore is piled a few meters high, and the heap is exhausted of recoverable copper in months.

In heap leaching, highly acidic water is pumped onto the top of the pile, where it percolates down, stimulating bacterial growth. The water exits at the bottom after approximately 2 weeks. Copper is removed from the leachate by electrolytic precipitation (a purely physical-chemical process). The waste fluid contains a bacterial inoculant, needed nutrients, and a favorable acidity. Therefore, after copper removal, the runoff waste is recycled by spraying onto the surface of the heap.

Gold. Until recently, the practice of microbial gold mining was only a theory. However, there are currently six large microbiological facilities on three continents, which together process perhaps 20% of the world's gold harvest from arsenopyrite ores. The first gold bioleaching factory was at the Fairview Mine in South Africa, where in 1986 a bacterial-based biooxidation process was developed, which involved a specific mine-adapted mixed bacterial culture. The biooxidation plant design includes a mixing tank in which finely pulverized gold ore is mixed with the bacterial seed culture and needed nutrients. The mixture is fed into primary aeration tanks, air is pumped in to provide both oxygen and carbon dioxide which are required for growth, and bioleaching begins. After a period of stirring and aeration, the slurry of mineralized ore, bacteria, and growth medium is transferred into secondary aeration tanks, where bioleaching continues until more than 70% of the sulfur in the ore has been solubilized, releasing more than 90% of the embedded gold. Less than 25% of the total ore mass is solubilized. The leached material is transferred to a settling tank, where physical processes separate the solubilized gold from the waste ore mass; a thickening process removes the ore. The addition of massive amounts of cyanide complexes the gold, retaining it in solution. The liquid and solid masses are separated, and gold is recovered from the cyanide complex by adsorption onto activated carbon. The carbon is recycled, but the spent cyanide is discarded.

About 30% of the world's gold reserves are high-sulfur- and high-arsenic-containing arsenopyrite ores, and thus candidates for bioleaching. The remaining 70% of gold ores are of different composition, often with the gold bound in carbonaceous complexes. Development of bacterial bioleaching of carbonaceous gold ores using heap-leaching technology has begun at the Newmont Mine site in Utah.

The Fairview Mine in South Africa currently operates on a scale of 39 tons (35 metric tons) of gold-bearing flotation concentrate per day. Four additional factories using the bioleaching process are now operating: the São Bento (beginning in 1990) in Brazil with a capacity of 165 tons (150 metric tons) per day; Harbour Lights (1992) in West Australia with a capacity of 44 tons (40 metric tons) per day; the Wiluna Mine (1993) in West Australia with a capacity of 127 tons (115 metric tons) per day; and the Ashanti Mine (1994) in Ghana with a capacity of 792 tons (720 metric tons) per day. A fifth factory uses a different but related process whereby the bacterial culture is adapted to the high-temperature conditions (sometimes 122°F or 50°C) of the West Australian summer, and minimizes the cost of cooling, which is substantial for bioleaching. The first commercial use of the high-temperature bioleaching process was at the Youanmi Mine in West Australia, which started production in 1994, with a capacity of 132 tons (120 metric tons) per day. Different processes for bioleaching of gold must be adapted to local ore compositions and environmental settings, because the requirements for low- or high-sulfur-containing arsenopyrite ores are quite different, as are those for carbonaceous ores.

Uranium. The use of microbial mining processes to extract soluble uranium(VI) from uranium ores was pioneered in northern Ontario, Canada, where an obvious advantage of such processes is the ability to work the subsurface where a more stable temperature can be maintained in the winter. Also, it appears economically cheaper and environmentally safer to pump water underground to subsurface ore that has been reduced to fine powder, and then to pump the water containing extracted minerals to the surface, rather than to move large volumes of ore to the surface for physical extraction. In northern Ontario, underground caverns were excavated, with the ore just moved to the side. The ore was reduced to fine powder by explosions and piled to form heaps called stopes, running at an approximately 45° angle up the face of the cavern wall. The bottom of the stope was walled off with concrete, and water containing bacterial nutrients (phosphate is especially important) and sometimes pregrown bacterial inoculants was pumped into the stope. However, in this flood-leaching arrangement, aeration was a problem because *Thiobacillus* and most other bioleaching bacteria are strictly aerobic; availability of air might be rate limiting, especially for an underground process. After a period of time (3 days in one test), the leachate water was pumped to the mine surface and uranium (exclusively at the high oxidation state of U_3O_8) was precipitated chemically with ammonium. After the stope had rested for 3 weeks, water was reintroduced and the process repeated, until the yield of uranium was no longer economically viable. In one test stope, 57% of the available uranium had been removed by bioleaching within 30 weeks. Overall recovery of uranium ranged from 69 to 86%. At the Denison Mines in Ontario, the ore body contains about 0.6 lb U_3O_8 per ton (0.5 kg per metric ton). In 1988, 484 tons (440 metric tons) of U_3O_8 were produced, with about 25% coming from bioleaching. *Thiobacillus ferrooxidans* appears to be the primary bacterial species responsible for uranium bioleaching, with psychrophilic isolates being important at surface sites.

Bioaccumulation. In bioleaching processes, the dissolved minerals must be removed from the liquid for further processing. Copper is electroplated onto ferrous metal materials, mostly old car bodies, and subsequently removed by physical processing. Soluble gold, stabilized as cyanide complexes, is adsorbed onto activated charcoal. The gold is harvested with the carbon, which is burned to yield semipure gold. Bioaccumulation onto microbial biomass, developed for removal of low-level toxic heavy metals from industrial effluents, may be an alternative to charcoal for gold bioaccumulation. Although small companies have used both bacterial and algal biomasses as bioabsorbants for mineral wastes from industrial effluents, bioaccumulation has not been tested on large-scale mining effluents.

Bioremediation. Two major uses of microbes for bioremediation of mining-associated organic wastes are the biodegradation of oil and petroleum spills (which are associated with all large-scale industrial processes) and the decomposition of waste cyanide after removal of soluble gold. This bioremediation technology has advanced to a practical scale. The best-characterized use of bacteria in cleaning up cyanide waste is at the Homestake Gold Mine in Lead, South Dakota. After absorption of the gold from gold-cyanide complex, the Homestake mine is left with a river of approximately millimolar cyanide. The cyanide-containing waters are passed over a series of rotating cylindrical plastic disks, each about 6.6 ft (2 m) in diameter and mounted on conventional sewage-system rotators. Bacterial growth on the disk surfaces forms masses of 20 tons (18 metric tons) per assembly and degrades the incoming cyanide to carbon dioxide and water. After processing, the effluent contains approximately 1 micromolar cyanide, low enough for discharge into a stream where trout thrive. The dominant microbe in the cyanide detoxification facility is a *Pseudomonas* species. However, this cyanide biodegradation process has not been adapted to other mine sites, which often have less abundant subsurface water.

For background information SEE BACTERIAL PHYSIOLOGY AND METABOLISM; BIOLEACHING; PSEUDOMONAS in the McGraw-Hill Encyclopedia of Science & Technology.

Simon Silver; Emilio Rodriguez

Bibliography. Australian Mineral Foundation, *Biomine '94: Proceedings of the International Conference and Workshop on Applications of Biotechnology to the Minerals Industry,* Perth, West Australia, 1994; J. Barrett et al., *Metal Extraction by Bacterial Oxidation of Minerals,* 1993; H. L. Ehrlich and C. L. Brierley (eds.), *Microbial Mineral Recovery,* 1990; D. E. Rawlings and S. Silver, Mining with microbes, *Bio/technology,* 13(8):773–778, 1995.

Microdialysis sampling

The use of microdialysis is a new approach to sampling biological systems. It provides the potential to sample the extracellular space of essentially any tissue or fluid compartment in the body. Continuous sampling can be performed for long periods with minimal perturbation to the experimental animal. Microdialysis provides a route to sample the extracellular fluid without removing fluid and to administer compounds without adding fluid. It provides a sample that is clean and amenable to direct analysis.

Microdialysis was initially developed to study neurochemical processes in the brain. The success of this technique in the study of neurotransmitter release has led to the development of microdialysis techniques for general pharmacokinetic and drug distribution studies.

Sampling system. Microdialysis sampling is performed by implanting a short length of hollow-fiber dialysis membrane at the site of interest. The fiber is slowly perfused with a sampling solution (the perfusate) whose ionic composition and pH closely match the extracellular fluid of the tissue being sampled. Low-molecular-weight compounds in the extracellular fluid diffuse into the fiber lumen, where they are swept to a collection vial for subsequent analysis. The system is analogous to an artificial blood vessel that can deliver compounds and remove the resulting metabolites. A microdialysis system for animals that are awake consists of a high-precision microinfusion pump connected to the microdialysis probe through a low-volume liquid swivel (a device that allows connection to be maintained without the tubing being twisted as the experimental animal moves in its container). The outlet of the microdialysis probe is then brought back through a second channel of the liquid swivel to a fraction collector.

The process. Microdialysis is a diffusion-controlled process. The perfusion rate through the probe is generally in the range of 0.5–5.0 microliters/min. At this rate there is no net flow of liquid across the dialysis membrane. The driving force for mass transport is then the concentration gradient existing between the extracellular fluid and the fluid in the probe lumen. If the concentration of a compound is higher in the extracellular fluid, some fraction will diffuse into the probe. This is known as a recovery experiment. Conversely, if the concentration of a compound is higher in the probe lumen, some fraction will diffuse out of the probe. This is known as a delivery experiment.

At typical perfusion rates, equilibrium is not established across the microdialysis membrane. The concentration of analyte collected in the dialyzate (the perfusate after it has gone through the microdialysis probe) is some fraction of the actual concentration in the extracellular fluid. The relationship between the fraction recovered and the actual concentration is known as recovery, and it is defined as the ratio of the concentration of analyte in the dialyzate to the true concentration in the tissue of interest. The recovery depends on a number of characteristics of the probe, the sample matrix, and the analyte. Probe characteristics that affect recovery include the type of membrane, the length of membrane, the geometry of the probe, and the perfusion flow rate. The recovery is influenced by matrix characteristics such as temperature, tortuosity (degree of twisting), and effective volume.

The size, hydrophobicity, charge, and diffusion coefficient of the analyte affect recovery. Finally, metabolic and active transport processes affect recovery.

The concentration recovery of a compound is independent of its concentration. A number of parameters affect the recovery of a given compound; they include perfusion flow rate, sample flow rate, blood flow rate, extent of tissue vascularization, metabolism rate, uptake into cells, physiological control of the concentration, diffusion rate in the sample matrix, diffusion rate in the dialysis membrane, surface area of the dialysis membrane, analyte size, and analyte charge. Under normal conditions of microdialysis sampling, these parameters remain constant; thus, although equilibrium is not established, a steady state is rapidly achieved.

Probe design. The major design consideration for a microdialysis probe is to provide a short length of hollow-fiber dialysis membrane through which a solution can be pumped, and in a geometry that can be implanted into an animal with minimal tissue damage. Microdialysis membranes made of cellulose, cellulose acetate, polyacrylonitrile, and polycarbonate ether fibers are available. These membranes are all hydrophilic and provide little difference in chemical selectivity. Membranes with nominal molecular-mass cutoffs between 5000 and 75,000 daltons are available, although membranes in the range of 10–20 kilodaltons are most commonly used. The dialysis efficiency decreases dramatically as the molecular mass of a compound increases. Compounds much above 1000 daltons typically have very low efficiencies and are not usually appropriate for microdialysis sampling. A variety of probe designs have been described that provide advantages for use in specific tissues. These designs can be divided into two classes: cannula-style probes with parallel inlet and outlet flow, and linear-style probes in which the inlet, membrane, and outlet are in series.

The most common design is the concentric cannula type, which consists of an inner and outer length of stainless-steel tubing. The inner cannula extends beyond the outer cannula and is covered with the dialysis membrane. This design is most appropriate for stereotaxic implantation in the brain. Its rigid nature provides good mechanical stability and allows for precise placement. The probe can be cemented to the skull to prohibit movement of the probe when the animal moves. Alternatively, a guide cannula can be implanted and glued in place. A concentric cannula microdialysis probe can then be implanted through the guide cannula. Thus the probe can be removed and precisely reinserted or another probe can be used later.

Microdialysis probes constructed from rigid materials are not appropriate for implantation in peripheral tissue. Unlike probes implanted in the brain, where the skull provides an excellent site to secure the probe and protect it from movement, peripherally implanted probes must be capable of moving as the animal moves without being damaged or causing damage. A modification of the concentric cannula design in which side-by-side pieces of fused silica are used has been found to be useful for intravenous implantation. In this design, one piece of fused silica extends beyond the other and is covered with the dialysis membrane. This flexible design is still needlelike and thus can be inserted into blood vessels. However, the fused silica bends as the animal moves, so that neither the dialysis membrane nor the blood vessel is punctured.

The most useful probe design for implantation in peripheral tissue is of a linear geometry. Several variations of this design are in use, but the general concept is that the hollow-fiber dialysis membrane is connected to small-bore tubing on both ends. One end is used as the inlet and the other as the outlet. A variety of tubing types can be used for the inlet and outlet, including Teflon, polyetherether ketone (PEEK), polyethylene (PE), and fused silica. Implantation is accomplished simply by pulling the probe through the tissue like a thread. It is often useful to use a longer piece of the dialysis fiber than desired for sampling, because the fiber is more flexible than the fused silica. In this case, most of the fiber is coated with an impermeable silicone resin, leaving a small uncoated window for sampling.

A fourth type of microdialysis probe is the flow-through or shunt design. This type is constructed by inserting the microdialysis fiber inside a length of polyethylene tubing. The microdialysis fiber is perfused with the sampling solution while the fluid sample flows through the polyethylene tubing. This design is useful for sampling flowing fluids from sites that are too small for implantation of a flexible microdialysis probe. The polyethylene tubing acts as a shunt to bring the biological fluid past the microdialysis membrane. This probe design has been most successfully used to sample the bile, but application to blood sampling from arteries or smaller veins may also be possible.

Applications. The greatest use of microdialysis sampling has been in the neurosciences. Microdialysis probes can be implanted in specific brain regions of conscious animals in order to correlate neurochemical activity with behavior. Most studies have focused on determining dopamine or other monoamine neurotransmitters. Microdialysis sampling has also proven to be a powerful technique for studying excitatory amino acids, such as glutamate, aspartate, and gamma-aminobutyric acid (GABA) in the brain. Recently, the use of microdialysis to sample neuropeptides has been explored. Another application has been in studying abnormal brain function in humans. Data on the neurochemical processes occurring prior to, during, and after an epileptic seizure have been obtained by using microdialysis in humans.

The use of microdialysis in the pharmaceutical sciences is growing rapidly. Because microdialysis provides continuous sampling without disruption of biological barriers, the technique is particularly well suited for studying the bioavailability of pharmaceutical compounds. Microdialysis probes have been implanted in the skin of experimental animals and humans to determine the transdermal delivery of drugs from ointments. Delivery of anticancer drugs to tumors has also been studied. Microdialysis may prove a useful technique for delivering toxic drugs, such as anticancer agents, to specific sites without systemic involvement. Although continuous sampling from tissues is possible only by microdialysis, sampling from the blood has been the most common use in pharmacokinetic studies.

Microdialysis sampling has also been used to study the metabolism of compounds in organisms. Metabolic organs such as the liver and kidneys have been studied. By also sampling the bile, complete metabolic profiles can be obtained from a single experimental animal. This approach dramatically decreases the number of experimental animals needed to assess the metabolism of a new drug.

For background information SEE DIALYSIS; PHARMACEUTICALS TESTING in the McGraw-Hill Encyclopedia of Science & Technology.

C. E. Lunte

Bibliography. D. E. Johnson (ed.), *Drug Toxicodynamics*, 1995; C. E. Lunte, D. O. Scott, and P. T. Kissinger, Sampling living systems using microdialysis probes, *Anal. Chem.*, 63:773A–780A, 1991; T. E. Robinson and J. B. Justice, Jr. (eds.), *Microdialysis in the Neurosciences*, 1991.

Microengineering

Microengineering is the design and production of small, three-dimensional objects, usually for manufacture in high volumes at low cost. Such designs have a wide range of applications, including automobiles, medicine, aircraft, printing, security, and insurance. The techniques employed come from as wide a range of disciplines, including biochemical, chemical, electrical, electronic, fluidic, mechanical, and optical engineering. Present designs occupy a volume about 1 mm^3 (40 mils^3), and are made to tolerances of a few micrometers (about 50 microinches).

Applications. The best-known application is the air-bag sensor, which detects crashes and sets off an air bag. These sensors are produced by using standard semiconductor processing, most often a conventional CMOS (complementary metal-oxide-semiconductor) process. The sensor has an accelerometer to measure shock, and fires the air bag if more than 50 g (500 m/s^2 or 1600 ft/s^2) is experienced for at least 10 ms.

Figure 1 shows a design which incorporates self-testing. The sensor is the flap around the outside, which is held still by electrostatic drives. Each time the ignition is turned on, the large bar in the middle of the sensor is heated to produce the equivalent of the pressure resulting from a crash for 10 ms; if no warning is given then the driver is informed that the air bag would not work in a crash. The driver can still

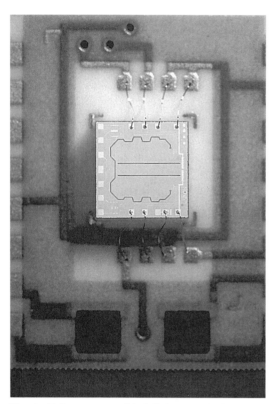

Fig. 1. Air-bag sensor with built-in auto-check (*Lucas-Novocasensor*)

drive off, but the warning is recorded and may be used as evidence in the event of a crash. The design must be sufficiently reliable to be accepted in a court of law. Microengineering is also used to provide automatic braking systems and intelligent suspension for smoother rides, and to optimize the flame profile in an engine to reduce fuel consumption.

LIGA process. This process uses synchrotron radiation to produce sharp cuts in thick-film photoresists, which may then be plated with plastics, metals, or other materials. The use of nickel produces strong metal parts in substantial quantities. In one design these are assembled to produce a very small motor (**Fig. 2***a*), which is suitable for use in microsurgery, watches, camcorders, and so forth. Components with high-precision requirements (the stator, rotor, and double-sided coils) are fabricated by means of microfabrication techniques, whereas the other parts are produced by precision mechanical methods (Fig. 2*b*).

Fiber clamp. Optical and other fibers are usually drawn in long lengths, and their terminations must be very precise. The fiber clamp holds fibers accurately in position over a length of 600 μm (25 mils), and the taper makes insertion easy to perform. It is machined directly from solid polycarbonate by using an excimer laser working through a reducing lens to make prototype quantities; when high-quantity production is needed a nickel mold is made.

Camera, watch, and communications industries. The use of aspheric lenses reduces the number of elements in a lens and makes telescopic lenses lighter and easier to use. However, such lenses are more difficult to make than spherical lenses and require very small motors to set them to the correct distance and aperture. Robots are used to assemble the fine parts used in the camera and watch industries.

Gyroscopes. Production gyroscopes whose smallest parts are 1 mm (40 mils) in size are made to an accuracy of 25 nanometers (1 microinch). They have been produced in the United Kingdom since the early 1960s, but were formerly security classified. Boron nitride, an unusual semiconductor, is used as the building material because it is easy to machine and provides a very fine surface. The overall gyroscope assembly is about 3 cm (1.2 in.) long. The central rotor has no imperfection in its crystalline structure over the whole length.

Intelligent fuses. These devices have been in production for use with weapons since the 1970s. They can withstand the shock due to protective armor and allow the shell to penetrate to the main body of a target before exploding. This equipment must be stored for many years and then work perfectly the first time it is used.

Water-quality monitoring. Up to 137 different properties of drinking water may need to be measured to ensure a high standard of water quality. The measuring units must be highly reliable because their failure could incur considerable legal penalties. The equipment may have to be situated where power supplies are unavailable and a small battery must be used. Also, the sensors can be polluted by the working environment, so self-cleaning is vital. The

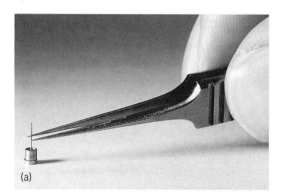

(a)

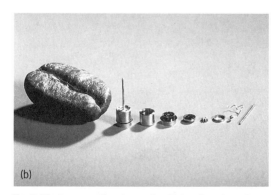

(b)

Fig. 2. Electromagnetic micromotor. Outer diameter = 2 mm (0.08 in.). (*a*) Motor being manipulated. (*b*) Components. (*IMM Institut für Microtechnik, GmbH*)

low power consumption and the self-cleaning require the use of microengineering.

Patient monitoring. This area provides many possible applications. Catheters a few hundred micrometers across can be made to pass within a vein to measure the properties of a patient's blood, including blood pressure, oxygen, carbon monoxide and carbon dioxide content, and pH.

Eye surgery. Microengineered motors are used for eye operations. Some of these are fluidic, with a harmless drive fluid turning the motor as well as keeping the eye lubricated.

Microsurgery. Within the body, minimally invasive surgery can require the use of microengineering. In remote prostate operations, surgical tools about 1 mm (40 mils) in diameter can carry out operations up to 30 cm (1 ft) into the body. Surgeons can use a remote scalpel to recognize cancerous material by its feel, exactly as they do with a hand-held scalpel.

Some observation techniques, such as magnetic resonance imaging, require remotely driven tools to be positioned where the image shows they are needed, to a very high degree of accuracy. These microengineered tools may have to be positioned in an unusual way, and are being developed. SEE MAGNETIC RESONANCE IMAGING.

Printing. Microengineering enables different textures and multicolors to be printed at very high speeds. Such printing can be used to enhance security in delicate situations. It can also be used to make art reproductions more realistic, which could increase the danger of forgery, and poses security problems.

Aviation. The robust nature of microengineered devices enables them to measure temperature and pressure on the leading edge of an aircraft wing or even within a jet engine. The packaging of these devices is very critical, and some work is being done on automatic self-sealing devices.

Production. In food production, artificial noses can smell the aromas of cooking and make sure that a properly finished product is produced, with no unpleasant taste. Perhaps the most common application of microengineering lies in production control in this and many other industries.

Other areas. In insurance, it is possible to give every item produced (and many older items of value) a unique identity that can be checked at any time. In the home, microengineering helps provide communication, interactive entertainment, and environmental controllers that act on humidity, smells, and so forth, as well as temperature.

Terminology. A wide variety of closely related terms are in use. In the United States, the term microelectronic and mechanical systems (MEMS) covers the electronic and mechanical parts of microengineering. In Japan, the term micromachines is used to cover a small specific subset that includes motors and robots. In Europe, the terms microsystems and microsystems technology cover systems and the technologies used to make them, but usually imply the use of semiconductorlike processing; they

also exclude the many small devices which operate independently.

The term nanoengineering is usually used to describe even smaller devices, nominally 1000 times smaller, which are usually much more expensive and are not normally mass produced. The term mechatronics covers the use of mechanical and electronic engineering irrespective of size. SEE MECHATRONICS.

For background information SEE INTEGRATED CIRCUITS; LASER; MICROMANIPULATION; OPTICAL SURFACES; SURGERY; SYNCHROTRON RADIATION in the McGraw-Hill Encyclopedia of Science & Technology.

Howard Dorey

Bibliography. A. D. Feinerman and S. R. Thodati, A millimeter-scale actuator with fibre optic roller bearings, *IEEE J. Micromech. Sys.*, 4(1):28–33, March 1995; F. Goodenough, Airbags boom when IC accelerometer sees 50 g, *Electron. Des.*, pp. 45–56, August 8, 1991; W. Qu, MST in China, *MST News*, 1(8):8–9, February 1994; G. Stemme, The progress of the Royal Institute of Technology on microengineering in Stockholm, Sweden, *MST News*, 1(12):18–19, April 1995; E. Yamamoto, Optical tactile sensor for medical application, *Micromachine*, 1(10):7, March 1995.

Microprocessor

With each new generation of microprocessors, the number of transistors increases as minimum feature size decreases, allowing for more complex architectures to be integrated on one chip. Supply voltage decreases in order to meet reliability and power requirements, and frequency increases to meet performance requirements. If advances continue at the same rate as they have since 1985, then by the year 2000 microprocessors in desktop computers should contain more than 20 million transistors, operate with a supply voltage of 1.8 V and at a frequency of 1 GHz, consume 20 W of power, and be fabricated by an inexpensive complementary metal-oxide-semiconductor (CMOS) process.

The microprocessor is the key component in the effort to develop more powerful personal computers. The performance of a microprocessor is measured by the clock frequency and the number of instructions per clock cycle that can be executed. In contemporary microprocessors, instructions are broken into smaller tasks which take several clock cycles to complete, and are pipelined to allow more instructions to be processed every clock cycle. Multiple units are added to be able to execute instructions in parallel. Memory size and speed are increased to match the performance of the microprocessor. Because of their regularity, memory arrays will typically have 10 times as many transistors as microprocessors for the same size of chip and process technology (**Fig. 1***a*).

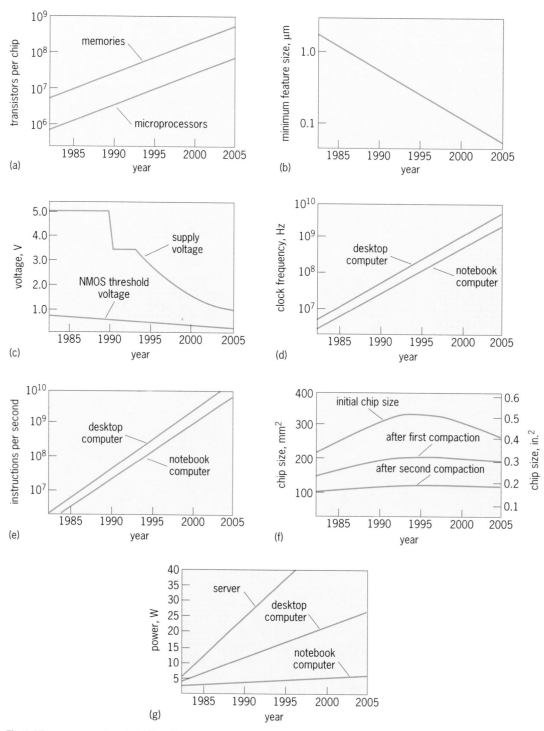

Fig. 1. Microprocessor trends. (*a*) Transistor per chip. (*b*) Minimum feature size. (*c*) Supply voltage and negative metal-oxide-semiconductor (NMOS) threshold voltage. (*d*) Clock frequency. (*e*) Instructions per second. (*f*) Chip size. (*g*) Power.

Transistors per chip. More transistors per chip allow for more advanced designs. Thus transistors are made smaller (Fig. 1*b*) and faster, and the chip size is increased. As silicon metal-oxide-semiconductor (MOS) transistors get smaller, the electric field increases at the drain, and resulting hot electrons can be injected into the gate oxide, causing degradation in performance over time. Supply voltages must be reduced to eliminate this reliability concern. Reducing the supply voltage slows down the part, because the amount of drive current is a function of the square of the voltage above the threshold of the device. To help offset the speed reduction at lower supply voltages, channel lengths and device thresholds are reduced, so tighter process control is required. Until 1990, the supply voltage was usually 5 V; afterward many designs dropped to 3.3 V (Fig. 1*c*).

Clock frequency. Clock frequency for silicon microprocessors has increased at a rate of about 36% per year as minimum feature sizes have been reduced, and is expected to reach 1 GHz for desktop computers by the year 2000 (Fig. 1*d*). Together with the increasing number of transistors per chip, this advance has resulted in a 47% increase per year in the number of instructions per second that a microprocessor executes (Fig. 1*e*). How long these rates of improvement continue will depend on scientific and technological innovation and the nature of the physical limits that are encountered. To an extent, architectural and circuit techniques can be used to make up for inabilities to scale the process up, but clock frequencies may eventually cease to increase at the same rate, or become less of a factor if asynchronous circuit design or other design methodologies are adopted.

One important consideration is the wire delay to get from one side of the chip to the other. A typical wire delay of 1 nanosecond is only 10% of the clock period for a 100-MHz processor, but it is 100% of the clock period for a 1-GHz processor. Designing with these constraints, and reducing the delay by finding lower-resistance and lower-capacitance materials, is a major area of research.

Notebook computers. Performance for notebook computers has lagged that of desktop computers by about 2 years because of power and cost constraints as well as the time required to design power-management circuits. Power is proportional to the capacitance times the square of the voltage times the frequency. It usually takes about 2 years to shrink the chip to the next generation process, in order to reduce the capacitance and manufacturing cost of the chip so that it can be designed into notebooks. The demand for notebook computers will strongly influence the chip size, frequency, and power for future designs.

Most microprocessors go through two compaction steps to reduce size, power, and cost (Fig. 1*f*). The die size is reduced by scaling down dimensions and by using more advanced equipment or processing steps, either by optically shrinking the mask or by digitally compressing the database used to make the mask. Frequency increases with each compaction. Desktop computers benefit from the increased frequency and lower power as well. Notebook computers will eventually become the dominant personal computer because of their portability and potentially lower cost.

Chip size. The initial chip size has not changed significantly since 1990 because of higher manufacturing costs, increased design complexity, and power constraints. This flat trend will probably continue because of the high-volume demand and the expense of raising chip production to a sufficiently high yield to remain competitive. When the die is compacted, two or more dies should fit on a reticle, thus reducing production costs. Power, interconnect delay, and assembly costs are other concerns limiting die size.

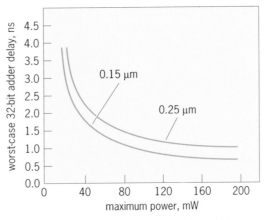

Fig. 2. Trade-off between speed and power in a 32-bit adder; numerical values express effective channel lengths.

Power. Since chip power is proportional to frequency, and the frequency of microprocessors is increasing exponentially with time, power might be expected to increase at the same rate. The reduction in supply voltage, combined with circuit techniques to stop unit clocks from switching, has helped to keep power under control. Because of the wide range of requirements for power between the notebook computers and servers (high-end computers that support large networks of smaller computers), design strategy must be carefully considered for each new microprocessor generation. Speed versus power becomes a key issue in designing a chip because speed and power can be traded off, as shown in **Fig. 2** for two different channel lengths. More efficient use of transistors can be expected to improve architectural performance (performance per number of transistors), and better circuit topologies can be expected to result in higher-frequency logic and memory. Quick analysis of the power-versus-delay trade-off for different circuits, in order to get the best circuit topology quickly (in minutes rather than days) and thereby reduce design time, will be a major development effort.

Circuit methodologies. Latch delay can be eliminated by combining gate and latch into one element by logically ANDing the clock with enables. The concept of combining latches and gates has been applied extensively in recent processors. Similar techniques were used previously in supercomputer designs. Disabling unit clocks has become a standard design methodology to reduce power for blocks that are not doing useful work, and still be able to meet clock skew requirements, which ensure that the clock arrives at all parts of a chip to within some small time domain.

The delayed precharge domino circuit (**Fig. 3**) is a faster, smaller, lower-power type of circuit for implementing logical functions. Commonly used in array designs, these circuits are gaining more widespread use in datapath because of their features of reduced delay, clock loading, area, and power. Such ideas will continue to address the ever-increasing requirements for speed.

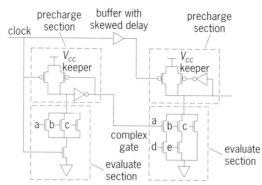

Fig. 3. Delayed precharge domino circuits. Elements a–e are inputs to the gate.

Interconnect delays. Interconnect delays, also known as RC delays, are the delays in getting a signal from one end of a wire to the other. As these delays become more dominant, techniques will be used to microarchitect around the delay problems. Extra pipe stages will be added to get signals from one side of the chip to the other. Buffers will be added in the middle of long wires to restore the very fast signal transitions needed for higher frequencies. Modeling and avoiding line-to-line capacitance effects due to smaller spacing between wires, and improved algorithms for routing the wires to increase density, along with modeling of on-chip inductance, resistance, and capacitance, will be important to better performance. Use of lower-resistance interconnects and insulators with lower dielectric constants to reduce wire delays will be major process development concerns.

Low thresholds. Development of low-threshold complementary metal-oxide-semiconductor processes which minimize leakage currents, or of innovative circuit methods of eliminating dc current leakage for these low-threshold devices, will be a focus for several generations of microprocessors. The scaling of supply voltages will depend on the ability to reduce device thresholds (Fig. 1c).

Soft error rates. An additional concern with scaling supply voltages is that memory cells become more susceptible to state changes due to impacts of alpha particles (soft errors). When an alpha particle (helium nucleus) passes through the depletion region of the drain of a metal-oxide-semiconductor device, it generates electron-hole pairs which cause a momentary current pulse that can switch the state of the memory cell. This phenomenon must be taken into account to minimize the probability of errors. Development of a new generation of devices beyond silicon complementary metal-oxide-semiconductor technology will eventually be required. *See* COMPUTER STORAGE TECHNOLOGY.

Phase-locked loops. These devices will be used to generate frequencies on the chip that may be 10–20 times higher than those at the board level, as compared to 2–4 times higher in current microprocessor systems. Controlling the jitter of these phase-locked loops will be difficult. Decoupling capacitors must be strategically placed on the power supplies to maintain quiet voltage references and reduce jitter on the clock. The higher chip frequencies will also create some memory management problems for architects, who must balance all aspects of a microprocessor carefully in order to arrive at optimum system solutions.

For background information *SEE* CONCURRENT PROCESSING; INTEGRATED CIRCUITS; MICROCOMPUTER; MICROPROCESSOR; PHASE-LOCKED LOOPS; SEMICONDUCTOR MEMORIES in the McGraw-Hill Encyclopedia of Science & Technology.

Thomas D. Fletcher

Bibliography. R. P. Collwell and R. L. Steck, A 0.6-μm BiCMOS processor with dynamic execution, *ISSCC*, pp. 176–177, February 1995; D. W. Dobberpuhl et al., A 200-MHz 64b dual-issue CMOS microprocessor, *IEEE J. Solid-State Circ.*, 27(11):1555–1567, November 1992; T. Fletcher, Microprocessor technology trends, *IEDM*, pp. 269–271, December 1994; Institute of Electrical and Electronics Engineers, *IEEE International Electron Devices Meeting, 1994, IEEE International Solid-State Circuits Conference, 1994, IEEE International Solid-State Circuits Conference, 1995.*

Mining

Environmental degradation due to contaminated mine drainage is one of the most challenging technical issues facing the hard-rock mining industry. In response, mine operators and regulators are placing increased emphasis on environmental management of mine wastes and on studies to determine mine waste geochemistry.

Environmental geochemical studies. These studies, typically called waste characterization studies, identify mineral system characteristics that may react with water or the atmosphere to produce acidic drainage or to leach heavy metals. Waste characterization studies provide mine operators and regulators with the data necessary to make an informed assessment of the short- and long-term geochemical behavior of mine wastes, and the potential environmental liability and risks involved in developing a specific mining project. These studies are also used to develop designs for mine waste containment facilities, monitoring systems, and closure measures to protect the environment.

Waste characterization studies involve collecting representative samples of project waste rocks, tailings, and spent leached ore, and subjecting these samples to a variety of tests to determine their geochemical behavior. There are two major types of waste characterization tests presently in use: acid generation tests that determine acid-generating and -neutralizing potential, and leach extraction tests that determine to what degree metals may be leached from project wastes when exposed to certain agents.

Acid generation tests. Mine wastes may be acid-generating because of the oxidation of sulfide minerals such as pyrite (FeS_2), marcasite (FeS_2), and pyrrhotite ($Fe_{1-x}S$), as these minerals react with air and water. The presence of ferric iron and biogeochemical reactions can also cause sulfide mineral oxidation. In some mineral deposits, oxidation of sulfide minerals produces low pH (acidic) mine drainage. Because the solubility of many metal sulfide and oxide minerals increases with decreased solution pH, acidic mine drainage typically contains elevated concentrations of metals. The contribution of metals such as iron, copper, and manganese to surface water and to ground-water systems draining mine sites is the principal environmental problem caused by acidic mine drainage.

The products of acid generation may be neutralized if a mine waste contains a sufficient quantity of acid-consuming or -neutralizing minerals, and if the rate of acid consumption exceeds the rate of acid generation. Carbonate minerals are the most effective acid-neutralizing minerals. However, clay and feldspar minerals may also consume acid. Carbonate, clay, and feldspar minerals are typically associated with many types of ore deposits as gangue minerals.

Acid generation tests can be divided into two main types: short-term, static tests (also called acid-base accounting), and long-term kinetic tests. Most acid generation testing programs begin by performing static acid-base accounting tests to provide an initial indication of whether the mine waste may be acid-generating or acid-consuming. These short-term tests are used as a screening tool to determine whether long-term kinetic acid generation tests are necessary.

If the acid-base accounting test indicates that the acid generation potential is equal to or greater than the acid consumption potential, the mine wastes could be acid-generating. These mine waste samples are then subjected to longer-term kinetic tests to determine the relative rate of the acid-generating and -neutralizing reactions. Kinetic acid generation tests thus estimate the time dependency and reaction kinetics of the waste.

Leach extraction tests. Mine operators typically augment acid generation test data with leach extraction tests to determine the geochemical mobility of metals in mine wastes. Because the prevailing geochemical conditions in a mine waste disposal facility are best simulated by acidic rain falling onto a mine waste pile, leach extraction tests use an inorganic acid leachant to simulate natural weathering conditions. Leachates produced from leach extraction tests are typically analyzed for the U.S. Environmental Protection Agency (EPA) primary and secondary drinking water parameters, and for any other chemical species that might be anticipated from the waste rock geochemistry or required by regulation.

If short-term leach extraction tests indicate that the mine wastes are leachable, long-term kinetic leachability tests may be used to evaluate the

chemical species that might leach from the mine waste with time. Long-term leaching tests consist of either laboratory tests, such as humidity cell tests, or field studies, such as pilot-scale waste piles or field columns.

Leach-extraction waste characterization tests are sometimes confused with leach-extraction waste classification tests developed by the EPA to determine whether industrial wastes should be classified as hazardous. The EPA has determined that most mine wastes, including waste rocks, tailings, and leached ore, are not hazardous. A much smaller group of mine wastes, known as mineral processing wastes, may be regulated as hazardous depending upon their characteristics. The EPA has developed a leach-extraction waste classification test to determine the leachability characteristics of mineral processing wastes. This test should not be used to evaluate the long-term geochemical behavior of waste rocks, tailings, or spent leached ore. Similarly, the EPA toxicity characteristic leaching procedure (TCLP) test, used to determine whether industrial wastes are hazardous, should not be used for any type of mine waste. The TCLP test uses an organic leachant, and thus it does not simulate natural weathering and leaching by an inorganic acid, the prevailing conditions in mine waste disposal facilities.

Regulatory agencies in some states have developed leach extraction test protocols specifically for mine wastes. These tests simulate leaching due to the infiltration of meteoric water through a mine waste pile.

Leachate generation tests. For projects in arid settings, it may also be appropriate to define the hydraulic controls on leachate generation in order to assess the potential for leachate to drain or seep from the site. Because conventional leach extraction test protocols simulate leaching under saturated conditions, they are not the best analog of the unsaturated conditions prevailing in waste rock piles in arid settings. Leachate generation tests determine the volume of seepage, if any, that would drain from mine wastes at arid mine sites because of both average and design storm-event precipitation levels.

Waste characterization data. State and federal regulatory environmental requirements for mining projects include mandates for waste characterization tests. Waste characterization data collected to satisfy regulatory requirements during project permitting are used by regulators and mine operators to determine the most appropriate design for waste containment facilities, to assess the need for special waste-handling measures, and to develop preliminary closure and reclamation plans. Waste characterization data collected subsequently during project operation are used to refine the ongoing waste management program, and to finalize mine closure and reclamation plans.

Project-specific waste characterization tests identify mine wastes that may need special waste management procedures and mitigation measures. The need for special mine waste management measures is site specific, and is highly dependent upon the mineralogy and geochemistry of project wastes, the amount of precipitation that falls on the site, the surface-water and ground-water hydrology at the site, and the potential risk to receptors at points of exposure.

As a general rule, contaminated mine drainage is less prevalent in arid settings. However, not all mine wastes in higher-precipitation areas are problematic. Mine wastes containing small amounts of or no sulfide minerals or unreactive sulfides are not acid-generating. Similarly, acid generation may not occur or may be significantly inhibited if sulfide-bearing mine wastes are deposited with a sufficient volume of wastes that possess a high acid neutralization potential.

Acid generation control. Depending upon mine site location, acidic drainage may create environmental problems requiring significant mitigation measures. Acidic drainage is generally of greater concern at sites with nearby potential receptors, such as aquatic life or surface- and ground-water users.

Adverse environmental impacts from acid-generating mine wastes at new and proposed mines can be controlled by measures designed to prevent or limit acid generation from the onset of project activities. Most measures focus on limiting the exposure of acid-generating wastes to water and air. Excluding water and air reduces acid generation in two ways. First, reducing the exposure of mine wastes to oxygen limits oxidation and slows the rate of acid generation. Second, low infiltration rates minimize the volume of leachate generated, thus limiting drainage of acidic or metals-bearing leachate from the mine waste pile.

The design of acid generation control measures is highly site specific and may include features to intercept and divert surface and ground water from acid-generating mine waste piles, or impervious caps or seals consisting of an engineered liner system to minimize infiltration of water and air. Depending upon the volume of acid-generating materials, it may be possible to encapsulate problematic mine waste with non-acid-generating or -neutralizing project wastes in a manner to minimize infiltration. At some sites, it may be preferable to blend acid-generating wastes with alkaline materials, such as lime, or with mine wastes having a high acid neutralization potential to inhibit acid generation and to neutralize acidic leachate. At other sites, it may be appropriate to dispose of acid-generating wastes in segregated facilities with an engineered system to collect and treat leachate during project operation. Subaqueous disposal of mine wastes below the water table in open-pit and underground mines may be an effective option for minimizing oxidation and controlling acid generation at some mine sites.

Preventing or limiting acid generation from occurring in the first place is far more effective than collecting and treating acidic mine drainage once a project is under way or after mining is completed. However, acidic drainage can be a problem at some historic mine sites and also at some sites that have been operating for many years. Natural outcrops and highway cuts that expose sulfide-bearing rocks may also be a source of problematic acidic drainage.

Controlling acid mine drainage at older operating sites, at abandoned historic mine sites, and at naturally occurring sites creates many technical and economic challenges. To prevent mine sites from becoming environmental problems, mine operators use data from mine waste characterization studies to design, operate, and close mine waste disposal facilities in a manner that is focused on preventing and minimizing environmental problems due to contaminated mine drainage.

For background information SEE CARBONATE MINERALS; CLAY MINERALS; MINING in the McGraw-Hill Encyclopedia of Science & Technology.

Debra W. Struhsacker

Bibliography. I. P. G. Hutchison and R. D. Ellison (eds.), *Mine Waste Management*, 1992.

Missile

The objective of the missile design process is to synthesize a balanced design based on trade-offs of the missile subsystems. The missile design process includes mission definition, specification of missile system requirements, integration of the missile with the launch platform, development of missile design concept alternatives, and evaluation of technology alternatives. An initial baseline design provides a starting point for the process of design convergence (**Fig. 1**). Characteristics such as the aerodynamic shape, propellant or fuel weight, flight trajectory range, time to intercept the target, maneuverability, seeker detection range, accuracy, lethality, and cost are evaluated; and the missile is resized and reconfigured in an iterative process. For example, the tail stabilizers and flight-control surfaces may be resized for improved stability or maneuverability. Another example is adding or subtracting propellant or fuel to match the flight range requirement. Typically, 3–10 design iterations are required before a synthesized missile converges to the performance requirements.

Figure 2 illustrates a typical arrangement of the missile subsystems; an asterisk indicates that the subsystem is sized by flight-performance requirements such as range, time to target, and maneuverability. The other subsystems are sized by considerations other than flight performance, and are less sensitive to flight performance. For example, the seeker is usually sized by the diameter required for target detection range and target resolution.

Missile versus aircraft design. Missile flight performance is usually more sensitive to changes in the size of the missile than to changes in the weight of the missile. A missile usually has 2–6 times the den-

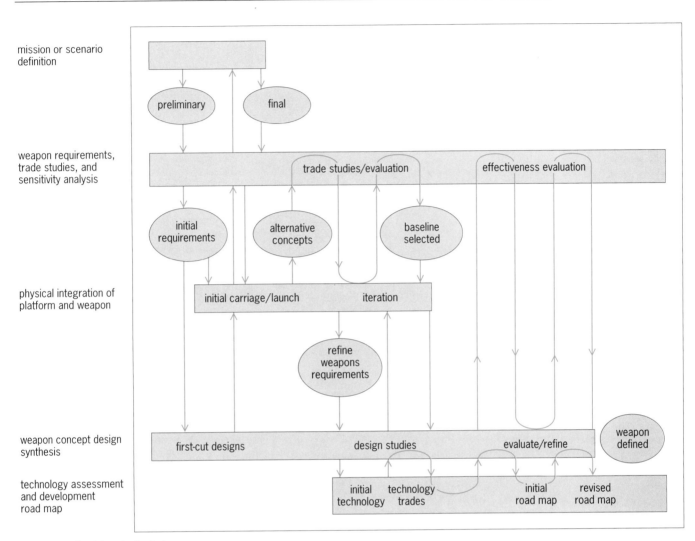

Fig. 1. Example of the missile design process.

sity of combat aircraft, and is more likely to be volume limited than weight limited. Design changes in nonpropulsion subsystem volume are often at the expense of the propellant or fuel volume.

Although the design process for missiles has much in common with the design process for combat aircraft, there are differences in the areas of design constraints, design boundaries, and subsystems. A major difference in missile design compared to crewed aircraft design is that the missile is not subject to the environmental limitations of the human pilot. For example, crewed combat aircraft are designed for maximum maneuverability less than 12 G's (1 G = 32.2 ft/s² = 9.8 m/s²), and missiles have been developed with maneuverability greater than 60 G's. A second example is thrust-to-weight ratio, or axial acceleration. Crewed combat aircraft use turbojet propulsion systems providing a thrust-to-weight ratio less than 2. Rocket-powered missiles have demonstrated thrust-to-weight ratios greater than 400. A third example is rotation rate. Crewed aircraft are limited in rotation rate (pitch, yaw, roll) to less than 90°/s, and missiles have demonstrated rotation rates of over 1000°/s.

Another major difference in the design process of missiles compared to combat aircraft is the constraint of cost: combat aircraft production cost is typically hundreds to thousands of times greater than missile production cost. Other design areas that are different for missiles include the seeker, guidance and control, warhead, and launch platform. The design process for the missile seeker and the missile guidance and control will be discussed.

Seeker. The missile seeker receives signals from the target, tracks the target, and sends commands to the flight-control system that are used to direct the flight trajectory. A driving concern in designing the seeker is the operational wavelength and the attenuation of the target signal due to weather. Infrared seekers and millimeter-wave seekers have acceptable performance in adverse weather with attenuation less than 3 dB/km.

The **table** gives an example ranking of alternative seeker concepts. The relative ranking of the seekers is based on the discriminants of technical maturity, cost, target resolution, robustness to countermeasures, weight, and performance in adverse weather. The best seeker for a given set of design require-

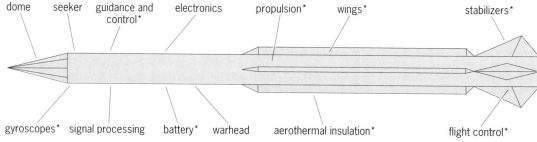

Fig. 2. Typical arrangement of missile subsystems. An asterisk indicates a subsystem that is sized by flight-performance requirements.

ments depends upon the weighting given to the discriminants, and is based on a specific mission. In this example, which assumes an equal weighting for each discriminant, the preferred seeker alternatives are passive-imaging infrared, active-imaging millimeter-wave, and imaging dual mode (millimeter-wave/infrared). Passive-imaging infrared seekers have the advantages of high technical maturity (they are currently in production), high resolution (less than 0.2-milliradian instantaneous field-of-view), and light weight. A disadvantage of passive-imaging infrared seekers is reduced target signal in adverse weather. Active-imaging millimeter-wave seekers have an advantage in their capability to compensate for loss of signal in adverse weather by increasing the transmitted power. A disadvantage of active millimeter-wave seekers is coarser resolution. Imaging dual-mode seekers have an advantage of robustness to countermeasures because they operate at more than one wavelength. A disadvantage is higher cost and weight because of the two different sensors.

All six types of seekers shown in the table, including passive-imaging millimeter-wave, active-imaging infrared or imaging laser-radar (LADAR), and synthetic aperture radar (SAR), are expected to be used in future missiles. The passive-imaging millimeter-wave seeker is rated lower because of the relative immaturity of the technology and limited resolution. The imaging laser-radar seeker is rated lower because of its limited resolution, lack of robustness to countermeasures, and lower performance in adverse weather. A laser-radar seeker operates in a narrow bandwidth and usually has less flexibility to make adjustments to countermeasures. Also, the laser-radar seeker operates at a relatively low pulse rate, limiting the field of regard and resolution that can be achieved in the short time interval of terminal flight. Finally, a synthetic aperture radar seeker has the disadvantages of greater weight due to the greater computation requirements, and higher cost due to the requirement for an accurate inertial navigation system and for the required electronics. Also, the squint (side-looking) flight trajectory of a synthetic-aperture-radar-guided weapon requires high maneuverability in terminal flight, and requires that the final portion of the flight trajectory use inertial guidance. Thus weapon accuracy is degraded.

Guidance and control. The typical guidance phases are launch, midcourse, and terminal. The launch phase is usually a programmed phase to safely separate from the launch platform. The midcourse phase begins when commands are initiated to direct the missile to the target. Some short-range missiles do not have a midcourse phase, but proceed immediately into a terminal phase. An inertial navigation system is typically used to direct the missile during midcourse guidance. A reference system may be used to update the missile inertial navigation system. Reference update systems include Global Positioning System (GPS) satellites, map matching, and loran.

One of the design trade-offs is the accuracy of the midcourse guidance system. An accurate midcourse guidance system reduces the performance requirements and cost of the terminal seeker.

Type of seeker	Technical maturity	Cost	Resolution	Robustness to countermeasures	Weight	Performance in adverse weather	Sum
Passive-imaging infrared	1	1	1	5	1	6	15
Active-imaging millimeter-wave	2	3	3	2	3	3	16
Imaging dual mode	5	4	4	1	5	2	21
Synthetic aperture radar (SAR)	4	6	2	3	6	1	22
Imaging laser radar (LADAR)	3	2	5	6	2	5	23
Passive-imaging millimeter-wave	6	5	6	4	4	4	29

Examples of ranking of seeker design alternatives*

* 1 = best.

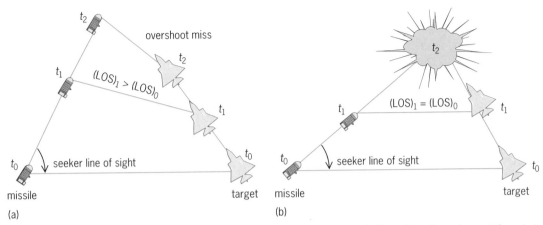

Fig. 3. Example of flight trajectories. (*a*) **Example of overshoot miss. Positions of missile and target are shown at times t_0, t_1, and t_2, with $t_2 > t_1 > t_0$. Line-of-sight angle increases. Thus, line-of-sight angle at time t_1, $(LOS)_1$, is greater than line-of-sight angle at time t_0, $(LOS)_0$.** (*b*) **Example of collision intercept. Line-of-sight angle is constant. Thus, $(LOS)_1 = (LOS)_0$.**

An efficient flight trajectory can be followed in terminal guidance by directing the missile with sufficient lead to a predicted intercept. A constant line-of-sight trajectory, or constant-bearing course, is known as proportional guidance. **Figure 3** compares an overshoot-miss flight trajectory to a constant-bearing intercept trajectory. In Fig. 3*a*, the seeker line-of-sight is increasing, resulting in an overshoot miss. In Fig. 3*b*, the line-of-sight between the missile and the target is constant, resulting in a collision intercept. Proportional-guidance terminal homing is based on adjusting the rate of change in the missile heading proportional to the rate of change in the seeker line-of-sight from the missile to the target. A missile turning-rate correction for heading error is usually designed to be 3–5 times the line-of-sight rotation rate.

Output. The output of the missile design process is the documentation of the evolved design. Specific information is developed for the following:

1. Justification of the evolved design includes reasons for the concept being selected, measures of merit, design trade-offs, criteria for selection, and advantages compared to alternative concepts.

2. Three-view drawings of the evolved design involve outboard profiles showing external features and dimensions, and inboard profiles showing subsystems and their dimensions.

3. Drawings of alternative concepts are included for comparison.

4. System measures of merit performance include maximum range, maximum speed, and accuracy. Accuracy is usually measured by circular error probable (CEP), the radius of the circle within which 50% of the intercepts occur.

5. Subsystem performance includes specific impulse of the propulsion system, thrust profile, aerodynamic stability, aerodynamic control effectiveness, lift, drag, subsystem response time, and structure strength-to-weight ratio.

6. The weight and balance statement includes component weight, component locations, missile center-of-gravity location at launch and burnout,

and missile moments of inertia. Also included are descriptions of the methods used to estimate weight and balance as well as the methods used to size aerodynamic stabilizers, wings, and flight-control surfaces.

7. Technical description of methods used in the design process: Methods must show traceability of the flowdown of requirements from system performance measures of merit to subsystem performance measures of merit to technology parameters. When practical, two or more methods are used to enhance confidence in the design.

8. Unit production cost includes units produced, the learning curve, and the basis for the learning curve. (The learning curve reflects the decrease in unit cost as more missiles are produced and the production process becomes more efficient. Typical learning curves range from 80 to 95%. An 80% learning curve means that the unit cost decreases by 20% with each doubling of the total units produced. The cost history of missile production substantiates the basis for learning-curve methodology.)

9. Technology road map: The missile design process identifies technologies that require development to demonstrate maturity and readiness for an operational application. Technology road maps provide the required schedule and milestones for timely insertion of technology into an operational system.

For background information SEE AIRCRAFT DESIGN; GUIDANCE SYSTEMS; GUIDED MISSILE; INERTIAL GUIDANCE SYSTEM; LASER; MISSILE; RADAR in the McGraw-Hill Encyclopedia of Science & Technology.

Eugene L. Fleeman

Bibliography. S. S. Chin, *Missile Configuration Design*, 1961; J. J. Jerger, *System Preliminary Design Principles of Guided Missile Design*, 1960; L. M. Nicolai, *Fundamentals of Aircraft Design*, 2d ed., 1984; J. N. Nielsen, *Missile Aerodynamics*, 1960, reprint 1988; D. P. Raymer, *Aircraft Design, A Conceptual Approach*, 2d ed., 1992.

Molecular nanotechnology

Molecular nanotechnology is an emerging interdisciplinary field combining principles of molecular chemistry and physics with the engineering principles of mechanical design, structural analysis, computer science, electrical engineering, and systems engineering. Molecular manufacturing is a method conceived for the processing and rearrangement of atoms to fabricate custom products. It would rely on the use of a large number of small manufacturing subsystems working in parallel and using commonly available chemicals. Built to atomic specifications, the products would exhibit order-of-magnitude improvements in strength, toughness, speed, and efficiency, and would be of high quality and low cost.

Distinctions from related disciplines. It is useful to illustrate some of the key points of the technology by drawing distinctions between it and some related fields. Molecular nanotechnology is distinguished from solution chemistry by the manner in which the chemical reactions occur: instead of the statistical process of molecules bumping together in random orientations and directions in solution until a reaction occurs, discrete molecules are brought together in individually controlled orientations and trajectories to cause a reaction at a specific site. Furthermore, this process is performed under programmable control.

In biological systems, ribosomes build proteins by "grabbing" onto transfer ribonucleic acid (tRNA) molecules and transferring their amino acids to a growing polypeptide chain, under the programming specified by messenger RNA (mRNA) from its deoxyribonucleic acid (DNA) template. Molecular manufacturing systems will have attributes different from those of biological systems. They will be able to transport raw materials and intermediate products more rapidly and accurately by conveyor belts and robotic arms (**Figs. 1** and **2**).

Molecular manufacturing systems will control all trajectories and orientations of all devices in the system, not just the relative orientations at points where reactions occur. In biological systems, ribosomes, tRNA, mRNA, amino acids, and DNA are suspended freely in the cell environment; they rely on random collisions and diffusion for the transport of raw materials and products. The molecular manufacturing systems will make heavy use of positional assembly (such as a blind robot thrusting a pin into the expected location of a hole) as opposed to matching assembly (a tRNA molecule bumping around a ribosome until it fits into the slot with the matching pattern of hills and valleys and positive and negative charges on its surface). Like auto factories and steel mills, the molecular manufacturing systems will lack the ability to independently evolve (a mutation in a molecular nanomachine would simply render it inoperable).

Microtechnology is also quite different: nanolithography is the patterning and selective etching of bulk material (often silicon) to create devices with features as small as a few nanometers at their narrowest point. Micromachines such as electrostatic motors and steam engines have been fashioned in this top-down manufacturing approach. Molecular manufacturing, by contrast, is bottom up, that is, structures are built by piecing together atoms and molecules.

Motivation for development. Based on calculations of theoretical strengths and experimental data on near-perfect whiskers, today's plastics, metals, and ceramics, as advanced as they are, could become 100 times stronger if made with molecular nanotechnology. Objects made from these materials could be up to 100 times lighter, using 100 times less material; and as much as 250 times lighter by substituting a composite made with stiff diamond fibers. Thus, ultralight cars, trucks, trains, and planes would use far less energy, especially with atomically smooth surfaces to reduce internal friction and air resistance losses.

Computer processors more powerful than the ones found in today's fastest engineering workstations could be made smaller than a typical human cell, as could microrobots. In medicine, the robots' capability to repair damaged cellular structures with molecular precision would provide cures for many illnesses.

One envisioned result of the technology is a portable manufacturing system able to build a wide range of useful products to atomic specification, including a copy of itself. A system capable of building a copy of itself to atomic specification—as well as a wide range of other products from food to computers—would have important economic consequences. Many goods would no longer need to be transported from remote locations but could be fabricated in the home, saving the time, energy, and

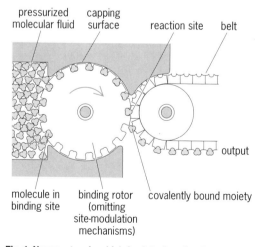

pressurized molecular fluid   capping surface   reaction site   belt

molecule in binding site   binding rotor (omitting site-modulation mechanisms)   covalently bound moiety

output

Fig. 1. Nanosystem in which feedstock molecules are transferred from a pipe onto a conveyor belt for transport to the assembly site where they will be bound to the finished structure. (*After K. E. Drexler, Nanosystems: Molecular Machinery, Manufacturing and Computation, John Wiley and Sons, 1992*)

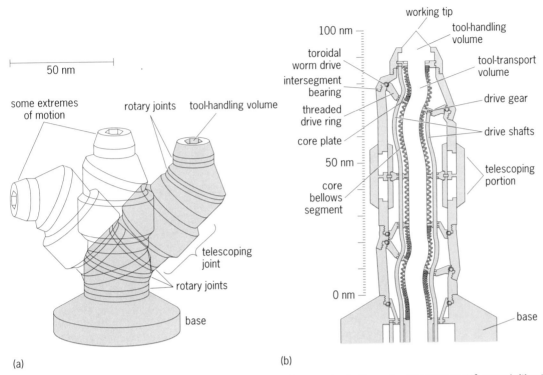

Fig. 2. Nanosystem manipulator arm for positioning molecules to form an atomically precise structure: 4×10^6 atoms (without base). (a) Diagram showing the wide range of motion. (b) Cross section showing the central core for transporting material to the tip. (After K. E. Drexler, Nanosystems: Molecular Machinery, Manufacturing, and Computation, John Wiley and Sons, 1992)

waste associated with transportation, packaging, and storage in warehouses and shops. Besides the waste reduction, the environment would benefit from a form of manufacturing that produces no toxic waste. In addition, the technology could be applied to both purify and mine existing waste sites.

Design studies and engineering analysis. Molecular mechanics computational techniques can be used to design reasonably accurate molecular-mechanical structures. For hydrocarbons, typical errors are about 0.1% for bond lengths and 0.6° for bond angles; energies are accurate to within a few times a milli-attojoule (maJ; a convenient unit of energy in molecular-mechanical systems equal to 10^{-21} joule). With this method, researchers have designed or outlined a variety of chemically stable molecular-mechanical components, subsystems, and computation systems, including gears, bearings, rods, springs, belt-and-roller systems, positioning devices, electrostatic nanomotors, and rod logic computers. A rod logic computer is a mechanical computer that transmits signals by pushing and pulling on carbon rods one atom in diameter. Molecular bumps on the rods can block the movement of other rods and serve as logic gates. An analysis of molecular gears showed that energy barriers to gear tooth slippage are large, greater than 500 maJ, so the teeth lock together without slipping; while energy barriers to corotation are small, less than 0.01 maJ, so the gears turn freely with very low friction. These gears could be highly efficient for transmitting power. For a gear system operating at a

shear force of 1 nanonewton (nN; a convenient unit of force in molecular mechanical systems equal to 10^{-9} newtons) phonon scattering and thermoelastic drag losses would account for only 0.003% of the transmitted power.

Similar exploratory engineering techniques were used to assign lower bounds to the performance of a variety of nanosystem components, with the conclusion that the following capabilities could be realized:

1. A 5×10^6-atom molecular robot arm could manipulate 10^6 atoms per second, each programmably positioned to within ~0.1 nanometer, thereby building a copy of itself in 5 s.

2. Compact parallel computing systems capable of 10^{15} millions of instructions per second (MIPS) [compared to supercomputers at ~10^5 MIPS or the human brain at ~10^9 MIPS] would be possible.

3. Computer memory would be compact enough to store the information content of the Library of Congress within the volume of a sheet of office paper.

4. The power dissipation in a mechanical nanocomputer (10^{16} instructions per second per watt) would be 500,000 times less per switching operation than in one of today's silicon chips. This would still require substantial cooling, but a fractal plumbing network would

make it possible to chill a cubic-centimeter, 10,000-W system to room temperature (the equivalent of cooling off 1000 ordinary lightbulbs shining in the space of a sugar cube).

5. Mechanochemical power conversion that approaches thermodynamic reversibility would be possible, for an estimated efficiency of greater than 99% (as compared to less than 40–60% for many macroscopic systems such as electric power plants and automotive engines).

6. Macroscopic components could be made with tensile strengths greater than 50 gigapascals (7×10^{16} lb/in.^{2}). With such material, 50 times stronger than ordinary steel, 10 strands only as thick as a human hair could support an average-sized person.

7. A 1-kg (2.2-lb) desktop assembler system could make a copy of itself (to atomic specification) in less than 3 h.

Although radiation damage is a concern, with sufficient numbers of backup subsystem devices the reliability can be quite high. After 200 years of terrestrial radiation, 4×10^{19} sets of components of 100 nm per side (10^{-18} kg or 2.2×10^{-18} lb) with a redundancy of $n = 25$ will have a failure probability of about 1 in 10^6.

Although a self-replicating molecular manufacturing system might be made from a large number of pieces ($>10^{27}$), it would not be impossibly complex. Most of the pieces would be small and could be made quickly. Many of the pieces can be identical, and it is not necessary to specify where every single atom on every atomically perfect part in the entire system goes, only where every atom of every different kind of part goes and how those parts fit together. Thus, the complexity (as represented by the number of bits of information needed to specify the entire system) of a replicating assembler system is on the order of 10^8 bits, larger than the 8×10^6 bits for an *Escherichia coli* bacterium but significantly less than the 10^{11} bits for a self-replicating lunar manufacturing facility proposed by the National Aeronautics and Space Administration.

Recent progress. Since the introduction of the idea in 1981, there have been major advances toward developing molecular nanotechnology. With scanning probe microscopes, researchers can now manipulate individual atoms to nanometer precision and catalyze reactions at specific points on surfaces; molecular biologists can design and synthesize proteins from scratch; synthetic chemists can design and create uncomplicated molecules to trap ions and molecules; researchers on molecular electronic devices have developed a working prototype and spurred commercial development; and the wide availability of scientific workstations and molecular modeling software is significantly reduc-

ing the time necessary to electronically construct and test new designs. Although much research is currently devoted to the related fields mentioned previously, a growing number of groups both in the United States and in Japan are working in the narrowly defined area of molecular nanotechnology.

Although new scientific developments are accelerating progress, no new scientific knowledge is needed to complete the development of this promising technology. The challenges are principally engineering and sociopolitical hurdles. Social and political issues are currently being addressed by at least two organizations in the United States: the Foresight Institute and the Center for Constitutional Issues in Technology, both of Palo Alto, California.

For background information SEE *MATERIALS SCIENCE AND ENGINEERING* in the McGraw-Hill Encyclopedia of Science & Technology.

David R. Forrest

Bibliography. B. C. Crandall and J. Lewis, *Nanotechnology: Research and Perspectives,* 1992; K. E. Drexler, *Engines of Creation,* 1986; K. E. Drexler, C. Peterson, and G. Pergamit, *Unbounding the Future,* 1991; R. P. Feynman, There's plenty of room at the bottom, *Eng. Sci.,* 23:22–36, 1960.

Multiple sclerosis

Multiple sclerosis is a chronic inflammatory disease of the central nervous system characterized by the damage and destruction of myelin sheaths in the brain parenchyma. As in other diseases with an assumed autoimmune etiology, multiple sclerosis often takes a relapsing-remitting course. There is increasing evidence that the immune system plays a role in the pathogenesis of this disease. In individuals with multiple sclerosis, an enhanced reactivity of lymphocytes, possibly directed against myelin antigens, can be detected in the peripheral blood as well as in the cerebrospinal fluid. Target antigens of T and B lymphocytes in multiple sclerosis include central nervous system molecules such as myelin basic protein, proteolipid protein, and myelin oligodendrocyte glycoprotein. Moreover, the relative risk for multiple sclerosis is correlated with the expression of certain human leukocyte antigen genes that play an important part in the normal and pathological function of the immune system. Finally, in some individuals with multiple sclerosis a beneficial effect of immunosuppressive therapy has been observed.

Experimental autoimmune encephalomyelitis. Additional evidence for a role of the immune system in multiple sclerosis is based on observations in animal models of demyelinating diseases, such as experimental autoimmune encephalomyelitis. This T-cell-mediated inflammatory autoimmune disease of the central nervous system serves as an experimental model for some aspects of multiple sclerosis, although its direct relevance is uncertain. In both diseases, circulating leukocytes enter the brain parenchyma and damage myelin sheaths, resulting in impaired nerve conduction and paralysis. Antigen-specific CD4$^+$-helper T cells are believed to play a major role in the initiation of experimental autoimmune encephalomyelitis and multiple sclerosis.

The inflammatory infiltrates characteristic for both diseases consist of lymphocytes and cells of the macrophage lineage. Each cell type contributes to the pathologic destruction of central nervous system tissue through various effector mechanisms, such as the release of cytokines and enzymes. At their final stages, the resulting intraparenchymal lesions are characterized histologically by a complete loss of myelin, an absence of the cells producing myelin sheaths (oligodendrocytes), and the relative sparing of axons.

Blood-brain barrier. The central nervous system has long been considered to be an immunologically privileged site, protected from immune surveillance by a specialized endothelium that is located in the endothelial cells of the capillaries in the brain (the blood-brain barrier). These cells differ from endothelial cells of systemic capillaries because they are rich in mitochondria, rarely have pinocytotic vesicles, and are fused by tight junctions of high electrical resistance. In addition, they make frequent contact on their abluminal (brain) side with projections from astrocytes. Through these contacts, astrocytes are thought to regulate capillary permeability.

The concept of a blood-brain barrier was developed early in the twentieth century when investigators noted that intravenously injected cationic dyes in experimental animals rapidly diffuse from capillaries and stained tissues in most organs but not in the brain. These observations promoted the notion that the two major central nervous system components, the brain and the cerebrospinal fluid, are separated or excluded from the blood-lymphatic compartment. It is now clear, however, that the blood-brain barrier is a two-way, regulatory interface between blood and nervous system rather than an impermeable fence. The endothelial cells selectively transport nutrients into the brain, such as sugars and amino acids down their concentration gradients. Further, they protect brain cells from extraneous fluctuations in extracellular ion concentrations, neurotransmitters, and growth factors. The vascular tight junctions, which normally prevent leukocytes from entering the central nervous system, may leak in a number of degenerative, infectious, epileptic, leukemic, and autoimmune diseases. For example, the inflammation during bacterial and aseptic meningitis is characterized by leukocytic infiltration and increased vascular permeability to water-soluble drugs such as penicillin.

The dynamics of the autoimmune T-cell infiltration into and out of the central nervous system parenchyma have proven difficult to investigate in humans with multiple sclerosis; however, a more complete picture is emerging from studies in animal models such as experimental autoimmune encephalomyelitis. The establishment of inflammatory experimental autoimmune encephalomyelitis lesions and clinical disease is a multistep event. It has been proposed that the first step requires that activated T cells cross the blood-brain barrier. Activated lymphocytes that bear a memory phenotype, suggesting previous activation by antigen and also expressing the adhesion molecule integrin beta 1 alpha 4, become attached to appropriate receptors on endothelial cells. This process has been shown to be dependent on the activation state of the T cells but independent of their antigen specificity. Next, central nervous system antigen-specific CD4+ T cells are reactivated by fragments of antigens such as myelin that are presented in the framework of major histocompatibility complex class II molecules on the surface of antigen-presenting cells (macrophages, microglia, and perhaps astrocytes). The result is a second wave of inflammatory recruitment and clinical onset of experimental autoimmune encephalomyelitis. Proinflammatory cytokines such as tumor necrosis factor alpha are probably key mediators of the full-blown inflammatory response. *SEE IMMUNO-LOGICAL MEMORY; INFLAMMATION.*

Encephalitogenic myelin-specific T cells may not be capable of mediating experimental autoimmune encephalomyelitis in the absence of this secondary leukocyte recruitment. In addition there are indications that T cells and antibodies may act synergistically to induce demyelination.

The role of infection. Autoreactive T cells are part of the normal lymphocyte repertoire of healthy individuals, but in the normal immune system, potent regulatory mechanisms control T- and B-cell tolerance to self antigens. For autoimmune disease to occur, autoreactive lymphocytes have to be activated, and this activation leads to clonal expansion of autoreactive cells. Only activated autoreactive lymphocytes are capable of migrating through the endothelial cell layer of the blood-brain barrier.

Although genetic factors influence susceptibility to autoimmunity in humans and experimental animals, environmental factors contribute to disease penetrance. Infectious pathogens are known to be the most potent activators of cells of the immune system, and bacteria and viruses have long been regarded as relevant exogenous factors in autoimmune diseases. Various infectious agents, including many viruses and bacteria, have been implicated in the pathogenesis of demyelinating diseases. Several mechanisms associated with infection could account for the activation of autoreactive lymphocytes leading to disease. One such mechanism is molecular mimicry: viruses and bacteria have determinants that are antigenic epitopes that cross-react with the self structures of the body. Under certain circumstances, these epitopes can lead to a T- or B-cell-mediated immunological attack on corresponding self components, such as myelin.

New mechanisms by which infectious pathogens affect autoreactive lymphocytes have been analyzed in detail. These mechanisms include the modulation of the cytokine phenotype of autoreactive cells during T-cell subset development and the enhancement of antigen spreading in central nervous system lesions during chronic inflammation. Superantigens might play a role during these pathological events by stimulating autoreactive T cells, both locally and systemically. SEE VIROLOGY.

Bacterial superantigens. Certain gram-positive cocci (*Staphylococcus aureus* and *S. pyogenes*) and *Mycoplasma arthritidis* secrete biologically active proteins that have been referred to as superantigens because of their powerful stimulatory activity for T lymphocytes. There are some similarities in the mechanism of T-cell activation by superantigens and conventional protein antigens, including autoantigens. Peptide antigens are presented to T cells bound to major histocompatibility complex class II molecules. Similarly, superantigens bind to class II molecules, forming a binary complex that stimulates T lymphocytes through defined V-region beta chains of the T-cell receptor. Recently, it has been demonstrated that superantigens are able to bind directly (without major histocompatibility complex presentation) with very low affinity to the T-cell receptor. In contrast to conventional protein antigens, superantigens do not need to undergo degradation prior to binding to major histocompatibility complex class II molecules.

Implications. Superantigens have been implicated in several pathological conditions, including acute responses such as severe fever, toxic shock, food poisoning, and multiple-organ failure. In addition, there is increasing evidence for a key role of superantigens in the pathogenesis of autoimmune diseases. Bacterial superantigens can specifically trigger classes of T cells through interaction with their T-cell receptor V beta region. Multiple sclerosis may arise from pathogenic T cells that somehow evaded mechanisms promoting self tolerance. These pathogenic T cells may rearrange a restricted number of T-cell receptor V alpha and V beta genes. This conclusion stems mainly from animal studies analyzing T-cell clones that recognize myelin basic protein, causing experimental autoimmune encephalomyelitis, and has recently been confirmed in studies of some T-cell clones from the peripheral blood in humans that recognize myelin basic protein. Experimental autoimmune encephalomyelitis disease could be induced or exacerbated through injection of low doses of the superantigen staphylococcal enterotoxin A and B. Recent research shows that exposure to various bacterial superantigens can induce myelin basic protein–specific T cells with different T-cell receptor V beta chain expression to mediate relapsing paralysis in experimental autoimmune encephalomyelitis. Similarly, because autoimmune conditions such as arthritis, diabetes, and multiple sclerosis are characterized by heterogeneous T-cell infiltrates during active disease, a broad spectrum of superantigens could precipitate exacerbation.

Recently, it was shown that bacterial superantigens play a role in the activation and expansion of T cells from individuals with multiple sclerosis. Two common superantigens, staphylococcal enterotoxin B and toxic shock syndrome toxin 1, were tested for activation of gamma-delta T cells. All the gamma-delta T cell clones tested showed a substantial reactivity to low amounts of the superantigens. Moreover, the T-cell clones lysed target cells pulsed with the superantigens, indicating a cytotoxic response elicited by these superantigens.

Although the evidence for a role of superantigens in autoimmune disease and multiple sclerosis is largely circumstantial, the potent stimulation of the immune response by superantigens can lead to acute disease and could reactivate autoreactive cells both systemically as well as inside organ lesions in autoimmune disease.

For background information SEE ANTIGEN; AUTOIMMUNITY; BRAIN; HISTOCOMPATIBILITY; NERVOUS SYSTEM DISORDERS; NEUROIMMUNOLOGY in the McGraw-Hill Encyclopedia of Science & Technology.

Stefan Brocke; Lawrence Steinman

Bibliography. S. Brocke et al., Infection and multiple sclerosis: A possible role for superantigens? *Trends Microbiol.*, 2(7):250–254, 1994; J. Burns et al., Isolation of myelin basic protein-reactive T cell lines from normal human blood, *Cell. Immunol.*, 81:435–440, 1983; P. Stinissen et al., Increased frequency of γδ T cells in cerebrospinal fluid and peripheral blood of patients with multiple sclerosis, *J. Immunol.*, 154:4883–4894, 1995; P. J. Vinken et al. (eds.), *Handbook of Clinical Neurology*, vol. 56, 1989.

Mycorrhizae

Plants do not grow alone. From the moment that roots develop in soil, they are penetrated and colonized extensively by filamentous fungi in fungus-root associations called mycorrhizae. These associations are mutualistic in that both host and fungus benefit. The plant receives inorganic nutrients and water from fungal hyphae foraging far beyond the root zone. The fungus, in turn, obtains a steady supply of carbon and energy directly from the plant with a minimum of competition from other soil microbes. The origin of mycorrhizae at least 400 million years ago indicates that they are critical to growth and reproduction of both plant and fungus. As a result of coevolution, mycorrhizae are found in most habitats worldwide and in approximately 95% of all plant species. Their universality is one reason that belowground microbial activity is being reevaluated from the perspective of a mycorrhizosphere. This mycorrhizosphere includes not only the root zone but a larger soil volume containing an integrative network of fungal hyphae, spores, and fruiting bodies (see **illus.**).

Mycorrhizal interactions between plants, fungi, and the environment are complex, interdependent, and often inseparable. Although much has yet to be learned about the dynamics of each association, it is clear that mycorrhizae are an essential belowground component for establishing and sustaining plant communities as the Earth's geography and climate undergo continual flux and change.

Mycorrhizal diversity. Seven types of mycorrhizae are known to occur in nature; all fit within two broad categories: (1) endomycorrhizae, where the fungal symbiont produces specialized structures for nutrient interchange with the host inside root cells and establishes a diffuse network of external fine hyphae in soil; and (2) ectomycorrhizae, where the fungus produces a Hartig net of hyphae between root cells together with an external mantle of hyphae encasing roots and penetrates surrounding soil with hyphae aggregated into thick shoestringlike rhizomorphs. Some fungi combine properties of each category. In ectoendomycorrhizae, the fungal symbionts form an ectomycorrhizal mantle, but hyphae in roots resemble endomycorrhizae by entering cortical cells.

Arbuscular endomycorrhizae are the most prevalent globally in more than 30,000 plant species, but fungal species currently number only 152 in the order Glomales of the phylum Zygomycota. Vegetative structural characters of fungi in roots define two broad groups: those forming finely branched arbuscules and lipid-rich intercellular vesicles, and those with coarser arbuscules and clusters of fragile, auxiliary cells outside roots. Species are defined by mode of spore formation and differences of microscopic structural characters inside broken spores. The simplicity of these characters is thought to be the cause of low number of species relative to the high diversity of habitats they coinhabit with plants.

Ectomycorrhizae also are widespread, but the range of hosts is limited to gymnosperms and some woody angiosperms. The fungi are much more taxonomically diverse, encompassing more than 4000 species spanning two phyla, Basidiomycota and Ascomycota. Species are classified mainly by structure and organization of fruiting bodies with diverse morphologies such as various mushrooms and puffballs. When sporulation is absent, some species can be identified by characteristics of the mantle covering the roots, such as color and branching patterns of component hyphae.

Less is known about other endomycorrhizal symbioses because they have narrow host ranges and are not as well studied. Ericoid mycorrhizae are formed by Ascomycete fungi that associate only with the roots of plants in Ericales. Hyphal coils penetrate in cortical cells that emerge and are arranged loosely on root surfaces. Orchidaceous mycorrhizae are restricted to orchids (Orchidaceae). The fungal symbionts are mostly Basidiomycetes that otherwise are

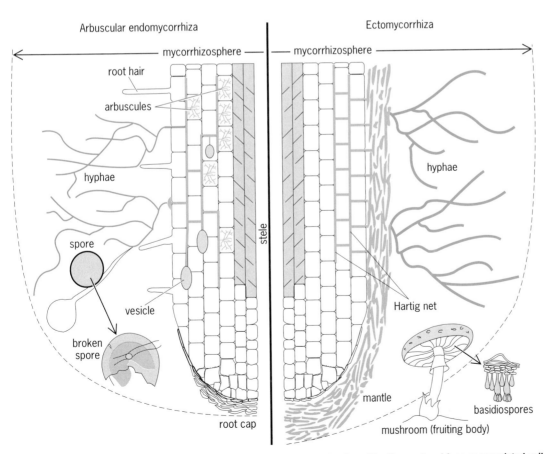

Fungal structures in an (arbuscular) endomycorrhiza and an ectomycorrhiza that with other root and fungus-associated soil microorganisms make up the mycorrhizosphere of individual plants.

pathogens, such as root and wood decay fungi. They establish hyphal coils in cells that degenerate and release nutrients during seed germination and early seedling development.

Host-fungus recognition. Arbuscular fungi have a wide host range, but plants in some families (for example, Cruciferae, Cyperaceae, and Polygonaceae) do not form mycorrhizae. They appear to have evolved away from the symbiosis, altering their genotypes to actively produce defense reactions against fungal colonization. Chemically induced mycorrhiza-resistant mutants of peas have been obtained to determine defense mechanisms. These plants release phenolic substances, callose, and pathogenesis-related proteins at the onset of fungal penetration into root cells, leading to suppression of colonization. Plants that are not mycorrhiza resistant lack such reactions, indicating that compatibility between host and fungus results from a suppression of genetic defense mechanisms. Mycorrhiza-resistant plants also do not form nodules, suggesting the presence of common plant symbiosis genes.

Similar mutants have not been obtained in ectomycorrhizal associations, but greater specificity between some fungi and their hosts provides model systems to determine the molecular basis for incompatible reactions. Nonhosts produce defense reactions similar to those observed by mycorrhiza-resistant legume mutants. However, hosts also produce some enzymes (such as chitinases and peroxidases) that may inhibit intracellular fungal colonization. Root cells normally undergo marked changes in shape and cell wall structure during mycorrhizal development, indicating strong host involvement. Host-encoded tubulin proteins increase with onset of mycorrhiza formation, and they are considered important determinants in host-fungus compatibility.

Nutrient exchange. The localized sites of nutrient exchange are intracellular in endomycorrhizal associations. The most intense activity occurs along a narrow interface zone between specialized fungal structures (arbuscules, hyphal coils) and the host cell plasma membrane. An arbuscule (or hyphal coil of ericaceous and orchidaceous fungi) develops after an intercellular hypha penetrates a root cell wall; it then is enveloped by the cell's invaginating plasma membrane. Ultrastructural studies indicate an electron-dense material is sandwiched between host and fungal boundaries, so there is no direct contact. Active bidirectional flow of carbon and inorganic ions occurs in this zone. The major nutrient exchanged, phosphorus, is released by the fungus in large amounts, and efficiency of the transfer to the host appears to be governed by arbuscule longevity, area of the interface, number of ion channels, and activity of transmembrane carriers. Complex interactions of source-sink relationships between shoots and roots also regulate the amount of mycorrhizal development. For example, increases in root-soluble carbohydrates tend to increase mycorrhizal colonization. Slight growth suppressions may occur from

carbon drain by the mycorrhizal fungus under conditions of high-soil nutrients, but usually disappear as soil nutrient levels decline and dependency on the symbiosis increases.

In ectomycorrhizal associations, host and fungal cell walls are in direct contact with each other. The fungal symbiont enzymatically induces cleavage of the middle lamella between root cells, followed by fusion of host and fungal cell walls into a common interfacial matrix. Elevated levels of enzymatic adenosinetriphosphate (ATPase) activity in fungal and host plasma membranes have been interpreted as indicators of bidirectional nutrient exchange.

Role of external hyphae. Arbuscular fungi produce a diverse range of external hyphae with different functions: (1) runner hyphae along roots for establishment of new sites of colonization to coincide with new root growth; (2) fertile hyphae in soil for production of spores individually, in aggregates, or in sporocarps; and (3) absorptive hyphae beyond the root zone for the uptake of nutrients and as a pipeline for translocation of nutrients back to host roots. External hyphae alter soil structure by increasing particle aggregation (important in sand dunes and disturbed soils) and by stimulating or inhibiting growth of other soil microbes. Ericoid endomycorrhizae in acidic heath soils produce external hyphae with a more active function, that is, an enzymatic release of nitrogen from predominant organic compounds that otherwise is unavailable to roots.

External hyphae of ectomycorrhizal fungi on tree species also excrete proteinases and phosphatases that break down organic matter and increase mineralization (and nutrient availability) of litter on the forest floor. Rhizomorphs greatly enhance the foraging potential of a fungus in organic matter and soil. Some fungi produce large mats of hyphae in litter to accelerate this process and also provide unique habitats for other soil microbes.

Ecological interactions. The capacity of plant communities to support a diverse fungal community is regulated partly by the genetics and physiology of the host plant and the aggressiveness of each fungal symbiont. Fungal colonization often is reduced under conditions of high soil nutrients (especially phosphorus) or foliar stresses on photosynthesis such as low light conditions. Some plant groups, such as alder and poplars, are able to support either arbuscular or ectomycorrhizal associations, depending on the environmental conditions and the kinds of fungal inoculum present. Arbuscular fungi have low host specificity, so as many as 8–10 species can coexist in the same plant root system. Species diversity seems to be similar in a wide range of habitats, from arid conditions to those characterized by a moderate amount of water. Higher degrees of host specificity regulate fungal species composition in ectomycorrhizal, ericoid, and orchidaceous associations.

Heterogeneous plant communities consist of plants with varying levels of dependency on the mycorrhizal association, which in turn impact on plant succession and other dynamic processes.

Plants may be obligately dependent (association required in all environments), facultatively dependent (associations required under some conditions such as low soil), or independent (associations not required). Some groups, such as ferns, range from total dependence (for example, Psilotopsida) to independence (such as aquatic Azollaceae) from mycorrhizae. Orchids rely completely on their fungal endophytes for seedling germination and establishment. In arid habitats, glacial moraines, and lava flows, early successional plants generally are mycorrhiza independent, and fungal communities are slow to develop. With the influx of mycorrhizal fungi, dependent hosts become more prevalent. In some tropical communities, plant diversity appears to increase when they are mycorrhizal.

Recent evidence indicates mycorrhizal fungi promote interplant communication, even among unrelated plants. Hyphal linkages have been found between roots of legumes and grasses by arbuscular fungi. Monotropoid mycorrhizae consist of hyphal linkages by ectomycorrhizal fungi between achlorophyllous parasitic plants and their tree hosts for nutrient exchange and other processes. The mycorrhizosphere includes many other organisms impacting on mycorrhizal processes, or vice versa. Helper bacteria appear to selectively promote the establishment and development of some ectomycorrhizal associations. Symbiotic nitrogen fixation by bacteria in legumes is strongly dependent on arbuscular mycorrhizae in order to meet the high demand for soil phosphorus. Soil pathogens may be inhibited or stimulated with changes in mycorrhizal root exudates, barriers created by ectomycorrhizal mantles, and many other processes.

Mycorrhizal manipulation. Many efforts have been made over the past 15 years to tap the benefits of mycorrhizae for agricultural and horticultural purposes. However, test results have not been consistently positive because of complications by many variables in fluctuating field situations. Success with inoculation of introduced organisms appears to be greatest in disturbed or polluted soils and in locations where the proper fungi for a given type of host association are absent.

Production of mass inoculum varies with mycorrhizal type because methods for culturing the fungal symbionts differ greatly. For example, arbuscular fungi must associate with living roots to grow and reproduce. Batch cultures are produced by growing whole plants in soil mixes, in a soilless growth media (such as sand), or in closed chambers where roots are bathed in a nutrient mist. Ericaceous or orchidaceous endomycorrhizal fungi can be cultured on synthetic media, but batch cultures are uneconomical because growth is too slow. Most ectomycorrhizal fungi also can be cultured rapidly on synthetic media. Commercial inoculum generally consists of fungal material grown or mixed with sterile vermiculite, expanded clay, or other inert and lightweight carriers.

In natural ecosystems, introduced fungi do not appear to offer a superior advantage to native species. Research has focused on developing strategies to optimize the ability of native fungi to colonize hosts in their natural habitat or to minimize loss of these fungi with disturbance. Practical efforts have been limited thus far to agroecosystems. For example, highly dependent crop hosts are selected over mycorrhizal independent hosts in crop rotations or in multiple cropping systems. Research suggests that traditional methods of breeding and producing crop plants in soils with high nutrients may select against the most efficient fungal communities or even against the mycorrhizal association, but preventive strategies have yet to be implemented. SEE FUNGI.

For background information SEE ECOSYSTEM; FOREST ECOLOGY; FUNGI; MYCORRHIZAE; PLANT PATHOLOGY; SOIL ECOLOGY in the McGraw-Hill Encyclopedia of Science & Technology.

Joseph B. Morton

Bibliography. A. D. Robson, L. K. Abbott and N. Malajczuk (eds.), *Management of Mycorrhizas in Agriculture, Horticulture and Forestry* 1994; A. Varma and B. Hock (eds.), *Mycorrhiza: Structure, Function, Molecular Biology, and Biotechnology*, 1995.

Nanostructure

Computer chip manufacturers are continually progressing in making smaller features on silicon, the dominant semiconductor used to fabricate such chips. With smaller transistors, diodes, and junctions, more devices can be packed onto a single computer chip, lowering manufacturing costs. How quickly circuits can be switched on and off depends on the speed of the electrons, the size of individual devices, and the length of connections between them. Packing circuits closer together allows them to run faster.

However, as this miniaturization continues, the characteristics of the underlying materials begin to change significantly. Of fundamental importance is the effect of reduced size and dimensionality on the electronic properties of materials. The motion of electrons within them is constrained as their dimensions are less than the wavelength of the electrons. Such low-dimensional solids display electronic properties that are significantly different from the bulk material; they are referred to as quantum size effects. Devices are beginning to be developed that utilize these effects and associated phenomena such as light emission from porous silicon. Among the techniques being explored for the fabrication of such devices is chemical vapor deposition (CVD) in a porous template.

Fabrication of small features. In the electronics industry, small devices and connections on computer chips are fabricated by photolithographic techniques. A photoresist is first applied to a silicon wafer. A light source shines through a mask onto the photoresist, causing a chemical change and thus patterning the photoresist. The wafer is placed in an etching solution which selectively removes the photoresist, and etches away silicon where the photoresist has been removed.

The size of features created by photolithography is limited by the wavelength of light used. The most advanced mass-produced chips have features on the order of 0.3 micrometer. To reduce this size further, light of shorter wavelength must be used, but this approach presents difficulties. Inexpensive, intense light sources in the deep ultraviolet have yet to be perfected, and good masks and photoresists with desired absorption properties need to be developed.

By using even shorter-wavelength sources, far smaller features can be created. Intense x-rays from a plasma discharge or synchrotron radiation have been proposed as sources. Electron and ion beams have also been used to create circuits. They have the advantage over x-rays of being easily focused to very small dimensions.

Scanning tunneling microscopy (STM) can be use to create features as small as a few atoms. By manipulating the tip voltage and current, a scanning tunneling microscope tip can be used to move atoms around, to carve channels and circuits into a surface with good precision and accuracy. Although a scanning tunneling microscope could be used to prepare nanostructures, it would take far too long to prepare a usable device with today's technology. Large, massively parallel arrays of scanning tunneling microscopes have been proposed to build nanostructures in a reasonable amount of time, but manufacturing a huge array of scanning tunneling microscopes has not yet been accomplished.

Quantum size effects. Circuits cannot be made smaller and run faster without limit. The speed at which electrons will travel in the material has a finite limit, and the material itself changes characteristics. As a circuit gets very small, it begins to show unusual properties. Isolated single atoms have well-characterized discrete electron energy levels, as shown on the left side of **Fig. 1**. If a second atom is added, forming a dimer, the energy levels of the atoms mix, forming new molecular orbitals higher and lower in energy than for the isolated atom. As more atoms are added and a crystal forms, the energy levels increase in number and continue to mix. When the crystal reaches a macroscopic size, as indicated on the right side of Fig. 1, individual discrete orbital levels are no longer discernible; a continuum of molecular orbitals is present, known as a band structure.

For certain crystal sizes between bulk solids and isolated atoms, on the order of 1 to 50 nanometers, called the mesoscopic size domain, the electronic properties can be described in terms of an electron-hole-in-a-box formalism. This endows the materials with unusual electronic properties known as quantum size effects (QSEs). A quantum dot is a crystal in which all dimensions are small enough to show

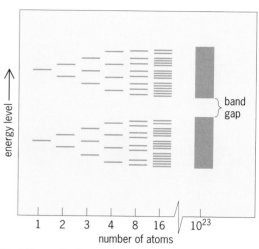

Fig. 1. Energy levels of a solid as a function of cluster size.

quantum size effects. A quantum wire has one long dimension, and shows quantum size effects in two dimensions. A quantum sheet has two long dimensions, and shows quantum size effects in one dimension. *SEE INFRARED RADIATION; LASER.*

Quantum size effects arise from confinement of electrons to small regions of space. The properties of a quantum dot can vary greatly with only a small shift in its size. Therefore, to make structures and devices in the mesoscopic size domain and take advantage of quantum size effects, strict control of size is crucial. Quantum size effects have been proposed to explain many unusual phenomena, such as light emission from porous and nanocrystalline forms of silicon.

In 1990, L. Canham reported the discovery of light emission from porous silicon. By anodizing silicon wafers in hydrofluoric acid solution, a porous network was created on the silicon surface. By adjusting the electrochemistry and acid concentration the pore size can be controlled over a wide range. Canham found his materials to be photoluminescent at room temperature. He immediately saw the potential for silicon-based optoelectronic devices to improve the interface between light-based fiber optics and electricity-based silicon chips.

Chemical vapor deposition in porous templates. Chemical vapor deposition (CVD) is a method of depositing a layer of material from a gaseous precursor. It is a versatile technique, and by choosing the proper gaseous precursor, a wide variety of materials have been synthesized, from insulators to superconductors. Chemical vapor deposition is commonly used industrially to synthesize layers of silicon and III-V semiconductors such as gallium arsenide for electronic devices. One method of preparing uniform quantum dots or wires is chemical vapor deposition in a porous template. By using a template, growth is constrained by the pore walls. Clusters can grow only as large as the pore volume, provided that the template remains intact.

Zeolite templates. Examples of such template materials include zeolites, a family of nanoporous aluminosilicate materials with well-defined pore shapes and volumes. Zeolites are used as catalysts, for gas separation, and as adsorbents. A variety of natural and synthetic zeolites are known, with large variations in pore sizes. Depending on the zeolite, the cavities can form one-, two-, or three-dimensional connected networks with known architecture. A periodic array or superlattice of uniformly sized clusters is more amenable to theoretical treatment and modeling than a random distribution of small crystals with differing sizes.

Zeolite Y has a porous adamantane structure, with a central 1.3-nm cavity, which is accessible through four tetrahedrally disposed windows of approximately 0.8-nm diameter. Chemical vapor deposition of disilane (Si_2H_6) in zeolite Y has been used to prepare an organized array of silicon quantum dots. Disilane initially anchors in the zeolite Y by reacting with the Brønsted acid sites of the zeolite Y. Reductive elimination of hydrogen from the anchored disilane can then be induced by controlled thermal treatment forming silicon nanoclusters within the zeolite Y. Further anchoring and heat treatment can increase the cluster size, up to a theoretical maximum of about 60 silicon atoms.

These stable silicon clusters are capped by the zeolite host and exhibit photoluminescence at room temperature. Similarly, germanium nanocluster arrays have been synthesized by chemical vapor deposition of digermane in zeolite Y, and composite silicon-germanium materials have also been prepared from disilane-digermane mixtures. This allows for compositional tuning of the energy levels in an analogous fashion to conventional semiconductors.

MCM materials. By using hosts with larger pore sizes, larger nanomaterials can be prepared. The MCM family of porous silica materials can be used to create clusters over a wide size and architecture range. For example, the MCM-41 materials consist of long parallel mesocylinders arranged hexagonally. Disilane can be anchored in MCM-41 and silicon clusters created by using similar conditions to those used with zeolite Y. **Figure 2** represents an idealized picture of silicon mesowires inside the mesoporous MCM-41 silica host. By choosing the synthesis conditions, the pore size in MCM-41 can be varied from 2 to 10 nm. Adjusting the pore size allows control over the dimensions of the resulting silicon clusters. As the silicon clusters increase in size, the optical absorption edge and light emission are red-shifted, both being manifestations of quantum size effects in silicon.

Anodic porous alumina templates. When aluminum is anodized in an acidic solution, a film of alumina with hexagonal pores can be formed, growing outward from the aluminum surface. The pores are parallel cylinders with diameters varying from 4 to 200 nm, depending on the reaction conditions used for the anodization. The pore size can be increased further by treating with phosphoric acid, which attacks the walls of the pores. These pores are in the size range

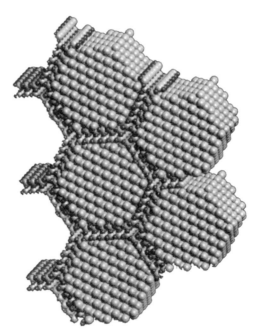

Fig. 2. Silicon mesowires formed in a hexagonal mesoporous silica host.

expected to show quantum size effects for many materials. They have a large aspect ratio, being on the order of 2 μm long. A number of materials have been deposited inside the porous alumina, such as gold and nickel, forming organized arrays of conductive nanowires in the insulating alumina substrate that exhibit novel optical and electrical properties.

Recently, replica materials were made from anodic porous alumina. A monomer, methyl methacrylate, was deposited into the porous alumina and polymerized to give a negative structure of polymethylmethacrylate (PMMA). The alumina was removed by using basic solution, leaving an array of parallel polymethylmethacrylate cylinders. By using electrochemical methods, platinum and gold were deposited around these cylinders. The polymethylmethacrylate can be removed with acetone, leaving the porous metal as a positive replica of the anodic porous alumina.

This replica method improves the mechanical and chemical stability of porous alumina. It may be possible to deposit other materials in the porous metal, forming composite materials with unique catalytic, membrane separation, electronic, and optical properties.

Luminescent quantum dots. Recently, luminescent indium arsenide (InAs) quantum dots have been prepared on a gallium arsenide (GaAs) surface by molecular beam epitaxy. Because the crystal lattice spacing in indium arsenide is different from gallium arsenide, indium arsenide settles randomly on the gallium arsenide surface, and acts as seeds for subsequent nucleation and growth, forming clusters that exhibit quantum size effects and luminescence.

For background information *SEE INTEGRATED CIRCUITS; NONRELATIVISTIC QUANTUM THEORY;*

SCANNING TUNNELING MICROSCOPE; SEMICONDUC-
TOR HETEROSTRUCTURES; ZEOLITE in the McGraw-
Hill Encyclopedia of Science & Technology.

Emmanuel Chomski; Geoffrey A. Ozin

Bibliography. L. T. Canham, Silicon quantum wire
array fabrication by electrochemical and chemical
dissolution of wafers, *Appl. Phys. Lett.,* 57:1046–
1048, 1990; R. P. Feynman, There is plenty of room
at the bottom, *Sci. Eng.,* 23: 22, 1960; C. N. R. Rao
and J. Gopalakrishnan, *New Directions in Solid
State Chemistry,* 1989; R. Turton, *The Quantum Dot:
A Journey into the Future of Microelectronics,* 1995.

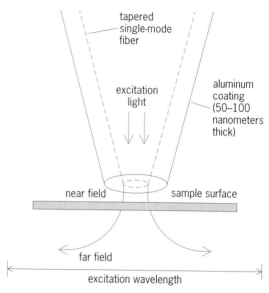

Near-field fiber optic probe and its position relative to the sample surface.

Near-field microscopy

The desire to conduct optical measurements with
subdiffraction limit spatial resolution has led to
the development of near-field scanning optical
microscopy. This technique circumvents the spatial
constraints normally encountered in optical microscopy, and has opened the way to some exciting
possibilities.

The spatial resolution limit in conventional optical microscopy (approximately one-half the wavelength of the light used) is not a fundamental limit
but rather arises from the position of the detection
element (lens) many wavelengths away from the
sample. It was recognized in the early 1900s that the
resolution in optical microscopy could therefore be
increased by scanning a nanometric probe (detector or radiation source) in close proximity to a sample surface. In this way, the resolution is a function
of only the probe diameter and probe-to-sample
distance, and not the wavelength of the light source.
Although conceptually simple, implementing the
optical equivalent to obtain superresolution has
only been realized in the last few years. The progress has been slowed by the formidable technical
challenges associated with (1) fabricating a nanometric light source and (2) positioning this light
source within nanometers of a sample surface.

Near-field fiber optic probe. In most applications
of near-field scanning optical microscopy, the nanometric spot of light is delivered by using a specially
fabricated, tapered, single-mode optical fiber that is
coated with aluminum around the sides (see **illus.**).
These probes are challenging to make, and their
quality and performance depend on a number of
factors, such as taper shape and aluminum-coating
grain size. The aluminum coating is used to confine
the light until it reaches the aperture at the end
(<100 nm). To obtain high spatial resolution, the tip
is positioned nanometers away from the sample surface to catch the emerging light before it diffracts
out. In this way, the spatial resolution is determined
by the tip diameter and tip-sample gap, and not by
the wavelength of the light used.

Typically probes with tip diameters of 80–100 nm
can deliver tens of nanowatts of light; although at
first this may sound prohibitively small, the power
density exiting the tip is on the order of 100 W/cm^2.

This provides an ample light source for most spectroscopic applications. Not surprisingly, reducing
the size of the aperture dramatically affects the
throughput of light, and research is focused on both
maximizing the existing tip geometry and exploring
alternative probe designs in hopes of pushing the
current spatial resolution limit.

Shear-force distance control. To obtain high resolution, the probe must be positioned and held within
nanometers of the sample surface during a scan (see
illus.). Although a number of schemes have been
introduced to accomplish this, a technique based on
shear-force microscopy is the most successful. In
this method, the fiber-optic probe is dithered laterally approximately 5–10 nm at its resonant frequency while the amplitude of the oscillation is
monitored. As the probe nears the sample surface,
there is a sharp decrease in the amplitude resulting
from interactions between the tip and the sample
surface. This drop in signal can be used in a feedback loop to regulate the tip-sample gap and maintain it at a fixed distance during a scan. Not only
does this provide a mechanism for keeping the tip
near the surface to obtain the high-resolution optical images, but it also provides a simultaneous force
mapping of the surface topology, similar to that
obtained in atomic force microscopy. This feature
allows for a unique comparison to be made in near-
field microscopy between surface morphology and
optical properties simultaneously.

Single-molecule detection. The detection of the
fluorescence from a single molecule at room temperature with near-field scanning optical microscopy was first demonstrated in 1993, and has since
been repeated by several groups. This extraordinary sensitivity in large part results from the dramatic reduction in background signal that results
from the use of the nanometric near-field light
source. The nanowatts of excitation light exiting
the near-field tip is easily blocked by using filters,

which results in a nearly background-free fluorescence measurement. The sensitivity is also enhanced by the evanescent wave components to the electric field present near the near-field aperture. These nonpropagating modes can excite the fluorophore, thereby increasing the fluorescence signal. Because these are nonpropagating modes, however, signal calculations based on the observed far-field excitation power and the absorption cross section of the molecule can give only a lower estimate on the true fluorescence signal in near-field scanning microscopy.

In the original single-molecule study, unusual features in the fluorescence image were noticed. The near-field fluorescence image of the individual molecules contained not only the expected round symmetric features but a distribution of rings, arcs, double arcs, and ellipsoids. These unusual observations are explained by considering the nature of the electric field exiting the near-field tip. Early treatments on the passage of light through a small aperture in a thin conducting screen predicted the presence of a large z component (perpendicular to the aperture) in the electric field near the edges of the aperture. This simple model turns out to give a good qualitative description of the electric field present at the near-field probe aperture, and therefore it predicts similar z components in the electric field near the tip edges. The electric field emerging from the near-field tip therefore contains both parallel (near the center of the tip) and perpendicular (near the edges of the tip) components. Molecules with dipoles oriented parallel to the tip aperture will be preferentially excited when the tip is centered above them, and molecules oriented perpendicular to the tip aperture will be preferentially excited when they are near the edges. Therefore, as the tip is scanned across a single molecule the observed fluorescence will depend on both the tip-molecule position and the orientation of the molecule, which explains the distribution in fluorescence shapes observed in the original single-molecule near-field scanning optical microscopy experiment. By analyzing the shapes observed in the near-field fluorescence image, it is possible to determine the orientations of the molecules.

Fluorescence lifetime measurements. The detected fluorescence count rate from a single molecule is impressive. Fluorescence count rates of thousands to tens of thousands of counts per second are routinely observed. Thus, not only the detection of single molecules but also spectroscopic measurements in either the time or frequency domains at the single-molecule level are possible. For example, several studies have reported fluorescence lifetime measurements of single-dye molecules, measured by incorporating an ultrafast laser system with near-field scanning optical microscopy. These measurements are carried out by accurately positioning the near-field tip above a single molecule and measuring the fluorescence decay using the time-correlated single-photon counting technique. Nominally, fluorescence decays with good signal-to-noise ratios can be collected in approximately 5 s; thus several measurements can be taken on the same molecule before it photobleaches. (Photobleaching is a light-induced process that leads to a change or disappearance of a molecule's spectral features.) Studies have concentrated on elucidating the effect that the aluminum coating of the near-field tip has on the observed fluorescence lifetime by measuring the lifetime as a function of tip-molecule position. The results indicate that this is a complicated problem.

Early reports showed that at tip-molecule gaps of 7 nm there was a decrease in the fluorescence lifetime as the tip was displaced laterally from being centered above the molecule. This decrease was attributed to quenching of the fluorescence by the nearby aluminum coating of the near-field tip. Later it was shown that the fluorescence decay could also increase under certain conditions as the tip was displaced laterally from the centered position. Therefore, both a lengthening and shortening of the fluorescence lifetime can be observed with near-field scanning optical microscopy as the tip is moved off center of the molecule. These seemingly contradictory results can be explained by examining the z dependence in the lifetime behavior. The observed single-molecule lifetimes are extremely sensitive to the tip-sample gap. This sensitivity results from a complicated competition between radiative and nonradiative energy transfer processes between the molecule and the nearby aluminum coating of the near-field probe. Theoretical treatments of these processes have been slowed by complications arising from the limited geometry of the tip. However, this complex problem is being studied, and a unified picture of the mechanisms involved has begun to emerge. Fortunately, however, for the measurement of fast processes (less than approximately 1–1.5 nanoseconds) there is minimal perturbation, and the true lifetime can be extracted as long as the tip is directly centered above the molecule.

Fluorescence spectra of single molecules. The emission and excitation spectra of individual molecules have also been studied with near-field scanning optical microscopy. As in the case of the lifetime measurements discussed above, the perturbative effect of the nearby aluminum-coated near-field tip needs to be considered before useful information can be extracted from these measurements. Recent calculations for a simplified near-field geometry, however, indicate that the perturbation to the measured spectral features will be on the order of 10^{10} Hz. At room temperature, this shift is negligible compared to the electronic spectral width, and essentially perturbation-free spectra can be obtained with near-field scanning optical microscopy. However, for low-temperature applications, where the spectral features become significantly narrowed, perturbations from the near-field tip may become important. This remains to be investigated.

The room-temperature fluorescence spectra of individual dye molecules measured with near-field scanning optical microscopy was first reported by J. K. Trautman and coworkers. In this study the single-molecule fluorescence was dispersed onto a sensitive charge-coupled device detector. The spectra of individual molecules were found to be typically narrower than that of the bulk, many-molecule spectrum. This result reflects a decrease in the inhomogeneous contribution to the line width that arises from the distribution in molecular environments in bulk measurements. The peak locations of the single-molecule spectra were distributed ±8 nm from that of the bulk spectrum. Again, these shifts reflect the different molecular environments of the individual molecules and illustrate the detailed, molecular-level information that can be obtained with near-field scanning optical microscopy. Interestingly, the spectra of some of the molecules were found to be dynamic, shifting as much as 10 nm on the time scale of the experiment (minutes). Thus, together with the observation that individual spectral widths varied a great deal, a model was developed in which there is a distribution in barrier heights to rearrangement in the molecular environment.

Excitation spectra of single molecules have also been measured with near-field scanning optical microscopy. In this arrangement, the total fluorescence is collected as the wavelength of a dye laser is scanned across the absorption feature. In agreement with the results from the above emission studies, the excitation spectra of single-dye molecules are found to be narrower than the bulk spectrum, and the location and shapes of the spectral features vary among the different molecules. This technique may have advantages in studying the dynamic processes responsible for the time-dependent spectral shifts observed for single molecules. In single-excitation wavelength studies, molecular events that shift the absorption peak significantly can result in the total loss of fluorescence signal. It is ambiguous in such experiments whether the loss in signal is due to spectral fluctuations or to photobleaching. By scanning the excitation source and collecting excitation spectra, further insight into these processes may be gained.

Biological applications. Up to this point this article has focused strictly on the high spatial resolution, sensitivity, and spectroscopic capabilities of near-field scanning optical microscopy. However, another advantage may turn out to be particularly informative in studies on biological specimens. As mentioned earlier, to obtain the high-resolution optical images, a feedback mechanism is implemented to keep the tip near the sample surface during a scan. A direct result of this is the simultaneous collection of a shear-force mapping of the surface topology, similar to that obtained in atomic force microscopy. Thus a unique comparison can be made in near-field scanning optical microscopy between surface contour features and optical properties.

Membrane fragments from photosynthetic systems have been studied with near-field scanning optical microscopy in which simultaneous shear-force and fluorescence images were collected. The vertical noise in the shear-force image was less than 0.5 nm, which allowed for the visualization of single-lipid bilayers (approximately 6–7 nm high) lying flat on the surface substrate. Also, the researchers could distinguish single bilayers (6–7 nm) from other more complicated stacked structures (12–14 nm, 18–21 nm, and so forth) that would have been difficult to discriminate against with only the fluorescence image. This feature removes many of the constraints on sample homogeneity because a region of interest can be found by using the shear-force image and simply zoomed in on for further study. In this particular study, the simultaneous shear-force and fluorescence images of single-membrane fragments were compared and used to study the distribution of embedded light-harvesting protein complex. These and similar studies demonstrate the feasibility of conducting simultaneous shear-force and fluorescence measurements on biological systems in a nonperturbative way, at the nanometric spatial dimension.

All the studies discussed above have been done on dry samples. An important next step for near-field scanning optical microscopy, especially for the biological applications, is the advancement to imaging in aqueous environments. Steps in this direction have been taken by using several approaches, such as one based on the use of bent-fiber optic probes. Instead of laterally dithering the near-field probe as in the shear-force technique, the bent-fiber tip is vibrated vertically to the sample surface. This method is analogous to tapping-mode atomic force microscopy, and shares many of its advantages such as gentleness to the sample being imaged. Preliminary work has shown that this form of near-field scanning optical microscopy tip control can be used to image in aqueous environments. The shear-force technique has also been successfully implemented in aqueous environments and used to image membrane fragments. Damping of the tip-dither amplitude by the intervening water layer was reduced in these experiments by using a thin, 250-micrometer water layer. Although researchers found no evidence of sample degradation due to interactions with the near-field tip, questions remain as to the gentleness of the shear-force technique for imaging softer specimens, such as the membrane of an intact cell.

Prospects. The major technical barriers of implementing near-field scanning optical microscopy have been overcome, and it is rapidly taking its place alongside other forms of microscopy. The single-molecule fluorescence detection limit, nanometric spatial resolution, noninvasiveness, and spectroscopic capabilities of near-field scanning optical microscopy provide a powerful new tool for researchers in a broad range of disciplines.

For background information *SEE FLUORESCENCE; OPTICAL MICROSCOPE; SCANNING TUNNELING MI-*

CROSCOPE in the McGraw-Hill Encyclopedia of Science & Technology.

Robert C. Dunn

Bibliography. E. Betzig and R. J. Chichester, Single molecules observed by near-field scanning optical microscopy, *Science,* 262:1422–1425, 1993; E. Betzig et al., Breaking the diffraction barrier: Optical microscopy on a nanometric scale, *Science,* 251: 1468–1470, 1991; D. W. Pohl, Scanning near-field optical microscopy (SNOM), *Adv. Opt. Electr. Micro.,* 12:243–312, 1991; J. K. Trautman et al., Near-field spectroscopy of single molecules at room temperature, *Nature,* 369:40–42, 1994; X. S. Xie and R. C. Dunn, Probing single molecule dynamics, *Science,* 265:361–364, 1994.

Neural network

In order to understand the application of neural networks to the control of aircraft, it is necessary to understand the mathematical description of a model of an aircraft, the use of the controller and the estimator to guide and direct the aircraft, and the neural-network implementation of the controller-estimator. **Figure 1** shows the architecture of a particular type of controller called the model-reference controller. Each subblock of the figure will be described in turn in order to arrive at an understanding both of this type of aircraft controller and of the advantages of a neural-network implementation. The analysis will begin with the block labeled neural-network controller. (It is also an estimator, as discussed below.)

Principles of neural networks. A neural network is a simple device capable of emulating complex models. Two neural networks are shown in **Fig. 2**. The lowest or input layer of the left neural network, labeled the feedforward network, is used to impress a set of numbers on the neural network. Each of the nodes of the layer (shown as circles) in this neural network is fully connected to the nodes of the next layer in the network successively until the output layer is reached. The output layer is shown as two nodes. Associated with each of the connecting lines is a weight which is adjusted to obtain a desired result.

All calculations in a neural network are done at the nodes. A typical node, as shown in the third layer of the feedforward network, is connected to all the nodes of the previous layer. The node sums all its inputs, and then by using a transfer function, which can be nonlinear, calculates an output to the nodes of the next layer.

Associated with each set of input numbers (vector) is a desired set of output numbers (vector). The key concept which distinguishes the neural-network approach to a control problem from other methods is the fact that the neural network is trained. The training is accomplished by presenting input-output data sets or vectors to the network and adjusting the connection weights so that with a given input vector the desired output vector is obtained. The method used in these calculations is called the back propagation algorithm. An analytical relationship can be developed between the input and output vectors in the case of an aircraft, although this is not necessary for the training of a neural network. Only experimental pairs of input and output vectors are needed. Any relationship between the input and output vectors can be emulated by a neural network. Returning to Fig. 1, an output vector from the controller drives the subblock labeled F/A-18 longitudinal dynamical model.

Longitudinal dynamical model. The F/A-18 is a current Navy fighter aircraft whose mathematical model is used in this article. The center block of the longitudinal dynamical model in Fig. 1 refers to the longitudinal dynamics of the F/A-18. This is the motion of the aircraft in a vertical plane through the axis of symmetry of the aircraft. A person standing in the center aisle of a commercial aircraft is standing in this plane. The forward and vertical velocities of the aircraft and the pitching motion, say, about the wings, are part of the longitudinal dynamics of the aircraft. The control surfaces of the aircraft such as the elevators are moved in order to cause the aircraft to climb or

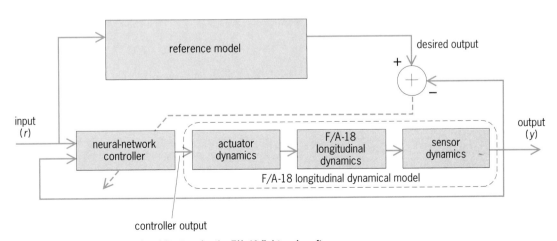

Fig. 1. Model-reference control architecture for the F/A-18 fighter aircraft.

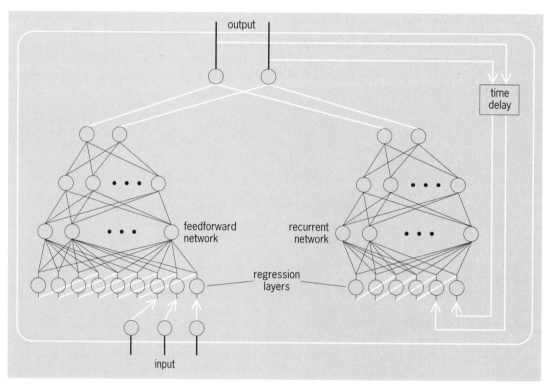

Fig. 2. Artificial neural-network emulator. Connections within the neural network are indicated by black lines. Information flow outside the neural network is indicated by white lines. In the regression layer, information is shifted to the left after each step, as indicated by diagonal lines.

descend in the vertical plane. Several control surfaces are used in the longitudinal dynamics of the F/A-18. These surfaces include the stabilators, the leading-edge flaps, and the trailing-edge flaps. The motion of the control surfaces is accomplished by means of actuators, whose dynamics are described, as with the airplane itself, by a set of equations. A further set of equations is used to describe the sensor dynamics.

A linear mathematical model of the F/A-18 including actuator and sensor dynamics is given by the set of equations below. Here, the vector z represents

$$z(k+1) = Fz(k) + Gr(k)$$

$$y(k) = Hz(k) + Dr(k)$$

sents the states (the variables needed to describe the system), the output is y, and the input is r. A discrete model is used, as shown by the discrete time variable k. The capital letters (F, G, H, and D) in the equations are matrices. The elements of the matrices are formed from the aerodynamic influence coefficients, which are a set of numbers that describe the particular aircraft under consideration, and depend on the geometry and the flight conditions (speed and air density) of the aircraft. In order to describe each of the flight regimes of the aircraft, multiple linear models, each having a different set of matrices F, G, H, and D, are needed. For the analysis discussed in this article, two different models for the F/A-18 were selected. The first model (called model 1) is that of normal straight and level flight. In the second model (model 2)

there is some damage to the control surfaces so that they are not as effective as in the normal case. The consideration of two models introduces a nonlinearity into the analysis.

In Fig. 1, the output (y) is subtracted from the desired output, and the difference is fed back into the neural-network controller. The desired output comes from the block labeled the reference model.

Reference model. As the aircraft goes through various flight regimes the response of the aircraft varies according to the control input generated by the pilot. It would be useful if the pilot always obtained the same response to the same input. This effect can be achieved by using a reference model and having the controller vary its inputs to the plant so that the output matches that of the reference model. The reference model selected for this controller is based on the military specification for the flying qualities of a piloted airplane. The same reference model was selected for both damaged and undamaged F/A-18 models.

The reference model can be described by specifying the modes associated with the longitudinal dynamics. These dynamic vibration modes are termed short period and phugoid. By giving the frequency and damping of these modes, the fourth-order dynamics (corresponding to the four variables: forward velocity, pitch rate, vertical velocity, and pitch angle) of the reference model can be completely specified.

In Fig. 1, the input (r) is fed into both the reference model and the neural-network controller. The

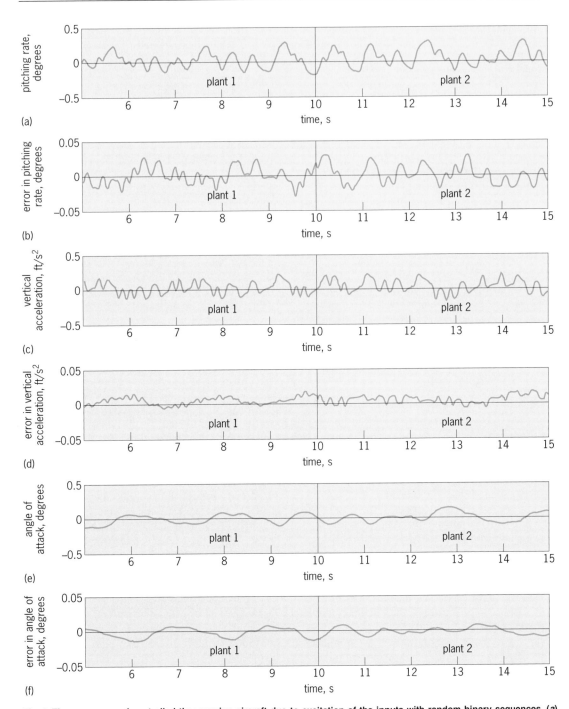

Fig. 3. Time response of controlled time-varying aircraft due to excitation of the inputs with random binary sequences. (*a*) Pitching rate. (*b*) Error in pitching rate. (*c*) Vertical acceleration. (*d*) Error in vertical acceleration. (*e*) Angle of attack. (*f*) Error in angle of attack.

output of the reference model (the desired output) and the output of the aircraft (*y*) are compared, and changes are made to the neural-network controller until outputs are the same. This is the significance of the broken line feeding back to (and through) the neural-network controller.

The description of the elements of a model reference neural-network controller is now complete. Thus, the definition of what is desired from the neural-network controller in a model reference architecture has been given. Given an input

(*r*), the controller should adjust its output so that the output (*y*) of the plant or aircraft matches the output of the reference model (the desired output). This matching should occur even with changes in the plant model. The question arises as to whether this is possible for a neural-network controller. A numerical experiment will now be described which shows that this desired result is indeed possible.

Numerical experiment. The neural-network architecture is given in Fig. 2. The network consists of two parts. The first part is a feedforward network, discussed above. The other network is a recurrent network, which involves direct feedback of the output as shown. The regression layer on the left of Fig. 2 (so called because it is involved in the evaluation of variables at previous times) requires classically 25 nodes for a 25-order plant (that is, a plant requiring 25 variables to model it). The connection weights for all but three nodes are very small in the trained network. This permits the use of a third-order regression (that is, the evaluation of variables at three previous times). This is an important result since it permits quick identification of aircraft model changes. The changes in the regression layer are effectively the estimator part of the neural network. For the F/A-18 there are 3 outputs and 6 inputs so the third-order regression layers contain 9 (3×3) and 18 (3×6) nodes, respectively. The layers between the input and the output contain 36 nodes.

Training of the neural network is based on excitation with random binary sequences of the inputs with further random selection of the model 1 and model 2 presented to the neural network. (For example, at the tenth step, model 1 was selected at the eleventh step, and so forth.) The mathematical models of both conditions give the needed input-output vectors. The neural-network controller is thus tasked to simultaneously learn the damaged and the undamaged models. Correct modeling of the two systems, which represent the nonlinearity in the modeling, is obtained.

The numerical experiment consists of inputting random binary sequences into model 1 and, at a time of 10 s, switching to model 2. The task of the neural-network controller is to identify that it is now dealing with model 2 and to properly control it. By proper control is meant that the reference model outputs are still matched after the model has been changed.

Figure 3 shows the time response of the controlled time-varying aircraft due to the random inputs. The change in models occurs at a time of 10 s, at which time the damaged aircraft is inserted in the simulation. Figure 3a has two curves plotted, the reference and plant pitching rates. The curves are so close to each other that it appears that only one curve is plotted. At the time of change of the model, 10 s, there are no large transients. Also shown in Fig. 3 are the vertical acceleration and the angle of attack. Below each of these curves is the error signal, that is, the difference of the reference model output and the plant or aircraft output. All the errors are small.

Thus, a neural-network controller can identify and control multiple models of an aircraft on which it has been trained. These models do not need to be damaged aircraft, as discussed in this article, but can be an expression of different flight regimes. The neural-network controller acts as an automatic estimator and reconfiguration controller.

For background information SEE CONTROL SYSTEMS; FLIGHT CHARACTERISTICS; FLIGHT CONTROLS; MILITARY AIRCRAFT; NEURAL NETWORK in the McGraw-Hill Encyclopedia of Science & Technology.

Daniel J. Collins

Bibliography. *Government Workshop on Neural Networks, NOSC, San Diego,* 1990; IEEE Neural Networks Council Staff, *IEEE-INNS International Joint Conference on Neural Networks, 1990,* 1992; W. L. Rogers and D. J. Collins, X-29 H∞ controller synthesis, *AIAA J. Guidance, Control Dyn.,* 15: 962–967, 1992; World Congress on Neural Networks, *Proceedings of the 1993 International Neural Network Society Annual Meeting,* 1993.

Nobel prizes

The Nobel prizes for 1995 included the following awards for scientific disciplines.

Physics. Frederick Reines, Professor Emeritus at the University of California, Irvine, and Martin L. Perl of the Stanford Linear Accelerator Center (SLAC) were awarded the prize for discoveries of a class of fundamental particles known as leptons. In 1956, Reines and the late Clyde Cowan, working at Los Alamos National Laboratory, detected the neutrino. Perl led a group at SLAC that discovered the tau lepton in 1975.

The neutrino, postulated in 1930 by W. Pauli to explain missing energy in beta decay, is an uncharged, apparently massless particle that interacts with matter so rarely that it was widely believed to be undetectable. To see it, Reines and Cowan placed a large tank of water with dissolved cadmium chloride next to a nuclear reactor, which should emit huge numbers of neutrinos. They observed a few events each hour in which a positron and a neutron were created when a neutrino collided with a hydrogen nucleus in the tank, resulting in characteristic emission of gamma rays. Besides raising the neutrino from a hypothetical entity to an observed particle, this discovery opened the field of neutrino physics, in which Reines has been deeply involved.

The electron and the muon are elementary particles which appear similar in most respects except that the muon is about 200 times more massive. They came to be called the first and second generations of the lepton family, but the duplication was puzzling. However, Perl decided to search for even more lepton generations by using the newly constructed Stanford Positron Electron Ring Accelerator (SPEAR) collider at SLAC. Events were observed in which only an electron and a muon were detected, and which appeared to be understandable only as the decay of a new, massive lepton. The 1975 paper reporting these observations met with initial skepticism, but by 1979 the existence of a new particle called the tau lepton, about 3500 times as massive as the electron but otherwise resembling the electron and muon, was confirmed. The discoveries of the neutrino and the tau lepton helped lay the basis for the currently accepted standard model of elementary particles.

Chemistry. The prize was awarded to three scientists for their pioneering work in atmospheric chemistry: Paul Crutzen of the Max Planck Institute for Chemistry in Mainz, Germany; Mario J. Molina of the Massachusetts Institute of Technology; and F. Sherwood Rowland of the University of California, Irvine. The Nobel committee honored them for their work that demonstrated the sensitivity of the stratospheric ozone layer to the influence of anthropogenic emissions of certain compounds. This was the first time that the Nobel Prize was awarded for significant work in the environmental sciences.

Atmospheric oxygen (dioxygen) consists of molecules in which two atoms of oxygen are chemically bonded to each other. Ozone (O_3) is created naturally in the stratosphere by sunlight in a photochemical process that converts the molecules of atmospheric oxygen, that is, dioxygen (O_2), to molecules in which three atoms of oxygen are bonded to each other. In 1970, Paul Crutzen's research demonstrated that the ozone layer also has a means of being destroyed naturally by nitrogen oxides, which react chemically with ozone. The ozone layer, when intact, serves to filter the ultraviolet radiation emanating from the Sun, thus protecting living organisms on Earth from exposure to harmful quantities.

Dr. Crutzen's work showed that naturally occurring sources of nitrogen oxides, for example, the nitrous oxide produced by soil-borne microbes, could drift upward and reach the stratosphere, where they react catalytically with the ozone, hastening its depletion.

In 1974 Rowland and Molina found that other chemicals could trigger destruction of ozone. Chemicals that are released to the environment as the result of industrial processes can follow the same route to the stratosphere as the nitrous oxide from the soil-microbe source. Their work indicated that the chlorine from a group of substances known as chlorofluorocarbons, with many and varied industrial uses, could migrate upward to the ozone layer, and because of their chemical stability remain in the stratosphere for a long time, almost a century. The Sun's ultraviolet radiation splits the chlorine out of the chlorofluorocarbons, setting off a chain reaction in which chlorine reacts with the ozone, destroying it. This work anticipated the later findings of ozone depletion in the stratosphere, the so-called ozone holes.

The work of Crutzen, Rowland, and Molina has led to a tremendous expansion of research in atmospheric chemistry. Although the precise chemical details of the ozone loss are still the focus of much research, the discovery of the recurring ozone destruction, particularly over Antarctica, has led to agreements by the industrialized nations to phase out the use of substances that deplete the ozone layer, especially the chlorofluorocarbons.

Physiology or medicine. Edward B. Lewis, Professor Emeritus of Developmental Genetics at the California Institute of Technology, Christiane Nüsslein-Volhard of the Max Planck Institute for Developmental Biology in Tübingen, Germany, and Eric F. Wieschaus, Squibb Professor of Molecular Biology at Princeton University, shared the prize for identifying the genes that are responsible for instructing the embryonic cells of the fruit fly, *Drosophila melanogaster,* to differentiate into segments, such as the thorax and head, and to develop body parts, such as the eyes, in the correct position.

During the 1940s, Lewis discovered that a distinct group of *Drosophila* mutation genes caused complete body parts to develop from uncharacteristic locations. By cross-breeding fruit flies with various mutations (such as an extra pair of wings or incorrectly placed antennae), Lewis located the affected genes, which he recognized as control genes, on chromosome 3, and named them homeotic selector genes. He found that these genes are positioned on the chromosome in the order that the developing segments appear in the embryo.

The homeotic selector genes are further directed by another series of genes classified by Nüsslein-Volhard and Wieschaus. In the late 1970s, these researchers examined the embryos of second-generation offspring of *Drosophilia* that were exposed to mutagenic chemicals and determined that three types of genes, the gap, pair-rule, and segment polarity genes, function in sequence to separate the embryo into segments. The gap genes are initially cued, which results in the basic partitioning of the embryo. Further segmentation is directed by pair-rule genes, and the anterior to posterior organization of each segment is determined by the segment polarity genes. The gap genes identify the bands in which the homeotic selector genes function, and the pair-rule and segment polarity genes further direct the homeotic genes.

The work done by Lewis, Nüsslein-Volhard, and Wieschaus has sparked research on corresponding regulatory genes that direct the arrangement of different body parts in other organisms, including humans. As the function of these genes is similar, regardless of the organism's complexity, it may be possible to relate characteristics of one organism's gene to that of another. The implications of these discoveries using pure genetic analytical techniques has established the general protocol that is allowing other geneticists to search for key genes responsible for specific congenital abnormalities in animals and humans. *SEE ANIMAL EVOLUTION; GENE.*

For background information *SEE ATMOSPHERIC CHEMISTRY; ATMOSPHERIC OZONE; CONGENITAL ANOMALIES; EMBRYOLOGY; GENE; GENETIC MAPPING; GENETICS; HALOGENATED HYDROCARBON; LEPTON; MUTATION; NEUTRINO* in the McGraw-Hill Encyclopedia of Science & Technology.

Nuclear fusion

Experiments on the Nova laser have recently demonstrated many of the key elements required for ensuring that the next proposed laser, the National Ignition Facility (NIF), will drive a fusion target to ignition.

Inertial confinement fusion. In the inertial confinement approach to fusion, spherical capsules containing deuterium and tritium (DT), the heavy isotopes of hydrogen, are imploded, creating conditions of high temperature and density similar to those in the cores of stars required for initiating the fusion reaction. When DT fuses, an alpha particle (the nucleus of a helium atom) and a neutron are created, releasing large amounts of energy. If the surrounding fuel is sufficiently dense, the alpha particles are stopped and can heat it, allowing a self-sustaining fusion burn to propagate radially outward; a high-gain fusion microexplosion ensues.

To create those conditions, the outer surface of the capsule is heated (either directly by a laser or indirectly by laser-produced x-rays) to cause rapid ablation and outward expansion of the capsule material. A rocketlike reaction to the outward-flowing heated material leads to an inward implosion of the remaining part of the capsule shell (called the pusher). The pressure generated on the outside of the capsule can reach nearly a gigabar (10^9 times atmospheric pressure or 10^{14} pascals), generating an acceleration of the shell of about 10^{13} times the acceleration of gravity, and causing that shell to reach, over the course of a few nanoseconds, an implosion velocity of 300 km/s. When the shell along with its contained fuel stagnates upon itself at the culmination of the implosion, most of the fuel is in a compressed shell which is at 1000 times solid density. That shell surrounds a hot spot of fuel with sufficient temperature (about 10 keV or 10^8 K) to ignite a fusion reaction.

The capsule must be uniformly heated over its entire surface to cause uniform compression of the fuel to the center. With direct drive, this uniform heating of the capsule is caused by simultaneously illuminating the capsule from all sides with many laser beams. With the recently declassified concept of indirect drive, the capsule is positioned in the center of a cylindrically symmetric container called a hohlraum. Laser beams enter the hohlraum through holes in the end caps and heat the walls of the cylinder, which then radiate soft x-rays, filling the hohlraum with a bath of radiant energy. This energy causes the fuel capsule to implode. Typically, 70–80% of the laser energy can be converted to x-rays. The hohlraum concept leads to a natural, geometric uniformity of x-ray flux on the capsule surface, since two points close to one another on the capsule surface "look out" at the heated hohlraum walls and see nearly identical sections of the walls, and hence a nearly identical heat environment.

National Ignition Facility. Targets have been designed for a new laser, the 1.8-megajoule, 500-terawatt glass laser. The hohlraum is made of a high-atomic-number material such as gold, which maximizes the production of x-rays. The hohlraum is roughly 1 cm (0.4 in.) long and 0.5 cm (0.2 in.) in diameter, and is predicted to contain an x-ray intensity of about 10^{15} W/cm^2 or an equivalent blackbody radiation temperature of about 300 eV (3×10^6 K). The capsule is composed of a low-atomic-number shell (such as beryllium or plastic), which maximizes the ablation pressure created by the absorbed x-rays. The DT is mainly in the form of a frozen layer on the inside of that shell. The capsule diameter is roughly 2 mm (0.08 in.). The target is predicted to give a yield of 10–20 MJ or a gain of 10–20. Similar gains are expected from direct-drive targets, which the NIF will also explore.

Because it requires only about 10 MJ/g to compress DT to 1000 times its solid density, and fusion yields about 100 gigajoules/g, gains of 10,000 might be expected. However, only about 10% of the x-ray energy is coupled to the capsule (the rest soaks into the hohlraum walls and escapes out of the laser entrance holes in the endcaps). The hydrodynamic rocket efficiency of absorbing heat on the capsule surface, ablating the material, and converting that energy to kinetic energy of the imploding shell turns out to be about 20%. On the NIF scale target, only about 10% of the fuel burns before it disassembles. These factors lead to the gain 10–20 result. For a larger-scale, reactor-size driver, the physics of hohlraums and the diffusive loss of x-rays to the walls allow for closer to 20% coupling to the capsule. Moreover, the larger-scale target has more inertia. Hence it will stay confined longer before it disassembles, and will burn 30% of its fuel. Gains in excess of 100 are expected, allowing the possibility of commercial reactors based on inertial confinement fusion (ICF).

Experiments on Nova. Recently, the Nova laser at Lawrence Livermore National Laboratory (LLNL) has been used to study the target physics issues that can most affect the NIF target performance. The results of these experiments have built confidence in the success of the NIF target.

Plasma physics issues. The first issue is laser-plasma coupling. Laser-driven parametric instabilities might result in scattering of laser light and production of high-energy electrons. The light scattering might degrade symmetry or represent a loss of potential drive energy, and the high-energy electrons can cause preheating of the capsule, which reduces the achievable compression.

To study these issues, 9 of Nova's 10 beams were used to create a large 2–3 mm-scale (0.1-in.) plasma, similar in conditions of temperature and density to a plasma that a typical beam of NIF might traverse. The tenth beam was configured as close as possible to NIF conditions, and used as a probe to test whether it would couple properly to the target. Acceptably low levels of scattering (about 5%) were measured, and the hot electrons created were of sufficiently low temperature so as not to cause any problems for the NIF target.

Drive and symmetry. The second issue, assuming the laser light has succeeded in entering the hohlraum and been absorbed on the walls, is x-ray drive and symmetry. The factors that determine the efficiency of converting laser light to x-rays, and the relative amounts of energy absorbed by the capsule versus those absorbed by the walls must be understood. The flux onto the capsules must be sufficiently uniform to achieve convergences (initial radius divided by final fuel radius) of 25–35 and remain nearly spherical. For this to occur, the time-integrated x-ray fluxes must be uniform to 1–2%.

On Nova, the temperatures reached by laser-heated hohlraums were measured by observing both the spectrum of x-rays emitted by the walls and the radiation-driven shock wave emerging through a plate of aluminum positioned on the hohlraum wall. Both methods gave consistent results. Temperatures up to 300 eV (3×10^6 K) were achieved. Computer simulations matched the data quite well. Because temperature is the result of a balance of x-ray sources (the conversion of laser light to x-rays) and x-ray sinks (the loss of x-rays absorbed by the walls of the hohlraum), this agreement could be the result of compensating errors. Therefore, the sinks were separately measured by measuring the time it took for x-rays to burn through thin patches of the hohlraum wall. The results of these measurements also turned out to be in excellent agreement with theoretical predictions. Thus, confidence in the predicted hohlraum temperatures on the NIF is quite high.

Since hohlraums naturally provide geometrically smooth drive for points relatively close to one another on the capsule surface, an important issue is to ensure uniformity between those points farthest from one another, typically a point on the pole of the spherical capsule compared to a point on the equator. Achieving this symmetry has been demonstrated by imploding capsules and imaging the imploded fuel volume. X-ray emission from the fuel is imaged by using an x-ray pinhole framing camera. By adjusting the pointing of the laser beams, either by varying the hohlraum length or by varying the beam focal position relative to the laser entrance hole, it is possible to control the imploded capsule shape to the 1–2% precision required by NIF and to achieve round implosions. Moreover, calculations accurately predict the optimal beam placement. Such data, combined with the fact that the NIF will have 192 beams (versus Nova's 10), give high confidence in the ability to provide good symmetric drive for the NIF capsule.

Capsule implosions. The third issue involves implosion physics. The Rayleigh-Taylor instability is prevalent in ICF implosions. An inverted glass of water is in principle in equilibrium (the atmosphere's pressure of 1 kgf/cm^2 or 14 lb/in.^{2} can keep the water in the glass) but it is in a Rayleigh-Taylor unstable equilibrium, wherein the dense water would "prefer" to lower the energy of the system by being lower in the gravitational potential than the lighter air it will replace on its way to the soon-to-be-wet floor. An ICF capsule is similar. The low-density ablated material accelerates the dense shell. The shell feels a huge effective gravity force, much like the force an astronaut feels at launch time. Thus, again there is dense matter in an effective gravity field wishing to exchange places with low-density matter. The target crinkles on its way toward implosion. The instability is mitigated somewhat by the ablative acceleration process, since the ablation tends to effectively burn off or smooth the perturbations. Upon deceleration at the culmination of the implosion, the low-density hot-spot DT gas holds up the dense DT shell, again in an effective gravity field, and again an unstable Rayleigh-Taylor situation arises and the cold shell mixes into the hot fuel. Understanding these phenomena quantitatively is required to ascertain just how smooth an initial target must be, since initial small perturbations will grow because of the Rayleigh-Taylor instability.

Many experiments have been performed on Nova with planar targets that confirm quantitative understanding of the acceleration phase of the Rayleigh-Taylor instability and the important mitigating effects of ablation. Moreover, significant progress has been made toward experimentally verifying the deceleration phase of the instability and modeling the mix of fuel and pusher material in ICF capsules. These latter experiments rely on spectroscopic measurements of emission from both dopants (such as argon) in the fuel and dopants (such as chlorine) in the pusher of ICF capsules. As the degree of mix or distortion at the pusher-fuel interface varies, the temperature and density of materials at that interface also vary. The increase in emission from the pusher dopants relative to that from the fuel dopants is observed as the mix increases because of the increase (in successive experiments) in the initial bumpiness (magnitude of the initial capsule surface finish) of the outside layer of the capsule. Quantitative agreement between model and experiment for the dependence of chlorine/argon line-intensity ratios on the initial surface perturbations once again raises confidence in NIF predictions.

Integral tests. Finally, all the aforementioned target physics knowledge was integrated, and capsules in hohlraums were successfully imploded with convergences of order 25, which is well into the NIF ignition target regime. These targets performed as predicted, increasing confidence in NIF target performance.

The experiments on Nova have thus addressed many of the key issues that can impact the performance of the NIF target. As a result of the success of these experiments and in the aftermath of the good agreement between theoretical predictions and the Nova data, the construction of the NIF is approached with increasing confidence in the ability to achieve, for the first time in the laboratory, fusion ignition and thermonuclear burn, thus creating a star on Earth.

For background information *SEE MAGNETOHY-DRODYNAMICS; NUCLEAR FUSION* in the McGraw-Hill Encyclopedia of Science & Technology.

Mordecai D. Rosen

Bibliography. W. J. Hogan, R. L. Bangerter, and G. L. Kulcinski, Energy from inertial fusion, *Phys. Today,* 45(9):42–50, September 1992; B. G. Levi, Veil of secrecy is lifted from parts of Livermore's laser fusion program, *Phys. Today,* 47(9):17–19, September 1994; J. D. Lindl, R. L. McCrory, and E. M. Campbell, Progress toward ignition and burn propagation in inertial confinement fusion, *Phys. Today,* 45(9):32–40, September 1992.

Nucleosynthesis

Nuclear astrophysics, at the interface between nuclear physics and astrophysics, has recently produced a significant advance in the understanding of the stellar process of explosive nucleosynthesis. This process is generally thought to occur in novae or in x-ray bursters, emissions from either of which are triggered by nuclear reactions operating at very high temperatures and densities. Important new information has been obtained from nuclear physics studies about the extremely proton rich nuclei that are involved in explosive nucleosynthesis. This information can be used in predicting the nucleosynthesis resulting from these astronomical events, and will have a profound effect on the theoretical-observational comparisons for any objects undergoing this mode of nucleosynthesis.

One site for explosive nucleosynthesis is a nova, literally new star. A nova is observed as a sudden brightening of a celestial object by several orders of magnitude during a time span of roughly a day. For some time, it has been thought that the nuclear reaction trigger for a nova occurs when material from the periphery of one star of a binary system is accreted onto its companion. The burst occurs at the surface of a collapsed star, a white dwarf, in tens of seconds, but the light from this event that ultimately is sent throughout the cosmos is processed in passing through stellar material, so it spreads in time to the observed day. Thus, the so-called new star really results from an exchange of matter between a pair of stars.

A second potential site for explosive nucleosynthesis is the surface of a neutron star (a much denser object than a white dwarf). Again, material is accreted from a companion star, but the resulting nuclear reactions now occur at higher temperatures in shorter times, and produce x-ray bursts.

Conditions for the rp-process. The star onto which the matter is accreted must have evolved to a collapsed star so that the accreted matter can achieve a temperature of $1–5 \times 10^8$ K upon falling into the deep gravitational potential well of a white dwarf, or in excess of 10^9 K if it is accreted by a neutron star. The infalling matter must also achieve a sufficiently high density to become degenerate, that is, to be constrained by the same Pauli principle that determines the distribution of electrons in atoms or of protons and neutrons in nuclei. At such densities the energy resulting from the nuclear reactions will increase the temperature of the matter in which it is produced without the concomitant expansion that would occur in a less dense environment (which would lower the temperature and slow the thermonuclear processes). Without the expansion, the matter undergoes thermonuclear runaway, that is, it simply heats up until it explodes. These are the conditions for a special mode of explosive nucleosynthesis known as the rapid proton capture process, or rp-process, by which it may be possible to synthesize many nuclides of the periodic table up to a mass of roughly 100 atomic mass units by the nuclear reactions produced in the accreted matter.

This process uses as seed nuclei the abundant nuclei lighter than iron that were produced in preceding generations of stars. Each of these nuclei captures many protons, populating the nuclides near the proton drip line. This drip line is defined as the sequence of the most proton-rich (radioactive) nuclei that ultimately beta decay, that is, decay by either electron capture or emission of a positron (a positively charged electron), but not by direct proton emission. The proton drip-line nuclei cannot capture another proton; the resulting nucleus would immediately proton decay. When the high temperatures and densities that cause the rp-process have abated, each nucleus that has been synthesized will undergo a series of beta decays back to one of the stable nuclides.

The pathway of the rp-process through a section of the periodic table of the nuclides is indicated in the **illustration**. The proton drip-line nuclei are those along the upper left of the nuclides shown. That the rp-process occurs well beyond the proton-rich side of the stable nuclei is seen by comparing the nuclei through which the rp-process path passes with the stable nuclei, indicated by a diagonal line in the lower right corner of their corresponding boxes. Vertical lines in the rp-process path result from proton captures, and diagonal lines are the result of beta decays (except for the nuclear reaction from potassium-37 to argon-34). The instability of the rp-process environment is thought to produce the short time scales of tens of seconds, and much less for the highest-temperature environments. Therefore, any of the nuclei along the rp-process path that cannot capture another proton, and that have a half-life much longer than 10 s, will slow or terminate the progression of the rp-process. Generally, the nuclei being processed would be expected to accumulate at the nuclei that decay slowly; high abundances of the stable nuclides to which those slowly decaying nuclides ultimately beta decay thus result.

Process termination point. In order to refine the abundance predictions of the rp-process models, the nuclides with masses around 64 atomic mass units were studied with the aim of, first, observing

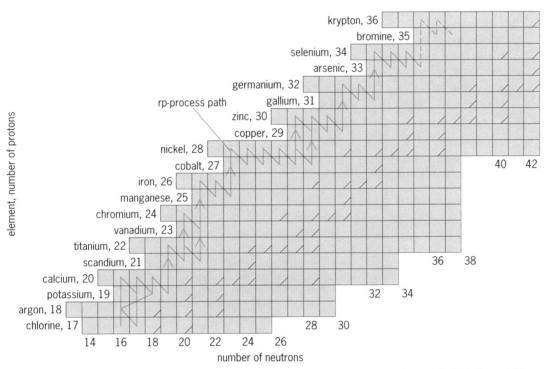

Chart of nuclides, showing the path of the rp-process. A stable nuclide is indicated by a short diagonal line in the lower right corner of its box. The environment for this calculation was assumed to have a temperature of 1.5×10^9 K and a density of 10^6 g/cm^3.

the most proton-rich nuclides in that mass region that are stable to proton emission, and then, in several cases, measuring their half-lives. The nuclei of interest are extremely proton-rich (or neutron-poor) nuclei such as arsenic-65. (For example, the only stable arsenic isotope, arsenic-75, has 10 more neutrons than the arsenic isotope of interest.) Theoretical predictions of nuclear masses had suggested that arsenic-65 would decay by proton emission, so that it could not be made by proton capture on germanium-64. Thus germanium-64, with its 1.06-min half-life, would block the rp-process from proceeding past it.

A radioactive-beam facility was utilized to produce the nuclei of interest. Such facilities have been a major area of growth in nuclear physics research, as several laboratories throughout the world have shifted a large fraction of their research emphasis to studies involving radioactive beams. These studies have been aimed not only at nuclear astrophysical questions, but also at fundamental questions about nuclei far from stability. The radioactive-beam facility first produces a beam of a nucleus somewhat more massive than those of interest, then bombards nuclei in a foil with the nuclei in that beam. Finally, it uses magnetic analyzers to separate out the recoiling nuclei of interest from the plethora of stable and unstable nuclides produced by the nuclear fragmentation reactions that occur in the foil. In some experiments of interest to the rp-process, the existence of the nuclei was the question of interest, whereas in others their subsequent decay products and half-lives were the important information.

It was found that arsenic-65 decayed not by proton emission but by beta decay. Thus, germanium-64 does not block the rp-process. However, a recent study at another radioactive beam facility showed that bromine-69 is unstable to proton emission, and so cannot be made by proton capture on selenium-68. Thus, selenium-68, with its half-life of 1.6 min, would effectively block the conventional rp-process. These results are included in the illustration, where the rp-process path is shown as proceeding through arsenic-65, but it is broken, indicating low probability, past selenium-68 (or actually arsenic-68, the nucleus resulting from the beta decay of selenium-68). These new results will clearly have a major effect on the abundances of the heaviest nuclei predicted by theoretical models of rp-process nucleosynthesis.

Possible higher-mass process. Calculations suggest that a higher-temperature rp-process can proceed to higher masses by using more massive (preexisting) seed nuclei. These seeds might have been produced in an earlier stage of nucleosynthesis within the star prior to its undergoing rp-processing. Recent work has shown that the rp-process could produce nuclei with masses around 90–100 atomic mass units. This finding might explain the isotopic abundances of the most neutron-poor molybdenum and ruthenium isotopes, synthesis of which has been a long-standing astrophysical puzzle, and might even provide a means of identifying Thorne-Żytkow objects. If such objects exist, they surely are among the most bizarre in nature. A Thorne-Żytkow object results when a star in its helium-burning, or red giant, phase swallows a companion neutron star which has resulted from the evolution of a very massive star. Such a composite object would produce conditions similar to those encountered by the matter infalling onto a neutron star; that is, the temperature could be in excess of 10^9 K. However, Thorne-Żytkow objects might be capable of expelling the products of their nucleosynthesis into the interstellar medium, a process thought to be difficult for accreting neutron stars. Thus, Thorne-Żytkow objects could produce copious amounts of the lightest molybdenum and ruthenium isotopes, thereby solving the question of the mechanism of their synthesis.

The calculations showed that in order to produce adequate amounts of those nuclides while not grossly overproducing others, the extremely proton rich nucleus silver-95 must be stable to proton emission. A radioactive-beam facility was used to search for silver-95. This nucleus was found to be stable to proton emission, confirming the first requirement of the rp-process scenario of molybdenum and ruthenium production. It is anticipated that further refinements of the nuclear data that provide the input for these models, specifically half-lives and decay modes, will provide sufficient predictive detail that astronomers will be provided with definitive observable signatures of Thorne-Żytkow objects, if they exist at all.

Thus, the combination of nuclear physics and astrophysics has provided the microscopic information necessary to elucidate the details of the events seen in such dramatic fashion by astronomers. Furthermore, the signatures provided by the theoretical predictions may be crucial for identifying the basic nature of a class of truly unusual astronomical objects.

For background information *SEE EXOTIC NUCLEI; NOVA; NUCLEOSYNTHESIS; RADIOACTIVITY; X-RAY ASTRONOMY* in the McGraw-Hill Encyclopedia of Science & Technology.

Richard N. Boyd

Bibliography. C. A. Barnes, D. D. Clayton, and D. N. Schramm (eds.), *Essays in Nuclear Astrophysics,* 1982; G. Biehle, Observational prospects for massive stars with degenerate neutron cores, *Astrophys. J.,* 420:364–372, 1994; A. E. Champagne and M. Wiescher, Explosive hydrogen burning, *Annu. Rev. Nucl. Part. Sci.,* 42:39–76, 1992; E. P. J. van den Heuvel and J. van Paradijs, X-ray binaries, *Sci. Amer.,* 269(5):62–71, November 1993.

Nutraceuticals

The term nutraceutical refers to any food or food ingredient that provides medical or health benefits, including the prevention and treatment of disease. In Japan, nutraceuticals are known as functional foods, that is, foods derived from naturally occurring substances that can be eaten as part of a daily diet and that regulate or otherwise affect particular body processes. Thus, nutraceuticals are regarded as something midway between a nutrient and a pharmaceutical. Broadly speaking, nutraceuticals include isolated nutrients, dietary supplements and diet plans, genetically engineered designer foods, herbal products, and processed foods. A nutraceutical can be taken in the form of a tablet, a capsule, a drink, a cereal, or some other type of preparation. Ideally, the formulation should be the most appropriate delivery system for the active ingredients.

Unlike foods, the claims for nutraceuticals must be supported by published clinical results. Nutraceuticals do not necessarily have to demonstrate definitive clinical activity; strongly suggestive data are sufficient. Unlike drugs, nutraceuticals are not currently subject to the U.S. Food and Drug Administration's strict regulations governing health claims.

Mushrooms. Among the various foods which qualify as nutraceuticals are mushrooms. Mushrooms have played an important role in traditional Chinese medicine for over 3000 years. Compared to modern Western medicine, such traditional medicine takes a preventive approach; food and medicine are not artificially separated. The aim of traditional medicines is to regulate well-being rather than to target specific diseases. Several types of mushrooms have been consumed through the centuries to maintain health, preserve youth, and increase longevity. The use of mushrooms for health purposes is gaining popularity in the West. Modern research has confirmed their effectiveness and, in some cases, found new applications for them.

Besides adding variety, flavor, and visual appeal to food, mushrooms are considered nutritious because of their protein, fiber, vitamin, and mineral content. A number of reliable laboratory analyses of edible mushrooms have been conducted, especially on commercially important species. Most assessments of the nutritional value have been based on the chemical composition of the fruit body.

Although mushroom protein ranks below most animal meats, it compares favorably with that of common vegetables, rice, and wheat. All essential amino acids are present, including leucine and lysine, which are lacking in most staple cereal foods. Mushrooms are a source of thiamin, riboflavin, niacin, biotin, ascorbic acid (vitamin C), provitamin A (measured as the beta-carotene equivalent), and ergosterol (vitamin D_2), which may be converted to vitamin D under ultraviolet light irradiation. They contain all the minerals present in their growth substrate, particularly magnesium, potassium, and calcium. The fiber constituent has a cholesterol-reducing effect and is important in the prevention of colonic cancer, coronary disease, and diabetes. Mushroom polysaccharides have been shown to have anticancer, antiviral, and anti-inflammatory activities, and are emerging as a new class of compounds in the treatment of immunocompromised individuals. Although mushrooms provide only small amounts of lipids, they are rich in unsaturated fatty acids, adding to their cholesterol-lowering food value.

More than 100 species of edible mushrooms can be considered to be nutraceuticals. Several examples are given below.

Shiitake. Lentinula edodes (shiitake in Japanese and hua-gu in Chinese) is the second most commonly produced edible mushroom in the world. In China, shiitake is used for conditions where the immune function needs strengthening. Considered the best of all plant foods in ancient Japan, shiitake was used for general health maintenance, flu and colds, measles in children, bronchial inflammation, stomachache, headache, high cholesterol, atherosclerosis, high blood pressure, liver ailments, diabetes, edema, smallpox, and mushroom poisoning. In modern Japanese traditional medicine, it is used to control blood pressure and lower plasma cholesterol level, and for ulcers, gout, constipation, myopia, allergies, hemorrhoids, neuralgia, poor complexion, and impotency. This mushroom has antiviral activity, and its polysaccharides have been shown to have antitumor and immune-enhancing effects. Shiitake can be found fresh or dried in markets. Formulations in capsule, tablet, or granular form, as well as liquid-extract products, are available in natural food stores and from Chinese herb dealers.

Hedgehog fungus. Hericium erinaceum, the hedgehog fungus, another gourmet mushroom, is known as yamabushi-take in Japanese and houton in Chinese. According to Chinese traditional medicine, this fungus is beneficial for internal organs (heart, liver, spleen, lung, kidney), and is used in the treatment of indigestion, gastric and duodenal ulcers, chronic gastritis, and fatigue. For medicinal purposes the fruit body is collected and sun-dried. It is then cooked in water or cut in thin slices and cooked with chicken or in chicken soup. In China, *H. erinaceus* pills are produced from dried mycelium. Each tablet is equivalent to 1 g of mycelium and contains polysaccharides and polypeptides reported to stimulate the immune response and expedite recovery from wounds or diseases. In addition, these pills exert a therapeutic effect on stomach and esophageal cancers.

Schizophyllum. The split-fold mushroom *Schizophyllum commune* grows on the branches of trees, and is one of the most widespread mushrooms in the world. Although edible, its fruiting bodies are relatively small and tough. In traditional medicine, it is used for general weakness and debility, chronic infections and immune suppression, and to cure gynecological diseases. A tonic can be prepared by simmering the fungus in water to make a tea. In China, it is stewed with eggs as a treatment for leukorrhea. In other countries it is regularly eaten as a food, being either fried or boiled, then seasoned before eating.

Polypores. A polypore is a bracket or shelf fungus that grows on the sides of trees. The panacea polypore *Ganoderma lucidum* is known as reishi in Japan and ling-zhi in China. Previously, *Ganoderma* was used to treat hepatitis, nephritis, high blood pressure, high cholesterol, arthritis, fatigue, insomnia, bronchitis, asthma, gastric ulcers, arteriosclerosis, low white blood cell count, and diabetes. The pharmacologic effect of *Ganoderma* has traditionally been associated with its color, which depends on the species of mushroom and the stage of growth used, as well as the environment in which it has grown. Bitter taste has long been identified with its therapeutic properties. Today *Ganoderma* is used for aging-related and degenerative conditions, cancer, allergies, chronic fatigue, insomnia, symptomatic relief of menopause and arthritis, and as an immune system stimulant. It has become the natural medicine of choice in North America, and is especially popular among high-risk groups with human immunodeficiency virus (HIV). This mushroom can be taken in a variety of forms—in syrups, soups, teas, tablets, tinctures, or in powdered form mixed with honey.

Grifola. Grifola frondosa, an edible polypore, called maitake in Japanese, has been shown to regulate blood pressure, cholesterol, blood sugar, and body weight. The polysaccharides and high-molecular-weight sugar polymers in the fruiting body appear to strengthen the immune system and inhibit tumors. Many researchers feel that maitake contains the most potent immunostimulant of any of the medicinal mushrooms and is useful for people with cancer, particularly those undergoing chemotherapy, and people infected with HIV. Maitake can be found in fresh or dried and packaged form in specialty stores and Chinese herb stores, and in prepared products, such as capsules or teabags.

Grifola umbellata is a species of mushroom closely related to *Gr. frondosa.* Its fruiting bodies arise from a tuberlike mass called a sclerotium. Extracts of the sclerotium have traditionally been taken as a diuretic and to treat urinary tract infections, jaundice, and diarrhea. Chinese physicians currently utilize these extracts to treat individuals with lung, cervical, esophageal, stomach, liver, intestinal, and breast cancers, and those with leukemia or lymphosarcoma. *Grifola umbellata* is a common component of many Chinese prepared products. Extracts are also available in tablet or capsule form.

Indian bread. Indian bread (*Poria cocos*), a fungus that grows underground on the roots of pine and other trees, is a common ingredient in herbal tonics. The sclerotium of this fungus is commercially gathered in Asian countries. After it is air-dried and the surface layer is removed, different portions of the sclerotium are used (depending on the desired effect) to lower blood sugar, as a diuretic, for diarrhea, to treat sterility or gynecological disorders, as a tranquilizer, or as a sedative. Although their clinical effectiveness in the treatment of autoimmune diseases, allergy, and inflammations has been demonstrated, the active principles, especially high-molecular-weight polysaccharides and triterpenoids, have only recently been isolated and identified. *Poria cocos* is available in Chinese herb shops and natural food stores and in a large number of prepared products.

Besides the use of single mushroom varieties in tonics, teas, and soups, many products combine a number of ingredients in one preparation. They are often made by freeze-drying a tea of herbs and mushrooms to obtain a powder, which is then made into pills. Two well-known products containing a combination of several mushrooms are manufactured in China. The first is Wu Ling San for the treatment of cirrhosis. It contains *Gr. umbellata, P. cocos,* and other ingredients. The other is a tendon-easing pill for relief of muscular aches and pains that is formulated with a very large variety of mushrooms.

Preparation. Since so many of the mushrooms mentioned as possible nutraceuticals can be eaten, it is interesting to consider the effect of cooking on their nutritional content. Some heat-labile nutrients, such as vitamin C, thiamine, and other B vitamins, are obviously reduced, often up to 50–70%, or are destroyed through cooking. Most minerals are unaffected by heat and can actually become more available. Fiber is generally broken down to a degree. Protein is affected, although its value is probably not reduced. Stir-frying or sautéing lightly, the traditional way to cook choice edible species, rather than boiling, preserves the more unstable nutrients. Generally, cooking mushrooms maximizes their nutritive value by increasing their digestibility, but overcooking removes some of the vitamins and most of the flavor.

The tough polypores are best boiled. Most of the active principles in medicinal species are associated with cell wall constituents and become more available after simmering for 45–60 min. Other active components such as terpenes are more soluble in very hot water and are relatively stable to heat. The immune-enhancing minerals, such as germanium and zinc, are also more accessible after cooking.

Implications. Mushroom nutraceuticals are a viable alternative to high-cost drugs with their frequent and severe side effects, and in some cases, even to surgery. The best way to take full advantage of the nutritional and medicinal benefits of mushroom nutraceuticals is to make them a regular part of the diet. They provide vital substances that the body needs for good health and immunity.

For background information *SEE ACQUIRED IMMUNE DEFICIENCY VIRUS (AIDS); AUTOIMMUNITY; FUNGI; IMMUNITY* in the McGraw-Hill Encyclopedia of Science & Technology.

Jeannette M. Birmingham

Bibliography. C. Hobbs, *Medicinal Mushrooms: An Exploration of Tradition, Healing and Culture,* 1995; K. Jones, *Shiitake: The Healing Mushroom,* 1995; T. Willard, *Reishi Mushroom: Herb of Spiritual Potency and Medical Wonder,* 1990.

Obesity

An estimated 68 million Americans are estimated to be overweight, that is, 10% or more above what is considered to be desirable or ideal weight for height, age, and body build. Of this number, 34 million are actually 20% or more above ideal weight, and are therefore considered to be obese. In many industrialized countries, obesity has reached epidemic proportions. Obesity is of major concern as it has been correlated with numerous health risk factors and diseases, including hyperinsulinemia, hypertension, hypertriglyceridemia, insulin resistance, impaired glucose tolerance, adult-onset diabetes, coronary heart disease, and the increased growth of certain forms of cancer, in particular breast and endometrial cancers.

A recent Harris Poll described the growth of obesity in the last decade as follows: the percentage of obese Americans in 1986 was 58%; in 1990, 64%; in 1994, 69%; and in 1995, 71%. A recent national weight survey that examined weight changes between 1962 and 1991 showed a steady increase in weight, with increases varying by ethnic group, age group, and gender. For ethnic groups, results for a 30-year-period varied from increases as low as 12.2% in black, nonhispanic women to a high of 37.3% in caucasian women. Weight gains for the 30-year-period ranged from a low of 31.7% in men to a high of 32.1% in women. Weight changes in men 20–34 years old showed a 22.2% increase, while there was a 42.9% increase in men 65–74 years old. Similar results were found in women: a 25.1% increase in 20–34-year-olds and a 48.5% increase in 55–64-year-olds. In real terms, the average American had gained 8–10 lb during the last decade. The current economic and health costs of obesity have been estimated to be in the double-digit billions of dollars, and this steady rise in weight and incidence of obesity continues in spite of all attempts to halt it.

For many years, obesity was believed to be a volitional disorder. The treatments of choice for obesity have included low-calorie diets, psychotherapy, behavior-modification programs, radical gastrointestinal bypass surgery, and even jaw wiring—all attempts to reduce what was assumed to be an excess intake of food. Although the rationale behind these treatments may appear sound, research has revealed that many of the obese do not consume more food than a normal-weight individual, and in some cases actually consume less. Thus, it is no surprise that the long-term success rates of current obesity treatments lie somewhere between 2–5%.

The role of genetics. Recent advances in science and technology have led to the current understanding that obesity is a symptom of real metabolic disease or physiological disorder in which there are genetic causes as well as environmental influences. Two fundamental procedures are being used to study the genetic basis of obesity: the macro-to-micro approach begins with the study of significant phenotypic expressions of obesity (body mass index or BMI) within families, and the micro-to-macro approach begins with genotypic linkages that connect specific genes known to be transmitted within families. Techniques used in the second approach are cloning, deoxyribonucleic acid (DNA) sequencing, genetically engineered animal models, and quantitative trait loci linkage.

The hunt for a genetic link to the cause of obesity began in 1962, when it was suggested that obesity is an expression of a so-called thrifty gene which becomes detrimental with cultural progress. New research into the interrelationship among activity patterns, food intake, human energy metabolism, and genetics is making it clear that many of the obese possess one or more thrifty or obesity genes that were important for survival during times when it was important to store surplus calories as fat during periods of abundance so as to use the stored calories during periods of paucity or famine.

Data from studies of body mass indices in families, fraternal and identical twins, and adopted children reveal a significant correlation in body mass indices of adoptees and their biological, but not their adoptive, parents and siblings. In addition, individual topographical distribution of fat has also been found to be related to genetics. These findings support the importance of the contribution of genetics to obesity. A gene can be viewed as a blueprint for life's structure and function. It often represents a range of expression for a characteristic, an environment selecting the appropriate expression for a gene. As an environment is altered, the expression of the gene may also be altered.

Geographical locations of early humans varied, but in general the Paleolithic diet consisted of roots, stems, leaves, fruits, and berries from plants, and lean animals. Many forms of food were available only seasonally. Specific foods were eaten infrequently, and were for the most part low in calories and high in fiber. In addition, these foods needed to be consumed within short periods of time of killing or harvesting because of the problems of preservation. With the arrival of agricultural science and technology and the development of food preservation and processing techniques, the Paleolithic diet (balanced by periods of famine) gradually shifted to a diet composed of highly processed and refined modern-day foods that became high in calories, low in fiber, easily maintained and preserved for long periods of time, and constantly available for consumption.

Survival genes, present in the Paleolithic era and conferring on early humans numerous benefits for survival in times of feast and famine, may actually be the obesity genes of today, contributing to the numerous detrimental health-related effects associated with obesity. Endocrinologists are now viewing obesity as a genetic disease, some even going so far as to say that 95% of obesity is genetic.

Identification of obesity genes. It would be expected that identification of an obesity gene would come from research into molecular genetics. Given the complex assortment of metabolic systems that undoubtedly contribute to obesity, it would be expected that more than one gene would be implicated. There is already a body of evidence that suggests that the genes for apolipoprotein B, apolipoprotein E, glucocorticoid receptor, insulin, and low-density lipoprotein are associated with obesity.

One promising area of research in the search for obesity genes is the dopaminergic system in the human brain. As with alcohol, amphetamines, cocaine, and other drugs that are abused, food for some people can become a reinforcing substance that can produce pleasure and euphoria or may be used to stave off withdrawal symptoms associated with physiological changes. Some scientists believe that individuals with obesity genes are at elevated risk of becoming addicted to carbohydrate-rich foods, alcohol, or other substances. Endogenous opioids, which are known to be regulated in part by dopamine, are elevated in the obese. Specificity and localization of the reinforcing properties of food are still unclear but appear to be in the brain's dopaminergic reward pathways, which makes the dopaminergic system a logical starting point in the hunt for an obesity gene. In initial studies, amphetaminelike drugs led to successful weight loss, but their safety and long-term efficacy were questionable.

DRD2/A1 allele. The first reports of a possible obesity gene came from research into the D_2 dopamine receptor gene (DRD2/A1 allele). In both animal and clinical studies in which neuroleptics (drugs producing symptoms of nervous system diseases) were used to block DRD2/A1 allele, there was significant weight gain. What appears to be a tiny alteration in this gene, located on human chromosome 11, leads to a reduction in the number of dopamine receptors in the brain. Fewer dopamine receptors cause a reduction in satisfaction from consuming food; therefore, cravings may be initiated for starches, snack foods, and sweets, which possess a high carbohydrate content. People with the DRD2/A1 allele exhibit obesity as well as strong desires for sugar-laden foods including cakes, candies, chocolates, ice cream, and pastries. The simple carbohydrates in these foods (glucose, fructose, and sucrose) quickly enter the blood stream and send a so-called sugar rush to the brain, which raises the levels of the neurotransmitter, dopamine. Elevated dopamine levels appear to overcome the effects of a reduction in the number of receptor sites on brain cells, thereby increasing feelings of happiness and satisfaction. One important consequence of the discovery of the DRD2/A1 allele is the understanding that the eating behavior in many of the obese is not volitional but rather has a significant biological contribution.

Leptin. Another possible obesity gene may be what has been termed the leptin gene. Leptin is a protein hormone found in mice that appears to regulate the amount of body fat. This hormone, synthesized by adipocytes (fat cells), enters the blood supply and signals the brain's appetite and metabolism control centers. The leptin signal may be thought of as a type of thermostat that regulates fat cells through a feedback system which affects the control of appetite and the rate of metabolism. In lean mice, the cycle of events appears to occur as follows: when more fat is being stored than is necessary, fat cells are stimulated to synthesize and release excess leptin. The excess circulating leptin reaches the brain, and causes appetite to decrease and metabolism to accelerate. Decreased appetite and accelerated metabolism will in turn decrease adipocyte fat storage, and thereby decrease leptin output. Decreased leptin output leads to increased appetite, decreased metabolism, and increased adipocyte fat storage.

In the 1960s, the first clues indicating the presence of leptin in the blood emerged from a series of studies using genetically obese mice that were linked to lean mice through a common blood supply. When a genetically fat mouse, suspected of producing inadequate amounts of a fat-signaling substance, was linked by a common blood supply to a lean mouse, the fat mouse, in spite of its genetic predisposition, acquired the thin mouse's hormonal balance and lost weight. Conversely, when a genetically enormously obese mouse, suspected of producing excessive amounts of this same fat-signaling substance, was linked by a common blood supply to a genetically lean mouse, the lean mouse's blood became flooded with the fat-signaling substance, and the lean mouse quickly stopped eating and starved to death. In both cases, the fat-signaling substance appeared to regulate food consumption and utilization.

In 1995 researchers reported the discovery of a leptinlike hormone in humans. If leptin in humans works in a similar manner as in mice, lowered levels of this protein hormone would help explain why some people are naturally fatter than others; that is, some obese people may be genetically predisposed to produce sparse amounts of leptin and would therefore naturally require greater numbers of fat cells than normal-weight people in order to release adequate levels of leptin into the bloodstream. Other obese individuals may be genetically predisposed to respond weakly to blood leptin levels. These individuals may possess lower numbers of leptin receptors in the brain, and a greater number of fat cells would be required in order to stimulate the mechanisms that result in the normal reduction of appetite and increases in metabolism.

By using the technologies developed in molecular biology and organic chemistry, copious quantities of leptin are now easily synthesized employing genetically engineered bacteria. With large quantities readily available, it is questioned whether leptin will be as effective in humans as it is in mice. Leptin levels appear to be 20–30 times higher in obese humans than in lean humans, and it is uncertain whether injecting leptin into obese humans will be an effective means of generating weight loss. Leptin may not be the total answer; many processes, both biological and environmental, influence obesity.

Supporters of the leptin hypothesis agree that leptin could have the same relationship for the obese as insulin has for diabetics. There are, however, serious concerns, similar to those related to insulin injections, regarding the use of leptin injections: the treatment would be life-long; injections would mimic the natural physiological releases of the hormone; the effectiveness of injected leptin might change with age or in the presence of other medications that the user may require with age. A great deal of research is still needed before the full potential of leptin is known, although leptin does appear to offer a hope for the future.

The future of obesity gene research. Obesity is beginning to receive attention as a symptom of a number of distinct disorders affecting weight. Obese individuals may be segregated into at least four different types: type I, individuals with excess body mass or percentage of body fat; type II or android obesity, those who exhibit an excess of truncal-abdominal subcutaneous fat; type III, those who exhibit an excess of abdominal visceral fat; and type IV, or gynoid obesity, those who exhibit an excess of glutofemoral fat. Most research into the genetic epidemiology of obesity has dealt with type I obesity. It is clear that the DRD2/A1 allele and the leptin gene both contribute to the obesity, but also that obesity is a symptom of a number of different conditions with multiple causes that are controlled by a number of different genes. Subsets of the obese may well require different treatments or corrections for the problem.

For background information *SEE ALLELE; CHROMOSOME; GENE; GENETICS; OBESITY* in the McGraw-Hill Encyclopedia of Science & Technology.

Richard F. Heller; Rachael F. Heller

Bibliography. D. E. Comings et al., The dopamine D$_2$ receptor (DRD2): A major gene in obesity and height, *Biochem. Med. Metabolic. Biol.*, 50:176–185, 1993; E. P. Noble et al., D$_2$ dopamine receptor gene and obesity, *Int. J. Eating Disord.*, 15:205–217, 1994; M. A. Pelleymounter et al., Effects of the obese gene product on body weight regulation in ob/ob mice, *Science*, 269:540–543, 1995; E. Ravussin and C. Bogardus, Energy expenditure in the obese: Is there a thrifty gene? *Infusionstherapie*, 17:108–112, 1990; T. I. A. Sorensen, The genetics of obesity, *Metabolism*, 44(3):4–6, 1995.

Occupational health and safety

Repetitive strain injury is a general term for a number of work-related conditions characterized by pain and discomfort in the soft tissues associated with sustained, repetitive, forceful movements usually performed in awkward or constrained postures. Similar terms include cumulative trauma disorder and upper extremity musculoskeletal disorder. Stress, poor conditioning and posture, work-style techniques, psychosocial factors, and genetic predisposition may contribute to the development of repetitive strain injury. Such intrinsic ergonomic factors are in contrast to the extrinsic ergonomic factors relating to workstation design and equipment. Because of the variety of conditions that can occur, many aspects of repetitive strain injury are often poorly understood, and the term is not a diagnosis in itself. Repetitive strain injury is not just carpal tunnel syndrome, which has been misused as a synonym for all repetitive strain injuries. The incidence of reported illness from repetitive strain injuries appears to be on the rise. In 1993, the Bureau of Labor Statistics in the United States reported a 7% increase over the 1992 total, accounting for nearly two-thirds of total work-related illness and almost 5% of all injuries and illness.

Repetitive strain injury is not a new disorder. In the nineteenth century, overuse syndromes were noted in writers and musicians and even in telegraphers, whose wrist and forearm pain diminished with the introduction of the Hughes perforator key (an early example of applied ergonomics). Repetitive strain injury is well known in the poultry, meat-packing, fishing, aircraft, automobile, and clothing industries; however, with the advent of the personal computer and the flat keyboard, much attention is now focusing on white-collar, computer-related injuries. Since the mid-1980s the occupational injuries of musicians have also received greater attention.

Characteristics. Occupational repetitive strain injury predominantly involves the soft tissues of the upper body, including the neck, shoulders, back, arms, and hands. These soft tissues include muscles, tendons, ligaments, nerves, connective tissue and cartilage, and blood vessels. Soft-tissue damage is a predominantly nonsurgical condition. The examining physician needs to combine the skills of the orthopedist, neurologist, physiatrist, and occupational medical specialist to make the correct diagnosis and prescribe treatment.

The symptoms and signs of repetitive strain injury can be divided into three stages: Early manifestations include muscle aches and fatigue that occur during work and may begin insidiously over weeks or months, but usually subside with rest. In later stages of repetitive strain injury aching and fatigue persist for longer periods, usually beyond the work cycle, causing diminished work capacity. Last-stage symptoms include aching and fatigue at rest, sleep disturbances, partial or total disability, and inability to perform many activities of daily living such as cooking, cleaning, driving, or even brushing one's teeth. As symptoms become more severe, treatment is more difficult and prolonged, especially if there is both proximal (head, neck, upper back, and shoulders) and peripheral (arms, forearms, hands, or fingers) involvement.

The common clinical manifestations of repetitive strain injury include postural misalignment, usually a round-shoulder posture; myofascitis, causing muscle tightness and tenderness in the forearms, resulting in diminished wrist range of motion; tendinitis; trigger finger and ganglion cysts; degenerative joint disease, often seen at the joints of the thumb; and neck pain, numbness, and weakness in the upper extremities related to poor posture. Thoracic outlet syndrome is a common yet often missed problem related to poor posture or previous neck injury. Nerve entrapment syndromes—associated with tight areas surrounded by muscle, fascia, ligament, or bone—cause numbness or pain as well as muscle wasting and weakness. Reflex sympathetic dysfunction and dystrophy can occur during the later stages of repetitive strain injury and can lead to dire consequences. Reflex sympathetic dystrophy can cause the unpleasant and uncomfortable form of chronic pain associated with anxiety and depression. Temperature changes make thermography the most sensitive test for early diagnosis. Bone scan, laser Doppler, and fluxmetry also show promise for confirming the diagnosis.

Diagnosis. This begins with a detailed work history, followed by a comprehensive physical examination. Too often, evaluation of work style and work conditions is neglected. Without treating both the intrinsic and extrinsic ergonomic factors, relapse is inevitable. A focused and orderly approach enables the examiner to sort out the multitude of factors related to repetitive strain injury and to prescribe appropriate treatment.

Occupational history. The occupational history should include the past occurrence of illness or injury; a description of the previous care given; itemization of the initial and subsequent symptoms, including pain, weakness, stiffness, and effect on the activities of daily living; details concerning the type of work, work conditions and tools, and the apparent relationship of work to symptoms; and other heavy use of upper extremities.

Clinical evaluation. The physical findings related to repetitive strain injury can be separated into proximal and distal disorders. Proximal disorders relate to the neck, shoulders, and back. Distal problems involve the arms, forearms, and hands. Evaluation should include both these areas, since they are links in a biomechanical chain.

Muscle. Skeletal muscle accounts for about 50% of the body's weight. Approximately 65% of individual muscles are located in the upper extremity. There is a balance between prime movers that contract (agonists) and opposing muscles which relax (antagonists), providing smooth movement, stability, and control. Muscle fatigue is a prime symptom of repetitive strain injury, for which several theories have been proposed. One theory attributes the condition to the localized reduction of blood supply to tissues (ischemia) from persistent tension, resulting in inflammation and fibrosis, and energy deficiency from glycogen depletion and lactate and neuropeptide accumulation. Preponderance of muscle fiber type has also been implicated as a cause of fatigue. An abundance of fast-twitch low-endurance muscle fibers and atrophy of slow-twitch high-endurance muscle fibers have been described in biopsy specimens of overuse patients.

Tendons and ligaments. There are three areas of vulnerability in the tendons. The first occurs at the tendon bone attachment (enthesis), where increased muscle loading and repetitive use cause local ischemia, microtears, and degenerative changes. A second area of vulnerability is the muscle tendon junction, where force vectors from the repeatedly contracting muscle are concentrated in a 10–100-mm (0.4–4 in.) swath of adhesive tissue subject to degenerative and inflammatory changes, causing symptoms of tendinitis. A third area of tendon vulnerability occurs where the tendons pass through guiding sheaths to enable them to change direction. Repetitive motion and awkward positioning create shearing forces that cause swelling and inflammation of both the lubricating sheath and the tendon, resulting in a painful condition known as tenosynovitis. Occasionally, this wear and tear can produce nodules on the tendon that become trapped in the tendon sheath, resulting in trigger finger. This condition locks the finger in a flexed position. Friction and shearing of tendons from excessive wrist movement may contribute to a formation of a bump on the tendon sheath or synovial lining of the joint (ganglion) and carpal tunnel syndrome. Ligaments that provide joint stability can indirectly contribute to symptoms of repetitive strain injury if they are too loose or too tight, resulting in biomechanical inefficiency.

Nervous system. There is increasing knowledge about the role that the nervous system plays in the syndrome of repetitive strain injury. Compression of nerve roots in the neck from cervical spondylosis (degeneration of the intervertebral discs of the neck) or thoracic outlet syndrome (caused by rib or scalene anomalies or postural misalignment) can contribute to peripheral nerve injury. A double-crush theory has been suggested to explain this phenomenon. Some studies suggest that central nerve fiber damage may result in damage to nerve fibers beyond the site of primary injury. Locally, swollen muscle or friction can irritate peripheral nerves in areas such as the carpal tunnel, causing entrapment. Traction of nerve fibers is another mechanism of injury.

The sympathetic nervous system plays a role in the genesis of many of the clinically observed changes in repetitive strain injury, ranging from dermographism (a form of skin allergy) to pain and reflex sympathetic dystrophy (a chronic pain disorder associated with thoracic outlet syndrome, immobilization surgery, or recent soft-tissue injury). Immobility, surgery, soft-tissue damage, or brachial plexopathy are known etiologic factors.

Arteries and veins. Blood supply to the musculoskeletal system is essential to muscle tendon and nerve viability. Excessive muscle contraction and damage resulting from overuse can diminish blood supply to all the soft tissues. Sympathetic nervous system hyperactivity can result in spasm of the blood vessels and shunting of blood to deep tissues, causing osteoporosis (decrease of bone tissue). Occasionally, larger arteries and veins can be occluded, as seen in the thoracic outlet syndrome.

Treatment. Most of the manifestations of repetitive strain injury involve soft-tissue injuries that, with a few exceptions, can be managed conservatively, unless tissue damage is severe. Physical and occupational therapy and changes in the intrinsic and extrinsic aspects of work are the critical components of treatment.

Thoracic outlet syndrome and reflex sympathetic dystrophy require specialized treatment approaches and protocols. In both cases, physical therapy is a pivotal aspect of treatment.

Ergonomics plays an important role in both treatment and prevention of repetitive strain injury. Symptoms of this condition begin when equipment is poorly set up or faulty. Faulty tools, a poorly setup workstation, or even a musical instrument that does not fit can predispose to injury. Recent studies have shown that devices such as wrist rests or splints can actually be harmful if incorrectly prescribed. Even good equipment can cause injury if work style or technique is faulty. A pianist who practices long hours with hyperextended fingers and who is loose jointed must be taught proper use of the hands. Computer users who fail to to take rest breaks, hit keys with force, hyperextend their fingers, or have poor posture need to have these intrinsic factors corrected.

It is becoming increasingly clear that repetitive strain injury has become a major work-site problem. The complexity of causative factors and clinical manifestations has made it difficult to fully understand. Prevention is the most desirable approach. When treatment is necessary, a complete evaluation including a comprehensive history and physical examination is the first step. A focused treatment program including correction of extrinsic and intrinsic ergonomic factors should follow.

For background information SEE INDUSTRIAL HEALTH AND SAFETY in the McGraw-Hill Encyclopedia of Science & Technology.

Emil F. Pascarelli

Bibliography. H. Hooshmand, *Chronic Pain: Reflex Sympathetic Dystrophy Prevention and Management,* 1993; E. F. Pascarelli and J. Kella, Soft tissue injuries related to use of the computer keyboard, *J. Occupat. Med.,* 35(5):522–532, May 1993; E. F. Pascarelli and D. Quilter, *Repetitive Strain Injury: A Computer User's Guide,* 1994; M. Pecina Marko and I. Bojanic, *Overuse Injuries of the Musculoskeletal System,* 1993.

Organometallic chemistry

Recent advances in organometallic chemistry include syntheses of compounds that have new forms of coordinated carbon and metal-silane complexes, and new water-soluble organometallic compounds.

New forms of coordinated carbon. Transition metals provide a unique means of stabilizing reactive molecules or fragments of molecules. Although graphite and diamond are the only two pure forms or allotropes of elemental carbon (C_x) that commonly occur in nature, chemists are finding that diverse types of C_x entities can be incorporated into isolable metal complexes. Some representative examples include solitary carbon atoms that are entombed within polymetallic clusters, fullerenes (buckyballs) such as C_{60} bound as pi ligands, and C_2 units that tether two metals. Such compounds have, in addition to their intrinsic appeal as basic research objectives, promising materials and catalytic properties. There is a methodology for preparing compounds in which chains of three or more sp carbons span two metals, $L_nM(C)_xM'L'_{n'}$ (M, M' = metals; L, L' = ligands). These linear, wirelike assemblies are attracting considerable attention from numerous research groups.

Complexes of odd-numbered sp carbon chains. This class of compounds was unknown until recently, and to date has been accessed only through species with M—$(C≡C)_z$—Li linkages (Li = lithium). These lithiated complexes are strong nucleophiles, similar to organic analogs R—C≡C—Li that have previously been used for many carbon-carbon bond-forming reactions. As shown in **Fig. 1**, they can attack carbon monoxide ligands in other metal complexes. The subsequent addition of a methylating agent [(Me$_3$O$^+$ BF$_4^-$); Me = CH$_3$] gives a type of species known as a Fischer carbene complex, in which a C_5OMe linkage spans two metals (carbene implies a double carbon-to-metal band). In a procedure closely modeled after earlier preparations of Fischer carbyne complexes, (carbyne implies a triple carbon-to-metal bond), an electrophile (BF$_3$) is added to remove the methoxy group (MeO). A complex in which a C_5 chain spans rhenium (Re) and manganese (Mn) can then be isolated (Fig. 1).

The ReC$_5$Mn complex has two possible resonance forms—one with alternating triple and single bonds and the other with double bonds up and down the chain. Spectroscopic and x-ray structural data show that the latter dominates. The ReC$_5$Mn complex is deeply colored and sensitive to room lighting. It exhibits some thermal decomposition after several hours at room temperature. An ultraviolet-visible spectrum shows an intense absorption at 480 nanometers that arises from rhenium-to-manganese charge transfer. Similar ReC$_3$Mn complexes are known, and are much more robust. Species with C_7 or higher odd-numbered chains have not been reported so far.

Complexes of even-numbered sp carbon chains. This class of compounds can exist in several types of redox states. For example, the sp carbon chain may consist of alkynyl units (containing carbon-carbon triple bonds) with metal-carbon single bonds at each terminus and a charge $n+$: $[M-(C≡C)_z M]^{n+}$. Removal of two electrons would give a cumulene-type chain with metal-carbon double bonds at each terminus and a charge $(n + 2)+$: $[M(=C=C)_z M]^{(n + 2)+}$. Removal of two more electrons would again give a chain consisting of alkynyl units, but with metal-carbon triple bonds at each terminus and a charge $(n + 4)+$: $[M≡C—C≡)_z M]^{(n + 4)+}$. Intermediate, radical-ion

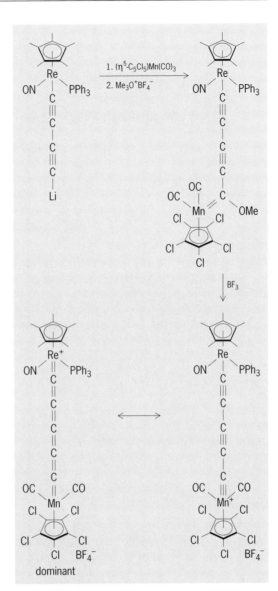

Fig. 1. Synthesis of a ReC$_5$Mn complex, Ph = C$_6$H$_5$ (phenyl). Me = CH$_3$ (methyl).

states are also possible. Several preparative strategies have been employed.

One approach that appears to have considerable generality is known as the Eglington coupling of ethynyl or higher $L_nM-(C≡C)_z$H complexes. This procedure simply involves a stoichiometric amount of the oxidizing agent Cu(OAc)$_2$, where Ac = acetate ion (CH$_3$COO$^-$). A complex with a Re—C≡C—C≡C—Re linkage forms in high yield. Subsequent one-electron oxidation with the silver tetrafluoroburate (Ag$^+$BF$_4^-$) gives an isolable radical cation. Data from both infrared and electron spin resonance spectroscopy show that the odd electron is rapidly delocalized between the two rhenium termini. In other words, there is no detectable resistance to conducting the electron from one metal to the other through the pi electron cloud of the chain. A second one-electron oxidation with Ag$^+$BF$_4^-$ gives an isolable dication with a $^+$Re=C=C=C=C=Re$^+$ linkage.

Special chain-extension procedures have allowed similar C_{12}, C_{16}, and C_{20} complexes to be prepared. The key step is known as the Cadiot-Chodkiewicz coupling in which a M$-(-C\equiv C-)_{\bar{z}}$Li species is first treated with copper iodide (CuI) to give an isolable alkynyl copper species. Then, an alkyne that is halogenated on one end and silylated on the other is added. Carbon-carbon bond formation occurs to give a longer alkynyl chain. The silyl group is subsequently replaced with a hydrogen, and an Eglington coupling is effected with $Cu(OAc)_2$. The Re*C_{20}Re*[Re*=(η^5-C_5Me_5)Re-(NO)(PPh$_3$)] species exhibits enormous visible absorptions, resulting in deeply colored solutions.

Future trends. Carbon normally has a valence requirement of four bonds. Thus, the edges of the three-dimensional and two-dimensional allotropes of carbon, diamond, and graphite must terminate in noncarbon end groups. Therefore, as methods are developed for the synthesis of complexes with still longer carbon chains, there will be a point at which such compounds can be regarded as genuine, one-dimensional carbon allotropes.

Another logical future research direction will involve switching from carbon chains to circles or loops. Only a few complexes have been prepared in which metals stabilize a simple monocyclic form of carbon; these include a triiron complex of cyclic C_3 and a hexacobalt complex of cyclic C_{18}. Other, topographically more complex, types of adducts are easily envisioned. These challenges are certain to occupy many synthetic chemists over the next few years.

Roman Dembinski; J. A. Gladysz

Metal-silane complexes. A large class of chemical compounds known generically as silanes consist of a silicon (Si) atom bonded to four other atoms or functional groups. These compounds are analogous to hydrocarbon compounds except that silicon is substituted for one or more carbon atoms. The simplest compound is silane (SiH_4), which is a close analog of methane (CH_4). Silanes are much more reactive to atmospheric oxygen and moisture than hydrocarbons, and SiH_4 gas spontaneously explodes in air. Silane is used in the electronics industry, and silane compounds are useful in chemical industry processes. Recently, SiH_4 has been discovered to bind to a transition metal to form a stable complex. In addition, the metal was found to break one of its Si-H bonds. This represents a model for the binding of methane to metals and subsequent C-H bond breaking, a long-sought goal of modern chemistry.

Chemical reactivity. Bonding and activation toward cleavage of simple molecules such as silanes, hydrocarbons, and hydrogen on transition metal complexes are crucial to their chemical reactivity, especially in catalytic conversions. However, SiH_4 and its organo-substituted derivatives such as SiH_3R (R = organic group) do not possess nonbonding electron pairs that typically "glue" molecules (known as ligands) to the metal center of a coordination complex. Most silanes have only

bonding electron pairs (sigma bonds as in Si-H) that until recently have been thought to be incapable of joining with a metal to give a stable species. Many such sigma bonds exist in chemical compounds, for example, the hydrogen molecule (H—H). However, the H—H molecule was found in 1983 to form stable chemical bonds to metal centers by approaching the metal side-on to form a T-shaped geometry. The isolation, structural characterization, and theoretical analyses of these unexpected dihydrogen complexes have shown that side-on (η^2) binding is the first step in breaking the strong H-H bond to give an atomically bound hydride [structure (I); W = tungsten]. This is neces-

$$
\begin{array}{c}
R_3 \\
P \quad\quad C{\equiv}O \\
| \quad\quad / \\
OC-W\!\!-\!\!\!\overset{\displaystyle H}{\underset{\displaystyle H}{|}} \\
O{\equiv}C \quad | \\
P \\
R_3
\end{array}
$$

(I)

sary for the controlled reaction of hydrogen long known to occur on metal catalysts. Most silanes possess Si-H bonds that can coordinate to the metal similarly to H-H bonds. It is important to note that this type of chemical bond is a nonclassical type, because it involves joining three atoms (Si, H, and metal) by essentially only the two electrons originally present in the Si-H bond.

Sigma-bond coordination. The bonding involving Si, H, and a metal is part of a new phenomenon in chemistry called sigma-bond coordination, analogous to coordination of pi bonds, for example, the C-C double bond in ethylene (**Fig. 2b**), which is critical in catalytic conversions of unsaturated petrochemicals. Any single two-electron bond (for example, H—H, Si—H, C—H, and C—C) is thus capable of binding to a metal by overlapping its electrons with a vacant metal d orbital. Such electron-deficient, 3-center bonding was initially thought too weak to give stable complexes. However, the bonding is strengthened by backbonding,

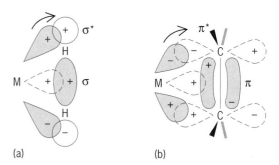

Fig. 2. Bonding pictures for (*a*) sigma-bond coordination as in metal-dihydrogen compared to (*b*) pi-bond coordination as in metal complexes of olefins such as ethylene. The empty, unshaded metal *d* orbitals are shown accepting electrons from the sigma (σ) or pi (π) orbitals, and the filled, shaded *d* orbitals backdonate electrons to the antibonding orbitals of H$_2$ (σ^*) or ethylene (π^*).

that is, donation of electrons from a filled metal d orbital to the sigma antibonding orbital (Fig. 2a), similar to metal donation to pi antibonding orbitals in ethylene coordination (Fig. 2b). It is now clear that this is the key component in binding silanes, H_2, or hydrocarbons to metals. If this backbond becomes too strong, for example, if more electron-donating coligands are put on the metal, the sigma bond breaks to give oxidative addition [that is, reaction of a metal center with a molecule such as H—H, where a bond is broken on the added molecule (H—H), and the fragments bind to the metal, increasing the metal's oxidation state by two]. This reaction is very important, particularly in catalysis. The bound silane (SiH_4) is fragmented to silyl (SiH_3) and hydride (H) ligands, and the oxidation state of the metal increases by two, for example, zerovalent (M^0) to divalent (M^{II}). Sigma-bond complexes can thus be looked upon as arrested oxidative addition, as in the reaction below.

The first metal-silane complexes, $CpMn(CO)_2$ ($SiHR_3$) [Cp = cyclopentadienyl; R = phenyl, chlorine] were prepared in the 1960s but were unrecognized as sigma-bond complexes and thought to be the oxidative-addition form, $CpMn(CO)_2$ (SiR_3)(H). Guided by comparison to metal-H_2 bonding, researchers eventually demonstrated similar 3-center bonding in manganese-silane complexes and synthesized new examples. The Si-H bond length determined by x-ray or neutron diffraction is a key parameter, typically 0.148 nm in unbound silanes. In metal-bound silanes it is elongated to 0.17–0.185 nm, much as H—H is elongated from 0.074 nm to greater than 0.082 nm in metal-H_2 complexes. In the latter case, a near continuum of H—H distances has been found, ranging up to the distance in separated hydrides (>0.16 nm). Thus, metal-mediated bond breaking can be arrested anywhere along the reaction coordinate, a notion that would have previously been difficult to believe. A logical question then arises as to the point at which a bond is broken on a metal; and the answer may be that it is never completely broken, at least not until one of its components is transferred away.

The metal complex that binds SiH_4 (and also $SiRH_3$ or SiR_2H_2) consists of molybdenum (Mo) in its zero-oxidation state surrounded by four phosphorus atoms of organophosphine ligands and a carbonyl (CO) ligand. The silane complexes are

synthesized in organic solvents by addition of SiH_4 gas to the metal complex, which is highly electron deficient and grasps one Si-H bond of the silane, as shown by x-ray crystallography and infrared and nuclear magnetic resonance (NMR) spectroscopy. The Mo—H—Si unit has a much longer distance for Mo—Si than for Mo—H, unlike the symmetrically bound H—H ligand. The silane complexes are crystalline yellow solids that decompose in air, and the silane can be removed from the metal by heating.

The Mo—SiH_4 complex in solution was discovered by NMR spectroscopy to be in chemical equilibrium with the oxidative-addition product, H—Mo—SiH_3. Both forms were found to be present at the same time in 1:1 ratio and rapidly undergo interchange, just as had been found for a tungsten-dihydrogen complex in equilibrium with its dihydride form. This is the first example of such an equilibrium bond-breaking and bond-making process other than for dihydrogen on metals. It is further direct evidence for η^2 sigma-bond coordination as the first step in sigma-bond cleavage on a metal. The cleaved fragments can then combine with other groups on the metal to form new compounds, the basis for catalytic conversion.

Gregory J. Kubas

Water-soluble compounds. The increasing demand for environmentally benign processes with high product selectivities and economically favorable reaction rates has renewed the interest in homogeneous organometallic catalysis. Simple and efficient separation of products from the catalyst under mild conditions is crucial to the application of homogeneous catalysts in industrial processes. The use of aqueous biphasic systems, in which the aqueous phase contains the dissolved organometallic catalyst, could allow particularly easy separation of the organic products. Because most biochemical reactions occur in aqueous environments, the reactions of organometallic compounds with biologically relevant ligands have recently attracted much interest. In general, aqueous organometallic chemistry is concerned with the synthesis, structure, reactivity, and catalytic activity of water-soluble organometallic compounds.

Although Zeise's salt [$PtCl_3(H_2C\!=\!CH_2)$; Pt = platinum], the first known organometallic compound, is soluble in water, most organometallic compounds have limited solubility in water. The most popular approach to increase the solubility of organometallic compounds in water has been the attachment of hydrophilic polar groups [(such as $-SO_3H$, $-SO_3Na$, $-COOH$, $-COONa$, $-OH$, $-NR_3^+$)(R = organic group)] to known ligands. Because phosphines are among the most frequently used ligands in organometallic catalysis, a variety of water-soluble phosphines bearing sulfonate, carboxylate, phosphonate, or ammonium groups have been prepared. The most studied water-soluble phosphine is the *meta*-trisulfonated-triphenyl phosphine [structure (II); Na = sodium].

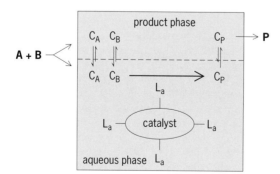

(II)

In certain cases nonionic ligands are also soluble in water such as 1,3,5-triaza-7-phosphaadamantane [P(CH₂OH)₃] and phosphines with polyethyleneglycol ether chains. Transition-metal complexes with highly basic ligands may also dissolve in water, usually because of the formation of ionic metal complex species. One of the most important structural implications of the presence of hydrophilic groups is the possible intra- and intermolecular interactions of the hydrophilic groups with each other and with other polar groups. The reactivity of aqueous organometallic compounds could also be affected by these interactions as well as by the possible interactions of the hydrophilic groups with water. Finally, the possible reactions of organometallic compounds with either proton (H^+) or hydroxyl ion (OH^-) should always be considered in aqueous organometallic chemistry.

Aqueous biphasic organometallic catalysis. There is an increasing interest in using water as a medium for transition-metal catalyzed organic reactions. The primary role of water as a solvent is the separation of the catalyst from the products (**Fig. 3**). An aqueous organometallic catalyst is designed to preferentially dissolve in the aqueous phase, although simply having a large partition toward the aqueous phase could also be appropriate for some applications. The required solubility can be achieved by attaching hydrophilic groups to the ligands of the organometallic catalyst. If the products (**P**) have limited solubility in water, a second, product phase could form during the conversion of the reactants (**A** and **B**), which can be easily separated. An aqueous organometallic catalyst could operate either in

the aqueous phase or at the interface of the two phases, depending on the solubilities of the reactants (**A** and **B**) in the aqueous phase. When the solubilities of the reactants are very low in water, an appropriate phase-transfer agent may be added to facilitate the reaction. Although the majority of investigations have been focused on hydrogenation, hydroformylation, carbonylation, hydrocyanation, and oxygenation, other types of reactions such as isomerization, alkylation, oligomerization, and polymerization are receiving increased attention. Catalytic asymmetric synthesis in aqueous media has been fundamentally advanced by successful modification of known chiral ligands via functionalization by sulfonate or quaternary ammonium groups. The hydrophilic analogs of well-known enantioselective rhodium and ruthenium catalysts were mainly used in the reduction of unsaturated amino acid precursors, imines, and hydrogenolysis of epoxides. In several cases very high enantioselectivities, close to the ones observed in the analogous hydrocarbon soluble systems, have been achieved. However, as a general rule, enantioselectivity was found to be lower in aqueous systems than in organic solutions. Addition of micelle-forming agents to aqueous systems increases the rate and enantioselectivity of hydrogenation of dehydro amino acids. An important feature of these biphasic reactions is that the chiral catalysts can be recycled in the aqueous phase with no or marginal loss in enantioselectivity.

Biological activity. A traditional field of organometallic catalysis is the activation of biologically important small molecules [for example oxygen (O_2), carbon dioxide (CO_2), and nitrogen (N_2)]. Recently further attempts have been made to perform such reactions in aqueous media under biologically acceptable conditions in order to construct working analogs of enzymatic systems. Catalytic hydrogenation of unsaturated lipids in biomembranes in model systems and in living cells has been demonstrated. Deuteration of membrane lipids by using either deuterium (D_2) or deuterium oxide (D_2O) offers a unique way of characterization of specific membrane domains. Such modifications take place in aqueous media, and a variety of investigations were carried out with water-soluble catalysts. Because of the enormously complex architecture of living systems, a very high degree of selectivity is required, including spatial (topological) selectivity.

Traditionally, the use of metal complex compounds as anticancer agents or as other therapeutic agents was studied by bioinorganic and biocoordination chemistry. Much is still left to be done in the area of bioorganometallic chemistry as exemplified by the binding of deoxyribonucleic acid (DNA) to surfaces by using (η^5-pentamethylcyclopentadienyl) rhodium complexes and by preparation of haptens for monoclonal antibody generation of catalysts.

Industrial application. The first large-scale industrial process utilizing a water-soluble organometallic cat-

Fig. 3 Aqueous biphasic organometallic catalysis. The required solubility of the catalyst in water is achieved by attaching hydrophilic groups (L_a = —SO₃H, —SO₃Na, —COOH, —COONa, —OH, —NR₃⁺) to the ligands of the organometallic catalyst. The catalyst converts reactants A and B to product P in the aqueous phase. Because product P has limited solubility in water, it forms the product phase that can be easily separated. The terms C_A, C_B, and C_P represent the concentrations of the species A, B, and P.

alyst has been operating in Oberhausen-Holton, Germany, since 1984. The aqueous biphasic hydroformylation of propylene is catalyzed by the water-soluble *meta*-trisulfonated triphenyl phosphine modified rhodium system, and it produces 300,000 metric tons per year of *n*-butanal (with about 4% isobutanal). Several smaller-scale biphasic processes have also been developed.

For background information SEE CHEMICAL BONDING; COORDINATION CHEMISTRY; COORDINATION COMPLEXES; HOMOGENEOUS CATALYSIS; MOLECULAR ORBITAL THEORY; ORGANOMETALLIC COMPOUND; PHASE-TRANSFER CATALYSIS; TRANSITION ELEMENTS in the McGraw-Hill Encyclopedia of Science & Technology.

István T. Horváth

Bibliography. T. Bartik et al., A step-growth approach to metal-capped one-dimensional carbon allotropes: Syntheses of C_{12}, C_{16}, and C_{20} μ-polyynediyl complexes via Cadiot-Chodkiewicz chain extension of $Re(C\equiv C)_n Cu$ species, *Angew. Chem. Int. Ed. Engl.*, 35: 414–417, 1996; R. H. Crabtree, Transition metal complexation of σ bonds, *Angew. Chem. Int. Ed. Engl.*, 32:789–805, 1993; I. T. Horváth and F. Joó, *Aqueous Organometallic Chemistry and Catalysis*, 1995; X. L. Luo et al., Synthesis of the first examples of transition metal η^2-SiH_4 complexes, *cis*-$Mo(\eta^2$-$SiH_4)(CO)(R_2PC_2H_4PR_2)_2$, and evidence for an unprecedented tautomeric equilibrium between an η^2-SiH_4 complex and a hydridosilyl species: A model for methane coordination and activation, *J. Amer. Chem. Soc.*, 117:1159–1160, 1995; U. Schubert, η^2-Coordination of Si-H σ bonds to transition metals, *Adv. Organometal. Chem.*, 30:151–187, 1990; J. W. Seyler et al., An isolable organometallic cation radical in which a C_4 chain conducts charge between two chiral and configurationally stable rhenium termini, *Organometallics*, 12:3802–3804, 1993; W. Weng, T. Bartik, and J. A. Gladysz, Towards one-dimensional carbon wires connecting single metal centers: A cumulenic C_5 chain that mediates charge transfer between rhenium and manganese termini, *Angew. Chem. Int. Ed. Engl.*, 33:2199–2202, 1994; Y. Zhou et al., New families of coordinated carbon: Oxidative coupling of an ethynyl complex to isolable and crystallographically characterized $MC\equiv CC\equiv CM$ and $^+M{=}C{=}C{=}C{=}C{=}M^+$ assemblies, *J. Amer. Chem. Soc.*, 115:8509–8510, 1993.

Paleontology

Fossilization results from the burial of organic remains in sediments. However, the processes of sedimentation rarely operate continuously in either space or time, and the preservation of organic remains within sediments is an uncommon event. Hence, the fossil records of most species are dominated by long gaps between fossil occurrences. Ever since Charles Darwin, workers have recognized that these gaps in the fossil record seriously compromise the ability of paleontologists to chart the course of

evolution. In particular, the incompleteness of the fossil record means that fossil ranges underestimate true longevity, and that the relative timing of evolutionary events is difficult to determine, especially for events that occurred within geologically short periods of time of each other. For example, it is often difficult to determine whether a group of species became extinct simultaneously or over a more extended period of time; this difficulty has been a major source of contention in understanding the nature of mass extinctions, particularly the Cretaceous-Tertiary boundary extinctions. It is also difficult to determine the order in which species originated if their first appearances in the fossil record occur in strata of approximately the same age, and to determine whether different lineages had different longevities. However, with recent developments in the statistical analysis of the gaps in the fossil record, the ability of paleontologists to interpret the history of life has been enhanced.

Gap analysis. Intuitively, the richer the fossil record of an evolutionary lineage the greater the proportion of that lineage's true longevity will be represented by fossils in the rock record. This intuition has now been quantified. From a statistical standpoint the stratigraphic positions correspond to the true times of origin and extinction of lineage fixed parameters; they are positioned at fixed points in time with probability 1.0, and everywhere else with probability 0.0. Hence, quantitative assessment of the incompleteness of the fossil record is based on confidence intervals. Specifically, the length of the confidence interval for a given confidence level, C, can be calculated such that there is a probability C that the true beginning (or end) of a lineage's stratigraphic range lies somewhere within that confidence interval.

The most reliable estimates can be given when the distribution of fossil horizons is statistically indistinguishable from a random distribution, a condition not infrequently met in the fossil record. More sophisticated methods that incorporate idiosyncrasies of the stratigraphic record are being developed, but all methods involve the computation of confidence intervals.

Rock versus time. Although paleontologists are primarily interested in the timing of evolutionary events, the primary data collected are the relative positions of fossils in the stratigraphic record. Hence, distances between fossiliferous strata are measured in meters of rock, rather than time, and thus confidence intervals are often given in units of stratigraphic distance rather than as temporal distances. Stratigraphic distances may be converted into intervals of time by calculating rates of sedimentation by, for example, measuring the stratigraphic distances between radiometrically dated strata (when available, which unfortunately is not often). However, this conversion process is not straightforward, and an area of active research focuses on ways of determining the completeness of the stratigraphic record. A recently developed method exploits the fact that the timing of the reversals in the Earth's magnetic field is independent of sedimentation events, but this approach may be used only to assess the stratigraphic completeness for relatively young rocks.

Sudden versus gradual extinctions. With mounting evidence of an enormous meteorite impact in Yucatán, Mexico (the Chicxulub crater), at the end of the Cretaceous Period, there has been increased interest in assessing the rapidity of the extinction of Cretaceous lineages, including the dinosaurs. However, because the fossil record is incomplete, the stratigraphic distribution of the last appearances of a group of species that became extinct simultaneously is expected to be distributed over a broad stratigraphic interval, giving the impression of a gradual extinction (the Signor Lipps effect). However, confidence intervals may be used to determine whether a pattern of gradual disappearances in the fossil record is in fact consistent with a sudden extinction scenario. For example, given that there is a 50% chance that the true time of extinction of a lineage occurs somewhere between the last observed fossil and the end of the 50% confidence interval placed on the top of its fossil range, for an ensemble of species that became extinct simultaneously, on average one-half of the end points of the 50% confidence intervals should lie above, and one-half below, the true extinction horizon. Thus, by plotting the distribution of the end points of the 50% confidence intervals on the top of the stratigraphic ranges of each of an ensemble of species, it is possible to determine if their fossil records are consistent with a sudden extinction; and if they are, to estimate the location of the extinction boundary.

This method has been applied to the fossil record of ammonites, one of the most important groups to become extinct at the end of the Cretaceous. Their fossil record is particularly well recorded on Seymour Island, Antarctica. It has been hypothesized, based on observed gradual declines of groups such as the Seymour Island ammonites, that high latitudes may have been somewhat protected from the consequences of the end Cretaceous impact. However, when 50% confidence intervals are applied to the tops of the stratigraphic ranges of the 10 uppermost Cretaceous ammonite species on Seymour Island, the pattern of disappearance of the ammonites is found to be consistent with the sudden extinction of all species at the Cretaceous-Tertiary boundary, even though the ammonites disappear from the fossil record over a 200-ft (60-m) stratigraphic interval below the boundary (identified by the stratum that contains the concentration of iridium derived from the meteorite impact). However, computer simulations show that the ammonite disappearances are also consistent with a gradual extinction that may have occurred over a period of time that corresponds to as much as 66 ft (20 m) of rock below the Cretaceous-Tertiary boundary. Thus,

although the use of confidence intervals shows that the ammonite fossil record is consistent with a mass extinction, despite appearances to the contrary, it does not prove there was a mass extinction of ammonites.

Order of appearance. It is generally reasonable to assume that the order in which lineages first appear in the fossil record is the order in which they really originated. However, if the fossil record of a later-appearing lineage is particularly poor, and the fossil record of an earlier-appearing lineage particularly rich, it is conceivable that the later-appearing taxon originated before the earlier-appearing one, especially if the gap between the times of first appearance is small. New methods have been developed, based on the assumption that the fossil distributions within each lineage are random, that make it possible for the first time to calculate the likelihood that a literal reading of the fossil record accurately reflects the true order of appearance of two lineages.

Determining relative longevities. The fossil record indicates that species of Cretaceous marine snail with larvae that fed in the plankton had greater longevities than species with nonplanktotrophic larvae. The gastropods with planktotrophic larvae also had greater geographic ranges than the other gastropods, and the correlation between species' longevity and geographic range has been used to support the hypothesis that natural selection may operate on species-level characteristics (for example, geographic range) as well as on individual-level characteristics. However, the observed correlation between geographic range and species' longevity may be an artifact of the incompleteness of the fossil record because species with smaller geographic ranges are less likely to be fossilized; they would be expected to have shorter observed longevities in the fossil record, even if they were actually as equally long lived. However, quantitative analyses of the fossil record, using both computer simulations and mathematical models, have shown that the gastropod fossil record is much too rich for the observed correlation to be an effect of the incompleteness of the fossil record. The gastropod fossil record does indeed support the hypothesis of species-level selection. With available analytic approaches this record can be used to explore the nature of the evolutionary process much more effectively than previously.

Limitations. The methods discussed above have two major limitations. First, the assumption of random distributions of fossil horizons underlies the statistics used. Yet, stratigraphic sequences with pronounced periodicities in the types of environment represented, or sequences that show secular trends in the sedimentary regime, will produce nonrandom distribution of fossil horizons. In these cases, the statistical calculations mentioned above will produce inaccurate confidence intervals. Further, many paleobiologic patterns may be artifacts of facies control, or of sequence architecture. For example, application of the concepts of sequence stratigraphy to an analysis of the fossil record suggests that first and last appearances are often artifactually concentrated at flooding surfaces caused by sea-level rises.

Second, most of these methods are designed for studies where fossils are recovered from a single section, or from limited geographic areas. The application of confidence intervals to the global fossil record is confounded by uncertainties in geographic ranges of species as well as major difficulties in correlating fossiliferous strata from remote areas.

The ability of paleobiologists to draw more conclusive deductions about the history of life from the fossil record will depend on an increased understanding of the periodic and secular forces that govern sedimentation, and on the spatial and temporal distribution of environments conducive to fossilization. Integration of this knowledge within a statistical framework will increase the scientific community's ability to understand the patterns and processes that have governed the history of life.

For background information SEE CRETACEOUS; EXTINCTION (BIOLOGY); FOSSIL; PALEONTOLOGY; STRATIGRAPHY; TERTIARY in the McGraw-Hill Encyclopedia of Science & Technology.

Charles Marshall

Bibliography. N. Gilinsky and P. W. Signor (eds.), *Analytical Paleobiology,* 4:19–38, 1991; K. A. Joysey and A. E. Friday (eds.), *Problems of Phylogenetic Reconstruction,* 1982; C. R. Marshall, Distinguishing between sudden and gradual extinctions in the fossil record: Predicting the position of the Cretaceous-Tertiary iridium anomaly using the ammonite fossil record on Seymour Island, Antarctica, *Geology,* 23(8):731–734, 1995; A. B. Shaw, *Time in Stratigraphy,* 1964.

Petroleum reservoir engineering

The petroleum industry is focusing on improving reservoir performance through the use of four-dimensional seismic data and dynamic reservoir characterization. Accurate and detailed seismic monitoring is critical to achieving economic success in reservoir production and development. Repeated three-dimensional seismic surveys that are shot for monitoring objectives are termed four dimensional, time lapse being the fourth dimension. A new wave is also being recorded that, combined with conventional seismic data, allows more accuracy in dynamic property sensing and determination.

Dynamic reservoir characterization. The relatively new concept known as reservoir characterization has provided an important tool that involves the pooling of information from different disciplines (geology, geophysics, and petroleum engineering) in order to provide a more complete understanding of the physical properties of a reservoir. Reservoir characterization initially takes data acquired from these disciplines to produce an integrated reservoir model. The purpose of this reservoir model is to predict the distribution of physical reservoir parameters and to tie these parameters into reservoir response. Generally, this process results in a static reservoir model with refined porosity and permeability distribution.

The development and production of hydrocarbons, however, change the physical dynamics of a reservoir. A logical extension would be to monitor these changes. If a change in the dynamic properties occurs that could negatively impact reservoir production, it makes economic sense to make an adjustment as soon as possible. An adjustment made early would cost far less than completion of a process that is not only inefficient but creates the need for additional processes in order to get the desired recovery. Periodic seismic surveys conducted over the reservoir would provide an instantaneous source of information. Comparative analyses and computer-simulated models would reveal changes in the dynamic properties of the reservoir. Thus, reservoir characterization would become truly dynamic and, as such, the ultimate reservoir management tool.

To the dynamic reservoir characterization team, a time-lapse or four-dimensional seismic survey is a tool that probes the reservoir and remotely senses its dynamic changes over time. In essence, the survey allows the team to hook up to a reservoir in a manner analogous to an annual physical at the doctor's office evaluating a person's general health. Monitoring vital signs on an individual or a reservoir are akin; it takes repeated observations to determine changes in condition and to prescribe remedial action.

Four-dimensional seismic. Computer systems enable three-dimensional seismic surveys to be acquired, processed, and interpreted routinely throughout the industry; in the 1980s the three-dimensional seismic survey was the domain of only a few large companies and was used only when deemed necessary. New seismic recording systems with increased sensitivity and channel capacity have also evolved. The high-tech systems of the 1990s have replaced their predecessors of the 1980s. The measurement system itself has changed, so that reservoir parameters and response can be determined more effectively. As a result, the applications of three-dimensional seismic technology are still under development.

Four-dimensional seismic methods compare one three-dimensional survey of a reservoir with a repeat survey taken in the same geographic location at a later time. In theory, there may be no limit to the number of repeat surveys; however, determination of the expected optimal timing of these surveys is linked to the reservoir model and to economics. Consistency between surveys is critical and is worth considerable effort. New seismic recording systems with increased fidelity and dynamic range now make this a reality. Part of the consistency or repeatability issue is resolved by using the same recording instruments from survey to survey and the same sources and procedures to balance effects of changes in source-receiver coupling that are near-surface induced.

Differences between surveys can indicate changes in the producing reservoir. For example, these changes can be calibrated through the reservoir model to monitor fluid movement. If determined accurately, the differences between surveys would assist in computer-aided production and the prediction and forecasting of reservoir performance. Early optimization of the development and production process would lead to improved recovery, extended reservoir life, and reduced environmental impact.

The four-dimensional seismic survey is more accurate, because a higher degree of resolution can be obtained by using differenced measurements (where one measurement is subtracted from the other). In the four-dimensional seismic method, attributes such as times and reflection amplitudes can be taken into account when differencing measurements. Rather than relying on absolute measurements that may have large margins of error, differencing similar measurements results in significantly reduced margins of error. It is important to determine those properties, or combinations thereof, that have the greatest sensitivity to reservoir response. These properties can be determined through the use of geostatistics and attribute analysis.

A breakthrough has occurred because of changes in the measurement system itself. A new method, multicomponent seismic recording, records two horizontal components and one vertical. This recording scheme is akin to the recording systems of earthquake seismologists and facilitates recording of combined compressional or primary (P) and shear or secondary (S) waves. Four-dimensional, multicomponent recording provides significantly more information about the rock and fluid properties of a reservoir than can be achieved from conventional P-wave seismic surveys alone. Because of technological innovations that have been developed since the mid-1980s, a reservoir can be probed remotely by recording P and S waves as they pass through it, and rock and fluid properties can be more accurately determined and monitored over time. Comparative travel time or velocity measurements and amplitudes of P and S waves enable the discrimination of rock and fluid properties, their characteristics, and their changes over time.

Field studies. Field studies using four-dimensional seismic methods are being undertaken in the North Sea. The Oseberg project has demonstrated movement of free gas in the Brent Sandstone reservoir. Water injection from a pressure maintenance program initiated in the early 1990s drove free gas back into solution with oil. The area and volume of free gas was monitored to optimize pressure maintenance and thus extend the life of the field. At Oseberg, for the first time, coupling four-dimensional seismic monitoring with reservoir simulation has improved petroleum recovery.

The first use of a four-dimensional seismic survey in deep water using permanently installed, ocean-bottom cables and sensors took place at Foinaven Field in the North Sea. The high cost of field development at a water depth of 1650 ft (500 m) contributes to the use of four-dimensional seismic technology, which reduces the number of wells needed to economically produce the field.

The first onshore four-dimensional, multicomponent survey took place at Vacuum Field, New Mexico, in 1995. The Colorado School of Mines Reservoir Characterization Project is monitoring a pilot carbon dioxide flood in a carbonate reservoir in the Permian Basin. Economic benefits of this industry-funded research include upscaling from pilot to maximum-efficiency, full-field flood through four-dimensional, multicomponent seismic monitoring and dynamic reservoir characterization.

Prospects. To develop a reservoir and manage it to achieve the maximum economic benefit requires the determination of rock and fluid characteristics and their changes over time. The addition of a four-dimensional seismic survey to reservoir characterization will refine the dynamic reservoir model. This advance will help produce structured economic decisions that will extend reservoir life and lead to improved recovery while reducing risk. The driving force behind the use of this technology is the profitable production of hydrocarbons. Careful calibration of these observations to measured changes is necessary, requiring a living or dynamic reservoir model and real-time processing of data from different disciplines and measurements systems. As a result, four-dimensional seismic surveying will enable dynamic reservoir characterization to extend the life spans of existing reservoirs. Improving recovery while minimizing environmental impact is critical to the future of the petroleum industry.

For background information SEE PETROLEUM RESERVOIR ENGINEERING; SEISMIC EXPLORATION FOR OIL AND GAS; SEISMOLOGY in the McGraw-Hill Encyclopedia of Science & Technology.

Thomas L. Davis

Bibliography. D. George, First deepwater 4-D seismic to be acquired over Foinaven, *Offshore*, pp. 31–32 July 1995; D. George, 4-D: The next seismic generation?, *Offshore*, pp. 21–22, October 1994; S. E. Johnstad, R. C. Uden, and K. N. B. Dunlip, Seismic reservoir monitoring over Oseberg Field, *First Break*, 11(5):177–185, 1993.

Phase transitions

An outstanding puzzle of modern cosmology is the structure of the universe on large distance scales. Current sky surveys have revealed striking patterns in the distribution of galaxies and galaxy clusters. The galaxies appear to form an interconnected spongelike network permeated with vast voids. Several theoretical models or mechanisms for generating this large-scale structure have been proposed. A fascinating example of such a mechanism arises out of the interplay between the physics of the very small (elementary particles) and the physics of the very large (cosmology).

Transitions in early universe. In the very successful big bang model, the universe evolves from a hot, homogeneous, structureless gas of elementary particles. For a tiny fraction of a second after the big bang, all the elementary particles interact according to the fundamental laws of a grand unified theory: the distinct strong, weak, and electromagnetic forces observed at low energies are united into one force. Grand unified theories also hypothesize the existence of a field or particle known as the Higgs particle, necessary to explain the observed masses of the particles detected in laboratory experiments.

As the universe expands and cools, it may undergo an abrupt transition to a state (phase) with very different physical properties and symmetries, much as water cooling below its freezing point turns to solid ice. Phase transitions are usually marked by a change in the symmetry of the physical system. This change may be abrupt in a first-order transition like freezing, which is accompanied by liberation or absorption of latent heat, or gradual, in a so-called continuous transition. Both cases are discussed in this article. Phase transitions may generate defects or irregularities in the fields describing the location of elementary particles in space, somewhat analogous to cracks in the ice. These objects, known technically as topological defects, prevent the system from being in its lowest possible energy state and, in the case of phase transitions in the very early universe, are extremely massive. Their mass is a seed for gravitational instability, which can ultimately lead to the clumping of matter on the enormous scales of galaxies and galaxy clusters.

A more precise description of topological defects requires consideration of the space of possible low-temperature ground states in a phase transition. The low-temperature phase, in almost every case, has less symmetry than the high-temperature phase; that is, the low-temperature phase is more ordered. If the high-temperature phase has a symmetry group, $\mathbf{G}$, then the low-temperature phase has a symmetry group, $\mathbf{H}$, which is some subgroup of $\mathbf{G}$. Now, for any ground state, ϕ, it is possible to construct another state of equal energy by applying a symmetry transformation from the group $\mathbf{G}$. (Although the full symmetry, $\mathbf{G}$, is not respected in any single ground state, it is still a symmetry of the laws governing the system.) Not all states obtained in this way are genuinely different. Any state obtained by acting on ϕ with an element of $\mathbf{H}$ is, by definition, equivalent to ϕ. The space of inequivalent ground states is the coset $\mathcal{M} = \mathbf{G}/\mathbf{H}$. When $\mathcal{M}$ is a space with nontrivial topology, topological defects are possible.

The most thoroughly investigated class of topological defects relevant to cosmology are linelike configurations of the Higgs field, known as cosmic strings. They occur whenever the space $\mathcal{M}$ admits loops that are not continuously contractible to points. Cosmic strings are thought to have formed about 10^{-37} s after the big bang, when the universe had a temperature of 10^{28} K, was no bigger than a mango. One centimeter of cosmic string weighs about 10^{19} kg, as much as the Rocky Mountains, and is 10^{15} times thinner than the atomic nucleus. In the cosmic-string model of the formation of large-scale structure, this huge energy density leads to gravitational perturbations that result in the clumping of the initially homogeneous universe into galaxies and galaxy clusters.

Cosmology in the laboratory. Although experimental tests of theories for the formation of large-scale structure are difficult, there are many examples of phase transitions, with associated topological defects, in more accessible condensed-matter systems such as liquid crystals and superfluid helium. Recently, both these systems have been investigated as laboratory models of certain aspects of the formation and dynamics of topological defects in the universe, and it has proven possible to verify some of the theoretical predictions of particle astrophysics.

Nematic liquid crystals. A simple and accessible laboratory analog of defect-producing phase transitions in the cooling universe may be observed in a heated sample of the simplest kind of liquid crystal, known as a nematic liquid crystal. Nematic liquid crystals are made of many individual rodlike organic molecules. A photograph of such a liquid crystal at high temperatures would reveal that the rods are randomly located and also randomly oriented. In other words, the rods have no preferred positions and no preferred axis of alignment. This high-temperature phase is known as the isotropic phase. As a liquid crystal cools, it undergoes an abrupt change of phase at a specific temperature, usually between 0 and 200°C (32 and 392°F), depending on the structure of the liquid crystal. Below this temperature the constituent rods are, on average, aligned with a particular axis in space, like a field of wild sunflowers inclined toward the setting Sun. The centers of mass of the rods, however, are still distributed randomly. Such a system is said to have orientational order, as opposed to the translational order of a crystal. This low-temperature, orientationally ordered phase of a liquid crystal is called the nematic phase. The isotropic-nematic transition is first order.

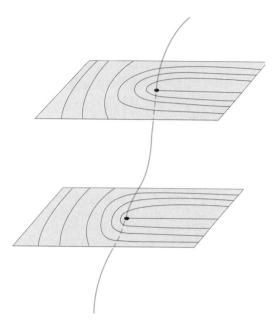

Fig. 1. Disclinations in a nematic liquid crystal. The vertical curved line is the disclination itself.

As a bulk sample of liquid crystal cools through the isotropic-to-nematic phase transition, nematic domains form spontaneously like droplets of early morning mist. The average orientation of molecules in each domain is roughly uniform. Different domains, however, orient independently. The average initial size of these nematic domains is called the correlation length, ξ, and is determined by the cooling rate and the microscopic physics of liquid crystals. In the cosmological analogy, a domain is akin to a causally connected region of space-time. Physical processes in two regions not causally connected are necessarily independent because communication can never have occurred between the regions.

With time the individual nematic domains grow and coalesce. Because it costs energy for molecular orientations to twist, bend, or splay significantly, it is preferable for the entire sample to orient uniformly. A striking consequence of the mathematical discipline of topology is that uniform orientation is impossible for a nematic liquid crystal. It is inevitable that regions of rapidly changing orientation, called singularities, will be trapped during the cooling to the nematic phase. At a singularity a liquid crystal is completely confused as to which orientation it should choose; it solves this dilemma by remaining in the high-temperature isotropic phase. These regions of isotropic phase are the topological defects. In nematic liquid crystals the most important topological defects are linelike, in exact analogy to cosmic strings, and are called disclinations. In **Fig. 1**, the field lines of nematic orientation have been illustrated for two planes cutting a disclination. **Figure 2** is a photograph of disclinations observed in the laboratory.

Probability of defect formation. With some further assumptions, it is even possible to calculate the probability that a disclination forms as randomly oriented nematic bubbles coalesce. First of all, the space of orientations in three dimensions is the space of all straight lines through the origin. Each line (orientation) can be imagined as a long toothpick completely piercing an orange through its center. Thus, the space of all orientations is equally well described mathematically as the space of all points on the surface of a sphere, with each point and its antipodal point being regarded as identical. This space is known mathematically as the projective plane RP^2.

A loop on RP^2 is either a loop on the two-sphere, S^2, or a path from a point on S^2 to its antipodal point. The first kind of loop may be continuously contracted to a point and is thus topologically trivial. The second kind of loop may not be continuously contracted to a point, because it has to be cut to contract and cutting is not a continuous operation; it is therefore topologically nontrivial. Superimposing two topologically nontrivial disclinations gives a configuration in which the lines of orientation wind 360° in a plane around the defect line. Such a configuration can relax into a topologically trivial configuration by a continuous bending of the lines into the third dimension. As a consequence, disclinations are described by the group $\{1, -1\} = \mathbf{Z}_2$.

Coalescing domains can be modeled by considering the meeting of three domains of size ξ. A random point on the manifold RP^2 can be assigned to each of the three domains, and orientations for points between domains may then be obtained by continuing the orientation from one domain to the next in as smooth a manner as possible. In this way it is possible to see if a disclination results within a correlation volume of order ξ^3. Repeating this process many times, typically numerically, yields a probability p, the expected number of disclinations per coalescing domain, or per correlation volume ξ^3. The total number of defects, N, in a volume V is then given by Eq. (1).

$$N = \frac{pV}{\xi^3} \qquad (1)$$

This theoretical number has been compared with the following experiment. A droplet of the nematic

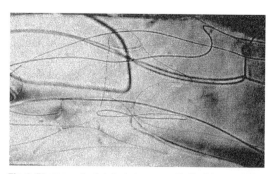

Fig. 2. Photograph of defects in a nematic liquid crystal. The thin lines are topological disclinations, and the thick lines are nontopological defects not discussed in this article. (*From L. Michel, Symmetry defects and broken symmetry, Rev. Mod. Phys., 52:617–651, 1980*)

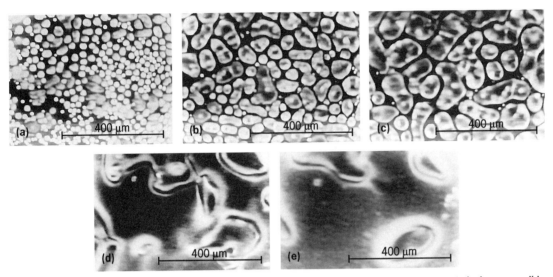

Fig. 3. Series of images showing the isotropic-to-nematic phase transition in a drop of 5CB on an untreated microscope slide. Stages of bubble nucleation and growth, bubble coalescence, and disclination formation and evolution are visible. The delay times for each image (referred to the first frame showing discernible bubbles) are (*a*) 2 s, (*b*) 3 s, (*c*) 5 s, (*d*) 11 s, and (*e*) 23 s. (*From M. Bowick et al., The cosmological Kibble mechanism in the laboratory: String formation in liquid crystals, Science, 263:943–945, 1994*)

liquid crystal 5CB was placed on a glass slide in a phase-contrast microscope equipped with a television camera and a videocassette recorder. The droplet was heated into the isotropic phase and allowed to cool through the isotropic-nematic transition at 35.3°C (95.5°F). A still photograph of nematic-domain formation was extracted from the video recording (**Fig. 3**) and used to count the total number of domains, N, and their effective linear size. This size also determines the average length, d, of disclination produced in a domain coalescence. The total length of disclination formed (L) is then given by Eq. (2). A second photograph of disclinations

$$L = pNd \qquad (2)$$

at the moment of formation was then used to measure the total length of disclination in the sample. It was found that theory is in reasonable agreement with the experimental result. This lends support to the mechanism described above for the generation of topological defects in phase transitions and gives a theory of the density of defects at the time of their formation.

Defect dynamics. The degree of large-scale structure generated by topological defects depends not just on their initial density but also on their evolution in time after formation. It is also possible to study directly the interactions of disclinations in a liquid crystal, as they evolve after formation, and to compare their behavior with that predicted for cosmic strings. There are remarkable similarities: disclinations shrink, straighten, cut, and reconnect, just as expected for cosmic strings. Because small loops of excised cosmic string are thought to ultimately seed single objects such as galaxies, this feature of the theory is important to verify experimentally, although in a dramatically different setting.

A fundamental idea of cosmic string theory that has been tested in this way is the scaling hypothesis. It is assumed that the dynamics of string interactions is governed by a single length scale, the correlation length, ξ. This scale corresponds to both the mean distance between strings and their mean radius of curvature. According to this hypothesis, the pattern of evolving defects remains the same statistically, with merely the overall scale of the structures growing. For disclinations in liquid crystals, it can be shown that ξ should grow as the square root of the elapsed time. By dimensional analysis the length per unit volume of disclination, ρ, should scale as $1/\xi^2$, and thus be expected to fall linearly with time. This result has been confirmed in experiments on rapid pressure-induced isotropic-nematic phase transitions in a cell of the nematic liquid crystal 5CB. The density of disclinations is measured by light transmission through the cell. It is seen quite graphically then that initially small defects produced by the phase transition lead to progressively larger structures at later times, just as required for the generation of large-scale structure in the universe. Analogous experiments have also been performed on two-dimensional systems. Here the expected result is also found, namely, that ρ scales as $1/\xi$, and thus decreases in proportion to the inverse of the square root of time.

Superfluid helium. Another laboratory system in which defects are produced in a phase transition is superfluid helium-4. In a rapid (few-millisecond) adiabatic expansion from high to low pressure, at a constant temperature of about 2 K (−456°F), normal liquid helium turns superfluid. This is a continuous phase transition. The decompression is performed with bellows mechanically linked to the top of the cryostat containing the liquid-helium sample. The defects in this case are superfluid vortices, resembling microscopic tornadoes, with normal fluid trapped inside and superfluid helium on the outside. Helium-4 is described phenomenologically by a many-body complex Bose-condensate wave function. In the superfluid state the magnitude of the wave function is nonzero, but its phase is arbitrary. The manifold of superfluid states is thus the set of all possible phases, namely the circle, S^1. The associated group is the symmetry group of the circle, $U(1)$, corresponding to rotations around a fixed axis. Around a vortex configuration the phase increases from zero to some integer multiple of 2π. Vortices are thus characterized by the integer winding number.

Superfluid helium vortices are detected in the experiment by their attenuation of heat pulses (more technically, second sound) propagated across a space of a few millimeters between a gold film heater and a carbon bolometer situated within the helium cell. Second sound is an entropy-temperature wave in which the normal fluid and the superfluid oscillate 180° out of phase. In contrast, ordinary, or first, sound is a pressure-density wave in which the normal fluid and superfluid oscillate in phase.

From the degree of attenuation of second sound, it is found that the length per unit volume of vortices formed in the transition is roughly $10^7/\text{cm}^2$. To produce the same density of vortices by spinning a vessel of liquid helium, the classic method of creating vortices, the vessel would have to be spun at the enormous rate of 630 revolutions per second. As with disclinations, vortices decay with time. Although a fascinating and fundamental system, superfluid helium suffers from the disadvantage that the vortices cannot be directly visualized. In this respect the liquid-crystal systems are clearly superior.

Prospects. There is a rich variety of directions toward which the developments described here may be extended. To begin with, there are many types of liquid crystals beyond the nematics discussed above. Some of the translation symmetry of the nematic may also be broken, in addition to the orientational symmetry, leading to chiral nematics and smectics. There may also be a second symmetry axis, giving rise to a biaxial nematic. Biaxial nematics are particularly fascinating, because they possess three fundamentally different types of disclination which should interact among themselves according to the algebra of the group of quaternions.

The liquid-helium experiment could be carried out with an annular geometry rather than in the bulk. As domains form statistically in a quench to the superfluid state, the phase of the wave function will vary randomly from one domain to the next. A spatial gradient of the phase will be generated which corresponds physically to a superfluid rotation around the annulus. Such spontaneously generated rotation would indeed be a graphic and dramatic consequence of domain formation giving rise to topological defects.

Finally, the defects discussed so far arise from the breaking of global symmetries. A global symmetry transformation is the same at all points in space. In both particle physics and condensed-matter systems, there also arise local (or gauge) symmetries—symmetry transformations that can change from point to point in space and yet still describe the identical physical system. The electromagnetic field itself is best understood as being a consequence of a local $[U(1)]$ symmetry.

Local cosmic strings could play as important a role in the early universe as global cosmic strings. Ideas about the formation of local cosmic strings might be tested in a rapid temperature quench of a type II superconductor in a vanishing external magnetic field. Spontaneously generated flux lines would be sought in this system, the analog of the local cosmic strings. These linelike defects contain a normal core surrounded by superconducting material and a spatially decaying magnetic field. Experiments with this system are hard to perform, because the quench to the superconducting state must be done by removing heat and the specific heat diverges at the phase transition. Thus it is difficult to perform the quench rapidly enough to generate a significant density of defects.

For background information SEE BIG BANG THEORY; COSMIC STRING; CRYSTAL DEFECTS; GAUGE THEORY; GRAND UNIFICATION THEORIES; GROUP THEORY; INFLATIONARY UNIVERSE COSMOLOGY; LIQUID CRYSTALS; LIQUID HELIUM; PHASE TRANSITIONS; QUANTIZED VORTICES; SECOND SOUND; SUPERCONDUCTIVITY; TOPOLOGY; UNIVERSE in the McGraw-Hill Encyclopedia of Science & Technology.

Mark J. Bowick

Bibliography. M. J. Bowick et al., The cosmological Kibble mechanism in the laboratory: String formation in liquid crystals, *Science*, 263:943–945, 1994; I. Chuang et al., Cosmology in the laboratory: Defect dynamics in liquid crystals, *Science*, 251:1336–1342, 1991; P. C. Hendry et al., Generation of defects in superfluid ^{4}He as an analogue of the formation of cosmic strings, *Nature*, 368:315–317, 1994; D. N. Spergel and N. G. Turok, Textures and cosmic structure, *Sci. Amer.*, 262(3):52–59, March 1992.

Photoionization

Studies of the interaction between photons and isolated atoms in the gas phase can provide detailed information on the internal structure of atoms and the fundamental interaction between charged particles and quantized electromagnetic radiation. Photons can scatter off atoms either elastically (Raleigh scattering) or inelastically (Compton scattering), or they can be absorbed by the atom, leading to transitions into excited states or to the ejection of electrons. The latter process, called photoionization, is the atomic analog of the photoelectron effect, in which electrons are ejected from solids following the absorption of electromagnetic radiation. Study

Table 1. Characteristic photon energies and wavelengths for photon-helium interaction

Region	Photon energy (E_{ph}), eV	Photon wavelength (λ), nm
First ionization potential of helium	24.6	50.4
Second (total) ionization potential	79	15.7
Photoionization and Compton cross sections are approximately equal	6000	0.2
Photon wavelength equals average distance of electron from nucleus in helium atom	25,000	0.05

of these processes with individual atoms has the advantage that the initial and final states of all participants in a reaction can, in principle, be determined with high precision. However, the small probability of the occurrence of this process for a low-intensity photon beam interacting with the low-density gas target poses a major obstacle to such a so-called complete experiment. With the recent availability of high-intensity photon beams from lasers (mostly in the infrared, visible, and near-ultraviolet spectrum) and from synchrotrons (in the vacuum ultraviolet and x-ray spectrum) some of these studies have become feasible. This article discusses recent advances in the study of one- and two-electron ejection from helium at high photon energies.

One- and two-electron emission. In quantum electrodynamics, the interaction between photons and charged particles is described by one-particle operators, so that one incident photon interacts with one electron at a time. Consequently, the photoionization of atoms by photons with energies exceeding the first ionization potential I_1 (the binding energy of the most loosely bound electron, $I_1 = 24.6$ eV for helium) results predominantly in the ejection of one electron, as in reaction (1). Here, $h\nu$

$$h\nu + A \longrightarrow A^+ + e_1 \qquad (1)$$

represents the photon energy (h is Planck's constant and ν is the photon frequency), A represents the atom, and e_1 represents the ejected electron. The kinetic energy of the ejected electron is the excess energy, that is, the photon energy, minus the ionization potential.

For photon energies above the second ionization threshold ($I_2 = 79$ eV for helium), the simultaneous ejection of a second electron (e_2), as in reaction (2), becomes energetically possible. Because of the

$$h\nu + A \longrightarrow A^{++} + e_1 + e_2 \qquad (2)$$

single-particle nature of the electron-photon interaction, this process occurs only as a result of strong electron-electron correlations in atoms. Correlation means that the electrons in the atom do not move independently of each other, as assumed in the traditional independent-particle picture, but are strongly coupled. Simply put, the first electron on its way out can cause a second electron to be ejected from the atom provided enough energy has been initially absorbed. Ejection of up to six electrons by a single energetic photon as a result of a complex rearrangement of the atomic charge cloud has been observed.

For the simplest case, two-electron ejection from the helium atom, an unprecedented look into the dynamics of three particles interacting with each other via Coulomb forces has become possible. The helium atom consists of an alpha particle nucleus and two electrons. This atom plays a prominent role because of the simplicity of its electron configuration. In addition, the high-energy region where the photon energy exceeds the total ionization potential can be reached at synchrotron facilities currently in operation. **Table 1** lists a few characteristic photon energies for the photon-helium interaction.

Photoionization dynamics. Although a photon can scatter off a free electron, absorption is forbidden because energy and momentum conservation cannot be simultaneously satisfied. Photoabsorption is possible only for bound electrons in atoms where the nucleus can take up the recoil. Because the mass ratio between the electron mass, m_e, and the mass, m_N, of the nucleus is very small ($m_e/m_N \lesssim 10^{-4}$), the absorbed energy must be shared between the electrons and the nucleus in a very asymmetric fashion. The electron carries away almost all of the excess energy, E_{kin}, with the remainder, $(m_e/m_N)E_{kin}$, taken up by the recoiling nucleus. The unfavored energy transfer to the nucleus is the origin of the precipitous decline of the photoionization cross section at high photon energies, where this cross section decreases in proportion to the $-7/2$ power of the photon energy.

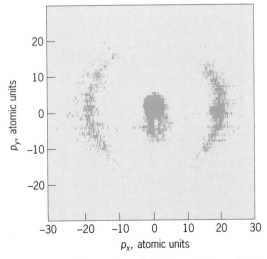

Two-dimensional momentum distribution of He$^+$ recoil ions produced in interactions of photons of energy equal to 9000 eV with helium. The direction of the photon beam is perpendicular to the plane of the figure.

Table 2. Ratios for double to single ionization at high photon energies

Initial-state wave function	Ratio by photoabsorption (R_{ph}), %	Initial-state wave function	Ratio by Compton scattering (R_C), %
Hartree-Fock	0.51	Screened hydrogenic	0.72
Screened hydrogenic	0.72	45-term configuration interaction wave function	0.815
18-parameter Hylleraas wave function	1.71	39-parameter Hylleraas wave function	0.835
39-parameter Hylleraas wave function	1.66		

The **illustration** displays the first observation of a two-dimensional momentum distribution of the recoiling helium nucleus following photoionization. The two lobes of the recoil distribution of the helium nucleus mirror the dipole angular distribution of the ejected electron. The coupling of the electromagnetic field to the electrons leads to the dipole selection rule for the angular momentum of the final state of the ejected electron: $|L_f - L_i| = 1$. Because the initial state has angular momentum $L_i = 0$, the final-state angular momentum, L_f, must be 1 for photoionization. The additional spherical distribution near zero momentum in the center of the figure results from ionization by Compton scattering.

Two-electron ejection probability. The probability of two-electron ejection can be determined by the ratio of the cross section for double photoionization to that for single photoionization, designated R_{ph}. Theoretically, it can be shown that R_{ph} becomes energy independent at high photon energies, as given by Eq. (3). Here, the sum in the numer-

$$R_{ph} = \frac{\sum_c | \int d^3 r \phi_c(\vec{r}) \psi_i(0, \vec{r})|^2}{\sum_b | \int d^3 r \phi_b(\vec{r}) \psi_i(0, \vec{r})|^2} \quad (3)$$

ator includes all final states of the ionized second electron in the continuum (labeled c, with wave function ϕ_c) and the sum in the denominator extends over all bound states (labeled b, with wave function ϕ_b) of the spectator electron (the electron that is not ejected) in the case of single ionization. The ratio R_{ph} is sensitive to correlations between the two electrons, because it is determined by the local behavior of the initial state, $\psi_i(\vec{r}_1, \vec{r}_2)$, of helium in its ground state. Loosely speaking, it is given by the probability of finding the spectator electron at a particular position, $\vec{r}$, while the photoabsorbing electron is in the vicinity of the nucleus ($\vec{r} \approx 0$). Because of the required momentum exchange, photoabsorption can take place only when the electron is localized in the vicinity of the nucleus. **Table 2** illustrates how strongly the predicted value of R_{ph} depends on the amount of correlation built into the initial state. The ratio ranges from 0.5% in the Hartree-Fock approximation to 1.66% when the wave function giving the most accurate value for the binding energy is used.

Measurements of R_{ph} have been successfully completed only since 1993. A major obstacle in the past has been the presence of an important competing process, single- and double-electron ejection

by an inelastic photon scattering process, called Compton scattering, as in reactions (4) and (5). The

$$h\nu + A \longrightarrow h\nu' + A^+ + e_1$$
$$\text{(one-electron ejection)} \quad (4)$$

$$h\nu + A \longrightarrow h\nu' + A^+ + e_1 + e_2$$
$$\text{(one-electron ejection)} \quad (5)$$

cross section for Compton scattering is of the order of r_0^2, where $r_0 = 2.8 \times 10^{-15}$ m is the so-called classical electron radius. This cross section is typically very small compared to photoabsorption cross sections and therefore very difficult to observe. However, because of the precipitous decline of the photoabsorption cross section, Compton scattering begins to dominate over ionization by photoabsorption in the case of helium at a photon energy of about 6000 eV. Photon scattering off the electron, unlike absorption, does not rely on the recoil of the nucleus to balance energy and momentum. Such events appear as the spherical spot in the center of the illustration, signifying near-zero recoil momentum.

Just as for photoabsorption, a single Compton-scattering process can eject more than one electron because of the correlations in the initial state. The ratio of double-to-single ionization, R_C, is given at high photon energies by Eq. (6).

$$R_C = \frac{\sum_c | \int d^3 r_2 \phi_c(\vec{r}_2) \psi(\vec{r}_1, \vec{r}_2)|^2}{\sum_b | \int d^3 r_2 \phi_b(\vec{r}_2) \psi(\vec{r}_1, \vec{r}_2)|^2} \quad (6)$$

This ratio provides a different measure of electron correlations present in the initial state. It probes the wave function at all positions of the primary Compton-scattered electron. Numerical values for R_C and for R_{ph} are predicted to be different. The first measurements do indeed show such differences. For photoionization, the asymptotic value of $R_{ph} = 1.66\%$ appears to be reached at a photon energy of about 8000 eV. The corresponding limit for two-electron Compton processes, which is predicted to be about 0.8% (Table 2), should be reached only at much higher energies and is not yet experimentally confirmed.

For background information SEE ATOMIC STRUCTURE AND SPECTRA; COMPTON EFFECT; QUANTUM ELECTRODYNAMICS; SCATTERING EXPERIMENTS (ATOMS AND MOLECULES); SYNCHROTRON RADIATION in the McGraw-Hill Encyclopedia of Science & Technology.

Joachim Burgdörfer

Bibliography. L. Andersson and J. Burgdörfer, Excitation-ionization and double ionization of helium by Compton scattering, *Phys. Rev.,* A50: R2810–2813, 1994; F. Byron and C. Joachain, Multiple ionization processes in helium, *Phys. Rev.,* 164:1–9, 1967; J. C. Levin et al., High-energy behavior of the double photoionization of helium from 2 to 12 keV, *Phys. Rev.,* A47:R16–19, 1993; L. Spielberger et al., Separation of photoabsorption and Compton scattering contributions to He single and double ionization, *Phys. Rev. Lett.,* 74:4615–4618, 1995.

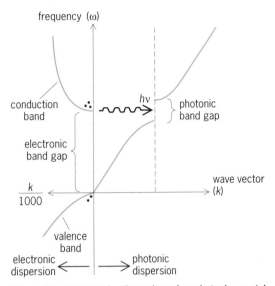

Fig. 1. Electromagnetic dispersion of a photonic crystal (right-hand side), together with the electron wave dispersion typical of a direct-gap semiconductor (left-hand side). The dots represent electrons and holes. (*After E. Yablonovitch, Photonic band-gap structures, J. Opt. Soc. Amer., B, 10: 283–295, 1993*)

Photonic crystals

As optical fiber has quickly replaced metal wire for electromagnetic transmission, the desire to use photons instead of, or in conjunction with, electrons has accelerated the field of photonics. Optical data storage, optical switching, and even optical computing are areas of intense activity. Although the potential benefits of these technologies are enormous, their development requires innovative ways to control light. A recent advance is the discovery of photonic crystals or photonic band-gap (PBG) materials, which are macroscopic periodic dielectric structures possessing spectral gaps (stop bands) for electromagnetic waves, in analogy with the energy bands and gaps in regular semiconductors. Although the periodicity in semiconductors is predetermined, the periodicity in the photonic crystals can be changed at will. Such structures have been built in the microwave regime (at frequencies from 10 to 500 GHz), and their potential applications continue to be examined. However, the greatest scientific challenge in the field of photonic crystals is to fabricate composite structures possessing spectral gaps at frequencies up to the optical region.

Photonic band gap. Electron waves traveling in the periodic potential of a crystal are arranged into energy bands separated by gaps where propagating states are prohibited. A semiconductor has a complete band gap between the valence and the conduction energy bands. It is now known that analogous band gaps can exist when electromagnetic waves propagate in a macroscopic periodic dielectric structure. Electromagnetic waves with frequencies inside such a gap cannot propagate in any direction inside the material. Such dielectric crystallike lattices have been referred to as photonic crystals or photonic band-gap materials.

Photonic crystals can have a profound impact on many areas in pure and applied physics. Because of the absence of optical modes in the gap, spontaneous emission is suppressed for atoms or molecules normally emitting photons with frequencies in the photonic gap. It has been suggested that, by tuning the photonic band gap to overlap with the electronic band edge, the electron-hole recombination process can be controlled, leading to enhanced efficiency and reduced noise in the oper-

ation of semiconductor lasers and other solid-state devices, such as solar cells.

The benefits of such a photonic band gap for direct-gap semiconductors are illustrated in **Fig. 1**. It shows a plot of frequency (ω) versus the wave vector (k) for the photon, with a forbidden gap at the wave vector corresponding to the periodicity of the dielectric structure. Sharing the frequency axis is a plot of the electron wave dispersion, showing conduction and valence bands appropriate for a direct-gap semiconductor. Because an optical wavelength is 1000 times larger than the atomic spacing, the electron wave vector must be divided by 1000 to fit the same graph with the photon wave vectors. If an electron were to recombine with a hole, it would produce a photon with energy $h\nu$ (where h is Planck's constant and ν is the photon frequency) equal to the electronic band-gap energy. As shown in Fig. 1, if the photonic band gap straddles the electronic conduction band edge, the recombination of electrons and holes is inhibited since the photons produced by electron-hole recombination have no place to go. The implications for solid-state devices are far reaching.

As in regular semiconductors where the material actually becomes more useful when defects are introduced, photonic crystals can be doped by removing or introducing extra dielectric material. Such defect modes are well suited to act as laser microresonator cavities. The electromagnetic interaction governs many properties of atoms, molecules, and solids. The absence of electromagnetic modes and zero-point fluctuations inside the photonic gap can lead to unusual physical phenomena. For example, atoms or molecules embedded in such materials can be locked in excited states if the photons emitted to release the excess energy have a frequency within the gap. The suppression of spontaneous emission can be used to prolong the lifetime of selected chemical species in catalytic processes.

Dielectric structures. To search for the appropriate structures, experimentalists employed a cut-and-try approach in which various periodic dielectric structures were fabricated in the microwave regime and the dispersion of electromagnetic waves was measured to see if a frequency gap existed. The process was time consuming and not very successful. After dozens of structures had been investigated over a period of 2 years, only one promising structure was found, but instead of a true photonic band gap it had a pseudogap (that is, a large depletion of the photon density of states). This structure consists of a periodic array of overlapping spherical holes arranged in a face-centered cubic lattice inside a dielectric block.

It was theoretically predicted that a diamond lattice of spheres exhibited a genuine photonic band gap. Shortly after this theoretical discovery, its experimental realization was given on a related structure, with the diamond-lattice symmetry. The fabrication of this crystal was carried out by drilling three sets of parallel cylinders into the top surface of a block of dielectric material, with each set angled 35° from normal and spread 120° on the azimuth. Photonic crystals with gaps in the microwave region are made by simply boring holes into a

block of dielectric material with a drill press. It was suggested that so-called three-cylinder structures with full three-dimensional band gaps at much higher frequencies might be made by using reactive ion-beam etching to drill the sets of very small diameter cylinders into semiconductors such as silicon (Si) or gallium arsenide (GaAs). However, so far there has been no report of a three-cylinder photonic lattice with a band gap above 20 GHz.

Layer-by-layer structure. Recently, a new dielectric structure was proposed that has a full three-dimensional photonic band gap. In such a structure, propagation of electromagnetic waves is forbidden for all directions. Thus, some of the roadblocks toward fabricating crystals with higher-frequency photonic band gaps may be lifted. The new photonic crystal (**Fig. 2**) is assembled by stacking layers of dielectric rods with each layer, consisting of parallel rods with a center-to-center separation of a. The rods are rotated by 90° in each successive layer. Starting at any reference layer, the rods of every second neighboring layer are parallel to the reference layer but shifted by a distance of 0.5 a perpendicular to the rod axes. The result is a stacking sequence along the z axis that repeats every four layers. This new structure has a sizable and robust photonic band gap over a range of structural parameters. The structure was fabricated by stacking layers of alumina rods, and confirmed the existence of a full photonic band gap at Ku-band frequencies of 12–14 GHz.

Smaller-scale structures with 100–500-GHz frequencies were fabricated by using semiconductor micromachining techniques. Here, the anisotropic etching of silicon by aqueous potassium hydroxide (KOH) is utilized, which etches the {110} planes of silicon rapidly while leaving the {111} planes relatively untouched. Thus, by using (110)-oriented silicon it is possible to etch arrays of parallel rods into wafers, and the patterned wafers can be stacked in the correct manner to make a photonic band-gap crystal (**Fig. 3**).

The excellent performance of the new layer-by-layer structure in the higher frequency range is such that a number of millimeter and submillimeter wave applications have been proposed, including microwave mirrors, substrates for planar antennas, resonators, waveguides, filters, and polarizers. The planar antenna application has already been realized. If such an antenna is placed on a photonic crystal instead of a dielectric substrate, it predominantly radiates into the air rather than into the substrate and therefore increases the directionality of the antenna.

In addition, defects or cavity modes can be easily introduced in the layer-by-layer structure by removing or adding rods. Another important result is that the average band-gap attenuation is roughly 16 dB per unit cell. This result is especially important for device considerations of these structures, since three unit cells (for a total of 12 wafers) will yield a photonic band gap with 45–50-dB attenuation,

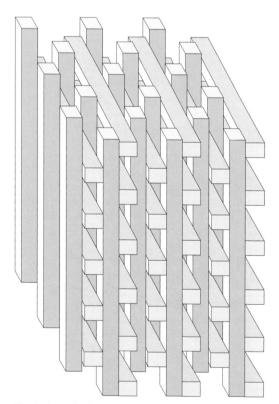

Fig. 2. Layer-by-layer photonic band-gap structure. The structure repeats every four layers in the stacking direction. (*After K. M. Ho et al., Photonic band gaps in 3D: New layer-by-layer periodic structures, Solid State Commun., 89: 413–416, 1994*)

Fig. 3. Fabrication of micromachined silicon photonic crystal. (*a*) Single etched silicon wafer. (*b*) Photonic crystal built by stacking silicon wafers. (*After E. Özbay et al., Micromachined millimeter-wave photonic band-gap crystals, Appl. Phys. Lett., 64:2059–2061, 1994*)

large enough for most applications. By using special silicon thinning methods and double-etching the wafers on both surfaces, the frequency range of this fabrication technology can probably be extended to build structures with photonic band gaps as high as 3 THz. However, the microfabrication of a three-dimensional photonic band gap at submicrometer wavelengths is the most important problem hindering the optoelectronic application of photonic crystals.

Two-dimensional structures. Concurrently with research on photonic crystals supporting three-dimensional band gaps, there has been development of structures with two-dimensional photonic band gaps, where transmission along the third dimension is allowed. Such two-dimensional structures have been fabricated by lithography at micrometer wavelengths, and their optical properties have been studied with the aim of developing high-efficiency miniature lasers and light-emitting diodes. The two-dimensional photonic crystals are easier to make than structures supporting three-dimensional band gaps, and might be immediately useful in a variety of optoelectronic applications in which light needs to be strongly confined in two directions. These applications include dielectric mirrors, resonant cavities, and waveguides, which reflect, trap, and transport light, respectively.

For background information SEE ANTENNA (ELECTROMAGNETISM); BAND THEORY OF SOLIDS; CAVITY RESONATOR; DIELECTRIC MATERIALS; SEMICONDUCTOR in the McGraw-Hill Encyclopedia of Science & Technology.

Costas M. Soukoulis

Bibliography. P. L. Gourley et al., Optical properties of 2D photonic lattices fabricated as honeycomb nanostructures in compound semiconductors, *Appl. Phys. Lett.,* 64:687–689, 1994; K. M. Ho, C. T. Chan, and C. M. Soukoulis, Existence of a photonic gap in periodic dielectric structures, *Phys. Rev. Lett.,* 65:3152–3155, 1990; E. Özbay et al., Measurement of 3D photonic band gap in a crystal structure made of dielectric rods, *Phys. Rev. B,* 50:1945–1948, 1994; C. M. Soukoulis (ed.), *Photonic Band Gaps and Localization,* 1993; E. Yablonovitch, Photonic band-gap structures, *J. Opt. Soc. Amer. B,* 10:283–295, 1993.

Photosynthesis

Photosynthesis is the process that efficiently converts solar energy into a form that can be used by living organisms. About eight times the world's energy consumption is fixed annually as high-energy carbohydrates. Photosynthetic organisms have found a unique solution for capturing the solar energy. The key steps are two ultrafast processes: excitation energy transfer in a light-harvesting antenna and electron transfer in a reaction center. A solar photon is absorbed by one of a number of pigments; chlorophyll *a* and *b* are found in plants and algae, and various forms of bacteriochlorophyll in bacteria. Furthermore, photosynthetic organisms contain carotenelike antenna pigments, which absorb light at wavelengths between 400 and 600 nanometers, where the solar spectrum peaks. After the absorption the excitation energy is rapidly transferred between many pigments, before it is trapped by a reaction center, where an ultrafast electron transfer converts the excitation energy into a stable charge separation. Because the fast electron transfer occurs across a membrane, an electrochemical gradient is formed that drives all subsequent biochemical reactions. More than 95% of the energy of the absorbed quanta is delivered to the reaction center, and the photosynthetic apparatus that mediates the highly efficient energy and electron transfer processes consists of a membrane-associated network of interconnected proteins that holds pigment molecules in an ordered state. In plants, two reaction centers, photosystems 1 and 2, each with its own light-harvesting system, operate in series. The electrochemical potential generated by photosystem 2 is sufficiently oxidizing to extract electrons from water, thereby producing oxygen, while photosystem 1 reduces $NADP^+$ (reduced nicotinamide adenine dinucleotide). Photosynthetic bacteria make do with one reaction center, but they cannot use water as a source for electrons.

This article discusses some of the structures of photosynthetic pigment proteins that have recently emerged. These structures carry out the initial steps of photosynthesis, and new techniques in femtosecond laser spectroscopy and genetic engineering are helping to understand how this takes place.

Reaction center. A great deal of progress has been made in understanding the function of reaction centers since the structure of the reaction center of the photosynthetic bacterium *Rhodopseudomonas viridis* was resolved to atomic resolution, followed by the structure of the *Rhodobacter sphaeroides* reaction center. The core of the reaction center is made of two homologous membrane proteins, called

L and M, while a third polypeptide, H, covers the cytosolic surface. Both L and M possess five membrane-crossing helices forming a cage around the four bacteriochlorophylls, two bacteriopheophytins, and two quinones noncovalently bound to the reaction center. The reaction center shows a remarkable C_2-symmetry with the pigments arranged in two branches, A and B. Surprisingly, electron transfer occurs only along the A (or active) branch, not the B (or inactive) branch. Following excitation of the centrally located so-called special pair of bacteriochlorophyll molecules, charge separation occurs in 3 picoseconds to yield the oxidized special pair, P^+, and a reduced bacteriopheophytin, H_A^-. The electron then hops to a quinone molecule, Q_A, in about 200 ps, and on to a second quinone, Q_B, in 100 milliseconds. P^+ is restored to neutrality by electron transfer from a cytochrome. This sequence of reactions is repeated after the absorption of a second photon. The doubly reduced Q_B molecule leaves the reaction center as Q_BH_2, and an electrochemical gradient has been created across the photosynthetic membrane in which the reaction center resides.

One major factor that has been debated since the reaction center structure became available concerns the role of the monomeric bacteriochlorophyll molecule, B_A, situated between P and H_A. Most scientists in the field support the proposal that B_A is transiently observable as an electron transfer intermediate, with the rate at which the electron arrives from the excited state of P (P^*) almost four times slower than the transfer away to H_A. The free energy of the initial radical pair $P^+B_A^-$ is only about 40 meV below P^*. However, the true situation is probably more complex and requires the further development of the spin-boson model with three electronic states. This incorporates the simpler sequential mechanism, but also allows for a fully coherent transfer in which B_A acts to mix the electronic states of P and H_A. Finally, the primary electron transfer events actually speed up as the temperature is lowered from room temperature to 10 K ($-442°F$), and much remains to be learned about the influence of temperature on the coherent and incoherent electron transfer.

In spite of the obvious symmetry of the reaction center, the electron transfer essentially occurs only along the A branch. Estimates of the coupling strengths along both paths do indeed favor the active branch, but not in the experimentally observed 200:1 ratio. One conspicuous feature in the reaction center structure is a tyrosine residue (Tyr-M208) strategically positioned between the cofactors in the active branch, while the symmetry-related residue in the inactive branch is a phenylalanine (PheL181). Reaction centers with mutations introduced at these positions were studied, and dramatic changes were observed in the rate of charge separation. For most mutations the rate slowed down, sometimes dramatically. Surprisingly, for a mutation that restored the symmetry with tyrosines at positions L181 and M208 the rate even increased

slightly. These observations were explained within the framework of the Marcus theory for electron transfer in the condensed phase, and ascribed to changes in the free energy of the initial radical pair due to the mutation. The free-energy dependence of the rate suggested that changes in the protein structure as electron transfer occurs (the reorganization energy) are very small. Nevertheless, in none of the mutants was electron transfer along the inactive branch observable. Many more mutants have since been generated. Reaction centers in which one of the bacteriochlorophylls of P is replaced by a bacteriopheophytin suggest the involvement of an intra-P charge transfer state before electron transfer occurs. Mutants in which hydrogen bonds of the protein were engineered around P produced dramatic changes in its the redox potential, again without affecting the asymmetry of electron transfer. Recently, a double mutant was studied with H_A replaced by a bacteriochlorophyll and $P^+B_A^-$ upshifted in free energy, and a measurable amount of $P^+B_B^-$ was detected. These results lead to doubts that the asymmetry can be due to only one or a few amino acids in the reaction-center structure. In fact, from the response of the reaction-center absorption to electric fields, it has been concluded that the major cause for asymmetric electron transfer is a collective one, involving many amino acids in the structure and resulting in a higher effective dielectric constant along the active branch.

The reaction center of photosystem 2 of green plants (or D_1D_2-particle) has been obtained in a purified form and shows a remarkable homology with the bacterial reaction center. Although D_1D_2 demonstrates photoactivity, it has lost the two quinones and no longer contains the manganese complex that is responsible for oxygen evolution in living organisms. The core of D_1D_2 binds four chlorophylls and two pheophytins, with the exception of the special pair, that seems to be more weakly coupled, a fact possibly related to the high oxidation potential generated by the complex in living organisms. D_1D_2 contains two additional chlorophylls that are located at its periphery. The response of D_1D_2 to a short laser pulse is complicated, largely because all the cofactors contribute to the absorption around 670–690 nm. As a consequence the dynamics appear to be more collective than in the bacterial reaction center, and so far the intrinsic rate of charge separation in D_1D_2 is not known, although many experiments indicate that it is in the range of a few picoseconds.

The reaction center of photosystem 1 is totally different from that of purple bacteria and photosystem 2. In its minimal form it consists of a heterodimer of two homologous polypeptides, each about 60 kilodaltons in mass, and the reaction center binds close to 100 chlorophyll a molecules. The 0.6-nm resolution structure reveals many of the transmembrane helices and suggests that the chlorophylls and other cofactors involved in primary electron transfer are again arranged in two symmetric branches. The reaction centers of green sulfur bacteria and the recently discovered heliobacteria are related to photosystem 1, with the very special additional feature that in these two photosynthetic systems the reaction center is truly symmetric and consists of a dimer of two identical proteins.

Light-harvesting antenna. The reaction centers of bacteria and plants are highly optimized devices that would be difficult to improve. However, on their own they would be of limited significance to the plant or bacterium, because the solar photon flux is too low to justify the efforts a cell must make for their synthesis. However, in nature the reaction centers are surrounded by a light-harvesting antenna, generally consisting of chlorophyll (or bacteriochlorophyll) and carotenoid molecules complexed to special, often membrane-bound, proteins. Photons are absorbed by the light-harvesting antenna and the excitation is efficiently transferred to the reaction center where a charge separation is initiated. The light-harvesting antenna greatly increases the absorption cross section of each reaction center and makes optimal use of their energy-converting capacity.

Recently the three-dimensional structure of two light-harvesting complexes has been resolved. In 1994, the structure of the major plant light-harvesting complex, LHC2, was discovered. This complex is the most abundant of the chlorophyll-binding proteins and binds about 50% of all chlorophyll. The elementary unit of the complex is a 27-kDa polypeptide with three transmembrane helices that binds 7 chlorophyll a pigments, arranged in the central part of the complex, and 5 chlorophyll b pigments, believed to be located more to the periphery. The pigments are arranged in two layers; the interpigment distances are of the order of 1 nm or less, allowing for ultrafast excitation transfer. In living organisms the complex is a trimer with C_3-symmetry and is located close to the core of photosystem 2.

In 1995 the structure of a bacterial light-harvesting complex was resolved, the peripheral light-harvesting complex of *Rhodop. acidophila*. In the heart of this structure are 18 strongly coupled bacteriochlorophyll a molecules bound to 9 αβ polypeptide heterodimers arranged in a C_9-symmetrical ring. The 9 bacteriochlorophyll dimers are non-covalently bound to pairs of highly conserved histidine residues located in the transmembrane region of the α and β polypeptides and are collectively responsible for the intense 850-nm absorption, which is strongly redshifted in comparison with that of free bacteriochlorophyll a. Within the αβ-dimer the bacteriochlorophylls are in van der Waals contact, while the dimer-dimer distance in the ring is about 1.7 nm. This complex also contains 9 weakly interacting bacteriochlorophyll molecules that lie close to the cytoplasmic surface and absorb at 800 nm. Energy transfer among all the pigments in this complex is in the subpicosecond time domain.

The **illustration** shows the result of an experiment in which the complex is excited at 825 nm, at the high-energy side of the major absorption band. After the excitation the spectrum dynamically shifts to the red on a 200-fs time scale. The dynamic redshift reflects energy transfer between pigments within a ring absorbing at slightly different energy, and in less than a picosecond an equilibrium distribution is reached. This experiment is representative of the ultrafast energy transfer that in fact causes this spectral equilibration, and the time scale at which this process takes place is a direct measure for energy-transfer dynamics.

In living organisms the excitation hops from ring to ring on a time scale of about a few picoseconds. Eventually, the light-harvesting antenna pigments neighboring the reaction center are reached and the excitation energy is transferred to one of the reaction center pigments. In fact, in these photosynthetic bacteria the light-harvesting antenna surrounding the reaction center is also arranged in a ringlike structure with a 16-fold symmetry axis. Again energy transfer along the ring proceeds on a subpicosecond timescale, but transfer to the reaction center is slow (20–40 ps) and the latter process may well be the rate-limiting step in the sequence of primary events.

The energy transfer is generally described by a well-known expression that assumes very weak electronic coupling between the participating pigments. Within the framework of this model excitation transfer is the incoherent hopping of fully localized excitations. However, from the available structural data it is evident that this model must be refined. The electronic couplings between the bacteriochlorophyll molecules are probably sufficient to cause a significant delocalization of the excitation. The decay of the coherence of the initially excited state and the interaction of the created exciton with low-frequency vibrations of the sur-

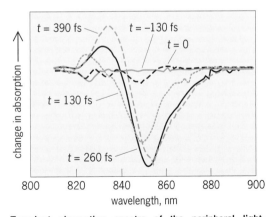

Transient absorption spectra of the peripheral light-harvesting complex of photosynthetic purple bacteria. At time $t = 0$ the complex is excited with a 250-femtosecond (full-width at half maximum) light pulse at 825 nanometers and the spectral response is recorded over a broad spectral region at 130-fs time intervals. (After R. Monshouwer, Faculty of Physics and Astronomy, Vrije Universiteit Amsterdam, The Netherlands)

rounding protein medium will be crucial parameters in the description of the dynamics of excitation transfer. However, a theory incorporating these elements has not been developed.

For background information *SEE CHLOROPHYLL; PHOTOSYNTHESIS* in the McGraw-Hill Encyclopedia of Science & Technology.

Rienk van Grondelle

Bibliography. J. Deisenhofer and J. R. Norris (eds.), *The Photosynthetic Reaction Centre*, vols. 1 and 2, 1993; G. R. Fleming and R. van Grondelle, The primary steps of photosynthesis, *Phys. Today*, 47(2):48–55, February 1994; B. A. Heller et al., Control of electron transfer between the L- and M-sides of photosynthetic reaction centres, *Science*, 269:940–945, 1995; W. Kühlbrandt et al., Atomic model of plant light-harvesting complex by electron crystallography, *Nature*, 367:614–621, 1994; G. McDermott et al., Crystal structure of an integral membrane light-harvesting complex from photosynthetic bacteria, *Nature*, 374:517–521, 1995; R. van Grondelle et al., Energy transfer and trapping in photosynthesis, *Biochim. Biophys. Acta*, 1187: 1–65, 1994.

Pinnipeds

The fossil record indicates that extant pinnipeds— Otariidae (fur seals and sea lions), Odobenidae (walruses), and Phocidae (true seals)—represent a small fraction of what was a much greater diversity. Only a single species of walrus exists today, whereas in the past no fewer than 10 genera and 13

species are known. New discoveries of fossil pinnipeds together with comparative study of living pinnipeds has enabled a more complete picture of their morphology as well as their origin and diversification. Evolutionary relationships among pinnipeds have remained uncertain until recently, when morphologic evidence was subject to reevaluation by modern systematic methods (**Fig. 1**).

Ancestry. Since the name Pinnipedia was first proposed for fin-footed carnivores more than a century ago, there has been debate on the relationships of pinnipeds to one another and to other mammals. Two hypotheses have been proposed. The monophyletic hypothesis proposes that pinnipeds share a single common evolutionary origin. The diphyletic view calls for an alliance of odobenids and otariids with ursids (bears) and a separate connection between the phocids and the mustelids (weasels, skunks, and kin). Until recently, morphologic evidence supported pinniped diphyly. Compelling evidence based on a reevaluation of cranial and postcranial anatomy among fossil and recent pinnipeds with rigorous phylogenetic methods of analysis (**Fig. 2**) supports a single or monophyletic origin for pinnipeds from arctoid carnivorans (bears and their relatives). Moreover, morphological studies also support a closer relationship between phocids and odobenids than with otariids. Biomolecular studies [mitochondrial deoxyribonucleic acid (DNA) sequence, DNA hybridization, amino acid sequences, and chromosomes] support pinniped monophyly, although there is disagreement regarding interrelationships among pinnipeds; that is, the molecular data favor a link between otariids and odobenids.

Origin and early evolution. The earliest pinnipeds, members of the Pinnipedimorpha clade, appear to have originated in the North Pacific during the late Oligocene or early Miocene [27–23 million years ago (Ma); Fig. 2]. The earliest known pinnipedimorph, *Enaliarctos*, is represented by five species. The primitive pinnipedimorph dental morphotype shows a flesh-eating dentition, cheek teeth with large blade-like cusps, broad protocone shelves (major cusp of upper molars), and elevated molar trigonids (triangular area on the lower molars). Other species of the genus show a trend toward decreasing the slicing function of the cheek teeth, heralding the development of the simple, peglike (homodont) dentition that is characteristic of living pinnipeds.

The pinnipedimorph *E. mealsi* is represented by a nearly complete skeleton from the Pyramid Sandstone Member of the Jewett Sand in central California (**Fig. 3**). The entire animal is estimated to be 4.6–4.9 ft (1.4–1.5 m) long and to weigh between 161 and 194 lb (73 and 88 kg), roughly the weight of a small male harbor seal. Considerable lateral and vertical movement of the vertebral column was possible in *E. mealsi*. Also, both the fore- and hindlimbs were modified as flippers and used in locomotion. Several features of the hindlimb suggest that this animal was highly capable of maneuvering on land

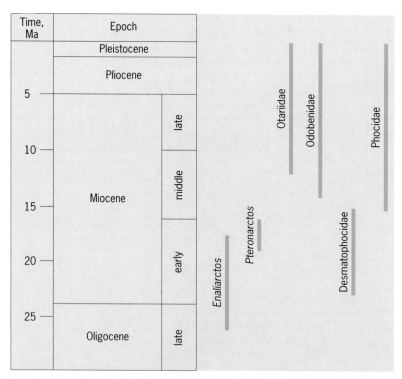

Fig. 1. Cenozoic time scale showing age ranges of major pinniped taxa.

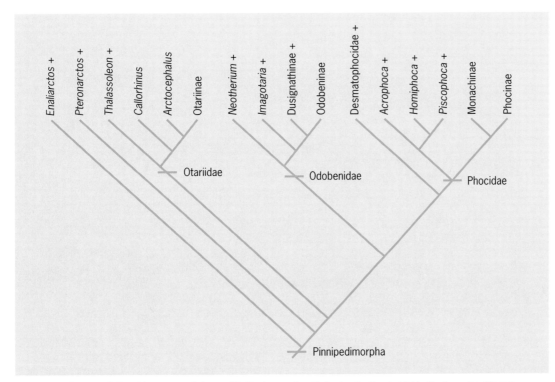

Fig. 2. Evolutionary relationships among living and better-known fossil pinnipeds. Symbol (+) indicates extinct taxon.

and probably spent more time near the shore than extant pinnipeds.

A later diverging lineage more closely allied with pinnipeds than with *Enaliarctos* is represented by *Pteronarctos* from the Miocene (19–16 Ma) of Oregon. *Pteronarctos* preserves the first evidence of the uniquely developed maxilla (cheek bone) seen in pinnipeds which makes a significant contribution to the orbital wall of the skull. In other mammals, the maxilla is usually excluded from contributing to the orbital wall by development of the jugal, lacrimal, palatine, or frontal bones. There is also evidence in this animal of a reduced shearing capability of the teeth.

Otariid radiation. Although otariids are often divided into two subfamilies, the Otariinae (sea lions) and the Arctocephalinae (fur seals), only the sea lions are believed to have descended from an exclusive common ancestor. Fossil otariids, which originated approximately 12 Ma, include the late Miocene taxon (8–6 Ma), *Thalassoleon*, represented by three species including the recently described *T. inouei* from central Japan. Prior records of otariid pinnipeds were confined mostly to the eastern North Pacific. The occurrence of *Thalassoleon* in Japan provides evidence for an earlier dispersal of otariid pinnipeds in the western North Pacific than was otherwise known. A single extinct species of the northern fur seal *Callorhinus gilmorei* has been recently described from the late Pliocene San Diego Formation (3 Ma) in southern California and Mexico. This is the earliest record of a modern otariid genus. Several species of the southern fur seal genus *Arctocephalus* have a fossil record that extends

back to the Pleistocene. The fossil record of sea lions is poorly known; only *Neophoca* and *Eumetopias* are represented.

Odobenid radiation. Arguably the most characteristic feature of the modern walrus *Odobenus* is a pair of elongated ever-growing upper canine teeth (tusks) found in adults of both sexes. A rapidly improving fossil record indicates that these unique structures evolved in a single lineage of walruses. Recent morphologic study of evolutionary relationships among walruses has identified two major radiations, the extinct Dusignathinae and the Odobeninae; the latter includes the modern walrus. Dusignathine walruses developed enlarged upper and lower canines, while odobenines evolved the enlarged upper tusks seen in the modern walrus. *Neotherium* and *Imagotaria* are identified as sequential sister taxa to the Dusignathinae and Odobeninae.

The earliest known walrus is *Neotherium* from the middle Miocene (13–14 Ma) of California. A newly described dusignathine walrus, *Dusignathus seftoni*, possessed enlarged upper and lower canines and a shortened rostrum. Another new species of walrus, possibly the most completely known fossil odobenine walrus, *Valenictus chulavistensis*, was recently described as closely related to modern *Odobenus* but distinguished in having a toothless mandible and lacking all postcanine teeth. The toothlessness of *V. chulavistensis* is unique among pinnipeds but parallels the condition seen in modern suction-feeding beaked whales and the narwhal. The co-occurrence of these new species of walrus in the San Diego Formation indicates a greater diver-

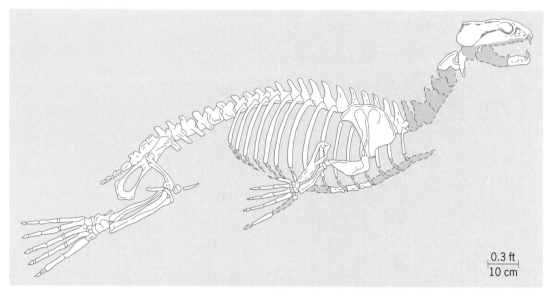

Fig. 3. Skeletal reconstruction of *Enaliarctos mealsi*, the oldest known pinnipedimorph. (*After A. Berta, C. E. Ray, and A. R. Wyss, Skeleton of the oldest known pinniped, Enaliarctos mealsi, Science, 244:60–62, 1989*)

sity of odobenids in the eastern Pacific during the late Pliocene than now.

Common ancestors of phocids-desmatophocids. A recent study of pinniped evolutionary relationships has identified a group of fossil pinnipeds including *Desmatophoca* and *Allodesmus* that are positioned as the common ancestors of phocid pinnipeds. This interpretation differs from previous work that recognized desmatophocids as otarioid pinnipeds, a grouping that includes walruses. Desmatophocids are known from the early and middle Miocene (23–15 Ma) of the western United States and Japan. Newly reported occurrences of *Desmatophoca* confirm the presence of sexual dimorphism and large body size in these pinnipeds.

Phocid radiation. Although traditionally phocids have been divided into two to four major subgroupings, recent work argues for the monophyly of only one of these groups, the Phocinae. Neither the Monachinae nor the genus *Monachus* is believed to be monophyletic, and they have been identified as sequential sister taxa to the remaining phocids. This phylogenetic framework has allowed the formulation of testable predictions of character evolution among phocids. Among the Phocinae, evolutionary reversals, those affecting the morphology of the flippers, have resulted in their more nearly resembling those of pinniped ancestors, that is, arctoid carnivorans. This pattern of widespread character reversal in the Phocinae coincides roughly with a decrease in body size from the ancestral condition of large body size in phocids. These evolutionary reversals have been at least partly attributed to changes in timing or the rate at which different body parts develop relative to one another. The result is an ancestral juvenile morphology. Previous interpretations of these character changes without benefit of a phylogenetic context attributed them to retrogressive evolution.

Both phocine and monachine seal lineages are recognized from the middle Miocene (15 Ma) of the North Atlantic. Abundant well-preserved cranial and postcranial material including several partial skeletons of the phocine seal *Leptophoca* from the middle Miocene Calvert Formation in Maryland are currently being described, as is material of the monachine seal *Callophoca* from the Pliocene Yorktown Formation at Lee Creek Mine in North Carolina. Recently reported monachines represented by well-preserved skeletal material include *Acrophoca* and *Piscophoca* from the early Pliocene of Peru and *Homiphoca* from South Africa. *Acrophoca* is unique among phocids with its extremely long slender skull, flexible neck, and elongated body.

For background information *SEE CARNIVORA; MIOCENE; OLIGOCENE; PINNIPEDS* in the McGraw-Hill Encyclopedia of Science & Technology.

Annalisa Berta

Bibliography. A. Berta and T. A. Deméré (eds.), Contributions in marine mammal paleontology honoring Frank C. Whitmore, Jr., *Proc. San Diego Soc. Nat. Hist.,* 29:33–56, 1994; A. Berta, C. E. Ray, and A. R. Wyss, Skeleton of the oldest known pinniped, *Enaliarctos mealsi, Science,* 244:60–62, 1989; A. R. Wyss, On "retrogression" in the evolution of the Phocinae and phylogenetic affinities of the monk seals, *Amer. Mus. Nov.,* 2924:1–38, 1987.

Plant development

The light environment affects many aspects of plant growth and development; for example, light is required by plants for photosynthesis to provide energy for growth. In addition, the quality and quantity of light perceived by plants affect the timing of developmental changes in the life of the plant, such as the transition to flowering.

In a process known as photomorphogenesis light exerts control over growth, development, and differentiation of plants independently of photosynthesis. One of the signaling systems by which plants detect differences in light quality and duration involves phytochrome, a chromoprotein. Phytochrome is synthesized in the red-light-absorbing form and undergoes conversion to the far-red-light-absorbing form in the presence of red light. The far-red absorbing form of phytochrome is thought to play an important role in the diversity of responses of plants to light quantity and quality. Rapid progress in understanding the phytochrome signaling system has been made by using molecular techniques.

In the light-grown plant, two processes affected by phytochrome are the ability to sense the ratio of red to far-red light and to perceive the length of the day. Differences in plant form obtained in different light environments involve differential expression of genes, and current research is focused on signal transduction, that is, the events which occur from the perception of light to the modification of gene expression. Previously, only two forms of phytochrome were thought to exist in plants, the light-labile phytochrome (phytochrome A) and the light-stabile phytochrome (phytochrome B). It is now know that more phytochrome species exist in some plants. In *Arabidopsis thaliana*, five phytochrome genes (*phyA* through *phyE*) have been identified. Studies with mutants of the different types of phytochrome have allowed researchers to assign specific functions to different phytochromes. For example, the growth and development of wild-type, phyA, phyB, and phyAphyB double-mutant plants lacking these phytochromes has been assessed in white-light, red-light, and far-red-light environments. It has been determined that phytochrome A and B act together in inhibiting elongation in the part of the stem known as the hypocotyl, and in promoting cotyledon expansion and chlorophyll accumulation. Different roles for phytochrome A and B have also been proposed with respect to flowering. Phytochrome A is important for sensing day length for flowering, and phytochrome B is important in sensing the growth environment with respect to the ratio of red to far-red light.

Plant hormones. One aim of photomorphogenesic research is to determine how differences in plant form are obtained. Past research has indicated that levels of plant growth substances (gibberellins, cytokinins, abscisic acid, ethylene, and auxin) change under different light environments. The level of biologically active gibberellin is a major factor in controlling plant height, and it is believed that phytochrome imposes some control on gibberellin biosynthesis and metabolism, thereby altering the levels of biologically active gibberellins. It has also been proposed that phytochrome alters the responsiveness of plant tissue to applied gibberellins. Researchers have studied the interaction of phytochrome and gibberellins by examining the gibberellin physiology of mutants which are defective in the phytochrome B signaling system. Mutants lacking the phytochrome B polypeptide have been identified in *Sorghum, Brassica,* and *Arabidopsis.* Although stems of wild-type plants are shorter in red light than white light, stem-elongation growth is not inhibited when phytochrome B mutants are grown in red light. In addition, these phytochrome mutants often contain higher levels of gibberellins than wild-type plants, and the mutants are also earlier flowering.

Detailed analyses of gibberellin levels determined at different times of the day and night have recently been measured in *Sorghum.* In normal plants, gibberellin levels are highest during the middle of the day, falling during the night. The phytochrome B mutant, thought to lack the phytochrome B polypeptide, continues to accumulate active gibberellins during the dark period. Thus, work with *Sorghum* indicates that the phytochrome B mutants do not simply overproduce gibberellins. It has been suggested that these plants possess an altered timing of gibberellin production, synthesizing gibberellins into the dark period when wild-type plants have decreased gibberellin production.

The link between phytochrome and gibberellin physiology has also been examined by genetically altering tobacco plants to overexpress the gene for phytochrome A. Three different promoters were used to drive expression of the phytochrome A gene. In the most extreme case, tobacco plants which constitutively overexpressed phytochrome near the vascular tissue in the plant stem were about one-fourth the height of control plants. Because the plants responded to gibberellin, it was suggested that responsiveness to gibberellin was not altered but that gibberellin levels were reduced. Measurements revealed that these plants possessed a fourfold reduction in gibberellin levels. Experimental results indicate that phytochrome status may affect gibberellin levels. Recently, several key genes in gibberellin biosynthesis have been cloned, enabling examination of where gibberellin-synthesis genes are expressed and how their expression is regulated in plants of different phytochrome status and grown under different light environments.

Transition to flowering. The transition from vegetative to reproductive development in plants has been studied extensively. The process of flowering is being dissected by using mutants and molecular biology techniques. The transition to flowering is dependent on both internal developmental programs and external environmental cues. Day length, temperature, and light quality can affect the timing of this transition. Growth of the plant proceeds through distinct stages prior to flowering. In the juvenile phase, the apical meristem is vegetative, initiating leaves and vegetative axillary buds. When the meristem becomes capable of initiating floral structures rather than vegetative buds, it has become an inflorescence meristem. This switch in apical meristem identity can be affected by a signal from the leaves as is the case for photoperiodic plants which flower when a critical night length is achieved. In cases where a plant species flowers following a cold period, the meristem may directly perceive the temperature. It is becoming evident that different pathways may lead to the transition of a vegetative meristem to an inflorescence meristem, and that these pathways may converge to affect the same genes involved in meristem identity.

Tremendous progress in understanding the internal developmental events which lead to flowering has been made by investigating species where genetic mutations in the flowering process exist. Meristem identity mutants are especially interesting because plant form and flowering behavior can be drastically altered depending on the activity of meristems. Meristem identity mutants have been studied in *A. thaliana* (a common weed), *Antirrhinum majus* (snapdragon), and *Pisum sativum* (the garden pea). In the garden pea, two recessive mutations illustrate how plant form and flowering behavior can be altered. The vegetative mutant (*veg-1*) is blocked in the ability to produce flowers, and instead produces leafy axillary branches. Another mutant of pea, determinant (*det*), ceases flowering after producing a limited number of flowers. The apical meristem appears to be converted to an axillary meristem in *det* plants, resulting in the formation of a final flower and cessation of growth. When the two recessive genes are combined in a double mutant, the block in flowering is removed and flowering eventually occurs.

In addition to internal developmental programs, the ways in which the environment regulates the transition to flowering is being addressed. In some species, such as *A. thaliana,* there is evidence that gibberellins might be involved. For example, it has been shown that some of the *Arabidopsis* meristem identity mutants respond differently under different photoperiods. The recessive *agamous* mutant is one example. The normal order of formation of floral organs is sepals, petals, stamens, and carpels. In the *agamous* mutant, sepals are followed by whorls of only petals resulting in a flower-within-a-flower phenotype. When *agamous* mutants are grown under conditions where the days are short, the floral meristem reverts back to producing an inflorescence. If gibberellins are applied to the *agamous* mutant, this reversion does not occur, indicating that gibberellin levels may play a role in regulating genes important for the transition of an inflorescence meristem to a floral meristem.

Future developments. The challenge will be to extend studies on factors, both internal and external, that regulate plant growth and development during the transition from the vegetative plant to the flowering plant. Results from the studies discussed above should allow various plant forms to be manipulated in new ways. For example, plants expressing a prolonged vegetative program would be beneficial in forage crops. Manipulation of light quality by using spectral filters also appears promising as a method of altering plant form. The ability to alter plant stature and flowering behavior would be especially beneficial to areas such as horticulture where precise control of plant form is highly desirable.

For background information SEE GENE; GENETIC ENGINEERING; MUTATION; PLANT GROWTH; PLANT HORMONES in the McGraw-Hill Encyclopedia of Science & Technology.

Sonja Maki

Bibliography. E. T. Jordan et al., Phytochrome A overexpression in transgenic tobacco: Correlation of dwarf phenotype with high concentration of phytochrome in tissue and attenuated gibberellin levels, *Plant Physiol.,* 107:797–805, 1995; N. C. Rajapakse and J. W. Kelly, Regulation of chrysanthemum growth by spectral filters, *J. Amer. Soc. Hort. Sci.,* 117:481–485, 1992; J. K. Okamuro, B. G. W. den Boer, and K. D. Jofuku, Regulation of *Arabidopsis* flower development, *Plant Cell,* 5:1183–1193, 1993; J. W. Reed et al., Phytochrome A and phytochrome B have overlapping but distinct functions in *Arabidopsis* development, *Plant Physiol.,* 104:1139–1149, 1994; S. R. Singer, S. L. Maki, and H. J. Mullen, Specification of meristem identity in *Pisum sativum* inflorescence development, *Flower. Newslett.,* 18:26–32, 1994.

Plant embryogenesis

Embryogeny is the sum of the progressive and highly coordinated processes that transform the single-celled zygote into a multicellular, miniature plant. Many of these developmental processes recur during the seedling, vegetative, and reproductive phases of the sporophyte's life cycle. Therefore, embryo development provides a useful model for studying many global problems in plant development, such as polarity, cell differentiation, and formation of the plant's basic architecture. Much of what is known about plant embryogeny has come from visual documentation of cell division patterns, formation of new anatomical features, and sequential changes in embryo shape, size, and color. These descriptive studies led to the formulation of hypotheses that were testable through the isolation and characterization of embryos with genetic defects. Mutant analysis is an especially powerful approach toward identifying developmentally important genes, highlighting interactions between plant parts, and unlinking growth, differentiation, and morphogenesis.

Seed plant embryogeny. Among seed plants there is a great diversity of embryonic patterns that lead to the stereotypical body organization of the seedling. However, the principles of embryogeny are similar between the angiosperms and the gymnosperms. Gymnosperm embryogeny differs in two notable ways: (1) Cellular division does not follow the first several mitotic divisions, leading to multinucleated cells. (2) Two or more embryos develop from each zygote in a process known as cleavage embryogeny.

In both angiosperms and gymnosperms, embryogeny passes through a series of distinguishable morphological shapes and biochemical differentiation without interruption until the end of embryogeny. However, it is convenient to divide embryo development into two sequential phases: embryogenesis and maturation (see **table**).

Embryogenesis. Embryogenesis encompasses that time period between fertilization of the ovule and formation of the embryonic axis. Within a few hours after union of the sperm and ovule, the resulting zygote divides into a small terminal cell and a larger basal cell, thereby establishing the polarity that is carried on throughout the life of the plant; the terminal cell develops into the embryo proper. This cell and its progeny rapidly divide to form a globular mass of apparently homogeneous cells that differentiate into the three primary tissue types: ground, vascular, and dermal. Organogenesis follows this cellular differentiation, leading to the formation of the cotyledon(s) and embryonic axis. The cotyledons become fully differentiated organs, storing photosynthates and serving as absorptive organs during food reserve mobilization from the endosperm during either the maturation phase or germination. Establishment of the root and shoot apical meristems delineates the embryonic axis. This bipolar axis embodies the fundamental architecture of the seedling. The initiation of the two apical meristems is critical to survival because, unlike other regions of the embryo, these tissues retain the capacity for postembryonic growth and development. If mature seeds are tiny, such as those produced by *Arabidopsis* and tobacco, the embryos cease morphogenesis at this time. The embryonic axes of large-seeded plants, such as maize (*Zea mays*) and pea (*Pisum sativum*), continue morphogenesis by producing a series of leaf and root primordia (rudimentary organs) before the end of the maturation phase. Primordium expansion through cell enlargement does not occur until germination.

The basal cell develops directly into a specialized organ known as the suspensor, a purely embryonic organ that rapidly disappears by midembryogeny. The suspensor promotes and regulates the growth and development of the immature embryo. It also orientates the developing embryo so that the cotyledon surface that functions to absorb nutrients is in physical contact with the endosperm, sup-

plies growth regulators, and carries nutrients from the maternal plant to the embryo.

Maturation phase. Following the formation of the apical meristems, maturation processes replace continued, rapid growth and account for the final two thirds of embryogeny. The embryo's developmental program switches from rapid cell division and novel morphogenesis toward the stockpiling of storage proteins, carbohydrates, and lipids needed for successful, rapid germination and reiterative morphogenesis (more of the same structures, such as leaf primordia, are initiated). In the midmaturation phase, the embryo gains tolerance to the gradual loss of water and begins to synthesize dehydration-associated late-embryo abundant proteins. The mature embryonic axis dehydrates and barely metabolizes until the proper environmental conditions (water, heat, and perhaps light) induce the resumption of growth and development in the process known as germination.

Seed drying plays a key role in the switch from embryo maturation to germination. These two very different developmental programs occur sequentially in the same cells. A mechanistic explanation of the effects of drying remains elusive. However, there is some evidence that drying changes cell-wall structure and decreases the embryo's sensitivity to the growth regulator, abscisic acid, which inhibits precocious germination in favor of maturation processes.

Immature embryos in many flowering plants acquire the capacity to germinate after completing cellular differentiation and meristem formation. Removal of embryos from maternal tissues followed by culture on nutrient media allows embryos to bypass the maturation phase and subsequent period of dormancy or quiescence.

Mutant analysis. One of many unresolved questions in plant embryogeny concerns the regulation of developmental processes. For example, very little is known about the regulation of meristem generation or the modulating effects of interacting seed components, such as the embryo proper and suspensor, and the embryo and maternal tissues. Clearly, gene activity plays a role in regulating development because morphological variation is heritable and specific gene mutations lead to developmental abnormalities. Therefore, in order to dissect the genetic regulation of embryonic development, muta-

Embryo development in seed plants

Characteristics	Embryogenesis	Maturation
Fraction of embryogeny	1/3	2/3
Cell division	Rapid	Slow
Cell enlargement	Little	Extensive (outside embryonic axis)
Morphogenesis	Novel	Reiterative
Major events	Establishment of basic body plan; first leaf primordia; three primary tissue types	Accumulation of storage reserves (lipids, starch, and proteins); accumulation of desiccation-associated late-embryo abundant proteins; acquisition of desiccation tolerance; drying; ability to germinate

tions that disrupt embryo development have been isolated and characterized.

Maize and *Ar. thaliana* are the premier model systems for the study of zygotic embryogeny in angiosperms, largely due to their well-studied genetics. The many advantages of maize include the vast store of biological and genetic information accrued about it during this century, its large embryo, and its long period of embryogeny (50–60 days). The *Arabidopsis* embryo is minuscule in comparison, but it is attractive as a model for genetic research because of its short regeneration time (seven weeks from seed to seed), small adult plant size, and relatively small genome.

Saturation mutagenesis studies, using agents that damage deoxyribonucleic acid (DNA; chemicals or radiation) or the insertion of jumping genes (transposable elements or transposons [T-DNA sequences] from *Agrobacterium*) have led to the isolation of hundreds of embryo mutants. The best-characterized mutants in maize are the *dek* (defective kernel) mutants, which have abnormal embryos and endosperm, and *emb* (embryo) mutants, which have defects largely confined to the embryo. Likewise, many *Arabidopsis* mutants, including the *emb* mutants and pattern mutants (mutants that suffer from specific morphological defects, such as the deletion or duplication of organs but otherwise are able to complete the embryogenic program), have been recovered following mutagenesis. The vast majority of these mutations are monogenetic, recessive, and embryolethal. However, there are a few dominant mutations, and some, including the pattern mutants of *Arabidopsis,* are lethal in the seedling stage. Both *dek* and *emb* mutants arrest uniformly at specific morphological stages or in a limited range of stages. Cell death and disorganized growth without morphogenesis are common. Finally, the expression of some *emb* and *dek* genes, particularly those that are active early in embryogenesis, may be active during development of the pollen and embryo sac (male and female gametophytes, respectively).

Characterization of mutants. Embryo mutants fall into three basic classes. The first class is the auxotrophs, which are unable to synthesize essential nutrients. A second class consists of cellular mutations that disrupt housekeeping functions, such as cell division and enlargement. The third class of mutants are unable to complete the embryogenic program because of dysfunction of developmentally important genes. Although all three classes lead to a variety of morphological abnormalities, mutants in the third class directly link morphology with genetic information.

Detailed characterization of some of these mutants has increased our understanding of plant development. For example, many of the same genes are expressed in the development of the gametophytes, embryo, and endosperm, as indicated by altered growth or development. Possibly, some developmentally important genes are active in different stages of the plant's life cycle as well as in different tissues.

Morphogenesis and differentiation. Morphogenesis and biochemical differentiation can be unlinked. Because of slow growth, the maize mutant *dek cp*-*1399A can generate only three of the usual six leaf primordia before the end of embryogeny. Time-course comparisons of mutant and wild-type embryo growth parameters, morphogenesis, germination potential, sensitivity to growth regulators (such as abscisic acid, auxin, cytokinin, and gibberellin), water relations, and accumulation of maturation products demonstrate that *cp*-*1399A mutant embryos grow and develop slowly. They appear and behave like immature wild-type embryos. The genetically disrupted timetables of mutant development alter the normal relationship between the developmental schedules of the embryo and maternal plant. Maternal signals that regulate embryogeny reach mutant embryos prematurely with respect to their morphological stage, thereby allowing the distinction between those maturation events that are controlled by seed-endogenous factors from those modulated by the maternal environment. Almost all aspects of embryo-specific biochemical differentiation follow the delayed morphological schedule of the *dek* embryos, including the synthesis of storage globulin proteins, pigments, and lipids. Also, developmental changes in sensitivity to water potential and growth regulators remain coupled to the morphological progression of the mutant embryos. However, the loss of water molecules and the appearance of desiccation-associated late-embryo abundant proteins occurs coincidentally in wild types and mutants, suggesting control by the maternal environment.

Independent development. Plant organs or compartments can develop independently of each other. Genetic and developmental analyses of the maize mutant, *dks*8 (defective kernel shootless mutation possibly caused by the insertion of the *Mu*8 transposon into a developmentally critical gene, *dks*8), provides an example of modular morphogenesis. The deletion of the shoot system does not interfere with the capacity of other components of the seed to develop to maturity. The *dks*8 embryos acquire the normal constellation of maturation products. The mutant weight profile parallels that of the wild type, indicating that their drying regimes are identical. Their responses to physiological cues (growth regulators and high sugar concentration) in culture are appropriate for their age. Finally, the mutant embryo is able to germinate its primary root from a dry seed. Therefore, the *dks*8 mutation demonstrates that morphogenesis of the shoot can be genetically separated from all other developmental events and that cell differentiation can proceed in the absence of normal morphogenesis.

Dissecting the regulatory pathways in plant embryogeny remains one of the greatest challenges for developmental botanists. However, the multifaceted characterization of embryo mutations through physiological, developmental, genetic, and molecular approaches is beginning to provide important insights into the ways and means of embryogeny and more global aspects of plant development.

For background information SEE EMBRYOGEN-ESIS; EMBRYOLOGY; GENE; MUTATION; PLANT GROWTH; PLANT MORPHOGENESIS in the McGraw-Hill Encyclopedia of Science & Technology.

John D. Sollinger

Bibliography. M. Freeling and V. Walbot (eds.), *The Maize Handbook,* 1994; U. Mayer et al., Mutations affecting body organization in the *Arabidopsis* embryo, *Nature,* 353:402–407, 1991; D. W. Meinke, Perspectives on genetic analysis of plant embryogenesis, *Plant Cell,* 3:857–866, 1991.

Plasma physics

In many applications the surface properties of a component are of paramount importance. Surface-modification techniques using plasmas can provide a component manufactured from a low-grade bulk material with a high-performance surface. In many instances the surface modification process effectively creates a new material at or on the surface. Surface modification can be used to improve the wear, corrosion, fatigue, or friction properties of the material, to change its electrical or optical properties, or to deposit a thin protective or biocompatible film. The material itself can be conducting or insulating, a metal, alloy, semiconductor, or even plastic. Similar plasmas can be used to produce the submicrometer-sized surface architectures required in the production of microelectronic devices. These physical and structural changes are possible because of the unique physical and chemical environments that can be created in a plasma.

Low-temperature plasmas. A plasma is a partially or fully ionized gas containing electrons, ions, and neutral atoms or molecules. The interactions of the electrically charged particles with each other, with the neutral gas, and with contact surfaces produce the unique physical and chemical properties of the plasma environment.

Plasmas exist in many different forms. The plasmas used for surface modification are generally created by passing an electric current through a gas. They can be created under a wide range of conditions. Plasmas used in materials processing generally fall into two groups. In thermal plasmas, for example, arc discharges, the constituent electrons, ions, and neutral atoms are all at about the same temperature, normally between 10^3 and 10^4 K. These plasmas operate over a range of gas pressures which are typically close to atmospheric pressure, and densities ranging between 10^{20} and 10^{25} electrons per cubic meter are attained. Thermal plasmas are used mainly to deliver heat to a surface, thereby increasing surface reaction rates or producing melting, sintering, or evaporation of the material. Such plasmas find application in, for example, welding, metal recovery, and waste treatment.

The second group comprises the nonthermal plasmas, which include glow discharges. In such plasmas the constituent electrons, ions, and neutral gas have widely differing temperatures. The electron temperature can range from 10^4 to 10^5 K and be several orders of magnitude higher than that of the ions or neutral constituents and thus the workpieces. The energetic electrons can be used to break up unreactive feedstock gases to produce positive ions and chemically reactive atoms or molecules, thus creating a unique plasma chemistry. The chemical species created in the volume of the plasma diffuse to the workpiece surface. Nonthermal plasmas are generally operated at low pressure, that is, between a millionth and thousandth of an atmosphere, with charged-particle densities ranging from 10^{14} to 10^{19} particles per cubic meter. Nonthermal plasmas find application in, for example, the etching of structures in semiconductor materials and the deposition of a wide range of surface coatings.

Plasmas have other unique physical properties which are important in surface modification. They can flow like liquids and so penetrate the features of even complex geometric structures. The difference in mass of the electrons and that of even the lightest ion means that the electrons tend to leave the plasma faster than the ions. This tendency is balanced by the net charge of the positive ions, and the plasma generally sits at a positive potential relative to any grounded surface in contact with the plasma. Energy is therefore delivered to the surface by plasma-produced ion bombardment. Generally, in nonthermal plasma processing this energy deposition is used to promote structural change or chemistry rather than surface heating. Indeed, an important feature of such processing is that the workpiece remains at a relatively low temperature.

Plasma processes. There are a number of processes by which workpiece surfaces can be modified by placing them in a plasma. Ion bombardment can remove material from the surface by a process called sputtering. The incident ions knock atoms off the surface and into the plasma. The amount of material removed can be precisely controlled to within a few atomic layers. The removal of material can also be made selective; for example, oxides can be removed more rapidly than the bulk element or vice versa. The removal rate and selectivity can be improved by using the plasma to produce chemically reactive species from chemically inert feedstock.

Sputtering can also be used to deposit target material on a workpiece. The sputtering rate at the target is optimized, and the liberated target material diffuses through the plasma, depositing as a film on the workpiece surface. This process can lay down, again to atomic-layer precision, thin films of, for example, metals to produce conducting films. Rather than deposit a single sputtered element directly, several targets, each made of different elements, can be used to deposit alloys. Recently there has been interest in producing layered structures on workpiece surfaces by sequentially exposing different targets to the plasma. In the reactive sputter deposition process the sputtered target material may react with the suitable plasma-generated species to deposit chemical compounds such as oxides or nitrides onto a surface. For example, in producing titanium nitride (TiN) films a titanium target is used in a plasma created in a gas mixture of argon and nitrogen.

In other approaches the plasma chemistry is used to create species which react at the surface and deposit thin films. This process is known as plasma-enhanced chemical vapor deposition (PECVD), and can be used in the growth of many different compounds. One dramatic example of this powerful technique is in diamond film growth. Quite complex materials, such as the high-temperature superconductor $YBa_2Cu_3O_{7-x}$, can be deposited on suitable materials by using the techniques described above. In many deposition processes initial, ongoing, or subsequent ion bombardment of the workpiece can also be used to alter the atomic structure of the surface and so enhance adhesion. *SEE METAL COATINGS.*

Applications. These plasma properties and processes are the basis for a large number of industrial applications of plasma-based surface modification. Some examples are listed in the **table.** A few specific examples are described below.

Plasma surface modification processes and applications

Process	Application
Anisotropic etching Surface oxidation Silicon nitride deposition Metal film deposition	Microelectronic chip fabrication
Amorphous silicon deposition	Solar-cell fabrication
Surface roughing Degreasing Stress relief	Improved adhesion
Nitriding Nitrocarbonizing	Enhanced wear, corrosion resistance, and fatigue strength of metals
Diamond film deposition	Surface hardness, chemical inertness, biocompatibility, enhanced heat conduction
Ceramic coatings	Wear and corrosion protection
Metal alloy coatings (for example, titanium carbide, titanium nitride, aluminum oxide)	Wear and corrosion protection
Ion implantation, (for example, nitrogen, boron, carbon, titanium)	Wear and corrosion protection; enhanced adhesion of films modified optical and electrical properties
Magnetic film deposition	Magnetic recording and memories

Plasma etching. Perhaps the most impressive examples of surface modification are found in the manufacture of microelectronic chips. Microelectronic devices are complex submicrometer-sized structures formed on the surface of semiconductor wafers. The structure is made by using subtractive processing. Thin films are first deposited over the entire water surface. Then the deposited material is selectively removed or etched from selected regions of the surface. In the production of a single device many films will be deposited and etched in the building of the final structure. Most of the steps are dependent on plasma processing to provide the required accuracy.

The unique combination of reactive and charged species in the plasma permits rapid, selective, and anisotropic etching. The process relies on the creation, in the plasma, of highly reactive species from inert feedstock gas. For example, fluorine is used in the etching of silicon but is introduced into the plasma in gases such as carbon tetrafluoride (CF_4), sulfur hexafluoride (SF_6), and nitrogen trifluoride (NF_3). The overall etch reaction in a carbon tetrafluoride plasma is given below. The products of

$$4CF_4 + Si \longrightarrow 2C_2F_6 + SiF_4$$

this reaction are volatile and desorb from the surface into the plasma from where they can be pumped away.

Energetic ions are accelerated by the potential between the plasma and workpiece. In striking the workpiece surface they enhance, through sputtering, the etching rate of the reactive species, stimulate the desorption of volatile products of the etching reaction, and clean the surface of unwanted byproducts. Because the ions strike the surface in a direction perpendicular to the surface, the holes and trenches are created with straight edges; that is, the etch is anisotropic.

The actual chemical pathways underlying processes such as those represented simply by the reaction above are complex. The use of specific gas and gas mixtures in the plasma allows selectivity so that the deposited material is removed more rapidly than the underlying material or the patterned photoresist mask which covers the surface to define the region to be etched. For example, adding a small amount of molecular hydrogen (H_2) to a carbon tetrafluoride plasma can provide a selectivity of silicon dioxide (SiO_2) to silicon sputtering of 15:1. Currently 0.2-micrometer-wide by 4-μm-deep trenches can be etched into silicon surfaces. It is expected that such structures will be routinely used in production lines with etch and deposition uniformities of better than 1% across wafers with diameters of 300 millimeters (12 in.).

Diamond deposition. When small amounts of hydrocarbon gases, such as acetylene, methane, or adamantine, are added to plasmas created in hydrogen gas it is possible to grow thin films of diamond onto the surfaces of workpieces placed in the plasma. In the plasma the hydrocarbon gas is dissociated into various carbon-containing radicals which migrate to the surface and deposit as solid carbon. Some will be in the form of diamond or diamondlike carbon, but most will be deposited as graphite. However, the hydrogen gas is also dissociated in the plasma, forming atomic hydrogen. The atomic hydrogen etches away the graphite from the surface at a rate approximately 20 times faster than it etches diamond. The atomic hydrogen also attaches to dangling bonds on the surface of the growing film, thus preventing the reconstruction of graphite but leaving the sites available for diamond growth. By balancing the atomic-hydrogen-to-carbon ratio and therefore the ratio of diamond growth to graphite etching, it is possible to grow synthetic diamond which is indistinguishable from naturally occurring diamond and in the form of a thin film. Cutting tools can routinely be covered with diamond layers a few tens of micrometers thick and with grain sizes in the region of 1 μm, and self-supporting films of diamond up to 135 mm (5.3 in.) in diameter are commercially available.

Plasma-source ion implantation (PSII). If ions of very high energy (tens of kiloelectronvolts) are incident on a surface they can penetrate and become embedded several hundred atomic layers below the surface. The penetration depth depends on the ion energy. The presence of the implanted atoms disturbs the bulk material, and essentially produces a new material in the surface region. This effect has been exploited for some time by using an external source of ions and an ion accelerator. Recently it has been recognized that the same effect can be achieved within the plasma environment. The workpiece, which is in a plasma containing the appropriate ion species, is repetitively pulse biased, for about 10 microseconds, to a high negative potential, generally about 50 keV. The positive plasma ions are thus accelerated to the surface and implant in the bulk material. The properties of the new surface material thus created can be selected by appropriate choice of the ion species and bias energy. The technique is particularly useful in producing corrosion-resistant hard coatings such as nitrides and carbides. Ion implantation offers many advantages over other surface-modification techniques. Because it is not a coating technique there are no film adhesion problems and no dimensional changes; therefore cutting edges retain their sharpness. Also, implantation is a nonequilibrium process; therefore a new range of alloys, not limited by classical thermodynamic properties and diffusion kinetics, can be produced. Conventional ion implantation is effective only on surfaces with direct line of sight for the incoming ions. By contrast, in plasma source ion implantation the plasma forms over the entire surface of the workpiece, allowing this important technique to be used on geometrically complex components.

Cleaning of archeological artifacts. The cleaning of many archeological artifacts requires the precise, preferential removal of material from often geometrically complex metal objects. Plasma processes are ideally suited for this task. Generally the material to be removed is an encrustation of hard ag-glomerate consisting of soil and migrated oxides. A few minutes of exposure to a hydrogen plasma can loosen the encrustation so that it can be readily removed by a simple tool, leaving the finer surface details preserved. Such hydrogen plasma exposure can also inhibit postcorrosion effects due to, for example, chlorine.

Environmental advantages. It is now possible in component design to specify the precise surface properties of a material. In addition to the new opportunities for cost savings, innovation, and the production of new materials, these techniques offer significant environmental advantages. For example, plasma surface modification eliminates the need to store, handle, and dispose of the large amounts of corrosive and hazardous liquids required in conventional, wet chemical techniques.

For background information SEE GAS DISCHARGE; INTEGRATED CIRCUITS; ION IMPLANTATION; ION-SOLID INTERACTIONS; PLASMA PHYSICS; SPUTTERING; SURFACE HARDENING OF STEEL; VAPOR DEPOSITION in the McGraw-Hill Encyclopedia of Science & Technology.

Bill Graham

Bibliography. M. J. de Graaf et al., Cleaning of archaeological artifacts by cascade arc plasma treatment, *Surface Coatings Tech.*, 74–75:351–354, 1995; M. Kenward, Why plasma science means business, *Phys. World,* 8(6):31–33, June 1995; J. Reece Roth, *Industrial Plasma Engineering*, 1995.

Population ecology

Numbers of most wild plants and animals fluctuate irregularly between limits that are extremely restricted compared with what their rate of increase will allow. In areas where conditions have not changed, the number of breeding birds is nearly the same each year. Many of the species present today were extant during the Renaissance, and the known changes are largely attributable to humans. Birds are certainly more stable in number than insects or shellfish, for example, but they also have a much lower intrinsic rate of population growth. The nature of the restriction of the potentiality of a population to increase and that of the coexistence of a variable number of populations that collect energy from a common pool at a well-defined trophic level are two related and central aspects of population ecology.

Potential limits to population abundance. A population can be established in an area only if the combination of occurring environmental conditions is included in its fundamental niche. The upper limit of a population's abundance will depend on its niche width, which sets the fraction of potentially available energy that the individuals can actually acquire. For a perfect generalist, the upper limit of population abundance is determined by the ratio between the overall amount of energy available and the population energy requirements per unit of body mass.

The energy requirements of individuals are known to increase with their body size. Therefore, in terms of number of individuals, the maximum potential limit of population abundance decreases proportionally as the mean individual body size increases. Consumers, however, cannot be too large because stochastic mortality factors and low chance of mating confer high risks of extinction on very sparse populations. Thus, the largest body size that a species can achieve is directly related to the amount of available energy.

Monospecific guilds (a guild being a group of species utilizing the same kinds of resources) are extremely rare or exceptional; each population co-occurs with others (from a few to a few hundreds) that obtain energy at the same distance from the ultimate source. The overall available energy is partitioned among a variable number of populations, the number and density of which should depend on the criteria of energy partitioning. Nevertheless, the expected patterns of covariation of individual body size with both population abundance and energy availability in energy-limited monospecific guilds are actually observed among real guilds. Body size represents one of the two criteria of energy partitioning allowing coexistence. The other is the conventional paradigm of resource partitioning and niche overlap.

Niche-mediated coexistence. In order to obtain a stable coexistence, populations must partition an axis of environmental heterogeneity, referred to as the resource axis, where resources consist of energy, space, or nutrients. The process is simplified as the one-species one-niche concept. Its basic assumption is that each species in a pair can actually acquire the control of a shared resource set if it is given enough time to realize its growth potential. Thus, according to a mathematical theorem relating to competition of two species for a single resource, the more efficient species drives the other to extinction. Coexistence is allowed when limiting resources are partitioned and niche overlap is less than the maximum value compatible with interspecific occurrence. Assuming size-related resource partitioning, differences in body size between competitors are commonly used as a morphological equivalent of niche overlap.

Plants are typically limited by essential resources such as nutrients and sunlight. Populations of plants can coexist if they are limited by different resources, each population being the most efficient in utilizing a different resource. In guilds of animals, resources are more likely to be substitutable because animals can obtain their essential nutrients from many different kinds of prey, and animal populations are limited by energy availability. Several models have been produced to explain interspecific coexistence and community structure on the basis of different mechanisms of resource apportionment, each niche portion supporting only a single species or species-equivalent assemblage of organisms.

The process of apportionment has economic and evolutionistic components. For example, an optimal niche is likely to be narrow when resources are abundant and wide when they are scarce; however, niche position (such as the location of the population niche on the resource axis) has genealogic and evolutionistic constraints. Population abundance follows from the energetic reward that the population gains from the control of the portion of resource spectrum determined according to its genealogic history and evolutionistic plasticity.

Interspecific interactions are also known to affect the size of the resource portion that a population acquires. However, they are the result of a random combination process of the competitive abilities of all potential colonists on all resource spectrum portions. In this sense, competitive abilities would be less relevant than the mechanism of resource apportionment, because they should affect neither number nor densities of coexisting competitors but only the kind of taxa occurring.

Body size, coexistence, and population abundance. A close analysis of the role of individual body size on interspecific coexistence has recently suggested that its influence on competitive interactions, coexistence, and guild structure cannot be reduced to resource overlap alone. Body size is likely to confer a competitive advantage to the larger competitor in a pair because of differences in resource acquisition rates or population interference ability, or both. When a body size–related competitive asymmetry occurs, the smaller competitor in a pair cannot acquire the control of a shared resource set. On these conditions, size-related constraints on resource acquisition at the individual level can determine coexistence conditions independently of the degree of niche overlap among competitors.

When the feeding behavior of an individual on a single patch in its own home range leads to a diminishing return of resources, individual ingestion rate will decrease until resources are no longer profitable in the patch and the animal must move. Ingestion rate is therefore expected to vary with time between two critical resource availabilities, that at which the rate starts to decrease and that at which the individual must leave the patch. The perception of resource availability is, however, relative to the individual size. On a shared resource set the same relationship between ingestion rate and relative resource availability can be repeated many times for many species, with the size of individuals and the absolute resource availability decreasing at each step. This represents a hierarchy, where all species are resource limited and the availability for each smaller species depends on an optimal inefficiency in resource exploitation of each larger species. The resource acquisition rate of a large species can limit the growth of a small, more efficient, species that from its higher efficiency will gain only a chance of coexistence.

Similarly with plants, when gradients of nutrient density are determined by the resource acquisition rates of an individual other individuals with lower requirements can find their optimal location on the gradient, independently of resource overlap. Generally, when differences among populations in the physiological size of individuals generate a size-related competitive asymmetry, niche apportionment is simply not required to achieve coexistence. A conceptual model has suggested that the difference in size required to obtain coexistence, measured as body mass ratios, is likely to be small enough to give a biological significance to this mechanism.

Integration of the coexistence mechanisms. The finding that competitive coexistence could occur even though the competitor niches overlap completely does not deny the role of niche apportionment at the population level in the field. It simply suggests that more species than niche portions can stably occur. Because coexistence among competitors of different size does not require any assumption about resource selection, the species of each of the allowed sizes can apportion resources in a way that is completely independent of the way they are apportioned at any other size. The upper limit to the number of coexisting populations is therefore determined by the product of number of allowed sizes and number of niche portions.

At the community level, the two coexistence mechanisms have, however, different implications. Body size–related coexistence should determine a hierarchy in resource and energy use, with a bias in energy flow toward large species that could also be commonly detected within guilds. This hierarchy is likely to constitute the ecological and evolutionary arena in which resource partitioning and species packing take place.

For background information SEE BIOLOGICAL PRODUCTIVITY; ECOLOGICAL COMMUNITIES; ECOLOGY; FOOD WEB; GUILD; POPULATION DISPERSION; POPULATION ECOLOGY in the McGraw-Hill Encyclopedia of Science & Technology.

Alberto Basset

Bibliography. A. Basset, Body size related coexistence: An approach through allometric constraints on home range use, *Ecology*, 76(4):1027–1035, 1995; A. P. Gutierrez, *Applied Population Ecology: A Supply-Demand Approach*, 1996; M. A. Huston and D. L. DeAngelis, Competition and coexistence: The effects of resource transport and supply rates, *Amer. Natur.*, 166(6):956–977, 1994; M. A. Leibold, The niche concept revisited: Mechanistic models and community context, *Ecology*, 76(5):1371–1382, 1995; P. D. Stiling, *Ecology: Theories and Applications*, 2d ed., 1995.

Potato, Irish

In the 1840s, the late blight disease of potatoes (caused by the fungus *Phytophthora infestans*) was introduced into Ireland. The disease destroyed potato plants in farmers' fields and rotted tubers stored after harvest. The result was the Irish famine of 1845–1849. Since then, potato farmers have had to combat this disease, which has until recently been controlled. Current destruction of potato and tomato crops is attributable to new migrations of *P. infestans*.

1840s migration. The center of diversity and probably the center of origin of *P. infestans* is located in the highlands of central Mexico. It coevolved with a large diversity of tuber-bearing *Solanum* species (wild potatoes). This fungus was probably sequestered in the highland valleys until the second quarter of the nineteenth century. The late blight of potato and tomatoes was not discovered until the 1840s, probably because potatoes (*S. tuberosum*) and tomatoes (*Lycopersicon esculentum*) evolved in the Andes of South America. Although yet unnamed, the potato late blight disease was probably first noticed in the early 1840s in the United States. *Phytophthora infestans* was probably carried from central Mexico to northeastern United States with plant material, but only a very small fraction of the genetic diversity in central Mexico was involved in this initial migration. Probably an even smaller portion of that genetic diversity (possibly a single genotype) was carried from the United States to Europe and initiated the Irish potato famine.

In Europe, the disease was first noticed at the end of June 1845, in Belgium. By October 1845 it had spread as far west as Ireland and as far east as Germany. Not only did the fungus destroy foliage, but it also infected tubers and rotted them in the ground or in storage. From western Europe, the fungus was subsequently carried in infected seed tubers to South America, Africa, the Middle East, and Asia. Even as late as the final quarter of the twentieth century, the populations of *P. infestans* in these locations were dominated by a single clonal lineage.

Twentieth-century migrations. Migrations have again increased the global importance of potato and tomato late blight. There have been several (probably independent) migrations. The first of the recent migrations probably occurred in the late 1970s, associated with a shipment of potatoes from Mexico to Europe. Both mating types (A1 and A2) were probably involved, so that sexual reproduction for the first time became possible in Europe. Previously most of the world contained only A1 mating type; both are required for sexual reproduction. The newly migrated population displaced the previous indigenous asexual population. As long as fungal populations remained asexual, the fungus was essentially an obligate parasite and could survive only in the presence of a living host cell. However, the products of sexual reproduction (oospores) could survive for long periods in soil, thus enabling the soil to serve as a source of the fungus. The new population subsequently appeared in South America, Africa, the Middle East, and Asia, and was probably carried in infected seed tubers from Europe.

During another migration, another genotype of the fungus was transmitted to Japan and Korea. The migrating population may have been a single genotype, and although of a different mating type (A2) from the previous indigenous population (A1) sexual matings did not give rise to viable progeny. Thus, the population in Japan and Korea persists as an asexual population of obligate parasites.

Migrations began impacting the United States in the 1980s but were especially important in 1992, 1993, and 1994. During these years, four clones (genotypes) became widely distributed throughout the United States and Canada. Some were A1 and others were A2. The immigrant strains of the fungus share several characteristics: (1) They are resistant to a fungicide that previously had been the most effective one. (2) Some of the immigrant strains of the fungus are especially pathogenic on tomatoes, whereas most of the previous population was primarily pathogenic on potatoes. Thus, infected tomato plants could easily contaminate neighboring potato fields, and vice versa. (3) The immigrant strains appear to cause severe disease more quickly than the previous population.

Pathways of migration. Although the pathways of migration are not known with certainty, technologies such as allozyme analysis and deoxyribonucleic acid (DNA) fingerprinting suggest that the origins of the immigrant strains are the potato and tomato production areas in Mexico. One possible migration mechanism is via infected tomato fruits. Young infections on tomato fruits are not visible, so infected tomatoes could be harvested and shipped to distant markets. During transit or after arrival, infections would become visible and the infected tomatoes discarded. If an infected tomato was discarded near a field of potatoes or tomatoes, the sporulating fungus could be dispersed via wind currents to susceptible potato or tomato foliage. This scenario could have occurred almost anywhere in the United States or Canada.

The occurrence of resistance to the most effective fungicide, metalaxyl, was especially problematic because growers had become accustomed to its efficacy, using it successfully to arrest developing epidemics. By the time growers realized that the new strains were resistant, an epidemic had consumed much of a particular crop. This scenario was repeated many times in the early 1990s.

Although not yet widespread in the United States, sexual reproduction in this fungus is now possible. Where sexual reproduction occurs, it will change the life history of the fungus and epidemiology of the disease. The resulting sexual population will be genetically diverse, a very different situation from the current homogeneous clonal populations.

Prospects. Late blight of potatoes and tomatoes is very likely to be more troublesome in the 1990s for home gardeners and commercial growers. Whenever the weather is favorable for the fungus (moderate temperatures and long periods of rain or dew) the disease is likely to be more prevalent. There will be an increasing need to rely on resistant plants, to use healthy seed tubers, and to employ fungicide judiciously.

For background information *SEE PLANT PATHOLOGY; POTATO, IRISH* in the McGraw-Hill Encyclopedia of Science & Technology.

William E. Fry

Bibliography. P. M. A. Bourke, Emergence of potato blight, 1843–46, *Nature,* 203:805–808, 1964; W. E. Fry et al., Historical and recent migrations of *Phytophthora infestans:* Chronology, pathways, and implications, *Plant Dis.,* 77:653–661, 1993; S. B. Goodwin, B. A. Cohen, and W. E. Fry, Panglobal distribution of a single clonal lineage of the Irish potato famine fungus, *Proc. Nat. Acad. Sci. USA,* 91:11591–11595, 1994; J. A. Lucas et al. (eds.), *Phytophtora,* 1991.

Prion disease

The transmissible spongiform encephalopathies (also referred to as prion diseases) constitute a biologically unique and fascinating group of disorders in both humans and animals. Scrapie is the most common form seen in animals, while in humans the most prevalent form is Creutzfeldt-Jakob disease. This group of disorders is characterized at a neuropathological level by vacuolation of the brain's gray matter (spongiform change). They were initially considered to be examples of slow virus infections. Experimental work has consistently failed to demonstrate detectable nucleic acids—both ribonucleic acid (RNA) and deoxyribonucleic acid (DNA)—as constituting part of the infectious agent. Contemporary understanding suggests that the infectious particles are composed predominantly, or perhaps even solely, of protein, and from this concept was derived the acronym prion (proteinaceous infectious particles). Also of great interest is the apparent paradox of how these disorders can be simultaneously infectious and yet inherited in an autosomal dominant fashion (from a gene on a chromosome other than a sex chromosome).

Disorders. Scrapie, which occurs naturally in sheep and goats, was the first of the spongiform encephalopathies to be described. Over the years, an increasing range of animal species have been recognized as occasional natural hosts of this type of disease. Examples of these disorders include transmissible mink encephalopathy, chronic wasting disease of mule deer and elk, and of greatest interest recently, an epidemic form in cattle in the United Kingdom called bovine spongiform encephalopathy. The last was believed to have been introduced by the accidental contamination of the food chain by meat and bone meal protein supplement derived from sheep presumably harboring scrapie. As a reaction, various measures including a feed ban were put in place, with recent surveys indicating the bovine spongiform encephalopathy epidemic is on the decline. The British surveillance unit and its detailed case-control study set up to monitor the incidence of Creutzfeldt-Jakob disease have noted no increase in the number of cases of this disorder, and no geographical clustering. However, the recent occurrence of Creutzfeldt-Jakob disease in younger persons (under the age of 42) has raised the possibility of the spread of the bovine disease to humans.

At this stage, the overall risk of transmission appears small. So far, animal models have indicated that only central nervous system tissue has been shown to transmit the disease after oral ingestion— a diverse range of other organs, including udder, skeletal muscle, lymph nodes, liver, and buffy coat of blood (that is, white blood cells) proving noninfectious.

The currently recognized spectrum of human disorders encompasses kuru, Creutzfeldt-Jakob disease, Gerstmann-Sträussler-Scheinker disease, and most recently, fatal familial insomnia. All, including familial cases, have been shown to be transmissible to animals and hence potentially infectious; all are invariably fatal with no effective treatments currently available.

Kuru has been almost eradicated since the cessation of ritualistic cannibalism, but occurred endemically among the Fore and some neighboring people of the Eastern Highlands of Papua New Guinea. The disorder was believed to be transmitted by the consumption of infected central nervous system tissues. In its typical form, the symptoms of kuru included a relentlessly progressive loss of muscular coordination (ataxia) associated with an unusual tremor; the duration of the entire illness spanned 1–2 years.

Creutzfeldt-Jakob disease is a rapidly progressive deterioration of intellectual abilities (dementia), which proceeds to death with a median duration of illness of around 4 months. As the disease evolves, it is typically accompanied by a range of additional clinical features which include ataxia, muscular rigidity, and spontaneous irregular limb jerks (myoclonus). For 90% of affected individuals Creutzfeldt-Jakob disease develops sporadically, without any clues as to causation. Approximately 10% of individuals have a genetic basis and occur within families. The final important epidemiological subgroup of Creutzfeld-Jakob disease comprises those cases arising after accidental person-to-person transmission (iatrogenic).

Gerstmann-Sträussler-Scheinker disease is relatively uncommon, and presents as a chronic progressive familial ataxia with somewhat distinctive neuropathological features. Fatal familial insomnia is the rarest and most recently recognized of the disorders, and is characterized by rapidly progressive and untreatable insomnia, ataxia, and myoclonus, with pathological examination revealing severe and selective degeneration of a restricted part of the brain (the thalamus). *SEE BRAIN.*

Molecular biology. Although not completely resolved, the prevailing view is that the transmissible agent of these disorders is an abnormal conformational form of the constitutively expressed prion protein molecule. The prion protein molecule normally is produced by a wide variety of cells throughout the body, but especially the central nervous system. Although initial investigations minimized its probable biological importance this normal, host-encoded glycoprotein probably plays a role in receptor-mediated signal transduction, and disordered function may be capable of explaining some of the clinical features observed in Creutzfeldt-Jakob disease.

It has been suggested that in sporadic cases of Creutzfeldt-Jakob disease, the change in the prion protein molecule occurs spontaneously as a random or chance event. However, once in this abnormal shape, the prion protein molecule becomes relatively resistant to enzymes usually capable of degrading proteins (proteases), and also serves as a template or nucleating factor for the conversion of other prion protein molecules through self-replicating polymerization. In some of the prion disorders, especially Gerstmann-Sträussler-Scheinker disease, the accumulation of the abnormal prion protein molecules can be detected as deposits of amyloid, which may also represent a mechanism of neurotoxicity inducing the spongiform change. However, the precise sequence of events subserving the translation of this subtle molecular change into a devastating fatal neurodegenerative disorder is unknown.

The study of the familial prion diseases has led to the recognition of a number of causal mutations within the prion protein gene located on chromosome 20. Each mutation is generally associated with a particular form of clinical illness, but considerable variability in clinical and neuropathological expression can occur within affected members of the same family. These mutations are thought to greatly increase the chance of the prion protein molecule adopting the abnormal configuration, leading to self-replicating polymerization. Similar to sporadic Creutzfeldt-Jakob disease, the consequent accumulation of abnormal prion protein molecules is thought to be linked to the eventual disruption of cellular function. If the abnormal form is introduced into an individual, it serves as the template for subsequent autocatalytic polymerization. Such a model attempts to reconcile the paradox of how this group of disorders can be transmissible and at the same time inherited in an apparent autosomal dominant fashion.

Although no causative mutation has been demonstrated in the prion protein gene to explain sporadic examples of Creutzfeldt-Jakob disease, a genetic component may still play a part in the susceptibility to developing this disorder, with normal variations in the gene possibly altering the likelihood of developing this and pituitary hormone-related Creutzfeldt-Jakob disease.

Human-to-human transmission. A variety of mechanisms of human-to-human transmission have been described, and include contaminated instruments such as electroencephalographic depth electrodes, dura mater homografts, corneal transplants, and cadaverically derived human pituitary growth hormone and gonadotrophins. Transmission is due in part to the ineffectiveness of conventional sterilization and disinfection procedures to control the infectivity of transmissible spongiform encephalopathies. Numerically, pituitary hormone–related Creutzfeldt-Jakob disease is the most important form of human-to-human transmission of disease. Although kuru and Creutzfeldt-Jakob disease are experimentally transmissible to nonhuman primates through the oral route, successful human-to-human transmission appears to require more invasive methods of inoculation of the infectious material, usually in the form of tissue transplants or injectable therapies or through the use of contaminated neurosurgical instruments.

Epidemiological evidence suggests that there is no increased risk of contracting Creutzfeldt-Jakob disease from exposure in the form of close personal contact during domestic and occupational activities. Nonetheless, a very small number of physicians, including pathologists and neurosurgeons, and a slightly larger number of allied health personnel (including histologists) have developed Creutzfeldt-Jakob disease. However, there is no clear evidence to suggest that occupational exposure poses a significant risk of contracting Creutzfeldt-Jakob disease, especially if precautionary and preventive guidelines are followed.

Incubation periods in cases involving human-to-human transmission appear to vary enormously, depending upon the mechanism of inoculation. In the case of pituitary hormone therapy and kuru, which presumably utilize peripheral methods of inoculation, incubation periods vary considerably from a few to more than 30 years, with infected individuals usually exhibiting an ataxic syndrome. However, when the infectious material is given more direct access to the central nervous system, as with intracerebral depth electrodes, the incubation periods are usually much shorter, generally of the order of 12–24 months, and cognitive decline similar to sporadic Creutzfeld-Jakob disease is the usual manifestation.

Current evidence suggests that transmission of Creutzfeld-Jakob disease from mother to child does not occur. This finding accords with the much larger epidemiological experience seen in the related disorder kuru. The evidence encompasses children born to mothers during both the presymptomatic and clinical phase of their illness. The explanation for the apparent transplacental and postpartum barriers to transmission remain to be elucidated.

Transmissibility. The transmissibility of human spongiform encephalopathies was first reported in 1966 after kuru had been successfully transmitted to three chimpanzees. Within 2 years, the successful transmission of Creutzfeldt-Jakob disease using the same animal model was announced. Using nonhuman primates (chimpanzees, squirrel and spider monkeys) transmission rates range from 90% for sporadic Creutzfeldt-Jakob disease, through 95% for kuru, and up to 100% for cases of human-to-human transmission of Creutzfeldt-Jakob disease. Two important factors pertaining to transmissibility are the method of inoculation and the dose of infectious material administered. A high dose of infectious material administered by direct intracerebral inoculation is clearly the most effective method of transmissibility and generally provides the shortest incubation time, which in the primate model is approximately 10 months. Peripheral methods of inoculation (for example, subcutaneous, intraperitoneal) are less effective routes for transmissibility which can be partly offset by the administration of higher doses of infectious material.

Smaller nonprimate host animals have been successfully utilized in transmissibility experiments. However, transmissibility rates are generally far lower, despite the use of similar inoculation techniques. Familial forms of disease are less transmissible, averaging around 70%. Within the familial group, the transmission rate also appears dependent on the underlying mutation in the prion gene.

There is some evidence of a species barrier to transmissibility. The prion protein gene appears to modulate both scrapie susceptibility and incubation times, and the more dissimilar the prion protein genes, the greater the species resistance to transmissibility.

Central nervous system tissue has been consistently shown to be the most infectious material in animal transmissibility experiments. However, other tissues have been shown to be infectious, although at lower rates of transmissibility, and include lung, liver, kidney, spleen, and lymph nodes. Although cerebrospinal fluid has been shown to be infectious, vaginal secretions, semen, sputum, saliva, nasal mucus, and tears are not. The testis, prostate, skeletal muscle, peripheral nerve, adipose tissue, adrenal gland, and thyroid gland have failed to transmit the disease; the status of urine and placental tissue is less clear.

The transmissibility status of blood and blood products is of some concern. White blood cells (buffy coat) and whole blood have been shown to be infectious and transmissible to nonprimate animals when taken from individuals during late stages of Creutzfeldt-Jakob disease. Clearly such findings raise some concern regarding the possible hematogenous transmission of disorders from individuals in the preclinical or clinical phases of their illness, but despite claims to the contrary there is no well-documented and verified example of a Creutzfeldt-Jakob disease being transmitted in this manner in humans.

For background information SEE *MUTATION; NERVOUS SYSTEM DISORDERS; SCRAPIE; VIRUS INFECTION, LATENT, PERSISTENT, SLOW* in the McGraw-Hill Encyclopedia of Science & Technology.

Colin L. Masters; Steven J. Collins

Bibliography. P. Brown, M. A. Preece, and R. G. Will, "Friendly fire" in medicine: Hormones, homografts, and Creutzfeldt-Jakob disease, *Lancet,* 340:24–27, 1992; B. N. Fields et al. (eds.), *Fields' Virology,* 3d ed., 1996; M. S. Palmer and J. Collinge, Human prion diseases, *Curr. Opin. Neurol. Neurosurg.,* 5:895–901, 1992; S. B. Prusiner, Natural and experimental prion diseases of humans and animals, *Cur. Opin. Neurobiol.,* 2:638–647, 1992.

Radio broadcasting

Recent developments in the field of radio broadcasting involve the introduction of digital audio broadcasting, the implementation of the Radio Broadcast Data System standard, and the replacement of the Emergency Broadcast System with the Emergency Alert System.

Terrestrial Digital Audio Broadcasting

The transmission of radio programs to the public by terrestrial broadcast in the United States is primarily by amplitude-modulated (AM) signals in the 535–1700-kHz radio-frequency band and by frequency-modulated (FM) signals in the 88–108-MHz radio-frequency band. Both modulations are analog. There is a desire to improve the quality of radio program transmission by use of digital modulation systems, commonly called digital audio broadcasting (DAB). Such systems have been developed, tested, and demonstrated, and will be soon implemented. Digital audio broadcasting will be the first major improvement in the quality of terrestrial radio transmission since the introduction of FM broadcasting in 1940, and will provide to homes and vehicles compact-disk-quality music, which is superior to FM in audio signal-to-noise ratio, bandwidth, and dynamic range.

System configurations. There are two basic methods for accomplishing terrestrial digital audio broadcasting transmissions. In one, the transmissions are broadcast in a new radio-frequency band. It requires both finding a sufficiently wide band of unused frequencies in a suitable radio propagation range (or requiring current users of such frequencies to move to other frequencies to avoid reception interference) and developing user radios that contain new antennas, radio-frequency elements, and demodulators. New broadcast transmitters are also required. A digital audio broadcasting system of this nature, termed Eureka 147, has been developed by the European Union and is in experimental operation using the L-band (1452–1492 MHz) in France, Canada, and Germany, and in full operation in the United Kingdom using the very high frequency (VHF) band (217–230 MHz).

The second method for accomplishing digital audio broadcasting terrestrial transmissions uses the existing radio-frequency bands. The digital modulation is transmitted either in combination with an analog AM or FM broadcast signal (in-band on-channel, or IBOC), or in the guard bands between the analog FM signals (in-band adjacent-

channel, or IBAC). Such in-band methods are favored in the United States because no new frequency spectrum is required (nor is new spectrum readily available), radio stations can add digital audio broadcasting capability when and if each individual radio station so desires, and equipment commonality exists with regard to broadcast transmitters, radio receiving antennas, and radio receiving electronics. Several in-band systems have been thoroughly tested and further evaluated in field trials, and one in-band on-channel AM and FM system has been publicly demonstrated.

Digital music compression. All digital audio broadcasting systems convert the audio analog program at its source to a digital replica and then compress the replica by using various techniques to remove frequency components that cannot be heard or are otherwise unnecessary to transmit. The two leading digital audio broadcasting compression systems are perceptual audio coding (PAC) and Musicam, which reduce the required transmission rate for compact-disk-quality stereo music by a factor of approximately 10 without impairing the perceived aural quality. Often, higher transmission rates than those required by the audio program are used so that ancillary channels can also be provided (facsimile, paging, data, or similar broadcast services). *SEE DATA COMPRESSION.*

Multipath. Because radio broadcasts are received by mobile vehicles, digital audio broadcasting transmissions have been designed to mitigate multipath, which is caused by reflected signals combining with the directly received signal so that, depending on their relative amplitudes and phases, fading or suppression of the desired signal results through cancellation. The difficulty in multipath mitigation can be compounded for moderate and rapidly moving vehicles by the Doppler shift in frequency, which at the L-band can be more than 100 Hz. The mitigation techniques in general involve spreading the transmitted waveform over space, frequency, and time, and include use of spatial diversity, forward error correction signal coding, resistive modulations such as spread spectrum, time interleaving of data, and adaptive equalizers in the receivers. Some receiver configurations are multichannel to allow the constructive combining of certain multipath components with the desired signal to enhance reception.

Networks. Radio broadcasts in other countries can be national in scope, and the implementation of digital audio broadcasting provides an opportunity to design a single-frequency network (SFN). With such a network, assuming enough terrestrial transmitters are built, a vehicle can drive through a large service area and receive the same radio program on one broadcast frequency. For very large countries, where the required number and cost of terrestrial transmitters for a single-frequency network may be prohibitive, direct-broadcast radio programs from satellites have been proposed to provide service coverage of the rural areas, with terrestrial transmitters providing urban service coverage. Single-frequency networks have been effected with appropriate digital audio broadcasting modulation schemes, including the Eureka 147 program. Normal analog and digital broadcast modulations cannot be used for a single-frequency network, since adjacent broadcast transmitters at the same frequency would be simultaneously received in at least the portions of the service area between them. The receivers could not distinguish between the multiple cofrequency transmissions and, because they were not in phase, aperiodic cancellations of the combined received signal would occur. The design of a single-frequency network is particularly difficult with mobile receivers: besides the multipath and Doppler effects previously mentioned, the relative signal amplitudes at the receivers from the several local broadcast transmitters will vary significantly over the service area.

Implementation examples. The two digital audio broadcasting systems that have been publicly demonstrated as of 1995 are the USA Digital Radio in-band on-channel systems and the Eureka 147 system.

USA Digital Radio. These in-band on-channel systems have been proposed as the digital audio broadcasting standard for existing AM and FM radio stations (**Fig. 1**). The FM in-band on-channel transmitting system employs Musicam compression to provide digital audio broadcasting using nominally 128–256 kilobits per second for achieving compact-disk-quality stereo music plus digital data of up to 64 kb/s along with the host analog FM broadcast. Both the digital music compression and the subsequent forward error correction (FEC) coding rates are variable depending on music complexity. The composite transmission rate is 384 kb/s. The modulation technique is unique and mitigates digital audio broadcasting signal interference into the analog FM signal, analog FM signal interference into the digital audio broadcasting signal, spillover of the digital audio broadcasting signal into adjacent FM channels, and multipath.

The digital audio broadcasting modulation scheme divides the 384-kb/s signal among 48 separate subchannels, each impressed with 8 kb/s of data by using biphase keying. These subchannels are modulated by using a spread-spectrum technique, which ensures they are transmitted wide-band (460 kHz) to mitigate multipath, mutually orthogonal to avoid self-interference, and noiselike to reduce host and adjacent-channel interference. Additionally, the transmitted waveform is shaped to suppress the center 220 kHz of the spectrum occupied by the analog FM signal and to avoid the need for complex modulation filtering. A special wide-band synchronization and reference signal is added at reduced power level to the composite 48 subchannels. The multipath mitigation requires that the digital audio broadcasting receivers have adaptive equalizers to reverse the effects of rapidly changing multipath over the transmission bandwidth. The reference signal allows rapid synchronization and training of the equalizer.

The AM in-band on-channel digital audio broadcasting system is similar to its FM counterpart, with differences in the modulation waveforms and capability. The AM system provides stereo 15-kHz audio bandwidth as compared with the stereo 20-kHz audio bandwidth and a 2.4-kb/s data channel in the FM system. The AM system occupies the 40-kHz AM broadcast emission mask and, after forward error correction coding, the composite digital audio broadcasting transmission rate is 128 kb/s.

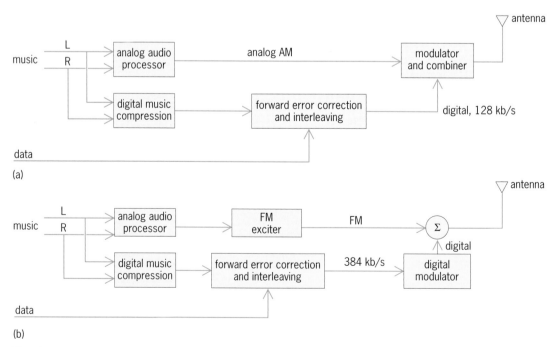

(a)

(b)

Fig. 1. In-band on-channel digital audio broadcasting systems. (*a*) AM system. (*b*) FM system. (*USA Digital Radio*)

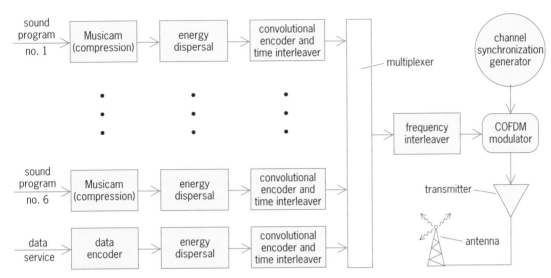

Fig. 2. Eureka 147 transmission system.

Eureka 147. Although many variants exist, the basic Eureka 147 system was designed to transmit six radio programs of compact-disk-quality stereo in 1.5 MHz of bandwidth at L-band frequencies. In many countries outside the United States, it is common for several different radio programs to be generated in a common studio and to be sent from the same transmitter and antenna tower. A simplified block diagram of such a transmitting system is depicted in **Fig. 2.** There are three main subsystems. The first is compression, where each of the six radio programs is digitized and compressed by using Musicam, which for independent encoding of stereo music provides a digital data stream at 256 kb/s. The six digital data streams are individually scrambled for energy dispersal, convolutionally encoded (that is, subjected to forward error correction), and time-interleaved before being multiplexed together. The data stream is frequency-interleaved and multiplexed by using coded orthogonal frequency-division multiplex (COFDM). The resultant signals are modulated by using differential quadrature phase-shift keying and are sent to the radio-frequency transmitting amplifier and antenna.

The previously described ability of Eureka 147 to form a single-frequency network and to mitigate multipath stems in part from COFDM, which consists of 384 individual orthogonal carriers (that is, subchannels), equispaced at 4-kHz intervals across the 1.5-MHz transmission bandwidth with each subchannel modulated so that a data throughput of 4 kb/s results. This data throughput is such that the corresponding symbol duration is larger than the delay spread of the transmission channel. Additionally, by inserting a guard time interval between successive symbols, the necessary rejection of self-interference in a single-frequency network and mitigation of multipath is achieved because of large reduction of intersymbol interference. For simplicity, Fig. 2 does not show various features of the Eureka 147 broadcast system for mode control, data transmission, and service channels.

Robert D. Briskman

Radio Broadcast Data System

The United States Radio Broadcast Data System (RBDS) refers to a technical standard developed by the National Radio Systems Committee (NRSC). The NRSC is a joint committee of the National Association of Broadcasters (representing broadcasters) and the Electronic Industries Association (representing receiver manufacturers). The RBDS standard defines a method by which FM broadcast stations, at each broadcaster's discretion, can transmit data to smart or intelligent receivers capable of performing a variety of automatic functions. These receivers may take such forms as consumer radios, pagers, and emergency alerting devices. The extensive geographical signal coverage of FM broadcast stations in the United States provides an extremely efficient infrastructure for the transmission of data via RBDS.

The RBDS standard is both an FM transmission standard and a standard for RBDS radio receivers. The RBDS is an extension of the Radio Data System (RDS) that has been in use in some European countries for a number of years. The NRSC released the RBDS standard on January 8, 1993.

Signal. FM broadcast stations are permitted by the Federal Communications Commission (FCC) to transmit subsidiary communications services, such as RBDS data, on a subcarrier within the FM baseband signal. The RBDS signal operates on a subcarrier frequency of 57 kHz, and is combined with the FM audio program signal by means of an RBDS encoder located at the input to the FM modulator or exciter. The RBDS subcarrier operates with an injection level of 3%, meaning its signal is set at 3% of the peak FM audio signal level into the exciter. No harmful interference to an FM station's main channel programming occurs.

Features. The RBDS provides a display (on radios so equipped) of station call letters, slogans, or logos, which may be more familiar to listeners than actual radio-frequency assignment numbers. The RBDS allows automatic receiver tuning of radio broadcast networks on individual stations. It also allows traffic or emergency bulletins to activate radios that are turned off or playing a cassette. Additionally, the RBDS allows broadcasters to transmit text information (weather reports, stock quotes, airline schedules, and so forth) for display immediately on special receivers or video monitors.

Functions. The RBDS standard consists of a number of codes used to implement specific functions in RBDS receivers.

Clock time (CT) and date code. This code is sent out by the RBDS encoder to update the time display in the radio receiver. The time and date information is sent out continuously and is available to the radio receiver whenever an RBDS station is tuned.

Program identification (PI) code. This code performs a variety of functions within the receiver. For example, the code allows the radio to know what area of the country it is in, and the code carries information relating to the individual station and its programs. One of the most important functions of the program identification code is to allow the receiver to search automatically for an alternative frequency carrying the same program in case of interference to the main frequency.

Broadcasters utilizing this function could have a radio automatically switch to a translator or other station carrying identical programming at an alternative radio frequency when the listener travels out of range of the main signal. The program identification code permits a cellular-type FM receiving environment whereby stations can fully utilize translators and not require listeners to tune manually to another frequency. Similarly, a station has the option to hand off its signal to another station with which it has a simulcast agreement, giving the traveler the illusion of listening to one particular frequency. The program identification code permits broadcasters effectively to set up networks for sports or other programming to allow the listener's radio to follow the network in and out of station coverage areas without retuning manually.

Traffic program (TP) code. This identification code permits the broadcaster to identify its station as one providing traffic information. If a listener desires to scan for stations providing traffic information, the radio will automatically stop at stations transmitting this code. Some receiver manufacturers include a special indicator on the radio showing that traffic information is available from the selected station.

Traffic announcement (TA) code. When a station transmits this identification code, the receiver can, at the listener's option, (1) switch automatically from any audio mode to the traffic announcement, or (2) switch on the traffic announcement automatically when the receiver is in a waiting reception mode and the audio signal is muted.

Program service (PS) code. This code represents actual alphanumeric characters which can be sent to the radio and displayed on a small screen. The typical RBDS radio will display up to eight characters and could provide the listener with the station call letters or slogan. For instance, a particular station may be known to its listeners as Z100. When the radio is tuned to this station, Z100 would be displayed on the small screen.

Radio text (RT) code. This code is primarily intended for home receivers. For safety reasons, it is not practical to display large amounts of text in a moving automobile. The radio text code transmits text for display on printers or computer-type monitors connected to the receiver. Any sort of one-way text can be sent to receivers specially equipped with the proper display device.

Program type (PTY) code. This code is used as a station program format identifier. The broadcaster selects the station's particular format from the available list. The RBDS receiver will then, during a format scan function, stop on the frequency of the station whose particular format has been selected. For instance, if a listener is interested in hearing only stations programming classical music, the radio can be set so that scanning stops only on stations offering a classical music format.

Another important function of the program type code is for emergency alerting. An alarm function can be implemented, which when activated by the broadcaster could "wake up" radios, similar to the traffic announcement code, and provide listeners with instructions during emergency conditions. Typical uses could range from severe weather alerts to nuclear power plant regional evacuation plans. The FCC is encouraging the use of RBDS emergency alerting as part of the United States Emergency Alert System, discussed below.

AM participation. The RBDS standard contains a provision to accommodate AM broadcasters and to allow AM stations to display call signs on RBDS radios and participate in the format-scanning feature. An in-receiver database system (I-RDS) technology is included within the RBDS standard. I-RDS technology employs an in-receiver database read-only memory (ROM) that contains a listing of all United States AM station call signs, frequencies, locations, and formats. A unique feature is its ability to accept update information from the RBDS signal transmitted by an FM station. By updating the database periodically, the AM station data will always be accurate. (For example, if an AM station changes its call sign, the new call-sign data will be updated in the receiver.)

Applications. Applications for RBDS data include sending information to display signs located in nearby hotels, restaurants, malls, moving vehicles, and so forth; uploading computer programs to home and business computer systems; sending data to in-car navigation systems, supplying them with updates on traffic problems and road closures; and digital paging and messaging.

With the establishment of the RBDS standard, opportunities exist to develop new uses for this form of data broadcasting. For instance, one company plans to use RBDS as the data carrier for coupon information transmitted by broadcasters directly to listeners. A specially equipped radio receives the so-called electronic coupons and stores them on a magnetic card. The card is then taken to the advertiser's store where the coupon can be printed and redeemed. Other companies are offering enhanced accuracy for civilian use of the military's Global Positioning System (GPS) by the transmission of local correction data via RBDS. Single-chip FM receivers with RBDS decoders can be designed into a variety of appliances and portable products, allowing data updates to be easily transmitted directly to these devices.

John D. Abel

Emergency Broadcast System and Emergency Alert System

The Emergency Broadcast System (EBS) was devised to provide the President and the federal government, as well as heads of state and local governments or their designated representatives, with a means of communicating with the general public during emergency situations. The Emergency Broadcast System has also been used extensively for the delivery of local and regional emergency information. The Emergency Broadcast System, which originated in the 1950s, was implemented in its present form in 1964. It is currently being modernized and replaced by the Emergency Alert System (EAS).

Emergency Broadcast System organization. The Emergency Broadcast System uses the facilities and personnel of the entire communications industry (broadcast stations and networks, telephone companies, national press services, and national cable programmers) on a voluntary, organized basis to establish an emergency broadcasting network. This network is operated by the industry under government regulations and procedures and in a manner consistent with national security requirements.

A national-level Emergency Action Notification (EAN) will be released only upon presidential authority. Release of the Emergency Action Notification constitutes the notice to all broadcast stations, participating radio and television networks, and national cable programmers of an emergency situation. Upon activation of the national-level Emergency Broadcast System (**Fig. 3***a*), the White House Communication Agency (WHCA), which is responsible for providing all communications for the President under all conditions, will deliver the presidential messages to selected originating points. From these points, the presidential messages or broadcast will be distributed to participating Emergency Broadcast System stations and cable systems via the Emergency Action Notification network.

The FCC has been assigned the overall responsibility for the development of the Emergency Broadcast System. The FCC ensures effective coordination between that agency, the Federal Emergency Management Agency (FEMA), and other government and nongovernment agencies concerned.

System operation. A dedicated network connecting the radio and television broadcasting networks and wire services transmits the Emergency Action Notification messages. Following the Emergency

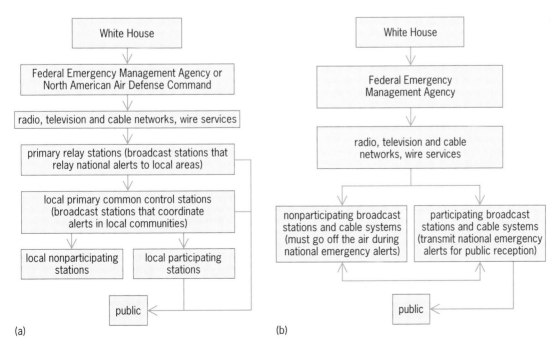

(a) (b)

Fig. 3. Message flow from the White House to the listener during a national alert. (*a***) Emergency Broadcast System. (***b***) Emergency Alert System.**

Action Notification, there is a pause to allow the broadcasting networks to transmit a message over their internal facilities, alerting stations that normal programming will be preempted by the Emergency Broadcast System. Simultaneously, the Emergency Action Notification is transmitted over the respective Radio Wire Teletype Networks to alert those stations equipped to receive this service. Broadcast stations not equipped to receive either network alerting information or Radio Wire Teletype Network transmissions must rely on receipt of the Emergency Action Notification via off-the-air Emergency Broadcast System monitoring of other stations.

Station responsibility. Upon receipt of the Emergency Action Notification, certain actions are taken by all stations: (1) authenticating the Emergency Action Notification; (2) discontinuing normal programming and broadcasting a special announcement alerting the public that important instructions are forthcoming; (3) transmitting the two-tone attention signal; and (4) broadcasting the message that an emergency situation exists, that some stations will remain on the air, and that additional news and information will follow. Those stations required to cease transmitting will so inform their listeners.

Weekly transmission tests. All radio and television stations are required to conduct a weekly test of the two-tone attention signal. The test must be done at a random day and time between 8:30 a.m. and local sunset.

Emergency Alert System. In 1991, the FCC began to consider technical improvements to the Emergency Broadcast System. In December 1994, the FCC announced the creation of the Emergency Alert System to replace the Emergency Broadcast System. Major new features of the new system include (1) a digital system architecture that allows broadcast, cable, satellite, and other services to send and receive alerting information; (2) multiple-source monitoring for emergency alerts; (3) a shortened attention tone; (4) automated and remote-control operations; (5) a weekly test that is unobtrusive to viewers and listeners, as well as a monthly on-air test; (6) the ability to issue alerts in languages other than English; (7) provisions for the hearing and visually impaired; (8) prohibition of the false use of the codes and the attention signal; and (9) a mandated standard protocol for sending messages.

Broadcasters are required to have decoding equipment in the Emergency Broadcast System and are likewise required to have such equipment in the Emergency Alert System. As in the case of the Emergency Broadcast System, stations may choose not to participate in the Emergency Alert System. Nonparticipating stations must still have Emergency Alert System coders and decoders, and must go off the air in the event of a national alert. In addition, cable operators are required to participate in the Emergency Alert System or go off the air.

The Emergency Broadcast System equipment relies on the broadcaster who receives the initial alert to alert other broadcasters in a "daisy chain," which has inherent reliability problems. The Emergency Alert System relies upon multiple-source monitoring (Fig. 3b), requiring broadcasters to monitor two information sources, which are less prone to break down from a single relay failure. The Emergency Alert System also allows automated and remote-control operation and involves only a monthly on-air test using a shortened (8 s as opposed to 20 s) attention signal. Weekly tests are unobtrusive to the listening or viewing public, consisting only of an 8.55-s burst of digital data. Under the transition rules for converting to the Emergency Alert System, broadcasters were required to modify existing Emergency Broadcast System equipment to decode a shortened, 8-s alerting tone. They are required to purchase Emergency Alert System encoders and decoders and place them in service by January 1, 1997. After a transition period during which both the old Emergency Broadcast System decoders and the new Emergency Alert System decoders must be operated, use of two-tone Emergency Broadcast System decoders may stop.

The implementation of the Emergency Alert System system is being phased in progressively until January 1, 1998, when use of the two-tone signal for system activation may be discontinued. FCC rules require Emergency Alert System encoders and decoders to be used as of January 1, 1997, but they also require use of the current two-tone attention signal decoder until January 1, 1998. This precautionary measure ensures that stations will have a backup system while any deficiencies are worked out in the new Emergency Alert System equipment.

Weekly tests under the new system last for 8.55 s and consist only of a digital data burst. Monthly tests consist of the weekly data bursts plus an 8-s two-tone attention signal, and some message text.

The FCC adopted a standard communications protocol for Emergency Alert System messages. This protocol is based on enhancements to the Weather Radio Specific Area Message Encoder (WRSAME) protocol used by the National Weather Service. It includes digitally encoded information about who originated an Emergency Alert System alert or test message, when it was originated, what type of message it is (for example, a tornado warning or a thunderstorm watch), the geographic area affected by the alert or test, and the period during which the alert or test is in effect. The message consists of a digital header, an attention signal, an audio or text message, and an end-of-message (EOM) signal. The digital codes are transmitted by using audio frequency-shift keying (FSK) tones.

For background information SEE AMPLITUDE-MODULATION RADIO; FREQUENCY-MODULATION RADIO; MODULATION; RADIO BROADCASTING; SPREAD SPECTRUM COMMUNICATION; TELECOMMUNICATIONS CIVIL DEFENSE SYSTEM in the McGraw-Hill Encyclopedia of Science & Technology.

Lynn D. Claudy

Bibliography. Federal Communications Commission, *Report and Order and Further Notice of Rule Making,* FO Docket 91-301: *In the Matter of Amendment of Part 73, Subpart G, of the Commission's Rules Regarding the Emergency Broadcast System,* December 9, 1994; National Association of Broadcasters, *NAB Engineering Handbook,* 8th ed., 1992; National Radio Systems Committee, *United States RBDS Standard,* January 8, 1993; *Proceedings of the 2d International Symposium on Digital Audio Broadcasting,* 1994; USA Digital Radio, *USA Digital Radio In-Band On-Channel Radio Description,* 1995.

Ram accelerator

The ram accelerator is a new launcher concept that uses chemical energy to accelerate projectiles to hypersonic speeds. Potential applications include weapons systems, hypersonic aerodynamic testing, and direct launch of space cargo to orbit. Although a ram accelerator outwardly resembles a conventional high-performance gun, its principle of operation is markedly different, being closely related to that of a supersonic air-breathing ramjet engine. A stationary tube, analogous to the cylindrical outer cowling of a ramjet engine (**Fig. 1**), is filled with a premixed gaseous fuel and oxidizer mixture (typically methane, oxygen, and a diluent such as nitrogen) at a pressure of 50–100 atmospheres (5–10 megapascals). Thin plastic diaphragms close off each end of the tube to contain the propellant gas. The projectile is similar in shape to the centerbody

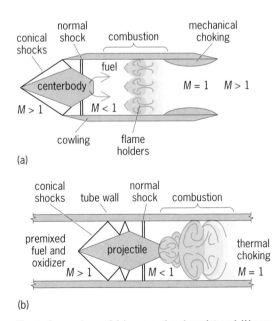

Fig. 1. Comparison of (*a*) conventional ramjet and (*b*) ram accelerator. The tube wall and projectile are analogous to the ramjet's cowling and centerbody, respectively. *M* denotes the Mach number of the gas flow relative to the ramjet or projectile. (*After AIAA Paper 92-3949, AIAA 17th Aerospace Ground Testing Conference, Nashville, Tennessee, July 6–8, 1992*)

of a ramjet, and has a diameter smaller than the bore of the accelerator tube. In the ramjet the centerbody is attached to the cowling; in the ram accelerator the projectile is free to move within the tube.

The operational sequence (**Fig. 2**) is initiated by firing the projectile into the ram accelerator at a speed of about 700–1000 m/s (2300–3300 ft/s) by means of a conventional powder gun or light-gas gun. A lightweight obturator, or piston, in contact with the base of the projectile seals the gun bore during this initial impulse. When the projectile-obturator combination enters the ram accelerator tube, the collision of the obturator with the gas that has been ram-compressed by the rapidly moving projectile generates a shock wave that ignites the gas directly behind the projectile. A stable combustion zone is thus formed that travels with the projectile, generating a wave of high pressure that propels the projectile forward, in a manner analogous to an ocean wave pushing a surfboard. The obturator rapidly falls behind, and does not participate in the acceleration process. To keep the projectile centered in the tube, either the projectile is fabricated with fins that span the bore of the tube or the tube is equipped with several internal rails that run its length.

What distinguishes the ram accelerator from a gun is that its source of energy (the combustible gas mixture) is uniformly distributed throughout the entire length of the accelerator tube, whereas in a gun the energy source is concentrated at the breech as either a charge of gunpowder or a high-pressure gas. During the ram acceleration process the highest pressure in the tube is always at the projectile's base, rather than at the breech as in a gun, and the bulk of the combustion products moves in a rearward direction. Only a small pocket of high-pressure gas exits the tube with the projectile. These characteristics of the ram accelerator result in much more uniform acceleration of the projectile, very high velocity capability, and very little muzzle blast and recoil. Furthermore, the acceleration and muzzle velocity can be easily tailored to specific needs by adjusting the gas composition and fill pressure.

Segmentation. The performance of the ram accelerator can be enhanced by segmenting the tube into several sections or stages, each separated from its neighbor by a thin plastic diaphragm and filled with a different propellant mixture. By selecting the sequence of mixtures in such a manner that the speed of sound and detonation speed increase toward the exit of the ram accelerator, the projectile Mach number can be kept within limits that maximize thrust and efficiency, resulting in high average acceleration. The predicted maximum velocity of the ram accelerator is in the range of 7–9 km/s (23,000–29,000 ft/s) for operation in hydrogen-based propellant mixtures. Although aerodynamic heating of the projectile at the associated high Mach numbers will be severe, its effects can be minimized through the judicious choice of refractory projectile materials.

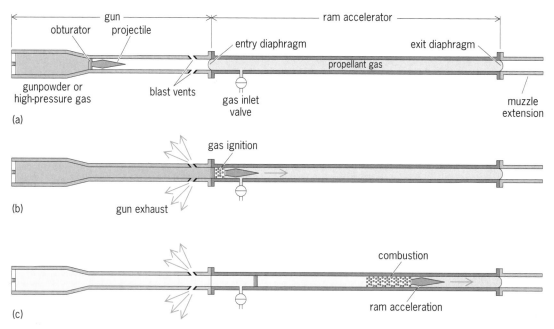

Fig. 2. Operational sequence of ram accelerator. (*a*) Gun is loaded with projectile and pistonlike obturator, and charge of gunpowder or high-pressure gas. The ram accelerator is loaded with combustible gas mixture at 50–100-atm pressure (5–10 MPa). (*b*) Gun fires obturator-projectile combination into ram accelerator. The collision of the obturator with the gas ram-compressed by the moving projectile causes ignition. (*c*) Combustion moves with projectile, sustaining traveling pressure wave that accelerates projectile to high velocity.

Facilities. The ram accelerator was conceived in 1983. An experimental 38-mm (1.5-in.) caliber ram accelerator has been in operation since 1985 and has propelled 75–100-gm (0.17–0.22-lb) projectiles to velocities approaching 3 km/s (9800 ft/s) in 50-atm (5-MPa) propellant mixtures, at average accelerations of 20,000–25,000 g (1 g = 9.8 m/s = 32 ft/s). This performance compares very favorably with that of well-developed advanced gun propulsion concepts, such as liquid propellant guns, electrothermal guns, and electromagnetic rail or coil guns. Several other ram accelerator facilities have also been constructed including a 120-mm (4.7-in.) caliber accelerator (the world's largest) and a 15 × 20-mm (0.78 × 0.6-in.) rectangular-bore installation. Projectile masses from 5 gm (0.18 oz) to 5 kg (11 lb) have been launched to velocities up to 2 km/s (6500 ft/s) in these facilities.

Velocity regimes. Recent work has focused on improving the understanding of the physical principles of ram acceleration, achieving higher velocities, developing robust projectile designs, and studying various near- and long-term applications. Experiments have revealed that acceleration is possible when the projectile is traveling both below and above the detonation speed of the propellant mixture; these velocity regimes are called subdetonative and superdetonative, respectively. The transition from subdetonative to superdetonative operation has been observed to occur smoothly, and is called the transdetonative velocity regime.

Considerable effort has been devoted to the understanding of the transdetonative and superdetonative regimes of operation. It has been observed that as the projectile approaches the detonation speed of the propellant gas, the combustion begins to move forward relative to the projectile, so that some of it takes place in the space between the projectile and the tube wall. As the projectile continues to accelerate to velocities above about 110% of the detonation speed, that is, into the superdetonative regime, the combustion appears to move almost entirely forward of the projectile's base. It is postulated that during this transition from subdetonative to superdetonative operation the combustion changes from purely subsonic to purely supersonic, and may even stabilize into an oblique detonation wave. During transdetonative operation, the velocity regime between approximately 90 and 110% of the detonation speed, it is believed that regions of both subsonic and supersonic combustion coexist.

Flow structure. High spatial resolution pressure measurements of the flow around the projectile have revealed a complex three-dimensional flow structure associated with the centering fins. These observations have been corroborated by high-speed in-bore photography of projectiles through transparent polycarbonate tube sections. Canting of the projectile in the tube has also been detected on a number of occasions, and is likely due to lateral forces and pitching moments generated by nonuniform pressure distributions around the nose and body of the projectile, coupled with erosion or bending of the projectile's centering fins. This problem has been circumvented through the use of titanium alloy as the projectile material, which is significantly stronger and more heat resistant than the magnesium and aluminum alloys commonly used previously.

Operational limits. Experiments performed with a variety of propellant mixtures have demonstrated the existence of operational limits governed by the heat release of the combustion, the Mach number of the projectile, and the projectile material. If the heat generated by the combustion is too small, the driving pressure wave is unable to remain coupled to the projectile and falls behind, resulting in a cessation of thrust. If the heat generated is too high, the driving pressure wave surges ahead of the projectile, causing a sudden deceleration (this phenomenon is called an unstart). Hence, selection of the appropriate propellant mixture is crucial to successful operation. The velocity limits, however, are believed to be related to projectile structural integrity. As the velocity increases, the pressure and aerodynamic heating increase markedly, and are capable of causing structural failure of the projectile. Computations of heat transfer to the nose cone, and to the leading edges and lateral surfaces of the centering fins, have shown that magnesium and aluminum alloys reach their melting points rapidly at these locations, resulting in potentially severe erosion by ablation and in loss of structural strength. Projectiles made of titanium alloy do not suffer these deleterious effects, and have been found to attain higher velocities in experiments.

Flow visualization and modeling. Experimental research has included flow visualization of the subdetonative mode in sacrificial transparent acrylic tubes. The results are recorded by using high-speed cinematography, and have yielded excellent correlations with theory. Computational fluid dynamics (CFD) has been used to develop both steady and unsteady numerical codes that model the subdetonative regime quite well.

Projectile materials and geometries. Aluminum, refractory-coated aluminum, and titanium projectiles have been used in an effort to identify the best projectile material for successful acceleration under the extreme pressures and temperatures characteristic of superdetonative operation. Research has begun on unusual projectile geometries, including hollow cylindrical projectiles and two-dimensional wedge-shaped projectiles. These efforts are aimed at developing capabilities to investigate supersonic reacting flow phenomena on a small scale.

Detonation ignition phenomena. The conditions under which a supersonic blunt body will initiate a detonation wave in a combustible gas at pressures of 5–10 atm are also being investigated. This work is relevant to the ram accelerator because of the similarity of the ignition processes believed to occur in the superdetonative regime. Parallel work with shock tubes is experimentally examining the detonability of various propellant mixtures at pressures up to several tens of atmospheres.

A high-pressure shock tube facility has been used to experimentally investigate the ignition kinetics of high-pressure ram accelerator propellant mixtures in an effort to obtain chemical reaction data at conditions that have not been heretofore explored. Expansion tube experiments have also been initiated to study the details of the reacting flow around stationary ram accelerator projectiles. Diagnostic techniques such as planar laser-induced fluorescence (PLIF) and Schlieren photography are being used.

Applications. Because the ram accelerator's operating principle is independent of dimension, the device can be scaled up or down over a broad range of sizes, as has already been experimentally demonstrated. Consequently, the ram accelerator is readily amenable to a variety of important applications, such as tactical and strategic weapons systems, hypersonic aerodynamic testing, and inexpensive direct launch of payloads into Earth orbit.

In the tactical weapons arena the ram accelerator is expected to be useful against a variety of targets, including fast-moving armor and low-flying supersonic cruise missiles and aircraft. Strategic applications include long-range bombardment and antiballistic missile defense. The flexibility and low mass of the ram accelerator allow it to be considered for installation on various launch platforms, such as land vehicles, ships, and even aircraft. For example, a mobile sea-borne defense system could be placed offshore to provide missile defense for land-based assets, as well as protection for a naval fleet.

The easily controllable acceleration level of the ram accelerator gives it the potential to soft launch, that is, gently accelerate, 30–60-cm (1–2-ft) instrumented models of hypersonic aircraft or atmospheric reentry vehicles to high Mach numbers for aerodynamic testing. Such tests conducted in the past with two-stage light-gas guns have been limited to typical model dimensions of the order of about 25 mm (1 in.). Because the propulsion principles of the ram accelerator are so similar to those of ramjets and supersonic combustion ramjets (scramjets), the device is also an excellent testbed for research on the reacting flow phenomena characteristic of these propulsive devices.

The potential use of the ram accelerator as a space launcher is perhaps the most intriguing. A ram accelerator with an inside diameter of 1 m (3.3 ft) and a length of 3.25 km (2 mi), installed up the side of a mountain at an inclination angle of approximately 15°, could launch a 2000-kg (4400-lb) projectile to 8 km/s (26,000 ft/s) at an average acceleration of about 1000 g (10^4 m/s^2 or 3×10^4 ft/s^2). Although such a high acceleration would preclude the launch of humans, it is acceptable for inanimate payloads such as water, liquid oxygen, rocket fuels, components of space structures, and other matériel necessary for the establishment of a space infrastructure. Only a few percent of the projectile's mass would be consumed by ablation during the atmospheric transit, and the projectile would lose less than 20% of its initial velocity to aerodynamic drag. An on-board rocket ignited at the peak of the projectile's trajectory would be used to provide the necessary additional impulse to place the projectile in a 500-km (300-mi) circular orbit. Because the payload mass fraction of the projectile would be nearly 50% (compared to 3–4% for conventional rocket launch vehicles) and because multiple launches per day would be possible, a ram accelerator would be capable of very low launch costs, nearly two orders of magnitude less than current costs.

Other applications of the ram accelerator, such as hypervelocity impact studies, testing of reentry thermal protection materials, and antiasteroid defense, have also been suggested. Progress in all application areas is predicated on the further development of the velocity capability of this innovative launcher concept.

For background information SEE RAMJET; SHOCK TUBE; SHOCK WAVE; SHOCK-WAVE DISPLAY in the McGraw-Hill Encyclopedia of Science & Technology.

Adam P. Bruckner

Bibliography. A. Hertzberg, A. P. Bruckner, and D. W. Bogdanoff, Ram accelerator: A new chemical method for accelerating projectiles to ultrahigh velocities, *AIAA J.,* 26:195–203, 1988; A. Hertzberg, A. P. Bruckner, and C. Knowlen, Experimental investigation of ram accelerator propulsion modes, *Shock Waves,* 1(1):17–25, January 1991; F. Kuznic, Battle of the big shots, *Air Space,* 8(3):54–61, August–September 1993; G. T. Pope, Ramming speed, *Discover,* 15(3):50–55, March 1994.

Relativity

Relativistic effects become important in applications requiring very accurate timing, time transfer, or synchronization. Many engineering systems are planning to rely on ultra-accurate, stable atomic clocks having fractional frequency stabilities in the range 10^{-12} to 10^{-13}. (Clocks with stabilities of 10^{-12}, or 1 part in 10^{12}, would be in disagreement by less than four-billionths of a second after an hour.) Future developments in technology may lead to clocks of stabilities better than that of the hydrogen maser, which is now less than 10^{-15} over periods of a few hours. The largest relativistic effects on clocks near the Earth's surface are of the order of a few parts in 10^{10}. Although the largest effects tend to cancel each other, residual relativistic effects of order 10^{-12} to 10^{-13} remain. Consequently, in order for engineering systems which depend on accurate timing to achieve the best performance, system design must consider many relativistic effects. In the Global Positioning System (GPS), for example, relativistic effects can be as large as 4 parts in 10^{10}. Three principal systematic relativistic effects impact clock rates and synchronization processes, and must be accounted for in order that such applications work properly: time dilation, gravitational frequency shifts, and the Sagnac effect.

Speed of light and metrology. Relativity enters metrology in a fundamental way through the principle of the constancy of the speed of light, c. This widely accepted principle asserts that the speed of light in free space has the same value in all inertial systems, independent of the motion of the source (or of the observer). The numerical value of c is defined by convention: $c = 299,792,458$ m/s. Much of the motivation for manufacture of more closely spaced transistor components in computer chips is to overcome the limitation in processing speed due to the finite signal propagation speed between processing elements.

With the adoption of a unit of time, the value of c defines the length of the meter. This implies that if a different time unit were chosen, the physical length of the new meter would change. Any physical quantity measured in such new meters would be numerically different. For example, the quantity GM_E/c^2 (G is the newtonian gravitational constant and M_E is the Earth's mass) has the units of length. Since c is fixed, the product GM_E would change if the time unit changed; this is significant in celestial mechanics.

Einstein synchronization. If a light signal leaves clock A at time t_A and arrives at clock B after traveling distance L in vacuum, then clock B can be synchronized with A by setting the arrival time, t_B, equal to $t_A + L/c$. Thus, the existence of the unique speed, c, provides a means for synchronizing clocks at different places; this is called Einstein synchronization. In an inertial frame such synchronization is path independent and can be used to set up a self-consistent network of synchronized clocks.

Applications. Networks of synchronized clocks have many practical applications, including precise positioning, fault detection, and communications.

Precise positioning. It is useful to consider a transmitter which sends out pulses at precisely timed intervals and a traveler who possesses a second clock, synchronized with the transmitter, and who moves directly away from the transmitter. By measuring the pulse-signal arrival times, the traveler's distance from the transmitter can be determined. With two transmitters at different locations, a moving clock can be located in two dimensions; with three transmitters appropriately placed, three position coordinates of the moving clock can be measured. With four transmitters, the moving clock can be located, and also synchronized with the transmitter clocks. In order to determine position to within 10 m (30 ft), synchronization must be maintained to better than 30 nanoseconds. Such a scheme is implemented in the GPS, with accurately synchronized atomic clocks placed in satellites. With a sufficiently dense network of transmitters on the Earth's surface and receivers in automobiles, existing roads might possibly handle much more traffic. Other positioning applications include search and rescue, mapping and surveying, and the study of motions of the Earth's crustal plates. SEE SATELLITE NAVIGATION SYSTEMS.

Fault detection. If a long power line has synchronized clocks at the ends, then the location of a fault in the line can be determined by monitoring the arrival times, at the ends of the line, of signals caused by the fault.

Communications. High-speed data transfer involves multiplexing of numerous signals, each fitting into a narrowly defined time slot. More efficient transfer can be achieved if clocks at the ends of the data link are precisely synchronized by means independent of the signals.

Time dilation. The phenomenon of time dilation causes a moving clock to beat more slowly as it moves through a network of synchronized clocks which are at rest in an inertial frame. If $\Delta\tau$ is the elapsed time on the moving clock, which is assumed to have uniform velocity v, and Δt is the elapsed time in the inertial frame, then these quantities are related by Eq. (1). Thus, a single moving clock appears

$$\Delta\tau = \sqrt{1 - v^2/c^2}\,\Delta t \qquad (1)$$

to beat more slowly than synchronized clocks at rest. This effect is sometimes also called the second-order Doppler shift. The factor $\sqrt{1 - v^2/c^2}$ in Eq. (1) differs from unity by approximately $-v^2/2c^2$. The Earth's rotation gives clocks at rest on the Equator a velocity of 465 m/s (1040 mi/h); relative to clocks in a nonrotating local inertial frame which is freely falling with the Earth (the Earth-centered inertial or ECI frame), Earth-fixed clocks will beat more slowly by about 1 part in 10^{12}. The velocity, v_s, of an atomic clock in a satellite may be greater than $10^{-5}c$, and its frequency may be slowed by as much as 3 parts in 10^{10}; for atomic clocks these are very large effects. For practical purposes it is the difference $(v^2 - v_s^2)/2c^2$ which is important. For a clock in a circular equatorial orbit, relativity theory predicts Doppler shifts varying from 1 part in 10^{12} to more than 3 parts in 10^{10}.

Gravitational frequency shifts. Two clocks whose gravitational potentials V differ by ΔV will beat at different frequencies; the fractional frequency difference is given by Eq. (2). Thus, the clock at the

$$\frac{\Delta f}{f} = \frac{\Delta V}{c^2} \qquad (2)$$

higher potential beats more rapidly. Near the Earth's surface the gravitational potential difference between two clocks, of altitude difference h, is approximately gh, where $g \simeq 9.8$ m/s^2 = 32.2 ft/s^2 is the acceleration of gravity. For example, the altitude of the frequency standards laboratory at the National Institute of Standards and Technology (NIST) in Boulder, Colorado, is 1654 m (5427 ft), causing a frequency shift of 1.8 parts in 10^{13}, or 15.5 ns per day, relative to mean sea level. Consequently the contribution of the NIST time standard to Universal Coordinated Time requires that a paper correction of −15.5 ns/day be applied to the NIST clock before it is compared to time standards at mean sea level.

A clock in a satellite orbiting the Earth at 100-km (62-mi) altitude compared to one at sea level has a fractional frequency shift $\Delta V/c^2 = 1.08 \times 10^{-11}$. For clocks in low Earth-orbiting satellites the quadrupole potential of the Earth, because of the Earth's oblateness, must also be considered for accurate prediction of the frequency shift.

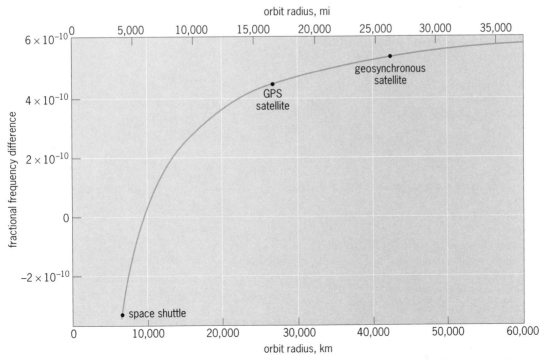

Fig. 1. Plot of the fractional frequency shift of an orbiting clock, relative to an Earth-fixed clock as a function of orbit radius, due to second-order Doppler and gravitational frequency shifts combined.

The combined time dilation and gravitational frequency shifts are plotted in **Fig. 1** versus orbit radius. The redshift due to time dilation tends to cancel the blueshift due to gravitation; at an orbit radius near 9547 km (5932 mi) these effects cancel precisely. The fractional frequency shift of a clock in a GPS satellite would be +4.46475 parts in 10^{10}, and if not compensated such clocks would drift ahead of Earth-fixed clocks by more than 38,000 ns/day. Therefore, to compensate for these relativistic effects the 10.23-MHz GPS satellite clock frequency is set lower before launch, to 10.229,999,999,543 MHz. GPS satellite clocks, at the same orbit altitude in circular orbits, all run at exactly the same rate as Earth-fixed clocks.

Calculations of combined frequency shifts are simplified as a result of a remarkable cancellation: All clocks at rest on the rotating Earth at mean sea level beat at the same rate. The Earth's surface has the approximate shape of a hydrostatic equipotential in the Earth-fixed rotating frame. Clocks near one pole will be closer to the Earth's center than clocks on the Equator, and will therefore be subject to a gravitational redshift. However, viewed from the nonrotating ECI frame, such clocks are closer to the rotation axis and are moving more slowly than clocks near the Equator, and so suffer less second-order Doppler shift. These effects cancel; thus it is possible to construct a network of standard clocks on the Earth's geoid, all beating at the same rate. However, to synchronize these clocks consistently it is necessary to correct for the Sagnac effect, due to the Earth's rotation.

If a satellite orbit has eccentricity, additional gravitational and Doppler-rate shifts occur as the satellite approaches and recedes from Earth. Thus an additional correction must be applied to orbiting clocks to synchronize them with Earth-fixed clocks. For a GPS orbit of eccentricity 0.01, the correction has an amplitude of 23 ns.

Sagnac effect. Another important relativistic effect occurs on the rotating Earth. If two clocks are fixed a small east-west distance x apart on the Earth, then, from a nonrotating frame, they will be moving with approximately equal speeds, $v = \omega r$, where ω is the Earth's angular rotation rate and r is the distance of the clocks from the rotation axis. If a clock synchronization process along x were carried out by Earth-fixed observers who ignored the Earth's rotation, then, as a consequence of the relativity of simultaneity, the two clocks would not be synchronous when viewed from the nonrotating frame. The magnitude of the synchronization discrepancy is approximately $vx/c^2 = (2\omega/c^2)(rx/2)$. In a rotating frame, this is called the Sagnac effect. The quantity $rx/2$ may be visualized as the equatorial projection of the area swept out by a vector from the Earth's center that follows the synchronization process. In general, the discrepancy depends on the path along which the light signals travel relative to the rotation axis. The same discrepancy occurs whether the synchronization is carried out by using slowly moving portable clocks, electromagnetic signals in a vacuum, or signals in an optical fiber. An acceptable way to avoid this problem is to use clocks synchronized in the underlying ECI frame.

Then, in synchronizing clocks on the rotating Earth, in order to avoid path-dependent discrepancies it is necessary to apply a correction, arising from the Sagnac effect, during the synchronization process. The correction term to be applied is given by Eq. (3), where A_E is the projected area on the

$$\Delta t = (2\omega/c^2) \times A_E = (1.6227 \times 10^{-21}\ \text{s/m}^2)A_E \quad (3)$$

equatorial plane swept out by a vector whose tail is at the center of the Earth and whose head follows the synchronizing signal or portable clock.

Figure 2 illustrates the relationship between the path of a portable clock or synchronizing signal, and the projected area from which the Sagnac correction may be calculated. For a synchronization path which traverses the Equator once from east to west, the correction is 207.3 ns. Along a meridian the correction of Eq. (3) vanishes. A synchronization link which goes from San Francisco directly to New York, for example, will disagree with a San Francisco–Miami–New York link by about 11 ns. This might become significant in an optical communications network operating at 10^{15} Hz. For a signal from a clock in a geosynchronous satellite to a ground station the correction can be larger than 210 ns.

Negligible effect of other bodies. The gravitational frequency shift of an Earth-orbiting clock due to the Sun's potential at first appears to be large: it is $-g_{\text{Sun}}r/c^2$, where r is the satellite orbit radius, and $g_{\text{Sun}} \simeq GM_S/R^2$, where M_S is the Sun's mass and $R \simeq 1.495 \times 10^{11}$ m $= 9.3 \times 10^8$ mi is the radius of the Earth's orbit about the Sun. This term by itself would give a frequency shift $\Delta f/f \simeq 1.7 \times 10^{-12}$ for a GPS clock. Einstein's principle of equivalence, however, implies that an observer in free fall in the grav-

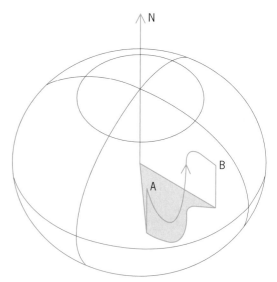

Fig. 2. Relationship between the path (from A to B on the Earth's surface) of a sychronizing signal and the projected area from which the Sagnac correction may be calculated. The Sagnac effect is proportional to the shaded area.

itational field of the solar system cannot sense the presence of external gravitational fields; these are transformed away by acceleration of the local system due to the external fields themselves. The observer's local clocks are synchronized differently from clocks at rest in the solar system. As a result, observable external gravitational potentials are reduced by an additional factor $\approx r/R$. For a clock in a GPS satellite the residual fractional frequency shift, $GM_S r^2/c^2 R^3$, is only a few parts in 10^{16} and is negligible. The principle of equivalence applies generally to a freely falling body under the influence of any number of external bodies. Therefore, at levels of stability and accuracy attained by presently existing clocks, other solar-system bodies have negligible influence on local clock-rate comparisons.

The GPS works well partly because relativistic effects have been carefully taken into account. In the GPS, about a dozen separate sources of relativistic effects must be considered. The success of the GPS provides continual confirmation of the principles of special and general relativity. With future improvements in clock technology, relativistic effects will become increasingly important in other engineering systems.

For background information SEE ATOMIC CLOCK; DOPPLER EFFECT; GEODESY; GRAVITATION; RELATIVITY; SATELLITE NAVIGATION SYSTEMS; TIME in the McGraw-Hill Encyclopedia of Science & Technology.

Neil Ashby

Bibliography. N. Ashby, Relativity in the future of engineering, *IEEE Trans. Instrument. Measur.,* 43:505–514, 1994; *The Global Positioning System: A Shared National Asset,* 1995; B. W. Parkinson, J. J. Spilker, Jr., and P. Axelrad, *The Global Positioning System: Theory and Application,* 1995.

Risk assessment

Risk assessment is a process used to predict the likelihood of unwanted events. The principles of risk assessment can be used to anticipate explosions, workplace injuries, failure of machine parts, natural disaster, injury or death due to voluntary activities (such as football, sky diving, or hunting), diseases caused by chemical exposure, death due to natural causes, and other events. Health risk assessment uses data from animal studies and human epidemiology, combined with information about an agent, to predict the likelihood of a particular adverse response (for example, liver cancer) in a specific human population.

Assessment of animal data to predict human health risk is not new. Regulatory agencies have used such information for almost 40 years, partially in response to public concern over the effects of chemical exposure. In addition, the industrial community has been concerned about ensuring that adequate scientific bases support regulation of their processes.

The health risk assessment process is separated into four distinct fields of analysis: hazard identification, dose-response assessment, exposure assessment, and risk characterization. Hazard identification is the first and most easily recognized step in risk assessment. It is the process of determining whether exposure to a given agent could increase the likelihood of an adverse health effect. Such adverse health effects include cancer and birth defects. Dose-response evaluations define the relationships between the dose of an agent and the probability of a specific adverse effect in laboratory animals. Exposure assessment addresses the probability of uptake from the environment by any combination of oral, inhalational, or dermal routes. Risk characterization summarizes and interprets the information collected during the previous activities and identifies the limitations and uncertainties in the risk estimates. It is the risk characterization upon which the decision maker, that is, the risk manager, relies.

Hazard identification. The purpose of hazard identification is to determine whether there is a causal relationship between a given agent and injury to humans or the environment. Information on the agent may come from laboratory studies in which test animals are deliberately exposed to toxic materials, or from other sources such as studies of chemical exposure in the workplace.

In recent years consideration has been given to whether all animal carcinogens ought to be considered human carcinogens. Past practice conservatively assumed that animal carcinogens were likely to be human carcinogens. Currently this assumption is being evaluated on the basis of several factors, such as tumor type, species tested, metabolism, pharmacokinetics, and epidemiological experience. These and other factors are weighed to predict whether a specific chemical poses a significant hazard to humans at doses to which they might reasonably be exposed.

An important realization in recent years is that not all studies may be appropriately compared. This recognition has resulted in a weight-of-evidence approach.

Dose-response assessment. Dose-response assessments require an extrapolation from the high doses typically administered to test animals in experiments to the low doses expected in human occupational or environmental exposures. Despite the great uncertainty involved, low-dose extrapolation models have become the backbone of the dose-response assessment for carcinogens.

Toxicologists acknowledge their limited ability to estimate risks associated with typical levels of environmental exposure based on the results of the standard rodent bioassay. Accordingly, they rely on a model or theory to estimate the response of humans to doses often 1000 times lower than the lowest animal dose tested. The most scientifically rigorous and likely the most valid models for bridging the gap between animal studies and estimating safe levels of human exposure are the physiologically based pharmacokinetic models.

Exposure assessment. The exposure assessment moves the analysis from the study of known populations to the task of identifying and characterizing exposure in hypothetical populations (for example, estimating exposure to children who might trespass on a hazardous waste site). Although understanding the transport and distribution of a chemical among a hypothetical population can be complicated, the concentration of a chemical in the environment and its resulting uptake by exposed persons can be adequately quantified. For example, the uptake of some chemicals can be measured by analysis of body fluids, excrement, or hair. Such analyses provide better estimates of uptake than the use of models alone.

However, direct measurement is not a substitute for predicting the frequency of exposure for a certain population and the severity of the exposure. Direct measurement may reveal only that the prediction was high or low for a specific time period. Further, the body of a person actually exposed may not retain the chemical long enough for measurements to detect it. Thus, although modeling has the appearance of overpredicting exposure, direct measurement may greatly underpredict exposure.

Risk assessors also determine the statistical confidence of risk assessments and the sensitivity of those estimates to important exposure variables. The Monte Carlo sensitivity analysis provides the kind of information that risk managers and the public need in order to form decisions and hold debate regarding the necessary response to a perceived health threat; the results of a Monte Carlo analysis indicate to what extent a given variable should be accurate. That is, it may be possible to allow a variable an order of magnitude without significantly affecting the accuracy of the calculation.

Risk characterization. Risk characterization reaches conclusions about the overall risk in a way that builds on the other three analyses. Characterizing risk has consistently been the weakest component of a risk assessment because it requires the assessor to draw on numerous aspects of science and regulatory policy to properly describe whether a significant human hazard exists in a specific setting. A thorough characterization will discuss background concentrations of the chemical in the environment and in human tissue. It should also discuss the pharmacokinetic differences between the animal test species and humans, the impact of using a physiologically based pharmacokinetic model or a biologically based model, the effect of selecting specific exposure parameters (sensitivity analysis), as well as other factors that can influence the estimated risks.

Other considerations. Experience in risk assessment has led to several major changes in the field. For example, the Environmental Protection Agency has published two new guidelines, one on exposure assessment and the other on interpreta-

tion of cancer bioassays for use in risk assessment. Low-dose extrapolation approaches for noncancer data have also been changed. The results utilize more rigorous scientific techniques and are therefore more precise than the previous EPA approach.

Another major development in risk assessment is the increased use of quantitative uncertainty analysis. Although the use of uncertainty modeling, such as with Monte Carlo or other mathematical tools, is not new to risk assessment, it is becoming more widely used.

Refinements are likely in all four fields of risk assessment. However, the biggest change will likely be in the use of risk assessment data by risk managers. Risk managers and the public will have access to information that is more precise and more reasonably characterized than the abstract notion additional cancer risk, common for the last several years. Characterizing risk assessment information not only in terms of increased likelihood of adverse health effects but also in terms of strength of the analysis and likelihood of an improvement in health will lead to better-reasoned decisions.

For background information SEE EPIDEMIOLOGY; MODEL THEORY; MONTE CARLO METHOD; RISK ANALYSIS; TOXICOLOGY in the McGraw-Hill Encyclopedia of Science & Technology.

Brian J. Pinkowski

Bibliography. B. Ames, Pesticides, risk and applesauce, *Science*, 244:755–757, 1989; D. Patton, The ABCs of risk assessment, *EPA J.*, pp. 10–22, Jan.–March 1993; D. Paustenbach, The practice of health risk assessment in the United States (1975–1995): How the U.S. and other countries can benefit from that experience, *Hum. Ecol. Risk Assess.*, 1:29–79, 1995.

River

Rivers and adjacent floodplains function as freshwater sources, productive ecosystems, and locations for economic development. These functions can suffer if natural forces or human activities change existing sedimentation patterns. Mining sediments can be used as tracers to correlate changes in river behavior with other watershed factors and to aid in the scientific evaluation of sedimentation problems.

Rationale. Age control is based on the assumption that the abundance of mining sediments in a river is highly correlated with upstream mine production. Thus, river sediments deposited during periods of peak mine production are expected to contain the highest concentrations of mining sediments.

Mine tailings produced during ore-milling operations provide the mining tracers used for sediment dating purposes. In the older mining districts, tailings dumps were poorly contained, with the result that large volumes of tailings were quickly mobilized by erosion and slope failure and directed into local drainage networks during periods of active mining. Tailing dumps usually contain sand- and gravel-sized materials of ore and gangue minerals and associated bedrock. Additionally, fine-grained clay and silt particles are introduced to water courses by mill and mine effluent discharges.

In places where large amounts of tailings have entered a river, gross differences in color, texture, and mineral composition can be used to identify the sedimentation patterns of mining sediments. Tailings composition usually differs from that of the sediments normally transported by the river, making it possible to identify mining sediment deposits by visual means. Geochemical detection methods must be used in rivers where mining sediment signals are highly diluted by sediments from other sources. These methods commonly involve the measurement of trace metal levels in river deposit samples. These metals are found in very high concentrations in the mill tailings and are related to the type of mineralization encountered at the mine. Unfortunately, these same metals are responsible for severe contamination problems in many mining districts.

Applications. Mining sediment tracers have been used to monitor long-term sediment transport rates, document channel position changes, and calculate floodplain sedimentation rates. Important information on how river sediment transport rates adjust to episodic sediment inputs has come from research in mining districts. For example, hydraulic gold mining activities in California's Sierra Nevada released very large quantities of tailings to rivers during the nineteenth century. The deposition of this sediment increased flooding hazards in downstream communities, and forced the government to construct extensive flood protection works and eliminate hydraulic mining practices by 1890. In this example, the hydraulic mining sediment was more rounded and contained larger percentages of white vein quartz than premining river sediments. Hence, the contribution of hydraulic mine tailings to channel deposits of different ages could be quantified. This study showed that while tailings contributions to channel sediment transport have generally decreased through time, they still remain high near upstream mining districts and represent about half of the channel sediment being transported today in downstream reaches. Clearly, this work focuses attention on the long-term effects of human activities on river systems.

Geochemical indicators of mining activities have also been used to chronicle the spatial patterns of channel adjustments to climate and anthropogenic factors. For example, the ages of channel and floodplain deposits along the River South Tyne, England, were estimated by correlating the lead and zinc concentrations in the deposit with upstream mining activities that occurred from 1700 to 1938. Interestingly, the age and textural composition of the deposits varied greatly across the valley floor. Results of the study showed that frequent episodes of channel erosion and deposition occur in spatially distinct sedimentation zones. These zones are separated by stable reaches that have changed little over the past century. This example stresses the discontinuous and episodic nature of sediment transport and channel adjustments in river systems.

Geochemical tracers can also be used to develop detailed records of floodplain sedimentation along rivers where sedimentary units are arranged in horizontally layered sequences. Overbank floodplain cores collected from along the Shullsburg Branch, Wisconsin, contain a good record of mining-related zinc contamination. The major changes in the zinc profile are related to the intensity of upstream ore production; therefore they can be dated by correlations with published government records. If these dates are superimposed on the profile trend, the sedimentation rate (in this case, centimeters per year) can be calculated for the time interval between the dated layers. The high floodplain deposition rates in the Shullsburg Branch between 1890 and 1925 are related to higher soil erosion rates and flood frequency caused by both land clearing for agricultural purposes by early settlers and above-average rainfall. Sedimentation rates have since decreased due to increased channel capacity and implementation of soil conservation practices. However, current rates are still almost 10 times greater than found in the presettlement period (prior to 1820). This study emphasizes the combined influence of both climate and human factors on changes in river behavior.

There is great potential for this method to be used in many parts of the world for which the scientific knowledge of the external controls on river behavior is scarce. Often this type of research is combined with investigations that focus on the toxic consequences of mining contamination. In this way, efforts to date river deposits can also be applied to understanding dispersal processes and spatial distributions of toxic substances in river systems. Interpretation of the results of mining sediment dating studies should be based on local conditions, since mining sediments do not always act as passive tracers. The introduction of large quantities of tailings may destabilize river channels, thus making it more difficult to isolate the effects of other environmental disturbances on river behavior.

Limitations. The usefulness of this dating technique depends largely on the degree to which mining sediment concentrations are related to the temporal fluctuations of upstream mine production. Hence, limitations of the method generally relate to the strength of the mining signal and the efficiency of mining sediment dispersal.

For adequate detection of the mining signal, tailings indicator concentrations (for example, mineral counts or metal levels) must far exceed levels found in premining sediments. The greater the difference in concentration between the two, the easier it is to measure minor variations in the mining signal. For example, geochemical mining indicators are usually found at concentrations more than 100 times greater than those found in premining sediments, whereas indicators based on texture or mineral properties tend to vary by less than 20 times. Hence, the resolution of geochemical detection methods will be about 10 times greater than those using the less variant physical properties.

For geochemical detection methods to work, the mining signal must be stable through time. Although most of the base metals released by mining activities tend to form very strong bonds with sediment particles, this relationship should be verified by independent means. Some elements, minerals, and organic materials may be somewhat mobile in the sediment column because of the ground-water seepage and weathering processes. The dissolution or redistribution of mining indicators could result in the reduction of the mine signal peaks and ambiguous age trends.

Mining sediments include a wide range of particle types that are transported in different ways. For example, silt particles tend to be transported downstream at a faster rate than gravels, because flows having sufficient velocity to transport the finer-grained particles occur more frequently. Variations in hydrologic conditions may change the composition of the mining sediments during transport, and so change the tracer behavior. Strong correlations between physical properties (for example, clay percentage or organic matter content) and trace-metal indicators may indicate that signal strength is being controlled more by the sorting process than by mining sources.

Recent studies suggest that most of the tailings materials discharged to a river are initially deposited in downstream reaches and stored in floodplains and channel bars for some period. However, subsequent erosion and geochemical releases may continue to redistribute mining indicators to new sedimentary locations long after the cessation of upstream mining activities. Hence, indicator profiles in deposits of postmining age may not be related to the mining history originally selected for age control, since the patterns of sediment reworking will also affect transport rates. This situation may explain why attempts to date laterally deposited floodplain features such as those found on the inside of meander bends have met with only limited success. These deposits tend to be inundated by floods more frequently than the higher-elevation floodplains and terraces, and they will act as sinks for recycled mining sediments.

For background information SEE DEPOSITIONAL SYSTEMS AND ENVIRONMENTS; RIVER in the McGraw-Hill Encyclopedia of Science & Technology.

Robert T. Pavlowsky

Bibliography. S. B. Bradley, Incorporation of metalliferous sediments from historic mining into river floodplains, *GeoJournal*, 19.1:5–14, 1989; L. A. James, Quartz concentration as an index of sediment mixing: Hydraulic mine-tailings in the Sierra Nevada, California, *Geomorphology*, 4:125–144, 1991; J. C. Knox, Historical valley floor sedimentation in the Upper Mississippi Valley, *Ann. Ass. Amer. Geog.*, 77(2):224–244, 1987; M. G. Macklin and J. Lewin, Sediment transfer and transformation of an alluvial valley floor: The River South Tyne, Northumbria, U.K., *Earth Surf. Proc. Landforms*, 14:233–246, 1989.

Road pavements

Highway managers have to decide when, where, and how to maintain and repair the road pavements under their care. In an era of rising costs, limited budgets, and increased use of aging highways, the manager must create a pavement preservation program that makes the best use of limited resources over the long term. Thus, individual projects are analyzed to determine what treatment schemes would be most cost effective, and then a particular scheme is selected for each project based on what is best for the network as a whole.

Program development. Individual projects are evaluated over a life-cycle period of at least 30 years. Following successful European experiences involving pavement design and construction, even longer-lasting designs (for example, 50 years) are becoming the trend in the United States. Whatever the selected analysis period, project life-cycle analysis requires determination and evaluation of existing pavement condition (cracking, faulting, rutting, and so forth), identification of suitable treatment repair options (for example, minor versus major rehabilitation), projection of the resulting future condition and subsequent deterioration, future treatments, and estimation of all associated costs. Numerous analytical methodologies are available to pursue each of these functions, and vary considerably in complexity and sophistication. Most agencies have their own customized approach to each step, including identification of specific treatment types and life expectancies appropriate for local conditions. As an example, the **table** lists selected treatments and lives for interstate service conditions in New York State. Treatment schemes are developed that identify types, magnitude, and frequency of maintenance during the analysis period. Variations can be developed by considering alternative combinations of treatments or by changing the timing of a treatment. Alternative treatment schemes are typically evaluated by using either cost-benefit or life-cycle cost analysis to determine which treatment scheme is the most cost effective over the entire serviceable life of the pavement project. Life-cycle cost analysis also facilitates evaluation of the effects of inflation and interest rates, which are significant for capital investment decisions. Recent advances in analytical models and computational tools employ optimization methods that facilitate life-cycle cost analysis of both individual pavement sections and groups of sections.

Optimization methods. Often what is most cost effective for a particular project may not be optimal

Typical maintenance and rehabilitation treatments and their life expectancy

Pavement type	Treatment category	Description	Life expectancy, years
Rigid	Preventive maintenance	Reseal joints	6
	Minor rehabilitation	Grind faults and reseal joints	5
		2.5-in. asphalt overlay	10
	Major rehabilitation	4-in. asphalt overlay	12
		5-in. asphalt overlay with crack and seal	14
Flexible	Preventive maintenance	Cleaning and sealing cracks	2
	Minor rehabilitation	Single overlay (shim course; 1.5-in. top course)	7
		Cold-mill single overlay* (1.5-in. top course)	7
	Major rehabilitation	Structural overlay (3-in. binder; 1.5-in. top course)	11
		Cold-mill structural overlay (3-in. binder; 1.25-in. top course)	11

* Pavement over existing road surface.

for the network as a whole. A simplified example of a hypothetical network consisting of two projects serves as an illustration of this point. Project A, representing 75% of the entire network, is currently in acceptable condition. The remainder of the network, Project B, is in very poor condition. Independent life-cycle cost analysis of the two projects indicates that this year Project A should have preventive maintenance that costs x dollars, and Project B should have major rehabilitation that costs $12x$ dollars. However, if the available annual budget is only $10x$ dollars, it is clear that the best treatment is not affordable for both projects. The notion that the optimal solution to the whole problem is different from the sum of the optimal solutions to each part (suboptimal solutions) is becoming better understood by practitioners. Optimization methods and associated operations research techniques are currently the state-of-the-art approach to simultaneously evaluating the trade-offs between condition, funding, and timing.

Three optimization approaches that have been successfully applied to pavement-treatment cost analysis and scheduling are Markov, semi-Markov, and state-increment methods. These methods use pavement states as an approach to characterizing condition. A state is a combination of specific levels of the variables (called the state variables) that completely describe pavement condition. Examples of state variables include pavement type, surface condition indexes, and traffic volume. The procedure of identifying treatment alternatives based on characterized conditions involves assigning several treatments for each state. Performance is predicted by using probabilistic models that describe the state transitions. Pavement state transition is an event that describes the change of state (value of at least one state variable) for the pavement as a consequence of a treatment action or deterioration over time. Subsequently, long-term treatment plans are established by combining a sequence of states and their assigned treatments. Economic analysis is performed by using a dynamic programming algorithm for the Markov and state-increment methods and a simple enumeration procedure for the semi-Markov method.

Most agencies use a computerized pavement management system to analyze highway needs and to allocate cost effectively the available resources among competing projects. A pavement management system combines detailed data on pavement conditions, treatment costs, and total life expectancy with a variety of computerized analytical models to predict service lives, evaluate life-cycle costs of alternative treatment schemes, and establish maintenance plans.

Life-cycle cost analysis. Long-term economic assessment of alternative treatment actions is an important capability of a pavement management system. The life-cycle cost analysis technique can be used to provide an economic assessment of several design (treatment) alternatives for specific projects. It involves reducing current and future costs of various treatment plans to an equivalent basis to indicate which alternative is most economical over the long term.

The recommended procedure for a life-cycle cost analysis involves three major steps: (1) identification of the project to be evaluated, (2) development of a treatment schedule for each alternative to be analyzed, and (3) conversion of the costs for each treatment schedule to equivalent dollars.

For the purposes of this analysis, engineers must review condition and other pavement characteristics, treatment requirements, expected future performance, and so forth, to define feasible construction projects. The treatment schedules developed for each project must depict the first year of analysis; the duration of the analysis period; and the type, timing, and cost of all treatments that are expected during the analysis period. Estimates of nontreatment costs such as salvage value, user costs, and design costs must also be specified wherever appropriate.

To make an economic evaluation of competing treatments, each of the treatment schedules must be analyzed to find the present worth and the equivalent uniform annual cost. This evaluation is done by converting the total cost associated with each year in the analysis period to either present worth or equivalent uniform annual cost. Appropriate choice of interest and inflation rates must be made in order to account for the so-called time

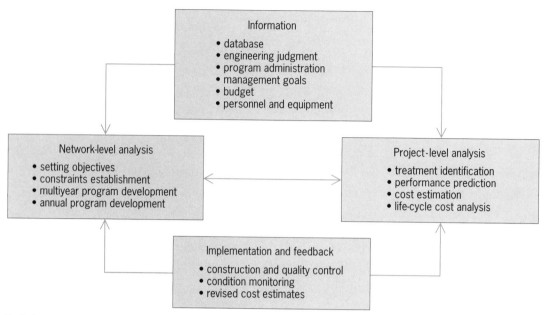

Typical pavement management system.

value of money while doing the cost conversions.

It is important to evaluate the effect of variability in factors such as treatment service lives and interest and inflation rates on life-cycle costs. This evaluation is made by systematically varying the values for these factors and observing changes in the life-cycle cost analysis results. A sensitivity analysis of various factors such as treatment service lives and interest and inflation rates is done by calculating life-cycle costs over a range of values for these factors. Finally, when interpreting the results of a life-cycle cost analysis, it is important to note that the obtained life-cycle costs can reflect only those quantitative factors that were included in the analysis. Other factors, such as availability of required resources in a given year and overall condition and safety improvement to achieve system goals, often significantly impact on the final choice of treatment. However, these factors cannot be satisfactorily captured by the life-cycle cost analysis. A more complete system-level analysis and optimization procedure is required to evaluate treatment-project trade-off scenarios with constraints in place.

Pavement management system. A pavement management system (see **illus.**) includes several interacting modules that can be grouped into an information system; models and computational tools to support project- and network-level analyses and decision making; and an implementation and feedback system that ensures continuous updating and improvement of both data and models. The core of the information system is typically a database of performance, history, cost, budget, geometry, environment, and related data items, which may be supplemented by a knowledge base (either explicit or implicit) of agency practices and policies. The information sys-

tem serves the two major analytical modules, which address network- and project-level analysis.

The network-level analysis module comprises tools for program planning and policy evaluation for an entire network. Optimization and other analytical models for long-term budget allocation and for selection and scheduling of projects are part of this module. The project-level analysis module includes tools and models for analyzing pavement condition, identifying suitable treatment alternatives, predicting performance, estimating treatment costs, analyzing life-cycle cost, detailing design, and implementing construction. Agencies may customize the models and specific data types used for pavement management systems to suit their needs and capabilities. For example, complex mathematical optimization models are available, but they are analytically and computationally demanding. Consequently, less rigorous techniques such as ranking, prioritization, and simple suboptimization models are in common use.

For background information SEE OPERATIONS RESEARCH; OPTIMIZATION; PAVEMENT; STOCHASTIC PROCESS in the McGraw-Hill Encyclopedia of Science & Technology.

Dimitri A. Grivas

Bibliography. American Association of State Highway and Transportation Officials, *AASHTO Guide for Pavement Management Systems*, 1990; R. Haas, W. R. Hudson, and J. Zaniewski, *Modern Pavement Management*, 1994; Transportation Research Board, *Methods of Cost-Effectiveness Analysis for Highway Projects*, National Cooperative Highway Research Program Synthesis of Highway Practice, vol. 142, 1988; Transportation Research Board, *Systematic Approach to Maintenance*, Trans. Res. Rec., vol. 1183, 1988.

Satellite navigation systems

The agricultural industry is increasingly employing new technologies in order to improve economic benefits as well as to enhance environmental stewardship. One such technology is the Global Positioning System (GPS), a satellite-based navigation system that has been deployed by the U.S. Department of Defense. Through the integration of GPS with farm machinery or crop-spraying aircraft, precise location information can be used to guide these vehicles during operations. Animals and livestock can also be outfitted with devices to monitor their movements and grazing patterns.

The GPS consists of 24 satellites at an altitude of 12,000 mi (20,000 km), and can currently provide a global navigation capability of approximately 300 ft (100 m) for 24 h each day in all types of weather. An improvement in accuracy can be achieved by using the differential GPS (DGPS) technique, in which error corrections are transmitted to the user (for example, a farm vehicle) from a reference GPS receiver situated at a known location. These corrections are then used to improve the user's accuracy to the few-meter level rather easily, and to the centimeter level in some cases. Various communications systems can be used for the transmission of the GPS errors with satellite links, very high frequency (VHF) data radios and frequency-modulation (FM) subcarriers being the most common.

Site-specific farming. In the past, farmers generally treated an entire field with the same quantity and type of fertilizer, without regard for possible variations in topography, salinity, and soil type within the field. The overall crop productivity can be improved through site-specific farming, in which a variable amount and type of fertilizer is applied throughout the area, as determined from a prescription map. As the fertilizer prescription is applied to smaller parcels of land, the overall economic gain is improved since very local conditions can be taken into account.

GPS plays a key role in site-specific farming since it provides the location information which links the prescription to the field. In order to generate a prescription map, data on the crop productivity must first be available. These data can be collected during harvest by adding a yield monitor and GPS receiver to the combine. As the crop is harvested, yield data are continuously recorded and tagged with the combine's position, which is available from DGPS with an accuracy of 3 ft (1 m) or better.

After the harvest is complete, a map can be generated which shows productivity as a function of location on the field. The productivity values can be color coded to clearly show the variability. Agriculturists can then use this information along with other soil data, such as salinity and soil type, to develop the prescription map. This map is stored in a computer database, and contains the type and amount of fertilizer to be applied the following spring. The fertilizing machine must be outfitted with a variable-rate fertilizer which receives prescription instructions from an on-board computer system connected to the GPS equipment. Using DGPS as the guidance system, the operator keeps the vehicle on a preprogrammed course to ensure that the fertilizer is spread according to the prescription map.

Several tests are being conducted of the feasibility of using DGPS for site-specific farming. In Alberta, for instance, four areas ranging from 80 to 200 acres (32 to 80 hectares) have been selected for study over a 4-year period. These areas vary in soil types as well as topography and salinity conditions. By analyzing changes in crop productivity through the introduction of site-specific farming, the expected increase in overall productivity can be derived.

As mentioned above, initial tests have shown that farm vehicles can be guided with an accuracy of better than 3 ft (1 m). Further postanalysis has demonstrated that accuracies better than 8 in. (20 cm) are feasible with improved GPS equipment and navigation software. This increase in accuracy will further expand the applications of DGPS to the agricultural industry. It is projected that in the future there will be unattended farm vehicles that can be remotely guided to within an 8-in. (20-cm) accuracy.

Airborne spraying and seeding. The use of aircraft for crop spraying is becoming increasingly important as a cost-effective means to apply pesticides to large areas. Although aircraft have been used for some time for this purpose, the use of DGPS is relatively new, and several systems using this technology for flight guidance are currently commercially available.

In order to cover the area of interest, the aircraft flies in swaths separated by 50–150 ft (15–50 m), depending on the aircraft altitude, spray-boom width, and the material to be sprayed. Traditionally, these swaths are marked by ground flaggers at each end of the line. The ground crew moves from field to field during the spray operation. With the advent of DGPS, these flaggers can be replaced by an accurate on-board navigation system. Similar to the fertilizing operations discussed above, the pilot simply follows the preprogrammed swath pattern by using a device (typically a light bar) to indicate if the aircraft is left or right of the centerline. The pilot can typically stay within 3 ft (1 m) of this line, which ensures that the coverage is even, no areas are without pesticide application, and no pesticide is applied twice to the same area.

GPS has significantly increased the cost-effectiveness of these aerial applicators since the time and labor to complete the work have been reduced significantly. Time may also be important in the context of the specific application. For example, a DGPS-based aircraft guidance system has been used to release millions of sterile Mediterranean fruit flies (medflies) into 1500 mi^2 (4000 km^2)

of the Los Angeles Basin in order to eradicate the flies, which had the potential to destroy the fruit industry there. Since medflies reach maturity in 4 days, the sterile flies had to be released quickly to mate and eliminate the local population.

Another significant benefit of having an accurate guidance system is that the exact flight path of the aircraft is recorded during the spraying operation. After spraying is complete, the pilot can overlay the aircraft path on a map to provide a permanent flight record. This flight documentation is important for environmental purposes since it is often required to show that only the targeted areas have been sprayed. This electronic record can also be used in case of litigation from neighboring land owners, for example.

Aerial application techniques have also been used to control the boll weevil in a 450,000-acre (180,000-hectare) cotton-growing area in Texas as well as for grasshopper eradication in a 6400-acre (2600-hectare) plot in North Dakota. Airborne seeding has also been used for planting rice, since it is often difficult to use ground vehicles because of moisture. Grain seeding has been done from the air when weather conditions limit the window to an extent that ground seeding is not possible. Airborne fertilization is also being used extensively.

Animal tracking. The use of GPS for animal tracking is relatively new but offers a tremendous gains over conventional techniques based on VHF transmitters. With the VHF-transmitter technique, a collar is attached to the animal, and signals are normally tracked from the air where the signal range is 10–15 mi (15–25 km). After the signal is received in the helicopter or aircraft, observers visually locate the animal and plot its position on a map. Errors of 300–1500 ft (100–500 m) have been reported with this technique, and in some cases visual location is impossible. Although accuracy is one of the disadvantages of this approach, the need for costly air time is an even bigger drawback.

In contrast, the DGPS system for animal tracking is efficient, accurate, and cost effective. Collars incorporating GPS receivers have been developed which can give 15–30 ft (5–10 m) position accuracy in postanalysis. These few-kilogram collars are mainly used for tracking wildlife such as caribou, moose, elk, and grizzly bears. In the case of the moose, which can travel in a 40-mi^2 (100-km^2) region, positions are typically recorded in computer memory in the collar every 3 h and saved until the user uploads the data via a communication link. These collars are designed to last approximately 12 months.

The specific application of animal tracking to livestock has been slowed by the size and weight of the collars. However, some experimental systems that do not use a collar have been tested. For example, in the United Kingdom sheep have been outfitted with DGPS-based systems in order to track their movements, which are then correlated to their radiation levels. Stemming from the Chernobyl nuclear plant accident, the large levels of radioactive cesium have motivated officials to screen sheep. The radiation levels varied greatly within a flock, indicating very local hot spots of radiation which could not be detected with land radiation monitors. The ability to track individual sheep to specific grazing locations would assist in the identification of these areas. The system consists of an inexpensive, low-power GPS receiver, a single-board computer, and a memory card for recording data for post-analysis. A mercury tilt switch to sense if the sheep is lying down, which triggers the GPS receiver to shut down and save power, and a jaw-movement sensor, which detects when the sheep is eating, are also included. The entire system weighs 5.5 lb (2.5 kg) and is installed in two pouches on either side of the sheep's back. The system is retrieved after 10 days, and the DGPS data are postanalyzed to give 15-ft (5-m) position information.

Another potential application is for tracking cattle on large ranches where the location and grazing patterns of the cattle can be monitored on a regular basis. This application is clearly important in the operational and environmental management of large ranching operations. It also has application in the isolation of cattle from predators such as wolves, since the whereabouts of cattle are known and any conflict can be quickly pinpointed. Collars of sufficiently small size and power have not yet been developed for these applications, but with improvements in GPS, computer, and battery technologies these systems will become available.

For background information *SEE AGRICULTURAL AIRCRAFT; SATELLITE NAVIGATION SYSTEMS* in the McGraw-Hill Encyclopedia of Science & Technology.

M. Elizabeth Cannon

Bibliography. Institute of Navigation, *Proceedings of GPS-93,* 1993; G. Lachapelle et al., GPS system integration and field approaches in precision farming, *Navigation,* 41(3):323–335, 1994; A. R. Rodgers and P. Anson, Animal-borne GPS: Tracking the habitat, *GPS World,* 5(7):21–32, 1994; M. W. Sampson, No small affair: DGPS battles the medfly, *GPS World,* 6(2):32–38, 1995.

Scattering experiments (atoms and molecules)

When electrons and ions collide, several processes can occur. With positively charged ions, the electron may be captured and become bound to the ion. This process is called recombination. When the ion carries electrons, excitation or further ionization may also take place if the initial energy of the free electron is sufficiently high. Collisions between free electrons and positive ions occur in many places such as in artificial plasmas, high-temperature astrophysical plasmas, and gaseous nebulae. Much of the early concern about electron-ion recombination was related to astrophysics; yet today detailed knowledge about electron-ion interactions is also required to govern the energy budget in thermonuclear fusion devices. Recombination as well as electron-impact excitation of atoms and ions result in the emission of one or more light quanta. These light quanta carry information which may be used to diagnose laboratory plasmas as well as plasmas in distant parts of the universe.

Precise knowledge about the structure of atoms and ions is required to accurately predict the rate at which recombination takes place since, as will be discussed later, doubly excited electronic states are often involved as well as singly excited states with highly excited electrons (Rydberg states). Some of these states are very sensitive to external fields. This sensitivity is reflected in the experimental data, and information on the recombination processes is an essential guide to the correct theoretical understanding of atomic structure and recombination processes.

Recombination. There has been a considerable increase in the understanding of the processes that lead to the formation of atoms from their constituents: free electrons and a positive nucleus. This increase is primarily due to the development of new experimental techniques and fast computers which can handle the complex programs used in theoretical descriptions.

There are three basic recombination mechanisms: radiative recombination, dielectronic recombination, and three-body recombination. The first theory of radiative recombination dates back to the early 1920s before quantum theory was developed. The first experimental test of this theory was not, however, performed until 1990. Dielectronic recombination was first discussed in the 1940s, and the first experimental results appeared in the mid-1980s. The theory for three-body recombination has been known since the 1960s, but up until now very few experiments have provided quantitative information about this process.

As an electron approaches a positively charged ion, it accelerates, and large kinetic energies may be obtained, especially in the vicinity of the nucleus. To become bound to the ion, the electron must get rid of some of its energy. It therefore has to exchange energy and momentum with a third particle. The different recombination reactions differ in the form of this exchange.

Radiative recombination. This recombination arises as the result of the coupling between matter and the electromagnetic field. During the large acceleration of the electron near the nucleus, a photon is created, and consequently the electron may no longer have sufficient energy to escape the Coulomb field of the positive ion and is thus captured. Predominantly, final states with low principal quantum number n (low energy) are populated. The radiative-recombination process may be writ-

ten as reaction (1), where X is the positive ion with

$$X^{Z+} + e^- \rightarrow X^{(Z-1)+} + h\nu \qquad (1)$$

total charge Z (in units of e, the elementary charge), e^- is the electron, and $h\nu$ is the energy of the emitted photon (where h is Planck's constant and ν is the photon frequency). The radiative-recombination process may take place for any initial energy of the free electron, but it is enhanced at low energies. It is clear from reaction (1) that the radiative-recombination process is the inverse of the photoionization process, in which a photon releases an electron from an atom or ion. For bare ions (no electrons), the cross section for radiative recombination can be calculated exactly since the problem then involves only one electron and one positive nucleus (the hydrogen problem). The simplest radiative-recombination process involves a proton (p^+) and a free electron which together may form hydrogen (H). As shown in reaction (2), the electron may be

$$
\begin{aligned}
p^+ + e^- &\rightarrow H(n) + h\nu \\
&\rightarrow H(n_1) + h\nu_1 \\
&\rightarrow H(n_2) + h\nu_2 \qquad (2) \\
&\rightarrow \dots \\
&\rightarrow \dots
\end{aligned}
$$

captured into a state (labeled n) that is not the lowest possible energy state (ground state). In this case, additional photons, with characteristic wavelengths ($h\nu_1$, $h\nu_2$, ...), are emitted when the electron makes a radiative transition to states of lower energy (labeled n_1, n_2, ...). The resulting characteristic wavelength pattern has been observed by telescope from regions of hydrogen gas both in the Milky Way Galaxy and in extragalactic space. Interest also focuses on the radiative-recombination reaction for nonbare ions. A comparison between experiment and theory can be used to test calculations of the photoionization of nonhydrogenic excited atoms. Such data are not readily available from other sources.

Dielectronic recombination. In this resonant process, the incoming electron excites a bound electron present on the ion to an excited electronic state. It thereby loses energy and becomes bound to the ion, provided its initial kinetic energy is less than the excitation energy. In this way, doubly excited ionic states are created. If these states decay by photon emission rather than by emission of an electron, the electron becomes bound to the ion. The dielectronic-recombination process can be represented by reaction (3), where the asterisk indicates

$$
\begin{aligned}
X^{Z+} + e^- \rightarrow (X^{(Z-1)+})^{**} &\rightarrow (X^{(Z-1)+})^* + h\nu_1 \\
&\rightarrow (X^{(Z-1)+})^* + h\nu_2 \\
&\rightarrow \dots \qquad (3) \\
&\rightarrow \dots \\
&\rightarrow X^{(Z-1)+} + h\nu_i
\end{aligned}
$$

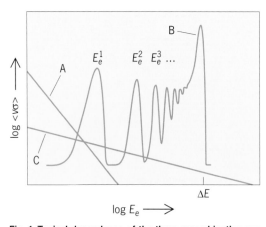

Fig. 1. Typical dependence of the three recombination processes. A certain finite experimental resolution is assumed for the dielectronic-recombination resonances, E_e^1, E_e^2, ... Curve A shows three-body recombination, B, dielectric recombination, and C, radiative recombination.

an excited atom or ion. If the electronic-excitation energy is denoted ΔE, then for capture to a state γ, conservation of energy yields the resonance condition, Eq. (4), where E_γ is the binding energy of the

$$E_e + E_\gamma = \Delta E \qquad (4)$$

state designated γ and E_e represents the energy of the initially free electron.

Three-body recombination. This process arises when the incoming free electron transfers energy and momentum to another free electron in the neighborhood of the ion rather than to a bound one. Three-body recombination may be written as reaction (5). Clearly, the three-body recombination process

$$X^{Z+} + e^- + e^- \rightarrow X^{((Z-1)+)*} + e^- \qquad (5)$$

is the inverse of the more familiar ionization process, where an electron hits an atom or ion and knocks off one of the bound electrons. It is more probable for the electron to transfer only a small amount of energy to a neighboring electron, and so the three-body recombination process greatly favors free electrons at low incident energy. As a result, highly excited electronic states become populated, and the process proceeds by stabilization of such states either collisionally or radiatively. The three-body recombination-reaction probability is

proportional to the square of the electron density since it involves two continuum electrons, in contrast to the linear response of the radiative-recombination and dielectronic-recombination reactions which involve only one active free electron. In plasmas with high electron density and low temperature, the three-body recombination process is the dominating recombination process.

It has been suggested that a possible way to produce antihydrogen is via the three-body recombination reaction (6), where an antiproton p^- captures

$$p^- + e^+ + e^+ \rightarrow \overline{H} + e^+ \qquad (6)$$

an antielectron (positron) e^+ to form an antihydrogen atom ($\overline{H}$). Researchers are now trying to produce the first antihydrogen atoms with antiprotons from storage rings.

In most electron-ion beam experiments in laboratories now, the electron densities and temperatures are such that the radiative-recombination and dielectronic-recombination contributions dominate the contribution due to three-body recombination. However, in the near future, experiments can be anticipated where densities and temperatures are such that the three-body recombination process can be studied thoroughly for the first time. **Figure 1** shows schematically the energy dependence of the rate coefficients (the products of cross section and incident electron velocity) for the three processes. Dielectronic-recombination resonances occur when reaction (4) is fulfilled ($E_e = \Delta E - E_\gamma$).

Molecular ions. The relationship to processes involving molecular ions should also be emphasized. Electron-molecule collisions are very important in the fields of chemistry, astrochemistry, astrophysics, and plasma physics. As with atomic ions, radiative recombination may take place in the case of molecular ions. However, when the incoming free electron approaches the charged molecule, it is more likely to excite the molecule both electronically and vibrationally, thereby losing energy and becoming bound to the molecule. The fate of the excited molecule then determines the final outcome of the process.

Dissociative recombination for a diatomic molecule AB^+ may be written as reaction (7). The

$$e^- + AB^+ \rightarrow AB^{**} \rightarrow A^* + B^* \qquad (7)$$

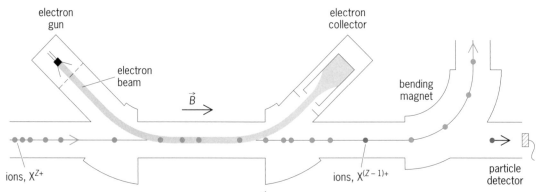

Fig. 2. Experimental merged-beams setup. A magnetic field ($\vec{B}$) is applied to confine the electrons.

excess energy is here dispersed into the new degree of freedom, which in this case is the internuclear separation. The dissociation, the second transition in reaction (8), is normally very rapid, and as a con-

$$e^- + ABC^+ \rightarrow A + BC$$
$$\rightarrow AB + C \qquad (8)$$
$$\rightarrow AC + B$$
$$\rightarrow A + B + C$$

sequence the dissociative recombination process takes place with a high probability. The molecules can separate into neutral atoms in their ground state as well as in excited states [indicated by the asterisk in reaction (7)]. For example, the green line at a wavelength of 557.7 nanometers seen in the night sky results from dissociative recombination with O_2^+ ions. One of the resulting oxygen atoms is produced in an electronically excited state which decays by emission of the characteristic green light.

Research in the field of dissociative recombination is concerned with the determination of reaction rates and branching ratios (ascertaining which atomic states get populated). For larger molecules, attention is paid to the formation of smaller neutral molecules. For triatomic molecules, for example, the reaction channels are as in (8). One important example of the creation of new molecules due to dissociative recombination is the formation of the water molecule (H_2O) from H_3O^+.

Experimental techniques. The main ingredients for a good experiment are an ensemble of ions in a well-defined state (electronic configuration and charge state), free electrons of a known density and possessing a well-controlled and narrow velocity distribution relative to the ions, and an effective detection scheme to monitor the products of the recombination reaction. Basically, two types of experiments are carried out. One uses ion traps, and the other merged or crossed beams of electrons and ions.

The merged-beams approach is utilized with electron coolers in ion-storage rings. This experimental arrangement has led to a great deal of progress in the study of electron-ion interactions. It is now possible to obtain high-energy resolution as the result of the careful transport of electrons in the cooler device and the relatively low density of electrons (typically 10^7 to 10^8 cm^{-3}). The electron density, however, is sufficiently high to provide a good signal, primarily because of the excellent vacuum conditions in the coolers. Thus electron capture from the background gas dominating the true recombination signal is prevented. An efficient detection scheme to monitor the recombination events is provided simply by counting the ions which have changed charge after the interaction with the electron target (**Fig. 2**). The ions are provided by accelerators or heavy-ion storage rings, which can produce intense beams of many different ions from the lightest to the very heaviest of the nuclear table.

Recombination measurements are carried out with energy resolutions of about 10^{-2} eV and with advanced atom-detector systems. Such measurements provide critical tests of the understanding of electronic interactions in atoms and ions and important information in the attempt to understand the physics of plasmas and the surrounding universe.

For background information SEE ATOMIC PHYSICS; ATOMIC STRUCTURE AND SPECTRA; MOLECULAR STRUCTURE AND SPECTRA; PARTICLE ACCELERATOR; PLASMA PHYSICS; SCATTERING EXPERIMENTS (ATOMS AND MOLECULES) in the McGraw-Hill Encyclopedia of Science & Technology.

Lars H. Andersen

Bibliography. W. G. Graham et al. (eds.), *Recombination of Atomic Ions*, NATO ASI Ser. B, Physics, vol. 296, 1992.

Sea-level change

Sea levels change continuously. Individuals who live near the coastlines have learned to adapt to and take advantage of these changes, despite occasional disastrous flooding events. Projected increases in atmospheric greenhouse gases, possibly leading to major increases of global sea levels in a future, warmer world, have focused public and scientific attention on the sea-level problem. Research is aimed at improving understanding the causes of past changes and at estimating the risks of more frequent, serious flooding in the future.

Global sea levels depend on the shape and size of the ocean basins and on the volume of water stored. Over geological time changes in the shape and size of the ocean basin are important, but not with regard to the time scales considered here. Also over geological time the redistribution of mass in the Earth can change the geoid, the surface adopted by equilibrium mean sea levels. For climate change the two major factors are the expansion of the ocean waters by heating and the addition of new water by the melting of grounded ice. The melting of floating ice has no effect on the sea levels, as Archimedes' principle applies.

Variation history. Sea levels change on a wide range of space and time scales, from local wind waves to global mean-sea-level variations over geological periods. During the most recent ice age some 20,000 years before present, sea levels were many tens of meters lower than at present. Locally, changes are dominated by tides and by the effects of weather. Extreme storms acting on high spring tides increase the risk of flooding. Compared with the daily tidal variations of sea level, the projected increases in the mean level due to global warming are small but will significantly affect the high flood levels when added to tides and storms.

Over the past 100 years for which a few long-term records of sea level have been made, the average increase globally is in the range 0.10 to 0.25 m (4 to

Table 1. Estimated contributions to sea-level rise from the 1890s to the present, cm (in.)

Component	Low	Middle	High
Thermal expansion	2 (0.8)	4 (1.6)	7 (3)
Glaciers and small ice caps	2 (0.8)	3.5 (1.4)	5 (2)
Greenland ice sheet	−4 (−1.6)	0	4 (1.6)
Antarctic ice sheet	−14 (−5.9)	0	14 (5.9)
Surface- and ground-water storage	−5 (−2)	0.5 (0.2)	7 (3)
TOTAL	−19 (−7.4)	8 (3)	37 (14)
OBSERVED	10 (3.9)	18 (7.0)	25 (9.7)

SOURCE: J. T. Houghton et al. (eds.), *Climate Change 1995: The Science of Climate Change,* 2d Assessment Report of the Intergovernmental Panel on Climate Change, 1996.

10 in.). Only a few major ports have maintained good-quality records over such a long period, and in all sea-level studies the lack of good data is a serious limitation. A global sea-level monitoring network (GLOSS) has been established by the Intergovernmental Oceanographic Commission, encouraging countries to measure sea levels to common standards and to make their data available for wider analysis.

Satellite measurements. Satellite technology is making fundamental contributions to sea-level measurements in two ways. First, altimetric satellites such as *TOPEX/POSEIDON* measure the sea levels below their orbit by timing the return of electromagnetic pulses, in the same way that an echo sounder works at sea. An accuracy of 0.02 m (1 in.) or better is possible, and the coverage is global and rapid.

The second application of satellites is to fix the zero baseline level at each tide gage into a standard system of global coordinates, so that comparisons are more exact; and local movements of the Earth's crust, which appear in the records as apparent sea-level changes, are eliminated. The most important of these crustal movements is the vertical recovery of land that was under an ice burden in the past ice age. In Scandinavia and Alaska, for example, the local sea levels are falling because the land is rising vertically more rapidly than the global sea levels appear to be increasing. The long records of sea level can be adjusted to allow for this effect.

Ocean thermal expansion. The volume of the oceans, at constant mass, depends on their density, which is controlled by the temperature and the salinity. Temperature is the most important factor, an increase in temperature expanding the water and increasing its volume. However, calculating this expansion is extremely complicated, because the oceans do not absorb additional heat in a simple way. The oceans are strongly stratified, with the coldest, most dense water, formed at high latitudes, filling the deep ocean basins. Typically the formation, sinking, and migration of these deep waters take place over hundreds of years. As a result, the oceans have a long time delay before they adjust to any changes in the mean atmospheric temperature.

Measuring ocean temperatures from research vessels is a slow and expensive process, and data collected over the past century are insufficient to show whether deep-sea warming is taking place. The

World Ocean Circulation Experiment (1990–1997) will provide a systematic baseline of measurements against which future changes may be detected. Coupled ocean atmosphere models driven by climates over the past century suggest that sea-level rise due to thermal expansion may be about a few centimeters, but the long time it takes for the oceans to respond makes the results of these models sensitive to initial conditions, as they neglect changes already under way at the start time. This initializing difficulty, known as the cold start problem, results in estimates of the sea-level rise due to thermal expansion in the past century, within a broad range of 0.02–0.07 m (1–3 in.).

Melting ice. The increase in ocean volume from melting ice comes potentially from two sources: the Antarctic and Greenland polar ice caps, and the melting of the wide range of mountain glaciers and ice caps in middle latitudes. The volume of water in the polar ice caps is much greater; but the role of the midlatitude ice is important, because the rates of accumulation and of loss are more intense.

Thinning of glaciers since the mid-nineteenth century has been observed in many parts of the world, although there are some examples of glaciers increasing in volume against the general trend. The coverage of systematic surveys is extensive but incomplete; for example, there is no monitoring of the glaciers feeding into the Gulf of Alaska or in the mountains of Patagonia.

Mass balance studies of the two major polar ice caps suffer from the vast areas to be covered and

Table 2. Estimated future contributions to global sea-level rise

Component	Amount, cm (in.)
Thermal expansion	28 (11)
Glaciers and ice caps	16 (6.2)
Greenland ice sheet	6 (2)
Antarctic ice sheet	−1 (−0.4)
BEST ESTIMATE*	49 (19)
RANGE	20–86 (7.8–34)

* The estimates use a climate sensitivity of 2.5°C (37°F) for the mid-projection and 1.5 and 4.5°C (35 and 40°F) for the low and high projections.
SOURCE: J. T. Houghton et al. (eds.), *Climate Change 1995: The Science of Climate Change,* 2d Assessment Report of the Intergovernmental Panel on Climate Change, 1996.

the need to determine a small imbalance between very large volumes of ice melting and vast volumes of new ice formation following precipitation. For the Antarctic ice sheet the imbalance is estimated at between 2 and 25% of the total input, and this uncertainty is one of the major limitations on water flux studies related to sea-level rise. Another major problem in monitoring the polar ice cap budgets is the use of recent short-period data to infer long-term changes. Some proposed improvements will use satellite altimetry and other geodetic techniques, but the present altimetric satellites are in orbits that do not reach the highest polar latitudes.

Other factors. Changes in ocean circulation patterns can change the levels of the sea surface as they adjust to balance the pressure forces with those due to the rotation of the Earth. These effects are unlikely to exceed a few tens of centimeters but are possible if sources and sinks of heat to the oceans adjust with climate change.

Long-term changes in the land-water budget, both natural and due to human activity, can also affect sea levels. Factors include ground-water depletion, surface-water storage, deforestation, loss of wetlands, and changes in levels of natural lakes such as the Caspian Sea. The uncertainties in these factors mean that their net contribution may be either negative or positive, but they cannot be ignored.

The range of estimated contributions to sea-level rise over the past century are summarized in **Table 1.** The highest and lowest estimates differ substantially, although they do encompass the range of estimated sea-level rise based on observations. However, the range of uncertainty, particularly in the contribution of the polar ice caps, illustrates the caution with which present understanding should be used as a basis for predicting future sea level changes.

Future changes. **Table 2** summarizes the best estimates of future sea-level changes provided by the Intergovernmental Panel on Climate Change. The range is wide, but excludes the more extreme rises of several meters proposed by many scientists in the 1980s. At the lower end a rise of 0.20 m (0.7 ft) is similar to that observed over the past century. The highest projected rise of 0.86 m (2.8 ft) is probably higher than anything experienced over the past two thousand years. Increased risks of coastal flooding in a warmer world will depend not only on increases in sea level but also on the changing frequency and intensity of the storms. For estimating the mean sea-level component of the total flood level, the major uncertainties are the mass balance of the polar ice caps and the extent to which greenhouse warming sets up an unavoidable commitment to ocean expansion as the oceans adjust on a long time scale to a warmer Earth.

For background information SEE ALTIMETER; APPLICATION SATELLITES; GREENHOUSE EFFECT; SEA-LEVEL FLUCTUATIONS in the McGraw-Hill Encyclopedia of Science & Technology.

David T. Pugh

Bibliography. S. Fankhauser, Protection versus retreat: The economic costs of sea level rise, *Environ. Plan. A*, 27:299–319, 1995; J. M. Gregory, Sea level changes under increasing atmospheric CO_2 in a transient coupled ocean-atmosphere GCM experiment, *J. Climat.*, 6:2247–2262, 1993; J. T. Houghton et al. (eds.), *Climate Change 1995: The Science of Climate Change*, 1996; D. T. Pugh, *Tides, Surges and Mean Sea Level: A Handbook for Engineers and Scientists*, 1987.

Serotonin

Serotonin is a simple molecule found throughout the body that is involved in a diverse range of functions. In neurons of the brain and gastrointestinal tract, it functions as a neurotransmitter. The neurons in the brain serotonin system play an important role in modulating several emotional states, including mood, appetite, sexual behavior, sleep, and aggression.

Brain serotonin neurons are organized in a pattern described as single-source divergent. Other neurotransmitter systems organized in a similar pattern include the noradrenergic, dopaminergic, and cholinergic systems. Single-source divergent neuronal circuits are unique because they arise from a focal area of the brain and diverge to project widely to many brain regions. They function in a modulatory capacity, influencing the rate and quality of transmission of information in other brain circuits (see **illus.**). SEE BRAIN.

Neuroanatomical organization. The serotonin system is the largest single-source divergent neurotransmitter system in the brain, and serotonin neurons project to all areas of the brain. Serotonin cell bodies originate from the brainstem (one of the most primitive brain regions) and are subdivided into nine groups. The overall neuroanatomical

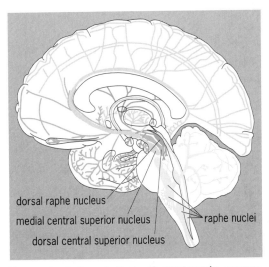

Neuroanatomical projections of serotonergic neurons. (*After P. M. Whitaker-Azmitia, Current Concepts, The Upjohn Company, 1992*)

organization of this system appears to be very similar across most mammalian species, suggesting evolutionary conservation of its physiological role. Two major subdivisions arise: an ascending and a descending arm. The descending arm projects to the spinal cord and is involved in pain perception. The ascending projections arising from the dorsal raphe and median raphe (also called the central superior nuclei) cell groups in the brain stem are the most relevant to emotional states.

Ascending projections from serotonin cell bodies go to distinct brain regions and are organized in a topographical fashion. There are two distinct classes of serotonin nerve terminals, each with different morphological, pharmacological, and functional properties. Nerve fibers from the dorsal raphe have a fine morphology with small granular varicosities and a greater degree of responsiveness to drugs that stimulate serotonin receptors; fibers from the median raphe are coarse, with large spherical varicosities, and demonstrate a reduced level of responsiveness to drugs that stimulate serotonin receptors.

Although serotonin fibers innervate all brain areas, some regions receive a more dense inervation. Limbic brain regions (hippocampus, amygdala, and temporal lobes) and those nuclei involved in sensory transmission (thalamus) are some of the most heavily innervated areas, and motor areas of the frontal cortex receive some of the lowest levels of innervation.

Serotonin neurons not only make classical synaptic connections but also end in nonsynaptic terminals. It can be inferred that some of the serotonin released functions like a hormone, diffusing in the area near the terminal. Serotonin receptors are found on both neuronal and nonneuronal tissues such as glial cells. The effects of serotonin on glial cells are not fully understood, but when serotonin stimulates receptors on glial cells, these cells produce peptide growth factors. Another hormone-like property is that serotonin is released from nerve terminals in more than one fashion, and some of the release may be independent of the rates of cell firing.

Synthesis. Serotonin is synthesized in the same way throughout the body. This synthesis involves the conversion of the amino acid tryptophan via two enzymatic steps. Because tryptophan is an essential amino acid that cannot be synthesized by humans, the synthesis of serotonin is entirely dependent on the dietary availability of tryptophan. Plasma tryptophan cannot passively cross the blood-brain barrier, but is transported across it by a carrier protein. This protein also carries other large neutral amino acids which compete with tryptophan for entry into the brain. The entry of tryptophan into the brain is therefore dependent on the ratio of tryptophan to other large neutral amino acids. The rate-limiting step in the synthesis of serotonin is the conversion of tryptophan to 5-hydroxytryptophan by the enzyme tryptophan hydroxylase. The tryptophan hydroxylase enzyme

is poorly saturated, and therefore increases and decreases in tryptophan availability as well as changes in the ratio of tryptophan to large neutral amino acids can influence the rate of brain serotonin synthesis. This fact has led to methods of experimentally increasing and decreasing brain serotonin in laboratory animals and humans.

Both increases and decreases in dietary tryptophan intake lead to corresponding changes in brain tryptophan, serotonin, and 5-hydroxyindoleacetic acid levels in laboratory animals. Ingestion of tryptophan-free amino acid mixtures in vervet monkeys decreases plasma tryptophan and cerebrospinal fluid tryptophan, and 5-hydroxyindoleacetic acid, with no change in cerebrospinal fluid markers of nonadrenergic and dopaminergic activity. Moreover, ingestion of tryptophan-free amino acid mixtures in laboratory animals leads to extremely rapid changes in both plasma tryptophan and brain serotonin, with maximal reductions of brain serotonin occurring within 2 h of ingestion of the tryptophan-free mixture.

Brain serotonin is metabolized primarily by the monoamine oxidase type A enzyme, which breaks it down into 5-hydroxyindoleacetic acid. Some studies have utilized levels of 5-hydroxyindoleacetic acid as an indirect marker of ongoing serotonin functional activity.

Serotonin receptors. As with most other neurotransmitters, serotonin released from the synaptic terminals of raphe neurons interacts with pre- and postsynaptic receptors. These receptors mediate the actions of the released serotonin, analogous to the way a key fits into a lock and causes it to open. Presynaptic receptors modulate the firing rate of serotonin neurons and the amount of serotonin released with each impulse. Postsynaptic receptors alter the firing rate of other neurons, the rate of release of glial growth factors, and the rate of synthesis of a variety of neuronal peptides and intraneuronal substances.

One of the most interesting aspects of the serotonin system is the large number of receptors that have been identified. There may be as many as 16 distinct subtypes of serotonin receptors in humans. The anatomic localization of serotonin receptor subtypes appears to follow a pattern closely related to the pattern of differing types of nerve terminals, and most likely also has functional significance. Several receptors are found presynaptically and serve to provide inhibitory feedback inhibition on firing rate and release of serotonin. Most of the subtypes also have been reported postsynaptically, and the physiological effects include both inhibitory and excitatory actions.

Hemispheric laterality and gender differences. Serotonin neurons show hemispheric laterality with increased markers of serotonin function in the left cerebral hemisphere. In laboratory rats, there is a greater amount of serotonin in the left striatum and accumbens, and greater concentration of 5-hydroxyindoleacetic acid in the left cortex. Similar findings have been reported in human autopsy specimens, with biochemical markers associated with increased serotinin neurons in the left cortex compared to the right. Interestingly, these changes are more pronounced in women than in men; women also have higher overall levels of serotonin markers. Increased serotonin activity in women compared to men is also demonstrated by studies of cerebrospinal fluid that show higher levels of 5-hydroxyindoleacetic acid in women. Laboratory animal studies confirm that females have increased levels of brain serotonin, 5-hydroxyindoleacetic acid, and one subtype of serotonin receptor as well as a greater serotonin turnover than males.

Physiological characteristics. Serotonin neurons are thought to play a major role in modulating motor activity, sensory information processing, and behavior. Preclinical studies in laboratory animals demonstrate that altering the function of the serotonin system alters many of the behaviors and somatic functions that form the core symptoms of clinical depression. These include changes in appetite, sleep, sexual function, pain sensitivity, body temperature, and circadian rhythms. Increasing serotonin neurotransmission above baseline decreases appetite, increases total sleep, decreases sexual activity, increases pain sensitivity, raises body temperature, and decreases aggressive behavior. In laboratory animals, lesioning the serotonin system or depleting serotonin causes opposite effects in these behaviors.

Whether direct influences on these behaviors are a normal part of the role of the serotonin system or whether these effects are seen with only dramatic, unnatural manipulations of this system is not known. When the activity of serotonin neurons is monitored in living animals, they are remarkably unperturbed by most environmental manipulations. These neurons have an intrinsic pacemaker-like activity, firing at nearly constant rates during the daytime and becoming quiescent during rapid eye movement sleep. If a freely moving cat is exposed to a mouse, a dog, or other environmental stresses, the firing rate of serotonin neurons does not change; in contrast, the rate of firing of nonadrenergic neurons dramatically increases in most of these same situations. Surprisingly, the most effective method of altering the firing rate of serotonin neurons is to cause the animal to move about. Physical activity seems to cause an increase in firing rate. This increase is especially true for motor activity involving chewing, licking, and grooming behaviors. The absence of rapid change in firing rate during exposure to environmental stresses suggests that the serotonin system may play a role in emotion by providing a stabilizing influence on those brain circuits that are activated by stress.

Serotonin also functions as a neurotrophic factor, maintaining the structural integrity of those neurons innervated by it. Astrocytes and glial cells release a neurotrophic factor in response to stimulation of a subtype of serotonin receptor. When serotonin innervation to the adult rat hippocampus is disrupted, there is a loss of neuronal synapses within 10–14 days. However, stimulation of the serotonin receptor leads to a restoration of lost synapses and to an increase in dendritic growth. Thus serotonin may have important effects on maintaining synaptic connections in several brain regions, such as the hippocampus. The role of these growth-factor-like effects in mental illness and in the mechanism of action of psychotherapeutic drugs is being considered.

Emotional effects. In spite of overwhelming evidence from studies involving manipulations of the serotonin system in laboratory animals, human studies manipulating this system have led to less robust effects. Drugs that increase the release of serotonin, such as fenfluramine and tryptophan, cause mild sedation, nausea, and a modest decrease in appetite in healthy humans. No significant effects on mood, sexual behavior, or aggression have been reported. Similar results are seen with drugs that block the reuptake of serotonin by nerve terminals. Both short-term (1–2-weeks) and long-term (4–6-weeks) administration of fluoxetine causes mild nausea, occasional disruption of sleep, and decreased appetite but has no effect on mood or aggression. Drugs that pharmacologically antagonize the effects of serotonin, such as cyproheptidine and methysergide, cause mild increases in appetite and sedation but effect no changes in mood, sexual behavior, or aggression. Similarly, depletion of serotonin by inhibiting its synthesis or by depleting the precursor amino acid tryptophan causes minimal decreases in mood, but causes no discernible changes in appetite, sexual behavior, or aggression.

The only drugs that increase serotonin function and cause dramatic behavioral effects in healthy humans are the psychedelic drugs [for example, lysergic acid diethylamide (LSD), mescaline, psilocin] and the designer drug 3,4-dioxymethylenemethamphetamine (MDMA; known as ecstasy). The psychedelics stimulate certain serotonin receptors and lead to a diverse and unpredictable array of changes in mood, appetite, sleep, sexual behavior, and aggression, as well as visual, auditory, and tactile hallucinations and distortions. Ecstasy causes profound release of serotonin and of the dopaminergic neurotransmitter. It causes euphoria, increased feelings of friendliness, and a desire to interact with others. For reasons that are not completely understood, ecstasy also irreversibly damages brain serotonin neurons.

Mental disorders. The functional state of the serotonin system in humans with a variety of mental disorders has been extensively investigated. These investigations have focused on individuals with major depression, obsessive-compulsive disorder, impulsive aggression, suicidal tendencies, panic disorder, and alcoholism. A variety of paradigms have been used, ranging from measurements of serotonin and its metabolite in the blood, cerebrospinal fluid, and the brain and of serotonin receptor number and affinity in blood cells and brain to more dynamic and challenging provocation of serotonin function. No method of directly measuring brain serotonin function is yet available.

Most studies have found that a subgroup of depressed individuals (35%) have low levels of cerebrospinal fluid 5-hydroxyindoleacetic acid group, and these individuals are more prone to impulsive, violent suicide. This finding is not restricted to individuals with depression but has also been reported in individuals with other mental illnesses (arsonists, some alcoholics, and some schizophrenics) with suicidal or impulsive features.

Data supporting the role of the serotonin system in the mechanisms underlying antidepressant action are very convincing. Most antidepressant drugs and electroconvulsive therapy enhance neurotransmission across serotonin synapses after long-term but not after short-term administration, corresponding with the delay in onset of therapeutic response in individuals. Antidepressants enhance serotonin function through different mechanisms. Tricyclic antidepressant drugs and electroconvulsive therapy appear to sensitize postsynaptic neurons to the effects of serotonin, monoamine oxidase inhibitors enhance its availability, and serotonin reuptake inhibitors (for example, fluoxetine) desensitize presynaptic inhibitory serotonin autoreceptors. Rapid depletion of serotonin by inhibition of synthesis or by depletion of the precursor tryptophan causes a rapid relapse of depression in most individuals being successfully treated with antidepressants (fluoxetine, fluvoxamine, and paroxetine) that increase serotonin function. This depressive relapse is transient, lasting only as long as the depletion.

Evidence for the involvement of serotonin in panic disorder, obsessive-compulsive disorder, and alcoholism is much more circumstantial in nature. Some drugs that mimic the effects of serotonin cause panic attacks in individuals with panic disorder, and can exacerbate symptoms in individuals with obsessive-compulsive disorders. However, drugs that cause serotonin release, such as fenfluramine and tryptophan, have no effect. The main reason that serotonin is suspected in these conditions is the finding that individuals with these disorders improve when treated with drugs that are exceptionally potent at blocking the reuptake of serotonin. In the case of obsessive-compulsive disorder, serotonin reuptake inhibitors are some of the only effective medications.

Implications. The brain serotonin system is an extremely large and diverse system, influencing the function of many different parts of the brain in complex ways. These influences include short-term effects common to most neurotransmitters, as well as other short- and long-term changes that are more consistent with the effects of hormonelike compounds. This diversity may explain why drugs increasing serotonin function affect many types of emotional disturbances and behaviors. Overall, the serotonin system seems to play a stabilizing and modulating role. The lack of immediate effects of manipulations of the system in some individuals with mental disorders suggests that the cause of some of these conditions does not lie in a disturbance of serotonin neurotransmission but may involve dysfunction of the brain regions modulated by serotonin.

For background information *SEE BRAIN; ENDOCRINE MECHANISMS; MONOAMINE OXIDASE; NEUROTIC DISORDERS; OBSESSIVE-COMPULSIVE DISORDERS; SEROTONIN; SYNAPTIC TRANSMISSION* in the McGraw-Hill Encyclopedia of Science & Technology.

Pedro L. Delgado

Bibliography. P. L. Delgado et al., Serotonin and the neurobiology of depression: Effects of tryptophan depletion in drug-free depressed patients, *Arch. Gen. Psychiat.*, 51:865–874, 1994; R. W. Fuller, Neural functions of serotonin, *Sci. Amer. Sci. Med.*, 7/8:48–57, 1995; B. L. Jacobs and E. C. Azmitia, Structure and function of the brain serotonin system, *Physiol. Rev.*, 72:165–229, 1992.

Simulation

Many phenomena in nature are driven by random processes. The global climate system, crystal growth, and brownian motion are simple examples. Many artificial phenomena are also driven by random processes. Examples include the stock market, computer networks, war games, and computer programs that use random numbers. In each case, macroscopic features result from many random events. Because of this randomness, it is desirable to be able to predict, with some degree of certainty, how future events may unfold.

Computer simulations are programs that generate scenarios based on models: descriptions of the physics of the problem, or relationships between key variables. When such programs use random numbers to generate scenarios involving random phenomena (for example, the stock market) or

Fig. 1. Simulation of traffic pattern on interstate highways in the United States.

solutions to deterministic problems (such as multi-dimensional integrals), they are called stochastic simulations. Cast in the latter form is a famous class of stochastic simulations, called Monte Carlo methods, pioneered by researchers at Los Alamos, New Mexico, near the end of World War II. They were motivated by the atomic bomb, in studying the diffusion of neutrons in fissionable material.

Parallelism. Because simulation runs can easily consume large amounts of computing time on even the fastest supercomputers, there is much interest in reducing run times. An obvious strategy is parallel simulation, where many processors cooperate on a single simulation to speed up the rate at which events are generated. When a model is too large to fit inside a single processor's memory, either data or functional decomposition may be used: the model's data are divided among processors, or distinct subtasks in the model are assigned to distinct processors. In such decomposed simulations, processors must coordinate their activities every so often to apprise each other of intermediate results, or to ascertain that they are working on subtasks correctly. This coordination of activities requires significant interprocessor synchronization. Alternatively, when a processor easily accommodates a model, or decomposition is not viable because interprocessor communication and synchronization times are large relative to times between simulation events, the model may simply be replicated across processors. During execution, such models require little interprocessor synchronization.

Replication. A single run of a sequential or decomposed stochastic simulation generates a single sample path of the process being studied. Replication readily yields statistically useful counterparts of a sample path when different random numbers are assigned to different processors. Variance-reduction techniques may be exploited across sample paths; these schemes aid a simulation analyst to reduce output variances, potentially also reducing the number of required runs. Decomposed simulations may also use replication to exploit available processors and improve estimation.

An example of application of replication is a traffic simulation. If the Department of Transportation is interested in traffic patterns between the cities shown in **Fig. 1**, a computer simulation model may be created. Real data on traffic patterns may be collected as an aid to generating synthetic data for a model. Each vehicle departs from one city and arrives in another. Vehicles departing from a given city choose a destination at random and take a random time, chosen from an appropriate probability distribution, to arrive there. On arriving, a vehicle may incur a random delay before moving on to another city.

In a simulation, arrival times and departure times are events. In a discrete-event simulation, time elapses nonuniformly, as the simulation moves from event to event. Simulation time is recorded in a variable known as the simulation clock. As the

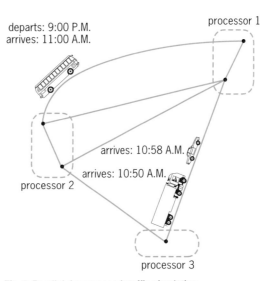

departs: 9:00 P.M.
arrives: 11:00 A.M.
processor 1
arrives: 10:58 A.M.
arrives: 10:50 A.M.
processor 2
processor 3

Fig. 2. Parallel decomposed traffic simulation.

simulation executes, events corresponding to vehicle arrivals and departures are generated. If randomness governing event occurrences is close to reality, computable quantities such as traffic density and average travel delays will also be close to reality. If the problem involves a few hundred cities and a few thousand vehicles, a single processor may require a week or more to generate useful data. To speed up the process, the same model may be replicated across many processors. By using different random numbers on each processor, different event sequences corresponding to different realizations are generated. Statistical results that would otherwise require weeks can now be obtained in hours or minutes.

Parallel replication is a conceptually simple but powerful technique. One notable application used this method in an experiment involving the simulation of polymer chains. With concerted replications on about 200 heterogeneous networked computers scattered across the country, it was shown that runs requiring 3 h on a CRAY Y/MP supercomputer could be done in less than 10 min. To effect sharp reductions in communication and data-combining overheads, cooperating processors were organized in a treelike structure over the Internet.

Decomposition. Clearly, a single processor may not be capable of hosting large models with 100 or 1000 cities, and perhaps 1,000,000 vehicles. Even if it is, a single simulation run may take too long. If events at different cities or sets of cities can be generated concurrently on different processors, a parallel run may be faster. The model from Fig. 1 can be decomposed, with subtasks assigned to three cooperating processors, as shown in **Fig. 2**. Processor 1 simulates traffic in Boston and New York, processor 2 simulates traffic in Chicago and Indianapolis, and processor 3 is responsible for Atlanta. In general, many such assignments are possible, with some better than others. Whereas a parallel replication run on three processors yields three distinct realizations, a parallel decomposition run on three processors yields a single realization.

Synchronization protocols. In a model, each vehicle traveling from one city to another is represented by a message sent between processors assigned to these cities. Each such message contains relevant information about the vehicle, and a time stamp (that is, its scheduled time of arrival at a destination). These time stamps represent times of event occurrences in the simulated system and occur along a simulation-time axis, distinct from real time. Each processor maintains its own notion of current simulation time through use of its own simulation clock. If it has no events to process between the current time on the clock and the occurrence time t of an imminent event, it may update its clock to time t and process the event.

As shown in Fig. 2, a bus that leaves Chicago at 9:00 P.M. is scheduled to arrive in Boston at 11:00 A.M. In the simulation, processor 2 sends a message containing this information to processor 1. Even if processors are separated by large geographic distances, such messages usually arrive reliably and within, at most, a few milliseconds. In the meantime, however, processor 1 may be busy simulating traffic within Boston and New York. It continues to process events, including the arrival of a truck from Atlanta in New York at 10:50 A.M., consequently updating its simulation clock to 10:50 A.M. If no other messages from processor 3 arrive, its next pending event will appear to be the arrival of the 11:00 A.M. Chicago Special in Boston. It now faces a dilemma. Not having received any event from Atlanta with a time stamp greater than 10:50 A.M., it has no way of knowing whether other inbound vehicles from Atlanta might arrive in Boston or New York between 10:50 and 11:00 A.M. How it chooses to handle the situation is determined by a synchronization protocol. Such a protocol is merely a set of rules encoded in the parallel program. Processors are forced to obey these rules to ensure that simulated events are ultimately processed in order of their occurrence in simulation time.

Conservative protocols. Processor 1 has a choice. It can be conservative and not process the arrival of the 11:00 A.M. Special. Instead it simply waits until some vehicle from Atlanta arrives in either Boston or New York. If the car from Atlanta arrives in Boston at 10:58 A.M., processor 1 must process the car arrival before processing the Special, because the car arrives earlier in simulation time. After processing the car's arrival and updating its clock to 10:58 A.M., it must wait to hear from processor 3 again, since its original dilemma has recurred. If some vehicle from Atlanta arrives exactly at 11:00 A.M., processor 1 may process either the Special or this vehicle first. Events that coincide in simulation time may be processed in arbitrary order. A model may also choose to process certain types of events (such as bus arrival events) last. If a message from processor 3 indicates a vehicle arrival with time

stamp greater than 11:00 A.M., processor 1 concludes that it is safe to process the 11:00 A.M. Special first, and does so without further delay.

The important point is that if a processor does not have at least one pending event from every processor that can send it events, it must wait until it does. Only then can it select the earliest pending event for processing. There are many variations on the basic theme. All seek to reduce the amount of time that a processor spends waiting for pending events despite work to be done.

Optimistic protocols. Faced with the choice described above, processor 1 can take a chance. It can be optimistic and process the event corresponding to the 11:00 A.M. Special, updating its clock to 11:00 A.M. After all, the next vehicle from Atlanta to arrive in Boston or in New York, after the 10:50 A.M. truck arrives in Boston, may roll in only after 11:00 A.M. If processor 1 gambles and wins, time otherwise spent idling while waiting to hear from processor 3 may gainfully be spent computing. Unfortunately, even the best gamblers may lose, and the inevitable occurs: processor 1 may allow the 11:00 A.M. Special to arrive in Boston, and perhaps even depart for another city before the car from Atlanta arrives in Boston. What is wrong with this scenario is that the car was scheduled to arrive in Boston at 10:58 A.M. Because of the randomness in processor loads and network delays, time-stamped messages arriving at a processor are never guaranteed to arrive in order of their time stamps.

With its simulation clock at 11:00 A.M., processor 1 finally receives and deciphers a message from processor 3, indicating a 10:58 A.M. car arrival in Boston. Now it must undo all the work it has done between 10:58 A.M. and 11:00, process the Atlanta car's arrival in Boston at 10:58 A.M., and and then repeat processing for any events occurring between then and 11:00 A.M. Undoing and redoing events in the right order is important because the car's arrival may potentially affect future events. For example, the car may bring passengers wanting to board the 11:00 A.M. Special for its next destination. In this protocol many variations on the basic theme can also be found. All seek to reduce the amount of time that a processor spends undoing work, or recording work it may have to undo later.

Prospects. Parallel simulation methodologies are still in their infancy. Hybrid schemes combining replication and decomposition are being developed, as are schemes that make long-running simulations resilient to processor failures and network outages. Parallel simulations hold great promise in supporting interactions between objects in virtual worlds, giving decision makers a significant edge over gamblers in complex economic situations involving chance, aiding engineers in designing sophisticated communication networks and in solving a variety of difficult problems. Both free and commercial software systems for sequential stochastic simulation have been available since the 1970s. The further development of enabling software technologies for fast parallel simulations can help solve otherwise intractable problems.

For background information SEE QUEUEING THEORY; SIMULATION; STOCHASTIC PROCESS in the McGraw-Hill Encyclopedia of Science & Technology.

Vernon Rego

Bibliography. A. Ferscha and S. Tripathi, *Parallel and Distributed Simulation of Discrete Event Systems*, CS-TR-3336, Department of Computer Science, University of Maryland, August 1994; H. Nakanishi, V. Rego, and V. Sunderam, On the effectiveness of superconcurrent computations on heterogeneous networks, *J. Parallel Distrib. Comput.*, 24(2):177–190, 1995; I. Peterson, Mix-and-match computing: Scientific supercomputing without supercomputers, *Sci. News*, 143:280–294, 1992.

Skeletal system

The shape of the human body and that of other vertebrates is to a large extent determined by the skeletal system, the bones of the skull in the head, the vertebral column along the back, the ribs in the chest, the pelvis, and the bones in the arms and legs, hands, and feet. Like the beams in a building, the skeleton serves as a scaffold, forming protected spaces for internal organs. Flexible connections between bones in joints, and muscles between different parts of the skeleton make a great variety of movements possible.

Hyaline cartilage tissue. In the adult, most of the skeleton is made up of bone tissue, composed of cells called osteocytes embedded within a hard substance produced by the cells. This extracellular matrix consists of complexes of organic molecules and densely packed crystals of calcium phosphate in the form of hydroxyapatite. The combination of hydroxyapatite crystals and organic molecules makes bone tissues hard, yet resilient and not brittle. The surfaces of bones where they meet in joints and portions of ribs, however, consist of hyaline cartilage tissue. Like bone tissue, hyaline cartilage contains cells (chondrocytes) within an extracellular matrix, but unlike bone this matrix does not contain hydroxyapatite crystals. The extracellular matrix of hyaline cartilage is therefore not hard, but firm and somewhat elastic. These properties make hyaline cartilage the ideal tissue for covering bone surfaces in joints, where it is termed articular cartilage. The lubricating properties of articular cartilage allow movements with little friction.

The properties of hyaline cartilage are primarily due to the unique composition and structure of its extracellular matrix. The most important components are fine threadlike protein fibers, and large complexes of proteins and polysaccharides. The fibers are made up of three types of collagen molecules and have a tensile strength equivalent to that of stainless steel. The protein-polysaccharide complexes, called proteoglycans, are highly charged and bind large amounts of water. This hydrophilic property allows them to swell and occupy a large amount of space as they take up water. Within cartilage, they are restricted in their swelling, however, by the collagen fibers. The fibers form a three-dimensional network, and the proteoglycans are squeezed into the spaces created by the network such that the volume they occupy is only about 25% of the volume they would have occupied had they been allowed to swell freely in water. The proteoglycans in cartilage therefore exert a strong swelling pressure and stretch the collagen fibers. The result is a tissue that resists compression because of its internal pressure and does not burst because of strong inextensible fibers that reinforce it.

Osteoarthritis. In most individuals, articular cartilage in joints holds up well during a lifetime of activities. Because the cartilage does not contain nerves or blood vessels, joints can be exposed to considerable force without an individual feeling pain or experiencing bleeding into the joint space. However, the lack of blood vessels incurs problems because it forces chondrocytes in articular cartilage to obtain nutrients and get rid of products of metabolism by diffusion from the joint fluid (synovial fluid) and from blood vessels in the underlying bone. Under normal circumstances, this diffusion is sufficient to allow chondrocytes to continuously degrade extracellular matrix molecules and replace them with new ones. Such turnover of matrix is important for the normal functioning of most tissues, including articular cartilage. However, following severe injury to articular cartilage or because of some underlying biochemical defect the chondrocytes are sometimes unable to replace matrix molecules as efficiently as they are degraded, and the result is net degradation of the extracellular matrix. Over time there is loss of articular cartilage, and the result is the clinical condition called osteoarthritis. As the cartilage disappears, the underlying bone tissue becomes directly exposed to the mechanical forces acting on a joint and movements become painful. At the periphery of the joint, spurs of cartilage and bone (osteophytes) are formed, limiting motion. Eventually, the condition may become so severe that the joint has to be surgically replaced with a metal substitute. Such joint replacement surgery is now common in the United States and other developed countries, especially for individuals with osteoarthritis of the hip and knee joints.

Despite the success of joint replacement surgery, a metal replacement is not as good as the original. The molecular causes of osteoarthritis are being studied in order to identify individuals at high risk for developing the disease, and to develop therapies to slow down, prevent, or even reverse the disease process once it has started. For example, genetic risk factors for osteoarthritis are being identified by studies of families in which osteoarthritis of knees and hips occurs early in life (late childhood, adolescence, or young adulthood). In families showing autosomal dominant or recessive inheritance of the disease, use of linkage analysis has allowed mapping of the gene defect to a specific region of a chromosome, identification of the gene, and even determination of the specific mutation within its deoxyribonucleic acid (DNA) sequence. The results show that mutations in genes coding for all the three collagen types that are components of cartilage collagen fibers can lead to cartilage abnormalities and early-onset osteoarthritis.

Molecular biology. The major collagen component in the fibers are molecules of type II collagen. These molecules, composed of three identical polypeptide chains encoded by the gene *COL2A1*, are rodlike structures that are packed together in a staggered, overlapping fashion within the fibers. This packing arrangement creates fibers of high tensile strength. It is therefore not surprising that mutations in *COL2A1* can dramatically affect the structure and properties of cartilage. In the worst cases, the mutations prevent the synthesis of collagen II and the formation of a normal skeleton, causing death of a fetus. Other mutations, however, appear only to cause early-onset osteoarthritis.

A minor collagen component in cartilage, representing only 5–10% of the total collagen, is collagen XI. Collagen XI molecules are localized close to the surface of cartilage collagen fibrils. They are composed of three polypeptide chains that are products of the genes *COL11A1*, *COL11A2*, and *COL2A1*. It is believed that collagen XI molecules are important for keeping the fibers in cartilage thin so that they can form a dense network to efficiently entrap the proteoglycan complexes.

In an autosomal recessive abnormality in mice called chondrodysplasia, the importance of this function is dramatically illustrated by a mutation in the *COL11A1* gene. A deletion of a single nucleotide in this gene causes the mutant messenger ribonucleic acid (mRNA) to be translated into a short peptide fragment instead of a normal size *COL11A1* product, and normal collagen XI molecules are not formed. In homozygous embryos, this abnormality leads to a few thick collagen fibers in cartilage instead of the fine network of many thin fibers seen in the normal tissue, loss of the ability to keep the proteoglycan complexes restricted in space, and a soft (almost fluid) cartilage matrix. Homozygous embryos die at birth, probably because the rings of hyaline cartilage that normally keep the trachea open are too soft, causing the trachea to collapse during breathing. Mice that are heterozygous for the mutation develop normally, but suffer from osteoarthritis in their knee joints a few months after birth.

Mutations affecting collagen XI are also associated with severe, early-onset osteoarthritis in humans. In one family a dominant mutation in *COL11A2* led to a syndrome characterized by deafness, osteoarthritis, and high risk of cleft palate. In a second family, a recessive mutation in the same gene caused deafness and osteoarthritis in individuals who inherited both mutated alleles. A mutation in the third type of collagen in cartilage fibers has also been shown to cause osteoarthritis in humans.

Collagen IX molecules are located on the surface of the fibers, and are encoded by three genes, *COL9A1*, *COL9A2*, and *COL9A3*. In one large Dutch family with early osteoarthritis of the knee, a mutation in *COL9A2* has been identified.

Growth and development. Hyaline cartilage is of primary importance not only in articular cartilage but also during the formation of the skeleton in the embryo. Except for the bones of the roof of the skull and a portion of the clavicle, all other bones in the skeleton are formed as hyaline cartilage models. These models are replaced by bone in a process called endochondral ossification. Cartilage is replaced by bone everywhere except at the end regions where it remains as articular cartilage and a disk of cartilage tissue separating the shaft of the bone from the end regions. Within these disks, called growth plates, proliferation of chondrocytes and endochondral ossification allow the long bones in the skeleton to grow in length. As long as growth continues, the growth plates remain cartilage, but sex hormones finally induce their ossification as an individual stops growing.

Because growth plates are composed of hyaline cartilage, it is not surprising that many mutations in the cartilage collagens have an effect on growth plate function, and can lead to short stature. The mutations in these collagens not only affect articular cartilage, and lead to osteoarthritis, but also affect skeletal development and growth plate function. Abnormalities caused by such mutations are for this reason called osteochondrodysplasias. Among osteochondrodysplasias with defects primarily in growth plate function are several forms of dwarfism in humans. The most common of these is achondroplasia. Achondroplasia is most frequently due to a single amino acid change in a molecule called fibroblast growth factor receptor 3. This receptor, bound to the surface of chondrocytes in growth plates, binds growth factors released by neighboring cells and signals the cells on which it is located to slow down proliferation, thus acting as a negative regulator of long-bone growth. The mutation causes the mutant receptor to become more active so that growth is slowed down too much. Thus the arms and legs of achondroplastic dwarfs are short. Another form of dwarfism is caused by mutations in collagen X, an extracellular matrix molecule that is made only in growth plates. Individuals with mutations in this collagen appear normal at birth, but become bow- and short-legged after they start walking.

Clinical benefits. One immediate benefit of the use of molecular biology and genetics in the study of the diseases that affect cartilage in joints and growth plates is the availability of reagents to detect mutations in several genes in families with a past history of osteochondrodysplasias. The more important and long-term benefits, however, are the novel insights that these studies provide into the molecular and cellular processes that lead to diseases of cartilage. Thus, scientists and clinicians are in a much better position to define targets for therapeutic intervention.

For background information SEE ARTHRITIS; BONE; COLLAGEN; DWARFISM AND GIGANTISM; GENE; GENE ACTION; GENETICS; MOLECULAR BIOLOGY; MUTATION; SKELETAL SYSTEM; SKELETAL SYSTEM DISORDERS in the McGraw-Hill Encyclopedia of Science & Technology.

Bjorn R. Olsen

Bibliography. L. Ala-Kokko et al., Single base mutation in the type II procollagen gene (COL2A1) as a cause of primary osteoarthritis associated with a mild chondrodysplasia, *Proc. Nat. Acad. Sci. USA,* 87:6565–6568, 1990; D. Chan et al., A COL2A1 mutation in achondrogenesis type II results in the replacement of type II collagen by type I and III collagens in cartilage, *J. Biol. Chem.,* 270:1747–1753, 1995; R. Shiang et al., Mutations in the transmembrane domain of FGFR3 cause the most common genetic form of dwarfism, achondroplasia, *Cell,* 78:335–342, 1994; M. Vikkula et al., Autosomal dominant and recessive osteochondrodysplasias associated with the COL11A2 locus, *Cell,* 80:431–437, 1995.

Snowpack monitoring

Snow-covered mountain landscapes are the source of an accumulation of moisture critical for human and wildlife populations and for the maintenance of vegetation. In the arid and semiarid western United States, for example, the mountain snowpack is a vital natural resource that provides up to 80% of the annual fresh-water supply. Spring snowmelt generates streamflow, recharges ground water, and fills reservoirs that are maintained to supply water for irrigation and municipal use, to generate hydroelectric power, and to provide flood control for downstream communities. By monitoring the seasonal accumulation of snow, estimates of the amount of water held in storage can be made. Hydrologists use this information to predict streamflow and to forecast annual water supply. In the American West, where populations continue to increase, the need to manage water resources is heightened by conflicting demands for water and by the long- and short-term climatic fluctuations. Snowpack forecasts, therefore, provide valuable information to facilitate watershed, reservoir, and fish and wildlife management goals.

Snow surveys. In order to forecast water supply, reliable and timely estimates of the snow water equivalent (the depth of water produced when a column of snow is melted) are needed. The basis for a systematic method of snowpack monitoring, called snow surveying, was developed in the American West in the early 1900s. Mountain snow water equivalent data were used to estimate the increase of lake level from spring melt water. The first snow course (a transect of permanent measurement points) was established on Mount Rose to predict the rise in Lake Tahoe water level. The technique made use of the high correlation that exists between the snow water equivalent and streamflow. In 1906, an instrument known as the Mount Rose snow sampler was developed for measuring snowpack depth and snow water equivalent. Subsequently, snow courses were rapidly established throughout the West and later spread around the world.

After a great Western drought in 1934, the Bureau of Agricultural Engineering of the U.S. Department of Agriculture developed a snow survey program for the purpose of maintaining an adequate irrigation water supply. In 1939, responsibilities for the federal-state-cooperative snow survey program were transferred to the Soil Conservation Service (now the Natural Resources Conservation Service). Under the management of the Soil Conservation Service snow measurement methods and instrumentation were standardized, and a schedule of surveys was established. Snow surveys occur on or about the first of the month, from January until June. Midmonth surveys are conducted when conditions require updated information, but more frequent updates are impractical because of the cost and labor involved in manual snow surveys. The measurement device used today, the federal snow sampler, has been modified very little from the original 1906 Mount Rose design. The snow survey network in the West reached its peak in the late 1980s, with more than 1800 active snow courses being used to predict streamflow at more than 500 forecast points.

Snow courses. A snow course consists of 5–10 snow-depth and snow-water-equivalent measurement points along a 1000-ft (300-m) transect. The locations of the transect end points are marked in the field to ensure that measurements are made at the same location each time a survey is conducted. The points along the transect are measured from the endpoints. The federal sampler consists of 30-in. (76.2-cm) sections of aluminum tube with a sharp-toothed cutter on one end. The tube sections are connected and inserted into the snowpack until the teeth contact the ground surface. The depth of the snowpack is measured by using the half-inch rule along the length of the tube. When the tube is removed, a snow core is extracted.

To determine snow water equivalent the core and tube are weighed on a spring scale. The result minus the weight of the empty tube equals inches of snow water equivalent. The diameter of the sampling tube cutter teeth at 1.485 in. (3.772 cm) makes it possible to measure the snow water equivalent directly: 1 in. (2.54 cm) of water at this diameter (1.73 in.^{3} or 7.359 cm^3) weighs 1 oz (28.34 g). Therefore, a scale calibrated to measure ounces also provides a direct reading in inches of water equivalent. The measurements for each transect point are averaged to produce one value for the site. The snow

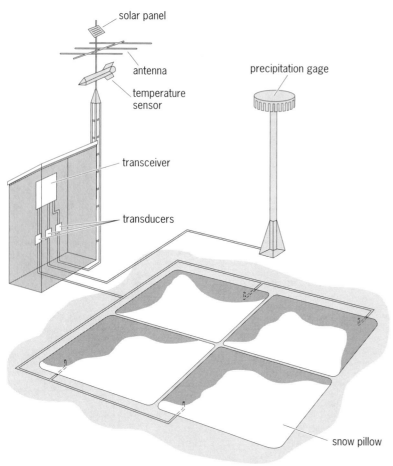

solar panel

antenna

temperature sensor

precipitation gage

transceiver

transducers

snow pillow

Configuration of a typical SNOTEL site for automated snow measurement. (*U.S. Department of Agriculture, Soil Conservation Service, Snow surveys and water supply forecasting, Agr. Inform. Bull. 536, 1988***)**

it is considered to be ripe. At this point the snowpack becomes isothermal; that is, it has a uniform temperature of 32°F (0°C) from top to bottom, and conditions are ready for the snow to melt. With knowledge of meteorologic conditions and snow density data hydrologists can determine if snow melt is imminent.

Automated snow measurement. As dependence on water supply forecasting continued to grow, there was a need for more frequent data on snowpack conditions. The Soil Conservation Service found it necessary to develop an automated snowwater-equivalent sensor and a system that could transmit the data to a central computer center, where they would be available to forecasters on a daily basis. By 1975 the snow pillow, a fluid-filled envelope that measures snow water equivalent by sensing the pressure of the snow load upon it, had been developed and was being installed along with radiotelemetry equipment (see **illus.**). The new snowpack telemetry (SNOTEL) sites were often co-located with snow courses to obtain comparative values of the two measurements methods. Other SNOTEL sites were located in remote areas of mountain watersheds that were inaccessible in winter. The use of a telemetry system with a technology based on meteor trails (meteor burst technology) to transmit data provided a low-cost, low-maintenance method of data retrieval because it operated without the use of satellites. SNOTEL improved water supply forecasts by providing daily updates of snow water equivalent and a rough indication of daily water loss from the snowpack. Many snow courses are being phased out in favor of SNOTEL sites. There are approximately 600 SNOTEL sites currently operational in the western United States and Alaska. Before a snow course is discontinued, however, the new SNOTEL site is co-located with it for several years to preserve the historical data record.

SNOTEL sites. A snowpack telemetry site consists of a snow pillow, radiotelemetry equipment (housed in a shelter), an antenna, and a solar panel. A typical site is also equipped with a storage precipitation gage and a temperature sensor. Early designs used an array of one to four 4-ft-square (1.22-m) stainless steel envelopes filled with an antifreeze solution. Later, a single 12-ft-diameter (3.66-m) butyl rubber pillow was substituted. When snow accumulates on the pillow, it exerts pressure on the antifreeze solution, which flows through tubing into a manometer. The system is corrected for the specific gravity of the antifreeze solution and calibrated so that the height of water in the manometer is equal to inches of snow water equivalent. The force exerted on the antifreeze is sensed by a pressure transducer and is converted into a voltage value. This electronic signal can then be stored on site in the memory of a microprocessor and later transmitted on demand to a central computer system. One drawback to the snow pillow is that it provides a measurement of only the snow water equivalent. Development of automated devices capable of measuring depth and density is in

course data provide a representative measure of snow conditions at a particular elevation within a basin or watershed. These data are incorporated into hydrologic models to assist in developing a forecast.

Another measure of snowpack condition, snow density, is determined from depth and water equivalent data as in the formula shown below. Snow

$$\text{density} = \frac{\text{snow water equivalent}}{\text{depth}}$$

density, the amount of water per unit volume of snow, increases with time through continuous metamorphic processes. When snow accumulates on the ground, its crystalline form immediately begins to change from that which fell through the atmosphere. The intricate shapes of snow crystals are transformed to rounded snow grains or angular, faceted crystals. The metamorphic processes are largely due to temperature and vapor fluxes within the snowpack and to wind action, resulting in increasing snow density as the season progresses. The snowpack has a connected network of air spaces where liquid water from melt or rain can be held by capillary action. As snowpack density approaches its seasonal maximum, its capacity to hold liquid water is reached, and

progress, but none are currently in use with SNO-TEL sites.

Meteor burst technology allows radio communication between two sites at a maximum distance of 1200 mi (1900 km). This telemetry system utilizes the trails left by billions of sand-sized meteorites, which enter the atmosphere daily, to reflect radio signals between a master station and the SNOTEL sites. The master station polls each site by its unique address and requests data. The site responds by transmitting the stored data. The communications are complete when the master station acknowledges receipt of the data and signals the site to stop transmitting. The entire procedure takes place in just a fraction of a second.

Areal measurements. The point measurements made at snow courses and SNOTEL sites provide input data for hydrologic models that estimate the total volume of water held as snow in a particular basin. The areal coverage of snow, however, is highly variable in depth and snow water equivalent. The recent development of techniques for the measurement of snow water equivalent over a larger area rather than a single point has the potential to improve forecast accuracy. One method currently being refined by the National Weather Service uses low-flying aircraft to monitor emissions of natural terrestrial gamma radiation along specified flight lines. The emission of gamma radiation from natural radioactive elements in the soil is measured with a gamma ray spectrometer. The presence of snow, water, or ice attenuates the gamma radiation signal; therefore, background gamma-level measurements are first made when the ground is free of snow. The flights are then repeated over snow-covered terrain. The attenuation of the gamma radiation signal due to the snowpack is used to determine the snow water equivalent. Corrections for soil moisture changes in the upper 8 in. (20 cm) of soil and for atmospheric moisture content must be applied. This technique provides real-time snow-cover water equivalent data for approximately 2.3 mi^2 (6 km^2) for each flight line. Other advances in snowpack monitoring technology include using satellite imagery to map snow cover, using digital photogrammetric terrain mapping techniques to calculate the volume of snow in a basin, and developing laser and acoustic sensors to automate snow-depth and -density measurements. The versatility of SNO-TEL continues to be realized; improvements in telemetry equipment now make it possible to incorporate up to 32 hydrometeorologic instruments at each site.

For background information SEE HYDROLOGY; SNOW; SNOW GAGE; SNOW SURVEYING in the McGraw-Hill Encyclopedia of Science & Technology.

Kristine E. Kosnik

Bibliography. S. S. Carroll and T. R. Carroll, Effect of uneven snow cover on airborne snow water equivalent estimates obtained by measuring terrestrial gamma radiation, *Water Resour. Res.,* 25:1505–1510, 1989; D. M. Gray and D. H. Male (eds.), *Handbook of Snow,* 1981; E. L. Peck, T. R. Carrol, and S. C. Van Demark, Operational aerial snow surveying in the United States, *Hydrol. Sci. Bull.,* 25:51–62, 1980; U.S. Department of Agriculture, Soil Conservation Service, Snow surveys and water supply forecasting, *Agr. Inform. Bull.* 536, 1988.

Software quality assurance

Software development is an error-prone process. Defects can be introduced during the development of requirements, design, and code. Much of the effort in software development is involved with finding and removing these defects. A number of approaches are used to detect defects, such as executing the code with specific test cases, using automated analysis tools, and performing a manual review of work in progress. A particular form of manual review, called software inspection, has been shown to be one of the most effective approaches for finding and removing defects.

Software inspection process. The inspection process involves a small group of software developers who review a work product with the objective of finding as many defects as possible. There are generally six steps in the inspection process: planning, overview, preparation, inspection meeting, rework, and followup.

1. The first step involves planning for the inspection of a particular work product, such as 10–20 pages of a requirements specification or 200–250 lines of code. Work products are inspected soon after they are completed. An inspection team consists of four to six software developers, each being assigned a role during the planning step. Most important is the moderator, who runs the inspection meeting and is responsible for keeping the inspection on track. The moderator ensures that the work product meets all entry criteria and is ready to be given to the inspection participants. The reader paraphrases the work product during the inspection meeting, and the author and other inspectors read along and comment on discrepancies. The recorder, also called the scribe, records the location and a brief description of all defects encountered during the inspection meeting (step 4).

2. The author provides the inspection team with an overview of the work product. This overview covers any information that will aid the inspectors in their review. For example, the author of a set of designs to be inspected may discuss the requirements from which the designs were drawn.

3. Inspectors prepare separately for the inspection meeting (step 4). Each inspector examines the work product in great detail with the objective of independently finding as many defects as possible. This examination may require several hours. Each inspector records a list of potential defects. Checklists of commonly occurring defects are often developed by the project and used to aid the inspector in finding defects.

4. A meeting is held to review the work product with all of the inspectors present. The moderator opens the meeting by verifying that all inspectors have individually reviewed the material in detail. If any individual has not sufficiently prepared for this meeting, the meeting is rescheduled. The reader paraphrases the work product, and defects that are discovered, whether during the individual preparation step or during the meeting, are briefly discussed and recorded. Solutions to the defects are not part of the agenda because such discussion would divert the meeting's focus away from finding defects. The duration of the inspection meeting is usually limited to a maximum of 2 h because the attention span of participants begins to dwindle after that time. Work products are sized appropriately during the planning step to fit this 2-h time limit.

At the conclusion of the meeting, the inspection team decides on whether a reinspection is necessary. If a large number of defects have been found, or if significant rework is necessary, the team may decide to reinspect the work product after corrections have been made.

5. All defects recorded during the inspection meeting are corrected. Corrections are the responsibility of the author of the work product. If the inspection team decides to hold a reinspection, the modified work product is scheduled for another review.

6. After the author has corrected all the defects that were detected, the moderator reviews the modified work product to ensure that the corrections were properly made and that no new defects have been introduced. The moderator then submits a summary of the inspection to project management. This summary includes information such as the number, type, and severity of the defects found; the amount of effort expended for the six inspection steps; and a list of inspection participants. This information is used to analyze defect trends, to characterize the return on investment for the effort involved, and to ensure that work products are being inspected according to the project plan.

Process variations. A number of variations and extensions to the inspection process described above have been successfully applied. Some organizations have found that the actual inspection meeting does not provide sufficient added value, since the majority of defects are found during the preparation step. Other organizations have added a step after the inspection meeting to discuss solutions to the defects recorded during that meeting. A step after the rework step might be included to discuss methods for preventing future occurrences of similar defects. For example, a postrework step would be useful if a particular mistake is frequently being made by several programmers.

Results. Software inspection can facilitate detection of a wide range of commonly occurring defects. These defects include missing or incorrect functionality, incorrect interfaces between modules, improper initialization or data validation, incorrect algorithms, and documentation problems. Inspection can reveal these defects during all phases of software development, not just the programming phase.

Organizations have reported extremely positive results from use of the inspection process. Although a variety of methods are used to detect defects, the inspection process often finds more than half of all defects discovered during a software development project. When compared to other forms of traditional testing, inspections are often 10–20 times more efficient at finding defects. Organizations that keep detailed records of the effort and results of inspection find that the return on investment for the process is on the order of 10 to 1.

The inspection process provides additional benefits. By participating on an inspection team, junior developers gain valuable exposure to the knowledge and practices of more experienced developers. Managers are better able to assess the status of a project because completed inspections provide a sound basis to judge progress.

State of practice. Although the benefits and effectiveness of inspection are well documented, not all software projects achieve the results described above for a number of reasons.

For example, the pressure of remaining on schedule is usually encountered in a software project; a project's schedule is often emphasized over the quality of the work products. Thus, the time required for inspection is sometimes spent completing work products. Too often, however, later efforts to fix defects in the finished work products are much more costly and time consuming than corrections made in response to the inspection process.

Effectiveness of inspections can be greatly reduced by not rigorously following the process. For example, allowing discussion of solutions to defects during the inspection meeting consumes valuable time that is better spent finding additional defects. Attempting to inspect work products that are too large or exceeding 2-h meeting limits similarly reduces inspection results. Many such potential pitfalls must be diligently avoided in order to maximize the benefits of inspection.

Software inspection is often not practiced because of a lack of awareness of the process. Software development is not a full-fledged engineering discipline with accepted standards of practice and methods for accrediting the use of such standards. Thus, many people involved in software development do not get exposed to the rigorous processes that would constitute a software engineering discipline.

Successfully introducing an inspection process into a software development project is not an easy task. It requires a highly motivated individual to guide project personnel in the use of the process. Training of the participants, and education for managers, who must understand and accept the value of the process, are also essential. Once the inspection process is in place, it must be periodically monitored to ensure that it is working properly.

For background information *SEE SOFTWARE ENGINEERING* in the McGraw-Hill Encyclopedia of Science & Technology.

Bill Brykczynski

Bibliography. T. Gilb and D. Graham, *Software Inspection*, 1993; G. W. Russell, Experience with inspection in ultralarge-scale developments, *IEEE Softw.*, 8(1):25–31, January 1991; E. F. Weller, Lessons from three years of inspection data, *IEEE Softw.*, 10(5):38–45, September 1993.

Soil

In regions of Africa, the Americas, Asia, and Europe, ancient agricultural land has been identified in archeological sites and in places still farmed by indigenous peoples. One such area occurs high in the western cordillera of the Peruvian Andes in the Colca Valley, known for a remarkable agricultural ecosystem created on its slopes more than 1500 years ago (**Fig. 1**). At the core of this terraced agroecosystem are soils, carefully built up to store and supply water and nutrients for crops that have nourished generations from ancient times to the present day. Besides its value in archeological and geographical understanding of human-environmental relationships, knowledge gained from studying these soils provides a unique and long-time perspective needed to further current efforts toward sustainable agriculture and the conservation of natural and cultural resources.

Environment and agriculture. Although the semiarid, cool climate and steep terrain make the Colca Valley a challenging environment for agriculture, the valley's potential productivity was realized by its ancient inhabitants. Environmental hazards throughout the more than 15 centuries of agriculture include limited water, frost, erosion (including landslides), earthquakes, and volcanism. The valley was carved by streams cutting deeply into the altiplano, or high plateau region, of the Andes, from which glaciated volcanic peaks reaching about 20,000 ft (6100 m) rise. Flights of agricultural terraces extend from the valley base to near the altiplano within a larger stepped topography of gently sloping fluvial surfaces and intervening steeper slopes, in an elevation range of roughly 9000–13,000 ft (3000–4000 m). Natural soils, sparsely vegetated with perennial grasses, shrubs, and cacti, are formed primarily in alluvium and colluvium derived from andesite and other volcanic rocks. Like the native grasslands of the Great Plains in the United States or of Ukraine, the valley contains Mollisols, mineral soils with thick dark surface horizons (layers) enriched in organic matter. The soils comprise a time sequence of development on progressively older and higher valley surfaces. Soils on older geomorphic surfaces dating back many millennia through the Pleistocene have had the time and the stability necessary to develop subsurface horizons with accumulations of clay, calcium carbonate, and cemented silica.

As in other Andean highlands, much of the Colca Valley has been transformed into a terraced landscape, with skillfully engineered rock walls up to 7–10 ft (2–3 m) high, and irrigation systems, originating near glaciers, that deliver good-quality water. The long history and continuity of Colca Valley agriculture, predating the Inca empire by at least 1000 years, is confirmed by radiocarbon ages from charcoal and buried organic matter in terraced soils and adjacent archeological sites. Buried walls and stratified sediments discovered beneath some agricultural terraces indicate a complex agricultural evolution. Although many agricultural terraces are still farmed, more than half are abandoned, probably since the early Spanish colonial period, when the indigenous population plummeted and social structure was disrupted. Today, a variety of native crops (maize, potatoes and other tubers, quinoa) and introduced crops (barley, wheat, oats, fava bean, alfalfa) are grown by descendants of the pre-Columbian farmers using traditional soil and crop management practices. Although production under traditional cultivation is difficult to quantify and to compare directly with modern intensive agriculture and crop varieties, Colca Valley yields reported in the 1970s and 1980s for crops such as barley and potatoes were within the midrange of yields in the United States.

Ancient agricultural soils. During centuries of terrace agriculture, land surfaces and the physical, chemical, and biological properties of soils have changed significantly from their natural state, altering conditions for crop growth. To evaluate soil change from long-term agriculture, nearby uncultivated soils in similar environmental settings and stages of natural development serve as references for comparison.

In constructing agricultural terraces, ancient Peruvian farmers modified steep terrain unsuitable for sustained agriculture into stepped, gently slop-

Fig. 1. Ancient agricultural terraces in the Colca Valley, Peru. The currently farmed and abandoned agricultural terraces are on the nearest slope. The volcanic peaks in the background (glacially faceted inactive volcano on right, active volcano on left) are about 20,000 ft (6100 m) in elevation. (*From R. R. Pawluk, J. A. Sandor, and J. A. Tabor, The role of indigenous soil knowledge in agricultural development, J. Soil Water Conserv., 47:298–302, 1992*)

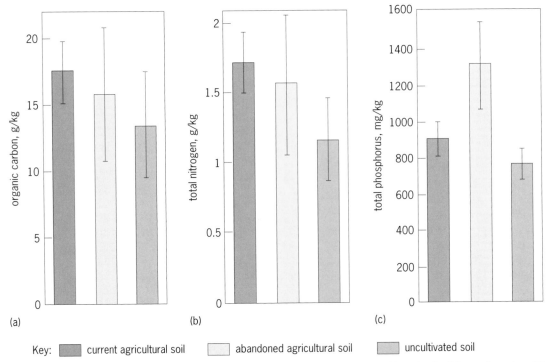

Key: current agricultural soil   abandoned agricultural soil   uncultivated soil

Fig. 2. Comparison of upper A horizon chemical properties among agricultural and uncultivated soils: (a) organic carbon, (b) nitrogen, and (c) phosphorus. Bars are means (vertical lines within bars show ±1 standard deviation) from 30–34 samples of each sample group. Mean differences are statistically significant minimally at the 0.05 probability level in all cases except between current and abandoned agricultural soils for organic carbon and total nitrogen.

ing segments with the stability required for crops and their irrigation. This transformation was accomplished by bench terracing, where stone walls were built along slope contours at intervals, with subsequent filling and retention of soil upslope of each wall. Wall bases were securely anchored into hard subsurface soil horizons. Although terrace walls require maintenance, the majority of them have endured intact, as attested to by today's farmers who climb between terraces along stepping-stones jutting from rock walls tightly fitted without mortar.

A primary consequence of terracing and long-term incorporation of organic fertilizers is dark, humus-rich topsoils (A horizons), commonly 1–4 ft (0.3–1.3 m) thicker than natural soils, which provide crops with a greater volume for root development and reservoir of moisture and nutrients. Other signs of improved soil tilth are strong aggregation of soil mineral and organic particles in granular shapes, many large pores, slightly lower density because of greater porosity, and earthworm activity. Deeper organic-matter-enriched horizons are also common; some are original A horizons buried by the filling process, while others may be old fertilizer additions.

Fertility of some ancient agricultural soils is also enhanced (**Fig. 2**). Both presently cultivated and abandoned agricultural A horizons contain significantly greater amounts of organic carbon and nitrogen than uncultivated soils. Levels of nitrate, a key form of plant-available nitrogen, are relatively high in presently farmed soils but lower in uncultivated and abandoned agricultural soils. Nearly all soils have optimal pH (slightly acid to neutral) for nutrient availability. Fertilization with manure and ashes, a long-standing practice in the Andes, has also enriched soils with phosphorus; available phosphorus for crops is more than sufficient. A surprising finding was the particularly high levels of phosphorus in some abandoned agricultural soils, possibly reflecting the prehistoric importance of guano (this phosphorus-rich seabird manure was especially abundant in Peru's coastal islands) as a fertilizer. Phosphorus in some soils reaches weight concentrations of 4000–10,000 mg/kg (versus typically 500–900 mg/kg in uncultivated soils), exceeding levels in many urban archeological sites where high phosphorus levels from human activity are expected. Substantial fertilizer-derived phosphorus has moved deeply into physically undisturbed subsurface horizons, which is notable because soil phosphorus is well known for low mobility. Relative to uncultivated soils, agricultural soils have steady or elevated levels of three enzymes (β-glucosidase, amidase, and phosphatase) associated with organic carbon, nitrogen, and phosphorus, in contrast to lower enzyme activities and related microbial biomass common in modern conventional agricultural systems in North America.

These soils exhibit similarities to old anthropogenic soils more commonly recognized in Europe and Asia. They include plaggen soils (thickened by long-term additions of organic or mineral amendments), anthropic horizons (enriched in phosphorus), and agric horizons (subsurface accumulation of fine mineral and organic particles translocated from cultivated topsoils).

Indigenous knowledge of soils. Evidence pointing toward careful management of Colca Valley soils by ancient and present populations led to an investigation by anthropologists and soil scientists into indigenous knowledge underlying the ancient agricultural system. The study found a complex soil classification system containing about 50 kinds of soil expressed in the native Quechua language (including Spanish loan words), and an in-depth awareness of soils, landscapes, and soil-crop relationships. The soil classification system has four hierarchical levels and incorporates several soil physical and chemical properties, with an emphasis on texture (proportions of sand, silt, and clay). Texture greatly influences soil water retention and availability for plants, an especially important consideration in semiarid land agriculture. One taxonomic class deals with geophagy (eating certain earths), which may be practiced to offset toxins found in potatoes and other tubers prevalent in Andean diet. Farmers also recognize depth and lateral soil variation, and make subtle distinctions among soils and other earth materials. Keen understanding of the physical environment among ancient farmers and their modern descendants is clearly evident in the astute placement of agricultural fields and features in different landscape settings. To illustrate, agricultural terraces were placed between parallel ridges of volcanic dikes oriented like a chute on a mountainside, and natural landslide depressions were modified to serve as reservoirs and wetlands for waterfowl. The way that agricultural terraces merge with the mountainous landscape, rather than being imposed upon it, reflects a cultural view in concert with its environment, a view characteristic of many indigenous societies living on the same land for a long time. A land ethic is also evident in religious customs associated with agriculture, such as pouring liquor on the soil as a toast prior to plowing and planting.

Sustainable agriculture. The Colca Valley agricultural system represents a fairly rare example of successful long-term soil conservation. Contributing factors are traditional management techniques such as terracing, minimum tillage, repeated inputs of organic residues, and crop rotation. More common is the decline of soil quality under intensive agriculture. Documented degradation of ancient agricultural land includes accelerated erosion in Europe, Africa, and Mesoamerica, salinization in Mesopotamia, and a case of compaction and decreased organic matter and nutrients persisting centuries after abandonment in New Mexico. Many instances of similar soil deterioration under modern intensive agriculture in the United States and other countries have been reported. Widespread environmental quality problems have spurred efforts to develop sustainable agriculture systems

that are both protective of land resources and productive. Ancient agricultural lands constitute a resource for studying soil processes and for learning which practices and conditions lead to conservation or degradation, across long time spans needed to evaluate sustainability.

For background information *SEE AGRICULTURAL SOIL AND CROP PRACTICES; SOIL; SOIL CHEMISTRY; SOIL CONSERVATION* in the McGraw-Hill Encyclopedia of Science & Technology.

Jonathan A. Sandor

Bibliography. D. Hillel, *Out of the Earth: Civilization and the Life of the Soil,* 1991; W. M. Denevan et al. (eds.), Pre-Hispanic agricultural fields in the Andean region, *Brit. Archaeol. Rep., Int. Ser.,* vol. 359, no. 1, 1987; J. A. Sandor and N. S. Eash, Ancient agricultural soils in the Andes of southern Peru, *Soil Sci. Soc. Amer. J.,* 59:170–179, 1995; K. D. Thomas (ed.), Soils and early agriculture, *World Archaeol.,* vol. 22, no. 1, 1990.

Soil chemistry

Soil chemistry is preeminently the study of natural particles and their reactions with substances that are dissolved in water or air (solutes). Foremost among these reactions is adsorption, the interaction between a solid particle and solutes that causes the latter to accumulate on the particle surface to form one or more layers. Soil chemists were the first to characterize adsorption reactions quantitatively, and they continue to bring forth new applications of basic concepts developed in chemical science.

A fundamental property of natural particles is their surface roughness. This facet of their structure reflects the unrelenting, etching attack of carbonic acid together with other agents of chemical weathering, as well as the sporadic deposition of highly heterogeneous, organic and inorganic products of weathering to form incomplete surface coatings. The interfacial features that evolve from these two multifarious processes are likely to be complex, but recent investigations have shown that some of the complexity is susceptible to description by fractal concepts. The term fractal refers to the limiting properties of mathematical objects that exhibit three essential attributes: (1) similar structure over a range of length scales (or magnifications); (2) intricate structure that is itself scale independent; and (3) irregular structure that cannot be described entirely by euclidean geometry, necessitating the use of a spatial dimension that is not an integer. Like the mathematical objects in euclidean geometry—spheres, cubes, circles, and squares—those in fractal geometry are idealizations that natural objects only approximate. Nonetheless, they are useful to the quantitative description of natural particles.

Fractal particle surfaces. When a molecule is adsorbed onto a roughened surface to form a single layer (termed a monolayer), it interacts with a broad array of surface sites exhibiting an irregular geometric structure. Some of this structure will be accessible to the molecule if the length scale over which irregularity occurs is considerably larger than the size of the molecule. In this case, the adsorbing molecule is said to resolve the irregular surface structure. If, however, the surface irregularity shows up only on a length scale that is considerably smaller than the size of the molecule, the molecule cannot resolve the surface structure except approximately, leaving undetected many shallow pits, impenetrable crevices, and grottos of excluded volume.

If the adsorbing molecule were replaced by a smaller molecular probe of surface structure, better resolution of the irregularity would be expected. An experimental signature of this change to a second, smaller probe is an observed increase in the number of molecules (n) required to form a complete monolayer on the surface. If the surface were smooth (that is, had no irregular structure inaccessible to either of the two probe molecules), n would increase in inverse proportion to the packing area covered in a monolayer by a single molecule; each type of probe molecule would adsorb simply to pave what for it is locally a flat surface. If the surface always exhibits roughness on a scale far too small for the two different probe molecules to detect, n will still increase as the probe size decreases, but now the increase will be larger (for a given drop in size) than what occurs for a smooth surface. The reason is that smaller probe molecules can in fact find more sites on which to adsorb than do larger probe molecules.

A mathematical representation of the inverse relationship between n and molecular size (expressed conveniently by the monolayer packing area, a_m) has the form shown in Eq. (1), where D_a is

$$n \propto 1/a_m^{D_a/2} \qquad (1)$$

a parameter known as the fractal dimension of a surface. The smallest possible value of D_a is 2, signifying a fully accessible (smooth) surface, and leading to the corresponding inverse proportionality between n and a_m as discussed above. The upper limit of D_a is 3, the euclidean dimension of the space into which the surface is embedded. As D_a approaches 3, the surface becomes so convoluted that the grottos of excluded volume are its predominant feature, and adsorbing molecules encounter an interfacial region that resembles a microporous sponge whose tortuous surface is effectively space filling. Between 2 and 3, D_a takes on noninteger values that characterize the degree of surface roughness or interfacial tortuosity. These properties, in turn, are important determinants of particle reactivity toward solutes. Surfaces with high fractal dimensions expose a broad variety of reactive sites whose accessibility depends strongly on the molecular size of the adsorbing solute, implying a high level of selectivity in the retention of nutrients, the trap-

ping of pollutants, and the susceptibility to acid attack in weathering.

Representative values of D_a for soil particles and for particulate matter composed of one type of soil constituent (for example, a metal oxide or a clay mineral) are listed in the **table**. These data were obtained by measuring n for a series of adsorbing molecules whose packing areas lay in the range 0.1–2.0 square nanometers. Alternatively, some of the data were the result of experiments in which particle size (and therefore adsorbing surface area) was varied while the same probe molecule was used to adsorb to form a monolayer. In this case, the surface area of the particle (A), calculated as the product, na_m, is assumed to be proportional to L^{D_a}, where L is the particle diameter, as shown in Eq. (2). This

$$A \propto L^{D_a} \qquad (2)$$

relationship is the conventional euclidean proportionality between area and diameter squared when $D_a = 2$, but a larger rate of increase in area with increasing particle diameter is expected if the probe molecule can explore more sites as a fractal particle surface becomes magnified by increasing the particle size. The table makes clear that natural particles found in soils have roughened surfaces that are conveniently described by fractal dimensions larger than 2.

Fractal coagulated particles. The particles most chemically reactive in soils fall into the diameter range of 0.01–10 micrometers. These particles usually do not dissolve readily in water but remain as identifiable particles in suspension. A suspension is stable (and the particles in it dispersed) if no particle growth to cause gravitational settling occurs over periods of hours to days. Stable suspensions of soil lead to erosion, because the particles so entrained remain highly mobile. Stable suspensions also affect the mobility of nutrients and pollutants that may become strongly adsorbed to soil particles. The process by which a suspension becomes unstable and its particles undergo settling is called

Surface fractal dimensions of soil particles and particles composing soil constituents

Particle composition	Fractal dimension of surface (D_a)
Soil	2.1–3.0
River alluvium	2.7–2.9
Sandstone	2.5–3.0
Shale	2.3–2.8
Aluminum oxide	2.0–3.0
Iron oxide	2.4–2.7
Clay minerals	2.1–2.4
Quartz	2.0
Silica	2.1–2.7
Carbonate	2.0–2.8
Feldspar	2.4
Organic matter	2.5

SOURCE: D. Avnir, D. Farin, and P. Pfeifer, A discussion of some aspects of surface fractality and its determination, *New J. Chem.*, 16:439–449, 1992.

coagulation. Coagulation that produces bulky particles with high porosity is called flocculation. Dense, organized masses of particles are formed during subsequent drying and wetting cycles in soil by the process of aggregation.

In quiescent water, the motions of soil particles are incessant and chaotic because of the particles' thermal energy. These brownian motions in a suspension are analogous to the diffusive motions of molecules in a solution. In the absence of long-range repulsive forces (Coulomb forces), the flocculation of soil particles is produced by collisions as a result of brownian motion. Flocculation caused in this way is said to exhibit transport control if the rate-limiting process is the movement of two (or more) particles toward one another prior to instantaneous combination into a larger particle. Reaction control occurs if particle combination instead of particle movement (toward collision) limits the rate of flocculation.

Microscope images of floccules (like a soil aggregate) have been examined to determine the relationship between the number of primary particles, N, and the spatial extent of the floccule, expressed conveniently as some length, L. In a simple approach, L can be estimated by the mean floccule diameter. The results of these studies can be described mathematically by a power-law equation [Eq. (3)], similar to the surface area–particle diameter relationship in Eq. (2). In Eq. (3), D_c is a positive parameter between

$$N \propto L^{D_c} \qquad (3)$$

0 and 3, termed the cluster fractal dimension. This power-law proportionality can be interpreted as a generalization of the relation between the number of primary particles in a cluster that is d-dimensional ($d = 1, 2,$ or 3) and the d-dimensional size of the cluster. For example, if a cluster is one-dimensional ($d = 1$), it can be portrayed as a straight chain of, say, circular primary particles of diameter d_0. The number of such particles in a chain of length L is proportional to L; that is, the relation shown in Eq. (3) holds, with $D_c = 1$. If the cluster is two-dimensional, it can be represented by a parquet of circular primary particles packed together so that they touch. The number of particles in the cluster then will be proportional to L^2, and $D_c = 2$.

In a more complex situation, a sequence of clusters is constructed by repeated additions of a unit comprising five disks, each of diameter d_0 (see **illus.**). The unit itself has a characteristic diameter equal to $3d_0$. If five units are combined to form a cluster with the same symmetry as one unit (that is, each unit in a cluster is arranged like a disk in one unit), the characteristic diameter grows to $9d_0$. If five clusters are then combined in a way that preserves the inherent symmetry, the diameter increases to $27d_0$, and so on. The clusters formed in this process exhibit similar structure (in the sense of their symmetry), intricate structure, and irregularity. Therefore, they qualify as fractal objects.

The number, N, of primary particles in each cluster of the sequence is 5, 25, 125, or 625 for the four

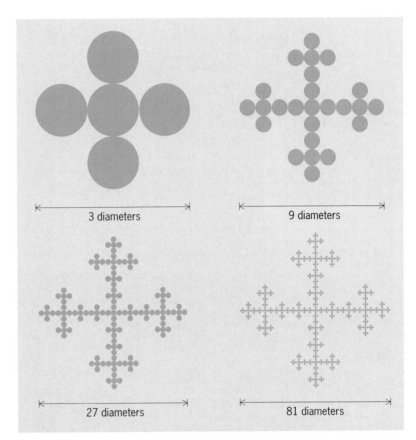

Sequence of clusters created by combining primary units comprising five particles to form larger structures that preserve the symmetry of the original unit. (*After P. Meakin, Fractal aggregates in geophysics, Rev. Geophys., 29:317–354, 1989*)

examples shown in the illustration. Thus, $N = 5^n$, where $n = 1, 2, \ldots$, denotes the stage of cluster growth in the sequence. In the same way, the characteristic diameter $L = 3^n d_0$, where $n = 1, 2, \ldots$, once again. The relationship between N and L in this example thus can be expressed by Eq. (4), where

$$5^n \propto (3^n)^{D_c} \qquad (4)$$

the value of D_c, given by Eq. (5), is the appropriate

$$D_c = \ln 5/\ln 3 \approx 1.456 \qquad (5)$$

cluster fractal dimension characterizing the sequence.

Like D_a for surfaces, the fractal dimension of a cluster, D_c, reflects its space-filling capability. The clusters in the illustration occupy space in a plane (euclidean dimension = 2). If they were disks arranged in a single row, the fractal dimension that characterizes them should be 1.0, because the cluster would be effectively one-dimensional objects. However, if they were compact structures comprising closely packed disks, their fractal dimension would be 2.0, indicating their space-filling nature. The fractal dimension of the clusters actually is near 1.5, because the clusters have a porous structure that is not entirely space filling.

Theoretical studies of particle coagulation based on computer models have shown that fractal clusters are formed in both transport-controlled and reaction-controlled flocculation processes. In the former case, $D_c = 1.8$, whereas for the latter, $D_c = 2$. The second, larger fractal dimension, as compared to that for floccules whose growth is transport controlled, is the result of the greater opportunity that the particles have to combine in close-packed arrangements when their coalescence is not instantaneous upon collision. Reaction-controlled flocculation thus is expected to be a slower process than transport-controlled flocculation because of the intervention of particle repulsion. This difference can be observed by following the time dependence of L. Measurements made on suspensions of silica, metal oxide, organic matter, and clay mineral particles all have confirmed these basic ideas, including the two predicted values of D_c.

For background information SEE ADSORPTION; BROWNIAN MOVEMENT; FRACTALS; SOIL CHEMISTRY; WEATHERING PROCESSES in the McGraw-Hill Encyclopedia of Science & Technology.

Garrison Sposito

Bibliography. D. Avnir (ed.), *The Fractal Approach to Heterogeneous Chemistry*, 1989; D. Avnir, D. Farin, and P. Pfeifer, A discussion of some aspects of surface fractality and of its determination, *New J. Chem.*, 16:439–449, 1992; N. Senesi and T. M. Miano (eds.), *Humic Substances in the Global Environment and Implications on Human Health*, 1992; H. van Damme, Scale invariance and hydric behavior of soils and clays, *C.R. Acad. Sci. Paris*, 320a: 665–681, 1995.

Solar cell

Solar cells are semiconductor devices which convert sunlight into electricity through the photovoltaic effect. The conversion process is quiet and produces no pollution.

Solar cells could in principle supply all the energy needs of humankind, but so far their application has been limited to special situations such as providing the energy needed on orbiting space satellites or space stations, providing small amounts of power in remote locations (for microwave repeaters, water pumping, pipelines, and so forth), and powering consumer products such as hand calculators and toys. The main impediment to large-scale generation of electricity for terrestrial applications by solar cells is their cost, which in turn is affected by their efficiency and long-term durability. Consequently, research on solar cells continues to focus on reducing costs, increasing efficiency, and enhancing durability.

Considerable progress has been made on all three fronts since the invention of the modern solar cell in 1953. The first cells had a solar energy conversion efficiency no higher than 6% and a life of at best several years. The commercially available 1995 solar cell costs about $\frac{1}{5}$ as much per peak watt as the 1953 cell, has an efficiency of about 14%, and can be expected to last for several decades. (A peak watt is the amount of solar cells which produce 1 W of electrical

power when exposed to terrestrial sunlight having an intensity of 1000 W/m². If the solar cell has a conversion efficiency of 10%, the area comprising 1 peak watt is 100 cm².) However, studies show that the cost per peak watt must drop an additional factor of 4 for 15%-efficient solar panels having a life of at least 20 years before large-scale electrical power generation by solar cells can become competitive economically with electrical power generated by fossil-fuel and nuclear power plants.

Early solar cells. Understanding current solar-cell research requires some knowledge of how it evolved historically. Two very different solar cells appeared in 1953: the single-crystal silicon (Si) cell (efficiency about 6%) and the single-crystal cadmium sulfide (CdS) cell (efficiency about 4%). Both were pn-junction devices. The silicon cell was a homojunction, that is, the p and n regions were parts of the same silicon crystal; but the cadmium sulfide cell was a heterojunction device which consisted of two different semiconductors, namely, a layer of p-type copper sulfide (Cu_xS) incorporated into an n-type cadmium sulfide substrate. The silicon homojunction device was amenable to analysis within the framework of the 1953 semiconductor device theory, and within a year its efficiency was increased to 10% by a small change in its design.

Analysis of the Cu_xS/CdS cell proved to be less successful, and so its efficiency remained around 4%. In 1956, a new Cu_xS/CdS cell which used a thin polycrystalline film of cadmium sulfide about 5 micrometers thick was invented. Its efficiency was comparable to that of cells made from single-crystal cadmium sulfide, but it promised to be much less expensive than the single-crystal silicon and cadmium sulfide cells because depositing thin films costs much less than growing single crystals. Research on solar cells ever since then has divided into two major streams: research on single-crystal cells and research on thin-film cells.

Solar-cell efficiency. In 1955, a more general analysis of the photovoltaic solar energy conversion process showed that silicon, cadmium sulfide, and copper sulfide were not the best choices as solar-cell semiconductors. The theory showed that the forbidden energy gap of a semiconductor is a

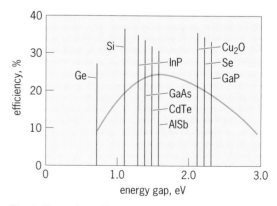

Fig. 1. Theoretical efficiency versus energy gap for solar cells exposed to sunlight.

first-order parameter in determining the ultimate efficiency of a solar cell made from that semiconductor. **Figure 1** shows how the ultimate efficiency depends on the energy gap. According to this theoretical curve, solar cells made from semiconductors having energy gaps between 1 and 2 eV have potential efficiencies in excess of 20%. Furthermore, the curve shows that the highest efficiency would be obtained from a semiconductor with an energy gap around 1.5 eV. Thus, the bandgaps of silicon (1.1 eV) and copper sulfide (1.0 eV) place these materials in the 1–2 eV range but not at the peak of the efficiency versus energy gap curve.

This theoretical result prompted research, which has continued ever since, on solar cells made from other semiconductors whose energy gaps are in the 1–2-eV range, and especially on solar cells from semiconductors with energy gaps close to the optimum value of 1.5 eV. Thus, cells were eventually based on materials such as indium phosphide (1.25 eV), gallium arsenide (1.35 eV), cadmium telluride (1.5 eV), tungsten diselenide (1.5 eV), copper indium sulfide (CISu; 1.5 eV), amorphous silicon hydride (a-Si:H; 1.7 eV), and copper indium selenide (CISe; 1.0 eV). Most research included both single-crystal and polycrystalline thin-film cells from the same semiconductor. In most cases, single-crystal cells had higher efficiencies than polycrystalline cells. Homojunction cells were usually more efficient than heterojunction cells. Cells with new designs based on sophisticated theoretical models have exhibited efficiencies close to the theoretical limits. Such cells, of single-crystal silicon, indium phosphide, and gallium arsenide, have achieved efficiencies in excess of 20%.

Thin-film cells. The Cu_xS/CdS cell is no longer a contender among thin-film cells because its efficiency never exceeded about 10% and this efficiency decayed with time. Thin-film research is now concentrated on copper indium selenide cells, which achieved a stable efficiency of 16% in 1994; on cadmium telluride cells, which achieved a stable efficiency of 15% in the 1990s; and on amorphous silicon hydride cells, which achieved a stable efficiency of about 10% in 1994.

The research on these thin-film cell semiconductors is aimed at elucidating the connection between the properties of these materials on an atomic scale and the macroscopic properties which determine the performance of solar cells made from them. For example, in the case of copper indium selenide, it has been found that high-efficiency solar cells require composition variations over the 2-µm thickness of the copper indium selenide layer. Some gallium is substituted for indium and some sulfur is substituted for selenium so that the actual composition changes from $CuInSe_2$ to $CuIn_yGa_{(1-y)}Se_2$ or $CuInSe_{2z}S_{2(1-z)}$. These alloys have larger bandgaps than copper indium selenide, but the best cells fabricated so far do not use any one of these materials to the exclusion of copper indium selenide. Rather, they are used to produce a thin-film semiconductor

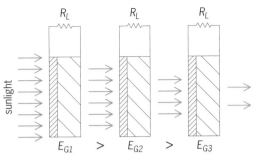

Fig. 2. Tandem-cell system with separate electric loads R_L. Cells are made from semiconductors with energy gaps $E_{G1} > E_{G2} > E_{G3}$.

in which the composition varies with position.

Another thin-film research area, still in its initial stages, is the fabrication of multijunction stacked solar cells (**Fig. 2**). The energy gap of the photovoltaic cell on the side receiving the sunlight is largest, and successive cells in the stack have successively smaller energy gaps. Theory shows that such tandem cells can achieve significantly higher efficiencies than single-junction cells. For example, an optimum-design three-junction tandem-cell structure has an efficiency limit of about 34%. As photovoltaic technology progresses, such tandem-cell systems will become the basic building blocks of large-scale photovoltaic power plants.

Cost reduction. As noted above, all research on thin-film cells can be classified as solar-cell cost-reduction research. A parallel effort at reducing the cost of single-crystal cells has focused on less expensive methods of making single-crystal silicon sheets. The current standard method of producing silicon single-crystal wafers is to grow large ingots, as long as 2 m (80 in.) and up to 20 cm (8 in.) in diameter, of silicon crystals, and to slice the crystal into wafers of the thickness required for solar cells (about 0.2 mm). The slicing by diamond saws or diamond-coated wires results in grinding almost 50% of the single-crystal ingot into dust. A number of techniques which result in the growth of single-crystal sheets have been developed. Laboratory or pilot-plant-scale production of solar cells based on these sheet-growing methods has been achieved. Analyses of these production methods show that if they were scaled up to achieve production levels required for large-scale solar-energy power plants, the cost of the solar cells would fall into the range needed to make such plants competitive with fossil-fuel and nuclear power plants. Very large investments would be necessary to determine whether the projected costs could be realized in an actual solar-cell plant.

World production. About 80% of solar cells now manufactured are either single-crystal or polycrystalline silicon. About 19% are made from amorphous silicon hydride; their efficiency is about 10%. About 1% are made from cadmium telluride, and have efficiencies around 12%. A significant fraction of the amorphous silicon hydride and cadmium telluride cells is used in indoor applications such as hand calculators.

The world production of photovoltaic solar-cell modules was around 60 peak megawatts in 1994, which represents double the production in 1987. Manufacturing capacity is expected to increase by about 100 peak megawatts per year over the next few years. These numbers represent steady progress in photovoltaic solar-cell demand but fall far short of the production needed to have a significant impact on the world's power needs. Current world installed electrical power generation capacity is around 4×10^6 MW. For photovoltaic solar power to make a significant contribution, solar-cell manufacturing capacity will need to increase a thousandfold.

Such an increase will require at least eight doublings of current production levels. Since each of the past three doublings have required about 4 years, eight doublings would require more than 30 years. It is not obvious that the growth level of the past decade will be sustained, but it is reasonable to expect that the world demand for energy will continue to grow, and that as the technical problems associated with photovoltaic power are resolved it will play an ever larger role in the world energy supply equation.

For background information SEE PHOTOVOL-TAIC CELL; PHOTOVOLTAIC EFFECT; SOLAR CELL in the McGraw-Hill Encyclopedia of Science & Technology.

Joseph J. Loferski

Bibliography. Institute of Electrical and Electronics Engineers, *IEEE 24th Photovoltaic Specialists' Conference, 1994,* 1995; Institute of Electrical and Electronics Engineers, *IEEE 25th Photovoltaic Specialists' Conference, 1996,* 1996; *Progress in Photovoltaics,* quarterly.

Solution mining

Conventional underground mining practices include excavating highly mineralized rock (ore) from a stope and transporting the ore to the surface for processing. Tons of solid rock are excavated and processed to produce kilograms, and sometimes only grams, of final product. Processing of ore on the surface results in most of the excavated material remaining on the surface as waste or tailings with exposure to long-term degradation from wind and water.

Surface disposal of processed rock can directly affect habitat by burial beneath the tailings from the processing plant. In some mines, certain types of rock contain sulfide minerals that can decompose with exposure to water and air to form an acidic runoff that may affect nearby streams.

Solution mining systems vary from in-place leaching, which circulates leaching solutions between wells drilled into the solid rock mass from the surface without excavation, to heap leaching, which transports excavated and crushed rock to prepared surface pads for the application of leaching solutions. Stope leaching, a variation of solution mining that applies to circumstances that exist between the in-place and heap-leaching applications of solution mining, involves the application of leaching solutions to mineralized rock that has been fragmented and left in place underground. Stope leaching modifies conventional mining practices to significantly reduce the amount of material brought to the surface from underground mines, or to remine existing surface mine dumps by returning processed rock (backfilling) to excavated underground voids (stopes).

Stope leaching. A stope is an underground working area in a mine. The underground stope-leaching mining system (see **illus.**) involves applying leach solutions to ore that has been fragmented by blasting in place or by backfilling empty stopes. Only enough material is removed from the stope to allow for adequate expansion of the rock during blasting. Thus, the amount of material brought to the surface is reduced by at least two-thirds. After blasting, a leach solution containing chemicals or bacteria (or both) is circulated through the fragmented ore to dissolve the target mineral. The resulting solution containing the dissolved mineral (pregnant leach solution) is pumped to the surface, where the product is removed; the leach solution is regenerated for recycling underground.

Solution mining systems, including stope leaching, can be applied to various geologic settings and types of orebodies. Stope leaching was successfully applied to uranium mining in Canada during the 1980s. Current interest in solution mining, particularly stope leaching, involves copper and gold mining. Conventional surface and underground mining of copper oxide orebodies in Arizona has been supplemented by in-place and heap-leaching production using dilute sulfuric acid as the leaching

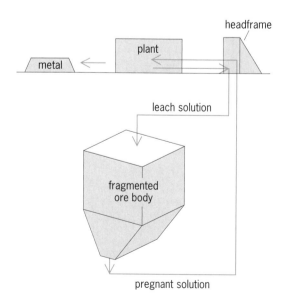

Stope leaching system in underground mines.

solution. After leaching, the copper is extracted from solution by using the solvent extraction/electrowinning (SX/EW) hydrometallurgical process.

Research requirements. Successful implementation of stope leaching in an underground mine requires economical methods for containing leach solutions at the stope and the development of economical leaching solutions that are environmentally compatible.

In stope leaching, the rock mass surrounding the stope becomes the leaching vessel. Detailed characterization of the rock mass prior to leaching is critical for ensuring solution containment and isolation. Preleach rock-mass characterization locates possible zones of high permeability, such as interconnected fractures in the rock, that can be sealed prior to introduction of leach solutions. Researchers have tested and compared rock-mass characterization procedures utilizing geologic mapping, various geophysical techniques, and hydrologic modeling at an underground test facility designed with boreholes serving as monitoring wells around a simulation leach stope in hard rock.

Researchers working on geochemical characterization of the rock matrix relative to various leach solutions and on leach solution chemistry have reported progress involving use of sodium thiosulfate for leaching gold from refractory carbonaceous ores, for leaching copper from chalcopyrite (a copper sulfide mineral) ores with ferric chloride solutions, and for the preleaching oxidation treatment of various minerals with biological agents.

Solution containment. In stope leaching, isolating the stope from surrounding ground water and the collection of metal-bearing solutions after leaching are critical. The goal is to design a system in which 100% of the solution is collected and pumped to the surface for processing. Although solution loss is the principal issue in solution management, ground water flowing into a stope can also be detrimental, because the water can render the leach solution ineffective with too much dilution; or the ground water can dilute the pregnant leach solution and make processing less effective.

Detailed rock-mass characterization prior to leaching is essential to ensuring minimal solution loss. It is important to collect data on the orientation, three-dimensional continuity, and condition of geologic features (that is, fractures, joints, faults, and foliation) that may be solution conduits. Standard geologic mapping and core logging practices are followed to obtain the required data on geologic features. Boreholes drilled in the rock mass to obtain core samples are used to conduct geophysical and hydrological testing. Much of the geophysical and hydrological equipment and techniques are adapted from water-well practices and from the petroleum industry.

Data evaluation includes statistical analysis to determine that the fracture sampling is adequate to represent the rock mass for computer modeling purposes. Fracture characteristics and results of permeability tests are used in a computer model to predict the direction and rate of fluid flow into the surrounding rock mass.

Leaching solution development. The type of leaching solution depends on the mineralogy and geochemical composition of the target mineral and the surrounding rock matrix. The rate that a mineral dissolves in a certain leaching solution is a major factor in applying solution mining to a specific orebody. The effect of these factors can be illustrated by copper heap leaching experience, where copper oxide minerals have recovery rates of 70–90% of the contained copper in a few weeks or months. Copper sulfide minerals may require 3 or more years to exceed 50% recovery.

Oxide copper mineralization in an orebody is generally the result of deposited sulfide minerals being exposed to air and water through natural processes over geologic time. Often an oxide orebody will overlie a sulfide orebody, both of which are part of the original mineral deposit. Breakthroughs in solution mining, including stope leaching, depend on the development of leaching solutions and systems that will treat sulfide orebodies more effectively.

The use of biological leaching to oxidize or to directly leach sulfide ore is an expanding area of research. Research on bacterial assistance in the leaching process has been focused on *Thiobacillus ferrooxidans*, which derives its life energy from the reaction between the iron(II) ion (ferrous) and the iron(III) ion (ferric). *Thiobacillus ferrooxidans* chemically attacks the ferrous ions (Fe^{2+}) to form ferric ions (Fe^{3+}). The bacteria catalyze the leaching reaction and accelerate the whole leaching process.

For background information SEE MICROBIAL DEGRADATION; SOLUTION MINING; UNDERGROUND MINING in the McGraw-Hill Encyclopedia of Science & Technology.

Carl H. Schmuck

Bibliography. N. C. Miller, *Predicting Flow Characteristics of a Lixiviant in a Fractured Crystalline Rock Mass,* U.S. Bureau of Mines, Rep. Investig. 9457, 1993; J. Salley, R. G. L. McCready, and P. L. Wichlacz (eds.), *Biohydrometallurgy,* Canada Centre for Mineral and Energy Technology, CANMET SP89-10, 1989; S. A. Swan and K. R. Coyne (eds.), *In Situ Recovery of Minerals II,* 1994.

Space flight

Space programs worldwide exhibited a number of significant developments in 1995, even while continuing to adjust to severe financial restrictions. As a consequence, developments increasingly tended to favor international consolidations and cooperative ventures between nations engaged in space scientifically, technologically, or commercially.

Significant activities, for example, included the progress made in the development of the International Space Station (ISS), the events around the operation of the Russian space station *Mir,* discoveries in the universe by the Hubble Space Telescope in its cosmic research, important findings from automated deep-space explorers such as *Ulysses* and the Jupiter-probe *Galileo,* and the growing attention by commercial firms to the potentials of space.

Commercial interest in worldwide space service markets has markedly increased. The advent of digital technology in telecommunications is a major force behind dozens of new advanced satellites being prepared for orbit, with most of the geosynchronous orbit growth expected to come from outside the United States, primarily from Europe and Asia. A new generation of commercial launchers for boosting these satellites is lining up, introducing competition not only in the many new voice, data, and video network markets but also in a worldwide contest to fill launch-vehicle payload manifests.

For the United States' space shuttle and Russia's

Earth-circling platform, the year featured the launching of phase I of the three-phase development process of the International Space Station, and, thereby, the end of decades of strict isolation and competition between these countries' spaceflight activities. Before actual construction begins in November 1997, the shuttle *Atlantis* is to link up with the more than 9-year-old *Mir* in a total of nine missions for the purpose of conducting joint onboard research in the interest of risk mitigation, gathering practical experience in joint operations activities, and testing assembly procedures for the two subsequent construction phases. Two of these linkups took place in 1995.

A total of nine crewed missions from the two major space-faring countries carried 48 humans into space, including 10 women, bringing the total number of humans launched into space since 1958 to 667 (counting repeaters), including 58 women; the actual number was 347 individuals (29 women).

Significant launches for 1995 are listed in **Table 1**, and the total number of successful launchings by

Table 1. Some significant space launches in 1995

Mission designation	Launch date	Country	Main payload or mission
STS 63 (*Discovery*)	February 3	United States	First rendezvous with and close approach and flyaround of Russia's *Mir* station, with second Russian cosmonaut on a United States spacecraft; beginning of phase I of International Space Station (ISS) development
STS 67 (*Endeavour*)	March 2	United States	Second flight of astronomical payload ASTRO; longest shuttle flight to date (16 days and 15 h); six-member crew
Soyuz TM 21	March 14	Russia	Mir 18 crew of three, with Norman Thagard, first United States astronaut on a *Soyuz*
H-2	March 17	Japan	Third flight of the new heavy lifter, carrying twin payloads (SFU; *Himawari 5*)
Ofeq 3	April 5	Israel	Observations technology research satellite in retrograde orbit
ERS 2	April 21	Europe	ESA's second operational environment, remote-sensing satellite; joined *ERS 1*
Spektr	May 20	Russia	Crewless Proton-launched scientific research module, docked to *Mir* on June 1
STS 71 (*Atlantis*)	June 27	United States	100th United States human space flight; second flight to and first linkup with *Mir;* carried crew of seven up, eight down (including two cosmonauts each way)
STS 70 (*Discovery*)	July 30	United States	Deployment of seventh Tracking and Data Relay Satellite (TDRS); crew of five
Mugunghwa (*Hibiscus*)	August 5	South Korea	South Korea's first communications satellite; Delta 2 launch partially successful
LLV 1	August 15	United States	First launch of Lockheed Launch Vehicle, a new small, low-cost launcher; failed
Soyuz TM 22	September 3	Russia	Mir 20 crew of three, with German ESA cosmonaut Thomas Reiter; did spacewalk
STS 69 (*Endeavour*)	September 7	United States	Crew of five; *SPARTAN 201* freeflyer and Wake Shield Facility; thirtieth shuttle spacewalk
STS 73 (*Columbia*)	October 20	United States	Seven-member crew with second United States Microgravity Laboratory
Radarsat	November 4	Canada	First Canadian remote-sensing (active radar) satellite; launched on United States Delta 2
STS 74 (*Atlantis*)	November 12	United States	Second *Mir* linkup of International Space Station phase I; delivered docking module
ISO	November 17	Europe	Large French-developed *Infrared Space Observatory* in highly elliptical orbit
Long March 2E (CZ)	November 28	China	Fifth (fourth successful) launch of new heavy lifter; deployed *Asiasat 2* communications satellite
SOHO	December 2	Europe/United States	French-built, United States–launched *Solar and Heliospheric Observatory;* orbiting around Earth-Sun lagrangian libration point L1 for sophisticated solar physics studies
XTE	December 30	United States	NASA's *X-Ray Timing Explorer;* largest x-ray telescope orbited so far

Table 2. Space launches and attempts in 1995

Country	Number of successful launches*	Number of attempts
Russia	32	33
United States (NASA, Department of Defense, Commercial)	27	30
Europe (European Space Agency, Arianespace)	11	11
People's Republic of China	2	3
Japan	1	2
Israel	1	1
TOTAL	74	80

* Launches that achieved Earth orbit or beyond.

the various countries, amounting to 74 (out of 80 attempts), is shown in **Table 2**.

United States Space Activity

The four-vehicle fleet of the United States space shuttle continued its operation of carrying people and payloads to and from Earth orbit for science, technology, and operational research. On the ground, the International Space Station program moved forward briskly with the finalization of the government-contractor development team, successful cooperation negotiations with the participating international partners, and accomplishment of critical review milestones.

Space shuttle. The National Aeronautics and Space Administration (NASA) successfully completed seven space shuttle missions into Earth orbit, three to the *Mir* space station, bringing the total number of shuttle flights since inception in 1981 to 73 and setting a new flight-duration record for shuttles. As in 1994, the shuttles carried 42 astronauts into space (including 10 women), with crew members from Russia and Canada.

STS 63. *Discovery* flew its twentieth mission from February 3 to 11 with the second Russian cosmonaut, Vladimir G. Titov, and the first woman shuttle pilot, Eileen M. Collins. As a precursor for the subsequent docking missions, the spacecraft performed a rendezvous with and flyaround of the space station *Mir.* Although no linkup was attempted, new flight techniques as well as the coordination between the mission control teams at Houston and Moscow required for such missions were validated, and the two 100-ton systems closed to a distance of 36 ft (11 m). After the 3-h joint flight, Titov used the shuttle's robot arm to release the free-flying astronomy satellite *SPARTAN-204* to make far-ultraviolet spectroscopic observations of the interstellar medium; it was retrieved 2 days later. The mission also included a 4-h 39-min spacewalk by mission specialists Bernard Harris and Michael Foale to evaluate spacesuit modifications and demonstrate large-object handling techniques.

STS 67. Setting a new shuttle mission record of 16 days 15 h in space as well as a distance mark of $6.9 \times$

10^6 mi (11.1×10^6 km), *Endeavour,* from March 2 to 18, carried the astronomical payload ASTRO on its second mission. ASTRO's three unique telescopes for ultraviolet astronomy mapped areas of space still largely uncharted in that spectrum, collected ultraviolet spectra of about 300 celestial objects including Jupiter and the Moon, and measured intergalactic gas abundances. Other experiments in weightlessness concerned large-structure control dynamics and materials processing.

STS 71. Originally planned for launch after STS 70, which was delayed for a month, *Atlantis* was launched to *Mir* on June 27 as the first phase I docking mission, 20 years after the historic joint Apollo-Soyuz Test Project (ASTP) between the United States and the former Soviet Union. Aboard the shuttle were the replacement crew (Mir 19) Anatoliy Solovyev and Nikolai Budarin. Steered by commander Robert Gibson, the orbiter docked to *Mir*'s *Kristall* module on June 29, forming the largest human-made structure in space so far (**Fig. 1**). Joint scientific (mostly medical) investigations were carried out for 5 days, with a record number of 10 individuals aboard a single space vehicle. *Atlantis* undocked on July 4 after unloading water and other supplies for *Mir* and taking on equipment no longer needed, returning to Earth on July 7 with the previous Mir 18 crew, Vladimir Dezhurov, Gennadiy Strekalov, and NASA astronaut Norman Thagard, who had flown to *Mir* on March 14 in *Soyuz TM 21* as a guest cosmonaut.

STS 70. Delayed from its June liftoff date by repair of minor damage to the cork insulation of its external fuel tank caused by a woodpecker, *Discovery* flew from July 13 to 22 to launch the seventh Tracking and Data Relay Satellite (TDRS) since 1983 into geosynchronous orbit, and to conduct research with a commercial protein growth facility, a bioreactor demonstrator, and various biological experiments. It was also the first flight of the new Block I space shuttle main engine with increased stability and safety.

STS 69. *Endeavour* at first had to be rolled back to the Kennedy Space Center's Vehicle Assembly Building on August 1 because of Hurricane Erin and then required some repair of its solid rocket boosters. It was launched on September 7, carrying the *SPARTAN 201* free-flyer satellite with instruments to observe the solar wind and the Sun's outer atmosphere; the Wake Shield Facility (WSF) on its second flight to experiment with the production of advanced, thin-film semiconductor materials; and various other payloads. Mission specialists James Voss and Michael Gernhardt conducted a 6-h 46-min spacewalk, the thirtieth of the shuttle program, to test construction tools, an arm-sleeve computer, and suit modifications for space station assembly. Return to Earth was on September 18.

STS 73. Flown from October 20 to November 5 after six delays (Hurricane Opal, a hydrogen leak, mechanical problems, and weather conditions), *Columbia* on its eighteenth flight carried the second United States microgravity laboratory with science

Fig. 1. Space shuttle *Atlantis* docked to the *Kristall* module of the Russian space station *Mir*, photographed by cosmonaut Nikolai Budarin on July 4, 1995. (*RKA/NASA*)

and technology experiments from government institutes, universities, and industry in areas such as fluid physics, materials science, biotechnology, combustion science, and commercial processing technologies.

STS 74. Performing its fifteenth mission from November 12 to 20, *Atlantis* was steered by commander Kenneth Cameron and pilot James Halsell to a perfect second docking with *Mir*, carrying a Russian-built 9000-lb (4000-kg) docking module, two new solar arrays, 992 lb (450 kg) of fresh water, and 1140 lb (517 kg) of equipment and food to the station. After three days of joint operations by the eight occupants from four of the five International Space Station partner countries—the United States, Russia, Canada (shuttle mission specialist Chris Hadfield), and Europe (Germany's Thomas Reiter)—*Atlantis* undocked, taking with it 816 lb (370 kg) of science samples, data, and equipment from previous *Mir* investigations and leaving the docking module attached for future shuttle visits.

International Space Station. In 1995, the development of the International Space Station, begun in 1994 after the redesign of the former *Freedom* concept and formal signings of the agreement between NASA and the Russian space agency RKA (Rossiyskoe Kosmicheskoe Agentstvo), made significant progress, remaining on schedule and within costs. A design and development contract with the prime contractor, Boeing Co., was definitized and signed. Boeing is responsible for integration and verification of the International Space Station system, as well as for design, analysis, manufacture, verification, and delivery of the United States onorbit segments of the station. Assembly is to begin in November 1997 with the Proton launch of the Russian FGB (Funktsionalya-gruzovod blok, or Functional Cargo Block) tug, purchased by the United States.

In March, the International Space Station program management, in the first of a series of incremental design reviews, successfully provided a comprehensive assessment of the design and technical feasibility for the first six United States and the first five Russian assembly flights as well as a forward-planning review of all assembly flights. In April, a major design review of the first construction element, the FGB, certified readiness to proceed with the manufacture of this Russian-built propulsion, guidance, and control module. Fabrication of the structure of NASA's first pressurized module, called Node 1, was completed in June, while Node 2, to be used as structural test article before its launch in 1999, was delivered in April. In September, Boeing also completed the main structure of the 28-ft (8.5-m) long, 14-ft (4.3-m) wide United States laboratory module, weighing about 6000 lb (2700 kg). Qualification testing of the so-called alpha joint for the space station's rotating solar arrays was begun, and construction of the first flight unit was started. Overall, United States contractors in 1995 delivered nearly 80,000 lb (36,300 kg) of hardware, including solar array panels, mast, truss segments, rack structures, hatch assemblies, and various mockups.

Phase I of the development, the joint Shuttle-*Mir* program, proceeded on schedule to meet its objectives of providing operations experience, risk mitigation, technology demonstrations, and early science opportunities. Major milestones during 1995 were three visits to *Mir* by United States space shuttles, including two linkups; ferrying of cosmonauts and supplies by *Atlantis;* and the first participation by a United States astronaut, Thagard, as a member of a Russian station crew, starting in March with the launch of *Soyuz TM 21* from Baikonur, Kazakstan, and ending in July with his return aboard *Atlantis*. Thagard's stay in space yielded the first long-duration medical data on an American astronaut since the Skylab program in 1973.

Development programs continued in other partner countries as well. In Canada, the Mobile Service System (MSS), which will provide external station robotics, progressed. Japan remained on schedule in developing the Japanese Experiment Module (JEM). In October, the European Space Agency (ESA) received final approval by the governments of the nine European countries involved in the International Space Station to proceed with the development of a pressurized laboratory called the Columbus Orbital Facility (COF) and the Ariane 5 launched the Automated Transfer Vehicle (ATV) for supplying logistics and reboosting the station.

Space sciences and astronomy. Numerous important and, in part, revolutionary discoveries in space from several automated or remotely directed missions enriched knowledge about the cosmos.

Hubble Space Telescope. A series of surprising observations from the perfectly functioning Hubble Space Telescope (HST) challenged prior concepts about cosmological phenomena and some of the most accepted theories about the form, buildup, structure, age, evolution, and future of the universe.

Particularly dramatic were images of about 100 new stars being born in the Eagle Nebula, M16 (**Fig. 2**), 7000 light-years away (1 light-year equals 5.9×10^{12} mi or 9.5×10^{12} km).

Other findings included detection of critical stars called cepheid variables in the remote Virgo cluster of galaxies, allowing better determination of the age of the universe; and compelling evidence for the existence of supermassive black holes in three galaxies: a 2.4×10^9 solar mass black hole in the core of the elliptical galaxy M87, a 4×10^7 solar mass black hole in the spiral galaxy NGC 4258, and an extremely puzzling black hole 10^8 light-years away in the direction of the constellation Virgo. Fueled from an 800-light-year-wide spiral-shaped disk of dust around it, it is offset by 20 light-years from the center of the host galaxy NGC 4261. Having presumably once been at the center, something must have pulled it outward, or else it is self-propelled by the rocketlike reaction from plasma jets expelled by temperatures of tens of millions of degrees. *SEE BLACK HOLE; UNIVERSE.*

Hubble photographed distant galaxies through the galaxy cluster Abell 2218, which acts as a spectacular gravitational lens, and detected mature, that is, fully developed, spiral galaxies that already existed when the universe was only 2×10^9 years old; bizarre, never-before-seen light structures in distant radio galaxies; blue dwarf galaxies; at least two (probably four) new satellites of Saturn; as well as signs of fresh volcanic activity on Jupiter's satellite Io and of oxygen on its satellites Europa and Ganymede.

Ulysses. The NASA-ESA solar-polar explorer *Ulysses,* having overflown the Sun's southern pole in November 1994, began its pass over the north pole in

Fig. 2. Columns of cool interstellar hydrogen gas and dust in the Eagle Nebula (M16), photographed by the Hubble Space Telescope. By photoevaporation, dense portions of the clouds are stripped free of surrounding matter by ultraviolet light from hot, massive newborn stars (off the top of the picture). Some of the uncovered globules of dense gas contain embryonic stars approaching birth as photoevaporation sets them free. (*Space Telescope Science Institute/NASA*)

June 1995, and concluded its main mission successfully in September, while already on its return journey to the orbit of Jupiter, to arrive there in April 1998. It will return in September 2000 for more polar overflights. The German-built probe, for the first time, detected periodic oscillations or wave motions in interplanetary space originating from deep within the Sun and, in effect, took the first snapshot of the spiral structure of the Sun's magnetic field extending past the orbit of Venus toward Earth's orbit. Some of its findings concerning the global differences in solar wind speed at different latitudes (up to twice as fast at high southern latitudes as near its equator, that is, 500 mi/s or 800 km/s versus 250 mi/s or 400 km/s), and the spiral magnetic field, could mean an upset of the current solar model, necessitating its revision.

Galileo. After taking a circuitous route lasting 6 years and involving one gravity-assist flyby of Venus and two of Earth, the Jupiter probe *Galileo* finally reached its target after traveling 2.3×10^9 mi (3.7×10^9 km). Throughout 1995, it had continued its long string of observations of targets of opportunity begun after its launch on October 18, 1989. On July 13, three explosive bolts connecting the entry probe to the main ship were detonated, sending the probe spinning to Jupiter. On July 27, *Galileo* fired its main propulsion system for the first time, changing its course to pass 133,000 mi (213,000 km) above Jupiter's clouds.

On December 7, after close approaches to the satellites Europa and Io, *Galileo* reached its closest point to Jupiter (perijove) of 134,000 mi (214,500 km). At 6:10 P.M. Eastern Standard Time, it began to receive and store 57 min of radio transmissions from the probe when it entered the atmosphere at latitude 6.5° N, longitude 4.4° W in the North-Equatorial Band. The data were subsequently relayed to Earth. Then, igniting its rocket engine on time at 8:20 P.M. and firing it for 49 min, *Galileo* established itself in a perfect orbit around Jupiter and began its 2-year mission of scientific studies of the Jovian system.

GOES 9. The second in a series of advanced weather satellites, *GOES 9* (Geostationary Operational Environmental Satellite) was launched by NASA on May 23 on an Atlas-2AS/Centaur rocket. After several months of testing, the new powerful observer was handed over to the National Oceanic and Atmospheric Administration (NOAA).

SOHO. The joint NASA-ESA solar observatory *SOHO* (*Solar and Heliospheric Observatory*) was launched on December 2 on an Atlas/Centaur. The 4100-lb (1860-kg) probe is the most sophisticated solar space observatory ever built, and promises to revolutionize solar physics. Science operations were scheduled to begin in mid-1996, after the spacecraft had reached its location about 1×10^6 mi (1.6×10^6 km) from Earth on the Earth-Sun line where it was to use its propulsion system to orbit the L1 lagrangian point.

XTE. NASA's *X-Ray Timing Explorer* (*XTE*) satellite, the largest x-ray telescope orbited so far,

was launched on December 30 on an upgraded two-stage Delta 7920-10 after six prior launch attempts, five of them prevented by high upper-atmospheric winds. The 6700-lb (3040-kg) spacecraft, measuring $6 \times 6 \times 18$ ft ($1.8 \times 1.8 \times 5.5$ m), was placed in a 360-mi (576-km) orbit inclined 23° to the Equator. Built by the Goddard Space Flight Center in Maryland, the *XTE* is equipped with three sophisticated telescopic instruments, the Proportional Counter Array (PCA), the High-Energy X-Ray Timing Experiment (HEXTE), and the All-Sky Monitor (ASM), to study dense objects such as white dwarf and neutron stars, binary systems, x-ray novae, black holes, active galactic nuclei, and quasars. The *XTE* will provide accurate timing and measurement of x-ray sources in the sky and can detect emissions as brief as 10–100 microseconds.

Voyager 1 and 2. Both *Voyager* deep-space probes, launched in 1977, were healthy at the end of 1995, continuing their departure from the solar system. As they travel farther from the Sun, they are returning data to characterize the environment of the outer solar system, searching for the heliopause, the boundary representing the outer limit of the Sun's magnetic field and outward flow of the solar wind. *Voyager 1,* cruising at 39,260 mi/h (17.45 km/s), was 5.78×10^9 mi (9.24×10^9 km) from Earth. *Voyager 2,* at a distance of 4.37×10^9 mi (7×10^9 km), was departing the solar system at 36,160 mi/h (16.07 km/s). Both spacecraft are expected to operate and send back valuable data until at least the year 2015.

Pioneer. On September 30, NASA ended operations of *Pioneer 11,* launched in 1973. Now far beyond the orbit of Pluto and more than 4×10^9 mi (6.5×10^9 km) from Earth, the probe is heading out into interstellar space. At that distance, faint signals traveling at the speed of light take more than 6 h to reach Earth. Because the spacecraft's power is too low to operate its instruments and transmit data, communications with it have been reduced from about 8–10 h a day to about 2 h every 2–4 weeks. Late in 1996, its transmitter is expected to fall silent altogether. Its sister ship, *Pioneer 10,* heading in the opposite direction, continues to return scientific data and may have enough power to last until 1999. It was launched in 1972, and, at a distance of almost 6×10^9 mi (9.5×10^9 km), is the most distant object launched from Earth. Another Pioneer, *Pioneer 6,* launched into a solar orbit in 1965, was contacted in December.

Department of Defense activities. Efforts continued to make space a routine part of military operations across all service lines. Joint initiatives are aimed at bringing launch and satellite operations increasingly on a level where they can be of maximum use in directly supporting military forces in the field.

At Cape Canaveral, continuing its buildup of space-based resources, the Air Force launched classified payloads on Titan 4/Centaur rockets on May 14 and July 10. On July 31, a DSCS 3 (Defense Satellite Communications System) spacecraft, the fifth in a constellation of upgraded military relay stations, was boosted into orbit by an Atlas 2/Centaur. During a ground test-firing of a Titan 4 second stage on July 31, a nozzle extension made of composite material failed, leading to launch delays of several months. The next Titan 4/Centaur launched the second MILSTAR DFS (Development Flight Satellite) communications satellite on November 6. The 10,000-lb (4500-kg) spacecraft will work in concert with MILSTAR *DFS 1,* deployed in 1994, to provide jam-proof, secure communications reaching from the Persian Gulf to the western Pacific Ocean. Four upgraded MILSTARs with higher data relay rates will be launched between 1999 and 2002 to provide a global military communications network. A fourth Titan 4 took off on December 5, carrying a classified reconnaissance satellite. It was the first launch from Vandenberg Air Force Base, California, since a Titan 4 failed after liftoff in August 1993, and the fifth from there since first launch in March 1991. Another ten Titan 4's have been launched from Cape Canaveral so far.

Commercial space activities. To assist in early space-shuttle missions of the International Space Station program, NASA in August contracted with Spacehab Inc. to lease a pressurized in-shuttle payload module, Spacehab, for ferrying cargo to *Mir.* In a first small step toward an eventual form of effective privatization of the shuttle, NASA in August announced plans to select a single contractor to operate the space shuttle fleet, replacing 85 separate shuttle contracts. Also in the commercial space sector in 1995, Boeing entered an agreement with the Russian firm Rocket Space Corp. (RSC) Energia to provide a turnkey service for placing and operating commercial and government payloads on the outside of the *Mir* space station.

In 1995, 11 Atlas 2AS rockets, 1 refurbished Atlas E (ICBM), 4 Titan 4's, and 3 Delta 2 vehicles were launched. To replace the current Atlas rockets, the Atlas 2AR is being developed, using Russian liquid-fueled engines, as a next-generation intermediate-class vehicle intended as the core of a new family of expendable launchers. A smaller rocket, the Lockheed Launch Vehicle (LLV), is being developed as an inexpensive booster for commercial payloads between 1 and 4 tons to low-Earth orbit. This rocket's first launch, however, failed when the vehicle pitched out of control shortly after liftoff on August 15.

The development of the Delta 3 was announced. It would be a derivative of the smaller but highly reliable Delta 2 rocket, using a larger payload fairing and a cryogenic upper stage. Its payload capability to geostationary transfer would be 8380 lb (3800 kg), similar to that of the current Atlas 2 and Europe's Ariane 4.

The first launch of the Conestoga 1620 rocket ended in failure on October 23, destroying its *Meteor* payload, a commercial microgravity recoverable capsule. The Pegasus XL, launched from an L 1011 airplane, succeeded on April 3 in launching three Orbcomm satellites, but its third flight on June 22 with a U.S. Air Force Space Test Experiment Platform (STEP) satellite as payload was a failure, the second in the program.

The Orbcomm system, comprising 36 satellites for two-way communications using handsets, is one of several competing space-based global communications systems in early stages of buildup. The large Iridium system, using 66 satellites to provide worldwide digitally switched point-to-point communications with multiple networks—terrestrial cellular, public telephone, and space-based satellites—is in full-scale development, the first spacecraft being prepared for launch in late 1996. The Globalstar system will employ 48 satellites on low-Earth orbits. The Odyssey system, an Atlas-launched, 12-satellite network, would provide voice, data, paging, and messaging to mobile subscribers beginning in 1999. ICO Global Communications is planning to provide mobile communications over 10 satellites. *SEE COMMUNICATIONS SATELLITE.*

Russian Space Activities

Represented by its space agency RKA, Russia in 1995 continued its robust space operations at a brisk pace, even if at a level considerable lower than in previous years. Again, it led the world in number of spacecraft launches: 32 successes in a total of 74 (43%; Table 2). The new partnership with the United States gained major substance in a number of crewed flight activities and began to play a major role in shaping the Russian space program. After cosmonaut Sergei Krikalev's participation in shuttle mission STS 60 in 1994, three more cosmonauts flew on the United States shuttles *Discovery* (STS 63) and *Atlantis* (STS 71) in 1995.

Space station Mir. By the end of 1995, *Mir* had been in operation for 3602 days, commencing in February 1986. In that time, it circled Earth approximately 56,380 times at 246 mi (393 km) altitude in an orbit inclined 51.65° to the Equator. Counting from its last brief period of nonoccupancy (September 1989) to the end of 1995, *Mir* has been inhabited continuously for 2308 days. Since its inception, it has been visited 24 times, including twice by a United States shuttle, by two- or three-person crews. To resupply the occupants during 1995, the space station was visited by five automated Progress M cargo ships, bringing the total of Progress and Progress M ships launched to *Mir* and the two preceding space stations *Salyut 7* and *Salyut 6* to 73, with no failure.

Soyuz TM 21. Launched on March 14, *TM 21* (spacecraft number 70) carried the Mir 18 crew, Dezhurov, Strekalov, and United States astronaut Thagard. Docking occurred on March 16. Thagard became the first American launched in a *Soyuz,* and during his stay on *Mir* he established a new United States record in space of 115 days, surpassing 84 days by the *Skylab 4* crew in 1973–1974. The Mir 18 crew returned to Earth on the shuttle *Atlantis* on July 7. On September 11, *TM 21* served as the return vehicle for the Mir 19 crew, Solovyev and Budarin, who had arrived with the *Atlantis* and stayed on *Mir* for 75 days.

Soyuz TM 20. In space since its launch on October 3, 1994, *TM 20* returned to Earth on March 22, bringing back the Mir 17 crew, Alexander Viktorenko, Elena Kondakova, and Valeriy Polyakov. Kondakova established a new women's space endurance record of 169 days, while Polyakov, who had been on *Mir* since January 1994, set a new overall record of 437½ days. Combined with an earlier stay on *Mir,* he has logged a total of 607 days in space.

Spektr. The 20-ton crewless *Spektr* module was launched on May 20 on a Proton rocket and docked to *Mir* on June 1, joining the similar *Kvant 2* and *Kristall* blocks. *Spektr,* based on a military spacecraft first tested in 1977, carries an array of remote-sensing instruments, a small manipulator arm, and a small science airlock.

Soyuz TM 22. The Mir 20 crew, Yuri Gidzenko, Sergei Avdeyev, and Reiter, was launched on September 3 and docked to *Mir* 2 days later. Reiter, a German, was the second astronaut sponsored by the ESA aboard *Mir.* In the EUROMIR 95 mission, undertaken by the ESA in cooperation with the RKA, he studied living and working conditions in space with an extensive program of 47 experiments in the life sciences, astrophysics, materials science, and technology. He also became the first Western European to perform a spacewalk, on October 20 for 5¼ h. During their stay, the crew were informed that their mission was to be extended through February 29, 1996. In a second spacewalk of 37 min, the fiftieth conducted from *Mir,* Gidzenko and Avdeyev prepared the remaining free port on the station's central transition node for the 1996 linkup of the *Priroda* module by attaching a docking cone.

Satellite launches. Russia launched about 24 military, scientific, and telecommunications satellites, among them the second Gals direct-broadcast television satellite, and the new large data-relay satellite *Luch 1,* a more powerful version of the earlier Luch (*Altair*). The proven heavy-lift carrier Proton was used in seven launches, three carrying geosynchronous payloads, three launching nine GLONASS (Globalnaya Navigatsionnaya Sputnikovaya Sistema) satellites in highly elliptical 12-h orbits (completing the GLONASS system), and one used for *Spektr.* GLONASS, the Russian satellite navigation system, has 25 satellites in orbital planes of 11,937 mi (19,100 km) altitude and 64.8° inclination. Originally developed for military use, it was available to civil users beginning in March, like its United States counterpart, the Global Positioning System (GPS).

Russian commercial activities. In its early stages of entering commercial space markets, Russian rocketry suffered a setback by a failure of a new Start booster on March 28 from Plesetsk Cosmodrome on its first commercial mission, carrying three spacecraft from Russia, Israel, and Mexico. To market its launch services, the Russian firm Khrunichev, builder of the Proton, entered a joint venture with Lockheed Martin and the Russian company RSC Energia to form International Launch Services (ILS) in 1993. A new heavy booster, Angara, larger than the Proton, is under development. Other high-quality space technology offered to Western companies includes rocket propulsion systems such as the NK 33 and the RD 180.

European Space Activities

The European Space Agency's participation in the International Space Station and in the EUROMIR 95 project aboard the Russian *Mir* station are discussed above. At Kourou, French Guyana, ESA and Arianespace launched 11 Ariane rockets carrying a variety of payloads, such as television satellites, Europe's first military reconnaissance satellite *Helios 1A,* the giant infrared space telescope *ISO (Infrared Space Observatory),* and the remote-sensing, environmental satellite *ERS 2.*

As compared to ESA's activities, national space programs in Europe were forced by economics to remain at an almost negligible level. In continuing its cooperation with Russia, Germany's space agency DARA entered an agreement with RKA for launching one of two German cosmonauts already in training in Russia to *Mir* toward the end of 1996.

Asian Space Activities

In 1995, space activities continued in Japan, the People's Republic of China, India, and South Korea.

Japan. After the successful introduction of the powerful H-2 heavy-lift launch vehicle with two flights in 1994, Japan's National Space Development Agency (NASDA) scored a third success from Tanegashima on March 17, 1995, launching twin payloads. The first, the SFU (Space Flyer Unit) experiment carrier, was deployed in Earth orbit at about 206 mi (330 km) altitude (from where it was retrieved by the United States shuttle *Endeavour* on January 13, 1996). The liquid oxygen–liquid hydrogen LE-5A second stage then reignited and injected the second payload, the Geostationary Meteorological Satellite 5, later named *Himawari 5 (Sunflower 5),* in a geostationary orbit at 22,500 mi (36,000 km) altitude.

On January 15, Japan's Institute of Space and Astronautical Science (ISAS) launched a four-stage *Mu-3S-II* from Kagoshima, carrying the German Experiment Reentry Space System (*EXPRESS*), a Russian-built reentry vehicle. It was due to land in Australia after a 5-day flight, but the thrust vector control system on the second stage malfunctioned during ascent, and the mission failed. Given up for lost in the Pacific Ocean, *EXPRESS* came safely down on its parachute 2.5 orbits later near a partially inhabited area of Ghana, West Africa.

Japan Satellite Systems' communications satellite *JCSAT 3* was launched by a United States Atlas on August 29. On the same day, the satellite *N-STARa* of Nippon Telephone and Telegraph (NTT) of Japan was carried into space by a European Ariane.

China. After its successful launches of the Long March (Chang Zheng) 2E in 1994, the People's Republic of China suffered a serious setback on January 25 with the explosion of a Long March 2E approximately 50 s after launch, which killed 6 persons near the Xichang Satellite Launch Center and injured 27 others by falling debris. A second Long March 2E was launched successfully on November 28, carrying the communications satellite *Asiasat 2* into geostationary orbit. On December 28, a third Long March 2E carried the television-broadcast satellite *EchoStar 1* successfully into orbit.

India. *Insat 2C*, a domestic communications, television, and radio-broadcast satellite built by the Indian Space Research Organization (ISRO), was launched on December 6 by an Ariane rocket from Kourou. A second Indian payload, the remote-sensing satellite *IRS 1C* was launched on December 28 on a Russian Molniya M four-stage rocket from Baikonur into a retrograde (99° inclination) orbit, a first for this carrier.

South Korea. South Korea got its first communications satellite with the (only partially successful) launch of *Mugunghwa* (*Hibiscus*) on a Delta 2 on August 5. South Korea also entered into negotiations with RKA for future participation in Russia's cosmonaut training program.

Space Activities of Other Countries

The heavy operational environmental satellite *Radarsat*, Canada's first venture in the field of Earth observation from space, was launched by NASA on November 4 on a Delta 2 from Vandenberg Air Force Base into a Sun-synchronous near-polar orbit of about 500 mi (800 km) altitude. Data from the spacecraft will be used by the Canadian Space Agency (CSA) for monitoring oceans, ice coverage, crop growth, forests, geology, and many other purposes.

The National Space Agency of Ukraine (NKAU) launched its first satellite, *Sich*, on a Tsiklon 3 rocket on August 31. It is a remote-sensing spacecraft based on the Okean satellites built by the Ukrainian firm NPO Yuzhnoe for the Soviet Union and later the Russian RKA. Attached to it was a microsatellite called *FASat Alfa* with imaging and communications experiments of the Chilean Air Force.

Argentina developed its first spacecraft, *SAC B* (Satélite de las Aplicacións Científicas), for launch in the near future on a Pegasus.

On April 5, Israel launched its most sophisticated satellite so far, the *Ofeq 3* (*Horizon 3*) three-axes-stabilized technology satellite. The 496-lb (225-kg) spacecraft was carried into retrograde (westward) orbit by a Shaviyt launcher. Earlier, on March 28, the technology research satellite *TechSat/Gurwin 1* from Israel was lost with two other payloads when its Russian Start rocket malfunctioned.

For background information *SEE COMMUNICATIONS SATELLITE; METEOROLOGICAL SATELLITES;* *MILITARY SATELLITES; SATELLITE ASTRONOMY; SATELLITE NAVIGATION SYSTEMS; SPACE BIOLOGY; SPACE FLIGHT; SPACE PROCESSING; SPACE SHUTTLE; SPACE STATION; X-RAY ASTRONOMY* in the McGraw-Hill Encyclopedia of Science & Technology.

Jesco von Puttkamer

Bibliography. Euroconsult 1994 world space markets survey, *SPACE Mag.*, 10(5):14–17, September–October 1995 and 10(6):12–17, November–December 1995; T. Furniss, The Proton threat, *SPACE Mag.*, 11(2):10–12, March–April 1995; *Jane's Space Directory, 1995–1996*; J. T. McKenna, Titan 4 lofts classified payload, *Aviat. Week Space Technol.*, 142(21):61, May 22, 1995; W. B. Scott, Major cultural change on tap in military space, *Aviat. Week Space Technol.*, 143(12):36–42, September 18, 1995.

Space navigation and guidance

Navigation for crewed space flight consists of on-board and ground-based navigation systems which together enable the accomplishment of the varied and complex mission objectives. This synergistic partnership between ground and on-board systems, which has existed since the start of crewed space flight and resulted in the achievements of the Apollo Lunar Landing Program, continues to provide the U.S. Space Shuttle Program with an accurate, robust, and adaptable navigation capability.

Navigation systems, whether ground based or on board the spacecraft, must satisfy four basic objectives: (1) determine in real time, through appropriate measurements or observations, the spacecraft state (position, velocity, and time); (2) provide these state estimates within the specified time period required for their use (the latency requirement); (3) propagate this state forward in time without the benefit of measurements with the required accuracy for the targeting and execution of spacecraft maneuvers to accomplish mission objectives; and (4) provide these state estimates with a specified degree of reliability (the redundancy and failure-mode requirement)

Mission phases. Missions are divided into phases or segments, each phase having different navigation requirements and navigation system configurations. The **illustration** shows a schematic representation of the shuttle mission profile, illustrating the mission phases from liftoff through touchdown. The shuttle ascent phase begins at liftoff with the firing of the solid rocket boosters and three main engines, for the first powered-flight portion of the mission. If no abort is declared, the powered-flight phase is followed by one or two on-orbit maneuvering system (OMS) burns, which place the orbiter in a safe, stable orbit about the Earth. If an abort has been declared during the ascent powered-flight phase, the orbiter is placed on a trajectory consistent with the abort profile—a transatlantic abort, return to launch site, abort to orbit, or abort once around

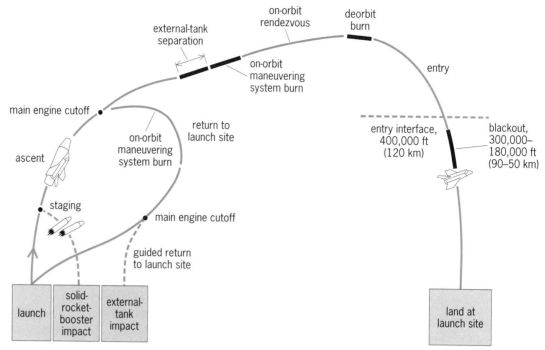

Space shuttle mission profile. (*C. S. Draper Laboratory*)

(one Earth orbit to landing). The on-orbit phase follows the ascent phase and consists of the on-orbit operations, which are normally the prime objectives of a shuttle flight. If a rendezvous with a target vehicle is required, the rendezvous mission phase is initiated shortly before the acquisition and tracking of the target with the on-board relative sensors, and it terminates with the grappling of or docking with the target vehicle. The entry-through-landing mission phase begins shortly before the deorbit burn, which takes the orbiter out of orbit and places it on the desired entry trajectory. The orbiter enters the atmosphere at an approximately 400,000-ft (120-km) altitude with a 40° angle of attack. Entry navigation is initiated and guidance and control executed to provide the desired downrange and crosstrack trajectory to touchdown at the landing site.

Synergistic use systems. During all mission phases, both the on-board and ground navigation systems provide vehicle position and velocity estimates which are used for independent targeting of maneuvers to accomplish the objectives of that mission phase. The primacy of either the on-board or ground solutions in providing the state estimates of the spacecraft are dependent on the navigation accuracies that each segment can bring to the particular mission phase. When the ground provides the more accurate state estimate, this estimate is used to reset the on-board state estimates required so that maneuver targeting and execution can be performed by using the on-board systems. When the on-board navigation state estimate is more accurate, its state is used and the ground solution is backup. This use of combined on-board and ground navigation system capabilities results in high-accuracy, robust, and backup navigation capabilities for all mission phases. **Table 1** shows the typical navigation accuracies for ground and on-board systems in crewed space flight.

Space shuttle program. Table 2 lists the ground and on-board navigation sensors used for each shuttle mission phase, and indicates the prime navigation system in a given mission phase.

Ground navigation. The shuttle ground navigation system uses range and Doppler measurements, and

Table 1. Typical accuracy of space shuttle navigation systems*

Mission phase	Ground	On-board
Ascent	200 ft (60 m), 0.5 ft/s (0.15 m/s)	200 ft (60 m), 0.5 ft/s (0.15 m/s)
Earth-orbit coast	3000 ft (1000 m), 3 ft/s (1 m/s)	Periodic ground uplink
Rendezvous: midcourse correction, targeting errors	1.5 ft/s (0.5 m/s)	0.3 ft/s (0.1 m/s)
Entry: at shuttle Tacan acquisition†	3000 ft (1000 m), 3 ft/s (1 m/s)	10,000 ft (3000 m), 10 ft/s (3 m/s)
Entry: at Microwave Scanning Beam Landing System acquisiton	—	400 ft (120 m), 2.0 ft/s (0.6 m/s)
Touchdown	—	15 ft (5 m), 0.5 ft/s (0.15 m/s)

* 1 standard deviation (1 sigma).
† Approximately 150,000 ft (45 km).

measurements of the spacecraft line of sight, from C-band (4–8-GHz) and S-band (2–4-GHz) tracking stations. Air Force Eastern Test Range radars are used during ascent, and the Manned Space Flight Network (MSFN) stations are used for the on-orbit, rendezvous, and entry phases. Since the launch of the Tracking Data and Relay Satellite System (TDRSS) satellites in the 1980s, tracking of the orbiter with two-way ranging through TDRSS, using the orbiter Ku-band (10–20-GHz) radar and ground stations at White Sands, New Mexico, has provided an enhanced ground navigation capability which reduces the time to obtain an orbiter navigation fix. Two 23-min TDRSS tracking passes over the course of an orbit are sufficient to provide accurate ground-based estimates of the orbiter state. During ascent, orbit, and entry operations, ground C-band and S-band radar and TDRSS tracking of the shuttle are used to provide measurements for the Mission Control Center High Speed Navigation Determination processor. This processor uses an extended Kalman-filter recursive navigation formulation for processing the measurements. The ground-state estimates are used to reset the on-board orbiter state when the on-board and ground states differ by values defined prior to the mission.

On-board general-purpose computers. The shuttle carries five general-purpose computers. Three are used by the primary avionics system, one for system management, and the one for the backup flight system. For each mission phase, a different set of computer software is loaded into computer memory. These mass memory loads are for ascent, aborts, and return to launch site; on-orbit, including coast and rendezvous; and entry through landing.

On-board systems and Kalman filters. The shuttle navigation systems are mission-phase dependent. During ascent and on-orbit coast, the estimate of the shuttle state is maintained by propagating the launch-site or ground-uplinked navigation state, respectively, by using selected acceleration measurements from the triply redundant inertial measurement units and appropriate drag and gravity models. For the entry-through-landing and rendezvous mission phases, an extended Kalman filter is used to sequentially process the measurements from the navigation sensors.

The shuttle rendezvous navigation filter is a 13-state extended Kalman filter. The filter states are the three components of position, velocity, and unmodeled acceleration, together with four sensor biases. The filter formulation assumes that one of the states, normally the target, is known "perfectly." Star-tracker optical measurements and rendezvous radar measurements of the range, range rate, and line-of-sight direction of the target relative to the orbiter are processed, and the filter updates the inertial state of the navigating vehicle relative to the target vehicle.

Table 2. Configurations of space shuttle navigation systems

| Mission phase | Navigation sensors and measurement types | | Prime navigation solution | |
	Ground	On-board*	Ground	On-board
Ascent	Eastern Test Range C-band and S-band radar tracking: range, Doppler, and line-of-sight angle	Triply redundant inertial measurement units (IMUs): velocity change in 3 axes of inertial frame		X
On-orbit coast	Tracking and Data Relay Satellite System (TDRSS): two-way range and Doppler	IMU and periodic resetting of on-board state to ground-determined navigation state	X	
Rendezvous	Manned Space Flight Network (MSFN) C-band radar tracking and TDRSS tracking: two-way range and Doppler	Automatic rendezvous radar: tracking of target, providing range, range-rate, line-of-sight angles; IMU sensed velocity change; Automatic star-tracker tracking of target in reflected sunlight		X
Entry through landing	C-band radar tracking and TDRSS tracking: two-way range and Doppler	Triply redundant IMUs: velocity change in three axes of inertial frame; Pseudodrag altitude measurements; Tacan range and bearing; Air date altitude measurements; Microwave Scanning Beam Landing System range, azimuth, and elevation measurements		X

* On-board systems also use drag and gravity models for advancing the state estimate in time. The complexity of the models is usually mission-phase dependent.

The shuttle entry filter is a six-state extended Kalman filter, consisting of the three components of position and velocity only. The six-state filter processes the following sets of measurements: pseudodrag altitude measurements from 400,000 ft (120 km) until barometric altimeter measurements are acquired at approximately 100,000 ft (30 km); Tacan range and bearing measurements (bearing measurements are inhibited when the elevation of the orbiter relative to the station is greater than 45°); barometric altimeter measurements; and Microwave Scanning Beam Landing System measurements from 15,000-ft (4.5-km) altitude and within 12° of the centerline to touchdown. This suite of sensors and the on-board navigation, guidance, and control system are capable of providing an automatic landing capability.

Operations. During the ascent powered-flight portion of the trajectory and the subsequent maneuvers of the on-orbit maneuvering system, the preliftoff state vector (set to the shuttle location on the launchpad) is integrated in an inertial coordinate system using the selected measurements of total velocity change from the triply redundant inertial measurement units. The navigation system maintains an estimate of position and velocity of the shuttle navigation base on which the inertial measurement units are mounted. During the coasting portions of this phase, a gravity model and an atmosphere model are used to propagate the state.

For the on-orbit coast mission phase, a gravity model and an atmosphere model (which uses orbiter attitude) are used in the propagation equations to maintain the current state estimate. Unlike the navigation system in the ascent mission phase, the on-orbit navigation system maintains an estimate of the orbiter center of mass. For thrusting periods, the redundant inertial measurement units are again used to provide selected acceleration measurements for state propagation. Sensed velocities from the inertial measurement units, which are located at the navigation base, are corrected to provide the sensed acceleration at the center of mass. Ground navigation provides periodic updates to the on-board computer position and velocity, according to premission flight rules which define difference limits between the ground and on-board states. The inertial measurement units are periodically aligned with star sightings from the two star trackers which are mounted on the navigation base.

Rendezvous operations are initiated by aligning the inertial measurement units and setting the on-board orbiter and target states to the states provided by ground navigation. The ground also provides maneuver solutions for the height-adjust and phasing maneuvers which place the orbiter on a trajectory to intercept a stable orbit point 8 mi (14 km) behind the target. As the orbiter closes on the target, optical measurements of the line of sight to the target are automatically taken by one of the two star trackers during the periods of reflected sunlight from the target. These measurements are processed in the rendezvous Kalman filter as discussed above at a rate of one every 8 s. Confirmation of the optical target track is made by the crew.

When the target is within the tracking range of the rendezvous radar, nominally 26 nautical miles (48 km), radar measurements of the range, range rate, and line-of-sight angles of the target with respect to the orbiter are processed by the rendezvous Kalman filter every 8 s. The rendezvous radar provides measurements down to a target range of 100 ft (30 m). Manual control of the trajectory is initiated following the last midcourse correction at a range of 4000 ft (1200 m). In the event of radar failure, the orbiter star trackers can be used until 8 min before the second midcourse correction (when the target enters darkness). The use of angle navigation alone accurately places the vehicle on an intercept trajectory.

The crew can also use navigation measurements made with the crew optical alignment sight, an instrument mounted in either the forward or overhead windows. This instrument is also used to confirm stars in the star tracker field of view, and for alignment of the inertial measurement units.

The entry mission phase is initiated by ground uplink transmission of the ground-determined orbiter state to the on-board general-purpose computers. By using atmosphere and gravity models, as well as the velocity sensed by the dedicated inertial measurement units during the deorbit burn, the state estimates maintained by the on-board system are propagated to the entry interface. When the drag acceleration on the orbiter is above the quantization level of the inertial measurement units, the acceleration (drag) sensed by these units and the gravity model are used to propagate the state. When the drag acceleration reaches 11 ft/s^2 (3.4 m/s^2), drag altitude measurements are processed. These measurements are obtained by using the sensed acceleration to compute atmospheric density (by using the standard drag equation) and by using this density in the on-board atmosphere model to compute an estimated altitude.

Triply redundant Tacan receivers provide range and bearing measurements to the general-purpose computers for processing. When the Microwave Scanning Beam Landing System is acquired, triply redundant receivers provide range, azimuth, and elevation measurements.

GPS navigation system. A Global Positioning System (GPS) navigation capability is being developed for the space shuttle. In addition to providing an enhanced on-board navigation capability for ascent and on-orbit operations, the GPS system will provide an entry navigation capability to replace the shuttle's current use of Tacan stations, which are planned to be eventually phased out. The shuttle will be an authorized user of GPS and hence have access to Precise Positioning Service accuracies, on the order of 48 ft (15 m) in position and 0.8 ft/s (0.24 m/s) in velocity.

A GPS receiver has been flown on several shuttle missions to obtain performance data in the operational space environment. GPS will be integrated as a separate navigation capability from the existing navigation capabilities (referred to as the baseline navigation system). Both systems will operate simultaneously with crew options for selecting which system provides inputs to guidance and for resetting the baseline navigation system states to the selected GPS state estimate.

This design provides three advantages: (1) it provides minimum disturbance of existing shuttle flight software; (2) it permits flight verification of the GPS system without affecting current shuttle navigation performance and therefore shuttle missions; and (3) it permits advances in GPS receiver navigation capabilities, such as differential navigation for landing, to be easily incorporated into the shuttle system.

Trajectory control sensor. A laser sensor for precise determination of the orbiter range, range rate, and attitude relative to a target vehicle during manual proximity operations following rendezvous (including docking and berthing with the target) is being flown on the orbiter. The sensor is mounted in the payload bay on the docking adapter. Sensor outputs are processed in a portable computer, and the resulting information is displayed to the crew for aiding proximity operations.

International Space Station Alpha. The U.S. On-Orbit Segment (USOS) of the International Space Station Alpha will use GPS navigation to provide position and velocity estimates, attitude determination, and precise time services. The station GPS receivers will be unclassified and will have accuracies comparable to those of the Standard Positioning Service, on the order of 300 ft (100 m) and 3 ft/s (1 m/s). Ground facilities will perform additional processing of the station GPS state estimates to obtain state estimate accuracies which will allow ground targeting of station maneuvers for orbital debris avoidance. To verify basic elements of the station GPS state and attitude determination system in a flight environment, an attitude and state determination flight experiment was scheduled for flight on the space shuttle in 1996.

For background information SEE ESTIMATION THEORY; INERTIAL GUIDANCE SYSTEM; MICROWAVE LANDING SYSTEM (MLS); SATELLITE NAVIGATION SYSTEMS; SPACE COMMUNICATIONS; SPACE FLIGHT; SPACE NAVIGATION AND GUIDANCE; SPACE SHUTTLE; SPACECRAFT GROUND INSTRUMENTATION; STAR TRACKER; TACAN in the McGraw-Hill Encyclopedia of Science & Technology.

Peter M. Kachmar

Bibliography. R. H. Battin and G. E. Levine, *Application of Kalman Filtering Techniques to the Apollo Program*, M.I.T. Instrumentation Laboratory, 1969; P. Kachmar and L. Wood, Space navigation applications, *Navigation*, 42(1):187–234, 1995; *Proceedings of the National Technical Meeting of the Institute of Navigation*, January 1995.

Space weather

Space weather refers to conditions on the Sun and in the solar wind, magnetosphere, ionosphere, and thermosphere that can influence the performance and reliability of space-borne and ground-based technological systems and endanger human life or health. Adverse conditions in the space environment can disrupt satellite operations, communications, navigation, and electric power distribution grids, leading to socioeconomic losses.

Throughout history, the wildly changing patterns of the northern lights, or the aurora, has been a source of awe. Normally seen only in the polar region, these lights mystified observers by occasionally appearing far south of that location. In 1934, J. Bartels noticed that periodic disturbances in the Earth's magnetic field corresponded with the Sun's rotation rate and postulated M regions on the sun as their cause. Once during World War II, radio operators in England were convinced that they were foiled by enemy jamming when all high-frequency radio communications ceased. These related events are early examples of the effects of space weather.

Space-Earth environment. At present, far more information of the space environment is available, from the turbulent surface of the Sun, with its continuous solar wind and periodic spewing of clouds of energetic ionized particles, to the protective boundary of the Earth's magnetic field, which provides a partial shield against deadly solar corpuscular radiation. The Earth's magnetic field is highly reactive to the onslaught of energy and pressure originating from the solar particles and fields. In a complex way, the Earth's magnetosphere redistributes its particle populations, often sending a rush of energetic particles along magnetic field lines into the atmosphere over the polar caps and creating the swirling red, green, and white auroras. Other particles pour into the Van Allen radiation belts and encircle the Earth with electric current. The Earth's magnetic field itself can distort to such an extent that compasses at the surface swing 10° away from the magnetic pole. The ionosphere (80–1000 km or 50–600 mi above the Earth's surface) changes in ways that affect radio transmissions: absorbing some radio frequencies, distorting others, and creating electric currents that affect systems on the ground.

This intricate picture of the connection between the Sun and the Earth's space environment has been uncovered in the last few decades. However, understanding of the physical processes that drive and couple this complex weather system in space is still rudimentary. In terms of the quantity of observations, basic understanding of processes, and physical models, current knowledge of space weather is about as advanced as that of tropospheric weather over a half-century ago. Meanwhile, in the United States the reliance on technological systems is growing exponentially, and many of these systems are susceptible to failure or unreliable performance because of extreme space weather conditions. The risks involved could be mitigated or avoided if reliable space weather forecasts were possible and available with sufficient lead time or, in some cases, if representative, quantitative models were available to systems designers.

Effects on aerospace engineering. Currently, space environmental support services in the United States are provided through the Space Environment Services Center operated by the National Oceanic and Atmospheric Administration in Boulder, Colorado, and the U.S. Air Force 50th Weather Squadron at Falcon Air Force Base in Colorado Springs, Colorado. Bulletins, forecasts, alerts, warnings, and data are routinely disseminated to a broad range of users, including satellite operators, power companies, telecommunications operators, navigational systems users, and research institutions.

Aerospace engineers use space environment information to specify the extent and types of protective measures that are to be designed into a system and to develop operating plans that minimize space weather effects. Engineers also use space environment information to determine the source of failures and develop corrective actions.

The impact of space weather on satellite systems is farther reaching than ever before, and the trend will almost certainly accelerate. Energetic particles that originate from the Sun, interplanetary space, and Earth's magnetosphere continually impact the surfaces of spacecraft. Highly energetic particles penetrate electronic components, causing spurious electronic signals that can result in wrong commands within the spacecraft or erroneous data from an instrument. Less energetic particles contribute to a variety of spacecraft surface charging problems, especially during periods of high geomagnetic activity. In addition, energetic electrons responsible for deep dielectric charging can degrade the useful lifetime of internal components.

Highly variable solar ultraviolet radiation continuously modifies terrestrial atmospheric density and temperature, affecting spacecraft orbits and lifetimes. Major geomagnetic storms result in heating and expansion of the atmosphere, causing significant perturbations in low-altitude satellite trajectories. At times, these effects may be severe enough to cause premature reentry of orbiting objects, such as occurred with *Skylab* in 1979. The space shuttle is also vulnerable to changes in atmospheric drag; reentry calculations for the orbiter are highly sensitive to atmospheric density. Besides being a threat to satellite systems, energetic particles present a hazard to astronauts on space missions. On Earth, protection from these particles is provided by the atmosphere, which absorbs all but the most energetic cosmic-ray particles. During space missions, astronauts performing extravehicular activities are relatively unprotected. The fluxes of energetic particles can increase hundreds of times, following an intense solar flare or during a large geomagnetic storm, to dangerous levels. Timely warnings are essential to give astronauts sufficient time to return to the spacecraft prior to the arrival of such energetic particles. Crews and passengers in high-altitude aircraft on polar routes [for example, the supersonic transport (SST) or the U-2] are also susceptible to radiation hazards during similar events.

The long power lines that traverse the landscape in many industrially advanced countries are susceptible to electric currents induced by the dramatic changes in high-altitude ionospheric currents that occur during geomagnetic storms. Surges in power lines from induced currents can cause massive network failures and permanent damage to multimillion-dollar equipment in power generation plants.

The Global Positioning System (GPS) operates by transmitting radio waves from satellites to receivers on the ground, aircraft, or other satellites. These signals are used to calculate location very accurately. However, significant errors in positioning can result when the signals are refracted and slowed by ionospheric conditions. Future high-resolution applications of GPS technology will require better space weather support to compensate for these induced errors.

Effects on communications. Communications at all frequencies are affected by space weather. High-frequency radio-wave communication is more routinely affected, because this frequency depends on reflection from the ionosphere to carry signals great distances. Ionospheric irregularities contribute to signal fading; highly disturbed conditions, usually near the aurora and across the polar cap, can absorb the signal completely and make high-frequency propagation impossible. Accurate forecasts of these effects can give operators more time to find an alternative means of communication.

Telecommunication companies increasingly depend on higher-frequency radio waves that penetrate the ionosphere and are relayed via satellite to other locations. Signal properties can be changed by ionospheric conditions so that the signals can no longer be accurately received at the Earth's surface. Degradation of signals may result, but more importantly critical communications, such as those used in search-and-rescue efforts and military operations, may be lost.

Research goals. Research in space weather is needed to advance state-of-the-art instruments and data-gathering techniques, to conduct future space missions, to understand the physical processes, to develop predictive models, to provide systems designers with input on conditions in the space environment, and to perform detailed analysis of data associated with past events that have caused significant impacts on space systems. New and creative experiments, employing present and planned space-based and ground-based sensors, are required.

The areas of space research that are relevant to space weather include studies of the Sun, the solar wind and interplanetary medium, the magnetosphere, the ionosphere, and the upper atmosphere. It is also important to understand the coupling of these regimes, which requires an interdisciplinary approach that merges observations, theory, and modeling. The research goal is to synthesize the scientific phenomenologies into a coherent and unified picture of the Sun-Earth system. Quantitative prediction models will be developed that are capable of assimilating data obtained by widely separated and disparate instruments on the ground and in space.

Today, space weather forecasting is in a situation similar to that of weather forecasting a half-century ago. Even with the present and planned instruments, the data are sometimes too sparse, and some critical data, such as in-place solar wind parameters, are not available at all. The gaps in ground-based observations are particularly acute at very high latitudes, where the magnetic field maps out to the distant regions of the magnetosphere. New ground- and space-based instruments, coupled with quantitative modeling, will enormously improve space weather specification and forecasting quality.

For background information SEE AURORA; GEO-MAGNETIC VARIATIONS; IONOSPHERE; MAGNETO-SPHERE; SATELLITE NAVIGATION SYSTEMS; SOLAR WIND; SUN; THERMOSPHERE; VAN ALLEN RADIATION in the McGraw-Hill Encyclopedia of Science & Technology.

Richard Behnke

Bibliography. D. Dooling, Stormy weather in space, *IEEE Spectr.*, pp. 64–72, June 1995; E. Lerner, Space weather, *Discover*, pp. 54–61, August 1995.

Spectroscopy

Advances in sensitive optical measurement schemes have led to the detection and the physical and chemical characterization of individual molecules. These technological achievements are much more than the ultimate milestone in low-level monitoring. Potential applications unique to single-molecule spectroscopy include sequencing of deoxyribonucleic acid (DNA) at high speeds, probing microscale environments, monitoring environmental pollution, studying the variability of molecular conformations, detecting disease infection at an early stage, and devising molecular-scale imaging probes.

Technical considerations. Even though single photons can be counted by photomultiplier tubes and avalanche photodiodes, many photons are needed to detect single molecules. The reason is that the photons cannot be collected with 100% efficiency. The simplest solution is to monitor molecular fluorescence, because the transition can be cycled through 10,000–100,000 times before photochemical destruction sets in.

Another approach is to cycle the molecule through a reaction repeatedly, that is, to monitor the products from a reaction catalyzed by the molecule of interest. The molecule is monitored indirectly, but substantial chemical amplification of the signal can be effected. If the reaction product is fluorescent, both the repeated excitation and the repeated reaction can be combined to gain extreme sensitivity.

The background (noise) level must also be low compared to the signal level in order to allow discrimination. Contributors to the background level include fluorescence or Rayleigh and Raman scattering of the bulk medium containing the molecule of interest. Careful rejection of stray light with the help of optical filters goes a long way toward background suppression. Also, each molecule can alternately absorb and emit light (cycle) many times, because of the large transition moment. Photons are therefore emitted in rapid succession, or are correlated, when the molecule is in the excitation region. In contrast, scattering and dark counts are not correlated.

Temporally, Rayleigh and Raman scattering are instantaneous. The use of gated detectors in combination with pulsed lasers can effectively block off scattered light in favor of fluorescence. Fluorescence of the medium can even be partially suppressed, as long as the lifetimes of the solvent and sample fluorescence are very different from each other.

Multiple fluorophores. A fluorophore is a fluorescent species or unit. Large molecules such as biopolymers can possess many fluorescent centers per molecule, increasing the signal in the measurement. Not surprisingly, one of the earliest demonstrations of single-molecule detection is based on multiple fluorescent labeling of DNA molecules deposited at a low concentration on a glass slide. The individual light spots were readily detectable through a microscope.

The ability to detect individual DNA molecules provides insight into the separation mechanisms of these molecules. Several research groups have produced so-called movies of DNA molecules migrating under an electric field. Motions due to sieving, snakelike movement of long molecules, or entanglement can thus be directly visualized, because the many fluorescent labels form an outline of the DNA strand that can be resolved under a microscope. Furthermore, the relative fluorescence intensity has been used to uniquely determine the size of the DNA strand.

Small volumes. Hydrodynamic focusing is a concentration effect due to compression in a flowing fluid. One approach is to adapt hydrodynamic focusing (as in flow cytometry, a cell-sorting technique) to restrict the volume of a stream of liquid containing the molecule of interest. Within the picoliter volume defined by the crossing of a micrometer-sized laser beam and a micrometer-sized flow stream, individual molecules pass through occasionally to generate bursts of fluorescent photons.

Even though the initial demonstration of single-molecule detection in picoliter volumes depended on having roughly 25 fluorescent groups in the protein β-phycoerythrin, further reduction of background fluorescence, improvements in light-collection efficiencies, and the selection of more stable fluorophores have led to the detection of single fluorescent entities in solution.

Small-volume monitoring is not limited to hydrodynamic focusing. A tightly focused laser beam has a beam waist (that is, the smallest part of the laser beam at the focal point) that is also in the picoliter range. If the focal region is placed in a solution contained in a much larger cuvette (transparent vessel), molecules will randomly diffuse in and out of the small volume. For a sufficiently dilute solution, bursts of fluorescence photons can be observed as single-molecule events.

An alternative scheme to miniaturize the observation volume is based on confocal microscopy. Either fluorescence correlation analysis or direct observation of photon bursts allows the counting of individual molecules within the femtoliter volume. In fact, the very small dimensions allow the use of molecular diffusion times to recognize single-molecule events. Conversely, such an approach can be used to determine molecular diffusion times of individual molecules.

Another approach for monitoring a small volume of liquid is the use of microdroplets. Microdroplets are generated at the end of a capillary tube carrying the sample solution. The micrometer-sized droplets can be suspended in a trapping electric field to allow interrogation by a laser beam.

A small volume can also be defined inside capillary tubes. By using a metal vapor cell and an excitation wavelength matching the absorption of the metal vapor, the additional stray light can be effectively filtered out to allow single-molecule detection. The use of a near-infrared wavelength further suppresses the background fluorescence signal from the solvent and from the capillary walls.

Small-volume observation is again the key to the study of single impurity molecules in a solid host. There, the line width can be extremely narrow, providing an unusually high transition probability for the excitation-emission process. Not only can single molecules be detected, but the spectral properties of each, which are influenced by the local environment, can also be recorded.

The smallest observation volume is provided by near-field microscopy, a new type of optical microscopy that allows imaging of subwavelength dimensions. The dimensions of the probe tip naturally restrict the area probed to below the diffraction limit. The excitation light diverges rapidly after the near-field region, and so the depth of observation is also limited. Near-field microscopy has been applied to the detection of individual dye molecules attached to solid surfaces. The sensitivity is high enough such that the emission lifetimes of single molecules can be measured. On irradiating for a longer period of time, it even becomes possible to see individual signals disappear because of fluorescence bleaching.

Finally, a relatively straightforward way to limit the observation volume is to excite the molecules through the evanescent wave generated at the surface by a light beam propagating behind it. The penetration depth is then of the order of only the wavelength of light. Fluorescence or scattering from solvent molecules becomes negligible. By imaging the surface onto a charge-coupled device camera, individual molecules can be registered as each migrates to the proximity of the surface (**Fig. 1**). The observation volume is defined by the penetration depth and the area of each pixel element in the image.

Catalytic amplification. In general, it is possible to design an appropriate nonfluorescent substrate for a given catalyst, so that a fluorescent product will be formed through reaction. Therefore it is possible to generate millions of fluorophores from each catalytic molecule, because the molecule itself does not become permanently altered during reaction, turning detection into a trivial problem.

Indeed, the very first report of single-molecule detection relied on catalytic amplification to generate easily detectable signals even from a nonlaser excitation source. It did take many hours to accumulate enough products for detection. To isolate individual enzyme molecules, a dilute solution was sprayed to produce micrometer-sized droplets in an inert oil. Here, the molecule itself need not be fluorescent, the molecule is not destroyed as a result of the measurement, and the chemical reactivity rather than the physical property of the species is being interrogated.

Individual enzyme molecules can also be isolated inside a narrow capillary tube. By incubation with the substrates, fluorescent product zones will be formed where each enzyme molecule resides. Single-molecule detection was confirmed by the fact that the sizes of the fluorescence signals and the number of discrete product zones observed were as predicted from the known activity and the prepared concentration of the sample solution (**Fig. 2**).

Implications. Single-molecule detection represents the final frontier in sensitive measurement. Some of the approaches allow counting every

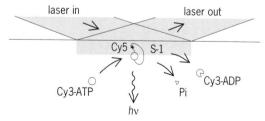

Fig. 1. Visualization of individual adenosinetriphosphate (ATP) turnovers by single myosin subfragments labeled with Cy5 dye to pinpoint their locations on a quartz slide. When Cy3 dye–labeled adenosinetriphosphate binds to myosin and reacts, additional fluorescence is recorded during the reaction period. The release of the products phosphate (Pi) and Cy3 dye–labeled adenosinediphosphate (ADP) completes the cycle. S-1 = single myosin molecule. (*From T. Funatsu et al., Imaging of single fluorescent molecules and individual ATP turnovers by single myosin molecules in aqueous solution, Nature, 374:555–559, 1995*)

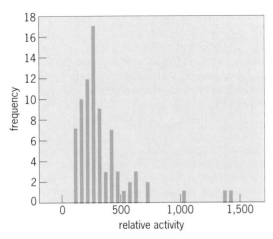

Fig. 2. Detection of single enzyme molecules. When an 8 × $10^{-17}M$ solution of the enzyme is incubated with nonfluorescent substrates inside a 20-micrometer-diameter capillary tube for 1 h, zones of fluorescent products are formed at the local regions where the enzyme resides. As these are migrated past a laser beam by electrophoresis, individual fluorescent peaks can be observed. (*After Q. Xue and E. S. Yeung, Differences in the chemical reactivity of individual molecules of an enzyme, Nature, 373:681–683, 1995*)

molecule in that small sample to provide measurement of the concentration down to $10^{-17}M$. If a target molecule [for example, a particle of the human immunodeficiency virus (HIV)] can be tagged with an enzyme, it can then be detected at the same low level, leading to early disease diagnosis.

Studying single molecules can reveal heterogeneities among them. Spectroscopic properties and reactivities are related to the microenvironment around the molecules. Thus, single-molecule spectroscopy provides a sensitive and specific probe of these micrometer- to nanometer-sized regions. Indeed, it was found that even when the microenvironments are identical, discrete molecular conformations can exist in otherwise identical molecules. Thus molecular modeling and drug design calculations need to become even more sophisticated to account for secondary energy minima in the molecules.

Finally, detection by amplification makes it possible to monitor an actual chemical reaction in progress. Although not all chemical reactions are expected to show microscopic variations beyond counting statistics, the ability to follow individual steps in a reaction should lead to new kinetic insights.

For background information SEE CONFOCAL MICROSCOPY; ELECTRON MICROSCOPE; FLUORESCENCE; RAMAN EFFECT; SCATTERING OF ELECTROMAGNETIC RADIATION; SPECTROSCOPY in the McGraw-Hill Encyclopedia of Science & Technology.

Edward S. Yeung

Bibliography. T. Funatsu et al., Imaging of single fluorescent molecules and individual ATP turnovers by single myosin molecules in aqueous solution, *Nature,* 374:555–559, 1995; B. Rotman, Measurement of activity of single molecules of β-D-galactosidase, *Proc. Nat. Acad. Sci. (USA),* 47:1981–1991, 1961; X. Shi, R. W. Hammond, and M. D. Morris, DNA conformational dynamics in polymer solutions above and below the entanglement limit, *Anal. Chem.,* 67:1132–1138, 1995; Q. Xue and E. S. Yeung, Differences in the chemical reactivity of individual molecules of an enzyme, *Nature,* 373:681–683, 1995.

Star clusters

Globular star clusters are remnants from the earliest phases of star formation in galaxies. Those clusters in the Milky Way Galaxy have been subjected to careful scrutiny for several reasons, including their primary importance in galactic chronology. The ages of the oldest clusters set a lower limit to the age of the Milky Way, as well as to the age of the universe. The age spread of the clusters provides crucial information on the speed with which the young galaxy evolved, both dynamically as its disk formed and chemically as the primordial gas left over from the big bang was polluted by the products of stellar evolution and supernova explosions. Globular cluster ages are determined by matching the theory of stellar structure and evolution with observations.

Stellar structure and evolution. Four differential equations are used to compute the structure of a star whose chemical composition is known as a function of radius. One equation defines the mass as a function of radius, the second defines the pressure gradient, the third the temperature gradient, and the fourth the energy-generation gradient. A variety of thermodynamic quantities must also be determined as functions of density and temperature, such as the opacity of matter to radiation, the heat capacities, the pressure, the energy generation, and the entropy. Any uncertainties in the physics that define these quantities will be reflected directly into uncertainties in stellar structure. The solution of these four differential equations also involves establishing four boundary conditions. Two of these conditions are simple ($L = 0$ and $M = 0$ at $r = 0$, where r is the radius parameter and L and M are the luminosity and mass of the portion of the star within radius r of the center). However, the other two, the outer boundary conditions for the pressure and temperature at $r = R$, where R is the star's radius, are difficult to apply. The definition of R is not simple, and uncertainties consequently arise in the solutions.

The evolution of stars is modeled by altering slightly, at a series of time steps, the chemical composition as a function of radius according to the processes of stellar energy generation via nucleosynthesis. A good test of such calculations is whether the models can reproduce the luminosity (L) and temperature (T), and hence the radius (since L depends on R^2 and T^4) of the Sun, since its age is known (from the radioactivity dating of meteoritic samples), as is its initial chemical composition (from analyses of the absorption-line spectrum of the solar atmosphere, which has retained the original mix of chemical elements). The model tests should improve in the near future as helioseismology studies that probe the Sun's core become available.

The primary result of such modeling is that the lifetime of a star is determined primarily by its mass and secondarily by its chemical composition. Thus, in a star cluster, where it is possible to assume (and test) that all cluster members came into existence at more or less the same time and with the same chemical composition, the more massive stars will expire first. Hydrogen-burning stars in a very young cluster will lie along the main sequence in a diagram of luminosity versus temperature (the Hertzsprung-Russell diagram) or the roughly equivalent diagram of magnitude versus color (a measure of temperature). At a later time, the cluster will be missing the shortest-lived massive stars, and the main sequence will be incomplete above a certain temperature and luminosity. This termination point of the main sequence, called the turnoff, moves down along the main sequence as the cluster ages, and successively less massive stars, which initially have lower temperature and luminosity, begin to evolve away from the main sequence. **Figure 1** shows isochrones (equal-age models) for two clusters of differing chemical compositions at three different stages in their lives, displaying the migration of the turnoff and the effect of composition. Globular-cluster ages may in principle be determined by comparing such models to observations, using, of course, models with the same chemical composition as determined ultimately from spectroscopy of cluster members.

The theory of convection is immature, and solar and globular-cluster main-sequence stars have convective envelopes. Uncertainties also remain in the

transformations between the temperatures from the models and the observed colors, and in the solutions to the differential equations because of the two outer boundary conditions. Thus, in Fig. 1 comparisons between models and observations along the horizontal axis are more uncertain than those along the vertical axis. Stellar luminosities are generated in the cores, and so are less sensitive to envelope effects.

Absolute ages. These ages are estimated most directly from the measurement of the apparent luminosities or magnitudes of cluster turnoffs, distances, and chemical compositions. The largest uncertainties in cluster ages arise from the remaining uncertainties in the distance scale. The clusters are all far too remote for direct-distance measurement via trigonometric parallax. An alternative is to obtain trigonometric parallaxes for nearby dwarf stars whose chemical compositions are like those of globular clusters. Comparing the apparent luminosities of the nearby star and stars of the same color (that is, temperature) in the clusters would give accurate relative distances, so knowing the distance to one gives the distance to the other. At the moment, the data are insufficient for this effort as well. With the availability of improved precision parallax measurements from ground-based telescopes, from the *Hipparcos* satellite, and from the Hubble Space Telescope, this main-sequence fitting will result in much-improved distances and, hence, ages for the clusters. Interstellar absorption in front of the clusters will remain a problem.

An alternative method, which is absorption independent, involves measuring in each cluster the gap in apparent luminosity (or magnitude) between the turnoff and the horizontal branch, which includes the RR Lyrae variable stars (**Fig. 2**). By calibrating the intrinsic luminosities of the RR Lyraes as a function of chemical composition, the gap in a cluster of known chemical composition yields the intrinsic turnoff luminosity and the cluster age.

The calibration of RR Lyrae luminosities proceeds in two steps. First, about two dozen such stars have been observed photometrically and spectroscopically throughout their pulsation cycle. Changes in brightness are changes in apparent luminosity. Changes in color indices may be transformed to changes in temperature. Since the luminosity (L) is proportional to R^2T^4, the change in the apparent luminosity is proportional to the change in θ^2T^4, where θ is the stellar angular diameter. The changing radial velocities throughout the cycle are obtained from the spectroscopy. The pulsational velocity of the stellar atmosphere versus time may then be derived, and by integration, the change in linear radius. Comparison with the change in angular diameter or radius yields a distance estimate. From such work the RR Lyrae luminosities have been found to depend on chemical composition.

This method provides good relative distances, but the zero point of the luminosity-metallicity relation must be set by more fundamental methods

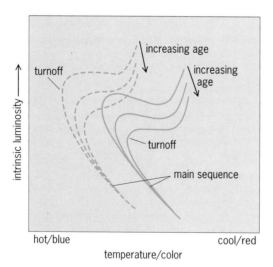

Fig. 1. Schematic representation of model isochrones (as would be observed in a star cluster) for two different chemical compositions (broken versus solid lines) at three different ages.

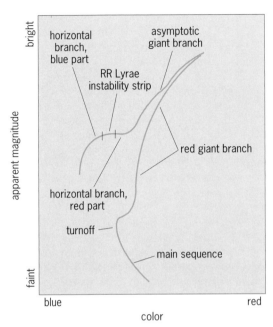

Fig. 2. Color-magnitude diagram of a globular cluster. The gap between the RR Lyrae instability strip and the turnoff, where stars leave the main sequence and move up the red giant branch, is an excellent age indicator.

because the derived distances depend on theoretical estimates of the relation between pulsational and radial velocity and between color and temperature. Currently, statistical parallax analyses of the motions of nearby RR Lyraes have been used with good success. Here, proper motions of RR Lyraes are converted into tangential velocities by using an adopted distance calibration. Combining these with radial velocities yields the space velocities for the stars. The ensemble average velocity across the direction of the Sun's motion should be zero, as should the average motion perpendicular to the galactic plane, and this result obtains for only a limited range of luminosity-metallicity zero points. Unfortunately, the samples are not large enough to test independently the slope of the luminosity-metallicity relation.

Most studies of RR Lyrae luminosities lead to an average age for the globular clusters of about 18×10^9 years. (The Galaxy and the universe would be even older since some clusters seem to be older than average and star formation did not occur immediately after the big bang.)

For two reasons, large uncertainties remain in this estimate. First, the distance derived for the RR Lyraes in the nearby Large Magellanic Cloud disagree with those derived for the Cepheid variables by about 12%, with the Cepheids' distance estimate being greater. The Cepheids are particularly important since they provide the primary calibration of the extragalactic distance scale, and hence the Hubble constant and the expansion age of the universe. If the Cepheid distance scale is correct, the RR Lyrae distance scale revision will result in an average globular cluster age of 14×10^9 years. If the RR Lyrae distance scale is correct, Hubble constant

estimates must be increased by 12% and the expansion age decreased by a like amount. A compromise between the two independent distance scales may be required. *See Universe.*

Second, a number of uncertainties in the physics of stellar evolution calculations remain, particularly in convection theory and in the diffusion of helium into the hydrogen-burning core. Neither effect is expected to produce age uncertainties as large as 10%, although diffusion, if present, can only decrease age estimates.

Relative ages. Most of the uncertainties in the distance scale zero point and in the physics become unimportant in differential age comparisons between clusters, so turnoff luminosities provide good relative age estimates. For clusters with similar chemical compositions, the color difference between the turnoff and the nearly vertical red giant sequence is predicted to be a good indicator of relative ages (Fig. 1). Further, for similar chemical compositions, the ratios of stars blueward of, within, and redward of the RR Lyrae instability strip are also predicted to be good indicators of relative ages (Fig. 2). All three methods reveal that globular clusters have a range in ages, with the scatter being typically 2–3×10^9 years and in a few cases up to 6×10^9 years, even at the same chemical compositions.

Two interpretations of these results have been offered. First, because most of the globular clusters are very deficient in heavy elements and occupy the low-density halo where star formation must have ceased long ago, the globular clusters are assumed to have formed during the Galaxy's earliest stages. The age spread observed among the clusters then implies that the Galaxy's youth was of very long duration, several billion years, and was also highly random, with different parts of it reaching similar levels of chemical enrichment at much different times. Alternatively, the Galaxy could have formed more rapidly and homogeneously. Globular clusters could have also formed in smaller, lower-density dwarf galaxies, such as those that still surround the Milky Way Galaxy, and at times even after the Milky Way had taken form and most of its own globular clusters had been formed. Some of the satellite galaxies and their globular clusters could then have been captured by the Milky Way Galaxy, leading to the age spread that is now observed. The Sagittarius dwarf galaxy, with its four associated globular clusters, appears to be an ongoing example of this phenomenon. Much more work remains to be done to determine which (or whether both) of these interpretations is correct.

For background information *see Big bang theory; Cepheids; Hertzsprung-Russell diagram; Hubble constant; Milky Way Galaxy; Star clusters; Stellar evolution; Variable star* in the McGraw-Hill Encyclopedia of Science & Technology.

Bruce W. Carney

Bibliography. C. Chiosi, G. Bertelli, and A. Bressan, New developments in understanding the H-R diagram, *Annu. Rev. Astron. Astrophys.*, 30: 235–285, 1992; I. Iben, Jr., Single and binary star evolution, *Astrophys. J. Supp.*, 76:55–114, 1991; A. C. Phillips, *The Ages of Stars*, 1995; A. Renzini and F. Fusi Pecci, Tests of evolutionary sequences using color-magnitude diagrams of globular clusters, *Annu. Rev. Astron. Astrophys.*, 26:199–244, 1988.

Superconductivity

High-transition temperature (high-T_c) superconductivity was discovered in many copper-oxide compounds, called cuprates, in 1986 and 1987 but has yet to be explained. In the superconducting state, electrons form pairs that undergo a collective condensation to a lower energy state at the transition or critical temperature, T_c. By using quantum mechanics, the properties of these paired electrons can be described by a wave function, which is also known as the order parameter. This wave function is characterized by a center-of-mass angular momentum, L, and a total spin, S. From Fermi-Dirac statistics, the wave function must change sign under exchange of spin and orbital angular momentum labels of the two electrons. It follows that electron pairs in a state with opposite spins ($S = 0$) may have any even L value; that is, $L = 0$ (s wave), 2 (d wave), and so on. Electrons in a state with parallel spins ($S = 1$) will have odd values of L.

Symmetry of the order parameter. These values of S and L define the symmetry properties of the wave function. For example, they determine how the wave function transforms under rotation, exchange of the electrons, or time reversal. The symmetry properties of the order parameter are worth knowing. Although they will not reveal the mechanism of high-T_c superconductivity, they will narrow the field by putting constraints on the possible mechanisms.

In the Bardeen-Cooper-Schrieffer (BCS) model for low-T_c superconductors, the electron pairing is mediated by a phonon exchange between the electrons. In this model, the electrons, called Cooper pairs, are paired in a spin singlet ($S = 0$) and have $L = 0$ (s wave). In high-T_c superconductors, there are good reasons to believe that the electrons (or holes) are also paired in a spin singlet ($S = 0$), and hence are also called Cooper pairs. However, the value of L is still unclear and is the subject of much controversy.

The cuprates have a layered structure, and every unit cell has at least one planar array of copper atoms connected by oxygen atoms, called the copper-oxide (CuO) layer (**Fig. 1**). The parent compounds of the cuprates are antiferromagnetically ordered, and as they are doped with either holes or electrons, they cross over from being insulating to being superconducting. For example, $YBa_2Cu_3O_6$ is

the parent compound of the most studied cuprate, $YBa_2Cu_3O_{7-\delta}$ (YBCO). As oxygen is added to $YBa_2Cu_3O_6$, the antiferromagnetic order is rapidly suppressed and the material becomes superconducting. However, in the superconducting phase antiferromagnetic correlations can still appear and disappear spontaneously within regions of the superconductor. The proximity of the antiferromagnetic phase to the superconducting phase has given rise to theories that the superconductivity is mediated by these antiferromagnetic spin fluctuations. The predicted symmetry for such unconventional pairing is the $d_{x^2-y^2}$ (d-wave) state (**Fig. 2a**). In this symmetry, the wave function has four lobes, with adjacent lobes having opposite phase, which is very different from the s-wave state (Fig. 2b).

Evidence for unconventional symmetry. The gap in the energy spectrum of electrons in superconductors has been much studied and depends on the properties of the order parameter. In conventional superconductors, the gap is uniform in momentum space. Hence, a minimum excitation energy is required to break electron pairs at low temperatures. This leads to an exponentially activated temperature dependence for properties such as the electronic specific heat.

For a d-wave superconductor, there are lines of nodes in the energy spectrum (between adjacent lobes of the order parameter). The existence of nodes means that there are always thermally excited quasiparticles (normal electrons) available

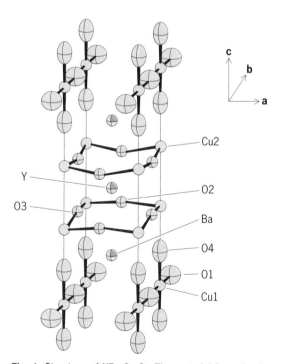

Fig. 1. Structure of $YBa_2Cu_3O_7$. The material is optimally doped by removing oxygen from some O1 sites to give $YBa_2Cu_3O_{7-\delta}$. The CuO planes are formed by the Cu2, O2, and O3 atoms, and the CuO chains by the Cu1 and O1 atoms. Unit cell parameters are $a = 0.382$ nanometer, $b = 0.388$ nm, and $c = 1.165$ nm. (*After W. E. Pickett, Electronic structure of the high-temperature oxide superconductors, Rev. Mod. Phys., 61:433–512, 1989*)

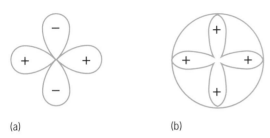

Fig. 2. Order parameter in wave-vector (or momentum) space. (*a*) *d*-wave symmetry. (*b*) Isotropic and anisotropic *s*-wave symmetries.

at low temperatures. Thus, there is a power-law temperature dependence for many properties. Such behavior has been observed in a number of measurements such as nuclear magnetic resonance relaxation rates, angle-resolved photoemission, penetration depth, and specific heat. These measurements are indirect evidence for nodes or near-nodes in the superconducting gap.

However, nodes or near-nodes in the gap could also arise from an order parameter that is highly anisotropic but does not change sign. Such an anisotropic *s*-wave gap has been predicted by a model that postulates BCS-type pairing of the electrons in each copper-oxide layer and Josephson tunneling between the layers (Fig. 2*b*). A more direct way, then, to test the symmetry of the order parameter is to look for sign changes in the order parameter.

Evidence for d wave. A new class of experiments using superconducting quantum interference devices (SQUIDs) and Josephson junctions was begun in 1993. These experiments are sensitive to both the magnitude of the order parameter and its phase.

The basic idea is to make a SQUID partially out of a conventional low-T_c *s*-wave superconductor such as lead, the barrier out of a normal metal such as gold or silver, and the rest of the loop out of the high-T_c superconductor to be studied, in this case YBCO. One such geometry is shown in **Fig. 3***a*. This type is called an *a-b* SQUID because it has one Josephson junction oriented normal to the YBCO crystalline **a** axis and the other normal to the **b** axis.

The two Josephson junctions allow tunneling of Cooper pairs between the superconductors. If no magnetic field B is present and YBCO has *d*-wave symmetry, pairs tunneling through the **a**-axis junction have a phase shift of π with respect to pairs tunneling through the **b**-axis direction. Thus, a pair which travels once around the SQUID loop acquires an intrinsic phase shift of π. Such a π shift will produce a circulating current, J, around the loop. If the SQUID has a substantial critical current, then $J \approx \Phi_0/(2L_s)$, where $\Phi_0 = h/2e$ is the flux quantum and L_s is the SQUID loop inductance. No such intrinsic phase shift or circulating current will be produced at $B = 0$ if YBCO has *s*-wave symmetry.

The most sensitive way to measure a circulating current, if any, is to measure the magnetic field it cre-

ates by using a SQUID-based magnetic sensor. High-resolution magnetic field microscopes have been built by using SQUIDs as the sensor and a platform which scans the sample to produce a magnetic image. Such a scanning SQUID microscope has been used to image an *a-b* SQUID at $B \approx 0$. Figure 3*b* shows a magnetic image of the *a-b* SQUID which reveals a circulating current producing about $-0.6\ \Phi_0$ of flux, consistent with *d*-wave pairing in YBCO.

Both the proposed *s*-wave and *d*-wave order parameters are time-reversal invariant. However, symmetries that break time-reversal symmetry, such as $s + id_{x^2 - y^2}$ and $d_{x^2 - y^2} + id_{xy}$, have also been proposed. A scanning SQUID microscope can also be used to test for the time-reversal symmetry of the superconductor, thus providing a powerful quantitative check on the pairing symmetry. For three *a-b* SQUIDs similar to that in Fig. 3, it was found that the intrinsic phase shift equals $(0.98 \pm 0.05)\pi$, where, as noted above, this phase shift equals 0 if YBCO has *s*-wave pairing, and is equal to π if YBCO has *d*-wave pairing. This result shows that the order parameter is time-reversal invariant, is consistent with *d*-wave pairing, and rules out such a symmetry as $s + id_{x^2 - y^2}$.

A scanning SQUID microscope has also been used to image superconducting YBCO rings with 0,

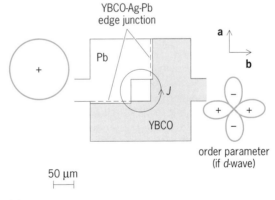

50 μm

(a)

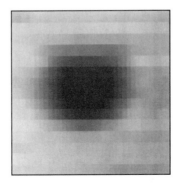

(b)

Fig. 3. Type *a-b* SQUID. (*a*) Geometry. (*b*) Magnetic image at $B \approx 0$, showing circulating current producing about $-0.6\ \Phi_0$ of flux. (*After A. Mathai et al., Experimental proof of a time-reversal invariant order parameter with a π-shift in YBa$_2$Cu$_3$O$_{7-\delta}$, Phys. Rev. Lett., 74:4523–4526, 1995*)

2, and 3 grain-boundary Josephson junctions. Spontaneous flux magnetization of $\Phi_0/2$ has been observed in the 3-junction ring but not in the 2-junction rings. This observation is also consistent with d-wave pairing.

Evidence for s wave. In contrast to the above experiments, a team of researchers have built and measured Josephson junctions that show tunneling currents in the direction perpendicular to the copper-oxygen planes in YBCO. If verified by other research efforts, this result is contrary to what is expected from pure d-wave pairing and may be evidence for s-wave pairing in YBCO.

Multicomponent order parameters. In general, the order parameter may have many angular momentum components; however, any combination of components must obey the underlying symmetry of the lattice. For example, an order parameter for a tetragonal lattice ($a = b \neq c$ in Fig. 1) may have four-fold symmetry, such as $L = 0$ (s wave) + $L = 4$ (g wave) + $\cdots$. It may also have twofold symmetry, such as $L = 2$ (d wave) + $L = 6 + \cdots$. However, a mixed combination of fourfold and twofold symmetries, such as $L = 0$ (s wave) + $L = 2$ (d wave) + $\cdots$, is not compatible with tetragonal lattice symmetry. Although YBCO has a slightly orthorhombic structure ($a \neq b \neq c$), most researchers have assumed that it is well described by a tetragonal lattice. However, penetration-depth measurements have revealed a large in-plane anisotropy associated with conducting copper-oxide chains running along one axis of the copper-oxide planes. Hence, YBCO must be treated as being orthorhombic, and its order parameter may be a combination of fourfold and twofold symmetric components. Thus, the order parameter may have both s-wave and d-wave symmetric components. Efforts are under way to reconcile the experimental results in YBCO.

Other cuprates. Alternatively, it would be easier to decipher the pairing mechanism by carrying out these experiments in tetragonal materials. However, YBCO samples are of much higher quality than those of the other cuprates. It has also been easier to make the needed samples (such as an a-b SQUID) by using YBCO. One such tetragonal material is $Nd_{2-x}Ce_xCuO_4$ (NCCO), which is unique among the cuprates in that it is an electron-doped superconductor. Penetration-depth measurements in this material are consistent with s-wave symmetry.

For background information SEE ANGULAR MOMENTUM; ANTIFERROMAGNETISM; CRYSTALLOGRAPHY; JOSEPHSON EFFECT; NONRELATIVISTIC QUANTUM THEORY; SQUID; SUPERCONDUCTIVITY in the McGraw-Hill Encyclopedia of Science & Technology.

Anna Mathai

Bibliography. D. L. Cox and M. B. Maple, Electronic pairing in exotic superconductors, *Phys. Today*, 48(2):32–40, February 1995; B. G. Levi, In high-T_c superconductors, is d-wave the new wave?, *Phys. Today*, 46(5):17–20, May 1993; A. Mathai et al., Experimental proof of a time-reversal invariant order parameter with a π-shift in $YBa_2Cu_3O_{7-\delta}$, *Phys. Rev. Lett.*, 74:4523–4526, 1995.

Supramolecular chemistry

Supramolecular chemistry may be defined as chemistry beyond the molecule. Individual molecules can recognize and associate with each other, forming aggregates between two or more single molecules. The kind of interactions that control the recognition processes are not those that have previously dominated chemistry (those interactions in which electrons are shared to form ordinary chemical or covalent bonds) but the much weaker noncovalent bonding interactions, such as hydrogen bonds, van der Waals forces, and hydrophobic interactions. Using these very weak interactions, chemists have managed to create a new artificial world populated by chemical systems displaying beautiful molecular architectures. However, one of the main goals of supramolecular chemistry is the production of molecular machines, for example, a molecular structure that can be switched reversibly between two different states (0 and 1) by an external stimulus. Control of the switching properties and the ability to address the two different states (that is, reading in, storing, and writing out information) are problems that have to be solved in order to achieve information storage at a molecular level. These challenges are some of the first that have to be met in the genesis of molecular computers.

From biological to artificial systems. Most biological events are regulated by noncovalent bonding interactions, enabling biochemical processes to take place with great efficiency and high selectivity. An example of the power inherent in the noncovalent bonding interactions is the form and function of deoxyribonucleic acid (DNA). A single strand of DNA will recognize its complementary strand in an act of so-called molecular recognition to form a dimeric structure that adopts the energetically most favored geometry, self-assembling into its characteristic double helix, which stores all the information contained in the genetic code and thus is essential for the support of life.

Many chemists are trying to transfer these concepts, rooted in the biological systems, to the artificial world of chemical synthesis, constructing unnatural molecular assemblies and supramolecular arrays designed on the basis of matching noncovalent bonds. Thus the self-assembly of novel molecular architectures in which two or more components are mechanically interlocked but not covalently bonded to each other has been made possible. Two classes of molecules (**Fig. 1a**) are becoming increasingly important: catenanes (I), which consist of two or more interlocked rings, and rotaxanes (II), in which one or more rings are threaded on to a dumbbell-shaped component. In

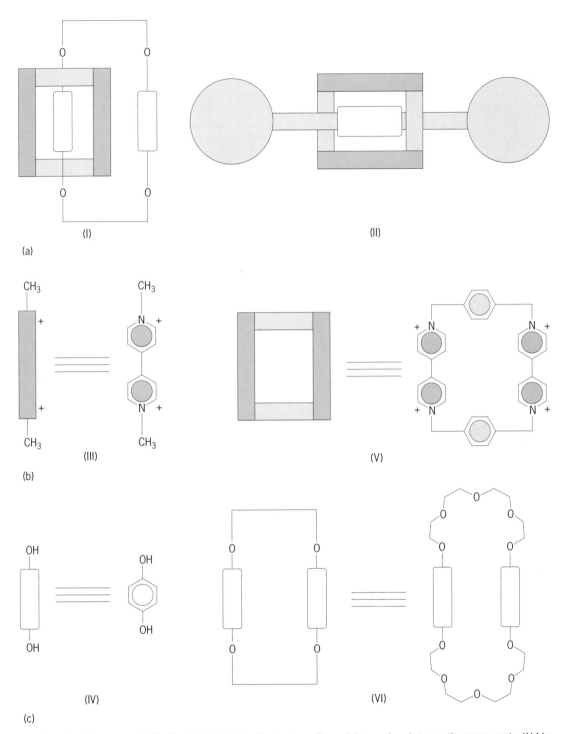

Fig. 1. Interlocked systems. (a) Mechanically interlocked molecules with weak interactions between the components. (b) Linear (left) and cyclic (right) electron acceptor units. (c) Linear (left) and cyclic (right) donor units.

rotaxanes, bulky assemblies of atoms at both termini of the dumbbell-shaped component act as stoppers, preventing the ring from slipping off the central position. In naming these interlocked molecular compounds, a number in brackets indicates the number of molecule components not covalently bonded to each other. Thus, a [2]catenane [structure (I) in Fig. 1a] comprises two ring components, and a [2]rotaxane [structure (II)] con-

tains a dumbbell-shaped component and a ring component.

Early attempts, in the 1960s and 1970s, to construct catenanes and rotaxanes relied upon the statistical chance of one molecule threading itself through a large ring. However, this threading is an unlikely event, unless molecular recognition at the behest of noncovalent bonds induces self-assembly. In order to produce catenanes and rotaxanes in an

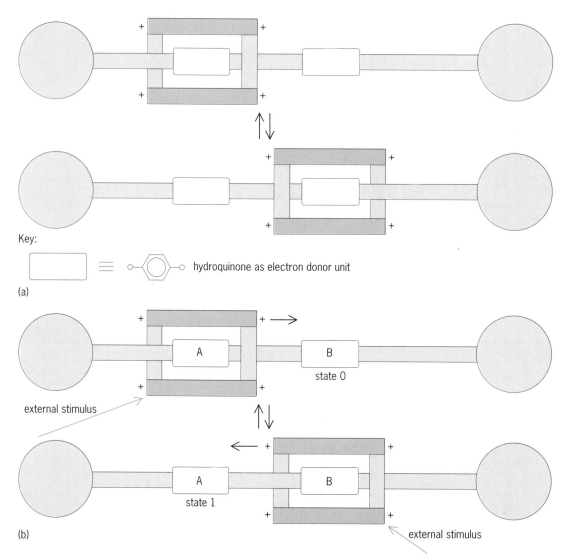

Key:

▭ ≡ o—◯—o hydroquinone as electron donor unit

(a)

external stimulus

state 0

state 1

external stimulus

(b)

Fig. 2. Rotaxane self-assembly. (*a*) Symmetrical molecular shuttle in which the electron acceptor ring moves back and forth between the two identical electron donor units. (*b*) The logic upon which the design of a binary molecular device based on the molecular shuttle is founded.

efficient manner, the noncovalent bonding interactions between the electron acceptor and donor units (Fig. 1*b* and *c*), have to be exploited. In an example that has worked well, the electron acceptors are derivatives of the nitrogen-containing aromatic compound, known as paraquat (III; Fig. 1*b*). These molecules carry two positive charges, which means that the aromatic rings are deficient in electrons; thus they tend to accept electrons from somewhere else. The matching electron donors are based on another aromatic compound, hydroquinone (IV; Fig. 1*c*). In this case, two hydrogen atoms of a benzene ring are substituted by two hydroxyl (OH) groups, and the oxygen atom of the OH group donates electrons toward the ring, making it more electron rich than it is in benzene. It is then possible to combine the donor and acceptor components in such a way that they interlock with each other spontaneously. In 1989, donor and acceptor ring components were self-assembled to produce a [2]catenane in a yield of 70% of the max-

imum possible. The ring components are connected not conventionally by chemical bonds but by mechanical bonds overlaid by weak noncovalent bonding interactions between the ring components (V) and (VI).

Although catenated structures can be modified to produce binary molecular devices, the rotaxanes with their abacuslike structures are better suited.

From shuttles to switches. By using the same synthetic strategy and building blocks as employed in the construction of the [2]catenane, the self-assembly of a [2]rotaxane that possesses two identical electron donor recognition sites in the dumbbell-shaped component has been achieved (**Fig. 2***a*).

The noncovalent bonding interactions between the electron-deficient ring and the electron-rich recognition units on the dumbbell are the stabilizing features of the components of this [2]rotaxane. Since both recognition units in the dumbbell component are identical, there is a dynamic equilibrium in which the positively charged ring moves back and forth like a shuttle (Fig. 2*a*) between the two identical stations in the form of hydroquinone residues. This dynamic molecular assembly was named the molecular shuttle.

It is not difficult to envisage the molecular shuttle as an ideal vehicle for the construction of a molecular device with binary characteristics. This objective might be achieved by designing a dumbbell component with two different donor recognition sites (A and B), each possessing particular electrochemical, photochemical, and chemical properties, such that the acceptor ring resides on one particular donor site until the system is perturbed by an external stimulus, forcing the ring to move from the originally preferred recognition site to the other donor site (Fig. 2*b*).

A series of desymmetrized molecular shuttles has been self-assembled in which the hydroquinone

ring is one of the donor recognition sites and the other ring is replaced by a different donor. Alternatively, both hydroquinone rings can be replaced by different recognition sites (**Fig. 3**). The electron donors in the linear dumbbell-shaped component were changed to a benzidine and a biphenol, with the acceptor-ring component being structure (V).

Self-assembly of the individual components produced the molecular shuttle in which the acceptor ring occupies preferentially the more donating benzidine unit. Furthermore, the position of the electron-deficient macrocyclic component can be switched from the benzidine to the biphenol recognition site by either chemical or electrochemical means.

The basic nitrogen atoms present in the benzidine unit can be protonated by acid in solution, forming a species that carries a double positive charge, which

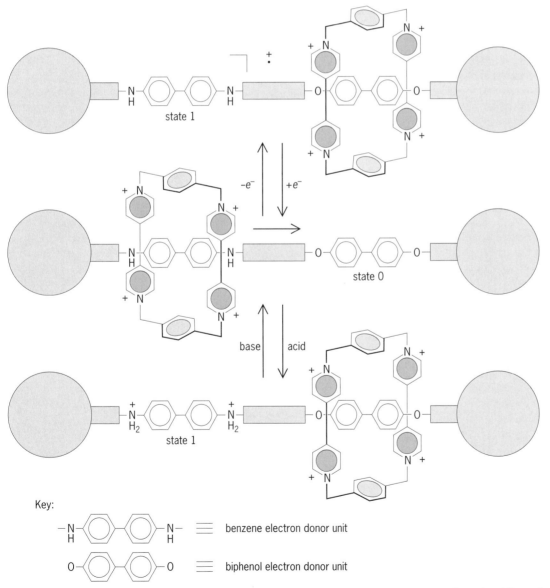

Fig. 3. Switchable molecular shuttle whose linear component comprises two different electron donor stations—benzidine and biphenol. The switching process is reversible, and can be controlled by either chemical or electrochemical means.

forces the positively charged acceptor ring to move to the biphenol donor unit as a consequence of repulsion between positive charges. Treatment of the molecular shuttle with base deprotonates the system, regenerating the original neutral benzidine donor unit. As a consequence of benzidine being a better donor than biphenol, the acceptor ring moves back to reside over the benzidine recognition site, rendering the cycle reversible (Fig. 3).

The benzidine unit also has a low and reversible oxidation potential. It can easily lose one electron in a reversible fashion, giving rise to the positively charged radical cation, where once again electrostatic repulsive forces repel the positively charged acceptor ring toward the biphenol donor recognition site. Since this oxidation is a reversible process, the radical cation can be reduced back to the neutral benzidine donor unit, switching the positively charged ring component back to its original location (Fig. 3).

By the use of noncovalent bonding interactions, it is possible to self-assemble matching components in order to construct a chemically interlocked molecule designed to function as a reversible switch between two different states.

From solution to the solid state. Molecules swimming around in solution behave in an incoherent manner. For molecular species to be able to support devicelike processes, the self-organization of molecules into larger macroscopic superstructures is necessary.

This ordering can be achieved in several ways, that is, as liquid crystals, in micelles, as vesicles, and in monolayers and multilayers. It has been possible to organize molecular compounds, such as the original [2]catenane, at a gas-water interface in both monolayers and multilayers.

The success obtained with relatively simple molecules indicates that molecular switches constructed on the basis of molecular recognition have the potential to be organized into macroscopic structures that in the future could be incorporated into nanoscale devices.

Prospects. Catenanes and rotaxanes can be designed and self-assembled to have specific microscopic properties such that they can be reversibly switched by stimuli from the macroscopic environment, thus constructing simple binary molecular devices. The problem that remains is the transformation of systems that can operate at the molecular level into devices and machines that function in the macroscopic world.

For background information SEE CHEMICAL BONDING; INTERMOLECULAR FORCES; MICELLE; MOLECULAR RECOGNITION; MONOMOLECULAR FILM in the McGraw-Hill Encyclopedia of Science & Technology.

Marcos Gómez-López; J. Fraser Stoddart

Bibliography. R. A. Bissell et al., A chemically and electrochemically switchable molecular shuttle, *Nature*, 369:133–137, 1994; J.-M. Lehn, *Supramolecular Chemistry: Concepts and Perspectives*, 1995; A. P. de Silva and C. P. McCoy, Switchable photonic molecules in information technology, *Chem. Ind.*, 992–996, 1994; G. M. Whitesides et al., Noncovalent synthesis: Using physical-organic chemistry to make aggregates, *Acc. Chem. Res.*, 28:37–44, 1995.

Television

From the time the television broadcasting systems first came into existence in the 1930s, the transmission of television signals over airwaves and cable has largely remained analog, unidirectional, and broadcast in nature. In particular, since the 1970s there has been a major expansion in the cable television networks. Typically, the cable distribution plants support 550-MHz, 750-MHz, or 1-GHz systems, and carry television channels in radio-frequency-modulated 6-MHz frequency bands. Moreover, sections of the overall spectrum are used to transport control and data channels. In addition to the basic television service with channels that do not require any descrambling, the premium encrypted channels are enabled either at the subscription time or on demand at the user's request. In the former case, the authorization is enabled by preprogramming the set top box, and in the latter case, by downloading a de-encryption key to the addressable set top box over a data or control channel. Scheduled and pay-per-view enabling are some of the variations of the analog broadcast service.

In the 1980s and 1990s there has been a major effort worldwide to define a digital television transmission standard, commonly referred to as a high-definition television (HDTV) standard. This standard will allow digital transmission of the video signal and high resolution on the television monitor. The digital transmission offers many advantages, namely, near error-free signal reception, improved bandwidth utilization, digital switching, interactive control, increased flexibility in signal manipulation, and enhanced features or functions in television systems. The high-definition television video resolution will increase about fourfold compared to present-day systems, and the aspect ratio will change from 4:3 to 16:9. Notwithstanding the standardization of high-definition television, the cable and telephone company providers have already begun to offer digital video of a quality comparable to the current analog standard (National Television Systems Committee; NTSC) and a host of other multimedia services over broadband terrestrial networks. One such service is called video-on-demand (VOD) digital interactive service. Several developments have contributed to the immense interest in this service. On the technical side, major advances have been made in server, switching, compression, transport, set top, and system control and management, making it possible to implement such systems.

Video dial-tone service. This service enables a subscriber to select a video information provider from among many such providers offering service in a neighborhood; some providers may be local and some may be thousands of miles away. From a selected video information provider, the subscriber accesses, on demand, a movie or other multimedia content for personal viewing from a library of contents prestored in high-speed servers with full videocassette-recorder (VCR) control features, such as play, stop, fast forward or reverse, and pause. This type of service includes other similar services, such as games and home shopping. Thus, a video dial tone is a common-carrier network transport service that provides a regulated, nondiscriminatory network enabling equal access to information providers (including the common carrier itself) and offering end users video programming with interactive capability. The same video dial-tone network can provide unregulated access to other interactive nonvideo programming services, such as home shopping and distance learning. The video dial tone enables consumers to, in effect, dial up on-demand video or multimedia applications offered by third parties. The current implementations use analog (NTSC) television signals that are digitized and compressed at the source. Such signals are transported over a video dial-tone network for delivery to the home. The set-top box decompresses the received digital stream and converts it back to analog form prior to the input to the television receiver. Compression makes it possible to transport an order of magnitude more video channels in the same analog or digital bandwidth.

Video dial-tone compression and switching. The digital video is typically compressed to a rate of 1.5–15 megabits per second depending on the application and the desired video quality, which can range from videocassette-recorder to high-definition television quality. At present, MPEG-2 is the preferred compression standard for video dial tone. (MPEG-2 is an international standard developed by the Moving Pictures Experts Group of the International Standards Organization.) MPEG-2 compressed bit streams for the video, audio, and data that make up a program are packaged into 188-byte MPEG-2 transport packets. (Other compression schemes, such as MPEG-1 and JPEG, can use the same MPEG-2 transport-packet structure for transmission.) To facilitate switching, MPEG-2 transport packets are converted to fixed-size packets, called asynchronous transfer mode (ATM) cells. An asynchronous transfer mode cell is a 53-byte packet with 5 bytes of header for addressing and cell routing and 48 bytes of data payload. The bandwidth of the asynchronous transfer mode virtual connection is typically about 10–12% more than the compressed video bandwidth because of the overheads of the cell structure. (Asynchronous transfer mode is an international multiplexing, switching, and transport standard developed by the International Telecommunication Union in conjunction with other national and international standards bodies, such as the American National Standards Institute.) SEE DATA COMMUNICATIONS; DATA COMPRESSION.

Video dial-tone system architecture. In video dial-tone service, a user typically selects an information provider from an on-screen interactive menu offered by the network provider. The selected information provider offers its own interactive and hierarchical on-screen menu of services. At the functional level, there are three networks: the information provider network, the asynchronous transfer mode network, and the access network. Typically, the asynchronous transfer mode and access networks together form the common-carrier video dial-tone network. The information provider network, which is a private network, connects to the asynchronous transfer mode backbone network that in turn connects to the access network, both via standard asynchronous transfer mode interfaces and transport. Synchronous optical network (SONET) transport is the preferred method to interconnect the video information providers, the asynchronous transfer mode network, and the access network, using ring or star architectures. The asynchronous transfer mode backbone network offers the switching and wide-area transport functions for video-on-demand connections between the video information provider and the access network. The digital broadcast video channels are either routed through the switches or more efficiently directly trunked from the video information provider to the access network. The access network is typically a local distribution network in the serving areas.

The four major competing access network (loop) technologies are hybrid-fiber/coax (HFC), fiber-to-the-curb (FTTC), fiber-to-the-home (FTTH), and asymmetric digital subscriber loop (ADSL). Hybrid-fiber/coax is fundamentally a bus architecture whereas the fiber-to-the-curb, fiber-to-the-home, and asymmetric digital subscriber loop are star architectures. Bus architectures are shared among multiple users, and therefore are economical but pose security and reliability problems. Subscribers on the same bus have access to all the broadcast video programs; however, only the authorized set top can tune in and decode the scrambled or encrypted broadcast or video-on-demand program that a user requests via signaling. Authorization is typically activated by loading an encryption key or a program identifier. In star architecture, a dedicated access link to each subscriber delivers a fixed bandwidth capable of carrying multiple sets of information (for example, broadcast video, video-on-demand, or multimedia) simultaneously. Only the information that is requested is routed to the subscriber's location. The switching or routing function is typically performed at the head end. In both architectures, signals are transported to optical network units (ONUs), also known as nodes, as close as possible to a cluster of subscribers. In general, the number of homes that can be served from a node decreases as the demand for bandwidth for interactive services increases.

The transport of the video-on-demand signal (in asynchronous transfer mode cell format) from the video information provider to the access network interface remains the same in all of the access network technologies. A video dial-tone session and call management system interacts with the resource and call management systems of both the asynchronous transfer mode network and the user access network to set up or release the end-to-end connections. However, the broadcast channels (analog or compressed digital), received over cable or satellite, are first verified for signal quality and then added in at the head end with the rest of the digital stream served by the same head end.

Hybrid-fiber/coax. In this star-bus architecture point-to-point optical links run from the head end to the optical network units. A so-called tree-and-branch bus network, similar to that in use by the cable companies, employs coaxial cable runs in a customer serving area or neighborhood. The access to the signals on the coaxial cable is shared among multiple users. Both the digital and analog channels are delivered over the same medium. To deliver a video-on-demand program, a virtual connection is made between a video information provider and a subscriber. The asymmetric transfer mode cells on that connection are routed to the appropriate broadband host digital terminal (BHDT) serving that neighborhood. (Typically, a broadband host digital terminal in the head end defines a functional block where asynchronous transfer mode cells terminate on the standard asynchronous transfer mode user-network interface.) Several such channels that are destined to the same neighborhood are demultiplexed from the incoming transport links and remultiplexed to fit within the 6-MHz bandwidth. The multiplexed bit stream, which may be made up of the asynchronous transfer mode cells or MPEG-2 packets, is encoded and modulated by using 64- or 256-quadrature amplitude-modulation (QAM) or 16-vestigial-sideband (VSB) modulation techniques. The 64 QAM enables a rate of about 28 megabits per second, and 256 QAM and 16 VSB enable a data rate of about 40 megabits per second to be accommodated within a 6-MHz bandwidth. Thus, each radio-frequency band carries multiple digital compressed video streams. For example, instead of carrying only one analog video program in a 6-MHz band, at an MPEG-2 rate of 3 megabits per second from 8 to 12 channels can be accommodated. The 6-MHz analog basebands, some carrying digital video and others analog video, are radio-frequency modulated to the appropriate nonoverlapping frequency bands, and combined with the rest of the analog spectrum in the distribution plant.

Fiber-to-the-curb and fiber-to-the-home. In these architectures, the delivery method of digital video (both video on demand and broadcast) from the video information provider to the access network is the same as in the hybrid-fiber/coax architecture, except that the analog broadcast programs are also delivered digitally in compressed format. The digitization, compression, and asynchronous transfer mode adaptation is done by either the video infor-

mation provider or the network as a value-added service. A broadband host digital terminal serves many optical network units, and one unit serves many homes. The physical transport from the optical network unit to the home is a dedicated coaxial cable or copper-twisted pair. The last drop into the home typically provides a downstream bandwidth of over 50 megabits per second and an upstream control channel of over 1.5 megabits per second, providing sufficient bandwidth in both directions for multiple set-top units in the same home to request different content or connections to different video information providers. All the broadcast programs are continuously available in the broadband host digital terminal. If a user requests a broadcast channel from a certain video information provider it is switched at the broadband host digital terminal or the optical network unit on a virtual channel in the physical link that connects the broadband host digital terminal with the network interface at the customer's residence.

Asymmetric digital subscriber loop. This architecture supports telephony and high-bit-rate asymmetric digital transport over the existing copper plant. It offers a transmission rate of 1.544–6.312 megabits per second downstream to transport video channels (a combination of 1.544- and 3.1252-megabit-per-second signals), and an upstream bandwidth of up to 640 kilobits per second. The actual transmission rate depends on the type of line encoding, the distance from the central office, and the noise environment. In asymmetric digital subscriber loop systems, the asynchronous transfer mode connections terminate either at the local central office or at a remote terminal over a standard 45- or 155-megabit-per-second asynchronous transfer mode interface, and switch to an asymmetric digital subscriber loop transmission system. From there, the connections are carried over a copper-twisted pair to the customer premises and delivered to the set-top box via a network interface.

Asymmetric digital subscriber loop is generally considered as an interim solution for serving sparsely populated areas, and lags in support in comparison to the hybrid-fiber/coax and fiber-to-the-curb technologies. Hybrid-fiber/coax and fiber-to-the-curb technologies offer the maximum flexibility with potential for evolution to fiber-to-the-home architectures as the demand for interactive services increases. Hybrid-fiber/coax, in particular, is well suited for the delivery of narrowband and broadband analog and digital signals over a single transport medium while offering compatibility with the legacy cable television systems and reliable power distribution to the home.

Control and management. The session and call management services are implemented in an external information management platform which can be centralized or distributed. These services interact with the call, session, and resource management entities in the various network elements of the video dial-tone network. Control and signaling information between the subscribers, the information providers, the access network, and the video dial-tone control and management system is transported over the same physical and logical video dial-tone network (in band) as the one used for video delivery, over a separate physical or logical network (out of band), or over a hybrid (combination of these two) control data network. In an in-band control network, the virtual connections are set up on demand between the various network elements, and in an out-of-band overlay control network, a separate data network shares the access network used for the downstream applications and services. In architectures where the asynchronous transfer mode layer terminates in the distribution loop at a standard asynchronous transfer mode user-network interface, the signaling and control messages are reassembled and transcoded, if necessary, to a new message set and encoded for physical transmission over a signaling and control channel between the head end and the set-top box. In the upstream direction, the reverse conversion takes place in the head end. In hybrid-fiber/coax architectures, finding enough bandwidth and using it for upstream traffic and signaling can be difficult since the efficiencies of the protocols used are heavily dependent on the upstream traffic load. Furthermore, the sophisticated timing technology required to avoid collisions is expensive.

Combining telephony and video dial tone. Existing telephone transmission systems use interfaces to connect switches in the central offices to the head ends. These 1.5-megabit-per-second lines carry up to a maximum of 24–64-kilobit-per-second voice circuits. In hybrid-fiber/coax architecture, part of the frequency spectrum is allocated to the upstream telephony signal, and part to the downstream telephony signal. In one approach the voice circuits are switched to the appropriate 6-MHz band; in another several upstream and downstream telephony carrier frequencies are allocated outside the video spectrum and are quadrature-phase-shift-keying (QPSK) modulated to carry multiple voice circuits. Depending upon the modulation technique, about 120–224 voice circuits can be multiplexed on a telephony carrier or in a 6-MHz band. The network interface, where the coaxial cable terminates at the home, filters out the telephony circuits and the rest of the coaxial cable spectrum. Inside the home the existing twisted pair wiring carries the telephony circuits and the existing cable infrastructure, the video dial-tone services. In other architectures, telephony and video dial-tone signals are kept on two physically separate loop distribution systems, similar to what exists today.

Video dial-tone performance. Performance requirements in a video dial-tone network are rigorous because sessions of different bandwidths are established and torn down on demand. Some network performance parameters that impact user-perceived quality of service are cell or packet delay jitter affecting end-to-end timing recovery, clock synchronization between the network elements, bit error rate and cell loss ratio, round-trip delays in the control and signaling network, session request rejection probability, connection request (or session set-up) response time, carrier-to-noise ratio, end-to-end reliability, fault-error monitoring and recovery, and overall operations, administration, management, and provisioning. Although reliable service and traffic forecasts are not available, the video dial-tone service providers are making major strides in implementing cost-effective and scalable networks to meet stringent end-to-end performance objectives.

For background information *SEE CLOSED-CIRCUIT TELEVISION; COAXIAL CABLE; DATA COMMUNICATIONS; MODULATION; OPTICAL COMMUNICATIONS; TELEPHONE SERVICE; TELEVISION* in the McGraw-Hill Encyclopedia of Science & Technology.

Sudhir S. Dixit

Bibliography. S. Dixit (ed.), Special issue on digital broadband video dial tone networks, *IEEE Netw.*, vol. 8, no. 5, September–October 1995; Feature topic on video-on-demand, *IEEE Commun.*, vol. 32, no. 5, May 1994; Special report on digital television, *IEEE Spectr.*, vol. 32, no. 4, April 1995.

Topography mapping

Global-scale topographic data are of fundamental importance to many earth science studies, and obtaining these data is a priority for the earth science community. Recent advances in the field of radar interferometry make feasible rapid generation of high-accuracy global map products. In addition, evidence of active deformation or motions of the surface on the order of centimeters can be retrieved from these data under certain conditions. The ability to characterize the entire Earth and study its changes with high precision will enable new geophysical studies on topics as varied as global warming and earthquakes.

Global-scale data requirements. Among the studies requiring continental topographic data are hydrology, ecology, glaciology, geomorphology, and atmospheric circulation. In hydrologic and terrestrial ecosystem studies, topography exerts significant control; examples include intercepted solar radiation, water runoff and subsurface water inventory, microclimate, vegetation type and distribution, and soil development. The topography of the polar ice caps and mountain glaciers is important, because it directly reflects ice-flow dynamics and is closely linked to global climate and sea-level change. Monitoring the amplitude of seasonal advance and retreat of mountain glaciers on a global basis and longer-term trends of the polar ice sheets can provide important information on the rate of global warming. Accurate mapping of the forms and slopes of young geomorphic features, such as glacial moraines and feature offsets and scarps due to recent geological faulting, can provide new information not only on underlying tectonic processes but also on climatic and paleoclimatic processes. Models of the present and past general circulation of the atmosphere require topography as a fundamental input.

Methods. There are at least three existing technologies for generation of future topographic data on a global scale: (1) optical-stereo photogrammetry, (2) laser profiling instruments, and (3) radar interferometry. Of these, the optical-stereo approach has the advantage that it utilizes existing or planned satellite systems used in a broad spectrum of applications. High accuracies cannot be achieved without suitable ground control point knowledge; thus this approach is logistically difficult to implement in a global system. Data also must be acquired in cloud-free image pairs, which is difficult for much of the world.

In laser profiling, one or more laser beams illuminate the Earth in a near-nadir direction to collect data directly beneath the satellite ground track. This approach has the advantage of very high vertical accuracy, perhaps less than 1 m (3.3 ft), but the disadvantage that for practical implementations only a very narrow swath may be acquired at one time. Although only a single pass is required over each region of the Earth's surface, the same atmospheric limitations that degrade optical imaging affect laser performance.

In radar interferometry, the required resolutions and accuracies are achieved in a reasonable mission lifetime without interference from clouds in the atmosphere. Since radar systems typically use wavelengths of several centimeters or longer, atmospheric water droplets of submillimeter dimension produce no consequential attenuation of the signal. If, for technological reasons, a very short radar wavelength is employed, the possibility of interference from severe storms remains; fortunately, such storms are much rarer than clouds in the sky.

Radar interferometry. A radar interferometer is formed by relating backscattered microwave signals from two spatially separated antennas; the separation of the two antennas is called the baseline. Existing radar interferometers have been realized in two ways. First, the two antennas may be mounted on a single platform. This is the usual implementation for aircraft systems; it has the advantage that simultaneous observation is allowed, and the disadvantage that the size of the airframe limits the achievable baseline. Second, synthetic interferometers have been formed by utilizing a single antenna on a satellite in a nearly exact repeating orbit: the interferometer baseline is formed by relating radar signals on two separate passes over the same site. Even though the antennas do not illuminate the same area at the same time, if the ground is essentially undisturbed between viewings, the signals will be very similar and a spatial baseline may be synthesized. Here the choice of a baseline is limited only by how accurately the position of the satellite orbit is known. A

third implementation utilizes two spacecraft flying in tandem orbit.

A radar system measures distances very accurately, and the topography may be deduced by geometrical construction from the distance measurements. Conventional radar systems are capable of obtaining distances to meter accuracies if the radar signal bandwidth is great enough. The use of two interferometric antennas permits, in addition, measurement of differential distances from two ray paths to the subwavelength level, yielding the fine precision necessary for high-resolution topographic mapping. Interferometer geometry for topographic mapping is shown in the **illustration**, where z is the elevation of the surface. Typically a geoid approximating mean sea level is used as a reference, but this varies with map projections and formats.

Deformation measurements. If the two observations forming the interferogram occur simultaneously, the observed phases depend solely on the topography of the surface. If, however, the measurements occur at different times, any motion of the surface will also add a phase signature. The motion phase signature is directly proportional to the displacement of the surface in the direction of radar illumination. Displacement of one-half wavelength along the radar line of sight will cause one full cycle of phase because of two-way propagation, and with a radar wavelength measured in centimeters the sensitivity of the deformation measurements will be on centimeter or millimeter scale. Therefore, the radar interferometer instrument can measure surface deformation to much finer precision than it can the topography itself.

In the usual case of nonsimultaneous observations, a radar interferogram contains the phase signatures of both topographic and surface displacement. It is necessary then to be able to separate these effects, which can be similar in magnitude. Two tech-

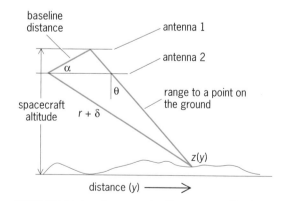

Interferometer imaging geometry. Two antennas illuminate the same patch of ground. Surface topography is given by $z(y)$, the spacecraft altitude is above a tangent plane at the point of interest, θ is the look angle, and α is the angle of the baseline with respect to horizontal. The radar signals transmitted from each antenna and received at the point of transmission form an interferogram where the phase at each point is proportional to the difference in path lengths (2δ), with the constant of proportionality $2\pi/\lambda$.

niques have been used for this. One compensates for the topographic term by using a digital elevation model of the surface derived independently and, using knowledge of the imaging geometry, calculates the phase field needed to compensate for the topographic phase. The second approach consists of using three rather than two radar passes of the surface and forming two (instead of one) interferograms. These interferograms contain different combinations of motion and topography signatures, and may be compared to remove the topographic component.

Both approaches have been demonstrated successfully, and each has its benefits and limitations. The former technique may be the only viable approach if only two radar passes are available, while the latter requires significant additional data processing to be useful. Obtaining a second interferogram of a region with favorable imaging geometry may be impossible; however, if available, it bypasses the need for the independent elevation model.

Applications. The ability to map surfaces rapidly is useful not only because it allows topographic data to be acquired quickly but also because the very change that occurs between mapping intervals becomes a topic of study. Characterizing changes in meter-scale topography is important in many areas, enabling monitoring of many of the geophysical processes noted above. Additional earth science investigations will follow from the precise surface deformation or motion measurements. Although many potential applications exist, to date the technique has been most successful when applied to seismological and ice motion problems.

Seismology. There have been several examples of coseismic earthquake deformation fields studied by radar interferometry, including the earthquakes at Landers, Eureka Valley, and Northridge, all in California, and the Kobe earthquake in Japan. In each case, centimeter-level deformation measurements were obtained for areas up to 100 by 100 km (60 by 60 mi), permitting characterization of the displacement fields from very near the fault zones to quite far away. At present these measurements have added only incrementally to what is known or inferred from the seismic record, but radar interferometry holds promise for those events occurring in remote areas or in less developed regions where in-place monitoring is not extensive or is nonexistent. Future refinements of the technique may lead to an increase in sensitivity such that creep or other preseismic motions may be detectable, thus impacting on earthquake forecasting. SEE EARTH CRUST.

Ice motion. Another area where interferometric measurement of motions has been applied is the flow of ice fields. Glacier dynamics are important observables in global climate studies. Important parameters that can be retrieved from the interferogram motion data are flow velocities, changes in glacier topography, and determination of terminus as well as the grounding line. The technique was used to generate a flow image of the Rutford ice stream in Antarctica. There is a bit of good fortune in the orbital requirements for polar ice mapping from existing satellites. Because of orbital convergence, interferometric pairs with small baselines close to the poles are much more likely to be found. The Rutford analysis was based on passes that were so close (4 m or 13 ft) that the sensitivity to topography was nearly zero, simplifying the interpretation. Thus, problems with the topographic component were completely avoided. The interferometric measurement of nearly instantaneous glacier velocities will be an important contribution to studies of glaciers because at present measurements are often average velocities derived from observations extending over a year.

For background information SEE APPLICATIONS SATELLITES; GEODESY; INTERFEROMETRY; RADAR; TOPOGRAPHIC SURVEYING AND MAPPING in the McGraw-Hill Encyclopedia of Science & Technology.

Howard A. Zebker

Bibliography. R. M. Goldstein et al., Satellite radar interferometry for monitoring ice sheet motion: Application to an Antarctic ice stream, *Science*, 262:1525–1530, 1993; H. A. Zebker et al., On the derivation of coseismic displacement fields using differential radar interferometry: The Landers earthquake, *J. Geophys. Res. Sol. Earth,* 99(B10): 19617–19634, 1994.

Transplantation biology

Skin, the largest organ of the body, serves to prevent fluid loss and protect the body from the environment. Thus, skin loss due to injury or disease can be life threatening and cause permanent disfigurement. Technology has now advanced to make grafting of artificial skin substitutes to heal wounds and replace lost skin.

Barrier properties. The important barrier property of skin is formed by the tough layers of the epidermal keratinocytes, cells that synthesize the protein keratin. Keratinocytes proliferate at the basal layer next to the dermis and gradually move up to the more superficial layers (see **illus.**). These cells are held together by numerous cell-cell connections called desmosomes. As keratinocytes migrate through the layers, they change in character or differentiate, becoming larger and flattened (squamous) and producing different keratin proteins and the lipids necessary to form the barrier of skin. They also contain other specialized proteins called envelope proteins, which are cross-linked by the enzyme transglutaminase K as the cell dies, and form a tough, flattened envelope of protein or squame. Squames are constantly sloughed and replenished. The specialized lipids made by the keratinocytes are extruded from the cells before envelope formation and spread between the squame layers, binding the squames together in a brick and mortar arrangement to form the stratum corneum. This cornified layer

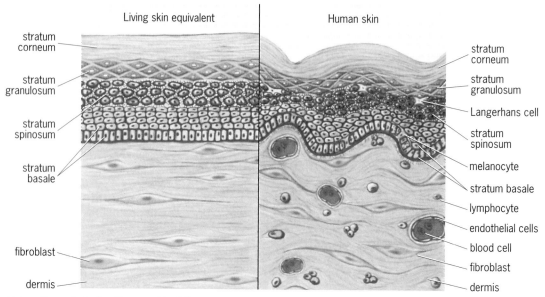

Living skin equivalent | Human skin

stratum corneum
stratum granulosum
stratum spinosum
stratum basale
fibroblast
dermis

stratum corneum
stratum granulosum
Langerhans cell
stratum spinosum
melanocyte
stratum basale
lymphocyte
endothelial cells
blood cell
fibroblast
dermis

Living skin equivalent and human skin. (*After Organogenesis Inc.*)

provides the barrier properties of skin. Thus, if keratinocytes are removed or lost because of injury the body is left unprotected.

The skin also consists of a second layer, the dermis, which lies beneath the epidermis and provides the structural foundation for skin. The dermis is a connective tissue formed by cells called fibroblasts. Dermal fibroblasts make collagen (types I and III), which is the main structural protein of the dermis. Collagen is the connective tissue of the dermis and provides the skin with strength, elasticity, and overall appearance, including wrinkles and scars. It provides an anchor for the epidermis and the other important structures of skin such as hair follicles, sweat glands, blood vessels, and nerves. When injured or lost, the dermis does not reform properly and scars. For example, scarring of burn victims is due to the body's inability to regenerate lost or damaged dermis.

Skin repair. Cells have a natural tendency to develop normally if given the proper environment. This characteristic has been used in the formation of aids to repair skin and of skin substitutes.

The structural framework of the dermis is difficult to reproduce when lost. The body quickly fills the space with wound tissue (granulation tissue) which contains many small blood vessels that transport immune cells to the wound to fight infection. As granulation tissue resolves, fibroblasts in the area replace it with additional collagen that is not organized in the same way as the original tissue. These cells do not have the appropriate three-dimensional framework and associated stimuli for regeneration of dermal connective tissue because the appropriate scaffold or precursor tissue once present in the fetal environment is no longer present. The adult is not normally able to recapitulate that environment for reasons that remain unclear.

Artificial scaffolds have been designed to help combat this problem. Dermis transplanted from another individual can provide a framework for dermal repair, if the tissue is available. However, one of the first artificial skin substitutes to be developed was a scaffold made of collagen and another extracellular matrix component called glycosaminoglycan. This material was made into a porous sponge material which provided a way of directing fibroblast ingrowth. The matrix consisted of a temporary impermeable silicone membrane which was substituted for the epidermis. The silicone membrane was eventually removed and replaced with the individual's own epidermis when sufficient tissue could be harvested from other areas of the body. A form of this collagen-glycosaminoglycan dermal substrate has also been combined with living cells to form a living skin substitute.

It is now possible to grow the two primary cell types of skin: keratinocytes and dermal fibroblasts. These cells can be separated from skin tissue and grown in large quantity in the laboratory. The epidermal keratinocyte is the more difficult of the two cell types to grow because of its inherent ability to differentiate, stop growing, and form a squame.

One of the most successful techniques developed for keratinocyte growth relies on the use of mouse fibroblast feeder cells to support the continued growth of the more immature keratinocytes while permitting other cells to differentiate. Keratinocytes grown in this way will spread to cover the dish, become multilayered and partially differentiate to form a cohesive sheet of rudimentary epidermis. This sheet can be released intact from the dish with enzymes such as dispase or thermolysin. These enzymes digest the matrix proteins attaching the cells to the culture dish without disturbing cell-cell adhesion. By using this technique, keratinocytes can be grown from a postage-stamp-sized sample of an individual's skin to produce enough epidermal sheets to cover the entire body in about 3 weeks. These sheets can be transplanted back to the individual to restore badly needed epidermal coverage. Once grafted, the cultured epidermis develops the features of normal epidermis. This technique has saved the lives of many severely burned individuals.

The dermal fibroblast is important for reconstitution of the dermal connective tissue. Fibroblasts naturally secrete collagen and other extracellular matrix molecules. When grown in the laboratory fibroblasts surround themselves with matrix. However, regulation of matrix production and the inability of the fibroblasts to efficiently stratify limit the amount of tissue deposition. Matrix deposition can be enhanced by providing a stimulus such as ascorbate (vitamin C) or growth factors which promote collagen synthesis such as tissue growth factor beta. Collagen deposition is also greatly aided by providing a three-dimensional environment. The fibroblasts are grown on a mesh made of resorbable suture material. The fibroblasts grow along the mesh fibers, eventually expanding to fill in the gaps of the mesh with cells and cell-produced matrix. A rudimentary connective tissue is thus formed on a resorbable scaffold which can be handled and implanted into the body to help repair of the dermis, providing an underpinning for epidermal regrowth and repair. This material may also be combined with an artificial silicone membrane outer layer to provide a temporary outer covering before epidermis becomes available.

Skin substitutes. The components of a matrix scaffold to help direct tissue formation, epidermal cells to form an epidermis, and dermal fibroblasts to help repair the dermis can also be combined to form a living full-thickness skin equivalent. Both keratinocytes and fibroblasts are grown from either the individual's own tissue or a piece of skin from another individual (such as tissue normally discarded after routine infant circumcision). Both cell types are grown in conditions which allow manifold culture expansion but minimal differentiation. Keratinocytes are dissuaded from differentiating by culturing them in a manner that potentiates the signal to cover while minimizing the signal to protect (differentiation) which will be needed later in the process.

Both self (autologous) and nonself (allogeneic) keratinocytes and fibroblasts may be used without distinction by the immune system. This may be due to the lack of other cell types normally found in skin grafts such as endothelial cells from blood vessels, pigment cells (melanocytes), and immune cells such as the Langerhan's cells normally found in the epidermis which could lead to classic rejection.

Dermal fibroblasts are incorporated into a three-dimensional tissue by combining them with a prepared collagen scaffold. The cellular dermal scaffold is then seeded with keratinocytes. Once the cultured keratinocytes have covered the surface of the matrix, conditions are changed to promote keratinocyte

differentiation (the protection mechanism). One important environmental factor that promotes this mechanism is exposure of the surface of the culture to air. Through this process, called organotypic culture, a tissue is formed with functional elements of both a dermis and an epidermis. Under these conditions, the cultured epidermis is able to stratify and differentiate normally as described to form a protective barrier. Thus immediate biological coverage can be provided to the open wound.

Clinical applications. Epidermal sheets continue to be used to save the lives of burn victims, and soon skin substitutes will be used routinely to aid the healing and reformation of skin tissue lost because of burns, cancer, disease, or trauma. Scientific advances in understanding the scarless wound healing of the developing fetus combined with the technology of skin substitutes will make the scarless reconstruction of skin tissue possible. SEE CELL (BIOLOGY).

For background information SEE BIOTECHNOLOGY; SKIN; TRANSPLANTATION BIOLOGY in the McGraw-Hill Encyclopedia of Science & Technology.

Nancy L. Parenteau

Bibliography. G. G. Gallico III et al., Permanent coverage of large burn wounds with autologous cultured human epithelium, *N. Eng. J. Med.,* 311:448–451, 1984; J. F. Hansbrough, C. Doré, and W. B. Hansbrough, Clinical trials of a living dermal tissue replacement placed beneath meshed, split-thickness skin grafts on excised burn wounds, *J. Burn Care Rehab.,* 13:519–528, 1992; L. M. Wilkins et al., Development of a bilayered living skin construct for clinical applications, *Biotech. Bioeng.,* 43:747–756, 1994; I. V. Yannas et al., Wound tissue can utilize a polymeric template to synthesize a functional extension of skin, *Science,* 215:174–176, 1982.

Tuberculosis

Tuberculosis has always been a serious threat to human health in much of the developing world. The World Health Organization estimates that fully one-third of the world's population has been infected with the bacterium (*Mycobacterium tuberculosis*) which causes tuberculosis. Each year, approximately 10 million new cases are diagnosed globally, and about 3 million people die of tuberculosis. After decades of steady decline in the United States, due principally to improved general health, case finding, and aggressive chemotherapy, tuberculosis reemerged in epidemic form during the mid-1980s. Compared to the expected incidence of tuberculosis, based upon the prior rate of decline, an estimated 75,000 excess cases have occurred in the United States since 1985.

Transmission and epidemiology. In the United States, tuberculosis is transmitted principally by the production of an aerosol containing the bacteria by an infected individual, usually by coughing or sneezing. The bacteria remain airborne and viable for minutes to hours, and are inhaled by the susceptible individual. Transmission can occur following a single, casual contact with a highly infectious person in a public place. The vast majority of otherwise healthy infected persons will not show any signs or symptoms of tuberculosis, other than the conversion of the tuberculin skin test to positive. However, if infected individuals are not detected and treated prophylactically with a drug which kills the mycobacteria, the bacteria will persist in a latent state in their tissues for many years, perhaps a lifetime. Any subsequent event which impairs the infected person's immune response [for example, infection with the human immunodeficiency virus (HIV), immunosuppressive chemotherapy, or aging] is likely to induce so-called reactivation tuberculosis.

Phases of intracellular survival. The principal mechanism by which *M. tuberculosis* produces disease in a susceptible host is by entering and surviving within the very host cells, the phagocytes, which are supposed to destroy it. It is not understood how the tubercle bacillus survives and multiplies within macrophages, but experimental evidence suggests that there are several distinct phases of the bacterium-macrophage interaction during which the tubercle bacilli or their products may modulate macrophage function and prevent their own demise: (1) receptor-mediated uptake of mycobacteria by the phagocytic cells; (2) trafficking of the ingested bacteria within the intracellular compartments (phagosomes) of the phagocyte; (3) modulation of the phagosome contents to enhance bacterial survival; and (4) production of immunoregulatory products, called cytokines, by the infected macrophage.

Invasion of the phagocyte. Tubercle bacilli enter phagocytes by binding to specific receptors on the surface of the cell. Although these receptors do not usually recognize the bacteria, they recognize host proteins which coat the bacteria. There appear to be multiple pathways (that is, multiple receptor types that mycobacteria can bind to), which result in internalization by host phagocytes. This redundancy is not surprising because to be a successful pathogen *M. tuberculosis* must get into the phagocytic cell.

A host who has already been exposed to mycobacteria may have specific antimycobacterial antibodies in serum and tissue fluids that will bind to newly inhaled tubercle bacilli and facilitate their uptake by receptors on the macrophage surface which bind to the antibodies (Fc receptors). Alternatively, the activation of the complement cascade (a series of interactive serum proteins), either by the complexes formed between mycobacteria and host antibodies or by interaction between bacterial cell wall constituents and complement proteins directly, may facilitate uptake of tubercle bacilli by complement receptors on macrophages. In particular, complement receptors CR3 and CR4 have been shown to play a major role.

In the absence of antibodies and complement activation, the mycobacteria will be coated with other host serum or secretory proteins for which macrophages also have receptors. These include fibronectin, laminin, and vitronectin. These substances are abundant, and their presence does not depend upon the previous exposure of the host to mycobacteria. It is clear that if one experimentally blocks the uptake of mycobacteria by antibody and complement receptor-mediated pathways (or both), significant ingestion will still occur by means of macrophage receptors for fibronectin and laminin. In the lung, which serves as the point of entry for mycobacteria in most human infections, an abundant protein in pulmonary secretions, called surfactant, may also facilitate entry of tubercle bacilli into alveolar macrophages.

Recently, a novel pathway of mycobacterial binding to phagocytic cells was described. This pathway involves direct binding of mycobacteria to the CD14 molecule on the macrophage surface which normally binds lipopolysaccharide, a prominent constituent of the cell walls of enteric bacteria such as *Escherichia coli*. The ligand for CD14 on the tubercle bacillus is lipoarabinomannan, a lipopolysaccharidelike molecule. Some investigators report that the lipoarabinomannan molecule may also participate in the binding of mycobacteria to mannose receptors.

Trafficking mycobacteria within the phagocyte. The significance of the precise pathway by which tubercle bacilli enter host phagocytes is that the intracellular fate of the bacteria may depend upon the mode of invasion. Mycobacteria which enter by means of FcR receptors end up in compartments (phagosomes) which are likely to fuse with enzyme-rich lysosomes, resulting in decreased intracellular survival. Tubercle bacilli taken up by other receptors may invade phagosomes which do not enter a lysosomal pathway (endosomal pathway), and thus are spared some of the antimicrobial effects of the macrophage.

There is additional evidence that mycobacteria produce enzymes called phospholipases, which can degrade lipid membranes like the ones which surround the phagosomes containing tubercle bacilli. Theoretically, these enzymes might allow the mycobacteria to escape into the cytoplasm of the macrophage, where they would not be subject to antibacterial influences. Other intracellular pathogens, such as *Listeria monocytogenes*, are known to utilize such a mechanism. However, the evidence supporting the escape of *M. tuberculosis* into the cytoplasm is not uniformly accepted.

Modulation of the phagosome environment. For many years, it has been known that one of the principal mechanisms by which mycobacteria escape intracellular destruction is by preventing the fusion of lysosomes laden with lethal enzymes with the bacteria-containing phagosomes. The precise manner by which fusion is prevented is unclear. In addition to the trafficking of mycobacteria into the endosomal (as opposed to lysosomal) pathway, the tubercle bacilli may produce substances which escape from the phagosome and interfere with the macrophage's signaling mechanisms. Furthermore, there is good experimental evidence that mycobacteria are able to prevent the pH within the phagosome from falling. An acidic environment is required for optimal activity of the degradative enzymes within the lysosomes. Therefore, even if fusion occurs, the high pH of the mycobacterial compartment would prevent the lysosomal enzymes from having their intended antimicrobial effect. Mycobacteria may maintain a high pH within the phagosome by at least two mechanisms: (1) they block a proton pump, which lowers the hydrogen ion content (and raises the pH) of the phagosome; and (2) they produce a urease enzyme which catalyzes the conversion of urea to ammonia, thus raising the pH. Together these experimental observations suggest that one of the most important strategies that are employed by mycobacteria to survive within phagocytic cells is to prevent exposure to, or inhibit the activity of, lysosomal enzymes which would normally kill the tubercle bacilli.

A second major strategy by which mycobacteria modulate the intracellular environment to ensure their own survival and replication involves access to critical nutrients, such as iron. Mycobacteria make iron-binding proteins, called exochelins, which compete very successfully for iron with host iron-binding proteins such as transferrin and lactoferrin. In addition, the *M. tuberculosis* phagosome is capable of receiving host transferrin from the extracellular environment by binding of the transferrin to macrophage-transferrin receptors and its trafficking to the early endosomal compartment where bound iron would conceivably be available for the mycobacteria. Additional evidence for the importance of iron availability to the intracellular survival of mycobacteria comes from the observation that macrophages activated by cytokines (such as interferon gamma) to suppress bacterial growth may do so, in part, by down-regulating their transferrin receptors and denying access to iron by the intracellular microbes.

Infected macrophage and immune cell interaction. Macrophages require the assistance of other cells, particularly T lymphocytes, to develop their full antimycobacterial capabilities. One of the strategies employed to resist infection with *M. tuberculosis* and other intracellular microbes is to have the infected macrophage signal other immune cells so that an effective antimicrobial response can develop. Some of these signals take the form of membrane-bound molecules on the surface of the macrophage which are required for cooperative binding to T lymphocytes. Molecules such as integrins are essential for the cell-to-cell contact which would normally result in activation of an antimycobacterial T-lymphocyte response. Recent studies have demonstrated that invasion of macrophages by *M. tuberculosis* down-regulates the expression of macrophage integrins, thus rendering the infected cells deficient in their ability to activate T lymphocytes.

The second major way that macrophages recruit other immune cells into an effective antimycobacterial response is by producing soluble messengers or cytokines. These cytokines act on lymphocytes and other macrophages to activate them to resist the infection. Infection of macrophages with *M. tuberculosis* has been shown to suppress the production of these activating cytokines, such as tumor necrosis factor alpha. The suppression of macrophage cytokine production may be mediated by the principal mycobacterial cell-wall component, lipoarabinomannan. Minor structural differences between the lipoarabinomannan molecules of virulent and avirulent mycobacteria have been postulated to explain the cytokine-suppressive activity of lipoarabinomannan from virulent strains of *M. tuberculosis.*

Thus, infection of macrophages by mycobacteria alters the ability of the host cell to cooperate productively with other cells in the immune network. The net effect of these alterations is to prevent effective macrophage-lymphocyte interactions which would normally lead to activation of macrophages and elimination of the tubercle bacilli. By down-regulating macrophage signaling, the mycobacteria improve their chances for survival within the host's phagocytic cells.

Prospects. The definition of a good microbial pathogen, in coevolutionary terms, is an organism which can alter its environment within the host in subtle ways which ensure the survival of both the microbe and the host. By this definition, *M. tuberculosis* is an excellent pathogen. Mycobacteria have evolved multiple mechanisms for gaining access to a protective niche within the host's phagocytic cells, and then altering their microenvironment and the functions of their host cell to prevent an effective immune response from developing. Scientists are just beginning to understand the myriad ways by which mycobacteria evade host defenses. This information will be crucial for the development of better vaccines, and new chemotherapeutic and immunotherapeutic approaches to preventing and controlling this ancient scourge.

For background information SEE IMMUNO-LOGICAL DEFICIENCY; MYCOBACTERIAL DISEASES; PHAGOCYTOSIS; TUBERCULOSIS in the McGraw-Hill Encyclopedia of Science & Technology.

David N. McMurray

Bibliography. B. R. Bloom (ed.), *Tuberculosis: Pathogenesis, Protection and Control* 1994; L. B. Reichman and E. S. Hershfield (eds.), *Tuberculosis: A Comprehensive International Approach,* 1993; W. N. Rom and S. Garay (eds.), *Tuberculosis,* 1995.

Turboramjet

The turboramjet is a high-speed, combined-cycle, air-breathing engine composed of a basic gas turbine engine and a ramjet. It is designed to perform efficiently at flight speeds from Mach 0 to 6–7, which spans the entire operating range of the two genera cycles. It is one of a very few candidate engine cycles which can be used in either aircraft or missile applications that must be capable of accelerating from brake release, or launch, to low hypersonic speeds. At present, there are no operational turboramjet-powered vehicles, but airframe design and ground test programs are in progress in several countries.

Missions and applications. Applications for which the turboramjet is a candidate engine cycle include a hypersonic cruise aircraft for military or civil use; the first stage of a two-stage horizontal takeoff-to-Earth-orbit transatmospheric accelerator; and a long-range surface-to-surface or air-to-surface missile. The preferred fuel for the accelerator is cryogenic liquid hydrogen. Missile applications use much higher density storable liquid hydrocarbon fuels. Cruise aircraft could use either, depending on the logistics of the operation including ground-support systems, required vehicle performance, and cooling requirements.

For a Mach 6 turboramjet-powered cruise aircraft that takes off from Washington, D.C., flight times of 2½ h or less provide access to nearly all population centers other than Australia and its environs. The attractiveness of transoceanic flights with greatly reduced travel times is apparent. Some of the great-circle routes from Washington, D.C., would require flying over population centers at hypersonic speeds to reach the destination in minimum times, which raises the issue of noise from sonic boom. Typically, cruise at Mach 6 will be at altitudes greater than 90,000 ft (27.4 km), which would reduce the noise signature on the ground to levels considerably below that of supersonic transports (SSTs). Furthermore, the engines would be designed to produce very low levels of nitrogen oxides (NO_x) and unburnt fuel in the exhaust gas and should be able to meet existing environmental standards. Nonetheless, arguments regarding potentially adverse effects on the environment in conjunction with the high cost of development have prevailed, and have precluded any commitment to proceed with development of a turboramjet Mach 6 cruiser.

For a two-stage space-access vehicle, the rocket-powered second stage would be silhouetted into the outline of the turboramjet-powered first stage.

Stage separation would occur at about Mach 6–7 with the first stage returning to a horizontal landing. A turboramjet-rocket two-stage-to-orbit vehicle is under study by the German government (**Fig. 1**). It is designed to take off from Germany, climb and accelerate to about Mach 4.5, and transit to the Equator, where it would resume accelerating with stage separation at Mach 6.2–6.8. The second-stage rocket would then carry a payload to Earth orbit. Transit to higher than equatorial latitudes for launch of the second stage would permit orbital capability at higher inclination angles. The development program was initiated in 1988 and named SÄNGER after the space pioneer E. Sänger. The first-stage vehicle has a total mass of 558,800 lb$_m$ (254 Mg) and carries 215,600 lb$_m$ (98 Mg) of liquid fuel. Two different second-stage vehicles are contemplated; a crewless version weighing 246,400 lb$_m$ (112 Mg) with a payload capability of 15,400 lb$_m$ (7000 kg) for a mission to a space station in a 28.5° orbital plane, and a crewed version with the same total mass and a payload of 6600 lb$_m$ (3000 kg). The respective payload-to-total-takeoff ratios are 0.019 and 0.008.

During the period 1988–1992 the development effort focused on establishing the structural design, determining the aerodynamic characteristics of the vehicle, and ground testing the turboramjet engine. High-temperature materials including carbon-titanium and titanium-aluminum alloys were selected for the various engine components. Computational-fluid-dynamic estimates of the lift, drag, and moment coefficients of the first-stage vehicle were verified in wind-tunnel tests. Similar agreement was obtained in determining the external flow field during stage separation. A liquid-hydrogen, regeneratively cooled turbojet engine operating in the scramjet mode was tested over the range from Mach 2.0 to 6.8 with good results. In 1992 the program was broadened and renamed as the German Hypersonics Technology Program. To proceed further, international cooperation and support is deemed essential. Intergovernmental agreements have been signed with Norway and Sweden, and several industrial agreements have been made with companies in Canada, Belgium, and Australia.

Engine configuration. **Figure 2** shows one of several possible configurations of the turboramjet engine. In this configuration, the core engine comprises the conventional turbojet components: the axial flow compressor, followed by a combustor and an axial flow turbine. The engine spool extends forward to a center body that supports the rotating machinery and provides the surfaces for the initial waves of the inlet compression process. The rotating components are surrounded by an annular duct which serves as the aft portion of the ramjet inlet. A single main combustor serves the dual functions of the afterburner, when operating as a turbojet, and as the ram burner, when operating in the ramjet mode. The exhaust nozzle has a variable-geometry,

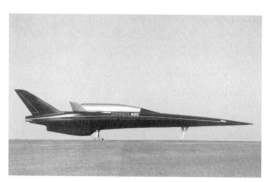

Fig. 1. SÄNGER, German two-stage-to-orbit aerospace plane. (*Deutsche Aerospace*)

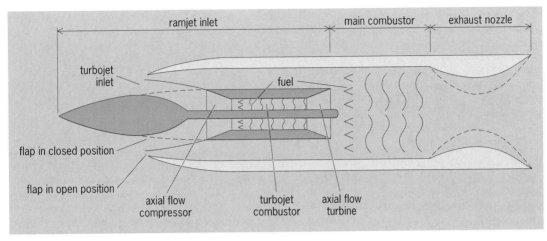

Fig. 2. Possible configuration of a turboramjet engine.

converging-diverging design which provides the large variation in throat area that is essential to obtain satisfactory engine performance over the broad range of flight speeds from the turboramjet cycle. A movable flap extends from the compressor face to the cowl lip. In its full open position, the ramjet duct is blocked and all of the flow captured in the inlet is directed to the turbojet compressor. In its closed position, the turbojet is closed off and all of the flow is channeled into the ramjet duct. Intermediate positions provide the capability of splitting the captured air and operating the two genera cycles in tandem. Translation of the forward portion of the inner body would be required to obtain optimum compression-wave structure to produce maximum performance. Details of the complex geometry of an inlet that could provide the required high-pressure recovery over the entire speed range are omitted.

Cycle diagrams. The temperature-entropy (T-S) and pressure-volume (P-V) diagrams of the modified Brayton cycle (**Fig. 3**) elucidate the factors that

limit the operating ranges of the genera cycles and those that lead to the attributes of the combined-cycle turboramjet. The curves shown are for engine operation at flight speeds of about Mach 3. At this speed the turboramjet could operate in the turbojet mode, the ramjet mode, or with a split of the inlet airflow in a tandem mode. Cycle A-B-C-D-E-F-G is for stoichiometric (the fuel that is added consumes all of the oxygen available in the captured air) operation of the engine in the turbojet mode. The extension E-G′ represents the completion of the turbojet cycle without the afterburner. This operation, with about ⅓ of the stoichiometric fuel rate and no afterburning, could represent cruise at constant speed, whereas stoichiometric operation would be typical of operation in an acceleration mode. The cycle A-B-F″-G″ depicts stoichiometric operation in the ramjet mode. Cruise operation in the ramjet mode is omitted for clarity. The pressure rise from point A in the free stream to B at the end of the inlet is through pressure waves on the compression surface followed by a terminal normal

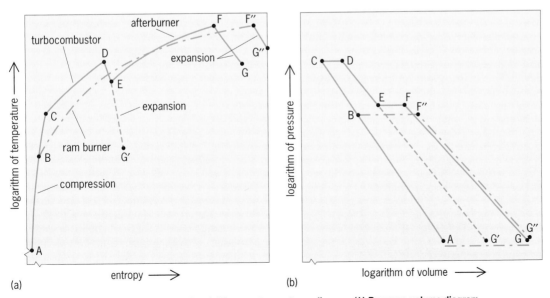

(a)   (b)

Fig. 3. Cycle diagrams for turboramjet engine. (a) Temperature-entropy diagram. (b) Pressure-volume diagram.

shock. From B to C the compression is primarily mechanical in the axial flow compressor. At much lower flight speeds points A and B and points F and G″ would be nearly superposed and the segments B-C and D-E would be considerably extended. At this low speed, the area enclosed in the pressure-volume diagram (Fig. 3b) for the closed path A-B-F-G″-A, which is directly related to engine thrust, would approach zero, which explains the futility of ramjet operation at low speeds. Conversely, the area B-C-D-E would be enlarged, which leads to the high thrust of the turbojet cycle at low speeds.

Paths C-D and E-F represent the turbojet combustor and afterburner, respectively, in the turbojet mode; path D-E is the turbine. In this example the turbine inlet temperature is 3000°R (1667 K). As depicted in Fig. 3a, segment B-C, which represents the work extracted in the compressor, is longer than segment D-E, the work provided by the turbine in the temperature-entropy diagram. Had a linear rather than the logarithmic scale of temperature been used the segment lengths would be nearly equal. Some of the work provided by the turbine is used to provide power for auxiliary systems, such as fuel pumps and electric power generators. Otherwise, the turbine work is matched to the compressor demand. Path B-F″ represents the combustion process in the ramjet cycle. The entropy at F″ is greater than that of F because the total pressure in the turbojet mode is higher than in the ramjet mode. Entropy rises (losses in total pressure) in the combustion processes are about equal in the two modes but the rise in total pressure in the compressor is greater than the loss in total pressure in the turbine. The work required to compress, or provided in expansion, is proportional to temperature difference whereas the total pressure increase, or loss, is proportional to temperature ratio.

In the speed range of Mach 3–4, the temperature at B begins to approach the maximum allowable turbine inlet temperature for structural reasons, and only a small amount of heat can be added in the turbocombustor. A small heat release combined with the losses in the compressor and turbine reduce the area B-C-D-E-B to near zero. Thus, the turbojet core becomes superfluous, and satisfactory operation is realized only in the ramjet mode.

Combining the turbojet and ramjet engines into the turboramjet provides the thrust capability and high fuel efficiency over the entire speed range that can be covered by the two genera cycles. The weight and complexity of the engine are increased. Nonetheless, it is the preferred cycle for several important applications, notably those discussed above.

For background information SEE AIRCRAFT ENGINE PERFORMANCE; BRAYTON CYCLE; JET PROPULSION; RAMJET; SONIC BOOM; TURBOJET; TURBORAMJET in the McGraw-Hill Encyclopedia of Science & Technology.

Frederick S. Billig

Bibliography. *AIAA 3d International Aerospace Planes Conference,* December 1991; *AIAA 6th International Aerospace Planes and Hypersonics Technologies Conference,* April 1995; E. T. Curran and S. N. Murthy (eds.), *High Speed Flight Propulsion Systems,* 1991.

Ulcer

Gastrointestinal ailments such as heartburn and upset stomach, collectively referred to as dyspepsia, have become a common symptom of the late twentieth century. If these symptoms persist ulcerations of the epithelial cell walls of the lower stomach or connecting intestine often develop. Until the 1980s, most physicians believed that these symptoms were the result of high stress and improper diet. A relatively small group of clinicians and biomedical researchers questioned this prevailing thought, and the result has been a new understanding of the etiology of these diseases.

Pioneering proof. Barry Marshall, working with J. R. Warren, identified in many ulcer patients spiral bacteria colonizing the stomach, an area previously thought too harsh to allow bacterial growth. Marshall hypothesized that these bacteria caused the ulcers, but to prove that these organisms were the etiologic agent, he needed to provide data to confirm Koch's four postulates. Pure cultures of viable organisms had been isolated from disease cases, fulfilling postulates 1 and 2. But he was unable to infect any laboratory animals to fulfill postulates 3 and 4 (that bacteria isolated in pure culture from ulcers could cause ulcers in animals and be reisolated). It was unlikely that human trials would be approved, so he grew a culture of these bacteria and drank it himself. He monitored his general health and underwent endoscopic examinations to monitor any disease progression. He eventually developed superficial inflammation, and organisms were isolated at the sites of inflammation. Fortunately for Marshall, the infection healed spontaneously. Reporting these results to other scientists began the process that has resulted in conclusive proof that these bacteria, later designated *Helicobacter pylori,* cause the majority of peptic and duodenal ulcers and likely are involved in the development of gastric carcinoma.

Host factors. One of the most troubling observations that required resolution in order to demonstrate that *H. pylori* causes gastrointestinal disease has been that most people who are persistently colonized are without apparent illness. For example, age is a clear predisposing factor in predicting who is infected. It has been estimated that over 60% of Americans beyond 60 years of age are colonized with *H. pylori,* yet very few display overt signs of illness. In some underdeveloped countries with poor sanitation, over 80% of the population over 12 years old have been infected. The explanation for this apparent paradox is provided in the proposed model of disease that can result from chronic infection. *H. pylori* bacteria colonize the stomach, and are primarily buried in the thick mucus layer that lines the gastric epithelium and protects these cells from the strong acid and digestive enzymes of the stomach. *Helicobacter pylori* also attach directly to the epithelial cells and release bacterial products that affect these cells. The systemic host response to these products is to send inflammatory cells to the site of colonization. These inflammatory cells produce factors that are designed to kill the bacterial parasite, but that also cause destruction of epithelial cells. In turn, a depletion of the protective mucus layer occurs, further destroying the gastric integrity and homeostasis.

The immune system also has the ability to control or suppress its response if it becomes too severe while trying to control the bacterial colonization. The balance between restricting the bacterial population and the severity of the inflammatory response determines whether an infected individual develops any of a wide spectrum of illnesses. Mild indigestion may, over time, develop into chronic superficial gastritis and eventually become a peptic ulceration. Additionally, it has been proposed that the proliferation of normally nondividing gastric epithelial cells increases their chances of becoming cancerous, since colonization with *H. pylori* is a significant predisposing factor for developing gastric cancers.

Bacterial factors. The host response to colonization by *H. pylori* and bacterial factors that cause human disease have been described by using several animal models, including mice, pigs, and primates. Bacterial factors have also been elucidated by using laboratory models and genetic manipulation techniques including cloning genes from *H. pylori* into more easily manipulated bacteria (*Escherichia coli*) and introducing mutated genes back into *H. pylori*. Mutating a suspected virulence gene and then testing the mutated bacteria for the ability to cause disease has become the molecular equivalent for fulfilling Koch's postulates for causality.

Probably the most important adaptation achieved by *H. pylori* is the copious production of a unique enzyme, urease. This protein literally coats the surface of the bacterium and metabolizes the minute quantities of urea available, generating ammonia as a by-product. Ammonia is a very basic small molecule that neutralizes the microenvironment of the *H. pylori* colony in the acidic stomach milieu. In fact, *H. pylori* organisms do not even grow below a pH of 6.0 which is 10,000 times less acidic than the normal stomach pH. The local ammonia may also be directly toxic to the gastric epithelium. Finally, strains of normally virulent *H. pylori*, which have a mutated, inactive gene for urease, are unable to colonize the stomach in animal models and therefore do not cause disease.

The ability to colonize the gastric mucus is enabled by additional bacterial factors. *Helicobacter* are spiral or helical rods, and this shape probably allows them to move within the very viscous medium of mucus. The movement is further facilitated by flagella, which are molecularly powered oars that propel organisms toward high concentrations of nutrients. *Helicobacter* that do not have flagella are unable to remain in the stomach, being carried through the digestive tract by the normal peristaltic movement of the gastrointestinal tract. Finally, *H. pylori* remain at sites of colonization by attaching to carbohydrate structures (receptors) found in the mucus and on the surface of gastric epithelial cells. Specific surface proteins on the bacteria (adhesins) interact with these host receptors through hydrogen bonds and hydrophobic forces.

The ability to mediate chronic inflammation is the result of several bacterial factors. A protein that is directly toxic to cells in culture is expressed by most strains of *H. pylori* that are isolated from individuals with severe ulcer disease. This cytotoxin causes cells to produce large vacuoles and inhibits normal functions. However, few gastric epithelial cells are seen in human biopsies from peptic ulcers, so the direct role in disease of this vacuolating cytotoxin is not clear. *Helicobacter pylori* are also able to produce fragments of their outer membranes (blebs) which are released and act on the host cells. Many of the constituents of these blebs elicit inflammatory mediators that participate in the disease process.

Treatment. With the recognition that most peptic and duodenal ulcers are caused by *H. pylori*, clinicians have been able to successfully treat these conditions with combined therapies. Diagnosing individuals who are colonized with *H. pylori* has become a routine test measuring the ability to metabolize labeled urea (actually the bacteria metabolize with urea, their unusual urease enzyme). Urease-test-positive individuals with clinical signs of preulcers or more severe disease may then be prescribed multiple therapeutics. Bismuth has been an effective treatment for years and controls the acid indigestion by coating the inflamed tissues. Long-term therapies have used histamine-2 receptor antagonists such as cimetidine or ranitidine to suppress acid secretion. Newer acid-suppressing drugs such as omeprazole, which inhibit proton pumps of the acid-secreting tissues, have proven extremely effective. One of these three types of drug is then combined with one or two antibiotics. Some combinations have proven to be over 95% effective in resolving *H. pylori* infection and eliminating disease. Most promising are data that strongly suggest that once individuals have been cleared of infection, reinfection is highly unlikely.

An important public health policy question is whether all individuals should be routinely screened to determine if they have an *H. pylori* infection and immediately treated if the bacteria are identified. Most infected people have no clinical signs of disease. Yet, the potential for progressing to ulcerative disease and the additional undefined risk of developing gastric cancer might warrant this drastic and expensive approach. Researchers are working on the alternative approach of developing a vaccine against *H. pylori*. For example, animals immunized with the urease protein are protected from infection with a subsequent challenge of *H. pylori*. A vaccine could be used either prophylactically to prevent infection or therapeutically to eliminate infection.

For background information *SEE BACTERIOLOGY; DIGESTIVE SYSTEM; GASTROINTESTINAL TRACT DISORDERS; INFLAMMATION; ULCER* in the McGraw-Hill Encyclopedia of Science & Technology.

James E. Samuel

Bibliography. M. J. Blaser, Hypothesis on the pathogenesis and natural history of *Helicobacter pylori*-induced inflammation, *Gastroenterology*, 102:720–727, 1992; A. Lee et al., Pathogenicity of *Helicobacter pylori:* A perspective, *Infect. Immun.*, 61:1601–1610, 1993; National Institutes of Health, *Helicobacter pylori in Peptic Ulcer Disease,* NIH Consen. Develop. Conf., February 7–9, 1994.

Underground mining

Mechanical mining of soft minerals and coal is common throughout the world. It is only recently that this technology has progressed to the point that some hard-rock mineral ores have had the benefit of these excavation techniques. Nearly all underground hard-rock mines use conventional drilling and blasting in patterns of small-diameter holes as the primary method of excavating the waste rock and ore. However, mechanical excavation systems have a number of advantages over conventional drilling and blasting. These methods result in safer working conditions, cause less damage to the surrounding rock structure, and therefore require less ground support. In addition, these systems, when properly applied, can excavate rock at less cost and with higher productivity. Mechanical mining systems are much easier to automate. Some of these systems can be used for selective mining of higher-grade ore without dilution from waste rock or low-grade ore. In addition, mechanical excavation is capable of breaking the average rock into smaller pieces; therefore it can be more easily transported and requires less crushing at the mill.

The major disadvantage of mechanical mining is associated with rock that is so hard, so abrasive, or so high in compressive strength that it will not allow effective penetration and breaking by the mechanical tools. In such a situation the cost will be high and the productivity low.

The various methods of mechanical excavation can be considered in terms of the tools used. These include full-face boring machines, partial-face excavating machines, and other tools that have limited application.

Full-face boring machines. This mechanical rock-cutting tool excavates the entire span of the opening at one time. For hard rock, raise boring and tunnel boring machines produce round openings.

Raise boring machines. In raise boring, a mechanically driven rock-cutting machine is used to advance an opening upward from a mine level to connect with a higher level or to explore the ground above one mine level. Raise boring is the most common form of mechanical excavation in underground metal mining. It has replaced nearly all of the conventional drilling and blasting methods of raise development, in which a vertical or inclined opening is advanced in the upward direction above an existing mine level. Raise boring machines have diameters in the 4–12-ft (1.2–3.7-m) range. There are three types of raise boring machines. The first, and by far the most common, type has traditionally been called a raise drill. **Figure 1** shows the operating cycle of a raise drill. At the beginning of the cycle, working above or below the surface, the machine drills a pilot hole (Fig. 1*a*). Then a reamer is attached in place of the drill bit (Fig. 1*b*). Finally the raise drill pulls the reamer toward itself; crushed rock falls down the shaft and is removed (Fig. 1*c*).

The second type of raise boring machine is known as a boxhole drill. The cycles of this drill are shown in **Fig. 2**. At the beginning of the cycle, the boxhole machine is positioned underground and drills a pilot hole (Fig. 2*a*). A reamer is then attached in place of the drill bit (Fig. 2*b*). Finally the reamer bores into the rock; crushed rock falls down the shaft and is deflected from the drill (Fig. 2*c*).

The third type of raise drill is shown in **Fig. 3**. It bores a hole from a transport tube by jacking against the sides of the tube while thrusting and turning the cutterhead against the rock. As the hole is made, the boring machine travels into and jacks against the hole and applies the thrust against the cutterhead of the machine within the hole. This type of raise drill was introduced in 1993. It has the advantage of being a mobile system that can be set into position within a few hours. In 1994, this

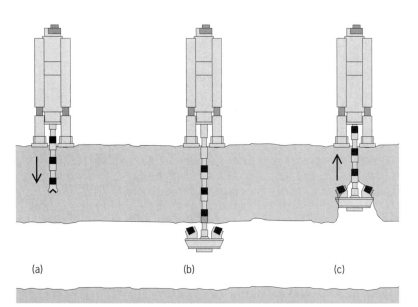

Fig. 1. Operating cycle of a raise drill. (*a*) The machine drills a pilot hole. (*b*) The drill bit is replaced by a reamer. (*c*) The raise drill lifts the reamer, and tailings fall down the shaft and are removed.

(a)   (b)   (c)

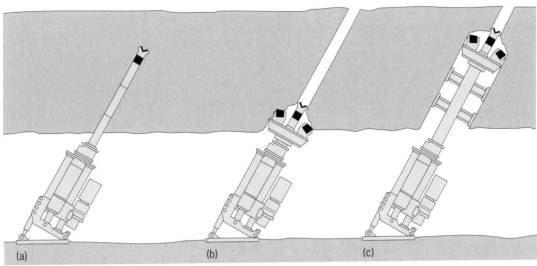

Fig. 2. Operating cycle of a boxhole drill. (*a*) The machine drills a pilot hole. (*b*) The drill bit is replaced by a reamer. (*c*) The reamer bores into rock, and tailings fall down the shaft and are deflected from the drill.

machine started drilling raises in the Coleman nickel mine of Sudbury, Ontario. A similar machine was introduced into BHP Minerals' Hartley platinum mine in Zimbabwe.

Tunnel boring machines. These machines are commonly used in civil construction works to drive long, straight tunnels. Full-face machines, they range in size from a few feet up to 35 ft (10.7 m) in diameter. Most of the early trials of tunnel boring machines in hard rock metal mines took place in the late 1960s and 1970s, where their success was somewhat marginal. Since that time, better and larger cutter assemblies that can utilize greater forces against the rock have been developed. These developments have allowed tunnel boring machines to cut very hard rock efficiently and economically. However, for application in the development of a mine, tunnel boring machines usually have lacked the flexibility to follow the curving nature of the exploration development of an orebody or to develop the curves required in the mine's production development.

In the late 1980s, at the Stillwater Mine, a platinum/palladium mine in Montana, a tunnel boring

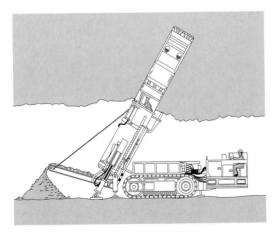

Fig. 3. Tube-launched raise boring machine.

machine 13.5 ft (4.1 m) in diameter was successfully applied to hard rock. This tunnel boring machine had a turning radius of 800 ft (244 m), which allowed the developed opening to nearly follow the curvature of the orebody. The main reason for using the tunnel boring machine was for rapid excavation. Other benefits were one-third less cost of development and only 10–12% of the ground support that would be required compared to drilling and blasting methods.

In another case, to expand production from the San Manuel copper mine complex in Arizona, a tunnel boring machine 15.2 ft (4.5 m) in diameter has been used in driving 34,000 ft (10,353 m) of mine exploitation development, beginning in 1994. This tunnel boring machine possesses a turning radius of 350 ft (106.7 m), which allows the machine to develop openings in a series of continuous loops on different levels; these serve as mining and haulage levels for a block-caving system of mining. A tunnel boring machine was chosen over drilling and blasting primarily because of the faster excavation.

Partial-face excavating machines. These include disc undercutting machines, disc cutters on revolving wheels, and roadheaders.

Disc undercutting machine. Several hard rock metal mining companies in Canada have joined with a German manufacturer to develop and test a new concept of a continuous mining machine. The continuous mining machine, a partial-face excavating machine, has four programmed circulating and ranging arms. Each arm has a large undercutting disc attached. Each disc is applied against the rock with great force, and it penetrates the rock at a low angle that causes the rock to break in tension. The machine cuts a circular, semisquare, or horseshoe-shaped opening to a maximum of 18.4 ft (5.6 m) across in hard rock. It has a very short turning radius. It is anticipated that this disc undercutting machine can be used for cutting development openings as well as for mining ore.

Disc cutters on a revolving wheel. The polymetal mines at Mount Isa, Queensland, and Broken Hill, New South Wales, Australia, have each acquired a mobile mining machine built in the United States. The cutting tool of this partial-face excavating machine is a large vertical rotating wheel equipped with steel discs 17 in. (432 mm) in diameter. This rotating wheel oscillates back and forth across the opening while thrusting the discs against the rock, and the indentations of the discs cause the rock to fracture. Recently, the machine at Broken Hill cut openings at about one-half to one-third the cost of similar work that uses drilling and blasting systems.

Roadheaders. Although originally designed to excavate softer rock in coal mines, these crawler-mounted machines for partial-face excavation are now being built in more robust models that can apply enough force to cut some of the hard rocks. All roadheaders have rotating drums laced with steel or tungsten-carbide tipped picks, and they are attached to the end of a sturdy boom. The machine forces the rotating cutterhead against the rock, and the pick's indentation causes the rock to break. A 100-ton (91-metric-ton) roadheader is excavating a dewatering tunnel in hard limestone for the copper-gold orebody at the Freeport mining operations in Indonesia. A roadheader has also been used in Western Australia to excavate a tunnel in a new nickel mine for the Western Mining Company.

Other tools. Two additional tools, the impact hammer and the rock splitter, have had limited application in hard-rock mining.

Impact hammer. This large mechanical hammer is fitted with a striking tool that impacts the rock when driven by a pneumatically or hydraulically actuated piston. An impact hammer can produce a blow of up to 20,000 ft-lb (27,116 joules). These tools have been tried in several narrow-vein hard rock metal mines for extracting ore, and in shaft-sinking applications. However, impact hammers have not yet proven productive enough to be considered economical.

Rock splitter. This machine first drills a hole in the rock and then, by inserting a hydraulically actuated wedging device into the hole, causes a cone of rock to break in tension from the solid rock. This system has been tried in several metal mines, but has not yet proven cost effective. It was tried in the Fletcher lead/zinc mine in Missouri for selectively mining the higher grade of ore in a mining area.

Minidisc cutter. This emerging technology is based on research undertaken at the Colorado School of Mines. Minidisc technology shows promise of allowing mechanical excavation machines to be developed that will be lighter and more maneuverable than older machines with larger discs. Thus, they may have greater application in underground hard-rock metal mining. The minidisc has been successfully tested on cutting drums that may be applied as drum miners. The drum miner is a revolving mechanically driven steel drum with some sort of rock-cutting tool mounted on the drum to excavate the rock or mineral. Minidiscs have also been successfully tested on roadheader cutterheads. Since these tests were completed, three firms have developed minidisc cutters that will be available commercially. They are being tested on boxhole drills in iron ore mines near Kiruna, Sweden.

For background information SEE MINING; UNDERGROUND MINING in the McGraw-Hill Encyclopedia of Science & Technology.

Richard L. Bullock

Bibliography. H. L. Hartman (ed.), *SME Mining Engineering Handbook*, 2d ed., vol. 1, 1992; W. A. Hustrulid (ed.), *Underground Mining Methods Handbook*, 1982; B. Stack, *Handbook of Mining and Tunnelling Machinery*, 1982.

Universe

E. P. Hubble's 1929 observation that the entire universe is expanding ranks as one of the most important astronomical discoveries of the twentieth century. The consequences of this result are remarkable. If the universe is now expanding, in the past it must have been smaller. Astronomers and physicists surmise that the universe began with expansion out of a state of immense pressure and density, and that its early moments were explosively hot. In 1965 the radio astronomers A. Penzias and R. W. Wilson discovered the now-cool (3 K) residual radiation from the fiery beginnings of what is now called the hot big bang. But these observations raise the questions of when this event occurred, how fast the universe is expanding, and what the size and age of the universe are. No conclusive answers have been obtained from more than a half-century of serious effort undertaken at major observatories around the world. Because of this observational impasse, the Hubble Space Telescope was designed and built. SEE COSMIC BACKGROUND RADIATION.

Hubble constant. Within the framework of general relativity, the expansion of the universe discovered by Hubble can be specified by the Friedmann equation. This equation relates the expansion rate, H_0, to the mean density of matter in the universe, the curvature of the universe, and a possible additional term, called the cosmological constant (Λ). A nonzero value for the cosmological constant would represent the gravitational effects of the energy density associated with the vacuum. In a uniform and isotropic universe, the relative expansion rate, v, is proportional to the relative distance, r, such that $v = H_0 \times r$. The Hubble constant, H_0, thus characterizes the expansion rate of the universe, and is instrumental in ultimately determining the age of the universe. A determination of the present-day value of the Hubble constant also provides constraints on the density of baryons produced in the big bang, the amount of dark matter in the universe, and the manner in which large-scale structure began forming in the early universe. SEE COSMOLOGY.

One of the simplest models for the big bang, the Einstein–de Sitter model, has become the standard working hypothesis. It makes a number of testable predictions about the density of the universe, its expansion rate, and the value of the cosmological constant (predicted to be equal to zero). Fortunately, these predictions can be tested observationally. A reliable measurement of the expansion rate (that is, the Hubble constant), an independent estimate of the ages of the oldest objects in the universe, and a further measurement of the average density of matter in the universe are separately required in order to make these tests, and ultimately provide constraints on cosmological models.

During a peer-review process in the mid-1980s, the determination of the extragalactic distance scale was designated as one of the Key (or highest-priority) Projects to be undertaken and completed by the Hubble Space Telescope. A team of about 20 astronomers from the United States, Canada, United Kingdom, and Australia has been actively involved in this effort. The Extragalactic Distance Scale Key Project involves determining accurate distances to about two dozen galaxies. At least two other groups have been awarded time on the Hubble Space Telescope to observe several additional galaxies. These galaxy distances will be used to establish an accurate and precise value for the Hubble constant.

Cepheid distance measurements. The method of obtaining distances with the Hubble Space Telescope is precisely that used by Hubble when he demonstrated that spiral galaxies such as the Milky Way Galaxy are major constituents of the universe. The key to this effort is in discovering a class of variable stars called Cepheids. These rhythmically pulsating supergiant stars are known to obey a tight relation between their period (or frequency) of oscillation and the total luminosity of the star. Discovered by H. Leavitt in 1912, this period-luminosity relation for Cepheids means that, given a measurement of the period of a Cepheid, its intrinsic brightness can be predicted. Measuring the observed brightness of the Cepheid and the difference between the predicted brightness and the measured brightness very simply yields the distance.

For decades, it has been recognized that the solution to the problem of the extragalactic distance scale would require observations made at very high spatial resolution. Although Cepheids are among the brightest stars found in galaxies, identifying these individual stars against the bright background of other stars in the galaxy in which they reside is feasible for only the very nearest systems. This difficulty is due to the turbulent motions in the Earth's atmosphere. Hence, the light that reaches telescopes on the ground is somewhat blurred or smeared out, and finding or recognizing individual Cepheids beyond a certain distance is not possible. The resolution of the Hubble Space Telescope is such that a volume of space a thousand times larger than is comparably available to most Earth-bound telescopes can now be surveyed.

Hubble Space Telescope. One primary motivation for building an optical telescope in space was to allow the discovery of Cepheids in galaxies and a measurement of an accurate value of the Hubble constant. The Hubble Space Telescope carries a suite of instrumentation, including high- and low-resolution spectrographs and imaging cameras. Originally launched in 1990, the space telescope was designed to be serviced and its instrumentation upgraded. A successful 1993 repair mission replaced the original wide-field imaging camera, and new optics were installed to correct for the unfortunate problem of spherical aberration in the primary mirror and to allow the telescope and other instruments to reach their original design specifications. Hubble is so efficient that in 2 years it will have more than doubled the number of galaxies that have accurate distances measured after more than a half-century of work by astronomers at the world's largest telescopes on the ground. The telescope is scheduled to be serviced again in 1997, at which time a new imaging spectrograph and a near-infrared camera will be installed. A mission in 1999 is scheduled to carry an imaging camera of higher resolution and sensitivity.

Key Project. The underlying basis of the Key Project is the measurement of accurate distances to galaxies by using the period-luminosity relation for Cepheid variables. There are three primary objectives. The first is the discovery of Cepheids and the measurement of distances to about 20 nearby spiral galaxies, with distances approximately from 4 to 20 megaparsecs. (1 Mpc equals 3.26×10^6 light-years, 1.9×10^{19} mi, or 3.1×10^{19} km). The distances of these primary galaxies will be used to calibrate several secondary distance techniques which extend to distances from 20 to 100 Mpc or more. (These methods include measuring the brightnesses of supernovae, the powerful, explosive deaths of stars; measuring the brightnesses and rotational velocities of entire galaxies; measuring the fluctuations in their light; and measuring another class of younger, more massive supernovae.) The second primary objective is the discovery of Cepheids and the measurement of the distances to galaxies in both the Virgo and Fornax clusters. The final objective is to test for systematic effects in the measurement of extragalactic distances, for example, to test the zero point of the Cepheid period-luminosity relation and to determine if there is a dependence of Cepheid luminosity on heavy-element abundance.

Progress is actively being made along all three of these lines. Prior to the refurbishment of the telescope, 30 Cepheids were discovered in two fields in the nearby galaxy M81. A field in the outer regions of the face-on spiral galaxy M101 was observed. This field is one of two chosen to allow a test of the sensitivity of the Cepheid period-luminosity relation to metallicity. Since the 1993 repair mission, the pace of the program has begun to increase rapidly. Data have been acquired for several galaxies: M101 (an inner field), the Virgo-cluster galaxy

M100, and three inclined spiral galaxies: NGC 925 (a member of the NGC 1023 group), NGC 7331, and NGC 3351 (a member of the Leo Group).

More recently a distance to the Virgo-cluster galaxy M100 has been measured. Given a measurement of the distance to M100 and the Hubble velocity of the Virgo cluster, a value of the Hubble constant (H_0) can be determined. However, at present the largest uncertainty in this determination is due to the fact that the distribution of spiral galaxies in the Virgo cluster is both extended and complex. Based on the new Cepheid distance to M100 of 17.1 Mpc, a value of $H_0 = 82$ km/(s)(Mpc), with uncertainties of ±6 km/(s)(Mpc) due to random errors and ±16 km/(s)(Mpc) due to systematic errors, is determined. An estimate of H_0 can also be made by using the measured relative distance between the Virgo cluster and the more distant Coma cluster. This determination avoids the uncertainty in the velocity of the Virgo cluster. In this case a value of $H_0 = 77$ km/(s)(Mpc), with uncertainties of ±6 km/(s)(Mpc) due to random errors and ±15 km/(s)(Mpc) due to systematic errors, is obtained. These results indicate that the value of the Hubble constant is 80 km/(s)(Mpc) out to a distance of 100 Mpc, with an accuracy of ±20%.

A value of $H_0 = 80 \pm 17$ km/(s)(Mpc) is consistent with a low-density universe (in which Ω, the ratio of the mass density of the universe to the critical density, lies between 0.1 and 0.3) and an age of 1.2×10^{10} years. An expansion age of 1.2×10^{10} years is consistent with other measured age estimates based on stellar evolution theory applied to globular clusters, white-dwarf cooling estimates for the galactic disk, and radioactive dating of elements, which give $14 \pm 2 \times 10^9$ years. However, for the standard cosmological model (the Einstein–de Sitter model, with $\Omega = 1$ and $\Lambda = 0$), the expansion age is $8 \pm 2 \times 10^9$ years for $H_0 = 80 \pm 17$ km/(s)(Mpc). This expansion age is well below the other age estimates listed above. This well-known age conflict highlights the importance of decreasing the uncertainties in all of these age estimates. *SEE STAR CLUSTERS.*

The remaining uncertainty in the value of H_0 is still dominated by systematic errors. An accuracy of 10% or better will be reached only when distances to a larger sample of galaxies have been measured so that the magnitude of these systematic errors can be assessed directly. Hence, the remaining observations of the Key Project will be critical. The results so far, however, suggest that the measurement of H_0 to an accuracy of 10% is a feasible goal.

For background information *SEE CEPHEIDS; COSMOLOGY; HUBBLE CONSTANT; SATELLITE ASTRONOMY; UNIVERSE* in the McGraw-Hill Encyclopedia of Science & Technology.

Wendy Freedman; Barry F. Madore

Bibliography. W. Freedman, Distance to the Virgo Cluster galaxy M100 from Hubble Space Telescope: Hubble Space Telescope observations of Cepheids, *Nature,* 371:757–762, 1994; S. W. Hawking, *A Brief History of Time,* 1988; E. Hubble, *The Realm of the Nebulae,* 1958; D. Overby, *Lonely Hearts of the Cosmos,* 1991.

Urban forestry

Urban forestry is the management of vegetation, particularly trees, to improve the environment and quality of life of people who live, work, and spend their leisure time in cities. Urban trees are found among high concentrations of people and within an intricate fabric of natural and artificial structures and processes. The complexity of this mosaic challenges urban foresters, who work to sustain healthy vegetation to meet the increasingly diverse needs of an urban society. Urban forest management emphasizes sustaining the physical, biological, economic, and social environment of the city.

Urban forests. A prerequisite to effective urban forest management is knowledge of the resource. More than just an assemblage of trees, urban forests are ecosystems that include other plants, as well as animals, water, air, soil, people, and the structures that people build. Attributes of the urban forest interact to affect the health and well-being of the city and its inhabitants.

The tree cover in cities varies across the United States. Cities in forested regions (for example, much of the eastern and western United States) have tree canopies that cover, on average, 31% of the city. In grassland regions (for example, central plain and intermountain areas), urban tree cover averages around 19%; in desert regions (southwestern United States), 10%. This variation is largely due to differences in the natural environment, particularly precipitation. Tree cover differs significantly among individual cities, ranging 15–55% in forested regions, 5–39% in grasslands, and 1–26% in deserts. These differences are mainly due to differences within a given city in the proportions of various land-use types. Each urban land-use type (for example, residential or commercial) has a characteristic pattern of buildings, roads, and associated functions that influence vegetation management, as well as the amount of space available for urban vegetation.

Management. Urban foresters, usually employed by a city government, guide urban forest management. They often conduct inventories of public trees, which may include information on tree species, health, location, size, maintenance needs, available planting locations, and other site information. Vegetation assessments of private lands may be conducted to provide additional resource information. With these data the forester can develop and implement management plans for trees in public areas, such as along streets and in parks, and guide tree management on private lands. These plans include designs for optimal species location, type, maintenance, and replacement.

Comprehensive plans and designs may be developed to increase desired benefits and reduce costs.

A substantial portion of the costs of urban forest management is associated with day-to-day tree maintenance activities such as planting, trimming, pruning, and removal, as well as controlling damage by insects and diseases. Other costs include repair of root damage to sidewalks and sewers, disposal of leaves and waste wood, and liability and hazard costs. In the United States, mean municipal tree management budgets range from around $10,000 in small towns to over $1 million in large cities. Mean total tree management expenditures per publicly owned tree in the United States average $4.64 per year. With proper management, which includes selecting and maintaining the appropriate tree species for each location, costs can be reduced while providing high levels of benefits.

Selecting and maintaining appropriate tree species for each location requires an understanding of site conditions and the contribution that the tree will make to the environment. Improper location of trees can reduce benefits and create significant problems, such as unhealthy or hazardous trees, or trees that interfere with utilities or other urban functions. With the introduction of exotic tree species into urban areas, selection and management issues encompass large numbers of tree species.

In many urban locations, individual trees have high monetary value, and often are managed by arboricultural practices, such as pruning, insect and disease control, and in some instances structural support. In other locations, such as in preserves or other natural areas, urban forests may be managed extensively, with little attention to individual trees or manipulation of the vegetative cover. Urban parks often fall in an intermediate category of management, with some attention to individual trees and extensive management of other areas.

In addition to direct management of public vegetation, urban foresters can use indirect means, such as ordinances and education, to influence vegetation management on private lands. Management of private holdings has a strong influence on the overall urban forest and the resulting benefits. Consequently, public assistance to private owners often is provided by urban foresters.

Benefits. Urban forest management can enhance particular attributes of the urban environment and provide benefits to people. Not all benefits can be realized at each location, but management can be targeted to provide those benefits that are most needed at a particular site, such as shading for energy conservation or esthetic enhancement of living environments.

Energy and carbon dioxide conservation. Trees reduce energy needs for heating or cooling buildings by shading buildings in the summer, reducing summer air temperatures (primarily through transpirational cooling), and blocking winter winds. However, trees also can increase heating needs by shading buildings in the winter. The energy effects of trees vary with regional climate and their location around the building. In northern climates, coniferous wind-breaks to block winter winds are generally most effective when planted to the north, northwest, and west of the building. Optimal deciduous shade tree locations are to the west, northwest, and east of the building. Trees planted to the south should be located or pruned to allow sunlight to reach south walls in midwinter. Properly establishing 100 million mature trees around residences in the United States could save $2 billion in energy costs annually.

Urban trees reduce carbon dioxide (CO_2), a major greenhouse gas, by directly removing it from the atmosphere and storing the carbon in the tree biomass. By reducing building energy use, trees can also reduce the emission of CO_2 from power plants. However, tree maintenance activities often require the use of fossil fuels that emit CO_2, and improperly located trees around buildings can increase energy demands and consequent emissions of CO_2.

Air quality. Trees influence air quality in a number of ways. Trees remove pollution from the air by intercepting airborne particles on their leaves and branches, and absorbing gaseous pollutants into their leaves via stomates. Trees also emit various volatile organic compounds that can contribute to the formation of ozone (O_3). By lowering air temperatures, trees lower the emission of volatile organic compounds from both vegetation and numerous anthropogenic sources (for example, gasoline), thus reducing the potential for ozone formation. Finally, by reducing building energy requirements, trees reduce pollutant emissions from power plants, thereby improving air quality. Sustaining widespread healthy forest cover through comprehensive urban forestry programs can lower local short-term levels of air pollution by 5% or more.

Urban hydrology. By intercepting and retaining precipitation or slowing its flow to the ground, urban forests can play an important role in hydrologic processes. They can reduce the rate and volume of storm-water runoff, flooding damage, and storm-water treatment costs; and they can enhance water quality. Estimates of runoff for an intensive storm in Dayton, Ohio, showed that the existing tree canopy reduced potential runoff by 7%; a modest increase in the canopy would have reduced it by nearly 12%.

Noise reduction. Properly designed plantings of trees and shrubs can reduce noise levels significantly. Wide belts (30 m or 100 ft) of tall dense trees combined with soft ground surfaces can reduce apparent loudness by 50% or more. Although the potential for noise reduction along roadsides in urbanized areas often is limited because of narrow planting space (less than 3 m or 10 ft), reductions of 3–5 dB can be achieved with dense vegetation belts of a row of shrubs roadside with a row of trees behind.

Quality of life. The presence of trees can make the urban environment a more pleasant place in which to live, work, and spend leisure time. Studies of people's preferences and behavior have confirmed the strong contribution of trees and forests to the quality of life in urban areas. Urban forests also provide significant outdoor leisure and recreation opportunities. The total annual contribution of trees in urban parks and recreation areas to the value of recreation experiences in the United States may exceed $2 billion. These benefits are often measured in terms of an individual's willingness to pay for experiences, as well as of increased property values.

Urban forest environments provide esthetically pleasing surroundings, increased enjoyment of everyday life, and a greater sense of connection between people and the natural environment. Trees are among the most important features that contribute to the esthetic quality of residential streets and community parks. Perceptions of such quality and of personal safety are highly sensitive to features of the urban forest, such as number of trees per acre and viewing distance.

Physical and mental health. Reduced stress and improved physical health for residents have been associated with the presence of urban trees and forests. Landscapes with trees and other vegetation have produced more relaxed physiological states in humans than landscapes without these natural features. A study of hospital patients determined that those with window views of trees recovered significantly faster and with fewer complications than comparable patients without such views.

Local economic development. Forest resources contribute to the economic vitality of a city, neighborhood, or subdivision. By improving the environment, trees contribute to increased property values, sales by businesses, and employment. Community action programs that begin with trees and forests often spread to other aspects of the community and result in substantial economic development.

Social development. A stronger sense of community, empowerment of inner-city residents to improve neighborhood conditions, and promotion of environmental responsibility and ethics can be attributed to involvement in urban forestry efforts. Active involvement in tree-planting programs enhances a community's sense of social identity, self-esteem, and territoriality; it teaches residents that they can work together to choose and control the condition of their environment. By improving the quality of the living environment, community planting programs also can help alleviate some of the hardships of inner-city living, especially for low-income groups.

Other benefits. Urban forests provide numerous other ecological, economic, and social benefits, for example, wildlife habitat, soil conservation, increased real estate values, and enhanced biodiversity. Such benefits are important to many urban dwellers and can contribute to the long-term well-being of urban ecosystems.

For background information SEE ARBORICULTURE; ECOSYSTEM; FORESTRY, URBAN; GREENHOUSE EFFECT; LAND-USE PLANNING in the McGraw-Hill Encyclopedia of Science & Technology.

David J. Nowak; John F. Dwyer

Bibliography. J. F. Dwyer et al., Assessing the benefits and costs of the urban forest, *J. Arboricult.,* 18:227–234, 1992; G. W. Grey and F. J. Deneke, *Urban Forestry,* 1986; R. W. Miller, *Urban Forestry,* 1988.

Urban infrastructure

Recent research in the field of urban infrastructure involves recycling and rehabilitation and the development of new approaches to infrastructure seismic hazard mitigation.

Recycling and Rehabilitation

As infrastructure ages and deteriorates, the urgency with which it must be repaired or replaced assumes greater importance. In a mature urban environment, rapid transit and railroad systems form an important part of the total transportation infrastructure, and the loss of any system increases the burden on the remaining portions of the network. Because the majority of rail and transit systems are in active use, the need to maintain a reasonable level of service must be balanced against the need to rehabilitate or improve. This may create significant problems, both in the repair or rehabilitation of existing infrastructure and in the upgrading of facilities to incorporate technological advances. Various rail and transit agencies throughout the United States have taken different approaches to satisfying these competing requirements.

Shutdown of rail system. The simplest approach to rehabilitation is to shut down a portion of the rail system totally and require riders to use an alternative method of transportation. This, of course, requires the existence of such an alternative with the capacity for the increased ridership generated by the closure. The Southeastern Pennsylvania Transportation Authority (SEPTA) was able to use this method to reconstruct a 4-mi (6.5-km), four-track elevated section of its regional rail mainline into Philadelphia. Since SEPTA's Broad Street subway paralleled the mainline in the area to be reconstructed, it was possible to terminate all train service at a rail station outside the construction area and transfer riders to the subway for the remainder of their commute. In order to minimize inconvenience and loss of ridership, SEPTA scheduled the closure and reconstruction project, named Railworks, to occur during two consecutive summers. During the first summer, approximately half of the 25 bridges within the project limits were rehabilitated or completely replaced; then the system was reopened until the following summer, when the remaining bridges were completed. The bridges rehabilitated during each shutdown were selected with input from the City of Philadelphia in order to minimize the effect of the reconstruction on vehicular traffic flow. At the end of the second summer, SEPTA had a completely new section of railroad including track, signal system, catenary traction power system, new or rebuilt through-girder and deck-girder bridges (the existing bridge abutments were reused with new superstructure installed in most cases), and a new elevated station. Because train service ceased during most of the heavy reconstruction, work did not have to be staged around rail operations. Thus, the contractors performing the work had much more latitude in their methods of construction than would have been possible had train service been maintained, and both the total cost and the overall duration of Railworks were significantly reduced.

Use of an abandoned line. When an agency has the opportunity to reuse an abandoned rail line to establish new service, there is no need to maintain rail operations during the reconstruction, and the contractor again has flexibility in construction methods. This is demonstrated by the Amtrak West Side Connection project in Manhattan, New York. In order to provide rail service to its Empire Division serving upstate New York from Pennsylvania Station in New York City, Amtrak utilized a long-abandoned 10-mi (16.5-km) freight rail line between the northern tip of Manhattan island at Spuyten Duyvil and the vicinity of 10th Avenue and 34th Street, where it was connected to a new cut-and-cover tunnel into Pennsylvania Station. Although most of the existing right-of-way required only grading, repair, or replacement of drainage structures and new track work, the existing movable bridge at Spuyten Duyvil, a historic structure, required major reconstruction. Because the swing bridge was not in service and there was difficulty in repairing it on site, the construction contractor elected to remove the bridge trusses from their abutments and piers, load them on barges, and move them to a more accessible site across the Hudson River in New Jersey. The removal of the three fixed and one movable span was accomplished by a high-capacity, barge-mounted crane. This method of construction had several benefits: first, the bridge trusses could be repaired safely in a controlled, accessible environment; second, the existing bridge abutment and piers could be more easily repaired with the bridge superstructure removed; and, third, the pier-mounted machinery used to swing the movable span could be readily replaced. After reconstruction, the bridge trusses were returned to Spuyten Duyvil and replaced on the renewed piers and abutments.

Maintenance of operations. Although a complete shutdown may be the most economical method of reconstructing a rail or transit line, it is not necessarily the most feasible, again because of the necessity of maintaining service during rehabilitation. Consequently agencies and design engineers have had to develop methods of maintaining rail operations during construction work.

Partial shutdown. For example, when Metro-North Commuter Railroad decided to reconstruct its Park Avenue viaduct in Manhattan, a detailed staging plan was developed to rehabilitate each of the four elevated tracks sequentially while maintaining train operations to and from Grand Central Terminal on the other three tracks. Thus, the contractor was permitted to perform the needed demolition and reconstruction work with minimal rail service disruptions. One restriction placed on the contractor's method of work was the requirement that pouring of concrete for the new viaduct deck be performed on Saturdays when the track adjacent to the one under reconstruction (on the same side of the center girder) could be shut down. This eliminated live load deflections in the new floor beams supporting the deck, and allowed the concrete to attain a significant portion of its design strength before rail traffic was reintroduced on the adjacent track on Monday morning. The staging method used on this project was dependent on the availability of sufficient excess capacity in the remaining in-service tracks in order to accommodate the loss of track under reconstruction without adversely impacting train operations.

Rerouting service. Because the conditions permitting long-term track shutdowns that do not impact train service are a rare occurrence, other methods of maintaining train operations have been developed. In order to replace the two-track Annsville Creek Bridge on the Metro-North Commuter Railroad Hudson Line in Peekskill, New York, another method was utilized. When the existing bridge construction did not lend itself to removal and replacement in stages while maintaining an operational track, a so-called runaround structure was designed. This two-track temporary structure had to be built parallel to the existing rail line, and train service was diverted prior to beginning replacement of the existing bridge spans, piers, and abutments. Upon completion of the reconstruction, train service was rerouted onto the new bridge, and the temporary runaround was removed. A similar method was used by the Connecticut Department of Transportation to replace the Peck Bridge, a bascule bridge (drawbridge) in Bridgeport, which carries Metro-North Commuter Railroad and Amtrak trains serving southern New England. Again, the condition and configuration of the existing structure did not permit staging the reconstruction while maintaining service. Use of a runaround structure is expensive and requires the availability of sufficient open space contiguous to the existing rail line, as well as modifications to signal systems and operating rules, but it is often the most viable option when balancing the need to reconstruct against the need to maintain operations.

Short-term shutdown. SEPTA has taken a unique approach in the reconstruction of its Frankford Elevated line serving northeast Philadelphia, Pennsylvania. This 5¼-mi (8.5-km) long, two-track viaduct required extensive rehabilitation because it was determined that only the existing columns and transverse girders could be reused. An entirely new superstructure, including new steel stringers (longitudinal steel beams), concrete deck, track, third rail, and signals had to be installed on these existing bents. Each bent consists of a transverse steel girder sitting on two steel columns and forms the primary support of the elevated structure. Because of the volume of ridership, SEPTA determined that no long-term shutdowns were acceptable, and the constricted urban environment did not permit construction of a runaround structure. Instead, the concept of short-term weekend shutdowns was developed by the design engineers. Service on the elevated line was terminated at 9:00 P.M. Friday night and did not resume until 6:00 A.M. the following Monday; bus service was provided as an alternative. During this 57-h period, the construction contractor, working on one-half of the structure, had to remove the existing track and ballast, demolish the existing steel and concrete jack arch deck, remove the steel trusses, replace the deck with either precast concrete track beams or a new precast deck, and install new, direct-fixation track. Much of the work was accomplished prior to the weekend shutdowns in order to minimize the amount to be done during the critical 57 h. For example, new steel stringers were installed between the transverse bents beneath the existing deck; steel shims were then placed between the existing deck and the new stringers so that the deck was supported by these stringers when it was cut away from the trusses that had previously supported it. This expedited the demolition, which had to occur during the weekend shutdown.

The sequence for the overall reconstruction was determined by the structural loads imposed by the new direct fixation track because portions of the existing structure required additional reinforcement. Although the original design specified precast concrete track beams to be installed during the first weekend, with a cast-in-place concrete deck poured during a later shutdown to complete the new deck, the design was refined over time and with contractor input, and later contracts specified large, precast concrete deck sections. This method accomplished total deck replacement during one weekend rather than the two or more originally envisioned, and permitted longer sections of structure to be replaced during each shutdown, with as many as 10 spans being completed during the 57-h period. The distance between existing steel bents varied, with the maximum span being almost 100 ft (30 m), but the average length was in the range of 50–60 ft (15–18 m). Although this use of short-term weekend shutdowns is a substantial restriction on the contractor's method of operation and has the effect of extending the project duration and increasing cost, it has provided SEPTA with an effective means of balancing the demand to provide unaffected service for the vast majority of its ridership with the need of the contractors to accomplish the reconstruction of this heavily used rail structure.

Douglas J. Fritz

Seismic Hazard Mitigation

Critical infrastructure systems in urban areas, such as highway and railroad transportation links, harbor facilities, and utility distribution networks, are severely affected by major earthquakes. Because infrastructure systems, or lifelines, play an essential role in earthquake rescue, reconnaissance, fire, medical, and general disaster-relief operations, special efforts need to be undertaken to keep these facilities functional following a major seismic event. For new facilities special design considerations are required that address not only a no-collapse scenario but also a functionality criterion ensuring serviceability through damage control or damage prevention. Older existing infrastructure systems need to be seismically strengthened or upgraded through seismic retrofitting, to meet the no-collapse or some specified functionality criteria. *SEE EARTH CRUST; EARTHQUAKE.*

Vulnerabilities of bridge structures. Recent earthquakes in urban areas worldwide have repeatedly demonstrated the vulnerability of bridge structures which form key components in lifeline and infrastructure networks. Older bridge structures designed and built 25–45 years ago have shown their susceptibility to seismic attack (**Fig. 1**). The majority of the existing bridge inventory in the United States, built in the late 1950s to early 1970s, is considered with respect to current seismic design criteria to have substandard designs and detailing. In California, a major turning point in seismic bridge design occurred following the 1971 San Fernando (magnitude on the Richter scale = 6.7) earthquake. Lessons learned and major problems identified from the collapse of bridge structures in the 1971 San Fernando earthquake led to a complete revision of seismic bridge design guidelines. The problems include the unseating of bridge spans from their supports at the end abutments or interior temperature movement joints, and the failure of bridge columns due to insufficient transverse or horizontal reinforcement to carry the seismic forces from the bridge superstructure to the foundation. The problem of unseating was addressed in California as well as in many seismic regions around the world by retrofitting bridge expansion joints with restrainer cables or similar deformation limiting devices (**Fig. 2**). The column problem required more research and development work. Actual retrofitting of bridge columns did not occur until the late 1980s. With an estimated 70–80% of all bridges built in the United States prior to the 1971 San Fernando earthquake, there are numerous bridge columns in need of seismic retrofit. These exist in areas where, even though the frequency of earthquakes may be smaller, the potential for a seismic event of large magnitude exists.

Bridge design. Because the source, location, and intensity of the next earthquake are unknown, bridge structures need to be designed and retrofitted to unknown force capacity levels. It is unrealistic to expect that a bridge structure be designed to

Fig. 1. Collapse of bridge structures caused by earthquakes. (*a*) SR-14/I-5 Southbound Connector in Los Angeles, California, during the 1994 Northridge earthquake. (*b*) Hanshin Expressway in Kobe, Japan, during the 1995 Hanshin earthquake.

perform completely elastically, without damage, under even the largest possible earthquake input. The current seismic bridge design philosophy assumes that local damage or inelastic structural action will occur during a major seismic event but that these regions of inelastic action are preselected and designed to provide large and controlled deformation capacities; can be easily inspected following an earthquake; and can be repaired without prolonged bridge closures or traffic interruptions. This general design philosophy has led to the design of bridges with potential inelastic flexural hinges formed at the column ends, keeping the superstructure, which carries the traffic, and the substructure or footings, which are typically below ground and difficult to inspect, in the essentially elastic or

undamaged range. Therefore, the bridge columns need to be designed to undergo large inelastic deformations during an earthquake without loss of load-carrying capacity (or danger of collapse). With this seismic design philosophy, the column in a bridge structure takes on an even more important role than just elevating and supporting the superstructure. Appropriate seismic column designs and retrofit concepts for new and existing bridge columns that will meet the above-stipulated performance criteria need to be developed and applied.

Bridge column retrofit. Three different types of failure modes can be observed under seismic load-deformation input in existing concrete bridge columns where insufficient transverse or horizontal and seismic detailing is provided.

Failure modes. The first and most critical failure mode is that of column shear failure, where inclined cracking, cover concrete spalling, and rupture or opening of the transverse reinforcement can lead to brittle or explosive column collapse (**Fig. 3**). The failure sequence consists of (1) the development of inclined cracks once the tensile strength of the concrete is exceeded, (2) the opening of inclined or diagonal cracks in the column and onset of cover concrete spalling (Fig. 3*a*), (3) rupture or opening of the transverse or horizontal reinforcement, (4) buckling of the longitudinal column reinforcement, and finally (5) complete disintegration of the column concrete core (Fig. 3*b*).

The second critical column failure mode consists of a confinement failure of the flexural plastic hinge region where, subsequent to flexural cracking, cover concrete crushing, and spalling, buckling of the longitudinal reinforcement or compression failure of the core concrete initiate plastic hinge capacity deterioration, associated with a shortening of the column in the plastic hinge zone.

Fig. 2. Movement joint restrainer retrofit concept. (*a*) Schematic bridge elevation. (*b*) Movement joint retrofit detail (circled in part *a*).

Fig. 3. Column failures of the I-10 Santa Monica Freeway caused by the 1994 Northridge earthquake. (*a*) Onset of shear failure in column. (*b*) Explosive column failure.

Finally, some existing bridge columns contain lap splices in the column reinforcement which for ease of construction are located at the lower column end to form the connection between the footing and the column. Starter bars for the column reinforcement are placed during the footing construction and lapped with the longitudinal column reinforcement in this region of maximum column moment demand, that is, the potential plastic hinge region. Lap-splice debonding occurs once vertical microcracks develop in the cover concrete and debonding gets progressively worse with increased vertical cracking and cover concrete spalling. This flexural capacity degradation can occur rapidly in cases

where short lap splices are present and little confinement from transverse or hoop reinforcement is provided.

None of the above failure modes and associated column retrofits can be viewed separately because retrofitting for one deficiency may only shift the seismic problem to another location and failure mode, without necessarily improving the overall deformation capacity.

Although all these seismic bridge column design issues can be addressed in new column designs through increased amounts of transverse or horizontal reinforcement in the form of circular hoops or spirals, even in rectangular columns through interlocking circular hoop or spiral reinforcement (**Fig. 4**), existing bridge columns can be retrofitted through external jacketing in the form of concrete, steel, or advanced composite jackets.

External jackets. The jacket retrofit concept is based on the simple principle that in order to confine the column core and to laterally support the vertical column reinforcement against buckling, transverse forces can be generated from an external jacket. The preferred jacket geometry is again circular, similar to the internal circular reinforcement hoops or spirals, because the radius of curvature provides a uniform lateral restraint around the entire column perimeter in case the column core dilates in shear or in flexural plastic hinging. Rectangular, oblong, or other column geometries can also be retrofitted with external jacketing by changing the column shape to a circular or elliptical cross section with concrete infills between the jacket and the original column (**Fig. 5**.) Rectangular or square column jackets are also possible, but are structurally less effective than circular or oval jackets and require significantly more jacket thickness.

Column jacket retrofitting with steel casings has been implemented on more than 500 bridges in California and on many bridge projects worldwide. During the Northridge earthquake in Los Angeles, California, in 1994, of the seven bridge structures which collapsed, six were due to column failures and all six failures could have been prevented with the available jacket retrofit technology. During this earthquake there were 24 bridges in regions of strong ground shaking, with horizontal ground

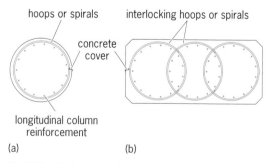

Fig. 4. Circular hoop or spiral reinforcement for confinement of the bridge column core. (*a*) Circular cross section. (*b*) Rectangular column cross section.

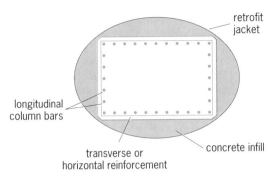

Fig. 5. Elliptical jacketing for rectangular bridge columns.

accelerations in excess of ≥0.5 **g** (**g** = the gravity vector), and 60 bridges in regions of moderate ground shaking, with estimated ground acceleration ≥0.25 **g**, all retrofitted with column steel jackets; no sign of damage to any of these structures was observed.

Recent development of a composite jacket utilizes new materials such as continuous carbon fiber wraps in epoxy matrix to provide column confinement and transverse reinforcement around the bridge column. This almost fully automated retrofit technology can significantly reduce retrofit installation time and through the automated application provide a high degree of quality assurance.

Seismic retrofit technology exists that can ensure safety and serviceability of critical urban infrastructure bridges and lifeline systems. Estimates are that in California alone seismic retrofit measures to the bridge infrastructure will cost two to three billion dollars over the next few years and that nationwide this amount may more than double.

For background information SEE BRIDGE; RAILROAD ENGINEERING; SEISMIC RISK in the McGraw-Hill Encyclopedia of Science & Technology.

Frieder Seible

Bibliography. M. J. N. Priestley, F. Seible, and M. Calvi, *Seismic Design and Retrofit of Bridges,* 1996; M. J. N. Priestley, F. Seible, and Y. H. Chai, *Design Guidelines for Assessment Retrofit and Repair of Bridges for Seismic Performance,* University of California, San Diego, Structural Systems Research Project, Rep. No. SSRP-92/01, August 1992.

Variational methods

Engineering science uses fundamental laws and equations to describe engineering systems. A prediction of a system's behavior is determined from the direct solution of these equations. In electrical engineering these equations are Kirchhoff's laws for circuits and Maxwell's equations for electromagnetic fields. In the case of Maxwell's equations, computational techniques are normally applied. One suitable and efficient approach is to use a variational formulation. With variational methods an energy or power expression is used to describe the

system. The unknowns (electromagnetic fields in this case) are varied until the power expression exhibits a stationary point. This is a unique minimum, maximum, or inflection point, which occurs at the true field solution. These stationary power methods can also be directly applied to electrical circuits. A power characteristic equation is constructed for the circuit, and the unknown voltages or currents are varied. When the power calculation yields a stationary turning point, the voltages or currents are at their correct values for the circuit.

The variational method provides unifying properties and is straightforward to implement. It can be used as the starting point to solve most electrical engineering problems, such as time-harmonic, time-transient, and nonlinear electric circuits, and nonlinear magnetic circuits. One characteristic energy equation describes an entire circuit no matter how complicated. Using variational methods for circuits allows the solution of hybrid problems where a system consists of field structures and lumped-element components. This type of system often occurs, for example, in very high frequency signal-processing applications. Variational methods are also suited to nonlinear systems, are easy to implement, can be graphically illustrated, and are applied throughout the physical sciences.

Pedagogical value. Apart from the research and design applications of this approach, it is also useful as a teaching tool. For example, the methods described can be used directly to replace conventional circuit analysis techniques at all levels of study. This energy-based approach has a number of advantages over conventional nodal and mesh circuit analysis techniques: (1) It is easy to implement. (2) One equation describes the entire circuit no matter how many circuit components are present. (3) The method avoids difficulties with voltage and current polarities. (4) The solution can be illustrated graphically. (5) The method uses simple calculus and the theory of maxima and minima. (6) The method is equally applicable to nonlinear components. (7) The method can be extended to include magnetic circuits and electromagnetic field devices.

This circuit application can also be used to introduce mathematical variational methods which are often used in advanced programs, usually involving electromagnetic field analysis. These variational procedures, energy expressions, and the minimum-energy principle are often difficult steps to introduce even at the advanced level. The simple circuit examples given here can be used to overcome these difficulties. The method includes all of the steps of the variational procedure in a very simple systematic form. Once circuits have been used to introduce and demonstrate the minimum-energy principle, applications to more conventional continuously varying field and potential problems are readily accepted by students.

Method of analysis. As an engineering analysis tool, the variational approach is well established and has a solid foundation in both physics and math-

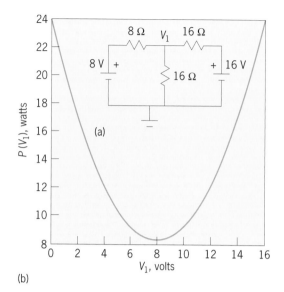

(b)

Fig. 1. Resistor circuit with dc supplies. (*a*) **Circuit diagram.** (*b*) **Variational power curve.**

ematics. A variational power equation for circuits can be constructed from the instantaneous power of an electrical network. Either unknown nodal voltages or branch currents are used as the unknown variables. As power is a scalar quantity, there are no restrictions on current and voltage polarities. Each circuit branch element is represented by an entry into the characteristic power equation. With nonlinear devices the voltage-current characteristic has to be integrated to form a power expression. One equation describes the entire circuit, and its stationary point coincides with the circuit's true solution. The stationary point can be found via trial-and-error graphical techniques (for simpler circuits) or a minimization procedure. This analytic approach is well known in mathematical physics and is widely applied to the calculation of electromagnetic fields. It is sometimes known as the principle of least action, and implies that physical systems naturally converge to a stationary energy state. A step-by-step guide to evaluating unknown circuit voltages is summarized as follows:

1. Identify unknown voltages in the circuit.
2. Form a power-energy characteristic equation for the entire circuit.
3. Solve graphically (when there are less than three unknowns).

As an alternative to step 3, the following steps can be carried out:

4a. Differentiate the characteristic equation with respect to the first unknown and set the result to zero.
4b. Carry out step 4a for each unknown in turn.
4c. Solve the generated equations.

Application to resistor circuits. The method of analysis is straightforward and is best introduced by considering a simple first-order example. Starting at step 1, there is one unknown voltage (V_1) at the junction of the three resistors in **Fig. 1***a*. In general, for a resistor (R) with a terminal voltage (V), the power (P) dissipated in that resistor element is defined by Eq. (1). By using this fundamental relationship, it

$$P = \frac{V^2}{R} \quad (1)$$

is possible to develop the characteristic equation for the entire circuit. The voltages across the three resistors from left to right are given by expression (2).

$$(V_1 - 8), (V_1 - 0), (V_1 - 16) \quad (2)$$

The power dissipated in the circuit is given by the linear addition of the power dissipated in each resistor. By using Eqs. (1) and (2), the characteristic power equation (3) is given by step 2. Following

$$P(V_1) = \frac{(V_1 - 8)^2}{8} + \frac{V_1^2}{16} + \frac{(V_1 - 16)^2}{16} \quad (3)$$

step 3, the solution for the circuit in Fig. 1 can be solved graphically by plotting the power $P(V_1)$ as a function of V_1. Clearly, a stationary minimum turning point is exhibited at the true solution of $V_1 = 8$ volts in the curve of Fig. 1*b*.

Alternatively, an analytical procedure can be adopted by following steps 4a, 4b, and 4c. The stationary point of Eq. (3) can be derived by differentiating $P(V_1)$ with respect to the unknown V_1 and by setting the result to zero. This operation (step 4a) yields Eq. (4). In this example, step 4b is unnecessary,

$$\frac{dP(V_1)}{dV_1} = \frac{2(V_1 - 8)}{8} + \frac{2V_1}{16} + \frac{2(V_1 - 16)}{16} = 0 \quad (4)$$

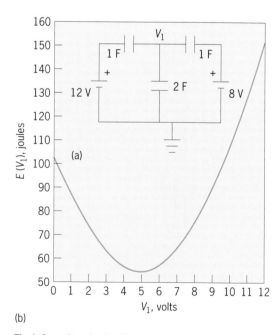

(b)

Fig. 2. Capacitor circuit with dc supplies. (*a*) **Circuit diagram.** (*b*) **Variational energy curve.**

and by step 4c the solution of Eq. (4) is $V_1 = 8$ volts, as before. The nature of this stationary point can also be investigated by taking the second-order derivative and by examining the sign, as in Eq. (5).

$$\frac{d^2 P(V_1)}{dV_1^2} = \frac{2}{8} + \frac{2}{16} + \frac{2}{16} \qquad (5)$$

The positive value is indicative of a minimum turning point, as shown in Fig. 1b.

Application to capacitor circuits. Electric lumped-element capacitor circuits also exhibit the minimum energy behavior. In the capacitor circuit in **Fig. 2a**, the potential V_1 is unknown (step 1). If V_1 is varied, the overall stored energy of the circuit will also vary. There is a unique value of V_1 which will give a stationary minimum value of energy for the circuit. The energy, E, stored by a capacitor, C, is written as $\frac{1}{2}CV^2$. In the circuit of Fig. 2a, there are three capacitors and there will be three terms in the characteristic energy equation. By taking each capacitor in turn, this gives an energy equation (6) for step 2.

$$E(V_1) = \frac{1}{2} 1 (V_1 - 12)^2 + \frac{1}{2} 2 V_1^2 + \frac{1}{2} 1 (V_1 - 8)^2 \qquad (6)$$

If V_1 is treated as a trial function and varied, $E(V_1)$ exhibits a minimum value when $V_1 = 5$ volts (step 3). This energy curve is plotted in Fig. 2b. As before, differential calculus can also be used to find the stationary point of Eq. (6). The quantity $E(V_1)$ is differentiated with respect to V_1, and the result is set to zero (step 4a), as in Eq. (7). The solution of

$$\frac{dE(V_1)}{dV_1} = (V_1 - 12) + 2V_1 + (V_1 - 8) = 0 \qquad (7)$$

this equation is 5 volts as before.

Application to nonlinear circuits. Analysis of networks containing nonlinear elements requires special attention. For passive nonlinear circuits, the solution may be calculated by using a variational approach so long as the current-voltage relationship in each component is single valued. In this case the resistance of the nonlinear component changes dynamically with its voltage and current. It then becomes necessary to cater for this dependency by using a mathematical integration technique. An example of a dc resistive circuit with a nonlinear component is illustrated in **Fig. 3a**. A powerlike characteristic equation can then be written as in Eq. (8).

$$P(V_1) = \frac{(V_1 - 8)^2}{1/2} + 2 \int_0^{V_1} V^2 \, dV$$

$$= 2(V_1 - 8)^2 + 2 \frac{V_1^3}{3} \qquad (8)$$

The stationary point of this equation gives Eq. (9).

$$\frac{\partial P(V_1)}{\partial V_1} = 4(V - 8_1) + 2V_1^2 = 0 \qquad (9)$$

This equation can be solved to give $V_1 = 3.12$ volts. The function $P(V_1)$ for the circuit is plotted against V_1 in Fig. 3b.

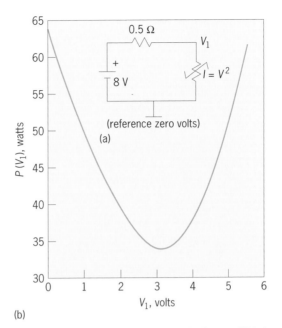

Fig. 3. Nonlinear resistive circuit. (*a*) Circuit diagram. (*b*) Variational power curve.

Application to hybrid problems. The variational solution of electrical circuits has certain unifying properties, some of which have been discussed. Since variational methods are universal, in principle unified lumped-element and electromagnetic field problems are possible. As a research tool, the variational method is of current interest as a technique for solving such hybrid structures. For example, a recent development in the area of broadbanding and miniaturizing microwave circuits is to use lumped-element capacitors and resistors with high-frequency distributed waveguides. The variational approach is particularly suited to this type of hybrid application. Variational finite elements are used to model the distributed region, and the circuit method discussed above caters for the lumped circuit elements. This approach provides a computationally efficient solution.

Another potential application area is in the modeling of nonlinear circuits and improving the convergence time to the correct circuit solution. Current work is concerned with developing these methods to include heat distribution and temperature profiles in high-power electrical systems. This work is being linked with numerical finite element solvers in order to cater for and analyze a broad range of distributed and lumped-element structures. In the future these methods will be developed to treat a wide range of electromechanical problems.

For background information *SEE ALTERNATING-CURRENT CIRCUIT THEORY; DIRECT-CURRENT CIRCUIT THEORY; FINITE ELEMENT METHOD; KIRCHHOFF'S LAWS OF ELECTRIC CIRCUITS; LEAST-ACTION PRINCIPLE; MAXWELL'S EQUATIONS; MICROWAVE* in the McGraw-Hill Encyclopedia of Science & Technology.

A. A. P. Gibson

Bibliography. A. A. P. Gibson and B. M. Dillon, The variational solution of electric and magnetic circuits, *IEE Eng. Sci. Educ. J.,* 4(1):5–10, February 1995; A. A. P. Gibson and B. M. Dillon, Variational solution of lumped element and distributed electrical circuits, *IEE Proc: Science Measurement Technology,* 141(5):423–428, September 1994.

Vertebrates

Miniaturization has evolved many times independently in many groups of animals. Among vertebrates, amphibians and bony fishes (teleosts) are the most miniaturized species. Among the smallest land vertebrates are the frogs *Psyllophryne didactyla* (family Brachycephalidae), *Sminthillus limbatus* (family Leptodactylidae), and members of the genus *Stumpffia* (family Microhylidae) with snout vent lengths of about 0.4 in. (10 mm); and the salamanders *Thorius pennatulus* and *T. narisovalis* (family Plethodontidae) with snout vent lengths of about 0.5–0.8 in. (13–20 mm). Among teleosts, miniaturized species are found in the clupeiform, characiform, siluriform, cyprinodontiform, and perciform orders. The smallest teleosts and vertebrates are found in the perciform family Gobiidae (gobies), with a minimum total length of about 0.3 in. (8 mm) [for example, *Trimmatom nanus*].

Gains and losses of miniaturization. The process of miniaturization is commonly viewed as an adaptive process, because miniaturization might enable animals to live in microhabitats with abundant food supplies that are inaccessible to both competitors and predators (such as leaf litter, bromeliads, or burrows of bark beetles as in the case of miniaturized tropical salamanders). Another advantage of miniaturization is the possibility of reaching sexual maturity very early in development. However, disadvantages of miniaturization are equally obvious with respect to morphology and function of at least some parts of the body. The most drastic consequences of miniaturization concern the sense organs (for example, the eyes and the brain, particularly with respect to visual acuity, object and depth perception, and visual learning) which are thought to heavily depend on the number of photoreceptors and visual neurons available for processing of visual information. Here, a series of processes might be expected to occur in miniaturized vertebrates to compensate for the otherwise drastic reduction in number of neural cells.

Visual system. Accordingly, in miniaturized vertebrates that heavily depend on vision, there is an increase in relative size of the eye; in the relative thickness of the retina, particularly the layer of retinal ganglion cells; in relative brain size, up to 15% of the size of the skull compared to 1% in large nonminiaturized species; in the size of brain parts containing visual centers, particularly the tectum of the midbrain (the most important visual center) and the dorsal thalamus; and in gray matter (containing nerve cell bodies) relative to white matter (containing nerve cell fibers) within these centers, up to 40% compared to 15% in nonminiaturized salamanders. These compensatory processes enlarge the relative space for sensory receptors and nerve cells, thereby ensuring a minimum number of cells required for visual perception and guidance of behavior. However, they occur in different degrees and different combinations among miniaturized taxa, and exceptions are found. For example, the minute leptodactylid frog *S. limbatus* has a remarkably small tectum in absolute as well as relative terms, as opposed to the general tendency of miniatures to increase the relative volume of the tectum.

However, more critical than the relative size of the eye, retina, brain, and visual centers are packing density (that is, the space occupied by nerve cell bodies inside the gray matter) and above all, the size of visual cells. Accordingly, miniatures with very small cells and high packing density generally have relatively many visual cells largely independent of relative eye and brain size. The average size of nerve cells varies enormously among miniaturized vertebrates. The smallest diameter of the average nerve cell is 2.4 micrometers in the gobiid boby fish *Pandaka lidvilli;* it is among the smallest vertebrate cells. The largest (11.3 μm) nerve cell is found in the plethodontid salamander *Batrachoseps attenuatus*—a cell size that is large even compared to nonminiaturized vertebrates. The smallest cell size found among miniaturized salamanders is 7.5 μm (*Desmognathus aeneus*), and among frogs, 5.1 μm (*S. limbatus*).

Cell size is among the most invariant features within a lineage. In miniaturized as well as nonminiaturized salamanders, cell size is correlated with neither brain size nor head size. As a consequence, miniaturized vertebrates have only slightly smaller cells than nonminiaturized close relatives. However, miniaturization has occurred predominantly in those taxa that phylogenetically already had small cells (for example, gobiid teleosts, leptodactylid frogs, or desmognathine salamanders) and therefore could tolerate a strong decrease in body and brain size.

Increases in nerve cell packing density relative to nonminiaturized species is typical of most miniaturized organisms compared to nonminiaturized amphibian taxa; this phenomenon is particularly strong among those species that have larger cells, where an increase up to 40% of gray matter is found (*Thorius*), while miniaturized taxa with small cells have packing densities around 18%. Thus, miniaturized taxa with small cells do not maximize their nerve cell number through maximum increase in packing density, but nonetheless have many tectal cells, while those with large cells generally have few cells. *Sminthillus limbatus,* because of its small cell size, has relatively many tectal cells, despite its relatively small tectum and submaximal packing density. Disproportional increase in relative brain size and in packing density can only partially compensate for large cell size.

Number of visual cells in miniatures. Cell counts for the entire visual system (brain and retinas) reach their minimum in the plethodontid salamander *T. narisovalis*, about 65,000 for the entire visual brain centers and about 60,000 for the retinas. These extraordinarily low numbers are the consequence of a combination of small head, eye, and brain and relatively large cells. However, the smallest extant salamander, *T. pennatulus* (which is much smaller than *T. narisovalis*) has 94,000 visual cells and a roughly equal number of retinal cells. This species has many more nerve cells than many larger salamanders with larger cells, a consequence of the combination of relatively large brain and visual centers, the highest nerve cell packing density among vertebrates, and medium cell size. The frog *S. limbatus* has about 400,000 cells in the visual centers of the brain, a consequence of small cell size.

Miniaturized and nonminiaturized taxa. Despite the enormous differences in head and brain size among large and miniaturized frogs and salamanders, the corresponding differences in number of visual neurons are remarkably small. For example, while the largest frogs and salamanders have heads that are 500 times and brains that are 40 times larger than those of the miniaturized taxa, they have only 4–10 times more visual neurons. Thus, the compensatory processes observed in miniaturized vertebrates appear to be highly effective in keeping the number of visual neurons as high as possible. No corresponding cell counts are available for miniaturized teleosts.

Compensation for loss in visual acuity. A reduction in the size of the eye and the visual centers does not necessarily lead to reduction in visual acuity provided that cell size is proportionally decreased. Under these circumstances, the angular distance between photoreceptors, which is critical for spatial resolution and consequently visual acuity, remains constant. However, in miniaturized amphibians, cell size remains constant or is only slightly reduced compared to nonminiaturized close relatives, and many miniaturized animals have remarkably large visual cells. Thus, the number of visual cells is decreased, and visual acuity is necessarily reduced.

Nevertheless, miniaturized salamanders and frogs exhibit good to excellent visual abilities. For example, the most miniaturized salamanders possess a very fast and precise projectile tongue that enables them to feed on fugitive prey (for example, collembolans) that is rather inaccessible to other salamanders. These salamanders apparently compensate the otherwise inevitable loss in visual acuity by restricting their foraging strategy to ambush feeding at very short distances, devoting a high portion of visual neurons to frontal vision (while giving up panoramic vision), and increasing the number of retinal ganglion cells relative to that of photoreceptors, thus making the entire eye a functional fovea.

In miniaturized salamanders with relatively large and consequently very few cells, the dendritic appendages of visual neurons in the brain are more widespread and consequently show greater overlap than in species with small and more numerous cells. Thus there might be an increase in the number of synaptic contacts between dendrites and visual afferents necessary for processing of visual information. Furthermore, theoretical considerations suggest that an increase in visual spatial resolution capabilities is achieved by the principle of coarse coding, based on a strong dendritic overlap of neurons that form a functional unity (a neuronal assembly). Here, features are coded in the distributed activity of the neurons belonging to the assembly. This principle requires fewer neurons than a coding based on the activity of specialized single neurons called detector cells.

Process of miniaturization. Animals cannot simply decide to become small in order to exploit the advantages of miniaturization. Rather, the process of miniaturization has to make use of particular developmental mechanisms, one of which is pedomorphosis, that is, the retention of juvenile traits into adulthood. In many ways, the brains of miniaturized vertebrates resemble early developmental stages of organisms in nonminiaturized taxa: features characteristic of embryonic and early larval brains are relatively large brains and tecta and relatively small endbrains, a large volume of gray compared to white matter, and a high cell packing density. Furthermore, anatomical complexity is generally low both in miniatures and in embryonic and larval brains. Accordingly, the tecta of the gobiid teleost (*P. lidvilli*), the leptodactylid frog, *S. limbatus,* and the salamander *T. narisovalis* uniformly exhibit a low degree of cell migration and consequently of cellular laminae and migrated nuclei compared to nonminiaturized close relatives, with *T. narisovalis* possessing the most larvalike and anatomically simple tectum among vertebrates, with no migrated cells and lamination at all. These features corroborate the assumption that a predominant way to achieve miniaturization phylogenetically is through a retardation and eventual arrest of somatic development at early stages, while becoming sexually mature.

For background information SEE AMPHIBIA; ANURA; CELL (BIOLOGY); NERVOUS SYSTEM (VERTEBRATE); NEURON; TELEOSTEI; URODELA; VERTEBRATA in the McGraw-Hill Encyclopedia of Science & Technology.

Gerhard Roth; Jens Blanke

Bibliography. J. Hanken, Miniaturization and its effects on cranial morphology in plethodontid salamanders, genus *Thorius* (Amphibia: Plethodontidae), II: The fate of the brain and sense organs and their role in skull morphogenesis and evolution, *J. Morphol.,* 177:255–268, 1983; J. Hanken and D. B. Wake, Miniaturization and evolution, *Annu. Rev. Ecol. Syst.,* 24:501–519, 1993; G. Roth, J. Blanke, and M. Ohle, Brain size and morphology in miniaturized plethodontid salamanders, *Brain Behav. Evol.*, 45:84–95, 1995.

Vibration

Fluid flow is a source of energy which can induce structural and mechanical oscillations. Flow-induced vibration denotes those phenomena associated with the dynamic response of structures immersed in or conveying fluid flow. The term covers those cases in which an interaction develops between fluid-dynamic forces and the inertia, damping, and elastic forces in structures. The study of these phenomena draws on three disciplines: structural mechanics, mechanical vibration, and fluid dynamics.

Flow-induced vibration is experienced in numerous fields, including aerospace engineering, such as in airplanes and the space shuttle; power generation and transmission, such as in steam generator tubes and valves; civil structures such as bridges and smoke stacks; and undersea technology, such as in periscopes and ships. Flow-induced vibration comprises complex and diverse phenomena; subcritical vibration of nuclear fuel assemblies, galloping of transmission lines, flutter of pipes conveying fluid, and whirling of heat exchanger tube banks are typical examples.

In general, all dynamic phenomena of fluid-structure interaction can be grossly divided into two groups: subcritical vibration for small oscillations and dynamic instability for large oscillations. For example, a typical response of a structure in a flow is given in **Fig. 1**. At small flow velocities, the response is small. As the flow velocity increases, there may be some peaks in the response curve; they may be due to vortex shedding or other excitation sources. As the flow is further increased, very drastic increase of the response is noted. This increase is due to fluid-elastic instability.

Turbulence-induced vibration. When the flow velocity is large enough, the flow becomes turbulent. Fluid turbulence is a continuous source of energy to excite structural oscillations. The response depends on various system parameters such as natural frequency, damping, and stiffness.

At present, the turbulence characteristics for different cases cannot be calculated with accuracy in the prediction of structural response. In addition, the turbulence in a flow may not be homogeneous and its characteristics depend on flow path and other system parameters. Thus, structural response to turbulence cannot be predicted accurately. Fortunately, small oscillations due to turbulence are acceptable in most cases. In certain cases in which the structural or mechanical components cannot be allowed to oscillate even with amplitudes no larger than several micrometers, special care must be exercised to reduce turbulence-induced vibration.

Vortex-induced lock-in resonance. The Kármán vortex street and other vortex patterns have been observed and studied for centuries. The earliest recorded observation of the phenomena of vortex shedding dates from the sixteenth century when Leonardo da Vinci made drawings of the surface pattern of the flow past an obstacle. Remarkable similarities have been found in these complex fluid-dynamic flow patterns over many orders of magnitude of the Reynolds number. The frequency of vortex shedding from a structure in a uniform flow is related to the structural dimension and flow velocity through the nondimensional Strouhal number. Specifically, the Strouhal number is equal to the product of this frequency (called the Strouhal frequency) and the structural dimension divided by the flow velocity. **Figure 2** shows the typical traces and a close-up view of a Kármán vortex street behind a cylinder at a Reynolds number (based on the diameter of the cylinder) of 80. Experiments show that, for flow past a cylinder with Reynolds numbers between 80 and 10^5, the Strouhal number is approximately 0.2.

Associated with the vortex shedding, there is a series of pulsating forces acting on the structure. One is perpendicular to the flow with its frequency equal to the Strouhal frequency, and another is in the flow direction with its frequency equal to twice the Strouhal frequency. When the natural frequency of the structure is the same as the vortex-shedding frequency in the lift direction or equal to twice the vortex-shedding frequency in the drag direction, structural vibration may lock in with the vortex shedding process. Thus the flow energy will transfer to structural motion through this process, a phenomenon called lock-in resonance. The vibration amplitude due to lock-in oscillations can be fairly large. For example, lock-in resonance of a single circular tube in cross-flow can increase to about 1–2 diameters of the tube in the lift direction and 0.1–0.2 diameters in the drag direction. Lock-in resonance for a single tube is fairly well characterized; however, for tube arrays, lock-in resonance is not well understood and much more research is needed.

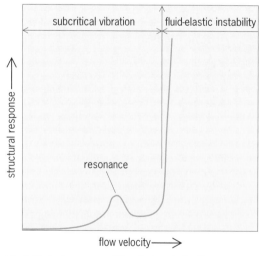

Fig. 1. Typical response of flow-induced vibration.

Axial flow-induced vibration. When vibration is induced by a flow parallel to the axis of a structure, such as pipes conveying fluid, it is called axial flow-induced vibration. At low flow velocity, the response, subcritical vibration, is similar to the response to cross-flow. However, as the flow velocity increases, fluid-elastic instabilities occur.

The response of the structure depends on the supports of the structure. For example, in a pipe supported at both ends, no dynamic instability, or flutter, will occur. However, in a cantilevered pipe fixed at the upstream end and free at the downstream end, such as a fire hose, large-amplitude oscillations, about 20 pipe diameters, can occur. **Figure 3** shows the flutter of a garden hose conveying fluid. This instability can also be demonstrated by blowing into a long balloon with an open end. Axial flow-induced vibration and instability have received considerable attention, in particular with regard to the postinstability region with large-amplitude chaotic motion.

Wake-induced galloping. A frequently noticed problem is the galloping of telephone wires in the wind. When one wire is in the wake of another wire, large-amplitude oscillations can be induced. This is a flutter phenomenon; the wind energy is transferred to the downstream wire through the wake of the upstream wire. The oscillation patterns depend on the location of the downstream wire with respect to the upstream one. Galloping can occur even with a single wire if ice is frozen on the wire, changing its shape. This often happens in northern areas with high winds. A submarine has two periscopes, and if they are not properly located, unacceptable vibration due to wake-induced galloping can occur.

Instability of tube arrays. The classical example of fluid-elastic instability is tube arrays in cross-flow. This instability was studied extensively because many components consist of a group of tubes subjected to cross-flow, such as steam generator tubes. Normally the spacing between the tubes can be as small as 0.1 tube diameter to several tube diameters. The flow energy is continuously transferred to the tube array, resulting in fluid-elastic instability. Once fluid-elastic instability (called flutter, fluid-elastic vibration, fluid-elastic coupling, fluid-elastic whirling, damping-controlling instability, or stiffness-controlled instability) develops, the tubes may impact one another.

Different tube arrangements have been used, including triangular and square arrays with different flow directions. The instability characteristics depend on tube arrangement, tube location, tube pitch, damping, and mass ratio. It is rather difficult to obtain a finite mathematical expression for the stability criterion. However, in many cases the following correlation can be used: the critical flow velocity at which a large-amplitude response develops is proportional to the square root of the mass-damping parameter. This relation is true for air flow in most cases but it is not valid for water flow

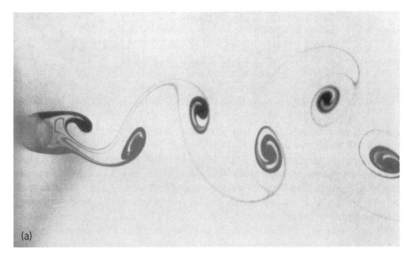

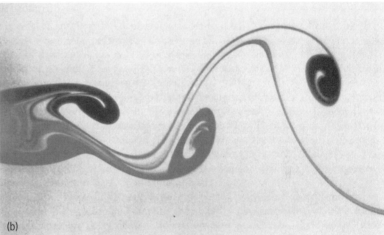

Fig. 2. Kármán vortex street behind a circular cylinder. (*a*) Typical dye traces. (*b*) Close-up view near the cylinder. (*From A. E. Perry, M. S. Chong, and T. T. Lim, The vortex-shedding process behind bluff bodies, J. Fluid Mech., 116:77–90, Cambridge University Press, 1982*)

in general. This topic is being studied extensively.

Chaotic vibration. When the structural oscillations become large, nonlinear effects become important. In some case, nonlinearity is an inherent property of the structure, such as in the case of loosely supported tubes in heat exchangers. In typical flow-induced vibrations with strong nonlinearity, chaotic vibration can occur. This is a classical example of mechanical systems which exhibit chaotic vibration. Pipes conveying fluid and tube arrays in cross-flow with loose supports have been studied analytically and experimentally to understand the characteristics of chaotic vibration. The techniques used include time histories, power spectra, phase planes, Poincaré planes, the Lyapunov exponent, correlation, and fractal dimension. These are simple mechanical systems but possess rich chaotic behavior.

Vibration control. Flow-induced vibration can be fun in a whirlpool but can be very detrimental to structural and mechanical components subjected to fluid flows. Therefore, all mechanical and structural components subjected to flow should be evaluated from the standpoint of flow-induced vibration and,

Fig. 3. Flutter of a polyethylene tube (garden hose) conveying fluid. The tube is illuminated by a stroboscope to show its motion.

as necessary, designed to avoid fluid-elastic instabilities and to limit flow-induced vibration to an acceptable level.

Some spectacular failures have been attributed to totally ignoring the potential for detrimental flow-induced vibration or otherwise insufficiently considering flow-induced vibration. Various design guides are available from professional societies. Even though there are significant gaps in codified knowledge, designers have been able to develop many system designs that have provided useful service without significant problems.

Design. Methods may be available for solving flow-induced vibration problems in the field, but it is always better to avoid such problems in the first place. Furthermore, the method used to reduce vibration frequently requires a compromise relative to the satisfaction of other design or operating requirements. However, designing with a large margin of safety generally leads to increased costs and often a reduction in performance and efficiency. Therefore, structural design should be optimized with respect to cost, performance, and safety. The current design procedure leaves much to be desired. The engineer charged with design must avoid detrimental vibrations by relying on existing information and sound judgment in modeling and testing.

For background information SEE CHAOS; KÁRMÁN VORTEX STREET; RESONANCE (ACOUSTICS AND MECHANICS); TURBULENT FLOW; VIBRATION; VORTEX in the McGraw-Hill Encyclopedia of Science & Technology.

Shoei-Sheng Chen

Bibliography. R. D. Blevins, *Flow-Induced Vibration,* 2d ed., 1994; S. S. Chen, *Flow-Induced Vibration of Circular Cylindrical Structures,* 1987; E. Naudascher and D. Rockwell (eds.), *Practical Experiences with Flow-Induced Vibrations,* 1980; T. Sarpkaya and M. Isaacson, *Mechanics of Wave Forces on Offshore Structures,* 1981.

Virology

Membranes have a defining role in eukaryotic cells, and their controlled fusion is essential for life processes. Fusion occurs in a variety of contexts, for example cell-cell fusion, the fusion of intracellular membranes, and liposome fusion. Furthermore, fusion of viral and host membranes is a key stage in the initiation of infection by enveloped viruses, allowing the release of the genetic material of the virus into the cytoplasm of the target cell. A central puzzle has been the mechanics of the process, that is, how the membranes are brought together in a controlled way. In this respect the simplicity of viral membranes makes them attractive model systems.

The paradigm for membrane fusion was provided by influenza virus, and recent work has provided fine structural detail of what happens. X-ray crystallography has allowed the visualization of unprecedented contortions in the key protein involved in the process of membrane fusion, hemagglutinin (HA), one of the two types of molecules located at the surface of the mature influenza virus. However, the structure of the surface protein of a flavivirus suggests that a different but somewhat analogous mechanism operates in a second family of viruses.

Influenza virus. Influenza virus hemagglutinin is a protein molecule, and each subunit starts life as a single polypeptide chain. This is then cleaved into two chains (HA_1 and HA_2,) to produce the mature molecule, which protrudes from the viral membrane that is ready to bind to a receptor molecule on the target cell. The receptors for influenza virus are the terminal sialic acid residues of the complex sugars attached to glycoproteins on the surface of susceptible cells; the hemagglutinin molecule recognizes these sugars specifically. A number of different viruses (for example, polyomavirus) recognize slightly different sialic acids, providing a mechanism by which these viruses modulate their tissue tropism and virulence. Although the recognition is specific, it is a relatively loose association, so that a number of points of attachment are required before the influenza virus is tightly enough anchored to be taken into the cell within a vesicle (endocytosis). In the cell, the vesicle undergoes changes in pH, and at a more acidic pH the hemagglutinin molecule undergoes extensive conformational rearrangements, initiating a process that leads to the fusion of the membrane of the virus with the membrane from the target cell.

The mature, extra-virion portion of hemagglutinin is composed of a globular head (comprising most of chain HA_1) mounted on a largely helical stalk (mainly a chain HA_2). It is now known how the hemagglutinin head recognizes the sialic acid group on the virus receptor; this knowledge is guiding the design of anti-influenza virus compounds. However, this structure alone does not explain membrane fusion. A hydrophobic region at the N terminus of the HA_2 chain, termed the fusion peptide, is known to insert into the target membrane; however, it is tucked into the stalk of the viral molecule, some 10 nanometers from the outermost tip of the hemagglutinin molecule (and therefore even further from the target membrane since the receptor on the cell must have some thickness).

It seems likely that the energy required to drive fusion is stored up during the folding of the hemagglutinin molecule. The molecule starts life as a single polypeptide chain. On cleavage of this chain to form the mature hemagglutinin the freed ends of the polypeptide chains snap apart 2 nm. Cleavage of the polypeptide chain provides, in principle, the chance of exploring new structures which might have more stable folds. This seems to be exactly what happens with hemagglutinin; the cleaved molecule is metastable, and is like a set trap: acidification of the endosome springs the trap, setting in motion the series of events that lead to fusion. A crystallographic structure of part of the hemagglutinin molecule at acid pH reveals numerous conformational changes, including a 10-nm movement of the fusion peptide. Thus, a model for fusion can be constructed: the fusion peptide is cast outward to engage the target cell membrane (the globular heads and hence the cellular receptor molecules separating to make way for this), while at the base of the molecule, tension is generated as a small globular domain tries to invert. These changes tend to flip the molecule onto its side, whereupon the viral and cell membranes are dragged into proximity.

Tick-borne encephalitis. Whereas in influenza virus the hemagglutinin stands up, bristling from the virus surface, the equivalent protein in the flavivirus tick-borne encephalitis virus (E protein) forms a dimer that lies flat across the surface of the viral membrane. This arrangement is deduced from the structure of a soluble form of the E protein, which was prepared by cleavage from the viral membrane. A little more than 100 of the approximately 500 residues of the intact polypeptide chain were thus removed, including almost 50 residues of the external domain leading into the membrane. The resulting protein fragment folds into an elongated structure, and individual molecules are formed by association of two of these fragments into a dimer which extends laterally in the plane of the membrane. The molecule has a slight curvature, as if it were made to fit onto the spherical surface of the virus.

The molecular architecture of the E protein is unusual. Each monomer consists of three domains arranged along its length. The domain containing the N terminus of the E protein chain lies at the center of the dimer. The second, elongated domain is principally responsible for holding the two subunits of the dimer together. This domain bears superficial similarity to other viral structural protein. The third domain, which resembles an immunoglobulin domain and has been implicated in receptor attachment, sits upright, projecting slightly above the others. Unlike influenza virus hemagglutinin, the E protein of tick-borne encephalitis is not cleaved dur-

ing maturation; however, it still undergoes a structural rearrangement at low pH, switching from dimeric to trimeric association as a prelude to membrane fusion. It is also believed that it is the precursor (prM) of the small viral transmembrane protein, M, which binds to E protein and modulates its properties. If this is considered as part of the E protein, then a situation analogous to that in influenza virus occurs: the prM-E complex is stable under acid conditions, protecting the E protein as it is transported through the acidic trans Golgi network during exocytosis of the immature virion. Then, during the virus maturation, the precursor of protein M is cleaved, releasing the dimeric E protein, now primed for the low pH conformational switch. Indeed, there is a peptide seen near the extremity of the E protein which is likely to act as a fusion peptide, although crystallography has not yet revealed if a low pH conformational switch exposes this peptide in an analogous fashion to influenza virus.

For background information SEE ANIMAL VIRUS; ARBOVIRAL ENCEPHALITIDES; INFLUENZA; VIRUS in the McGraw-Hill Encyclopedia of Science & Technology.

David Stuart; Patrice Gouet

Bibliography. P. A. Bullough et al., Structure of influenza haemagglutinin at the pH of membrane fusion, *Nature*, 371:37–43, 1994; F. A. Rey et al., The envelope glycoprotein from tick-borne encephalitis virus at 2 Å resolution, *Nature*, 375:291–298, 1995; I. A. Wilson, J. J. Shekel, and D. C. Wiley, Structure of the haemagglutinin membrane glycoprotein of influenza virus at 3 Å resolution, *Nature*, 289:366–373, 1981.

Vortex

Vortex breakdown may be defined as an abrupt change in the structure of the core of a swirling flow. Its occurrence is often (although not always) marked by the presence of a free stagnation point (where the fluid flow has zero velocity) on the axis of the vortex followed by a region of reversed flow of finite extent, which may be axisymmetric, and a corresponding divergence of the stream surfaces near the axis. With increasing swirl velocity ratio (ratio of swirl to axial velocity) three basic types of breakdown are generally observed: (1) for mild swirl, a helix or double helix with a slight departure of the central filament from the axis; (2) for intermediate swirl, a spiral characterized by an abrupt departure of the central filament from the axis, which then follows a corkscrew or spiral path (with sense the same as or opposite to that of the vortex); and (3) for large swirl, a bubble, a (nearly) axisymmetric envelope surrounding a region of reversed flow. (The interior flow may be asymmetric and unsteady, with one or two vortex rings embedded within the bubble.)

In all cases, there is a sudden decrease of axial and angular velocities, and generally the flow in the wake of the breakdowns is turbulent. There may also be a succession of such breakdowns, for example, a bubble followed by a spiral. For flows with sufficient swirl, vortex breakdown can be expected if the vortex persists for a sufficiently long distance without otherwise being attenuated (such as by diffusion, other instabilities, and so forth).

Figure 1 shows vortex breakdown on a delta wing, with a spiral breakdown on one of the leading-edge vortices, and a bubble breakdown on the other, symmetrically located vortex. **Figure 2** shows breakdowns in a diverging channel: a spiral breakdown in Fig. 2*a* and a bubble followed by a spiral breakdown in Fig. 2*b*. Figure 1 illustrates the occurrence of breakdown in unbounded external flows, and Fig. 2, breakdown in internal flows, the divergence of the channel providing the adverse pressure gradient leading to breakdown. (In an adverse pressure gradient the pressure increases in the direction of flow, called adverse, or unfavorable, because it acts to decrease the axial momentum of the flow.)

Applications of vortex breakdown. Vortex breakdown has major technological, aerodynamic and nonaerodynamic, applications. In the last, use is made of the closed recirculating flow in a bubble-type breakdown. Because the fluid in the bubble is isolated from containing walls and undergoes extensive mixing, processes and operations involving fluids that can damage or be damaged by such walls, or that require efficient or extensive mixing, are made possible or enhanced. Aerodynamically, vortex breakdown can be beneficial or destructive. If it occurred naturally, or could be artificially induced, in the trailing vortices in the wake of large-lift aircraft, it would hasten the dissipation and demise of these vortices, and hence lessen the threat to following, smaller aircraft. This possible contribution of vortex breakdown to the wake alleviation problem, and its consequent potential impact on airport congestion, has been a focus of recent research. Contributing also to the recent interest is the harmful effect of breakdown when, in

the case of high-speed aircraft flying or maneuvering at large angles of attack, it occurs over the top surface of the wing or near the tail surfaces, causing degradation of aerodynamic performance, buffeting, and early structural failure.

Breakdown and vortex dynamics. Of less immediate technological application but great importance is the role of vortex breakdown in vortex dynamics in general, and in the transition to turbulence. Flow simulations with vortex filaments show that vortices have a tendency to form local isolated concentrations, which may then exhibit behavior reminiscent of breakdown flows. Streamwise vortices are a feature of the intermediate stages of transition from laminar to turbulent flow. Waviness in these streamwise vortices may be associated with spiral or helical breakdown, and similarly it is possible that breakdown, or bursting of vortices, is a factor in the final stages of transition of a laminar boundary layer, in the eruption of the streamwise vortices from the solid surface, or in the formation of turbulent spots (isolated, sporadic patches of intense turbulent motion, characteristic of the last stages of transition from laminar to turbulent flow of a wall-bounded boundary layer).

Mechanism of breakdown. In spite of the importance of vortex breakdowns and decades of research, the basic underlying mechanism leading to breakdown is not yet established. Several explanations have been proposed, and variously assert that vortex breakdown (1) is similar to the separation of a two-dimensional boundary layer; (2) is a consequence of hydrodynamic instability; (3) is dependent on the existence of a critical state, and is a finite transition from a supercritical to a subcritical state, analogous to a hydraulic jump; and (4) resembles a solitary wave, or soliton, the result of the trapping of long, weakly nonlinear waves propagating in nearly critical swirling flows.

To a greater or lesser degree, each of these theories fails to explain fully, or adequately, breakdown in all its aspects. Experiments, particularly those on swirling flows in pipes, have elucidated certain aspects of breakdown, flow visualization being especially useful in illuminating the various forms of breakdown and their characteristics. These visualizations, together with flow measurements, have to some extent enabled evaluation of the theories of breakdown. However, experiments and numerical simulations, discussed below, have been less successful in establishing a universally accepted general physical mechanism(s) leading to the breakdowns usually observed in aerodynamic and swirling internal flows. There are a number of reasons for this. Measurements, whether invasive (for example, using hot wires) or noninvasive (for example, laser Doppler velocimetry), and flow visualization are difficult to obtain and to interpret in these three-dimensional, and often unsteady, flows, and there are often other problems, one of the most important being the tendency of the breakdown to move back and forth in the test sec-

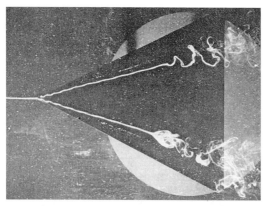

Fig. 1. Vortex breakdown over a delta wing, with spiral breakdown on one leading-edge vortex and bubble breakdown on the other. (*From N. C. Lambourne and D. W. Bryer, The bursting of leading-edge vortices: Some observations and discussion of the phenomenon. Aeronaut. Res. Coun., R & M 3282, 1961*)

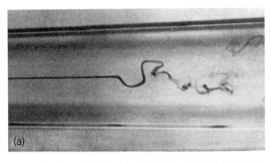

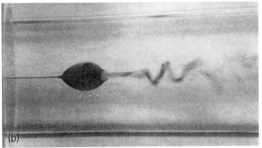

Fig. 2. Vortex breakdowns in a divergent channel. (*a*) Spiral form. (*b*) Bubble form followed by a spiral. (*From S. Leibovich, The structure of vortex breakdown, Annu. Rev. Fluid Mech., 10:221–246, 1978*)

tion or on the aerodynamic surface, often quite dramatically.

Some of the seemingly different postulated physical mechanisms of vortex breakdown are equivalent or related, leading to the same or similar criteria for breakdown. To the extent that the various theories predict a criterion for vortex breakdown, the predictions are quite similar, namely, that the onset of breakdown occurs when the ratio of the swirl to the axial velocity is approximately 1.5. Therefore, this criterion cannot be used to distinguish between the theories.

Numerical simulations. Full numerical simulations of breakdown are usually based on the complete Navier-Stokes equations, appropriate for viscous, rotational flows (flows with vorticity), or the Euler equations, for inviscid, rotational flows. Until recently, because of computer limitations, it was necessary to assume that the flow was both axisymmetric and steady, but more recent calculations solve the unsteady, three-dimensional, full viscous Navier-Stokes equations. Common characteristics of many of the earlier and more recent numerical simulations, as well as many of the experiments, are: (1) extreme sensitivity to flow parameters (such as swirl velocity ratio, external axial velocity variation, or pressure gradient); (2) suddenness of breakdown (that is, breakdown with no evidence of the prior growth of an instability of the basic swirling flow); and (3) a tendency for the breakdowns to migrate upstream to the initial station, that is, the upstream boundary of the computational domain or test section (unless some means is employed to prevent this). This tendency of breakdowns to move to the initial station has

caused the validity of such numerical simulations to be questioned. Some of these issues are particularly troubling because, to some extent, the difficulties with the numerical simulations, including the slow rates of convergence to final steady states often encountered, may be intrinsic aspects of the physics of breakdown and not numerical artifacts.

Upstream migration of the vortex breakdown occurs mainly for unbounded, or unconfined, flows, such as aerodynamic flows and flows in diverging channels. Numerically simulated breakdowns, usually in the form of steady axisymmetric bubbles, that occur in bounded or confined swirling flows, for example, in finite closed cylinders with one end wall rotating, or flow in the gap between rotating spheres, do not exhibit this behavior, nor do experiments on these flows. Because of the robustness of the breakdown of the swirling flow in such cylinders, with respect to both the type of breakdown (invariably a symmetric steady bubble) and its location (unmoving and far from the end walls), this configuration has been the focus of much experimental and numerical work. The newest breakdown criterion, based on azimuthal vorticity considerations, was advanced with some success in connection with this problem, and has been applied to other physical situations as well.

Universality. It is not known whether there is a single underlying mechanism of vortex breakdown with universal application, although this is implicitly assumed in much of the above discussion. There may be, also, no fundamental etiological difference between bubble, helical, and spiral breakdowns, or combinations thereof, the particular form being perhaps dependent upon initial and boundary conditions. It is not known whether the same is true for swirling flows in geometries as different as those discussed above.

For background information SEE BOUNDARY-LAYER FLOW; FLOW MEASUREMENT; FLUID FLOW; FLUID-FLOW PRINCIPLES; HYDRAULIC JUMP; SIMULATION; TURBULENT FLOW; VORTEX in the McGraw-Hill Encyclopedia of Science & Technology.

Stanley A. Berger

Bibliography. J. M. Delery, Aspects of vortex breakdown, *Prog. Aerosp. Sci.*, 30:1–59, 1994; M. Escudier, Vortex breakdown: Observations and explanations, *Prog. Aerosp. Sci.*, 25:189–229, 1988; S. Leibovich, Vortex stability and breakdown: Survey and extension, *AIAA J.*, 22(9):1192–1206, 1984.

Wind power

Recent developments involving generation of electricity with machines utilizing wind energy have provided a new potential for pumping water in remote areas. For example, lack of an adequate year-round water supply is still a major impediment to livestock grazing in many arid regions. Cattle tend to graze up to about 1 km (0.6 mi) from a water supply; therefore, several water supplies are needed in most large pastures. Ranchers have found that if sufficient watering places are not provided, livestock do not move to areas of the pasture where grass may be abundant. For this reason, many ranchers continue to haul water for livestock in remote areas.

Another application of wind-powered systems is to provide a safe and dependable source of water to about one-third of the world's population. Many people depend on surface waters that are polluted and harmful to their health. Water cannot be pumped, because energy and labor for servicing engine-driven pumps are usually unavailable. Thousands of mechanical water pumping systems have been installed over the years to meet the water requirements of people and livestock. However, because of high maintenance requirements and aging equipment, many water users must seek other energy sources to power their pumps. The availability and cost for new electrical grid service are often prohibitive.

System equipment. A wind-electric water pumping system consists of a wind turbine that produces alternating-current electric power at variable voltage, variable frequency; a pump controller; and a standard utility-grade electric motor and pump. The wind-electric water pumping system allows the wind turbine to operate at variable speed, thus producing a variable-voltage, variable-frequency system that can supply electric power directly to a standard electric motor. The direct-drive, permanent-magnet alternators nominally produce three-phase, 240-V alternating-current power at 60 Hz at 325 revolutions per minute (rpm). The frequency varies between 0 and 90 Hz, and the corresponding voltage varies between 0 and 330 V. At a wind speed of 6 m/s (13 mi/h), the system stabilizes, and the voltage and frequency increase rapidly together until the rotor furls (turns out of the wind) at a wind speed of 13.5 m/s (30 mi/h). The benefit of this type of system is seen in **Fig. 1**, which shows the voltage-frequency ratio. The voltage-frequency ratio exceeds 3 at a wind speed of 6 m/s (13 mi/h) and remains almost constant until furling at 13.5

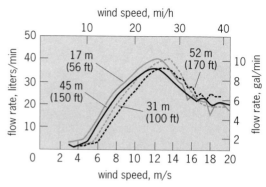

Fig. 2. Water-flow rates for four pumping depths, using a submersible pump and a 1500-W wind turbine.

m/s (30 mi/h). The electric motor, rated at 240 V and 60 Hz, will operate best at a voltage-frequency ratio near 4. Although the voltage-frequency ratio varies from 3 to 4, this range is acceptable for most motors. When the voltage-frequency ratio is constant or nearly constant, the current draw to the motor is proportional to the power provided, and it is always equal or below the design current. Motor overheating will not occur as long as design current is not exceeded.

Operation. Each wind turbine system has a mechanical rotor overspeed control, which allows the unit to run unloaded. The units furl and slow the rotor by turning sideways out of the wind flow. The rotor blades are usually constructed of fiber-reinforced plastic or epoxy-coated wood, and they operate at rotor speeds between 100 and 500 rpm. Rotor diameters for small water pumping systems range from 2 to 7 m (7 to 23 ft).

The wind pumping system is controlled by an electronic circuit that senses the frequency output of the wind turbine generator; when a preset cut-in frequency is reached, a standard motor solenoid connects the electric power from the wind turbine to the standard electric pump motor. The controller performs four control functions by starting the electric pump motor at the low-speed cut-in, stopping the motor at a low-speed cut-out, stopping the motor at a high-speed cut-out, and restarting the motor after a high-speed cut-out.

The pumps used in wind-electric systems are multistage submersible pumps powered by three-phase, 240-V standard submersible electric pump motors. Pumps and motors operate at 3450 rpm when powered at a constant 60 Hz (utility power). Systems have been tested at several pumping heads to determine the effect of pumping head on the wind speed at which pumping is initiated, and to develop the pumping curves under the different pumping heads as a function of wind speed. (Pumping head is the specific energy that is required to move water from a beginning point to a discharge point.)

Performance. A 1500-W wind-electric water pumping system that operates independently of the electric utility was operated at seven different pumping heads ranging from 17 to 59 m (56 to 190

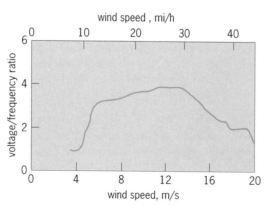

Fig. 1. Voltage-frequency ratio for a 1500-W wind turbine pumping from a depth of 45 m (150 ft).

ft). Performance data were collected for more than 700 h at each pumping head. During all these tests, the wind turbine, pump controller, electric submersible motor, and pump required no maintenance. These systems experienced wind speeds in excess of 30 m/s (65 mi/h). The water-flow rates for four pumping heads are given in **Fig. 2**. For the 17-m (56-ft) pumping head, flow was initiated at a wind speed of 3 m/s (7 mi/h), and a peak flow of 40 liters/min (10 gal/min) was recorded at a wind speed of 12 m/s (27 mi/h) when furling occurred. The peak flows varied from 36 to 41 liters/min (9 to 10 gal/min) for all heads tested.

The flow curve for a pumping head of 45 m (150 ft) was selected for conducting a prediction of yearly pumping. Monthly wind-speed histograms from 10 years of wind-speed data collected at a height of 10 m (33 ft) at Bushland, Texas, were used to calculate an average daily pumping volume for each month. **Figure 3** shows the average daily water pumped when the pumping head was 45 m (150 ft). The highest daily average water pumped was in March with a volume of 16,139 liters/day (4264 gal/day), and the lowest was in August with 7349 liters/day (1942 gal/day). The average for the year was 12,534 liters/day (3311 gal/day); all months, except August, exceeded 10,000 liters/day (2642 gal/day). A beef cow requires 40–50 liters/day (10–15 gal/day); therefore, this pumping system would provide for well over 100 head. A rancher should plan for storage of a 5-day supply and size the herd for the lowest daily amount available. However, in this case a rancher might choose to select the average of July, August, and September, or 9680 liters/day (2560 gal/day), as the available water supply.

Since the multibladed windmill has been used for many years to provide water for livestock, its performance was compared to this electrical water pumping system. A month-by-month comparison of the two pumping systems using the average daily water volume is given in Fig. 3. The average daily water volume for the wind-electric system exceeds the wind-mechanical system by almost 4000 liters/day (1050 gal/day), or 45% more water. The wind-electric pump provided more water in all months except August, when the average wind speed is significantly lower than in the other months. These data clearly show that electrical wind pumps operate better than mechanical systems when the average wind speed is above 5 m/s (11 mi/h), and operate about the same as mechanical systems when the wind speed is 4–5 m/s (9–11 mi/h). Comparisons made between mechanical and electrical wind pumps for pumping heads of 17–30 m (56–100 ft) show that the electrical wind pump will pump about twice as much water.

In the two systems the electric wind turbine rotor at 3.05 m (10 ft) is larger than the mechanical windmill rotor at 2.44 m (8 ft). The mechanical windmill starts pumping at a lower wind speed, but the difference is less than 1 m/s (2 mi/h) and is dependent

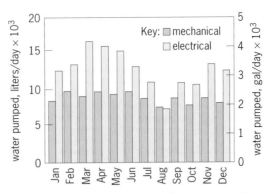

Fig. 3. Comparison of the average daily water pumped by a mechanical windmill and a wind-electric pumping system. The pumping depth was 45 m (150 ft). Wind data are from 1983–1992.

upon the pumping head. Probably the most important comparison is between the costs of the two systems. The turbines and towers cost about the same, but the costs of the controls and pumps differ greatly. The higher cost of steel pipe and the requirement for a pump rod for the mechanical system more than offset the cost of the pump controller for the electrical system. These small-sized submersible pumps are often supported by a hanger wire, and polyethylene pipe is used to transport the water to the surface, thus reducing the cost of the pump installation. The overall costs are almost identical for the two systems.

These machines have proved to be reliable and robust enough to be installed in remote areas where the greatest need for pumping water for livestock and domestic use occurs. This wind-electric water pumping system has consistently performed better than the wind-mechanical system. Although data are presented for one pump and four pumping heads, several pumps using three different wind turbines have been tested, and all perform better than mechanical pumps. Much of the improved performance of wind-electric systems is a result of using submersible pumps that have a low starting torque and a flow that is proportional to the speed of the pump. In contrast, the piston pump used with wind-mechanical water pumping systems has a high starting torque, with flow proportional to the stroke length and stroke speed. Mechanical wind systems furl and reduce the pump speed when the wind speed exceeds 10 m/s (22 mi/h), thus wasting significant amounts of energy.

For background information SEE ALTERNATING-CURRENT MOTOR; PUMP; WIND POWER in the McGraw-Hill Encyclopedia of Science & Technology.

R. Nolan Clark

Bibliography. R. N. Clark and B. Vick, Determining the proper motor size for two wind turbines used in water pumping, *Wind Energy 1995*, ASME Publ. SED-16, pp. 65–72, 1995; F. C. Vosper and R. N. Clark, Autonomous wind-generated electricity for induction motors, *Trans. ASME J. Solar Energy Eng.*, 110:198–201, 1988.

Wood

In the past, most foresters believed that variation in wood properties among trees within a species was primarily the result of the differing environments under which the trees had been grown. However, it is now known that wood properties have about the highest level of genetic control of any important characteristic of forest trees. Most wood, growth, and form characteristics of forest trees are not related genetically, except those with obvious relationships such as cell wall thickness or latewood percentage, both of which determine wood density. This independence enables the geneticist to breed trees with the suitable adaptability, pest resistance traits, tree form, and growth characteristics combined with the desired wood properties. Improvements through selection and breeding can be relatively rapid and substantial because of the magnitude of variability and intensity of genetic control of wood properties. Many tree improvement programs have at least one wood property (such as wood density) as a major characteristic requiring genetic improvement.

As the world's supply of wood becomes more restricted and of lower quality, and especially as intensive forestry with shortened rotations becomes general practice, wood properties have become more important. The supply of high-quality wood from old-growth indigenous forests is rapidly disappearing, and must be replaced by plantation-grown trees. This is especially true for species with highly desirable wood from the tropical rainforests and those with short rotations. In some conifer species, such as the pines, the percentage of juvenile wood, with properties that confer only minimal strength and instability for solid wood products, increases rapidly as rotation ages are shortened. If the quality of the finished products made from wood is to be maintained, action must be taken to counter this trend by changing wood properties of younger trees.

A great deal is known about the genetics of wood properties in a few widely used and widely planted coniferous species such as *Pinus taeda* (loblolly pine), *P. elliottii* (slash pine), *P. radiata* (radiata pine), and Douglas-fir (*Pseudotsuga menziesii*). Also well understood is the genetics of wood in some species of eucalypts, poplars, and other diffuse porous hardwoods. However, less is known about the ring porous species, and essentially nothing is known about the genetics of wood of most tropical hardwoods.

Genetic control. Wood is affected by tree growth and form, by the environment, and by internal physiological characteristics which have strong genetic control. Often, the fastest and the best changes in wood can be obtained by improving the straightness of the tree's stem and to a lesser extent, by developing smaller and flatter angled limbs. If a tree is not straight, it will produce reaction wood, that is, compression wood in the conifers (on the underside of the lean) or tension wood in the hardwoods (on the upper side of the lean). Compression wood has as much as 9% more lignin than normal wood and short cells with fracture cracks; it is hard to pulp and bleach, and is totally unstable in solid wood products such as boards. Tension wood has a low lignin content and short cells. Its wood is unstable, giving low pulp yields and poor-quality finished lumber. Heritability of straightness is moderately high and variability is large; therefore significant and quick gains in wood properties can result from developing the straighter trees. One generation of intensive selection is usually enough. Limb angle has strong genetic control but the genetic control of limb size is weak and fewer gains can be obtained from breeding.

Changes in the cambium (the layer of cells responsible for secondary growth and generating new cells) caused either by the environment or by internal physiological controls determine wood properties. Anything that affects the growth of a tree may also influence the kind of wood produced. However, the determination of wood properties is not simple. All trees have juvenile wood near the tree center (the pith) and mature wood toward the bark. In many conifers, juvenile wood, in comparison to mature wood, has low wood density, short tracheids (xylem cells), low cellulose yields, and cell walls that are unstable upon drying. Most diffuse porous hardwoods have fewer differences between the juvenile and mature wood than do the conifers. Only a few studies have been made relative to the control of juvenile wood quality and quantity by genetic manipulation. When successful, such manipulation results in much more uniformity of wood within the tree.

Internal wood property changes. The most important change in wood that can result from genetic manipulation is the uniformity of wood within a tree as well as uniformity of wood among trees in a plantation. Nonuniformity in genetic makeup allows tolerance to pests, varied sites, and extreme environments. However, because of their genetic independence, it is possible to breed trees that have uniform wood but at the same time show tolerance to and grow well under adverse conditions. Uniformity of wood obtained in this way is widely utilized in operational forestry programs. Wood uniformity within a tree is difficult to obtain because of the juvenile-mature growth pattern and the inherent differences between earlywood and latewood within the annual ring of the tree. A few studies have indicated that genetic modification for internal uniformity of wood is possible, but this has not been generally applied operationally.

Wood density. Along with uniformity, the most important wood property included in a tree improvement program is wood density because gains in product quality are large and rapid and the effect on both fiber and solid wood products is outstanding. Wood density is the ratio of dry weight of wood to its green volume. When expressed as grams per cubic centimeter, the measure is often called specific gravity. Wood density has a very high narrow sense heritability, ranging from 0.6 to 0.8, and a large amount of variability (for example, *Eucalyptus grandis* varies in specific gravity from tree to tree from 0.35 to 0.80). Heritability is a common measure of the strength of inheritance; it is the ratio of genetic variation to total variation (genetic and environmental variation). This ratio ranges from 0 to 1, and indicates how much of a characteristic is inherited and how much is the result of environmental differences. Because of its genetic pattern and high variability, and its utilitarian value, the bulk of genetic studies on wood have been made on wood density.

Individual tree heritabilities obtained for 11 species of pines are presented in the **table**. Although great significance cannot be given to differences in heritability values because of the varied ages and environments in the studies, what is remarkable is how high the heritabilities are (about 0.60) for all but one species. These can be compared with heritabilities of characteristics such as volume growth, which are around 0.15. Inheritance patterns of wood density are equally strong for other conifers as well as in the hardwoods, despite the complexity of wood density, which is determined by the combination of a number of factors such as earlywood-to-latewood ratio, wall thickness, and cell size, each of which has its own definitive inheritance pattern.

Cell length. The length of both tracheids in conifers and fibers in the hardwoods has a very high inheritance, nearly as high as wood density. However, the importance of genetic changes in cell length is not as great because the effect of cell length on the quality of the final product is not as important. In the conifers, tracheid length is satisfactory for all products except those derived from the juvenile wood. Even though genetic control of fiber length in the hardwoods is strong, the variability within species is small and the gains in length obtained from breeding are so small that they have very little effect on product quality. Many studies have been made on the genetics of cell length; but even though the inheritance pattern is strong, cell length is rarely used in operational tree improvement programs.

Spiral grain. Spiral grain is defined as the alignment of wood fibers to the long axis of the tree stem. This characteristic of wood is of vital importance for the quality of solid wood products, and its inheritance pattern is moderately strong. It has essentially no importance for fiber products. The measurement of spiral grain is difficult and the grain angle changes with the age of the cambium from the pith, often starting as a left-hand spiral, then changing to no spiral and sometimes even to a right-hand spiral near the bark. Spiral grain is always most severe near the center of the tree. For conifers, primarily used for solid wood products, with shortened rotations, spiral grain has become of major consideration in breeding programs of some species such as radiata pine. Improvement of spiral grain by genetic manipulation can have great utility.

Fibrillar angle. Although sometimes confused with spiral grain, with which there is no relationship, fibrillar angle (the orientation of the fibrous elements that make up the cell wall) has become of major importance in both fiber and solid wood production because of its effects on stability. Only a few genetic studies have investigated fibrillar angle, and they indicate strong genetic control. Measurement of fibrillar angle is time consuming and requires specialized equipment; it is only now beginning to be used in operational programs.

Miscellaneous properties. Many other wood properties, such as chemical characteristics, pulpability, wood color, rot resistance, reaction wood, internal defects (cracks, shake), wood extractives, moisture content, bark characteristics, and heartwood production, can be described as having small to moderate inheritance patterns. Some of these are used in specialized breeding programs but are not in general use. There has been some recent special interest in breeding for wood color and heartwood formation for high-value hardwoods.

For background information *SEE* FOREST GENETICS AND BREEDING; FOREST MANAGEMENT; TREE; WOOD in the McGraw-Hill Encyclopedia of Science & Technology.

Bruce J. Zobel; Jackson B. Jett

Bibliography. B. J. Zobel and J. B. Jett, *Genetics of Wood Production,* 1995.

Wood specific gravity: range in narrow-sense heritability for 11 pine species	
Species	Range
Pinus banksiania	0.55–0.73
Pinus caribaea	0.62–0.71
Pinus contorta	0.33–0.69
Pinus elliottii	0.43–0.69
Pinus nigra	0.60–0.70
Pinus oocarpa	0.58–0.82
Pinus pinaster	0.44–0.75
Pinus radiata	0.20 to nearly 1.00 (mostly 0.55–0.65)
Pinus sylvestris	0.46–0.57
Pinus taeda	0.42–0.90 (mostly 0.45–0.60)
Pinus virginiana	0.13–0.41

Xiphosurida

Each spring large numbers of horseshoe crabs nest on beaches along the Atlantic and Gulf coasts of the United States. The high tides during the new and full moons are the highest tides of the month (spring tides); at this time, horseshoe crabs crawl to the top of the tide line to nest in the high intertidal region of the beach. Intertidal nesting is a surprising phenomenon because it is associated with several hazards: crabs may be stranded by the receding tide and die from desiccation or predation; eggs laid in this area are more likely to be lost through predation, erosion, or extremes of temperature and moisture; breeding only during the highest tides of a month necessitates strongly synchronized crab nesting, which increases competition for nesting sites and mates.

Horseshoe crabs that are ready to spawn arrive on the beach in pairs, the female trailed by the smaller male holding onto her carapace. Upon reaching the high tide line, she buries herself in the sand. While buried, she lays eggs that the attached male fertilizes externally with aquatic, free-swimming sperm. After laying several thousand eggs, she moves forward in the sand about 4 in. (10 cm) and lays another batch of eggs; thus a nest consists of 2–20 discrete egg clusters buried 3–8 in. (7–20 cm) beneath the sand surface. Males that are not attached to females also come ashore, crowd around the nesting couples, and eject sperm. Recent research using deoxyribonucleic acid (DNA) fingerprinting techniques has shown that unattached, satellite males often fertilize many of the eggs. After nesting for 10–60 min, the couple returns to the sea. They may reposition themselves lower on the beach as the tide recedes and continue nesting, or they may return to nest again on the following tide. After 2–4 weeks, the eggs hatch into nonfeeding larvae, which remain in the sand for a few more weeks until a tidal inundation occurs that is high enough to reach the nest. The now free-swimming trilobite larvae move to the surface and enter the ocean, where they soon molt into juvenile horseshoe crabs.

The advantages of nesting in the high intertidal were evaluated in an experimental manipulation. Clutches of eggs were dug up and systematically reburied (at a normal depth) just above, just below, and in the region of the beach in which crabs normally nested. Temperature, salinity, beach erosion, and predation by birds did not affect egg survival and development differentially in the various regions of the beach. Rather, oxygen levels in the sand and sand moisture were the two principal factors affecting egg success: eggs buried high on the beach became desiccated; eggs placed at the bottom of the beach developed slowly or died because of low oxygen levels. Thus, horseshoe crabs nest in the high intertidal region, because this is where the eggs survive and develop best. Other studies also demonstrate that female horseshoe crabs do not nest in areas with low oxygen levels, such as on sand covering peat beds or near sewer outflows, because nesting on these beaches results in reduced egg development. Females are thought to have oxygen-sensitive sensory receptors that help them avoid beaches unfavorable for their eggs.

The pattern of horseshoe crab nesting differs between the Delaware Bay and the Gulf Coast of Florida. In the Delaware Bay, crabs nest over the entire beach, whereas in Florida they nest only along a narrow strip of high intertidal sand. In part, this pattern can be explained by the higher tides in Delaware, which make it possible for crabs to reach the upper parts of the beach. More important, however, is the fact that a wider portion of the Delaware beach is conducive to egg development as compared with Florida. In Delaware, egg development can occur much lower on the beach since the oxygen concentration in the sand exceeds 1 part per million at 3 ft (1 m) from the bottom of the beach. In Florida, this minimal oxygen content occurs only in areas that are more than 10 ft (3 m) above the bottom of the beach. Differences in oxygen concentrations in the sand are caused by differences in beach morphology. The fine-grained sand on Florida beaches can hold more water and provide more surface area for microbial growth than the coarse-grained sand on Delaware beaches. Thus, differences in the pattern of nesting by crabs from one area to another reflect differences in their beach environments.

Delaware Bay and Florida Gulf Coast horseshoe crabs also differ in their pattern of nesting synchrony, which can also be explained in part by differences in beach morphology. In Florida, horseshoe crabs rarely spawn on neap tides (tides occurring between the new and full moons) because these tides remain below the aerobic zone of the beach. However, in Delaware, neap-tide nesting is common because these tides often inundate favorable regions of the beach. Crabs in Florida nest on both tides of a day (during spring tides), whereas in Delaware crabs prefer the higher of the two tides, which usually occurs at night. The two tides of a day are nearly equal in Florida, and both allow crabs to nest above the anoxic zone; but in Delaware only the higher of the two tides allows the full use of the beach. Crabs in Delaware generally nest during the receding portion of a tide, whereas those in Florida nest both as the tide comes in and as it goes out. Tidal inundations are smaller in Florida, and less of the beach is available for nesting; thus selection may favor crabs that arrive early. In Delaware the higher tidal amplitudes and greater availability of beach allow crabs to delay moving until the tide begins to go out; this strategy minimizes the chance that females will dig up each other's nests. Nonetheless, during the height of the nesting season the beaches of the Delaware Bay are strewn with horseshoe crab eggs. In fact, the migration of some shorebirds is timed to take advantage of the mass quantities of eggs available on the beach, and the birds gorge on them. In general, these eggs have not been eroded from the sand by tidal or wave action but have been dug up by female crabs nesting in the same stretch of beach on successive nights.

For background information *SEE TIDE; XIPHOSURIDA* in the McGraw-Hill Encyclopedia of Science & Technology.

H. Jane Brockmann

Bibliography. R. B. Barlow et al., Migration of *Limulus* for mating in relation to lunar phase, tide height, and sunlight, *Biol. Bull.,* 171:310–329, 1986;

M. L. Botton, R. E. Loveland, and T. R. Jacobsen, Beach erosion and geochemical factors: Influence on spawning success of horseshoe crabs (*Limulus polyphemus*) in Delaware Bay, *Mar. Biol.*, 99: 325–332, 1988; G. Castro and G. P. Myers, Shorebird predation on eggs of horseshoe crabs during spring stopover on Delaware Bay, *Auk*, 110:927–930, 1993; D. Penn and H. J. Brockmann, Nest-site selection in the horseshoe crab, *Limulus polyphemus, Biol. Bull.*, 187:373–384, 1994.

Zircon

Zircon ($ZrSiO_4$) is an extremely durable accessory mineral in igneous and metamorphic rocks, as well as a heavy mineral typically found in placer deposits of streams. For centuries zircons were mined from the river gravels of Sri Lanka and southeast Asia, providing gems of a wide variety of color and high brilliance. Zircon was studied because of changes in color, birefringence, and specific gravity on heating. The variation in physical and optical properties is the result of radiation damage caused by the alpha decay of nuclides in the uranium-238 (^{238}U), uranium-235 (^{235}U), and thorium-232 (^{232}Th) decay chains; and heating anneals this damage. Zircon can typically contain up to 5000 parts per million (ppm) uranium and lesser amounts of thorium. At saturation alpha-decay doses, zircon becomes amorphous, the metamict state, which was recognized 3 years before the discovery of radioactivity by H. Becquerel in 1896. The presence of uranium and thorium in zircon has led to numerous studies, as zircon is the most commonly used mineral in geologic age dating techniques, which measure the uranium/thorium/lead (Pb) nuclide ratios. Additionally, radiation effects in zircon have been extensively researched because of applications in the development of high-technology materials (such as optic waveguides in optoelectronic devices) and the development of solids for the immobilization of nuclear waste.

Geological dating. The widespread distribution of zircon in the continental crust, its tendency to concentrate trace elements (lanthanoids as well as actinoids), the low diffusion rates that slow loss of impurity nuclides (such as U, Pb, and rare-earth elements), and resistance to chemical and physical degradation have made zircon one of the most useful accessory minerals in geologic studies, providing a remarkable record of crustal processes. The most recent progress has come from the use of the sensitive high-mass-resolution ion microprobe (SHRIMP), a method that allows the measurement of isotopic ratios on areas as small as 20–30 micrometers, thus providing age dates on separate zones within single crystals of zircon. Detrital zircons in the metamorphosed sandstone, a quartzite, at Mount Narryer, Western Australia, have been dated at 4100–4300 million years ago (Ma), the

oldest terrestrial minerals yet found. The trace-element ratios and the rare-earth element distribution patterns are consistent with granitic parent rocks, although the parent rocks have not been located (perhaps completely removed by erosion). In Western Australia, the Jack Hills contain slightly younger (3900–4270 Ma) detrital zircons. The zircons in Australia are individual, recycled grains in a younger (3500 Ma) sequence of metamorphosed sedimentary rocks, as are similarly dated zircons from the Sino-Korean craton in northeast China ($\geq$ 3800 Ma). The oldest so-called intact crust (that is, the dated zircons are in the original rock) is found in the early Archean (3800–3960 Ma) granitoids (granitelike materials) in northwest Canada and western Greenland. These intact sequences confirm the presence of a normal, granitic crust as early as 4000 Ma. The accepted age of the Earth is 4600 Ma; thus, these results demonstrate that the continental crust had already begun to form very early in the Earth's history. The oldest zircons in the solar system are found as rare inclusions in meteorites, and by using the SHRIMP ion probe are dated at 4560 Ma.

Resistance to high pressure. Zircons that have been deformed by high-pressure impact, such as meteorite impact, are known as shocked zircons. The extraordinary ability of zircon to retain its U-Pb systematics has been recently demonstrated in studies of zircons from the Chicxulub impact structure (crater) of the Yucatán Peninsula. The shocked zircons were exhumed from the Chicxulub basement rock during meteorite impact and dispersed in the fine dust of the impact cloud. Discordancies in the U-Pb systematics (for example, Pb loss) are proportional to the extent of impact-induced shock textures, and isotopic resetting is consistent with partial lead loss at the time of impact (65 Ma), thus providing convincing support for the meteorite impact origin of the Chicxulub crater and its being the source of ejected material found at the Cretaceous-Tertiary (K-T) boundary in North America. Uranium-lead dating studies of shocked zircons in the fine-grained ejecta, deposited in areas as wide apart as Colorado and Saskatchewan, at the Cretaceous-Tertiary boundary have a predominant age of 545 Ma. These zircon ages are in agreement with dates for shocked zircons from the Chicxulub crater and from Beloc, Haiti.

The remarkable resistance of zircon to high-pressure events and zircon's ability to retain its U-Pb nuclide systematics have been confirmed experimentally and empirically. Zircons experimentally shocked to extremely high pressures (up to 59 gigapascals or 590 kilobars) show no evidence for fractionation or loss of lead. SHRIMP analysis of zircons from the Kokchetav massif in northern Kazakhstan indicates that they have retained their original isotopic compositions, which yield ages of 2000 Ma (although diffusive lead loss had commenced). This is despite a later ultrahigh-pressure metamorphic event that occurred at 530 Ma at depths as great as 125 km (75 mi; at 4 GPa or 40 Kbar, and 900–1000°C or (1650–1830°F)) and resulted in the formation of diamond-bearing gneisses.

Immobilization of plutonium. One of the most recent applications of zircon is to the disposition and disposal of plutonium recovered from dismantled nuclear weapons. Under the first and second Strategic Arms Reduction treaties, as well as unilateral pledges made by the United States and Russia, several thousand nuclear weapons will be dismantled. Thus, an estimated 100 metric tons of excess weapons plutonium (essentially ^{239}Pu) will require long-term disposition. The disposal strategy should be designed not only to protect the public and the environment but also to ensure that the plutonium is not readily recoverable for remanufacture of weapons. Any option that utilizes geologic disposal will require that the plutonium be incorporated into a highly durable solid that will prevent its release [because of its high toxicity and long half-life (24,500 years for ^{239}Pu)] for periods of at least 250,000 years, and will also prevent the remobilization and concentration of plutonium to a critical mass that could sustain spontaneous nuclear fission reactions.

Plutonium is readily incorporated into the zircon structure, which consists of independent SiO_4 tetrahedral monomers linked across edges to triangular dodecahedral ZrO_8 groups (see **illus.**). The actinide elements (U, Th, Pu) can substitute for the zirconium (Zr). Compositions of $XSiO_4$ (X^{4+} = zirconium, hafnium, thorium, protactinium, uranium, neptunium, plutonium, and americium) were synthesized in the early 1960s; most recently, radiation effects in natural zircons and in synthetic zircon with 10 wt % Pu have been thoroughly studied. The fact that a pure, end-member composition, $PuSiO_4$, exists suggests extensive substitution of plutonium for zirconium is possible. Additionally, neutron absorbers, such as hafnium and gadolinium, are easily incorporated into the zircon structure; thus, the possibility of criticality events (in which nuclear fission reactions are self-sustaining) is reduced. The affinity of the zircon structure for actinides has been demonstrated by Russian scientists, who have identified zircon [with up to 10 wt % uranium dioxide (UO_2)] as a primary crystalline phase in the Chernobyl lavas, the melted core of the reactor. The zircon formed during the reaction between the melting nuclear fuel and construction materials (such as zirconium metal, concrete, and sand).

Based on the previous, extensive studies of zircon, a number of the issues inevitably raised in the evaluation of a nuclear waste form are already understood. Radiation damage studies demonstrate that zircon undergoes a radiation-induced transformation from the periodic to aperiodic state (metamict state) at doses over the range of 10^{18}–10^{19} alpha-decay events per gram [0.2–0.6 displacement per atom (dpa)], with a density decrease and a corresponding volume expansion of 18%.

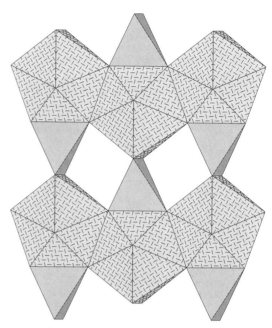

Atomic structure of zircon. The large zirconium (Zr) ion is surrounded by eight oxygens (O) at the corners of a triangular dodecahedron. The smaller silicon (Si) atoms are surrounded by four oxygen atoms (shaded areas) at the corners of a tetrahedron. The ZrO_8 and SiO_4 polyhedra are joined across shared edges to form the sheet shown.

The displacements per atom value is a measure of the number of atoms actually displaced (moved) from their structural positions by nuclear collisions. These studies are unique, as they include the analysis of natural zircons that have accumulated alpha-decay-event damage up to doses of nearly 0.7 dpa over 550 million years; ^{238}Pu-doped zircons, up to doses of 0.7 dpa in 6.5 years; and heavy-ion irradiations, to doses of 0.2–2.3 dpa in less than 1 h. These experiments cover a dose rate range of $>10^8$ and provide a firm basis for predicting the microstructure of the radiation-damaged zircon as a function of dose and temperature. The chemical durability has been confirmed by laboratory studies, as well as by zircon's natural occurrence in rocks of great age. At low temperatures (<80°C or 180°F) and near-neutral pH values (conditions pertinent to nuclear waste disposal conditions), zircon is extremely insoluble. Estimated leach rates may be as low as 10^{-14} g/(cm^2)(day). Such a low leach rate results in essentially no loss of plutonium for periods of up to 250,000 years (10 half-lives of ^{239}Pu).

Criticality concerns can be addressed by reducing the fissile plutonium loading in the zircon or by including neutron absorbers, such as hafnium and gadolinium, which are readily substituted for zirconium in the zircon structure. A fissile atom undergoes fission; that is, it splits into two fragments plus a few neutrons. Not all elements are fissile, but ^{235}U and ^{239}Pu are. Processing and production technologies have already been demonstrated in Japan, and zircon can be synthesized by sintering (1200–1300°C or 2200–2400°F). Zircon yields of nearly 90% are obtained in less than 1 day. In the case of weapons plutonium, the high purity of weapons-grade ^{239}Pu may be used to advantage in engineering the synthesis and properties of the waste form. Thus, zircon is a unique phase whose natural occurrence as an actinide-bearing mineral combined with extensive laboratory studies of natural and synthetic zircons supports the use of zircon for the long-term, safe immobilization of plutonium.

For background information SEE METAMICT STATE; METEORITE; PLUTONIUM; RADIOACTIVITY; THORIUM; ZIRCON; ZIRCONIUM in the McGraw-Hill Encyclopedia of Science & Technology.

Rodney C. Ewing

Bibliography. E. B. Anderson, B. E. Burakov, and E. M. Pazukhin, High-uranium zircon from "Chernobyl lavas," *Radiochim. Acta,* 60:149–151, 1993; B. F. Bohor, W. J. Betterton, and T. E. Krogh, Impact-shocked zircons: Discovery of shock-induced textures reflecting increasing degrees of shock metamorphism, *Earth Planet. Sci. Lett.,* 119:419–424, 1993; R. C. Ewing, W. Lutze, and W. J. Weber, Zircon: A host-phase for the disposal of weapons plutonium, *J. Mater. Res.,* 10(2):243–246, 1995; S. L. Kamo and T. E. Krogh, Chicxulub crater source for shocked zircon crystals from the Cretaceous-Tertiary boundary layer, Saskatchewan: Evidence from new U-Pb data, *Geology,* 23:281–284, 1995; A. C. McLaren, J. D. Fitz Gerald, and I. S. Williams, The microstructure of zircon and its influence on the age determination from Pb/U isotopic ratios measured by an ion microprobe, *Geochim. Cosmochim. Acta,* 58:993–1005, 1994; M. Roland et al., The Earth's oldest known crust: A geochronological and geochemical study of 3900–4200 Ma old detrital zircons from Mt. Narryer and Jack Hills, Western Australia, *Geochim. Cosmochim. Acta,* 56:1281–1300, 1992; W. J. Weber, R. C. Ewing, and L.-M. Wang, The radiation-induced crystalline-to-amorphous transition in zircon, *J. Mater. Res.,* 9(3):688–698, 1994.

Contributors

The affiliation of each Yearbook contributor is given, followed by the title of his or her article. An article title with the notation "in part" indicates that the author independently prepared a section of an article; "coauthored" indicates that two or more authors jointly prepared an article or section.

A

Abbott, Dr. Andrew P. *Department of Physical Chemistry, University of Leicester, England.* ELECTROOPTIC MATERIALS.

Abel, John D. *National Association of Broadcasters, Washington, D.C.* RADIO BROADCASTING—in part.

Agee, Prof. James K. *Division of Ecosystem Science and Conservation, College of Forest Resources, University of Washington, Seattle.* FOREST FIRES.

Andersen, Dr. Lars H. *Institute of Physics and Astronomy, University of Århus, Denmark.* SCATTERING EXPERIMENTS (ATOMS AND MOLECULES).

Ansell, Dr. Alan D. *The Scottish Association for Marine Science, Argyll, Scotland.* DECAPODA (CRUSTACEA).

Ashby, Prof. Neil. *Department of Physics, University of Colorado, Boulder.* RELATIVITY.

B

Barbosa-Cánovas, Prof. Gustavo V. *Department of Biological Systems Engineering, Washington State University, Pullman.* FOOD MICROBIOLOGY—coauthored.

Basset, Dr. Alberto. *Department of Animal Biology and Ecology, University of Cagliari, Italy.* POPULATION ECOLOGY.

Behnke, Dr. Richard. *Division of Atmospheric Sciences, National Science Foundation, Arlington, Virginia.* SPACE WEATHER.

Bendinelli, Prof. Mauro. *Director, Retrovirus Center, Department of Biomedicine, University of Pisa, Italy.* FELINE IMMUNODEFICIENCY VIRUS.

Berger, Prof. Stanley A. *Department of Mechanical Engineering, University of California, Berkeley.* VORTEX.

Berta, Prof. Annalisa. *Department of Biology, San Diego State University, California.* PINNIPEDS.

Beusse, Dr. Deke O. *Avian, Wildlife, Exotic Medicine, Oviedo, Florida.* MARINE MAMMAL MEDICINE.

Bhattacharjee, Prof. Amitava. *Department of Physics and Astronomy, The University of Iowa, Iowa City.* LASER—in part.

Billet, Dr. Michael L. *Head, Fluid Dynamics and Turbomachinery Department, Pennsylvania State University, State College.* CAVITATION.

Billig, Dr. Frederick S. *Applied Physics Laboratory, The Johns Hopkins University, Laurel, Maryland.* TURBORAMJET.

Birge, Dr. Robert R. *W. M. Keck Center for Molecular Electronics, Center for Science and Technology, Syracuse University, New York.* BIOELECTRONICS.

Birmingham, Dr. Jeannette M. *Mycology and Protistology Program, American Type Culture Collection, Rockville, Maryland.* NUTRACEUTICALS.

Blanke, Dr. Jens. *Brain Research Institute, University of Bremen, Germany.* VERTEBRATES—coauthored.

Bowick, Prof. Mark J. *Department of Physics, Syracuse University, New York.* PHASE TRANSITIONS.

Boyd, Prof. Richard N. *Associate Dean, College of Mathematical and Physical Sciences, Ohio State University, Columbus.* NUCLEOSYNTHESIS.

Bright, Prof. Frank V. *Department of Chemistry, University of Buffalo, State University of New York.* BIOSENSOR—coauthored.

Briskman, Robert D. *President, CD Radio, Inc., Washington, D.C.* RADIO BROADCASTING—in part.

Brocke, Dr. Stefan. *National Institutes of Health, National Institute of Neurological Disorders and Stroke, Bethesda, Maryland.* MULTIPLE SCLEROSIS—coauthored.

Brockmann, Prof. Jane H. *Department of Zoology, University of Florida, Gainesville.* XIPHOSURIDA.

Brody, Dr. Aaron L. *Managing Director, Rubbright Brody, Inc., Duluth, Georgia.* FOOD ENGINEERING—in part.

Bruckner, Prof. Adam P. *Aerospace and Energetics Research Program, University of Washington, Seattle.* RAM ACCELERATOR.

Brykczynski, Bill. *Computer and Software Engineering Division, Institute for Defense Analyses. Alexandria, Virginia.* SOFTWARE QUALITY ASSURANCE.

Bucher, Dr. Martin A. *Department of Physics, Princeton University, New Jersey.* COSMOLOGY—in part.

Buchleiter, Dr. Gerald W. *Agricultural Engineer, U.S. Department of Agriculture, Agricultural Research Service, Colorado State University, Fort Collins.* IRRIGATION (AGRICULTURE)—in part.

Buede, Dr. Dennis M. *Department of Systems Engineering, George Mason University, Fairfax, Virginia.* FUNCTIONAL ANALYSIS DIAGRAMS.

Bullock, Richard L. *Richard L. Bullock Associates, Las Vegas, Nevada.* UNDERGROUND MINING.

Burgdörfer, Dr. Joachim. *Department of Physics, University of Tennessee, Knoxville, and Oak Ridge National Laboratory.* PHOTOIONIZATION.

Bürgmann, Prof. Roland. *Department of Geology, University of California, Davis.* EARTH CRUST.

Butler, Prof. P. J. *The University of Birmingham, School of Biological Sciences, Edgbaston, United Kingdom.* BIRD MIGRATION.

C

Cannon, Prof. M. Elizabeth. *Department of Geomatics Engineering, University of Calgary, Alberta, Canada.* SATELLITE NAVIGATION SYSTEMS.

Capasso, Dr. Federico. *AT&T Laboratories, Murray Hill, New Jersey.* LASER—coauthored.

Carney, Dr. Bruce W. *Samuel Baron Professor of Astronomy, Department of Physics and Astronomy, University of North Carolina, Chapel Hill.* STAR CLUSTERS.

Catalan, Kemal V. *Department of Chemistry, Michigan State University, East Lansing.* BIOINORGANIC CHEMISTRY—coauthored.

Chase, Scott. *Publisher, Via Satellite, Potomac, Maryland.* COMMUNICATIONS SATELLITE.

Chen, Dr. Shoei-Sheng. *Argonne National Laboratory, U.S. Department of Energy, The University of Chicago, Illinois.* VIBRATION.

Chomski, Emmanuel. *Materials Chemistry Research Group, Lash Miller Chemical Laboratories, University of Toronto, Canada.* NANOSTRUCTURE—coauthored.

Clark, Dr. R. Nolan. *Laboratory Director, U.S. Department of Agriculture, Conservation and Production Research Laboratory, Bushland, Texas.* WIND POWER.

Clark, Dr. William R. *Professor of Immunology, Department of Biology, University of California, Los Angeles.* CYTOLYSIS.

Claudy, Lynn D. *National Association of Broadcasters, Washington, D.C.* RADIO BROADCASTING—in part.

Collins, Prof. Daniel J. *Aeronautics and Astronautics Department, Naval Postgraduate School, Monterey, California.* NEURAL NETWORK.

Collins, Dr. Steven J. *Department of Clinical Neurosciences, St. Vincent's Hospital, Melbourne and Department of Pathology, The University of Melbourne, Australia.* PRION DISEASE—coauthored.

Cook, Stephen. *Marshall Space Flight Center, Alabama.* LAUNCH VEHICLE—in part.

Cory, Dr. Jenny S. *Natural Environment Research Council, Institute of Virology and Environmental Microbiology, Oxford, United Kingdom.* INSECT CONTROL, BIOLOGICAL—in part.

Courchesne, Prof. Eric. *Department of Neurosciences, School of Medicine, University of California, San Diego, and Director, Autism and Brain Development Research Laboratory, San Diego Children's Hospital.* BRAIN—in part.

Crawmer, Dr. Daryl E. *Chief Scientist, Miller Thermal, Inc., Appleton, Wisconsin.* METAL COATINGS.

Cressey, Prof. Barbara A. *Department of Geology, University of Southhampton, United Kingdom.* CHRYSOTILE.

Crossley, Dr. D. A., Jr. *Research Professor, Institute of Ecology, The University of Georgia, Athens.* FOREST SOIL.

Currey, John D. *Department of Biology, The University of York, United Kingdom.* BONE.

Cutler, Dr. Alan H. *Visiting Scientist, Department of Paleobiology, National Museum of Natural History, Smithsonian Institution. Washington, D.C.* EXTINCTION (BIOLOGY).

D

Davis, Dr. Thomas L. *Department of Geophysics, Colorado School of Mines, Golden.* PETROLEUM RESERVOIR ENGINEERING.

Delgado, Dr. Pedro L. *Director of Research, Department of Psychiatry, The University of Arizona, College of Medicine, Tucson.* SEROTONIN.

Dembinski, Dr. Roman. *Department of Chemistry, The University of Utah, Salt Lake City.* ORGANOMETALLIC CHEMISTRY—coauthored.

Dent, Dr. David R. *School of Pure and Applied Biology, University of Wales, Cardiff.* INSECT CONTROL, BIOLOGICAL—in part.

Di Rosa, Dr. Francesca. *Department of Health and Human Services, National Institutes of Health, Bethesda, Maryland.* IMMUNOLOGICAL MEMORY—coauthored.

DiVincenzo, Dr. David P. *IBM Research Division, Thomas J. Watson Research Center, Yorktown Heights, New York.* COMPUTER.

Dixit, Dr. Suhrir S. *NYNEX Science & Technology, Inc., Framingham, Massachusetts.* TELEVISION.

Dodds, Dr. W. Jean. *President, HemoPet, Santa Monica, California.* GENETIC DISEASE.

Dorey, Prof. A. P. *Department of Engineering, University of Lancaster, United Kingdom.* MECHATRONICS.

Dorey, Prof. Howard. *Department of Electrical and Electronic Engineering, Imperial College of Science, Technology and Medicine, London, United Kingdom.* MICROENGINEERING.

Dugan, Dr. Frank M. *Mycology and Botany, American Type Culture Collection, Rockville, Maryland.* BIOLOGICAL PEST CONTROL.

Dunbar, Kim R. *Department of Chemistry, Michigan State University, East Lansing.* BIOINORGANIC CHEMISTRY—coauthored.

Dunn, Prof. Robert C. *Department of Chemistry, The University of Kansas, Lawrence.* NEAR-FIELD MICROSCOPY.

Dushay, Dr. Mitchell. *Department of Molecular Biology, Stockholm University, Sweden.* INSECT PHYSIOLOGY.

Dwyer, Dr. John F. *U.S. Department of Agriculture, Forest Service, Evanston, Illinois.* URBAN FORESTRY—coauthored.

E

Eberts, Dr. Ray E. *School of Industrial Engineering, Purdue University, West Lafayette, Indiana.* INDUSTRIAL ENGINEERING.

Elias, Dr. Antonio L. *Orbital Sciences Corporation, Dulles, Virginia.* LAUNCH VEHICLE—in part.

Ellman, Dr. Jonathan A. *Department of Chemistry, University of California, Berkeley.* COMBINATORIAL CHEMISTRY—coauthored.

Ewing, Dr. Rodney C. *Department of Earth and Planetary Sciences, University of New Mexico, Albuquerque.* ZIRCON.

F

Faghri, Prof. Amir. *Head, Department of Mechanical Engineering, University of Connecticut, Storrs.* FLUID FLOW—in part.

Faist, Dr. Jerome. *AT&T Bell Laboratories, Murray Hill, New Jersey.* LASER—coauthored.

Faulkner, Dr. D. John. *Scripps Institution of Oceanography, La Jolla, California.* MARINE NATURAL PRODUCTS.

Federline, Gregory E. *Hughes Network Systems, Germantown, Maryland.* DATA COMMUNICATIONS.

Feduccia, Dr. Alan. *Department of Biology, University of North Carolina, Chapel Hill.* AVES.

Flammang, Dr. Patrick. *Laboratoire de Biologie Marine, Université de Mons, Hainaut, Belgium.* ECHINODERMATA.

Fleeman, Eugene L. *Rockwell International Corporation, Tactical Systems Division, Duluth, Georgia.* MISSILE.

Fletcher Dr. Thomas D. *Intel Corporation, Hillsboro, Oregon.* MICROPROCESSOR.

Forrest, Dr. David R. *Research Specialist, Allegheny Ludlum Corporation, Technical Center, Brackenridge, Pennsylvania.* MOLECULAR NANOTECHNOLOGY.

Freedman, Dr. Wendy. *The Observatories of the Carnegie Institution of Washington, Pasadena, California.* UNIVERSE—coauthored.

Freeman, Dr. Richard R. *Department Head, Research Division, AT&T Bell Laboratories, Murray Hill, New Jersey.* MATTER IN STRONG FIELDS.

Fritz, Douglas J. *O'Brien-Kreitzberg & Associates, Inc., Professional Construction Managers, Pennsauken, New Jersey.* URBAN INFRASTRUCTURE—in part.

Fry, Prof. William E. *Department of Plant Pathology, Cornell University, College of Agriculture and Life Sciences, Ithaca, New York.* POTATO, IRISH.

Fryrear, Dr. Donald W. *Location Leader, U.S. Department of Agriculture, Agricultural Research Service, The Wind Erosion Management Unit, Big Spring, Texas.* AIR POLLUTION.

G

German, Dr. Randall M. *Brush Chair Professor in Materials, Engineering Science and Mechanics Department, The Pennsylvania State University, University Park.* METAL INJECTION MOLDING.

Gibson, Dr. A. A. P. *Department of Electrical Engineering and Electronics, University of Manchester Institute of Science and Technology, United Kingdom.* VARIATIONAL METHODS.

Gladysz, Dr. John A. *Department of Chemistry, The University of Utah, Salt Lake City.* ORGANOMETALLIC CHEMISTRY—coauthored.

Gomez-López, Marcos. *Department of Chemistry, University of Birmingham, Edgbaston, United Kingdom.* SUPRAMOLECULAR CHEMISTRY—coauthored.

Gouet, Dr. Patrice. *Laboratory for Molecular Biophysics, Oxford, United Kingdom.* VIROLOGY—coauthored.

Graham, Dr. Bill. *Department of Pure and Applied Physics, Queen's University of Belfast, United Kingdom.* PLASMA PHYSICS.

Graves, Dr. Jennifer A. Marshall. *Department of Genetics and Human Variation, La Trobe University, Bundoora, Victoria, Australia.* CHROMOSOME—in part.

Gray, Prof. Dennis J. *Institute of Food and Agricultural Sciences, Central Florida Research and Education Center, University of Florida, Leesburg.* BREEDING (PLANT).

Green, Dr. Paul J. *Center for Astrophysics, Harvard College Observatory and Smithsonian Astrophysical Observatory, Cambridge, Massachusetts.* COSMOLOGY—in part.

Grivas, Prof. Dimitri A. *Department of Civil and Electrical Engineering, Center for Infrastructure and Transportation Studies, Rensselaer Polytechnic Institute, Troy, New York.* ROAD PAVEMENTS.

Gunapala, Dr. Sarath D. *Jet Propulsion Laboratory, California Institute of Technology, Pasadena.* INFRARED RADIATION.

H

Hamed, Prof. Awatef. *Department of Aerospace Engineering and Engineerng Mechanics, University of Cincinnati, Ohio.* EROSION.

Harrison, Prof. D. Jed. *Department of Chemistry, University of Alberta, Edmonton, Canada.* CAPILLARY ELECTROPHORESIS—in part.

Hartley, Prof. Alan A. *Department of Psychology, Scripps College, Claremont, California.* BRAIN—in part.

Heller, Dr. Rachael F. *Department of Pathology, Mount Sinai School of Medicine, New York.* OBESITY—coauthored.

Heller, Dr. Richard F. *Department of Pathology, Mount Sinai School of Medicine, New York.* OBESITY—coauthored.

Hensler, Dr. Patrick J. *Division of Molecular Virology, Baylor College of Medicine, Houston, Texas.* CELL (BIOLOGY).

Hewell, James. *Director of Sales and Marketing, Cantrell Machine Company, Inc., Gainsville, Georgia.* FOOD ENGINEERING—coauthored.

Hillenius, Dr. Willem J. *Department of Biology, University of California, Los Angeles.* MAMMALIA.

Hines, Dr. Anson H. *Smithsonian Environmental Research Center, Edgewater, Maryland.* CRAB.

Hoeppner, Prof. David W. *Department of Mechanical Engineering, The University of Utah, Salt Lake City.* CORROSION.

Holland, Dr. Linda Z. *Scripps Institute of Oceanography, University of California, San Diego, La Jolla.* ANIMAL EVOLUTION.

Hopp, Dr. Wallace J. *Department of Industrial Engineering and Management Sciences, Northwestern University, Evanston, Illinois.* MANUFACTURING ENGINEERING—coauthored.

Hoppe, Walter H. *Laser Perfect, Inc., Warrensville Heights, Ohio.* MANUFACTURING ENGINEERING—in part.

Horvath, Dr. Istvan T. *Exxon Research and Engineering Company, Corporate Research Chemical Sciences Laboratory, Annandale, New Jersey.* ORGANOMETALLIC CHEMISTRY—in part.

I

Inagaki, Dr. M. *Department of Applied Chemistry, Faculty of Engineering, Hokkaido University, Sapporo, Japan.* COMPOSITE MATERIAL—in part.

Ingersoll, Dr. Christine M. *Department of Chemistry, University of Buffalo, State University of New York.* BIOSENSOR—coauthored.

J

Jackson, Dr. Charles L. *Strategic Policy Research, Bethesda, Maryland.* INFORMATION SUPERHIGHWAY.

Jacquez, Dr. Geoffrey M. *President, BioMedWare, Ann Arbor, Michigan.* MEDICAL GEOGRAPHY.

Jaouen, Prof. Gerard. *National School of Chemistry of Paris.* BIOINORGANIC CHEMISTRY—coauthored.

Jett, Dr. Jackson B. *Associate Dean for Research, College of Forest Resources, North Carolina State University, Raleigh.* WOOD—coauthored.

Jorden, Dr. Paul R. *Royal Greenwich Observatory, Cambridge, United Kingdom.* CHARGED-COUPLED DEVICES.

Joshi, Prof. Chandrashekhar. *Electrical Engineering Department, University of California, Los Angeles.* LASER—in part.

Jurgen, Ronald K. *Fort Lauderdale, Florida.* AUTOMOBILE.

K

Kachmar, Peter M. *C.S. Draper Laboratory, Cambridge, Massachusetts.* SPACE NAVIGATION AND GUIDANCE.

Kaiser, Dr. Walter J. *Research Plant Pathologist, U.S. Department of Agriculture, Agricultural Research Service, Regional Plant Introduction Station, Washington State University, Pullman.* LENTIL—coauthored.

Karlen, Dr. Douglas L. *Research Soil Scientist, U.S. Department of Agriculture, Agricultural Research Service, National Soil Tilth Laboratory, Ames, Iowa.* CROP ROTATION.

Kartha, Dr. Sivan. *Center for Engineering and Environmental Studies, Princeton University, New Jersey.* FUEL CELL.

Kear, Dr. Amanda J. *Department of Geology, University of Bristol, United Kingdom.* CEPHALOPODA.

Kelly, Dr. Robert J. *Baltimore, Maryland.* AIR NAVIGATION.

Kennedy, Prof. Robert T. *Department of Chemistry, University of Florida, Gainesville.* CYTOCHEMISTRY.

Klein, Dr. George. *Karolinska Institute, Microbiology and Tumorbiology Center, Stockholm, Sweden.* CANCER (MEDICINE).

Ko, Prof. Edmond I. *Department of Chemical Engineering, Carnegie Mellon University, Pittsburgh.* AEROGEL.

Korobko, Prof. E. V. *Luikov Heat and Mass Transfer Institute, Belarus Academy of Sciences, Minsk, Belarus, Russia.* FLUID FLOW—coauthored.

Kosnik, Kristine E. *Environmental Scientist/Cartographer, Applied Environmental Consultants, Inc., Tempe, Arizona.* SNOWPACK MONITORING.

Kraus, Per. *Department of Physics, Princeton University, New Jersey.* BLACK HOLE—in part.

Krieg, Dr. Arthur M. *Department of Internal Medicine, The University of Iowa College of Medicine, Iowa City.* INFECTION.

Kronfeld, Dr. David S. *The Paul Mellon Distinguished Professor of Agriculture and Professor of Veterinary Medicine, Department of Animal and Poultry Sciences, Virginia Polytechnic Institute and State University, Blacksburg.* CATTLE.

Kubas, Dr. Gregory J. *Department of Chemistry, Los Alamos National Laboratory, New Mexico.* ORGANOMETALLIC CHEMISTRY—in part.

Kutas, Dr. Marta. *Department of Cognitive Science, University of California, San Diego.* LANGUAGE PROCESSING.

L

Larish, John. *Eastman Kodak, Rochester, New York.* DIGITAL CAMERA.

Lehrer, Dr. Robert I. *Department of Medicine, University of California, Los Angeles.* ANTIBIOTIC.

Lieber, Prof. Charles M. *Department of Chemistry, Harvard University, Cambridge, Massachusetts.* CHEMICAL FORCE MICROSCOPY.

Liou, Prof. Ming. *Department of Electrical and Electronic Engineering, Hong Kong University of Science and Technology, Kowloon.* DATA COMPRESSION—in part.

Lloyd, Prof. David K. *Department of Oncology, McGill University and Meakins-Christie Laboratories, Montreal, Canada.* CAPILLARY ELECTROPHORESIS—in part.

Lockley, Dr. Martin G. *Department of Geology, University of Colorado at Denver.* DINOSAUR.

Lofereski, Prof. Joseph J. *Engineering Division, Brown University, Providence, Rhode Island.* SOLAR CELL.

Loftus, Elizabeth F. *Department of Psychology, University of Washington, Seattle.* MEMORY—coauthored.

Lowman, Dr. Margaret. *Director of Research, Stark Research Center, The Marie Selby Botanical Gardens, Sarasota, Florida.* FOREST CANOPY.

Lucchesi, Dr. John C. *Asa G. Candler Professor and Chair, Department of Biology, Emory University, Atlanta, Georgia.* CHROMOSOME—in part.

Lunte, Prof. Craig E. *Department of Chemistry, University of Kansas, Lawrence.* MICRODIALYSIS SAMPLING.

M

McCully, Dr. Kilmer S. *Department of Veterans' Affairs, Medical Center, Providence, Rhode Island.* ARTERIOSCLEROSIS.

McMenamin, Prof. Mark A. S. *Department of Geography and Geology, Mount Holyoke College, South Hadley, Massachusetts.* FOSSIL.

McMurray, Prof. David N. *Department of Medical Microbiology and Immunology, Texas A&M University, College Station.* TUBERCULOSIS.

Madore, Dr. Barry F. *Infrared Processing and Analysis Center, California Institute of Technology, Pasadena.* UNIVERSE—coauthored.

Maki, Dr. Sonja L. *Department of Horticulture, College of Agricultural Sciences, Clemson University, South Carolina.* PLANT DEVELOPMENT.

Manning, Dr. Charles G. *Department of Psychology, University of Washington, Seattle.* MEMORY—coauthored.

Marshall, Prof. Charles. *Department of Earth and Space Sciences, University of California, Los Angeles.* PALEONTOLOGY.

Masters, Dr. Colin L. *Head of the National Creutzfeldt-Jakob Disease Registry, Department of Pathology, The University of Melbourne, Australia.* PRION DISEASE—coauthored.

Mathai, Dr. Anna. *Department of Physics, Center for Superconductivity Research, University of Maryland at College Park.* SUPERCONDUCTIVITY.

Matzinger, Dr. Polly. *Department of Health and Human Services, National Institutes of Health, Bethesda, Maryland.* IMMUNOLOGICAL MEMORY—coauthored.

Miller, Prof. Arnold I. *Department of Geology, University of Cincinnati, Ohio.* EVOLUTION.

Moran, Prof. James. *Center for Astrophysics, Harvard College Observatory, Smithsonian Astrophysical Observatory, Cambridge, Massachusetts.* BLACK HOLE—in part.

Morgan, Dr. Vincent T. *Division of Applied Physics, National Measurement Laboratory, Sydney, Australia.* CONDUCTOR (ELECTRICITY).

Morton, Prof. Joseph B. *Chairman, Division of Plant and Environmental Microbiology, West Virginia University, Morgantown.* MYCORRHIZAE.

Moseley, Dr. Michael E. *Department of Radiology, Lucas Center for MRS and MRI, Stanford University, California.* MAGNETIC RESONANCE IMAGING.

Muehlbauer, Dr. Fred J. *U.S. Department of Agriculture, Agricultural Research Service, Regional Plant Introduction Station, Washington State University, Pullman.* LENTIL—coauthored.

Murray Dr. Keith. *Head, CSIRO Australian Animal Health Laboratory, Victoria.* EQUINE MORBILLIVIRUS.

N

Nilsson, Stig L. *Technical and Business Consultant, Los Gatos, California.* ELECTRIC POWER TRANSMISSION.

Nowak, Dr. David J. *Research Forester, U.S. Department of Agriculture, Forest Service, SUNY College of Environmental Science and Forestry, Syracuse, New York.* URBAN FORESTRY—coauthored.

O

Olhoeft, Prof. Gary R. *Department of Geophysics, Colorado School of Mines, Golden.* ENVIRONMENTAL GEOPHYSICS.

Olsen, Dr. Bjorn R. *Hershey Professor of Cell Biology, Department of Biology, Harvard Medical School, Boston, Massachusetts.* SKELETAL SYSTEM.

Olsen, Prof. Robert G. *School of Electrical Engineering and Computer Science, Washington State University, Pullman.* ELECTROMAGNETIC FIELD.

Ozin, Prof. Geoffrey A. *Materials Chemistry Research Group, Lash Miller Chemical Laboratories, University of Toronto, Canada.* NANOSTRUCTURE—coauthored.

Rom, Prof. Josef. *Department of Aerospace Engineering, Technion, Israel Institute of Technology, Haifa.* AERO-DYNAMICS.

Rosen, Dr. Mordecai D. *X-Division Leader, Department of Physics, Lawrence Livermore National Laboratory, University of California, Livermore.* NUCLEAR FUSION.

Roth, Dr. Gerhard. *Brain Research Institute, University of Bremen, Germany.* VERTEBRATES—coauthored.

P

Palmer, Dr. Fred L. *Research Laboratory of Electronics, Massachusetts Institute of Technology, Cambridge.* MASS SPECTROMETRY—coauthored.

Parenteau, Dr. Nancy L. *Organogenesis, Canton, Massachusetts.* TRANSPLANTATION BIOLOGY.

Pascarelli, Dr. Emil F. *Miller Institute, St. Lukes-Roosevelt Hospital, New York, New York.* OCCUPATIONAL HEALTH AND SAFETY.

Pavlowsky, Dr. Robert T. *Department of Geography, Carthage College, Kenosha, Wisconsin.* RIVER.

Pepper, Dr. Darrell W. *Department of Mechanical Engineering, University of Nevada, Las Vegas.* ENVIRON-MENTAL FLUID DYNAMICS.

Pinkowshi, Brian J. *Denver, Colorado.* RISK ASSESSMENT.

Plunkett, Dr. Matthew J. *Department of Chemistry, University of California, Berkeley.* COMBINATORIAL CHEM-ISTRY—coauthored.

Power, Dr. J. F. *Research Leader, U.S. Department of Agriculture, Agricultural Research Service, Soil and Water Conservation Research Unit, University of Nebraska, Lincoln.* COVER CROPS.

Pritchard, Prof. David E. *Research Laboratory of Electronics, Massachusetts Institute of Technology, Cambridge.* MASS SPECTROMETRY—coauthored.

Pugh, Dr. David T. *Natural Environment Research Council, Southhampton Oceanography Center, United Kingdom.* SEA-LEVEL CHANGE.

Q

Qin, Bai-Lin. *Department of Biological Systems Engineering, Washington State University, Pullman.* FOOD MICROBIOLOGY—coauthored.

Quarles, Dr. John M. *Department of Medical Microbiology and Immunology, Texas A&M University, Health Science Center, College Station.* HEPATITIS.

Quarles, Dr. William. *Bio-Integral Resource Center, Berkeley, California.* INSECT CONTROL, BIOLOGICAL—in part.

R

Raymer, Daniel P. *Conceptual Research Corporation, Sylmar, California.* AIRCRAFT DESIGN.

Rego, Dr. Vernon. *Department of Computer Sciences, Purdue University, West Lafayette, Indiana.* SIMULA-TION.

Rodriquez, Dr. Emilio, Jr. *Department of Microbiology and Immunology, University of Illinois at Chicago.* MICROBIAL MINING—coauthored.

S

Samuel, Prof. James E. *Department of Medical Microbiology and Immunology, Texas A&M University, Health Science Center, College Station.* ULCER.

Sandor, Jonathan A. *Department of Agronomy, Iowa State University of Science and Technology, Ames.* SOIL.

Saykally, Prof. Richard J. *Department of Chemistry, University of California, Berkeley.* CARBON—coauthored.

Schmuck, Carl H. *Mining Engineer, Department of Labor, Mines Safety and Health Administration, Denver, Colorado.* SOLUTION MINING.

Schneider, Dr. Dieter H. *Lawrence Livermore National Laboratory, University of California, Livermore.* ATOMIC PHYSICS.

Schwartz, Mel. *Consultant, Madison, Connecticut.* COM-POSITE MATERIAL—in part; HYBRID MATERIALS.

Scott, Prof. James F. *Dean of Science, University of New South Wales, Sydney, Australia.* COMPUTER STORAGE TECHNOLOGY.

Seible, Prof. Douglas J. *Division of Structural Engineering, University of California, San Diego.* ULTRA INFRA-STRUCTURE—in part.

Senter, Dr. Peter. *Bristol-Myers Squibb, Pharmaceutical Research Institute, Seattle, Washington.* BIOCONJUGA-TION.

Serikawa, Dr. Kyle A. *Department of Plant Biology, University of California, Berkeley.* GENE.

Sheldon, Ray W. *Director of Development, SynCoal, Rosebud SynCoal Partnership, Billings, Montana.* COAL, LOW-RANK.

Siginer, Prof. Dennis A. *Department of Mechanical Engineering, Auburn University, Alabama.* FLUID FLOW—coauthored.

Silver, Dr. Simon. *Department of Microbiology and Immunology, University of Illinois at Chicago.* MICRO-BIAL MINING—coauthored.

Simister, Dr. Neil E. *Rosenstiel Center for Basic BioMedical Sciences and Biology Department, Brandeis University, Waltham, Massachusetts.* IMMUNOLOGY—coauthored.

Sleep, Dr. Norman H. *Department of Geophysics, Stanford University, California.* EARTHQUAKE.

Sojka, Dr. R. E. *U.S. Department of Agriculture, Agricultural Research Service, Kimberly, Idaho.* IRRIGATION (AGRICULTURE)—in part.

Sollinger, Dr. John D. *Department of Biology, Carleton College, Northfield, Minnesota.* PLANT EMBRYO-GENESIS.

Soukoulis, Prof. Costas M. *Department of Physics, Iowa State University, Ames.* PHOTONIC CRYSTALS.

Spearman, Dr. Mark L. *Department of Industrial and Systems Engineering, Georgia Institute of Technology, Atlanta.* MANUFACTURING ENGINEERING—coauthored.

Sposito, Prof. Garrison. *Department of Environmental Science, Policy, and Management, Division of Ecosystem Sciences, University of California, Berkeley.* SOIL CHEMISTRY.

Steinhardt, Prof. Paul J. *Mary Amanda Wood Professor of Physics, Department of Physics and Astronomy, University of Pennsylvania, Philadelphia.* COSMIC BACK-GROUND RADIATION.

Steinman, Dr. Lawrence. *Department of Neurology and Neurological Sciences, Stanford University Medical Center, and Beckman Center for Molecular and Genetic Medicine, Stanford, California.* MULTIPLE SCLEROSIS—coauthored.

Stoddart, Prof. J. Fraser. *Department of Chemistry, University of Birmingham, Edgbaston, United Kingdom.* SUPRAMOLECULAR CHEMISTRY—coauthored.

Story, Dr. Craig M. *Rosenstiel Center for Basic BioMedical Sciences and Biology Department, Brandeis University, Waltham, Massachusetts.* IMMUNOLOGY—coauthored.

Struhsacker, Debra W. *Environmental Permitting and Government Relations Consultant, Reno, Nevada.* MINING.

Stuart, Dr. David *Laboratory for Molecular Biophysics, Oxford, United Kingdom.* VIROLOGY—coauthored.

Stute, Dr. Martin. *Lamont-Doherty Earth Observatory of Columbia University, Palisades, New York.* GLACIAL EPOCH.

Sugiyama, Prof. Junta. *Institute of Molecular and Cellular Biosciences, The University of Tokyo, Japan.* FUNGI—in part.

Swanson, Barry G. *Department of Biological Systems Engineering, Washington State University, Pullman.* FOOD MICROBIOLOGY—coauthored.

T

Tang, Dr. Man-Chung. *DRC Consultants, Inc., Flushing, New York.* BRIDGE.

Taylor, Dr. Carson W. *Carson Taylor Seminars, Portland, Oregon.* ELECTRIC POWER SYSTEMS.

Thewissen, Dr. J. G. M. *Department of Anatomy, College of Medicine, Northeastern Ohio Universities, Rootstown.* CETACEA.

Toledo, Prof. Romeo T. *Department of Agricultural and Environmental Sciences, The University of Georgia, Athens.* FOOD ENGINEERING—coauthored.

Tuite, Prof. Mick F. *Department of Biosciences, University of Kent at Centerbury, United Kingdom.* FUNGAL GENETICS.

U

Unger, Dr. Paul W. *Soil Scientist, U.S. Department of Agriculture, Agricultural Research Service, Southern Plains Area, Bushland, Texas.* AGRICULTURAL SOIL AND CROP PRACTICES.

V

Van, Dyck, Dr. Robert E. *GEC-Marconi Systems, Wayne, New Jersey.* DATA COMPRESSION—in part.

Van Grondelle, Prof. Rienk. *Department of Biophysics, Vrije Universiteit, Amsterdam, The Netherlands.* PHOTOSYNTHESIS.

Van Orden, Dr. Alan. *Department of Chemistry, University of California, Berkeley.* CARBON—coauthored.

von Puttkamer, Dr. Jesco. *NASA Headquarters, Office of Space Flight, Washington, D.C.* SPACE FLIGHT.

Vessières, Dr. A. *Director of Research, National Center of Scientific Research, Paris, France.* BIOINORGANIC CHEMISTRY—coauthored.

W

Wainwright, Dr. Milton. *Department of Molecular Biology and Biotechnology, The University of Sheffield, United Kingdom.* FUNGI—in part.

Ward, Dr. Peter A. *Godfrey D. Stobbe Professor, Department of Pathology, University of Michigan Medical School, Ann Arbor.* INFLAMMATION.

Webster, Dr. Robert G. *Rose Marie Thomas Chair, Department of Virology and Molecular Biology, St. Jude Children's Research Hospital, Memphis, Tennessee.* INFLUENZA.

White, Dr. Stephanie M. *Advanced Technology and Development Center, Northrop Grumman, Bethpage, New York.* COMPUTER-BASED SYSTEMS ENGINEERING.

Willmer, Dr. Pat. *School of Biological and Medical Sciences, University of St. Andrews, United Kingdom.* ANIMAL PHYLOGENY.

Y

Yeung, Prof. Edward S. *Department of Chemistry, Iowa State University, Ames.* SPECTROSCOPY.

Young, Dr. Craig M. *Senior Scientist, Department of Larval Ecology, Harbor Branch Oceanographic Institution, Fort Pierce, Florida.* CRINOIDEA.

Z

Zebker, Dr. Howard A. *Department of Geophysics, Stanford University, California.* TOPOGRAPHY MAPPING.

Zobel, Dr. Bruce J. *Department of Forestry, North Carolina State University, Raleigh.* WOOD—coauthored.

Index

Asterisks indicate page references to article titles.

For Reference

Not to be taken from this room

Hillsborough Community
College LRC